S_pringer_ **M**_onographs_ in **M**_athematics_

For further volumes:
www.springer.com/series/3733

Władysław Narkiewicz

Rational Number Theory in the 20th Century

From PNT to FLT

Springer

Władysław Narkiewicz
Institute of Mathematics
Wrocław University
Plac Grunwaldzki 2-4
50-384 Wrocław
Poland
narkiew@math.uni.wroc.pl

ISSN 1439-7382 Springer Monographs in Mathematics
ISBN 978-1-4471-2715-4 ISBN 978-0-85729-532-3 (eBook)
DOI 10.1007/978-0-85729-532-3
Springer London Dordrecht Heidelberg New York

British Library Cataloguing in Publication Data
A catalogue record for this book is available from the British Library

Mathematics Subject Classification: 11-00, 11-03, 01A60

Softcover reprint of the hardcover 1st edition 2012

Cover design: VTeX UAB, Lithuania

Printed on acid-free paper

Springer is part of Springer Science+Business Media (www.springer.com)

Preface

1. The beginning of a new century provides a good moment for looking back. Number theory has changed its appearance during the last hundred years. At the end of the 19th century it was regarded as a collection of dispersed results dealing with various old and newer problems, obtained by people who were mostly specializing in other subjects. After one hundred years number theory became a well-established part of mathematical sciences, having close relations to commutative algebra, homological algebra, algebraic geometry, function theory, real analysis, functional analysis, group theory and topology.

2. The aim of this book is to give a short survey of the development of the classical part of number theory between the proof of the Prime Number Theorem (PNT) and the proof of Fermat's Last Theorem (FLT), covering thus the twentieth century. Results obtained earlier or later will be also quoted, as far as they are connected with our main topics.

Actually it is now difficult to indicate the borders of number theory, as it tends to acquire grounds reserved earlier to analysis, algebra or geometry. It seems that A. Weil thought about limiting the possessions of number theory, when he wrote: "To the best of my understanding, analytic number theory is not number theory," [6630, p. 8] but nowadays it is fashionable to believe that number theory encompasses more and more of mathematical research.

The word "rational" in the title indicates that we shall concentrate on that part of number theory which deals with properties of integers and rational numbers, hence the theory of algebraic numbers will be excluded. This is motivated by the fact that its inclusion would enormously increase the size of the book, and, moreover, a large bibliography covering this part of number theory is available in my previous book [4543]. Nevertheless, some exceptions will be made, as we shall consider the class-number problems for quadratic and cyclotomic fields. The first of them coincides with the class-number problem for binary quadratic forms, and the second is intimately connected with the earlier approach to Fermat's Last Theorem. We shall also comment on the generalization of the Waring problem to algebraic number fields and describe the creation of class-field theory because of its influence on the reciprocity laws.

The history of the theory of modular forms which played a decisive role in the proof of Fermat's Last Theorem, and which underwent great progress in the last century, deserves a book of its own. Therefore we shall describe only those parts of its development which had a direct influence on number theory proper. This applies also to other branches of mathematics providing tools for arithmetical research. In particular we will not touch the more advanced topics in Diophantine geometry.

In consecutive chapters we shall present the main achievements of the relevant period, accompanied by comments about the development occurring in the next periods. An exception will be made for Fermat's Last Theorem, to which the last chapter is devoted. Our exposition will be concise, sometimes imitating the style of the celebrated Dickson's *History of the Theory of Numbers* [1545], although there is neither the possibility nor need to comment on all number-theoretical production. We have tried to list all the main achievements, quote many important papers, but restrain from including technical details in order to make the text available to non-specialists also. More attention will be paid to earlier work, in the hope that this will help to save it from falling into oblivion.

3. The first chapter contains a very short summary of the development of number theory in the 19th century, starting with Gauss's book *Disquisitiones Arithmeticae* [2208], and ending with the proof of the Prime Number Theorem by Hadamard and de la Vallée-Poussin and Hilbert's talk at the 1900 Congress of Mathematicians in Paris. The second chapter begins with a survey of some famous old problems (perfect numbers, Mersenne and Fermat primes, primality, ...), and then brings the story of our subject at the begin of the century (solution of the Waring problem, Brun's sieve, theorem of Thue, ...). In the next chapter the development up to 1930 will be covered (the inventing of the circle method by Hardy and Ramanujan, progress in the theory of Diophantine equations starting with Siegel's thesis, Mordell's finite basis theorem in the theory of elliptic curves). The most important events in the thirties, covered in Chap. 4, were Vinogradov's proof of the ternary Goldbach conjecture for large numbers, the solution of Hilbert's problem about transcendence of numbers α^β (with algebraic $\alpha \neq 0, 1$ and algebraic irrational β) obtained by Gelfond and Schneider, and the revival of the theory of modular forms by Hecke. The next two chapters report on later development, including the creation of the large sieve, and Chen's theorem on the binary Goldbach problem. The last chapter is devoted to Fermat's Last Theorem.

Information about results obtained after the period considered in each particular chapter is set in a smaller font.

Acknowledgements In writing this book I used the important collections of old books and journals available in the library of the Faculty of Mathematics and Computer Sciences of Wrocław University. I am very grateful to the librarians for their extraordinary patience. I have also used available reference journals, as well as various databases on the web, in particular those of Jahrbuch, Zentralblatt and MathSciNet.

I would like to express my sincere thanks to numerous friends and colleagues, who helped me in the search for information. In particular I would like to thank

Jerzy Browkin, Kalman Győry, Franz Lemmermeyer, Tauno Metsänkylä, Andrzej Schinzel and Michel Waldschmidt who read preliminary versions of my manuscript and suggested several improvements.

I am also very grateful to the Springer copyeditors for their remarkable job. Particular thanks go to Ms Karen Borthwick and Ms Lauren Stoney for their cooperation. I would like also to thank the Springer team of TeX experts for helpful suggestions which removed typesetting problems.

Wrocław, Poland
August 15, 2011

Władysław Narkiewicz

Contents

Notation

We will use the standard notation utilized in modern texts on number theory. In particular we shall denote by **Z**, **Q**, **R** and **C** the ring of rational integers, the fields of rational, real and complex numbers, respectively. The field of p-adic numbers and its ring of integers will be denoted by $\mathbf{Q}_p$ and $\mathbf{Z}_p$, respectively.

The number of divisors, Euler's phi-function, the sum of divisors and the sum of kth powers of divisors of an integer n will be denoted by $d(n)$, $\varphi(n)$, $\sigma(n)$ and $\sigma_k(n)$, respectively. By $\omega(n)$ we shall denote the number of prime divisors of n, and by $\Omega(n)$ the number of prime factors of n, counted with their multiplicities. The symbol $\left(\frac{n}{p}\right)$ will denote the quadratic residue symbol of Legendre.

By $\mu(n)$ we shall denote the Möbius function, defined by

$$\mu(n) = \begin{cases} (-1)^{\omega(n)} & \text{if } n \text{ is square-free,} \\ 0 & \text{otherwise.} \end{cases}$$

The number of representations of n as a product of k factors > 1 will be denoted by $d_k(n)$. By ζ_n we shall denote the primitive nth root of unity $\exp(2\pi i/n)$.

The letter p in formulas will always denote a prime. By P_k we shall denote a number having at most k prime factors (i.e., $\Omega(P_k) \le k$), and $\pi(x)$ will be the number of primes $p \le x$.

By $\mathrm{li}(x)$ we shall denote the logarithmic integral defined by

$$\mathrm{li}(x) = \int_2^x \frac{dt}{\log t},$$

and $\Gamma(z)$ will be the usual Gamma-function.

We shall use Landau's o-notation, writing

$$f(x) = o(g(x))$$

when the ratio $\frac{|f(x)|}{g(x)}$ tends to 0, when x tends to infinity, and the O-notation

$$f(x) = O(g(x)),$$

introduced on p. 225 in [2520], to mean the existence of a constant C with

$$|f(x)| \le Cg(x)$$

holding for large x. In particular $f(x) = o(1)$ means that $f(x)$ tends to 0, and $f(x) = O(1)$ implies that the function $f(x)$ is bounded. Instead of $f(x) = O(g(x))$ we shall also use Vinogradov's notation $f \ll g$ and $g \gg f$. If the implied constant depends on some parameters $a, b, \ldots$, then we shall write $O_{a,b,\ldots}$, $\ll_{a,b,\ldots}$, or $\gg_{a,b,\ldots}$, respectively.

The notation

$$f(x) = \Omega(g(x))$$

stands for the falsity of the relation $f(x) = o(g(x))$, and

$$f(x) = \Omega_+(g(x)), \qquad f(x) = \Omega_-(g(x))$$

means that for a sequence x_n tending to infinity one has

$$f(x_n) \geq Cg(x_n), \qquad f(x_n) < -Cg(x_n)$$

respectively, with a suitable positive constant C, assuming $g(x)$ to be positive.

The letter ε will usually be reserved for arbitrarily small positive numbers.

By $\Re z$, $\Im z$ we denote the real and imaginary part of the complex number z. The distance of a real number x from the nearest integers will be denoted by $\|x\|$, and by $\{x\}$ we shall denote the fractional part of x. The number of elements of a set A will be denoted by $\#A$.

For an algebraic integer a we shall denote by $\overline{|a|}$ the house of a, defined as the product of all conjugates of a, lying outside the unit circle.

Chapter 1
The Heritage

The 19th century brought essential progress in all branches of mathematics and number theory was no exception. The biggest steps in its development are connected with the names of three eminent mathematicians: C.F. Gauss[1], P.G.L. Dirichlet[2] and B. Riemann[3].

The book *Disquisitiones Arithmeticae* [2208] by the young Gauss, published in 1801, gave a solid basis for the subsequent development, presenting for the first time proofs of such fundamental results as the unique factorization property of positive integers and the quadratic reciprocity law. Its main subject was the theory of binary quadratic forms with integral coefficients. Gauss considered the action of the group $\Gamma = SL_2(\mathbf{Z})$ of unimodular 2×2 matrices with integral entries on the set of primitive binary forms $f(X, Y) = aX^2 + bXY + cY^2$ with even middle coefficient (this restriction turned out later to be unnecessary) and a fixed discriminant, defining the action of

$$M = \begin{bmatrix} \alpha & \beta \\ \gamma & \delta \end{bmatrix} \in \Gamma$$

by

$$M \cdot f = f(\alpha X + \beta Y, \gamma X + \delta Y). \tag{1.1}$$

He introduced a composition in the set of resulting orbits inducing the structure of a finite Abelian group. He did not have yet the notion of a group which arose later, but several of his results in [2208] have a group-theoretical meaning. In particular his comments at the end of art. 306 indicate that he guessed the decomposition of the group formed by classes in the principal genus into a product of cyclic factors.

P.G.L. Dirichlet was the first who, in the thirties, applied analytical tools to arithmetical questions. This brilliant idea allowed him [1584, 1585] to prove the infini-

[1]Carl Friedrich Gauss (1777–1855), professor in Göttingen. See [1653, 3285].

[2]Peter Gustav Lejeune Dirichlet (1805–1859), professor in Breslau, Berlin and Göttingen. See [4327].

[3]Bernhard Riemann (1826–1866), professor in Göttingen. See [3722].

W. Narkiewicz, *Rational Number Theory in the 20th Century*,
Springer Monographs in Mathematics,
DOI 10.1007/978-0-85729-532-3_1,

tude of primes in arithmetical progressions $ax + b$ with co-prime integers a, b, and to give an analytical formula for the number of classes of binary quadratic forms [1586, 1587] studied by Gauss. For this aim Dirichlet introduced L-functions by the formula

$$L(s, \chi) = \sum_{n=1}^{\infty} \frac{\chi(n)}{n^s},$$

where χ is a character of the group $G(k) = \{a \bmod k : (a, k) = 1\}$ of reduced residue classes mod k, and s is a positive real number.

One should point out that Dirichlet followed Gauss in the style of presentation, giving very precise and flawless arguments, which was rather a rarity at that time.

Another revolutionary idea came from B. Riemann who in 1860 related in [5224] analytical properties of the series

$$\zeta(s) = \sum_{n=1}^{\infty} \frac{1}{n^s} \tag{1.2}$$

to the properties of prime numbers[4]. It had been known since L. Euler[5] that this series converges for $s > 1$ and its sum is connected with prime numbers due to the formula

$$\zeta(s) = \prod_{p} \frac{1}{1 - p^{-s}}, \tag{1.3}$$

but Riemann was the first to consider $\zeta(s)$ for complex arguments. He obtained its analytical continuation to a meromorphic function in the plane with a single simple pole at $s = 1$ and showed that it satisfies the functional equation

$$\Gamma\left(\frac{s}{2}\right)\pi^{-s/2}\zeta(s) = \Gamma\left(\frac{1-s}{2}\right)\pi^{-(1-s)/2}\zeta(1-s) \tag{1.4}$$

for all complex $s \neq 0, 1$. He stated several properties of $\zeta(s)$ and the counting function of primes,

$$\pi(x) = \sum_{p \le x} 1,$$

for which complete proofs were later obtained, except for the assertion that every non-real zero of $\zeta(s)$ lies on the line $\Re s = 1/2$. This is the famous *Riemann Hypothesis* (RH).

The only previous result about $\pi(x)$ was obtained by P.L. Čebyšev[6], who proved in 1850 [970] that for large x one has

$$0.92129 \frac{x}{\log x} < \pi(x) < 1.1055 \frac{x}{\log x}. \tag{1.5}$$

---

[4]For an analysis of Riemann's memoir see the book by H.M. Edwards [1691].

[5]Leonhard Euler (1752–1833), professor in St. Petersburg and Berlin. See [852, 1986].

[6]Pafnutiĭ Lvovič Čebyšev (1821–1894), professor in St. Petersburg. See [5015].

To the list of outstanding personalities of that time one should also add C.G.J. Jacobi[7], whose main achievement in number theory consisted of applying the theory of elliptic functions to various arithmetical questions. In that way he obtained several identities in the theory of partitions and gave explicit formulas for the determination of $r_k(n)$, the number of representations of an integer n as the sum of k squares for $k = 2, 4, 6$ and 8 [3075, 3077].

The theory of algebraic numbers, which started with Gauss's study [2211] of the complex integers $a + bi$ $(a, b \in \mathbf{Z})$, was further developed by P.G. Dirichlet (see [1593, 1594]) and E.E. Kummer[8] (see [3583]), who utilized it in his work on Fermat's Last Theorem, and reached adult status in the hands of R. Dedekind[9] (see [1423]).

The leading French mathematicians interested in number theory at that time were J. Liouville[10], who constructed the first transcendental numbers and produced several important results in the theory of Diophantine equations, and C. Hermite[11] who, among other deep results, established the transcendence of the number e, the basis of natural logarithms.

In the second half of the century two great personalities, J. Hadamard[12] and H. Poincaré[13], became interested in number theory. Hadamard [2426] confirmed a conjecture of Gauss's by proving in 1896 the Prime Number Theorem as a by-product of his theory of entire functions, and Poincaré, whose main interest was rather far away from number theory, published two papers in our subject. In the first [4935] he generalized Čebyšev's bound (1.5) to the case of primes congruent to unity mod 4, and in the second [4936] he considered rational points on elliptic curves defined by an equation of the form $y^2 = f(x)$, where f is a cubic polynomial without multiple roots.

One should also mention here E. Cahen[14], E. Maillet[15], H. Padé[16] and T. Pépin[17]. Maillet showed [4108] that for sufficiently large k, depending on the prime p, Fermat's equation

$$X^{p^k} + Y^{p^k} = Z^{p^k}$$

[7]Carl Gustav Jacob Jacobi (1804–1851), professor in Königsberg and Berlin. See [3438, 3439].

[8]Ernst Eduard Kummer (1810–1893), professor in Breslau and Berlin.

[9]Richard Dedekind (1831–1916), professor in Zürich and Braunschweig.

[10]Joseph Liouville (1809–1882), professor in Paris. See [4035].

[11]Charles Hermite (1822–1901), professor in Paris. See [4624, 4849].

[12]Jacques Salomon Hadamard (1865–1963), professor in Bordeaux and Paris. See [3868, 4221].

[13]Jules Henri Poincaré (1854–1912), professor in Paris. See [4850].

[14]Eugéne Cahen (1865–1941), teacher at Collège Rolin in Paris.

[15]Edmond Maillet (1865–1938), engineer, president of the Société Mathématique de France in 1918.

[16]Henri Padé (1863–1953), professor in Poitiers and Bordeaux.

[17]Jean François Théophile Pépin (1826–1904), catholic priest, Jesuit, teacher of mathematics.

has no solution in positive integers not divisible by p, and in [4107] studied representations of integers as sums of values of polynomials of small degree.

Cahen [878] made the first systematic study of general Dirichlet series and published in 1900 a textbook on number theory [879]. Padé introduced in [4714] a new kind of continued fractions which later played an important role in analysis and number theory, and Pépin [4775] studied various Diophantine equations, providing in particular a description of all integral solutions of $X^4 + 35Y^4 = Z^2$.

At the end of the 19th century several bright young mathematicians interested in number theory started their careers, Most of them came from Germany. One should list here first of all A. Hurwitz[18], a student of F. Klein[19]. He got his degree in 1881 in Leipzig, next year made his habilitation in Göttingen[20], became in 1884 extraordinary professor at the University of Königsberg[21], and stayed there until 1892 when he switched to ETH[22] in Zürich. In Königsberg he had two extremely bright students, D. Hilbert[23] and H. Minkowski[24]. Hilbert obtained his doctorate in 1886 at Königsberg, got an extraordinary professorship, and in 1893 became ordinarius there. In 1895 he left Königsberg for Göttingen which at that time was, in the hands of Klein, the center of mathematical life. Minkowski, a close friend of Hilbert, got his doctorate in 1885, in 1892 became professor in Bonn, returned in 1895 to Königsberg University, in 1896 went to ETH, and in 1902 settled in Göttingen. In 1899 E. Landau[25], a student of G. Frobenius[26], got his doctorate in Berlin, and after the early death of Minkowski became in 1909 his follower in Göttingen. To this list one should also add K. Hensel[27] who studied in Berlin and obtained his degree in 1884 under L. Kronecker[28]. Two years later he made his habilitation there, became professor in Berlin, and in 1901 switched to Marburg. He invented the p-adic numbers, which would play an important role in algebra and number theory in the coming century. Most of the work in number theory in Germany at the beginning of the 20th century was done by these men and their students.

---

[18] Adolf Hurwitz (1859–1919). See [2791, 6794].

[19] Christian Felix Klein (1849–1925), studied in Bonn under Plücker, passed his doctorate at the age of 19, professor in Erlangen, Munich, Leipzig and Göttingen. See [1259].

[20] He was unable to do this in Leipzig, since he pursued his secondary education at a Realgymnasium, considered unacceptable by Leipzig University [2791].

[21] Called now Kaliningrad, in honor of a Soviet official Mikhail Kalinin.

[22] Eidgenössische Technische Hochschule in Zürich.

[23] David Hilbert (1862–1943), professor in Königsberg and Göttingen. See [5149].

[24] Hermann Minkowski (1864–1909), professor in Bonn, Königsberg, Zürich and Göttingen. See [2790].

[25] Edmund Landau (1877–1938), professor in Göttingen. See [2518, 3418].

[26] Ferdinand Georg Frobenius (1849–1917), professor in Zürich and Berlin.

[27] Kurt Hensel (1861–1941), edited the *Journal für reine und angewandte Mathematik* from 1901 on. See [2603, 5533].

[28] Leopold Kronecker (1823–1891), professor in Berlin. See [3403].

The English mathematicians at the brink of the 20th century were not particularly keen on number theory. Among the few who produced papers on our subject one should point out H.J.S. Smith[29], who for the proof of his formula for $r_5(n)$ shared with Minkowski the prize of the Paris Academy in 1887, J.J. Sylvester[30], who after his return from America considered the problem of odd perfect numbers (see [2239]), J.W.L. Glaisher[31], who at that time was the editor of the journal *Messenger of Mathematics*, and A. Cunningham[32], whose interest encompassed the elementary theory of numbers, in particular factorization, and primality tests of large integers. Knowledge of foreign literature was not particularly high, so, for example, Glaisher devoted two papers to the series $\sum_{n=0}^{\infty}(-1)^n/(2n)^2$, not noticing its relation to $\zeta(2)$, and rediscovered [2245] Dirichlet's class-number formula in the case of discriminant equal to 7, writing "I do not know whether the following series for $\frac{\pi}{\sqrt{7}}$ has been remarked before." This was followed by a paper by N.M. Ferrers[33] [1996], who found Dirichlet's formula for discriminants 11 and 19, and only in [2246] Glaisher acknowledged Dirichlet's priority. Glaisher also wrote a series of papers [2242–2244, 2247] dealing with Bernoulli, Eulerian and related numbers, and later studied representations of integers by sums of squares (see Sect. 2.4.1).

The situation changed drastically when G.H. Hardy[34] became interested in arithmetical problems. In his first paper [2504] on this subject, published in 1906, he presented an analytical formula giving the maximal prime divisor $\theta(N)$ of a positive integer N:

$$\theta(N) = \lim_{r\to\infty}\lim_{m\to\infty}\lim_{n\to\infty}\sum_{j=0}^{m}\left(1-\cos\left(\frac{\pi(j!)^r}{N}\right)\right)^{2n}.$$

This was not a very serious result, but soon Hardy found out that the analytical tools at his possession could be used for the proof of important arithmetical applications, and this resulted in a series of path-breaking papers, co-authored in a later period by J.E. Littlewood[35] and S. Ramanujan[36].

In America there was increased interest in number theory at the time when Sylvester was professor at Johns Hopkins University from 1877 until 1883. During that time he published several papers in the *American Journal of Mathematics*,

---

[29]Henry John Stephen Smith (1826–1883), professor in Oxford. See [5659].

[30]James Joseph Sylvester (1814–1897), professor in Baltimore and Oxford. See [4748].

[31]James Whitbread Lee Glaisher (1848–1928), Fellow of Trinity College, Cambridge. See [2042].

[32]Allan Joseph Champneys Cunningham (1842–1928), military engineer, Fellow of King's College, London. In his obituary [6638] A.E. Western wrote: "It is probably true that no single person has ever before calculated and printed so large an amount of numerical work in this subject."

[33]Norman Macleod Ferrers (1829–1903), Fellow of Gonville and Caius College, Cambridge.

[34]Godfrey Harold Hardy (1877–1947), professor in Oxford and Cambridge. See [6181, 6370, 6663].

[35]John Edensor Littlewood (1885–1977), professor in Cambridge. See [865].

[36]Srinivasa Aiyangar Ramanujan (1887–1920), Fellow of Trinity College, Cambridge. See [84, 2512, 2516, 3237]. Cf. also [456, 457].

which he founded and edited until 1884. Among them was a paper [6008] in which he improved Čebyšev's bounds for the number of primes in the interval $[2, x]$. He published also an important treatise on partitions [6009]. Browsing through early issues of Sylvester's journal one also finds other papers written by J.C. Fields[37] [1998], A.S. Hathaway [2613], O.H. Mitchell[38] [4339] and others, in which various elementary and algebraic aspects of number theory were treated. After the return of Sylvester to Britain that interest weakened.

The list of prominent mathematicians doing number theory in other countries at the end of the 19th century is rather short. One should mention N.V. Bugaev[39], A.A. Markov[40], G.F. Voronoĭ[41], A.N. Korkin[42] and E.I. Zolotarev[43] in Russia[44], L. Gegenbauer[45] and F. Mertens[46] in Austria, and E. Césaro[47] and A. Genocchi[48] in Italy.

There were very few books on number theory at that time. After Gauss's [2208] came two editions of A.M. Legendre's[49] [3767] book (in 1808 and 1830), and in 1863 there appeared the first edition of Dirichlet's lectures [1592], edited and provided with several appendices by Dedekind. These lectures covered divisibility properties, congruences, the quadratic reciprocity law and the theory of binary quadratic forms. In Dedekind's appendices one finds i.a. the proof of Dirichlet's theorem on the infinitude of primes in progressions, the theory of Pell's equation, composition of binary quadratic forms and the principal results of the theory of algebraic numbers. Surprisingly this book aged quite well and can be read even today.

---

[37] John Charles Fields (1863–1932), professor in Toronto. Initiator of the Fields Medal. See [6015].

[38] Oscar Howard Mitchell (1851–1889), teacher of mathematics in Marietta College, Springfield, student of C.S. Peirce, worked mainly in mathematical logic. He introduced the English term *power residue*, writing in a footnote in [4338]: "Power residues is a term not used, I believe, but a needed translation of 'Potenz-Reste'." See [1580].

[39] Nikolaĭ Vasilievič Bugaev (1837–1903), professor in Moscow. See [5667].

[40] Andreĭ Andreevič Markov (1856–1922), professor in St. Petersburg. See [2349].

[41] Georgiĭ Fedoseevič Voronoĭ (1868–1908), professor in Warsaw. His name is sometimes spelt "Voronoï". See [5940].

[42] Aleksandr Nikolaevič Korkin (Korkine) (1837–1908), professor in St. Petersburg. See [4998].

[43] Egor Ivanovič Zolotarev (1847–1878), professor in St. Petersburg. See [3594].

[44] For a very detailed survey of the development of number theory in Russia before 1918 see the book by E.P. Ožigova [4711].

[45] Leopold Gegenbauer (1849–1903), professor in Czernowitz, Innsbruck and Vienna. See [5960].

[46] Franz Carl Josef Mertens (1840–1927), professor in Cracow, Graz and Vienna. See [1533].

[47] Ernesto Césaro (1859–1906), professor in Palermo and Naples. See [42].

[48] Angelo Genocchi (1817–1889), professor in Turin.

[49] Adrien-Marie Legendre (1752–1833), worked in Paris.

The exposition [198] of number theory written by P. Bachmann[50] also played an important role. In five volumes, appearing between 1892 and 1905, he treated elementary, analytical and algebraic number theory. Bachmann wrote also a survey of Gauss's achievements in number theory [200].

An introduction to the theory of algebraic numbers formed a part of the monumental treatise on algebra, published in 1894–1908 by H. Weber[51] [6602].

The list of other books dealing with our subject is rather short: one has to mention two books by V.A. Lebesgue[52] [3754, 3755], a textbook by P.L. Čebyšev [969], and the book [4030] by É. Lucas[53], published in 1891. In the last book one finds a simple primality test, which works fine for numbers which are not too large.

The end of the century brought certain important events, which may even suggest the idea that for number theory the 20th century actually began a few years earlier.

In 1891 Minkowski published his first paper [4321] in a subject which later acquired the name *Geometry of Numbers*. He showed there that geometrical considerations essentially simplify the study of reduction and extremal values of quadratic forms.

Two years later two independent proofs were given of the Prime Number Theorem, a statement conjectured already by Legendre [3767] and Gauss, asserting that for the number $\pi(x)$ of primes in the interval $[2, x]$ the asymptotic formula

$$\pi(x) = (1 + o(1))\frac{x}{\log x} \tag{1.6}$$

holds. These proofs were discovered in 1896 by Hadamard [2426] and C. de la Vallée-Poussin[54] [6263]. They both utilized the non-vanishing of Riemann's zeta-function $\zeta(s)$ on the line $\Re s = 1$. Hadamard established the inequality $\zeta(1+it) \neq 0$ by a short argument based on the behavior of the series

$$S(\sigma + it) = \sum_p \frac{\cos(t \log p)}{p^\sigma}$$

to the right of the presumed zero of the zeta-function. He then deduced the Prime Number Theorem by considering the integral

$$\frac{1}{2\pi i}\int_{a-i\infty}^{a+i\infty} \frac{\zeta'(s)}{\zeta(s)}\frac{x^s}{s^2}\,ds \tag{1.7}$$

[50] Paul Gustav Heinrich Bachmann (1837–1920), student of Kummer, professor in Breslau and Münster. See [2749].

[51] Heinrich Weber (1842–1913), professor in Heidelberg, Zürich, Königsberg, Charlottenburg, Marburg, Göttingen and Strassburg. See [6477].

[52] Victor Amédée Lebesgue (1791–1875), professor in Bordeaux.

[53] François Édouard Anatole Lucas (1842–1891), teacher in Moulins and Paris. See [1419].

[54] Charles de la Vallée-Poussin (1866–1962), professor in Louvain.

for $a > 1$, and applying the formula

$$\sum_{n \le x} a_n \log(x/n) = \frac{1}{2\pi i} \int_{a-i\infty}^{a+i\infty} \left(\sum_{n=1}^{\infty} \frac{a_n}{n^s} \right) \frac{x^s}{s^2} \, ds, \tag{1.8}$$

for non-integral x. To be valid, it suffices to have the absolute convergence[55] of the series

$$\sum_{n=1}^{\infty} \frac{a_n}{n^s}$$

on the line $\Re s = a$. This resulted in the formula

$$\sum_{p \le x} \log p \cdot \log(x/p) = (1 + o(1))x,$$

from which an easy elementary argument led to

$$\theta(x) := \sum_{p \le x} \log p = (1 + o(1))x. \tag{1.9}$$

The usual form of the Prime Number Theorem, given by (1.6), is obtainable from (1.9) by a simple argument.

De la Vallée-Poussin, who at that time was interested in a class of real functions which later became known as almost periodic functions, based his proof on the uniqueness of the Fourier expansion of functions from this class. His arguments[56] as well as the deduction of (1.9) from the non-vanishing of $\zeta(1 + it)$ were rather cumbersome. At the end of his paper he gave a very simple proof of the last fact, whose modification, due to Mertens [4259], found its way into most textbooks.

The third big event was the publication in 1897 of Hilbert's report on the theory of algebraic numbers, the famous *Zahlbericht* [2783]. In it Hilbert recapitulated and partially simplified the results of previous research. The presented topics included the theory of quadratic and Abelian fields, and Kummer extensions, i.e., extensions of the form $k(\sqrt[p]{a})/k$, where p is an odd prime, k is the field generated by pth roots of unity, and $a \in k$ is not a pth power. Hilbert's work had a great influence on subsequent researchers, although in later times some aspects of his approach underwent severe criticism (see the highly interesting introduction to the English translation of [2783] written by F. Lemmermeyer and N. Schappacher). An exposition of Hilbert's results was presented a few years later in the book [5846] by J. Sommer.

Shortly afterwards Hilbert published an important paper [2786] containing his theory of quadratic extensions of arbitrary algebraic number fields. In it the Hilbert norm-residue symbol was introduced, which led to the quadratic reciprocity law in number fields, and later played an important role in several problems of algebraic number theory. A generalization of these results to Abelian extensions led to the

---

[55]Later O. Perron [4784] showed that this formula can also be used under much weaker assumptions.

[56]For details of the proofs of de la Vallée-Poussin and Hadamard see, e.g., [4542].

notion of the class-field (H. Weber [6603–6605], D. Hilbert [2785]; for a modern description of Weber's ideas see G. Frei [2072]). This notion led later, due to the work of T. Takagi[57], E. Artin[58], N.G. Čebotarev[59], P. Furtwängler[60] and H. Hasse[61] to class-number theory, which dominated algebraic number theory in the first half of the coming century.

The *Dedekind zeta-function*

$$\zeta_K(s) = \sum_I \frac{1}{N(I)^s} \tag{1.10}$$

(where I runs over all non-zero ideals of the ring of integers of a fixed algebraic number field K) was studied in 1900 by Dedekind [1422] in the case when $K = \mathbf{Q}(\sqrt[3]{D})$ is a pure cubic extension of the rationals. He showed that in this case one can write

$$\zeta_K(s) = \zeta(s)H(s),$$

where $\zeta(s)$ is the Riemann zeta-function, and $H(s)$ is a certain well-behaved Dirichlet series. A similar result in the case of an arbitrary algebraic number field was later established by E. Landau [3624].

The main mathematical event of the last year of the 19th century was certainly the International Congress of Mathematicians held in Paris. On that occasion Hilbert [2788] gave his famous talk on open mathematical problems. Among the 23 problems presented by him, six (7–12) were devoted entirely to number theory, and the 18th problem also contained a question related to the geometry of numbers (Kepler's conjecture).

The seventh problem dealt with transcendence proofs. Hilbert stated here his belief that in general a transcendental entire function should assume transcendent values at algebraic arguments, although he knew of examples of such functions assuming rational values at all algebraic arguments. In particular he asked for a proof of transcendence of values of the exponential function $e^{i\pi x}$ at irrational algebraic arguments x, and posed the question of transcendence, or at least irrationality, of numbers of the form α^β with algebraic[62] α, and algebraic irrational β, like $2^{\sqrt{2}}$ and $e^\pi = i^{-2i}$. Both problems were solved in the nineteen thirties (see Sect. 4.3.1).

The eighth problem dealt with prime numbers and contained a long list of questions commencing with the celebrated *Riemann Hypothesis* (RH), also called the *conjecture of Riemann*.

---

[57]Teiji Takagi (1875–1960), professor in Tokyo. See [2852, 4345].

[58]Emil Artin (1898–1962), student of Herglotz, professor in Hamburg, Princeton and at Notre Dame University and Indiana University. See [1051].

[59]Nikolai Grigorievič Čebotarev (Tschebotareff) (1894–1947), professor in Kazan.

[60]Philipp Furtwängler (1869–1940), professor in Bonn, Aachen and Vienna. See [2846, 2942].

[61]Helmut Hasse (1898–1979), studied in Göttingen and Marburg and got his doctorate in 1920 under K. Hensel. Professor in Halle (1925–1930), Marburg (1930–1934), Göttingen (1934–1939), Berlin (1949–1950) and Hamburg (1950–1966). See [2071].

[62]He did not state the obvious necessary condition $\alpha \neq 0, 1$.

The Riemann Hypothesis *Prove that all non-real zeros of the Riemann zeta-function lie on the line* $\Re s = 1/2$.

The next question asked for the evaluation of the difference $R(x)$ between the number $\pi(x)$ of primes $p \le x$ and the integral logarithm $\mathrm{li}(x)$, defined by

$$\mathrm{li}(x) = \lim_{\varepsilon \to 0} \left(\int_0^{1-\varepsilon} \frac{dt}{\log t} + \int_{1+\varepsilon}^{x} \frac{dt}{\log t} \right). \tag{1.11}$$

In particular Hilbert asked whether this difference is not of higher order than $\sqrt{x}$. This can be interpreted either as $R(x) \ll \sqrt{x}$ or as $R(x) = O(x^{1/2+\varepsilon})$ for every positive ε. It is now known that the second bound is equivalent to the Riemann Hypothesis (H. von Koch[63] [3433]), and the fate of the first is still undecided.

Next came two old questions: the first was the binary *Goldbach conjecture*, which goes back to an exchange of letters in June 1742 between Euler and Goldbach[64] (see P.H. Fuss[65] [1909; 2168, I, letter 43]), and states that every even integer ≥ 6 is the sum of two primes. The second dealt with *twin primes*, asking whether there are infinitely many primes p, p' with $p - p' = 2$. Next came a generalization of the last question.

Show that if $(a, b, c) = 1$, *then the equation*[66] $ax + by + c = 0$ *is solvable with prime* x, y.

The last problem of this sequence asked for a generalization of results on the distribution of prime numbers to the case of prime ideals in algebraic number fields.

In his ninth problem Hilbert asked for reciprocity laws of power residues modulo prime powers in arbitrary algebraic number fields. He expressed the belief that its solution would follow from a generalization of the theory of cyclotomic fields (i.e., fields of the form $\mathbf{Q}(\zeta)$, where ζ is a root of unity) and quadratic extensions of arbitrary algebraic number fields developed earlier by him in [2783, 2786]. This problem later found a solution as a consequence of class-field theory.

The tenth problem dealt with the question of existence of a finite algorithm for checking the solvability of Diophantine equations in rational integers. A negative solution was found in the second half of the next century (see Sect. 6.6).

In the eleventh problem Hilbert asked for a theory of quadratic forms having coefficients in an arbitrary algebraic number field. In particular he proposed finding a method of solving quadratic Diophantine equation in several variables. This problem was solved later by H. Hasse and C.L. Siegel[67].

[63]Niels Fabian Helge von Koch (1870–1924), professor in Stockholm.

[64]Christian Goldbach (1690–1764), lived in St. Petersburg, where he served as an official responsible for code breaking.

[65]Paul Heinrich Fuss (1796–1855), great-grandson of Euler, worked in St. Petersburg.

[66]Hilbert omitted the necessary assumption $2 | a + b + c$.

[67]Carl Ludwig Siegel (1896–1981), professor in Frankfurt, Göttingen and Princeton. See [2835, 5543].

The twelfth problem was the last concerning number theory. Here Hilbert asked for a description of finite Abelian extensions of arbitrary algebraic number fields. In the case of the field of rational numbers such a description is provided by the *Kronecker–Weber theorem* stating that every such extension is contained in a cyclotomic field. It was stated by Kronecker [3524] and the first proof (by an incomplete argument[68]) was presented by Weber [6599, 6600]. The first correct proof appeared in Hilbert's paper [2782] and in his report [2783]. Hilbert asked for a similar result in the case of an arbitrary algebraic number base field, and expressed the belief that a proof of Kronecker's conjecture asserting that Abelian extensions of an imaginary quadratic field are generated by certain values of elliptic functions could be obtained on the basis of the theory of complex multiplication[69], developed by Weber [6601]. He pointed out that the key to the solution may lie in the construction of reciprocity laws governing power residues in algebraic number fields.

In the eighteenth problem, devoted to geometry, Hilbert mentions the question of the maximally dense arrangement of spheres and tetrahedrons in three-dimensional space.

For surveys of Hilbert's problems and the following development see [46, 744, 3250]. For the seventh problem see N.I. Feldman[70] [1982], and for the twelfth see R.-P. Holzapfel [2849] and N. Schappacher [5422].

A survey of the main achievements of the 19th century in number theory has been presented by H. Opolka and W. Scharlau [4686]. It is noteworthy that in the two volumes of the classical work of Klein [3354, 3355] devoted to the history of mathematics in the 19th century one finds only a few mentions of arithmetical results.

A list of all papers dealing with the theory of numbers in that period, except those related to various reciprocity laws, can be found in L.E. Dickson's[71] *History of the Theory of Numbers* [1545] and the *Report on Algebraic Numbers* [1569]. An early survey was published in 1859–1865 by H.J.S. Smith [5831].

---

[68]See O. Neumann [4579].

[69]For the history of complex multiplication see the book [6450] of S.G. Vlăduţ.

[70]Naum Il'ič Feldman (1918–1994), professor in Moscow. See [3489].

[71]Leonard Eugene Dickson (1874–1954), professor in Chicago. See [43].

Chapter 2
The First Years

2.1 Elementary Problems

2.1.1 Perfect Numbers

1. One of the oldest mathematical problems concerns *perfect numbers*. A positive integer N is called *perfect*, if it equals the sum of its proper divisors, i.e., the equality $\sigma(N) = 2N$ holds[1]. It had been noted already by Euclid that if the numbers $2^p - 1$ and p are both prime, then $2^{p-1}(2^p - 1)$ is perfect. After 2000 years Euler [1907] proved that every even perfect number is of this form. Therefore the problem of the existence of infinitely many even perfect numbers is equivalent to the question of whether there are infinitely many *Mersenne primes*, i.e., primes of the form $M_p = 2^p - 1$. The first four such primes, corresponding to $p = 2, 3, 5$ and 7, were known already to the ancient Greeks, and the next three, M_{13}, M_{17} and M_{19}, were found, according to L.E. Dickson [1545], in the 15th and 16th centuries. To this list M. Mersenne[2] (see [1545, pp. 12–13]) added M_{31} and M_{127}, and asserted incorrectly the primality of M_{67} and M_{257}.

A factorization of M_{67} was given by F.N. Cole[3] in 1903 [1174], and the fact that M_{257} is composite was established in 1932 by D.H. Lehmer[4] [3774].

Mersenne also stated that for every other prime $p \le 257$ the number M_p is composite. Mersenne did not indicate any proofs of his assertions, and the first proofs of the primality of M_{31} and M_{127} were obtained by Euler [1902] and É. Lucas [4025, 4028], respectively. Lucas formulated two primality tests of M_p (the first working only for $p \equiv 3 \bmod 4$) but it seems that he never published complete proofs of them.

---

[1] As pointed out by F. Acerbi [11], the equality $6 = 1 + 2 + 3$ can be found in Plato's *Theaetetus*, which may be the first occurrence of a perfect number.

[2] Marin Mersenne (1588–1648), French monk, friend of Descartes.

[3] Frank Nelson Cole (1861–1926), professor at Columbia University. See [2009].

[4] Derrick Henry Lehmer (1905–1991), son of D.N. Lehmer, student of J. Tamarkin, professor at Berkeley. See [729].

W. Narkiewicz, *Rational Number Theory in the 20th Century*, Springer Monographs in Mathematics,
DOI 10.1007/978-0-85729-532-3_2,

They were provided much later by D.H. Lehmer [3773, 3779] and A.E. Western[5] [6639]. The second test, which later has been widely used, runs as follows.

Define a sequence S_n *by putting* $S_1 = 4$ *and* $S_{k+1} = S_k^2 - 2$ *for* $k \geq 1$. *If* $p \neq 2$ *is prime, then* M_p *is prime if and only if* S_{p-1} *is divisible by* M_p.

A simple proof was later provided by M.I. Rosen [5288], and an extremely simple proof of the sufficiency part was given by J.W. Bruce [763]. For a description of the extension of Lucas' ideas see the book [6673] by H.C. Williams.

The number M_{61} was asserted to be prime by I.P. Pervušin[6] (see [4711, p. 277]), J. Hudelot (see [4029]) and P. Seelhoff[7] [5597]. Seelhoff's argument was later shown to be incomplete by F.N. Cole [1174] (cf. D.H. Lehmer [3771]). The primality of M_{89} and M_{107} was proved by R.E. Powers [5003, 5004] in 1911 and 1914, respectively.

The results of the first use of computers in the study of Mersenne primes were presented by R.M. Robinson[8] [5243] in 1954. With the advent of computers the Lucas–Lehmer test led to the discovery of several new Mersenne primes, and a special program, called GIMPS (*Great Internet Mersenne Prime Search*[9]), was created to find them. At the moment of writing, 47 Mersenne primes are known, the largest being M_p with $p = 43\,112\,609$ comprising almost 13 million digits, found in August 2008. This is actually the largest known prime number. In fact, after 1951 every largest known prime has been a Mersenne prime, the last other record being $189 \cdot M_{127}^2 + 1$ found by J.C.P. Miller[10] and D.J. Wheeler [4309] in 1951.

The problem of the existence of infinitely many even perfect numbers can be stated in group-theoretical terms. It was shown in 1997 by M.P.F. du Sautoy [5417] that there are only finitely many such numbers if and only if the sum of the series

$$\sum_{n=1}^{\infty} \frac{a(2^n)}{2^{ns}}$$

is a rational function, $a(m)$ denoting the number subgroups of the product $\prod_p PSL_2(\mathbf{F}_p)$ having index m.

2. It is still unknown whether an odd perfect number exists and there is a strong belief that this is not the case. It seems that R. Descartes[11] was unique in his belief in its existence when he wrote to B. Frénicle de Bessy[12] on December, 20th 1638: "... je[13] juge qu'on peut trouver des nombres impairs véritablement parfaits."

---

[5] Alfred Edward Western (1873–1961), worked as a solicitor. See [4308].

[6] Ivan Mikheevich Pervušin (1827–1900), orthodox priest. He spent forty years preparing a table of all primes below 10^7.

[7] Paul Peter Heinrich Seelhoff (1829–1896), teacher in Mannheim.

[8] Raphael Mitchell Robinson (1911–1995), professor at Berkeley. See [2728].

[9] Homepage: http://www.mersenne.org.

[10] Jeffrey Charles Percy Miller (1906–1981), worked at Cambridge University.

[11] René Descartes (1596–1650).

[12] Bernard Frénicle de Bessy (1605–1675).

[13] "I believe that one can find truly perfect odd numbers."

Certain necessary conditions for odd N to be perfect had already been given by Euler, and in 1832 B. Peirce[14] [4764] showed that an odd perfect number has at least four distinct prime divisors. J.J. Sylvester [6012] stated later that it must have at least five such divisors. The first correct proof of this assertion was provided in 1913 by L.E. Dickson [1543].

It was later shown that an odd perfect number N must have at least 6 (I.S. Gradštein [2297], U. Kühnel [3566]), 7 (C. Pomerance [4966]), 8 (E.Z. Chein [1009], P. Hagis, Jr. [2435]) and 9 (P. Nielsen [4613]) distinct prime divisors. If N is not divisible by 3, then it must have at least eleven prime divisors (M. Kishore [3343], P. Hagis, Jr. [2436]), and one of them must exceed 10^8 (T. Goto, Y. Ohno [2286]). Previous lower bounds for the largest prime divisor were 60 (H.-J. Kanold [3239]), 11 200 and 10^5 (P. Hagis, Jr., W.L. McDaniel [2439, 2440]), 10^6 (P. Hagis, Jr., G.L. Cohen [2438]) and 10^7 (P.M. Jenkins [3120]). Moreover N has to exceed 10^{300} (R.P. Brent, G.L. Cohen, H.J.J. te Riele [711]), and must satisfy $\Omega(N) \geq 75$ (K.G. Hare [2547]). Previous lower bounds were 29 (M. Sayers [5418]), 37 (D.E. Iannucci, M. Sorli [2999]) and 47 (K.G. Hare [2546]). The maximal prime-power divisor of N must exceed 10^{30} (G.L. Cohen [1135]). Several congruences which odd perfect numbers must satisfy were found by J.A. Ewell [1945], L.H. Gallardo [2186] and L.H. Gallardo, O. Rahavandrainy [2187].

Let $A(x)$ be the number of odd perfect numbers $\leq x$. In a letter to Mersenne in 1638 Descartes observed that an odd perfect number must have the form pa^2 with prime p, and this leads, with the use of Čebyšev's bound $\pi(x) = O(x/\log x)$, to

$$A(x) = O\left(\frac{x}{\log x}\right).$$

In 1955 B. Hornfeck[15] [2906] established $A(x) = O(\sqrt{x})$, and this bound was later reduced to $A(x) = o(\sqrt{x})$ and $O(x^{1/4}\log x/\log\log x)$ (H.-J. Kanold [3240, 3242]), and

$$A(x) = O\left(\exp\left(c\frac{\log x \log\log\log x}{\log\log x}\right)\right)$$

with certain $c > 0$ (B. Hornfeck, E. Wirsing [2908]). In 1959 E. Wirsing [6692] eliminated the triple logarithm in the last formula.

L.E. Dickson proved in [1543] that there can be at most finitely many odd perfect numbers with a given number of prime divisors, and in fact he established the same assertion for odd numbers N which satisfy the inequality $\sigma(N) \geq 2N$ and for every proper factor $M > 1$ of N one has $\sigma(M) < 2M$ (cf. [1544]).

Dickson's proof utilized algebraic tools and a simple elementary proof was much later found by H.N. Shapiro [5676]. In 1977 an effective proof was provided by C. Pomerance [4967], leading to the exorbitant bound

$$\log\log N \ll 2^{k^2}\log k$$

for odd perfect N with k prime divisors, improved later by D.R. Heath-Brown [2651] to $\log N < 4^k \log 4$ and by P. Nielsen [4612] to $\log N < 4^k \log 2$.

---

[14] Benjamin Peirce (1809–1880), professor at Harvard. See [1146].

[15] Bernhard Hornfeck (1929–2006), professor in Clausthal-Zellerfeld.

3. A number N is called *multi-perfect* if it divides its sum of divisors but is not perfect i.e., $\sigma(N) = kN$ holds with an integer $k \geq 3$. Several such numbers had already been found in the 17th century by Descartes, P. Fermat[16] and A. Jumeau (see [1545]), the first few being 120, 672, 30 240, 32 760, 523 776, 23 569 920, 33 550 336 and 45 532 800. In 1901 D.N. Lehmer[17] [3801] noted that also 2 178 540 is multi-perfect, and R.D. Carmichael[18] [910] showed that this list exhausts all such numbers below 10^9. He also extended an earlier result of J. Westlund [6641] by proving that 120 and 672 are the only multi-perfect numbers having three prime divisors [908], and later [911, 912] listed all those with four and five prime divisors (in the last case restricting himself to even numbers).

Now more than 5000 multi-perfect numbers are known, all even, and this leads to the conjecture that there are no odd multi-perfect numbers. It was proved by E.A. Bugulov [825] in 1966 that such a number must have at least 11 distinct prime divisors. Later G.L. Cohen and M.D. Hendy [1138] showed that if $k = \sigma(n)/n \geq 3$ and n is odd, then n has at least $(k^5 + 387)/70$ prime divisors, hence $\omega(n) \geq 20$ holds for $k \geq 4$ (for $k = 3$, H. Reidlinger [5153] proved $\omega(n) \geq 12$ for odd n).

In 1985 G.L. Cohen and P. Hagis, Jr. [1137] proved that an odd multi-perfect number has to exceed 10^{70} and to have a prime factor $> 10^5$. Dickson's result in [1543], quoted above, has been extended by H.-J. Kanold [3240] to multi-perfect numbers with fixed ratio $\sigma(n)/n$, which are not multiples of an even perfect number, and an effective proof has been provided by C. Pomerance [4967].

2.1.2 Pseudoprimes and Carmichael Numbers

1. Fermat's theorem states that if p is a prime and $p \nmid a$, then the number $a^{p-1} - 1$ is divisible by p. In particular p divides $2^{p-1} - 1$. This necessary condition for primality is not sufficient as there exist composite numbers n satisfying the congruence

$$2^{n-1} \equiv 1 \pmod{n}.$$

Such composites are called *pseudoprimes*. It seems that the first pseudoprime appeared in a paper by F. Sarrus[19] [5408] in 1819, who observed that $341 = 11 \cdot 31$ divides $2^{170} - 1$. This answered a question posed anonymously in [5022] asking if one can test an integer n for primality by checking whether the congruence

$$2^n \equiv \pm 1 \pmod{2n + 1}$$

[16]Pierre Fermat (1601–1665), lawyer in Toulouse and Bordeaux. See [3035, 4096].

[17]Derrick Norman Lehmer (1867–1938), student of E. Moore, father of D.H. Lehmer, professor at Berkeley.

[18]Robert Daniel Carmichael (1879–1967), professor at the University of Illinois. He wrote two textbooks on number theory: [917, 918].

[19]Pierre Frédéric Sarrus (1798–1861), professor in Strasbourg.

holds. In view of the fact that $2^{170} - 1 | 2^{340} - 1$ this was also a counterexample to the converse of Fermat's theorem, but this fact had not been noted by Sarrus.

It is not difficult to see that there are infinitely many pseudoprimes and, more generally, it was shown in 1904 by M. Cipolla[20] [1114] that for every $a \geq 2$ there exist infinitely many composite n with $a^{n-1} \equiv 1 \pmod{n}$.

Denoting by $P(x)$ the number of pseudoprimes below x, P. Erdős[21] [1802] showed first

$$P(x) \leq x \exp\left(-\frac{1}{3}\sqrt[4]{\log x}\right),$$

and then [1815]

$$P(x) \leq x \exp(-c\sqrt{\log x \log\log x})$$

with some $c > 0$. This was later improved to

$$P(x) \leq x \exp\left(-\frac{1}{2}\log x \frac{\log\log\log x}{\log\log x}\right)$$

for large x by C. Pomerance [4974], who also obtained in [4975] the lower bound

$$P(x) \gg \exp(\log^{5/14} x).$$

This was improved in 1994 to $P(x) \gg x^{\alpha}$ with $\alpha = 2/7$ by W.R. Alford, A. Granville and C. Pomerance [53], and consecutive improvements were obtained by R.C. Baker and G. Harman [266] ($\alpha = 0.2932 > 2/7$) and G. Harman ($\alpha = 0.3322$ [2564], $\alpha = 1/3$ [2566]).

All pseudoprimes below 10^{13} have been computed (R.G.E. Pinch [4878]). Earlier this had been done up to $2.5 \cdot 10^{10}$ (C. Pomerance, J.L. Selfridge, S.S. Wagstaff, Jr. [4981]).

It was proved by A. Rotkiewicz [5317–5319] in 1963 that every progression $aX + b$ with co-prime a, b contains infinitely many pseudoprimes. A bound for the distance between consecutive pseudoprimes in a progression was given by H. Halberstam and A. Rotkiewicz [2458] in 1968. It is also known that every primitive binary quadratic form in the principal genus having a fundamental discriminant[22] and not negative definite represents infinitely many pseudoprimes (A. Rotkiewicz, A. Schinzel [5321]).

A survey of the theory of pseudoprimes was given in 1972 by A. Rotkiewicz [5320].

2. It was observed in 1899 by A. Korselt[23] [3491] that there exist composite integers n, e.g., $n = 561 = 3 \cdot 11 \cdot 17$, satisfying $a^{n-1} \equiv 1 \pmod{n}$ for all a prime to n. He showed also that this happens if and only if $n = p_1 p_2 \cdots p_r$ is square-free and $n - 1$ is divisible by the least common multiple of the numbers $p_1 - 1, \dots, p_r - 1$. Numbers having this property were later studied by R.D. Carmichael [914, 915] and are now called *Carmichael numbers*.

---

[20] Michele Cipolla (1880–1947), professor in Catania and Palermo. See [4290].

[21] Paul Erdős (1913–1996), student of L. Fejér, professor in Budapest, published more than 1200 papers. See [188, 189, 2446, 5351].

[22] An integer d is called a *fundamental discriminant* if it is either square-free and congruent to unity mod 4, or is of the form $d = 4D$, where D is square-free and congruent to 2 or 3 mod 4.

[23] Alwin Reinhold Korselt (1864–1947), schoolteacher, got his Ph.D. in 1902 in Leipzig.

Denote by $C(x)$ the number of Carmichael numbers less than x. The first upper bound for $C(x)$ was given by W. Knödel in 1953, who first got [3411]

$$C(x) \ll x \exp\left(-\log 2\sqrt{\log x}\right),$$

and then [3412]

$$C(x) \ll x \exp\left(-c\sqrt{\log x \log\log x}\right)$$

for every $c < 1/\sqrt{2}$. This was improved three years later by P. Erdős [1815] who proved

$$C(x) \le x \exp\left(-c\frac{\log x \log\log\log x}{\log\log x}\right)$$

with some $c > 0$ and conjectured $C(x) \gg x^{1-\varepsilon}$ for every $\varepsilon > 0$. Some arguments against Erdős's conjecture were given by A. Granville and C. Pomerance [2321].

Much later, in 1994, W.R. Alford, A. Granville and C. Pomerance [53] proved that there are infinitely many Carmichael numbers; more precisely, one has

$$C(x) \gg cx^{2/7}$$

with a certain $c > 0$. The exponent $2/7 = 0.2857\ldots$ was replaced four years later by $0.2932\ldots$ (R.C. Baker, G. Harman [266]), and later G. Harman increased it first to 0.3322 [2564] and then to $1/3$ [2566]. It was conjectured by C. Pomerance [4980] that for $k \ge 3$ there are $x^{1/k+o(1)}$ Carmichael numbers $\le x$ having exactly k prime factors. In 1980 C. Pomerance, J.L. Selfridge[24] and S.S. Wagstaff, Jr. [4981] gave in the case $k = 3$ the bound $O(x^{c+\varepsilon})$ with $c = 2/3$ and any $\varepsilon > 0$. This was later improved to $c = 1/2$ (I.B. Damgård, P. Landrock, C. Pomerance [1319]), to $c = 5/14$ (R. Balasubramanian, S.V. Nagaraj [280]), and to $c = 7/20$ (D.R. Heath-Brown [2660]).

2.1.3 Primality

1. Testing of the primality of Fermat numbers $F_n = 2^{2^n} + 1$ goes back to Fermat, who in several letters (listed in Dickson's *History* [1545, p. 375]) asserted that all numbers F_n are prime. This is true for $1 \le n \le 4$ but fails already for $n = 5$ in view of the factorization $F_5 = 641 \cdot 6700417$ found by Euler [1897]. In 1877 T. Pépin [4772] stated the following test.

The number F_n ($n \ge 1$) is prime if and only if it divides $a^{(F_n-1)/2} + 1$, where a is a quadratic non-residue of F_n.

In the last quarter of the 19th century, using this test and other elementary tools, it was possible to show that F_n is composite for $n = 6, 11, 12, 23, 32$, and 36. In the new century this list has been quickly enhanced due to the efforts of A. Cunningham, J.C. Morehead and A.E. Western who showed that also for $n = 7, 8, 9, 38$ and 73 one gets composite F_n [1298, 4418, 4419, 4421].

---

[24] John Selfridge (1927–2010), professor at Northern Illinois University.

The factorization of Fermat numbers forms a difficult task which for F_7 was done successfully only in 1971 by M.A. Morrison and J. Brillhart [4436, 4437]. Now one knows factorizations of F_n for $n \leq 11$ (R.P. Brent [709, 710], R.P. Brent, J.M. Pollard [712], A.K. Lenstra, H.W. Lenstra, Jr., M.S. Manasse, J.M. Pollard [3820]).

There is a polynomial in seven variables, whose positive values at non-negative integers coincide with Fermat primes, but its practical importance is minimal. The same applies also to Mersenne primes. This was established in 1979 by J.P. Jones [3154].

Now over 200 composite Fermat numbers are known and the smallest Fermat numbers of unknown status are F_{33}, F_{34} and F_{35}. A wealth of information about Fermat numbers is contained in a recent book by M. Křížek, F. Luca and L. Somer [3523]. The actual status is given on the web page http://www.prothsearch.net/fermat.htm.

2. Various elementary methods of primality testing were developed by A. Cunningham and H.J. Woodall (see, e.g., [1299]), who were able to find several large primes, the largest lying in some neighborhood of 3^{15}. They initiated the *Cunningham Project*[25] [1301], consisting of factoring numbers of the form $a^n \pm 1$. D.N. Lehmer also dealt with factorizations, and published lists of the smallest factors of integers up to 10^7 [3803, 3804].

These simple methods could not be used to test very large numbers for primality. The first real progress in this matter was made by D.H. Lehmer [3771, 3772] who in 1927 modified the Lucas test so that it could be applied to numbers like $(10^{24}+1)/(10^8+1)$ of 16 decimal digits (cf. J. Brillhart [730]). Later D.H. Lehmer [3773] formulated a test which used Lucas sequences, and which, in particular, leads to the modern form of the test for Mersenne primes.

In 1983 a new primality test, based on Gauss and Jacobi sums, was found by L.M. Adleman, C. Pomerance and R.S. Rumely [21]. It needed

$$O(\exp(c \log\log n \log\log\log n))$$

steps to test an integer n. This test has been simplified by H. Cohen and H.W. Lenstra, Jr. [1143], who also provided an implementation [1144].

A test based on the theory of elliptic curves was invented in 1993 by A.O.L. Atkin[26] and F. Morain [165].

The question of the existence of a polynomial time algorithm for primality testing obtained a positive answer due to the work of M. Agrawal, N. Kayal and N. Saxena [24]. The new algorithm uses the elementary fact that an integer n is a prime if and only if for some a not divisible by n the polynomial

$$(X-a)^n - X^n + a$$

has all its coefficients divisible by n. The original algorithm was later modified by H.W. Lenstra, Jr. and C. Pomerance [3825], and this modification uses $O(\log^6 n)$ operations

[25] See S.S. Wagstaff, Jr. [6490] for the current standing of this project.

[26] Arthur Oliver Lonsdale Atkin (1925–2008), professor at the University of Illinois in Chicago. See [6101].

to test the primality of n. For expositions of this algorithm see A. Granville [2318] and F. Morain [4372].

3. The old method of Fermat based on the identity $a^2 - b^2 = (a-b)(a+b)$ remained for a long time the only non-trivial way to factorize large numbers. To find a factor of N one takes $m = [\sqrt{N}]$ and tries to find a square in the sequence $m^2 - N, (m+1)^2 - N, (m+2)^2 - N$, and if for some a one gets $(m+a)^2 - N = b^2$, then $N = (m+a-b)(m+a+b)$. This elementary method has been applied by A. Cunningham (see, e.g., [1294]), who also used various tricks to factorize several integers of specific form (for binomials $a^k + b$ see [1297]). In the *Jahrbuch* one finds a long list of papers by Cunningham and other authors in which large numbers were factorized at the beginning of the century.

Modern versions of Fermat's method were given by R.S. Lehman [3770] in 1974 and J. McKee [4234] in 1999.

A method based on the continued fraction expansion of $\sqrt{N}$, which works if the denominator of a complete quotient is a square, was used by D.N. Lehmer [3802, 3805].

Later this method was extended to the case when the product of two or more denominators of complete quotients is a square (D.H. Lehmer, R.E. Powers [3800]). It was further enhanced by M.A. Morrison and J. Brillhart in 1975 [4437], who applied it to a complete factorization of the seventh Fermat number.

A factorization algorithm, using $O(n^{1/4}\log^2 n)$ operations was proposed in 1974 by J.M. Pollard [4944], however its practical importance is minimal in view of the large implied constant. The following year he presented [4946] a more practical method (*Pollard's* ϱ *method*) based on the iteration of a polynomial f mod n. If f_k denotes the kth iterate of f, then one looks at the greatest common divisor D of n and $f_r(a) - f_s(a)$ for $r \neq s$ and suitable a. If D happens to be distinct from n, then it is a proper divisor. A variant of this method was applied by R.P. Brent and J.M. Pollard [712] to factorize F_8. J.M. Pollard also gave [4944] another algorithm (*Pollard's* $p-1$ *method*) in which one computes $\mathrm{GCD}(n, 2^m - 1)$, where m runs over integers having many factors of the form $p-1$ with prime p.

At the beginning of the eighties C. Pomerance [4976, 4978] observed that the problem of finding a proper factor of N is equivalent to the question of finding a non-trivial square root of unity mod N, and that in certain cases the second task can be more feasible. This led to the *quadratic sieve factorization method*. Later development brought to life the *number field sieve*, in which elementary algebraic number theory is used to find a proper factor (see [3818]).

A test using elliptic curves (the *ECM algorithm*) was proposed in 1987 by H.W. Lenstra, Jr. [3822] (see Chap. 7 in the book [1277] by R. Crandall and C. Pomerance).

All these methods are heuristical, hence not quite rigorous, but in many cases lead to the desired result. There are also rigorous algorithms, but their practical value seems to be rather limited (see J.D. Dixon [1598], H.W. Lenstra, Jr., C. Pomerance [3824], C.-P. Schnorr [5545] and B. Vallée [6262]).

For surveys of primality tests and factorization methods see the books of H.C. Williams [6673] and R. Crandall, C. Pomerance [1277].

2.1.4 Other Questions

1. In 1907 R.D. Carmichael [909] asserted that there are no integers m such that the equation $\varphi(x) = m$ has only one solution, but later he recognized [920] that his proof was defective. It is still unknown whether his assertion, known as *Carmichael's conjecture*, is true.

Due to the much later work of V.L. Klee, Jr.[27] [3349] it is now known that any counterexample to Carmichael's conjecture must have many large prime divisors, and this has been used to infer that it should exceed $10^{10^{10}}$ (K. Ford [2028]). Previous bounds were 10^{10^4} (P. Masai, A. Valette [4165]) and 10^{10^7} (A. Schlafly, S. Wagon [5456]). If Carmichael's conjecture is false, then the set of counterexamples has a positive lower density (K. Ford [2028]).

It was conjectured by W. Sierpiński[28] (see [5430]) that for every $m \geq 2$ there exists an integer k such that the equation $\varphi(x) = k$ has exactly m solutions, and A. Schinzel [5433] deduced this from a conjectured assertion about prime numbers (conjecture H, see Sect. 6.1.3). At the end of the century the conjecture of Sierpiński was settled for even m by K. Ford and S. Konyagin [2034], and K. Ford [2029] established it in the general case.

2. At the meeting[29] of the London Mathematical Society on 13 June 1901 A. Cunningham stated that there are no *idoneal numbers* between 1849 and 50 000. These numbers were originally defined by Euler [1903–1905] as positive integers N with the property that any odd number which has a unique representation in the form $x^2 + Ny^2$ with $(x, Ny) = 1$ is necessarily either a prime or a square of a prime, or a double of a prime or, finally, a power of 2 (see [1545, p. 361]). This definition later obtained the following simpler form: a positive integer N is idoneal if any odd number which has a unique representation in the form $x^2 + Ny^2$, and in this representation the numbers x and Ny are co-prime, is necessarily a prime (see, e.g., the book [5784, 2nd ed., p. 229] by W. Sierpiński).

Euler gave a method of finding idoneal numbers (a precise version of it was presented by F. Grube [2369] in 1874), found 65 of them, the largest being 1848, and in [1905] checked that there are no others below 10 000. It is an open question whether there are any larger ones. Later Gauss [2208, Sect. 303] interpreted idoneal numbers in the language of quadratic forms: it turned out that N is idoneal if and only if every genus of classes of positive definite quadratic forms of discriminant $-N$ contains only one class, hence the class-group is isomorphic to C_2^N for some N.

Much later it was proved by S. Chowla[30] [1074] that there are only finitely many idoneal numbers, and in 1954 Chowla and W.E. Briggs [1096] showed that under the *General Rie-*

---

[27] Victor LaRue Klee (1925–2007), professor at the University of Washington. See [2348].

[28] Wacław Sierpiński (1882–1969), student of Voronoĭ, professor in Lwów and Warsaw. See [5442].

[29] See Proc. London Math. Soc., 34 (1902), p. 541.

[30] Sarvadaman Chowla (1907–1995), student of J.E. Littlewood, professor in Benares, Waltair, Lahore, at the University of Kansas, University of Colorado in Boulder and Pennsylvania State University. See [185].

mann Hypothesis[31] (GRH) Euler's list is complete, and in any case there can be only one idoneal number larger than 10^{60}. Earlier J.D. Swift [5999] showed that there are no new idoneal numbers below 10^7. Cf. also E. Grosswald[32] [2361].

The idoneal numbers are related to the Diophantine equation

$$n = xy + yz + zx,$$

with positive integers x, y, z, as pointed out by J. Borwein and K.K.S. Choi [649] (see also R. Crandall [1275] and M. Peters [4803]). The connection of this equation with class-groups of quadratic forms had already been noted in 1862 by J. Liouville [3935] (see also E.T. Bell[33] [396], L.J. Mordell[34] [4381] and W.G. Gage [2175]).

There is also a relation between idoneal numbers and sums of three squares (see E. Grosswald, A. and J. Calloway [2367], A. Schinzel [5432]).

2.2 Analytic Number Theory

2.2.1 Dirichlet Series

1. The first years of the 20th century were marked by a powerful entrance of methods of complex analysis into number theory. After the work of Dirichlet and Riemann it slowly became clear that the study of the analytical properties of suitable Dirichlet series may lead to essential progress in various arithmetical problems. The fundamental properties of the Riemann zeta-function, touched upon by Riemann in [5224], were developed further at the end of 19th century by H. von Mangoldt[35] [4125] and J. Hadamard [2424]. These results led to the proof of the Prime Number Theorem by J. Hadamard [2426] and C.J. de la Vallée-Poussin [6263, 6264] in 1896.

The corresponding studies of Dirichlet's L-functions were made by H. Kinkelin[36] [3338], A. Hurwitz [2955] and R. Lipschitz[37] [3937], and the first general treatment of functions defined by Dirichlet series

$$f(s) = \sum_{n=1}^{\infty} \frac{a_n}{n^s} \tag{2.1}$$

[31]This hypothesis, called sometimes the Great Riemann Hypothesis, states that the non-trivial zeros of a large class of functions, encompassing Dirichlet L-functions and Dedekind zeta-functions lie on the line $\Re s = 1/2$.

[32]Emil Grosswald (1912–1989), professor at the University of Pennsylvania and Temple University. See [3419].

[33]Eric Temple Bell (1883–1960), professor at the University of Washington and California Institute of Technology. See [5151].

[34]Louis Joel Mordell (1888–1972), professor in Manchester and Cambridge. See [1375].

[35]Hans von Mangoldt (1854–1925), professor in Hannover, Aachen and Danzig. See [3417].

[36]Hermann Kinkelin (1832–1913), professor in Basel. See [5425].

[37]Rudolf Lipschitz (1832–1903), professor in Breslau and Bonn. See [3493].

appeared in the thesis of E. Cahen [878]. It contained a wealth of results, but it was soon pointed out (e.g., in [6263] and [4784]) that some of Cahen's arguments are fallacious. Several years later E. Landau wrote in his book [3636, p. 724]: "... vierzehn[38] Jahre erforderlich waren, bis es möglich wurde bei jedem einzelnen der Cahenscher Resultate festzustellen, ob es richtig oder falsch ist."

2. Over the course of years several new kinds of functions defined by Dirichlet series were defined. So in his proof of the functional equation for Dirichlet's L-functions with real characters [2955] A. Hurwitz utilized the partial zeta-functions

$$\zeta_{k,l}(s) = \sum_{n \equiv l \pmod{k}} \frac{1}{n^s},$$

and proved that they are meromorphic, the only singularity being a simple pole with residue $1/k$ at $s = 1$. For rational $0 \le u < 1$ (the case of arbitrary $u \in (0, 1)$ was treated later by H. Mellin[39] [4238]) he also considered the series

$$\zeta(s, u) = \sum_{n=0}^{\infty} \frac{1}{(n+u)^s}, \tag{2.2}$$

now called the *Hurwitz zeta-function*. The behavior of these functions for large $|s|$ was studied by G.N. Watson[40] [6589] in 1913.

A class of functions encompassing the Hurwitz zeta-functions was studied by C.J. Malmstén[41] [4116], R. Lipschitz [3937] and M. Lerch[42] [3834]. They are defined by

$$\mathfrak{K}_{u,x}(s) = \sum_{n=0}^{\infty} \frac{e^{2n\pi x i}}{(n+u)^s}, \tag{2.3}$$

and are usually called *Lerch zeta-functions*[43].

A modern exposition of the theory of these functions has been given recently by A. Laurinčikas and R. Garunkštis [3729].

---

[38] "... it took fourteen years until it was possible to determine whether a particular Cahen's result is true or false."

[39] Robert Hjalmar Mellin (1854–1933), professor in Helsinki. See [3894].

[40] George Neville Watson (1886–1965), professor in Birmingham. See [5121].

[41] Carl Johan Malmstén (1814–1886), professor in Uppsala, later minister and governor of a province.

[42] Matiaš Lerch (1860–1922), professor in Prague, Fribourg and Brno. See [1302, 4996].

[43] Lerch himself called them *Lipschitz functions*.

In 1900 H. Mellin [4240] attached a class of zeta-functions (*Mellin zeta-functions*) to every non-constant polynomial $F(X)$ with non-negative coefficients, by putting

$$\zeta_F(s;a) = \sum_{n=1}^{\infty} \frac{1}{F(n+a)^s}$$

for positive a. This series converges in the half-plane $\Re s > 1/\deg F$. He showed also that if $P(x_1, \dots, x_n)$ is a complex polynomial whose coefficients have positive real parts, then the function

$$\sum_{k_1,\dots,k_n} \frac{1}{P(k_1,\dots,k_n)^s}$$

can be continued to a meromorphic function in the plane with all its poles on the real line (cf. E. Landau [3653], K. Mahler[44] [4060]).

These functions were later generalized (see T. Shintani[45] [5708], P. Cassou-Noguès [958]) and found important applications in the analytical theory of algebraic numbers.

3. In a huge paper [3619], published in 1903, E. Landau proved that the Dedekind zeta-function $\zeta_K(s)$ can be extended across the line $\Re s = 1$ to a regular function in the half-plane $\Re s > 1 - 1/n$ (n being the degree of the field K) with a simple pole at $s = 1$, and does not have zeros on that line as well as in a small region to the left of it. He noted also that in the case of quadratic or cyclotomic fields the zeta-function is meromorphic. Earlier only the existence of the limit

$$\lim_{s\to 1} (s-1)\zeta_K(s)$$

was known. Landau used these results to obtain lower and upper bounds for the number $\pi_K(x)$ of prime ideals in the ring of integers of the field K, both having the form $cx/\log x$ with certain $c > 0$, generalizing thus the classical result of Čebyšev [970] concerning $\pi(x)$. In the case of the Gaussian field $Q(i)$ this result had been proved earlier by Poincaré [4935]

The next step was taken by E. Hecke[46] [2677], who in a series of papers published in 1917 and 1918 succeeded in continuing $\zeta_K(s)$ to a meromorphic function in the whole plane with a single pole at $s = 1$. Hecke proved also that $\zeta_K(s)$ obeys a functional equation similar to that of the Riemann zeta-function. Previously this was known only in the case of Abelian fields, when $\zeta_K(s)$ can be written as the product of suitable Dirichlet L-functions, and also for some other particular fields.

Hecke's proof was based on the theory of theta-functions and details were given only in the case of a real cubic field, having two imaginary conjugated fields.

---

[44] Kurt Mahler (1903–1988), studied in Göttingen, lecturer in Manchester, professor in Canberra. See [947, 1129].

[45] Takuro Shintani (1943–1980), professor in Tokyo. See [3007].

[46] Erich Hecke (1887–1947), professor in Göttingen and Hamburg. See [4818, 5550].

A modern exposition of Hecke's method was presented in the book by J. Neukirch[47] [4576]. A proof based on ideas going back to Riemann was provided by C.L. Siegel in 1922 [5744, 5745], and another proof was furnished by Ch.H. Müntz[48] [4480] two years later. In 1950 K. Iwasawa (unpublished) and J. Tate [6057] found a proof based on integration in the group of ideles (see e.g., [3690, 4543]). This method was applied later to other kinds of zeta-functions (see e.g., G. Fujisaki [2135, 2136], S. Raghavan, S.S. Rangachari [5049], G. Shimura [5694], A. Weil[49] [6624]).

4. In his next paper [2678] E. Hecke proved the analytical continuation and functional equation for a class of Dirichlet series associated with characters of certain groups of ideal classes in algebraic number fields. Again theta-functions were used, but this time all details were given. At the end of his paper Hecke noted that analogous results can be proved[50] also for L-functions associated with partition of ideals in narrower classes, and his idea was later realized by E. Landau [3657]. This led to the generalization of Dirichlet's theorem on primes in progressions to arbitrary algebraic number fields. This generalization had been known to be true earlier in the case of the Gaussian field $\mathbf{Q}(i)$ (F. Mertens [4260], H. Weber [6606]). A broad exposition of Hecke's results in [2677] with numerous applications was given in Landau's book [3654] which appeared in 1918.

5. A new class of L-functions was brought to life by E. Hecke in [2679, 2680], who introduced a new family of ideal characters, called by him *Größencharaktere*, i.e., *characters of the magnitude*. Nowadays one defines these characters with the use of the idele group[51] and calls them *Hecke characters*. With every Hecke character χ the function

$$L(s,\chi)=\sum_I \frac{\chi(I)}{N(I)^s}$$

was associated and E. Hecke showed that if $\chi \neq 1$, then this function can be extended to an entire function satisfying a functional equation.

Utilizing Kummer's ideal numbers Hecke used this result to study the distribution of ideal prime numbers in cones, which in the case of imaginary quadratic fields can be interpreted as the distribution of primes represented by a quadratic form $f(x, y)$ with the point (x, y) lying in an angle.

Hecke's result in the case of real quadratic fields was later improved by H. Rademacher[52] [5033], who provided an error term for Hecke's formula (cf. [5034, 5035], where the distribution of primes of a real quadratic field in rectangles has been studied).

---

[47] Jürgen Neukirch (1937–1997), professor in Regensburg.

[48] Chaim Hermann Müntz (1884–1965), worked in Göttingen, Berlin and Leningrad. See [4700].

[49] André Weil (1906–1998), professor in Chicago and Princeton. See the special issue of the *Notices AMS*, vol. 46/4, 1999.

[50] He wrote: "Ich habe die Rechnung nicht durchgeführt." ["I did not perform the calculations."]

[51] See, e.g., [3690, 4543].

[52] Hans Rademacher (1892–1969), professor in Breslau and at the University of Pennsylvania. See [89].

In the fifties Hecke's method was extended by J. Kubilius [3542, 3543, 3545] who proved in particular that there are infinitely many primes of the form $p = x^2 + y^2$ with $x \le p^a$ for $a > 25/64 = 0.3906\dots$. Under the General Riemann Hypothesis this holds even with $x \le C \log p$ with a suitable C (N.C. Ankeny[53] [100], J. Kubilius [3545]). Later this was shown to hold for $a > 1/3$ (K. Bulota [838]), $a > 1/4$ (F.B. Kovalčik [3501]), $a \ge 0.1631$ (M.D. Coleman [1178]) and $a \ge 0.119$ (G. Harman, P. Lewis [2568]).

6. A few years later the *Epstein zeta-functions* were introduced by P. Epstein[54] [1768, 1769]. They were associated with n-ary quadratic forms $Q(X_1, \dots, X_n)$ with complex coefficients whose values at real arguments lie in the half-plane $\Re s > 0$, and were defined by the formula

$$Z(s, Q) = \sum_{\substack{m_1,\dots,m_n=-\infty \\ (m_1,\dots,m_n)\neq(0,\dots,0)}}^{\infty} \frac{1}{Q(m_1, \dots, m_n)^s}. \tag{2.4}$$

Epstein showed that in the case when the form Q is positive definite the function $Z(s, Q)$ can be continued to a meromorphic function in the plane having a single simple pole at $s = n/2$ and satisfying a functional equation (in the case of two variables a part of this assertion had been shown earlier by H. Mellin [4240]).

Later M. Deuring[55] [1498] gave fresh proofs in the case $n = 2$ and obtained for the zeros of $Z(s, Q)$ an asymptotical formula analogous to Mangoldt's formula (see (2.20)) concerning $\zeta(s)$ (cf. H.M. Stark [5898]). He found also [1502] a quickly convergent expansion of $Z(s, Q)$.

It turned out that the distribution of zeros of Epstein's zeta differs essentially from that of Riemann's zeta or Dirichlet's L-functions. Indeed, M. Deuring showed in [1498] that if the absolute value $D(Q)$ of the discriminant of $Q(X, Y)$ is sufficiently large, then $Z(s, Q)$ has a real zero between 1/2 and 1 (cf. Chowla [1086], Chowla, A. Selberg[56] [1100, 5627], P.T. Bateman, E. Grosswald [356]), and one year later H. Davenport[57] and H. Heilbronn[58] [1386, 1387] proved the existence of zeros in the half-plane $\Re s > 1$. It was shown in 1935 by H.S.A. Potter and E.C. Titchmarsh[59] [5001] that $Z(s, Q)$ has infinitely many zeros on the line $\Re s = 1/2$.

The paper [1498] of M. Deuring contains the first example of the *Deuring–Heilbronn phenomenon* asserting that a real zero in $[1/2, 1)$ pushes away complex zeros ϱ with $1/2 < \Re\varrho < 1$. In the case of $Z(s, Q)$, M. Deuring showed that for large $D(Q)$ one must have $|\Im\varrho| > D^c$ for every such zero, with the exponent c being independent of the form Q.

---

[53] Nesmith Cornett Ankeny (1927–1993), professor at MIT.

[54] Paul Epstein (1871–1939), professor in Frankfurt.

[55] Max Deuring (1907–1984), professor in Jena, Marburg, Hamburg and Göttingen. See [1704, 5277].

[56] Atle Selberg (1917–2007), professor in Princeton, Fields Medal 1950. See [187].

[57] Harold Davenport (1907–1969), professor in Bangor, London and Cambridge. See [4406, 4407, 5256].

[58] Hans Arnold Heilbronn (1908–1975), student of Landau, professor in Bristol and Toronto. See [951].

[59] Edward Charles Titchmarsh (1899–1963), professor in Liverpool and Oxford. See [924].

Zeta-functions for indefinite quadratic forms were treated by C.L. Siegel [5755] (over the rationals) and K. Ramanathan[60] [5071] (over algebraic number fields).

L-functions for quadratic forms were introduced in 1968 by H.M. Stark [5899, 5900, 5904–5906] in connection with the class-number problem for quadratic forms (see Sect. 6.5).

7. In 1899 H. Mellin [4238, 4240, 4241] introduced the *Mellin transform*, and applied it in the theory of differential equations [4239]. This transform became later, in hands of E. Hecke, a powerful tool in the theory of modular forms. It is defined by

$$g(s) = \int_0^\infty f(x)x^{s-1}dx, \tag{2.5}$$

and its inverse is given by

$$f(x) = \frac{1}{2\pi i}\int_{a-i\infty}^{a+i\infty} \frac{f(s)}{x^s}ds.$$

W.L. Ferrar[61] [1993] in 1937 and N.S. Košliakov[62] [3495] in 1941 used Mellin transform to prove various summation formulas.

An important step forward in the theory of Dirichlet series was made in 1908 by O. Perron[63] [4784], who put right several assertions of E. Cahen [878] and showed that if the function f is defined by the Dirichlet series (2.1) convergent for $\Re s > c$, then for positive $a > c$ and $x > 1$ one has the following formula (*Perron's formula*):

$$\sum_{n\le x} a_n = \frac{1}{2\pi i}\int_{a-i\infty}^{a+i\infty} f(s)\frac{x^s}{s}ds.$$

(See also H. Mellin [4241] and W. Schnee [5534].)

An introduction to the theory of Dirichlet series was published in 1915 by G.H. Hardy and M. Riesz[64] [2543].

For a survey of early results in the theory of Dirichlet series see the article by H. Bohr[65] and H. Cramér[66] [580], published in 1923.

---

[60] Kollagunta Gopalaiyer Ramanathan (1920–1992), professor at the Tata Institute. See [5048].

[61] William Leonard Ferrar (1893–1990), professor in Oxford. See [4580].

[62] Nikolai Sergeevič Košliakov (1891–1958), professor in Leningrad.

[63] Oskar Perron (1880–1975), professor in Tübingen, Heidelberg and Munich. See [2068, 2719].

[64] Marcel Riesz (1886–1969), professor in Stockholm and Lund. See [2195, 2909, 2910].

[65] Harald Bohr (1887–1951), professor in Copenhagen. He seems to be the only mathematician who won a medal at the Olympic Games. He did it in 1908, playing soccer for the Danish team. See [3127].

[66] Carl Harald Cramér (1893–1935), student of Marcel Riesz, professor in Stockholm, worked in number theory and probability theory. See [2501].

2.2.2 Prime Number Distribution

1. Already Riemann had formulated a relation between zeros of the zeta-function $\zeta(s)$ and the distribution of primes. A simple-looking formula of this type (the *explicit formula for* $\psi(x)$) was proved by H. von Mangoldt [4125]. He considered the function

$$\psi(x) = \sum_{p^k \le x} \log p$$

introduced in 1850 by Čebyšev [970] to be closely related to

$$\vartheta(x) = \sum_{p \le x} \log p,$$

due to

$$\psi(x) = \vartheta(x) + \sum_{k=2}^{[\log x/\log 2]} \sum_{p^k \le x} \log p = \vartheta(x) + O\left(\sqrt{x}\log^2 x\right).$$

The function $\vartheta(x)$ is related to $\pi(x)$ by the inequalities

$$\vartheta(x) \le \pi(x)\log x,$$

and

$$\begin{aligned}\vartheta(x) \ge \sum_{x^{1-\varepsilon} \le p \le x} \log p &\ge (1-\varepsilon)\left(\pi(x) - \pi(x^{1-\varepsilon})\right)\log x \\ &= (1-\varepsilon)\pi(x) + O(x^{1-\varepsilon}),\end{aligned}$$

valid for every positive ε. Therefore the Prime Number Theorem can be also stated in the form

$$\psi(x) = (1+o(1))x$$

and

$$\vartheta(x) = (1+o(1))x.$$

In 1895 H. von Mangoldt established the following relation between $\psi(x)$ and zeros of $\zeta(s)$.

If $x > 1$ is not a prime power, then

$$\psi(x) = x - \sum_{\varrho} \frac{x^{\varrho}}{\varrho} - \frac{1}{2}\log\left(1 - \frac{1}{x^2}\right) - \log(2\pi), \tag{2.6}$$

ϱ running over all non-real zeros of the zeta-function, arranged according to their absolute value.

A similar formula utilizing roots of $\zeta(s)$ from a bounded region was obtained in 1910 by H. von Koch [3435], who established

$$\psi(x) = x - \sum_{|\Im\varrho|<x^c} \frac{x^\varrho}{\varrho}\Gamma\left(1 - \frac{c\varrho\log x}{kx^c}\right) + O\left(x^{1-c}\log^2 x\right)$$

for any $0 < c < 1$ and arbitrary fixed $0 < k < \pi/2$. Two years later E. Landau [3646] deduced from (2.6) the equality

$$\psi(x) = x - \sum_{|\Im\varrho|\le T} \frac{x^\varrho}{\varrho} + O\left(\frac{x}{T}\log^2 x + \frac{x\log T}{T} + \log x\right)$$

for $T \ge 3$.

Landau's result has been improved in 1982 by D.A. Goldston [2268, 2269] who replaced the error term by

$$O\left(\frac{x\log x\log\log x}{T} + \frac{x\log T}{T} + \log x\right),$$

and showed that under the Riemann Hypothesis the error term for large T does not exceed

$$\frac{x}{2T} + 2\sqrt{x}\log^2 T.$$

2. The first proofs of the Prime Number Theorem found by Hadamard and de la Vallée-Poussin did not lead to any evaluation of the error term. The first such evaluation appeared in a later paper of de la Vallée-Poussin [6264], who established

$$\pi(x) = \mathrm{li}(x) + O\left(x\exp\left(-c\log^{1/2} x\right)\right), \tag{2.7}$$

with some positive c. He recognized the influence of the existence of a zero-free region of $\zeta(s)$ to the left of the line $\Re s = 1$ on the size of the error term, and based his proof of (2.7) on the observation that $\zeta(\sigma + it)$ does not vanish in the region

$$\sigma > \begin{cases} 1 - 0.0328214/\log(t - 3.806) & \text{if } |t| > 574, \\ 0.9872 & \text{if } |t| \le 574. \end{cases}$$

To evaluate the error term in the Prime Number Theorem one usually does not deal directly with the Dirichlet series

$$P(s) = \sum_p \frac{1}{p^s},$$

whose sum of the first N coefficients coincides with $\pi(N)$, because $P(s)$ has a logarithmic singularity at $s = 1$. Therefore usually one considers the series

$$\sum_{n=1}^{\infty} \frac{\Lambda(n)}{n^s} = -\frac{\zeta'(s)}{\zeta(s)}, \tag{2.8}$$

where $\Lambda(n)$ is the *von Mangoldt function*, defined by

$$\Lambda(n) = \begin{cases} \log p & \text{if } n \text{ is a power of a prime } p, \\ 0 & \text{otherwise.} \end{cases} \tag{2.9}$$

The series (2.8) converges absolutely in the half-plane $\Re s > 1$, and since the right-hand side can be extended over the line $\Re s = 1$ to a function having a simple pole at $s = 1$ it is easier to deal with it than with $P(s)$.

Because of the equality $\psi(x) = \sum_{n \le x} \Lambda(n)$ the asymptotical behavior of $\psi(x)$ can be obtained by applying the formula (1.8) to (2.8), provided one is able to evaluate the integral occurring there. This can be done by shifting the integration path to the left, encompassing the point $s = 1$, where the integrand has a pole, but remaining in a region of non-vanishing of $\zeta(s)$. Since the residue of the integrand at $s = 1$ equals x, it remains to obtain a good bound for the part of the integral taken outside the line $\Re s = 1$.

A zero-free region for $\zeta(s)$ is usually obtained by using bounds for $\zeta'(s)/\zeta(s)$. These bounds are easily obtainable in the region to the right of the line $\Re s = 1$, as there $\zeta(s)$ and its derivative are represented by sums of absolutely convergent Dirichlet series. To obtain such bounds to the left of that line one uses the following formula, which is a consequence of Hadamard's theory of entire functions [2425]:

$$\frac{\zeta'(s)}{\zeta(s)} = \frac{1}{1-s} - \frac{1}{2}\frac{\Gamma'}{\Gamma}\left(1 + \frac{s}{2}\right) + \sum_{\varrho}\left(\frac{1}{s-\varrho} + \frac{1}{\varrho}\right) + C,$$

where ϱ runs over all non-real zeros of $\zeta(s)$, and C is a constant.

Later, when powerful methods of evaluation of exponential sums were invented, one could use the approximation of $\zeta(s)$ by the sums $\sum_{n \le N} 1/n^s$, as it turned out that although the series $\sum_{n=1}^{\infty} 1/n^s$ diverges to the left of the line $\Re s = 1$, nevertheless its partial sums can be used to approximate $\zeta(s)$, due to the equality

$$\zeta(s) = \sum_{n \le N} \frac{1}{n^s} + \frac{N^{1-s}}{s-1} + O(N^{-\Re s}), \tag{2.10}$$

valid for $\Re s \ge 1/2$, $1 \le \Im s \le N$. This reduces the problem of bounding $|\zeta(s)|$ to the evaluation of the exponential sum

$$\sum_{n \le N} \exp(-s \log n).$$

An explicit form of the relation between zero-free regions and the error term in formula (2.7) was presented by A.E. Ingham[67] in 1932 in his book [3018]. The converse of Ingham's theorem was proved in an important particular case by P. Turán[68] [6219, 6223] and in the general case by J. Pintz [4894] (see also W. Staś [5911]). A similar result for the error term in the asymptotic formula for

$$\psi(x; k, l) = \sum_{\substack{n \le x \\ n \equiv l \pmod{k}}} \Lambda(n)$$

gave K. Wiertelak [6664].

[67] Albert Edward Ingham (1900–1967), worked in Leeds and Cambridge. See [864].

[68] Paul Turán (1910–1976), professor in Budapest. See [1827, 2444].

3. C.J. de la Vallée-Poussin obtained in [6264] the evaluation

$$\psi(x) = x + O\big(x\exp(-c_1\log^{1/2}x)\big), \tag{2.11}$$

(with a certain $c_1 > 0$), and then deduced (2.7) by a rather complicated argument. Later E. Landau [3617] observed that this can be obtained by elementary partial summation, utilizing the equality

$$\mathrm{li}(x) = \sum_{n=2}^{[x]} \frac{1}{\log n} + O(1).$$

De la Vallée-Poussin used his approach also to obtain a quantitative form of Dirichlet's theorem on primes in progressions. He showed that if for co-prime k, l we denote by $\pi(x; k, l)$ the number of primes $p \le x$ congruent to l mod k, then

$$\pi(x; k, l) = \left(\frac{1}{\varphi(k)} + o(1)\right)\mathrm{li}(x). \tag{2.12}$$

4. The next step was taken by H. von Koch, who proved in 1901 [3433, 3434] that the equality

$$\pi(x) = \mathrm{li}(x) + O\big(\sqrt{x}\log x\big) \tag{2.13}$$

is a consequence of the Riemann Hypothesis. On the other hand it was shown by E. Schmidt[69] [5474] that one cannot have

$$\pi(x) = \mathrm{li}(x) + O\left(\frac{\sqrt{x}}{\log x}\right)$$

or

$$\psi(x) = x + o\big(\sqrt{x}\big).$$

A simpler proof of Schmidt's result can be found in J.E. Littlewood's paper [3944]. Much later P. Turán [6218] proved an effective result, showing that for large x one has

$$\max_{y \le x} |\psi(y) - y| > \sqrt{x}\exp(-c\log x \log\log\log x / \log\log x)$$

with some explicit $c > 0$ (cf. W. Staś [5910]). These results were made more precise in 1980 by J. Pintz [4893, 4896] who showed that for a sequence x_n tending to infinity the difference $|\psi(x_n) - x_n|$ may be close to

$$\sup_\rho \left|\frac{x^{\Re\rho}}{\rho}\right|,$$

as suggested by formula (2.6). See also S.G. Révész [5169].

[69] Erhard Schmidt (1876–1959), student of Hilbert, professor in Zürich, Erlangen, Breslau and Berlin. See [1579, 5264].

5. In 1903 E. Landau [3620] presented an important simplification of Hadamard's proof of the Prime Number Theorem eliminating the use of the theory of entire functions, and basing his arguments on simple analytical methods. His main tool was the inequality

$$\left|\frac{\zeta'(\sigma+it)}{\zeta(\sigma+it)}\right| \ll \log^c |t|,$$

with a certain c, valid in a region of the form

$$\sigma > 1 - \log^{-b} |t| \quad (|t| > \delta > 0)$$

which he used to evaluate the integral (1.7) in the case $a = 2$.

The error term obtained had the form $O(x\exp(-\log^{1/12} x))$, hence was worse than that given by (2.7), but later modifications [3630, 3634] permitted Landau to obtain the same order of the error term as in de la Vallée-Poussin's formula. He used the same approach [3623] to obtain an error term of the form $O(x\exp(-\log^c x))$ in (2.12), with $c > 0$ depending on k, and in [3631] he showed that c may be taken to be independent of k. A further simplification [3629] led to a proof which is now regarded as the standard proof of the Prime Number Theorem. In [3632] he obtained asymptotics for the number of zeros ϱ of L-functions, satisfying $0 < \Im\varrho < T$, and this led to the non-vanishing of these functions in the region

$$\left\{s = \sigma + it : \sigma \geq 1 - \frac{1}{a\log t},\ t \geq 2\right\},$$

for certain a, depending on k, and to the equality

$$\pi(x;k,l) = \frac{\mathrm{li}(x)}{\varphi(k)} + O\big(x\exp(-b\log^{1/2} x)\big) \tag{2.14}$$

with positive b depending on k.

Landau's simplification [3620] of Hadamard's proof of the Prime Number Theorem enabled him to produce the first proof of the asymptotic equality (the *Prime Ideal Theorem*)

$$\pi_K(x) = \mathrm{li}(x) + O\big(x\exp(-\log^{1/13} x)\big).$$

Later he was able to reduce the error term to $O(\exp(-c\sqrt{\log x}))$ [3654].

The best known evaluation of it, obtained in 1968 by T. Mitsui[70] [4344] and A.V. Sokolovskiĭ [5844], is

$$O\big(\exp(-c\log^{3/5} x(\log\log x)^{-1/5})\big).$$

(See also J. Hinz [2816]).

The proof of the Prime Ideal Theorem came as a great surprise even for its author, who stated earlier in [3619] that a proof of the existence of the limit

$$\lim_{x\to\infty} \frac{\pi_K(x)}{x/\log x}$$

[70]Takayoshi Mitsui (1929–1997), professor at Gakushuin University.

is not possible “at the contemporary state of the theory of ideals.” It turned out that no achievements in that theory were needed.

6. In 1905 E. Landau [3625] proved a very useful observation: if a Dirichlet series

$$f(s)=\sum_{n=1}^{\infty}\frac{a_n}{n^s}$$

with non-negative coefficients converges for $\Re s > a$ and diverges for $\Re s < a$, then $f(s)$ has a singularity at $s=a$. He utilized this result for a new proof of Čebyšev's [972] assertion[71] that the ratio

$$\frac{\pi(x;4,3)-\pi(x;4,1)}{\sqrt{x}/\log x}$$

can attain values arbitrarily close to unity. Čebyšev stated also that

$$\lim_{x\to+0}\sum_{p>2}(-1)^{(p-1)/2}e^{-px}=-\infty, \tag{2.15}$$

hence there are, in some sense, more primes congruent to 3 mod 4 than to 1 mod 4.

It was later proved by G.H. Hardy and J.E. Littlewood [2523] that (2.15) follows from the hypothetical non-vanishing of the series $L(s,\chi_4)$ (where χ_4 is the unique non-principal character mod 4) in the half-plane $\Re s > 1/2$. The same result was obtained independently by E. Landau [3655, 3656] who also established the converse implication.

7. J.E. Littlewood recalled in [3946] that his supervisor E.W. Barnes[72] proposed, as a subject for his Ph.D. thesis, the proof of the Riemann Hypothesis. This turned his interests toward Riemann's zeta-function and although no proof was found, nevertheless some results of primary rank emerged. One of them [2523, 3940] implied that the old conjecture based on numerical calculations for $x < 10^9$, stating that for all $x \geq 2$ the inequality $\pi(x) < \mathrm{li}(x)$ holds, is incorrect. He obtained this by showing the existence of a positive constant c such that each of the inequalities

$$\pi(x)-\mathrm{li}(x) < -c\frac{\log\log\log x}{\log x}x^{1/2}$$

and

$$\pi(x)-\mathrm{li}(x) > c\frac{\log\log\log x}{\log x}x^{1/2}$$

occurs infinitely often, i.e.,

$$\pi(x)-\mathrm{li}(x)=\Omega_{\pm}\left(\frac{\log\log\log x}{\log x}x^{1/2}\right),$$

---

[71] The first proof of this assertion was given by E. Phragmén [4840]. In his paper Landau lists several errors of earlier authors touching this subject.

[72] Ernest William Barnes (1874–1953), Fellow of Trinity College in Cambridge, bishop of Birmingham (1924–1952). See [219].

hence the difference $\pi(x) - \mathrm{li}(x)$ changes its sign infinitely often. Since this had already been deduced by E. Schmidt [5474] from the falsity of the Riemann Hypothesis, so Littlewood could assume its truth. Among his principal tools were the theorem of Phragmén-Lindelöf (proved in 1908 [4841]) and an old result of Kronecker from the theory of Diophantine approximations[73].

The use of the Phragmén-Lindelöf theorem in Littlewood's proof was later eliminated by A.E. Ingham [3020]. Much later, in 1975, H.G. Diamond [1523] also eliminated the use of the explicit formula for $\psi(x)$, used in previous proofs.

Littlewood's proof did not give any bound for the smallest integer N with $\pi(N) > \mathrm{li}(N)$, and this question was pursued by S. Skewes[74] in 1933. In [5801] he showed that the Riemann Hypothesis implies $N < 10^{10^a}$ with $a = 10^{34}$, and in [5802], more than twenty years later, he got $N < \exp\exp\exp(7.703)$, if the Riemann Hypothesis is true, and $N < 10^{10^b}$ with $b = 10^{1000}$, if it is false. This was reduced in 1966 by R.S. Lehman [3768] to $N < 1.65 \cdot 10^{1165}$, by H.J.J. te Riele [6116] to $N < 6.69 \cdot 10^{370}$, by C. Bays and R.H. Hudson [368] to $N < 1.4 \cdot 10^{316}$, by K.F. Chao and R. Plymen [999] to $1.3984 \cdot 10^{316}$ and to $N < 1.3972 \cdot 10^{316}$ by Y. Saouter and P. Demichel [5394]. On the other hand the inequality $N > 10^{14}$ was established by T. Kotnik [3497].

Nevertheless the inequality $\pi(N) < \mathrm{li}(N)$ is true on average, as the integral

$$\int_1^\infty (\pi(t) - \mathrm{li}(t)) \exp(-\log^2 t/x) dt$$

is negative for all large x (Pintz [4903]).

Similar questions have also been posed for error terms in other asymptotical formulas. It seems that the first result of this type is due to E. Phragmén[75], who obtained in 1891 [4840] a sufficient condition for the infinitude of sign changes for a large class of functions, and applied this to the differences $\psi(x) - x$ and $\Pi(x) + \log 2 - \mathrm{li}(x)$, where

$$\Pi(x) = \sum_{n \geq 1} \frac{\pi(x^{1/n})}{n} = \pi(x) + O(\sqrt{x}). \tag{2.16}$$

Littlewood's paper [3940] contains also the proof of

$$\psi(x) - x = \Omega_\pm(\sqrt{x} \log\log\log x).$$

Later E. Schmidt [5474] showed also that the difference $\Pi(x) - \mathrm{li}(x)$ changes its sign infinitely often (cf. also E. Landau [3625]). See Sect. 4.1.2 for further development in the question of sign changes.

73 The importance of Kronecker's result in the theory of Dirichlet series had been demonstrated earlier by H. Bohr [576] who used it to show that $|\zeta(s)|$ attains arbitrarily small values in the half-plane $\Re s > 1$.

74 Stanley Skewes (1899–1988), professor in Capetown.

75 Lars Edvard Phragmén (1863–1937), professor in Stockholm (1892–1903), worked later in an insurance company. See [900].

8. In 1917 a new proof of the Prime Number Theorem was presented by G.H. Hardy and J.E. Littlewood [2523], who showed that it is an easy consequence of their Tauberian theorem, established earlier in [2522]. This theorem states, in its simplest form, that if λ_n is an increasing sequence tending to infinity and satisfying $\lambda_{n+1}/\lambda_n \to 1$, the series

$$f(y) = \sum_{n=1}^{\infty} a_n \exp(-\lambda_n y)$$

converges for positive y, the coefficients a_n satisfy

$$a_n > -c\lambda_n^{a-1}(\lambda_n - \lambda_{n-1})$$

with a certain positive c and $a \geq 0$, and

$$\lim_{y\to 0+0} f(y)y^{-a} = A,$$

then

$$\sum_{n\leq x} a_n = \left(\frac{A}{\Gamma(1+a)} + o(1)\right)\lambda_n^a.$$

There are various similar results available with arithmetical applications. The interested reader should consult the book [2517] by G.H. Hardy, published in 1949.

9. De la Vallée-Poussin considered in [6263] also the number $\pi(x; f)$ of primes $p \leq x$ represented by a primitive binary quadratic form[76] $f(X,Y) = aX^2 + bXY + cY^2$ of discriminant $D = b^2 - 4ac$. It had been known since Dirichlet (the *Dirichlet–Weber theorem* [1588, 6598]) that $\pi(x; f)$ tends to infinity with x, and C.J. de la Vallée-Poussin established the asymptotic equality

$$\pi(x; f) = \left(\frac{c(D)}{H(D)} + o(1)\right)\mathrm{li}(x), \tag{2.17}$$

where $H(D)$ is the number of classes of forms of discriminant D, and $c(D)$ is equal either to 1 or to $1/2$, depending on algebraic properties of the class to which f belongs.

The first evaluation of the error term in this formula appeared in 1906 in a paper by E. Landau [3627], who showed that it is $O(x\exp(-\log^c x))$ for some unspecified $c > 0$, independent of the form in question. Later P. Bernays[77] in his thesis [443] used Landau's method to show that one can take $c = 1/8$. For positive definite forms this was later improved by E. Landau [3648] to $O(x\exp(-a\sqrt{\log x}))$, with a certain $a > 0$.

---

[76] Actually he followed Gauss in considering only forms with even b, however there is no problem in applying his approach to the general case.

[77] Paul Isaac Bernays (1888–1977), student of Landau, professor in Göttingen, Zürich and Princeton. His main work concerned logic and foundations of mathematics. See [4470].

It is now known that the error term in (2.17) is $O(x\exp(-c(\log x)^{3/5}(\log\log x)^{-1/5}))$. This follows from the bound for the error term in the asymptotical formula for the Prime Ideal Theorem for ideal classes in quadratic fields (see J. Hinz [2816]).

10. In 1911 a note by Gauss was discovered (see F. Klein [3353]) in which he stated that the number $\pi_k(x)$ of integers $n \le x$ with $\omega(n) = k$ is asymptotically equal to

$$\frac{1}{(k-1)!}\frac{x(\log\log x)^{k-1}}{\log x}.$$

Although a proof of this assertion was at that time already known (occurring in Landau's book [3636]), nevertheless Landau, after hearing of this discovery, returned to the subject and proved in [3639] an asymptotic expansion of $\pi_k(x)$. Later a uniform upper bound for $\pi_k(x)$ was obtained by G.H. Hardy and S. Ramanujan [2540].

Asymptotics for $\pi_k(x)$ which is uniform for $k < c\log\log x$ (for all $c < e$) were obtained in 1953 by L.G. Sathe [5411–5414], and A. Selberg [5617] presented a simplification. The range for k was later extended by D. Hensley [2750], C. Pomerance [4977] and A. Hildebrand, G. Tenenbaum [2808].

11. An important conjecture concerning prime numbers was formulated in 1908 by L.E. Dickson, who wrote in [1537]:

> "In order that m forms $a_i n + b_i$ shall give m prime numbers for at least one integer n, it is necessary for every prime $p \le m$ and for every set of the a_i chosen from those not divisible by p, that at least two of the b_i/a_i shall be[78] congruent mod p. The sufficiency of these conditions is proposed as a problem worthy of an investigation ...."

(Dickson omitted here the obviously necessary condition $(a_i, b_i) = 1$.) Nowadays one states this conjecture in a stronger form.

Dickson's conjecture *If*

$$f_i(X) = a_i X + b_i \in \mathbf{Z}[X] \quad (i = 1, 2, \dots, m;\ a_i > 0)$$

are given linear polynomials with the property that their product does not have a constant divisor > 1, *then for infinitely many positive integers* n *the values* $f_1(n), \dots, f_m(n)$ *are prime.*

For $m = 1$ this is equivalent to Dirichlet's theorem on primes in progressions, but already in the case $m = 2$ its proof seems to be beyond reach. Indeed, the truth of this conjecture for the pair X, $X + 2$ would imply the infinitude of twin primes, which forms one of the greatest open problems in number theory.

[78] The ratio here should be understood mod p.

Sieve methods provided a way to satisfy the assertion of Dickson's conjecture by almost primes, i.e., by numbers having a bounded number of prime factors. The first such results are presented in the book [2455] by H. Halberstam and H.-E. Richert[79], and subsequent improvements were made by D.R. Heath-Brown [2654] and K.-H. Ho, K.-M. Kwang [2838].

In 1973 D. Hensley and I. Richards [2755, 2756] showed that Dickson's conjecture contradicts the old conjecture asserting the inequality $\pi(x+y) \le \pi(x) + \pi(y)$, but it is now commonly believed that this inequality may fail for some large x, y.

12. In 1909 E. Landau published his monumental treatise [3636]. In almost 1000 pages he presented a detailed survey of the actual knowledge of the distribution of prime numbers and included many of his own improvements. G.H. Hardy and H. Heilbronn wrote in [2518] about Landau's book:

> "In it the analytic theory of numbers is presented for the first time, not as a collection of few beautiful scattered theorems, but as a systematic science. The book transformed the subject, hitherto the hunting ground of a few adventurous heroes, into one of the most fruitful fields of research...."

It was pointed out by T.H. Gronwall[80] in his long review of Landau's book [2351] that the exposition in it "is a model of clearness and rigor."

This was one of the first steps towards recognizing number theory as a serious research subject, as most mathematicians at the begin of the century regarded it as a kind of mathematical recreation. As late as in 1911 A. Châtelet[81] wrote in his thesis: "Quoique les[82] méthodes de la théorie des nombres paraissant encore bien vague et imprécises, on peut néanmoins signaler dans cette partie de la Science l'existence d'un petit nombre d'idées générales ..." [1001, p. 105].

In August 1912 the International Congress of Mathematicians was held in Cambridge and E. Landau was invited to give a plenary talk. He chose to speak about solved and unsolved problems in the distribution of primes and the zeta-function [3641]. This fact can be regarded as the recognition of number theory as an important branch of mathematics (note that at the previous congresses in Heidelberg (1904) and Rome (1908) there were no plenary talks devoted to number theory). After sketching the main results concerning these topics, Landau stated four open problems. One of them was the Goldbach conjecture and the other three also dealt with primes.

(a) *Does the polynomial $X^2 + 1$ represent infinitely many primes at integral arguments?*

---

[79] Hans-Egon Richert (1924–1993), professor in Marburg and Ulm.

[80] Thomas Haakon Gronwall (Grönwall) (1877–1932), professor of physics at Columbia University. See [2251].

[81] Albert Châtelet (1883–1960), father of F. Châtelet, professor in Lille and Paris.

[82] "... Although the methods of the theory of numbers appear rather vague and unprecise, one can point out the existence of a small number of general ideas in this part of science...."

This question is usually called *Landau's conjecture*, although it is a special case of a problem posed in 1857 by V. Bouniakowsky[83] [659], who conjectured that if $f(X)$ is an irreducible polynomial with integral coefficients and d is the maximal integer dividing all numbers $f(1), f(2), \ldots$, then the polynomial $f(X)/d$ represents infinitely many primes at integral arguments.

Landau's conjecture (a) is equivalent to the following assertion.

For infinitely many primes one has

$$\{\sqrt{p}\} < \frac{1}{p^c} \tag{2.18}$$

with $c = 1/2$.

Now one knows that (2.18) holds for every $c < 0.262$, as shown by G. Harman and P. Lewis [2568]. Earlier I.M. Vinogradov[84] [6448] had this for $c < 0.1$, R.M. Kaufman [3284] proved this for $c < 0.1631\ldots$, and A. Balog [297] and G. Harman [2553] got $c < 1/4$. Earlier N.C. Ankeny [100] and J. Kubilius [3545] deduced (2.18) for any $c < 1/2$ from the Riemann Hypothesis for Hecke L-functions, and R.M. Kaufman [3284] has shown that this follows already from the usual Riemann Hypothesis. There are similar results in the case when the primes p are restricted to an arithmetic progression (see A. Balog [298], D.I. Tolev [6187]).

The related question of bounding $\|\alpha p\|$ for irrational α for infinitely many prime p has been considered by I.M. Vinogradov [6436], who proved that the bound $p^{-\tau}$ with any $\tau < 1/5$ is possible, and this was later improved by R.C. Vaughan [6351] ($\tau < 1/4$), G. Harman [2554] ($\tau = 0.3$), C.H. Jia [3136, 3137] ($\tau = 4/13 = 0.3076\ldots$), G. Harman ([2555]; $\tau = 7/22 = 0.3191\ldots$), C.H. Jia ([3137]; $\tau = 9/28 = 0.3214\ldots$), and D.R. Heath-Brown and C.H. Jia [2666] ($\tau < 16/49 = 0.3265\ldots$). Recently K. Matomäki [4195] showed that one can take any $\tau < 1/3$ (earlier A. Balog [299] showed this to be a consequence of the General Riemann Hypothesis). On the other hand it was shown by G. Harman [2560] that there exist irrational numbers θ with

$$\|\theta p\| \geq \frac{0.002 \log p}{p \log\log p}$$

for all primes p.

The same question for simultaneous approximations was treated by A. Balog and J.B. Friedlander [302] and G. Harman [2558].

It was noted in 1972 by P. and S. Chowla [1066] that Landau's conjecture would follow if for every k there would exist prime numbers p such that the length of the period of the continued fraction of $\sqrt{p}$ equals k, and in 1988 C. Friesen [2113] showed that there exist infinitely many square-free integers n with a given period length of $\sqrt{n}$ (cf. F. Halter-Koch [2484]).

It should be pointed out that although no non-linear univariate polynomial is known to represent infinitely many primes, there exist polynomials representing many primes. The classical example is given by the polynomial $X^2 - X + 41$ which represents primes for $X = 0, 1, \ldots, 40$. This was observed by Euler in 1772 and an explanation was given in 1913 by G. Rabinowitsch[85] [5023], who showed that

[83] Victor Bouniakowsky (1804–1889), professor in St. Petersburg.

[84] Ivan Matveevič Vinogradov (1891–1983), professor in Moscow. See [954].

[85] Georg Rabinowitsch (Rainich) (1886–1968), professor at Johns Hopkins University and the University of Michigan.

the polynomial $X^2 - X + m$ represents primes for $X = 0, 1, \ldots, m - 1$ if and only if the ring of integers of the quadratic field generated by $\sqrt{1-4m}$ has the unique factorization property.

Today we know that this holds only for $m = 2, 3, 5, 7, 11$ and 41 (theorem of Heegner–Stark, see Sect. 6.5). It was shown in 1992 by S. Louboutin, R.A. Mollin and H.C. Williams [4003] that Dickson's conjecture implies that for every N there is a negative m such that the polynomial $X^2 - X + m$ represents primes for $X = 1, 2, \ldots, N$.

It seems that the actual record for a polynomial representing the maximal number of distinct primes at consecutive integers is held by the polynomial

$$f(X) = 3X^5 + 7X^4 - 340X^3 - 122X^2 + 3876X + 997,$$

found in 2001 by F. Dress and B. Landreau (see [5181]), whose absolute value represents 49 distinct primes for $X \in [-24, 24]$. A heuristic approach was presented by F. Dress and M. Olivier [1624] in 1999.

On the other hand there are irreducible polynomials without a fixed divisor which do not attain prime values in long intervals (see K.S. McCurley [4227–4229]).

For a study of quadratic polynomials representing many primes see the book [4348] by R.A. Mollin.

(b) *Are there infinitely many primes p, p' satisfying $p - p' = 2$?*

Such prime pairs are called *twin primes*. It seems that this problem was first stated by A. de Polignac[86] [4940, 4941] in 1849. For bounds of the number of twin primes in an initial interval see the next section.

Several large twin primes have been found in the computer era, the largest having more that 100 000 digits[87].

(c) *Does every interval $[n^2, (n+1)^2]$ contain a prime number?*

This conjecture was first stated in 1882 by L. Oppermann[88] [4697] and forms a part of the more general problem of the behavior of prime differences. See Sect. 3.1.3 for a survey of this question.

The contemporary status of Landau's conjectures (b) and (c) has been described in a recent paper by J. Pintz [4908].

2.2.3 *Riemann Zeta-Function and L-Functions*

1. In his celebrated memoir [5224] Riemann asked whether all non-trivial[89] zeros of the zeta-function lie on the line $\Re s = 1/2$, usually called the *critical line*. The

---

[86] Alphonse de Polignac (1817–1890).

[87] For the largest twin primes see the Prime Page of C. Caldwell: http://primes.utm.edu/top20.

[88] Ludvig Henrik Ferdinand Oppermann (1817–1883), professor in Copenhagen. See [2308].

[89] The so-called trivial zeros are $s = -2, -4, -6, \ldots$. They are poles of the function $\Gamma(s)$ and by the functional equation for Riemann's zeta-function they are zeros of $\zeta(s)$.

first 15 such zeros were computed by J.P. Gram[90] [2309, 2310] and E. Lindelöf[91] [3892]. Later R.J. Backlund[92] [205] determined a further 64 zeros and it turned out that they all lie on the critical line. It is highly remarkable that the question of zeros of the zeta-function is so difficult in contrast to its non-zero values. It was shown in 1915 by H. Bohr [577] that for every complex $w \neq 0$ the equation $\zeta(z) = w$ has a solution in every strip $1/2 < \alpha < z < \beta < 1$ (this was established earlier under the Riemann Hypothesis by H. Bohr and E. Landau [585]). The density of the set of values of $\zeta(s)$ at every vertical line $\Re s = \theta$ for $\theta \in (1/2, 1]$ was evaluated by H. Bohr and R. Courant[93] [579] in 1914.

There are several assertions equivalent to the Riemann Hypothesis. One of the first results of this type was proved by J.E. Littlewood [3939] in 1912. He considered the Farey series[94] $0 < \varrho_1 < \cdots < \varrho_M$ of order n, consisting of all rationals in $(0, 1)$ having denominators not exceeding n, and proved that the evaluation

$$\sum_{j=1}^{M} \cos(2\pi \varrho_j) \ll n^{1/2+\varepsilon}$$

for every $\varepsilon > 0$ is equivalent to the Riemann Hypothesis.

An elementary statement equivalent to the Riemann Hypothesis, similar to Littlewood's criterion, was presented by J. Franel[95] [2067] in 1924, who showed that the Riemann Hypothesis is equivalent to the validity of the bound

$$\sum_{j=1}^{M} \left(\varrho_j - \frac{j}{n}\right)^2 \leq \frac{B(\varepsilon)}{n^{1-\varepsilon}}$$

for every $\varepsilon > 0$ with a suitable $B(\varepsilon)$. For interesting early variants of Franel's result see the papers [3669, 3677] by E. Landau.

Much later several related equivalences involving Farey series were found by J. Kopřiva [3475, 3476], and a simpler proof given by A. Zulauf [6849] in 1977. Other relations between Farey series and the Riemann Hypothesis were studied by M. Mikolás, K.I. Sato, S. Kanemitsu and M. Yoshimoto [3233–3235, 4305–4307, 6790–6792].

---

[90] Jorgen Pedersen Gram (1850–1916), worked in an insurance company.

[91] Ernst Leonard Lindelöf (1870–1946), professor in Helsinki. See [4499].

[92] Ralf Josef Backlund (1888–1949), student of Lindelöf, worked in the insurance company Kaleva. One of the founders of the Actuarial Society of Finland.

[93] Richard Courant (1888–1972), professor in Münster, Göttingen and at New York University. See [48, 5150].

[94] The Farey series is named after John Farey (1766–1826), a geologist working in London, who in 1816 [1957] established certain simple properties of this series. It was pointed out by G.H. Hardy [2515] that actually Farey's results were anticipated by Haros [2569] in 1802 (his first name seems to be unknown, and the initial "C." appearing in Dickson's *History* stems from "citoyen" [= citizen], occurring in Haros's paper in the form "C^{en}". See [115].

[95] Jérôme Franel (1859–1939), professor in Zürich.

Another elementary condition equivalent to the Riemann Hypothesis was found in 1984 by G. Robin [5239], who showed that the Riemann Hypothesis is equivalent to the assertion that for every $n \geq 5041$ one has

$$\sigma(n) < e^{\gamma} n \log\log n, \tag{2.19}$$

where γ is Euler's constant. It was shown by Y.J. Choie, N. Lichiardopol, P. Moree and P. Solé [1058] that (2.19) holds for all square-free $n > 30$, and M.Wójtowicz [6710] established (2.19) for almost all n.

In 2002 J.C. Lagarias [3605] eliminated Euler's constant from this statement, showing that the Riemann Hypothesis is equivalent to the following assertion.

If we put $H_n = \sum_{j=1}^{n} 1/j$, *then for all* $n \geq 1$ *one has*

$$\sigma(n) \leq H_n + \exp(H_n)\log(H_n),$$

with equality occurring only for $n = 1$.

In 1922 H. Cramér [1268] showed the equivalence of the Riemann Hypothesis with the evaluation

$$\int_2^x \left(\frac{\psi(t) - t}{t}\right)^2 dt = O(\log x).$$

It was proved in 1948 by P. Turán [6217] that the Riemann Hypothesis would follow from the non-vanishing of partial sums $\sum_{n=1}^{N} n^{-s}$ of the series for $\zeta(s)$ in the half-plane $\Re s > 1$. However, it was shown by C.B. Haselgrove[96] [2574] that this assumption may fail, and R. Spira [5865, 5867] showed that this happens already for $N = 19$.

Another statement equivalent to the Riemann Hypothesis was formulated by B. Nyman [4646] and A. Beurling[97] [495] (cf. H. Bercovici, C. Foias [423], J.C. Carey [899]). It states that if we put $\varrho_a(t) = \{1/t\}$, then the Riemann Hypothesis will be equivalent to the density of the linear space spanned by the set $\{a\varrho_a(t) - \varrho_1(t) : a > 0\}$ in the L^2 space on the positive real half-line (the *Nyman–Beurling criterion*). For a refinement see L. Báez-Duarte [213].

The idea that zeros of $\zeta(s)$ may have an interpretation as eigenvalues of a linear operator in a suitable Hilbert space was used by A. Connes [1196, 1197] (see also P.B. Cohen [1150]) to prove the equivalence of the Riemann Hypothesis with the trace formula for a certain Hilbert space operator.

See J.B. Conrey [1203] for a survey.

In 1916 M. Riesz [5226] showed that the Riemann Hypothesis is equivalent to the bound

$$\left|\sum_{k=1}^{\infty} \frac{(-1)^{k+1}}{\Gamma(k)\zeta(2k)} x^k\right| \ll x^{1/4+\varepsilon}$$

for every $\varepsilon > 0$ and large x.

Early unsuccessful attempts to prove the Riemann Hypothesis are described in Chap. 4 of [4542]. Later there were more such attempts, but without attaining the number of fruitless efforts to establish Fermat's Last Theorem.

---

[96] Colin Brian Haselgrove (1926–1964), lecturer in Manchester.

[97] Arne Karl August Beurling (1905–1986), professor in Uppsala and Princeton. See [26].

2. The fact that the strip $0 < \Re s < 1/2$ contains infinitely many zeros of the zeta-function follows from the formula for the number $N(T)$ of these zeros lying in the rectangle $0 < \Re s < 1/2$, $0 < \Im s < T$, conjectured by Riemann [5224] and established by H. von Mangoldt [4125] in 1895:

$$N(T) = \frac{1}{2\pi} T \log\left(\frac{T}{2\pi}\right) - \frac{T}{2\pi} + R(T), \tag{2.20}$$

with $R(T) = O(\log^2 T)$.

The error term in this formula was later reduced by von Mangoldt to $O(\log T)$ [4127], and a simpler proof was provided by R.J. Backlund [206, 208]. The same formula was obtained by H. Bohr, E. Landau and J.E. Littlewood [588] for the number of solutions of the equation $\zeta(s) = a$ in the strip $1 \le \Im s \le T$ for $a \ne 1$ (in the case $a = 1$ the term $\log(T/2\pi)$ must be replaced by $\log(T/4\pi)$). One conjectures that actually one has $R(T) = O(\log T / \log\log T)$, but it is only known that this is a consequence of the Riemann Hypothesis (J.E. Littlewood [3942], E.C. Titchmarsh [6165]). (Earlier H. Bohr [588] proved that the Riemann Hypothesis implies the bound $o(\log T)$.) On the other hand E. Landau [3637] proved that the Riemann Hypothesis implies the falsity of $R(T) = o(\log\log T)$, and later showed with H. Bohr [585] the existence of a positive c such that even $R(T) = o(\log^c T)$ is incompatible with the Riemann Hypothesis.

The integral

$$\int_0^T \frac{R(t)}{t} dt$$

was evaluated in 1924 by F. and R. Nevanlinna[98] [4581], who showed that it equals

$$\frac{7}{8} \log T + \kappa + O\left(\frac{\log T}{T}\right),$$

with a certain constant κ.

In 1946 A. Selberg [5610] obtained unconditionally

$$R(T) = \Omega_{\pm}\left(\frac{\log^{1/3} T}{(\log\log T)^{7/3}}\right).$$

The number of sign changes of $R(T)$ in $[0, T]$ was evaluated by A. Selberg [5609, 5610]. For a survey see A.A. Karatsuba, M.E. Korolev [3254].

The analogue of Mangoldt's formula (2.20) for zeros of Dirichlet's L-functions was proved by E. Landau [3632].

According to [580, p. 800] it was A. Piltz [4877] who first conjectured that all zeros of L-functions in the half-plane $\Re s > 0$ lie on the line $\Re s = 1/2$. In any case the first computations of small zeros of a sample of these functions performed by J. Großmann [2359] in 1913 showed that they seem to obey this law. This conjecture

[98] Frithiof Nevanlinna (1894–1977) and Rolf Herman Nevanlinna (1895–1980), brothers, professors in Helsinki. See [2621].

of Piltz forms a part of the General Riemann Hypothesis. Of particular importance are the possible real zeros of L-functions.

It was checked by J.B. Rosser[99] [5300, 5301] in 1949 that L-functions associated with real characters modulo d do not have real zeros in the interval (0, 1) for all $d \le 227$. Rosser's result has been extended to $d \le 593\,000$ (with one possible exception) by M. Low [4005], and to $d \le 800\,000$ by G. Purdy [5017]. Low's approach to this question was much later used by M. Watkins [6576] to prove that for real odd (i.e., satisfying $\chi(-1) = -1$) Dirichlet characters χ mod k the corresponding L-function does not have zeros in (1/2, 1) for $k \le 3 \cdot 10^8$, and for even χ this has been shown for $k \le 2 \cdot 10^5$ by K.S. Chua [1103]. At about the same time J.B. Conrey and K. Soundararajan [1220] established that for at least 20% of odd square-free integers k the L-function corresponding to the character

$$\chi(n) = \left(\frac{-8k}{n}\right)$$

does not have a real positive zero.

It was observed in 1912 by M. Fekete[100] (see [1970]) that if for a prime p and

$$\chi(n) = \left(\frac{n}{p}\right)$$

the polynomial

$$f_p(X) = \sum_{n=0}^{p-1} \chi(n) X^n$$

(*Fekete polynomial*) does not have zeros in (0, 1), then the function $L(s, \chi)$ has no real positive zeros.

Although for most small primes $f_p(X)$ does not have zeros in (0, 1), nevertheless it was shown much later by R.C. Baker and H.L. Montgomery [268] that for almost all primes p the interval (0, 1) contains many zeros of $f_p(X)$, hence Fekete's criterion can be applied rather rarely. For a study of complex zeros of Fekete polynomials see J.B. Conrey, A. Granville, B. Poonen, K. Soundararajan [1217].

3. The question of the size of $|\zeta(s)|$ in the critical strip was considered by E. Lindelöf [3893] in 1908. He showed that the function $\mu(\sigma)$, defined as the greatest lower bound of numbers a for which one has

$$|\zeta(\sigma + it)| \ll t^a \tag{2.21}$$

for large t, is continuous and convex and conjectured (*Lindelöf's conjecture*) that one has

$$\mu(\sigma) = \begin{cases} 0 & \text{if } \sigma \ge 1/2 \\ 1/2 - \sigma & \text{if } \sigma < 1/2. \end{cases}$$

---

[99] John Barkley Rosser (1907–1989), professor at Princeton, Harvard, Cornell and the University of Wisconsin. See [5361].

[100] Michael Fekete (1886–1957), professor at the Hebrew University of Jerusalem. See [5262].

This is equivalent to the assertion

$$\zeta(1/2+it)=O(t^{\varepsilon})$$

for all $t>0$ and $\varepsilon>0$, as well as to the following two properties of the moments of $\zeta(s)$:

$$\int_1^T |\zeta(\sigma+it)|^{2k}dt=(T+o(T))\sum_{n=1}^{\infty}\frac{d_k^2(n)}{n^{2\sigma}},\quad \sigma>1/2,\ k=1,2,\dots, \tag{2.22}$$

where $d_k(n)$ is the number of decompositions of n into k factors, and

$$\int_1^T |\zeta(1/2+it)|^{2k}dt \ll T^{1+\varepsilon}$$

for every $\varepsilon>0$.

These equivalences were established by G.H. Hardy and J.E. Littlewood in [2533], where also other equivalent conditions are presented. Lindelöf's conjecture is also equivalent to the bound $o(\log T)$ for the number of zeros of the zeta-function in the region

$$\{\sigma+it: 1/2+\delta\le\sigma\le 1,\ T\le t\le T+1\}$$

for every $\delta>0$ (R.J. Backlund [207]), thus it is a consequence of the Riemann Hypothesis[101] (cf. [3939]).

On the other hand Lindelöf's conjecture implies the still unproved *density conjecture*, which asserts that if $N(\alpha,T)$ denotes the number of zeros $\sigma+it$ of Riemann's zeta-function in the region $0<t\le T$, $\sigma\ge\alpha$, then the bound

$$N(\alpha,T)\ll T^{2(1-\alpha)(1+\varepsilon)} \tag{2.23}$$

holds for every $\varepsilon>0$ and sufficiently large T, uniformly in $\alpha\in[1/2,1]$. Let $\varrho(\alpha)$ be the largest lower bound of exponents ϱ such that one has $N(\alpha,T)\ll T^{\varrho}$. In 1914 H. Bohr and E. Landau proved for $a>1/2$ first that $\varrho(\alpha)\le 1$ [586], and then $N(\alpha,T)=o(T)$ [587], showing thus that the majority of zeros of $\zeta(s)$ lie close to the line $\Re s=1/2$. Six years later S. Wennberg [6636] established $N(\alpha,T)=O(T/\log^c T)$ with $c>0$ depending on α, and in the same year F. Carlson[102] [905] proved $\varrho(\alpha)\le 4\alpha(1-\alpha)$.

Later E.C. Titchmarsh [6168] established $\varrho(\alpha)\le 4(1-\alpha)/(3-2\alpha)$. In 1937 A.E. Ingham [3021] related $\varrho(\alpha)$ to $\mu(\alpha)$ by proving

$$\varrho(a)\le 2+4\mu(1/2),$$

which shows that (2.23) follows from the Lindelöf conjecture, and three years later he established that $\varrho(\alpha)\le 3(1-\alpha)/(2-\alpha)$ [3023].

Note that G. Halász and P. Turán [2447] showed that for fixed $\Re s>3/4$ the Lindelöf conjecture implies the stronger estimate

$$N(\sigma,T)\ll T^{\varepsilon}$$

for every $\varepsilon>0$.

[101] The question of whether these two conjectures are equivalent is still open.

[102] Fritz David Carlson (1888–1952), professor in Stockholm. See [2122].

The value of $\varrho(\alpha)$ was reduced to $\varrho(\alpha) \le 2.5(1-\alpha)$ by H.L. Montgomery [4355] in 1969, and to $\varrho(\alpha) \le 2.4(1-\alpha)$ by M.N. Huxley [2972] in 1972. This led to improvements in bounds for the difference of consecutive primes (see Sect. 3.1.3).

In a paper published in 1941 P. Turán [6215] deduced the density conjecture from a conjectured lower bound for sums of powers of complex numbers. Unfortunately this conjectured bound has recently been shown to be false by J. Andersson, who in his thesis [74, 75] provided a counterexample.

The book [6223] by P. Turán, published in 1953, presented his new method of studying sums of the form

$$1 + \sum_{j=1}^{k} z_j^t$$

with complex z_j and integral t, which led to several applications in number theory and function theory. It permitted to show that one can obtain a bound for the number of zeros of the zeta-function in a strip close to the line $\Re s = 1$ which does not differ much from that given by the density conjecture [6221]. The same approach gave certain necessary and sufficient conditions for the non-existence of roots of $\zeta(s)$ in some half-plane $\Re s > c$ with $c < 1$ (the *quasi-Riemannian hypothesis*) [6216]. The proof of the converse of Ingham's theorem concerning the relation between zero-free regions of $\zeta(s)$ and the error in the Prime Number Theorem was also achieved with this method [6219].

The density conjecture was shown to be true for σ close to 1 by P. Turán [6224], and this was made more precise in 1969 by H.L. Montgomery [4355] who established it for $\sigma \ge 0.9$. Later improvements were made by M.N. Huxley ($\sigma \ge 5/6 = 0.8333\ldots$ [2972], $\sigma \ge 189/230 = 0.8217\ldots$ [2976]), K. Ramachandra ($\sigma \ge 21/26 = 0.8076\ldots$ [5062]), M. Forti and C. Viola ($\sigma \ge 0.8059\ldots$ [2043]), M.N. Huxley ($\sigma \ge 0.8011\ldots$ and $\sigma \ge 0.8$ [2977, 2978]), M. Jutila ($\sigma \ge 43/54 = 0.7962\ldots$ and $\sigma \ge 11/14 = 0.7857\ldots$ [3170, 3171]), D.R. Heath-Brown ($\sigma \ge 15/19 = 0.7894\ldots$ [2627]) and J. Bourgain ($\sigma \ge 25/32 = 0.78125$ [661]).

It is now known that the equality (2.22) holds for all real $k > 0$ and $\sigma > \sigma_k$ with certain $\sigma_k < 1$. Early results are quoted in Landau's book [3636], and for later development see A.E. Ingham [3019], H. Bohr, B. Jessen[103] [581–583], H. Davenport [1345], R.T. Turganaliev [6228] and the books of A. Ivić [3041, 3042].

4. In 1915 G.H. Hardy [2505] succeeded in obtaining a breakthrough in the search for zeros of Riemann's zeta-function. Earlier it had been known only that most of these zeros lie close to the critical line (Bohr, E. Landau [586, 587]) and Hardy showed that infinitely many zeros actually lie on that line. His main tool was a formula, due to H. Mellin [4242], expressing the theta-function

$$\Theta(y) = \sum_{n=-\infty}^{\infty} e^{-n^2 y} \tag{2.24}$$

by the zeta-function:

$$\Theta(y) = 1 + \sqrt{\pi/y} + \frac{1}{2\pi i} \int_{1/2-i\infty}^{1/2+i\infty} \Gamma(s/2) y^{-s/2} \zeta(s) ds,$$

valid for $\Re y > 0$.

[103] Børge Jessen (1907–1993), professor in Copenhagen. See [431].

The theta-function, defined by (2.24), was considered first in 1823 by S.D. Poisson[104] [4937] and made its first application to number theory in hands of C.G.J. Jacobi [3075, 3077]. He used it to obtain formulas for the number of representations of a positive integer as the sum of 2, 3, 6 and 8 squares. Riemann [5224] utilized it in one of his proofs of the functional equation for $\zeta(s)$.

For modern expositions of the theory of theta-functions see the books of J. Fay [1963], J.-I. Igusa [3003] and D. Mumford [4477–4479].

Another proof of Hardy's theorem dealing with the existence of infinitely many zeros of $\zeta(s)$ on the critical line was given by H. Mellin in 1917 [4243].

E. Landau [3650] simplified Hardy's argument and proved that if $N_0(T)$ denotes the number of zeros $\varrho = 1/2 + it$ $(0 < t \le T)$ of the zeta-function lying on the critical line, then

$$N_0(T) \gg \log\log T. \tag{2.25}$$

Landau's method was modified in 1937 by E. Hecke [2699], who studied a class of Dirichlet series $f(s)$ extendable to the complex plane so that the product $(s-k)f(s)$ (with a certain $k > 0$) is entire, and which with certain $\lambda > 0$ and $\varepsilon = \pm 1$ satisfy the functional equation

$$R(k-s) = \varepsilon R(s),$$

where

$$R(s) = \left(\frac{2\pi}{\lambda}\right)^{-s} \Gamma(s) f(s).$$

He showed that $f(s)$ has infinitely many zeros on the line $\Re s = k/2$.

E. Landau also obtained the same assertion for Dirichlet L-functions. The role of the function Θ was played in this case by the series

$$\sum_{n=1}^{\infty} \chi(n) e^{-n^2 y},$$

χ being a Dirichlet character.

The inequality (2.25) has been consecutively improved to $N_0(T) \gg T^{\alpha}$, with $\alpha = 1/2$ (C.J. de la Vallée-Poussin [6265]), any $\alpha < 3/4$ (G.H. Hardy, J.E. Littlewood [2523]), and $\alpha = 1$ (G.H. Hardy, J.E. Littlewood [2527]).

A simple proof of the weaker bound

$$N_0(T) \gg \left(\frac{T}{\log T}\right)^{1/2}$$

was given in 1926 by M. Fekete [1969].

An explicit value for the constant c in the bound $N_0(T) \ge cT$ was found much later by C.L. Siegel [5749]. See Sect. 4.1.2, where also further results on $N_0(T)$ are quoted.

[104] Siméon Denis Poisson (1781–1840), professor in Paris.

5. The first lower bound for the value at $s = 1$ of an L-function with a non-real character χ mod k was given by E. Landau [3640] who in 1911 showed

$$|L(1,\chi)| \gg \log^{-5} k.$$

This was improved two years later by T.H. Gronwall [2350] to

$$|L(1,\chi)| \gg \frac{1}{\log k(\log\log k)^{3/8}}.$$

and in 1926 by E. Landau [3673] to

$$|L(1,\chi)| \gg \frac{1}{\log k}.$$

Now for non-real characters modulo k bounds

$$|L(1,\chi)| \ge \frac{c}{\log k}$$

with explicit c are known. Such a result with $c = 1/58$ was obtained by T. Metsänkylä [4268] in 1970, and larger values of c were later given by S. Louboutin [3995] and P. Barrucand and S. Louboutin [338].

The elementary upper bound

$$|L(1,\chi)| \le \log k + 2 \tag{2.26}$$

can be obtained by partial summation from the trivial bound

$$\left|\sum_{n\le x}\chi(n)\right| < k.$$

For improvements see S. Louboutin [3994, 3996–4002], O. Ramaré [5091, 5092] and A. Granville and K. Soundararajan [2325, 2326]. The last authors obtained for non-real characters χ mod k the bound

$$|L(1,\chi)| \le \lambda(34/35 + o(1))\log k,$$

with

$$\lambda = \begin{cases} 1/4 & \text{if } k \text{ is 3-free},\\ 1/3 & \text{otherwise}.\end{cases}$$

For corresponding bounds in the case of real characters see Sect. 4.1.2.

2.2.4 Character Sums

1. Sums involving a quadratic character χ mod p, with prime $p \ge 3$, appear for the first time in the work of Gauss [2208, Sect. 356; 2209], who gave an explicit formula for the value of the sum

$$\tau_p(\chi) = \sum_{j=1}^{p-1}\chi(j)\zeta_p^j = \begin{cases}\sqrt{p} & \text{if } p \equiv 1 \pmod 4,\\ i\sqrt{p} & \text{if } p \equiv 3 \pmod 4.\end{cases} \tag{2.27}$$

There are several proofs known of this formula and a survey of them was given by B.C. Berndt and R. Evans [452]. See also the book [453] by B.C. Berndt, R. Evans and K.S. Williams.

Sums (2.27) with arbitrary characters χ, as well as the sums

$$\tau_N(\chi) = \sum_{j=1}^{N-1} \chi(j)\zeta_N^j$$

with arbitrary characters modulo N are now called *Gauss sums*. They appear in a natural way in many branches of number theory and for prime p they are related to sums of the form

$$G_k(p) = \sum_{n=0}^{p-1} \exp\left(\frac{2\pi i n^k}{p}\right).$$

(For quadratic characters χ one has $\tau_p(\chi) = G_2(p)$.)

For small k explicit formulas for $G_k(p)$ are known (see B.C. Berndt, R. Evans [451, 452] and the book [453]).

An old problem, proposed by E.E. Kummer in 1842 [3576, 3577], found its solution in 1979. Kummer considered for prime $p \equiv 1 \pmod 3$ the cubic Gauss sums

$$\tau(p) = \sum_{n=1}^{p-1} \chi_p(n)\zeta_p^n,$$

χ_p being a cubic character mod p, and conjectured on the basis of calculations that the arguments of $\tau(p)$ are not uniformly distributed on the unit circle. This seemed not to be supported by numerical experiments performed by J. von Neumann[105] and H.H. Goldstine [4578] in 1953 and E. Lehmer [3806] three years later, and was finally disproved by D.R. Heath-Brown and S.J. Patterson [2671]. For an exposition of the proof see A.B. Venkov, A.B. Proskurin [6381]. It turned out later (S.J. Patterson [4759]) that Gaussian sums corresponding to characters of larger orders also have a similar behavior.

In the case of an odd prime power modulus p^k with $k \geq 2$ the Gaussian sums can be evaluated explicitly (R.W.K. Odoni [4658], J.-L. Mauclaire [4206, 4207]). Explicit formulas for cubic and quartic sums were given by C.R. Matthews [4199, 4200] in terms of Weierstrass $\wp$-functions and Jacobian elliptic functions, respectively, confirming earlier conjectures stated by J.W.S. Cassels [945], A.D. McGettrick [4231] and J.H. Loxton [4006, 4007].

A monograph on Gaussian sums has been written by B.C. Berndt, R. Evans and K.S. Williams [453].

2. The first non-trivial evaluation of the sum of character values in an interval was obtained for arbitrary primitive (i.e., not induced by a character belonging to a

[105] John von Neumann (1903–1957), professor in Princeton. See [2249, 6251].

proper divisor of k) characters χ mod k by G. Pólya[106] [4957] and I.M. Vinogradov [6406] in 1918, who proved for

$$S(\chi) = \max_{x\geq 1}\left|\sum_{n\leq x}\chi(n)\right|,$$

the bound

$$S(\chi) \leq c\sqrt{k}\log k \tag{2.28}$$

with an absolute constant c. A simpler proof was provided by I. Schur[107] [5574]. Pólya's proof gave the value $1/\pi + o(1)$ for c, and this was replaced by $1/(\pi\sqrt{2})$ by E. Landau [3658], who also removed the assumption of primitivity.

For a generalization see E. Dobrowolski, K.S. Williams [1604].

In 1977 H.L. Montgomery and R.C. Vaughan [4366] deduced $S(\chi) \ll \sqrt{k}\log\log k$ from the General Riemann Hypothesis, and this bound cannot be improved, as it was shown by R.E.A.C. Paley[108] [4725] (cf. E. Landau [3678]) that there exists an infinite sequence χ_i mod k_i of quadratic characters satisfying

$$S(\chi_i) \geq \frac{1}{7}\sqrt{k_i}\log\log k_i.$$

In Paley's result the numbers k_i may be assumed to be prime, as shown by P.T. Bateman, S. Chowla and P. Erdős [354] in 1950.

The value of the constant c in (2.28) has been reduced several times, and the best known results are due to A. Hildebrand [2804] and A. Granville and K. Soundararajan [2326]. In the last paper the bound

$$S(\chi) \ll_k \sqrt{k}\log^{1-a(r)} k,$$

with $a(r) > 0$, r denoting the order of χ, was established.

A generalization of (2.28) to algebraic number fields was provided by J. Hinz [2817] (cf. P. Söhne [5842]).

3. Bounds for character sums turned out to be useful in the study of the distribution of primitive roots. The first result of this type was obtained in 1918 by I.M. Vinogradov [6406], who showed that $g(p)$, the least primitive root mod p, does not exceed

$$4^{\omega(p-1)}\sqrt{p}\log p,$$

$\omega(n)$ being the number of distinct primes dividing n.

A small modification in Vinogradov's argument allows replacement of the factor $4^{\omega(p-1)}$ in this inequality by $2^{\omega(p-1)}$, but it took 12 years before I.M. Vinogradov [6415] could improve his result to

$$g(p) \leq 2^{\omega(p-1)}\sqrt{p}\log\log p.$$

---

[106]George Pólya (1887–1985), professor at ETH in Zürich, Brown University and Stanford. See [52].

[107]Issai Schur (1875–1941), professor in Bonn and Berlin. See [3162].

[108]Raymond Edward Allan Christopher Paley (1907–1933), worked in Cambridge and at MIT. See [6662].

The tables of $g(p)$ for $p < 25\,410$ prepared by A. Cunningham, H.J. Woodall and T.G. Creak [1300] indicated that this bound is much larger than it should be and after the next 12 years Vinogradov's bound was improved to

$$g(p) < 2^{1+\omega(p-1)}\sqrt{p}$$

by L.K. Hua[109] [2929]. In 1945 P. Erdős [1794] established for large p the bound

$$g(p) < \sqrt{p}\log^{17} p,$$

and in 1950 P. Erdős and H.N. Shapiro [1859] showed

$$g(p) \ll \omega(p-1)^c\sqrt{p}$$

with a certain constant c. The bounds for character sums obtained in 1962 by D.A. Burgess led to the evaluation

$$g(p) \ll p^{1/4+\varepsilon}$$

for every $\varepsilon > 0$ (D.A. Burgess [855], Y. Wang [6553]), which still holds the record. This bound seems to be rather far from optimal, as it was proved by E. Bach [190] in 1990 that the General Riemann Hypothesis implies

$$g(p) < 3\log^2 p.$$

Explicit bounds were provided by E. Grosswald [2364], who showed that for $p > \exp\exp(24)$ one has $g(p) < p^{0.449}$.

In 1969 P.D.T.A. Elliott showed [1728] that infinitely often one has $g(p) < 475\log^{8/5} p$, and this was superseded in 1984 by R. Gupta and M.R. Murty [2393] who obtained the bound $g(p) < 2250$ for infinitely many p. Two years later this bound was reduced to $g(p) \le 7$ for infinitely many p by D.R. Heath-Brown [2640].

On the other hand it was shown by S.S. Pillai[110] [4871] in 1944 that $g(p) \gg \log\log p$ holds infinitely often, and now it is known that infinitely often $g(p)$ can exceed $c\log p\log\log\log p$ for a certain $c > 0$ (S.W. Graham, C.J. Ringrose [2307]). It follows from the General Riemann Hypothesis that the bound $g(p) > c\log p\log\log p$ holds for infinitely many primes p (H.L. Montgomery, in [4357, Theorem 13.5]).

The mean value of $g(p)$ was considered by D.A. Burgess [861], who established

$$\sum_{p\le x} g(p) \ll x\log^A x$$

with some unspecified A, and this was made more precise by him and P.D.T.A. Elliott [863] to

$$\sum_{p\le x} g(p) \ll x\log x(\log\log x)^4.$$

In 1991 L. Murata [4482] showed that the General Riemann Hypothesis implies

$$\sum_{p\le x} g(p) \ll x(\log\log x)^7,$$

[109] Loo Keng Hua (1910–1985), professor in Beijing and at the University of Illinois. See [2451, 6558].

[110] S. Sivasankaranarayana Pillai (1901–1950), worked in Annamulai, Travancore and Calcutta. See [991].

and the exponent 7 has been replaced by any number > 4 by P.D.T.A. Elliott and L. Murata [1750]. For a numerical study see A. Paszkiewicz, A. Schinzel [4758].

Let $h(p)$ be the least primitive root mod p^2. For small primes p one usually has $g(p) = h(p)$, and in fact one knows only two primes for which this equality fails: $p = 40\,487$ (E.L. Litver, G.E. Yudina [3949]) and $p = 6\,692\,367\,337$ (A. Paszkiewicz [4757]). For the mean value of $h(p)$ S.D. Cohen, R.W.K. Odoni[111] and W.W. Stothers [1153] obtained

$$\sum_{p \le x} h(p) \le x \log x (\log\log x)^4,$$

improving by the factor $\log^3 x (\log\log x)^2$ an earlier bound of D.A. Burgess [862]. They also proved $h(p) \ll p^c$ for every $c > 1/4$.

Denote by $g^*(p)$ the least prime primitive root mod p. In 1969 P.D.T.A. Elliott [1728] showed that for almost all primes p one has

$$g^*(p) \ll \exp(c \log\log p \log\log\log p)$$

for certain $c > 0$, and this was improved by A. Nongkynrih [4625] in 1995 and G. Martin [4160] two years later. Under the Riemann Hypothesis, L. Murata [4482] obtained $g^*(p) = O(p^\varepsilon)$ for all $\varepsilon > 0$ and almost all primes p.

4. G. Pólya used (2.28) to show that if $0 \le \alpha < \beta < 1$ are given, then for large primes p the interval $[\alpha p, \beta p]$ contains approximately the same number of quadratic residues and non-residues.

Another application of the Pólya–Vinogradov inequality was given by I.M. Vinogradov [6407, 6409–6411], who evaluated $n_2(p)$, the smallest quadratic non-residue mod p. He showed that for large p one has

$$n_2(p) < p^c \log^2 p$$

with $c = e^{-1/2}/2 = 0.303\ldots$, and also obtained a similar bound for the smallest kth non-residue mod p.

The exponent of the logarithm in the last inequality was reduced by H. Davenport and P. Erdős [1381]. Elementary methods allow $n_2(p) = O(p^{2/5})$ to be obtained, as shown by A. Brauer[112] [676].

It took 30 years to halve the exponent in Vinogradov's bound, which was done by D.A. Burgess [854] in 1957. One expects that actually the bound $n_2(p) = O(\log^2 p)$ is true, which is known to be a consequence of the General Riemann Hypothesis (N.C. Ankeny [99], K.A. Rodosskiĭ[113] [5245]). On the other hand it was shown by S.W. Graham and C.J. Ringrose [2307] that the inequality

$$n_2(p) \ge c \log p \log\log\log p$$

(with a certain positive c) holds for infinitely many primes p. This is an improvement upon the earlier lower bound

$$n_2(p) \ge c \log p$$

[111] Robert Winston Keith Odoni (1947–2002), professor in Exeter and Glasgow. See [1152].

[112] Alfred Brauer (1894–1985), elder brother of Richard Brauer. Student of Schur, worked at Berlin University until 1935. Professor at the University of North Carolina. See [5266].

[113] Kirill Andreevič Rodosskiĭ (1913–2004), professor in Moscow. See [126].

for infinitely many p, obtained independently by V.R. Fridlender [2094], H. Salié [5380] and P. Turán [6220]. For most primes one has $n_2(p) \leq C(\varepsilon)p^{\varepsilon}$ for every $\varepsilon > 0$, as shown in 1942 by Yu.V. Linnik[114] [3903], and the inequality $n_2(p) \leq 7$ for infinitely many p follows from the result of D.R. Heath-Brown quoted above.

I.M. Vinogradov considered in [6407] also the question of the size of $n'(p)$, the smallest prime quadratic residue mod p, and was able to show that for large p one has

$$n'(p) \leq p^{1/2} \exp(-\log p/\log\log p).$$

He conjectured also that for every $\varepsilon > o$ one has $n'(p) = o(p^{\varepsilon})$.

This conjecture is still open, but in 1967 P.D.T.A. Elliott [1727] showed it to be a consequence of the General Riemann Hypothesis. The best known unconditional result is due to Yu.V. Linnik and A.I. Vinogradov[115] [3932], who in 1966 obtained $n'(p) = O(p^a)$ for every $a > 1/4$, using Burgess's method.

2.2.5 Möbius Function and Mertens Conjecture

1. The leading person promoting the use of analytical methods in number theory at the beginning of the century was E. Landau, who considered in his thesis [3615] the Möbius function $\mu(n)$ and presented a simple proof of the equality[116]

$$\sum_{n=1}^{\infty} \frac{\mu(n)}{n} = 0,$$

proved first by H. von Mangoldt [4126]. Later [3618] he established the bound

$$\sum_{n=1}^{N} \frac{\mu(n)}{n} = o\left(\frac{1}{\log N}\right).$$

Another property of the Möbius function, equivalent to the Prime Number Theorem, is given by

$$M(x) = \sum_{n \leq x} \mu(n) = o(x).$$

[114] Juri Vladimirovič Linnik (1915–1972), professor in Leningrad. See [3000, 4123].

[115] Askold Ivanovič Vinogradov (1929–2005), professor at the Steklov Institute.

[116] Later E. Landau [3638] proved that this equality is equivalent to the Prime Number Theorem, i.e., each of these results implies the other in a simple way.

It was asserted in 1885 by T.J. Stieltjes[117] [5952] that with a certain constant B one has $|M(x)| \leq B\sqrt{x}$. Since this bound implies the convergence of the series

$$\frac{1}{\zeta(s)} = \sum_{n=1}^{\infty} \frac{\mu(n)}{n^s}$$

in the half-plane $\Re s > 1/2$, Stieltjes' assertion, if true, would imply the non-vanishing of $\zeta(s)$ in that half-plane and hence the truth of the Riemann Hypothesis.

2. The stronger inequality

$$|M(x)| \leq \sqrt{x}$$

is usually called the *Mertens conjecture*, since F. Mertens stated it in 1897 [4258] and checked it for $x < 10000$. Later R.D. von Sterneck[118] [5935, 5936, 5938] pursued the checking up to 500000 and noted that apart from some numbers in the vicinity of 200 one even has

$$|M(x)| \leq \frac{1}{2}\sqrt{x}. \tag{2.29}$$

In 1912 J.E. Littlewood [3939] showed that the evaluation

$$M(x) = \sum_{n \leq x} \mu(n) = O\left(x^{1/2+\varepsilon}\right) \tag{2.30}$$

is equivalent to the Riemann Hypothesis.

It was shown later that the Riemann Hypothesis allows replacement of the right-hand side of (2.30) by

$$O\left(\sqrt{x}\exp\left(\frac{c\log x \log\log\log x}{\log\log x}\right)\right)$$

(E. Landau [3670]) and by

$$O\left(\sqrt{x}\exp\left(\frac{c\log x}{\log\log x}\right)\right)$$

(E.C. Titchmarsh [6165]). Recently H. Maier and H.L. Montgomery [4100] reduced this bound to

$$O(\sqrt{x}\exp(c\log^{39/61} x)),$$

K. Soundararajan [5857] obtained

$$O(\sqrt{x}\exp(c\log^{1/2} x(\log\log x)^{14})),$$

and M. Balazard and A. de Roton [288] showed that the exponent 14 can be replaced by any number exceeding $5/2$. It was conjectured by S.M. Gonek (see N. Ng [4590]) that, still under the Riemann Hypothesis, one has

$$M(x) = O(\sqrt{x}(\log\log\log x)^{5/4})$$

and this bound is best possible.

---

[117] Thomas Jan Stieltjes (1856–1894), professor in Toulouse. See [6335].

[118] Robert Daublebsky von Sterneck (1871–1928), professor in Czernowitz, Graz and Vienna.

The first proved result concerning the Mertens conjecture is due to A.E. Ingham who showed in 1942 [3024] that if the Riemann Hypothesis is true and there are only finitely many non-trivial vanishing linear relations between imaginary parts of complex zeros of the zeta-functions, then Mertens conjecture is false, as in that case one has

$$\liminf_{x\to\infty} \frac{M(x)}{\sqrt{x}} = -\infty,$$

and

$$\limsup_{x\to\infty} \frac{M(x)}{\sqrt{x}} = \infty.$$

He established also the same result for the related sum

$$L(x) = \sum_{n\le x} \lambda(n),$$

where $\lambda(n) = (-1)^{\Omega(n)}$ is the *Liouville function*. This showed that the conjecture of G. Pólya, who in 1919 [4959] predicted the truth of the inequality

$$L(x) \le 0$$

for all $x \ge 2$, contradicts the assumed properties of $\zeta(s)$. Pólya's conjecture was checked by H. Gupta[119] [2387] up to 20 000, but it was disproved in 1958 by C.B. Haselgrove [2574]. Pólya observed also that the equality $L(n) = 0$ is related to discriminants of binary quadratic forms with class-number one.

The oscillations of $M(x)$ were studied by S. Knapowski[120] [3379, 3380], who showed under the Riemann Hypothesis that $M(x)$ assumes both positive and negative values satisfying

$$|M(x)| > x^{1/2} \exp\left(\frac{-15 \log x \log\log\log x}{\log\log x}\right).$$

This was improved by I. Kátai [3268]. See also J. Pintz [4898].

The fact that Sterneck's inequality (2.29) does not persist indefinitely had been noticed in 1963 by G. Neubauer [4574] who showed that for a certain $x \le 8 \cdot 10^9$ one has $M(x) > 0.53\sqrt{x}$, and W.B. Jurkat [3165] obtained

$$\liminf_{x\to\infty} \frac{M(x)}{\sqrt{x}} < -0.505.$$

In 1981 R.J. Anderson and H.M. Stark [73] established

$$\limsup_{x\to\infty} \frac{M(x)}{\sqrt{x}} > 0.557$$

(cf. J. Pintz [4897, 4899]), and a few years later Mertens conjecture was disproved by A.M. Odlyzko and H.J.J. te Riele [4657], who showed that infinitely often the ratio $M(x)/\sqrt{x}$ attains positive and negative values lying outside the interval $[-1, 1]$ (cf. H.J.J. te Riele [6115]). Two years later J. Pintz [4902] managed to show that this happens for some $x \le \exp(3.21 \cdot 10^{64})$ and in 2006 T. Kotnik and H.J.J. te Riele [3498] replaced this bound by the still exorbitant $x \le \exp(1.59 \cdot 10^{40})$.

[119] Hansraj Gupta (1902–1988), professor at Panjab University. See [1059, 1651].

[120] Stanisław Knapowski (1931–1967), docent in Poznań and professor in Miami. See [6226].

3. The Möbius function plays a decisive role in the distribution of square-free integers, because the number $M_2(x)$ of square-free integers below x equals $\sum_{n\le x}\mu^2(n)$.

In the 19th century it had already been shown (see, e.g., L. Gegenbauer [2218]) that one has

$$M_2(x)=\frac{6}{\pi^2}x+R(x)$$

with $R(x)=O(\sqrt{x})$. In 1905 E. Landau [3626] improved this to

$$R(x)=O(\sqrt{x}/\log\log x),$$

and the results of A. Axer[121] [181] show that the Riemann Hypothesis implies $R(x)=O(x^\lambda)$ for every $\lambda>\lambda_0=2/5$.

More generally, if $M_k(x)$ denotes the number of k-free integers below x, then

$$M_k(x)=\frac{x}{\zeta(k)}+O(x^{1/k}) \tag{2.31}$$

as was known already to L. Gegenbauer [2219], and it was shown in 1929 by C.J.A. Evelyn and E.H. Linfoot [1914–1918, I] that the error term in (2.31) can be improved to

$$O\big(x^{1/k}\exp\big(-c\sqrt{\log x\log\log x}\big)\big)$$

with certain $c>0$. They proved also [1914–1918, IV] that for any $a<1/2k$ the error term here is $\Omega(x^a)$.

Much later A. Walfisz[122] [6535] reduced the error term in (2.31) to

$$O\big(x^{1/k}\exp\big(-a_k\log^{3/5}x(\log\log x)^{-1/5}\big)\big)$$

with certain $a_k>0$.

Assuming the Riemann Hypothesis, the number λ_0 in Axer's result has been consecutively reduced to $9/28=0.3214\ldots$ (H.L. Montgomery, R.C. Vaughan [4367]), $8/25=0.32$ (S.W. Graham [2302]), $7/22=0.3181\ldots$ (R.C. Baker, J. Pintz [269], S.W. Graham [2303], C.H. Jia [3130]) and $17/54=0.3148\ldots$ (C.H. Jia [3134]). For larger k the best known bounds under the Riemann Hypothesis were obtained by S.W. Graham and J. Pintz [2306]: for sufficiently large k one has $O(x^{d_k})$ with $d_k=1/(k+bk^{1/3})$ with constant b.

On the other hand the error term in (2.31) is $\Omega_\pm(x^{1/2k})$ (H.M. Stark [5896]), hence in particular $R(x)=\Omega(x^{1/4})$. See also R. Balasubramanian, K. Ramachandra [283, 284], B. Saffari [5368].

Sign changes of $R(x)$ have been studied by R. Balasubramanian and K. Ramachandra [284] and A. Sankaranarayanan [5391].

2.2.6 *Ramanujan*

1. S. Ramanujan was a self-educated person with an extraordinary talent. After discovering several amazing identities involving integrals, continued fractions and

[121] Alexander Axer (1880–1948), teacher in Zürich.

[122] Arnold Walfisz (1892–1962), worked first in an insurance company in Warsaw, becoming later professor in Tbilisi. Co-founder of *Acta Arithmetica*. See [3986].

arithmetical functions, he wrote in 1913 to G.H. Hardy asking him for advice. Ramanujan's formulas clearly impressed Hardy who invited the young man to Cambridge, initiating in this way a fruitful collaboration which ended abruptly when Ramanujan's health deteriorated and he died in 1920 shortly after returning to India.

2. In a letter, sent in February 1913 to G.H. Hardy (see [5088]) S. Ramanujan formulated certain formulas, which are particular cases of the identities

$$1+\sum_{n=1}^{\infty}\frac{x^{n^2}}{\prod_{j=1}^{n}(1-x^j)}=\prod_{n=0}^{\infty}\frac{1}{(1-x^{5n+1})(1-x^{5n+4})},$$

and

$$1+\sum_{n=1}^{\infty}\frac{x^{n(n+1)}}{\prod_{j=1}^{n}(1-x^j)}=\prod_{n=0}^{\infty}\frac{1}{(1-x^{5n+2})(1-x^{5n+3})}.$$

His proof appeared in [5082].

Actually these identities were found earlier by L.J. Rogers [5259, 5260] and therefore they are now called the *Rogers–Ramanujan identities*.

Later I. Schur [5573] noted that these identities conceal information on partitions. If we define the *difference of the partition* $n = a_1 + \cdots + a_s$ as the minimum of $a_i - a_j$ for $i \neq j$, then the Rogers–Ramanujan identities show that the number of partitions of n with difference ≥ 2 is equal to the number of partitions of n into parts congruent to 1 or 4 mod 5, and the number of partitions of n with difference equal to 2 is equal to the number of partitions of n into parts congruent to 2 or 3 mod 5.

A combinatorial construction of the bijections leading to a proof of Schur's result was given by A.M. Garsia and S.C. Milne [2198, 2199] in 1981. Their method turned out to be applicable also to other similar problems. A survey was given in 1986 by P. Paule [4760]. For later results see, e.g., K. Alladi, B. Gordon [54], C. Boule, I. Pak [657], D. Stockhofe [5954], J.A. Sellers, A.V. Sills, G.L. Mullin [5631].

For other proofs of the Rogers–Ramanujan identities see, e.g., G.E. Andrews [83], G.E. Andrews, R.J. Baxter [85], D.M. Bressoud [714], D.M. Bressoud, D. Zeilberger [715], F.J. Dyson [1674], A. Selberg [5605].

In 1961 B. Gordon [2282] obtained an extension of the Rogers–Ramanujan identities extending Schur's interpretation to other moduli. Several other generalizations of these identities were obtained much later by H.L. Alder [44], G.E. Andrews (see [78, 79]), W.N. Bailey [218], W. Connor [1198], B. Gordon [2283] and L.J. Slater [5819, 5820]. A survey was presented in 1969 by H.L. Alder [45] (see also Chap. 7 in Andrews' book [80]). In 2006 a survey of bijections between various classes of partitions was given by I. Pak [4723].

A problem related to the Rogers–Ramanujan identities was proposed in 1979 by G.E. Andrews [81], who asked for the determination of all pairs of sequences $\{a_n\}$, $\{b_n\}$ of positive integers satisfying the equality

$$\prod_{i=1}^{\infty}(1-x^{a_i})^{-1}=1+\sum_{n=1}^{\infty}\frac{x^{b_n}}{\prod_{j=1}^{n}(1-x^j)}$$

(the *Ramanujan pairs*). Several such pairs were found later (see D. Acreman, J.H. Loxton [12], M.D. Hirschhorn [2824], R. Blecksmith, J. Brillhart, I. Gerst [553]).

3. The first important paper by S. Ramanujan appeared[123] in 1915 [5076]. He considered there the sequence of *highly composite numbers*, defined as positive integers having more divisors than all smaller integers, and established the lower bound

$$Q(x) \gg \frac{\log x(\log\log x)^{1/2}}{(\log\log\log x)^{3/2}}$$

for the number $Q(x)$ of such numbers $\leq x$.

Much later P. Erdős [1793] established

$$\log^a x \ll Q(x) \ll \exp\bigl(b(\log\log x)^2\bigr)$$

with certain $b > 0$ and $a > 1$. J.-L. Nicolas proved $1.136 \leq a \leq 1.71$ [4594, 4595] and presented a survey [4596] in 1988.

In 1943 S.S. Pillai [4870] introduced the analogue of highly composite numbers, called *highly abundant numbers*, replacing $d(n)$ by $\sigma(n)$ in their definition. This notion was studied in 1944 by L. Alaoglu[124] and P. Erdős [41]. A variant, in which $\sigma(n)$ is replaced by $\sigma(n)/n$ leads to *superabundant numbers* (see [41], P. Erdős, J.-L. Nicolas [1845], J.-L. Nicolas [4593]).

4. In Ramanujan's paper [5078], devoted to the study of the sum

$$\Sigma_{r,s}(n) = \sum_{j=0}^{n} \sigma_r(j)\sigma_s(n-j)$$

(with $\sigma_k(0) = -\zeta(-k)/2$), the function $\tau(n)$ appeared, which is defined by

$$\Delta(q) = q\prod_{n=1}^{\infty}(1-q^n)^{24} = \sum_{n=1}^{\infty}\tau(n)q^n, \tag{2.32}$$

$\Delta(\exp(2\pi i z))$ being the classical discriminant function in the theory of elliptic functions. The function $\tau(n)$ is called the *Ramanujan function*, although it had been considered already by Jacobi, and the formula (2.32) is called *Jacobi's formula*. There are several proofs of (2.32), see, e.g., the books [3427, 5641, 5698]. The function $\tau(n)$ appears in [5078] in the formula for the number of representations of an integer as the sum of 24 squares. S. Ramanujan computed its values up to $n = 30$, and stated several conjectures. He asserted that $\tau(n)$ is multiplicative and for prime p and $n = 2, 3, \ldots$ one has

$$\tau(p^n) = \tau(p)\tau(p^{n-1}) - p^{11}\tau(p^{n-2}),$$

thus

$$\Phi_\tau(s) = \sum_{n=1}^{\infty}\frac{\tau(n)}{n^s} = \prod_p \frac{1}{1-\tau(p)p^{-s}+p^{11-2s}} \tag{2.33}$$

holds for sufficiently large $\Re s$.

[123] For some reason only a part of Ramanujan's manuscript was printed, and the missing sections appeared in print as late as 1997 [5087].

[124] Leonidas Alaoglu (1914–1981), worked at the Lockheed Corporation.

Both of these assertions were proved[125] by L.J. Mordell [4376] in 1917.

Moreover S. Ramanujan established the bound $\tau(n) = O(n^7)$, showed $\tau(n) = \Omega(n^5)$, stated the inequality

$$|\tau(p)| \le 2p^{11/2} \tag{2.34}$$

for prime p and asserted that for infinitely many n the inequality

$$|\tau(n)| \ge n^{11/2} \tag{2.35}$$

holds.

In 1929 J.R. Wilton[126] [6681] established a functional equation for the function $\Phi_\tau(s)$ and showed that it has infinitely many zeros on the line $\Re s = 6$. The last result was later generalized by E. Hecke [2699] to Dirichlet series associated with a class of modular forms. Over 12 000 smallest zeros $\Phi_\tau(s)$ have been shown to be simple and to lie on the line $\Re s = 6$ by J.B. Keiper [3294].

In 1927 G.H. Hardy [2514] showed $\tau(n) = O(n^6)$ and this bound was reduced in the same year to $O(n^{47/8})$ by H.D. Kloosterman [3371]. In 1933 Kloosterman's bound was improved to $O(n^{35/6})$ by H. Davenport [1343] and H. Salié [5379], and in 1939 R.A. Rankin [5111] replaced it by $O(n^{29/5})$. The bounds for exponential sums established by A. Weil in [6617] imply $O(n^c)$ for every $c > 23/4$ (R.A. Rankin [5114]). In 1967 Y. Ihara [3006] (cf. Y. Morita [4431]) deduced (2.34) from Weil conjectures, concerning relations between modular forms and representations of the Galois group of $\overline{\mathbf{Q}}/\mathbf{Q}$, which were established in 1974 by P. Deligne [1447, 1448]. Deligne's result implied, more generally, the bound

$$|a_p| \le 2p^{(k-1)/2}$$

for the coefficients of the Fourier expansion

$$f(q) = q + \sum_{n=1}^{\infty} a_n q^n \quad (q = e^{2\pi i z})$$

of any normalized primitive cusp form f of weight ≥ 2 and any level, and this led to

$$|a_n| \le d(n) n^{(k-1)/2}.$$

For an exposition see J.-P. Serre [5649]. A table of values of $\tau(n)$ for prime $n < 300$ was prepared by D.H. Lehmer [3783] in 1943, H. Gupta [2390] gave values of $\tau(n)$ for $n < 400$, and G.N. Watson [6592] extended it up to 1000.

On the other hand G.H. Hardy [2514] established

$$x^{12} \ll \sum_{n \le x} \tau^2(n) \ll x^{12}, \tag{2.36}$$

and this implies that for infinitely many n one has

$$\tau(n) \ge c n^{11/2} \tag{2.37}$$

with a certain $c > 0$ which is close to (2.35).

---

[125] R. Fueter wrote in his review of Mordell's paper in the *Jahrbuch* that the multiplicativity of τ was also established by J.W.L. Glaisher. I was unable to confirm this.

[126] John Raymond Wilton (1884–1944), professor in Adelaide. See [923].

In 1939 R.A. Rankin [5111] proved asymptotics for the sum in (2.36), and several years later improved the bound (2.37) by showing that the ratio $|\tau(n)|/n^{11/2}$ is unbounded [5123]. It was conjectured that

$$\tau(n) = \Omega\left(n^{11/2} \exp\left(\frac{c_1 \log n}{\log\log n}\right)\right) \tag{2.38}$$

holds with certain positive c_1. H. Joris [3161] obtained in 1975

$$\tau(n) = \Omega\big(n^{11/2} \exp(c_2 \log^a n)\big)$$

for every $a < 1/22$, and ten years later R. Balasubramanian and M.R. Murty [278] showed this for every $a < 2/3$ (cf. M.R. Murty [4483]). Finally (2.38) was established in 1983 by M.R. Murty [4484], who also obtained its analogue for the coefficients of cusp forms of weight k, with the exponent $11/2$ being replaced by $(k-1)/2$.

In 1972 H. Joris [3160] established

$$A(x) = \sum_{n \le x} \tau(n) = \Omega_{\pm}\big(x^{23/4} \log\log\log x\big),$$

and this was improved to

$$A(x) = \Omega_{\pm}\big(x^{23/4} \exp\big(c \log\log^{1/4} x \log\log\log^{-3/4} x\big)\big)$$

by J.L. Hafner and A. Ivić [2433] in 1989 as a particular case of their bound for the sum of coefficients for a class of cusp forms.

It was conjectured by A.O.L. Atkin and J.-P. Serre (see [5650]) that for prime p one has

$$|\tau(p)| \gg p^{\alpha}$$

for every $\alpha < 9/2$, with the implied constant depending on a. In this direction M.R. Murty, V.K. Murty and T.N. Shorey showed in 1987 [4491] that if $\tau(n)$ is odd (which happens if and only if n is an odd square, as shown in 1943 by H. Gupta [2389]), then

$$|\tau(n)| \ge \log^c n,$$

holds with an absolute constant $c > 0$.

The Atkin–Serre conjecture implies the non-vanishing of $\tau(p)$. This was earlier conjectured by D.H. Lehmer [3783], who showed in 1947 [3785] that $\tau(n)$ is non-zero for $n \le 3\,316\,798$ and later extended this to $n \le 21\,492\,639\,999$ (unpublished, see Lehmer's review of [2391] in Math. Reviews, 10, p. 514). Much later this was extended to $n \le 10^{15}$ (see J.-P. Serre [5654]) and $n \le 2 \cdot 10^{19}$ (J. Bosman [655]).

It is also not known whether there exist infinitely many primes p with $p|\tau(p)$. Only six such primes are known: $2, 3, 5, 7, 2411$ and $7\,758\,337\,633$, the last found in 2010 by N. Lygeros and O. Rozier [4036].

Hardy's conjecture [2516]

$$\sum_{p \le x} \tau(p) \log p = O\big(x^{13/2}\big)$$

was established by R.A. Rankin [5110–5112] in 1939 in the stronger form

$$\sum_{p \le x} \tau(p) \log p = o\big(x^{13/2}\big),$$

and in 1972 C.J. Moreno [4423] improved this to

$$\sum_{p \le x} \tau(p) \log p = O\big(x^{13/2} \exp(-A\sqrt{\log x})\big)$$

for some $A > 0$.

It is not known whether $\tau(n)$ assumes infinitely many prime values. The smallest n with prime $\tau(n)$ is $n = 63\,001$, as shown by D.H. Lehmer [3790] in 1965.

It was conjectured by J.-P. Serre in [5639] that if one writes

$$\tau(p) = 2p^{11/2}\cos(\theta_p)$$

with $0 \le \theta_p \le \pi$, then the sequence $\{\theta_p\}$ is uniformly distributed with respect to the measure

$$\frac{2}{\pi}\sin^2 t\,dt.$$

This conjecture, which resembles the *Sato–Tate conjecture* for elliptic curves, formulated by J. Tate [6060, 6061] in 1965 (see Sect. 6.7), is still open (cf. P. Solé [5845]).

A survey covering, more generally, evaluations of coefficients of arbitrary modular forms was given by R.A. Rankin [5128] in 1986. Another survey of problems around Ramanujan's τ-function was prepared by M.R. Murty [4487].

At the end of the short note [5083], S. Ramanujan asserted the following divisibility properties of $\tau(n)$: $5|\tau(5n)$, $7|\tau(n)$ if $\left(\frac{n}{7}\right) \ne 1$, and $23|\tau(n)$ if $\left(\frac{n}{23}\right) = -1$. Several other congruences satisfied by the values of $\tau(n)$ modulo certain powers of the primes $2, 3, 5, 7, 23$ and 691 appear in Ramanujan's manuscript, published in 1999 by B.C. Berndt and K. Ono [455] (cf. R.A. Rankin [5124]). One of them,

$$\tau(n) \equiv \sigma_{11}(n) \pmod{691},$$

appears there with a proof (the first published proof of this congruence is due to Wilton [6682]).

The conjectured congruences for $\tau(n)$ were later proved by M.H. Ashworth [152], R.P. Bambah [307], O. Kolberg[127] [3449, 3450], D.B. Lahiri [3612] and J.R. Wilton [6683], and a uniform proof was given by J.-P. Serre and H.P.F. Swinnerton-Dyer [6002, 6003]. (See also [5638, 5644].)

Since

$$\sum_{n=1}^{\infty} \tau(n) e^{2\pi i n z}$$

is the Fourier expansion of the unique normalized cusp form of weight 12, the question could be asked whether similar congruences hold also for Fourier coefficients of other cusp forms. In the case, when the space of cusp forms of a fixed weight k is of dimension one, which happens for $k = 12, 16, 18, 20, 22$ and 26. S. Ramanujan stated certain congruences in an unpublished manuscript, whose contents were described by R.A. Rankin in [5124].

In these cases all congruences mod p for prime p have been listed by J.-P. Serre and H.P.F. Swinner ton-Dyer [6002], with one possible exception, which concerned the unique normalized cusp form of weight 16 in the case $p = 59$. This case was settled later by K. Haberland [2418–2420] in 1983. Congruences modulo odd prime powers were considered in [6003] (see also B. Gordon [2284], E. Papier [4739]). The more general question,

[127] The congruence for $\tau(n)$ mod 49 proved in [3450] is attributed in [6002, 6003] to some unpublished notes of D.H. Lehmer.

regarding congruences modulo prime powers between coefficients of modular forms was answered by N.M. Katz [3278] (see also F. Diamond [1517], K.A. Ribet [5187]).

An explicit formula for $\tau(n)$ was found in 1984 by J.A. Ewell [1943]:

$$\tau(n) = \sum_{k=1}^{n} (-1)^{n-k} r_{16}(n-k) 2^{3e(k)} \sigma_3(o(k)),$$

where $o(k)$ denotes the maximal odd factor of k, $e(k) = k/o(k)$ and $r_{16}(k)$ is the number of representations of k as the sum of 16 squares.

5. The paper [5079] by S. Ramanujan settled the problem when a diagonal quaternary quadratic form

$$\sum_{j=1}^{4} a_j X_j^2 \tag{2.39}$$

with positive integral coefficients represents all positive integers. Particular cases of this question were considered earlier by J. Liouville, H.J.S. Smith and T. Pepin (for an account of these results see [1545, Chap. 10]). Ramanujan asserted that there are exactly 55 such forms, and listed all of them, but made an error[128], stating that the form $x^2 + 2y^2 + 5(z^2 + t^2)$ has the desired property, although it does not represent the number 15. He was particularly interested in the form $x^2 + y^2 + 10z^2$, noted that odd numbers which are not represented by it "do not seem to obey any simple law," and listed sixteen such numbers, the largest being 391.

Later two more such numbers (679 and 2719) were discovered (B.W. Jones, G. Pall [3152], H. Gupta [2388]), and it was shown in 1997 by K. Ono and K. Soundararajan [4685] under the General Riemann Hypothesis that 2719 is the largest number having this property.

For later results on representations of integers by quadratic forms see Sect. 3.2.2.

6. S. Ramanujan stated a wealth of formulas, most of which were later proved. He published some of them, but the majority was hidden in his notebooks [5085, 5086], analyzed later by G.E. Andrews and B.C. Berndt [86–88, 447, 449] (see also the proceedings of a conference devoted to the work of S. Ramanujan [84]).

Several formulas, related to the number $d(n)$ of divisors of n were presented in [5077]. Among them one finds the following asymptotic formula, which in the case $k = 1, l = 0$ was established by G.F. Voronoĭ [6469] (see Sect. 2.6.2):

$$\sum_{n \le x} d(kn + l) = a(k,l)x(\log x + 2\gamma - 1) + b(k,l)x + O\left(x^{1/3} \log x\right), \tag{2.40}$$

where k, l are fixed positive integers, and $a(k,l)$, $b(k,l)$ are appropriate coefficients of the Dirichlet series of certain explicitly given functions.

This formula with a weaker error term of order $O(\sqrt{x})$ later got an elementary proof by T. Estermann[129] [1877], and in 1932 A. Page [4716] established (2.40).

---

[128] This error was pointed out by L.E. Dickson [1546], who also gave a fresh proof of Ramanujan's result.

[129] Theodor Estermann (1902–1991), professor at University College, London. See [5315].

The paper [5077] also contains asymptotic expressions for the sums

$$\sum_{n\le x} d^a(n), \quad \sum_{n\le x} 1/d(n), \quad \sum_{n\le x} \sigma^2(n),$$

as well as closed formulas for various Dirichlet series, like

$$\sum_{n=1}^{\infty} \frac{\sigma_a(n)\sigma_b(n)}{n^s} = \frac{\zeta(s)\zeta(s-a)\zeta(s-b)\zeta(s-a-b)}{\zeta(2s-a-b)}.$$

These assertions were later proved by B.M. Wilson[130] [6677] and similar sums involving the divisor function were treated by A.E. Ingham [3016] in 1927, who obtained asymptotics for the sums

$$\sum_{m\le n} d(m)d(m+k) \quad (k\ne 0), \qquad \sum_{m<n} d(m)d(n-m).$$

The error terms were later improved by T. Estermann [1880–1882], R.A. Smith [5835], J.-M. Deshouillers, H. Iwaniec [1489, 1490], M. Ishibashi, S. Kanemitsu [3029], S. Kanemitsu [3232], Y. Motohashi [4460] and Y.-F.S. Pétermann, J. Wu [4801].

Not every assertion made by Ramanujan turned out to be true. Considering the number $B(x)$ of integers $\le x$ which are sums of two squares, he stated in an unpublished manuscript the formula

$$B(x) = c\int_1^x \frac{dt}{\log^{1/2} t} + O\left(\frac{x}{\log x}\right)^{1/2},$$

with a certain positive c and an arbitrary N. At that time only the evaluation

$$B(x) = (c+o(1))\frac{x}{\sqrt{\log x}}$$

with

$$c = \frac{1}{\sqrt{2}} \prod_{p\equiv 3 \pmod 4} \frac{1}{\sqrt{1-p^{-2}}}$$

was known (E. Landau [3633]).

In 1928 a paper by G.K. Stanley[131] [5888] appeared in which she claimed that Landau's evaluation cannot be improved, hence Ramanujan's formula fails for $N > 3/2$. She was right, but there were errors in her calculation, pointed out by D. Shanks [5674], who established

$$B(x) = \frac{cx}{\log^{1/2} x} + \frac{c_1 x}{\log^{3/2} x} + O\left(\frac{x}{\log^{5/2} x}\right),$$

with positive c_1, showing definitely the falsity of Ramanujan's statement.

A quick way of computing $B(x)$ was proposed by P. Shiu [5715] who adapted an old method of counting primes in large intervals developed by E. Meissel[132] in 1870 [4236], modified in 1959 by D.H. Lehmer [3789].

---

[130] Bertram Martin Wilson (1896–1935), professor in Dundee. See [2576].

[131] Gertrude Katherine Stanley (1898–1974), professor at Westfield College, London. See [5126].

[132] Daniel Friedrich Ernst Meissel (1826–1895), teacher in Kiel. See [4763].

For a discussion of other assertions by S. Ramanujan which turned out to be incorrect see P. Moree [4415].

The corresponding question for numbers which are sums of three squares was considered in 1908 by E. Landau [3633]. He showed that the number of such integers $\leq x$ equals

$$\frac{5}{6}x + \Delta(x),$$

with $\Delta(x) = O(\log x)$.

Much later, in 1988, the mean value of $\Delta(x)$ was determined by P. Shiu [5716]:

$$\sum_{n \leq x} \Delta(n) = \left(\frac{3}{16\log 2} + o(1)\right)x \log x.$$

7. The paper [2540] by G.H. Hardy and S. Ramanujan gave birth to probabilistic number theory. Roughly, its main result states that with probability one every natural number n has $\log\log n$ distinct prime divisors. More precisely, evaluating the number of $n \leq x$ with a given $\omega(n)$, they established the following assertion.

If $f(x)$ is a function tending to infinity, then for the number $N_f(x)$ of $n \leq x$ satisfying

$$\frac{|\omega(n) - \log\log n|}{\sqrt{\log\log n}} < f(n)$$

one has $N_f(x) = o(x)$.

Later a much simpler proof based on the inequality

$$\sum_{n \leq x} (\omega(n) - \log\log n)^2 = O(x \log\log x)$$

was found by P. Turán [6212]. See Sect. 5.5 for further results.

8. In 1916 G.H. Hardy and S. Ramanujan [2539] proved a Tauberian theorem[133], which in the case of the usual Dirichlet series takes the following form

If the series $f(s) = \sum_{n=1}^{\infty} a_n n^{-s}$ has positive coefficients, converges for $s > 0$ and for $s \to 0$ satisfies

$$\log f(s) = (A + o(1))s^{-\alpha}\left(\log\frac{1}{s}\right)^{-\beta},$$

then for x tending to infinity one has

$$\log \sum_{n \leq x} a_n = (B + o(1)) \log^{\alpha/(1+\alpha)} x (\log\log x)^{-\beta/(1+\alpha)},$$

with B depending explicitly on A, α, β.

[133] A monograph devoted to Tauberian theorems has been written by H.R. Pitt [4918]. For their history see the recent book by J. Korevaar [3477].

They used it to obtain an asymptotical formula for the number of integers $n \le x$ having the form

$$n = 2^{\alpha_2} 3^{\alpha_3} \cdots p^{\alpha_p}$$

with $\alpha_2 \ge \alpha_3 \ge \cdots \ge \alpha_p$, considered earlier by S. Ramanujan in [5076], and applied it also to a study of the *partition function* $p(n)$, counting all representations of a positive integer n as sums of positive integers. This function had been considered already by Euler [1899], who proved the identity

$$1 + \sum_{n=1}^{\infty} p(n)x^n = \prod_{k=1}^{\infty} (1 - x^k)^{-1} \quad (|x| < 1). \tag{2.41}$$

At Hardy's request all values of $p(n)$ for $n \le 200$ were calculated by P.A. MacMahon[134] (see [2541, pp. 114–115]). This function grows rapidly, its value at $n = 200$ being 3 972 999 029 388.

Later H. Gupta [2385, 2386] extended MacMahon's table up to $n = 600$.

The Tauberian theorem of G.H. Hardy and S. Ramanujan led to the equality

$$\log p(n) = (c + o(1))\sqrt{n}$$

with $c = \pi\sqrt{2/3}$, and in the next year they observed [2541] that a more precise result can be obtained with the use of Cauchy's integral theorem.

It soon turned out that the same approach can be used to solve various additive problems, and so the *circle method* was born. They considered the sum $f(z)$ of the power series occurring in (2.41), and noted that if Γ is a path lying inside the unit circle, and encompassing the origin, then one has

$$p(n) = \frac{1}{2\pi i} \oint_{\Gamma} \frac{f(z)}{z^{n+1}} dz.$$

One reads in their paper:

> "The idea which dominates this paper is that of obtaining asymptotic formulae for $p(n)$ by a detailed study of the integral .... This idea is an extremely obvious one; it is the idea which has dominated nine-tenths of modern research in the analytic theory of numbers: and it may seem very strange that it should never have been applied to this particular problem before."

They explained this situation by "the extreme complexity of the behavior of the generating function $f(x)$ near a point of the unit circle." In fact, they noted that every point of the unit circle is an essential singularity for f.

Taking for Γ the circle $z = Re^{2\pi i\theta}$ $(0 \le \theta < 1,\ 0 < R < 1)$ they dissected Γ into arcs $\gamma_{p,q}$, corresponding to θ lying in a small interval centered at the rational number $p/q \in [0, 1)$ with co-prime p, q, belonging to the *Farey series* of order N, which consists of all rational numbers from the unit interval, having denominator $\le N$, and

---

[134] Percy Alexander MacMahon (1854–1929), worked at military schools. He wrote a large monograph on combinatorics [4055]. See [246].

arranged according to their value. They also introduced an auxiliary function $F_t(z)$, defined for positive t by

$$F_t(z) = \sum_{n=1}^{\infty} \psi_t(n) z^n,$$

where

$$\psi_t(x) = \frac{d}{dx} \frac{\cosh(t\sqrt{x - 1/24}) - 1}{\sqrt{x - 1/24}}.$$

Putting

$$\Psi(z) = \frac{1}{\pi\sqrt{2}} \sum_{q \le a\sqrt{n}} \sqrt{q} \sum_{\substack{p<q \\ (p,q)=1}} \omega_{p,q} F_{C/q}\left(z \exp\left(\frac{-2p\pi i}{q}\right)\right) \tag{2.42}$$

(where a is a suitable positive constant, $C = \pi\sqrt{2/3}$, and the numbers $\omega_{p,q}$ are certain 24th roots of unity), and

$$\Phi(z) = f(z) - \Psi(z)$$

they succeeded in showing that on the one hand all integrals

$$\frac{1}{2\pi i} \int_{\gamma_{p,q}} \frac{\Phi(z)}{z^{n+1}} dz$$

are small, and on the other hand the integral

$$\frac{1}{2\pi i} \int_{\gamma_{p,q}} \frac{\Psi(z)}{z^{n+1}} dz$$

can be swiftly evaluated. This led to an explicit expression of $p(n)$ as a finite sum of $\sqrt{n}$ terms with an error of the order $O(n^{-1/4})$, implying in particular the equality

$$p(n) = \left(\frac{1}{4\sqrt{3}} + o(1)\right) \frac{\exp(\pi\sqrt{2n/3})}{n}.$$

A simpler proof of the last result was given in 1920 by J.V. Uspensky[135] [6254]. It is reproduced in the book [184] by R. Ayoub.

An elementary proof of this formula (but without identifying the leading constant) was provided by P. Erdős [1792] in 1942.

In 1937 H. Rademacher [5036, 5037] observed that a modification of the Hardy–Ramanujan approach leads to an infinite series converging rapidly to $p(n)$, whose tail may be effectively estimated. The first 18 terms evaluated at $n = 599$ approximate $p(599)$ with an error smaller than $1/2$, leading to

$$p(599) = 435\,350\,207\,840\,317\,348\,270\,000,$$

justifying a conjecture by D.H. Lehmer [3780]. An analogue of his result for the number of partitions into unequal parts was obtained in 1942 by L.K. Hua [2930].

---

[135] Jacob Victorovič Uspensky (1883–1947), professor in St. Petersburg and Stanford.

For the theory of partitions see the books [80, 82] by G.E. Andrews.

The roots of unity $\omega_{p,q}$ occurring in (2.42) are made explicit by the formula

$$\omega_{p,q} = \exp(s(p,q)),$$

where

$$s(p,q) = \pi i \sum_{j=1}^{q} \frac{j}{q}\left(\left\{\frac{pj}{q}\right\} - \frac{1}{2}\right) \tag{2.43}$$

(see H. Rademacher [5032]). The sums $s(p,q)$ were introduced by R. Dedekind in his comments to Riemann [1420], and are called *Dedekind sums*. They were later studied by H. Rademacher [5031] and H. Rademacher and A. Whiteman [5042].

Later several arithmetical applications of Dedekind sums were given, and there were even applications outside number theory (see the book [2830] by F. Hirzebruch and D. Zagier). A book on Dedekind sums was written by H. Rademacher and E. Grosswald [5041] in 1972, and generalizations were given by T.M. Apostol [111], L.J. Goldstein [2267], G.J. Rieger [5221], K.H. Rosen, W.M. Snyder [5287], and D. Zagier [6808].

The method applied by G.H. Hardy and S. Ramanujan [2541] to study the partition function $p(n)$ was applied to related problems by E.M. Wright[136] in three papers [6739–6741]. In particular, he obtained asymptotics for the number of partitions of an integer into kth powers, improving upon an earlier result by G.H. Hardy and S. Ramanujan [2541].

9. The *Ramanujan sums*, defined by

$$c_r(n) = \sum_{\substack{1\le a<r \\ (a,r)=1}} \exp\left(\frac{2\pi i a n}{r}\right),$$

were used by Ramanujan[137] in [5080] to expand arithmetical functions f in a series of the form

$$f(n) = \sum_{r=1}^{\infty} a_r c_r(n).$$

He proved several interesting examples of these expansions, such as

$$\frac{\sigma(n)}{n} = \frac{\pi^2}{6}\sum_{r=1}^{\infty}\frac{c_r(n)}{r^2},$$

and

$$d(n) = -\sum_{r=1}^{\infty}\frac{\log r}{r}c_r(n),$$

and obtained similar formulas for the number of representations of an integer n as the sum of an even number of squares studied earlier by him in [5078] (for an exposition see [2516, Chap. 9]).

---

[136] Edward Maitland Wright (1906–2005), professor in Aberdeen. See [4761].

[137] It has been pointed out in [2545] that these sums already appear in Landau's book [3636].

Simpler proofs of Ramanujan's results were given in 1921 by G.H. Hardy [2513].

It was shown by R.D. Carmichael [922] in 1932 that Ramanujan sums satisfy

$$\lim_{N\to\infty}\frac{1}{N}\sum_{n\le N} c_r(n)c_s(n) = \begin{cases}\varphi(r) & \text{if } r=s,\\ 0 & \text{otherwise,}\end{cases}$$

and hence can be regarded in some sense as an orthogonal system. The theory of Ramanujan expansions was later systematically developed by several authors (see, e.g., A. Hildebrand [2796], W. Schwarz [5581], L.G. Lucht [4031]). For connections with the theory of almost periodic functions see M. Kac[138], E.R. van Kampen, A. Wintner[139] [3178][140]. A survey with ample bibliography was given in 1988 by W. Schwarz [5582].

10. A few years later S. Ramanujan [5081, 5084[141]], considered again the partition function, and proved that for $(a,b)\in\{(5,4),(7,5),(11,6)\}$ it satisfies the congruence

$$p(an+b)\equiv 0 \pmod{a}. \tag{2.44}$$

He conjectured also that there exist similar congruences with the modulus being a power of 5, 7 or 11. More precisely, he asserted that if q is one of these three primes then for $n\ge 1$ the congruence $24m\equiv 1 \pmod{q^n}$ implies the divisibility of $p(m)$ by q^n. In an unpublished manuscript (see [5124]) he considered also the residues of $p(n)$ modulo powers of 13.

Ramanujan's assertion was not quite accurate. This was noted by S. Chowla [1072] who pointed out (using a table of values of $p(n)$ prepared by H. Gupta [2385, 2386]) that it fails for the modulus 7^3. It turned out, however, that a modified form of this assertion is true. This was shown for $q=5,7$ by G.N. Watson in 1938 [6591] (for some special cases see H.B.C. Darling [1328, 1329], V. Krečmar [3521]) and for $q=11,13$ by J. Lehner [3808] and A.O.L. Atkin [161]. For related results see A.O.L. Atkin [163], F.G. Garvan [2200], B. Gordon [2284], K. Hughes [2946].

Some of Ramanujan's congruences have a combinatorial interpretation discovered by F.J. Dyson [1675] and proved by A.O.L. Atkin and H.P.F. Swinnerton-Dyer [167]. For further developments see F.G. Garvan [2201, 2202], G.E. Andrews, F.G. Garvan [91], F.G. Garvan, D. Kim, D. Stanton [2203].

In 1992 I. Kiming and J.B. Olsøn [3336] showed that if (2.44) holds for a prime a, then $24b\equiv 1 \pmod{a}$, and in 2003 S. Ahlgren and M. Boylan [28] showed that in this case the only congruences of the form (2.44) are those found by Ramanujan.

Other congruences for $p(an+b)$ were found later by M. Newman[142], S. Ahlgren, G.E. Andrews, A.O.L. Atkin, J.N. O'Brien, D. Eichhorn, K. Mahlburg, K. Ono and R.L. Weaver [27, 30, 92, 166, 1698, 4059, 4585, 4683, 6596].

---

[138] Mark Kac (1914–1984), professor at Cornell University and Rockefeller University. See [3310].

[139] Aurel Friedrich Wintner (1903–1958), professor in Baltimore. See [2573].

[140] Essentially the same paper, with a different proof of one of the lemmas and an author (P. Erdős) added, appeared in 1940 [1836].

[141] This paper was published posthumously by G.H. Hardy.

[142] Morris Newman (1924–2007), worked until 1977 at the National Bureau of Standards, later became professor at the University of California in Santa Barbara. See [2256].

It was conjectured by P. Erdős that for every prime p there is an integer n such that p divides $p(n)$, and in 1989 A. Schinzel and E. Wirsing [5454] showed that the number of primes dividing at least one of the numbers $p(n)$ with $n \le x$ is $\gg \log x$. K. Ono [4683] established Erdős's conjecture for primes $p \ge 5$, and showed that in this case the set of such numbers n has a positive lower density. It is still unknown whether $p(n)$ is divisible by 3 for infinitely many n.

Still open is a stronger conjecture, posed in 1960 by M. Newman [4587], asserting that every arithmetic progression contains values of the partition function. A sufficient condition for its validity was given by K. Ono [4683], implying its correctness for all primes moduli $3 < p < 1000$ (for certain small values of p this was shown earlier by A.O.L. Atkin [162], T. Kløve [3375, 3376], O. Kolberg [3448] and M. Newman [4587]). This was extended to $3 < p < 2 \cdot 10^5$ by J.H. Bruinier and K. Ono [794], and in 2005 S. Ahlgren and M. Boylan [29] established Newman's conjecture for prime powers moduli $\neq 3^n$.

These results were obtained with the use of the theory of modular forms, which is not surprising, as already S. Ramanujan in his *Lost Notebook* [5086] gave congruences relating the generating functions of partitions to modular forms, e.g.,

$$\sum_{n=0}^{\infty} p(13n+6)q^n \equiv p(6) \prod_{n=1}^{\infty} (1-q^n)^{11} \pmod{13}.$$

Similar congruences modulo arbitrary prime powers were later given by S. Ahlgren and M. Boylan [28].

In 1966 M.V. Subbarao[143] [5983] conjectured that every arithmetical progression $Ax+B$ contains infinitely many integers n with $p(n)$ even and this was established in 1996 by K. Ono [4681]. M.V. Subbarao conjectured that the same holds also for odd values of $p(n)$. See also K. Ono [4682].

In 2000 J. Getz, a high school student, proved the truth of Subbarao's conjecture for infinitely many progressions [2236]. Both conjectures were established earlier for several small values of A (F.G. Garvan, D. Stanton [2204], M.D. Hirschhorn [2826], M.D. Hirschhorn, M.V. Subbarao [2828]).

11. The paper [5084] by Ramanujan contained only part of the results which were described in the manuscript of 43 pages found in his papers. In 1928 the collection of Ramanujan's notebooks was given by G.H. Hardy to G.N. Watson, who in 1935 presented in [6590] complete proofs of three theorems contained in that manuscript. The first deals with positive integers n having the property that for primes lying in a fixed residue class mod k and dividing n their maximal power dividing n is a square. If $B(x)$ denotes the number of such $n \le x$, then one has

$$B(x) = (c(k) + o(1)) \frac{x}{\log^{1-1/\varphi(k)} x}$$

with some positive $c(k)$. The second asserts that for every odd k, every d and almost all n, the number $\sigma_k(n)$ is divisible by d, more precisely,

$$A_{k,d} = \#\{n \le x : d \nmid \sigma_k(n)\} \ll \frac{x}{\log^{1/\varphi(k)} x}.$$

[143] Matukumalli Venkata Subbarao (1921–2006), professor at the University of Alberta. See [3273].

Much later this was made more precise by R.A. Rankin [5118] and E.J. Scourfield [5591] (cf. [4535]).

The third assertion states that $\tau(n)$ is divisible by 691 for almost all n.

Similar results were established later for larger moduli (R.P. Bambah and S. Chowla [309, 1088]), and it was shown by J.-P. Serre [5646, 5647, 5654] that for every integer k and almost all integers n one has $k|\tau(n)$, and the same applies to coefficients of modular forms of integral weight.

12. In [5075] S. Ramanujan formulated the following question: "$2^n - 7$ is a perfect square for the values $3, 4, 5, 7, 15$ of n. Find other values." It turned out much later that there are no other solutions[144]. This was proved by T. Nagell[145] in 1948 in a Norwegian paper, and an English translation appeared in 1961 [4517].

The same result was proved independently in 1956 by J. Browkin and A. Schinzel [749] and in 1959 by T. Skolem, S. Chowla and D.J. Lewis [5810]. The more general equation

$$x^2 + A = p^y, \tag{2.45}$$

where p is a fixed prime and A is a fixed integer not divisible by p, has been considered by R. Apéry[146] [109], who showed that for $p > 2$ and positive A there are at most two solutions, and the same holds for $p = 2$ and $15 \le A \equiv 7 \pmod 8$. For $p = 2$ and $A \not\equiv 0, 4, 7 \pmod 8$ there is at most one solution (J. Browkin, A. Schinzel [750]). Much later F. Beukers [485] confirmed the conjecture posed in [750] that for $p = 2$ and positive A the equation (2.45) has at most one solution, except when $A = 7, 23$ or $2^k - 1$ with a certain $k \ge 4$. M.H. Le [3746] showed that if p is odd and $\max\{D, p\}$ is sufficiently large, then two solutions exist only if for some a one has $p = 4a^2 + 1$ and $D = 3a^2 + 1$. For the case $A = 7$ see J.-L. Lesage [3838], S. Siksek, J.E. Cremona [5788]. In the case of A negative, F. Beukers [485] showed that for $p = 2$ one has at most 4 solutions, and in [486] he extended this to the case of odd p. Later M.H. Le [3742] showed that for large $|A|$ there are at most three solutions and determined in [3743, 3745] all cases for $p = 2$ with four solutions. Now it is known that for $p \ne 2$ this equation has at most three solutions (M.L. Bauer, M.A. Bennett [361]).

In paper [823] by Y. Bugeaud, M. Mignotte and S. Siksek all solutions of the equation $x^2 + D = y^n$ (with $n \ge 3$) were found for all $1 \le D \le 100$ for which this had not been done earlier, in particular for $D = 7$.

The equation

$$x^2 + y^m = 2^n$$

(with $(x, y) = 1$, $y \ge 2$, $n \ge 3$) was considered in 1995 by Y.-D. Guo and M.H. Le [2384], who showed that it has finitely many solutions and in all solutions one has $m \le 10^9$. Later Y. Bugeaud [808] reduced this bound to $m \le 7.3 \cdot 10^5$, and M.H. Le showed [3748] that for large y it does not have solutions with $m \ge 2$. Later M.A. Bennett and C.M. Skinner [419] showed that the only solutions with $m \ge 2$ are $(x, m, n, y) = (\pm 13, 3, 9, 7)$, and for odd negative $y \ne -1$ with $m \ge 3, n \ge 1$ one has $(x, m, n, y) = (\pm 71, 3, 7, 17)$.

---

[144] On p. 327 of [5088] it is noted that this question was answered by K.J. Sanjana and T.P. Trivedi in 1913, but actually they noticed only that there are no other small solutions.

[145] Trygve Nagell (1895–1988), docent in Oslo and professor in Uppsala. See [946].

[146] Roger Apéry (1916–1994), professor in Caen. See [108].

For the equation

$$x^2 \pm y^m = p^n$$

with odd prime p see Y. Bugeaud [809], A. Bérczes, I. Pink [427].

It was shown by M.A. Bennett, M. Filaseta and O. Trifonov [418] that if with positive m, n one has

$$x^2 + 7 = 2^n m,$$

then either $x < m^2$ or $m \in \{1, 3, 5, 11, 181\}$.

For a survey of the developments up to 1992 see J.H.E. Cohn [1164].

2.2.7 *Modular Forms*

1. The theory of modular forms in one variable forms a part of the theory of automorphic functions, defined as functions on a complex space, which behave in a special way under the action of a discrete group of transformations. This subject developed quickly in the second half of the 19th century, mainly in hands of C. Hermite, F. Klein, H. Poincaré and their students. (The main results being contained in the papers of F. Klein [3351, 3352], H. Poincaré [4930–4934], A. Hurwitz [2954] and in the lectures of F. Klein [3356, 3357], published in 1890. See also the early survey by R. Fricke[147] [2092].) It is interesting to observe how Klein, who often used various transformation groups in his research, avoided the use of properties of abstract groups. This can be clearly seen in his lectures, where in several places he meticulously defines factor groups in each individual case although the general notion was at that time already known. His attitude towards abstract groups never did change, as can be seen from the following sentence in his book [3354, p. 338] published in 1926:

> "Die[148] Lehre von den Vertauschungsgruppen hat sich ... zu einer selbständigen Disziplin entwickelt. Wir begegnen da Namen wie Cayley, Sylow, Dyck, Hölder, Frobenius, Burnside und in neuerer Zeit vielfach auch Amerikanern. Für viele Gemüter ist es ein besonderer Reiz, daß man auch hier wieder arbeiten kann, ohne von sonstiger Mathematik viel zu wissen ... ."

2. We recall quickly some properties of modular forms. Let $\Gamma = SL_2(\mathbf{Z})$ be the group of invertible 2×2 matrices with integral entries and unit discriminant. It acts on the upper complex half-plane $\mathfrak{H}$ by

$$A \cdot z = \frac{az+b}{cz+d},$$

[147] Robert Fricke (1861–1930), professor in Braunschweig.

[148] "The theory of permutation groups developed itself ... to an independent discipline. We meet there the names of Cayley, Sylow, Dyck, Hölder, Frobenius, Burnside, and in recent time also many Americans. For many minds it is a particular attraction that one can work here without knowing much of the rest of mathematics ... ."

where

$$A = \begin{bmatrix} a & b \\ c & d \end{bmatrix} \in \Gamma. \tag{2.46}$$

A complex-valued function defined and regular on $\mathfrak{H}$ is called a *modular form* of weight k, if for $A \in \Gamma$ one has

$$f(A \cdot z) = (cz + d)^k f(z). \tag{2.47}$$

Since the last equality implies $f(z+1) = f(z)$, one can expand f in a series of the form

$$f(z) = c_0 + \sum_{n=1}^{\infty} c_n \exp(2\pi i n z), \tag{2.48}$$

with complex coefficients c_n. If $c_0 = 0$, then f is called a *cusp form.*

Now let $N \geq 1$ be an integer. Denote by $\Gamma(N) \subset \Gamma$ the group of matrices (2.46) with $a \equiv d \equiv 1 \pmod N$ and $b \equiv c \equiv 0 \pmod N$. If f is regular on $\mathfrak{H}$, satisfies (2.47) for $A \in \Gamma(N)$, and for every $A \in \Gamma$ one has

$$f(A \cdot z)(cz+d)^{-k} = c_0(A) + \sum_{n=1}^{\infty} c_n(A) \exp\left(\frac{2\pi i n z}{N}\right),$$

then f is called a *modular form of weight k and level N* (note that in older literature these forms were said to be of weight $-k$). If for all $A \in \Gamma$ one has $c_0(A) = 0$, then f is called a cusp form. Modular forms of weight zero are called *modular functions.*

In the same way one defines modular forms associated to the subgroups $\Gamma_0(N)$ and $\Gamma_1(N)$ of Γ, which contain $\Gamma(N)$ and are defined by

$$\Gamma_0(N) = \left\{ \begin{bmatrix} a & b \\ c & d \end{bmatrix} : N | c \right\},$$

and

$$\Gamma_1(N) = \left\{ \begin{bmatrix} a & b \\ c & d \end{bmatrix} : a \equiv d \equiv 1 \pmod N,\ N | c \right\}.$$

Every subgroup of Γ containing $\Gamma(N)$ for some N is called a *congruence subgroup* and is of finite index in Γ. There exist subgroups of finite index in Γ which are not congruence subgroups, and the first such example was given in 1880 by F. Klein [3351] without a proof. First proofs were provided six years later by R. Fricke [2091] and G. Pick[149] [4854].

The groups $\Gamma(N)$, $\Gamma_0(N)$ and $\Gamma_1(N)$ are finitely generated. In the case of prime N a set of generators and their relations for $\Gamma_0(N)$ was given in 1929 by H. Rademacher [5030], and for $\Gamma(N)$ by H. Frasch [2069] in 1933. For the group

$$\Gamma_0^0(N) = \left\{ \begin{bmatrix} a & b \\ c & d \end{bmatrix} : N | (b, c) \right\}$$

this was done in 1950 by E. Grosswald [2360].

[149] Georg Alexander Pick (1859–1942), professor in Prague.

Now it is known that Γ contains infinitely many subgroups of finite index which are not congruence subgroups (see, e.g., I. Reiner[150] [5154], K.K. Wohlfahrt [6706, 6708], M. Newman [4588]). Remarkably, this phenomenon happens only in dimension $n = 2$, as it was shown in 1962 by H. Bass, M. Lazard and J.-P. Serre [348] and J.L. Mennicke [4252] that for $n \geq 3$ any subgroup of finite index in $SL_n(\mathbf{Z})$ contains a group of matrices, congruent to the unit matrix modulo a certain integer. This was extended in 1967 to $SL_n(R)$, where R is a ring of integers in a not totally complex field by H. Bass, J. Milnor, J.-P. Serre [349]. The case $n = 2$ for arbitrary global fields was worked out by J.-P. Serre [5640]. In 1969 H. Matsumoto [4197] generalized these results to a large class of algebraic groups. For surveys of the theory of congruence subgroups of algebraic groups see A.S. Rapinchuk [5129] and M.S. Raghunathan [5050].

One considers also, more generally, modular forms with characters[151], i.e., functions f satisfying for $A \in \Gamma_0(N)$, instead of (2.47), the equality

$$f(A \cdot z) = \chi(d)(cz + d)^k f(z), \tag{2.49}$$

where χ is a character $\bmod N$. The simplest examples are provided by theta-functions

$$\theta(z, F) = \sum_{n_1,\dots,n_k \in \mathbf{Z}} \exp(2\pi i z F(n_1, \dots, n_k)),$$

where F is a positive-definite quadratic form in an even number of variables (see, e.g., E. Hecke [2700, 2701]).

The set of all modular forms of fixed weight k, level N and character χ forms a finite-dimensional linear space $\mathfrak{M}_k(N, \chi)$ over the complex field, and the cusp forms form a subspace $\mathfrak{C}_k(N, \chi)$ of $\mathfrak{M}_k(N, \chi)$. Similarly one defines modular forms and cusp forms associated with other congruence subgroups.

3. At the end of the 19th century Hilbert considered a generalization of the theory of modular forms to several variables. He did not publish his results and gave his notes to O. Blumenthal[152] who used them in his habilitation thesis [565].

Hilbert modular forms are associated with the *Hilbert modular group*

$$\Gamma(K) = GL_2(\mathbf{Z}_K)/\{\pm E\},$$

where K is a totally real algebraic number field of degree $n \geq 2$, $\mathbf{Z}_K$ is its ring of integers and E denotes the unit matrix. This group acts on the product $\mathfrak{H}_n$ of n upper complex half-planes by

$$A(z_1, \dots, z_n) = \left(\frac{a^{(1)}z + b^{(1)}}{c^{(1)}z + d^{(1)}}, \dots, \frac{a^{(n)}z + b^{(n)}}{c^{(n)}z + d^{(n)}} \right)$$

---

[150] Irving Reiner (1924–1986), professor at the University of Illinois. See [3104].

[151] Sometimes called, after E. Hecke, modular forms of *Nebentypus*.

[152] Otto Blumenthal (1876–1944), professor in Aachen (1905–1933), executive editor of *Mathematische Annalen* (1906–1938). See [385].

for

$$A = \begin{pmatrix} a & b \\ c & d \end{pmatrix}, \tag{2.50}$$

$x^{(j)}$ denoting the jth conjugate of $x \in K$.

A regular function $f : \mathfrak{H}_n \to \mathbf{C}$ is called a *Hilbert modular form* of weight k if for A as in (2.50) one has

$$f(A\bar{z}) = N(cz + d)^k f(z), \tag{2.51}$$

$N(\cdot)$ denoting the norm $K \to \mathbf{Q}$.

Originally an extra regularity condition appeared in the definition, but it was shown in 1954 by M. Koecher [3436, 3437] that it is redundant. In the special case $K = \mathbf{Q}(\sqrt{5})$ this was shown earlier (in 1928) by F. Götzky [2289].

In 1903 O. Blumenthal [565] presented a construction of the fundamental domain for $\Gamma(K)$, applying a method of E. Picard[153] [4846–4848], who used it to construct functions of two variables satisfying (2.51) with $k \geq 4$.

Blumenthal's construction of the fundamental domain was not quite correct. This was pointed out in 1940 by H. Maass[154] [4037], who gave an explicit construction of the fundamental domain for a class of groups generalizing $\Gamma(K)$ (cf. O. Herrmann [2768], T. Tamagawa [6041] and Chap. 3 in Siegel's lectures [5776]).

In 1928 H.D. Kloosterman[155] [3372] introduced modular forms associated to congruence subgroups of the Hilbert group. Later [3373] he constructed a family of modular forms, generalizing the classical θ-series.

The theory of Hilbert modular forms was presented in the books by E. Freitag [2085], G. Shimura [5697] and C.L. Siegel [5776].

2.3 The First Sieves

1. The first arithmetical sieve appeared in the second edition of Legendre's book [3767]. Legendre formalized the ancient method of Eratosthenes for counting prime numbers in an interval, which can be described as follows.

If an integer N is given, and $2 = p_1 < p_2 < \cdots < p_k$ are all primes smaller than $\sqrt{N}$, then deleting from the set of all integers in $I_N = [\sqrt{N}, N]$ all numbers divisible by at least one of the p_j's we obtain a list of primes in I_N. To count these primes one first forms a list of all integers in the interval $[\sqrt{N}, N]$ and deletes from it all multiples of the primes $p_1, p_2, \dots, p_k$. This gives $\sum_{i=1}^{k}[N/p_i]$ deletions. Numbers divisible by two primes p_i were deleted twice, hence one has to subtract their number, equal to $\sum_{p_i \neq p_j}[N/p_i p_j]$. One realizes now that numbers divisible by three primes are left undeleted, so one has to

[153] Charles Émile Picard (1856–1941), professor in Toulouse and Paris. See [2427].

[154] Hans Maass (1911–1992), professor in Heidelberg. See [867].

[155] Hendrik Douwe Kloosterman (1900–1968), professor in Leiden. See [5880].

delete them, and proceeding in this way one is led to the following formula for the number $\pi(N)$ of primes not exceeding N.

$$\pi(N) = \pi(\sqrt{N}) + \sum_{i=1}^{k}(-1)^i \sum_{p_{j_1}<\cdots<p_{j_i}<\sqrt{N}} \left[\frac{N}{p_{j_1}\cdots p_{j_i}}\right] = \sum_{d|D}\mu(d)\left[\frac{N}{d}\right],$$

where D denotes the product of all primes $< \sqrt{N}$ and $\mu(d)$ is the Möbius function.

This formula does not lead to a non-trivial evaluation of $\pi(N)$, since the error obtained by the replacement of $[N/p_i]$ by $N/p_i + O(1)$ is larger than the main term, due to the $2^{\pi(\sqrt{N})}$ terms in the considered sum.

The include-exclude procedure described above is applicable also in a more general setting. Let Ω be a given finite set, let $X_1, \dots, X_k \subset \Omega$, $Y = \Omega \setminus \bigcup_{i=1}^{k} X_i$, and let f be a function on Ω having non-negative values. If, for $X \subset \Omega$, one puts

$$\Phi(X) = \sum_{x\in X} f(x),$$

and

$$A_r(f) = \Phi(\Omega) + \sum_{m=1}^{r}(-1)^m \sum_{1\le i_1<i_2<\cdots<i_m\le k} \Phi(X_{i_1}\cap\cdots\cap X_{i_m})$$

for $r = 1, 2, \dots, k$, then this procedure leads to the formula

$$\Phi(Y) = A_k(f). \tag{2.52}$$

2. There are rather few results which can be achieved with the use of this simple sieve. A variant of it, dealing with arithmetic progressions, was utilized in 1909 by J.C. Morehead [4420] in his search for Fermat primes.

An attempt to modify the Eratosthenian sieve method appeared in the paper [4256] by J. Merlin, edited posthumously by J. Hadamard; however the arguments appearing there are unconvincing. This paper gave an impulse to V. Brun[156] [798, 800–802], who presented the first non-trivial applications of the sieve. His approach was based on the observation that to obtain a reasonably small error term one should at some point stop the procedure leading to (2.52). In fact, a short calculation utilizing the simple fact that the sum

$$\sum_{j=0}^{r}(-1)^j\binom{k}{j}$$

is non-negative for even r, and non-positive for odd r, leads to

$$A_r(f) \le \Phi(Y) \le A_s(f), \tag{2.53}$$

for odd r and even s, satisfying $1 \le r, s \le k$.

[156] Viggo Brun (1885–1978), professor in Oslo. See [5594].

Although this idea is simple, its application involves cumbersome combinatorial arguments. Nevertheless V. Brun succeeded in proving the following important results.

(a) *The number* $\pi_2(x)$ *of twin primes below* x *does not exceed* $100x/\log^2 x$ *for large* x, *hence the series of their inverses converges.*

The coefficient 100 was later reduced to 19.43 by A.A. Buhštab[157] [833]. One expects that for large x one has

$$\pi_2(x) \le (c+\varepsilon)B\frac{x}{\log^2 x}, \tag{2.54}$$

with $c = 1$ and

$$B = 2\prod_{p>2}\left(1 - \frac{1}{(p-1)^2}\right).$$

In 1947 A. Selberg [5611] proved this with $c = 10.6$, and in [5614] stated that one can have $c = 8$. Later this was improved to $c = 6$ (C.D. Pan[158] [4734]), $c = 4$ (E. Bombieri, H. Davenport [608]), $c = 3.9171$ (J.R. Chen[159] [1026]), $c = 34/9 = 3.77\ldots$ (É. Fouvry, H. Iwaniec [2056]), $c = 64/17 = 3.76\ldots$ (É. Fouvry [2047, 2048]), $c = 3.5$ (E. Bombieri, J.B. Friedlander, H. Iwaniec [611–613]), $c = 3.454$ (É. Fouvry, F. Grupp [2053]), $c = 3.418$ (J. Wu [6755]), $c = 3.406$ (Y. Cai, M.G. Lu [886]) and $c = 3.3996$ (J. Wu [6759, 6760]), which is the current record.

Brun's method is also applicable to prime pairs with any fixed even difference. See B.I. Segal [5598].

(b) *Every large even integer is the sum of two positive integers, each of which has at most* 9 *prime divisors.*

For later progress, culminating in the replacement of 9 by 2, see Sect. 5.2.

(c) *In every interval of the form* $[x, x+\sqrt{x}]$ *there exists an integer, having at most* 11 *prime divisors.*

Later the number 11 was replaced by 9 (W.E. Mientka [4288]), by 4 (S. Uchiyama [6249]) and by 2 (J.R. Chen [1023, 1024]).

It took some time to recognize the importance of the new sieve, although V. Brun gave a short and readable exposition of his idea in [803]. It seems that the first who used the new method was T. Nagell [4504], who proved in 1922 that if f is a polynomial with integral coefficients, and $P(x)$ denotes the number of primes represented by $f(n)$ with $n \le x$, then $P(x) = o(x)$ (for $P(x) = x^2 + 1$ this had been shown already by Brun in [799]). This re-

---

[157] Aleksandr Adolfovič Buhštab (1905–1990), professor in Baku and Moscow. See [3541].

[158] Pan Cheng Dong (1934–1997), professor in Shandong.

[159] Chen Jing Run (1933–1996), professor at the Academia Sinica in Beijing. See [4736].

sult was strengthened in 1931 by H. Heilbronn [2706] to $P(x) = O(x/\log x)$ using Brun's sieve. In his result H. Heilbronn allowed f to be an integer-valued polynomial with rational coefficients. It was shown by G. Pólya [4953] that every such polynomial is of the form

$$\sum_{j=0}^{n} a_j \binom{x}{j}$$

with $a_j \in \mathbf{Z}$. The analogue of the last result for polynomials in algebraic number fields was obtained in 1919 by A. Ostrowski[160] [4707] and Pólya [4958] (for the theory of integer-valued polynomials see the book by P.-J. Cahen and J.-L. Chabert [880]; see also [4541]).

3. Let $F(X)$ be a polynomial with integral coefficients. We have already mentioned Bouniakowsky's old question of whether an irreducible polynomial $F \in \mathbf{Z}[X]$ without constant factors represents infinitely many numbers $\pm p$ with p prime. Since this seems to be out of reach, one tried to show that such a polynomial represents at least infinitely many numbers with a bounded number of prime divisors. The result by V. Brun [802] implied that the polynomial $x(x+2)$ represents infinitely many integers having at most 18 prime factors, and in 1923 H. Rademacher [5024] gave a simplified exposition of Brun's sieve utilizing it to show that an irreducible polynomial of degree d having integral coefficients and without fixed divisors represents infinitely many integers having at most $4d-1$ prime divisors.

This bound was later reduced by G. Ricci[161] [5194, 5195, 5199, 5200] first to $[cd]$ with $c = 3.0243$, and then to $3d-1$. It took some time before this bound was reduced to 3 for $d = 2$ and $d + c\log d$ with a certain constant c for $d \geq 3$ by P. Kuhn [3563, 3564], and to $d+1$ by A.A. Buhštab [837] and H.-E. Richert [5209] (for $3 \leq d \leq 5$ and $d = 6, 7$ this had been shown earlier by Y. Wang [6552] and B.V. Levin [3856], respectively). For $d = 2$ this bound was reduced in 1978 to 2 by H. Iwaniec [3055].

A.A. Buhštab's result was generalized to the case of polynomials in algebraic number fields by H. Sarges [5397] (cf. J. Hinz [2823]).

For analogues of these results for binary forms $F \in \mathbf{Z}[X, Y]$ see G. Greaves[162] [2331, 2332].

In another paper [5025] H. Rademacher extended Brun's method to algebraic number fields, and this enabled him to settle certain additive questions in this domain. An exposition of the early version of Brun's method (V. Brun [801]) was given in the first volume of Landau's textbook [3674].

Various classes of sieve, including the sieve of Brun, are described in the book by H. Halberstam and H.-E. Richert [2455].

---

160 Alexander Ostrowski (1893–1986), professor in Basel. See [1705, 3119].

161 Giovanni Ricci (1904–1973), professor in Pisa and Milano. See [1292].

162 George Richard Herbert Greaves (1941–2008), professor in Cardiff.

2.4 Additive Problems

2.4.1 Sums of Squares

1. It was asserted by Fermat and proved by Euler that any prime congruent to unity mod 4 can be represented as the sum of two squares. An explicit formula (using character sums) for the summands occurring in such a representation was given in 1907 by E. Jacobsthal[163] [3084], and a similar result for the representation of primes $p \equiv 1 \pmod 6$ by the form $x^2 + 3y^2$ was obtained by L. Schrutka von Rechtenstamm[164] [5569] in 1911. Earlier results giving explicit solutions of $x^2 + dy^2 = p$ for various values of d and primes p are described in [1545, Chap. 2].

Later similar results for other quadratic forms were obtained by S. Chowla ($x^2 + 27y^2 = 4p$ [1085]), A.L. Whiteman ($x^2 + 2y^2 = p, x^2 + 7y^2 = 4p$ [6651, 6652]) and B.W. Brewer ($x^2 + 5y^2 = p$ [723, 724]). The last author used the sums

$$\Lambda_n(p) = \sum_{x=0}^{p-1} \left(\frac{V_n(x)}{p} \right),$$

where the polynomials V_n are defined by

$$V_1(X) = X, \qquad V_2(X) = X^2 - 2, \qquad V_{n+2}(X) = xV_{n+1}(X) - V_n(X),$$

and are special cases of Dickson's polynomials of the first kind, defined by L.E. Dickson in his thesis [1535]. For the properties and applications of Dickson's polynomials see the monograph by R. Lidl, G.F. Mullen and G. Turnwald [3887].

The sum $\Lambda_n(p)$ is called the *Brewer sum* and was later investigated by several authors (see, e.g., A.L. Whiteman [6653, 6654], R.E. Giudici, J.B. Muskat, S.F. Robinson [2241], B.C. Berndt, R. Evans [450] and the book by B.C. Berndt, R. Evans and K.S. Williams [453]).

2. In the 19th century much work was done to find explicit formulas for $r_k(n)$, the number of representations of an integer n as the sum of k squares. Such formulas were provided for various small k by C.F. Gauss [2208, Sect. 291], C.G.J. Jacobi [3075, 3077, 3078], P.G. Dirichlet [1587], G. Eisenstein[165] [1709, 1710], L. Kronecker [3526, 3529], J. Liouville [3936], H. Minkowski [4318] and H.J.S. Smith [5833].

Elementary proofs of Jacobi's formulas were given much later by M.D. Hirschhorn [2827] for $k = 2$, and by J.A. Ewell [1942], M.D. Hirschhorn [2825], G.E. Andrews, S.B. Ekhad, D. Zeilberger [90] for $k = 4$.

This classical question was revived in the first decades of the 20th century. E.T. Bell [394] wrote that this was possibly due "to its connection with X-ray analyses of crystal structure," but this is a rather doubtful assertion. In 1907

---

[163] Ernst Jacobsthal (1882–1965), professor in Berlin and Trondheim. See [5628].

[164] Lothar Wolfgang Schrutka von Rechtenstamm (1881–1945), professor in Brno and Vienna.

[165] Gotthold Eisenstein (1823–1852), worked in Berlin.

J.W.L. Glaisher [2248] presented his results concerning explicit formulas for $r_k(n)$ ($k = 2, 4, \ldots, 18$) obtained with the use of modular forms[166], the details appearing in a series of papers in the *Quarterly Journal of Mathematics*. Similar formulas in the cases of $k = 3, 5, 7, 9$ and 11 were found in 1920 by E.T. Bell [394, 395]. Formulas for the number of representations of even numbers as sums of ten and twelve squares, stated by Liouville, were proved in 1907 by G. Humbert[167] [2947] and K. Petr[168] [4828] with the use of theta-functions. In the same year A. Hurwitz [2962] expressed $r_3(n^2)$ in terms of divisor functions, the case of $r_5(n^2)$ having been treated by him already in 1884 [2956]. In 1914 B. Boulyguine[169] [658, 839] gave formulas for $r_{2k}(n)$. A few years later L.J. Mordell [4375, 4377] applied modular forms to a study of $r_k(n)$ in the case when k is either a power of 2 or odd.

Modular forms were also later used to obtain explicit formulas for $r_k(n)$ (see, e.g., R.A. Rankin[170] [5119]). In 1965 R.A. Rankin [5120] proved that for $k \geq 9$, $r_k(n)$ cannot be expressed by divisor functions. The analogue of this result for representations of integers by an arbitrary quadratic form was obtained by G.P. Gogišvili [2254, 2255].

An elementary proof of Hurwitz's formula for $r_3(n^2)$ was given much later by J. Lagrange [3609]. Similar formulas for $r_k(n^2)$ for $k = 7, 9, 11$ and 13 were established by H.F. Sandham [5387, 5388]. In 1950 P.T. Bateman [351] proved that the function $r_k(n)/2k$ is multiplicative only for $k = 1, 2, 4$ and J. Lagrange [3609] showed in 1972 that $r_k(n^2)/2k$ is multiplicative if and only if $1 \leq k \leq 8$.

Exact formulas for an infinite sequence of k ($k = 4m^2$ and $k = 4m(m + 1)$) were given much later by S.C. Milne [4314, 4315] (cf. D. Zagier [6812], K. Ono [4684]).

2.4.2 The Waring Problem

1. The first great old problem in number theory which was settled in the 20th century was the celebrated *Waring problem*, which got its name from an English mathematician of the 18th century, E. Waring[171]. In 1770 he published a book *Meditationes Algebraicae* [6565] in which, among other questions, he considered the possibility of representing positive integers as sums of powers. He did not prove anything in this respect, but formulated an assertion which later became known as the Waring Problem. He wrote[172]:

---

[166] Later elementary proofs of these formulas have been found. See J.A. Ewell [1944], J.G. Huard, K.S. Williams [2941] and the book [4548] by M.B. Nathanson.

[167] Marie Georges Humbert (1859–1921), professor in Paris, father of P. Humbert.

[168] Karel Petr (1868–1950), professor in Prague.

[169] Vasiliĭ Vasilevič Bulygin [Boulyguine] (1888–1918).

[170] Robert Alexander Rankin (1915–2001), professor in Birmingham and Glasgow. See [454].

[171] Edward Waring (1734–1798), professor in Cambridge.

[172] We use the English translation by D. Weela [6565, Theorem 47, part 9].

"Every integer is a sum of two, three, ..., nine cubes; every integer is also the square of a square, or the sum of up to nineteen such; and so forth. Similar laws may be affirmed for the correspondingly defined numbers or quantities of any degree."

2. One denotes by $g(k)$ the smallest number of terms which are sufficient to represent any positive integer as the sum of kth powers of non-negative integers. Thus Waring's assertion states that $g(3) = 9$, $g(4) = 19$ and usually one extrapolates his statement to the following form:

The number $g(k)$ is finite for every $k \geq 2$.

In the case $k = 2$ this question had been considered already by C. Bachet[173] and Fermat [1989, p. 307], and the corresponding theorem ($g(2) = 4$) had been established by J.L. Lagrange[174] [3610] in 1770. A similar problem was considered by Descartes, who in two letters to Mersenne [1477, 1478] stated that the only numbers which are not sums of four positive squares are $a \cdot 4^n$ for $a = 2, 6, 14$, $n = 0, 1, 2\ldots$ as well as $1, 3, 5, 9, 11, 17, 29$ and 41. This was shown to be true as late as in 1911 by E. Dubouis [1633] and rediscovered by G. Pall [4727] in 1933 (Pall's proof is reproduced in the book [5784] by W. Sierpiński). Dubouis also described positive integers which are not sums of a given number of positive squares.

Waring [6565] himself checked that every number up to 3 000 is the sum of nine cubes, and C.G.J. Jacobi arranged the checking at first up to 3 000 [6835], and then to 12 000 [3080]. This led him to conjecture that all large integers are sums of at most five positive cubes and every integer larger than 8 042 is the sum of at most six positive cubes. These conjectures are still open.

During the 19th century the values of $g(3)$, $g(4)$ and $g(5)$ were shown to be finite, and explicit bounds $g(3) \leq 21$, $g(4) \leq 41$ and $g(5) \leq 192$ were found by E. Maillet [4106], É. Lucas [4027] and E. Maillet [4107], respectively. They proved this by establishing various polynomial identities. For example, to prove the finiteness of $g(4)$ the following easily verified identity can be used:

$$6\left(x_1^2 + x_2^2 + x_3^2 + x_4^2\right)^2 = \sum_{1 \leq i < j \leq 4} (x_i + x_j)^4 + \sum_{1 \leq i < j \leq 4} (x_i - x_j)^4. \tag{2.55}$$

As, according to Lagrange, every positive integer is the sum of at most 4 squares, this identity shows that every number of the form $6n^2$ is the sum of at most 12 fourth powers, and, again using Lagrange's result, one obtains that every multiple of 6 is the sum of at most 48 fourth powers. This implies the bound $g(4) \leq 53$. This argument is due to J. Liouville (see [3754]). Using essentially the same approach S. Realis[175] [5136] and É. Lucas [4026, 4027] reduced Liouville's upper bound to

---

[173] Claude Gaspar Bachet de Méziriac (1581–1638), lived in Bourg-en-Bresse.

[174] Joseph Louis Lagrange (1736–1813), worked in Turin, Berlin and Paris.

[175] Savino Realis (1818–1886), engineer in Turin. See [962].

47 and 41, respectively. In 1906 A. Fleck[176] [2012] modified this method and obtained $g(4) \leq 39$, and one year later E. Landau [3628] combined formula (2.55) with a careful investigation of 48 cases to get $g(4) \leq 38$. This was slightly improved by A. Wieferich[177] [6658] to $g(4) \leq 37$. In the same year he also considered [6659] the fifth and seventh powers, and established the bounds $g(5) \leq 59$ and $g(7) \leq 3\,806$. Later $g(5) \leq 58$ was proved by W.S. Baer [211].

For the next step one had to wait several years, when in 1933 L.E. Dickson [1556] proved $g(4) \leq 35$, earlier giving [1546] a new proof of Wieferich's bound.

Going in the opposite direction, A.J. Kempner[178] showed in his thesis [3298], that infinitely many integers require at least 16 biquadrates.

3. In the cubic case R.D. von Sterneck [5939] in 1903 extended Jacobi's check up to 40 000, and pointed out that every number in the interval $(8\,042, 40\,000)$ can be written as the sum of six cubes. In 1906 A. Fleck [2012] reduced Maillet's bound to 13, and three years later A. Wieferich [6656] was able to confirm Waring's assertion in the cubic case, proving the equality $g(3) = 9$. His proof followed the ideas developed in the previous century, as it was based on the identity

$$\sum_{i=1}^{3}\left((y + x_i)^3 + (y - x_i)^3\right) = 6y\left(y^2 + x_1^2 + x_2^2 + x_3^2\right). \tag{2.56}$$

Actually Wieferich's proof was incomplete, as it omitted a rather long interval. This was pointed out by P. Bachmann [199] and repaired by A.J. Kempner [3298].

See [4538] for an exposition of Wieferich's argument modified by B. Scholz [5559]. Simpler proofs eliminating the use of numerical computations were later given by L.E. Dickson [1551] and E. Herzog [2772].

In [2013] A. Fleck gave the first proof of the finiteness of $g(6)$, establishing $g(6) \leq 2\,451$, and this was reduced to 970 by A.J. Kempner in 1912 [3298], and to 488 by W.S. Baer in 1913 [211]. The inequality $g(8) < \infty$ was established in 1908 by E. Maillet [4110], and the first explicit bound for this number ($g(8) \leq 36\,119$) was obtained by A. Hurwitz [2963] in the same year.

The next improvement came in 1927 when L.E. Dickson [1548] got $g(6) \leq 190$ and $g(8) \leq 2690$.

4. The year 1909 turned out to be rather good for the Waring problem as in that year E. Landau [3635] proved that all large integers can be represented as the sum

---

[176] Albert Fleck (1861–1943), medical doctor, known for finding errors in purported proofs of Fermat's Last Theorem. See [5980].

[177] Arthur Josef Alwin Wieferich (1884–1954), student of M. Dehn in Münster, schoolteacher.

[178] Aubrey Kempner (1880–1973), professor at the University of Colorado in Boulder. Student of Minkowski, obtained his Ph.D. in Göttingen in 1909.

of 8 non-negative cubes[179] (another proof was later found by L.K. Hua [2915]), and D. Hilbert [2789] succeeded in establishing the essential part of Waring's assertion, proving the finiteness of $g(k)$ for every k.

Hilbert's proof was the culmination of previously applied methods being based on a rather complicated identity, found earlier by F. Hausdorff[180] [2614], involving an integral in 25-dimensional real space. Later work by F. Hausdorff [2615], E. Schmidt [5475], E. Stridsberg[181] [5969], G. Frobenius [2117] and R. Remak[182] [5157] simplified Hilbert's reasoning.

Hilbert's approach to the Waring problem is now only of historic interest since newer methods are much simpler and lead to better bounds for the number of summands. It was later shown by G.J. Rieger [5216] that Hilbert's proof gives the bound

$$g(k) \le (2k+1)^{260^{A(k)}}$$

with

$$A(k) = (k+3)^{3k+8},$$

whereas it is now known that for large k one has

$$g(k) = \left[(3/2)^k\right] + 2^k - 2,$$

which is of much smaller order (K. Mahler [4088]). For expositions of Hilbert's approach see [1754] and [4538].

5. The minimal number of non-negative kth powers needed to represent every sufficiently large integer is denoted by $G(k)$. Already in 1908 A. Hurwitz [2963] and E. Maillet [4110] had proved the lower bound $G(k) \ge 1 + k$ which has not been improved since, except for certain classes of k (see Sect. 3.2.1). As no number congruent to 7 mod 8 is the sum of three squares, the theorem of Lagrange implies $G(2) = 4$, and Landau's result in [3635] gives $G(3) \le 8$. On the other hand one has $G(3) \ge 4$, as no number congruent to ± 4 mod 9 is the sum of three cubes.

Landau's bound has not been improved for several years. In 1943 Yu.V. Linnik ([3906], cf. [3931]) established $G(3) \le 7$, which still holds the record. The proof was based on his previous results on representations of integers by ternary quadratic forms [3901]. It was pointed out by G. Pall [4729] that the argument in [3901] is incomplete, but can be corrected. A simpler proof, using the Siegel–Walfisz theorem on primes in progressions, was presented by G.L. Watson[183] [6578]. Another proof, due to R.C. Vaughan, can be found in [6360, 6361]. Later, effective proofs were found by R.J. Cook [1242], K.S. McCurley [4223], H. Kadiri [3210] and O. Ramaré [5093, 5094]. In the last paper it is shown that every integer $n > \exp(524)$ is the sum of at most 7 cubes. One expects that this happens already for

---

[179] Actually 23 and 239 are the only positive integers which are not of that form (L.E. Dickson [1567]).

[180] Felix Hausdorff (1869–1942), professor in Leipzig and Bonn. See [727].

[181] Eric Stridsberg (1871–1950), worked in an insurance company.

[182] Robert Remak (1888–1942), student of Frobenius, worked in Berlin. See [4262].

[183] George Leo Watson (1909–1988), professor at University College, London.

$n > 455$. It was proved by J. Brüdern [773] that the summands could be taken to have at most 69 prime divisors, and later 69 was replaced by 4 by K. Kawada [3286].

Vaughan's paper [6360, 6361] also contains an asymptotic formula for the number of representations of an integer N as the sum of 8 positive cubes, its main term being $cN^{5/3}$, and also a lower bound of order $N^{4/3}$ in the case of 7 cubes.

In 1926 A.E. Western [6637] after doing some numerical experiments conjectured that all integers $\geq 10^{14}$ are sums of at most 4 cubes, thus $G(3) = 4$, and his conjecture was made more precise by J.-M. Deshouillers, F. Hennecart and B. Landreau [1493], who believe that there are exactly 113 936 676 integers which are not sums of four cubes, the largest being 7 373 170 279 850.

It was shown in 1939 by H. Davenport [1356] that the number of $n \leq x$ which are not sums of 4 cubes is $O(x^c)$ for every $c > 29/30$. Denote by α the largest lower bound for the value of c in this result. In 1986 R.C. Vaughan [6360, 6361] proved $\alpha \leq 103/115 = 0.895\ldots$, then J. Brüdern [769, 772] improved this first to $\alpha \leq 131/147 = 0.891\ldots$, and then to $\alpha \leq 37/42 = 0.8809\ldots$, and T.D. Wooley [6728] got $\alpha \leq 0.8808$.

R.C. Vaughan [6365, 6366] proved in 1989 that the number of integers $n \leq x$ which are sums of four positive cubes is $\gg x^{\beta}$ for every $\beta < 11/12 = 0.9167$, and this was later improved by T.D. Wooley [6728, 6735] to any $\beta < 0.91681$ and $\beta < 0.91686$.

It has been conjectured that the number $N_k(x)$ of integers $\leq x$ which are sums of three positive kth powers is of order $x^{3/k}$ for $k \geq 3$. The best known lower bound,

$$N_k(x) \gg x^{3/k - \exp(-k/17)}$$

is due to T.D. Wooley [6728], who improved earlier results by R.C. Vaughan [6365, 6366].

In the case of biquadrates, the inequality $G(4) \geq 16$ obtained by A.J. Kempner [3298] was turned into equality by H. Davenport [1354] in 1939. In 2000 J.-M. Deshouillers, F. Hennecart and B. Landreau [1488] showed that every integer between 13 793 and 10^{245} is the sum of at most 16 fourth powers. Five years later J.-M. Deshouillers, K. Kawada and T.D. Wooley [1494] proved that the maximal integer which is not a sum of 16 fourth powers equals 13 792.

2.5 Diophantine Approximations

2.5.1 Approximation by Rationals, Theorem of Thue

1. The first result dealing with approximations of real numbers by rationals goes back to Dirichlet who proved in 1842, in a paper concerning continued fractions [1589], that for every real α and $N = 1, 2, \ldots$ one can find a rational p/q with $q \leq N$ and

$$\left|\alpha - \frac{p}{q}\right| \leq \frac{1}{qN}.$$

He showed, more generally, that for any given reals $\alpha_1, \ldots, \alpha_n$ and $N \geq 1$, one can find integers $q_1, \ldots, q_n$ satisfying

$$1 \leq \max\{|q_1|, \ldots, |q_n|\} \leq N,$$

and

$$\left\|\sum_{j=1}^{n} \alpha_j q_j\right\| < \frac{1}{N^n}.$$

In 1850 C. Hermite [2761] showed the existence of a constant c depending only on n such that if $\alpha_1, \dots, \alpha_n$ are distinct real numbers, then for infinitely many positive integers q there exist $b_1, \dots, b_n \in \mathbf{Z}$ such that

$$\left|\alpha_j - \frac{b_j}{q}\right| \leq \frac{c}{q^{1+1/n}}. \tag{2.57}$$

This was also proved by L. Kronecker [3530] in 1884, and in 1903 É. Borel[184] [636] showed that the exponent $1 + 1/n$ is best possible. Borel's paper contains the first explicit use of the notion of the height of a rational number.

An effective version of the last result was given by A.K. Lenstra, H.W. Lenstra, Jr. and L. Lovász [3819] in 1982, as a corollary to their polynomial-time algorithm to find a nearly orthogonal basis of a lattice in $\mathbf{R}^n$. For other applications of this algorithm to Diophantine approximations see R. Tijdeman [6161].

2. Let $c(n)$ denote the greatest lower bound of possible values of c in (2.57). H. Minkowski proved in his book [4324] that $c(n) \leq n/(n+1)$ and this was slightly strengthened by H.F. Blichfeldt[185] [557]. The equality $c(1) = \sqrt{5}$ had been established already in 1891 by A. Hurwitz [2958], but for $n \geq 2$ the exact value of $c(n)$ is not known. Minkowski in [4325] established the bound $c(2) \leq \sqrt{8/19} = 0.6488\dots$.

Later the inequality $c(2) \geq 1/\sqrt[4]{23} = 0.4566\dots$ was proved by P. Furtwängler[186] [2165, 2166], who confirmed an earlier assertion by H.F. Blichfeldt [558]. It is conjectured that $c(2) = \sqrt{2/7} = 0.5345\dots$. For numerical investigation in the case $n = 3$ see G. Szekeres[187] [6020].

In 1921 O. Perron [4787] obtained the first lower bound for $c(n)$ of the order $1/(n2^n)$. Later P. Furtwängler [2165, 2166] strengthened Perron's bound and his result was improved in 1985 by S. Krass [3513].

3. The result proved in 1851 by J. Liouville in [3934] went in another direction. He proved the following assertion, which allowed him to construct several transcendental numbers.

If α is a real algebraic number of degree $n \geq 2$, then for every rational p/q one has

$$\left|\alpha - \frac{p}{q}\right| > \frac{C}{q^n}, \tag{2.58}$$

where C is a positive constant, depending only on α.

The proof is a simple consequence of the mean value theorem. The analogue of Liouville's result for approximations of algebraic numbers by algebraic numbers of lower degree was given in 1910 by É. Borel in his book [638].

---

[184] Émile Borel (1871–1956), professor in Paris. See [1183, 2070].

[185] Hans Frederick Blichfeldt (1873–1945), professor at Stanford. See [1568].

[186] Philipp Furtwängler (1869–1940), professor in Bonn, Aachen and Vienna. See [2846, 2942].

[187] György Szekeres (1911–2005), worked in chemical industry in Budapest and Shanghai, professor at Macquarie University. See [2238].

Liouville's theorem has subsequently been improved. The first such result is due to A. Thue[188] [6142], who proved[189] that if α is algebraic of degree $n \geq 2$, then for all rational integers $q > 0$, p one has

$$\left|\alpha - \frac{p}{q}\right| \geq \frac{c(\alpha, \varepsilon)}{q^t} \tag{2.59}$$

for $t = n/2 + \varepsilon$ with arbitrary $\varepsilon > 0$ and positive $c(\alpha, \varepsilon)$. In the special case when α is an nth root of a rational number, Thue had established this result earlier [6141].

4. Thue's result had important applications to the theory of Diophantine equations and to multiplicative properties of polynomials. It implied in particular that the maximal prime divisor of $(ax + b)(cx + d)$ (where $a, b, c, d \in \mathbf{Z}$, $ad \neq bc$ and $ac \neq 0$) tends to infinity with x. Later G. Pólya [4954] showed that the same holds for all irreducible quadratic polynomials. This was a rather far generalization of an old result of C. Størmer[190] [5961], who established this in 1898 for the polynomials $x^2 + x$, $x^2 + 2x$ and $x^2 + 1$. For later developments see Sect. 3.1.4.

In his thesis [5738], C.L. Siegel gave a strengthening of Thue's result, replacing in (2.59) the exponent $t = d/2 + \varepsilon$ by $2\sqrt{d}$. He used this to prove Pólya's assertion for every polynomial over $\mathbf{Z}$ with at least two distinct zeros, and also gave an extension to algebraic number fields. Siegel [5741] also discovered another consequence of Thue's theorem by proving that if u_n is a recurring sequence of rationals of order 3, defined by

$$u_{n+3} = Pu_{n+2} - Qu_{n+1} + u_n,$$

with integral P, Q, then apart from six explicitly given cases the equation

$$u_n = 0 \tag{2.60}$$

can have at most a finite number of solutions. A similar result was proved by M. Ward[191] [6561] for linear recurrences of order 4. Much later it was shown by S.J. Scott [5588] that if the polynomial $X^3 - PX^2 + QX - 1$ has a negative discriminant and non-zero roots, then there are at most three solutions (under stronger assumptions this was proved earlier by M. Ward [6564] and M.F. Smiley [5830]).

In 1934 T. Skolem[192] [5806] showed that if u_n is a linear recurrence in $\mathbf{Q}$, then the set of integers n satisfying (2.60) forms a union of finitely many (possibly zero) arithmetical progressions and a finite set. This was generalized in 1935 to algebraic number fields by K. Mahler [4078] and to arbitrary fields of zero characteristic by C. Lech [3756] in 1953 (the *Skolem–Mahler–Lech theorem*). This result fails for fields of positive characteristics (C. Lech [3756]), but an analogue of it was established in this case in 2007 by H. Derksen [1475].

---

[188] Axel Thue (1863–1922), professor in Oslo.

[189] For comments on Thue's work on Diophantine approximations and Diophantine equations see [5773].

[190] Frederik Carl Størmer (1874–1957), professor in Oslo. See [804].

[191] Morgan Ward (1901–1963), professor at Caltech. See [3792].

[192] Thoralf Albert Skolem (1887–1963), professor in Oslo and Bergen. See [4518].

In the same paper [4078] K. Mahler proved that if

$$u_{n+k} = \sum_{j=0}^{k-1} A_j u_{n+j}$$

is a linear recurrence with coefficients in an algebraic number field K, which is non-degenerate, i.e., the roots and ratios of roots of its companion polynomial

$$P(X) = X^k - \sum_{j=0}^{k-1} A_j X^j \tag{2.61}$$

are not roots of unity, then $|u_n|$ tends to infinity, hence the equation

$$u_n = c \tag{2.62}$$

can have only finitely many solutions.

In the case $k = 2$, M. Ward conjectured that for a non-degenerate linear recurrence with integral terms equation (2.62) can have at most five solutions. In 1973 R. Alter and K.K. Kubota [61] proved the conjecture in the case when $(M, N) = 1$, and four years later K.K. Kubota [3553–3555] established it in all cases. Later F. Beukers [483] showed that apart from finitely many cases, which can be explicitly listed, the equation $u_N = \pm c$ can have at most 3 solutions. Kubota [3555] also obtained an upper bound depending only on the degree N of the field K, containing all elements u_n. This was made more precise by F. Beukers and R. Tijdeman [493], who showed that in this case there are at most $100\max\{300, N\}$ solutions (for the rational field they got the bound 29).

In 1984 J.-H. Evertse [1923] showed that if u_n is a non-degenerated linear recurrence of any order with algebraic terms and not of the form $u_n = a\zeta^n$ with a root of unity ζ, then there can be only finitely many pairs m, n with $u_m = u_n$. For sequences lying in an algebraic number field, explicit bounds for the number of solutions of (2.62) were given by A.J. van der Poorten[193] and H.-P. Schlickewei [6305] (for $c = 0$) and Schlickewei [5468] (any c).

It was conjectured that for a non-degenerate recurrent complex sequence of order k the number of solutions of (2.62) with $c = 0$ is bounded by a value depending only on k. For rational sequences this was established in 1996 by H.P. Schlickewei [5469], who in the case $k = 3$ proved this also for complex sequences [5470]. In this case the bound does not exceed 61 (F. Beukers, H.P. Schlickewei [492]). In the case when the companion polynomial has only simple zeros the conjecture was established by J.-H. Evertse, H.P. Schlickewei and W.M. Schmidt [1940] with the explicit bound $\exp((6k)^{3k})$. They also obtained an effective version of the Skolem–Mahler–Lech theorem in this case. In the general case the conjecture was proved in 1999 by W.M. Schmidt [5526] with the bound $\exp\exp\exp(3k\log k)$. This bound has subsequently been reduced to somewhat smaller numbers $\exp\exp\exp(20k)$ (W.M. Schmidt [5527]) and $\exp\exp(k^{\sqrt{11k}})$ (P.B. Allen [55]). See also H.P. Schlickewei [5468–5470] and A.J. van der Poorten, H.P. Schlickewei [6305].

5. It was shown by Čebyšev [971] in 1868 that if α is irrational and β is not an integer, then for infinitely many integers q one has

$$\|\alpha q - \beta\| < \frac{c}{|q|}$$

[193] Alfred Jacobus van der Poorten (1942–2010), professor at Macquarie University.

with $c = 1/2$. Hermite [2765] in 1879 proved this with $c = \sqrt{2/27} = 0.272\ldots$, and in 1896 Minkowski [4323] showed that one can take $c = 1/4$. It was later proved by J.H. Grace[194] [2296] that this is best possible.

In 1946 A.J. Khintchine[195] [3326] considered the function $M(\alpha)$, defined as the least upper bound (taken over $\beta \in \mathbf{R}$, not of the form $\beta = m\alpha + n$ with integral m, n) of

$$\liminf_{q\to\infty} q\,\|q\alpha + \beta\|.$$

He proved the bound

$$M(\alpha) \le \frac{1}{4}\left(1 - 2\lambda(\alpha)^2\right)^{1/2},$$

where

$$\lambda(\alpha) = \liminf_{q\to\infty} q\,\|q\alpha\|.$$

6. For irrational $\alpha > 0$ let p_n/q_n be the nth convergent of the continued fraction of α, and define the sequence λ_n by

$$\left|\alpha - \frac{p_n}{q_n}\right| = \frac{1}{\lambda_n q_n^2}. \tag{2.63}$$

It was proved by A. Hurwitz [2958] in 1891 that the inequality $\lambda_n > \sqrt{5}$ holds for infinitely many n. This result was made more precise by É. Borel [635, 636], who in 1903 showed that the inequality

$$\max\{\lambda_{n-2}, \lambda_{n-1}, \lambda_n\} > \sqrt{5} \tag{2.64}$$

holds for every n. The inequality

$$\max\{\lambda_{n-1}, \lambda_n\} > 2$$

had been known earlier (T. Vahlen [6257]).

Borel's result was generalized in 1950 by N. Obreškov[196] [4652], who showed that if $[a_0; a_1, a_2, \ldots]$ is the continued fraction of α, then the left-hand side of (2.64) exceeds $\sqrt{a_n^2 + 4}$ (cf. F. Bagemihl and R.C. McLaughlin [216]). For $s = 2$ this had been proved earlier by M. Fujiwara[197] [2139, 2140] and G. Humbert [2948]. A very simple proof of Borel's theorem was provided in 1963 by H.G. Forder [2040], and one year later A.L. Schmidt [5473] gave simple proofs for the theorems of Borel and Fujiwara–Humbert. The mean value $M_N(\alpha) = \frac{1}{N}\sum_{j=1}^{N} \lambda_j$ was studied in 1949 by A. Brauer and N. Macon [683, 684], who established $M_N(\alpha) \ge \sqrt{5}$.

---

[194] John Hilton Grace (1873–1922), worked in Cambridge. See [6182].

[195] Aleksandr Jakovlevič Khintchine [Hinčin] (1894–1959), professor in Moscow. See [2252].

[196] Nikola Obreškov (1896–1963), professor in Sofia. See [6195].

[197] Matsusabûro Fujiwara (1881–1946), professor at Tôhoku University. See [3556].

For real α put $M(\alpha) = \limsup \lambda_n$, with λ_n defined in (2.63). It was shown in 1921 by P.J. Heawood[198] [2673] that the inequality $M(\alpha) < 3$ holds only for certain quadratic irrationalities, whereas there exist uncountably many α's with $M(\alpha) = 3$. In the same year O. Perron [4788] showed that the distinct values of $M(\alpha) < 3$ form a sequence and determined the first 11 of its elements[199]: $\sqrt{5}$, $\sqrt{8}$, $\sqrt{211}/5$, $\sqrt{1517}/13, \ldots$. This has been also established independently by K. Shibata [5688]. This sequence arises also in A.A. Markov's [4153, 4154] theory of indefinite quadratic forms (see the monograph [1310] by T.W. Cusick and M.E. Flahive).

7. In a paper by Jacobi [3081] (published posthumously by E. Heine[200] in 1868) a generalization of the continued fraction algorithm was proposed. This algorithm was later studied by P. Bachmann [197], and in 1907 O. Perron [4783] published a large paper in which he essentially generalized Jacobi's approach. The algorithm consists of iterations of the mapping T defined in the following way: if $u = (x_1, \ldots, x_n)$ is an n-tuple of positive real numbers, and x_s is the smallest of them, then for $i = 1, 2, \ldots, n$ write $x_i = A_i x_s + y_i$ with integral A_i and $0 \le y_i < x_s$, and put $T : u \mapsto (y_1, \ldots, y_n)$. This algorithm can be used to determine the greatest common divisor of n given positive integers and to find units in algebraic number fields (see L. Bernstein, H. Hasse [468] and the book by L. Bernstein [467].

A broad exposition of the theory of continued fractions was published in 1913 by O. Perron [4785].

The history of continued fractions up to 1939 has been written by C. Brezinski [725], who also prepared a complete bibliography [726] on that subject. The theory of numbers with bounded partial quotients (with a bibliography of over 300 items) was surveyed in 1992 by J. Shallit [5670].

8. The first results dealing with the approximation of complex numbers by ratios of integers from a fixed imaginary quadratic field appear in the papers of C. Hermite [2762] and A. Hurwitz [2957] as well as in Minkowski's book [4324, Sect. 39]. In particular, Hermite showed that for every complex z there are infinitely many distinct pairs of co-prime integers a, b of the field $\mathbf{Q}(i)$ with

$$\left| z - \frac{a}{b} \right| < \frac{1}{c|b|^2}$$

with $c = \sqrt{2}$, and Hurwitz proved the analogous result for the field $\mathbf{Q}(\sqrt{-3})$. Minkowski showed that in the case of the field $\mathbf{Q}(i)$ one can replace here $\sqrt{2}$ by $\pi/\sqrt{6}$, but left open the question of whether this constant is optimal. In 1925 L.R. Ford[201] [2039] showed that this is not the case, the optimal value being $\sqrt{3}$ (cf. [2038]). Ford's result was rediscovered a few years later in 1930 by O. Perron [4789, 4790].

---

[198] Percy John Heawood (1861–1955), professor in Durham. See [1581].

[199] The first two terms were known already to Hurwitz [2958].

[200] Heinrich Eduard Heine (1821–1881), professor in Bonn and Halle.

[201] Lester Randolph Ford (1886–1967), professor at the Illinois Institute of Technology.

Most of these results were obtained with the use of various kinds of continued fractions for complex numbers.

Later Perron [4792] extended Hurwitz's result to all imaginary quadratic fields $\mathbf{Q}(\sqrt{-D}\,)$. The least upper bound of possible values of the corresponding constant c, say $c(D)$, is called the *Hurwitz constant* of the field K, thus Ford's result gives $c(1)=\sqrt{3}$. The value of $c(D)$ is known only in the following cases: $c(2)=\sqrt{2}$ (O. Perron [4792]), $c(3)=\sqrt[4]{13}$ (O. Perron [4791]), $c(5)=c(6)=1$ (L.J. Vulakh [6485]), $c(7)=\sqrt[4]{8}$ (N. Hofreiter[202] [2844]), $c(11)=\sqrt{5}/2$ (R. Descombes, G. Poitou[203] [1479]), $c(15)=\sqrt{2}$ (L.J. Vulakh [6486]), $c(19)=1$ (G. Poitou [4939]). This constant is related to the geometry of fundamental domains of Fuchsian groups (see, e.g., R.A. Rankin [5116], L.J. Vulakh [6484, 6485]).

2.5.2 Uniform Distribution

1. At the International Congress of Mathematicians held in Cambridge in 1912, G.H. Hardy and J.E. Littlewood [2519] announced a far reaching generalization of the following result, due essentially to Kronecker, concerning the distribution of fractional parts of sequences of the form $u_n = n\alpha$ for irrational α.

If the real numbers $1,\alpha_1,\dots,\alpha_N$ are linearly independent over the rationals, and $\beta_1,\dots,\beta_N$ are arbitrary reals, then for every $\varepsilon>0$ the system of inequalities

$$|y\alpha_j+\beta_j-x_j|<\varepsilon \quad (j=1,2,\dots,N)$$

is solvable with integral $y,x_1,\dots,x_N$.

They applied the results of A. Pringsheim[204] [5011] and F. London [3987], dealing with limits of double sequences, to prove that if $\theta_1,\dots,\theta_n$ are irrational real numbers, linearly independent over the rationals, then for any given numbers $0\le\alpha_{ij}<1$ $(i,j=1,2,\dots,n)$ one can find a sequence $n_1,n_2,\dots$ of integers such that

$$\lim_{r\to\infty}\{n_r^i\theta_j\}=\alpha_{ij}$$

holds for $i,j=1,2,\dots,n$.

The proof appeared in [2520]. There is also a proof there of the assertion that if λ_n is any sequence tending to infinity, then for almost all real θ the set $\{\{\theta\lambda_n\}:n=1,2,\dots\}$ is dense in the interval $(0,1)$. This may fail for some irrational θ. Indeed, Hardy and Littlewood showed in [2519] that there exist irrational numbers θ such that the sequence $\{2^n\theta\}$ is not dense in the unit interval. This led to the question of existence of numbers α for which the sequence $\{q^n\alpha\}$ (with a certain integral $q>1$) tends to zero. The answer was given by Hardy, who proved in [2510] that if

[202] Nikolaus Hofreiter (1904–1990), professor in Vienna. See [2836].

[203] Georges Poitou (1926–1989), professor in Paris. See [3215].

[204] Alfred Pringsheim (1850–1941), professor in Munich. See [4795].

$\alpha > 1$ is an algebraic integer then it has this property if and only if all its remaining conjugates lie in the interior of the unit circle. Such numbers have several peculiar properties, and are known under the name of *Pisot–Vijayaraghavan numbers*[205], or *PV-numbers*, since Hardy's result was later rediscovered by C. Pisot[206] [4916, 4917] and T. Vijayaraghavan[207] [6391–6394]. Actually this characterization of PV-numbers was proved for the first time in 1912 by A. Thue [6143] in a paper unknown to Hardy, Pisot and Vijayaraghavan.

C. Pisot [4916] and T. Vijayaraghavan [6391–6394] independently obtained certain characterizations of PV-numbers. One of them states that $a > 1$ is a PV-number if and only if for some real $\lambda \neq 0$ the series

$$\sum_{n=1}^{\infty} \|\lambda a^n\|^2$$

is convergent. R. Salem[208] [5373] proved that the set S of all PV-numbers is closed and nowhere dense, as conjectured in [6392]. In the same paper he studied algebraic integers $a > 1$ all of whose remaining conjugates lie in the closed unit circle and which are not PV-numbers. They are now called *Salem numbers*. It turned out later that PV- and Salem numbers play an important role in harmonic analysis (see R. Salem [5371, 5372] and the book by Y. Meyer [4282]).

The two smallest PV-numbers were found by C.L. Siegel [5765] (they are the positive roots of $X^3 - X - 1$ and $X^4 - X^3 - 1$). He conjectured that $\theta = (1 + \sqrt{5})/2$ is the smallest limit point of the set of PV-numbers, and this was later established by J. Dufresnoy and C. Pisot [1635], who in [1636–1638] found all PV-numbers smaller than θ.

The book by M.J. Bertin, A. Decomps-Guilloux, M. Grandet-Hugot, M. Pathiaux-Delefosse and J.-P. Schreiber [470] gives a broad account of the theory of PV-numbers and Salem numbers.

2. In 1914 H. Weyl[209] [6646, 6647] introduced the notion of uniform distribution mod 1 of real sequences.

A sequence $x_1, x_2, \dots$ of elements in the interval $[0, 1)$ is called *uniformly distributed*[210] mod 1, provided for every $0 \le a < b \le 1$ one has

$$\lim_{N\to\infty} \frac{\#\{n \le N : a \le x_n < b\}}{N} = b - a. \tag{2.65}$$

---

[205] This name was introduced by R. Salem [5371, 5372].

[206] Charles Pisot (1910–1984), professor in Bordeaux and Paris. See [66].

[207] Tirukkannapuram Vijayaraghavan (1902–1955), professor at Andrha University and Director of the Ramanujan Institute in Madras. See [1369].

[208] Raphael Salem (1898–1963), professor at MIT and in Paris.

[209] Hermann Weyl (1885–1955), professor at ETH in Zürich, Göttingen and Princeton. See [1053, 4540, 4589].

[210] In [6646] Weyl did not use the term *uniform distribution*, calling this property *gleichmäßig dicht*, i.e., *uniformly dense*.

Weyl proved that a sequence (x_n) has this property if and only if for every non-zero integer m one has

$$\sum_{n=1}^{N} \exp(2\pi i m x_n) = o(N). \tag{2.66}$$

This is the original *Weyl criterion*. This name is nowadays usually attached to uniform distribution in compact Abelian groups G, where it takes the form

$$\sum_{n=1}^{N} \chi(x_n) = o(N)$$

for every character $\chi \neq 1$ of G.

In 1948 P. Erdős and P. Turán [1866, 1867] obtained a bound (the *Erdős–Turán inequality*) for the difference

$$\frac{\#\{n \leq N : a \leq x_n < b\}}{N} - (b - a)$$

in the case of a finite sequence x_n (cf. J.D. Vaaler [6256]). This result was later generalized to sequences in $\mathbf{R}^n$ (see J.F. Koksma[211] [3446], T. Cochrane [1131], P.J. Grabner, R.F. Tichy [2295]).

Weyl's criterion implies in particular that for every irrational real number α the sequence $\{n\alpha\}$ is uniformly distributed mod 1, a fact earlier proved independently by P. Bohl[212] [575], H. Weyl [6645] and W. Sierpiński [5782]. This strengthened an old result by Kronecker [3530] which implied that for irrational α the sequence $\{n\alpha\}$ is dense in [0, 1].

A simple heuristical argument shows that for irrational α the sum $\sum_{k\leq N}\{k\alpha\}$ should be close to $N/2$. The difference

$$C_\alpha(N) = \sum_{k\leq N}\{k\alpha\} - \frac{N}{2} \tag{2.67}$$

was considered by C. Størmer [5962] and M. Lerch [3836] in 1904, who showed that if α is irrational and has bounded partial quotients, then $C_\alpha(N) = O(\log N)$ (cf. A. Ostrowski [4708]).

G.H. Hardy and J.E. Littlewood [2528, 2529] showed later that for any irrational α one has $|C_\alpha(t)| \geq c \log t$ infinitely often, and A. Ostrowski [4708] obtained this with $c = 1/720$. In 1995 T.C. Brown and P.J.-S. Shiue [755] showed that one can take $c = 1/256$. In 1957 V.T. Sós [5848] showed that $C_\alpha(t)$ may be bounded from one side. For the use of Dirichlet series in this question see E. Hecke [2682].

---

[211] Jurjen Ferdinand Koksma (1904–1964), professor in Amsterdam. See [3570].

[212] Piers Bohl (1865–1921), professor in Riga. See [3404].

3. To apply his criterion to polynomial sequences H. Weyl needed evaluations of exponential sums of the form

$$S_f(N) = \sum_{n=1}^{N} \exp(2\pi i f(n)), \tag{2.68}$$

where $f(X) = a_M X^M + \cdots + a_0$ is a polynomial with real coefficients, of which at least one, distinct from a_0, is irrational. Such sums are now called *Weyl sums.*

Assuming that the leading coefficient a_M of f is irrational, writing

$$|S_f(N)|^2 = \sum_{m=1}^{N}\sum_{n=1}^{N} \exp\bigl(2\pi i\bigl(f(m) - f(n)\bigr)\bigr),$$

and putting $m = n + r$, Weyl got

$$f(m) - f(n) = rMa_M r^{M-1} + \cdots = rg(r,n),$$

where g is a polynomial in r with integral coefficients of degree $M - 1$, and this allowed him to reduce the problem to the case of linear polynomials, which can be settled elementarily. This led to

$$|S_f(N)| \ll N^{1+\varepsilon}\left(N^{-t} + q^{-t} + \left(\frac{N^M}{q}\right)^{-t}\right),$$

where $t = 1/2^{M-1}$ and q is the denominator of a rational approximation r to the leading coefficient a_M of f, satisfying the condition

$$|a_M - r| \le 1/q^2.$$

This bound implies the asserted uniform distribution mod 1 of the sequence of consecutive polynomial values. Independently the same result was obtained by G.H. Hardy and J.E. Littlewood [2520, 2521].

In 1928 I. Schoenberg[213] [5546] gave a generalization considering arbitrary continuous distributions of sequences of real numbers in the unit interval. He obtained an analogue of Weyl's criterion, applied it to prove the existence of the limit

$$F(t) = \lim_{x\to\infty} \frac{1}{x}\#\left\{n \le x : \frac{\varphi(n)}{n} \le t\right\}$$

(φ being Euler's function), and this implied the existence of a distribution function for the additive function $\log(\varphi(n)/n)$.

H. Weyl proved also in [6647] that for every sequence $0 < \lambda_1 < \lambda_2 < \cdots$ of integers the sequence $\lambda_n x$ is uniformly distributed mod 1 for almost all x.

[213] Isaac Jacob Schoenberg (1903–1990), professor at the Universities of Pennsylvania and Wisconsin. See [153].

It is not always easy to find a number x having this property. The first effective example in the case $a_n = n!$ was given in 1950 by N.M. Korobov[214] [3485]:

$$x = x_c = \sum_{k=1}^{\infty} \frac{[k^c]}{k!}$$

for every $c \in (1, 2)$.

If λ_n is a lacunary sequence of positive reals (i.e., $\lambda_{n+1}/\lambda_n \geq c > 1$), then there are irrationals θ such that the sequence $\{\lambda_n \theta\}$ is not dense in $(0, 1)$, hence not uniformly distributed (we have noted already that this has been observed for $\lambda_n = 2^n$ by G.H. Hardy and J.E. Littlewood [2519]). This was shown in 1979 by A.D. Pollington [4947], who answered a question from P. Erdős [1824] (for sequences with $c \geq 5^{1/3}$ this had been proved earlier by E. Strzelecki [5978]). B. de Mathan [4186] proved later that the set of such θ's has Hausdorff dimension one. The *Hausdorff dimension* of a set A of reals was introduced by Hausdorff in [2617] in the following way.

For $r > 0$ let $\mu^r(A)$ be the largest lower bound of the sums

$$\sum_n |I_n|^r,$$

where I_n is a sequence of intervals whose union contains A. The number ρ with the property

$$\mu^r(A) = \begin{cases} 0 & \text{if } r > \rho, \\ \infty & \text{if } r < \rho \end{cases}$$

is called the Hausdorff dimension of A. Cf. A.S. Besicovitch[215] [475] and the book [5255] by C.A. Rogers[216].

Weyl's sums turned out to be of importance in several arithmetical problems. In particular they were applied to evaluations of the zeta-function, the distribution of primes, the Waring problem and Diophantine approximations.

Stronger bounds for Weyl sums were later obtained by I.M. Vinogradov [6422, 6428, 6436, 6447] and Yu.V. Linnik [3908]. See L.K. Hua [2938] and G.I. Arkhipov, A.A. Karatsuba[217], V.N. Čubarikov [123, 124] for surveys. For later improvements see E. Bombieri, H. Iwaniec [615, 616], E. Bombieri [604], L.D. Pustylnikov [5018], M.N. Huxley, G. Kolesnik [2991], M.N. Huxley [2984].

4. H. Weyl considered also sums of the form

$$S_f(p) = \sum_{j=0}^{p-1} \exp\left(\frac{2\pi i f(x)}{p}\right)$$

[214] Nikolai Mikhailovič Korobov (1917–2004), professor in Moscow. See [4616].

[215] Abram Samoilovič Besicovitch (1891–1970), professor in Perm, Petrograd (Leningrad) and Cambridge. See [6090].

[216] Claude Ambrose Rogers (1920–2005), professor in Birmingham and at University College London.

[217] Anatolij Alekseevič Karatsuba (1937–2008), professor in Moscow. See [120].

with prime p, where $f \in \mathbf{Z}[X]$ is of degree n and not all of its coefficients are divisible by p. He obtained the bound

$$\left|S_f(p)\right| = O(p^c), \tag{2.69}$$

for any $c > 1 - 1/2^{n-1}$. In the case $n = 2$ one has $|S_f(p)| = \sqrt{p}$, by reduction to quadratic Gauss's sums. This bound was later improved by L.J. Mordell [4382] (for further development see Sect. 4.1.1).

5. A real number a is called *simply normal* with respect to an integer $q \geq 2$ if in the q-expansion of a every digit occurs with the same frequency. A number which is normal with respect to every q is called *absolutely simply normal*. Note that in earlier literature these numbers were called *normal* and *absolutely normal*, respectively. Now a number θ is said to be normal with respect to q if every combination of k digits in the q-expansion of θ occurs with frequency q^{-k}.

One of the first applications of probability theory to arithmetical questions occurred in a paper by É. Borel, who asserted in 1909 [637] that almost all real numbers are simply normal with respect to any $q \geq 2$. His arguments were rather heuristical but soon other proofs of this assertion were given by G. Faber[218] [1948] and F. Hausdorff [2616]. Borel also stated in [637] that almost every number is absolutely simply normal (cf. [640]), and an elementary proof of this result was provided in 1917 by W. Sierpiński [5783] who also gave the first effective example.

Later A.M. Turing[219] gave an effective way of constructing such numbers (see V. Becher, S. Figueira, R. Picchi [371]).

Borel's result implies in particular that if for real $x \in (0, 1)$ one denotes by $S_n(x)$ the number of occurrences of 1 in the first n binary digits of x, and $\mu_n(x) = |S_n(x) - n/2|$, then for almost all x one has

$$\mu_n(x) = o(n).$$

Hausdorff's argument in [2616] improved this result to

$$\mu_n(x) = O(n^\alpha)$$

for every $\alpha > 1/2$, and G.H. Hardy and J.E. Littlewood [2520, 2521] in 1914 established

$$\mu_n(x) = O(\sqrt{n \log n}),$$

and showed that the bound

$$\mu_n(x) = O(\sqrt{n})$$

fails for a set of x of positive measure.

---

[218]Georg Faber (1877–1966), professor in Tübingen, Stuttgart, Königsberg, Strasbourg and Munich.

[219]Alan Mathison Turing (1912–1954), worked in Cambridge and Manchester. During Second World War, a member of the Bletchley Park code-breaking team. Forerunner of computer science. See [2839, 6230].

Almost ten years later A.J. Khintchine [3317] showed that for almost all $x \in (0,1)$ one has

$$\limsup_{n\to\infty} \frac{|\mu_n(x)|}{\sqrt{n \log\log n}} \leq 1.$$

The Hausdorff dimension of the set

$$\left\{x \in (0,1) : \limsup_{n\to\infty} \frac{S_n(x)}{n} \leq t\right\}$$

was determined for $t < 1/2$ by A.S. Besicovitch [478] and V. Knichal[220] [3408, 3409].

The more general case of q-ary expansions was considered by I.J. Good[221] in [2280]. He conjectured a formula for the Hausdorff dimension of sets of real numbers whose digits appear with a prescribed frequency, and it was later established by H.G. Eggleston [1692] (*theorem of Besicovitch–Eggleston*) (cf. L. Barreira, B. Saussol, J. Schmeling [335], J. Cigler [1111], L. Olsen [4676], F. Schweiger [5586], B. Volkmann [6457–6462]).

A characterization of normal numbers in terms of Rademacher functions was given later by M. Mendès France [4245].

In 1932 D.G. Champernowne[222] [989] proved that the number 0.123456789101112... is normal, and conjectured that the same happens for the number whose decimal digits are formed by the juxtaposition of the sequence of primes. This was established in 1946 by A.H. Copeland and P. Erdős [1246], who showed that the same happens if the primes are replaced by any quickly increasing sequence of integers. Later H. Davenport and P. Erdős [1382] extended Champernowne's result to all numbers with decimal expansion $0.f(1)f(2)f(3)\ldots$, where $f(x)$ is an integer-valued polynomial attaining positive values for $x = 1, 2, \ldots$. As shown later by K. Mahler [4081] all these numbers are transcendental. For further variants and generalizations see P. Bundschuh [844], P. Bundschuh, P.J.-S. Shiue, X. Yu [849], M.G. Madritsch, J. Thuswaldner, R.F. Tichy [4057], K. Mahler [4094], H. Niederreiter [4600], J.W. Sander [5386], Z. Shan, E.T.H. Wan [5672], I. Shiokawa [5710], Y. Nakai, I. Shiokawa [4528–4530], P. Szűsz[223], B. Volkmann [6032].

Denote by $B(q)$ the set of all numbers normal with respect to an integer $q \geq 2$. It was shown in 1960 by W.M. Schmidt [5484] that the equality $B(r) = B(s)$ holds if and only if the ratio $\log r/\log s$ is rational. For other results on relations between sets $B(q)$ for various q see J.W.S. Cassels [941], W.M. Schmidt [5485], C.M. Colebrook, J.H.B. Kemperman [1175], A.D. Pollington [4948, 4949]. For rational r one has $B(q) + \{r\} = B(q)$. Examples of irrational r with this property were given by N. Spears and J.E. Maxfield [5860], and a description of the set of such r's was provided by G. Rauzy [5135].

The notion of normality was generalized to the case of non-integral bases. One says that a real number a is normal with respect to $\theta > 1$ if the sequence $\{a\theta^n\}$ is uniformly distributed mod 1. Let $B(\theta)$ be the set of all real a satisfying this condition. M. Mendès France [4249] asked whether the above result by W.M. Schmidt [5484] holds also for non-integral bases, and this got a negative answer in 1993, when G. Brown, W. Moran and A.D. Pollington established $B(10) \neq B(\sqrt{10})$ [751] (cf. [752] and W. Moran, A.D. Pollington [4373]).

---

[220] Vladimír Knichal (1908–1974), professor in Brno. See [3587].

[221] Irving John Good (1917–2009), professor at Virginia Tech. During Second World War, a member of the Bletchley Park code-breaking team.

[222] David Gawen Champernowne (1912–2000), professor of economics in Oxford and Cambridge.

[223] Peter Szűsz (1924–2008), professor at SUNY.

For a historical survey of the theory of normal numbers see G. Harman [2563].

M. Mendès France defined (in a footnote in [4246]) a set B of reals to be *normal* if there exists a sequence λ_n of reals such that the sequence $\lambda_n x$ is uniformly distributed modulo 1 if and only if x lies in B. Various examples, including real number fields of finite degree or the set of all real transcendental numbers, were given by M. Mendès France [4247] and Y. Meyer [4281, 4282]. Characterizations of normal sets were given by F. Dress and M. Mendès France [1623] and G. Rauzy [5134]. See also F. Dress [1621], J. Lesca, M. Mendès France [3839], M. Mendès France [4248].

G. Rauzy's question of whether a finite union of normal sets is normal was answered in the positive by T.C. Watson [6593] (this had been shown earlier for subsets of $\mathbf{Z} \setminus \{0\}$ by F. Dress and M. Mendès France [1623]). Watson's paper characterizes normal sets B as $F_{\sigma\delta}$-sets satisfying $0 \notin B$ and $-B = B$.

Normal sets associated with sequences which are recognizable by automatons have been studied by C. Mauduit [4210].

2.6 Geometry of Numbers

2.6.1 Lattice Points

1. Although geometrical arguments in arithmetical questions may also be found in earlier papers, one usually associates the name *Geometry of Numbers* with the research started by H. Minkowski in 1891, who in his first paper on this subject [4322] applied geometrical methods to the study of minima of positive-definite quadratic forms with real coefficients. He later published two books [4324, 4328] in which he showed that the utilization of geometrical methods can provide insight in various arithmetical problems.

Recall that a subgroup Λ of the real n-space $\mathbf{R}^n$ is called an *n-dimensional lattice*, if it is isomorphic to $\mathbf{Z}^n$. Every such lattice can be written in the form

$$\Lambda = \left\{ \sum_{j=1}^{n} x_j \omega_j : x_1, \ldots, x_n \in \mathbf{Z} \right\},$$

where $\omega_1, \ldots, \omega_n$ are suitably chosen elements of Λ, linearly independent over the rationals. The volume $d(\Lambda)$ of the set

$$\left\{ \sum_{j=1}^{n} r_j \omega_j : 0 \le r < 1 \right\}$$

is called the *discriminant of* Λ. Lattices in small dimensions had already been considered by C.F. Gauss [2212] and A. Bravais[224] [693].

Minkowski used lattices very efficiently in various problems. The central result of the book [4324] is the *convex body theorem* (a *convex body* is defined as a compact

[224] Auguste Bravais (1811–1863), professor of physics and mineralogy in Paris.

convex subset of $\mathbf{R}^n$, symmetrical about the origin, and containing the origin in its interior).

Let X be a convex body in $\mathbf{R}^n$, and let $\Lambda \in \mathbf{R}^n$ be a lattice of discriminant $d(\Lambda)$. If $V(X)$ exceeds $2^n d(\Lambda)$, then X contains at least one non-zero point of Λ. If X is compact, then the same assertion holds under the weaker assumption $V(X) \leq 2^n d(\Lambda)$.

For simple proofs of this theorem see H.F. Blichfeldt [557], L.J. Mordell [4384], R. Remak [5161].

2. The convex body theorem has several important applications, most of them already appearing in [4324]. One of them is the theorem on linear forms (*Linearformensatz*), which in its simplest form runs as follows.

Let for $i = 1, 2, \dots, n$

$$L_i(X_1, \dots, X_n) = \sum_{j=1}^{n} a_{ij} X_j$$

be real linear forms in n variables with non-zero discriminant $D = \det[a_{ij}]$. If $t_1, \dots, t_n$ are positive numbers satisfying $t_1 t_2 \cdots t_n > |D|$, then the system of inequalities

$$|L_j(x_1, \dots, x_n)| \leq t_j \quad (j = 1, 2, \dots, n) \tag{2.70}$$

has a non-zero integral solution.

The proof is based on the observation that the volume of the set defined by the inequalities (2.70) equals $\det[a_{ij}]$. There also exist proofs avoiding the use of geometry. Several of them are listed in the book by J.F. Koksma [3444].

From several other applications of the convex body theorem we mention here only two simple proofs of the old result stating that an integer is the sum of three squares if and only if it is not of the form $4^a(8b+7)$, given by N.C. Ankeny [102] and L.J. Mordell [4398].

3. In his book [4324] Minkowski established the following assertion in the case $n = 2$.

Let

$$L_i(X_1, \dots, X_n) = \sum_{j=1}^{n} a_{ij} X_j \quad (i = 1, 2, \dots, n)$$

be linear forms with real coefficients and non-vanishing determinant $D = \det[a_{ij}]$. Then for every real $y_1, \dots, y_n$ there exist rational integers $x_1, \dots, x_n$ satisfying

$$\prod_{i=1}^{n} |L_i(x_1, \dots, x_n) - y_i| \leq \frac{D}{2^n}. \tag{2.71}$$

The assertion (2.71) in the general case is usually quoted as a conjecture of Minkowski, although, as noted by F.J. Dyson [1678], there is no evidence that Minkowski considered the case $n > 2$.

The inequality (2.71) was established for $n = 3$ by R. Remak [5159] in 1923, and other proofs were provided later by H. Davenport [1353] and B.J. Birch and H.P.F. Swinnerton-Dyer [536]. The case $n = 4$ was handled successfully in 1948 by F.J. Dyson [1678], and in the case $n = 5$ a sketch of the proof was given in 1973 by B.F. Skubenko[225] [5816]. A complete proof on lines indicated by Skubenko was given by R.P. Bambah and A.C. Woods [311]. A proof for $n = 6$ was found by C.T. McMullen [4235] in 2005, and recently the conjecture was established for $n = 7$ (R.J. Hans-Gill, R. Madhu, S. Ranjeet [2499]).

It was shown by N.G. Čebotarev [968] that on the right-hand side of (2.71) one can put $D/2^{n/2}$ (for a small improvement see L.J. Mordell [4389]), and later this was replaced by $c_n D/2^{n/2}$ with $c_n = 1/(2e-1+o(1))$ (H. Davenport [1361]), and $c_n = 1/(3(2e-1)+o(1))$ (E. Bombieri [594]). This was later improved to $c_n \ll (\log n/n)^{1/3}$ (B.F. Skubenko [5817]), $c_n \ll n^{-3/7}\log^{4/7} n$ (Kh.N. Narzullaev, Skubenko [4545]), $c_n \ll \log n/\sqrt{n}$ (A.K. Andriyasyan, I.V. Ilin, A.V. Malyšev[226] [97]), and $c_n \ll \sqrt{\log n/n}$ (A.V. Malyšev [4124]).

4. In [4324] one also finds the first occurrence of an abstractly defined metric in the set of all n-term sequences of reals, defined as a function $f(p,q)$ satisfying the triangle inequality and the condition: $f(p,q) = 0$ is equivalent to $p = q$. Note that Minkowski explicitly avoided the symmetry condition.

An important result established by Minkowski in [4324] is the *Brunn–Minkowski theorem*, which gives a lower bound for the volume of the set

$$A + B = \{a + b : a \in A, b \in B\},$$

where A, B are convex sets in $\mathbf{R}^n$, in terms of volumes of A and B:

$$V(A+B)^{1/n} \geq V(A)^{1/n} + V(B)^{1/n}. \tag{2.72}$$

It was shown by L.A. Lusternik[227] [4033] in 1935 that (2.72) also holds for arbitrary non-empty compact sets A, B, the volume being replaced by the Lebesgue measure. For further developments see R.J. Gardner [2196].

Minkowski's book also contains applications of his methods to the theory of algebraic numbers. In particular, there he proved a lower bound for the absolute value of the discriminant of an algebraic number field, which implied the non-existence of a non-trivial extension of the rationals with discriminant equal to 1, and led to a simple proof of Hermite's [2763] assertion that there can be only finitely many algebraic number fields having a fixed discriminant. One also finds in [4324] a new proof of Dirichlet's theorem on the structure of units in algebraic number fields as well as a geometrical interpretation of the theory of continued fractions.

---

[225] Boris Faddeevič Skubenko (1929–1993). See [96].

[226] Aleksandr Vasilevič Malyšev (1928–1993), professor in Leningrad. See [643].

[227] Lazar Aronovič Lusternik (1899–1981), professor in Moscow. See [47].

5. In [4324, Sect. 40] Minkowski considered systems $S=\{L_1,\dots,L_n\}$ of n real linear forms in n variables of unit discriminant, and defined

$$M_n=\sup_S \min_{P\neq 0}|L_1(P)\cdots L_n(P)|, \tag{2.73}$$

$P=(x_1,\dots,x_n)\in \mathbf{Z}^n$. He proved the inequality $M_n\le n!/n^n$ implying

$$M=\limsup_{n\to\infty}\sqrt[n]{M_n}\le\frac{1}{e}=0.367\dots.$$

The value of M_n is related to $D(n)$, the minimal absolute value of discriminants of algebraic number fields of degree n, because of the inequality

$$M_n\ge\frac{1}{\sqrt{D(n)}}$$

(N. Hofreiter [2845]).

Minkowski's bound was improved in 1936 by H.F. Blichfeldt [563], who obtained $M\le\sqrt{2/3\pi e}=0.279\dots$, and, as noted by C.A. Rogers in [5251], the inequality $M\le 0.188\dots$ is hidden in Blichfeldt's paper [564]. In [5251] Rogers obtained $M\le\pi/4\exp(1.5)=0.175\dots$.

The equality $M_3=1/7$ was established in 1938 by H. Davenport [1350]. Simpler proofs were later provided by H. Davenport [1359] and L.J. Mordell [4391]. In [1352] Davenport determined forms with

$$\min_{P\neq 0}|L_1(P)L_2(P)L_3(P)|\ge\frac{1}{9.1}$$

(cf. H.P.F. Swinnerton-Dyer [6001]). The corresponding constant in the case of three forms, one real and the other two complex conjugate, equals $1/\sqrt{23}$ (H. Davenport [1351]).

If the forms L_i in (2.73) have complex coefficients and are pairwise conjugated, then the corresponding value of M does not exceed $2/\sqrt{\pi e^3}=0.2517\dots$ (H.P. Mulholland [4469]).

6. The second book by Minkowski [4328] is, contrary to its title, devoted mostly to a new development of the theory of algebraic numbers, in which the main role is played by lattices in real n-spaces. Nevertheless it contains some important results concerning lattice points in convex bodies such as, for example, the following *theorem on successive minima*, which extends the convex body theorem.

Let X be a convex body of unit volume, put for $\lambda>0$

$$\lambda X=\{\lambda x: x\in X\},$$

and for $j=1,2,\dots,n$ let λ_j be the minimal value of λ with the property that λX contains j linearly independent points of the integral lattice. Then

$$\frac{2^n}{n!}\le\prod_{j=1}^{n}\lambda_j\le 2^n.$$

For simpler proofs of this result see H. Davenport [1357], T. Estermann [1892] and R.P. Bambah, A.C. Woods, H. Zassenhaus[228] [312].

[228] Hans Zassenhaus (1912–1991), professor in Hamburg and at McGill University, Notre Dame University and Ohio State University. See [4923, 4928].

In the same book a conjecture is stated concerning the minimal area $\Delta(p)$ of a parallelogram having one of its vertices at the origin and the remaining three lying on the line $|x|^p + |y|^p = 1$, with $p \geq 1$ being fixed. Minkowski conjectured that for $1 < p < 2$ one has

$$\Delta(p) = (1 - 2^{-p})^{1/p},$$

and that for $p > 2$ the value of $\Delta(p)$ is a root of the polynomial

$$\left(4^{1/p}x + 1\right)^p + \left(4^{1/p}x - 1\right)^p - 2^{p+1}.$$

It turned out later that this conjecture is correct for large p (H. Cohn [1154]), as well as for certain small p (L.J. Mordell [4390], V.G. Kuharev [3560]), but incorrect for several other small values of p (C.S. Davis[229] [1411], H. Cohn [1154], G.L. Watson [6580, 6581]).

7. A subset X of $\mathbf{R}^n$ is called a *star body* if it contains the origin and every radius vector from the origin meets the boundary of X in one point. The following assertion was stated without proof by Minkowski in [4321].

If $X \subset \mathbf{R}^n$ is a bounded star body of volume $V < \zeta(n)$, then there exists a lattice of unit determinant, not containing a non-zero point of X. If X is, in addition, symmetrical about the origin, then this holds under the assumption $V < 2\zeta(n)$.

The first proof of his assertion was given by E. Hlawka[230] in 1943 [2832], and therefore it is usually called the *Minkowski–Hlawka theorem*. Other proofs were later found by C.A. Rogers [5250] and C.L. Siegel [5767].

A stronger result for symmetric convex bodies was shown by K. Mahler [4084]. Let K be such a body in $\mathbf{R}^n$ of volume $V(K)$, let $\Delta(K)$ be the largest lower bound for the discriminant of a lattice having only the origin in common with K, and put

$$Q(K) = \frac{V(K)}{\Delta(K)}$$

and

$$c_n = \inf_{K \subset R^n} Q(K).$$

Mahler's result gives $c_2 \geq \sqrt{12} = 3.4641\ldots$, and

$$c_n \geq 2\zeta(n) + \frac{1}{6}$$

for every $n \geq 2$. Later H. Davenport and C.A. Rogers [1400] improved upon this and obtained

$$\liminf_{n \to \infty} c_n \geq 4.921\ldots.$$

Cf. C.G. Lekkerkerker [3813], C.A. Rogers [5252], W.M. Schmidt [5483].

In 1961 V. Ennola [1766] proved $c_2 \geq 3.5252\ldots$, and in 1970 P.P. Tammela [6042] got $c_2 \geq 3.5706\ldots$, which is not very far from the upper bound $c_2 \leq 3.6096\ldots$, established by K. Reinhardt [5155] and K. Mahler [4085], conjectured to be optimal.

---

[229] Clive Selwyn Davis (1916–2009), professor at the University of Queensland.

[230] Edmund Hlawka (1916–2009), professor in Vienna. See [5529, 6153].

8. Denote by $m(f)$ the minimal value attained by a positive definite real quadratic form f at a non-zero point of the lattice $\mathbf{Z}^n$, and define the *Hermite constant* γ_n by

$$\gamma_n = \max \frac{m(f)}{D^{1/n}(f)}, \tag{2.74}$$

the maximum being taken over all forms f in n variables, $D(f)$ being the discriminant of f.

A.N. Korkin and E.I. Zolotarev introduced and studied the important notion of *extreme forms* [3478–3480], presented an algorithm to check whether a form is extreme and described all such forms in dimensions ≤ 5. Extreme forms are positive definite quadratic forms f having the property that for every form g of the same discriminant, whose coefficients are sufficiently close to the corresponding coefficients of f one has $m(g) < m(f)$.

H. Minkowski proved the bound

$$\gamma_n \leq \frac{4}{\pi}\Gamma\left(1+\frac{n}{2}\right)^{2/n},$$

improving for $n \geq 9$ the inequality

$$\gamma_n \leq \left(\frac{4}{3}\right)^{(n-1)/2},$$

due to C. Hermite [2761]. For large n a better evaluation was provided in 1914 by H.F. Blichfeldt [557], who established

$$\gamma_n < \frac{2}{\pi}\Gamma\left(2+\frac{n}{2}\right)^{2/n} \tag{2.75}$$

(cf. R. Remak [5161]).

An important step forward was made by G.F. Voronoĭ [6471–6473] who in 1908 presented an algorithm for γ_n, which is however extremely cumbersome in higher dimensions. This algorithm also gives a way of constructing all *perfect quadratic forms*, defined in the following way: let $f(\bar{x}) = f(X_1, \ldots, X_n)$ be a positive definite quadratic form, put

$$M(f) = \min_{\bar{x} \neq \bar{0}} f(\bar{x}),$$

and let this minimum be attained at points $\bar{u}_1, \ldots, \bar{u}_s \in \mathbf{R}^n$. The form f is said to be perfect if the conditions $f(\bar{u}_i) = M(f)$ $(i = 1, 2, \ldots, s)$ determine f uniquely. Every *extreme form* is perfect.

Voronoĭ's algorithm allowed a complete list of perfect forms for $n = 6$ to be obtained later (see H.S.M. Coxeter[231] [1261], E.S. Barnes[232] [330]). There are seven such forms. A simpler method of producing perfect and extreme forms was presented by Barnes in 1958 [331, 332]. Much later, Voronoĭ's method allowed the determination of all 33 perfect forms for $n = 7$

[231] Harold Scott MacDonald Coxeter (1907–2003), professor in Toronto, See [5237].

[232] Eric Stephen Barnes (1924–2000), professor in Adelaide.

(K. Stacey [5883], D.-O. Jacquet-Chiffelle [3091]). Other criteria for an extreme form appear in E.S. Barnes [333]. In the book [4164] by J. Martinet 1096 perfect forms of dimension 8 are described, and it was proved later by M.D. Skirič, A. Schürmann and F. Vallentin [5786] that this list is complete.

The lattices associated with perfect forms are called *perfect lattices*, and a monograph dealing with them has been written by J. Martinet [4164].

The analogue of γ_n for forms over number fields was introduced in 1997 by M.I. Icaza [3001] and it was determined for binary forms in certain particular fields by R. Baeza, R. Coulangeon, M.I. Icaza, M. O'Ryan [212], R. Coulangeon, M.I. Icaza, M. O'Ryan [1258] and M. Pohst, M. Wagner [4929].

9. The evaluation of γ_n is closely related to the problem of arranging disjoint spheres of the same radius in an n-dimensional space (the *sphere packing problem*), with centers located in grids of a lattice (the so-called *regular arrangement*).

It had been observed already by Gauss [2212] that the density ϱ_n of the closest such arrangement (i.e., the limit of the ratio of the volume occupied by the spheres to the volume of the smallest cube containing them) is related to γ_n by the formula

$$\varrho_n = \frac{1}{\Gamma(1+n/2)}\left(\frac{\pi\gamma_n}{4}\right)^{n/2}. \tag{2.76}$$

Thus the old result $\gamma_2 = 2/\sqrt{3}$ of Lagrange [3611] implies $\varrho_2 = \pi/\sqrt{12}$, and the equalities $\gamma_3 = \sqrt[3]{2}$ and $\varrho_3 = \pi/\sqrt{18}$ follow from Gauss's comments [2212] on the book by L.A. Seeber[233] [5596]. The cases $n = 4, 5$ were treated by A.N. Korkin and E.I. Zolotarev [3478–3480], who obtained $\gamma_4 = \sqrt{2}$, and $\gamma_5 = \sqrt[5]{8}$, implying $\varrho_4 = \pi^2/16$ and $\varrho_5 = \pi^2/\sqrt{450}$.

In 1905 the lower bound

$$\varrho_n \geq \frac{\zeta(n)}{2^{n-1}} \tag{2.77}$$

was obtained by H. Minkowski in his paper [4326], where he constructed a reduction theory for positive definite quadratic forms in several variables. An approach to the reduction theory based on a study of form matrices was proposed in 1911 by A. Châtelet [1001].

A simpler presentation of Minkowski's reduction theory was given in 1928 by L. Bieberbach[234] and I. Schur [503]. A generalization to quadratic forms over arbitrary fields was made much later by P. Humbert[235] [2949, 2950].

In the twenties H.F. Blichfeldt [559, 560] obtained $\varrho_6 = \pi^3/48\sqrt{3}$ and $\varrho_7 = \pi^3/105$. The result in dimension 6 has been independently obtained by N. Hofreiter [2843]. Blichfeldt published his proofs only in 1935 [562], when he also established the equality $\varrho_8 = \pi^4/384$. These results imply

$$\gamma_6 = 2/\sqrt[6]{3}, \qquad \gamma_7 = \sqrt[7]{64}, \qquad \gamma_8 = 2$$

for the constants γ_n defined by (2.74).

---

[233] Ludwig August Seeber (1793–1855), professor of physics in Freiburg.

[234] Ludwig Bieberbach (1886–1982), professor in Basel, Frankfurt and Berlin. See [2375].

[235] Pierre Humbert (1891–1953), professor in Montpellier, son of G. Humbert.

For $9 \leq n \leq 12$ and $n = 15$ lower bounds for ϱ_n were obtained later by T.W. Chaundy [1006], E.S. Barnes [331, 332], E.S. Barnes and G.E. Wall [334] and H.S.M. Coxeter and J.A. Todd[236] [1264] by explicit constructions. They are listed in the book by C.A. Rogers [5254], where also the opinion is expressed that these bounds are optimal. A construction of packings for $n = 2^k$ $(k \geq 4)$ as well as for $n \leq 24$ was given in 1964 by J. Leech [3760, 3761] (cf. J.H. Conway [1223]). The lattice Λ found by Leech in the case $n = 24$ (the *Leech lattice*) plays an important role in coding theory (see I.F. Blake [545], W. Ebeling [1681]). It turned out later that the factor group of the group of rotations of Λ by its center is a new sporadic simple group of more than $8 \cdot 10^{15}$ elements, containing two other new sporadic groups (J.H. Conway [1222]). Various constructions of the Leech lattice were given later (J.H. Conway, N.J.A. Sloane [1226, 1227], N.D. Elkies [1719], N.D. Elkies, B.H. Gross [1725], J. Lepowsky, A. Meurman [3832], M. Craig [1265]).

The Leech lattice is an example of a *unimodular lattice*, i.e., having discriminant 1. All such lattices in $\mathbf{R}^N$ for $N \leq 23$ were listed by J.H. Conway and N.J.A. Sloane [1228, 1229], for $N = 24$ this has been done by H.-V. Niemeier [4615], and for $N = 25$ by R.E. Borcherds [627]. For partial results in larger dimensions see, e.g., R. Bacher, B.B. Venkov [192], C. Bachoc, G. Nebe, B.B. Venkov [203], P. Gaborit [2172], H. Koch, G. Nebe [3430], H. Koch, B.B. Venkov [3431, 3432], M. Kervaire[237] [3306]. See also the book by J.H. Conway and N.J.A. Sloane [1230].

The case of higher dimensions is still open, and only some bounds are known. Upper bounds result from (2.75) as well as from the inequality

$$\gamma_n \leq \gamma_{n-1}^{(n-1)(n-2)}$$

proved by L.J. Mordell [4392] (cf. A. Oppenheim[238] [4694]). The lower bound

$$\gamma_n \geq (1 + o(1)) \frac{8}{\pi^2 e} \frac{n\zeta(n)}{2^{n-1}}$$

follows from Minkowski's bound (2.77) for ϱ_n. It was improved much later by C.A. Rogers [5250], H. Davenport and C.A. Rogers [1400], and K. Ball [291]. The upper bound was made smaller (also in the non-regular case) by H.F. Blichfeldt [561], R.A. Rankin [5113], C.A. Rogers [5253], V.M. Sidelnikov [5733], G.A. Kabatiansky and V.I. Levenštein [3176] (for $n \geq 43$), V.I. Levenštein [3846] and K. Bezdek [496] (for $n \geq 8$). See the books by C.A. Rogers [5254], J.H. Conway and N.J.A. Sloane [1230] and K. Bőrőczky [646], as well as a survey by G. Fejes Tóth [1964]. The best known upper bounds for $4 \leq n \leq 36$ were given by H. Cohn and N.D. Elkies [1159].

In the three-dimensional non-regular case upper bounds were given by J.H. Lindsey [3899] and D.J. Muder [4464, 4465], and the final step was taken by T. Hales in a series of papers, one with S.P. Ferguson, published between 1997 and 2006 ([2460–2465, 2467], cf. S.P. Ferguson [1988]) who confirmed an old conjecture[239] of Kepler stating that the maximal density equals $\pi/\sqrt{18}$ (for corrections see [2468]). This concluded the realization of the program put forward by Hales in 1992 [2459]. For the history of Kepler's conjecture see T. Hales [2466].

---

[236] John Arthur Todd (1908–1994), reader in Cambridge. See [160].

[237] Michel Kervaire (1927–2007), professor at the Courant Institute and in Geneva. See [1712].

[238] Alexander Oppenheim (1903–1997), professor in Singapore, Reading and at the Universities of Ghana and Benin.

[239] It formed a part of Hilbert's eighteenth problem.

A related conjecture was proposed in 1943 by L. Fejes Tóth [1966] (cf. his books [1967, 1968]). Let Λ be the set of centers of spheres realizing a packing of spheres in $\mathbf{R}^3$, and for any $P \in \Lambda$ define its *Voronoĭ cell* as the set of all points of $\mathbf{R}^3$ which are closer to P than to any other point in Λ. The conjecture asserts that the volume of every Voronoĭ cell is at least equal to the volume of a regular dodecahedron of unit inradius. It has been known for some time that it implies Kepler's conjecture. Recently the dodecahedron conjecture has been established by T. Hales and S. McLaughlin [2469].

10. Dense regular arrangements of disjoint congruent convex bodies other than spheres have also been considered, and it seems that this question in $\mathbf{R}^3$ has been first considered by H. Minkowski [4325] in 1904. In the case of congruent tetrahedrons the problem is equivalent to the question of minimal positive value attained by $|x|+|y|+|z|$ at points of a lattice of given determinant D. Minkowski showed that this minimum is $\leq cD$, with $c = (108/19)^{1/3}$, and deduced the possibility of the approximation of two real numbers a_1, a_2 by rationals $r_1 = p_1/q, r_2 = p_2/q$ with the same denominator, so that both differences $|r_i - a_i|$ are $\leq \sqrt{8/19}/q^{3/2}$.

The methods used by H.F. Blichfeldt in [562] were later adapted by G. Fejes Tóth and W. Kuperberg [1965] to obtain upper bounds for the packing density for a large class of convex bodies.

A problem which is in some sense dual to the packing problem deals with a covering of $\mathbf{R}^n$ by overlapping equal spheres whose centers form a lattice. The main question is to find the sparsest such covering, but the answer is known only for $n=2$ (R. Kershner [3304]), $n=3$ (R.P. Bambah [308]), $n=4$ (B.N. Delone[240], S.S. Ryškov [1456]) and $n=5$ (S.S. Ryškov, E.P. Baranovskiĭ [5359, 5360]). In higher dimensions only bounds are known (see H. Davenport [1366], R.P. Bambah, H. Davenport [310], P. Erdős, C.A. Rogers [1852], C.A. Rogers [5253], H.S.M. Coxeter, L. Few, C.A. Rogers [1263], M.N. Bleicher [554], P. Erdős, L. Few, C.A. Rogers [1828], S.S. Ryškov [5358]).

11. The problem of minimas attained by indefinite binary quadratic forms at integral points had already been considered by A.A. Markov in two papers [4153, 4154] published in 1879 and 1884. If $f(x,y)=ax^2+bxy+cy^2$ is such a form with real coefficients, $D(f)=b^2-4ac$ is its discriminant, and $m(f)$ is the minimal absolute value of $f(x,y)$ attained at points $(x,y)\neq(0,0)$ with integral coordinates, then Markov showed that the maximal value of the ratio

$$\delta(f) = \frac{m(f)}{\sqrt{D(f)}}$$

equals $2/\sqrt{5}$, and there are only denumerably many non-equivalent forms f with $\delta(f) > 2/3$. His cumbersome proof used continued fractions. He found a connection between this problem and the Diophantine equation

$$x^2+y^2+z^2=3xyz, \tag{2.78}$$

and showed that solutions of it could be arranged in a binary tree (the numbers x occurring in this equation are called *Markov numbers*).

---

[240] Boris Nikolaevič Delone [Delaunay] (1890–1980), professor in Leningrad and Moscow. See [1950].

Similar results for solutions of the equation $x^2 + y^2 + z^2 = nxyz$ were obtained in 1907 by A. Hurwitz [2961].

Solutions of the more general equation

$$x_1^2 + x_2^2 + \cdots + x_n^2 = ax_1x_2\cdots x_n$$

were investigated in a series of papers of A. Baragar [316, 317, 319].

Simpler proofs of Markov's results were given by G. Frobenius [2118] in 1913 (see also R. Remak [5160] and J.W.S. Cassels [927]).

It was noted by H. Cohn [1155, 1156] that Markov numbers are of importance in the theory of geodesics on Riemann surfaces (cf. A. Haas [2417]). In another paper [1157] Cohn showed that they are also related to primitive words of the free group with two generators.

In 1982 D. Zagier [6811] showed that the number $M(x)$ of Markov numbers below x satisfies

$$M(x) = c\log^2 x + O\big(\log x(\log\log x)^2\big)$$

with some explicitly given $c > 0$. He also noted that the existence of

$$\lim_{x\to\infty} \frac{M(x)}{\log^2 x}$$

was earlier established in C. Gurwood's Ph.D. thesis [2396].

The problem of whether there are two solutions of (2.78) with the same value of z was stated by G. Frobenius [2118] in 1913 and is still open. Only some partial results are known (see A. Baragar [318], P. Schmutz [5531], J.O. Button [870], Y. Zhang [6829], A. Srinivasan [5881]).

A large part of the book [1310] by T.W. Cusick and M.E. Flahive is devoted to Markov numbers.

In [4156] A.A. Markov treated the analogous problem for indefinite ternary quadratic forms f, and gave a proof for the inequality

$$\frac{M(f)^3}{|D(f)|} \geq \frac{2}{3}$$

(the discriminant $D(f)$ being defined as the determinant of the matrix of f), which, as he wrote in a footnote, had been communicated to him by A.N. Korkin around 1880. He also determined the next two possible values of the ratio $M(f)^3/|D(f)|$, namely $2/5$ and $1/3$.

In the case of indefinite quaternary quadratic forms of signature zero the first seven possible values of $M(f)^4/|D(f)|$ were listed later by A. Oppenheim [4692, 4693], who noted also that the largest value, equal to 4/9 had been found much earlier by A.A. Markov. The next four values were found in 1945 by B.A. Venkov[241] [6384].

12. Packing and coverings were treated in the books by K. Bőrőczky, Jr. [646], J.H. Conway and N.J.A. Sloane [1230] and C.A. Rogers [5254].

---

[241] Boris Alekseevič Venkov (1900–1962), professor in Leningrad. See [6386].

An introduction to the geometry of numbers can be found in the books by J.W.S. Cassels [934], C.G. Lekkerkerker [3814] and C.L. Siegel [5777]. For a survey see [2371] and the recent monograph by P.M. Gruber on convex geometry [2370] contains also much information on that subject.

2.6.2 Integral Points in Regions

1. For a bounded region $\Omega \subset \mathbf{R}^2$ denote by $N(\Omega)$ the number of points with integral coordinates lying in Ω. Already C.F. Gauss [2213] had shown that if Ω is a plane convex body of area $V(\Omega)$ and for positive x,

$$x\Omega = \{xP : P \in \Omega\}$$

denotes its dilatation, then

$$N(x\Omega) = x^2 V(\Omega) + R(x) \tag{2.79}$$

with $R(x) = O(x)$.

The next general result in this topic was obtained in 1917 by I.M. Vinogradov [6403] who showed that if $\Omega \subset \mathbf{R}^2$ is a region bounded by a closed curve Γ, the curvature $\kappa(P)$ exists at each point $P \in \Gamma$ and satisfies $0 < r \leq \kappa(P) \leq R$, then

$$N(\Omega) = V(\Omega) + O\left(\frac{R^2 \log^{2/3} r}{r^{4/3}}\right).$$

An exposition of Vinogradov's proof, leading to a better error term (R^2 being replaced by $R^{4/3}$) can be found in the book by A.O. Gelfond[242] and Yu.V. Linnik [2234, Chap. 8].

The next results were obtained by J.G. van der Corput[243], who in his thesis [6274, 6275] developed a method to evaluate the number $N(\Omega)$ in the case when the boundary of Ω consists of convex arcs, satisfying certain regularity conditions. This allowed him, in a joint paper with E. Landau [3682], to improve Vinogradov's result to

$$N(\Omega) = V(\Omega) + O(R^{2/3}). \tag{2.80}$$

The exponent $2/3$ in (2.80) was later reduced by M.N. Huxley [2981–2983] to $0.636\ldots$, $0.630\ldots$ and $0.629\ldots$.

The error term $R(x)$ in (2.79) was shown by V. Jarník [3106] in 1924 to be $\Omega(\sqrt{x})$ and in 1985 W.G. Nowak [4635] proved

$$R(x) = \Omega\left(\sqrt{x}\log^4 x\right)$$

for regions with boundary of class C^∞ and finite non-zero curvature. He also obtained [4638–4640] a corresponding result in higher dimensions. A broad survey of these questions in

[242] Aleksandr Osipovič Gelfond (1906–1968), professor in Moscow. See [3858].

[243] Johannes Gualtherus van der Corput (1890–1973), student of J.C. Kluyver, professor in Groningen, Amsterdam and Berkeley. See [787].

arbitrary dimension was given in 2006 by A. Ivić, E. Krätzel, M.Kühleitner and W.G. Nowak [3043].

In his next papers [6276, 6278–6281] J.G. van der Corput produced bounds for exponential sums

$$\sum_{a \le n \le b} \exp(\pi i f(n)),$$

depending on bounds for the derivatives of f, which can be used in certain important cases to obtain good bounds for the difference between $N(\Omega)$ and the volume of Ω. In particular he improved the bounds of the error terms in the classical circle and divisor problems (see below).

It was observed by V. Jarník[244] (see H. Steinhaus[245] [5920]) that this method implies that if Γ is a closed rectifiable Jordan curve of length $L > 1$ in the plane, encompassing a region of area V, then for the number N of points with integral coordinates lying in that region one has $|N - V| < L$. For convex regions M. Nosarzewska [4627] obtained the inequalities

$$-1 - L/2 < V - N < L/2,$$

and showed that the left inequality is best possible. An analogous result in 3-space was obtained by W.M. Schmidt [5504] and J. Bokowski and J.M. Wills [591], and in $\mathbf{R}^n$ by J. Bokowski, H. Hadwiger and J.M. Wills [590]. For a survey see J.M. Wills [6676].

In 1972 H. Chaix [981] proved that one can take 16400 for the implied constant in (2.80).

See also B. Randol [5102], Y. Colin de Verdière [1182], W.G. Nowak [4632, 4633], A. Iosevich [3027] and the book by M.N. Huxley [2986].

The books by F. Fricker [2093] and by E. Krätzel [3517] published in 1982 and 1988, respectively, give a good presentation of problems concerning the number of lattice points in various regions. A survey of later development was given by W.G. Nowak [4637].

2.6.2.1 The Circle Problem

1. In the *circle problem* one asks for the number $F(x)$ of points with integral coordinates lying in the circle of radius $\sqrt{x}$, centered at the origin. Clearly

$$F(x) = \sum_{n \le x} r_2(n) = \sum_{a^2+b^2 \le x} 1 = 1 + 4 \sum_{0 \le n < \sqrt{x}} \left[\sqrt{x - n^2}\right],$$

$r_2(n)$ being the number of representations of n as the sum of two squares of integers.

Gauss's result (2.79) implies

$$F(x) = \pi x + O(\sqrt{x}),$$

and the first improvement was obtained in 1906 by W. Sierpiński [5780]. He used the method applied a few years earlier by his teacher G.F. Voronoĭ [6469] to the divisor problem to obtain

$$F(x) = \pi x + A(x) \tag{2.81}$$

[244] Vojtěch Jarník (1897–1970), student of Landau, professor in Prague. See [4628].

[245] Hugo Steinhaus (1887–1972), professor in Lwów and Wrocław. See [3177].

with $A(x) \ll x^{1/3}$. A simpler, but nevertheless rather involved, proof was later provided by E. Landau [3645].

A really simple proof was found much later by W.G. Nowak [4629]. See also the book [3064, Sect. 4.4] by H. Iwaniec and E. Kowalski.

Denote by a the greatest lower bound for numbers c, for which one has $A(x) \ll x^c$. There were several improvements of Sierpiński's result $a \le 1/3$. The first was obtained by J.G. van der Corput [6277], who got $a < 0.33$. (It was pointed out by A. Walfisz [6526] that his argument in fact gives $a \le 163/494 = 0.3299\ldots$.) Independently, a slightly weaker result ($a \le 37/112 = 0.3303\ldots$) was obtained by J.E. Littlewood and A. Walfisz [3948].

Here is the list of subsequent improvements:

- $a \le 27/82 = 0.3292\ldots$ (L.W. Nieland [4611], using the method of evaluations of exponential sums, introduced by J.G. van der Corput in [6282]; cf. E.C. Titchmarsh [6171]),
- $a \le 15/46 = 0.32608\ldots$ (E.C. Titchmarsh [6175]),
- $a \le 13/40 = 0.325$ (L.H. Hua [2928]),
- $a \le 12/37 = 0.3243\ldots$ (W.L. Yin [6780], J.R. Chen [1015]),
- $a \le 35/108 = 0.32407\ldots$ (W.G. Nowak [4631]),
- $a \le 139/429 = 0.32400\ldots$ (K. Kolesnik [3457], W.G. Nowak [4641]),
- $a \le 7/22 = 0.3181\ldots$ (H. Iwaniec, C.J. Mozzochi [3067]; cf. W. Müller, W.G. Nowak [4475]),
- $a \le 23/73 = 0.31506\ldots$ (M.N. Huxley [2982]).

The best known result is due to M.N. Huxley [2983] who in 2003 obtained

$$A(x) = O\big(x^{131/416} \log^{18637/8320} x\big).$$

(Note that $131/146 = 0.3149\ldots$.)

For a long time it has been conjectured that $a = 1/4$, and G.H. Hardy [2509] showed that this holds in the mean, by proving

$$\frac{1}{x}\int_1^x |A(t)|\,dt \ll x^{1/4+\varepsilon}$$

for every $\varepsilon > 0$. On the other hand the inequality $a \ge 1/4$ was established by G.H. Hardy [2506] and E. Landau [3652], and in [2508] Hardy proved that $A(x) = \Omega((x \log x)^{1/4})$.

This was improved in 1940 by A.E. Ingham [3022], who showed that one cannot have $A(x) = O((x \log x)^{1/4})$. Much later J.L. Hafner [2430] established

$$A(x) = \Omega_-\big(\big(x \log x \sqrt{\log\log x}\big)^{1/4}\big),$$

and the current record holder is K. Soundararajan [5852], who in 2003 got

$$A(x) = \Omega\big((x \log x)^{1/4} (\log\log x)^a (\log\log\log x)^{-5/8}\big)$$

with $a = 3(\sqrt[3]{2} - 1)/4 = 0.1949\ldots$.

The integral $I_2(x) = \int_0^x A^2(t)\,dt$ was considered in 1922 by H. Cramér [1269], who obtained

$$I_2(x) = cx^{3/2} + O\big(x^{5/4+\varepsilon}\big)$$

for every $\varepsilon > 0$ with a certain $c > 0$. Later E. Landau [3665] and A. Walfisz [6526] replaced the error term by $O(x^{1+\varepsilon})$ and $O(x\log^3 x)$, respectively.

After forty years I. Kátai [3268] proved the error term here to be $O(x\log^2 x)$, and in 2004 W.G. Nowak [4644] replaced this by $O(x\log^{3/2} x\log\log x)$.

Asymptotics of higher moments $\int_0^x A^n(x)dx$ for small n were found by K.M. Tsang ($n = 3, 4$) [6202] and D.R. Heath-Brown ($n \le 9$) [2648].

The related problem of evaluating the sum $S_2(x) = \sum_{n\le x} r_2^2(n)$ was considered by W. Sierpiński who, in his dissertation [5781, Sect. 6], obtained

$$S_2(x) = 4x\log x + ax + R_2(x) \tag{2.82}$$

with a certain constant a and $R_2(x) = O(x^{3/4}\log x)$.

S. Ramanujan stated that the error in (2.82) is $O(x^c)$ for every $c > 3/5$, and B.M. Wilson [6677] conjectured that this holds for every $c > 1/2$. This was accomplished in 1989 by W. Recknagel [5137], who showed

$$R_2(x) = O\big(\sqrt{x}\log^6 x\big).$$

Now the bound $R_2(x) = O(x^{1/2}\log^{11/3} x\,(\log\log x)^{1/3})$ is known (M. Kühleitner [3561]). On the other hand A. Schinzel [5447] obtained $R_2(x) = \Omega(x^{3/8})$.

2. The corresponding problem in the 3-space concerns the evaluation of

$$F_3(x) = \sum_{a^2+b^2+c^2\le x} 1,$$

the number of lattice points in the sphere of radius $\sqrt{x}$, centered at the origin.

It is not difficult to show by an argument similar to that used by Gauss that one has

$$F_3(x) = \frac{4\pi}{3}x^{3/2} + r(x),$$

with $r(x) = O(x)$, and in 1914 D. Cauer [967] obtained $r(x) = O(x^{3/4})$.

In 1926 A. Walfisz [6525] showed that if c denotes the greatest lower bound for numbers a with $r(x) = O(x^a)$, then $c \le 43/58 = 0.7413\ldots$, and for the next step one had to wait until 1949, when I.M. Vinogradov [6438] obtained $c \le 113/162 = 0.6975\ldots$, six years later [6439] improved this to $c \le 11/16 = 0.6875$, and stated in 1960 [6444] that he could show $c \le 19/28 = 0.6785\ldots$. Later J.R. Chen [1014] got $c \le 35/52 = 0.6730\ldots$, and I.M. Vinogradov [6445] and J.R. Chen [1016, 1017] showed $c \le 2/3$. This was made more precise by I.M. Vinogradov [6446] (see also his book [6448]), who proved

$$r(x) \ll x^{2/3}\log^6 x.$$

The next improvement came from F. Chamizo and H. Iwaniec [987], who in 1995 got $c \le 29/44 = 0.6590$. The best known evaluation of the error term is due to D.R. Heath-Brown [2655], who four years later proved $c \le 21/32 = 0.65625$.

On the other hand it is known that one has $r(x) = \Omega_-((x\log x)^{1/2})$, as shown by G. Szegö[246] [6017] in 1926, and in 2000 K.M. Tsang [6204] obtained $r(x) = \Omega_+((x\log x)^{1/2})$, improving upon a previous result by W.G. Nowak [4636].

[246]Gábor Szegő (1895–1985), professor in Königsberg, St. Louis and at Stanford. See [154].

This problem is closely related to a question in the theory of equivalence classes of binary quadratic forms. This notion goes back to Lagrange, who in a long paper [3611] considered binary quadratic forms

$$f(x, y) = ax^2 + bxy + cy^2 \tag{2.83}$$

with rational integral coefficients a, b, c and fixed discriminant $d(f) = b^2 - 4ac$. A form is called *primitive* if $(a, b, c) = 1$. In the case of a negative discriminant only positive definite forms (i.e., with $a > 0$) were considered. Lagrange defined two forms f, g to be equivalent if one could transform f into g by a linear substitution

$$x' = ax + by, \qquad y' = cx + dy,$$

assuming that the matrix

$$\mathfrak{M} = \begin{bmatrix} a & b \\ c & d \end{bmatrix}$$

lies in the group $GL_2(\mathbf{Z})$, i.e., $a, b, c, d \in \mathbf{Z}$ and $\det(\mathfrak{M}) = \pm 1$. The *class-number*, i.e., the cardinality of the set of equivalence classes, turned out to be finite. Later Gauss [2208] built an extensive theory of quadratic forms, however he distinguished between *proper* and *improper equivalence*, depending on the sign of the determinant (proper equivalence of two forms meaning their equivalence under the action of $SL_2(\mathbf{Z})$). Moreover he considered only forms with even middle coefficient in which case the discriminant is a multiple of 4. The number of equivalence classes of primitive forms of discriminant d is usually denoted by $h(d)$, and in the older literature the number of classes of Gaussian forms $aX^2 + 2bXY + cY^2$ with the determinant $D = b^2 - ac$ was denoted by $H(D)$, thus $H(D) = h(4D)$.

Much later it was shown by B.J. Birch and J.R. Merriman [535] that if one defines, in the same way, the equivalence of two binary forms with integral coefficients in a fixed algebraic number field, having the same degree and discriminant, then the number of resulting classes is finite. An effective proof of this result was given by K. Győry [2400] for forms over $\mathbf{Z}$ and by J.-H. Evertse and K. Győry [1934] in the general case (cf. A.Bérczes, J.-H. Evertse, K. Győry [425], J.-H. Evertse [1925]). This has been generalized to the case of decomposable forms in several variables by J.-H. Evertse and K. Győry [1935].

It was shown by Gauss ([2208, Sect. 291], cf. also [3526]) that if $r_3(n)$ is the number of representations of the number n as the sum of three squares, then r_3 and H are related by the formula

$$r_3(n) = \begin{cases} 12H(n) & \text{if } n \equiv 1 \pmod 4 \text{ and } n \text{ is not a square,} \\ 8H(n) & \text{if } n \equiv 3 \pmod 8, \\ 12H(n) & \text{if } n \equiv 2 \pmod 4, \\ 12H(n) - 6 & \text{if } n \text{ is an odd square.} \end{cases}$$

Since $r_3(n) = r_3(n/4)$ in the case $4|n$, and

$$H(4n) = \begin{cases} 2H(n) - 1 & \text{if } n \text{ is an odd square,} \\ 2H(n) & \text{otherwise,} \end{cases}$$

then, as shown by E. Landau [3642–3644], the asymptotics of

$$\mathcal{H}(x) = \sum_{d \le x} H(-d),$$

can be deduced from evaluations of the sum $\sum_{a^2+b^2+c^2 \le x} 1$, in which a, b, c are restricted to fixed residue classes mod 2.

The study of $\mathcal{H}(x)$ has its origin in Gauss's book [2208], where it is stated in Sect. 302 that $\mathcal{H}(x)$ is asymptotically equal to

$$\frac{4\pi}{21\zeta(3)} x^{3/2} - \frac{2}{\pi^2} x + o(x). \tag{2.84}$$

Although Gauss asserted that he found that formula "through theoretical investigations," his book does not contain the proof of this assertion. If one also admits forms with b odd, then Gauss's formula has to be modified to the form,

$$\mathcal{H}'(x) = \sum_{d \le x} h(-d) = \frac{\pi}{18\zeta(3)} x^{3/2} - \frac{3}{2\pi^2} x. \tag{2.85}$$

The first proof of a result about $\mathcal{H}(x)$ is due to F. Mertens [4257], who established that in fact $\mathcal{H}(x)$ is asymptotic to $(4\pi/21\zeta(3))x^{3/2}$. Later E. Landau considered this question, and in 1912 proved [3644] Gauss's assertion (2.84) with the error term $R(x) = O(x^{5/6} \log x)$. The first improvement of this error term was achieved by I.M. Vinogradov who in [6403] slightly reduced the exponent of the logarithm and in [6404] obtained $R(x) = O(x^{3/4} \log^2 x)$.

Subsequent improvements in the bound for the error term in the problem of lattice points in three-dimensional spheres, quoted above, led to corresponding improvements for $\mathcal{H}(x)$ and $\mathcal{H}'(x)$ (see F. Chamizo, H. Iwaniec [988]).

Asymptotic formulas for the sums $\sum_{d \le x} h^k(-d)$ for integers $k \ge 2$, with $-d$ running over fundamental discriminants, were given by M.B. Barban[247] [321] (cf. M.B. Barban, G. Gordover [326] and A.F. Lavrik [3736]). The error terms in these formulas were improved by D. Wolke [6711].

In [6405] I.M. Vinogradov also considered the analogue of formula (2.84) for the case of indefinite binary quadratic forms. Its origin goes back to Gauss, who in [2208, Sect. 304] asserted that the sum

$$\mathcal{H}^+(x) = \sum_{0<d \le x} H(d) \log \varepsilon_{4d}$$

(with $\varepsilon_m = x + y\sqrt{m}$, where x, y are the smallest positive integers satisfying $x^2 - my^2 = 4$) is asymptotic to $cx^{3/2}$ with a certain constant c. The value of this constant given in [2208] is not correct, however the true value, $c = \pi^2/18\zeta(3)$ can be found in Gauss's manuscripts.

The equality

$$\mathcal{H}^+(x) = cx^{3/2} + O(x \log x)$$

[247] Mark Borisovič Barban (1935–1968), professor in Taškent. See [6402].

has been established as late as in 1944 by C.L. Siegel [5764]. Later, in 1975, T. Shintani [5706] proved the equality

$$\mathcal{H}^{+}(x) = cx^{3/2} + c_1 x \log x + c_2 x + O(x^{\alpha}),$$

with explicitly given constants c_1, c_2 and any $a > 3/4$. In 2006 F. Chamizo and A. Ubis [986] obtained a similar asymptotic formula for the sum

$$\sum_{0<d\leq x} H(d) \log \varepsilon_d .$$

The case when the summation runs only over fundamental discriminants (in which case the corresponding sums give mean values for the class-number of imaginary quadratic fields and the product of the class-number by the regulator of real quadratic fields) was treated in 1985 by D. Goldfeld and J. Hoffstein [2260]. This was later extended by B. Datskovsky [1338], who considered similar sums with conditions on ramification of given finitely many primes. Cf. P. Sarnak [5405, 5406] and N. Raulf [5132].

The asymptotic behavior of the related sum $\mathcal{H}_0(x) = \sum_{0<d\leq x} H(d)$ is still unknown. In 1984 C. Hooley [2898] conjectured the equality

$$\mathcal{H}_0(x) = \left(\frac{25}{12\pi^2} + o(1)\right) x \log^2 x,$$

which remains unproved.

Surprisingly, a similar question for class-numbers of irreducible binary cubic forms of given discriminant found its answer in 1951 (H. Davenport [1364, 1365]). In the case of higher degree, only partial results are known (J. Nakagawa [4526]).

3. Similar questions have also been considered in higher dimensions. Denote by $P_k(x)$ the number of points with integral coordinates in a sphere of radius $\sqrt{x}$ in k-dimensional real space, i.e.,

$$P_k(x) = \sum_{a_1^2+\cdots+a_k^2 \leq x} 1,$$

and let $V_k(x) = c_k x^{k/2}$ with

$$c_k = \frac{\pi^{k/2}}{\Gamma(1+k/2)}$$

be the volume of that sphere. The problem consists of the evaluation of the difference

$$D_k(x) = |P_k(x) - V_k(x)|,$$

and the first result in the case $k \geq 3$ seems to be due to H. Minkowski [4326], who proved in 1905 the evaluation

$$D_k(x) = O\big(x^{(k-1)/2}\big). \tag{2.86}$$

In the case $k = 4$ the bound (2.86) was improved by E. Landau [3642, 3643] to $D_4(x) \ll x^{1+\varepsilon}$ for every positive ε, but his proof, as noted in [3667, 3668], actually gives the stronger bound $D_4(x) = O(x \log x)$.

The next improvement for $k=4$ was obtained in 1927 by A. Walfisz [6526], who deduced $D_4(x) \ll x\log x/\log\log x$ from his formula[248]

$$\sum_{n\le x}\frac{\sigma(n)}{n}=\frac{\pi^2}{6}x-\frac{1}{2}\log x+O\left(\frac{\log x}{\log\log x}\right).$$

In [6527], published in 1931, he considered the second moment of $D_4(x)$ and showed

$$\int_1^x D_4^2(t)\,dt=\frac{2\pi^2}{3}x^3+O\big(x^{5/2}\log x\big),$$

improving upon a result of V. Jarník [3112–3114], who had the bound $O(x^3)$. Much later L.K. Hua [2932] proved $D_4(x)\ll x\log^{3/4}x(\log\log x)^{1/2}$, and the best known result is due to A. Walfisz, who in the Russian edition of his book [6532] gave the bound

$$D_4(x)\ll x\log^{2/3}x.$$

On the other hand, in 1927 A. Walfisz [6526] proved $D_4(x)=\Omega(x\log\log x)$, and the more precise results $D_4(x)=\Omega_+(x\log\log x)$ and $D_4(x)=\Omega_-(x\log\log x)$ were obtained in the nineties by S.D. Adhikari, R. Balasubramanian and A. Sankaranarayanan [16] and S.D. Adhikari and Y.-F.S. Pétermann [17], respectively.

The case $k\ge 5$ turned out to be the simplest. In this case it was observed by V. Jarník (see [3668]) that the bound

$$D_k(x)\ll x^{k/2-1},$$

proved in 1924 in the thesis of A. Walfisz [6516], cannot be improved.

In 1927, for $P_k(n)$ ($k\ge 5$, n integral) H. Petersson[249] [4804] obtained the expansion

$$P_k(n)=\sum_{j=0}^{[(n-1)/4]}\alpha_{k,j}(n)(n+1/2)^{k/2-j}+O\big(n^{k/4}\log n\big),$$

with bounded $\alpha_{k,j}(n)$, using the theory of theta-functions and complex integration. For $5\le k\le 8$ he got a better error term, namely $O(n)$ for $k=5,7$ and $O(n\log n)$ for $k=6,8$. A very detailed exposition of these results for $k\ge 4$ is contained in the book by A. Walfisz [6532].

4. A natural generalization of the circle problem concerns the number of integral points in an ellipse, or, more generally, in a k-dimensional ellipsoid $Ell(x)$, defined by the equation

$$F(x_1,\dots,x_k)\le x,$$

where F is a positive definite quadratic form with real coefficients. Let $D=D(F)$ be the discriminant of F. There is a difference of behavior between forms which are constant multiples of forms with rational coefficients (such forms are called *rational*), and all other forms, called *irrational*.

---

[248] Much later he showed the error term to be $O(x\log^{4/5}x\log\log x)$ [6521].

[249] Hans Petersson (1902–1984), student of Hecke, professor in Hamburg and Münster. See [6709].

E. Landau showed in [3642, 3643] that if $N(x)$ is the number of integral points in *Ell*(x), and

$$V = \frac{\pi^{k/2}}{\Gamma(1+\frac{k}{2})\sqrt{D}}$$

is the volume of *Ell*(1), then

$$P(x) = N(x) - Vx^{k/2} = O(x^{\lambda+\varepsilon}),$$

holds for every $\varepsilon > 0$ with $\lambda = k(k-1)/2(k+1)$ (later [3651] he was able to remove ε from this formula).

In the case of rational f and $k = 4$, E. Landau [3667, 3668] proved $P(x) \ll x\log^2 x$, and this was improved in the case of diagonal forms to $O(x\log x)$ by H.D. Kloosterman [3370]. In 1932 A. Walfisz [6519] utilized the connection between lattice points in ellipsoids and modular functions discovered by E. Hecke [2687] to get the bound $P(x) = O(x\log x/\log\log x)$. Later [6521] he got $P(x) = O(x\log^{4/5} x\log\log x)$, and in 1960 reduced the bound to $P(x) \ll x\log^{2/3} x$ [6534]. The second moment of $P(x)$ was considered in 1935 by A. Walfisz [6519], who used modular forms to achieve

$$\int_0^x P^2(t)\,dt = cx^3 + O\big(x^{5/2}\log^2 x\big)$$

with $c > 0$. Later [6520] he found a simpler proof.

For $5 \le k \le 7$ the bound

$$P(x) \ll x^{k/2-1}$$

was established in 1924 by E. Landau [3667] and A. Walfisz [6516] proved the same evaluation for all $k \ge 8$. A. Walfisz also showed [6517] that for $k \ge 5$ this bound is best possible.

The case of irrational f turned out to be much harder. For diagonal forms with exactly one irrational coefficient, say α, A. Walfisz [6518] proved in 1927 the bound

$$P(x) = o\big(x^{k/2-1}\big)$$

for $k \ge 10$, showed that this evaluation cannot be improved and obtained a slightly stronger result holding for almost all α. This last result was superseded in the following year by V. Jarník who in [3108] proved that for almost all diagonal forms in $k \ge 4$ variables one has

$$P(x) \ll x^{k/4+\varepsilon}$$

for every $\varepsilon > 0$ (cf. [3115]). He also obtained, for such forms, the evaluation

$$P(x) \ll \begin{cases} x^{k/2-1}\log x & \text{if } k \ge 5, \\ x\log^2 x & \text{if } k = 4. \end{cases}$$

In [3109] he proved that for every irrational diagonal form in at least six variables one has

$$P(x) = o\big(x^{k/2-1}\big)$$

and

$$P(x) = \Omega\big(x^{(k-1)/4}\big).$$

For $k = 5$, V. Jarník and A. Walfisz [3117] showed later that $P(x) = o(x^{3/2})$ and in the case $k = 4$ obtained $P(x) = \Omega(x\log\log x)$.

Optimal lower and upper bounds for the first and second moment of $P(x)$ in the case of rational diagonal forms were given by V. Jarník in the last two parts of [3112–3114].

Analogous questions for shifted ellipsoids, defined by

$$\sum_{i,j=1}^{n} a_{ij}(X_i - b_i)(X_j - b_j) \leq x$$

with rational a_{ij}, b_i were studied by E. Landau [3668] and Ch.H. Müntz [4481].

5. In 1914 D. Cauer [967] considered in his thesis the number $A_\lambda(t)$ of lattice points in the region bounded by the curve defined by

$$|x|^\lambda + |y|^\lambda = t,$$

and showed that for every real $\lambda \geq 2$ the equality

$$A_\lambda(t) = \frac{2\Gamma^2(1/\lambda)}{\lambda\Gamma(2/\lambda)} t^{2/\lambda} + O(t^c)$$

with $c = c(\lambda) = 1/\lambda - 1/(2\lambda^2 - \lambda)$.

The value of c was improved for $\lambda > 3$ to $c \leq 1/\lambda - 1/\lambda^2$ in the thesis of J.G. van der Corput [6274]. For $\lambda = 3$ this was done by E. Krätzel [3514], W.G. Nowak [4630] did this for $\lambda > 41/14 = 2.928\ldots$, and W. Müller and W.G. Nowak [4474] extended it to $\lambda > 38/13 = 2.923\ldots$. The bound obtained is best possible. See also B. Randol [5102] and S.B. Ablialimov [6].

2.6.2.2 Dirichlet's Divisor Problem

1. In 1849 Dirichlet [1590] considered the mean value of the number $d(n)$ (earlier denoted mostly by $\tau(n)$) of positive divisors of an integer n and proved the following asymptotic formula:

$$D(x) = \sum_{n \leq x} d(n) = x \log x + (2\gamma - 1)x + \Delta(x), \tag{2.87}$$

where $\Delta(x) = O(x^{1/2})$ and $\gamma = 0.577215\ldots$ is *Euler's constant*[250], defined by

$$\gamma = \lim_{n\to\infty} \left(\sum_{k=1}^{n} \frac{1}{k} - \log n \right).$$

Since

$$\sum_{n \leq x} d(n) = \sum_{n \leq x} \sum_{d|n} 1 = \sum_{d \leq x} \sum_{\substack{n \leq x \\ d|n}} 1 = \sum_{d \leq x} \left[\frac{x}{d} \right],$$

[250]Whether γ is irrational is an old unsolved problem.

the function $D(x)$ counts lattice points (a,b) with $1 \le a \le x, b \ge 1$, lying under the hyperbola $y(t) = x/t$. Dirichlet's idea consisted of separately considering points (a,b) with $a \le \sqrt{x}$ and $b \le \sqrt{x}$, which leads to

$$D(x) = 2\sum_{a\le\sqrt{x}}\sum_{b\le x/a} 1 - \sum_{a,b\le\sqrt{x}} 1 = 2x\sum_{a\le\sqrt{x}}\frac{1}{a} - x + O(\sqrt{x}),$$

and so his result is a consequence of the formula

$$\sum_{n\le t}\frac{1}{n} = \log t + \gamma + O\left(\frac{1}{t}\right).$$

In a letter to Kronecker [1591], Dirichlet asserted that he can reduce the exponent in the error term in (2.87) but gave no details and the first published improvement of Dirichlet's evaluation was obtained by G.F. Voronoĭ [6469] in 1903. He used a special partition of the region $\{(x,y) : x, y \ge 1, xy \le t\}$ in the plane to deduce that the error term in (2.87) is of the order $O(x^{1/3}\log x)$. In the next year [6470] he proved a Bessel function expansion of the sum

$$\sum_{n=a}^{b} d(n) f(n),$$

where f is a continuous function, having only finitely many maxima and minima in the interval (a,b). This gave, in particular, an expansion of $D(x)$ (cf. G.H. Hardy [2508], H. Cramér [1270], W.W. Rogosinski[251] [5261], A. Walfisz [6515]).

Simpler proofs were later provided by N.S. Košliakov [3494, 3495], A.L. Dixon[252] and W.L. Ferrar [1596], J.R. Wilton [6684], T. Meurman [4275].

Similar expansions for sums $\sum_{n\le x} F(n)$ were later obtained for various arithmetic functions F. For $F = \sigma_k$ this was done in 1927 by A. Oppenheim [4689] (cf. J.R. Wilton [6685]), and for $F = r_2$ by J.R. Wilton [6680] in 1928.

Voronoĭ's formula formed a special case of a more general conjectured expansion (*Voronoĭ's summation formula*), generalizing the classical Poisson formula. In the simplest case it has the following form.

For every arithmetic function $A(n)$ there exist analytic functions $\alpha(x)$ and $\delta(x)$ such that for every function f continuous in $[a,b]$ and having there finitely many maxima and minima, one has

$$\frac{A(a)f(a) + A(b)f(b)}{2} + \sum_{n=a+1}^{b-1} A(n)f(n)$$
$$= \int_a^b f(t)\delta(t)\,dt + \sum_{n=1}^{\infty} A(n)\int_a^b f(t)\alpha(nt)\,dt.$$

---

[251] Werner Wolfgang Rogosinski (1894–1946), professor in Königsberg and at the University of Durham.

[252] Arthur Lee Dixon (1867–1955), professor in Oxford. See [1994].

For certain classes of arithmetical functions this conjecture has been established by A. Walfisz [6523, 6524] in 1925, N.S. Košliakov [3494] in 1928, W.L. Ferrar [1992, 1993] in 1935, A.P. Guinand [2379] in 1937, and B.C. Berndt [444–446] in 1969–1972.

E. Landau [3644, 3649] gave fresh proofs of Voronoĭ's bound for $\Delta(x)$, and in [3642, 3643] proved a very general theorem concerning the sum of coefficients $A(x) = \sum_{n\le x} a_n$ of a Dirichlet series $\sum_{n=1}^{\infty} a_n n^{-s}$. This theorem asserted that if the series converges in a half-plane $\Re s > \alpha$, its coefficients satisfy, for every positive ε, the inequality $|a_n| \ll n^{\alpha+\varepsilon}$, the function $f(s)$ defined by this series satisfies a kind of functional equation and does not grow too quickly when $|\Im s|$ tends to infinity, then the sum $A(x)$ is well approximated by the sum of residues of $x^s f(s)/s$ in a certain well-defined perpendicular strip. This allowed him to make important progress in the problem of evaluation of the number of points with integral coordinates in various regions. In particular, this led to a new proof of Voronoĭ's result.

2. It is not difficult to show (see, e.g., [3661]) that the error term in the divisor problem equals

$$\Delta(x) = -2 \sum_{n\le\sqrt{x}} \left(\left\{ \frac{x}{n} \right\} - \frac{1}{2} \right) + O(1),$$

so all further work was aimed at the evaluation of the sum occurring here. Since the Fourier series of the function $\alpha(t) = \{t\} - 1/2$ has the form

$$\alpha(t) = -\frac{1}{\pi} \sum_{n=1}^{\infty} \frac{\sin(2\pi n t)}{n} = \frac{i}{2\pi} \sum_{\substack{n\in\mathbf{Z}\\ n\ne 0}} \frac{\exp(2\pi n x t i)}{n},$$

one can approximate $\alpha(t)$ by a convenient partial sum of this series, and the whole question is reduced to a good evaluation of the resulting exponential sum. This type of argument is applicable to a wealth of number-theoretical problems, and this explains why evaluations of exponential sums became a very important tool.

A lower bound for $\Delta(x)$ was furnished by G.H. Hardy [2507, 2508], who obtained

$$\limsup_{x\to\infty} \frac{\Delta(x)}{(x\log x)^{1/4}\log\log x} > 0,$$

and

$$\liminf_{x\to\infty} \frac{\Delta(x)}{x^{1/4}} < 0.$$

Much later A.E. Ingham [3022] showed

$$\liminf_{x\to\infty} \frac{\Delta(x)}{x^{1/4}} = -\infty.$$

It is conjectured that $\Delta(x)$ is of the order $O(x^{1/4+\varepsilon})$ for every $\varepsilon > 0$, and this is supported by a result by G.H. Hardy [2509], who proved

$$\frac{1}{x}\int_1^x |\Delta(t)|\,dt \ll x^{1/4+\varepsilon}$$

for every positive ε.

The best known lower bound for $\Delta(x)$ is due to K. Soundararajan [5852], who obtained

$$\Delta(x) = \Omega\big((x\log x)^{1/4}(\log\log x)^{a}(\log\log\log x)^{-5/8}\big)$$

with $a = 1.139\ldots$. The previous record belonged to J.L. Hafner [2429], who in 1981 got

$$\Delta(x) = \Omega_{+}\big((x\log x)^{1/4}(\log\log x)^{(3+\log 4)/4}\exp(-c(\log\log\log x)^{1/2})\big),$$

with a certain $c > 0$, improving upon previous results of K.S. Gangadharan [2190] and K. Corrádi, I. Kátai [1250].

The first improvement of Voronoĭ's upper bound was obtained by J.G. van der Corput [6277] who in 1922 obtained

$$\Delta(x) \ll x^{a}$$

with $a \le 163/494 = 0.32995\ldots$. He applied his method of dealing with exponential sums developed in [6276], as well as Weyl's evaluations of such sums [6647]. In 1923 he presented a survey of various methods of dealing with the divisor and circle problems and their generalizations in a talk in Geneva [6280], and five years later [6282] improved the bound to

$$\Delta(x) \ll x^{a}\log^{b} x$$

with $a = 27/82 = 0.3292\ldots$ and $b = 11/41$, using evaluations of sums of the form

$$\sum_{n=a}^{b} \exp(4\pi i\sqrt{un}),$$

with a suitable parameter u, which are special cases of the main result of [6279].

Later the following evaluations of the form $\Delta(x) \ll x^{a+\varepsilon}$ for every positive ε were obtained:

- $a = 15/46 = 0.3260\ldots$ (T. Chih [1054], H.-E. Richert [5204]),
- $a = 13/40 = 0.325$ (W.L. Yin [6778]),
- $a = 12/37 = 0.3243\ldots$ (G. Kolesnik [3453]),
- $a = 346/1067 = 0.3242\ldots$ (G. Kolesnik [3454]),
- $a = 35/108 = 0.3240\ldots$ (G. Kolesnik [3456]),
- $a = 7/22 = 0.3181\ldots$ (H. Iwaniec, C.J. Mozzochi [3067]; cf. W.Müller, W.G. Nowak [4475]),
- $a = 23/73 = 0.315\ldots$ (M.N. Huxley [2982]).

Now the record is held by M.N. Huxley [2983], who in 2003 got $a = 131/416 = 0.31498\ldots$.

The behavior of higher moments

$$M_k(x) = \int_1^x \left|\Delta^k(t)\right| dt$$

and

$$M_k^*(x) = \int_1^x \Delta^k(t)\,dt \quad (2\nmid k)$$

of the error term in the divisor problem is, for large k, still unknown. The evaluation

$$M_1^*(x) = o\big(x^{5/4}\big)$$

follows from results by G.F. Voronoĭ [6470], and H. Cramér [1269] established

$$M_2(x) = cx^{3/2} + \Delta_2(x)$$

with $\Delta_2(x) \ll x^{5/4+\varepsilon}$ for every $\varepsilon > 0$ (earlier G.H. Hardy [2509] proved $M_2(x) \ll x^{3/2+\varepsilon}$ for all $\varepsilon > 0$). Cramér's result supports the conjecture about the true order of $\Delta(x)$.

In 1956 K.C. Tong [6191, 6192] reduced the last bound to $O(x \log^5 x)$, and in 1988 E. Preissmann [5010] diminished it to $O(x \log^4 x)$. This seems to be fairly close to the optimal bound, as this error term is certainly not $o(x \log^2 x)$ (Y.K. Lau, K.M. Tsang [3733]). It was conjectured by Tsang [6203] that

$$\Delta_2(x) = -\frac{1}{4\pi^2} x \log^2 x + cx \log x + O(x)$$

holds with a certain constant c.

One conjectures also that $M_k(x)$ equals $(c_k + o(1))x^{1+k/4}$ with a certain $c_k > 0$, but this has been established only for $k \leq 9$ by D.R. Heath-Brown [2648], who showed also the existence of the limits

$$\lim_{x \to \infty} M_k^*(x) x^{-1-k/4}$$

for these k. For the values of c_k and bounds of the corresponding error terms see K.M. Tsang [6202], A. Ivić, P. Sargos [3047] for $k = 3, 4$, and W. Zhai [6821–6823] for $4 \leq k \leq 9$.

In 1952 P. Erdős [1807] proved that if $f \in \mathbf{Z}[X]$ is irreducible, then the ratio

$$\frac{1}{x \log x} \sum_{n \leq x} d(f(n))$$

lies between two positive constants depending on f (for a correction see F. Delmer [1453]). For reducible polynomials a corresponding result was obtained in 1968 by V. Ennola [1767].

If f is linear the problem reduces to the evaluation of the sum

$$D(x; q, a) = \sum_{\substack{n \leq x \\ n \equiv a \pmod{q}}} d(n).$$

Such sums were considered in 1957 by A.I. Vinogradov and Yu.V. Linnik [6401], who showed that for $c < 1$, $q \leq x^c$ and $(a, q) = 1$ one has

$$|D(x, q, a)| \ll_c \frac{x}{q} \prod_{p \mid q} \left(1 - \frac{1}{p}\right) \tag{2.88}$$

(see M.G. Qi [5020] for a correction). Also in 1957 C. Hooley [2855], applied Weil's evaluations of Kloosterman sums to show that for $q < x^{2/3}$ one has, with suitable $A > 0$,

$$D(x; q, a) = \left(1 + O\left(\frac{1}{\log^c x}\right)\right) \sum_{\substack{n \leq x \\ (a,q)=1}} d(n),$$

uniformly in $q \leq x^a$. H.G. Kopetzky [3474] gave explicit formulas for the coefficients α, β in the equality

$$D(x; q, a) = \alpha(q, a) x \log x + \beta(q, a) x + O(\sqrt{x}),$$

valid for fixed q. The error term was later diminished to $O(x^c)$ for every $c > 35/108 = 0.3240\ldots$ by W.G. Nowak [4634].

For further progress see É. Fouvry [2049] and É. Fouvry, H. Iwaniec [2057].

In the quadratic case there is an asymptotical equality

$$\sum_{n\leq x} d(\alpha n^2 + \beta n + \gamma) = (A + o(1))x \log x,$$

proved (according to P. Erdős [1807]) by R. Bellman and H.N. Shapiro. More precise results were later obtained by C. Hooley [2859], N. Gafurov [2174] and J. McKee [4232, 4233]. See also E.J. Scourfield [5590].

3. The question of the asymptotic behavior of the sum

$$S_a(x) = \sum_{n\leq x} \sigma_\alpha(n)$$

can be regarded as a generalization of the divisor problem, to which it reduces in the case $\alpha = 0$. Already Dirichlet [1590] had considered the sum $S_1(x)$ and proved that it equals $cx^2 + O(x \log x)$ with $c = \pi^2/12$. In 1914 S. Wigert[253] [6666] established

$$S_{-1}(x) = \frac{\pi^2}{6}x + O(\log x),$$

and elementary methods lead, for $0 < |\alpha| < 1$, to

$$S_\alpha(x) = \zeta(1-\alpha)x + \frac{\zeta(1+\alpha)}{1+\alpha}x^{1-\alpha} + T_\alpha(x),$$

with $T_\alpha(x) = o(x^{1-\alpha})$.

The maximal order of σ_a was determined in 1913 by T.H. Gronwall [2350]:

$$\limsup_{n\to\infty} \frac{\sigma_a(n)}{n^a} = \zeta(a) \quad (a > 1),$$

$$\limsup_{n\to\infty} \frac{\sigma(n)}{n \log\log n} = e^\gamma,$$

γ being Euler's constant, and

$$\limsup_{n\to\infty} \frac{\log(\sigma_a(n)/n^a)\log\log n}{\log^{1-a} n} = \frac{1}{1-a}, \qquad 0 < a < 1.$$

The equality

$$\limsup_{n\to\infty} \frac{\log d(n) \log\log n}{\log n} = \log 2$$

had earlier been proved by S. Wigert [6665].

In 1927 A. Walfisz [6526] established

$$S_{-1}(x) = \frac{\pi^2}{6}x - \frac{\log x}{2} + O\left(\frac{\log x}{\log\log x}\right)$$

[253] Severin Wigert (1871–1941), worked in Stockholm.

and

$$S_1(x) = \frac{\pi^2}{12}x + O\left(\frac{x\log x}{\log\log x}\right),$$

and now it is known that

$$S_1(x) = \frac{\pi^2}{12}x + O\big(x\log^{2/3+\varepsilon} x\big)$$

holds for every $\varepsilon > 0$ (N.M. Korobov [3487]). On the other hand, it has been shown by Y.-F.S. Pétermann [4799] that the error term is $\Omega(x\log\log x)$ (for the case $25/38 < |a| < 1$ see [4800]).

In [6527], which appeared in 1931, A. Walfisz obtained evaluations of the second moment of the error terms T_α in the case $\alpha = \pm 1$, showing

$$\int_1^x T_\alpha^2(t)\,dt = \begin{cases} c_{-1}x + O(x^{1/2}\log x) & \text{if } \alpha = -1, \\ c_1x^3 + O(x^{5/2}\log x) & \text{if } \alpha = 1. \end{cases}$$

Similar results for $\alpha \in (-1,1)$, $|\alpha| \neq 1/2$ were later obtained by H. Cramér [1269, 1271] and S. Chowla [1071]. The case $|\alpha| = 1/2$ was treated in Walfisz's paper [6528]. For integral α see R.A. MacLeod [4053, 4054].

4. There is another generalization of the divisor problem, going back to the dissertation of A. Piltz [4876] who considered the summatory function $D_k(x)$ of coefficients of the Dirichlet series for $\zeta^k(s)$. These coefficients have a simple arithmetic interpretation as it can easily be seen that for $\Re s > 1$ we have

$$\zeta^k(s) = \sum_{n=1}^{\infty} \frac{d_k(n)}{n^s}, \tag{2.89}$$

where $d_k(n)$ denotes the number of representations of n as a product of k factors. In particular $d_2(n)$ coincides with the number of divisors of n. Piltz proved the asymptotic formula

$$D_k(x) = xA_k(\log x) + \Delta_k(x), \tag{2.90}$$

where $A_k(x)$ is a polynomial of degree $k-1$, and

$$\Delta_k(x) = O\big(x^{1-1/k}\log^{k-2} x\big) \quad (k \geq 2).$$

A simpler proof was given by E. Landau [3647] in 1912.

If we denote by α_k the lower bound of the numbers α for which the bound $\Delta_k(x) = O(x^\alpha)$ holds, then Piltz's result gives $\alpha_k \leq 1 - 1/k$. The first improvements of this bound were given by E. Landau [3642, 3643], who used his general theorem about sums of coefficients of Dirichlet series to obtain $\alpha_k \leq (k-1)/(k+1)$ for $k \geq 3$, and in [3649] proved

$$\Delta_k(x) \ll x^{(k-1)/(k+1)}\log^{k-1} x.$$

At the same time he observed [3642, 3643] that the Riemann Hypothesis implies $\alpha_k \leq 1/2$. On the other hand one has $\alpha_k \geq 1/2 - 1/2k$ (G.H. Hardy [2508]).

The next improvement was obtained by G.H. Hardy and J.E. Littlewood [2530], who proved $\alpha_k \le 1 - 2/k$ for $k \ge 4$ and in [2527] obtained[254] $\alpha_k \le 1 - 3/(k+2)$. Moreover they showed in [2533] that the evaluation $\alpha_k \le 1/2$, as well as

$$\int_1^x \Delta_k^2(t)\,dt \ll x^{1-1/k+\varepsilon}$$

for every $\varepsilon > 0$ and $k = 1, 2, \ldots$, are both equivalent to the Lindelöf conjecture.

The lower bound for α_k had been made more precise first by K.C. Tong [6192] in 1956, then by J.L. Hafner [2430], but the best known result is due to K. Soundararajan [5852], who in 2003 showed

$$\Delta_k(x) = \Omega\big((x\log x)^{1/2-1/2k}(\log\log x)^a(\log\log\log x)^{-b}\big),$$

with $a = (1+1/2k)(k^{2k/(k+1)} - 1)$, $b = 1/2 + (k-1)/4k$.

The mean value of $|\Delta_k(x)|$ was first evaluated in 1922 by H. Cramér [1270], who for $k \ge 3$ got

$$\int_1^x |\Delta_k(t)|\,dt \ll x^{2-1/2k+\varepsilon}$$

for every $\varepsilon > 0$.

In the case $k = 3$ A. Walfisz [6525] got $\alpha_3 \le 43/87 = 0.4942\ldots$ in 1926, and several years later F.V. Atkinson[255] [168] proved $\alpha_3 \le 37/75 = 0.4933\ldots$. This was slightly improved by R.A. Rankin [5115] and further improvements followed:

- $\alpha_3 \le 14/29 = 0.4827\ldots$ (M.I. Yüh [6804]),
- $\alpha_3 \le 25/52 = 0.4807\ldots$ (W.L. Yin [6779]),
- $\alpha_3 \le 8/17 = 0.4705\ldots$ (M.I. Yüh, F. Wu [6805]),
- $\alpha_3 \le 5/11 = 0.4545\ldots$ (J.R. Chen [1019]),
- $\alpha_3 \le 43/96 = 0.4479\ldots$ (G. Kolesnik [3455]).

For larger values of k, upper bounds were given by H.-E. Richert [5207] who showed that for a_k one can take any number exceeding $1 - ck^{-2/3}$ with a certain constant $c > 0$. Later A.A. Karatsuba [3251] made this more precise by establishing

$$\alpha_k \le 1 - \frac{1}{2(200k)^{2/3}}$$

for large k. Improvements were later provided by A. Ivić [3037], D.R. Heath-Brown [2631], A. Ivić and M. Ouellet [3039, 3046] and K. Ford [2032], who obtained

$$\alpha_k \le 1 - \frac{0.195\ldots}{k^{2/3}}.$$

The mean value of the error term was evaluated by I. Kiuchi [3347] and T. Meurman [4276].

The Piltz problem in arithmetical progressions was considered in 1966 by A.F. Lavrik [3734], who obtained an asymptotical formula for the sum

$$\sum d_k(n)$$

extended of $n \le x$, congruent to a mod q.

---

[254] This was slightly improved for $k \ge 5$ by A. Walfisz [6522].

[255] Frederick Valentine Atkinson (1916–2002), professor in Ibadan, Canberra and Toronto. See [4317].

The error term was later improved by A.F. Lavrik and Ž. Edgorov [3738], Ž. Edgorov [1686], M.M. Petečuk [4796] (for prime power q), R.A. Smith [5837], K. Matsumoto [4198], J.B. Friedlander and H. Iwaniec [2105, 2106], D.R. Heath-Brown [2641], C.E. Chace [977] and H. Li [3876].

One also considers the analogue of the Piltz problem in algebraic number fields, consisting of the evaluation of the difference between the sum of the coefficients a_n (with $n \leq x$) of an mth power of Dedekind zeta-function $\zeta_K(s)$ and the sum of residues at $s = 0$ and $s = 1$ of $\zeta_K^m(s)x^s/s$. This problem was studied by G. Szegö and A. Walfisz [6018, 6019, 6523, 6524]. The strongest known lower bounds for the error term are due to K. Girstmair, M. Kühleitner, W. Müller and W.G. Nowak [2240], who proved analogues of K. Soundararajan's [5852] bounds in the divisor problem. Upper bounds were provided by A. Ivić [3041] and W.G. Nowak [4642].

A recent survey of various questions connected with the divisor problem has been prepared by A. Ivić and E. Krätzel [3043].

A related problem was considered in 1932 by O. Hölder[256] [2848], who considered the mean value of $t_k(n)$, the number of k-free divisors of n.

2.7 Diophantine Equations and Congruences

1. In 1887 a paper by C. Runge[257] [5341] appeared in which certain necessary conditions were given for an equation of the form

$$F(x, y) = 0 \tag{2.91}$$

(where $F \in \mathbf{Z}[X, Y]$ is irreducible) to have an infinite number of integral solutions.

A new proof of Runge's result has been given by T. Skolem [5804]. For improvements see A. Schinzel [5440] and M. Ayad [182].

A quantitative version was given by D.L. Hilliker and E.G. Straus[258] [2811] in 1983, and their bounds were later improved by P.G. Walsh [6536]. See also A. Grytczuk, A. Schinzel [2377], S. Tengely [6111], A. Sankaranarayanan, N. Saradha [5392].

Four years later D. Hilbert and A. Hurwitz published a joint paper [2793] in which they considered equations of the form

$$F(x, y, z) = 0,$$

where F is a homogeneous polynomial with integral coefficients, and the corresponding curve Γ in the projective plane is of genus[259] zero. They showed that if

[256] Otto Ludwig Hölder (1859–1937), professor in Tübingen, Königsberg and Leipzig. See [6310].

[257] Carl David Tolmé Runge (1856–1927), professor in Hannover and Göttingen. See [1260].

[258] Ernst Gabor Straus (1922–1983), professor at the University of California in Los Angeles. See [889, 1826].

[259] The simplest way to define the genus of a projective curve Γ over a field k is to say that it is the $\overline{k}$-dimension of the linear space of holomorphic differential forms of Γ (see, e.g., [5791, Chap. 2]), $\overline{k}$ denoting the algebraic closure of k.

such an equation has infinitely many rational solutions, then Γ is birationally equivalent to a line or a conic. The same result was also obtained by H. Poincaré [4936].

A fresh proof of the Hilbert–Hurwitz theorem was given by D. Poulakis [5002] in 1998.

2. The next important step forward occurred in 1909, when A. Thue [6142] proved his theorem on approximations by rationals and obtained as a corollary the following assertion.

If $F(X,Y) \in \mathbf{Z}[X,Y]$ is an irreducible form of degree $N \geq 3$ with rational integral coefficients, then for every fixed $a \in \mathbf{Z}$ the equation

$$F(x,y) = a \tag{2.92}$$

can have only finitely many integral solutions x, y.

(Such equations are called now *Thue equations*.)

Shortly afterwards E. Maillet [4112] filled some lacuna in Thue's proof and used Thue's theorem to show that if $F(X,Y)$ is an irreducible form of degree n with integral coefficients, and $G(X,Y)$ is a polynomial prime to F, of degree $m < n/2 - 1$, then the equation

$$F(x,y) = G(x,y) \tag{2.93}$$

can have only finitely many integral solutions.

Another consequence of Thue's theorem was presented by its author in 1917 [6144], when he showed that the Diophantine equation

$$ax^2 + bx + c = dy^n$$

can have, for fixed $n \geq 3$, a, b, c, d, only finitely many integral solutions, provided $ad \neq 0$ and $b^2 \neq 4ac$ (cf. L.J. Mordell [4378] and E. Landau, A. Ostrowski [3681]).

An analogous result, with the quadratic polynomial replaced by a cubic one, was obtained by L.J. Mordell [4380], who conjectured in [4379] that the same happens also for polynomials of larger degrees. This was established in the anonymous paper [5746] (signed by "X") of C.L. Siegel, who deduced from his strengthening of Thue's theorem given in [5738] that if the polynomial f has no repeated roots, then the equation $y^n = f(x)$ has, for fixed $n \geq 3$, at most finitely many integral solutions. For effective results see Sect. 6.6.

Siegel's thesis [5738] brought several applications to Diophantine equations. First of all he was able to show that equation (2.93) has finitely many solutions under less stringent assumptions about the degrees, and at the same time he showed the truth of the analogue of this assertion in the case that the rational integers are replaced by integers of a fixed algebraic number field K containing the coefficients of both F and G. He proved also that a polynomial over K, which has at least two different zeros, can represent only finitely many units (i.e., invertible integers of K) at algebraic integral arguments.

3. For a long time it has been known that the minimal solutions of the Pellian equation

$$x^2 - Dy^2 = 1 \tag{2.94}$$

(where D is positive and square-free) may be very large, even for relatively small values of D. The first such example seems to be that presented by J. Wallis[260] in 1658, who showed that the minimal solution of the equation $x^2 - 151y^2 = 1$ is $x = 1\,728\,148\,040$, $y = 140\,634\,693$, and another resulted from the solution of the cattle problem attributed to Archimedes (see B. Krummhiebel, A. Amthor [3537], H.W. Lenstra, Jr. [3823], H.L. Nelson [4556]). The first upper bound for the minimal solution was given by R. Remak in 1913 [5158], who showed that in such a solution one has

$$\log y \ll D^{3/2} \log D.$$

In the following year Perron [4786] improved this to

$$\log y \le (1 + o(1)) D \log D,$$

using the observation that if $p(D)$ denotes the length of the period of the continued fraction of $\sqrt{D}$, then $p(D) \le D + \sqrt{D}$ and

$$\log y \le (p(D) + o(1)) \log D. \tag{2.95}$$

In 1916 T. Schmitz [5530] proved that $\log y \le (8 + o(1))D$, and two years later I. Schur [5574] got

$$\log y \le \frac{1}{2}\sqrt{D} \log D.$$

Note that minimal solutions of the equations $x^2 - Dy^2 = \pm 1$ and $x^2 - Dy^2 = \pm 4$ are closely related to fundamental units $\varepsilon > 1$ of the quadratic field $K = \mathbf{Q}(\sqrt{D})$, hence bounds for such solutions imply corresponding bounds for the regulator $R(K) = \log \varepsilon$ of K.

For integers D with large values of y in (2.94) see C. Reiter [5156], F. Halter-Koch [2483], Y. Yamamoto [6773].

Later stronger bounds for $p(D)$ were obtained. In 1927 T. Vijayaraghavan [6390] got $p(D) \ll D^{1/2+\varepsilon}$ and in 1942 L.K. Hua [2928] obtained $p(D) \ll \sqrt{D} \log D$. Thirty years later K. Hirst [2829] showed that for square-free D

$$p(D) < 2\sqrt{D} \log D + O(\sqrt{D}),$$

and D.R. Hickerson [2778] obtained for non-square D

$$p(D) \le D^{1/2} \exp\left(\log 2 \frac{\log D}{\log\log D} + O\left(\frac{\log D \log\log\log D}{(\log\log D)^2}\right)\right).$$

Next, R.G. Stanton, C. Sudler, Jr. and H.C. Williams [5894] proved

$$p(D) < 0.72\sqrt{D} \log D$$

for every square-free $D > 7$, and for non-square D's they showed

$$p(D) < 3.76\sqrt{D} \log\left(\frac{D}{s^2}\right),$$

260 John Wallis (1616–1703), professor in Oxford. See [5587].

where s^2 is the maximal square factor of D. This estimate was improved in 1977 by J.H.E. Cohn [1163] to

$$p(D) < \frac{7}{2\pi^2}\sqrt{D}\log D + O(\sqrt{D})$$

for every non-square D. He showed also that $p(D) = \Omega(\sqrt{D}/\log\log D)$ (cf. J.C. Lagarias [3601]), and in 1986 H.W. Lu [4014] replaced the constant in Cohn's bound by 0.24 in the case of square-free D. The evaluation $p(D) \ll \sqrt{D\log D}$ was conjectured in 1962 by D. Shanks[261], and numerical support for it was provided by B.D. Beach and H.C. Williams [369]. Since for square-free D

$$p(D) < \frac{\log(x + y\sqrt{D})}{\log\alpha}$$

with $\alpha = (1+\sqrt{5})/2$ (see [4767, 5894]), hence using Dirichlet's class-number formula jointly with a result of J.E. Littlewood [3945] one sees (E.V. Podsypanin [4925]) that the General Riemann Hypothesis implies

$$p(D) = O(\sqrt{D}\log\log D).$$

A survey of computational aspects was given by H.C. Williams [6674].

4. A formula for the number of solutions of a polynomial congruence modulo a prime was given in 1903 by A. Hurwitz [2959]. It is simpler than the formulas established earlier by J. König (see G. Rados[262] [5045] and L. Kronecker [3531, p. 363]), but nevertheless its applicability is limited. Generalizations to the case of equations in finite fields were later obtained by L.E. Dickson [1536] and H.S. Vandiver [6328].

Denote by $S_f(N)$ the number of solutions of the congruence

$$f(X) \equiv 0 \pmod{N},$$

where f is an irreducible polynomial with integral coefficients. If $k \geq 2$ is the degree of f, and $D(f)$ is its discriminant, then the bound

$$S_f(N) \leq k^{\omega(N)} D(f)^2$$

was established in 1921 by O. Ore[263] [4699] in the case when $D(f) \neq 0$, and the leading coefficient of f is prime to N. This was later improved by G. Sándor [5389], and in 1979 M.N. Huxley [2979] established the bound $S_f(N) \leq k^{\omega(N)}\sqrt{|D(f)|}$.

The bound $S_f(N) \leq d^{k-1}(N)N^{1-1/k}$, which does not depend on the discriminant, was established in 1924 by E. Kamke [3230], and much later this was improved by S.V. Konyagin and S.B. Stečkin[264] [3473] to

$$S_f(N) \leq \left(\frac{k}{e} + o(\log^2 k)\right)N^{1-1/k}.$$

[261] Daniel Shanks (1917–1996), professor at the University of Maryland. See [6672].

[262] Gusztáv Rados (1862–1942), professor in Budapest.

[263] Oystein Ore (1899–1968), professor at Yale. See [4710].

[264] Sergei Borisovič Stečkin (1920–1995), professor in Moscow. See [480, 6252].

Later O.M. Fomenko [2023] obtained

$$\sum_{N\le x} S_f(N) = C(f)x + O\left(\frac{x}{\log^c x}\right)$$

for every $c < 1/2$, and the error term was shown to be $O(x^{a(d)})$ with explicit $a(d) < 1$ by H.H. Kim [3333] and G. Lü [4013].

Fix a prime p. The *Poincaré series* $P_f(T)$ of a polynomial $f(X_1, \dots, X_N)$ with integral (or rational p-integral, i.e., with denominators not divisible by p) coefficients is defined by

$$P_f(T) = \sum_{n=0}^{\infty} a_n T^n,$$

where $a_0 = 1$ and a_n denotes the number of solutions of the congruence

$$f(x_1, \dots, x_N) \equiv 0 \pmod{p^n}.$$

It seems that this series appeared for the first time in 1964 in the book by Z.I. Borevič[265] and I.R. Šafarevič [642], who conjectured that $P_f(T)$ is a rational function. This was confirmed later by J.-I. Igusa [3004, 3005], and a generalization to algebraic varieties was obtained by D. Meuser [4277]. Another proof, based on the elimination of quantifiers, was given by J. Denef [1466, 1467]. On this topic see also J. Denef [1468], A. Macintyre [4051], D. Meuser [4278], J. Pas [4754–4756].

5. The equation

$$\frac{x^m - 1}{x - 1} = \frac{y^n - 1}{y - 1} \quad (y > x \ge 2,\ m, n \ge 3) \tag{2.96}$$

has been considered in 1917 by R. Goormaghtigh[266] [2281], who showed that the only solutions with $(x^m - 1)/(x - 1) \le 10^6$ are $(m, n, x, y) = (5, 3, 2, 5)$ and $(13, 3, 2, 90)$. It is conjectured that there are no other solutions.

It was observed in 1956 by H.-J. Kanold [3241] that for fixed x, y there are at most finitely many solutions m, n of (2.96), and in 1986 T.N. Shorey [5722] gave the bound 17 for their number. In 2002 this bound was reduced to 2 by Y. Bugeaud and T.N. Shorey [824]. They also proved that if $y > 10^{11}$, then there is at most one solution and six years later B. He and A. Togbé [2623] proved that this holds for every y.

An effective bound for the solutions in the case when the maximal prime divisor of xy is bounded was found by R. Balasubramanian and T.N. Shorey [287] in 1980.

In 1961 H. Davenport, D.J. Lewis and A. Schinzel [1398] also showed that for fixed m, n there are only finitely many solutions x, y.

In the case $m = 3$ and n odd, P.Z. Yuan [6803] has shown that there are no new solutions. This had been known earlier for prime power y (M.H. Le [3749]). Yu.V. Nesterenko and T.N. Shorey [4573] got an explicit bound for the solutions of (2.96) in the case $d = (m - 1, n - 1) \ge 2$. This bound depends on $(m - 1)/d$ and $(n - 1)/d$.

---

[265] Zenon Ivanovič Borevič (1922–1995), professor in Leningrad.

[266] René Victor Goormaghtigh (1893–1960), director of a steel plant in Belgium. See [1418].

2.8 *p*-Adic Numbers

1. Right at the beginning of the new century K. Hensel had the brilliant idea of considering formal power series in which the role of the variable was played by a fixed prime number and the coefficients were digits $0, 1, \dots, p-1$. These entities were introduced in [2731, 2732] and although at the beginning were regarded rather as a curiosity, they were later to play an overwhelming role in the development of number theory and algebra. Hensel wrote in [2736] that the idea of this new class of numbers has its roots in the theory of algebraic functions, where one studies their behavior through expansions at points of the complex plane. In his new approach the role of points is played by prime numbers. His hope, expressed on page 3 of [2736], that his method would lead to a transcendence test for complex numbers was never fulfilled.

Hensel recapitulated his method in the book [2736], of which only the first volume appeared. He defined in it, for every prime p, the p-adic integers as finite or infinite sequences $C = (c_0, c_1, \dots)$ of digits $0, 1, \dots, p-1$ with properly defined arithmetical operations. He chose the form

$$C = c_0 + c_1 p + c_2 p^2 + \cdots,$$

to represent these sequences, pointing out that these series do not converge in the usual sense, but have to be understood as a symbolical way to show that the sequence $c_0, c_0 + c_1 p, c_0 + c_1 p + c_2 p^2, \dots$ approximates the p-adic integer C. He wrote: "Die[267] ... p-adischen Zahlen sind reine Symbole, mit denen nach bestimmten Vorschriften zu rechnen ist ...." Extending this definition by considering series of the form

$$c_{-n} p^{-n} + c_{-n+1} p^{-n+1} + \cdots + c_0 + c_1 p + c_2 p^2 + \cdots,$$

K. Hensel arrived at the field $\mathbf{Q}_p$ of p-adic numbers[268]. The set $\mathbf{Z}_p$, consisting of series with $c_n = 0$ for negative n, forms a subring of $\mathbf{Q}_p$, its ring of integers.

Considering finite extensions of $\mathbf{Q}_p$, Hensel developed in a series of papers [2732, 2734, 2735, 2737, 2739–2748] all the main results of $\mathfrak{p}$-adic fields, as they are now called.

One of the first followers of Hensel's new approach was G.E. Wahlin [6491–6494], who studied the structure of units in $\mathfrak{p}$-adic fields and applied Hensel's theory to the factorization of prime ideals in extensions of the rationals.

An important tool in the elementary theory of p-adic and $\mathfrak{p}$-adic fields is *Hensel's lemma*, proved by K. Hensel first [2733] for p-adic numbers (a simpler proof was given by L.E. Dickson [1541] in 1910), and then for their finite extensions [2734]. This lemma can be formulated in several ways. In its simplest form (which we state in the case of the field $\mathbf{Q}_p$) it asserts that if $F(X)$ is a polynomial with integral

[267] "The ... p-adic numbers are pure symbols, on which one has to perform calculations according to particular laws" [2736, p. 39].

[268] The second book [2738] by Hensel gave an exposition of elementary number theory based on the theory of p-adic numbers.

p-adic coefficients, $f \in \mathbf{F}_p[X]$ is its reduction mod p, and one has $f = gh$, where g, h are relatively prime non-constant polynomials, then one can write

$$F(X) = G(X)H(X),$$

where G, H are relatively prime polynomials with coefficients in $\mathbf{Z}_p$ and g, h are their reductions mod p.

2. Hensel's method was not easily accepted, because its foundations were considered as being rather vague. This imposed the necessity of giving a sound basis for this theory. The first result in this direction came from A. Fraenkel[269], who gave in [2065] an axiomatic definition of p-adic numbers, but his approach did not have any influence on the later development of number theory. The second approach to this problem, which turned out to be of utmost importance and exerted a big influence on a large part of commutative algebra, was due to J. Kűrschak[270] [3586], who in 1913 introduced *valuations*, as functions $v(x)$, defined in a field K, with non-negative real values, obeying the following conditions:

(a) $v(x) = 0$ holds if and only if $x = 0$,
(b) $v(xy) = v(x)v(y)$,
(c) $v(x \pm y) \leq v(x) + v(y)$.

If v satisfies

(d) $v(x \pm y) \leq \max\{v(x), v(y)\}$,

then it is called a *non-Archimedean valuation*.

This paper, which marks the start of valuation theory, which later became an important part of algebra, was followed by a sequence of three papers [4704–4706] by A. Ostrowski in which he determined all valuations of the rational field as well as of its finite extensions (a much simpler proof of this result was given later by E. Artin [143]). He also observed that the algebraic closure of the p-adic field $\mathbf{Q}_p$ is not complete and showed that if a field K is complete under a non-archimedean valuation, L/K is a finite extension, β is algebraic and separable over K, and its distance from an element of L under the metrics $d(x, y) = w(x - y)$ (where w extends v to its separable closure) is sufficiently small, then β lies in K. This result is usually called *Krasner's lemma*, after M. Krasner[271], who rediscovered it in [3512]. One should also mention here the paper [5356] by K. Rychlik[272], where Hensel's lemma is extended to arbitrary complete fields.

The theory of valuated fields found its culmination in the work of H. Hasse and F.K. Schmidt [2609], published in 1934, where the structure of fields complete under a discrete valuation was determined. The rather complicated proof was later simplified by

[269] Adolf Abraham Halevi Fraenkel (1891–1965), professor in Marburg, Kiel and Jerusalem.

[270] József Kűrschak (1864–1933), professor at the Technical University in Budapest.

[271] Marc Krasner (1912–1985), professor in Clermont-Ferrand and Paris.

[272] Karel Rychlik (1885–1968), professor at the Technical University in Prague. See [2997].

O. Teichmüller[273] [6098, 6099]. In the first of these papers, a mapping $\mathbf{Z}_p^* \mapsto \mathbf{Z}_p^*$ was defined, known today as the *Teichmüller character*, which turned out to be extremely useful.

The ideas of K. Hensel, J.Kűrschak and A. Ostrowski were developed by M. Deuring, W. Krull[274] [3536], A. Ostrowski [4709], F.K. Schmidt[275] [5480] and others to form a sovereign branch of mathematics, the theory of valuations (see O.F.G. Schilling[276] [5429], P. Ribenboim [5177, 5180], O. Endler[277] [1763]). A broad exposition of its history was given by P. Roquette [5278].

[273] Oswald Teichmüller (1913–1943), student of Hasse, worked in Göttingen and Berlin. See [5423].

[274] Wolfgang Krull (1899–1971), professor in Freiburg, Erlangen and Bonn. See [5560].

[275] Friedrich Karl Schmidt (1901–1977), professor in Jena, Münster and Heidelberg. See [3584].

[276] Otto Schilling (1911–1973), professor at Purdue University.

[277] Otto Endler (1929–1988), professor in Rio de Janeiro.

Chapter 3
The Twenties

3.1 Analytic Number Theory

3.1.1 Exponential Sums

1. Sums of the form

$$S_f(A,B)=\sum_{n=A}^{B} f(n),$$

where f is a well-behaved function had already been considered in the 18th century, when Euler and C. Maclaurin[1] proposed a formula[2] [4052], now bearing their names, relating this sum to the integral

$$I_f(A,B)=\int_A^B f(t)\,dt.$$

If f is a function having $2M+1$ continuous derivatives in the interval (A,B), then this formula in its modern form gives, for the difference $S_f(A,B)-I_f(A,B)$, the expression

$$\sum_{k=1}^{M}\frac{B_{2k}}{(2k)!}\left(f^{2k-1}(A)-f^{2k-1}(B)\right)-\frac{f(A)+f(B)}{2}+R_f(A,B),$$

where

$$\left|R_f(A,B)\right|\le\frac{2}{(2\pi)^{2M}}\int_A^B\left|f^{2M+1}(t)\right|\,dt.$$

---

[1]Colin Maclaurin (1698–1746), professor in Aberdeen and Edinburgh. See [6235].

[2]Hardy devoted a chapter of his book [2517] to this formula. He pointed out that the first rigorous proof of it was given by Jacobi [3078] in 1834, and the first discussion of the error term occurs in Poisson's paper [4938] in 1823.

W. Narkiewicz, *Rational Number Theory in the 20th Century*, Springer Monographs in Mathematics,
DOI 10.1007/978-0-85729-532-3_3, © Springer-Verlag London Limited 2012

This formula is not applicable to functions having discontinuities and the first attempt to deal with such a case appears to occur in 1917 in a paper by I.M. Vinogradov [6403]. He used completely elementary methods to deal with sums of the form

$$\sum_{n=A}^{A+B} \{f(n)\},$$

where f is assumed to have a second derivative in $[A, B]$. If f satisfies the inequalities

$$\frac{1}{kM} < f''(x) < \frac{1}{M}, \quad x \in [A, B]$$

with certain positive k, M, then Vinogradov's result gives

$$\sum_{n=A}^{A+B} \{f(n)\} = \frac{B}{2} + O\left(\left(\frac{B}{kM} + 1\right)(M \log M)^{2/3}\right),$$

the implied constant being less than 2.

This allowed him to obtain the result (2.85) on the mean value of class-numbers of positive-definite binary quadratic forms.

2. In [2521] G.H. Hardy and J.E. Littlewood considered the sum

$$S_n(x) = \sum_{k \le n} \exp\left(2\pi i k^2 x\right)$$

(with real x), as well as two other similar sums, and proved for irrational x the evaluation $S_n(x) = o(n)$ which cannot be, in general, improved. In the case when the partial quotients of the continued fraction for x are bounded which happens, for example, if x is a quadratic irrationality, they obtained $S_n(x) = o(\sqrt{n})$. They showed also that

$$S_n(x) \ll n^{1/2} \log^{1/2+\varepsilon} n$$

holds for almost all x, but on the other hand there is no irrational x for which one has $S_n(x) = o(\sqrt{n})$.

From the results of a later paper [2530] one can infer using Roth's theorem that if x is algebraic, then for every $\varepsilon > 0$ one has

$$S_n(x) = O\left(n^{1/2+\varepsilon}\right).$$

3. The next step was made by J.G. van der Corput [6276], who in his dissertation [6274, 6275] simplified Voronoĭ's method of calculating the number of lattice points in a given region, and in 1921 made the important observation that if one controls the behavior of the derivatives $f', f'', \ldots, f^{(r)}$ of a well-behaved function f, then the sum

$$S_f(I) = \sum_{n \in I} \exp(2\pi i f(n)), \tag{3.1}$$

where I is an interval, can be approximated by the integral

$$\int_I \exp(2\pi i f(t))\,dt. \tag{3.2}$$

He used the Euler–Maclaurin formula to obtain new evaluations of exponential sums (3.1), when f is a real-valued function with monotonic derivative, for which some good bounds are available. This formula allowed him to replace $S_f(I)$ by the integral (3.2), the error term being bounded by a small constant[3], provided the inequality $0 \le f'(t) \le 1/2$ holds and f' is non-decreasing. The evaluation of the integral depended on certain assumptions concerning the derivatives of f. If, for example, one would have $|f''(t)| \ge \varrho$ in $I = [a, b]$ (with a certain $0 < \varrho \le 1$), then the bound

$$|S_f(I)| \le \frac{11}{4\sqrt{\varrho}} + \frac{9}{4} \tag{3.3}$$

would follow.

This method permitted new evaluations to be obtained of error terms in formulas giving the number of lattice points in various regions.

In [6277] J.G. van der Corput introduced the notion of exponent pairs, which still plays a very important role. To define them in the simplest case consider (3.1), with $I \subset (N, 2N)$ and a well-behaved real-valued function f, such that $f'(x)$ is approximated on I by yx^{-s} with certain positive s, y satisfying $y \ge N^s$. If

$$S \ll (yN^{-s})^k N^l,$$

then the pair (k, l) is called an *exponent pair*[4]. E. Phillips [4838] gave a method to produce such pairs.

For further development of the theory of exponent pairs see R.A. Rankin [5115], G. Kolesnik [3457, 3458], S.W. Graham [2303] and the book by S.W. Graham and G. Kolesnik [2305].

4. A few years later J.G. van der Corput [6282] made his approach more precise, which allowed him to improve previous bounds in Dirichlet's divisor problem [6279]. His method was applicable to sums of the form (3.1) in the case when the function f has its kth derivative (with $k \ge 2$) of constant sign, and is bounded away from zero in the considered interval. More precisely, if $f^{(k)}(x) \ge 1/R$ and

$$L = R\left|f^{(k)}(a) - f^{(k)}(b)\right|,$$

then for $\kappa = 2^{k-1}$ one gets

$$|S_f(x)| \ll L\left(R^{-1/(2\kappa-2)} + L^{-1/\kappa} + \left(\frac{R}{L^k}\right)^{2/(3\kappa-2)}\right).$$

---

[3] It was later proved by V. Jarník and E. Landau [3116] that the optimal value of this constant equals $1/2 + 1/\pi + (1/4 + 1/\pi^2)^{1/2} = 1.4110\ldots$.

[4] Note that sometimes (e.g., in [3064]) the exponent β in this definition is replaced by $\beta + 1/2$. Here we follow the definition given in [2305].

In van der Corput's paper [6279], published in 1929, one finds an important inequality reducing the evaluation of the sum

$$\sum_{n\le x}\exp(2\pi i f(n))$$

to the study of upper bounds of sums $\sum_{n\le x}\exp(2\pi i(f(n+k)-f(k)))$ for $k=1,2,\dots,$ f being a real-valued function. This inequality implies immediately that a sequence a_n of real numbers is uniformly distributed modulo one if this happens for all sequences $a_{n+h}-a_h$ $(h=1,2,\dots)$.

Far-reaching generalizations with several applications were proved much later by E. Hlawka [2833] and J. Cigler [1112].

A modern exposition of van der Corput's method is given in the book by S.W. Graham and G. Kolesnik [2305]. See also M. Radouaby, P. Sargos [5145], H.Q. Liu [3958] and M. Radouaby [5144] for recent progress.

3.1.2 The Zeta-Function

1. The method of evaluation of exponential sums (2.68) developed in 1916 by H. Weyl [6647] showed its power when its author applied it in [6648] to bound the zeta-function on the line $\Re s=1$. He obtained the evaluation

$$\zeta(1+it)=O\left(\frac{\log t}{\log\log t}\right),$$

using the equality

$$\zeta(1+it)=\sum_{n=1}^{[t]}\frac{\exp(-ti\log n)}{n}+O(1),$$

and then dealing with the obtained exponential sum. This improved a result by G.H. Hardy and J.E. Littlewood [2519] who established that $\zeta(1+it)=o(\log t)$. Earlier H. Bohr and E. Landau [584] proved that the evaluation $\zeta(1+it)=o(\log\log t)$ does not hold.

Much later N. Levinson [3863] made the result of [584] explicit by establishing

$$\limsup_{t\to\infty}\frac{|\zeta(1+it)|}{\log\log t}\ge e^{\gamma},$$

γ denoting the Euler's constant.

In 1922 J.E. Littlewood [3941, 3943] applied Weyl's method to the sum

$$\sum_{n\le x}\exp(it\log n),$$

which led to the bound

$$\zeta(\sigma+it)=O\left(\log^5 t\right)$$

in the region $t > c$, $\sigma \geq 1-(\log\log t)^2/\log t$. He used it to obtain the non-vanishing of $\zeta(s)$ for $\sigma \geq 1 - c_1(\log\log t)/\log t$, and this led to

$$\pi(x) = \mathrm{li}(x) + O\left(x \exp\left(-c_2 (\log x \log\log x)^{1/2}\right)\right) \tag{3.4}$$

with suitable positive constants c_1, c_2. He also got for $\mu(\sigma)$ defined by (2.21) the bound

$$\mu(\sigma) \ll \frac{1-\sigma}{\log(1/(1-\sigma))} \quad (1/2 < \sigma < 1) \tag{3.5}$$

and proved that the Riemann Hypothesis implies the bound $\zeta(1+it) = O(\log\log t)$, which showed that the result of [584] cannot be improved if the Riemann Hypothesis is true. In 1930 J.G. van der Corput and J.F. Koksma [6297] showed that for the implied constant in (3.5) one can take $\log 2$.

The bound $\zeta(1+it) = O(\log^{3/4} t\,(\log\log t)^{1/2+\varepsilon})$ with any $\varepsilon > 0$ was obtained in 1950 by T.M. Flett[5] [2014] and in 1958 A. Walfisz [6533] removed ε from the exponent. In the same year N.M. Korobov [3486] obtained

$$\zeta(1+it) = O\left(\log^c t\right)$$

with any $c > 4/7$, and now it is known (I.M. Vinogradov [6442], N.M. Korobov [3488]) that this holds with $c = 2/3$.

The Riemann Hypothesis implies the evaluation

$$\zeta(\sigma+it) \ll \exp\left(\frac{C\log t}{\log\log t}\right)$$

for $\sigma \geq 1/2$ and it has been shown by K. Ramachandra, A. Sankaranarayanan [5067] that one can take $C = 0.4665\ldots$. In a close neighborhood of the line $\sigma = 1$ the bound

$$\zeta(\sigma+it) \ll t^{2^{15}(1-\sigma)^{3/2}} \log^{5/2} t$$

was established by W. Staś [5912], improving an earlier bound of L. Schoenfeld [5551], and later H.-E. Richert [5208] obtained

$$\zeta(\sigma+it) \ll t^{100(1-\sigma)^{3/2}} \log^{2/3} t \tag{3.6}$$

for $1/2 \leq \sigma \leq 1$, $|t| \geq 4$. A similar result for Dedekind's zeta-functions was obtained by W. Staś [5913]. For an analogue in the case of Hecke–Landau zeta-functions see K. Bartz [342, 343], who used it to prove an effective version of Čebotarev's theorem in [343]. The factor 100 in the exponent in (3.6) was reduced to 20.9973 by Bartz [344], to 18.4974 by M. Kulas [3572], and to 4.45 by K. Ford [2032] in 2002. K. Ford also showed that the implied constant may be taken to be 76.2, which leads to the following explicit estimate for the zeros of $\zeta(s)$:

$$N(\sigma, T) \ll T^{58.05(1-\sigma)^{3/2}} \log^{15} T.$$

[5] Thomas Muirhead Flett (1923–1977). See [6092].

2. In the preceding chapter we mentioned that the Lindelöf conjecture is equivalent to the bound $\zeta(1/2+it) \ll |t|^{\varepsilon}$ for every $\varepsilon > 0$. This created interest in evaluations of Riemann's zeta-function on the critical line. Denote by c the smallest number such that the bound

$$\zeta\left(\frac{1}{2}+it\right) \ll |t|^{c+\varepsilon} \tag{3.7}$$

holds for every $\varepsilon > 0$. Thus the Lindelöf conjecture is equivalent to $c = 0$. It seems that the first evaluation of c, namely $c \le 1/6$, occurred in an unpublished paper by G.H. Hardy and J.E. Littlewood, and a proof was given by E. Landau [3666], who replaced the factor t^{ε} by $\log^{3/2} t$. In the same year A. Walfisz [6522] utilized van der Corput's method of exponent pairs to get $c \le 163/988 = 0.1649\ldots$. Evaluations of c are closely connected with these pairs, as one can show that if (p,q) is an exponent pair, then

$$c \le \frac{1}{4}\frac{p+q}{p+1}$$

(see, e.g., A. Ivić [3041]).

Later improvements were obtained by E.C. Titchmarsh [6172] ($c = 27/164 = 0.1646\ldots$), E. Phillips [4838] in 1933 ($c \le 229/1392 = 0.1645\ldots$), E.C. Titchmarsh [6179] in 1942 ($c \le 19/116 = 0.1637\ldots$) (he generalized van der Corput's idea, introducing two-dimensional exponent pairs), S.H. Min in 1949 [4316] ($c \le 15/92 = 0.1630\ldots$), W. Haneke [2497] in 1962 ($c \le 6/37 = 0.1621\ldots$), G. Kolesnik [3456, 3458] in 1982 and 1985 ($c \le 35/216 = 0.16203\ldots$, $c \le 139/858 = 0.16200\ldots$), E. Bombieri and H. Iwaniec [615] in 1986 ($c \le 9/56 = 0.1607\ldots$). The new way of handling exponential sums presented in the last paper and its sequel [616] also permitted improvement of the bound on the error term in Dirichlet's divisor problem (see [3067]). Refinements of this method (M.N. Huxley, N. Watt [2994]) led soon to further progress. So in 1989 N. Watt [6594] obtained $c \le 89/560 = 0.1589\ldots$, M.N. Huxley and G. Kolesnik [2991] two years later got $c \le 17/108 = 0.1574\ldots$, and M.N. Huxley [2984] proved in 1993 the inequality $c \le 89/570 = 0.15614\ldots$, reducing this after 12 years to $c \le 32/205 = 0.15609\ldots$ [2985], which is the best bound at the time of writing. The progress of 0.01 in 80 years indicates the difficulty of the problem. In his review of [2497] in Mathematical Reviews [1092] S. Chowla wrote: "If the proof of the exponent 6/37 needs 74 pages, how many pages will be required to achieve the exponent ε of Lindelöf?"

The constants in these evaluations are not explicit, with the unique exception of the bound

$$\left|\zeta\left(\frac{1}{2}+it\right)\right| \le 3t^{1/6}\log t,$$

valid for $t \ge e$, which was proved in 2004 by Y.F. Cheng and S.W. Graham [1043].

On the other hand it was proved by E.C. Titchmarsh [6167] that for every $a < 1/2$ there are arbitrarily large values of t with

$$\left|\zeta\left(\frac{1}{2}+it\right)\right| > \exp\left(\log^{a} t\right),$$

and this has been improved by N. Levinson[6] [3863] to

$$\left|\zeta\left(\frac{1}{2}+it\right)\right| > \exp\left(\frac{b\sqrt{\log t}}{\log\log t}\right),$$

and by R. Balasubramanian and K. Ramachandra [282] to

$$\left|\zeta\left(\frac{1}{2}+it\right)\right| > \exp\left(b\sqrt{\frac{\log t}{\log\log t}}\right),$$

(with a certain $b > 0$) for infinitely many t. Recently K. Soundararajan [5854] showed that in the last inequality one can take for b any number not exceeding 1.

3. The study of moments

$$M_k(T) = \int_0^T \left|\zeta\left(\frac{1}{2}+it\right)\right|^{2k} dt,$$

of Riemann's zeta-function started in 1917, when G.H. Hardy and J.E. Littlewood [2523] established

$$M_1(T) = (1+o(1))T\log T.$$

A few years later J.E. Littlewood [3941] refined this to

$$M_1(T) = T\log T - cT + E(T)$$

with

$$E(T) = O(T^{a+\varepsilon}) \tag{3.8}$$

for $a = 3/4$ with $c = 1+\log(2\pi) - 2\gamma$ and any $\varepsilon > 0$ (cf. E.C. Titchmarsh [6166]). An analogous result for L-functions was obtained by Z. Suetuna[7] [5985].

The error term $E(T)$ was later reduced by E.C. Titchmarsh first to $O(T^{1/2}\log T)$ [6173], and then to $O(T^{5/12}\log^5 T)$ [6174]. Much later R. Balasubramanian [273] and D.R. Heath-Brown [2628] proved $E(T) = O(T^{1/3+\varepsilon})$ for every $\varepsilon > 0$.

It was shown in 1977 by A. Good [2279] that a in (3.8) must be $\geq 1/4$, and later J.L. Hafner and A. Ivić [2432] established a result slightly stronger than

$$E(T) = \Omega\left((T\log T)^{1/4}\right).$$

A relation between the error terms in this problem and the divisor problem was established by M. Jutila [3174], who in [3175] proved that if the error term in the divisor problem is $O(T^{b+\varepsilon})$ for every $\varepsilon > 0$, then one can take $a = (1+2b)/5$, hence using the best known value of b [2983] one gets $a \leq 339/1040 = 0.3259\ldots$. The evaluation $a \leq 35/108 = 0.3240\ldots$ appears in Ivić's book [3041], and the best known result, $a \leq 7/22 = 0.3181\ldots$, has been established by D.R. Heath-Brown and M.N. Huxley [2663].

Second moments for a large class of functions defined by Dirichlet series were studied in 1940 by H.S.A. Potter [4999, 5000], who extended earlier results by F. Carlson [906, 907].

---

[6] Norman Levinson (1912–1975), professor at MIT.

[7] Zyoiti Suetuna (1898–1970), professor in Tokyo.

An analogue for $M_1(T)$ of Voronoĭ's summation formula was established in 1949 by F.V. Atkinson [169].

The case $k = 2$ was considered by G.H. Hardy and J.E. Littlewood [2530, 2533], who showed

$$M_2(T) \ll T \log^4 T.$$

Later A.E. Ingham [3017] improved this to

$$M_2(T) = \frac{1}{2\pi^2} T \log^4 T + O\left(T \log^3 T\right).$$

A simpler proof of the last result was given much later by K. Ramachandra [5061], and D.R. Heath-Brown [2628] made it more precise, by showing

$$M_2(T) = T \sum_{j=0}^{4} c_j \log_j T + O(T^a)$$

with certain constants c_j and every $a > 7/8$. The error term was reduced to $O(T^{2/3+\varepsilon})$ by N.I. Zavorotnyĭ [6820], and to $O(T^{2/3} \log^8 T)$ by A. Ivić and Y. Motohashi [3044, 3045], who also showed that it is not $o(\sqrt{T})$.

For larger k the first evaluation of $M_k(T)$ was provided by K. Ramachandra [5064] in 1978, who showed that for integral $k > 0$, the Riemann Hypothesis implies

$$M_k(T) \gg T \log^{k^2} T,$$

and three years later D.R. Heath-Brown [2630] established this for all rational $k > 0$. They showed also [2630, 5063] that the Riemann Hypothesis implies this bound for every positive k. J.B. Conrey and A. Ghosh [1208–1210] made this result effective, and in 1990 R. Balasubramanian, K. Ramachandra [285] showed that for integral k one can eliminate the assumption of the Riemann Hypothesis. For an improvement see K. Soundararajan [5850].

A recent result by K. Soundararajan [5856] shows that the upper bound

$$M_k(T) \ll T \log^{k^2+\varepsilon} T$$

(for every $\varepsilon > 0$) follows from the Riemann Hypothesis.

It was conjectured that for integral $k \geq 1$ one has

$$M_k(T) = (c_k + o(1)) T \log^{k^2} T$$

with an explicitly given constant c_k (see J.B. Conrey, A. Ghosh [1210] for $k = 3$, J.B. Conrey, S.M. Gonek [1216] for $k = 4$, J.P. Keating, N.C. Snaith [3292], A. Diaconu, D. Goldfeld, J. Hoffstein [1515], J.B. Conrey, D.W. Farmer, J.P. Keating, M. Rubinstein and N.C. Snaith, [1205, 1206]). A similar conjecture for functions from a much larger class (the so-called Selberg class, introduced by A. Selberg [5623] in 1989; see Sect. 6.1.4) was formulated in [1205].

In 1991 A. Ivić published a book [3042] devoted to $M_k(x)$. See also K. Ramachandra [5066].

4. In 1921 G.H. Hardy and J.E. Littlewood [2527, 2530, 2537] proved the following *approximate functional equation* for $\zeta(s)$.

If $x, y > H > 0$ and $xy = |t|/2\pi$, then for $s = \sigma + it$ one has

$$\zeta(s) = \sum_{n\le x} \frac{1}{n^s} + \pi^{s-1/2}\frac{\Gamma((1-s)/2)}{\Gamma(s/2)} \sum_{n\le y} \frac{1}{n^{1-s}} + O\left(y^{\sigma-1}|t|^{1/2-\sigma}\right),$$

the implied constant depending on H.

They obtained also a similar equation for $\zeta^2(s)$.

It was later shown by C.L. Siegel [5749] that the idea of this equation could already be found in Riemann's notes.

A similar equation was later established for Dirichlet L-functions by Z. Suetuna [5986, 5987] (see also N.G. Čudakov [6097], Yu.V. Linnik [3914], T. Tatuzawa[8] [6068], A.F. Lavrik [3735], V.V. Rane [5103]), and for a large class of Dirichlet series, including Dedekind zeta-functions, by K. Chandrasekharan and R. Narasimhan [995].

5. A characterization of the Riemann zeta-function among meromorphic functions by its functional equation was given in 1921 by H. Hamburger[9] [2489, 2491–2493], who also proved a parallel result for L-functions [2490, 2493]. Later he gave in [2494] a simpler proof in the case of $\zeta(s)$ (cf. E. Hecke [2683]).

Later, in 1936, E. Hecke [2695] considered Dirichlet series $f(s)$, for which the product $(s-k)f(s)$ (with some k) can be extended to an entire function in the plane satisfying certain growth conditions and for which the function

$$R(s) = \left(\frac{2\pi}{\lambda}\right)^{-s} \Gamma(s) f(s)$$

satisfies the functional equation

$$R(s) = \gamma R(k-s)$$

for $\gamma = \pm 1$, and determined the dimension of the linear space consisting of such f's. These series turned out to be closely connected with modular forms. Cf. P.W. Marke [4151].

6. In 1925 J.I. Hutchinson [2968] computed all zeros $\sigma + it$ of Riemann's zeta with $200 \le t \le 300$ and found that the first 138 non-trivial zeros lie on the critical line.

Ten years later E.C. Titchmarsh [6176] continued the search for non-trivial zeros of $\zeta(s)$ and showed that the first 195 lie on the critical line.

For further progress one had to wait for the advent of computers, and the first who applied them to this task was A.M. Turing [6229] in 1953, who extended Titchmarsh's search. The first large scale computation of zeros of $\zeta(s)$ was made by D.H. Lehmer [3787, 3788] in 1956. He used a computer to find that the first 25 000 non-trivial zeros lie on the critical line.

---

[8] Tikao Tatuzawa (1915–1997), professor in Nagoya and Tokyo.

[9] Hans Ludwig Hamburger (1889–1956), professor in Berlin, Ankara and Köln. See [2346].

In 1960 C.B. Haselgrove and Miller [2575] listed the first 1600 zeros (this book also contains tables of values of $\zeta(s)$ on the critical line and at $\Re s = 1$), and in 1966 R.S. Lehman [3769] showed that the first 250 000 non-trivial zeros lie where they should. This was superseded in 1969 by J.B. Rosser, J.M. Yohe and L. Schoenfeld [5304], who reached 3 500 000 zeros, all on the critical line. The development of high-speed computers led to a quick extension of that search. In 1979 R.P. Brent [707] showed that the first 75 million non-trivial zeros are simple and lie on the critical line; three years later this was extended to $2 \cdot 10^8$ (R.P. Brent J. van de Lune, H.J.J. te Riele, D.T. Winter [713]), and in the next years this limit has been raised to $1.5 \cdot 10^9$ (J. van de Lune, H.J.J. te Riele [6269–6271]), 10^{11} (S. Wedeniwski [6607]) and 10^{13} (X. Gourdon [2292]).

3.1.3 Prime Numbers

1. Denote by p_n the nth consecutive prime, and put $d_n = p_{n+1} - p_n$. The truth of Oppermann's conjecture, mentioned in the preceding chapter, would give the bound

$$d_n = O\left(p_n^{1/2}\right).$$

In 1884 A. Piltz [4877] put forward the stronger conjecture

$$d_n = O(p_n^{\varepsilon})$$

for every positive $\varepsilon > 0$, and the results of H. von Koch [3433, 3434], quoted in the preceding chapter, show that the Riemann Hypothesis implies

$$d_n = O\left(p_n^{1/2} \log^2 p_n\right). \tag{3.9}$$

The Prime Number Theorem implies the bound $d_n = o(p_n)$, and Landau's evaluation of the error term in that theorem leads to the slightly better bound

$$d_n = O\left(p_n \exp\left(-c\sqrt{\log p_n}\right)\right)$$

with a certain constant $c > 0$.

In 1921 H. Cramér [1266] showed that the Riemann Hypothesis permits replacement in (3.9) of the factor $\log^2 p_n$ by $\log p_n$. He also showed, still under the Riemann Hypothesis, that if $\Phi(x)$ is non-decreasing and satisfies $\Phi(x) = o(x)$ and $\sqrt{x}\log^2 x = o(\Phi(x))$, then one has

$$\pi(x + \Phi(x)) - \pi(x) = (1 + o(1))\frac{\Phi(x)}{\log x}. \tag{3.10}$$

Another consequence of the Riemann Hypothesis is the bound $d_n = O(p_n^{\varepsilon})$ valid for every $\varepsilon > 0$ and almost all primes p_n, established by H. Cramér in [1267].

Later A. Selberg [5608] showed that the equality (3.10) holds for almost all x provided

$$\liminf_{x\to\infty} \frac{\log \Phi(x)}{\log x} > \frac{19}{77},$$

and proved under the Riemann Hypothesis that if $\log^2 x = o(\Phi(x))$, then (3.10) holds for almost all integers x. In 1972 M.N. Huxley [2972] proved (3.10) for almost all x with $\Phi(x) =$

x^a with any $a > 1/6$, and the exponent a was later reduced to $1/10$ (G. Harman [2552]), $1/14$ (N. Watt [6595], C.H. Jia [3141]), and $1/20$ (C.H. Jia [3144]). One expects $a = 0$, and it was shown by D.R. Heath-Brown [2633] that this follows from the Elliott–Halberstam conjecture (see Sect. 6.1.1) on the distribution of primes in arithmetical progressions. He showed also [2632] that this is also a consequence of Montgomery's Pair Correlation Conjecture (see Sect. 6.1.2) and the Riemann Hypothesis.

J. Mueller [4466] deduced from the Riemann Hypothesis and the Pair Correlation Conjecture the bound

$$d_n = O\left(\sqrt{p_n} \log^{3/4} p_n\right).$$

In 1982 D.R. Heath-Brown [2633] improved this to

$$d_n = O\left(\sqrt{p_n \log p_n}\right),$$

and using a stronger form of Montgomery's conjecture obtained

$$d_n = o\left(\sqrt{p_n \log p_n}\right)$$

in a paper with D.A. Goldston [2662].

An important step forward was taken in 1930 by G. Hoheisel[10] [2847], who used F. Carlson's [905] bound for $N(\alpha, T)$ and Littlewood's zero-free region for $\zeta(s)$ to prove (3.10) for $\Phi(x) = x^\theta$ with $\theta = 1 - 1/33\,000 < 1$. This implied

$$d_n = O\left(p_n^\theta\right). \tag{3.11}$$

Hoheisel's argument was simplified in 1933 by H. Heilbronn [2707] who proved (3.11) with $\theta = 1 - 1/250$, and obtained an analogous result for primes in arithmetic progressions.

Denote by A the greatest lower bound for numbers a such that (3.10) holds for $\Phi(x) = x^a$. It was shown in 1937 by A.E. Ingham that the bounds for c in (3.7) have influence on the values of A by proving

$$A \le (1+4c)/(2+4c),$$

and this led to $A \le 5/8$ in view of the bound $c \le 1/6$ obtained by G.H. Hardy and J.E. Littlewood. This implied the existence of a prime between two sufficiently large consecutive cubes, and the first effective version of this assertion was obtained recently by Y.F. Cheng [1042] who showed that this holds for cubes exceeding $\exp\exp(15)$.

Further progress was made by N.G. Čudakov[11] who got $A \le 3/4$ [6094, 6095] (already H. Heilbronn had noted in his paper that $A \le 3/4$ would follow from the non-vanishing of $\zeta(s)$ in $\Re s > (1 - B(t) \log\log t / \log t)$ for some $B(t)$ tending to infinity with t, and this was established by Čudakov).

Consecutive improvements to the value of c led directly to upper bounds for A, starting with $A = 48/77 = 0.6233\ldots$ (E.C. Titchmarsh [6179] in 1942). The best known result achieved in this way, $A = 333/538 = 0.6189\ldots$ is due to M.N. Huxley [2985]. On the other hand it is known that (3.10) does not hold for $\Phi(x) = \log^a x$ with $a > 1$ (H. Maier [4098]).

---

[10] Guido Hoheisel (1894–1968), professor in Breslau and Köln.

[11] Nikolai Grigorevič Čudakov (1905–1986), professor in Saratov, Moscow and Leningrad. See [3817, 6475].

Any progress in the evaluation of the constant A implies a bound for the difference d_n of consecutive primes, however it turned out that other methods may give better results. So in 1969 H.L. Montgomery [4355] used his improvement of the bound for $N(\sigma, T)$ to prove

$$d_n = O\left(p_n^a\right) \tag{3.12}$$

with $a = 3/5$.

Later the value of a was reduced several times, and the actual record stands at $a \le 0.525$, achieved in 2001 by R.C. Baker, G. Harman and J. Pintz [267]. Previous steps were made by M.N. Huxley ($7/12 = 0.5833\ldots$) [2972] in 1972, H. Iwaniec and M. Jutila ($13/23 = 0.5652\ldots$) [3063] in 1979, D.R. Heath-Brown and H. Iwaniec ($11/20 = 0.55$) [2664] in 1979, J. Pintz ($17/31 = 0.5483\ldots$) [4900, 4901] in 1981, H. Iwaniec and J. Pintz ($23/42 = 0.5476\ldots$) [3068] in 1984, C.J. Mozzochi ($11/20 - 1/384 = 0.5473\ldots$) [4462] in 1986, S.T. Lou and Q. Yao ($6/11 = 0.5454\ldots$ and $7/13 = 0.5384\ldots$) [3991, 3992] in 1992 and 1993 and R.C. Baker and G. Harman ($107/200 = 0.535$) [265] in 1996.

It is conjectured that $A = 0$, thus (3.10) holds for $\Phi(x) = x^\varepsilon$ for every positive $\varepsilon > 0$, but even the much weaker conjecture, stating that for every $\varepsilon > 0$ the interval $[x, x + x^\varepsilon]$ contains a square-free integer is still open. Denote by B the greatest lower bound for numbers a such that for all large x the interval $[x, x + x^a]$ contains a square-free number. In 1941 E. Fogels [2017] established $B \le 2/5$, and in 1951 K.F. Roth [5307] proved $B \le 3/13 = 0.2307\ldots$. The next improvements were rather small: $2/9 = 0.2222\ldots$ (H.-E. Richert [5205]), $0.2212\ldots$ (R.A. Rankin [5115]), $1057/4785 = 0.2208\ldots$ (S.W. Graham, G. Kolesnik [2304], note the advance by 0.0004 in 34 years), $17/77 = 0.2207\ldots$ (O. Trifonov [6199]), $47/217 = 0.2165\ldots$ (M. Filaseta [2001]), $3/14 = 0.2142\ldots$ and $1/5 = 0.2$ (M. Filaseta, O. Trifonov [2002, 2003]). Later Filaseta [2001] showed that $B = 0$ is equivalent to the existence of the limit

$$\lim_{x\to\infty} \frac{1}{x} \sum_{n\le x} (s_{n+1} - s_n)^\gamma$$

for every $\gamma > 0$, s_n denoting the nth consecutive square-free integer. The existence of this limit was established in 1951 by P. Erdős [1805] for $\gamma \le 2$, by C. Hooley [2867] for $\gamma \le 250/79 = 3.1645\ldots$, by M. Filaseta [2001] for $\gamma < 29/9 = 3.2222\ldots$, by M. Filaseta, O. Trifonov ([2005]) for $\gamma < 43/13 = 3.3076\ldots$, and by M.N. Huxley for $\gamma < 11/3 = 3.6666\ldots$ [2987] and $\gamma < 59/16 = 3.6875$ [2988].

2. It has been observed in 1929 by I. Schur [5576, 5577] that Čebyšev's result on primes [970] implies that for every $n \ge 24$ there is a prime in the interval $[n, 5n/4]$. The Prime Number Theorem implies that for every $\varepsilon > 0$ and $n \ge n_0(\varepsilon)$ the interval $[n, (1+\varepsilon)n]$ contains a prime, so Schur's observation lead to $n_0(1/4) = 24$. Later explicit values of $n_0(\varepsilon)$ for various ε were obtained by R. Breusch [720], L. Gatteschi [2206] and H. Rohrbach[12], J. Weis [5268].

In 1976 L. Schoenfeld [5552] computed $n_0(1/16597) \le 2010760$, and in 2003 O. Ramaré and Y. Saouter [5096] gave upper bounds for $n_0(\varepsilon)$ for several small values of ε. In the last paper it was also deduced from the Riemann Hypothesis that for $x \ge 2$ there is a prime in the interval $(x - \frac{8}{5}\sqrt{x}\log x, x)$. This improved the result of L. Schoenfeld [5552], who obtained this for the interval $(x - (\sqrt{x}\log^2 x)/4\pi, x)$.

Similar results for primes in progressions were obtained by H. Kadiri [3210].

[12]Hans Rohrbach (1903–1993), professor in Göttingen and Mainz. See [5584].

3. The Prime Number Theorem implies

$$\lambda = \limsup_{n\to\infty} \frac{d_n}{\log p_n} \geq 1.$$

This was improved to $\lambda \geq 2$ in 1929 by R.J. Backlund [209] and the next year A. Brauer and H. Zeitz[13] [687] obtained $\lambda \geq 4$.

In 1931 E. Westzynthius[14] [6643, 6644] proved λ to be infinite by establishing

$$\limsup_{n\to\infty} \frac{d_n}{\log p_n} \frac{\log_4 p_n}{\log_3 p_n} \geq 2e^{\gamma},$$

$\log_k x$ denoting the kth iterate of the logarithm. This was improved first by P. Erdős [1776] to

$$\limsup_{n\to\infty} \frac{d_n}{\log p_n} \frac{\log_3^2 p_n}{\log_2 p_n} > 0,$$

and then by R.A. Rankin who in [5105] showed

$$\limsup_{n\to\infty} \frac{d_n}{\log p_n} \frac{\log_3^2 p_n}{\log_2 p_n \log_4 p_n} \geq c$$

with $c = 1/3$. The value of c has subsequently been replaced by $e^{\gamma}/2$ (A. Schönhage [5561]), e^{γ} (Rankin [5109]), $1.31\ldots \cdot e^{\gamma}$ (H. Maier, C. Pomerance [4102]), and $2e^{\gamma}$ (J. Pintz [4904]).

Large differences between consecutive primes below 10^7 were listed by A.E. Western [6640].

In 1961 F. Gruenberger and G. Armerding [2372] checked maximal prime differences for the first six million primes, with the largest gap having length 220. Denote by $M(n)$ the largest gap for primes below n. In 1967 L.J. Lander and T.R. Parkin [3684] computed $M(1.46 \cdot 10^9) = 382$. This computation was extended by R.P. Brent [706, 708], J. Young, A. Potler [6793], T.R. Nicely [4591] and P.A. Cutter [1312].

The paper [687] also settled a question stemming from a statement by Legendre which appeared in the second edition of [3767]. It implied that if $p_1, p_2, \ldots, p_r, p_{r+1}$ are the first $r+1$ odd primes then any sequence of p_{r+1} consecutive integers contains at least one which has no odd prime divisor $\leq p_r$. A. Brauer and H. Zeitz showed in [687] that this fails for every $r \geq 13$, confirming a conjecture of A. Piltz [4877]. It was known earlier (A. Dupré [1655]) that Legendre's assertion fails for certain small values of r.

For a discussion of Legendre's assertion see also the paper [5195] by G. Ricci.

Denote by $C_0(k)$ the maximal number m such that there exist m consecutive integers, each having a prime divisor not exceeding the kth consecutive prime. The truth of Legendre's assertion would imply $C_0(k) \ll k \log k$. The true order of $C_0(k)$ is not known and the best known upper bound, $C_0(k) \ll k^2 \log^2 k$, was obtained by H. Iwaniec [3051]. A lower bound was given by R.A. Rankin [5109]:

$$C_0(k) \gg \frac{k \log^2 k \log\log\log k}{(\log\log k)^2}.$$

[13] Hermann Zeitz (1870–1939), worked in a bank. See [678].

[14] Erik Johan Westzynthius (1901–1980), worked as actuary in an insurance company.

A similar function was considered in 1961 by E. Jacobsthal [3085–3089], who denoted by $g(k)$ the smallest number m such that every interval of length m contains a number prime to k. Putting

$$C(r) = \max\{g(k) - 1 : \omega(k) = r\},$$

he showed $C(r) \leq 2^r$, and conjectured $C(r) = O(r^2)$ and $C(r) = C_0(r)$. One year later P. Erdős [1821] showed that for almost all r one has

$$C(r) = (1 + o(1))\frac{r \log\log r}{\varphi(r)}.$$

Later H.-J. Kanold [3244] got $C(r) \leq 2^{\sqrt{r}}$ for large r, and this has been improved to

$$C(r) \leq 2r^{2+2e\log r}$$

for $r \geq 15$ by H. Stevens [5944]. The best known upper bound is due to H. Iwaniec [3056], who obtained $C(r) \ll r^2 \log^2 r$, saving $\log^2 r$ upon an earlier bound of R.C. Vaughan [6348]. Exact values of $C(r)$ are known for $r \leq 50$ (T.R. Hagedorn [2434]).

On the other hand J. Pintz [4904] obtained

$$g(n) \geq (2e^\gamma + o(1))\frac{\log n \log\log n \log_4 n}{\log_3^2 n}$$

for infinitely many n.

4. In the last part of [2531] G.H. Hardy and J.E. Littlewood considered the problem of existence of infinitely many sets of primes of the form $n, n + a_1, n + a_2, \ldots, n + a_m$, where the numbers $a_1, a_2, \ldots, a_m$ are assumed to satisfy certain necessary conditions. No definite results in this particular case of Dickson's conjecture were achieved but some conjectural asymptotical formulas are deduced from a *Hypothesis X*, concerning the behavior of the function

$$\sum_p \Lambda(p)\Lambda(p + a_1) \cdots \Lambda(p + a_m) X^p$$

to the right of $X = 1$.

They also studied the function $\varrho(x)$, giving the maximal number of primes which occurs in an infinite number of sequences of $[x]$ consecutive integers, and using the exclude-include principle of Legendre they proved the bound

$$\varrho(x) < (1 + o(1))e^{-\gamma}\frac{x}{\log\log x}.$$

They suggested, based on an examination of primes below 200, that the inequality

$$\varrho(x) \leq \pi(x) \tag{3.13}$$

might be true for all $x \geq 2$.

Conjecture (3.13) is related to the inequality

$$\pi(x + y) \leq \pi(x) + \pi(y), \tag{3.14}$$

expected to hold for all $x \geq 2$, $y \geq 2$.

It was shown by A. Schinzel and W. Sierpinski [5452] that these two conjectures are equivalent if the conjecture H, formulated in [5452] (see Sect. 6.1.3), is true. The inequality (3.14) has been shown to hold for $x, y \leq 10^5$ by S.L. Segal [5602], and later P. Dusart [1662] established it for $2 \leq x \leq y \leq \frac{7}{5} \log x \log\log x$. See also L. Panaitopol [4738].

A very surprising result was obtained in 1972 by D. Hensley and I. Richards [2755, 2756] who showed that (3.14) is incompatible with the conjecture of Dickson. Indeed, assuming the truth of Dickson's conjecture they proved that for every given $\varepsilon > 0$ and sufficiently large x the inequality

$$\max_{y \geq x} (\pi(x+y) - \pi(y)) \geq \pi(x) + (\log 2 - \varepsilon)\frac{x}{\log^2 x}$$

holds, which obviously contradicts (3.14).

The truth of (3.14) would imply that the interval $(x, x+y]$ contains at most $(1 + o(1))y/\log y$ primes. In 1950 A. Selberg [5616] obtained the bound

$$\pi(x+y) - \pi(x) \leq \frac{2y}{\log y} + O\left(\frac{y \log\log y}{\log^2 y}\right),$$

and J.H. van Lint[15] and H.-E. Richert [6334] removed the iterated logarithm from the numerator, proving

$$\pi(x+y) - \pi(x) \leq \frac{2y}{\log y + c} \tag{3.15}$$

with a certain $c < 0$. They obtained also an analogous result for the difference $\pi(x+y; k, l) - \pi(x; k, l)$. In 1971 E. Bombieri [598] proved (3.15) for $x > \sqrt{y} > e^{3/2}$ with $c = -3$, and H.L. Montgomery and R.C. Vaughan [4364] showed that this holds with $c = 5/6$, hence there are at most $2y/\log y$ primes in an interval of length y. A. Selberg ([5624, Sect. 22]) increased c up to 2.81 and recently O. Ramaré and J.-C. Schlage-Puchta [5097] got (3.15) with $c = 3.53$.

3.1.4 Multiplicative Problems

1. It seems that the first results dealing with consecutive quadratic residues were obtained by R.D. von Sterneck [5937] in 1898 and E. Jacobsthal [3083] in 1906. Jacobsthal proved that for every large prime p there exist three consecutive quadratic residues mod p, and this was made more precise in 1925 by A.A. Bennett[16] [403], who showed that this happens for all $p \geq 19$ (for a certain class of primes this was proved also by H.S. Vandiver [6319]). Later A.A. Bennett showed [404, 405] that for $p > 53$ there are four consecutive quadratic residues and for $p > 193$ there are five of them. Cf. K. Dörge [1613] and H. Hopf[17] [2905].

---

[15] Jacobus Hendricus van Lint (1932–2004), professor in Eindhoven. See [788].

[16] Albert Arnold Bennett (1888–1971), professor at the University of Texas, Lehigh University and Brown University.

[17] Heinz Hopf (1894–1971), professor at ETH Zürich, worked in algebraic topology. See [2812].

Essential progress in this topic occurred after B.L. van der Waerden ([6309], see also [6311]), answering a question by P.J.H. Baudet[18], established in 1927 the existence of a number $W(k,l)$ having the property that if the integers not exceeding $W(k,l)$ are divided into k classes, then at least one of these classes contains an arithmetic progression of length l. This implies that if the set of all positive integers is divided into finitely many classes, then at least one class contains arbitrarily long arithmetic progressions.

Other proofs were given later by R. Rado[19] [5044], R.L. Graham, B.L. Rothschild [2299] and A.D. Taylor [6076]. A generalization to binary trees was obtained by H. Furstenberg and B. Weiss [2149] in 2003.

It was proved in 1952 by P. Erdős and R. Rado [1848] that $W(k,l)$ exceeds $\sqrt{2lk^l}$, and this has been superseded for large k by $W(k,l) > lk^{c\log k}$ with a certain positive c (L. Moser[20] [4441]) and $W(k,l) > k^{l-c\sqrt{l\log l}}$ (W.M. Schmidt [5486]). For small k, E.R. Berlekamp [442] gave stronger bounds. Upper bounds, which are rather large, were provided by S. Shelah [5685] and W.T. Gowers [2294]. A table of known values and bounds was given in [1619].

Several problems related to van der Waerden's theorem have been listed in the book [1831] by P. Erdős and R.L. Graham. One of them asked for a constant $b > 1$ such that if the integers from the interval $[2, b^k]$ are divided into k classes, then one of these classes contains a subset S with

$$\sum_{n\in S} \frac{1}{n} = 1.$$

This problem was solved in 2003 by E.S. Croot, III [1283], who showed that $b = \exp(167\,000)$ will do.

Van der Waerden's theorem was used by A. Brauer in [674] to prove, as conjectured by A.A. Bennett [404], the existence of arbitrarily long chains of consecutive kth power residues or non-residues mod p, provided the prime $p \equiv 1 \pmod{k}$ is sufficiently large. H. Davenport [1342] showed that for $p > 100k^{12}$ there exists a triple of kth power residues.

More precise results have been obtained in the case $k = 2$. Denote by $R(p)$, $N(p)$ the length of the biggest chain of consecutive quadratic residues, and non-residues mod p, respectively. It was conjectured by I. Schur that for large p one has $N(p) < \sqrt{p}$ (note that the inequality $N(p) < 2\sqrt{p}$ is trivial, as the difference of consecutive squares smaller than p is less than $2\sqrt{p}$). This conjecture has been established for p congruent to 3 mod 4 by A. Brauer [677] in 1932, who obtained also $R(p) < \sqrt{p}$ in this case. The bound $R(p) = O(\sqrt{p})$ was proved in 1950 by H. Davenport and P. Erdős [1381]. The best known lower bound, $R(p) \gg \log p$ follows from Weil's bounds for character sums.

Progress next came in 1957, when D.A. Burgess [854] found a new method of evaluating character sums which in the quadratic case for large p led to $\max\{N(p), R(p)\} < p^c$ with

[18] Pierre Joseph Henry Baudet (1891–1921), professor at the Technical School at Delft. See [5843].

[19] Richard Rado (1906–1989), professor in Exeter. See [3826, 5257].

[20] Leo Moser (1921–1970), professor at the University of Alberta. See [6764].

c being any number larger than 1/4. Later [855] he showed that $R(p) \ll p^{1/4}\log^{3/2} p$ and in [858] replaced the exponent 3/2 by 1. This is still the best known bound. Much later R.H. Hudson [2943, 2944] proved $N(p) < \sqrt{p}$ for all primes $p \not\equiv 13 \pmod{24}$ and expressed the opinion that this holds for all primes $p \neq 13$. This turned out to be true due to the work of P. Hummel [2951], hence now we know that Schur's conjecture holds for all primes $p \neq 13$. Bounds for the smallest pair of consecutive quadratic non-residues mod p were given by P.D.T.A. Elliott [1731, 1732] and A. Hildebrand [2803].

In 1962 D.H. Lehmer and E. Lehmer[21] [3795] considered consecutive higher residues and defined $r(k,l,p)$ as the minimal integer r with the property that the interval $[r, r+l-1]$ consists of kth power residues mod p. They conjectured that if one puts

$$\Lambda(k,l) = \limsup_{p\to\infty} r(k,l,p),$$

then $\Lambda(k,4) = \infty$, proved this for small k and established $\Lambda(2k,3) = \infty$. In a joint paper with W.H. Mills [3796] they showed $\Lambda(k,2) < \infty$ for $5 \le k \le 6$ and asked whether this inequality persists for every k. For $k = 4$ this was proved by R.G. Bierstedt and Mills [504] and for $k = 7$ by J. Brillhart, D.H. Lehmer, E. Lehmer [732]. In 1986 A. Hildebrand [2800, 2801] showed that $\Lambda(k,2)$ is finite for prime k and five years later [2801] proved this for every k.

The first of Lehmer's conjectures ($\Lambda(k,4) = \infty$) was proved in 1964 by R.L. Graham [2298]. The status of $\Lambda(k,3)$ for odd $k \ge 5$ is still unknown, except for $\Lambda(3,3) = 23\,532$ (D.H. Lehmer, E. Lehmer, W.H. Mills, J.L. Selfridge [3797]).

2. In 1930 K. Dickmann [1534] studied the number $\phi(u,x)$ of integers $n \le x$ which have a prime divisor larger than n^u (for fixed $0 < u < 1$) and showed the existence of the limit

$$\varrho(u) = \lim_{x\to\infty} \frac{\phi(u,x)}{x},$$

where $\varrho(u) = 1$ for $0 \le u \le 1$, and $u\varrho'(u) = -\varrho(u-1)$ for $u > 1$ (the *Dickmann function*).

This was later rediscovered by S. Chowla and T. Vijayaraghavan [1102].

The related function $\Psi(x,y)$ giving the number of integers $n \le x$ having all its prime divisors $\le y$ (note that Dickmann's function $\phi(u,x)$ is close to $x - \Psi(x,x^u)$) appears in R.A. Rankin's paper [5105], where an upper bound for it has been given. Bounds for the error term $\Psi(x,x^u) - \varrho(u)x$ were given by A.A. Buhštab [834] and V. Ramaswami [5098]. Asymptotics for $\Psi(x,y)$ with a slightly different main term were obtained by N.G. de Bruijn [785, 786] (cf. E. Saias [5369]), who in [784] made a thorough study of Dickmann's function $\varrho(u)$. Much later A. Hildebrand [2799] established the equality

$$\Psi(x,y) = x\varrho(u)\left(1 + O\left(\frac{\log(u+1)}{\log y}\right)\right)$$

(where $u = \log x/\log y$) valid uniformly in the range $\exp(\log^c \log x) \le y \le x$ with any $c > 5/3$. A more complicated asymptotic formula for $\Psi(x,y)$, uniform for $2 \le y \le x$,

[21] Emma Lehmer (1906–2007). See [731].

was given in 1986 by A. Hildebrand and G. Tenenbaum [2807]. For further improvements see G. Tenenbaum [6109] and T.Z. Xuan [6769]. For a probabilistic approach see R. de la Brèteche, G. Tenenbaum [1424, 1425].

Similar questions for integers in residue classes were considered by É. Fouvry, G. Tenenbaum [2064], A. Granville [2315, 2316], G. Harman [2562] and K. Soundararajan [5855].

Surveys were given by K.K. Norton [4626] in 1971 and by A. Hildebrand and G. Tenenbaum [2809] in 1993.

The number $\Phi(x, y)$ of integers $n \leq x$ having all its prime divisors $\geq y$ forms a counterpart of $\Psi(x, y)$. It was first considered by A.A. Buhštab [831] in 1937, who showed that for fixed $u = \log x / \log y$ the value $\Phi(x, y)$ is asymptotic to $u\omega(u)x/\log x$, where $\omega(u) = 1/u$ for $1 \leq u \leq 2$ and $(u\omega(u))' = \omega(u-1)$ for $u \geq 2$ (the *Buhštab function*). For improvements see N.G. de Bruijn [785, 786], A.A. Buhštab [835].

This topic has been presented in the last two chapters of the book [6107] by G. Tenenbaum.

3. Denote by $P(n)$ the maximal prime divisor of n. In 1895 A.A. Markov [4155] considered

$$P_f(x) = P\left(\prod_{n \leq x} f(n)\right)$$

and showed that for $f(X) = X^2 + 1$ this product tends to infinity quicker than x, confirming an earlier statement by Čebyšev. In 1917 G. Pólya [4954] proved the same assertion for cyclotomic polynomials, and in 1921–1922 T. Nagell [4503, 4504] first established this for every irreducible polynomial in $\mathbf{Z}[X]$ and then for all polynomials having at least one irrational root. More precisely, he showed that for every $\varepsilon > 0$ and x sufficiently large one has

$$P_f(x) \geq x \log^{1-\varepsilon} x. \tag{3.16}$$

In 1929 S. Chowla [1068] made Markov's result more precise by showing that for large x and every $c < 2$ one has

$$P_{X^2+1}(x) > cx \log x.$$

The ε in (3.16) was removed in 1934 by G. Ricci [5197]. It was shown much later by P. Erdős [1808] that if f is not a product of linear factors with integral coefficients, then

$$P_f(x) \gg x \exp(c \log\log x \log\log\log x)$$

holds with a certain $c > 0$. This was improved in 1990 by P. Erdős and A. Schinzel [1856] to

$$P_f(x) \gg x \exp\exp\left(c (\log\log x)^{1/3}\right),$$

and the best known result is due to G. Tenenbaum [6108], who proved

$$P_f(x) \gg x \exp\left(\log^c x\right)$$

for every $c < 2 - \log 4 = 0.6137\ldots$. For quadratic binomials $f(x) = x^2 + a$ a better lower bound, $P_f(x) \gg x^{11/10}$ was obtained by C. Hooley [2863] (cf. [2868]) and this was improved by J.-M. Deshouillers and H. Iwaniec [1490, 1491] first to $\gg x^{6/5}$ and then to $\gg x^{1.202}$. For the polynomial $f(X) = X^3 - 2$ the bound $P_f(x) \gg x^{31/30}$ was established by

C. Hooley [2892] and for $f(X) = X^3 + 2$ D.R. Heath-Brown [2656] obtained $P_f(x) \gg x^{1+\delta}$ with some $\delta > 0$.

The paper [1068] by S. Chowla also contains the evaluation $P(n^2 + 1) \gg \log\log n$.

For certain other quadratic and cubic polynomials f the lower bound $P(f(n)) \gg \log\log n$ was obtained later by K. Mahler [4071, 4077], T. Nagell [4512], A. Schinzel [5439] and M. Keates [3291]. S.V. Kotov [3499] did this later for all non-linear polynomials, improving upon the results of J. Coates [1124] ($P(f(n)) \gg (\log\log n)^{1/4}$) and V.G. Sprindžuk[22] [5876], who obtained $P(f(n)) \gg \log\log n / \log\log\log n$. It was conjectured by T.N. Shorey, A.J. van der Poorten, R. Tijdeman and A. Schinzel [5731] that for all such polynomials f the maximal prime divisor of $f(n)$ is, for large n, larger than $B \log^c n$ with some B, c depending on f. On the other hand it has been shown by C. Dartyge [1336] that for a set of integers of positive density one has $P(n^2 + 1) \ll n^a$ for every $a > 149/179 = 0.8324\ldots$. See also C. Dartyge, G. Martin, G. Tenenbaum [1337].

4. T. Nagell proved in [4504] that irreducible polynomials in $\mathbf{Z}[X]$ of degree d and without fixed divisors represent infinitely many d-free integers. There is an old conjecture that every such polynomial represents infinitely many square-free integers.

Nagell's result was made more precise in 1935 by G. Ricci [5194, 5195] who used Brun's sieve to show that the set of integers n with d-free $f(n)$ has a positive density.

In 1953 P. Erdős [1809] showed that $(d-1)$-free values occur infinitely often, and in 1967 C. Hooley [2864] (cf. [2868]) proved Ricci's result for $(d-1)$-free polynomial values. Further improvement was made by M. Nair [4519, 4520] who showed that for $k \geq (\sqrt{2} - 1/2)d$ infinitely many k-free values occur, and in part II obtained the same for prime arguments, improving a previous result by C. Hooley [2890, 2891]. In [2992] M.N. Huxley and M. Nair showed that for $d \geq 15$ infinitely many $(d-2)$-free numbers occur.

It was shown in 1998 by A. Granville [2317] that the conjecture on square-free values is a consequence of the ABC conjecture (see Sect. 4.4.3).

5. The multiplicative analogue $F(n)$ of the partition function, giving the number of solutions of

$$x_1 \cdot x_2 \cdots x_k = n \quad (k = 1, 2, \ldots;\ x_i \geq 2), \tag{3.17}$$

(solutions differing in ordering being identified) was introduced in 1923 by P.A. MacMahon [4056]. Using the generating series

$$\sum_{n=1}^{\infty} \frac{F(n)}{n^s} = \prod_{m=2}^{\infty} \frac{1}{1 - m^{-s}},$$

A. Oppenheim [4687, 4688] succeeded in obtaining the asymptotic expansion

$$\sum_{n \leq x} F(n) = \frac{1}{2\sqrt{\pi}} \frac{x}{\log^{3/4} x} \exp\left(2\sqrt{\log x}\left(1 + \sum_{r=1}^{N-1} \frac{c_r}{\log^{r/2} x} + O\left(\frac{1}{\log^{N/2} x}\right)\right)\right),$$

[22] Vladimir Gennadievič Sprindžuk (1936–1987), professor in Minsk. See [3255].

where N is arbitrary, and $c_1, c_2, \ldots$ are constants.

Oppenheim's result was rediscovered later by G. Szekeres and P. Turán [6021]. In 2010 F. Luca, A. Mukhopadhyay and K. Srinivas [4024] considered the behavior of $F(n)$ in an interval $I = (x, x + y)$ (with $y \geq 2$) and proved

$$\sum_{n \in I} F(n) \gg \frac{y}{\log^{3/4} y} \exp(2\sqrt{y}).$$

They also obtained an upper bound for the number of values of $F(n)$ in $[1, x]$.

In 1983 J.F. Hughes and J.O. Shallit [2945] proved $f(n) \leq 2n^{\sqrt{2}}$, and this was improved by E.R. Canfield, P. Erdős and C. Pomerance [888] to

$$f(n) \leq n \exp\left(-(1 + o(1)) \frac{\log n \log_3 n}{\log_2 n}\right),$$

which is best possible.

For a similar function, in which the order of the factors in (3.17) is taken into account see Sect. 4.1.3.

3.2 Additive Problems

3.2.1 The Waring Problem

1. The idea applied by G.H. Hardy and S. Ramanujan in [2541] to study the number of partitions was later exploited in other additive problems, based on the following observation: let $a_1 < a_2 < \cdots$ be a sequence of non-negative integers, and put

$$F(z) = \sum_{m=1}^{\infty} z^{a_m},$$

the series converging in the open unit disc $\{z : |z| < 1\}$. Then for $k = 2, 3, \ldots$ the nth coefficient of the power series for $F^k(z)$ equals the number $f_k(n)$ of representations of n as the sum of k elements of the sequence $\{a_m\}$.

For $r \in (0, 1)$ Cauchy's integral formula now leads to the equality

$$f_k(n) = \frac{n!}{2\pi i} \oint_{|z|=r} \frac{F^k(z)}{z^{n+1}} dz, \tag{3.18}$$

hence in order to show that all large integers can be represented as the sum of k elements of the sequence $\{a_m\}$ it suffices to show that for a certain $0 < r < 1$ the integral in formula (3.18) is non-zero for large n.

This method, called *the circle method*, found a wealth of applications, some of which we shall describe.

2. In 1920 G.H. Hardy [2511] used this method to obtain asymptotic formulas for the number $r_k(n)$ of representations of an integer n as the sum of k squares, and this led to a new proof of the classical exact formula for $k=5$. The same method works also for $k=6,7$ and 8 (details for $k=7$ were described by G.K. Stanley [5886] in 1927).

For $k=3,4$ this method needs amendments due to the convergence problems of some series (P.T. Bateman [352]). The pure circle method was successfully applied to the case $k=3$ in 1959 by T. Estermann [1894].

A survey of asymptotical results concerning $r_k(n)$ was published in 1951 by A. Walfisz [6260].

The distribution of solutions of the equation $\sum_{j=1}^{n} x_j^2 = m$ on the sphere of radius $\sqrt{m}$ in $\mathbf{R}^n$ has been shown in the case $n=3$ to be uniform in a certain sense by Yu.V. Linnik, A.V. Malyšev [3930] and Yu.V. Linnik [3918, 3919] (see also the book [3927] by Yu.V. Linnik devoted to the application of ergodic theory to this type of problem). The case $n \ge 4$ was later treated, also for integer points on ellipsoids, by Ch. Pommerenke [4982] and A.V. Malyšev [4122].

Later, explicit formulas in the case of sums of $9 \le k \le 32$ squares were given by G.A. Lomadze[23] [3984, 3985]. Books dealing with sums of squares were published by E. Grosswald [2366] in 1985 and by C.J. Moreno and S.S. Wagstaff, Jr. [4424] in 2002.

3. In the hands of G.H. Hardy and J.E. Littlewood [2524, 2525], the circle method led to a new solution of the Waring problem. The paper [2525] was the first of the cycle *Partitio Numerorum*, which had a tremendous influence on the further development of additive number theory. In it the same approach as in [2541] was applied to the problem of representing an integer n as the sum of s kth powers but this time some additional complications arose. It was not enough to divide the circumference of the integration circle into similar arcs, but one had to consider two types of arcs: those with $q \le n^{1/k}$, called *major arcs*, and those with $n^{1/k} < q \le n^{1-1/k}$, called *minor arcs*. The integrals over the major arcs could be evaluated with a rather good error term, thanks to the use of Weyl's method[24] [6647] of dealing with exponential sums, and the same approach was applied to obtain upper bounds for the integral on minor arcs. For some reason the authors of [2525] used the function $f(z) = 1 + 2\sum_{m=1}^{\infty} z^{m^k}$ instead of the more natural looking series $\sum_{m=0}^{\infty} z^{m^k}$, so the coefficients of

$$f^s(z) = 1 + \sum_{n=1}^{\infty} r_{k,s}(n) z^n$$

were equal to the number of solutions $R_{k,s}(n)$ of the equation

$$n = x_1^k + x_2^k + \cdots + x_s^k \tag{3.19}$$

[23] George Lomadze (1912–2005), professor in Tbilisi.

[24] The treatment of major arcs in [2525] was soon simplified by E. Landau [3662] and H. Weyl [6649].

(also taking in account negative bases x_i, as well as permutations) only in the case of k even. For odd k one has the relation

$$r_{k,s}(n) = \sum_{j=1}^{s} \binom{s}{j} 2^j R_{k,j}(n).$$

An asymptotic formula for $r_{k,s}(n)$ in the case of $s \geq 2^k + 1$ had the form

$$r_{k,s}(n) = \left(\frac{\Gamma(1+1/k)^k}{\Gamma(s/k)} + o(1) \right) \mathfrak{S}(n) n^{s/k-1}, \tag{3.20}$$

where $\mathfrak{S}(n)$ is the *singular series*, defined by

$$\mathfrak{S}(n) = \sum_{q=1}^{\infty} \sum_{\substack{1 \leq p < q \\ (p,q)=1}} \left(\frac{S_{p,q}}{q} \right)^s \exp\left(-\frac{2\pi i n p}{q} \right), \tag{3.21}$$

with

$$S_{p,q} = \sum_{h=0}^{q-1} \exp\left(\frac{2\pi i h^k p}{q} \right).$$

This formula is useful only in the case when $\mathfrak{S}(n)$ exceeds a positive constant depending on k, and the question of which s this happens for was answered later in [2532].

4. In the second paper [2526] of their series they presented a sharpening of their argument[25] in the special case $k = 4$, which allowed them to obtain the inequality $G(4) \leq 21$. This improved upon the previous bound $G(4) \leq 37$, due to A. Wieferich [6658].

It took thirty years before this bound was improved to $G(4) \leq 17$ by T. Estermann [1887] and H. Davenport, H. Heilbronn [1388]. The final step was taken by H. Davenport [1354] who established the equality $G(4) = 16$.

The paper [2532] contains a detailed study of the singular series (3.21) in the case of an arbitrary k, which shows that for $s \geq 1 + 2^k$ this series has a product representation

$$S = \prod_{\pi} \left(1 + \sum_{r=1}^{\infty} A_{\pi^r}(n) \right),$$

with π running over all primes and

$$A_q(n) = \sum_{p=1}^{\infty} \left(\frac{S_{p,q}}{q} \right)^s \exp\left(-\frac{2\pi i n p}{q} \right).$$

[25] They used the bound $O(n^{\varepsilon})$ for the number of representations of an integer n as the sum of two kth powers, proved by D. Cauer in [967].

It turned out that the singular series is intimately related to the number $M(\pi^k, n)$ of solutions of the congruence

$$\sum_{j=1}^{s} x_j^k \equiv n \pmod{\pi^k}$$

for prime powers π^k. Indeed, [2532, Lemma 2] gives the formula

$$M(\pi^k, n) = \frac{1}{\pi^{k(s-1)}} \sum_{j=0}^{k} A_{\pi^j}(n).$$

This lemma hid a crucial observation, and H. Heilbronn wrote in his comments to Hardy's *Collected Papers* (vol. I, p. 378): "It is historically interesting that the important relation between the singular series and the congruence was formulated by the authors as a lemma only ...."

This result implies that the singular series $\mathfrak{S}(n)$ is related to the Waring problem in p-adic fields and its positivity is equivalent to the representability of n as the sum of sth powers of p-adic integers.

5. In [2532] the asymptotical formula[26] for $r_{k,s}(n)$ was shown to hold for $s \geq (k-2)2^k + 5$, which implied the effective bound $G(k) \leq (k-2)2^{k-1} + 5$, and in the sixth paper of the series [2535] the order of this bound was essentially halved. In particular, the inequality $G(4) \leq 19$ was obtained there, which was of particular importance in view of the inequality $g(4) \geq 19$, resulting from the observation that the number 79 requires 19 biquadrates. For other small exponents the inequalities $G(5) \leq 41$, $G(6) \leq 87$, $G(7) \leq 193$, $G(8) \leq 425$, $G(9) \leq 949$ and $G(10) \leq 2113$ resulted.

Denote by $\tilde{G}(k)$ the minimal number s with the property that the asymptotic formula holds for sufficiently large n. The bound $\tilde{G}(k) \leq (k-2)2^k + 5$ was consecutively reduced by I.M. Vinogradov in a series of papers in 1935–1936 [6421, 6425–6427]. In 1942 Vinogradov [6435] proved $\tilde{G}(k) \leq 20k^2 \log k$ for $k \geq 10$, and a few years later L.K. Hua [2933, 2934] got

$$\tilde{G}(k) \leq 4k^2 \left(\log k + O\left(\sqrt{\log k}\right)\right).$$

In 1992 T.D. Wooley [6725, 6726] improved this to

$$\tilde{G}(k) \leq (2 + o(1))k^2 \log k,$$

using his refinement of Vinogradov's mean value theorem. This bound was halved by K. Ford [2027]. Recently T.D. Wooley [6738] established

$$\tilde{G}(k) \leq 2k^2 + 2k - 3.$$

For small k R.C. Vaughan [6360, 6361, 6363] proved in 1986 for $k \geq 3$ the bound $\tilde{G}(k) \leq 2^k$, which implies the equality $\tilde{G}(4) = 16$, and two years later D.R. Heath-Brown [2643] got

$$\tilde{G}(k) \leq 7 \cdot 2^{k-3} + 1 \quad (k \geq 6).$$

The unit summand in this inequality was removed by K.D. Boklan [589].

---

[26] A simplified proof of this formula was given by E. Landau [3663] and found its way into Landau's book [3674].

6. A part of [2535] is devoted to consequences of the conjecture (called by the authors *Hypothesis K*), asserting that the number $F_k(n)$ of solutions of the equation

$$x_1^k + \cdots + x_k^k = n$$

is $O(n^\varepsilon)$ for every $\varepsilon > 0$. G.H. Hardy and J.E. Littlewood showed that this conjecture implies the bounds

$$G(k) \le \begin{cases} 4k & \text{if } k \text{ is a power of 2,} \\ 2k+1 & \text{otherwise,} \end{cases}$$

but, unfortunately, it was noted in 1936 by K. Mahler [4080] that it fails for $k = 3$ as a consequence of the identity

$$\left(9x^4\right)^3 + \left(3xy^3 - 9x^4\right)^3 + \left(y^4 - 9x^3y\right)^3 = y^{12}.$$

Its status in the case $k \ge 4$ is unknown.

It was proved by S. Chowla and S.S. Pillai in 1936 [1099] that for $k \ge 5$ one has $F_k(n) = \Omega(\log\log n)$ and P. Erdős [1782] improved this to

$$F_k(n) = \Omega\left(\exp\left(c\frac{\log n}{\log\log n}\right)\right)$$

for a suitable $c > 0$.

The paper [2535] also contains results concerning representations of almost all numbers as sums of powers. If $G_1(k)$ denotes the minimal number of kth powers of non-negative integers needed to represent almost all integers, then the bound $G_1(k) \le (k-2)2^{k-2} + 3$ was proved for $k \ne 4$, which implied, in particular, that almost all numbers are sums of five cubes. Moreover the equality $G_1(4) = 15$ has been established. The only previous result for cubes was obtained by W.S. Baer [210] who showed that the set of integers which are sums of seven cubes has a positive density.

In 1939 H. Davenport [1356] obtained $G_1(3) = 4$, in 1986 R.C. Vaughan [6362] established $G_1(8) = 32$, and four years later T.D. Wooley [6724] proved $G_1(16) = 64$.

7. The eighth[27] and last paper of the *Partitio Numerorum* series [2536], was devoted to the study of the number $\Gamma(k)$, defined as the minimal number s such that every arithmetical progression contains infinitely many sums of at most s positive kth powers. This number can be also defined as the smallest s such that every congruence

$$\sum_{j=1}^{s} x_j^k \equiv n \pmod{p^m}$$

[27] The seventh paper of that series never appeared. It had to contain the proof that under a certain generalization of the Riemann Hypothesis one has $\liminf(p_{n+1} - p_n)/\log p_n \le 2/3$, p_n being the nth consecutive prime.

with prime p, $m = 1, 2, \ldots$ and integral n has a solution with not all x_j's divisible by p. Therefore $\Gamma(k)$ gives a bound for the Waring constant for the exponent k in the ring $\mathbf{Z}_p$ of p-adic integers. Clearly one has $\Gamma(k) \le G(k)$, and in [2536] it was shown that one has $\Gamma(k) \le k$, except when k has one of the following forms:

$$2^m, \quad 3 \cdot 2^m \quad (m \ge 2), \quad p^m(p-1), \quad p^m(p-1)/2 \quad (m \ge 0),$$

p being an odd prime. In all these exceptional cases the value of $\Gamma(k)$ was determined in [2536], e.g., one has $\Gamma(2^m) = 2^{m+2}$. If p and $2p+1$ are both primes, then $\Gamma(p) = p$. The conjecture

$$\lim_{k \to \infty} \Gamma(k) = \infty$$

stated there is still open. It is related to the question of the smallest prime in an arithmetical progression (see C. Elsholtz [1760]).

If p is a prime such that $2p+1$ is composite, then $\Gamma(p) \le 2 + 2[p/3]$ has been shown in 1937 by I. Chowla [1064]. In 1973 M. Dodson [1608] established the inequality $\Gamma(k) < k^{7/8}$ for infinitely many k. Earlier he evaluated the mean value of $\Gamma(k)$ [1606]:

$$\sum_{k \le x} \Gamma(k) = \left(\frac{5\pi^2}{24} + o(1) \right) \frac{x^2}{\log x}.$$

The value of $\Gamma(k)$ is now known for every $k \le 200$ (H. Sekigawa, K. Koyama [5604]), and C. Elsholtz [1760] established the inequality $\Gamma(k) \ge 5$ for $20 \le k < 5 \cdot 10^8$.

8. An extension of the Waring problem, in which the kth power is replaced by a fixed polynomial (the *Waring–Kamke problem*) was obtained by E. Kamke[28] [3226, 3229], who generalized the old result of Maillet [4107] to polynomials of arbitrary degree. He also considered [3227, 3228] the analogous problem of representing rational numbers as sums of polynomial values at rational arguments and its generalization to algebraic number fields.

Such a result had been conjectured already by E. Waring [6565] and the first theorem of this type goes back to A. Cauchy[29] [966] who established Fermat's assertion made in [1990] that every positive integer is the sum of N N-gonal numbers, which are the values of the quadratic polynomial

$$(N-2)\frac{X^2 - X}{2} + X$$

at integers[30].

It was later shown by L.E. Dickson [1558] that for $N \ge 5$ only a few integers need N N-gonal summands.

---

[28] Erich Kamke (1890–1961), student of Landau. Professor in Tübingen (1926–1937 and 1946–1958). See [6540].

[29] Augustin Louis Cauchy (1789–1857), professor in Paris. See [393, 6266].

[30] Other proofs were later given by T. Pépin [4774] and L.E. Dickson [1549].

Kamke's proof was elementary, based on a lemma concerning the Hilbert–Kamke problem (see (3.22) below) but the circle method is also applicable here, as shown later by E. Landau [3663] and E. Kamke [3230].

An elementary proof based on Schnirelman's theorem was given in 1956 by A.V. Kuzel [3588].

For quadratic polynomials a bound for the number of needed summands was given by L.E. Dickson [1552] (cf. [1553, 4726]).

The case of cubic polynomials was treated in 1934 by R.D. James[31] [3101, 3102] who showed that in absence of a non-trivial constant divisor all large integers are sums of at most 9 polynomial values, and six years later L.K. Hua [2926] showed that already eight summands are sufficient. In the case of *pyramidal numbers* $(x^3 - x)/6$ this had earlier been proved by R.D. James [3098].

The number of summands needed in the case of polynomials of small degree ≥ 4 was bounded in 1935 by L.E. Dickson [1561] and M.G. Humphreys [2952]. For the quartic case see also H.B. Yu [6798, 6799].

An asymptotic formula for the number of representations of an integer as the sum of $s \geq 2^{k-1}(k-2)+5$ values of a polynomial of degree $k \geq 3$ was proved by L.K. Hua [2916] in 1936, who later [2921] obtained it for $s \geq 2^k + 1$. He also found [2917–2919] the first upper bound for the minimal number $G(f)$ of summands $f(x_j)$ needed to represent every large integer by showing that if f is an integer-valued polynomial of degree $k \geq 21$ and positive leading term and for every integral c the polynomial $f(X) - c$ does not have a constant factor > 1, then $O(2^k k^3)$ summands are sufficient for this task. In [2919] he reduced this to $G(f) \leq (k-1)2^{k+1}$. He conjectured also that

$$G(f) \leq \begin{cases} 2^k & \text{if } 2|k, \\ 2^k - 1 & \text{if } 2 \nmid k, \end{cases}$$

and presented examples to show that these inequalities cannot be improved.

Hua's conjecture was established in 1998 by H.B. Yu [6797]. In 1997 T.D. Wooley [6731] established an asymptotic formula for the number of representations for more than $ck \log k$ summands, provided the corresponding singular series is bounded away from zero. Further progress was obtained by Ford [2030].

Representations of integers in the form

$$N = \sum_{j=1}^{k} f(x_j) + g(y)$$

with polynomials f, g were considered in the case of a quintic f by H.B. Yu [6796] and K. Kawada, T.D. Wooley [3287]. They showed that for any integer-valued g this is possible for all large N, satisfying the necessary congruence conditions with $k = 24$ and $k = 20$, respectively. An early survey of the Waring–Kamke problem was published in 1951 by V.I. Nečaev [4552].

[31] Ralph Duncan James (1909–1979), professor at the University of Saskatchewan and University of British Columbia.

9. E. Kamke's paper [3226] also contains another generalization of Waring's problem (the *Hilbert–Kamke problem*). It was shown there that the system of Diophantine equations

$$\sum_{j=1}^{s} x_j^k = N_k \quad (k = 1, 2, \ldots, n) \tag{3.22}$$

is solvable provided the integers $N_1, \ldots, N_n$ are divisible by a certain $A = A(n)$ and s is sufficiently large. It was conjectured by Hilbert in a seminar around 1910 that this system is solvable if s is large and the N_k's satisfy the necessary local conditions (i.e., (3.22) has real solutions, and for every $M > 1$ the corresponding congruence is solvable mod M).

The first result of this type had been obtained already by A. Cauchy [966], who considered the case $n = 2$ and showed that (3.22) is solvable with $s = 5$, provided it has real solutions and $N_1 \equiv N_2 \pmod 2$.

It was shown later by K.K. Mardžanišvili[32] first [4143] that Hilbert's conjecture holds for $s \geq n^4 2^{n^2-n-2}$, and then [4144] that it holds already for $s \geq 5n(n+1)(n+2)\log n$. He found also for $s > 4^n(n+1)^3 n!$ an asymptotical formula for the number of solutions. Later, in 1940, L.K. Hua [2927] proved solvability for $s \geq (7 + o(1))n^2 \log n$, G.I. Arkhipov [116] obtained in 1984 asymptotics for $s > 3n^3 2^n$, and finally in 1986 D.A. Mitkin [4340, 4341] determined for $n \geq 12$ the minimal possible number of summands needed for solvability. An exposition of the Hilbert–Kamke problem has been given in the book [123] by G.I. Arkhipov, A.A. Karatsuba and V.N. Čubarikov.

The analogous problem in which the x_j's in (3.22) have to be primes has been considered by Mardžanišvili [4145, 4147], L.K. Hua [2932] and V.N. Čubarikov [1284]. The minimal possible number of summands in the case $n \geq 17$ was determined by D.A. Mitkin [4343] in 1987. See also S. Turjányi [6231].

10. Important progress in the Waring problem was made a few years after the work of G.H. Hardy and J.E. Littlewood by I.M. Vinogradov [6408, 6413]. He eliminated the use of power series, replacing them with finite exponential sums. His idea consisted of the simple observation, overlooked by earlier writers, that to apply Cauchy's integral formula to obtain the number of representations of an integer N as the sum of a fixed quantity of kth powers, it suffices to deal with the polynomial $P_N(z) = 1 + \sum_{j=1}^{N} z^{j^k}$ instead of the power series $1 + \sum_{j=1}^{\infty} z^{j^k}$, and this allowed him to move the integration path to the unit circle, where the polynomial P_N is transformed in a natural way into the exponential sum

$$P_N\left(e^{it}\right) = 1 + \sum_{j=1}^{N} \exp\left(itj^k\right).$$

This enormously simplified the proof without changing its essential idea. In particular major and minor arcs appear again but their treatment becomes less technical. This approach led to the same bound for $G(k)$ as in [2532].

---

[32] Konstantin Konstantinovič Mardžanišvili (1903–1981), professor in Tbilisi. See [4617].

The next paper by Vinogradov [6414] applied the same technique to the Waring–Kamke problem. An exposition of Vinogradov's result was given by E. Landau [3671, 3675]. In [3671] one finds the first upper bound for $g(k)$:

$$g(k) \leq \left(\frac{1}{2} + o(1)\right) k2^k. \tag{3.23}$$

For essential improvements of these bounds one had to wait until I.M. Vinogradov in 1934 [6416, 6417, 6419] presented a modified version of his approach (see Sect. 4.2.1).

11. The analogue of the Waring problem in number fields had been considered already by Hilbert, who in [2787] asserted that every totally positive element of an algebraic number field can be represented as the sum of four squares (an algebraic number α lying in a field K is called *totally positive* if the conjugates of α lying in real embeddings of K are all positive). A proof of this assertion for quadratic fields was given by E. Landau [3660] in 1919 and later C.L. Siegel [5740] extended this result to all algebraic number fields. The question of whether integers can be represented as sums of four squares of integers was left open, although in [5742] C.L. Siegel was able to show that in a real quadratic field every totally positive integer is the sum of five squares having bounded denominators, and in 1923 he reduced [5743] the number of summands to four, obtaining this assertion for all algebraic number fields. He applied the Hardy–Littlewood method to obtain asymptotics for the number of representations of totally positive quadratic integers, satisfying certain necessary conditions, as the sum of $s \geq 5$ squares. Six years later F. Götzky [2289] showed that if K is a real quadratic number field, then every totally positive integer of K is the sum of four squares of integers if and only if $K = \mathbf{Q}(\sqrt{5})$, and obtained a simple formula for the number of such representations in this case, analogous to the classical formula of Jacobi in the case of rational integers (see J. Kirmse [3340] for a particular case). For further progress see Sect. 5.2.

Later it was shown by H. Maass [4038] that in $\mathbf{Q}(\sqrt{5})$ already three square summands suffice.

These questions are related to Hilbert's seventeenth problem in which D. Hilbert asked whether every real polynomial assuming non-negative values is the sum of squares of rational functions. Earlier he showed in [2781] that it must not be the sum of squares of polynomials (as the simple example $z^6 + x^4z^2 + x^2y^4 - 3(xyz)^2$, due to T.S. Motzkin[33], shows). This problem got the final positive answer by E. Artin [140] in 1927, as a corollary to the theory of formally real fields built by E. Artin and O. Schreier[34] in [146].

[33] Theodore Samuel Motzkin (1908–1970), professor in Jerusalem and at the University of California in Los Angeles.

[34] Otto Schreier (1901–1929), worked in Hamburg.

3.2.2 Quadratic Forms

1. The problem of representing integers by a diagonal quadratic form

$$\sum_{j=1}^{s} a_j X_j^2, \quad (a_1, \ldots, a_s) = 1 \tag{3.24}$$

was pursued in the thesis of H.D. Kloosterman [3367], where an asymptotical expression for the number of representations is given in the case of positive a_j's and $s \geq 5$. Later he did the same for $s = 4$ [3369] and showed that this number equals

$$\frac{\pi^2}{\sqrt{\Delta}} n S(n) + O\left(n^{c+\varepsilon}\right)$$

for every $\varepsilon > 0$ with $c = 17/18$, $\Delta = a_1 a_2 a_3 a_4$ and

$$S(n) = \sum_{d|n} \frac{1}{d} \left(\frac{\Delta}{d}\right).$$

He gave also a sufficient condition for the equality

$$S(n) \geq \frac{c}{\log\log n} > 0,$$

implying that the involved form represents all sufficiently large integers, and disproved a conjecture of E. Waring [6565, 2nd ed., p. 349], asserting that every form (3.24) represents all large integers.

Later M. Eichler[35] [1699] showed that the last inequality holds with $c = 1/2$.

In 1926 Kloosterman [3368, 3369] dealt with diagonal quaternary forms representing all sufficiently large integers, but he could not decide all cases, leaving the status of four forms undecided.

A complete list was given in 1946 by G. Pall [4730], who showed that the missing four forms represent all large integers. As shown by P. Halmos[36] [2481] there are 88 such forms representing all but one positive integers. The last such form was found by G. Pall [4728] in 1940. In 1946 A.E. Ross [5293] showed that the discriminants of quaternary forms representing all positive integers do not exceed 112. Using his results M.F. Willerding [6671] found 178 such inequivalent forms. Her list turned out to be incomplete and in fact there are 204 such forms (see M. Bhargava [499]).

In [3369] the *Kloosterman sums*

$$S(u, v; \lambda, \Lambda, q) = \sum_{x \equiv \lambda \pmod{\Lambda}} \exp\left(\frac{2\pi i(xu + x'v)}{q}\right)$$

[35] Martin Eichler (1912–1992), professor in Münster, Marburg and Basel. See [3407].

[36] Paul Richard Halmos (1916–2006), professor at the University of Chicago, the University of Michigan, Indiana University and Santa Clara University. See [1946].

(where x is restricted to residues mod q and x' is the inverse of x mod q) appear for the first time. They were later to play an important role in several problems. In particular the bound

$$|S(u, v; \lambda, \Lambda, q)| \ll q^{3/4+\varepsilon}(u, q)^{1/4},$$

with $s = u, v$, and $\Lambda|q$ played a decisive role in [3369]. A simpler proof of the last bound provided T. Estermann [1878]. Note that due to the multiplicative property of Kloosterman sums it suffices to treat the case when q is a prime power.

2. Positive definite quaternary forms in four variables representing all positive integers were considered in 1927 by L.E. Dickson [1547], who in his book [1554], published in 1930, gave a presentation of the theory of quadratic forms in 3 and 4 variables. In [1550] he introduced the notion of a regular form. A quadratic form f in n variables with integral coefficients is called *regular*, if for every $a \neq 0$ the equation $f(x_1, \dots, x_n) = a$ is solvable in integers if and only if it is solvable in $\mathbf{R}$ and in every $\mathbf{Z}_p$, i.e., the integral Hasse principle holds[37] for f (Dickson used an equivalent definition of the Hasse principle in terms of congruences, not utilizing p-adic numbers). He characterized all ternary regular forms of type $x^2 + ay^2 + bz^2$. See also A. Oppenheim [4691], who characterized ternary quadratic forms representing all integers, and A.E. Ross[38] [5292]. A criterion for regularity for positive definite ternary quadratic forms was given by B.W. Jones [3151] and it was used by him and G. Pall [3152] to list all such primitive regular forms.

More than 70 years later W.C. Jagy, I. Kaplansky[39] and A. Schiemann [3097] proved that there are 913 regular primitive ternary forms. The finiteness of this set had been established by G.L. Watson [6579] in 1953.

In 1993 J.H. Conway and W.A. Schneeberger discovered a surprising theorem (the *fifteen theorem*) concerning the representation of integers by positive definite quadratic forms $\sum_{i \leq j} a_{ij} X_i X_j$ with all even coefficients a_{ij} $(i \neq j)$. It states that if such a form represents all positive integers ≤ 15, then it represents every positive integer. A sketch of the proof appeared in Schneeberger's thesis [5535]. Later a simpler proof was given by M. Bhargava [499] who also proved a far reaching generalization [500] asserting that every set S of positive integers has a finite subset S_0 with the property that if a positive definite quadratic form with integer coefficients represents every integer in S_0, then it represents every integer from S. In [1224] J.H. Conway conjectured that if S is the set of all positive integers, then one can take for S_0 the interval $[1, 290]$ and this was confirmed by M. Bhargava and J. Hanke [501]. See [1224] for the history of these questions.

Ternary positive quadratic forms representing all odd integers were studied by I. Kaplansky [3248], who showed that there are at most 23 of them (cf. W.C. Jagy [3096]). Indefinite ternary forms with this property were listed by J. Bureau and J. Morales [853] in 2005.

---

[37] For the Hasse principle see Sect. 3.4 below.

[38] Arnold Ephraim Ross (1906–2002), professor in St.Louis, at Notre Dame University and Ohio State University.

[39] Irving Kaplansky (1917–2006), professor in Chicago and Berkeley. See [347].

3. A quadratic form f is called *almost regular*, if the integral Hasse principle holds for the equation $f = a$ for every sufficiently large integer a. It is called p-isotropic if the equation $f = 0$ has a non-trivial solution in $\mathbf{Q}_p$. For a large class of positive definite forms f in $n \geq 4$ variables it was shown by V.A. Tartakovskiĭ [6054, 6055] that if f is p-isotropic for all primes p then it is almost regular.

This last result was completed in 1946 by A.E. Ross and G. Pall [5294], who showed that this is true for all positive definite quadratic forms. This implies that every such form in $n \geq 5$ variables is almost regular. There are infinitely many inequivalent almost regular primitive forms in 3 (W.K. Chan and B.-K. Oh [990]) and in 4 variables (J. Bochnak and B.-K. Oh [567]).

For effective versions of Tartakovskiĭ's result see G.L. Watson [6587], J.S. Hsia, M.I. Icaza [2911] and the book [3346] by Y. Kitaoka.

An almost regular form is called *exceptional*, if for at least one prime p it is not p-isotropic. Such forms exist only for $n \leq 4$ variables, and in the case $n = 4$ it was shown by J. Bochnak and B.-K. Oh [567] that this can happen only for a single prime not exceeding 37. In 2009 the same authors described all quaternary positive definite quadratic forms which represent all sufficiently large integers [568].

3.2.3 Primes

1. In 1923, G.H. Hardy and J.E. Littlewood [2531] applied the circle method to the *Goldbach conjecture*. This conjecture has its origin in two letters (dated 7 and 30 June 1742; see [1909]) exchanged between Goldbach and Euler, and its usually accepted form stated that every even integer $\neq 2$ is the sum of two primes and every odd integer ≥ 5 is the sum of three primes. For a recent discussion of various wordings of this conjecture see the survey paper of J. Pintz [4905].

For a long time only numerical and heuristical results concerning these assertions were available (see [2619, 5884, 6006]), and the first correct result dealing with representations of integers as sums of primes is due to E. Landau [3616], who used his results concerning sums of values of arithmetical functions at primes to deduce the equality

$$\sum_{j=1}^{n} P_2(2j) = (2 + o(1))\frac{n^2}{\log^2 n}$$

for the mean value at even integers of the function $P_2(n)$, giving the number of representations of n as the sum of two primes.

The circle method turned out to be insufficient to solve the problem, but it led to many new insights, which prepared the way to the partial solution achieved later by I.M. Vinogradov. At the beginning of [2531] the authors wrote about Goldbach's problem:

> "We do not solve it; we do not even prove that any number is a sum of 1 000 000 primes. In order to prove anything we have to assume the truth of an unproved hypothesis, and, even on this hypothesis, we are unable to prove Goldbach's theorem itself. We show, however, that the problem is not 'unan-

greifbar', and bring it into contact with the recognized methods of the Analytic Theory of Numbers."

In fact, Hardy and Littlewood showed that if no L-function vanishes in the half-plane $\Re s > c$ for some $c < 3/4$, then every sufficiently large odd integer is the sum of three primes[40] and the number of representations of n in such a form is asymptotically equal to

$$C\frac{n^2}{\log^3 n}\prod_{\substack{p|n\\p\neq 2}}\frac{(p-1)(p-2)}{p^2-3p+3},$$

where

$$C=\prod_{p>2}\left(1+\frac{1}{(p-1)^3}\right).$$

This result was shortly afterwards generalized to algebraic number fields by H. Rademacher [5026–5028], who also succeeded in showing [5029] that the assumption on zeros of L-functions implies that every sufficiently large odd integer is the sum of three primes lying in prescribed residue classes, which have to satisfy certain compatibility conditions.

The three primes theorem for large odd integers was later established unconditionally by I.M. Vinogradov (see Sect. 4.2.2). The dependence of Rademacher's assertion on unproved hypotheses was later removed (see K. Iseki [3028], R. Ayoub [183], A. Zulauf [6846–6848], D.I. Tolev [6188], K. Halupczok [2487, 2488]).

2. Paper [2531] also contains a list of conjectures, all concerning primes. We now state these conjectures in a slightly modified form. The first is a version of the binary Goldbach conjecture.

A: *Every large even integer n is the sum of two odd primes, and for the number $P_2(n)$ of such representations one has*

$$P_2(n)=(2C_2+o(1))\frac{n}{\log^2 n}\prod_{\substack{p|n\\p\neq 2}}\frac{p-1}{p-2}, \tag{3.25}$$

where

$$C_2=\prod_{p>2}\left(1-\frac{1}{(p-1)^2}\right). \tag{3.26}$$

For a method of computing C_2 and similar constants see P. Moree [4413].

The next conjecture concerned a generalization of the problem of twin primes, and the next pair generalized both A and B (the conjectures C and D differ in the sign of b).

[40] A simpler proof was later given by E. Landau [3664].

B: *For every even k there are infinitely many prime pairs $p_1 < p_2$ with $p_2 - p_1 = k$. The number of such pairs less than x is asymptotically equal to $P_2(n)$, defined by* (3.25).

C, D: *If $(a,b) = (ab,n) = 1$, and exactly one of the numbers a, b, n is even, then the number of solutions of the equation*

$$n = ap_1 + bp_2$$

in prime numbers p_1, p_2 is asymptotically equal to $P_2(n)/ab$.

The solvability of this equation for negative b with p_2 being either a prime or a product of two primes was established in 1990 by M.D. Coleman [1176].

3. The next four conjectures are related to squares. The first of them gives a quantitative version of Landau's conjecture.

E: *There are infinitely many primes of the form $n^2 + 1$. The number of such primes less than x equals*

$$(C + o(1))\frac{\sqrt{x}}{\log x},$$

where

$$C = \prod_{p>2}\left(1 - \frac{1}{p-1}\left(\frac{-1}{p}\right)\right).$$

F: *If $a > 0$ and the polynomial $f(X) = aX^2 + bX + c \in \mathbf{Z}[X]$ has no fixed factor > 1 and its discriminant $D = b^2 - 4ac$ is not a square, then f represents infinitely many prime numbers, and the number of such primes less than x equals*

$$\left(\frac{C(a)}{\sqrt{a}} + o(1)\right)\prod_{\substack{p\mid(a,b)\\ p>2}}\frac{p}{p-1}\frac{\sqrt{x}}{\log x}, \tag{3.27}$$

where

$$C(a) = \varepsilon\prod_{\substack{p\nmid a\\ p>2}}\left(1 - \frac{1}{p-1}\left(\frac{D}{p}\right)\right),$$

and

$$\varepsilon = \begin{cases} 2 & \text{if } a \equiv b \pmod 2, \\ 1 & \text{otherwise.}\end{cases}$$

G: *If $a > 0$, $(a,b,n) = 1$, $2 \nmid (n, a+b)$ and $D = b^2 + 4an$ is not a square, then the number of representations*

$$n = am^2 + bm + p$$

with $m \in \mathbf{Z}$ and prime p equals the expression given by (3.27) *with x replaced by n.*

Two special cases of G were pointed out as separate conjectures.

H: *Every large number n is either a square or the sum of a prime and a square. The number of such representations equals*

$$(1+o(1))\frac{\sqrt{n}}{\log n}\prod_{p>2}\left(1-\frac{1}{p-1}\left(\frac{n}{p}\right)\right).$$

It was deduced by G.K. Stanley in 1928 [5887] from General Riemann Hypothesis that the set of integers of the form $p+a^2$ with prime p has positive density. Later, in 1934, N.P. Romanov [5272] showed this unconditionally. For generalizations see Sect. 4.2.3.

I: *Every large number n can be written in the form $n=p+2a^2$ with prime p. The number of such representations equals*

$$(1+o(1))\frac{\sqrt{2n}}{\log n}\prod_{p>2}\left(1-\frac{1}{p-1}\left(\frac{2n}{p}\right)\right).$$

The last conjecture dealing with squares concerns representations of integers as sums of two or four squares and a prime.

J: *If $N_k(n)$ denotes the number of representations of n as the sum of k squares and a prime, then*

$$N_2(n)=(C+o(1))n\prod_{\substack{p>2\\ p|n}}\frac{A_p}{p^2-p+1},$$

with

$$C=\pi\prod_{p>2}\left(1+\frac{1}{p^2-p}\left(\frac{-1}{p}\right)\right),$$

$$A_p=\begin{cases}(p-1)^2 & \text{if } p\equiv 1 \bmod 4,\\ p^2-1 & \text{if } p\equiv 3 \bmod 4,\end{cases}$$

and

$$N_4(n)=(C_1+o(1))n^2\prod_{\substack{p>2\\ p|n}}\frac{(p-1)^2(p+1)}{p^3-p^2+1},$$

where

$$C_1=\frac{\pi^2}{2}\prod_{p>2}\left(1+\frac{1}{p^3-p^2}\right).$$

In the case $k=4$ this was established in 1935 by S. Chowla [1080]. The formula for $N_2(n)$ turned out to be a consequence of the General Riemann Hypothesis (C. Hooley [2854]), and a proof free of any unproved assumptions was given in 1960 by Yu.V. Linnik [3921–3923] (see Sect. 6.2).

4. The next three conjectures concern primes and cubes.

K: *If a is not a cube, then there are infinitely many primes of the form $n^3 + a$ and the number of such primes less than x equals*

$$(1 + o(1))\frac{\sqrt[3]{x}}{\log x}\prod\left(1 - \frac{2}{p-1}\lambda(-a, p)\right),$$

where p runs over prime divisors of a congruent to unity mod 3 *and*

$$\lambda(a, p) = \begin{cases} 1 & \textit{if } a \textit{ is a cubic residue} \bmod p, \\ -1/2 & \textit{otherwise.} \end{cases}$$

L: *Every large integer n which is not a cube is the sum of a positive cube and a prime, and the number of such representations equals*

$$(1 + o(1))\frac{\sqrt[3]{n}}{\log n}\prod\left(1 - \frac{2}{p-1}\lambda(n, p)\right),$$

where p runs over prime divisors of n congruent to unity mod 3.

M: *If $a \neq 0$, then there are infinitely many primes of the form $m^3 + n^3 + a$ with m, n positive. If $f(p)$ is the number of such representations of a prime p, then*

$$\sum_{p\le x} f(p) = \left(\frac{\Gamma^2(4/3)}{\Gamma(5/3)} + o(1)\right)\frac{x^{2/3}}{\log x}\prod\left(1 - \frac{2}{p}\right)\prod_{\substack{p\nmid n\\ p\equiv 1 \bmod 3}}(1 + A_p),$$

where p runs over prime divisors of n congruent to unity mod 3 *and A_p is defined in a rather complicated way, depending on the prime factors of p in the ring of integers of the field* $\mathbf{Q}(\sqrt{-3})$.

N: *There are infinitely many primes p which are sums of three positive cubes, and if $g(p)$ denotes the number of such representations, then*

$$\sum_{p\le x} g(p) = (C + o(1))\frac{x}{\log x}$$

with

$$C = \Gamma^3(4/3)\frac{x}{\log x}\prod_{p\equiv 1 \bmod 3}\left(1 - \frac{B_p}{p^3}\right).$$

To define B_p let $a + b\zeta_3$ be that prime factor of p in the ring of integers of $\mathbf{Q}(\sqrt{-3})$ *for which $3|a + 1$ and $3|b$. Then $B_p = 2a - b$.*

The first part of N has been established in 2001 by D.R. Heath-Brown [2657], who showed the existence of infinitely many primes of the form $x^3 + 2y^3$. Earlier C. Hooley [2903] proved

$$\sum_{p\le x} g(p) \le (4C + o(1))\frac{x}{\log x},$$

and under the General Riemann Hypothesis got

$$\sum_{p\le x} g(p) \le (3C + o(1))\frac{x}{\log x}.$$

A short survey of the status of the Hardy–Littlewood conjectures before 2000 is given in [4542].

3.3 Creation of the Class-Field Theory

1. The work of D. Hilbert [2783, 2785, 2786] and H. Weber[41] [6602–6605] paved the way towards the creation of class-field theory, which describes Abelian extensions (i.e., extensions with Abelian Galois group) of algebraic number fields. In the case of the rational base field such a description is contained in the *Kronecker–Weber theorem* stating that every Abelian extension of the rationals is contained in a suitable *cyclotomic field*, i.e., a field generated by a root of unity. In [2785] Hilbert proposed certain conjectures dealing with the Abelian extension of an algebraic number field.

(a) *Every algebraic number field K has a unique maximal unramified Abelian extension L/K.*

(A finite extension L/K is called *unramified*, if for every prime ideal $\mathfrak{p}$ of the ring $\mathbf{Z}_K$ of integers of K its lifting $\mathfrak{p}\mathbf{Z}_L$ to the ring $\mathbf{Z}_L$ of integers of L is a product of distinct prime ideals.)

(b) *The extension L/K is of finite degree and its Galois group is isomorphic to the narrow class-group*[42] *$H^*(K)$ of K.*

(c) *Two prime ideals $\mathfrak{p}$, $\mathfrak{q}$ of $\mathbf{Z}_K$ have the same factorization pattern in L if and only if they lie in the same class in $H^*(K)$.*

(d) (The Principal Ideal Theorem.) *For every ideal I of $\mathbf{Z}_K$ its lifting $I\mathbf{Z}_L$ is a principal ideal.*

A field L satisfying (a) is called the *class-field* of K.

2. An important step was made by P. Furtwängler who in a sequence of papers [2151–2153, 2155]) proved conjectures (a), (b) and (c). He presented his proof again in [2156], and in [2161] described the factorization of prime ideals in the class-field.

Hilbert's conjecture (d) is related to a question considered earlier by L. Kronecker (see [3528, pp. 66–67]), who proposed to study extensions L/K in which every ideal of $\mathbf{Z}_K$ becomes principal and did this in the case of imaginary quadratic K. In the case when the class-group of K is either cyclic or isomorphic to $C_p \oplus C_p$ with

---

[41] For a modern description of Weber's ideas see G. Frei [2072].

[42] The narrow class-group of $\mathbf{Z}_K$ consists of equivalence classes of ideals of $\mathbf{Z}_K$, two ideals I, J being equivalent if there exist totally positive elements $\alpha, \beta \in \mathbf{Z}_K$, i.e., all of whose conjugates in real embeddings of k are positive, such that $\alpha I = \beta J$.

prime p, conjecture (d) was established by P. Furtwängler in 1907 and 1916 ([2156] and [2163], respectively).

Hilbert's conjecture (d) was settled in 1930 by Furtwängler [2167] in 1930. In fact, he established a group-theoretic statement involving metabelian groups to which the proof of (d) was reduced by E. Artin [141]. Later, several proofs of this assertion were given (Z.I. Borevič [641], S. Iyanaga[43] [3072], W. Magnus[44] [4058], H.G. Schumann [5571], K. Taketa [6038]).

Early research on Abelian extensions of imaginary quadratic fields was presented in R. Fueter's[45] [2129] report, published in 1911. Earlier R. Fueter [2125–2128] considered imaginary quadratic base fields, found a formula for the class-number of Abelian extensions of such fields, and in [2130, 2132] solved a particular case of Hilbert's twelfth problem.

3. The quest for a description of all finite Abelian extensions of an algebraic number field K ended in 1920, when T. Takagi [6037] showed that to every Abelian extension L/K there corresponds an ideal $\mathfrak{f}$ in the ring $\mathbf{Z}_K$ of integers of K such that a factor group $H(L/K)$ of the group $H_{\mathfrak{f}}^*(K)$ of narrow classes of ideals prime to $\mathfrak{f}$ is isomorphic to the Galois group $Gal(L/K)$ of the extension L/K.

This correspondence has the following properties.

(i) *A prime ideal of* $\mathbf{Z}_K$ *ramifies*[46] *in* L/K *if and only if it divides the ideal* $\mathfrak{f}$.
(ii) *All prime ideals of* $\mathbf{Z}_K$, *unramified in* L/K *and lying in the same class of* $H_{L/K}$ *factorize in* $\mathbf{Z}_L$ *in the same manner.*

These results generalize the results of P. Furtwängler [2156] who dealt with the case $\mathfrak{f} = \mathbf{Z}_K$. They form the essence of *class-field theory* and solve a major part of Hilbert's twelfth problem.

A detailed exposition of Takagi's theory was given by H. Hasse in [2582].

It has been observed in 1936 by C. Chevalley ([1049]) that class-field theory can be reformulated with the use of infinite algebraic extensions of algebraic number fields. These extensions appeared for the first time in a paper by E. Stiemke[47] [5953], published posthumously in 1926. An extensive theory of these extensions was created by W. Krull [3533–3535] and extended by J. Herbrand[48] [2757, 2758] and A. Scholz [5558].

To achieve his goal C. Chevalley introduced *ideal elements* (which are now called *ideles*), which were elements of the *idele group* I_K, defined as the restricted direct product[49] of the

---

[43] Shokichi Iyanaga (1906–2006), professor in Tokyo.

[44] Wilhelm Magnus (1907–1990), professor in Göttingen, at the Courant Institute and Polytechnical Institute in New York. See [5].

[45] Rudolf Fueter (1880–1950), professor in Zürich. See [851, 5862].

[46] A prime ideal ramifies in the extension L/K if it divides the relative discriminant of that extension.

[47] Erich Stiemke (1892–1915).

[48] Jacques Herbrand (1908–1931). See [1571].

[49] The restricted direct product of groups G_n with respect to subgroups $H_n < G_n$ is the subgroup of the direct product $\prod_n G_n$ consisting of elements (g_n) with almost all g_n lying in H_n.

multiplicative groups of all completions K_v of K with respect to their unit groups $U(K_v)$, i.e., the groups of invertible elements of the closure R_v in K_v of the ring $\mathbf{Z}_K$ of integers of K. In the case of archimedean K_v (i.e., $K_v = \mathbf{R}$ or $\mathbf{C}$) one puts $U(K_v^*) = K_v^*$. Chevalley gave I_K a certain non-Hausdorff topology. His main result stated that finite Abelian extensions of K are in a one-to-one correspondence with certain subgroups of the factor group I_K/P_K, P_K being the group of *principal ideles* consisting of ideles of the form (a_v) with $a_v = a \in K^*$.

The introduction of the restricted direct product topology in I_K, (now in normal use), is hidden in a paper by A. Weil [6611] who reformulated Chevalley's result in terms of group characters.

A complete presentation of his approach was presented by C. Chevalley a few years later [1050].

Class-field theory can also be constructed in the case of algebraic function fields over a finite ground-field. This was done by F.K. Schmidt [5478] and the main elements of class-field theory were carried over to the case of arbitrary algebraic function fields in the thesis of M. Deuring [1496].

For expositions of class-field theory in a modern manner see the books by Artin, J. Tate [147], S. Lang [3693] and A. Weil [6629].

4. An explicit form of the isomorphism between *Gal*(L/K) and $H(L/K)$ was given in 1927 by E. Artin [139]. To state it one has to define the *Frobenius symbol*. We shall do it for arbitrary normal extensions, not necessarily Abelian. So let L/K be such an extension, and let $\mathfrak{P}$ be a prime ideal in $\mathbf{Z}_L$, put $\mathfrak{p} = \mathfrak{P} \cap \mathbf{Z}_K$, and assume that $\mathfrak{p}$ is unramified in L/K. The Frobenius symbol corresponding to $\mathfrak{P}$ is the unique element $s_{\mathfrak{P}}$ of *Gal*(L/K), such that

$$s_{\mathfrak{P}}(x) - x^{N(\mathfrak{p})} \in \mathfrak{P}$$

holds for all $x \in \mathbf{Z}_L$. It was introduced by G. Frobenius [2114] in 1896.

If L/K is Abelian, then $s_{\mathfrak{P}}$ depends only on $\mathfrak{p}$, so we may denote it by $F_{L/K}(\mathfrak{p})$. *Artin's reciprocity law* (E. Artin [139]) states that if L/K is Abelian, then $F_{L/K}(\mathfrak{p})$ depends only on the class of $\mathfrak{p}$ in $H_{\mathfrak{f}}^*(K)$, and the induced map

$$H_{\mathfrak{f}}^*(K) \to Gal(L/K)$$

is surjective.

5. Artin's reciprocity law can be used to generalize the quadratic reciprocity law to higher exponents and larger fields. The quartic and cubic cases had already been treated by C.F. Gauss [2211] and G. Eisenstein [1708], respectively, and Eisenstein [1711] also treated odd prime exponents p in the pth cyclotomic field $\mathbf{Q}(\zeta_p)$. After further work by E.E. Kummer [3581] D. Hilbert [2783, 2786] stated a general reciprocity law for prime exponents p in fields containing the pth roots of unity and a proof of it was given by P. Furtwängler [2157–2159], who earlier did this under certain additional assumptions [2150, 2154]. This gave a solution of Hilbert's ninth problem in an important special case. The case of prime powers was dealt with by Furtwängler [2164]. A clarified form of this theory was presented in a cycle of papers by H. Hasse, culminating in [2585] and [2586], where Artin's reciprocity law is used to obtain a general reciprocity law for power residues for any exponent m in algebraic number fields containing the mth roots of unity. A more explicit form of it appears in [2589]. The Artin–Hasse theory was described in the second volume of Hasse's report [2582–2584].

In 1919 E. Hecke [2681] introduced Gaussian sums in algebraic number fields and utilized them in his proof of the analogue of the quadratic reciprocity law in quadratic fields. He proved this law for arbitrary algebraic number fields in his book[50] [2685].

The new Gaussian sums, introduced by E. Hecke, turned out to be of fundamental importance in the creation of the modern theory of algebraic number fields (see, e.g., the thesis of J. Tate [6057]). They were also very useful in the theory of modular forms in several variables, as demonstrated by H.D. Kloosterman [3373].

An explicit form of the reciprocity law was given by I.R. Šafarevič [5362] in 1950 (for an exposition see M. Kneser[51] [3405]).

A historical account of Artin's reciprocity law is given in a forthcoming book by P. Roquette [5282].

The history of reciprocity laws was described in the book [3816] by Lemmermeyer.

6. An important result was established in 1926 by N.G. Čebotarev [6205]. He showed that for every normal extension L/K of algebraic number fields the set of prime ideals $\mathfrak{p}$ of $\mathbf{Z}_K$ for which the Frobenius symbol $s_{\mathfrak{P}}$ lies in a fixed class X of conjugated elements of $Gal(L/K)$ has a density equal to $\#X/[L:K]$ (*Tschebotareff*[52] *density theorem*). This result made more precise the density results for prime ideals established earlier by L. Kronecker [3527] and G. Frobenius [2114]. The rather complicated proof was later simplified by O. Schreier [5568], A. Scholz[53] [5554] and M. Deuring [1499].

The Čebotarev density theorem has found many important applications in number theory. J.-P. Serre [5654] applied it to the theory of modular forms, and J.C. Lagarias [3602] showed that it can be used to describe sets of primes determined by systems of polynomial congruences.

Effective versions of Čebotarev's theorem under the assumption of the General Riemann Hypothesis have been given by J. Oesterlé [4660], J.C. Lagarias, H.L. Montgomery, A.M. Odlyzko [3608] and E. Bach, J. Sorenson [191].

7. In the particular case, when the ground field k is imaginary quadratic, a description of all Abelian extensions can be done using the theory of modular functions (R. Fueter [2130, 2132]). A new presentation of this fact was given in 1927 by H. Hasse [2587, 2588]. He showed in particular that these extensions can be generated by values of a single analytic function at points of k.

---

[50] This book, along with the third volume of Landau's book [3674], served as the main introduction to the theory of algebraic numbers for years to come.

[51] Martin Kneser (1928–2004), professor in Göttingen. See [5979].

[52] Čebotarev himself used this spelling of his name in papers published outside Russia.

[53] Arnold Scholz (1904–1942), worked in Freiburg and Kiel. See [6075].

An exposition of the analytical method was given in a Princeton seminar in 1957/1958 [632].

A purely algebraic approach without the use of analytical concepts was given in 1949–1952 by M. Deuring [1504, 1505, 1510]. For further development of this subject up to 1961 in the more general setting of Abelian varieties the reader may consult the books by G. Shimura [5705] and G. Shimura and Y. Taniyama[54] [5705].

The case of a real quadratic base field was treated by T. Shintani [5709].

8. A new class of L-functions was introduced by E. Artin in 1924 [138, 142]. Let K/k be a finite normal extension of an algebraic number field k with Galois group G and let ϱ be a finite-dimensional representation of G. The *Artin L-function* corresponding to ϱ is defined for $\Re s > 1$ by the product

$$\prod_{\mathfrak{p}} (\det(E - A_{\mathfrak{p}} N(\mathfrak{p})^{-s})^{-1}), \tag{3.28}$$

where $\mathfrak{p}$ runs over unramified prime ideals of $\mathbf{Z}_k$,

$$A_{\mathfrak{p}} = \varrho(s_{\mathfrak{P}}),$$

with $\mathfrak{P}$ being any prime ideal in K dividing $\mathfrak{p}$, and $s_{\mathfrak{P}}$ being the corresponding Frobenius symbol. The correctness of this definition follows from the observation that this product does not depend on the choice of $\mathfrak{P}$, dividing $\mathfrak{p}$. Moreover it depends only on the character χ of ϱ and therefore one denotes the function defined by (3.28) by $L(s, \chi)$. If G is Abelian, then the corresponding Artin L-functions coincide with L-functions associated with characters of suitable groups $H_{\mathfrak{f}}^*(k)$, and in the general case every Artin L-function is a product of such functions with suitable rational exponents.

It was later proved by R. Brauer[55] [691] that these exponents can be taken to be integral, and this showed that Artin L-functions can be continued to meromorphic functions in the whole plane.

Artin established also a functional equation for his L-functions.

In this equation a factor $W(\chi)$ of absolute value appears (*Artin root number*), and it was shown in 1956 by B. Dwork[56] [1666] that $W(\chi)$ can be factorized into local factors associated with prime ideals of the relevant field. The question of the signs of these factors, left open in [1666], has been settled by R.P. Langlands [3713] and P. Deligne [1446] (cf. J. Tate [6064]).

The Artin root numbers are of importance for the Galois module structure in algebraic number fields (see Ph. Cassou-Noguès, T. Chinburg, A. Fröhlich[57] [959], M.J. Taylor, J. Cougnard [1257], A. Fröhlich [2121]).

---

[54] Yukata Taniyama (1927–1958), professor at the University of Tokyo. See [5704].

[55] Richard Dagobert Brauer (1901–1977), brother of Alfred Brauer, student of Schur, assistant in Königsberg, professor in Toronto, Ann Arbor and at Harvard University. See [5265].

[56] Bernard Dwork (1923–1998), professor at Johns Hopkins University, Princeton and Padua. See [3283].

[57] Albrecht Fröhlich (1916–2001), professor at King's College, London. See [6077].

E. Artin conjectured [138] that all his L-functions are entire, with the obvious exception of the case when ϱ is the trivial representation. This is true for fields with Abelian Galois group, and was later shown to be true in the case when the Galois group is a p-group, or has an Abelian commutator subgroup (see A. Speiser [5861, 2nd ed.]). Artin himself established this for the Galois group A_4.

Artin's conjecture has been established for two-dimensional representations ϱ by Langlands [3712] and J. Tunnell [6210]), except for the case when the image $\varrho(G)$ is isomorphic to the alternating group A_5. For further progress in this case see J.P. Buhler [826], K. Buzzard, M. Dickinson, N. Shepherd-Barron, R. Taylor [871], K. Buzzard, W.A. Stein [872], K. Buzzard, R. Taylor [873], T. Kiming, X.D. Wang [3337], N. Shepherd-Barron, R. Taylor [5687] and R. Taylor [6085, 6086].

For surveys see S. Gelbart [2220], D. Prasad, C.S. Yogananda [5008], R. Taylor [6087].

Artin's conjecture implies, in particular, the truth of a conjecture posed in 1900 by Dedekind [1422], stating that if $k \subset K$ are two algebraic number fields, then the ratio ζ_K/ζ_k is entire. For Abelian extensions of the rationals this conjecture follows from the entirety of Dirichlet's L-functions, and Dedekind established its truth for pure cubic extensions $K/\mathbf{Q}$. In 1923 E. Artin [135] established the truth of Dedekind's conjecture in the case when K/k is normal with the icosahedral Galois group and then showed [142] that if $k \subset K$, then ζ_K/ζ_k is meromorphic.

For all normal extensions Dedekind's conjecture was established by H. Aramata [113, 114] in the thirties and later also by R. Brauer [689, 690]. For certain classes of solvable extensions a proof was given by M. Ishida [3030], and the next step was made independently in 1975 by K. Uchida [6248] and R.W. van der Waal [6308], who proved it in the case when the Galois group of the normal closure of K/k is solvable.

3.4 The Hasse Principle

1. The first important application of Hensel's p-adic method came from the hands of H. Hasse, who in his thesis[58] tackled Hilbert's eleventh problem. He proved that a quadratic form $f(X_1,\dots,X_n)$ with rational coefficients represents a given number $m \in \mathbf{Q}$, i.e., the equation

$$f(x_1,\dots,x_n) = m$$

has rational solutions, if and only if it represents m in every p-adic field $\mathbf{Q}_p$, and in $\mathbf{R}$ [2577]. This result in the case $n = 3$ is hidden in the third edition of Legendre's book [3767] and for $n \geq 5$ follows from a theorem by A. Meyer [4279], but H. Hasse was the first to give a uniform proof. Hasse also proved that two quadratic forms

[58] He obtained his doctorate in 1920 in Marburg under supervision of K. Hensel.

f, g in n variables with rational coefficients are equivalent, i.e., there exist rational numbers a_{ij} $(i, j = 1, 2, \ldots, n)$ with $\det[a_{ij}] \neq 0$ such that

$$g(X_1, \ldots, X_n) = f\left(\sum_{i=1}^{n} a_{i1} X_i, \ldots, \sum_{i=1}^{n} a_{in} X_i\right)$$

holds if and only if they are equivalent over $\mathbf{R}$, and over every p-adic field[59] [2578]. Later he also showed [2580, 2581] that a similar situation holds for quadratic forms with coefficients in an algebraic number field.

These results were later obtained by E. Witt[60] [6703] as corollaries of his general theory of quadratic forms in arbitrary fields.

If two quadratic forms with integral coefficients are equivalent over the reals and over the rings $\mathbf{Z}_p$ of integers of $\mathbf{Q}_p$, then they may not be equivalent over $\mathbf{Z}$. H. Minkowski [4319] asserted that for every prime p each form could be transformed into the other by means of a linear mapping with rational coefficients having denominators not divisible by p. In three dimensions this had already been shown by H.J.S. Smith [5832] in 1867, but the general case had to wait till 1941, when C.L. Siegel [5759] produced a proof.

2. H. Hasse's results are examples of a principle, called now the *Hasse principle* (also known as the *local–global principle*), which states that a property holds in an algebraic number field K if and only if it holds in all completions of K. There are several known properties obeying this principle. It seems that the first result of this type is due to G. Rados [5046], who used Kronecker's density theorem [3527] to deduce that a monic polynomial $f \in \mathbf{Z}[X]$ splits into linear factors if and only if its reduction mod p does the same for every prime p. This implies that a monic $f \in \mathbf{Z}[X]$ splits over $\mathbf{Q}$ if and only if it splits over every p-adic field. The example $f(X) = X^4 + 13X^2 + 81$ shows that there exist polynomials irreducible over $\mathbf{Z}$ whose reductions mod p^k are reducible for every prime power p^k (D. Hilbert [2784]), hence over every p-adic field.

An important example of the Hasse principle was given in 1931 by H. Hasse [2592], who showed that if K is an algebraic number field and L/K is its extension with cyclic Galois group, then every element $a \in K$ which is a local norm is also a global norm. Recall that if L/K is a field extension of finite degree, then the norm $N_{L/K}(\alpha)$ of an element $\alpha \in L$ is defined as the determinant of the L-linear map $x \mapsto \alpha x$ in L, regarded as a linear space over K. In the case of characteristic zero $N_{L/K}(\alpha)$ coincides with the product of conjugates of α over K. In the case of quadratic extensions of the rationals Hasse's result already appears in Hilbert's *Zahlbericht* [2783] and for cyclic extensions of prime degree it had earlier been established by P. Furtwängler [2157–2159].

---

[59] Hasse pointed out that essentially the same result can be deduced from a theorem by Minkowski [4320] which gives a complete set of invariants of quadratic forms under the action of invertible linear maps.

[60] Ernst Witt (1911–1991), professor in Hamburg. See [3305].

This result may fail for non-cyclic extensions (H. Hasse [2592], A. Scholz [5555], H. Reichardt[61] [5148]). The factor group

$$\left(\bigcap_{\mathfrak{p}}\bigcap_{\mathfrak{P}|\mathfrak{p}} N_{L_{\mathfrak{P}}/K_{\mathfrak{p}}}(L^*_{\mathfrak{P}})\right) \Big/ N_{L/K}(L^*),$$

was studied by A. Scholz [5556], who called it the *knot of the extension L/K*. A modern version of Scholz's theory was given by W. Jehne [3118].

The Hasse principle for representations of integers by an indefinite quadratic form in $n \geq 4$ integer variables was established in 1951 by C.L. Siegel [5769, 5770] (cf. G.L. Watson [6583, 6586]). The corresponding result for positive definite quadratic forms in $n \geq 5$ variables (and for $n = 4$ with certain exceptions) was proved in 1959 by Ch. Pommerenke [4982]. In 1977 M. Peters [4802] deduced the local–global principle for $n = 3$ (with certain exceptions) from the General Riemann Hypothesis, and the unproved assumption was removed in 1990 by W. Duke and R. Schulze-Pillot [1650]. This form of the local–global principle had been conjectured in 1946 by G. Pall and A.E. Ross [5294], who established a weaker form of it.

H. Reichardt [5148] used his results on quartic equations [5147] to give examples of such equations (e.g., $X^4 - 2Y^2 = 17$) having solutions in all completions of the rationals, but being without rational solutions. Actually the first examples of this type could already be found in a paper by T. Pépin [4773] from 1879, without proof however, which was only supplied in 1999 by F. Lemmermeyer [3815]. Similar examples exist also in the case of cubic and sextic polynomials, e.g., for equations $(X^2 - 2)(X^2 + 7)(X^2 + 14) = 0$, $X^3 + 3Y^3 - 22 = 0$ (T. Skolem [5808]) and $3X^3 + 4Y^3 + 5Z^3 = 0$ (E.S. Selmer[62] [5632]). Cf. J.W.S. Cassels [935].

L.J. Mordell [4396] conjectured in 1949 that the Hasse principle also applies to equations defining cubic surfaces other than cones in the projective space, but this was shown to be false by H.P.F. Swinnerton-Dyer [6000] (cf. L.J. Mordell [4402]). Later J.W.S. Cassels and M.J.T. Guy [952] proved that the diagonal form $5X^3 + 12Y^3 + 9Z^3 + 10T^3$ is also a counterexample. A family of cubic forms in 3 variables violating Hasse's principle was presented by B. Poonen [4985] in 2001.

On the other hand E.S. Selmer [5633] established Hasse's principle for the equations

$$\sum_{j=1}^{4} a_j x_j^3 = 0$$

in the case when the ratio a_3a_4/a_1a_2 is a cube of a rational number, and T. Skolem [5809] proved it for cubic surfaces having at least one singular point. A. Schinzel [5446] observed that Hasse's principle fails for rational quartic forms in $n \geq 3$ variables.

Yu.I. Manin [4134, 4136] formulated a condition in terms of the Brauer–Grothendieck group preventing the Hasse principle from being applicable for a variety. A method for checking this condition for cubic diagonal forms was presented by J.-L. Colliot-Thèlene, D. Kanevsky and J.-J. Sansuc [1186] in 1985. They conjectured that the Manin condition is the only obstruction in this problem. It has been shown in 1999 by A.N. Skorobogatov [5811] that this is not so in the case of quartic surfaces. See also J.L. Colliot-Thèlene, H.P.F. Swinnerton-Dyer [1189], T. Fisher [2008], P. Salberger [5370], A.N. Skorobogatov [5812], H.P.F. Swinnerton-Dyer [6005], and the survey by E. Peyre [4830].

---

[61] Hans Reichardt (1908–1991), professor in Berlin. See [3429].

[62] Ernst Sejerstedt Selmer (1920–2006), professor in Bergen.

It was shown in 2009 by T.D. Browning and D.R. Heath-Brown [762] that the Hasse principle holds for rational points on smooth projective quartic hypersurfaces of dimension $n \geq 39$, improving upon B.J. Birch [523], who proved this for $n \geq 47$. For diagonal quartics this holds already for $n \geq 17$ (Davenport [1372]).

In 1971 V.A. Iskovskih[63] [3033] showed that Hasse's principle fails for common zeros of a pair of rational quadratic forms in 5 variables. It holds however for any finite system of quadratic forms $f_i(X, Y, Z)$, as shown by A. Schinzel [5446].

M. Fujiwara [2141]) proved that Hasse's principle fails for forms of degree 5, and later he obtained, with M. Sudo [2143], the same result for forms of degree congruent to 5 mod 10. See also B.J. Birch, H.P.F. Swinnerton-Dyer [539] and N.Q. Thang [6124] for further examples. On the other hand it has been shown by W.C. Waterhouse [6572] that the Hasse principle holds for the equivalence of two pairs of quadratic forms over a global field of odd characteristics (cf. [6573], D.B. Leep, L.M. Schueller [3764]). It also holds for the equivalence of bilinear forms (W.C. Waterhouse [6574]).

Various forms of the Hasse principle were given by J.L. Colliot-Thélène and J.J. Sansuc [1188]. A weakened form of this principle for systems of forms was studied in 1985 by R. Danset [1326].

The Hasse principle for irreducible varieties over the ring Ω of all algebraic integers was established in 1986 by R.S. Rumely [5339], who used it to show that Hilbert's tenth problem has a positive solution in Ω (see D.C. Cantor, P. Roquette [890] for an earlier special case). A model-theoretic proof was given in 1988 by L. van den Dries [6272] (cf. van den Dries, A. Macintyre [6273]).

For various versions of this principle for elliptic curves and, more generally, for Abelian varieties see R. Dvornicich, U. Zannier [1663–1665], W. Gajda, K.Górnisiewicz [2176], E. Kowalski [3503], T. Weston [6642], S. Wong [6719]. A family of curves of genus 1 violating the Hasse principle was constructed by Colliot-Thélène and Poonen [1187].

The Hasse principle also holds for exponential equations

$$a_1^{x_1} \cdots a_m^{x_m} = b$$

in algebraic number fields, as shown by A. Schinzel [5441, 5444], but fails for systems of such equations.

A kind of Hasse principle in polynomial dynamics was established by T. Pezda [4831].

A broad survey with good bibliography was given in 1992 by Colliot-Thélène [1184]. A report on the Hasse principle on the border of number theory and algebraic geometry was presented by B. Mazur [4217] in 1993.

3. In his proof of local-global principle for quadratic forms H. Hasse was led to a reformulation of the quadratic reciprocity law in algebraic number fields [2579]. The extension of this idea to higher powers (see [2608]) resulted in the creation by H. Hasse [2590, 2591, 2597] and F.K. Schmidt [5477] in 1930 of *local class-field theory*, which describes Abelian extensions of $\mathfrak{p}$-adic fields.

Hasse's idea was to associate with every finite Abelian extension L/K of a $\mathfrak{p}$-adic field K the group $H(L) = N_{L/K}(L^*)$ of non-zero elements of K which are norms of suitable elements of L. Then he applied Takagi's theory to obtain the following assertions.

---

[63] Vasiliĭ Alekseevič Iskovskih (1939–2009), professor in Moscow. See [574].

(i) *There exists an explicit isomorphism* τ *between the Galois group* $G(L/K)$ *and the factor group* $K^*/H(L)$.
(ii) *The isomorphism* τ *agrees with the Galois correspondence between subgroups of* $G(L/K)$ *and subfields of* L *containing* K.
(iii) *If* L, M *are Abelian extensions of* K *then one has* $L \subset M$ *if and only if* $H(M) \subset H(L)$, *thus, in particular, to distinct fields there correspond distinct groups* $H(\cdot)$.

This implies that the groups $H(L)$ are groups of finite index in the multiplicative group of K, and the question arising of whether every subgroup of finite index of K^* is of the form $H(L)$ for some Abelian extension L/K was settled in affirmative by F.K. Schmidt [5477].

The use of global class-field theory to prove the main assertions of the local theory was eliminated in the thesis of C. Chevalley[64] [1047], who based his approach on pure algebraic constructions. H. Hasse succeeded in [2594] in proving algebraically almost all the main theorems of local class-field theory and obtained a new proof of the reciprocity law for power residues. The isomorphism between the Galois group $G(L/K)$ and factor group $L^*/N_{L/K}(L^*)$ found an algebraic proof in Chevalley's paper [1046]. Also F.K. Schmidt eliminated the use of analytical tools in this aim, but he did not publish it. In 1935 C. Chevalley and H. Nehrkorn [1052] presented a purely algebraic construction of local class-field theory and their idea was later used by M. Moriya [4434] who did the same in the case of algebraic function fields over a finite field.

A simple proof of main results of local class-field theory was given in 1975 by M. Hazewinkel [2622].

A very general approach to class-field theory encompassing both global and local theories was presented by J. Neukirch in his book [4575].

3.5 Geometry of Numbers and Diophantine Approximations

1. In 1925 V. Jarník [3107] studied the number of integer points lying on plane convex curves and showed that under certain rather weak regularity assumptions a curve of length N can have at most $cN^{2/3}$ such points, where c is an absolute constant. He showed also that this result is best possible.

H.P.F. Swinnerton-Dyer [6004] showed in 1974 that if the curve Γ is sufficiently regular, then the number of integral points on $x\Gamma$ is $O(x^c)$ for every $c > 3/5$ (cf. W.M. Schmidt [5517, 5519]). Later E. Bombieri and J. Pila [620] reduced the exponent to $1/2$ for curves of the class C^∞ (see also J. Pila [4857, 4858], D.R. Heath-Brown [2658]).

The number of integer points on the boundary of a convex body in $\mathbf{R}^n$ was bounded in the sixties by G.E. Andrews [76, 77].

For a uniform version of Járnik's result see A. Plagne [4919] (cf. also F.V. Petrov [4829]).

M.N. Huxley devoted a part of his book [2986] to the questions of lattice points lying close to a curve.

64 Claude Chevalley (1909–1984), professor in Paris. See [1572, 3073].

2. In 1921 C.L. Siegel [5738, 5739] obtained an essential strengthening of the theorem of Thue (2.59). He was able to replace Thue's bound for t by $t \geq 2\sqrt{N}$.

The coefficient 2 in this result was later replaced by $\sqrt{2}$ (F.J. Dyson [1677], A.O. Gelfond [2227]), and finally K.F. Roth [5311] succeeded in showing that for irrational algebraic α and any $t > 2$ the inequality

$$\left|\alpha - \frac{p}{q}\right| < \frac{C(\alpha, t)}{q^t}$$

does not have solutions in integers p, q for sufficiently small $C(\alpha, t)$. This brought him the Fields Medal in 1958. An exposition of Roth's proof is given in the book [933] by J.W.S. Cassels.

Unfortunately, the theorems of Thue, Siegel, Dyson, Gelfond and Roth are not effective, in contrast to Liouville's theorem. The first effective strengthening of Liouville's theorem was obtained in 1971 by N.I. Feldman [1981], who showed that if α is an algebraic number of degree $n \geq 3$, then for all integers p, q one has

$$\left|\alpha - \frac{p}{q}\right| > cq^{b-n}$$

with certain effective positive b, c, depending on α. (Other proofs were given much later by E. Bombieri [605] and Y.F. Bilu, Y. Bugeaud [514]).

Siegel's main theorem in [5738] also covers approximations of algebraic numbers by algebraic numbers of smaller degrees. To state it one needs the notion of the *height* $H(\alpha)$ of an algebraic number α, defined as the maximal modulus of the coefficients of the minimal polynomial of α over $\mathbf{Z}$ with co-prime coefficients.

If α is an algebraic number of degree $n \geq 2$, then there is a constant $C(\alpha) > 0$ such that for all algebraic numbers β of degree smaller than n one has

$$|\alpha - \beta| > \frac{C(\alpha)}{H(\beta)^{2\sqrt{n}}}. \tag{3.29}$$

If one restricts the number β in this result to a fixed field, then the exponent $2\sqrt{n}$ can be replaced by any number greater than 2. This extension of Roth's theorem was proved by W.J. LeVeque[65] [3850] in 1956.

In certain cases a bound stronger than (3.29) can be obtained from a result by A. Brauer [675] who showed in his thesis that if $\deg \alpha = n$, $\deg \beta = m$, then

$$|\alpha - \beta| \geq \frac{c(\alpha, m)}{H(\beta)^r},$$

where r denotes the degree of α over the field $\mathbf{Q}(\beta)$.

---

[65] William J. LeVeque (1923–2007), professor at the University of Michigan and Claremont Graduate University. See [4212].

3. In 1924 an important paper [3319] by A.J. Khintchine appeared, dealing with rational approximations to almost all real numbers. The main result of it states that if $f(t)$ is a positive continuous function and $tf(t)$ decreases, then the inequality

$$\left|\alpha - \frac{p}{q}\right| < \frac{f(q)}{q}$$

has for almost all real α infinitely many relatively prime integral solutions $q > 0, p$ if and only if the integral

$$\int_1^\infty f(t)\,dt \tag{3.30}$$

diverges.

A.J. Khintchine showed moreover that if $a_1, a_2, \dots$ are consecutive partial quotients of the continued fraction of α, then for almost all α the geometric mean $G(a_1, \dots, a_n)$ satisfies

$$\limsup_{n\to\infty} G(a_1, \dots, a_n) \le \exp\left(\sqrt{2\log 2}\right).$$

He also considered the arithmetic mean $A(a_1, \dots, a_n)$ and showed that if $f(t)$ is positive, $f(t)/t$ increases and the series $\sum_{n=1}^{\infty} 1/f(n)$ converges, then for almost all α one has

$$A(a_1, \dots, a_n) = O(f(n)).$$

(Earlier he showed this for $f(n) = n^\varepsilon$ with positive ε [3318]).

This strengthened a previous result by É. Borel [637, 639] and F. Bernstein[66] [466], who proved that the following conditions for a given positive function f are equivalent.

(a) *The series* $\sum_{q=1}^{\infty} 1/f(q)$ *diverges,*
and
(b) *The set of real numbers for which the inequality* $a_n < f(n)$ *holds for all n is of measure zero.*

This implies in particular that the set of all numbers having bounded partial quotients is of zero measure.

Two years later A.J. Khintchine [3320] obtained a generalization of his result to higher dimensions, proving the following assertion.

If $f(x) > 0$ *is continuous and* $xf^n(x)$ *decreases to zero, then the inequality*

$$\max_{i=1,\dots,n} |\alpha_i q - p_i| < f(q) \tag{3.31}$$

has infinitely many solutions in integers $q > 0$, $p_1, \dots, p_n$ *for a set of* $(a_i)_{i=1}^n \in \mathbf{R}^n$ *of full measure if and only if the integral* $\int_c^\infty f^n(t)\,dt$ *diverges. If this integral converges, then the inequality* (3.31) *has infinitely many solutions only for a set of n-tuples* $(\alpha_1, \alpha_2, \dots, \alpha_n)$ *of measure zero.*

---

[66]Felix Bernstein (1878–1956), professor in Göttingen, New York and Syracuse. See [2087].

Another generalization was made in 1938 by A.V. Grošev [2352] who replaced in the last theorem the terms $\alpha_i q$ by linear forms $L_1(q_1, \dots, q_k)$ in k variables, and $f(q)$ by $f(\max_i |q_i|)$.

Much later R.J. Duffin and A.C. Schaeffer [1634] replaced, in A.J. Khintchine's first theorem, the assumption on the monotonicity of $tf(t)$ by a weaker condition, and asked whether the assumptions on f could be waived and the condition on the integral (3.30) could be replaced by the divergence of the series

$$\sum_{q=1}^{\infty} \frac{f(q)\varphi(q)}{q}.$$

Under certain additional assumptions this has been shown to be true (P.A. Catlin [963, 964], P. Erdős [1823], G. Harman [2559], O. Strauch [5966–5968], V.T. Vilčinskiĭ [6395]; see also P.X. Gallagher [2178]), but the general question is still open. Its higher-dimensional analogue, formulated in the book by V.G. Sprindžuk [5877], was established by A.D. Pollington and R.C. Vaughan [4950, 4951].

The books by V.G. Sprindžuk [5877] and G. Harman [2561] contain expositions of the main results in the metric theory of Diophantine approximations.

Khintchine's conjecture, stated in [3318], asserts that for any measurable set $A \subset (0, 1)$ and any increasing sequence m_j of integers one has

$$\#\{j \le n : \{m_j\theta\} \in A\} = (|A| + o(1))n$$

(with $|A|$ denoting the measure of A) was shown to be false by J.M. Marstrand [4158] (cf. R. Nair [4522–4525]).

4. Approximations of real numbers by rationals with restrictions on their denominators had been considered already in 1914 by G.H. Hardy and J.E. Littlewood [2520, 2521], who conjectured that there exists a constant c with the property that for every irrational θ and $N = 1, 2, \dots$ one can find a positive integer $n \le N$ and an integer m with

$$\left|\theta - \frac{m}{n^2}\right| \le \frac{c}{n^2 N} \le \frac{c}{n^3}.$$

This inequality can also be equivalently stated in the form

$$\|n^2\theta\| \le \frac{c}{N}. \tag{3.32}$$

The first result concerning this conjecture was obtained in 1927 by I.M. Vinogradov [6412] who showed that the left-hand side of (3.32) can be made to be smaller than $c(\varepsilon)/N^{2/5-\varepsilon}$ for every $\varepsilon > 0$.

For an improvement one had to wait till 1948, when H. Heilbronn [2711] replaced Vinogradov's bound by $c(\varepsilon)/N^{1/2-\varepsilon}$ for any $\varepsilon > 0$. In 1970 M.C. Liu [3967] showed that one can take $1/\log\log N$ instead of the exponent $1/2 - \varepsilon$ with $c(\varepsilon) = 1$ for large N.

Heilbronn's result was improved in 1995 by A. Zaharescu [6813], who showed that one can replace the number $1/2$ by $4/7$ for $N > N(\varepsilon)$, and moreover the inequality

$$\|n^2\theta\| < \frac{1}{n^c}$$

has infinitely many solutions for every $c < 2/3$.

For later results concerning the more general question of evaluating $\|n^k\theta\|$ see D.R. Heath-Brown [2645], R.C. Baker, J. Brüdern, G. Harman [259], T.D. Wooley [6727].

This question has been generalized to approximations of several reals by I. Danicic [1324], R.C. Baker, J. Gajraj [261] and R.C. Baker [253].

Another generalization, in which n^2 is replaced by an arbitrary polynomial, was made by H. Davenport [1378] (cf. R.J. Cook [1235–1239], J. Gajraj [2177], M.C. Liu [3968]). In 1977 W.M. Schmidt's book [5508] appeared, containing a study of the fractional part of polynomials and introducing new methods. They were later used by R.C. Baker [248, 249, 251, 252] (see also the book [254]) for important improvements, culminating in the proof that for any polynomial f of degree k, vanishing at 0 and for every $N > N_0(k,\varepsilon)$ there is some $n \le N$ with

$$\|f(n)\| \le \frac{1}{N^{1/k-\varepsilon}}.$$

A recent result by N.G. Moshchevitin [4443] implies that if $\deg f = k$, then the set of real θ such that

$$\liminf_{n\to\infty} n \log n \|f(n)\theta\| > 0$$

has Hausdorff dimension $\ge k/(k+1)$.

Similar problems were considered also for polynomials in several variables. In 1958 I. Danicic [1323] showed that if $f(X_1,\dots,X_k)$ is a real quadratic form, then for $N \ge 2$ and every $\varepsilon > 0$ one can find integers n_i with $|n_i| \le N$ such that

$$\|f(n_1,\dots,n_k)\| \le \frac{c(k,\varepsilon)}{N^{a-\varepsilon}},$$

where $a = k/(k+1)$. It was shown later by R.J. Cook [1241] that this holds already with $a = 2$.

Approximations of irrational numbers by rationals with square-free numerators and denominators were treated by R.C. Baker, J. Brüdern, G. Harman [260], A. Balog, A. Perelli [306], R. Dietmann [1570], G. Harman [2556], I and D.R. Heath-Brown [2638]. The case when the denominator and the numerator are both primes was considered by G. Harman in [2557], where it was shown that if $\psi(n)$ is non-increasing, then for almost all α there are infinitely many solutions of

$$|p\alpha - q| < \psi(p)$$

with prime p, q if and only if the series

$$\sum_{n=1}^{\infty} \frac{\psi(n)}{\log^2 n}$$

diverges. This result has been generalized to simultaneous approximations by H. Jones [3153].

5. In 1928 R.O. Kuzmin[67] [3589, 3591] solved a problem posed by Gauss in a letter to Laplace dated 30 January 1812 [2210]. Let $[a_0; a_1, a_2, \dots]$ be the continued fraction of a real number x and let $\xi_n(x)$ be the fractional part of $[a_n; a_{n+1}, \dots]$. Gauss asserted that the probability $P_n(t)$ of the inequality $\xi_n(x) < t$ $(0 \le t \le 1)$

[67] Rodion Osievič Kuzmin (1891–1949), professor in Perm and Leningrad. See [6385].

tends to $\log_2(1+t)$ for n tending to infinity, and asked Laplace to evaluate the difference $|P_n(t) - \log_2(1+t)|$. In modern formulation $P_n(t)$ is the measure of the set

$$\{x \in [0,1] : \xi_n(x) < t\}.$$

Kuzmin proved Gauss's assertion and established the equality

$$P_n(t) = \log_2(1+t) + O\left(c^{\sqrt{n}}\right)$$

with a certain $0 < c < 1$. An independent proof was provided by P. Lévy[68] [3866] in 1929, who obtained a stronger evaluation of the error term, namely $O(a^n)$ with $a = 0.7$.

Much later P. Szüsz [6031] showed that some $a < 0.49$ will also do, and E. Wirsing [6700] obtained this with $a = 0.303663\ldots$. He also showed that $a \le 0.302$ is impossible. A proof of Kuzmin's result based on ergodic theory was found in 1951 by C. Ryll-Nardzewski [5357].

P. Lévy also showed in [3866] that if $q_n(x)$ denotes the denominator of the nth convergent of x, then for almost all x one has

$$\lim_{n\to\infty} \frac{\log q_n(x)}{n} = \frac{\pi^2}{12\log 2}.$$

This limit (the *Lévy constant*) does not always exist and it was proved in 1999 by C. Baxa [364] that if $(1+\sqrt{5})/2 \le a \le b$, then there exist non-denumerably many numbers x, non-equivalent under the action of $SL_2(\mathbf{Z})$, such that

$$\liminf_{n\to\infty} \frac{\log q_n(x)}{n} = a, \qquad \limsup_{n\to\infty} \frac{\log q_n(x)}{n} = b.$$

On the other hand the Lévy constant exists for all algebraic numbers of degree 2 (H. Jager, P. Liardet [3095]), and its mean value for such numbers was found by C. Faivre [1953].

Recently C. Baxa showed ([365]) that every number $\ge (1+\sqrt{5})/2$ is the Lévy constant of a transcendental number.

6. In 1929 V. Jarník [3110, 3111] considered the Hausdorff dimension of the set of reals $\theta \in (0,1)$ for which there exist infinitely many co-prime integers p, q with

$$\left|\theta - \frac{p}{q}\right| < \frac{1}{q^c}$$

and showed that it equals $2/c$ for $c > 2$ (another proof was given five years later by A.S. Besicovitch [477]). Jarník considered also the set M_N of irrational numbers $\theta \in (0,1)$ having their partial quotients bounded by N and obtained lower and upper bounds for their Hausdorff dimension.

Jarník's result was generalized in 1970 by A. Baker and W.M. Schmidt [243] who considered sets of $\xi \in \mathbf{R}$ for which there exist infinitely many polynomials $P \in \mathbf{Z}[X]$ of degree n satisfying

$$|P(\xi)| < H(P)^{-\alpha},$$

[68] Paul Pierre Lévy (1886–1971), professor in Paris. See [6091].

with $H(P)$ denoting the height of P. They stated also a conjecture about the Hausdorff dimension of these sets proved later by V.I. Bernik [459, 460]. For $n = 3$ this conjecture had been established earlier by R.C. Baker [247], and the case $n = 2$ is a consequence of results by F. Kasch, B. Volkmann [3266] and A. Baker, W.M. Schmidt [243]. Cf. the book by V.I. Bernik and Yu.V. Melničuk [463].

The Hausdorff dimension of sets defined by the behavior of continued fractions was later considered by Good [2280], who showed that the set of numbers whose partial quotients a_n tend to infinity has dimension $1/2$ and the same applies to the set of numbers with unbounded $a_n^{1/n}$. The Hausdorff dimension of such sets was also studied by J.R. Kinney, T.S. Pitcher [3339], Y. Kifer, Y. Peres, B. Weiss [3332]. For a study of Jarník's sets M_N and, more generally, sets of $\theta \in (0, 1)$ whose partial quotients lie in a fixed set of positive integers see T.W. Cusick [1307–1309], D. Hensley [2751, 2752, 2754], O. Jenkinson [3121], M. Kesseböhmer, S. Zhu [3307] and G. Ramharter [5099].

A historical survey of the metric theory of continued fractions was given in 1977 by G.J. Rieger [5223].

7. The problem of minimal values attained at integral points of indefinite quaternary quadratic forms f was considered in 1929 by A. Oppenheim [4690] who showed that

$$\inf_{P \neq 0} |f(P)| \leq \sqrt{\frac{2}{3}}\, D^{1/4},$$

D being the discriminant of f. At the end of the paper he wrote: "It is very likely that $L(f)$ must be zero when the coefficients are incommensurable. But this has not yet been proved." This became known as *Oppenheim's conjecture* and was formulated again, in a more general form, by Oppenheim in 1953 [4695, 4696]. A special case of this conjecture was settled in 1934 by S. Chowla [1075].

Oppenheim's conjecture found its final solution in 1987, when G.A. Margulis [4148–4150] found its proof utilizing ergodic theory. He showed that if f is a real indefinite quadratic form in $n \geq 3$ variables which is not a multiple of a form with rational coefficients, then it attains, at integral non-zero arguments, arbitrarily small values. This was earlier shown to be true for diagonal forms in $n \geq 5$ variables (H. Davenport, H. Heilbronn [1391], G.L. Watson [6582]), and for forms with $n \geq 21$ through the work of B.J. Birch, H. Davenport, D. Ridout and G.L. Watson [530, 1367, 1368, 1399, 5215, 6584]. See R.C. Baker, H.P. Schlickewei [270] for certain other cases.

G.L. Watson [6585] proved in 1960 that the truth of Oppenheim's conjecture implies that the values of f at integral points lie dense on the real line. This was strengthened in 1989 by S.G. Dani and G.A. Margulis [1321] who showed that already the values of f at primitive integral points (i.e., points with relatively prime coordinates) lie dense. A simpler proof was given by Dani and Margulis in [1322] (cf. Dani [1320]). A quantitative version of Oppenheim's conjecture was proved by A. Eskin, G.A. Margulis and S. Mozes [1874, 1875]. A further proof was provided in 1999 by V. Bentkus and F. Götze [422].

For expositions see J.C. Sikorav [5787] and A. Borel[69] [630].

---

[69] Armand Borel (1923–2003), professor in Princeton. See [134].

8. Let $N(t)$ be the number of lattice points (x, y) (with $xy \neq 0$) lying inside a rectangular triangle with two sides lying on the coordinate axes, and the third lying on the line $Ax + By = t$, with fixed A, B and t growing to infinity. It is easy to see that $N(t)$ is asymptotically equal to $t^2/2AB$, and the question arises about the size of the error term. This is of interest only if the ratio $\theta = A/B$ is irrational as otherwise the trivial error term $O(t)$ cannot be improved.

This problem was attempted for the first time in 1922 by G.H. Hardy and J.E. Littlewood in [2528, 2529]. They proved the equality

$$N(t) = \frac{t^2}{2AB} + \frac{t}{2}\left(\frac{1}{A} + \frac{1}{B}\right) + R(t)$$

with $R(t) = o(t)$, and showed that the error term is related to the sum $C_\alpha(t) = \sum_{k \le N}\{k\alpha\}$, which we considered in Sect. 2.5.2. They showed also that this bound cannot be improved in general, although in some special cases one can get better results. So if θ is algebraic, then $o(t)$ could be replaced by t^c with some $c < 1$. The proof used complex integration, but A. Ostrowski [4708] discovered an elementary proof.

For irrational algebraic numbers Roth's theorem implies the evaluation $R(t) = O(t^\varepsilon)$ for every $\varepsilon > 0$.

The paper [2529] brought a representation of $R(t)$ as a series for which another proof, not using complex integration, was later furnished by Wilson [6678]

Asymptotic formulas for the number of lattice points in an n-dimensional tetrahedron

$$\left\{(x_1, \dots, x_n\} : x_i \ge 0, \quad \sum_{i=1}^{n} a_i x_i \le T\right\}$$

were obtained later by D.H. Lehmer [3781], D.C. Spencer [5864] (see F. Beukers [481] for an elementary proof; cf. also D.C. Spencer [5863] and G. Lochs [3981, 3982]). In the case when $T = 1$ and each a_i is an inverse of an integer, explicit formulas for $n = 3$ were obtained by L.J. Mordell [4397] in terms of Dedekind sums, and J. Kallies [3218] generalized this to arbitrary n. Sharp upper bounds in dimensions 3, 4 and 5 were given by K.P. Lin, Y. Xu and S.S.-T. Yau [3890, 6767, 6768].

It was shown in 1923 by A.J. Khintchine [3318] that for almost all plane polygons P one has

$$\#\{tP \cap \mathbf{Z}^2\} = t^2 \operatorname{vol}(P) + O\left(\log^{1+\varepsilon}\right). \tag{3.33}$$

In 1962 E. Ehrhart [1694–1697] showed that if P is a convex polyhedron in $\mathbf{R}^N$ whose vertices lie in $\mathbf{Z}^N$, then the number of lattice points in tP is a polynomial in t (*Ehrhart polynomial*); see also I.G. Macdonald [4049, 4050]. For the study of Ehrhart polynomials see C.A. Athanasiadis [159], B. Chen [1010], J.-M. Kantor [3245], R. Diaz, S. Robins [1532], J.-M. Kantor, A. Khovanskii [3246], R.P. Stanley [5892, 5893]. This is closely related to the Euler characteristic of algebraic varieties (see, e.g., S.E. Cappell, J.L. Shaneson [895], J.E. Pommersheim [4983]).

An algorithm for counting the lattice points in a polyhedron was proposed by A. Barvinok [345].

Another formula for $\#\{tP \cap \mathbf{Z}^N\}$ has been given by M. Brion [739–741] (cf. M. Brion, M. Vergne [742], M.-N. Ishida [3031], J. Agapito [23]). R. Morelli [4422] found an N-dimensional generalization of G. Pick's old formula [4854], who showed in 1886 that if P is a plane polygon with vertices in $\mathbf{Z}^2$, then its volume equals $I + B/2 - 1$, where I, B denote the number of lattice points inside, and on the border of P, respectively.

In 1993 M.M. Skriganov [5813] showed that if the slopes of sides of P are algebraic irrationals, then the error term in (3.33) is $O(t^\varepsilon)$ for every $\varepsilon > 0$. He later considered [5814] the case of polyhedrons P in $\mathbf{R}^N$ and used ergodic theory to deduce that for almost all unimodular lattices $\Lambda \subset \mathbf{R}^N$ one has

$$\#\{tP \cap \Lambda\} = t^N \operatorname{vol}(P) + O\left(\log^{N-1+\varepsilon}\right)$$

for every positive ε. See also M.M. Skriganov, A.N. Starkov [5815].

3.6 Transcendental Numbers

1. The first examples of transcendental numbers occur in a paper by J. Liouville [3933, 3934], who proved that any real number whose continued fraction has quickly growing partial quotients is transcendental.

Liouville's result on approximations of algebraic numbers by rationals, established in [3934], may be used for the construction of several examples of transcendental numbers, however the numbers so obtained (called *Liouville numbers*) form a set of Lebesgue measure zero, thus the majority of transcendental real numbers is not of this form. Moreover, this approach is unsuitable for the proof of transcendence of numbers naturally appearing in mathematics, like e or π. The transcendence of these two numbers was demonstrated in the last part of the 19th century, the transcendence of e in 1873 by C. Hermite [2764] and the transcendence of π by F. Lindemann[70] in 1882 [3895].

Details of Hermite's argument with a simplification due to H. Greminger [2343] may be found in the book by N.I. Feldman [1982]. A good bibliography of several variants of Hermite's proof is given in the book by A.B. Šidlovskiĭ[71] [5736].

Later several proofs of the transcendence of e and π were found. The book by J.F. Koksma [3444], published in 1936, quotes 27 of them and later years added several more to this list (see, e.g., J. Hančl [2495], Yu.V. Nesterenko [4566], and the book by K. Mahler [4093]).

At the end of his paper Lindemann sketched the proof of the following assertion.

If $\{a_1, \dots, a_n\}$ *and* $\{b_1, \dots, b_n\}$ *are algebraic numbers, the* a_i*'s are pairwise distinct and none of the* b_i*'s vanishes, then the linear combination* $\sum_{j=1}^n b_j e^{a_j}$ *does not vanish.*

He wrote also: "Eine[72] genauere Darlegung der hier nur angedeuteten Beweise behalte ich mir für eine spätere Veröffentlichung vor." Unfortunately he never pub-

---

[70] Ferdinand Lindemann (1852–1939), professor in Freiburg, Königsberg and Munich.

[71] Andreĭ Borisovič Šidlovskiĭ (1915–2007), student of Gelfond, professor in Moscow. See [3507].

[72] "I reserve a more precise presentation of the proofs sketched here for a later publication."

lished it as his efforts turned towards Fermat's Last Theorem[73], and a complete proof was later provided by Weierstrass [6608]. Therefore this result is usually called the *Lindemann–Weierstrass theorem*. It implies in particular that if a is a non-zero algebraic number, then e^a is transcendental, and so, in particular, are logarithms of algebraic numbers distinct from unity.

A modern proof, due to K. Mahler, can be found, e.g., in the book by P. Bundschuh [845], and another proof appears in F. Beukers, J.-P. Bézivin, P. Robba [490].

There was no real progress in the theory of transcendental numbers in the first two decades of the 20th century. Several new such numbers, defined as sums of suitable series, were constructed by G. Faber [1947], E. Maillet [4109] and A.J. Kempner [3299].

The first book devoted to transcendental numbers, written by E. Maillet [4109], appeared in 1906.

2. In 1929 J. Popken[74] [4989, 4990] considered approximations of the numbers e and π by algebraic numbers and showed that if θ is an algebraic number of degree n and height H, then for every $\varepsilon > 0$ one has

$$|\pi - \theta| > \frac{1}{2^{H^\nu}}, \tag{3.34}$$

and

$$|e - \theta| > \frac{B(n, \varepsilon)}{H^{n+\varepsilon}}, \tag{3.35}$$

where $B(n, \varepsilon)$ is positive, and ν depends on n. The result concerning e improved a previous result by Borel [634], who was the first to consider this type of problem.

The bounds in (3.34) and (3.35) were improved for rational θ first by K. Mahler in 1932 [4064, 4065], and the best known lower bound for approximation of π by rationals is due to V.Kh. Salikhov [5382], who in 2008 established

$$\left|\pi - \frac{p}{q}\right| \geq q^{-7.6063}.$$

Earlier K. Mahler [4086] proved

$$\left|\pi - \frac{p}{q}\right| \geq q^{-42},$$

G. Rhin [5173] had for large q the exponent -16.2 in the last inequality and M. Hata [2610] increased the exponent to -8.016045.

Approximations by algebraic numbers have been treated by K. Mahler [4064, 4065], N.I. Feldman [1973, 1974], P.L. Cijsouw [1113] and Yu.M. Aleksentsev [49].

In 1971 P. Bundschuh [843] obtained

$$\left|e - \frac{p}{q}\right| \geq c\frac{\log\log q}{q^2 \log q}$$

[73] He published three incorrect proofs of this theorem [3896–3898].

[74] Jan Popken (1905–1970), professor in Utrecht and Amsterdam. See [3094].

for $c = 1/18$ and large q, and the value of c was increased to $c = 1/3$ (for all q) by T. Okano [4673] and in 1998 H. Alzer [62] got $c = 0.386\ldots$, with equality for the 19th convergent of the continued fraction for e.

Let θ be a transcendental number. A function $\Phi(n, H) > 0$ is called a *transcendence measure* of θ if for every polynomial $f \in \mathbf{Z}[X]$ of degree n and height H one has

$$|f(\theta)| \geq \Phi(n, H).$$

Transcendence measures for several classes of transcendental numbers were evaluated by K. Mahler [4064, 4065], N.I. Feldman [1973–1976], S. Lang[75] [3689].

3. In 1929 C.L. Siegel [5747] modified the method of Hermite, and this permitted him to obtain transcendence results for values at algebraic arguments of certain entire functions, belonging to the family of E-functions, defined in the following way.

A function

$$f(z) = \sum_{n=0}^{\infty} \frac{a_n}{n!} z^n$$

is called an *E-function* if the coefficients a_n are algebraic numbers lying in a fixed algebraic number field, the absolute value of every conjugate of a_n is $\ll n^{\varepsilon n}$ for every $\varepsilon > 0$, and for every n one can write $a_n = A_n/q_n$ with an algebraic integer A_n and $q_n \in \mathbf{N}$ such that for every $\varepsilon > 0$ one has $q_n \ll n^{\varepsilon n}$.

Siegel was able to obtain transcendence results for values of an E-function f at non-zero algebraic numbers in the case when f was a solution of a certain linear differential equation. As a corollary he proved transcendence results for values of the Bessel function $J_0(z)$, as well as for $K_\lambda(z)$, differing from the Bessel function $J_\lambda(z)$ by the factor $(z/2)^{2\lambda}\Gamma^{-1}(1+\lambda)$, in the case when λ is rational, distinct from a negative integer and the half of an odd integer. In this case he established algebraic independence of the values of K_λ and its derivative at a non-zero algebraic number. Siegel's method is presented in his book [5768]. An important tool in his investigation was the following elementary result, concerning small integral solutions of a system of linear equations, known now as *Siegel's lemma*[76].

Siegel's lemma *Let $n > m$ and let $L_j(X_1, \ldots, X_n) = \sum_{i=1}^{n} a_{ij} X_j$ $(j = 1, 2, \ldots, m)$ be linear forms with integral coefficients, and put $H = \max_{ij} |a_{ij}|$. Then there exist integers $x_1, \ldots, x_n$, satisfying*

$$0 < \max_i |x_i| \leq 1 + (nH)^{m/(n-m)}$$

with

$$L_j(x_1, \ldots, x_n) = 0 \quad (j = 1, 2, \ldots, m). \tag{3.36}$$

[75] Serge Lang (1927–2005), professor at Columbia University and Yale. See [3158, 3159, 6511].

[76] This lemma later found several different applications in number theory.

It is not an easy task to deduce results concerning particular E-functions from the main theorems of C.L. Siegel. In 1955 A.B. Šidlovskiĭ [5734, 5735] used a modification of Siegel's approach to produce a rather simple criterion for algebraic independence of values of E-functions. He showed, namely, that if $f_1(z), \dots, f_n(z)$ are E-functions, algebraically independent over the field $\mathbf{C}(z)$ of rational functions over the complex field, and satisfying a system of linear differential equations of the form

$$y_i' = V_{i0} + \sum_{j=1}^{n} V_{ij}(z) y_i \quad (i = 1, 2, \dots, n)$$

with $V_{ij} \in \mathbf{C}(z)$, then for every non-zero algebraic number α which is not a pole of some V_{ij} the numbers $f_1(\alpha), \dots, f_n(\alpha)$ are algebraically independent over $\mathbf{Q}$. In particular, each of them is transcendental. For a quantitative version see Yu.V. Nesterenko [4562].

A modern approach to the theory of E-functions was presented by F. Beukers, W.D. Brownawell and G. Heckman [491] in 1988. Cf. also V.Kh. Salikhov [5381].

A generalization of Siegel's lemma to algebraic number fields was given by E. Bombieri and J. Vaaler [623] in 1983, who established the existence of a basis of solutions of (3.36) with a small product of heights of its elements, and it was shown by D. Roy, J.L. Thunder [5324] that the bounds obtained are sharp. Cf. D. Roy, J.L. Thunder [5325]. See also D.W. Masser [4176]. For a generalization to function fields see J.L. Thunder [6148].

3.7 Diophantine Equations

1. In 1923 A. Arwin [148] considered the system of two Pellian equations

$$x^2 - ay^2 = 1, \quad y^2 - bz^2 = 1 \quad (a \neq b) \tag{3.37}$$

in the case $a = 2, b = 3$, and conjectured that its only solution in positive integers is $(x, y, z) = (3, 2, 1)$. He was able to reduce this question to a finite check (it follows from the Siegel–Thue theorem that this equation has only finitely many solutions) and his conjecture was established in 1941 by W. Ljunggren[77] [3971] who also gave an upper bound for solutions of (3.37) in the general case.

The similar system

$$x^2 - ay^2 = 1, \qquad z^2 - by^2 = 1 \tag{3.38}$$

had already been considered in 1904 by A. Boutin and P.F. Teilhet [667], who showed that in the case $a = 6$, $b = 3$ there are no solutions. The same assertion was obtained for $a = 2$, $b = 3$ by M. Rignaux [5229] in 1918.

The first explicit although very large bound for the number of positive solutions of (3.38) can be extracted from the results of H.P. Schlickewei [5465] on the number of solutions of unit equations. This enormous bound was reduced in 1996 to 16 by D.W. Masser and J.H. Rickert [4177], and to 3 by M.A. Bennett [408] in 1998. The ultimate bound was obtained by Bennett, M. Cipu, M. Mignotte and R. Okazaki [416] who showed that there are at most two solutions. It follows from an earlier result by M.A. Bennett [407] that this cannot be improved. Earlier

[77] Wilhelm Ljunggren (1905–1973), professor in Oslo and Bergen.

the bound 2 had been obtained in the case $\max\{a,b\} > 1.4 \cdot 10^{57}$ by P.Z. Yuan [6801]. If a has at most seven prime factors, and $b = 2a$, then (3.38) can have at most one solution and such a solution exists only for seven values of a (X. Dong, W.C. Shiu, C.I. Chu, Z. Cao [1612]). A list of all solutions for $2 \le a < b \le 200$ was given by W.S. Anglin [98].

The third system of a similar form,

$$x^2 - ay^2 = 1, \qquad z^2 - bx^2 = 1 \tag{3.39}$$

was considered by M. Cipu and M. Mignotte [1116], who showed that it can have at most two solutions. It is conjectured that there is at most one solution. In the case $a = 4m(m+1)$ this conjecture has been proved by P.Z. Yuan [6802] in 2004, and P.G. Walsh [6539] established it for $a = m^2 - 1$. In both cases the results are definitive, as in the cases $a = 8, b = 35$ and $a = 3, b = 15$, corresponding to $m = 1$ and $m = 2$ respectively, this system is solvable.

An algorithm to solve the more general system

$$a_1x^2 + b_1y^2 = c_1, \qquad a_2x^2 + b_2z^2 = c_2,$$

working for not too large coefficients can be found in the paper by L. Szalay [6016] where also a list of particular equations of this type which were solved earlier is given.

2. In [5747] C.L. Siegel used his lemma as well as ideas stemming from the work of L.J. Mordell [4379] and A. Weil [6609] for an important breakthrough in the theory of Diophantine equations in two variables, showing the following result.

If $F(X,Y)$ is an irreducible polynomial with integral coefficients, then the equation $F(x,y) = 0$ has infinitely many rational solutions with bounded denominators if and only if the corresponding curve $\Gamma : F(X,Y) = 0$ has a parametrization of the form

$$x = \sum_{j=-m}^{m} a_j t^j, \qquad y = \sum_{j=-n}^{n} b_j t^j,$$

with suitable m, n and integral coefficients a_j, b_j. The same assertion holds also in the case when rational integers are replaced by algebraic integers from a fixed algebraic number field.

This result implies that this equation can have infinitely many such solutions only if the genus $g(\Gamma)$ of Γ is zero. The sufficiency of this condition follows from an older result of Maillet [4113, 4114].

In 1934 K. Mahler [4074] showed the truth of Siegel's assertion for genus 1 also in the case when rational integers are replaced by rational numbers with denominators divisible only by primes from a fixed finite set. In an appendix to Mahler's paper H. Hasse gave a simpler proof based on the theory of algebraic functions (see also [2604]). A proof based on non-standard analysis was found in 1975 by A. Robinson[78] and P. Roquette [5240]. In 2002 P. Corvaja and U. Zannier [1254] presented a proof of Siegel's theorem, based on the subspace theorem of W.M. Schmidt (see Sect. 6.4.1).

[78] Abraham Robinson (1918–1974), professor in Toronto, Jerusalem and at the University of California and Yale. See [4446].

In 1960 S. Lang [3687] extended Siegel's result to the case of finitely generated integral domains containing **Z**.

In 1998 D. Poulakis [5002] established upper bounds for the smallest integral solution lying in an algebraic number field K of a Diophantine equation of genus zero with integral coefficients from K. Such bounds were known earlier for the equation

$$ax^2 + by^2 + cz^2 = 0,$$

due to the work of L. Holzer [2851] and L.J. Mordell [4404] in the rational case, and C.L. Siegel [5775] in the general case.

3. The cubic equation $x^3 + dy^3 = 1$ was treated by B. Delaunay, who had announced already in 1916 [1436] that if d is not a cube, then there can be at most one integral solution apart from the trivial one ($x = 1, \ y = 0$), with the exception of $d = 2$, when there is one more solution ($x = -1, \ y = 1$), but published his proof in an obscure local journal [1454] so it only became known in 1928, when the paper [1438] appeared. This result was independently obtained by T. Nagell [4505, 4507, 4508]. They both used the theory of units in cubic number fields and applied the same approach when dealing with binary cubic forms $f(x, y)$ with integral coefficients and negative discriminant, assumed to be irreducible over the rationals. In [4508] it was proved that the more general equation $ax^3 + by^3 = c$ with $c = \pm 1, \pm 3$ has at most one solution, with the unique exception of $x^3 + 2y^3 = 1$ which has two solutions.

B. Delaunay showed later [1437, 1439] that the equation

$$f(x, y) = 1, \tag{3.40}$$

where f is a cubic form of negative discriminant can have at most five integral solutions, and T. Nagell [4509] obtained the same bound and proved that apart from three explicitly given forms (and forms linearly equivalent to them) there are at most three solutions. The more general equation $f(x, y) = k$, where f is an arbitrary cubic form was shown by C.L. Siegel [5747] to have at most eighteen solutions.

Delaunay's method used in [1454] was applied in 1926 by V.A. Tartakovskiĭ to show [6053] that if $d \neq 15$ is not a square, then the equation $x^4 + dy^4 = 1$ can have at most one integral solution.

In the omitted case $d = 15$ this was shown later by D.K. Faddeev[79] [1949].

The number of solutions of (3.40) in the case when f is a cubic form of positive discriminant has been shown to be ≤ 12 by J.-H. Evertse [1921, 1922] in 1983, and ≤ 10 by M.A. Bennett [411]. For forms with sufficiently large discriminants this had been established much earlier by A.E. Gelman (see [1455]), and in this case R. Okazaki [4674] later got the bound 7 (cf. I. Wakabayashi [6495]).

Siegel's method leads, in the case of $g(\Gamma) \geq 1$, to an explicit upper bound for the number of integral points on Γ, however it does not give bounds for the size of them. As an example C.L. Siegel [5754] showed in 1937, using Thue's [6145] approach to this question, that the equation

$$ax^n - by^n = c$$

[79] Dmitriĭ Konstantinovič Faddeev (1907–1989), professor in Leningrad. See [5367].

has for a fixed $n \geq 3$ at most one integral solution in co-prime integers x, y, provided $|ab|$ exceeds a value, depending on c and n. It was established in 1954 by Y. Domar [1611] that the equation

$$|ax^n - by^n| = 1 \tag{3.41}$$

cannot have more than two solutions for $n \geq 5$, and S. Hyyrö [2998] showed in 1964 that for fixed $n \geq 7$ this equation can have at most one solution with $x \neq \pm 1$. Much later it was shown by M.A. Bennett and B.M.M. de Weger [417] that it has at most one solution, except possibly in the case when $n \leq 347$ and $b = a + 1 \leq 84$, and finally Bennett [409] established that in all cases (3.41) can have at most one solution. A bound for the number of solutions of the more general equation $|ax^n - by^n| = m$ was given by J.-H. Evertse [1920] in 1982.

Siegel's approach is not applicable to the case of rational solutions. Already L.J. Mordell conjectured in [4379] that in the case $g(\Gamma) \geq 2$ there can be only finitely many such solutions, and C.L. Siegel expressed the same opinion, as he wrote:

"Doch dürfte wohl der Beweis der Vermutung, daß jede solche Gleichung, wenn ihr Geschlecht größer als 1 ist, nur endlich viele Lösungen in rationalen Zahlen besitzt, noch die Überwindung erheblicher Schwierigkeiten erfordern[80]."

For the fulfillment of these expectations one had to wait several years up to 1983, when G. Faltings [1954] established Mordell's conjecture (see Sect. 7.2).

For certain special classes of curves the finiteness of rational points had been established earlier (see, e.g., V.A. Demyanenko [1458, 1459]).

4. Let $f(X) = \sum_{n=0}^{\infty} a_n X^n$ be a power series with coefficients in a $\mathfrak{p}$-adic field K and $v(a_n) \to 0$. It had been shown in 1928 by R. Straßmann [5965] that if f is not identically zero, then the number of solutions of the equation $f(x) = 0$ in integers of K is finite. This result can be used to solve Diophantine equations in the rational field (see, e.g., the book by H. Cohen [1142]).

3.8 Elliptic Curves

1. The study of Diophantine equations of the form $y^2 = f(x)$, where f is a cubic polynomial, commenced in the middle of the 17th century when Fermat [1989, pp. 333, 345] asserted that the only positive integral solution of the equation

$$y^2 = x^3 - 2$$

is formed by the pair $x = 3$, $y = 5$, and the equation

$$x^2 = y^3 - 4$$

[80] "But the proof of the conjecture that every such equation, when its genus is larger than 1, has only finitely many solutions in rational numbers, will necessitate to overcome considerable difficulties" [5747, p. 34].

has two solutions ($x = 2, 11$). Proofs of these assertions were given later by Euler [1901], who tacitly assumed unique factorization in the ring of numbers generated by $\sqrt{2}$, and the first complete proof is due to T. Pépin [4771]. Euler [1898] showed also that the equation $y^2 = x^3 + 1$ has $x = 2$, $y = 3$ for its only integral solution. Later several other equations of the form $y^2 = x^3 + a$ were considered by various authors (for details see L.E. Dickson's book [1545, Chap. 20]), but no attempt to construct a general theory was made in the 19th century.

Nowadays one considers, more generally, equations of the form

$$E : y^2 + a_1xy + a_3y = x^3 + a_2x^2 + a_4x + a_6, \tag{3.42}$$

with x, y lying in a field K. In the case when the characteristic of K is neither 2 nor 3, a linear transformation can be applied to obtain an equation of the form

$$y^2 = x^3 + ax + b. \tag{3.43}$$

In this case one defines the *discriminant* $\Delta(E)$ as the discriminant of the cubic polynomial on the right-hand side of (3.43), i.e.,

$$\Delta(E) = -(4a^3 + 27b^2).$$

In the general case the discriminant is defined by a rather unpleasant formula, which the interested reader may find in the paper by J. Tate [6063] or in J.H. Silverman's book [5791], which jointly with [5794] serves as an introduction to the theory of elliptic curves.

It is convenient to consider the set $E(L)$ of all solutions (x, y) of (3.43) with x, y lying in a fixed extension L/K, as a curve in a two-dimensional L-space. It follows from the classical theory of elliptic functions that they can be used to parametrize this curve in the case when L is the field of complex numbers and the discriminant $\Delta(E)$ does not vanish. Indeed, in this case one has

$$x = \wp(z), \qquad y = \wp'(z),$$

where $\wp(z)$ is the Weierstrass elliptic function associated with a suitable lattice $\Lambda \subset \mathbf{C}$, and z runs over points of the fundamental parallelogram of this lattice. Since the points of that parallelogram are in a bi-unique correspondence with elements of the factor group $\mathbf{C}/\Lambda$, this fact can be used to induce a group structure in the set of all complex solutions of (3.43) with an added point at infinity. It took, however, some time before this fact was utilized in number theory and the first explicit mention of this group structure appears in a paper by C.S. Juel [3164], although this idea, as pointed out in D.H. Husemöller's book [2966], can be traced back to Jacobi [3079]. From the geometric point of view this group structure is very simple: if 0 is the point at infinity, then three points A, B, C lying on the curve E defined by (3.43) satisfy

$$A + B + C = 0$$

if and only if they are collinear. This property permits a group structure to be defined on $E(K)$ in the case of an arbitrary field K. Of particular importance is the case, when K is the field $\mathbf{Q}$ of rational numbers or one of its finite extensions.

Curves obtained in this manner are called *elliptic curves*. With every such curve E defined over a field k one associates its field of functions $K(E) = k(X, Y)/I_E$, where I_E is the ideal of all rational functions $F \in k(X, Y)$ with $F(x, y) = 0$ for $(x, y) \in E$.

If the curve E is defined over a field K, which is either the field of rational numbers or its finite extension, and $\mathfrak{p}$ is a prime ideal of the corresponding ring of integers $\mathbf{Z}_K$, then by a linear change of variables (which usually changes the discriminant) one can assume that all coefficients in (3.42) are $\mathfrak{p}$-integral, and the additive valuation corresponding to $\mathfrak{p}$ of the discriminant $\Delta(E)$ is minimal. Such a form of (3.42) is called the *minimal equation* at $\mathfrak{p}$. Reducing mod $\mathfrak{p}$ the coefficients of such an equation one obtains a curve over the finite field $k = \mathbf{Z}_K/\mathfrak{p}$. If $\mathfrak{p} \nmid \Delta(E)$, then E is said to have *good reduction* mod $\mathfrak{p}$. In this case the reduction E mod $\mathfrak{p}$ is an elliptic curve over k. If the discriminant is divisible by $\mathfrak{p}$, then one distinguishes between *multiplicative reduction* at $\mathfrak{p}$, when E mod $\mathfrak{p}$ has a node at the singular point, which happens if in (3.42) one has

$$\mathfrak{p} | (a_1^2 + 4a_2)^2 - 48a_4 - 24a_1a_3,$$

and *additive reduction* at $\mathfrak{p}$ in the remaining case (geometrically, in the last case E mod $\mathfrak{p}$ has a cusp at the singular point). The non-singular points of E mod $\mathfrak{p}$ form a group which is isomorphic in the first case to the multiplicative group of the field $\mathbf{F}_\mathfrak{p} = \mathbf{Z}_K/\mathfrak{p}$, and in the second case to the additive group of $\mathbf{F}_\mathfrak{p}$.

If E_1, E_2 are elliptic curves with zero elements O_1 and O_2 respectively, then a rational map $f : E_1 \to E_2$ carrying O_1 in O_2 is called an *isogeny*, and one shows that every isogeny is at the same time a group homomorphism. The set of all isogenies $E \to E$ forms the endomorphism ring $\mathrm{End}(E)$ of E with addition induced by the addition in E and composition as multiplication. Two curves E_1, E_2 are *isogenous* if there exists a surjective isogeny $f : E_1 \to E_2$. In this case the field $K(E_2)$ is isomorphic to a subfield K of $K(E_1)$, and the degree of the extension $K(E_1)/K$ is called the *degree* $\deg f$ of f. If E_1 is defined over a field of zero characteristic, then $\deg f$ equals the cardinality of the kernel of f.

With any curve E defined by (3.42) one associates the j-invariant $j(E)$ by putting

$$j(E) = \frac{(b_2^2 - 24b_4)^3}{\Delta(E)}. \tag{3.44}$$

Since for curves E defined over the complex field one has $E = \mathbf{C}/\Lambda$, where $\Lambda = \omega_1\mathbf{Z} + \omega_2\mathbf{Z}$ ($\Im(\omega_1/\omega_2) > 0$) is a lattice, then the equation

$$j(z) = j(E) \tag{3.45}$$

with $z = \omega_1/\omega_2$ defines a function regular in the upper half-plane, earlier called *Klein's modular function* (see, e.g., R. Dedekind [1421], F. Klein [3350], A. Hurwitz [2960]).

The isomorphism classes of elliptic curves over the complex field correspond to orbits of the upper complex half-plane $\mathfrak{H}$ under the action of $SL_2(\mathbf{Z})$, and are classified by the value of the j-invariant.

If E is defined over a number field and $\mathrm{End}(E(\mathbf{C}))$ is larger than $\mathbf{Z}$ (in which case this ring is an order in an imaginary quadratic field), then E is said to have *complex multiplication* (CM-curve).

For a quick introduction to the theory of CM-curves see B.H. Gross [2353].

2. In the first years of the century there was no great interest in elliptic curves, which were to play an important role in the future development of number theory. The first important general result concerning $E(\mathbf{Q})$ was obtained by L.J. Mordell [4379] in 1922, who proved, confirming a statement attributed[81] to H. Poincaré [4936], that this group is finitely generated, hence is isomorphic to

$$A \oplus \mathbf{Z}^r, \tag{3.46}$$

where A is finite, and $r \geq 0$. This result was later extended by A. Weil[82] [6609] to Abelian varieties over algebraic number fields (*Mordell–Weil theorem*).

Another proof of the Mordell–Weil theorem was given in 1942 by H. Hasse [2601], and an arithmetical proof of the finiteness of the torsion part of $E(K)$ for extensions $K/\mathbf{Q}$ of finite degree was given by G. Bergman [437].

Later the Mordell–Weil theorem was generalized by A. Néron [4558] to Abelian varieties over arbitrary finitely generated fields (see also S. Lang, A. Neron [3704]). The same question was considered (sometimes in a more general setting) in the case of infinite Abelian extensions of $\mathbf{Q}$ by B. Mazur [4213], D.E. Rohrlich [5269, 5270] and K. Rubin, A. Wiles [5333].

A survey of developments related to Mordell's finite basis theorem was prepared by J.W.S. Cassels [948] in 1986. See also K. Rubin, A. Silverberg [5332] and J.-P. Serre [5658].

3. Of special interest were curves of the form

$$E(k): y^2 = x^3 - k, \tag{3.47}$$

and the corresponding Diophantine equations. We noted already that Fermat considered the case $k = 2$. See Dickson's [1545, Chap. 20] for a survey of early papers on that subject.

It was shown in 1907 by A.S. Verebrusov [6387] that equation (3.47) cannot have non-constant polynomial solutions. In 1919 L.J. Mordell proved [4378] that it can have at most finitely many integral solutions, although a few years earlier he had believed (see [4374]) the contrary. In 1926 A. Brauer [673] developed a method which enabled, in certain cases, complete solution of these equations, and he applied it to the case $k = -2$, which was attempted previously by several writers without success. In 1930 R. Fueter [2133] showed that if $k \neq -1, 2^4 \cdot 3^3$, and there is a point $(x, y) \in E(k)$ with $xy \neq 0$, then $E(k)$ is infinite. He showed also that if $k > 0$ satisfies certain congruences mod $3^2 \cdot 2^4$ and the class-number of the field $\mathbf{Q}(\sqrt{-k})$ is not divisible by 3, then $E(k)$ is trivial (for a simpler proof see L.J. Mordell [4403]).

---

[81] Actually Poincaré described a geometrical procedure for generating new points of a curve from a finite number of points given, and defined the rank of the curve as the minimal number of its generators. He tacitly assumed this number to be finite, and on p. 173 formulated the question of which numbers are ranks of rational elliptic curves.

[82] The paper [6610], which is sometimes quoted as the source of the Mordell–Weil theorem, contains only a fresh proof of Mordell's result for the base field $\mathbf{Q}$.

R. Fueter also gave examples of elliptic curves over **Q** having no finite rational points.

Generators of the group $E(k)$ for $|k| \leq 25$ were listed in 1938 by G. Billing [506]. Later this list was extended by V.G. Podsypanin [4926] in 1949, J.W.S. Cassels [929] in 1950, O. Hemer [2726, 2727] in 1952–1954, E.S. Selmer [5634] in 1956, and R. Finkelstein and H. London [2006]. All integral solutions of (3.47) for $|k| \leq 10^4$ were found by J. Gebel, A. Pethő and H.G. Zimmer [2216].

More information on the equation (3.47) may be found in the books by L.J. Mordell [4393] and H. London and R. Finkelstein [3989].

Several examples of elliptic curves over **Q** without finite points were given later by L.J. Mordell [4394, 4395] and K.L. Chang [996], but the existence of infinitely many such curves was only proved as late as 1988 by J. Nakagawa and K. Horie [4527].

4. An early exposition of the arithmetic theory of elliptic curves, illustrated by several examples, was presented by T. Nagell [4510] in 1928. He posed the question of whether for every $k = 1, 2, \ldots$ there exists an elliptic curve over **Q** having exactly k rational points, and was able to answer it positively for $k \leq 4$ and $k = 6$, the case $k = 5$ going back to Euler [1901] who showed that the equation $y^2 = x^3 + 1$ has exactly 5 rational solutions (cf. T. Nagell [4506]).

The final answer was given in 1977 by B. Mazur [4215], who described the torsion subgroups of $E(\mathbf{Q})$ (see Sect. 6.7).

Chapter 4
The Thirties

4.1 Analytic Number Theory

4.1.1 Exponential and Character Sums

1. The problem of evaluating sums of the form

$$T_{f,\chi}(p) = \sum_{x=1}^{p-1} \chi(f(x)), \tag{4.1}$$

where p is a prime, $\chi(a)$ is the Legendre symbol mod p and f is a monic cubic polynomial with integral zeros, was proposed by J.E. Littlewood to H. Davenport at the beginning of the thirties. For quadratic f such sums had been evaluated earlier by E. Jacobsthal [3084], and E. Artin conjectured in [137] that for cubic f one has

$$|T_{f,\chi}(p)| \le 2\sqrt{p}.$$

H. Davenport [1339, 1340] established the bound

$$|T_{f,\chi}(p)| \le Bp^{3/4}$$

for cubic f, obtained the same bound in the case when f is a monic quartic polynomial with integral zeros, and also proved bounds of order $O(p^{1-c})$ for such sums with polynomials of degrees ≤ 8.

The sum (4.1) is closely related to the number $N_p(f)$ of solutions of the congruence

$$y^2 \equiv f(x) \pmod{p},$$

due to the equality

$$N_p(f) = T_{f,\chi}(p) + p - r - 1,$$

with r denoting the number of non-zero solutions of the congruence $f(x) \equiv 0 \pmod{p}$. Thus Davenport's result implied

$$N_p(f) = p + O\left(p^{3/4}\right)$$

W. Narkiewicz, *Rational Number Theory in the 20th Century*,
Springer Monographs in Mathematics,
DOI 10.1007/978-0-85729-532-3_4,

for the polynomials considered.

Later L.J. Mordell [4382, 4383] obtained bounds of similar shape for the number of solutions of the congruence

$$y^m \equiv f(x) \pmod{p}$$

for certain small values of m and $\deg f$, and showed also that if $f(x, y)$ is a cubic polynomial, irreducible mod p and irreducible over $\mathbf{C}$, then the congruence

$$f(x, y) \equiv 0 \pmod{p}$$

has $p + O(p^{2/3})$ solutions.

Later developments in algebraic geometry, in particular Weil's proof [6615] of the analogue of the Riemann Hypothesis for curves, led to the bound

$$|T_{f,\chi}(p)| \leq (\deg f - 1)\sqrt{p}.$$

2. The first treatment of sums (4.1) with an arbitrary character χ appeared in the paper [1343] by H. Davenport, who showed that if $f \in \mathbf{Z}[X]$ is a quadratic polynomial whose leading term and the discriminant are not divisible by the prime p, and χ is a non-principal character mod p, then $T_{f,\chi}(p) \ll p^{2/3} \log p$. Later he treated, more generally, sums of the form

$$S(f_1, \dots, f_r, \chi_1, \dots, \chi_r) = \sum_{x \in \mathbf{F}_q} \prod_{j=1}^{r} \chi_j(f_j(x)),$$

where χ_j are non-principal characters of the multiplicative group of the field $\mathbf{F}_q$, and the polynomials $f_j \in \mathbf{F}_q[X]$ are monic [1355]. To bound such sums he introduced a kind of L-function defined in the half-plane $\Re s > 1$ by

$$L(f_1, \dots, f_r, \chi_1, \dots, \chi_r) = \sum_{n=0}^{\infty} \frac{\sigma_n}{n^s},$$

where

$$\sigma_n = \sum_{g} \prod_{j=1}^{r} \chi_j(R(f_j, g)),$$

g running over all monic polynomials over $\mathbf{F}_q$ of degree n and $R(f, g)$ denoting the resultant of the polynomials f, g. In the case $r = 1$ this L-function coincides with the L-function occurring in the theory of function fields generated by the equation $Y^n = f(X)$, considered by H. Hasse [2597]. H. Davenport showed that the zeros $\varrho \neq 0$ of that L-function satisfy

$$\theta_K \leq \Re \varrho \leq 1 - \theta_K,$$

with $K = \sum_{j=1}^{r} \deg f_j \geq 4$ and $\theta_K = 3/(2K + 8)$ (in the case $K = 3$ and quadratic characters χ_j the equality $\Re \varrho = 1/2$ was known already at that time, due to H. Hasse [2596]). This led to

$$|S(f_1, \dots, f_r, \chi_1, \dots, \chi_r)| \leq (K - 1)q^{1-\theta_K}.$$

3. The first significant improvement of Weyl's bound (2.69) for the sum

$$S_f(n) = \sum_{x=0}^{n-1} \exp\left(\frac{2\pi i f(x)}{n}\right)$$

was obtained in 1932 by L.J. Mordell [4382], who established for prime p the evaluation

$$|S_f(p)| \le C p^{1-1/d}$$

(with an absolute constant C) for polynomials $f \in \mathbf{Z}[X]$ of degree $d < p-1$, having not all coefficients divisible by p. The next year H. Davenport [1343] showed that if $d \ge 4$, then

$$|S_f(p)| \ll p^{1-1/m},$$

where m is the maximal number of the form 2^k or $3 \cdot 2^k$ not exceeding d, and obtained

$$|S_f(p)| \ll p^{5/8}$$

for $d = 3$.

Since for co-prime m, n one has

$$S_f(mn) = S_g(m) S_h(n)$$

with $g(X) = f(nX)/n$, $h(X) = f(mX)/m$, and moreover $\omega(n) \ll \log n / \log\log n$, Mordell's result implied the bound

$$|S_f(n)| \le C^{\omega(n)} n^{1-1/d} \ll n^{1-1/d+\varepsilon} \tag{4.2}$$

for square-free n and every $\varepsilon > 0$.

In 1940 L.K. Hua [2925] proved for prime powers p^a the bound

$$|S_f(p^a)| \le d^3 p^{1-1/d},$$

implying (4.2) for all n, and the ε in the exponent was removed in 1953 by Nečaev [4553], who obtained for $d \ge 12$ the explicit inequality

$$|S_f(n)| \le B(d) n^{1-1/d},$$

with $B(d) = \exp(2^{1+d})$. The value of $B(d)$ was later reduced by J.R. Chen [1013] to $\exp(cd^2)$ for some c, by V.I. Nečaev [4554] to $\exp(5d^2/\log d)$, by J.R. Chen [1025] to $\exp(cd)$ with $c \le 6.1$ (and $c \le 4$ for $d \ge 10$), by S.B. Stečkin [5917] to $\exp(d + O(d/\log d))$, by M.H. Lu [4017] to $\exp(1.85d)$, and the best known bound, $B(d) \le \exp(1.74d)$ was obtained in 1989 by P. Ding and M.G. Qi [1577]. Cf. also S.B. Stečkin [5915] (who treated the polynomial $f(t) = at^k$), O. Körner and H. Stähle [3484], R.A. Smith [5836], V.I. Nečaev, V.L. Topunov [4555], J.H. Loxton, R.A. Smith [4009], J.H.H. Chalk[1] [983], P. Ding [1574–1576], M.V. Kudryavtsev [3559], W.K.A. Loh [3983], J.H. Loxton, R.C. Vaughan [4011], J.H. Loxton [4008], T. Cochrane, Z. Zheng [1133].

[1] John H.H. Chalk (1923–2004), professor in Toronto.

The analogue of Hua's result for polynomials over an algebraic number field was proved by L.K. Hua [2936] in 1951.

In 1957 L. Carlitz[2] and S. Uchiyama [904] deduced from Weil's results in [6615] the bound

$$|S_f(p)| \leq (d-1)\sqrt{p}$$

for prime p. An elementary proof of this bound was given in 1986 by D.A. Mitkin [4342] using S.A. Stepanov's method [5925]. In the special case $f(X) = aX^d$ an improvement was obtained by H.L. Montgomery, R.C. Vaughan and T.D. Wooley [4370]. For large d stronger bounds were given by I.E. Šparlinskiĭ [5859] and D.R. Heath-Brown, S.V. Konyagin [2667].

Bounds depending on the number of non-vanishing coefficients of $f(X)$ were found by Shparlinski [5732], T. Cochrane, C. Pinner, J. Rosenhouse [1132] and J. Bourgain [663, 665].

4. The Kloosterman sums

$$S(u,v;q) = \sum_{\substack{x \bmod q \\ (x,q)=1}} \exp\left(\frac{2\pi i(xu + x'v)}{q}\right)$$

were treated by H. Salié [5378] in 1932. He observed that in the case when $q = p^m$ is a prime power with exponent $m \geq 2$ they are reducible to quadratic Gauss sums, and in that case this leads to the bound $|S(u,v;q)| \leq C\sqrt{q}$ with an absolute constant C.

Other proofs were later given by A.L. Whiteman [6650] and K.S. Williams [6675].

H. Salié conjectured also that for all q and $\varepsilon > 0$ one has

$$|S(u,v;q)| \ll q^{1/2+\varepsilon}(u,q)^{1/2},$$

and was able to prove this for square-full q. One year later H. Davenport [1343] obtained the bound

$$|S(u,v;p)| \ll p^{2/3}$$

for prime p improving upon the previous bound $O(p^{3/4})$ due to H.D. Kloosterman [3369] (see also [5379]).

Weil's results in [6615] imply the bound $O(\sqrt{p})$ (Weil [6617]). At the end of the seventies N.V. Kuznetsov [3595] and R.W. Bruggeman [781, 782] considered certain weighted mean values of Kloosterman sums and Kuznetsov used them to establish the bound

$$\left|\sum_{q=1}^{x} \frac{S(u,v;q)}{q}\right| \leq x^{1/6+\varepsilon} \tag{4.3}$$

for every $\varepsilon > 0$ (a simpler proof due to D. Zagier appears in H. Iwaniec's [3060] lectures). It was conjectured earlier by Yu.V. Linnik [3926] and A. Selberg [5618] that this sum is $O(x^\varepsilon)$ (the *Linnik–Selberg conjecture*). Kuznetsov's bound showed its power in several arithmetical problems. In particular J.-M. Deshouillers and H. Iwaniec used it in [1489, 1490] to obtain

$$\sum_{n=1}^{x} d(n)d(n+1) = xP(\log x) + O(x^c)$$

[2]Leonard Carlitz (1907–1999), professor at Duke University. See [2620].

for every $c > 2/3$, $P(X)$ being a quadratic polynomial, and to show that for infinitely many n the number n^2+1 has a prime factor exceeding $n^{6/5}$ and $n^{1.202}$.

For generalizations see J.-M. Deshouillers, H. Iwaniec [1490], D. Goldfeld, P. Sarnak [2262] (cf. also P. Sarnak's book [5407]), N.V. Proskurin [5013] (the preprint of the last paper appeared in 1980, see the review [3574] in Math. Reviews).

A survey covering evaluations of exponential sums and their applications in number theory was given by L.K. Hua in 1959 [2938].

4.1.2 Zeta-Function, L-Functions and Primes

1. The exceptional role played by the zero values of the zeta-function has been underlined by the result by H. Bohr and B. Jessen [581, 582] who showed in 1930–1932 that for every a every sufficiently large rectangle of the form

$$\{\sigma + it : \alpha \le \sigma \le \beta,\ |t| \le T\}$$

contains points s with $\log \zeta(s) = a$, and the number of such points is asymptotic to $c(a)T$ with some positive $c(a)$.

For polynomials of large degrees the evaluations of the sums (2.68) made by H. Weyl and J.G. van der Corput were essentially improved by I.M. Vinogradov [6420, 6424] in 1935–1936, and this led to an enlargement of the zero-free region of Riemann's zeta-function.

First N.G. Čudakov [6094] showed in 1936 that $\zeta(s)$ does not vanish in the region

$$\left\{\sigma + it : \sigma > 1 - \frac{a}{|\log t|^c}\right\} \tag{4.4}$$

for every $c > 10/11$, and this implied

$$|\pi(x) - \mathrm{li}(x)| \ll x \exp\left(-c \log^d x\right)$$

for any $d < 8/15$, and then [1285] improved his result to any $d < 4/7$. In 1938 E.C. Titchmarsh [6177] simplified Čudakov's argument and replaced $10/11$ in (4.4) by $4/5$ which allowed d to be any number below $5/9$.

Further development of Vinogradov's method led in 1958 to the non-vanishing of $\zeta(s)$ first for

$$\sigma > 1 - \frac{C}{(\log t \log\log t)^{2/3}}$$

(I.M. Vinogradov [6441]), and then for

$$\sigma > 1 - \frac{C}{\log^{2/3} t (\log\log t)^{1/3}}, \quad t \ge 2,$$

(N.M. Korobov [3487], I.M. Vinogradov [6442]; for a simpler proof see Y. Motohashi [4458]), and this implied

$$|\pi(x) - \mathrm{li}(x)| \ll x \exp\left(-c \log^a x (\log\log x)^{-b}\right) \tag{4.5}$$

valid with $a = 3/5$, $b = 2/5$ [3486] and $a = 3/5$, $b = 1/5$ [3487, 3488]; the asserted $b = 0$ has never been confirmed. This forms the best known unconditional evaluation of the error term in the Prime Number Theorem.

The best known evaluation of C is $C \geq 0.0173\ldots$ obtained in 2002 by K. Ford [2031, 2032], and this leads to the value $c = 0.2098$ in (4.5). Earlier $c \geq 0.001$ had been obtained by Y.F. Cheng [1041].

A very detailed exposition of Vinogradov's method can be found in his books [6428, 6436, 6447, 6448].

2. Phragmén's [4840] result on the sign changes of the difference

$$\Delta(x) = \psi(x) - x$$

obtained a quantitative form in 1930, when G. Pólya [4961] used his results on singularities of power series obtained in [4960] to prove evaluations of the number of sign changes for a large class of functions. His results implied in particular that the number $\omega(T)$ of sign changes of $\Delta(x)$ in the interval $[2, T]$ satisfies

$$\limsup_{T\to\infty} \frac{\omega(T)}{\log T} \geq \frac{\gamma_0}{\pi},$$

with $\gamma_0 = 14.134\ldots$ being the imaginary part of the first zero of the zeta-function on the critical line. Assuming the Riemann Hypothesis , G. Pólya showed[3] later [4962] that lim sup in this inequality can be replaced by lim inf (with another constant in place of γ_0).

Improvements to the value of this constant were provided later by B. Szydło [6033–6036].

It was shown much later by H.L. Montgomery and U.M.A. Vorhauer [4371] that the Riemann Hypothesis implies the existence of sign changes of $\Delta(x)$ in every interval $(x, 19x)$, which is best possible, as in the interval $[1, 19]$ this difference has a constant sign.

Some generalizations of Pólya's method were proved later by E. Grosswald [2362, 2363].

In 1935 A.E. Ingham [3020] considered the function $\nu(T)$, giving the number of sign changes of the difference $\pi(x) - \mathrm{li}(x)$ in the interval $[2, T]$, and proved that if there is a number σ_0 such that $\zeta(s)$ does not vanish in the half-plane $\Re s > \sigma_0$, and there is a zero on the line $\Re s = \sigma_0$, then for large T one has

$$\nu(T) \geq c \log T, \tag{4.6}$$

with a certain $c = c(\sigma_0) > 0$. He also deduced from the Riemann Hypothesis the existence of a constant B such that $\pi(x) - \mathrm{li}(x)$ changes its sign in every interval (x, Bx).

It was pointed out by J. Kaczorowski [3179–3182] that Ingham's approach leads to the same result for the differences

$$\Delta(x), \quad \Pi(x) - \mathrm{li}(x), \quad \text{and} \quad \theta(x) - x,$$

where $\Pi(x)$ is defined by (2.16).

[3] A gap in Pólya's proof was filled by J. Steinig [5921] in 1969.

For the first unconditional lower bound for $\nu(T)$ there was a 25 year wait, until S. Knapowski [3377, 3378] obtained for large T the ineffective bound

$$\nu(T) \gg \log\log T,$$

as well as the effective bound

$$\nu(T) \geq e^{-35} \log\log\log\log T.$$

He also proved [3377]

$$\omega(T) > \frac{\log\log T}{\log 3} + O(1),$$

After further steps made by N. Levinson [3865], S. Knapowski, P. Turán [3397, 3398] and J. Pintz [4892–4896], finally J. Kaczorowski obtained in [3180] the inequality

$$\nu(T) \geq c \log T$$

for large T, with a certain ineffective constant $c > 0$, as well as the same assertion for the difference $\theta(x) - x$. In [3179] the inequality

$$\omega(T) \geq \frac{\gamma_0}{4\pi} \log T$$

was shown to hold for large T and the same assertion was proved for the number of sign changes of $\Pi(x) - \mathrm{li}(x)$. Later Kaczorowski [3187] obtained similar results for primes in residue classes, using his earlier study of functions

$$k(z,\chi) = \sum_{\Im\varrho>0} \exp(\varrho z), \qquad K(z,\chi) = \sum_{\Im\varrho>0} \frac{\exp(\varrho z)}{\varrho},$$

ϱ running over zeros of $L(s,\chi)$ [3183–3186]. In [3181] J. Kaczorowski considered the number $V(T)$ of sign changes in $[2,T]$ of the function

$$\sum_{n=1}^{\infty} (\Lambda(n) - 1) \exp\left(-\frac{n}{x}\right),$$

and obtained $V(T) \geq c \log T$ with some explicit c and showed that the Riemann Hypothesis implies $V(T) = 4c\log + O(1)$. In [3182] he established $V(T) = o(\log^2 T)$ unconditionally.

Similar results for the error term in the Prime Ideal Theorem were obtained by J. Kaczorowski and W. Staś [3208, 3209, 5914].

For oscillatory properties of other classes of functions see J. Kaczorowski, J. Pintz [3206, 3207], M. Radziejewski [5047].

3. In 1931 S. Ikehara [3008] applied the new method of constructing Tauberian theorems invented two years earlier by N. Wiener[4] [6660, 6661] to a proof of an important strengthening of Landau's Tauberian theorem [3630]. Landau's theorem in its simplest form asserted that if the series

$$F(s) = \sum_{n=1}^{\infty} a_n n^{-s} \tag{4.7}$$

[4]Norbert Wiener (1894–1964), professor at MIT. See [3862, 4166].

with non-negative coefficients converges in the half-plane $\Re s > 1$ and can be extended to a function regular in $\Re s \geq 1$, except for a simple pole at $s = 1$ with residue A, and moreover there is a constant k such that the bound

$$|F(s)| \ll |\Im s|^k$$

holds uniformly for $|\Im s| \to \infty$, then

$$\lim_{x\to\infty} \frac{1}{x} \sum_{n=1}^{\infty} a_n = A.$$

S. Ikehara showed that the growth condition in this theorem can be omitted. This enabled a rather simple proof of the Prime Number Theorem, applying Ikehara's theorem to the series

$$-\frac{\zeta'(s)}{\zeta(s)} = \sum_{n=1}^{\infty} \frac{\Lambda(n)}{n^s}.$$

Ikehara's proof of his result was rather involved, but soon much simpler proofs were found (S. Bochner[5] [569], H. Heilbronn and E. Landau [2716]).

The theorems of Landau and Ikehara did not allow evaluation of the error terms arising, but in 1939 A. Kienast[6] [3331] presented a version of Landau's theorem which did.

The problem of evaluating error terms in various Tauberian theorems was considered in the books by T.H. Ganelius [2189] and M.A. Subkhankulov [5984]. See also G. Tenenbaum [6105].

In 1945 A.E. Ingham [3025] deduced from Wiener's Tauberian theorem a new kind of Tauberian result which led to another proof of the Prime Number Theorem.

A generalization of Ikehara's theorem was obtained in 1954 by H. Delange[7] [1426]. It allows the function F defined by the series (4.7) with non-negative coefficients to have a more complicated singularity at $s = 1$. This result allows the main term to be obtained quickly in various arithmetic asymptotic formulas. Several applications were presented by H. Delange in [1427].

4. In all previous treatments of the Prime Number Theorem for progressions the evaluations of $\pi(x; k, l)$ were non-uniform with respect to the difference k of the progression considered. The first attempt to obtain uniform results was made by E.C. Titchmarsh in 1930 [6169] who observed that Brun's sieve leads to the inequality

$$\pi(x; k, l) \leq \frac{c(a)}{\varphi(k)} \frac{x}{\log x}, \tag{4.8}$$

[5] Salomon Bochner (1899–1982), professor in Princeton.

[6] Alfred Kienast (1879–1969), professor in Zürich.

[7] Hubert Delange (1913–2003), professor in Clermont-Ferrand and Paris-Sud.

with some positive constant $c(a)$, valid for every fixed a and $k < x^a$. The same method leads to a slightly more general inequality, valid for all $k < x$,

$$\pi(x;k,l) \le \frac{c}{\varphi(k)} \frac{x}{\log(x/k)}, \tag{4.9}$$

in which c is an absolute constant (the *Brun–Titchmarsh theorem*).

In the same paper [6169] E.C. Titchmarsh showed that for characters χ mod k the corresponding L-functions do not vanish in the region $|t| \ge 3$, $\sigma > 1 - c/\log(k|t|)$ (with some $c > 0$), and if χ is non-real, then the condition on t can be waived. He deduced from this the bound

$$\pi(x;k,l) = \frac{\mathrm{li}(x)}{\varphi(k)} + O\left(x \exp\left(-c\sqrt{\log x}\right)\right),$$

valid for $k \le \exp(\sqrt{\log x})$ with at most one exception, and deduced from the General Riemann Hypothesis the equality

$$\pi(x;k,l) = \frac{\mathrm{li}(x)}{\varphi(k)} + O\left(\sqrt{x}\log x\right).$$

The analogue of the Brun–Titchmarsh theorem in algebraic number fields was obtained in 1959 by G.J. Rieger [5218]. For stronger forms of the Brun–Titchmarsh theorem see Sect. 6.1.3.

5. Titchmarsh's paper [6169] also contains the proof of the bound

$$A(x) = \sum_{a<p\le x} d(p-a) = O\left(x \log\log x\right)$$

for $a \neq 0$. He deduced from the General Riemann Hypothesis the asymptotic equality

$$A(x) = E(a)x + O\left(\frac{x \log\log x}{\log x}\right) \tag{4.10}$$

with

$$E(a) = \frac{\varphi(a)}{a} \prod_{p\nmid a} \left(1 + \frac{1}{p(p-1)}\right),$$

and asserted that $A(x) \gg x/\sqrt{\log x}$ holds unconditionally, but his argument had to be corrected.

Much later Yu.V. Linnik [3924] used his dispersion method (see Sect. 6.2) to get an unconditional proof of (4.10) with the error term of the order $O(x/\log^a x)$ for every $a < 1$ (cf. B.M. Bredikhin [694]). A simpler deduction of this result (with the bound given in (4.10)) from the Bombieri–Vinogradov theorem (see Sect. 6.1.1) was provided by G. Rodriquez [5247]. A more general sum, with $d(n)$ replaced by $d_k(n)$ was treated by J.W. Porter [4993]. Asymptotics for the similar sum

$$\sum_{p\le n} d(n-p)$$

were found in 1966 by P.D.T.A. Elliott and H. Halberstam [1749].

It follows from Titchmarsh's results that if $p(k,l)$ denotes the least prime congruent to l mod k ($(k,l)=1$), then the bound $p(k,l) \ll k^{2+\varepsilon}$ for every $\varepsilon > 0$ is a consequence of the General Riemann Hypothesis. Slightly later S. Chowla [1076] observed that one can replace k here by $\varphi(k)$, and established unconditionally the inequality

$$p(k,l) < \exp\left(Ak^{3/2}\log^6 k\right)$$

with certain fixed A, in the case when $k \equiv 3 \pmod 4$ is a prime. In [1079] he obtained for all k the bound

$$p(k,l) \ll \exp\left(\log^B k\right)$$

with a certain $B > 1$.

The stronger bound

$$p(k,l) \ll \varphi(k)\log^{2+\varepsilon} k$$

holding for every $\varepsilon > 0$ and for $(1+o(1))\varphi(k)$ residues mod k was deduced from the General Riemann Hypothesis by P. Turán [6213, 6214] in 1937, who in [6214] showed that a weaker bound, $p(k,l) \ll \varphi(k)^c$, follows from the hypothetical nonexistence of zeros of the L-function in a fixed parallelogram

$$\{\sigma + it : 1 - \delta < \sigma < 1, |t| \le c\}$$

($0 < \delta < 1/2, c > 0$). For important further progress see Sect. 5.1.

6. In 1935 A. Page [4720] enlarged the zero-free region for L-functions to

$$|t| > 0, \qquad \sigma > 1 - \frac{c}{|t|\log k}$$

and one year later Čudakov [6093] showed that for large $|t|$ there are no zeros in

$$\sigma > 1 - \frac{c(k)}{\log^{\gamma}|t|} \tag{4.11}$$

with $c(k) > 0$ and a certain $\gamma < 1$, not depending on k. Page showed also that if χ is real, then $L(s,\chi)$ does not have real zeros in the interval $(1 - c_1/(\sqrt{k}\log^2 k), 1]$, with some absolute constant $c_1 > 0$. This led to the first uniform evaluation of the error term in the Prime Number Theorem for progressions. The bound obtained depends on the real zeros of L-functions.

Let $k \ge 3$ be an integer and $(k,l) = 1$. If $1/2 \le \sigma < 1$ and the product of L-functions associated with characters mod k *does not have real zeros $\varrho > \sigma$, then with a suitable $C > 0$ one has*

$$\left|\pi(x;k,l) - \frac{1}{\varphi(k)}\mathrm{li}(x)\right| \ll x\exp\left(-C\sqrt{\log x}\right) + \frac{x^{\sigma}}{\varphi(k)\log x}. \tag{4.12}$$

Later N.G. Čudakov [1288] showed that the number $c(k)$ in (4.11) does not depend on k, provided $|t|$ exceeds $\exp(\log^{1/\gamma} k)$, and T. Tatuzawa [6067] established the non-vanishing of $L(s,\chi)$ in

$$\sigma > 1 - \frac{c}{\log k + (\log|t|\log\log|t|)^{3/4}},$$

improving the error term in (4.12).

The search for positive real zeros of $L(s,\chi)$ was reduced to the case of real characters by H. Heilbronn [2714] who showed, more generally, that every real simple zero of a Dedekind zeta-function is a zero of $L(s,\chi)$ with quadratic χ.

7. The size of $L(1,\chi)$ for real χ is important in the study of the class-number $h(d)$ of the quadratic field K_d with discriminant d, due to the following relation found by Dirichlet [1586, 1587]:

$$h(d) = \begin{cases} \frac{w|d|^{1/2}L(1,\chi_d)}{2\pi} & \text{if } d<0, \\ \frac{d^{1/2}L(1,\chi_d)}{2\log\varepsilon} & \text{if } d>0, \end{cases} \tag{4.13}$$

where w is the number of roots of unity contained in K_d,

$$\chi_d(x) = \left(\frac{d}{x}\right) \tag{4.14}$$

is Kronecker's extension of the Legendre symbol, and $\varepsilon > 1$ is the fundamental unit of K_d. Note that every primitive real character mod k is of the form χ_d with $d=k$ or $d=-k$ (see, e.g., A. Walfisz [6531, Lemma 1]).

It was shown by J.E. Littlewood in 1928 [3945] that if χ is a non-principal real character mod k such that $L(s,\chi)$ does not vanish in the half-plane $\Re s > 1/2$, then

$$(1+o(1))\frac{\pi^2}{12e^\gamma\log\log k} < L(1,\chi) < (1+o(1))2e^\gamma\log\log k, \tag{4.15}$$

where γ is Euler's constant, and in the case of complex χ the right-hand side forms an upper bound for $|L(1,\chi)|$. In 1934 S. Chowla [1078] established unconditionally that for a non-principal real character mod k one has

$$L(1,\chi) = \Omega(\log\log k),$$

and in 1935 A. Page [4720] established the lower bound

$$L(1,\chi) \gg \frac{1}{\sqrt{k}} \tag{4.16}$$

for primitive real characters.

The trivial bound

$$|L(1,\chi)| < \log k + 2$$

which one gets by partial summation of $\sum_{n=1}^{N}\chi(n)/n$ was improved in 1942 for real characters by L.K. Hua [2931] who obtained

$$L(1,\chi) < 1 + \frac{\log k}{2}.$$

In 1942 Yu.V. Linnik [3905] and A. Walfisz [6259] proved

$$\limsup_{k\to\infty} \frac{L(1,\chi)}{\log\log k} \ge e^{\gamma},$$

(for a simpler proof see S. Chowla [1087]). They showed also that for infinitely many k the product $L(1,\chi)\sqrt{\log\log k}$ remains bounded, and this was made more precise by S. Chowla [1089] who proved that infinitely often one has

$$L(1,\chi) < \left(\frac{\pi^2}{6e^{\gamma}} + o(1)\right)\frac{1}{\log\log k}.$$

In 1964 S. Chowla [1093] proved for large primes p the bound

$$L(1,\chi) < c\log p$$

for every $c > 1/4$. Three years later D.A. Burgess [860] replaced $1/4$ by 0.2456, and in 1972 P.J. Stephens [5933] showed that this bound holds for every $c > 0.1968$. A simpler proof, holding also for all real primitive characters $\bmod k$ for composite k, was later given by J. Pintz [4889].

For most primes p one can get better bounds. So P.D.T.A. Elliott [1729] showed in 1969 that for all except $O(x^{\varepsilon})$ primes $p \le x$ and all non-principal characters $\chi \bmod p$ one has

$$\frac{1}{\log\log p} \ll |L(1,\chi)| \ll \log\log p.$$

8. A decisive step forward was taken by C.L. Siegel [5753] in 1935. Namely, he showed that for every $\varepsilon > 0$ and any non-principal real character $\chi \bmod k$ one has

$$L(1,\chi) > \frac{c(\varepsilon)}{k^{\varepsilon}} \tag{4.17}$$

for every $\varepsilon > 0$ with a certain positive $c(\varepsilon)$ (unfortunately the value of $c(\varepsilon)$ is ineffective). This inequality implies, in view of the class-number formula (4.13), the relation

$$\lim_{d\to-\infty} \frac{\log h(d)}{\log|d|} = \frac{1}{2},$$

$h(d)$ denoting the class-number of the quadratic field with discriminant d. For positive d Siegel's result implied

$$\lim_{d\to\infty} \frac{\log(R_d h(d))}{\log|d|} = \frac{1}{2},$$

where $R_d = \log\varepsilon_d$, $\varepsilon_d > 1$ denoting the fundamental unit of K_d.

A. Page noted in [4720] that a theorem by E. Landau [3659] (for which he gave a simplified proof) implies that among all L-functions associated to real primitive characters $\bmod k$ for $k \le x$ there is at most one which has a zero in the interval $(1 - c/\log k, 1)$ for any fixed $c > 0$. Such zeros, whose existence would contradict the General Riemann Hypothesis, are now called *Siegel zeros*. The inequality (4.17) was used by A. Walfisz [6531, Lemma 2], to show that for any Siegel zero ϱ the inequality

$$1 - \varrho > \frac{C(\varepsilon)}{k^{\varepsilon}} \tag{4.18}$$

holds with some $C(\varepsilon) > 0$. His argument shows, more generally, that with a certain absolute constant $A > 0$ one has

$$1 - \varrho > \frac{L(1,\chi)}{A \log^2 k}.$$

A. Walfisz utilized (4.18) [6531] to prove the equality

$$\pi(x;k,l) = \frac{\mathrm{li}(x)}{\varphi(k)} + O\left(x \exp\left(-c\sqrt{\log x}\right)\right),$$

valid uniformly for all $k \le \log^q x$, with q being arbitrary and the implied constants depending on q only. This result is usually called the *Siegel–Walfisz theorem.* Soon a simpler proof was found by H. Heilbronn [2710].

Later, other proofs of Siegel's theorem were found (see, e.g., S. Chowla [1090], N.G. Čudakov [1287], T. Estermann [1893], D. Goldfeld [2257], Yu.V. Linnik [3915], J. Pintz [4882, 4890], K. Ramachandra [5065] and K.A. Rodosskiĭ [5246]), but none of them is effective.

It was proved in 1951 by T. Tatuzawa [6066] that for small ε and $k > \exp(1/\varepsilon)$ one has

$$L(1,\chi) > \frac{0.655\varepsilon}{k^{\varepsilon}}$$

with the exception of at most one k (cf. T. Tatuzawa [6073]). This result was later consecutively improved by J. Hoffstein [2840], C.G. Ji, H.W. Lu [3129] and Y.G. Chen [1038], who obtained for $0 < \varepsilon < 1/(6\log 10)$ and $k > \exp(1/\varepsilon)$ the inequality

$$L(1,\chi) > \min\left\{\frac{1}{7.732\log k}, \frac{1.5 \cdot 10^6 \varepsilon}{k^{\varepsilon}}\right\}$$

with the exception of at most one k.

In 1983 D.R. Heath-Brown [2634] proved that if there are only finitely many twin primes, then for every k and every χ mod k one has $L(s,\chi) \neq 0$ for real $s \ge 1 - C/\log k$, hence the existence of Siegel zeros implies the infinitude of twin primes.

It was later shown by J.B. Friedlander and H. Iwaniec [2110–2112] that in certain questions concerning primes in progressions, primes in short intervals and primes in polynomials, the existence of Siegel zeros allows results to be obtained which are stronger than those obtained under the Riemann Hypothesis.

In 1966 H. Davenport [1376] showed

$$1 - \varrho > \begin{cases} c/(\sqrt{k}\log\log k) & \text{if } \chi \text{ is odd,} \\ c\log k/(\sqrt{k}\log\log k) & \text{if } \chi \text{ is even.} \end{cases}$$

The factor $\log\log k$ was later removed by D. Goldfeld, A. Schinzel [2263] and J. Pintz [4884].

9. In 1932 C.L. Siegel [5749] published an analysis of Riemann's notes concerning the zeta-function, preserved in the archive of Göttingen University, and provided proofs for several of Riemann's statements. In particular he showed that the number $N_0(T)$ of zeros ϱ of $\zeta(s)$ lying on the critical line and satisfying $0 < \Im\varrho \le T$ satisfies

$$N_0(T) \ge (c + o(1))T,$$

with $c = \frac{3}{8\pi}\exp(-3/2) = 0.026\ldots$. In 1934 R.O. Kuzmin [3592] replaced the 8π here by 2π, and in [3593] obtained a similar evaluation for the zeros of Dirichlet L-functions.

The constant in the last result was later improved by C.L. Siegel [5760], who also proved a similar result for Dirichlet L-functions and the Epstein zeta-function.

In 1942 A. Selberg [5606] obtained $N_0(T) \gg T \log\log\log T$ and in [5607] improved this to $N_0(T) \gg T \log T$, showing that a positive fraction of the roots of $\zeta(s)$ lie on the critical line. His proof was soon simplified by E.C. Titchmarsh [6180]. N. Levinson [3864] later showed that more than $1/3$ of zeros of $\zeta(s)$ lie on the critical line (for a simpler proof see J.B. Conrey, A. Ghosh [1211]), and this was pushed to 36.58% by J.B. Conrey [1201]. Levinson's method was used by R.J. Anderson [72] in 1983 to show that more than 35% of zeros lie on the critical line and are simple. Today it is known (J.B. Conrey [1202]) that this happens for at least 40.1% of zeros. H.L. Montgomery [4359] deduced from the Riemann Hypothesis that at least $2/3$ of zeros are simple (cf. A.Y. Cheer, D.A. Goldston [1007]) and, using both RH and a generalization of the Lindelöf conjecture to L-functions, J.B. Conrey, A. Ghosh and S.M. Gonek [1215] obtained this for at least 70% of zeros.

An analogue of Selberg's result for L-functions was obtained by Š.S. Parmankulov [4744] and T. Hilano [2780] generalized Levinson's result to this case. Later P.J. Bauer [362] proved that more than 35% of their zeros lie on the critical line and are simple.

Selberg's result was also extended by J.L. Hafner [2431] to a class of Dirichlet series associated with cusp forms.

The first to consider mean values of Dirichlet L-functions on the critical line was R.E.A.C. Paley [4724], who studied the sum

$$S(q,t) = \sum_{\chi \bmod q} |L(1/2+it,\chi)|^4$$

taken over non-principal characters χ and showed that it is bounded by $Cq^{1+\varepsilon}$ (with suitable constant $C = C(\varepsilon, t)$).

Further work on this topic has been done by Yu.V. Linnik [3922], P.X. Gallagher [2179] and M.N. Huxley [2970], who also considered higher even powers of $L(1/2+it,\chi)$ as well as of Hecke zeta-functions.

10. The first paper by P. Erdős [1771], who had a very important role to play in the development of number theory in years to come, contained a very simple elementary proof of Čebyšev's theorem stating that for certain $0 < A < B$ one has for $x \geq 2$

$$\frac{Ax}{\log x} < \pi(x) < \frac{Bx}{\log x}.$$

In 1936 A.O. Gelfond and L.G. Schnirelman[8] (see [973, pp. 285–288], cf. also [4362]) gave a proof based on polynomial approximations.

The hope that this approach may lead to an elementary proof of the Prime Number Theorem was destroyed in 1956 by D.S. Gorškov [2285] who showed that the level of approximation needed in the interval [0, 1] of the zero polynomial by non-zero polynomials with

[8]Lev Genrikhovič Schnirelman (1905–1938), professor in Moscow. See [6163]. Actually the modern transcription of his name is 'Šnirelman' but we shall adhere to the traditional one.

integral coefficients cannot be attained. An extension of this approach to the case of several variables was given by I.E. Pritsker [5012].

The analogous question for primes in certain arithmetical progressions was treated by R. Breusch [720], G. Ricci [5193, 5196] and P. Erdős [1774].

Explicit bounds for the functions $\pi(x)$, $\vartheta(x)$ and the nth prime were given by J.B. Rosser [5299] in 1941. These bounds were later made more precise by J.B. Rosser and L. Schoenfeld [5302, 5303, 5552], G. Robin [5238], J.-P. Massias and G. Robin [4185] and P. Dusart [1659, 1660]. See also H.G. Diamond, P. Erdős [1526] and N. Costa Pereira [1256] for elementary estimates.

Similar results for primes in arithmetical progressions with a small difference were provided by K.S. McCurley [4225, 4226], O. Ramaré and R.S. Rumely [5095] and P. Dusart [1661].

11. In 1934 H. Cramér [1272, 1273] proposed a probabilistic model of the behavior of primes and formulated a series of conjectures based on it. In particular his model led to

$$\limsup_{n\to\infty} \frac{p_{n+1} - p_n}{\log^2 p_n} = 1,$$

(p_n denoting the nth consecutive prime) and

$$\limsup_{x\to\infty} \frac{\pi(x) - \mathrm{li}(x)}{(x \log\log x / \log x)^{1/2}} = \sqrt{2}.$$

The results obtained in 1985 by H. Maier [4098] show that the application of Cramér's model to the distribution of primes in short intervals may lead to incorrect results. Other discrepancies between the model and reality were found in 2007 by J. Pintz [4907].

12. In 1939 J.G. van der Corput [6292] established that there are infinitely many triplets of primes forming an arithmetic progression.

Much later E. Grosswald [2365] proved that the number of such triplets $p_1 < p_2 < p_3 \le x$ equals

$$\frac{1}{2} \prod_{p\neq 2} \left(1 + \frac{1}{(p-1)^2}\right) \frac{x^2}{\log^3 x} + O\left(\frac{x^2}{\log^4 x}\right).$$

It was shown later by A. Balog [300, 301] that for all $m \ge 2$ the number of m-tuples $p_1, \dots, p_m$ of primes such that $(p_i + p_j)/2$ is prime for $1 \le i < j \le m$ exceeds

$$B \frac{x^m}{\log^{m(m+1)/2} x}$$

with some positive B, and for $m = 3$ equals

$$(c + o(1)) \frac{x^3}{\log^6 x}$$

with

$$c = \frac{27}{16} \prod_{p\ge 5} \frac{p^3(p^2 - 5p + 7)}{(p-1)^5}.$$

The question of the existence of arbitrarily long progressions of primes remained open for many years, until it got a positive answer in the work by B. Green and T. Tao [2334, 2335] in 2004 (for an exposition of the proof see [3508]). They also obtained an analogue of Grosswald's result for progressions of four terms and a lower bound for the number of k-term progressions formed by primes $\leq x$.

Later B. Green [2333] went one step further by showing that every subset of primes with a positive upper $(x/\log x)$-density contains an arithmetical progression of three elements, and in a paper with Tao [2336] showed that the same applies to the set of primes p such that $p+2$ is either prime or a product of two primes.

Recently T. Tao and T. Ziegler [6050] made a far reaching generalization, proving that if $P_1, \dots, P_k$ are polynomials with integral coefficients, vanishing at 0, then there exist infinitely many $m, n > 0$ such that all numbers $P_i(n) + m$ are primes.

13. The Prime Number Theorem was put into a more general context by A. Beurling [494]. He considered multiplicative semigroups S of real numbers in $[1, \infty)$ (their elements are called *generalized primes*) and obtained asymptotical results for the number $P(x)$ of generators $\leq x$, provided the asymptotic behavior of the number $N(x)$ of elements of S below x is known.

A further generalization was made by W. Forman and H.N. Shapiro [2041], who considered abstract semigroups partitioned into a finite number of classes and obtained analogues of the Prime Number Theorem for progressions. Another approach to this question was made by B. Nyman [4645], who showed that the conditions

$$N(x) = cx + O\left(\frac{x}{\log^n x}\right) \quad \text{for all } n \quad (c > 0)$$

and

$$P(x) = \mathrm{li}(x) + O\left(\frac{x}{\log^n x}\right) \quad \text{for all } n$$

are equivalent.

For further results concerning the relation between $N(x)$ and $P(x)$ see S.A. Amitsur[9] [67], A.E.P. Balanzario [272], H.G. Diamond [1521, 1522, 1524], A.S. Faĭnleĭb [1952], R.S. Hall [2480], J.-P. Kahane [3216, 3217], P. Malliavin [4115] and H. Müller [4471].

Many results concerning additive and multiplicative functions were generalized to arithmetical semigroups. See, e.g., F. Halter-Koch, R. Warlimont [2485], K.-H. Indlekofer, E. Manstavičius [3014, 3015], L.G. Lucht, K. Reifenrath [4032], R. Warlimont [6566], W.-B. Zhang [6826–6828].

The analogue of Linnik's theorem on the smallest prime in progressions in this context was established by E. Fogels[10] [2022]. An exposition of the analytic theory of these semigroups, usually called *arithmetic semigroups*, was given in the books by J. Knopfmacher[11] [3414–3416], the last written with W.-B. Zhang.

---

[9] Shimshon Avraham Amitsur (1921–1994), professor in Jerusalem. See [1167].

[10] Ernests Fogels (1910–1985), professor in Riga. See [3551].

[11] John Knopfmacher (1937–1999), professor in Johannesburg. See [5583, 6567].

4.1.3 Other Questions

1. The first progress in a question concerning primitive roots, posed by E. Artin in 1927, was made by H. Bilharz [505] in 1937. Artin conjectured[12] that if a is an integer which is neither -1 nor a square, then there are infinitely many primes p such that a is a primitive root mod p, and the number $N_a(x)$ of such $p \le x$ equals

$$(c(a)+o(1))\frac{x}{\log x}, \tag{4.19}$$

with a certain positive constant $c(a)$.

Artin also gave a heuristical argument in support of his conjecture, reproduced in Bilharz's paper. This argument leads to the assertion that $a \neq \pm 1$ is a primitive root mod p if and only if for every prime q the ideal generated by p in the field $K = \mathbf{Q}(\zeta_q, a^{1/q})$ is not a product of $[K:\mathbf{Q}]$ distinct prime ideals in the ring of integers of K. A similar statement holds also for primitive roots in rings of integers of finite extensions of $\mathbf{Q}$ and $\mathbf{F}_q(X)$. In the last case Bilharz showed that the analogue of Artin's conjecture is a consequence of the Riemann Hypothesis for zeta-functions of curves, which was later established by A. Weil [6615]. Actually some weaker bounds for the zeros of zeta-functions are sufficient, and as they were proved in 1939 by H. Davenport [1355], Bilharz's assertion became a theorem.

The value of $c(a)$, proposed by Artin, was

$$A = \prod_p \left(1 - \frac{1}{p^2 - p}\right),$$

but it was later modified as it did not conform with numerical experiments (see D.H. Lehmer, E. Lehmer [3794], P. Stevenhagen [5943]). Now it is believed that if a is not a proper power, and $a = a_1 a_2^2$ with a_1 square-free, then

$$c(a) = \begin{cases} A & \text{if } 4 \nmid a_1 - 1, \\ A(1-\mu(|a_1|))\prod_{p|a_1} \frac{1}{p^2-p-1} & \text{otherwise.} \end{cases}$$

When a is a proper power, the formula is more complicated.

In 1967 C. Hooley [2862] succeeded in deducing Artin's conjecture from the General Riemann Hypothesis. His proof gave the error term in (4.19) as $O(x \log\log x / \log^2 x)$. It was later proved by A.I. Vinogradov [6399] that the inequality

$$N_a(x) \le (c(a)+o(1))\frac{x}{\log x}$$

can be deduced without the use of any unproved assumptions.

An approximation to Artin's conjecture was obtained in 1967 by P.X. Gallagher [2179] who applied the large sieve to show that the number of integers lying in an interval of length T which are not primitive roots for any prime is $O(\sqrt{T}\log T)$, and the exponent of the logarithm was later reduced by R.C. Vaughan [6344]. This has been carried over to algebraic number fields by J. Hinz [2819].

---

[12] According to [505] Artin communicated this conjecture to H. Hasse on 13 September 1927.

It was proved in 1969 by P.J. Stephens [5932] that Artin's conjecture is true in mean by showing that if M exceeds $\exp(4(\log x \log\log x)^{1/2})$, then

$$\frac{1}{M}\sum_{a\le M} N_a(x) = \prod_p \left(1 - \frac{1}{p^2-p}\right)\mathrm{li}(x) + O\left(\frac{x}{\log^c x}\right)$$

with some $c > 1$.

Much later, in 1984, R. Gupta and M.R. Murty [2393] invented an approach which led to the assertion that for any three distinct primes p, q, r the lower bound

$$N_a(x) \gg \frac{x}{\log^2 x} \tag{4.20}$$

holds for some $a = p^\alpha q^\beta r^\gamma$ with $\alpha, \beta, \gamma \le 3$. This was improved in 1986 by D.R. Heath-Brown [2640], whose result implies in particular that the bound (4.20) can fail for at most two prime values of a. This led to a peculiar situation, as now we know that at least one of the primes 2, 3, 5 is a primitive root for infinitely many primes but we do not know which.

A generalization of Artin's conjecture for algebraic number fields was presented in 1977 by H.W. Lenstra, Jr. [3821], who also showed that it is a consequence of the General Riemann Hypothesis (cf. H. Roskam [5289–5291]). For analogues of the Gupta–Murty result see J. Cohen [1147, 1148] and W. Narkiewicz [4539].

2. A positive integer n is called *k-full* if for every prime $p|n$ one has $p^k|n$. The number $A_k(x)$ of k-full integers $n \le x$ was studied in 1934 by P. Erdős and G. Szekeres [1861], who proved

$$A_k(x) = c_k x^{1/k} + O\left(x^{1/(k+1)}\right)$$

with certain $c_k > 0$.

In 1958 P.T. Bateman, E. Grosswald [355] added one term to this equality by establishing

$$A_k(x) = c_k x^{1/k} + d_k x^{1/(k+1)} + O(\Delta_k(x)) \tag{4.21}$$

with $\Delta_k(x) = O(x^{1/(k+2)})$ for $k \ge 3$ and

$$\Delta_2(x) = O\left(x^{1/6}\exp\left(-c\log^{4/7} x(\log\log x)^{-3/7}\right)\right). \tag{4.22}$$

They showed also that a bound $\Delta_2(x) = O(x^a)$ with $a < 1/6$ would imply the non-vanishing of $\zeta(s)$ in a half-plane $\Re s > c$ with some $c < 1$. Progress in the evaluation of the error term in the Prime Number Theorem allowed replacement of the exponents 4/7 and $-3/7$ in (4.22) by 3/5, and $-1/5$, respectively (D. Suryanarayana, R. Sitaramachandra Rao [5994]). On the other hand R. Balasubramanian, K. Ramachandra and M.V. Subbarao obtained $\Delta_2(x) = \Omega(x^{1/10})$ [286].

Assuming the Riemann Hypothesis one can improve the bound (4.22) and the best known evaluation, $\Delta_2(x) = O(x^c)$ with $c = 12/85 = 0.1411\ldots$, is due to J. Wu [6758], who improved upon the earlier bounds by H.Q. Liu [3955], X. Cao [892, 893], W. Zhu and K. Yu [6830] and Y. Cai [881]. It has been conjectured that the Riemann Hypothesis implies $\Delta_2(x) = O(x^c)$ for every $c > 1/10$. This is true in mean, as D. Suryanarayana's [5993] bound

$$\int_1^x |\Delta(t)|dt = O\left(x^{1/10}\right)$$

shows.

In the case of $k \geq 3$ it is expected that

$$A_k(x) = \sum_{j=0}^{k-1} c_{j,k} x^{1/(k+j)} + O\left(x^{1/2k}\right). \tag{4.23}$$

In 1972 E. Krätzel [3515] added more terms to (4.21) without reaching (4.23), and six years later A. Ivić [3036] established the conjectured equality for $k = 3, 4$, and showed that the Lindelöf conjecture implies (4.23) for all k with a slightly worse error term $O(x^{1/2k+\varepsilon})$ for every $\varepsilon > 0$. In 1982 A. Ivić and P. Shiu [3048] established (4.23) for $k \leq 7$, obtaining even better bounds for $k = 3, 4$. Later A. Ivić [3039] proved the conjecture for $k = 8, 9, 10$. For improvements in the cases $k = 5, 6, 7$ see H. Menzer [4254].

For Ω-results see R. Balasubramanian, K. Ramachandra, M.V. Subbarao [286], T. Zhan [6824].

In the case $k = 4$ stronger results were later obtained by E. Krätzel [3516, 3518], H. Menzer [4253], G. Yu [6795], H.Q. Liu [3956], P. Sargos and J. Wu [5398]. The last authors proved the bound $O(x^{0.1}\log^9 x)$.

Let q_n be the sequence of all square-full numbers and put $\delta_n = q_{n+1} - q_n$. The result by P.T. Bateman, E. Grosswald [355] implies $\delta_n = O(q_n^{2/3})$, and one expects to have $\delta_n = O(n^{1/2+c})$ for every $c > 0$, which would be best possible. In 1984 P. Shiu [5714] showed that this holds with $c = 0.1526\ldots$, and the value of c has consecutively been reduced to $0.1507\ldots$ (P.G. Schmidt [5482]), $0.1490\ldots$ (C.H. Jia [3131]), $0.1425\ldots$ (H.Q. Liu [3951]), $0.1318\ldots$ (D.R. Heath-Brown [2647]), $14/107 = 0.1308\ldots$ (H.Q. Liu [3954]), $5/39 = 0.1282\ldots$ (M. Filaseta, O. Trifonov [2004]), $5/8$ (M.N. Huxley, O. Trifonov [2993]) and $0.1233\ldots$ (O. Trifonov [6200]).

P. Erdős's question of whether every large integer is the sum of three square-full numbers was answered in the positive by D.R. Heath-Brown [2644] in 1988.

3. The *distribution function* $F(t)$ of a real-valued arithmetic function f is defined by

$$F(t) = \lim_{x\to\infty} \frac{\#\{n \leq x : f(n) < t\}}{x}.$$

In Sect. 2.5.2 we have already noted I. Schoenberg's result [5546], implying the existence of a distribution function for Euler's φ-function and its logarithm, and in 1933 H. Davenport [1344] proved that $\sigma(n)/n$ also has a distribution function. This implied in particular that the set of *abundant numbers*, i.e., numbers n with $(\sigma(n)/n) > 2$, has a density. According to [1735] this answered a question from E. Bessel-Hagen[13] (cf. S. Chowla [1077] and P. Erdős [1772]).

It was shown in 1998 by M. Deléglise [1440] that the density of abundant numbers equals $0.247\ldots$.

An integer is called *primitive abundant* if it as abundant without having abundant proper divisors. Let $Ab(x)$ denote the number of such integers in $[2, x]$. The bounds

$$x\exp\left(-c_1\sqrt{\log x \log\log x}\right) \leq Ab(x) \leq x\exp\left(-c_2\sqrt{\log x \log\log x}\right),$$

with $c_1 = 8$, $c_2 = 1/25$ were obtained in 1935 by P. Erdős [1777].

---

[13] Erich Bessel-Hagen (1898–1949), worked at the universities of Halle and Bonn.

In 1985 A. Ivić [3040] showed that for large x one can take $c_1 = \sqrt{6}+\varepsilon$, $c_2 = 1/\sqrt{12}-\varepsilon$ with arbitrary $\varepsilon > 0$, and in 1996 M.R. Avidon [174] replaced these constants by $\sqrt{2}+\varepsilon$ and $1-\varepsilon$, respectively.

In 1935 P. Erdős [1778] proved the existence of distribution functions for all non-negative additive functions which assume distinct values at distinct primes, and in 1936 I. Schoenberg [5547] showed that this happens if for a real-valued additive function f the series

$$\sum_p \frac{(\log f(p))^+}{p}$$

converges absolutely, where for any function g one puts

$$g^+(n) = \begin{cases} g(n) & \text{if } |g(n)| \le 1, \\ 1 & \text{otherwise.} \end{cases}$$

Two years later P. Erdős [1779] showed that if both series

$$\sum_p \frac{f^+(p)}{p}, \qquad \sum_p \frac{(f^+(p))^2}{p}$$

converge, then the distribution function exists, and later in a joint work with A. Wintner [1869] showed that this condition is also necessary. In [1788] he produced examples of additive f with absolutely continuous or purely singular distribution functions.

A similar sufficient condition for the existence of a distribution function for $f(p+1)$, where f is a real-valued additive function and p runs over primes was obtained by I. Kátai [3269] in 1968, and later A. Hildebrand [2805] and N.M. Timofeev[14] [6164] established its necessity.

Necessary and sufficient conditions for additive functions f such that there exists a limit distribution of

$$\frac{1}{x}\#\{n \le x : f(n) - g(x) \le t\}$$

(with a suitable function $g(x)$) were given independently by H. Delange [1433], P.D.T.A. Elliott and C. Ryavec [1751] (for strongly additive functions, i.e., satisfying $f(p^k) = f(p)$ for prime p and $k \ge 1$), and B.V. Levin and N.M. Timofeev [3859].

4. The sum of values of Euler's function in an interval has been considered already by F. Mertens [4257] in 1874, who obtained for the difference

$$E(x) = \sum_{n\le x} \varphi(n) - \frac{3}{\pi^2}x^2$$

the bound

$$E(x) \ll x \log x.$$

[14] Nikolai Mikhailovič Timofeev (1944–2002), professor in Vladimir. See [118].

In 1930 S.S. Pillai and S. Chowla [4873, 4874] established $E(x) = \Omega(x \log\log\log x)$ and

$$\sum_{n \le x} E(n) = cx^2 + R(x)$$

with $c = 3/2\pi^2$ and $R(x) = o(x^2)$. Later S. Chowla [1071] proved that the integral $\int_2^x E^2(t)\,dt$ is asymptotic to $x^3/6\pi^2$.

Much later P. Erdős and H.N. Shapiro [1858] obtained $E(x) = \Omega_{\pm}(x \log\log\log\log x)$, disproving a conjecture by H. Mellin, who in [4241] expressed the belief that $E(x) = O(x)$. In 1987 H.L. Montgomery [4361] showed

$$E(x) = \Omega_{\pm}\left(x\sqrt{\log\log x}\right).$$

His method was later applied to other arithmetic functions by J. Herzog, R.P. Smith [2773].

An elementary proof was later provided by S.D. Adhikari, R. Balasubramanian, A. Sankaranarayanan [16]. These results imply that $E(x)$ changes its sign infinitely often, and it was shown by Y.-F.S. Pétermann [4797, 4798] in 1986, confirming a conjecture of P. Erdős, that the number of these changes in $[2, x]$ equals $(C + o(1))x$ with a certain $C > 1.5$. In [4798] the same question was considered for other arithmetical functions.

In 1963 A. Walfisz [6535] established the bound

$$E(x) \ll x \log^{2/3} x \log\log^{4/3} x,$$

improving upon his previous bounds

$$E(x) \ll x \log^{3/4} x \log\log^{a} x$$

with $a = 2$ and $a = 3/2$ [6261, 6533].

The next improvement was obtained in 1972 by D. Suryanarayana and R. Sitaramachandra Rao [5995]:

$$R(x) \ll x^2 \exp\left(-c \log^{3/5} x \log\log^{-1/5} x\right),$$

with a certain positive c. They proved also that the Riemann Hypothesis implies $R(x) \ll x^a$ for every $a > 9/5$, and in 1981 P. Codecá [1134] showed that the bound $R(x) \ll x^{3/2+\varepsilon}$ for every $\varepsilon > 0$ is equivalent to RH.

Sylvester's conjecture [6010], asserting the positivity of $E(n)$, was shown to be false by M.L.N. Sarma [5404] in 1936, who noted that $E(820) < 0$.

The result by P. Erdős and H.N. Shapiro [1858] shows that $E(n)$ is negative for infinitely many n.

5. In 1932 D.H. Lehmer [3776] posed the question of the existence of a composite number N with $\varphi(N) | N - 1$ (*Lehmer conjecture*) and showed that such a number must have at least 7 distinct prime divisors.

The first improvement of Lehmer's result occurred in 1970, when M. Lieuwens [3889] showed that if L denotes the set of all integers violating Lehmer's conjecture, then any member of L should have at least 11 prime divisors, and this was improved to 13 by M. Kishore [3342] in 1977 and to 14 by G.L. Cohen and P. Hagis, Jr. [1136] in 1980. If the ratio $(N-1)/\varphi(N)$ is $\neq 1, 2$, then $\omega(N) \ge 1991$ and $N > 10^{8171}$ (P. Hagis, Jr. [2437]).

For the number $L(x)$ of elements of L not exceeding x the results of P. Erdős [1815] imply

$$L(x) \ll x \exp\left(-c\frac{\log x \log\log\log x}{\log\log x}\right).$$

This was later improved by C. Pomerance [4968] to

$$L(x) \ll x^{2/3} \log^{1/3} x$$

and ([4969])

$$L(x) \ll \frac{x^{1/2} \log^{3/4} x}{(\log\log x)^{1/2}}.$$

The exponent of the logarithm was replaced in 1985 by $1/2$ (Z. Shan [5671]), and recently W.D. Banks and F. Luca [315] established

$$L(x) \ll \sqrt{x \log\log x}.$$

There can only be finitely many integers in L with a given number of prime divisors (C. Pomerance [4968, 4969]).

6. In 1935 P. Erdős [1781] considered the number $N(x)$ of values attained in the interval $[2, x]$ by Euler's φ-function and showed

$$\frac{x \log\log\log x}{\log x} \ll N(x) \ll \frac{x}{\log^a x}$$

for every $a < 1$, improving an earlier upper bound of S.S. Pillai [4860], who had $a = \log 2/e = 0.2549\ldots$.

Ten years later P. Erdős [1795] obtained the lower bound

$$N(x) \gg \frac{x}{\log x} \log\log x.$$

Much later he returned to this subject and in a joint paper with R.R. Hall [1832] improved his previous upper bound to

$$N(x) \ll \frac{x}{\log x} \exp\left(B\sqrt{\log\log x}\right)$$

with a certain $B < 4.1$. The lower bound was subsequently improved by them [1833] to

$$N(x) \gg \frac{x}{\log x} \exp\left(A(\log\log\log x)^2\right)$$

with some $A > 0.36$, and C. Pomerance [4979] showed in 1986 that this expression (with another constant A) also provides a lower bound. Two years later H. Maier and C. Pomerance [4101] obtained the equality

$$N(x) = \frac{x}{\log x} \exp\left((c + o(1))(\log\log\log x)^2\right)$$

with an explicit c, and the exponent was made more precise by K. Ford [2028].

P. Erdős showed also in [1781] that for the number $N(m)$ of solutions of the equation $\varphi(x) = m$ one has infinitely often $N(m) > m^c$ with a certain positive c. Cf. P. Erdős [1783, 1784], S. Chowla [1082].

It was shown in 1979 by K. Wooldridge [6720] that here c can take any value less than $(3-2\sqrt{2})=0.1715\ldots$, C. Pomerance [4971] showed the next year that any $c \le 0.55655\ldots$ will do, and in 1998 R.C. Baker and G. Harman [266] showed this for $c \le 0.7039$. On the other hand one has

$$N(m) \le m \exp\left(-(1+o(1))\frac{\log m \log\log\log m}{\log\log m}\right)$$

(C. Pomerance [4971]).

An algorithm for the determination of all solutions of the equation $\varphi(x)=m$ was given by S. Contini, E. Croot and I.E. Shparlinski [1221].

7. The question of the mean value of the number $R(n)$ of representations of n as a product of factors ≥ 2 has been considered by L. Kalmár [3219], who obtained

$$\sum_{n\le x} R(n) = cx^{\varrho} + O\left(x^{\varrho}\exp(-a\log\log x\log\log\log x)\right)$$

with $c=-1/\varrho\zeta'(\varrho)$ and some $a>0$, $\varrho>0$ being defined by $\zeta(\varrho)=2$.

A more general result was obtained in 1937 by E. Hille[15] [2810], who showed that if $\mathcal{P}$ is a sequence of primes, and $A=\{a_1<a_2<\cdots\}$ is the sequence of all integers having their prime factors in $\mathcal{P}$, then for the number $R_A(n)$ of representations

$$n = a_{i_1}\cdots a_{i_k}$$

one has

$$\sum_{n\le x} R_A(n) = (c+o(1))n_P^{\varrho},$$

where ϱ_P is defined by the equality

$$\sum_{n=1}^{\infty}\frac{1}{a_n^{\varrho}} = 2.$$

He showed also $R(n) \ll n^{\varrho}$.

In 1941 P. Erdős [1790] showed that Hille's assertion holds for any integer sequence a_j for which the series

$$\sum_{j=1}^{\infty}\frac{1}{a_i^{1+\varepsilon}}$$

converges for every $\varepsilon>0$, provided not all a_i's are powers of a_1.

In 1972 P. Erdős and P. Turán [1868] showed that almost all factorizations of integers $n\le x$ into distinct factors consist of

$$\left(\frac{2\sqrt{3}\log 2}{\pi}+o(1)\right)\sqrt{x}$$

[15] Einar Carl Hille (1894–1980), professor in Princeton and at Yale. See [1652].

terms. This is a special case of a more general result dealing with the number of solutions of

$$\sum_{i=1}^{k} \lambda_{j_i} \leq x \quad (k = 1, 2, \ldots)$$

with distinct λ_{j_i}, where the λ_j form a sequence of positive real numbers tending to infinity. For $\lambda_j = \log j$ one gets the result of P. Erdős and P. Turán, and for $\lambda_j = j$ one regains an earlier result of P. Erdős and J. Lehner [1841].

The bound for $R(n)$ proved by Hille was improved first to $R(n) < n^{\rho}$ (B. Chor, P. Lemke, Z. Mador [1060]) and then to

$$R(n) < n^{\varrho} \exp\left(-\frac{\log^{1/\varrho} n}{(\log\log n)^{c}}\right)$$

for every $c > 1$ and large n (M. Klazar, F. Luca [3348]). The last paper also contains a lower bound holding for infinitely many n, improved later to

$$R(n) > n^{\varrho} \exp\left(-a\frac{\log^{1/\varrho} n}{\log\log n}\right)$$

(with some $a > 0$) by M. Deléglise, M.-O. Hernane and J.-L. Nicolas [1442], who also improved the upper bound to

$$R(n) < n^{\varrho} \exp\left(-b\frac{\log^{1/\varrho} n}{\log\log n}\right)$$

(with some $b > 0$). See also M.-O. Hernane, J.-L. Nicolas [2767].

For a survey see A. Knopfmacher, M.E. Mays [3413].

4.2 Additive Problems

4.2.1 The Waring Problem

1. The year 1934 brought an important leap forward for Waring's problem. First a simplification in the study of major arcs was made by M. Gelbcke [2221], allowing R.D. James [3100] to halve Landau's evaluation (3.23) for the value of $g(k)$, then James [3099] improved the bounds for $G(k)$ in the case of odd k (showing in particular $G(5) \leq 35$, $G(7) \leq 164$ and $G(9) \leq 824$), and finally as a great surprise came the achievement of I.M. Vinogradov [6416–6419], who introduced a new idea, permitting him to obtain the bound $G(k) = O(k \log k)$, which compares well with previous evaluations, which were exponential in k. Instead of evaluating the integral counting all representations of an integer as the sum of s kth powers he considered the number of solutions of the equation

$$N = \sum_{j=1}^{4m-2} x_j^k + u_1 + u_2 + y^k u_3,$$

where $m = [2k\log k + k\log 6]$, and the u_i's are sums of at most $O(k\log k)$ kth powers lying in suitable intervals. This permitted him to get

$$G(k) \le 6k\log k + (\log 216 + 4)k < 6k\log k + 10k,$$

which essentially superseded the bound obtained by G.H. Hardy and J.E. Littlewood. In 1936 H.S. Zuckerman [6838] gave the explicit bound

$$\exp\left(18k2^{ck^2}\right)$$

(with $c = 14/3$) for the maximal number requiring more than $(6k\log k + 10k)$ kth powers.

An improvement to Vinogradov's bound was made in 1936 by L.E. Dickson [1562] (e.g., he got $G(13) \le 310$, whereas Vinogradov's bound gave $G(13) \le 321$, and the best previously known result was $G(13) \le 22\,639$). For small k the bounds for $G(k)$ were improved in 1937 by T. Estermann [1890], who, in particular, obtained $G(5) \le 29$ and $G(6) \le 42$.

A successive important step was taken in 1936 by I.M. Vinogradov, who first essentially improved known bounds for Weyl's sum in [6422], and then applied them to show [6427] that the asymptotic formula for the number of representations of an integer as the sum of s kth powers holds for $s > 10k^3\log k$.

The exponent 3 in the last expression was reduced to 2 in Vinogradov's book [6436, Chap. 7], and in his next book [6447, Chap. 6] Vinogradov showed the validity of the asymptotical formula for $s > k^2(2\log k + \log\log k + 2.6)$.

A careful exposition of Vinogradov's proof was presented by H. Heilbronn[16] [2709], who introduced important simplifications. His proof implies[17] that if $k \ge 9$, then every integer exceeding $n(k) = \exp(18k^6)$ is the sum of at most

$$6k\log k + (4 + 3\log(3 + 2/k))k + 3 \tag{4.24}$$

kth powers. In 1938 Vinogradov [6433] reduced his bound to

$$G(k) \le 4k(\log k + 2\log\log k + 3) \quad (k \ge 800).$$

Later I.M. Vinogradov [6436] obtained

$$G(k) < 3k\log k + 11k \quad (k \ge 3),$$

and

$$G(k) < 2k\log k + 4k\log\log k + 2k\log\log\log k + 13k \quad (k > 170000),$$

[6443, 6448]. This has been improved for small k by J.R. Chen [1012] to

$$G(k) < 3k\log k + 5.2k \quad (k \ge 3)$$

and by K. Thanigasalam [6131] to

$$G(k) < 3k\log k + 4.7k \quad (k \ge 3).$$

---

[16] See also the books by H. Davenport [1372] and R.C. Vaughan [6357].

[17] See, e.g., [4538, Chap. 4.2].

Later A.A. Karatsuba [3253] obtained

$$G(k) < 2k\log k + 2k\log\log k + 12k \quad (k > 4000),$$

and the current record holder is T.D. Wooley [6724], who in 1992 got

$$G(k) \le k(\log k + \log\log k + O(1)).$$

An elementary deduction of the bound $G(k) = O(k\log k)$ using Linnik's dispersion method was given by B.M. Bredikhin and T.I. Grišina [700].

2. The progress in the Waring problem achieved by Vinogradov resulted from a good bound for the integral

$$J_s^{(k)} = \int_{I^k} \left| \sum_{x\le T} \exp\left(a_1x + a_2x^2 + \cdots + a_kx^k\right) \right|^{2s} da_1\,da_2 \cdots da_k,$$

I denoting the unit interval.

This result is commonly called *Vinogradov's mean value theorem.*

For simplifications and improvements of Vinogradov's proof see L.K. Hua [2934], Yu.V. Linnik [3908], A.A. Karatsuba [3252] and S.B. Stečkin [5916].

Stronger evaluations of the integral $J_s^{(k)}$ were obtained later by T.D. Wooley [6725, 6726, 6730], and this was made explicit by K. Ford [2032] in 2002. The best known bounds were proved in 2011 by T.D. Wooley [6738].

The integral $J_s^{(k)}$ counts the number of solutions of the Diophantine system

$$x_1^j + x_2^j + \cdots + x_n^j = y_1^j + y_2^j + \cdots + y_n^j \quad (j = 1, 2, \ldots, k)$$

with $1 \le x_j, y_j \le T$, and hence is related to the old *Prouhet–Tarry–Escott problem*, which in its modern formulation asks for the determination of the minimal value $P(k,s)$ of n for which the system

$$\sum_{i=1}^n x_{i1}^m = \sum_{i=1}^n x_{i2}^m = \cdots = \sum_{i=1}^n x_{is}^m \quad (m = 1, 2, \ldots, k) \tag{4.25}$$

has a solution with the sets $\{x_{ir}\}$ being distinct. One defines also the number $L(k,s)$ as the smallest value of n for which (4.25) has a solution for which not all sums

$$\sum_{i=1}^n x_{ij}^{k+1} \tag{4.26}$$

are distinct. The number $W(k,s)$ is defined similarly, with all sums (4.26) being distinct.

This question had already appeared for the first time (in the case $k = s = 2$, $n = 4$) in the letter of 18 July 1750 from Goldbach to Euler (see [2168, letter 131] and [1909]). In 1851 E. Prouhet [5014] asserted $P(k,s) \le s^k$ and proved this for $k = 2$, $s = 3$. Proofs of this assertion were given later by D.H. Lehmer [3784] and E.M. Wright [6753, 6754]. Several authors produced examples of solutions of (4.25) and E.B. Escott in 1908 presented [1873] a complete solution in the case $k = 3$, $s = 2, n = 3$.

E.M. Wright [6746–6748] showed $L(k, 2) = O(k^4)$, and L.K. Hua [2922] improved this to $L(k, 2) = O(k^2 \log k)$.

For an early survey see H.L. Dorwart, O.E. Brown [1616].

In 1948 E.M. Wright [6751] proved $P(k, s) \le (k^2 + k + 2)/2$ and conjectured $P(k, s) = k + 1$. This has been established for $k = 2, 3, 5$ and every s (see A. Gloden [2250]) as well as for $s = 2$ and $k \le 9$ (cf. C.J. Smyth [5839]). Wright [6752] also proved $W(k, s) = O(k^2 \log(ks))$, and gave a bound for $L(k, s)$ of the order $O(k^2 \log k)$, improving upon the earlier bound $L(k, s) \le s^k$ proved by D.H. Lehmer [3784]. In the next year L.K. Hua [2935] got a bound of the order of $k^2 \log k$ for $W(k, s)$, not depending on s. Almost fifty years later T.D. Wooley [6730] improved Hua's bound in the case $s = 2$, proving

$$W(k, 2) \le \frac{k^2}{2}(\log k + \log\log k + O(1)),$$

and showing that infinitely often one has

$$W(k, 2) \le \frac{k^2 + k}{2} + 1.$$

In January 2011 T.D. Wooley [6738] obtained

$$W(k, s) \le k^2 + k - 2$$

for all $k, s \ge 2$.

It is an old question whether (4.25) has solutions (called *ideal solutions*) for $s = 2$, every n and $k = n - 1$. This was shown for $n = 7$ by G. Tarry [6052] in 1913 and J. Chernick [1044] gave parametric solutions for $n \le 7$.

It is now known that ideal solutions exist for every $n \le 12$ (see P. Borwein, P. Lisonék, C. Percival [654], A. Choudhry, J. Wróblewski [1062], A. Gloden [2250], C.J. Smyth [5839]).

The numbers $P(k, 2)$ also occur in certain questions related to polynomials with integral coefficients. See H.L. Dorwart [1615] and Z.A. Melzak [4244] on that topic. For a survey see P. Borwein and C. Ingalls [653].

3. A lower bound for the Waring constant $g(k)$ had already been provided by J.A. Euler[18] [1896], who observed that for every $k \ge 2$ one has

$$g(k) \ge I(k) := \left[\left(\frac{3}{2}\right)^k\right] + 2^k - 2,$$

because the number $2^k[(3/2)^k] - 1$ requires $I(k)$ kth powers, and it was conjectured in 1853 by C.A. Bretschneider [716] that for all k one has

$$g(k) = I(k). \tag{4.27}$$

[18] Johann Albrecht Euler (1734–1800), the eldest son of Leonhard Euler, professor of physics in St. Petersburg.

This assertion is called the *ideal Waring theorem*. The results of J.L. Lagrange [3610] and A. Wieferich [6656] confirmed Bretschneider's conjecture for $k = 2$ and 3.

In 1927 L.E. Dickson [1548] presented for the first time his method of ascent, which allowed him to show in [1556] that every integer below $21 \cdot 10^{10}$ is the sum of $I(15) = 33\,203$ 15th powers. He was also able to improve previous bounds on $g(k)$ for several small values of k. In particular he obtained $g(4) \le 35$ [1556], $g(5) \le 54$ [1555] and $g(6) \le 160$ [1556], the previous bounds being 37 (Wieferich [6658]), 58 (W.S. Baer [211]) and 183 (R.D. James, quoted in [1556]), respectively. The same approach led him in [1557, 1559, 1560] and [1566] to the inequalities $g(6) \le 110$, $g(7) \le 258$, $g(9) \le 981$ and $g(11) \le 4425$. Developing his method he was able to establish three years later the truth of (4.27) for $11 \le k \le 15$ and $k = 17$ [1562, 1563]. Later he [1564, 1565] also reduced the determination of the value of $g(k)$ for most k to a finite check by establishing the following theorem.

If $k \ge 35$ *and we put* $q = [(3/2)^k]$, $r = 3^k - 2^k q$ *and* $A = [(4/3)^k]$, *then*

$$g(k) = \begin{cases} I(k) & \text{if } r \le 2^k - q - 3, \\ I(k) + A & \text{if } r > 2^k - q - 1,\ qA + q + A = 2^k, \\ I(k) + A - 1 & \text{if } r > 2^k - q - 1,\ qA + q + A \ne 2^k. \end{cases} \tag{4.28}$$

This implied the truth of (4.27) for $7 \le k \le 180$. For $15 \le k \le 20$ this was independently achieved by H.S. Zuckerman [6838]. The first case of (4.28) was independently established by S.S. Pillai [4863–4865]. This was used by S. Chowla [1083] to show that the ideal Waring theorem holds for a set of integers of upper density 1.

The cases $r = 2^k - q$, $r = 2^k - q - 1$ and $r = 2^k - q - 2$ are not covered by Dickson's result, however the first case never occurs (R. Rubugunday [5335]), the second can be easily eliminated by congruence considerations, and in the last case the equality $g(k) = I(k)$ was later proved by I. Niven[19] [4620].

The equality $g(6) = I(6) = 73$ was established in 1940 by S.S. Pillai [4868], who earlier [4865] proved $g(6) \le 104$.

The remaining cases of small integers, $k = 4$ and $k = 5$, were settled much later. In 1964 J.R. Chen [1018] proved $g(5) = I(5) = 37$, and then $g(4) = I(4) = 19$ was shown to be true in 1986 by R. Balasubramanian, J.-M. Deshouillers and F. Dress [276, 277, 1485, 1486].

It is now known that the ideal Waring theorem holds for all sufficiently large k. This was established in 1957 by K. Mahler [4088], who used Ridout's extension [5213] of Roth's theorem [5311] to deduce that the inequality $r > 2^k - q - 2$ appearing in the statement of Dickson's theorem can happen only for finitely many k. Unfortunately Mahler's result does not give a limit after which the ideal Waring theorem would hold. It would be effective, if one would have

$$\left\| \left(\frac{3}{2}\right)^k \right\| \ge 2c^k \tag{4.29}$$

[19] Ivan Niven (1915–1999), professor at the University of Oregon.

with $c = 3/4$ for $k \geq k_0$ with an effective k_0, $\|\alpha\|$ denoting the distance of α to the nearest integer. In 1975 A. Baker and J. Coates [241] established this with c slightly larger than $1/2$, and in 1981 F. Beukers [487] proved that for $k \geq 5000$ the left-hand side of (4.29) exceeds c^k with $c = 0.5358$, and for large k got this with $c = 0.5637\ldots$. The last result was improved to $c = 0.5769$ by A. Dubickas [1627], to $c = 0.5770$ by L. Habsieger [2422] (he showed also that for $k \geq 5$ $c = 0.5743\ldots$ does the job), and recently V. Zudilin [6845] obtained $c = 0.5803$ for $k \geq K$ with an effective constant K (cf. Yu.A. Pupyrev [5016]).

Mahler's main result in [4088] showed that if $\Lambda(\alpha)$ denotes the set of limit points of the sequence $\|\alpha^n\|^{1/n}$, then for rational α one has $\Lambda(\alpha) = \{1\}$. Later Mahler and G. Szekeres [4095] proved that the last equality holds for almost all real α. See also P. Corvaja, U. Zannier [1255] and Y. Bugeaud, A. Dubickas [814].

Numerical checks, based on Dickson's theorem established the truth of (4.27) for $7 \leq k \leq 471\,600\,000$ (R.M. Stemmler [5923], J.M. Kubina, M.C. Wunderlich [3552]).

4. By 1922 H.W. Richmond[20] [5211] had already proved that every positive integer is the sum of three positive rational cubes (cf. H. Davenport, E. Landau [1392]). In 1938 H. Davenport [1349] considered such representations with positive integral cubes, and showed that the number of integers $\leq x$ which are sums of three such cubes exceeds x^a for every $a < c$ with $c = 13/15 = 0.8666\ldots$.

In 1950 H. Davenport [1363] returned to this subject and obtained $c = 47/54 = 0.8703,\ldots$, and in 1985 R.C. Vaughan [6359] got $c = 8/9$. The next steps were taken by C.J. Ringrose [5230] ($c = 113/127$) and again by R.C. Vaughan [6360, 6361] ($c = 19/21 = 0.9047\ldots$).

C. Hooley studied the number of representations of integers as sums of 4, 7 or 8 cubes [2888, 2889, 2899] (cf. also R.C. Vaughan [6358]). Upper bounds for the number of representations of an integer as the sum of four kth powers were obtained by C. Hooley [2888] for $k = 3$ (in [2889] an Ω-result was obtained), [2904] for $k = 4$ and J.M. Wisdom [6701] for $k = 5$.

5. Several variants of the Waring problem were also considered. We have already mentioned the Waring–Kamke problem, concerned with representations of integers as sums of polynomial values. Another problem concerns representations of integers n as a given number of kth powers which are close to each other. This has been considered by S. Chowla [1067, 1069] in the case of 4 squares and 8 cubes, and a far reaching generalization of this problem was treated in 1933 by E.M. Wright [6743]. He showed, using the Hardy–Littlewood circle method, that if $k \geq 3$, $s \geq (k-2)2^{k-1} + 5$, and λ_i $(i = 1, 2, \ldots, s)$ are positive numbers with unit sum, then every large integer n can be written in the form

$$n = \sum_{j=1}^{s} x_j^k, \tag{4.30}$$

with

$$\left|\lambda_j n - x_j^k\right| = o(n) \quad (j = 1, 2, \ldots, s).$$

---

[20] Herbert William Richmond (1863–1948), Fellow of King's College, Cambridge.

Earlier [6742] he considered the case $k = 2$, obtaining the same assertion for all integers having a large odd factor. In a later paper [6744] Wright proved that in the case $k \ge 3$ under the same assumptions the stronger condition

$$\left|\lambda_j n - x_j^k\right| \ll x^{1-\beta} \quad (j = 1, 2, \ldots, s)$$

holds with a certain positive β, and stated that in the case $k = 2$ and $s \ge 4$ this can be proved to hold for almost all integers n. He later presented a proof in [6749]. Cf. F.C. Auluck, S. Chowla [173], E.M. Wright [6750].

Another kind of distribution of solutions of (4.30) was studied by M. Laborde [3596].

The question of representing large integers n in the form $n = \sum_{j=1}^{s} a_j x_j^k$ with given non-negative integers a_j was considered in 1935 by R.E. Huston [2967], who showed that this is possible with $s \ge (k-2)2^{k-1} + 5$, provided n satisfies the necessary congruence conditions.

6. It seems that the first result concerning representations of integers as sums of distinct powers with fixed exponents is due to G.H. Hardy and J.E. Littlewood who showed in [2534] that almost all positive integers are sums of two squares and a cube. This was followed by the investigations of G.K. Stanley [5890, 5891] concerning sums of a square and s kth powers. She showed that every large integer is the sum of a square and at most 6 cubes and almost all integers are sums of a square and at most three cubes. She also obtained similar results for sums of one or two squares and a certain number of higher powers. Later H. Davenport and H. Heilbronn [1389, 1390] proved that almost all integers are of the form $x^2 + y^2 + z^k$ $(k \ge 3$, odd) and $x^2 + y^3 + z^3$.

G.L. Watson [6588] showed in 1972 that every large integer is the sum of a square and five cubes. Later R.C. Vaughan [6364] obtained a lower bound for the number of such representations of the expected order of magnitude, and T.D. Wooley [6736] proved an asymptotical formula valid for all except $O(x^{\varepsilon})$ integers $n \le x$. It was conjectured that every large integer is the sum of a square and two cubes, and, more generally, every large integer n satisfying the necessary congruence conditions, can be written in the form

$$n = \sum_{j=1}^{m} x_j^{k_j}, \tag{4.31}$$

provided $k \ge 2$, and

$$\sum_{j=1}^{m} \frac{1}{k_j} > 1.$$

This conjecture has been established only in the rather old case of three squares, but there are several results implying in various particular cases that (4.31) holds for almost all integers n, subject to the necessary conditions. In particular this holds for $m = 3$, which follows from the results of H. Davenport and H. Heilbronn quoted above and the later work of K.F. Roth [5305] and R.C. Vaughan [6354, 6357]. For $m = 4$ the same assertion was established for the following 4-tuples of exponents: $(2, 3, 3, 6)$ (S. Sastry, R. Singh [5409]),

(3, 3, 3, 3) (H. Davenport [1356]), (3, 3, 3, 4) (J. Brüdern [766, 767]), (3, 3, 3, 5) (M.G. Lu [4018]), and (2, 3, 5, k), (2, 3, 6, k) for any k (R.J. Cook [1240]).

In the case of two squares and three cubes stronger results are known. In 1972 Yu.V. Linnik [3928] proved that every large integer can be so represented, and in 1981 C. Hooley [2894] described the asymptotical behavior of the number $\nu(n)$ of representations, proving

$$\nu(n) = \frac{\pi}{27}\Gamma^3\left(\frac{1}{3}\right)\mathfrak{S}(n)n + O\left(\frac{n}{\log^c n}\right)$$

with some $c > 0$ and the singular series $\mathfrak{S}(n)$ being larger than a positive constant.

Several cases with a larger number of summands were considered by J. Brüdern [764], J. Brüdern and T.D. Wooley [778, 779], C. Hooley [2897, 2899], K. Kawada and T.D. Wooley [3288], H.W. Lu [4015], M.G. Lu and H.B. Yu [4019], K. Thanigasalam [6127, 6128] and G.L. Watson [6588].

A related problem consists in finding the smallest s with the property that every large integer can be written as

$$\sum_{j=1}^{s} x_j^{1+j}.$$

The first bound, $s \le 50$, was proved by K.F. Roth [5306] in 1951 and through the efforts of several authors it has been consecutively diminished. The current record is $s \le 14$, due to K. Ford [2026], previous results being due to K. Thanigasalam [6127] ($s \le 35$), R.C. Vaughan [6340, 6342] ($s \le 30$ and $s \le 26$), K. Thanigasalam [6130, 6132] ($s \le 22$ and $s \le 20$), J. Brüdern ([765], $s \le 18$; [768], $s \le 17$) and K. Ford [2025] ($s \le 15$). K.F. Roth proved also in [5306] that for almost all integers n one has

$$n = \sum_{j=2}^{5} a_j^j,$$

and the number of exceptional $n \le x$ is $O(x^c)$ for every $c > 9/10$. Later M.G. Lu [4016] got this for $c > 53/60$. In 1953 K. Prachar [5005] showed that for almost all even n this holds also with prime a_j's. This has been strengthened recently by X.M. Ren and K.M. Tsang [5162], who obtained for the number of exceptional $2n \le x$ the bound $O(x^a)$ with $a = 65/66$.

Another variant consists in finding $g_2(k)$, defined as the smallest s such that every positive integer can be written as $\sum_{j=1}^{s} x_j^{n_j}$ with $n_j \ge k$. In certain cases $g_2(k)$ was determined by M. Haberzetle [2421] in 1939, and one year later S.S. Pillai [4869] proved

$$g_2(k) = 2^k + \left[\frac{\log A}{\log 2}\right] - 1$$

for $k \ge 32$ with $A = [(3/2)^k]$.

7. Numbers which are sums of two kth powers were studied by P. Erdős and K. Mahler in 1938. They obtained [1843] the bound

$$N_k(x) := \sum_{a^k+b^k \le x} 1 \gg x^{2/k},$$

which improved bounds obtained by G.H. Hardy and J.E. Littlewood [2535, Lemma 22], E. Landau [3672] and S.S. Pillai [4859].

Much later C. Hooley [2860, 2893, 2895, 2896] obtained

$$N_k(x) = c(k)x^{2/k} + O\left(x^a\right)$$

for odd k and every $a > 5/3k$. He had treated the case of cubes earlier ([2856]; cf. T.D. Wooley [6729]). The case $k = 4$ was treated by G. Greaves [2330]. In 1990 W. Müller and W.G. Nowak [4475] obtained

$$N_k(x) = c_1(k)x^{2/k} + c_2(k)x^{1/k-1/k^2} + O\left(x^a \log^b x\right)$$

with $a = 7/11k$ and $b = 45/22$.

Sums of two kth powers of primes with prime k were treated by P. Shields [5689].

There is an old conjecture that for $k \geq 5$ there are no integers having more than one representation as the sum of two kth powers. Let $\nu_k(x)$ be the number of such integers $\leq x$. C. Hooley [2856] proved $\nu_3(x) \ll x^{2/3}$, later [2893] got $\nu_3(x) \ll x^c$ for all $c > 5/9$, and D.R. Heath-Brown [2653] obtained this for every $c > 4/9$. For larger k, C. Hooley [2895, 2896] showed $\upsilon_k(x) \leq x^c$ for $c > 5/3k$, and T.D. Browning [761] improved this to $c > 2/3k$ for $k \geq 27$ (cf. C.M. Skinner, T.D. Wooley [5803], D.R. Heath-Brown [2658]).

For the mean value of $\nu_k(x)$ one has in the case $k \geq 3$ the rather complicated formula (W.G. Nowak [4630])

$$\sum_{n \leq x^{k/2}} \nu_k(n) = \frac{2}{k}\frac{\Gamma^2(1/k)}{\Gamma(2/k)}x + \frac{8k^{1/k}}{\pi(2\pi)^{1/k}}\Gamma\left(1+\frac{1}{k}\right)\Phi_k(x)x^{1/2-1/2k} + P_k(x),$$

where $\Phi_k(x)$ is the sum of an explicit series with

$$|\Phi_k(x)| \leq \sum_{n=1}^{\infty} \frac{1}{n^{1+1/k}},$$

and the best known upper bound for the error term is

$$P_k(x) \ll x^{23/73} \log^{315/146} x,$$

obtained in 1990 by G. Kuba [3538, 3539]. The first lower bound was obtained in 1982 by L. Schnabel [5532] who got $P_k(x) = \Omega(x^{1/6})$, and this was improved to $P_k(x) = \Omega(x^{1/4})$ (E. Krätzel [3517, Chap. 3]) and $P_k(x) = \Omega_-((x\log x)^{1/4})$ (W.G. Nowak [4643]). For $k = 3$ one has

$$P_3(x) = \Omega_+\left((x \log\log x)^{1/4}\right)$$

(M. Kühleitner, W.G. Nowak, J. Schoissengeier, T.D. Wooley [3562]).

The old conjecture of Sylvester which states that every prime congruent to 4, 7 or 8 mod 9 is the sum of two rational cubes is still open. The only approximation to this assertion was obtained by P. Satgé [5410] who showed that if p is a prime congruent to 5 mod 9, then $2p^2$ is the sum of two rational cubes, and if $p \equiv 2 \bmod 9$, then $2p$ is such a sum.

8. In the *easier Waring problem* one looks for representations of integers N in the form

$$N = \pm x_1^k \pm x_2^k \pm \cdots \pm x_s^k$$

with integral x_i. The smallest value of s such that every N can be written in that form is usually denoted by $v(k)$, and we have clearly $v(k) \le g(k)$; however, the finiteness of $v(k)$ can be proved[21] directly, as noted by V. Veselý [6388] in 1933. A very simple argument was presented later by E.M. Wright [6745], who used the equality

$$\sum_{j=0}^{k-1}(-1)^{k-1-j}\binom{k-1}{j}(X+j)^k = k!X + a,$$

where $a = a(k)$ is an integer, not depending on X (see, e.g., [2545, Theorem 402]), to get

$$v(k) \le 2^{k-1} + 4k.$$

This has been improved by K. Subba Rao [5981] to

$$v(k) \ll 16^{k/7},$$

and S. Chowla [1081] showed that for infinitely many k one has $v(k) = O(k^2)$.

It is a simple exercise to prove the inequality $v(k) \le 1 + G(k)$, hence T.D. Wooley's result [6724], mentioned above, implies

$$v(k) \le (1 + o(1))k \log k.$$

The equality $v(2) = 3$ is trivial. The first result concerning cubes goes back to a paper by G. Oltramare[22] [4679] in 1894, where the inequality $v(3) \le 5$ was established. It was conjectured that every integer n is the sum of at most four cubes, and this was established for $n \equiv a \pmod 9$ with $a \ne \pm 2, \pm 4$ by H.W. Richmond [5210], L.J. Mordell [4386], and W. Sierpiński and A. Schinzel [5451], using representations of linear polynomials as sums of cubes of linear polynomials.

The case $a = \pm 2$ was added to that list by V.A. Demyanenko [1457], who considered sums of 4 cubes of quadratic polynomials (cf. P. Revoy [5170]). The case $a = \pm 4$ is still open.

It was shown in 1937 by H. Davenport and H. Heilbronn [1389] that for almost all integers four cubes are sufficient. On the other hand one has $v(3) \ge 4$, as no sum of three cubes is congruent to ± 4 mod 9.

It is believed that all numbers not congruent to ± 4 mod 9 are sums of three cubes and D.R. Heath-Brown [2648] conjectured that in this case there are infinitely many solutions. It is usually a difficult task to determine whether a particular integer is the sum of three cubes. The representation of the integer 30 in such a form was established only in 2007 (M. Beck, E. Pine, W. Tarrant, J. Yarbrough [380]). For other numerical results see A. Bremner [702], W. Conn, L.N. Vaserstein [1195], A.-S. Elsenhals, J. Jahnel [1758], V.L. Gardiner, R.B. Lazarus, P.R. Stein [2194], D.R. Heath-Brown [2649], D.R. Heath-Brown, W.M. Lioen

---

[21] It was pointed out by E.T. Bell [398] that the finiteness of $v(k)$ is an immediate consequence of an identity established in 1851 by P. Tardy [6051], but possibly known earlier.

[22] Gabriel Oltramare (1816–1906), professor in Geneva. See [2173].

and H.J.J. te Riele [2668], K. Koyama [3504, 3505], K. Koyama, Y. Tsuruoka, H. Sekigawa [3506], J.C.P. Miller and M.F.C. Woollett [4310]. The smallest integers for which no such representation is known are 33, 42 and 74.

In the case of biquadrates one has $9 \leq v(4) \leq 10$, as shown by W. Hunter [2953] (earlier H. Davenport [1358] proved $v(4) \leq 11$). Moreover one has $5 \leq v(5) \leq 9$ and the bounds in the cases $6 \leq k \leq 20$ and $21 \leq k \leq 30$ were given by W.H.J. Fuchs, E.M. Wright [2124] and D.P. Banerjee [313], respectively.

For subsequent improvements see A. Choudhry ($v(7) \leq 12$) [1061], A. Choudhry and J. Wróblewski ($v(13) \leq 27$) [1062], T. Rai [5051], L.N. Vaserstein ($v(8) \leq 28$) [6338]. This problem is related to the Tarry–Escott problem (see P. Revoy [5171]).

A generalization to algebraic number fields was obtained by R.M. Stemmler [5922].

The analogues of the Waring problem and the "easier" Waring problem in arbitrary commutative rings R with identity were considered by J.-R. Joly [3149], T. Chinburg [1055], and L.N. Vaserstein [6337].

The easier Waring problem with prime-power summands was considered by P. Erdős who in [1786] showed that for every k there is a constant $h(k)$ such that every integer can be written in the form $\sum_{j=1}^{r} \pm p_j^k$ with $r \leq c(k)$ and prime p_j (in [1785] he established this for $k = 2$).

9. Still another variant of the Waring problem has been considered by B.I. Segal [5599, 5601] who proved that if $c > 1$ is not an integer, then there exists $k = k(c)$ such that every large integer N can be written as

$$N = \sum_{j=1}^{k} \left[x_j^c\right] + a(N)$$

with integral x_j and uniformly bounded $a(N)$. Later [5600] he obtained $k(c) = O(c^2 2^c)$. For $1 < c < 1.5$ he showed that 3 terms are sufficient.

This was improved by J.-M. Deshouillers to $k(c) < c(4\log c + 2\log\log c + 14)$ for $c > 12$, and $k(c) \leq 2$ for $c \in (1, 4/3)$ [1481, 1482]. The last result was extended to $1 < c < 55/41 = 1.3414\ldots$ by S.A. Gritsenko [2347], and to $1 < c < 1.5$ by S.V. Konyagin [3472]. Asymptotics for the number of such representations were given by G.I. Arkhipov and A.N. Zhitkov [125] for $c > 12$ and $k \gg c^2 \log c$. Later Yu.R. Listratenko [3938] established this for all $c > 0$.

4.2.2 The Goldbach Conjecture

1. A great step forward was achieved in 1930 by L.G. Schnirelman [5544, 5840], who applied Brun's sieve to obtain the first unconditional result towards the Goldbach conjecture.

There is a constant C with the property that every integer ≥ 4 is the sum of at most C primes.

The proof of this result was based on a new notion of density, now called *Schnirelman density*. If A is a set of non-negative integers and $A(x)$ denotes the number of its non-zero elements not exceeding x, then the Schnirelman density $\delta(A)$ of A is defined by

$$\delta(A) = \inf_{x\geq 1} \frac{A(x)}{x}.$$

One proves by a simple counting argument that if A, B are sets of non-negative integers such that $1 \in A$ and $0 \in B$, then for the sumset

$$A + B = \{a + b : a \in A,\ b \in B\}$$

one has

$$\delta(A + B) \geq \delta(A) + \delta(B) - \delta(A)\delta(B). \tag{4.32}$$

This implies that if for a set A we have $\delta(A) > 0$, then there exists a number h with the property that every positive integer is the sum of at most h members of the set A, i.e., A is a *basis* of order $\leq h$, or a h-basis. Schnirelman's theorem is a consequence of the fact, that if $\mathbf{P}_1$ is the set of all primes to which the number 1 is adjoined, then the set $\mathbf{P}_1 + \mathbf{P}_1$ has positive density. This in turn follows from the bound

$$O\left(\frac{n}{\log^2 n} \prod_{p|n}\left(1 + \frac{2}{p}\right)\right)$$

for the number of representations of an integer n as the sum of two primes, obtained via Brun's sieve. For an early exposition of Schnirelman's proof see E. Landau [3676].

It has been conjectured ($\alpha + \beta$-*hypothesis*) that the inequality (4.32) could be improved to

$$\delta(A + B) \geq \min\{1, \delta(A) + \delta(B)\},$$

and this was shown to be true in 1942 by H.B. Mann[23] [4139] (see Sect. 5.2).

2. The constant C in Schnirelman's theorem is called the *Schnirelman constant*. In 1935 N.P. Romanov[24] [5273] claimed that every integer is the sum of at most 1104 primes, however it was pointed out by H. Heilbronn, E. Landau and P. Scherk[25] in [2717] that Romanov's argument is not quite convincing, and they used his method to show that every large integer is the sum of at most 73 primes. In 1937 G. Ricci [5199, 5200] reduced this bound to 67, but in the same year this was overshadowed

[23] Henry Berthold Mann (1905–2000), professor at Ohio State University, the University of Wisconsin and the University of Arizona. See [4678].

[24] Nikolai Pavlovič Romanov (born 1907), professor in Tomsk, Samarkand and Tashkent.

[25] Peter Scherk (1910–1985), professor at the University of Saskatchewan and the University of Toronto. See [5274].

by Vinogradov's [6429, 6430] proof of the Goldbach ternary conjecture for large odd integers, which implies that every large integer is the sum of at most 4 primes.

The question remained of how far one can reduce Schnirelman's constant for large integers without using heavy analytical tools. Ricci's bound was reduced to 20 by H.N. Shapiro and J. Warga [5684] in 1950 (C.L. Siegel noted in his review of that paper in *Zentralblatt* that the same result was also announced by A. Selberg). Much later it was proved by R.C. Vaughan [6347] that Schnirelman's method combined with the Siegel–Walfisz theorem and the large sieve can be used to show that every large integer is the sum of at most six primes.

The number C of primes needed to represent additively every integer > 2, obtained by Schnirelman's method, was bounded first by $2 \cdot 10^4$ (A.A. Šanin [5390]) and then by 159 (J.-M. Deshouillers [1480]), 115 (N.I. Klimov, G.Z. Pilt'jai, T.A. Šeptickaĭa [3362]), 113 (N.I. Klimov [3363]), 27 (R.C. Vaughan [6349]), 26 (J.-M. Deshouillers [1483]), 24 (M.Y. Zhang, P. Ding [6825]), 19 (H. Riesel, R.C. Vaughan [5225]), and the best known bound, $C \leq 7$, was obtained by O. Ramaré [5090]. It has been shown by L. Kaniecki [3236] that the Riemann Hypothesis implies $C \leq 5$.

Schnirelman's method was applied in 1943 by Yu.V. Linnik [3907] to an elementary proof of Waring's theorem.

3. A fresh presentation of Brun's method was given by T. Estermann [1886] in 1932, who used it to show that every sufficiently large even integer N is the sum of two numbers, each having at most six divisors, i.e., $N = P_6 + P_6$. He proved also that the General Riemann Hypothesis allows one of these summands to be a prime. An important modification of Brun's sieve was introduced by A.A. Buhštab [831–833], and this allowed him to replace the number 6 in Estermann's result first by 5 and then by 4. He showed also that the equality $P_4 - P_4 = 2$ holds infinitely often.

Buhštab's result was not improved until 1956, when A.I. Vinogradov [6397] showed that $2N = P_3 + P_3$ holds for every large N. Earlier, in 1950, A. Selberg [5616] announced that $2N = P_2 + P_3$ holds for large N, and the first proof of this assertion was published by Y. Wang [6552] in 1959. For further progress see Sect. 5.2.

4. The decisive step towards the solution of Goldbach's assertion for odd numbers was made in 1937 by I.M. Vinogradov [6429, 6430], who showed that every sufficiently large odd integer is the sum of three primes. He used the Hardy–Littlewood circle method in which the necessary evaluation of the integral over major arcs was made possible by an evaluation of the exponential sum

$$\sum_{p \leq x} \exp(2\pi i \alpha p) \tag{4.33}$$

for real α having a good approximation by rationals. An exposition of the proof was given later in Vinogradov's book [6436]. Stronger bounds for the sum (4.33) were obtained in [6431].

More general sums

$$\sum_{p \leq x} \exp\left(2\pi i F(p)\right), \tag{4.34}$$

where $F(X)$ is a polynomial with real coefficients, were considered by Vinogradov in a series of papers ([6432, 6434, 6437, 6440]; see also his books [6436, 6447, 6448]). For improvements and variants see G. Harman [2549–2551], H. Maier, A. Sankaranarayanan [4104].

The sum (4.34) was considered by É. Fouvry and P. Michel [2060] in the case when $F(X) = P(X)/q$ with a fixed prime q and a rational function $P(X)$. They obtained the bound $O(q^a x^b)$ with $b = 25/32$ and any $a > 3/16$.

Vinogradov's proof of his three primes theorem was simplified by K.K. Mardžanišvili [4146], and later Yu.V. Linnik [3912, 3914] gave two other proofs, based on properties of L-series (see also N.G. Čudakov [6097]). Another proof was given by B.M. Bredikhin [699]. In 1977 R.C. Vaughan found [6350, 6356] a simple elementary identity which allowed him to obtain a rather short proof of the Vinogradov–Goldbach three primes theorem (see [6357]) and the Bombieri–Vinogradov theorem [6355].

Vinogradov's result was made effective in 1956, when K.G. Borozdkin [648] showed that every odd integer exceeding exp(exp(16.038)) is the sum of three primes. Now it is known that every odd integer $n > \exp(3100)$ is the sum of three primes (M.C. Liu, T.Z. Wang [3969]), the previous records being $n > \exp(\exp(13.465))$ (A.M. Kasimov [3267]), $n > \exp(\exp(11.503))$ and $n > \exp(\exp(9.715))$ (J.R. Chen and T.Z. Wang [1034, 1035]). The General Riemann Hypothesis implies that this enormous bound can be reduced to exp(114), (T.Z. Wang, J.R. Chen [6546]), exp(94) (T.Z. Wang [6545]), and 10^{20} (D. Zinoviev [6833]). This led the way to the proof that the General Riemann Hypothesis implies the Goldbach ternary conjecture, which was achieved by J.-M. Deshouillers, G. Effinger, H.J.J. te Riele and D. Zinoviev [1487]. The use of computers enabled checking of the Goldbach ternary conjecture for large integers, up to $1.13256 \cdot 10^{22}$ (O. Ramaré, Y. Saouter [5096]). The previous record was 10^{20} (Y. Saouter [5393]).

5. Around 1937 N.G. Čudakov [1286, 6096] and T. Estermann [1891] used Vinogradov's method to show independently that almost all even integers are sums of two primes. They proved that if $E(x)$ denotes the number of exceptions, then for every M one has

$$E(x) \ll \frac{x}{\log^M x}.$$

The same result was obtained by a different method by J.G. van der Corput [6287], who also got the same assertion for even numbers which are not differences of two primes. This had been known earlier, with the bound $E(x) \ll x^{1/2+\varepsilon}$ (for every $\varepsilon > 0$), to be a consequence of the General Riemann Hypothesis (G.H. Hardy, J.E. Littlewood [2534]).

An elementary proof was later provided by B.M. Bredikhin and N.A. Jakovleva [701]. The upper bound for $E(x)$ was lowered first in 1972 by R.C. Vaughan [6343] who showed

$$E(x) \ll x \exp\left(-c\sqrt{\log x}\right)$$

with a certain $c > 0$, and then by H.L. Montgomery and R.C. Vaughan [4365] who proved $E(x) \ll x^a$ with some unspecified $a > 0$. Later J.R. Chen and C.D. Pan obtained $a = 0.99$ [1033], J.R. Chen got $a = 0.96$ [1028] and with J. Liu obtained $a = 0.95$ [1032], and this was reduced by H. Li to $a = 0.921$ and $a = 0.914$ [3879, 3880]. Recently W.C. Lu [4021] obtained $a = 0.879$.

It was shown in 1974 by K. Ramachandra [5060] that if $h(x) \gg x^a$ with $a > \alpha = 3/5$, then almost all even integers in the interval $(x, x + h(x))$ are sums of two primes. The admissible value of α was later consecutively reduced to $0.5476\ldots$ (C.H. Jia [3133]), $1/2$, $1/3$

and $7/36 = 0.1944\ldots$ (A. Perelli, J. Pintz [4778, 4779]), $7/48 = 0.1458\ldots$ (H.H. Mikawa [4304]), $7/78 = 0.0897\ldots$ (C.H. Jia [3142, 3143]), $7/81 = 0.0864\ldots$ (H. Li [3877]). Assuming the truth of the General Riemann Hypothesis one can take for $h(x)$ any function satisfying $\lim_{x\to\infty} h(x)/\log^6 x = \infty$ (J. Kaczorowski, A. Perelli, J. Pintz [3205]; for a correction see A. Languasco, A. Perelli [3715]).

In 1940 the binary Goldbach conjecture was checked by classical means for all integers $2n \leq 99998$ by N. Pipping[26] [4914, 4915], but subsequently, the use of computers enabled checking to be pursued up to $2 \cdot 10^{10}$ (A. Granville, J. van de Lune, H.J.J. te Riele [2328]), $4 \cdot 10^{11}$ (M. Sinisalo [5799]), $4 \cdot 10^{14}$ (J. Richstein [5212]) and $1.6 \cdot 10^{18}$ (T. Oliveira e Silva; see his web page http://www.ieeta.pt/~tos/goldbach.html).

6. Several open problems surrounding the Goldbach conjecture have been formulated. One of them appeared in 1938 in a paper by L.K. Hua [2920], who considered integers which are sums of three squares of primes. He showed that if n satisfies certain necessary congruence conditions, then at most $O(x/\log^c x)$ integers $n \leq x$ do not have such a representation, with c being a certain positive constant.

Much later W. Schwarz [5580] showed that Hua's bound holds for every c, and as late as in 1993 M.-C. Leung and M.C. Liu [3841] showed that the number of exceptional $n \leq x$ is $O(x^{1-\delta})$ with a certain $\delta > 0$. Now it is known (see J.Y. Liu, T. Zhan [3965]) that this holds with every $\delta < 1/12$ (previously $\delta < 3/50$ was known (J.Y. Liu, T. Zhan [3964])).

Similar results are also known for sums of four squares of primes (see, e.g., G. Harman, A. Kumchev [2567]).

L.K. Hua proved also that every large integer $n \equiv 5 \pmod{24}$ is the sum of five squares of primes. He considered also representations of large integers as sums of a prime, one or two squares of primes and a prime power with fixed exponent. Later [2923] he studied the more general question of representing integers as the sum of kth powers of primes for $k \geq 4$. Let $k = \prod_p p^{\theta_p}$ be the canonical factorization of k, put

$$\gamma_p = \begin{cases} \theta_2 + 2 & \text{if } p = 2 \text{ and } 2|k, \\ \theta_p + 1 & \text{if } p \neq 2, \end{cases}$$

and $K = \prod p^{\gamma_p}$, the product extended over primes p with $p - 1|k$. Denote by $H(k)$ the smallest integer s such that every large integer congruent to s mod K is the sum of s kth powers of primes. Hua showed that for odd k one has $H(k) = (6 + o(1))k\log k$, and obtained $H(3) \leq 11$, $H(4) \leq 19$ and $H(5) \leq 31$.

Later H. Davenport [1354] proved $H(4) \leq 15$. The book by L.K. Hua [2932], written in 1941, but published only in 1947, contains a broad exposition of these questions, leading in particular to the bounds $H(3) \leq 9$, $H(5) \leq 25$, $H(6) \leq 37$. Later K. Thanigasalam [6133, 6135] improved Hua's bounds for $5 \leq k \leq 10$, showing in particular $H(5) \leq 23$. In 2001 K. Kawada and T.D. Wooley [3289] obtained $H(4) \leq 14$, and $H(5) \leq 21$.

---

[26] Nils Pipping (1890–1982), professor in Turku.

4.2.3 Other Additive Questions

1. The circle method of Hardy and Littlewood, as well as more elementary methods, has been successfully applied by several authors to study various additive problems. In 1929 C.J.A. Evelyn and E.H. Linfoot [1914–1918] considered the number of representations of a large integer n as the sum of s k-free integers, i.e., integers not divisible by a kth power of a prime. They applied the circle method in the case $s \geq 3$ to show that the said number is asymptotic to $c(n)n^{s-1}$, with $c(n)$ lying between two positive constants, and in [1915] their paper they extended this result to the case $s = 2$, this time using an elementary approach (a simpler proof was given by T. Estermann [1883]). In [1916] they obtained an elementary proof also for $s \geq 3$. The error term in their formula was reduced in [1917, 1918], and a further reduction was made in [1895].

Further improvements (some under the General Riemann Hypothesis) were later provided by L. Mirsky[27] [4336, 4337], J. Brüdern, A. Perelli [777], J.-C. Schlage-Puchta [5458] and D.I. Tolev [6189].

The number of representations of large integers as sums of a k-free and an l-free integer was evaluated by A. Page [4715], who later also considered representations of numbers as sums of squares and products [4717–4719].

T. Estermann also successfully applied this method. In 1929 he found asymptotics for the number of representations of an integer as the sum of three products [1879], one year later he did the same for the case of two products [1880, 1881], and in 1931 he obtained an asymptotic formula for the number of representations of integers as sums of a prime and a square-free number [1884, 1885]. The error term in this formula was later provided by A. Page [4720] and improved by A. Walfisz in [6531].

This result was later generalized by C. Hooley [2890, 2891] who showed that for $r \geq 2$ all large integers are sums of an rth power and an $(r-1)$-free integer.

In the next years T. Estermann obtained a series of similar results. In [1889] he showed that every large integer is the sum of two primes and a product of two primes, and in [1888] he proved that every large integer is of the form $p_1 + p_2 + x^2$ with prime p_i.

The more general problem of representing large integers as the sum of r squares and s primes (the special case $s = 1, r = 2, 4$, forming problem **J** of [2531]) was settled by S. Chowla [1080] for $s = 1$, $r = 4$ and by A. Walfisz [6530, 6531] for $s \geq 1$, $r \geq 5$. They confirmed the asymptotical formulas for the number of representations, obtained earlier by G.K. Stanley [5889] under the assumption of the General Riemann Hypothesis (under the same assumption Stanley [5887] showed earlier that all large integers are sums of three squares and a prime, and of two primes and a square).

---

[27] Leon Mirsky (1918–1983), professor in Sheffield. See [866].

2. A set A of positive integers is called an *essential component* if its Schnirelman's density $\delta(A)$ equals zero, but for every set B with $\delta(B) > 0$ one has $\delta(A+B) > \delta(B)$. The first essential component was found by A.J. Khintchine [3322], who showed that the sequence of squares has this property. A large class of such sets was discovered in 1935 by P. Erdős [1775], who showed that if A is a basis of order h and $\delta = \delta(B)$, then one has

$$\delta(A+B) \geq \delta + \frac{(1-\delta)\delta}{2h}, \tag{4.35}$$

hence every basis is an essential component.

The first example of an essential component which is not a basis was constructed in 1942 by Yu.V. Linnik [3904] (it had $O(x^{\varepsilon})$ elements $\leq x$ for every $\varepsilon > 0$), and a simpler example was furnished in 1956 by A. Stöhr and E. Wirsing [5958]. It was shown much later by I.Z. Ruzsa [5343] that for every $c > 1$ there exist essential components having $O(\log^c x)$ elements below x.

The inequality (4.35) was later improved, first by F. Kasch [3262–3264] in 1955, and then by H. Plünnecke [4924] in 1970.

The question of the existence of small bases of the integers was tackled first by H. Rohrbach, who showed in 1937 that for every $h \geq 2$ and $\varepsilon > 0$ there exists an h-basis of natural numbers having $O(x^{1/h+\varepsilon})$ elements less than x [5263]. The ε in the exponent was subsequently removed by A. Stöhr [5955] and D. Raĭkov [5052]. H. Rohrbach showed also that for the counting function $A(x)$ of a 2-basis one has

$$\liminf_{x\to\infty} \frac{A(x)}{\sqrt{x}} \geq \sqrt{2}.$$

It was shown later by W. Klotz [3374] that in the last result one can replace $\sqrt{2}$ by 1.441, and a basis with

$$\liminf_{x\to\infty} \frac{A(x)}{\sqrt{x}} \leq \sqrt{7/2} = 1.807\ldots$$

was constructed in 2001 by G. Hofmeister [2841].

An analogous result for *multiplicative bases*, which are sequences $a_1 = 1 < a_2 < \cdots$ with the property that every positive integer n can be written in the form $n = a_{i_1} \cdots a_{i_h}$ was obtained by D. Raĭkov [5053].

Several results about bases of integers can be found in the paper [5956, 5957] by A. Stöhr.

3. A rather unexpected result was established in 1934 by N.P. Romanov [5272] who showed that for fixed $k \geq 2$ numbers of the form $p + x^k$ with prime p form a set of positive density, and the same applies to numbers $p + k^x$ (for $k = 2$ this was deduced earlier from the General Riemann Hypothesis by G.K. Stanley [5887]).

In a letter to P. Erdős Romanov asked[28] whether every large odd integer is of the form $2^k + p$ with prime p, and Erdős proved that the set A of odd integers not having

[28] This question was first considered by Polignac in 1849 [4941].

such a form contains an infinite arithmetic progression, but published his result only in 1950 ([1801], cf. [5784]). At the same time J.G. van der Corput [6296] showed that the set A has a positive lower density.

Evaluations of lower and upper densities of A were given by Y.G. Chen and X.G. Sun [1039], L. Habsieger and X.-F. Roblot [2423], G. Lü [4012] and J. Pintz [4906].

Later H. Davenport and H. Heilbronn [1390] strengthened Romanov's result by showing that almost all integers are of the form $p + x^k$.

In the case $k = 2$ the number of exceptions $\leq x$ was shown to be $\ll x\exp(-c\sqrt{\log x})$ (with $c > 0$) by I.V. Polyakov [4964] in 1981 and this bound was improved by A.I. Vinogradov [6400] to $O(x^a)$ with a certain $a < 1$ (see also R. Brünner, A. Perelli, J. Pintz [805], I.V. Polyakov [4965]). Later T.Z. Wang [6544] showed that this holds with $a = 0.99$ and H. Li [3882] improved this to $a = 0.914$. A.I. Vinogradov [6400] obtained a similar result for sums of a prime and a cube, and A. Zaccagnini [6806] extended this to the case of an arbitrary power (see also J. Brüdern, A. Perelli [776], A. Perelli, A. Zaccagnini [4780]).

A more general question was considered by J.G. van der Corput [6288–6291], who used Vinogradov's approach to the Goldbach problem to show that if $F(X) \in \mathbf{Z}[X]$ has no fixed divisor, then almost all integers n satisfying the necessary congruence conditions can be represented in the form $p + F(x)$ with a prime p, the number of exceptions $\leq T$ being $O(T/\log^m T)$ for all m. Later [6292] he obtained similar results for the representation of integers in the forms $a_1p_1 + a_2p_2$, $a_1p_1 + a_2p_2^2 + a_3p_3^2$ and $\sum_{j=1}^{4} a_jp_j^2$ (with prime p_j's and fixed co-prime coefficients a_j).

It was shown later by A. Zaccagnini [6807] that the number of exceptions $\leq x$ to van der Corput's result is $O(x\exp(-c\log x/\log\log x))$.

4.3 Transcendence and Diophantine Approximations

4.3.1 Transcendence

1. Around 1930 K. Mahler [4061–4063] developed a new method, which allowed him to prove transcendence of values at algebraic arguments of transcendental functions $f(z_1, \ldots, z_n)$, satisfying certain functional equations. In particular he showed that if $\alpha \neq 0$ is algebraic and lies inside the unit circle, then the numbers $\prod_{j=0}^{\infty}(1 - \alpha^{2^j})$ and $\sum_{j=0}^{\infty}\alpha^{2^j}$ are both transcendental, and for every positive quadratic irrationality β the sum of the series

$$\sum_{n=1}^{\infty}[n\beta]\alpha^n$$

is also transcendental.

An exposition of Mahler's method and its extensions was given by J.H. Loxton and A.J. van der Poorten [4010] and K. Nishioka [4618]. This method was much later used to prove the transcendence of several numbers, including

$$\sum_{n=0}^{\infty} \frac{1}{u_{d^n} + a}$$

for a class of linear recurrences u_n and $d \geq 2$, $a \in \mathbf{Z}$ (P.G. Becker, T. Töpfer [381]).

2. A classification of transcendental numbers was proposed in 1932 by K. Mahler [4064, 4065]. It depends on the size of values attained at such a number by polynomials $f \in \mathbf{Z}[X]$ of given height $H(f)$.

If z is a complex number, then $\omega_n(z)$ is defined as the least upper bound for numbers $\alpha > 0$ with the property that there exist infinitely many polynomials $f \in \mathbf{Z}[X]$ of degree n satisfying

$$0 < |P(z)| < H(P)^{-\alpha},$$

and one puts

$$\omega(z) = \limsup_n \frac{\omega_n(z)}{n}.$$

Algebraic numbers are characterized by the property $\omega(z) = 0$ and $\omega_n(z) < \infty$ for all n. If $0 < \omega(z) < \infty$, then z is called an S-number. If $\omega(z)$ is infinite but all $\omega_n(z)$'s are finite, then z is called a T-number. Finally, if $\omega(z)$ is infinite but there is a number m such that $\omega_n(z)$ is finite for all $n < m$ and $\mu = \mu(z)$ is the least value of m having this property, then z is called a U-number, more precisely, a U_μ-number.

Mahler showed in his paper that for every non-zero rational r the number e^r is an S-number, π and logarithms of rational numbers $\neq 0, 1$ are not U-numbers, and noted that the set U_1 coincides with that of Liouville numbers. He proved also the existence of a positive constant c such that for almost all z and every $P \in \mathbf{Z}[X]$ of degree n and sufficiently large height one has

$$|P(z)| > \frac{1}{H(P)^{cn}}. \tag{4.36}$$

K. Mahler proved that for c one can take any number larger than 4, and if z is not real, then 4 can be replaced by 3/2, and conjectured that instead of 4 and 7/2 one can take 1 and 1/2.

He showed also that if $a_1, a_2, \ldots, a_n$ are algebraic numbers linearly independent over the rationals, and λ is a Liouville transcendental number, then the set $\{e^{a_1}, \ldots, e^{a_n}, \lambda\}$ is algebraically independent over the field of all algebraic numbers, and the same applies also to the pair π, λ. In [4066] he showed that the set of S-numbers is of full measure.

Later K. Mahler [4087] showed that logarithms of algebraic numbers $\neq 0, 1$ are not U-numbers.

For a long time it was unknown whether T-numbers exist at all, and only much later their existence was established by W.M. Schmidt [5497, 5499]. The existence of U_n-numbers for every n was proved in 1953 by W.J. LeVeque [3849].

It was shown by J.F. Koksma [3445] in 1939 that Mahler's classification is strongly related to the degree of approximation by algebraic numbers. For a complex z and $n = 1, 2, \dots$ denote by $\omega_n^*(z)$ the least upper bound for positive reals α for which there exist infinitely many algebraic numbers ξ of degree n, satisfying

$$|z - \xi| < \frac{1}{h(\xi)^{1+\alpha}},$$

where the height $h(\alpha)$ of α, is defined as the height of the minimal polynomial of α with rational integral and co-prime coefficients. Moreover, let

$$\omega^*(z) = \limsup \frac{\omega_n^*(z)}{n},$$

and denote by $\mu^*(z)$ the minimal index n with $\omega_n^*(z)$ infinite, if such an index exists, otherwise put $\mu^*(z) = \infty$. This leads to a classification of transcendental numbers which is formally different from Mahler's, but, as shown later by E. Wirsing [6693] these differences are unimportant. J.F. Koksma showed that z is an S-number if $\omega^*(z) < \infty$, it is a T-number if both $\omega^*(z)$ and $\mu^*(z)$ are infinite, and if $\omega^*(z)$ is infinite, but $\mu^*(z)$ is finite, then z is a U-number. Using these properties J.F. Koksma was able to replace the pair $4, 7/2$ occurring in Mahler's bound in (4.36) by $3, 5/2$.

This result was later consecutively improved by W.J. LeVeque [3849], B. Volkmann [6463–6465], F. Kasch [3265] and V.G. Sprindžuk [5871]. Finally it was V.G. Sprindžuk [5872, 5874] who in 1964 established the truth of Mahler's conjecture on (4.36). He gave an exposition of the proof in his book [5875]. There one also finds a p-adic analogue.

In 1966 A. Baker [227] generalized Sprindžuk's result by showing that if $f(h)$ is a decreasing function for which the series $\sum_{h=1}^{\infty} f(h)$ converges, then for almost all real θ the inequality

$$|P(\theta)| < f(H)^n$$

can hold only for finitely many polynomials $P \in \mathbf{Z}[X]$ of degree n and height H. In 1989 V.I. Bernik [461] extended this by showing that under the same assumptions for almost all real θ the inequality

$$|P(\theta)| < \frac{f(H)}{H^{n+1}}$$

can hold only for finitely many $P \in \mathbf{Z}[X]$ of degree n and height H. This result is best possible (V. Beresnevich [428]). Later Beresnevich [429] showed that the assumption of monotonicity of f in Bernik's result can be omitted. A p-adic version of Bernik's theorem was obtained by E.I. Kovalevskaya [3502].

In 1971 K. Mahler [4092] introduced another classification of transcendental numbers and posed several questions which, with one exception, were later answered by A. Durand [1656] and Yu.V. Nesterenko [4561].

Still another classification was given by V.G. Sprindžuk [5870] in 1962 (see M. Amou [69]). Generalizations of the Mahler and Koksma classifications to larger dimensions were considered by W.M. Schmidt [5528] and K.R. Yu [6800].

In [4066] K. Mahler formulated a conjecture which can be stated in the following form: for every n, and almost no real t there exists $\varepsilon > 0$ such that the inequality

$$\max\{\|qt\|^n, \dots, \|qt^n\|^n\} \le \frac{1}{|q|^{1+\varepsilon}}$$

holds for infinitely many integers q.

This conjecture was proved in 1964 by V.G. Sprindžuk [5873, 5875], who later [5878] proposed its generalization, in which the sequence $t, t^2, \ldots, t^n$ is replaced by a sequence of well-behaved functions. In the case $n = 2$ it follows from a result by W.M. Schmidt [5488], and a proof for $n = 3$ was given in 1996 by V. Beresnevich and V.I. Bernik [430] (see also the book by V.I. Bernik and M. Dodson [462]). For arbitrary n the conjecture was established in 1998 by D.Y. Kleinbock and G.A. Margulis [3360] (cf. D.Y. Kleinbock [3358, 3359]).

3. The year 1934 brought a solution of Hilbert's seventh problem. This was achieved independently by A.O. Gelfond [2224, 2225] and T. Schneider[29] [5537, 5538], who established the following theorem.

If $\alpha, \beta \neq 0, 1$ are algebraic numbers, then the ratio $\log\alpha / \log\beta$ is either rational or transcendental.

This implies that if $\beta \neq 0, 1$ is algebraic, and $\alpha \notin \mathbf{Q}$ is algebraic, then α^β is a transcendental number.

A.O. Gelfond, who earlier established the transcendence of e^π and of the numbers $\alpha^{i\sqrt{n}}$ with algebraic $\alpha \neq 0, 1$ and non-square positive integer n ([2222, 2223], cf. [3447, 3590]) considered complex functions of the form

$$f(z) = \sum_{k=0}^{m} \sum_{l=0}^{m} C_{k,l} \left(\alpha^k \beta^l\right)^z,$$

with rational integral coefficients $C_{k,l}$, and showed that if $\log\alpha / \log\beta$ is algebraic and irrational, then there is a choice of the coefficients $C_{k,l}$ such that $f \neq 0$, and for some large N the function f has high order zeros at $z = 0, 1, 2, \ldots, N$. A clever utilization of Cauchy's integration formula enabled him to show that f must in fact be the zero function, thus giving a contradiction.

T. Schneider applied in [5537, 5538] another method based on the approach used earlier by C.L. Siegel [5747] in his study of E-functions. He utilized, in particular, a version of Siegel's lemma.

A.O. Gelfond's method was used by G. Ricci [5198] to show that α^β is transcendental if $\alpha \neq 0, 1$ and $\beta \notin \mathbf{Q}$ are both well approximated by algebraic numbers.

4. A.O. Gelfond's early method was utilized in 1932 by C.L. Siegel [5748] to study the periods of elliptic functions, and he was able to show that if the coefficients g_2, g_3 in the differential equation

$$\wp'(z)^2 = 4\wp(z)^3 - g_2\wp(z) - g_3$$

for the Weierstrass function $\wp(z)$ are both algebraic, then at least one of the periods of $\wp(z)$ must be transcendental. He showed also the transcendence of the number

$$\int_{-1}^{1} \frac{dx}{\sqrt{1 - x^4}},$$

[29] Theodor Schneider (1911–1988), professor in Erlangen and Freiburg. See [850].

which equals the ratio of the length of the lemniscate to its diameter, and obtained the same assertion for the integral

$$\int_{-1}^{1} \frac{dx}{\sqrt{1-x^6}}.$$

Note that these results are parallel to Lindemann's theorem on the transcendence of π, since

$$\int_{-1}^{1} \frac{dx}{\sqrt{1-x^2}} = 2\pi.$$

An extension of Siegel's result was given by T. Schneider in [5538].

C.L. Siegel also pointed out how to obtain a generalization of his results to periods of Abelian integrals, which for $n = 2, 3, \dots$ leads to the proof of transcendence of at least one of the numbers

$$\Gamma(m/n)\pi^{-m/n} \quad (m = 1, 2, \dots, [(n-1)/2]).$$

For further results on the Weierstrass function $\wp(z)$ see D.W. Masser, G.Wüstholz [4180, 4182] and R. Tubbs [6208, 6209].

5. One of the first results dealing with transcendence questions in p-adic number fields was obtained in 1932 by K. Mahler [4067] who showed that the p-adic exponential function defined by the usual series

$$\exp(x) = \sum_{j=0}^{\infty} \frac{x^j}{j!}$$

in its convergence disc, assumes transcendental values for algebraic non-zero arguments. Three years later he proved also [4076] an analogue of the Gelfond–Schneider result, showing that if α, β are algebraic elements of the algebraic closure of $\mathbf{Q}_p$, both distinct from 1 and lying sufficiently close to 1, then the ratio $\log\beta/\log\alpha$ is either transcendental or lies in $\mathbf{Q}_p$.

Another proof, based on Schneider's approach, was given in 1940 by G.R. Veldkamp [6377]. Cf. A. Günther [2383], W.W. Adams [15].

K. Mahler [4075] also proposed a classification of transcendental numbers in the p-adic case, introducing the analogues of S- and T-numbers.

In 1967 V.G. Sprindžuk [5875] established in this case a result corresponding to Mahler's conjecture in the complex case, and in 1981 H.P. Schlickewei [5463] established the existence of p-adic T-numbers.

4.3.2 Uniform Distribution and Diophantine Approximations

1. In 1931 J.G. van der Corput [6283, 6284] considered the extension of the notion of uniform distribution mod 1 to several variables, also obtaining new results in the one-dimensional case. In particular he found a simple proof of his earlier theorem [6279] stating that if for every $k = 1, 2, \ldots$ the sequence $u_{n+k} - u_n$ is uniformly distributed modulo one, then so is the sequence u_n.

In 1952 N.M. Korobov and A.G. Postnikov [3490] generalized van der Corput's theorem by showing that its assumption implies uniform distribution mod 1 of the sequence u_{an+b} for all integers $a \neq 0, b$.

The sequence θ^n was considered in 1935 by J.F. Koksma [3443], who showed that for almost every $\theta > 1$ it is uniformly distributed mod 1.

2. Let $A = \{a_n\}$ be a sequence of real numbers in the unit interval and for any interval $I \subset [0, 1)$ put

$$\Delta_A(N, I) = |\#\{n \leq N : a_n \in I\} - n|I||,$$

$|I|$ denoting the length of I.

Restricting I to initial intervals one defines similarly

$$\Delta_A^*(n, x) = |\#\{n \leq N : a_n \leq x\} - nx|.$$

One defines two notions of *discrepancy* $d_A(N)$ and *star-discrepancy* $d_A^*(N)$ of A by putting

$$d_A(N) = \sup_I \Delta_A(N, I),$$

and

$$d_A^*(N) = \sup_x \Delta_A^*(n, x).$$

These notions are connected by the inequalities

$$\Delta_A^*(N) \leq \Delta_A(N) \leq 2\Delta_A^*(N).$$

Vinogradov's bounds for Weyl sums [6422] have been used by him in [6423, 6427] to bound the discrepancy of polynomial sequences. His bounds were later improved by J.G. van der Corput and C. Pisot [6298–6300].

For later improvements see J.W.S. Cassels [932].

J.G. van der Corput [6285, 6286] conjectured in 1935 that for every infinite sequence A the discrepancy $d_A(N)$ tends to infinity with n.

This conjecture was confirmed in 1945 by T. van Aardenne-Ehrenfest [6267], who later established [6268]

$$\limsup_{n \to \infty} \frac{d_A(n) \log\log\log n}{\log\log n} > 0,$$

and K.F. Roth [5310] improved this to

$$\limsup_{n \to \infty} \frac{d_A(n)}{\sqrt{\log n}} > 0.$$

See also W.M. Schmidt [5490, 5495].

In 1972 W.M. Schmidt [5496] obtained

$$\limsup_{n\to\infty} \frac{d_A(n)}{\log n} \geq c$$

(with $c = 0.01$), confirming a conjecture of P. Erdős [1822]. Later R.Béjian [391, 392] showed that one can take $c = 0.06$ and $c = 0.12$. The example given by J.G. van der Corput [6285, 6286] implied the upper bound $c \leq 1/3 \log 2 = 0.4\ldots$, and this bound was later reduced by L. Ramshaw [5101] ($c \leq 0.416$), H. Faure [1961] ($c \leq 4828/(5181 \log 12) = 0.3750\ldots$; for the star-discrepancy he got the bound 0.223), and A. Thomas [6136], who diminished Faure's value by $2 \cdot 10^{-5}$. Cf. also V.T. Sós [5849] and R. Tijdeman, G. Wagner [6162].

The discrepancy of sequences $A = (\{n\theta\})$ for irrational θ is closely related to the sum (2.67), considered in Sect. 2.5.2. Bounds for the discrepancy in this case, depending on the degree of approximations of θ by rational numbers were obtained by H. Behnke [384], E. Hecke [2682] and A. Ostrowski [4708].

In 1964 H. Kesten [3309] showed that in this case the ratio

$$\frac{d_A(n)}{\log n \log\log n}$$

converges in measure to $2/\pi^2$, i.e., for every $\varepsilon > 0$

$$\lim_{n\to\infty} \mu\left(\left\{\theta \in (0,1) : \left|\frac{d_A(n)}{\log n \log\log n}\right| > \varepsilon\right\}\right) = 0,$$

μ denoting the Lebesgue measure. The speed of this convergence was evaluated by J. Schoissengeier [5553] in 2000.

The discrepancy of sequences of the form $a_n = \{c_n x\}$ was studied by P. Erdős and J.F. Koksma [1838] in 1949. They showed that if the sequence c_n is increasing, then for almost all x one has

$$\Delta_A(n,x) \ll \sqrt{n} \log^c n$$

with any $c > 5/2$, and in 1981 R.C. Baker [250] replaced $5/2$ here by $3/2$. The optimal value of c is still unknown, but I. Berkes and W. Phillip [439] showed in 1994 that one cannot have $c = 0$. If the sequence c_n is lacunary (i.e., $c_{n+1}/c_n \geq q > 1$), then for almost all x one has

$$\frac{1}{16} \leq \limsup_{n\to\infty} \frac{\Delta_A(n,x)}{\sqrt{n}\log\log n} \leq 166 + \frac{664}{\sqrt{q}-1},$$

as shown by W. Phillip [4834] in 1975, improving an earlier bound given by P. Erdős and J.F. Koksma [1837]. This result was later extended to a broader class of sequences (C. Aitsleitner, I. Berkes [35], I. Berkes, W. Phillip, R.F. Tichy [440]).

The discrepancy of the sequence $n\overline{a}$ for $\overline{a} = (a_1, \ldots, a_r) \in \mathbf{R}^r$ was considered by W.M. Schmidt [5487] in 1964. He showed that for almost all sequences it is $O(\log^c n)$ for every $c > r + 1$, and J. Beck [376, 377] showed that this bound holds for all $c > r$. More precisely, he obtained the bound $O(\log^r n f(\log\log n))$ for every function $f(t) > 0$ for which the series $\sum_{n=1}^{\infty} 1/f(n)$ converges.

Sequences with discrepancy not exceeding $c_r \log^r n + O(\log^{r-1} n)$ have been constructed by H. Faure [1962], who improved earlier results by J.H. Halton [2486] and I.M. Sobol [5841] (cf. H. Niederreiter [4599, 4601, 4602]). For lower bounds see W.M. Schmidt [5491–5494], J. Beck [376] and R.C. Baker [257] (cf. R. Alexander [51], J. Beck [372–375]).

These results are of importance in construction of random numbers and also in evaluations of higher-dimensional integrals (see the papers by H. Niedereiter and C. Xing [4606–4610] as well as the books by E. Hlawka, F. Firneis and P. Zinterhof [2837], L.K. Hua and Y. Wang [2940] and H. Niederreiter [4603]).

For surveys of discrepancy theory see J. Beck and V.T. Sós [379] and the books by J. Beck and W.W.L. Chen [378] and M. Drmota and R.F. Tichy [1626].

3. In 1935 A.J. Khintchine [3323] showed that if a_n denotes the nth partial quotient of the continued fraction for a real number x, then for almost all x the geometric mean of $a_1, a_2, \dots, a_n$ tends to a constant. He obtained this as a special case of the following result.

If a function $f(t)$ satisfies $f(t) = O(t^c)$ for some $c < 1$ and $t \to \infty$, then for almost all x one has

$$\lim_{n\to\infty} \frac{1}{n} \sum_{j=1}^{n} f(a_j) = \sum_{r=1}^{\infty} f(r) \log_2 \left(1 + \frac{1}{r(r+2)}\right).$$

In a later paper [3324] he extended this result to functions of several variables. For a probabilistic approach to these questions see the paper by P. Lévy [3867].

Later A.J. Khintchine [3325] considered the question of simultaneous inhomogeneous approximations. Earlier [3319] he showed that the inequality

$$\|x\theta - \alpha\| < \frac{1}{t}$$

has for every real α and $t > 1$ a solution $x = O(t)$ if and only if the continued fraction of θ has bounded partial quotients. Now he extended this and obtained a necessary and sufficient condition for a set $\theta_1, \dots, \theta_n$ of $\mathbf{Q}$-linearly independent reals to have the following property.

For all real $\alpha_1, \dots, \alpha_n$ and $t > 1$ the system of inequalities

$$\|x\theta_i - \alpha_i\| < \frac{1}{t} \quad (i = 1, 2, \dots, n)$$

has a solution $x = O(t^n)$.

4. In 1934 K. Mahler [4072] studied several problems of Diophantine approximations in the p-adic domain and proved the analogue of Minkowski's theorem on linear forms. In [4073] he considered the p-adic version of continued fractions and in [4082] he dealt with inhomogeneous approximations.

C.L. Siegel's result [5738] on the approximation of algebraic numbers by rationals had been transferred to the p-adic case by K. Mahler [4068] where it took the following form.

Let the polynomial $f \in \mathbf{Z}[X]$ of degree $n \geq 3$ be irreducible, and let $\{p_1, \dots, p_t\}$ be a finite set of primes. Let α be a root of f in the complex plane, and let α_i be the *root of f in the algebraic closure of the p-adic field $\mathbf{Q}_{p_i}$ ($i = 1, 2, \ldots, t$). Then for every $M \geq 1$ and $\varepsilon > 0$ the inequality*

$$\min\left\{1, \left|\alpha - \frac{p}{q}\right|\right\} \prod_{j=1}^{t} \min\left\{1, |p - q\alpha_i|_{p_i}\right\} \leq \frac{M}{(\max\{|p|, |q|\})^{\beta+\varepsilon}},$$

where

$$\beta = \min_{j=1,\ldots,n-1} \left\{\frac{n}{j+1} + j\right\}, \tag{4.37}$$

and $|\cdot|_q$ denotes the q-adic norm, has only finitely many solutions $p, q \in \mathbf{Z}$.

This result was later extended by C.J. Parry [4745, 4746] to cover approximations by algebraic numbers.

In 1936 J.F. Koksma published a survey of the theory of Diophantine approximations [3444], containing a large bibliography, in the series *Ergebnisse der Mathematik*.

4.4 Diophantine Equations and Congruences

4.4.1 Polynomial Equations

1. The equation

$$x^2 - dy^2 = -1, \tag{4.38}$$

with positive d, usually called the *non-Pellian equation*, has a long story. Already A.M. Legendre [3766] had established that it has integral solutions for prime $d \equiv 1 \pmod 4$. It had been known for a long time that a solution exists if and only if the length of the period of the continued fraction for $\sqrt{d}$ is odd, but this condition is rather tedious to check. A criterion utilizing another type of continued fraction was given by A. Hurwitz [2957] in 1887.

A way of constructing all positive integers d for which the equation has integral solutions was presented in 1934 by P. Epstein [1770] (cf. L. Rédei[30] [5138]), who showed that (4.38) is solvable if and only if d can be written in the form

$$d = (\alpha m + \beta)^2 + (\gamma m + \delta)^2,$$

where $\alpha^2 + \gamma^2$ is a square and $\alpha\delta - \beta\gamma = \pm 1$.

An algorithm for checking the solvability of (4.38) which involves $O(d^c)$ operations for every $c > 1/4$ was constructed in 1980 by J.C. Lagarias [3601]. Later P. Morton [4439] proved that there is no criterion for the solvability of this equation using congruences. For various sufficient conditions for solvability see J.E. Cremona, R.W.K. Odoni [1280], C.U. Jensen

[30] László Rédei (1900–1980), professor in Szeged and Budapest. See [4152].

[3122]. It was conjectured by P. Stevenhagen [5942] that the number of integers $d \le x$ for which (4.38) is solvable equals

$$(c + o(1))\frac{x}{\sqrt{\log x}}$$

with $c = 0.2813\ldots$.

2. In 1933 K. Mahler [4068] deduced from (4.37) that if $F \in \mathbf{Z}(X, Y)$ is an irreducible form of degree $n \ge 3$ and P_0 is a finite set of primes, then there can be only finitely many co-prime pairs x, y such that $F(x, y)$ is composed of these primes. Therefore for any finite set of primes $\{p_1, \ldots, p_t\}$ and non-zero $a \in \mathbf{Z}$ there are only finitely many integral solutions $x, y, u_1, \ldots, u_t$ of the equation

$$F(x, y) = a p_1^{u_1} \cdots p_t^{u_t}. \tag{4.39}$$

(Such equations are called *Thue–Mahler equations*.)

In [4069] K. Mahler obtained upper bounds for the number of solutions and this led to the bound $O(|m|^{\varepsilon})$ ($\varepsilon > 0$, arbitrary) for the number $N_F(m)$ of integral solutions of the equation $F(x, y) = m$ for given non-zero m. In [4070] he established the asymptotical formula

$$\sum_{|m| \le x} N_F(m) = c(F)m^{2/n} + O\left(m^{1/(n-1)}\right), \tag{4.40}$$

using an unpublished result by C.L. Siegel from 1925 stating that if $F \in \mathbf{Z}[X, Y]$ is a binary irreducible form of degree $n \ge 3$, then the function

$$Z_F(s) = \sum_{(a,b) \ne (0,0)} \frac{1}{|F(a, b)|^s},$$

defined for $\Re s > 2/n$ can be continued to a meromorphic function in $\Re s > 1/(n-1)$ with a simple pole at $s = 2/n$.

In 1955 H. Davenport and K.F. Roth [1401] gave the first explicit upper bound for the number $N_F(m)$ in the case of an irreducible form F. This bound was improved in 1960 by D.J. Lewis and K. Mahler [3873], who obtained for sufficiently large m

$$N_F(m) \le (cn)^{\omega(m)}$$

where c is an absolute constant c.

A bound of another type was given by J. Silverman [5790].

An important step forward was made in 1983 in the thesis of J.-H. Evertse [1921], who showed that $N_F(m)$ is bounded by a quantity which depends exponentially on n and $\omega(m)$. This answered positively a question from C.L. Siegel. Evertse's bound was replaced in 1986 (by E. Bombieri and W.M. Schmidt [621]) by $O(n^{1+\omega(m)})$ with the implied constant being absolute. For $n \ge 6$ and large m this bound was later reduced by B. Brindza[31] [735].

In 1988 J. Mueller and W.M. Schmidt [4468] confirmed C.L. Siegel's conjecture (stated in [5747, Sect. 7]) which asserted that the number of solutions is bounded by a number depending only on the number $k(F)$ of non-zero coefficients of F (cf. W.M. Schmidt [5520]).

[31] Béla Brindza (1958–2003), professor in Debrecen. See [2415].

Earlier they did this [4467] for forms with $k(F) = 3$, in which case sharp bounds for the number of solutions were given by E. Thomas[32] [6138]. Note that Siegel's conjecture fails to hold in the case of function fields, as shown by A.I. Lapin [3719] in 1971. It fails also in the case of reducible F (S. Chowla [1094]).

Upper bounds for $\sum_{|m|\le x} N_F(m)$ were obtained in 1987 by W.M. Schmidt [5520].

For cubic forms of non-vanishing discriminant K. Mahler [4079] proved in 1935 the relation

$$N_F(m) = \Omega\left(\log^{1/4}(|m|)\right).$$

After almost 50 years J.H. Silverman [5789] replaced the exponent $1/4$ by $1/3$, and C. Stewart [5947, 5948] increased that exponent first to $1/2$, and then to $6/7$. An explicit bound for the error term in (4.40) for cubic forms was given by J.L. Thunder [6149].

A generalization for forms of more variables was given much later by G. Lachaud [3599].

Explicit bounds for the solutions of the Thue–Mahler equations were later provided by Baker's method (see K. Győry [2402], Y. Bugeaud and K. Győry [815] and Y. Bugeaud [810]). Most of these equations have few solutions (E. Bombieri [607], J.-H. Evertse [1924, 1928], J.-H. Evertse, K. Győry [1933]). A method of finding all solutions was proposed by N. Tzanakis and B.M.M. de Weger [6244]. For a generalization to rings of algebraic integers see N.P. Smart [5825].

3. The equation

$$x^2 - Dy^4 = 1 \tag{4.41}$$

was studied by W. Ljunggren [3970, 3974, 3977] who showed that it can have at most two positive solutions.

The equation

$$x^4 - Dy^2 = 1 \tag{4.42}$$

also has at most two natural solutions (W. Ljunggren [3973]), and this is best possible, as the example $D = 1785$, with solutions $x = 13$ and 239 shows (see also W. Ljunggren [3977]). In 1989 M.H. Le [3741] proved that for large D there is at most one solution of (4.42), and in 1997 J.H.E. Cohn [1162] and Q. Sun and P.Z. Yuan [5989] showed that two solutions occur only for $D = 1785$.

Later W. Ljunggren [3976] established explicit bounds for certain other quartic equations (cf. M.A. Bennett, P.G. Walsh [421], J.H.E. Cohn [1166], P.G. Walsh [6537]).

The equation

$$ax^4 - by^2 = 1$$

was considered in 1967 by W. Ljunggren [3978, 3979], who showed that in certain cases it can have at most one positive solution. Much later S. Akhtari [38] established the bound 2 for the number of solutions for arbitrary $a, b > 0$, and it is now known that for $b = 1$ and $a \ge 3$ there is at most one solution (J.H. Chen, P.M. Voutier [1011]). Cf. M.A. Bennett, A. Togbé, P.G. Walsh [420], A. Togbé, P.M. Voutier, P.G. Walsh [6186].

[32] Paul Emery Thomas (1927–2005), professor in Berkeley.

The similar equation

$$ax^4 - by^2 = 2$$

has been considered by S. Akhtari, A. Togbé and P.G. Walsh [39, 40] and P.M. Voutier [6483] who proved that it has at most two solutions. It was conjectured in [40] that for odd ab there is at most one solution.

4. Although certain special cases of the assertion that the product of consecutive integers cannot be a proper power were considered by earlier authors (see [1545, p. 679]), it was stated first in 1857 by O. Terquem [6114] and E. Prouhet (editor's comment at the end of [22]). Using elementary methods it was shown by S. Narumi [4544] in 1917 that a product of at most 202 consecutive integers cannot be a square, and from a result in Siegel's paper [5746] on power values of polynomials it follows that if $m \geq 3$ and k are fixed, then for sufficiently large n the product

$$n(n+1)\cdots(n+k) \tag{4.43}$$

cannot be an mth power. In 1933 R. Obláth[33] [4648] dealt with several special cases of the problem, but the first important step towards a complete proof was made independently in 1939 by P. Erdős [1787] and O. Rigge [5227], who showed that the product (4.43) with $k \geq 2$ cannot be a square. Their arguments were based on an old result by Sylvester [6013], stating that a product of more than k consecutive integers is always divisible by a prime larger than k. This result is called usually the *Sylvester–Schur theorem*, since it was rediscovered in 1929 by I. Schur [5576, 5577]. A new proof was provided by P. Erdős [1773].

Consecutive improvements of a lower bound for the maximal prime divisor $P(n,k)$ of the product (4.43) were given by D. Hanson [2500] and M. Langevin [3709, 3710] (who also dealt with the same problem for products of terms of an arithmetic progression), and now it is known, that if $k > 200$ and $n > 279k/262$, then $P(n,k) > 2k$ (S. Laishram, T.N. Shorey [3614]).

Denote by $P(x)$ the maximal prime divisor of the product of all integers from the interval $[x, x+\sqrt{x}]$. The Sylvester–Schur theorem implies $P(x) \geq x^{1/2}$, and the first improvement was obtained in 1970 by K. Ramachandra [5058], who proved

$$P(x) \geq x^{\beta} \tag{4.44}$$

with $\beta = 15/26 = 0.5726\ldots$, and in [5059] got $\beta = 5/8 = 0.625$. In 1981 S.W. Graham [2301] obtained $\beta = 0.66$, and this was improved to $\beta = 0.7$ (R.C. Baker [255]), $\beta = 0.71$ (C.H. Jia [3132]), $\beta = 0.723$ (C.H. Jia [3138], H.Q. Liu [3952]), $\beta = 0.728$ (C.H. Jia [3139]), $\beta = 0.732$ (R.C. Baker, G. Harman [263]), and $\beta = 0.738$ (H.Q. Liu, J. Wu [3960]). The best known result, $\beta = 0.74$ is due to G. Harman [2565].

Stronger results have been achieved for the maximal prime divisor $P_{\varepsilon}(x)$ of the product of integers from the slightly larger interval $(x, x+x^{1/2+\varepsilon})$ with $\varepsilon > 0$. One can deduce from the Riemann Hypothesis that for large x one has $P_{\varepsilon}(x) > x$. In 1973 M. Jutila [3169] showed $P_{\varepsilon}(x) \gg x^{\gamma}$ for every $\gamma < 2/3$, seven years later A. Balog obtained $\gamma = 0.7388$ [295, 296] and $\gamma = 0.77$ [296]. This was improved to $\gamma = 0.82$ (A. Balog, G. Harman, J. Pintz [304]).

[33] Richárd Obláth (1882–1959), worked in an insurance company.

In 1996 D.R. Heath-Brown [2652] showed that one can take any $\gamma < 11/12 = 0.9166\ldots$, and this was increased to $\gamma < 17/18 = 0.9444\ldots$ by D.R. Heath-Brown and C.H. Jia [2665]. The best known result is due to Jia and M.C. Liu [3145], who in 2000 obtained (4.44) with any $\gamma < 25/26 = 0.9615\ldots$.

If $f(m)$ denotes the smallest integer k such that for $n \geq m$ the maximal prime divisor of $(n+1)(n+2)\cdots(n+k)$ exceeds k, then the Sylvester–Schur theorem implies $f(m) \leq m$. P. Erdős [1825] conjectured $f(m) = O(\log^c m)$ for some c, and the best known bound is due to T.N. Shorey [5721]:

$$f(m) \ll \frac{m \log\log\log m}{\log m \log\log m}.$$

Soon P. Erdős [1787] was able to extend his result by showing that for any fixed $m \geq 2$ and sufficiently large k, say for $k > k_0(m)$, the product (4.43) cannot be an mth power, and O. Rigge [5228] obtained this for $m \geq 17$.

It was stated in [1812] that P. Erdős and C.L. Siegel had proved that one can replace $k_0(m)$ by an absolute constant, but this result remained unpublished until P. Erdős [1812] found an elementary proof of this assertion. Finally, in the paper by P. Erdős and J.L. Selfridge [1857] it was shown that one has $k_0(m) = 2$ for all $m \geq 2$, and this settled the old problem. At the end of their paper they wrote: "No doubt our method would suffice to show that the product of consecutive odd integers is never a power.... In fact, the proof would probably be simpler." They also conjectured that the product of consecutive elements of a sufficiently long arithmetic progression $ax + b$, with $(a, b) = 1$ cannot be a power, and this was shown to be true by R. Marszałek [4159] in 1985. He modified the approach of P. Erdős and J.L. Selfridge to show that the equation

$$x(x+d)(x+2d)\cdots(x+(k-1)d) = y^n \tag{4.45}$$

has no solutions if $(x, d) = 1$ and $k > c\exp(d^{3/2})$ with some c. For $k \geq 5$ it suffices to have $k > cd$. Later these bounds were improved by T.N. Shorey and R. Tijdeman [5728–5730].

Note that there are infinitely many triples $a, b = a + d, c = a + 2d$ with a square product, and it is widely believed that this is the only possibility for the solution of (4.45).

It was proved by K. Győry [2408] that (4.45) has no solutions in the case $k = 3, n \geq 3$, and, as shown later, the same happens also for $k = 4, 5$ (K. Győry, L. Hajdu, N. Saradha [2413]), as well as for $6 \leq k \leq 11$ (M.A. Bennett, N. Bruin, K. Győry, L. Hajdu [415]), and $7 \leq k \leq 34$ (K. Győry, L. Hajdu, Á. Pintér [2412]).

In 1998 N. Saradha [5395] showed that in the case $n = 2$ and $d \leq 22$ there is only one solution: $x = 18, d = 7, k = 3$, and if $d \geq 23$, then $k < cd\log^2 d$ with $c = 1.3$ if $2|d$, and $c = 4$ otherwise. An algorithm to find all solutions for a fixed d has been produced by P. Filakovszky and L. Hajdu [2000].

H. Darmon and A. Granville [1334] showed that for fixed $k \geq 2$ and $n \geq 3$, with $k + n \geq 6$ there are at most finitely many solutions of (4.45).

The result of P. Erdős and J.L. Selfridge has been strengthened by J. Turk [6232], who proved in 1983 the existence of a constant c such that if $s \geq 2$, then the product of s distinct integers, all lying in an interval of the form $[N, N + c\log\log\log N]$ cannot be a proper power.

The related question of whether the binomial coefficients are proper powers has been considered by P. Erdős in [1787], who showed that the equation

$$\binom{n}{k} = x^j \tag{4.46}$$

has no solutions with $n \geq 2k$, $k \geq 2$ and $j = 3$.

It was later noted by R. Obláth [4649] that the same holds in the cases $j = 4, 5$ and in 1951 P. Erdős [1804] extended this to all $k \geq 4$, $j \geq 2$. The final step was made in 1997 by H. Darmon and L. Merel [1335] and K. Győry [2406], and now it is known that the only solution of (4.46) apart from the case $k = j = 2$ is $n = 50$, $k = 3$, $j = 2$, $x = 140$.

A conjecture by P. Erdős, asserting that the middle binomial coefficient $\binom{2n}{n}$ is not square-free for $n \geq 5$, was established in the nineties independently by A. Granville and O. Ramaré [2322] and G. Velammal [6376]. Earlier A. Sárkőzy [5402] had shown its truth for sufficiently large n.

For studies of prime divisors of other binomial coefficients see E.F. Ecklund, Jr., P. Erdős, J.L. Selfridge [1682], P. Erdős, G. Kolesnik [1839], P. Erdős, C.B. Lacampagne, J.L. Selfridge [1840], S.V. Konyagin [3471], J.W. Sander [5384, 5385].

It has been conjectured that the only integers which can be written at least twice in the form $\binom{n}{k}$ with $2 \leq k \leq n/2$ are 120, 210, 1540, 3003, 7140, 11628, 24310 and

$$c_n = \binom{F_{2n+2}F_{2n+3}}{F_{2n}F_{2n+3}},$$

F_n being the nth Fibonacci number. It is known that there no others below 10^{30} (B.M.M. de Weger [1513]). The corresponding equation

$$\binom{x}{k} = \binom{y}{l}$$

reduces for certain small k, l to the equation of an elliptic curve and this case has been resolved in 1999 by R.J. Stroeker and B.M.M. de Weger [5977].

5. A new method in the study of Diophantine equations was invented by T. Skolem [5805, 5807]. It involved a study of exponential equations in the p-adic setting, and led to a new proof of Thue's theorem on the equation (2.92) in the case when the polynomial $F(X, 1)$ has at least one non-real zero.

This method works also in the case, when $F(X, 1)$ has only real zeros, as shown in 1941 by C. Chabauty[34] [975], who also showed that Skolem's method leads to a proof of Mahler's theorem on prime divisors of form values. W. Ljunggren [3972] successfully applied Skolem's method to equations involving cubic forms.

4.4.2 Representations of Integers by Forms

1. In 1935 C.L. Siegel [5750] published the first of his three important papers on quadratic forms in $n \geq 2$ variables. Its main idea came from the observation that the Hasse principle holds neither for the problem of representability of integers by quadratic forms at integral arguments, nor for the equivalence (over $\mathbf{Z}$) of forms with integral coefficients. Siegel defined two forms with integral coefficients

[34] Claude Chabauty (1910–1990), professor in Grenoble.

and the same discriminant to lie in the same *genus* if they are equivalent over all completions of **Q**. Denote by A_f the matrix of a form f, thus $f = x'A_f x$ (with $x = (x_1, \dots, x_n)$ and x' being the transpose of x). A form f is said to be integrally transformable in a form g, if there exists a matrix C with integral entries such that $A_g = C'A_f C$, and it is locally transformable if this happens with **Z** replaced by $\mathbf{Z}_p$ (for all primes p) and **R**. Siegel discovered a relation between the number of solutions $a_p(f, g)$ of $X'A_f X = A_g$ with X being a matrix over $\mathbf{Z}_p$ and the numbers of solutions of $X'A_\phi X$ with an integral matrix X for all forms ϕ lying in the genus of f. In special cases this relation implies Dirichlet's formula for the class-number of binary quadratic forms. It explains also why it is only possible to obtain explicit formulas for the number of representations of an integer by a quadratic form in the case of at most 8 variables. He also studied representations of a quadratic form by another form (see A. Weil [6625] for another approach to this problem).

In the first paper of the series Siegel presented his theory in the case of positive definite forms over **Z**, and in the next two ([5751, 5752], see also [5763, 5769, 5770]) extended it first to indefinite forms over **Z**, and then to forms with integral algebraic coefficients. He also gave an interpretation of his results in the language of modular forms.

A further three of his papers from that period dealt with quadratic forms: in the first two [5755, 5756] the zeta-functions of indefinite forms were introduced and studied, and in the third [5758] Siegel showed that the group of integral automorphisms of an indefinite form with integral coefficients is finitely generated.

The last result was generalized by P. Humbert [2950] to forms having integral coefficients in an algebraic number field. Cf. K. Ramanathan [5069, 5070].

In [5758] Siegel also showed that for certain linear algebraic groups G over complex numbers the invariant measure of $G_\mathbf{R}/G_\mathbf{Z}$ (where $G_\mathbf{R}, G_\mathbf{Z}$ denote the sets of elements of G with real or integral entries, respectively) is finite.

The last assertion was later established for all complex linear groups by A. Borel and Harish-Chandra [633].

2. In [3900, 3901] Yu.V. Linnik applied quaternions to study the representation of integers by positive definite ternary quadratic forms. This method was first applied by B.A. Venkov [6382] in the case of the form $x^2 + y^2 + z^2$. See also L.J. Mordell [4388] where quaternions have been applied to problems of representing a quadratic form as the sum of squares.

For further development of this method see A.V. Malyšev [4118–4122], Yu.V. Linnik, A.V. Malyšev [3931].

In 1985 Yu.G. Teterin [6122] used Linnik's ergodic method to obtain an asymptotic formula for the number of representations by a positive definite ternary quadratic form (see also E.P. Golubeva [2277]).

3. E. Artin conjectured that a form $f(X_1, \dots, X_n)$ of degree d with p-adic coefficients and $n > d^2$ represents zero non-trivially in every p-adic field. This assertion, if true, would be optimal as for every $n \geq 1$ and prime p L.J. Mordell [4387] constructed forms f of degree n in n^2 variables over $\mathbf{Q}_p$, for which the equation $f(x_1, \ldots, x_n) = 0$ has only the zero solution. The truth of Artin's conjecture for $d = 2$ follows from a theorem in H. Hasse's thesis [2577].

A related question of Artin's was answered by C. Chevalley [1048], who showed that if $f_1, \ldots, f_m$ are polynomials in n variables over a finite field k having a common zero in k, and $n > \sum_{j=1}^{m} \deg f_j$, then there is also another common zero in k. This was strengthened by E. Warning [6568], who proved that under these conditions the number of common zeros in k is divisible by the characteristics of k.

The Chevalley–Warning theorem was generalized by J. Ax[35] [175], N.M. Katz [3277] and S.H. Schanuel [5421] (cf. also D.Q. Wan [6541]). For a survey see J.-R. Joly [3150].

In 1936 C.C. Tsen[36] [6206] defined a field K to be of level α if every finite set of forms $f_1, \ldots, f_m$ in n variables over K satisfying $n > \sum_{j=1}^{m} \deg^{\alpha} f_j$ has a common non-trivial zero in K. This notion is closely related to that of *C_α-fields*, which are fields in which every form of degree r in more than r^α variables has a non-trivial zero (S. Lang [3685]; cf. G. Terjanian [6118]). Chevalley–Warning's result shows that finite fields are C_1-fields. Artin's conjecture on forms can be reformulated as follows: *p-adic fields are C_2-fields.*

The case $d = 3$ of Artin's conjecture was settled for $p \neq 3$ by V.B. Demyanov [1462] in 1950, and for every p by D.J. Lewis [3870] in 1952.

In 1960 B.J. Birch and D.J. Lewis [532] showed that a form of degree $d \in \{1, 2, 3, 5\}$ over $\mathbf{Q}_p$ in $1 + d^2$ variables represents zero non-trivially, provided the prime p is sufficiently large, and the same is true for all finite extensions of the p-adic fields. The same holds also for degrees 7 and 11, as shown by R.R. Laxton and D.J. Lewis [3740] in 1965.

In 1963 H. Davenport and D.J. Lewis [1395] succeeded in establishing Artin's conjecture for diagonal forms

$$\sum_{j=1}^{s} a_j X_j^d \tag{4.47}$$

of degree $d \geq 18$. This was later extended by R.C. Vaughan [6352] to the case $11 \leq d \leq 17$.

If one denotes by $\Gamma^*(d)$ the minimal value of s such that every form (4.47) has a non-trivial zero in every p-adic field, then this gives $\Gamma^*(d) \leq 1 + d^2$. It was also shown in [1395] that if $1 + d$ is prime, then $\Gamma^*(d) = 1 + d^2$. For odd values of d S. Chowla and G. Shimura [1101] established

$$\Gamma^*(d) < \left(\frac{2}{\log 2} + o(1)\right) d \log d,$$

and showed that for infinitely many odd d this bound cannot essentially be improved. Later A. Tietäväinen [6156] halved this estimate and proved that this bound is best possible. The values $\Gamma^*(3) = 7$ and $\Gamma^*(5) = 16$ were determined earlier by D.J. Lewis [3872] and

---

[35] James Burton Ax (1937–2006), professor at Cornell University and at Stony Brook.

[36] Chiungtze C. Tsen (1898–1940), student of E. Noether, professor at Beniyang University and the National Northwestern United University. See [1578].

S. Chowla [1091], respectively. For even $d \neq 8, 32$ with composite $1+d$ M. Dodson [1605] obtained

$$\Gamma^*(d) \le \frac{d^2}{2} + O(d),$$

and showed that for all d one has $\Gamma^*(d) \ge 1+d$. The same author considered in [1606] the mean value of $\Gamma^*(d)$, establishing

$$\sum_{d=1}^{N} \Gamma^*(d) = \left(\frac{\pi^2}{18} + o(1)\right)\frac{N^3}{\log N}.$$

Later the equality $\Gamma^*(8) = 39$ was established by J.D. Bovey [668]. All values of $\Gamma^*(k)$ for $k \le 12$ are listed in this paper.

An important result was obtained in 1966 by G. Terjanian [6117] who produced a biquadratic form in 18 variables over $\mathbf{Q}_2$ without a non-trivial zero. This disproved Artin's conjecture in its original form. A family of similar counterexamples in every field $\mathbf{Q}_p$ was presented by J. Browkin [745, 746] (see also G. Terjanian [6120, 6121]). These examples show that even the change in Artin's assumption from $n > d^2$ to $n > d^{3-\delta}$ (n being the number of variables and d the degree) does not save the conjecture. The question of whether Artin's assertion holds when the assumption $n > d^2$ is replaced by $n > d^m$ for some fixed m got a negative answer in 1981 (G.I. Arkhipov and A.A. Karatsuba [121, 122]; cf. W.D. Brownawell [759], D.J. Lewis, H.L. Montgomery [3874]). An analogous result also holds for finite extensions of $\mathbf{Q}_p$ (Y. Alemu [50], V.T. Vilčinskiĭ [6396]). Note that for all known counterexamples to Artin's conjecture in $\mathbf{Q}_p$ one has $d(d-1)|p$, and G. Terjanian [6121] asked whether this is always so (cf. L. Chakri, M. El Hanine [982]).

In 1965 J. Ax and S. Kochen [179] showed that Artin's conjecture holds in $\mathbf{Q}_p$ for forms of every degree d provided $p \ge M(d)$ with a certain $M(d)$. Their result was the first successful application of model theory to arithmetical problems.

The first explicit although enormous bound for $M(d)$ was given by S. Brown [754]. For certain small d better bounds are known: $M(2) = 2$ (H. Hasse [2580]), $M(3) = 2$ (D.J. Lewis [3870]), $M(5) \le 41$ (D.B. Leep, C.C. Yeomans [3765]), $M(7) \le 883$, $M(11) \le 8053$ (T.D. Wooley [6737]; earlier M.P. Knapp [3401] had bounds of orders 10^{17} and 10^{19}, respectively).

It was shown by D.B. Leep and W.M. Schmidt [3763] that it suffices to have

$$n > B(c)\exp\left((d!)^2 c^d\right)$$

for every $c > 1$, and this bound was reduced by W.M. Schmidt [5516] to $d > B\exp(2^d d!)$ with a suitable B.

The first general result concerning systems of forms was obtained in 1945 by R. Brauer [688], who showed that if K is an arbitrary field having the property that there exists a constant $\Phi(d, K)$ such that every diagonal form f of degree d in at least $\Phi(d, K)$ variables has a non-trivial zero in K, then the system

$$f_j(x_1, \dots, x_n) = 0 \quad (j = 1, 2, \dots, r), \tag{4.48}$$

(where f_i are forms over K with $\deg f_j = d_j$) has a non-trivial zero for $n \ge \psi(K, d_1, \dots, d_r)$. (For an explicit dependence of $\psi(K, d_1, \dots, d_r)$ on $\Phi(d, K)$ see D.B. Leep and W.M. Schmidt [3763] and T.D. Wooley [6733, 6734].)

This result implied that if $f_1, \dots, f_r$ are forms in n variables over $\mathbf{Q}_p$ and n is large, say $n \ge \psi(p, d_1, \dots, d_r)$ (where $d_j = \deg f_j$), then $\mathbf{Q}_p$ contains a non-trivial solution of the system (4.48).

R. Brauer also proved that if the forms f_i have coefficients in an algebraic number field K, and n is sufficiently large, then (4.48) has a solution in a suitable solvable extension of K. This has also been shown by B. Segre[37] [5603] in a geometrical setting. In the case of totally complex K it has been shown by L.G. Peck [4762] that for large n (4.48) has solutions already in K. In 1957 B.J. Birch [521] extended Peck's result by showing that any system of forms over an algebraic number field with odd degrees has a common non-trivial zero, provided the number of variables exceeds a value depending on the field and on the degrees of the forms. The case of cubic forms was treated earlier by D.J. Lewis [3871]. An explicit version of Birch's result was given by T.D. Wooley [6732, 6733].

Later B.J. Birch [525] extended Brauer's result to finite extensions of p-adic fields and gave explicit bounds for $\Phi(d, K)$, independent of p. Later M. Dodson [1609] found smaller bounds, of order $O((dN \log d)^2)$ with N being the degree of the extension $K/\mathbf{Q}_p$.

In 1952 S. Lang conjectured in his thesis [3685] the inequality $\psi(d_1, \dots, d_r) \leq \sum_{j=1}^r d_j^2$. He was able to prove the corresponding assertion for the field of formal power series over a finite field and V.B. Demyanov [1463] proved Lang's conjecture for pairs of quadratic forms (cf. [534]). For triples of quadratic forms and $p \geq 53$ this conjecture was established in 1964 by B.J. Birch and D.J. Lewis [533], and F. Ellison [1752] did this for diagonal f and odd[38] p. For systems of 3 diagonal cubic forms in the case $p \neq 3, 7$ this was done in 1982 by E. Stevenson [5945].

An approximation to Lang's conjecture in the cases $d_1 = \dots = d_r = 2$ or 3 was obtained by W.M. Schmidt [5512, 5514], who got

$$\psi(2, 2, \dots, 2) \leq 4r(r+1), \qquad \psi(3, 3, \dots, 3) \leq 5300r(3r+1)^2$$

(Lang's conjecture would give the bounds $4r$ and $9r$, respectively). In the first inequality the right-hand side was later replaced by $2r^2 + 2r - 4$ (D.B. Leep [3762]).

G. Terjanian's result, quoted above, disproved Lang's conjecture, but, as in the case of Artin's conjecture, it turned out that it holds in $\mathbf{Q}_p$ for large p (J. Ax, S. Kochen [179]).

Systems of diagonal equations

$$\sum_{j=1}^{n} a_{ij} X_j^k = 0 \quad (i = 1, 2, \dots, r) \tag{4.49}$$

with integral coefficients were investigated by H. Davenport and D.J. Lewis [1397] in 1969. They showed that if n exceeds an explicitly given value, depending on k and r, then (4.49) has a non-trivial p-adic zero, and in the case of odd k this happens in $\mathbf{Q}$. Earlier [1396] they considered the case of pairs of diagonal cubic forms, showing the existence of a p-adic zero for $n \geq 16$ (the case $p = 7$ shows that this is best possible) and a rational zero for $n \geq 18$.

The result for cubic forms was consecutively improved by R.J. Cook [1231] and R.C. Vaughan [6353].

In the case $R = 2$ and odd $k \geq 19$ there are integral solutions provided $N \geq 2k^2 + 1$. For $k = 5$ it suffices to have $N \geq 51$ (R.J. Cook [1232, 1233]) and for $k = 6$ some sufficient conditions for the existence of solutions for $N \geq 73$ were given in [1234]. It was shown later by J. Brüdern [771] that in the quintic case it suffices to have $N \geq 37$, provided there are zeros in $\mathbf{Q}_{11}$, and S.T. Parsell and T.D. Wooley reduced this to $N \geq 34$

---

[37] Beniamino Segre (1903–1977), professor in Bologna and Rome. See [6040].

[38] She wrote: "The author has verified Artin's conjecture for the case $p = 2$ as well, but the proof is prohibitively long for inclusion here."

[4747]. In the cubic case 14 variables suffice if there is a solution in $\mathbf{Q}_7$ (J. Brüdern [770]).

In 1985 R.J. Cook showed [1243] that a p-adic zero already exists for $n \geq 13$, with the exception of $p = 7$. This was used in 1988 by R.C. Baker and J. Brüdern [258] to show that if two cubic diagonal forms in 15 variables have a common non-trivial zero in $\mathbf{Q}_7$, then they have such a zero in $\mathbf{Q}$. Later J. Brüdern [770] replaced 15 here by 14. Finally J. Brüdern and T.D. Wooley [780] established the Hasse principle for pairs of cubic diagonal forms in $n \geq 13$ variables. Pairs of quintic forms were considered by R.J. Cook [1244, 1245].

In 1988 L. Low, J. Pitman and A. Wolff [4004] improved the result in [1397] by showing that any system of $R \geq 3$ diagonal forms of degree k in $N \geq 48Rk^3 \log(3Rk^2)$ variables has a non-trivial zero in every $\mathbf{Q}_p$. This was in turn improved in 1999 by J. Brüdern and H. Godinho [774], who relaxed the condition to $N \geq R^3k^2$, unless $R = 3$ and k is a power of two, in which case one must have $N \geq 36k^2$. In [775] they considered the case $R = 2$ and showed that $N \geq 8k^2$ suffices. In this case for $p > k^6$ just $N > 4k$ is enough, as shown in 1989 by O.D. Atkinson, and R.J. Cook [171]. The last improvement in the case $R \geq 3$ is due to M.P. Knapp [3400], who showed that $N \geq 4R^2k^2$ is sufficient. For primes $p > p_0(k, R)$ then $N > 2kR$ is sufficient, as shown in 1990 by E. Dörner [1614], and it was shown by Atkinson, J. Brüdern, R.J. Cook [170] that one can take $p_0(k, R) = k^{2R+2}$, confirming a conjecture by T.D. Wooley, who in a cycle of three papers [6721–6723] treated systems of diagonal forms having different degrees.

The book [2338] by M.J. Greenberg, published in 1969, surveys the problems arising from Artin's conjecture.

Small zeros of diagonal forms with integral coefficients and odd degree were considered by B.J. Birch [528] and W.M. Schmidt [5510, 5511]. See also J.S. Hwang [2995, 2996], Y. Wang [6555, 6556].

The number $N_\Phi(T)$ of solutions $(x_1, \dots, x_s)$ with $|x_i| \leq T$ of the system

$$\Phi : F_i(x_1, \dots, x_s) = 0, \quad \dots, \quad F_r(x_1, \dots, x_s) = 0$$

of forms of odd degrees $\leq k$ was evaluated by W.M. Schmidt [5518], who obtained

$$N_\Phi(T) \gg T^{s-c},$$

where $c > 0$ depends on k and r, and the implied constant depends on Φ.

4.4.3 Exponential Equations

In 1931 S.S. Pillai [4861] considered exponential equations of the form

$$ra^x - sb^y = c \tag{4.50}$$

with given a, b, r, s and non-zero c, and showed that for every $\lambda < 1$ and sufficiently large x one has

$$|ra^x - sb^y| > a^{\lambda x}.$$

An effective version of this result was given by W.J. Ellison [1753] in 1970.

The finiteness of the set of solutions of (4.50) had been known earlier (A. Thue [6141], G. Pólya [4956]).

In the case $r = s = 1$ S.S. Pillai [4862] showed in 1936 that for $c > c_0(a, b)$ the equation (4.50) has at most one solution. For $a = 3, b = 2$ this had been established earlier by A. Herschfeld [2771]. S.S. Pillai also showed that the number of positive integers $c \leq N$ for which this equation has solutions is asymptotically equal to $\log^2 N/(2 \log a \log b)$. He conjectured (*Pillai conjecture*) that (4.50) can have for $r = s = 1$ and every fixed $c > 0$ only finitely many positive solutions a, b, x, y. The case $c = 1$ forms the *Catalan conjecture* (see Sect. 6.6).

Later S.S. Pillai conjectured [4872] that $c_0(3, 2) = 13$, and this was established in 1982 by R.J. Stroeker and R. Tijdeman [5972] who used Baker's method. This result was later extended by M.A. Bennett [413] who showed $c_0(N + 1, N) = 13$ for every $N \geq 2$. Bennett showed also [410] that for all $a, b \geq 2$ the equation (4.50) has at most two solutions ≥ 1.

The Pillai conjecture can be deduced from a conjecture, which was stated by D.W. Masser (in the problem section of [1037, p. 25]) and J. Oesterlé [4663] at the end of the century, and is called the *ABC conjecture*. To formulate it denote by $R(n)$ the product of all prime divisors of n.

The ABC conjecture *If a, b, c are relatively prime positive rational integers satisfying $a + b = c > 2$, then for every $\varepsilon > 0$ one has*

$$c \leq B(\varepsilon) R(abc)^{1+\varepsilon}, \tag{4.51}$$

with a certain $B(\varepsilon)$.

The first result of this type was established in 1986 by C.L. Stewart and R. Tijdeman [5949], who showed that the above assumptions imply the bound

$$\max\{|a|, |b|, |c|\} < \exp\left(B R(abc)^{15}\right)$$

with a certain B. Later this was improved by C.L. Stewart and K. Yu [5950, 5951] to

$$c \ll \exp\left(B(\varepsilon) R(abc)^{2/3+\varepsilon}\right).$$

The ABC conjecture can also be stated in the following way.

If a, b are co-prime positive integers, $c = a + b$, and we put

$$L(a, b) = \frac{\log c}{\log(R(abc))},$$

then the set of values of $L(a, b)$ is bounded, and its maximal limit point equals 1.

The largest known value of $L(a, b)$ is $1.62991\ldots$, and corresponds to $a = 2, b = 109 \cdot 3^{10}$ in view of the equality $2 + 109 \cdot 3^{10} = 23^5$ (E. Reyssat).

This conjecture has many important consequences. It easily implies, for example, Fermat's Last Theorem for sufficiently large exponents. In 1991 N.D. Elkies [1717] showed that it can be also used to obtain a simple proof of Mordell's conjecture stating that there can be only finitely many rational points on a plane curve of genus exceeding 1. This is a special case of the much more general theorem of G. Faltings [1954] (see Sect. 7.2).

Several further consequences of the ABC conjecture are listed together with appropriate references on the web page maintained by A. Nitaj: http://www.math.unicaen.fr/~nitaj. An account of the ABC conjecture and its consequences was given by J. Browkin [747].

In 2000 A. Granville and H.M. Stark [2327] showed that a generalization of the ABC conjecture to algebraic number fields (due to N.D. Elkies [1717] and P. Vojta [6451, 6456], cf. J. Browkin [748]) implies the non-existence of Siegel zeros for Dirichlet's L-functions corresponding to negative discriminants. In 2007 A. Surroca [5991] proved that the ABC conjecture makes Siegel's theorem on real zeros of L-functions effective, and in [5992] established a kind of converse (cf. L. Moret-Bailly [4425]). The best known approximation to the ABC conjecture in number fields was obtained in 2008 by K. Győry [2411].

Note that the analogue of the ABC conjecture for polynomials is now a theorem, established by R.C. Mason [4168] and W.W. Stothers [5963]. Another proof was found by K. Yamanoi [6776] (see G. Gasbarri [2205] for an exposition of Yamanoi's proof as well as of a proof found by M. McQuillan).

Another analogue in this case was given by L.N. Vaserstein [6339]. An analogous result for function fields was proved by P.-C. Hu and C.-C. Yang [2913, 2914].

4.4.4 Other Equations

1. Mahler's theorem on equation (4.39) in [4068–4070] implies in particular that the equation

$$x_1 + x_2 = x_3 \tag{4.52}$$

has only finitely many solutions in co-prime integers x_i, having all their prime factors in a fixed finite set, since one can write $x_1 = ay_1^5$, $x_2 = by_2^5$ with a, b in a finite set, and apply Mahler's result to $ax^5 + by^5$.

This implies in particular that for fixed integers $a, b, c \geq 2$ the equation

$$a^x + b^y = c^z \tag{4.53}$$

has only finitely many solutions.

It was conjectured by M.H. Le [3750] that (4.53) can have at most one solution with $x, y, z \geq 2$ (for an earlier version see N. Terai [6113]). This extends an earlier conjecture by L. Jeśmanowicz [3126] stating that if $a^2 + b^2 = c^2$, then (4.53) has no other solutions. This conjecture has been established in several particular cases (see, e.g., M. Cipu and M. Mignotte [1117] and M.H. Le [3750–3752]).

An effective method to find all solutions of (4.52) was given by J.W.S. Cassels [942] in 1961.

Mahler's result was later extended to a much more general setting. First S. Lang [3687] and D.J. Lewis and K. Mahler [3873] showed in 1960 that the equation

$$u_1 + u_2 = 1$$

can have only finitely many solutions in units of $\mathbf{Z}_K$, the ring of integers of a finite extension of the rationals. All solutions of this equation in the case when K is either quadratic or cubic with a negative discriminant were found in 1959 by T. Nagell [4515].

A solution of the more general equation

$$\sum_{j=1}^{n} x_j = 0 \tag{4.54}$$

is called non-trivial, if none of the proper subsums vanishes.

The analogue of Mahler's assertion on the equation (4.52) has been established for non-trivial solutions of (4.54) by E. Dubois and G. Rhin [1632], and independently by H.P. Schlickewei [5462], the case $n = 3$ having been treated earlier by T. Schneider [5542]. The next step was made in the eighties, when J.-H. Evertse [1923] proved that the equation

$$\sum_{j=1}^{n} u_j = 1$$

with u_j lying in a finitely generated subgroup of the multiplicative group of an algebraic number field can have at most finitely many non-trivial solutions, and A.J. van der Poorten and H.P. Schlickewei [6304] extended this to cover the case of any finitely generated field of characteristic 0. Another proof was given by J.-H. Evertse and K. Győry [1932].

The first explicit bound for the number of solutions of linear unit equations was given in 1979 by K. Győry [2401], who considered the equation

$$au_1 + bu_2 = c \tag{4.55}$$

(with $abc \neq 0$) in S-units of an algebraic number field K (S-*units* are invertible elements of the ring K_S of S-*integers* of an algebraic number field K, defined as elements $a \in K$ such that the $\mathfrak{p}$-adic valuation of a is non-negative for all prime ideals of $\mathbf{Z}_K$ lying outside a fixed finite set S). He obtained an upper bound for the number of solutions in the case when c has a sufficiently large norm. A similar result without the last assumption was obtained in 1984 by J.-H. Evertse [1924], who showed that (4.55) can have at most $3 \cdot 7^{d+2s}$ solutions in S-units of a field K, with d being the degree of $K/\mathbf{Q}$ and $s = \#S$ (note that in most cases there are at most two solutions, as shown by J.-H. Evertse, K. Győry, C.L. Stewart and R. Tijdeman [1937]).

In the following year J.-H. Evertse and K. Győry [1930] proved that a similar bound applies when the group of S-units is replaced by any finitely generated multiplicative group contained in K. Later these results were extended to non-trivial solutions of equations of the form

$$\sum_{j=1}^{n} a_j u_j = 1 \tag{4.56}$$

(J.-H. Evertse, K. Győry [1931], H.P. Schlickewei [5465, 5466], A.J. van der Poorten, H.P. Schlickewei [6306]). A general result of this type was achieved in 2002 by J.-H. Evertse, H.P. Schlickewei and W.M. Schmidt [1940], who proved (cf. also [1939]) that if K is a field of characteristic 0, and $G \subset K$ is a finitely generated multiplicative group of rank r, then the equation (4.56) has at most $\exp((r+1)(6n)^{3n})$ non-trivial solutions in G (for $n = 2$ a stronger bound, $\exp(8\log 2(r+1))$, was earlier established by F. Beukers and H.P. Schlickewei [492]; cf. also E. Bombieri, J. Mueller and M. Poe [619]). They conjectured that this bound may be replaced by

$$\exp\left(c(n)r^{n/(n+1)}\right)$$

with a certain $c(n)$. On the other hand they observed that for each $r \geq 2$ there are equations (4.56) with at least

$$\exp\left(a(n)(r/\log r)^{n/(n-1)}\right)$$

solutions with a suitable $a(n) > 0$ (for $n = 2$ this was proved earlier by P. Erdős, C.L. Stewart and R. Tijdeman [1860]). Recently F. Amoroso and E. Viada [68] showed that (4.56) has at most $\exp(4n^4(n+r+1)\log n)$ solutions. Bounds for the number of solutions of (4.55) in S-units of $\mathbf{Q}$ were given by M. Poe [4927].

Effective bounds for all solutions of (4.55) in S-units of an algebraic number field were provided by Y. Bugeaud, K. Győry [815], Y. Bugeaud [810] and K. Győry, K. Yu [2416].

The finiteness of the set of non-trivial solutions of (4.55) was later applied to the arithmetic of dynamical systems (see, e.g., [4541]).

Surveys of the theory of unit equations were given by J.-H. Evertse, K. Győry, C.L. Stewart and R. Tijdeman [1936] in 1988, K. Győry [2404] in 1992 and J.-H. Evertse and H.P. Schlickewei [1938] in 1999.

2. It has been known for a long time that if an integer a is a quadratic residue for almost all primes (i.e., for all primes except finitely many), then a is a square (see, e.g., G. Rados [5046]). The analogue of this for higher exponents fails in certain cases, and the resulting question was settled in 1934 by E. Trost [6201] who showed that if the congruence

$$X^n \equiv a \pmod p$$

is solvable for almost all primes p, then either a is an nth power, or n is divisible by 8 and a has the form

$$a = 2^{n/2}b^n$$

with some integer b.

Trost's result has been rediscovered in 1951 by N.C. Ankeny and Rogers [106] and generalized to algebraic number fields by H. Flanders [2011] and H.B. Mann [4140]. For another generalization see I. Gerst [2235] (cf. A. Schinzel [5441]).

4.5 Elliptic Curves

1. An important step towards the understanding of the torsion subgroup of an elliptic curve over $\mathbf{Q}$ was made by É. Lutz[39] [4034] and T. Nagell [4511] who proposed a simple algorithm for its computation. They proved (the *Nagell–Lutz theorem*) that rational points of finite order on $E(Q)$ defined by an equation of the form $y^2 = x^3 + ax + b$ with integral a, b must have integral coordinates. This may fail for general elliptic curves, as shown by the example

$$y^2 + xy = x^3 + 4x + 1,$$

(see the book [5791, p. 178] by J.H. Silverman), but is true for points whose order is not a prime power (J.W.S. Cassels [928], G. Bergman [436]).

A quicker algorithm was proposed by D. Doud [1617] in 1998. An algorithm for curves over an algebraic number field has been given by F. Châtelet[40] [1002, 1005].

[39] Élisabeth Lutz (1914–2008), professor in Grenoble.

[40] François Châtelet (1912–1987), son of A. Châtelet, professor in Besançon. See [1185].

2. The number $r(E(k))$ of infinite summands in (3.46), called the *rank* of $E(k)$, is still a mystery. Even in the simplest case $k = \mathbf{Q}$, it is not known whether it can be arbitrarily large. Computation of the rank is a difficult task, which can be seen from the fact that the first example of a curve of rank larger than 2 was found only in 1938 by G. Billing [506]. He also gave an upper bound for the rank of $E(Q)$ in the case when the right-hand side of (3.43) is irreducible.

In 1944 A. Wiman[41] [6686–6688] produced examples of curves of rank ≥ 4 (note that the rank definition used by Wiman differs from ours as he counted all independent generators, including those of finite order). In 1952 A. Néron [4558] proved the existence of a curve of rank ≥ 10, and in [4559] sketched a construction of a curve of rank 11 (the details were provided later by M. Fried [2097]).

The first explicit examples of curves having rank ≥ 6 and ≥ 7 were found by D.E. Penney and C. Pomerance [4768, 4769] in 1974 and 1975, respectively. Curves of ranks ≥ 8 and ≥ 9 were later found by F.J. Grunewald[42] and R. Zimmert [2374] and A. Brumer and K. Kramer [795]. J.-F. Mestre presented a method of finding curves of high rank [4263–4265] which has been used in the construction of curves up to rank ≥ 28 (N.D. Elkies [1724]). See also T.J. Kretschmer [3522] and the survey paper by K. Rubin and A. Silverberg [5332].

A table of consecutive records can be found on A. Dujella's web page: http://web.math.hr/~duje/tors/rankhist.html.

It was proved by D. Clark [1118] that if the number of solutions of the system

$$xy = n, \qquad x + y = z^2$$

is unbounded on the set of square-free n, then there exist curves over $\mathbf{Q}$ with arbitrarily large rank.

In the case when k is the field of rational functions over a finite field J. Tate and I.R. Šafarevič [6065] proved the existence of elliptic curves having arbitrarily large rank. The same is true also for fields of algebraic functions over $\mathbf{C}(x)$ (I.A. Lapin [3717, 3718]).

3. It turned out later that the zeta-functions introduced in 1931 by F.K. Schmidt [5479] play an important role in the theory of elliptic curves. Schmidt defined them for every finite extension F of the field $\mathbf{F}_q(X)$ of rational functions over the finite field $\mathbf{F}_q$ by the equality

$$\zeta_F(s) = \prod_{\mathfrak{p}} \frac{1}{1 - |\mathfrak{p}|^{-s}}, \qquad \Re s > 1,$$

the product running over all prime divisors $\mathfrak{p}$ of F. The prime divisors $\mathfrak{p}$ of F were defined as the unique maximal ideals of integrally closed local subrings of R, i.e., proper subrings R of K in which the non-invertible elements form an ideal (which is necessarily maximal). Schmidt put $|\mathfrak{p}| = p^f$, where f is the degree of the extension $(R/\mathfrak{p})/\mathbf{F}_q$ and called his functions the *congruence zeta-functions*

In the case of a quadratic extension this function differs only by a finite number of factors from the zeta-function introduced by E. Artin in [136, 137] and from

[41] Anders Wiman (1865–1959), professor in Lund and Uppsala. See [4516].

[42] Fritz Grunewald (1949–2010), professor in Düsseldorf.

the zeta-function considered earlier by F.K. Schmidt in [5476]. Schmidt proved that $\zeta_F(s)$ can be extended to a meromorphic function with infinitely many poles at lines $\Re s = 0$ and $\Re s = 1$. He showed moreover that it satisfies a functional equation of the form

$$p^{(g-1)s}\zeta_F(s) = p^{(g-1)(1-s)}\zeta_F(1-s),$$

with some integer g, the *genus* of F, and after introducing the new variable $t = p^{-s}$ $\zeta_F(s)$ turned out to be a rational function of t of the form

$$\frac{L(t)}{(1-qt)(1-t)}, \tag{4.57}$$

with

$$L(t) = q^g t^{2g} + \cdots + (N - q - 1)t + 1,$$

where N denotes the number of prime divisors of degree 1. In the case when $f(X,Y)$ is irreducible over K and $F = K(X,Y)$ with $f(X,Y) = 0$, then, as pointed out by H. Hasse [2595], the number N is related to the number N_f of solutions of the equation $f(x,y) = 0$ in K by

$$0 \le N - N_f \le C,$$

where C depends only on the degree of f. In particular, if f is the reduction mod q of a polynomial $g \in \mathbf{Z}[X,Y]$, then N differs from the number N_g of solutions of the congruence

$$g(x,y) \equiv 0 \pmod q$$

by a constant.

Applying Schmidt's construction of functions $L(t)$ to elliptic curves over the rationals one is led to L-functions of an elliptic curve E by putting

$$L_E(s) = \prod_p \frac{1}{L_p(p^{-s})},$$

where $L_p(t)$ is defined for primes p at which E has good reduction as the function $L(t)$ occurring in the numerator in (4.57) in the case of the field $\mathbf{F}_p$, and in other cases $L_p(t)$ equals either $1 \pm t$ or 1, according to the form of bad reduction at p. For curves over other algebraic number fields one proceeds in the same way, the role of primes being taken by prime ideals of rings of integers.

In [2595] H. Hasse formulated the analogue of the Riemann Hypothesis for zeta-functions of algebraic function fields[43] in the following form.

If one writes

$$L(t) = \prod_{i=1}^{2g}(1 - \omega_i t),$$

[43] Now it is customary to call it the Riemann Hypothesis for curves.

then for all i one has

$$|\omega_i| = q^{1/2}. \tag{4.58}$$

Note that its truth implies the bound $|N_g - q| \ll q^{1/2}$.

A special case of (4.58) was proved by G. Herglotz[44] [2759] in 1921 as a statement about solutions of the congruence

$$x^2y^2 + x^2 + y^2 \equiv 1 \mod p,$$

and some further cases were established by E. Artin in his thesis [136, 137] and by H. Hasse [2593]. In the elliptic case (i.e., $g = 1$) a proof was found in 1933 by H. Hasse [2593, 2596, 2598–2600], followed by another proof given by M. Deuring [1500, 1501].

In 1943 H. Hasse [2602] presented a new proof of his theorem, and simpler proofs were later found by Yu.I. Manin [4130] and H.G. Zimmer [6832].

In 1935 H. Hasse, in a joint paper with H. Davenport [1385], gave a proof for some special classes of function fields, and the general case was settled[45] finally in 1940 by A. Weil[46] [6612, 6613], who applied deep methods of algebraic geometry, built by him in [6614]. It was pointed out by Weil in [6617, p. 497] (and [6618]) that a special case of the Riemann Hypothesis is equivalent to a result proved by Gauss in [2211].

Another proof was given by J.-I. Igusa [3002], and a proof without the use of algebraic geometry is contained in P. Roquette's dissertation [5275, 5276]. Later other proofs were given by S.A. Stepanov [5928, 5929] (cf. E. Bombieri [600], W.M. Schmidt [5505]) and K.-O. Stöhr, J.F. Voloch [5959].

In 1938 J. Weissinger [6635] studied the analogue of L-functions for function fields and, in particular, proved that they satisfy a functional equation[47].

Another proof was given by H.L. Schmid[48] and O. Teichmüller [5471] in 1947.

An exposition of the classical theory of algebraic function fields in one variable was given in 1973 by M. Deuring [1512], and the history of the proof of the Riemann Hypothesis for curves was described by P. Roquette [5279–5281].

In 1949 A. Weil [6618] defined the more general zeta-function $\zeta_V(t)$ of an arbitrary non-singular projective variety V of degree n defined over a finite field $k = \mathbf{F}_q$ by the formula

$$\zeta_V(t) = \exp\left(\sum_{m=1}^{\infty} \frac{N_m t^m}{m}\right),$$

[44] Gustav Herglotz (1881–1953), professor in Leipzig. See [6158].

[45] According to P. Roquette [5279–5281], H. Hasse was very close to the proof in [2597].

[46] The full account of the proof appeared only in 1945 [6615].

[47] According to [5471] this was also achieved in 1936 by E. Witt, who, however, did not publish his result.

[48] Hermann Ludwig Schmid (1908–1956), professor in Berlin and Würzburg. See [2605, 5481].

where N_m denotes the number of points of V lying in the extension of degree m of k. A. Weil formulated four conjectures concerning the behavior of this function.

I: *$\zeta_V(t)$ is a rational function.*
II: *One has*

$$\zeta_V(t) = \frac{P_1(t)P_3(t)\cdots P_{2n-1}(t)}{P_2(t)P_4(t)\cdots P_{2n}(t)}, \tag{4.59}$$

where the P_i's are polynomials. Moreover

$$P_i(t) = \prod_{j=1}^{b_i}(1-\alpha_{ij}t), \tag{4.60}$$

with $|\alpha_{ij}| = q^{i/2}$. (The Riemann–Weil conjecture.)
III: *The function $\zeta_V(t)$ satisfies a functional equation, relating $\zeta_V(t)$ to $\zeta_V(1/(q^n t))$.*
IV: *If X is the reduction* $\bmod q$ *of a variety Y in zero characteristic, then the degrees b_i occurring in* (4.60) *are equal to Betti numbers of Y considered as a complex manifold.*

In a later paper [6621] A. Weil pointed out that for projective varieties having singular points some of these conjectures may fail, and gave an example in which there is no functional equation.

The first of these conjectures was established in 1960 by B. Dwork [1667–1671] and another proof was given by A. Grothendieck [2368].

A proof of (4.59) was found by B. Dwork two years later [1668], except for the case of even q and n. The excluded cases were later covered by the results by P. Deligne [1447, 1448]. B. Dwork's paper [1668] also contains the proof of the functional equation (conjecture III) (cf. S. Lubkin [4022, 4023]).

The final step was made in 1974 by P. Deligne [1447, 1448], who proved the Riemann–Weil conjecture (4.60), as well as conjecture IV. For expositions see N.M. Katz [3279] and J.-P. Serre [5649].

For certain classes of varieties, (4.60) was established earlier by E. Bombieri, H.P.F. Swinnerton-Dyer [622], P. Deligne [1445], G. Harder [2502], V.A. Iskovskih [3034], Yu.I. Manin [4132], G.I. Perelmuter [4781] and B.R. Tennison [6112].

Dwork's results in [1667] were utilized by E. Bombieri [596, 597] in his study of exponential sums with polynomials in several variables over a finite field (cf. J.H.H. Chalk, R.A. Smith [984] and É. Fouvry, N.M. Katz [2059]). The case of a cubic polynomial over $\mathbf{F}_p$ was considered earlier by H. Davenport, D.J. Lewis [1393] and L.J. Mordell [4400].

4.6 Hecke's Revival of Modular Forms

1. The modern theory of modular forms was created by E. Hecke, who discovered remarkable proprieties of the associated Dirichlet series. In his dissertation, published in 1912 [2675], Hecke had already considered Hilbert modular forms in two variables and used them in his habilitation thesis [2676] to construct Abelian extensions of quadratic number fields, and later [2684] constructed a class of Hilbert modular forms in two variables.

In a series of papers [2686–2693], E. Hecke completed the classical approach to modular forms. In [2686] he constructed new families of modular forms, using θ-functions attached to indefinite quadratic forms in two variables (see also B. Schoeneberg[49] [5548], who used forms in four variables), and in [2687] constructed an *Eisenstein series* of higher levels. After showing that every modular form is a unique linear combination of an Eisenstein series and a cusp form, he applied this to give bounds for the Fourier coefficients c_n of modular forms, as well as their sums $C(x) = \sum_{n \leq x} c_n$. In particular he obtained the bound $O(n^{k/2})$ for the nth coefficient of a cusp form of weight k, and for $k \geq 3$ deduced the evaluation

$$c_n = n^{k-1} C(n, f) + O\left(n^{k/2}\right) \tag{4.61}$$

with bounded $C(n, f) \geq 0$ (in the case $k = 2$ he had $c_n = O(n^{k-1+\varepsilon})$ for every $\varepsilon > 0$ [2697, 2698, Theorem 6]). He also obtained the bound

$$C(x) = O\left(x^{k/2} \log x\right)$$

in the case of cusp forms.

The papers [2689–2693] were devoted to the study of modular forms of higher levels.

In [2688] Hecke put forward the question of how one can characterize functions which can be expanded in a half-plane in a Dirichlet series

$$\sum_{n=1}^{\infty} \frac{a_n}{n^s}.$$

This natural question still does not have a complete answer.

The exponent in the error term in (4.61) was refined by H.D. Kloosterman [3371] to $k/2 - 1/8 + \varepsilon$ for every $\varepsilon > 0$ (a simpler proof was provided later by T. Estermann [1878]), and the bound for $C(x)$ was later improved to $O(x^a)$ for every $a > k/2 - 1/24$ by A. Walfisz [6529], who used bounds for Kloosterman sums obtained by H.D. Kloosterman in [3369]. A. Walfisz pointed out in a footnote that using H. Salié's stronger bounds [5378] one can replace $1/24$ here by $1/12$. He also proved the equality

$$\int_0^x |C(t)|^2 dt = \lambda(f) x^{k+1/2} + O\left(x^k \log^2 x\right).$$

This result implies, in particular, the relation

$$\sum_{n \leq x} \tau(n) = \Omega\left(x^{23/4}\right)$$

for Ramanujan's τ-function.

W.B. Pennington[50] [4770] showed in 1951 that one can replace Ω here by Ω_+ and Ω_- (this is a particular case of a more general result proved later by K. Chandrasekharan and R. Narasimhan [994]).

---

[49] Bruno Schoeneberg (1906–1995), professor in Hamburg. See [448].

[50] William Barry Pennington (1926–1968), professor in Aberystwyth. See [5122].

2. The main novelty in the two papers [2697, 2698] (the first paper dealing only with level 1) published by E. Hecke in 1937 and preceded by their summary [2694] in 1935 (cf. also [2696]), consisted in the association of a Dirichlet series

$$\Phi_f(s) = \sum_{n=1}^{\infty} \frac{a_n}{n^s} \tag{4.62}$$

with every modular form

$$f(z) = \sum_{n=0}^{\infty} a_n \exp(2\pi i a_n z/N) \in \mathfrak{M}(k, N). \tag{4.63}$$

This series is actually the Mellin transform (2.5) of f.

Hecke showed that $\Phi_f(s)$ converges absolutely in the half-plane $\Re s > k$, and can be extended to a meromorphic function in the plane. In the case of a cusp form this function is entire and in other cases it has a single simple pole. Moreover, it satisfies a functional equation which in the case $N = 1$ has the following simple form: if we put

$$R(s) = \left(\frac{\sqrt{N}}{2\pi}\right)^s \Gamma(s)\Phi_f(s),$$

then one has

$$R(s) = CR(k-s) \tag{4.64}$$

with $C = \pm 1$.

To obtain his result Hecke introduced in [2697, 2698] a family of linear operators acting in $\mathfrak{M}_k(N)$, noting that in a special case these operators were used already by Hurwitz. In the simplest case $N = 1$ they were defined by the formula

$$T_n(F) = n^{k-1} \sum_{\substack{a\in\mathbf{Z}, d>0, 0\le b<d \\ ad=n}} F\left(\frac{az+b}{d} d^{-k}\right) \quad (n = 1, 2, \ldots). \tag{4.65}$$

These operators generate a commutative ring, which is the predecessor of various *Hecke rings* and *Hecke algebras*, occurring in the theory of group representations.

It was noted by H. Rademacher [5039] that for prime p and $n = 2, 3, \ldots$ the operator $T(p^n)$ can be written as a Čebyšev polynomial in $T(p)$.

For $N = 1$ and certain small weights k ($k = 12, 26$ and $16 \le k \le 22$) Hecke was able to show that the space $\mathfrak{C}_k(1)$ has a particular basis consisting of eigenfunctions of all operators T_n. The elements of this basis have multiplicative coefficients c_n, i.e., the condition $c_{mn} = c_m c_n$ holds for co-prime m, n, and have an Euler-type factorization of the form

$$\prod_p \frac{1}{1 - c_p p^{-s} + p^{k-1-2s}}. \tag{4.66}$$

This was only known earlier (due to S. Ramanujan [5078]) in the case of the space $\mathfrak{C}_{12}(1)$ which is of dimension 1 and is generated by the classical discriminant function Δ (see (2.32)).

In the general case E. Hecke could only show that if f is an eigenfunction of all operators T_n, then the series $\Phi_f(s)$ has the Euler product (4.66), but was unable to provide a basis for $\mathfrak{C}_k(1)$ consisting of such eigenforms. The case of higher levels was considered by E. Hecke in [2698]. The theory became more complicated, because the operators T_p for primes p dividing the level had to be defined in different ways in certain subspaces of the space $\mathfrak{M}(N,k)$.

3. The restrictions on k in Hecke's result concerning the Euler product (4.66) were removed by H. Petersson [4813–4817] in 1939. He showed the existence of a basis of $\mathfrak{C}_k(1)$ consisting of eigenvalues of Hecke operators, which have multiplicative coefficients in all cases when this space is non-zero[51]. His main tool was the introduction of the structure of a Hilbert space under the scalar product

$$\iint_D f(z)\overline{g(z)}y^{k-2}dx\,dy \quad (z = x+iy)$$

in $\mathfrak{C}_k(1)$, and the observation that Hecke operators are commuting Hermitian operators, hence $\mathfrak{C}_k(1)$ has a basis consisting of common eigenfunctions of these operators. In [4814] he obtained a corresponding extension of Hecke's results to the level $N>1$, and also considered analogous questions for other subgroups of Γ. In particular he dealt with modular forms attached to the groups $\Gamma_0(N)$.

It was later proved by H.J. Hsiao and H.C. Lee [2912] that a modular form of level one and a fixed even weight is a common eigenfunction of all Hecke operators T_n if and only if its Fourier coefficients satisfy a finite set of relations.

For a survey of consequences of Petersson's introduction of the Hilbert space structure see J. Elstrodt and F.J. Grunewald [1761].

A new exposition of his theory, utilizing Petersson's result, was presented by E. Hecke [2701] in 1940. This paper also contains applications of the theory of modular functions to the problem of representing integers by positive definite quadratic forms.

An explicit expression for the scalar product in $C_k(1)$ was provided in 1952 by R.A. Rankin [5114].

The Hecke–Petersson theory for groups $\Gamma_0(N)$ was simplified much later by A.O.L. Atkin and J. Lehner[52] [164]. They introduced the important distinction between *oldforms* and *newforms* in the following way: if $M|N$, f is a cusp form of level M, and $dM|N$, then the function $f^*(z) = f(dz)$ is a cusp form of level N. Every such function f^* is called an oldform of level N, and the subspace of $S(N,k)$ spanned by them is denoted by $S^-(N,k)$. Atkin and Lehner showed that the orthogonal complement $S^+(N,k)$ of $S^-(N,k)$ (the members of $S^+(N,k)$ are the newforms) has a basis consisting of eigenvalues of Hecke operators T_p for primes p, allowing p to be a divisor of N.

[51] In the case $N=1$ this happens for $k=12$ and for all even $k \geq 16$.

[52] This paper contains an elegant exposition of the theory of Hecke operators.

For a generalization of the Atkin–Lehner theory to a more general class of modular forms see W.-C.W. Li [3883].

In 1955 M. Eichler [1700] confirmed a conjecture by Hecke [2701], which asserted that the space of cusp forms with respect to $\Gamma_0(q)$ with prime q has a set of generators consisting of theta-functions

$$\theta(z; F) = \sum_{\overline{x} \in \mathbf{Z}^4} \exp(2\pi i F(\overline{x}))$$

associated with quaternary quadratic forms F of discriminant q^2, and obtained a similar result in the case of arbitrary even weight and sufficiently large q. He showed also that for composite q this assertion may fail. For the case $q = 2, 3, 5, 11$ see Y. Kitaoka [3345]. For the analogue of Eichler's result for forms mod p see M. Ohta [4672]. Generating sets in the case of square-free N were given by M. Eichler [1702], and the general case was treated by H. Hijikata, A. Pizer and T. Shemanske [2779]. G. Shimura's [5701, 5702] conjecture that forms of weight $1/2$ are a linear combination of theta series

$$\sum_{n=-\infty}^{\infty} \chi(n) q^{tn^2}$$

(where χ is a Dirichlet character and $t \geq 1$ is an integer) was established by J.-P. Serre and H.M. Stark [5662]. In 1978 J.-L. Waldspurger [6513] showed that every newform of level N is a linear combination of theta series of the same level, associated with quadratic forms of square discriminant.

Tables of dimensions of the spaces of modular and cusp forms for $\Gamma_0(N)$ for $N \leq 200$ were given by H. Cohen and J. Oesterlé [1145].

Hecke operators were generalized to Hilbert modular forms in N.G. de Bruijn's thesis [783], and the extension of the Hecke–Petersson theory to this case was realized in 1954 by O. Herrmann [2769] and K.-B. Gundlach [2381], following a suggestion by C.L. Siegel [5769, p. 38]. K.-B. Gundlach also treated modular forms associated with congruence subgroups of the Hilbert group.

4. The function $J(q)$ is defined as the Fourier expansion at infinity of the modular function $j(z)$, defined by (3.45). More precisely, for z in the upper half-plane $J(q) = j(e^{2\pi i z})$ holds. One has

$$J(q) = \frac{1}{q} + 744 + 196884q + \sum_{n=2}^{\infty} c(n) q^n, \tag{4.67}$$

and the question arises of the size of the coefficients $c(n)$. This was first considered in 1932 by H. Petersson [4806] who expressed $c(n)$ as the sum of convergent series involving Kloosterman sums and Bessel functions (the same was independently obtained by H. Rademacher [5038]), and deduced that $c(n)$ is asymptotic to

$$F(n) = \frac{\exp(4\pi\sqrt{n})}{\sqrt{2}n^{3/4}}.$$

It was proved by N. Brisebarre and G. Philibert [743] in 2005 that for $n \geq 1$ one actually has $c(n) \leq F(n)$. Earlier (in 1975) the upper bound

$$c(n) \leq 6 \exp\left(4\pi\sqrt{n}\right)$$

was obtained by O. Herrmann [2770].

The first 25 coefficients $c(n)$ were computed in 1939 by H.S. Zuckerman [6839].

This list was extended up to $n = 100$ by A. van Wijngaarden [6336] in 1953, and to $n = 6002$ by O. Herrmann [2770] in 1975. Various congruences satisfied by the coefficients $c(n)$ were found by D.H. Lehmer [3782] (see also H.-F. Aas [2, 3], A.O.L. Atkin, J.N.O'Brien [166], O. Kolberg [3451], J. Lehner [3807], M. Newman [4586], A. van Wijngaarden [6336]).

There is a remarkable connection between the coefficients $c(n)$ in (4.67) and the degrees of irreducible representations of the largest sporadic simple group (the *Monster*), whose existence was conjectured in the seventies by B. Fischer and R. Griess, and finally constructed by Griess [2344] in 1980. This connection was pointed out in 1979 by J.H. Conway and S.P. Norton [1225] and led to an extensive theory (the *Moonshine*), culminating in the paper [626] by R.E. Borcherds where a conjecture stated in [1225] was established (cf. C.J. Cummins, T. Gannon [1293]). See T. Gannon [2191, 2192] for broad surveys.

Series expansions of Fourier coefficients of modular forms in terms of Bessel functions were given by H. Petersson [4807] and H. Rademacher, H.S. Zuckerman [5043, 6840, 6841].

5. Modular forms of half-integral weight appeared first in the work of G.H. Hardy [2511] and L.J. Mordell [4377], who dealt with sums of an odd number of squares. The first general results were obtained later by H. Petersson [4805, 4808–4812] who considered arbitrary complex weights. His approach in [4808–4812] also essentially simplified the theory in the classical case, eliminating the use of the uniformization theory, used in earlier methods.

In his last paper [2702] E. Hecke considered certain particular modular forms of half-integral weight, namely the classical η-function

$$\eta(z) = \exp\left(\frac{\pi i z}{12}\right) \prod_{n=1}^{\infty} (1 - \exp(2\pi i n z)),$$

the simplest theta-function

$$\theta(z) = \sum_{n=-\infty}^{\infty} \exp\left(2\pi i z n^2\right),$$

and the products $\eta^m(z)\theta^n(z)$ with integral m, n. He studied the question of the existence of analogues of Hecke operators, which would imply the existence of associated Dirichlet series. The answer turned out to be negative in general, as it turned out that this is possible only in four particular cases:

$$(m, n) \in \{(1, 0), (0, 1), (3, 0), (4, -1)\}.$$

Although E. Hecke pointed out in [2696, 2702] that the Hecke–Petersson theory cannot be extended to modular forms of half-integral weights, G. Shimura [5701, 5702] was able to show in 1973 that important parts of that theory can be carried over to this case. See also W. Kohnen [3441, 3442], M. Manickam, B. Ramakrishnan and T.C. Vasudevan [4128], S. Niwa [4622], T. Shintani [5707], and J.-L. Waldspurger [6514].

Modular forms of any real weight had already been considered by G.H. Hardy and S. Ramanujan [2542] in 1919, and by H.S. Zuckerman [6841] in 1940, but the first modern approach to their theory appears in the papers of H. Petersson [4819, 4820] in 1950. Later Hecke operators for these forms were constructed by K.K. Wohlfahrt [6705].

6. Hecke's theory of modular forms has been generalized to other situations, and the first such step was taken by C.L. Siegel [5757, 5761] in 1939. Earlier, in his first paper on the theory of quadratic forms [5750], he introduced *Siegel modular forms* associated with symplectic groups acting on *Siegel's upper half-space*, defined as the set $\mathcal{H}_g$ of complex symmetric matrices of dimension g, whose elements have positive imaginary parts. Let us recall that the symplectic group $\Gamma_g = Sp_{2g}(\mathbf{Z})$ (also called the *Siegel modular group of degree g*) is the automorphism group of the lattice $\mathbf{Z}^{2g}$ equipped with the symplectic form $\sum_{1\leq i\neq j\leq g} x_i y_j$.

This group is finitely generated, and a system of generators was found by L.K. Hua and I. Reiner [2939]. Complete systems of defining relations were found by E. Gottschling [2287, 2288] for $g = 2$, and by H. Klingen [3364, 3365] for all g (cf. H. Behr [386], J.S. Birman [542]).

The group Γ_g, whose elements can be written in the form

$$\gamma = \begin{pmatrix} A & B \\ C & D \end{pmatrix}$$

(where A, B, C, D are $g \times g$ matrices), acts on $\mathcal{H}_g$ in a natural way. One fixes a representation ρ of $GL_g(\mathbf{C})$ in a finite-dimensional space V, and calls a vector-valued function $f \to \mathcal{H}_g$ a *Siegel modular form* of weight ρ, if for $\tau \in \mathcal{H}_g$ and $\gamma \in \Gamma_g$ one has

$$f(\gamma(\tau)) = \rho(C\tau + D)f(\tau).$$

(In the case $g = 1$ the function f should be holomorphic at infinity. M. Koecher [3436] proved in 1954 that for $g \geq 2$ this condition is automatically satisfied.) For $g = 1$ Siegel modular forms have values in the complex field and are called *classical Siegel modular forms*. They satisfy the condition

$$f(\gamma(\tau)) = (\det(C\tau + D))^k f(\tau)$$

for some k, and are said to be of weight k.

For $g = 1$ one has to assume that the function f is holomorphic at infinity. For $g \geq 2$ M. Koecher [3436] proved in 1954 that this condition is automatically satisfied.

In his papers [5750–5752] C.L. Siegel introduced generalized Eisenstein series, and the question of their domains of convergence was answered in 1938 by H. Braun[53] [692]. In the same year M. Sugawara [5988] introduced Hecke operators $T(m)$ in the spaces of Siegel modular forms and showed that generalized Eisenstein series are eigenfunctions of each $T(m)$.

M. Sugawara also asserted that the last property characterizes Eisenstein series, but in 1951 H. Maass [4042] constructed other Siegel modular forms having this property, and in a series of papers [4040–4043] generalized important fragments of the Hecke–Petersson theory of modular forms to Siegel modular forms. In [4040] he dealt with the case $g = 2$ and constructed the associated Dirichlet series, and in [4041] the associated Poincaré series to Siegel forms, generalizing the theory built by H. Petersson [4816] for ordinary modular forms.

[53] Helene Braun (1914–1986), professor in Hamburg. See [5964].

In 1956 I.I. Piateckiĭ-Šapiro[54] [4843–4845] presented a general theory of automorphic functions in one and several variables. He showed in particular that in most cases the field of automorphic functions is generated by two elements, generalizing several classical results.

The theory of Siegel modular forms is covered in the books by E. Freitag [2084], H. Klingen [3366], H. Maass [4045] and C.L. Siegel [5776].

7. Hecke's method related, in particular, the zeta-functions of imaginary quadratic fields with modular forms, but was not applicable to zeta-functions of real quadratic fields. This case was settled in 1949 by H. Maass [4039], who introduced a class of functions, called *automorphic wave functions*, having a prescribed behavior under the action of the group generated by matrices

$$\begin{pmatrix} 1 & a \\ 0 & 1 \end{pmatrix}, \quad \begin{pmatrix} 0 & -1 \\ 1 & 0 \end{pmatrix},$$

for a fixed real a. Maass also generalized Hecke operators to his space of functions obtaining results analogous to Hecke's.

Expositions of the theory of modular forms, including the Hecke–Petersson theory were provided in the sixties by R.C. Gunning [2382], J. Lehner [3809, 3810], H. Maass [4044] and A. Ogg [4667].

4.7 Other Questions

1. Let $A = \{a_1 < a_2 < \cdots\}$ be an infinite sequence of integers with the property that for $i \neq j$ one has $a_i \nmid a_j$. Such sequences are called *primitive*. H. Davenport [1344] and S. Chowla [1077] asked whether every such sequence has a density, and a negative answer was provided by A.S. Besicovitch [476]. It was proved in 1935 by F. Behrend[55] [387] that for every such sequence one has

$$\sum_{a_i \leq n} \frac{1}{a_i} = O\left(\frac{\log n}{\sqrt{\log\log n}}\right) \tag{4.68}$$

and S.S. Pillai [4867] showed this to be best possible (cf. P. Erdős, A. Sárkőzy, E. Szemerédi [1854]). F. Behrend also obtained that the lower density of A vanishes, A.S. Besicovitch [479] gave an example with positive upper density, and P. Erdős [1780] proved the convergence of the series

$$\sum_{i=1}^{\infty} \frac{1}{a_i \log a_i}.$$

[54] Ilya Piateckiĭ-Šapiro (1929–2009), professor in Tel Aviv and at Yale.

[55] Felix Adalbert Behrend (1911–1962), professor in Melbourne. See [4577].

This implies the bound

$$\sum_{a_i \le x} 1 = O\left(\frac{x}{\log\log x \log\log\log x}\right). \tag{4.69}$$

In 1967 P. Erdős, Sárkőzy and E. Szemerédi proved in [1855] that in (4.68) the implied constant may be taken to be independent of A (cf. I. Anderson [71]). For a survey see Sárkőzy [5403].

In 1999 R. Ahlswede, L.H. Khachatrian and A. Sárkőzy [31] constructed a primitive sequence whose counting function differs from the upper bound (4.69) only by an arbitrarily small power of $\log\log x$ (cf. [32]).

A.S. Besicovitch's solution of S. Chowla and H. Davenport's question was based on his results concerning integers having factors in an interval. Let $H(x, y, z)$ be the number of integers $n \le x$ having a divisor in the interval $[y, z]$, and put

$$\varepsilon(y, z) = \lim_{x\to\infty} \frac{H(x, y, z)}{x}.$$

Besicovitch's argument implied

$$\liminf_{x\to\infty} \varepsilon(x, 2x) = 0,$$

and it was soon proved by P. Erdős [1780] that one can replace lim inf here by lim, and, more generally, one has

$$\lim_{x\to\infty} \varepsilon\left(x, x^{1+f(x)}\right) = 0$$

for every function $f(x)$ tending to 0.

25 years later P. Erdős returned to that subject and showed [1819]

$$\varepsilon(y, 2y) = \log^{-c+o(1)} a,$$

with a constant $c = 0.086\ldots$ (cf. G. Tenenbaum [6102]). Later G. Tenenbaum [6103] showed for $\lambda \in [0, 1]$ the existence of the limit

$$h(\lambda) = \lim_{y\to\infty} \varepsilon\left(y^\lambda, y\right)$$

with continuous $h(\lambda)$. The asymptotical behavior of $H(x, y, z)$ was studied in certain cases by G. Tenenbaum [6104, 6106] and recently K. Ford [2033] determined the order of magnitude of $H(x, y, z)$.

In 1948 P. Erdős [1798] conjectured that almost all integers have a pair of divisors $d_1 < d_2$ satisfying $d_2 < 2d_1$, and this was proved in 1984 in a stronger form (which permitted replacement of the coefficient 2 by any number exceeding 1) by H. Maier and G. Tenenbaum [4105].

See also the book by R.R. Hall and G. Tenenbaum [2479].

2. In 1936 P. Erdős and P. Turán [1864] asked whether a sequence of integers having positive upper density must contain arbitrarily long arithmetic progressions. Recall that the upper density $\bar{d}(\alpha)$ of a sequence $\alpha = (a_n)$ of integers is defined by

$$\bar{d}(\alpha) = \limsup_{x\to\infty} \frac{1}{x} \sum_{a_n \le x} 1.$$

The origin of this problem lies in van der Waerden's [6309] theorem on the existence of arbitrarily long arithmetic progressions. In [1864] the authors introduced the function $r_k(n)$, defined as the least number r with the property that any sequence of r integers in $[1,n]$ contains an arithmetic progression of k terms, and two years later F. Behrend [388] showed that for all k the limit

$$\gamma_k = \lim_{n\to\infty} \frac{r_k(n)}{n}$$

exists, and either all γ_k vanish, or one has $\lim_{k\to\infty} \gamma_k = 1$.

The conjecture stated in [1864] that $r_3(n) = O(n^{\delta})$ holds with some $\delta < 1$ turned out to be incorrect since R. Salem and D.C. Spencer [5375] proved in 1942

$$r_3(n) \geq n^{1-c/\log\log n}$$

for every $c > \log 2$, and four years later Behrend [389] improved this to

$$r_3(n) \geq n^{1-c/\sqrt{n}}$$

for a certain $c > 0$. Soon Salem and Spencer [5376] extended their method to obtain, for every k, the lower bound

$$r_k(n) > n^{1-\varepsilon}$$

for every $\varepsilon > 0$ and sufficiently large n, and R.A. Rankin [5117] obtained an analogue of Behrend's result for $r_k(n)$ with any k.

In 1952 K.F. Roth [5308] used the circle method to establish the vanishing of γ_3, and an improvement of this approach lead him later [5309] to the bound

$$r_3(n) \ll \frac{n}{\log\log n}.$$

This was improved in 1987 by D.R. Heath-Brown [2642] to $O(x/\log^c x)$ with some positive c, and E. Szemerédi [6024] showed that one can take $c = 1/4$.

Much later J. Bourgain [660] first obtained

$$r_3(n) \ll n\left(\frac{\log\log n}{\log n}\right)^{1/2},$$

and then [664]

$$r_3(n) \ll n\frac{(\log\log n)^2}{\log^{2/3} n}.$$

The case $k = 4$ was settled in 1969 by E. Szemerédi [6022], who used purely combinatorial methods to show $\gamma_4 = 0$ and another proof was later found by K.F. Roth [5313, 5314]. In 1975 E. Szemerédi [6023] succeeded in showing $\gamma_k = 0$ for all k using an extremely complicated combinatorial argument.

Two years later it turned out that the question can be attacked with the tools of ergodic theory, when H. Furstenberg [2146] reformulated Szemerédi's result in the language of measure-preserving transformations in a measure space, and provided a proof of it using ergodic machinery. A very clear exposition of Furstenberg's proof was given in the paper by H. Furstenberg, Y. Katznelson and D. Ornstein [2148]. Neither Szemerédi's nor Furstenberg's proof provided any estimate for $r_k(n)$, and the first such estimate emerged only in the new century, when W.T. Gowers [2294] produced a third proof of Szemerédi's theorem, based on completely new ideas, the origin of which can be traced to the first paper by K.F. Roth on that subject. His proof leads to

$$r_k(n) \ll \frac{n}{(\log\log n)^{c_k}}$$

with an explicit positive but very small c_k. Earlier [2293] he did this in the case $k = 4$. In this case the best known evaluation is due to B. Green and T. Tao [2337] who in 2009 got

$$r_4(n) \ll n \exp\left(-c\sqrt{\log\log n}\right)$$

with $c > 0$.

The ergodic method was used later to obtain higher-dimensional generalizations (H. Furstenberg, Y. Katznelson [2147]), or to replace progressions by values of suitable polynomials (A.Sárkőzy [5399–5401], A. Balog, J. Pelikán, J. Pintz, E. Szemerédi [305], V. Bergelson, A. Leibman [432], S. Slijepčević [5821]). For a further generalization see V. Bergelson and R. McCutcheon [433]. An account of the development around Szeméredi's theorem is given by T. Tao [6048].

The question of the existence of an infinite arithmetical progression in a set of positive integers was considered by E. Szemerédi and V.H. Vu [6026]. They showed that if the counting function $A(x)$ of a set $A \subset \mathbf{N}$ satisfies $A(x) \gg \sqrt{x}$, and the sumset $S(A)$ is defined by

$$S(A) = \{a_1 + \cdots + a_k : a_i \in A,\ k = 1, 2, \ldots\}$$

then $S(A)$ contains an infinite arithmetical progression. They also studied long arithmetical progressions in sumsets $S(A)$ for finite sets ([6025–6027]; cf. the book by T. Tao and V.H. Vu [6049]).

3. In 1937 A. Scholz [5557] introduced *addition chains*, defined as finite sequences of positive integers $a_0 = 1,\ a_1 = 2, \ldots,$ having the property that every element is the sum of two earlier elements. He defined the function $l(n)$ as the shortest addition chain having n as its last element. He stated three properties of $l(n)$ in the problem section of the *Jahresbericht*, however no solution was published there, and the first proof of two of the properties

$$l(mn) \le l(m) + l(n), \quad m + 1 \le l(n) \le 2m \quad \text{for } n \in (2^m, 2^{m+1}]$$

was presented by A. Brauer [681], who also established the equality

$$l(n) = \log_2 n + O\left(\frac{\log n}{\log\log n}\right).$$

Several years later P. Erdős [1818] showed that for almost every n one has

$$l(n) = \log_2 n + (1 + o(1))\frac{\log n}{\log\log n}.$$

The third proposed property,

$$l\left(1 + 2^m\right) \le m - 1 + l(m),$$

was established only in certain cases.

4. Although various types of linear recurrences u_n were studied earlier (e.g., by É. Lucas [4028, 4030]), it seems that the question of the size of periods of the sequence u_n mod m obtained its first treatment in the hands of R.D. Carmichael [919, 921], who was able to resolve this question in certain cases. His research was continued later by H.T. Engstrom[56] [1765]. Their methods did not permit the length of the minimal period to be determined and this was done later by M. Ward [6560], and M. Hall[57] [2470].

See the book by G. Everest, A.J. van der Poorten, I. Shparlinski and T. Ward [1919, Chap. 3].

5. A function f of N variables is called multiplicative, if one has

$$f(m_1n_1,\dots,m_Nn_N)=f(m_1,\dots,m_n)f(n_1,\dots,n_N)$$

provided the products $\prod_i m_i$ and $\prod_i n_i$ are co-prime. In 1931 a thorough study of such functions was done by R. Vaidyanathaswamy[58] [6258], who described all multiplicative functions f satisfying an identity of the form

$$f(mn)=\sum_{d|(m,n)} f\left(\frac{m}{d}\right)f\left(\frac{n}{d}\right)F(d)$$

with a suitable function F and obtained similar results for functions in several variables. Identities of this type are called the *Busche–Ramanujan identities*, as they were discovered for divisor functions by E. Busche[59] [868] and S. Ramanujan [5077].

6. Denote by $W(n)$ the numerator of the sum

$$\sum_{\substack{m<n\\(m,n)=1}} \frac{1}{m}.$$

An old theorem usually attributed to J. Wolstenholme[60] [6718] (although it had appeared already in Waring's book [6565], see [2545, 4th ed., p. 93]), stating that for prime $p\ge 5$ one has $p^2|W(p)$, was generalized in 1889 by C. Leudersdorf [3840] to cover composite integers. In the case $(n,6)=1$ one gets $n^2|W(n)$ and in other cases $n^2|c_nW(n)$ with $c_n\le 6$. Fresh proofs of these results were given by G.H. Hardy and E.M. Wright [2544] and S. Chowla [1070] (who dealt with the case $(n,6)=1$), and an extension has been provided by N. Rama Rao [5089].

---

[56]Howard T. Engstrom (1902–1962), professor at Yale.

[57]Marshall Hall, Jr. (1910–1990), professor at Ohio State University and Caltech. See [6819].

[58]Ramaswamy Vaidyanathaswamy (1894–1960), professor in Madras. See [4533].

[59]Edmund Busche (1861–1916), teacher in Bergedorf and Hamburg.

[60]Joseph Wolstenholme (1829–1891), professor at an Engineering College near London.

7. A curious elementary problem arose in the thirties, associated with the names[61] of L. Collatz[62], H. Hasse and S. Kakutani[63] (the *Collatz–Hasse–Kakutani problem*).

For a positive integer N define

$$T(N) = \begin{cases} N/2 & \text{if } N \text{ is even,} \\ 3N+1 & \text{if } N \text{ is odd.} \end{cases}$$

Prove that the sequence of consecutive iterations $N, T(N), T(T(N)), \ldots$ *eventually reaches* 1.

There are very few proven results about this assertion. It has been checked for every $N < 3 \cdot 2^{53}$ by T. Oliveira e Silva [4675] in 1999, who later extended his search up to $2 \cdot 10^{18}$ (according to J.C. Lagarias [3606]). An excellent bibliography can be found in [3606]. For earlier surveys see J.C. Lagarias [3603] and the book by G.J. Wirsching [6691].

---

[61] It is also called the *Syracuse problem* as well as the *Ulam problem*.

[62] Lothar Collatz (1910–1990), professor in Hannover and Hamburg. See [2378].

[63] Shizuo Kakutani (1911–2004), professor at Yale.

Chapter 5
The Forties and Fifties

5.1 Analytic Number Theory

1. Brun's sieve enabled evaluation of the number of integers $\le x$ which remain after removing k residue classes mod p (k being a fixed integer) for every prime $p \le \sqrt{x}$. The question of what happens in the case when the number of deleted residue classes grows with p was first considered by Yu.V. Linnik [3902] in 1941. He showed that if $\{p_1, \dots, p_m\}$ is a set of primes not exceeding $\sqrt{N}$, and for each i one deletes $0 < f(p_i) < p_i$ residue classes mod p_i from the set $\{1, 2, \dots, N\}$, then the number E of integers which remain satisfies

$$E \le \frac{20\pi N}{\tau^2 m}, \tag{5.1}$$

where $\tau = \min_i f(p_i)/p_i$. He used this to show [3903] that for every $\varepsilon > 0$ the number of primes $p \le x$ for which the minimal quadratic non-residue exceeds p^ε is $O(x^\varepsilon)$.

Later A. Rényi[1] reformulated Linnik's method using an analogue of the probabilistic notion of variance. If $A \subset \{1, 2, \dots, N\}$ is a set of integers having Z elements and for prime p one denotes by $Z(p, h)$ the number of elements of Z congruent to h mod p, then the variance $D(p)$ is defined by

$$D(p) = \sum_{h=0}^{p-1} \left(Z(p,h) - \frac{Z}{p} \right)^2. \tag{5.2}$$

Rényi [5163] obtained a bound for $D(p)$, which may be stated, as in [2457], in the form

$$\sum_{p \le x} p D_p \ll Z^{2/3} N^{4/3} x^{1/3}, \tag{5.3}$$

provided $x \le N^{3/5}$. This allowed him to show that every even integer is the sum of a prime and an *almost prime*, i.e., a number having at most k prime divisors with a

[1] Alfréd Rényi (1921–1970), student of Linnik, professor in Budapest. See [520, 2399].

W. Narkiewicz, *Rational Number Theory in the 20th Century*,
Springer Monographs in Mathematics,
DOI 10.1007/978-0-85729-532-3_5,

certain fixed k. It had been known earlier (T. Estermann [1886]) that this assertion, with $k = 6$, can be deduced from the General Riemann Hypothesis with the use of Brun's sieve, and A.A. Buhštab [833] and V.A. Tartakovskiĭ [6056] showed that every large even integer is the sum of two integers each having at most four prime divisors, improving upon previous results by V. Brun, which we quoted in Sect. 2.3.

Rényi followed Estermann's method, but replaced the General Riemann Hypothesis by mean value theorems about zeros of L-functions.

Later Y. Wang [6549] and A.I. Vinogradov [6397] showed that the General Riemann Hypothesis implies that every large even integer can be written as $p + P_k$ with $k = 4$, and Y. Wang [6551, 6554] obtained $k = 3$, again under the General Riemann Hypothesis.

The first explicit value of k, free of any unproven assumption, was obtained by C.D. Pan [4732], who proved in 1962 that one can take $k = 5$. Soon M.B. Barban [322], B.V. Levin [3855, 3856], Y. Wang [6554] and C.D. Pan [4733] showed that $k = 4$ is sufficient, and in 1965 A.I. Vinogradov [6398] and A.A. Buhštab [836, 837] got $k = 3$. Another proof, obtained by an improvement to Selberg's sieve, was sketched by H. Halberstam, W.B. Jurkat and H.-E. Richert [2453] in 1967, and a complete proof appeared in the book [2455].

The penultimate step towards the proof of the binary Goldbach conjecture (one still waits for the definitive step) was taken in 1973 by J.R. Chen [1021], who proved that every large even integer is the sum of a prime and a number which is either prime or a product of two primes. His proof was soon simplified[2] by P.M. Ross [5295]. For other modifications see A. Fujii [2134] and C.D. Pan, X.X. Ding and Y. Wang [4735].

For the number $P_{1,2}(N)$ of representations $N = p + P_2$ for even N, Chen's result gave the lower bound

$$P_{1,2}(N) \geq cA(N) \tag{5.4}$$

where

$$A(N) = \prod_{\substack{p|N \\ p\neq 2}} \frac{p-1}{p-2} \prod_{p\neq 2} \left(1 - \frac{1}{(p-1)^2}\right) \frac{N}{\log^2 N}$$

and $c = 0.67$. The value of c was increased to 0.7544 and 0.81 in [1022].

Much later Y. Cai and M.G. Lu [885] got $c = 0.8285$, and soon further improvements followed with $c = 0.836$ (J. Wu [6759]), $c = 0.867$ (Y. Cai [884]) and $c = 0.899$ (J. Wu [6760]).

Another proof of Chen's result was given by É. Fouvry and F. Grupp [2054].

A. Rényi proved also in [5163] that there are infinitely many primes p such that $p + 2$ is an almost prime. He noted that his approach can be applied also to show that for any given k there are infinitely many primes p such that $p + k$ is an almost prime and every odd integer ≥ 5 can be written as $p + P$ with prime p and an almost prime P. One of his principal lemmas shows that for almost all primes p the corresponding L-functions do not have Siegel zeros.

[2] The first version of Ross's argument can be found in the book by H. Halberstam and H.-E. Richert [2455].

In 1964 N.C. Ankeny and H. Onishi [105] showed that the General Riemann Hypothesis implies the existence of infinitely many primes p with $p+2=P_3$ and this was later established unconditionally by A.A. Buhštab [837], H.-E. Richert [5209], H. Halberstam, W.B. Jurkat, H.-E. Richert [2453].

Now it is known (J.R. Chen [1021, 1022]) that one has $p+2=P_2$ for infinitely many primes p, which is the best known approximation to the problem of twin primes.

For the number $\Pi_{1,2}(x)$ of primes $p\le x$ with $p+2=P_2$, Chen's result gives the lower bound

$$\Pi_{1,2}(x)\ge 2c\prod_{p\ne 2}\left(1-\frac{1}{(p-1)^2}\right)$$

with $c=0.335$ for large x, and obtained a similar result for primes $p\le x$ with $ap+b=P_2$ for given a,b. The value of the constant c was later increased to 0.3445 (H. Halberstam [2450]), 0.3772 and 0.405 (J.R. Chen [1022]), 0.71 (É. Fouvry, F. Grupp [2053]), 1.015 (H.Q. Liu [3950]), 1.05 (J. Wu [6755]), 1.0974 (Y. Cai [882, 883]), 1.104 (J. Wu [6759, 6760]) and 1.13 (Y. Cai [883]).

Later A. Rényi [5164–5166] improved (5.3) by deducing the bound

$$\sum_{p\le x^{1/3}} pD_p\le 9xZ$$

from one of his theorems in probability theory. He also succeeded in improving Linnik's original statement by showing that the number E in (5.1) does not exceed $2N/\tau m$, provided that the considered primes are less than $(N/12)^{1/3}$. As an application, in [5165] he proved the inequality

$$\sum_{p<x^{\alpha}} p\sum_{l=1}^{p-1}\left(\psi(x;p,l)-\frac{\psi(x)}{p}\right)^2\ll x^{3\alpha-1},$$

where

$$\psi(x;p,l)=\sum_{\substack{n\le x\\ n\equiv l \pmod p}}\Lambda(n)$$

valid for $1/3\le\alpha<1/2$. In 1958 he strengthened [5167] the bound in (5.3) to $O((N+X^3)Z)$.

2. In 1944 Yu.V. Linnik [3910, 3911] essentially improved the upper bound for $p(k,l)$, the minimal prime number in the progression l mod k ($(k,l)=1$). Using his previous results on the distribution of zeros of Dirichlet L-functions he obtained[3] $p(k,l)\le k^C$, where C is an absolute constant. Linnik's proof was essentially simplified in 1954 by K.A. Rodosskiĭ [5244]. The lower bound of possible values for C is called the *Linnik constant*, and the first explicit bound for C was obtained in 1957 by C.D. Pan [4731], who got $C\le 5448$.

---

[3] Earlier [3909] he had obtained this for a part of progressions mod k.

A few years later P. Erdős [1799] showed that the bound $p(k,l) \ll \varphi(k)\log k$ holds for a positive proportion of residues $l \bmod k$, but on the other hand, for a positive proportion of such residues one has $p(k,l) > C\varphi(k)\log k$ for any fixed C.

In 1961 K. Prachar [5007] showed that for any fixed l there exist infinitely many k with

$$p(k,l) \geq c(l)\frac{k\log k \log\log k \log\log\log\log k}{(\log\log\log k)^2},$$

the constant $c(l) > 0$ being independent of k, and A. Schinzel [5435] observed that this constant can be made absolute. Later C. Pomerance [4970] established the bound

$$\max_l p(k,l) \geq (e^\gamma + o(1))\varphi(k)\frac{\log k \log\log k \log\log\log\log k}{(\log\log\log k)^2}$$

for almost all k.

A new proof of Linnik's theorem was given by P.D.T.A. Elliott [1747].

The value of Linnik's constant was later reduced to 777 (J.R. Chen [1020]), 550 and 80 (M. Jutila [3167, 3172]), 36 and 20 (S.W. Graham [2300, 2301]) 17 (J.R. Chen [1027]), 16 (W. Wang [6547]), 13.5 and 11.5 (J.R. Chen, J.M. Liu [1029–1031]) and 8 (W. Wang [6548]). In 1991 D.R. Heath-Brown [2650] obtained $C \leq 11/2$, which holds the actual record. Earlier he had shown [2646] that stronger bounds would result from the existence of Siegel zeros.

It was shown by M.B. Barban, Yu.V. Linnik and N.G. Čudakov [327] that for prime p and sufficiently large n one has $p(p^n,l) \ll_p p^{cn}$ with any $c > 8/3$. In 1974 H. Iwaniec [3053] showed that this holds with $c = 2.4601\ldots$. For infinitely many primes p one can have $c \leq 1.638$ (Y. Motohashi [4449]).

Much later E. Bach and J. Sorenson [191] showed that the General Riemann Hypothesis implies the bound $p(k,l) \leq 2k^2\log^2 k$, and for large k the factor $2k^2$ can be replaced by $(1+\varepsilon)\varphi^2(k)$ for every $\varepsilon > 0$. For an improvement see A. Granville, C. Pomerance [2320].

An analogue of Linnik's result for prime ideals in an algebraic number field was established by E. Fogels [2021], who used his work on the zeros of L-functions [2018–2020] associated with characters of the class-groups $H_{\mathfrak{f}}^*$ to obtain upper bounds for the smallest norm of a prime ideal of first degree lying in a prescribed class $\bmod \mathfrak{f}$.

3. In 1946 Yu.V. Linnik [3913] considered the number $N(\sigma,T,\chi)$ of zeros ϱ of L-functions with primitive characters χ in the region $\sigma \leq \Re\varrho < 1$, $|\Im\varrho| \leq T$, and obtained for the sum

$$\Sigma_1(q,\sigma,T) := \sum_{\chi \bmod q} N(\sigma,T,\chi),$$

the evaluation

$$\Sigma_1(q,\sigma,T) \ll q^{2\sigma-1}T^{4(1-\sigma)/(3-2\sigma)} + q^{30},$$

which for $q = 1$ corresponds to Titchmarsh's result ([6168]) concerning $\zeta(s)$. Yu.V. Linnik followed Titchmarsh's method and used the approximative functional equation for L-functions.

Another proof was given later by P. Turán [6225].

It is conjectured (the *density conjecture for L-functions* mod q) that for every $\varepsilon > 0$ one has

$$\Sigma_1(q,\sigma,T) \ll (qT)^{2(1-\sigma)+\varepsilon}. \tag{5.5}$$

For further progress see Sect. 6.1.2.

4. In 1947 A. Selberg [5611, 5616] invented a new sieve method, which now bears his name. To explain it, observe that if $n_1, n_2, \dots, n_N$ is a sequence of integers, and $S(z)$ denotes the number of n_i's without prime divisors below z, then Legendre's include-exclude procedure leads to the formula

$$S(z) = \sum_{j=1}^{N} \sum_{d|(D,n_j)} \mu(d) = \sum_{d|D} \mu(d) S_d,$$

where D is the product of all primes $p \le z$, and S_d is the number of n_i's divisible by d. In many cases one can write

$$S_d = \frac{f(d)}{d} N + R_d,$$

with a certain multiplicative function f and an error term R_d, but in the resulting formula for $S(z)$ the final error term $\sum_{d|D} |R_d|$ is usually larger than the main term.

Selberg noted that if one can find a sequence ϱ_d which for every n satisfies

$$\sum_{d|n} \varrho_d \ge \sum_{d|n} \mu(d), \tag{5.6}$$

then one is lead to

$$S(z) \le \sum_{d|D} \varrho_d S_d,$$

and if many ϱ_d's vanish, then the resulting error term can have a good upper bound. The inequality (5.6) can be achieved with the help of an auxiliary sequence λ_d, satisfying $\lambda_1 = 1$, by putting

$$\varrho_d = \sum_{[a,b]=d} \lambda_a \lambda_b,$$

and to get a reasonable bound for the error term in the evaluation of $S(z)$ one has to choose the λ's minimizing a quadratic form with one extra condition.

This idea also works in algebraic number fields as shown by G.J. Rieger [5217–5219].

In 1964 N.C. Ankeny and H. Onishi [105], applied Selberg's sieve to improve upon earlier results in several questions. So they were able to show that there are infinitely many integers n with $\omega(n) \le 2$, $\omega(n+2) \le 3$, and deduced from the Riemann Hypothesis for L-functions that for infinitely many primes p one has $\Omega(p+2) \le 3$.

A variant of Selberg's sieve based on an idea introduced in 1937 by A.A. Buhštab [831] appears in the paper by W.B. Jurkat and H.-E. Richert [3166] and leads to rather strong upper and lower bounds. The authors applied this to obtain short intervals containing integers having a prescribed number of prime factors and lying in a given arithmetic progression. For example, if $a > 14/25$ and x is large, then the interval $[x - x^a, x]$ contains at least two integers with at most two prime factors, and if $a > 25/11$, then there is an integer $n \equiv l \pmod{k}$ in that interval with $\Omega(n) \le 2$, provided k is sufficiently large and $(k, l) = 1$. Previously this was known with $a = 10/17$ (Wang [6550]) and $a = 7/2$ (S. Uchiyama [6250]), cf. also P. Kuhn [3563], W. Fluch [2016].

Further improvements are contained in the book by H. Halberstam and H.-E. Richert [2455] (cf. [5209]).

5. It had been believed for a long time that the Prime Number Theorem is an essentially analytical result, and there was no way of proving it without utilizing some complex analysis. It came as a great surprise when in 1948 P. Erdős [1800] and A. Selberg [5613] provided elementary proofs of the equality

$$\psi(x) = (1 + o(1))x,$$

equivalent to the Prime Number Theorem.

Both proofs were based on the following formula known as *Selberg's identity*:

$$\sum_{p\le x} \log^2 p + \sum_{pq\le x} \log p \log q = 2x \log x + O(x), \tag{5.7}$$

where p and q run over primes.

Later K. Iseki and T. Tatuzawa [6074] showed that the following identity may also be used to obtain an elementary proof of the Prime Number Theorem:

$$\psi(x) \log x + \sum_{n\le x} \psi(x/n)\Lambda(n) = 2x \log x + O(x). \tag{5.8}$$

Note that in this formula the function ψ can be replaced by θ.

These proofs did not lead to any evaluation of the error term but some time later P. Kuhn [3565] modified Selberg's method, and this yielded the equality

$$\vartheta(x) = x + O\left(\frac{x}{\log^a x}\right) \tag{5.9}$$

with $a = 1/10$.

In 1960 R. Breusch [722] showed that (5.9) holds for every $a < 1/6$, and two years later E. Bombieri [593] and E. Wirsing [6696, 6697] proved that a can be taken arbitrarily large. For an exposition of Wirsing's method see A.F. Lavrik [3737].

A modification of Selberg's method was used by H.G. Diamond and J. Steinig [1527], leading to

$$\psi(x) = x + O\left(x \exp\left(-c \log^a x \,(\log\log x)^{-b}\right)\right), \tag{5.10}$$

with $a = 1/7$, $b = 2$, $c = 1$. Pushing forward this approach, A.F. Lavrik and A.Š. Sobirov [3739] proved this with $a = 1/6$, $b = 3$, $c = 1$, the same evaluation being true also for the difference $\pi(x) - \mathrm{li}(x)$. Later B.R. Srinivasan and A. Sampath [5882] eliminated the iterated logarithm from the last result. Now it is known that for every $\varepsilon > 0$ one can have

$$\pi(x) = \mathrm{li}(x) + O\left(x \exp\left(-c \log^a x\right)\right)$$

with any $a < 1/2$ (W.C. Lu [4020]).

Elementary proofs unrelated to the Selberg–Erdős method were given by H. Daboussi [1314] in 1984 and A. Hildebrand [2797] in 1986. They used sieve methods to deduce

$$\sum_{n\le x}\mu(n)=o(x)$$

which is known to be equivalent to the Prime Number Theorem.

6. Selberg's method can also be used for an elementary proof of the existence of infinitely many primes in arithmetic progressions $an+b$ with co-prime a,b. The first such proofs were given in 1949 by A. Selberg [5612, 5615] and H. Zassenhaus [6818]. Selberg used the following analogue of his identity (5.7):

$$\sum_{\substack{p\le x\\ p\equiv l \pmod k}}\log^2 p+\sum_{\substack{pq\le x\\ pq\equiv l \pmod k}}\log p\log q=\frac{2}{\varphi(k)}\frac{x}{\log x}+O(x),$$

as well as an inequality involving the Jacobi symbol, equivalent to the non-vanishing at $s=1$ of Dirichlet L-functions associated with real characters. This led to the formula

$$\sum_{\substack{p\le x\\ p\equiv l \pmod k}}\log p=(1+o(1))\frac{x}{\varphi(k)},$$

without any evaluation of the error term. Other elementary proofs of this result were given by H.N. Shapiro [5678, 5679] and K. Yamamoto [6771].

In 1989 H. Daboussi [1315] gave an elementary proof of the Prime Number Theorem for progressions based on the approach in [1314].

An elementary proof of the Dirichlet–Weber theorem on primes represented by binary quadratic forms was given by W.E. Briggs [728] and H. Ehlich [1693], and H.N. Shapiro [5677] gave the first elementary proof of the Prime Ideal Theorem.

An elementary proof of the Prime Ideal Theorem with an evaluation of the error term was later provided by Y. Eda and N. Nakagoshi [1684]. Other proofs were given by C. Touibi and H.S. Zargouni [6197] using Daboussi's method and by J. Hinz [2822] who used Hildebrand's approach.

A survey of elementary methods in the theory of primes was given by H.G. Diamond [1525].

7. In 1949 it was asked by G.H. Hardy and J.E. Littlewood [2538] whether the lower bound

$$\int_{-\pi}^{\pi}\left|\sum_{j=1}^{N}\cos(n_jx)\,dx\right|\ge C\log N$$

holds for every sequence of distinct positive integers n_j with an absolute constant C, and it became customary to call *Littlewood's conjecture* the inequality

$$\int_{-\pi}^{\pi}\left|\sum_{j=1}^{N}\exp(in_jx)\,dx\right|\ge C\log N. \tag{5.11}$$

The first result concerning this bound was obtained by R. Salem [5374] who showed in 1955 that if the sequence n_j has polynomial growth, then the left-hand side of (5.11) is $\gg \sqrt{\log N}$. After five years the first general lower bound was obtained by P.J. Cohen[4] [1151], who showed that the sum in (5.11) exceeds $c(\log N/\log\log N)^{1/8}$ with some $c > 0$. The exponent $1/8$ was soon replaced by $1/4$ (H. Davenport [1371] in 1960) and by $1/2$ (S.K. Pichorides [4851] in 1974). Later S.K. Pichorides [4852, 4853] replaced the right-hand side of (5.11) first by $\sqrt{\log N}$, and then by $\log N/\log\log N$. Finally, Littlewood's conjecture was settled in 1981 independently by O.C. McGehee, L. Pigno and B. Smith [4230] and S.V. Konyagin [3469].

8. In 1953 I.I. Piateckiĭ-Šapiro [4842] used Vinogradov's method of trigonometric sums to show that if $1 \le c < 12/11$, then the sequence $u_n(c) = [n^c]$ contains infinitely many primes, or more precisely, for the number $\pi_c(x)$ of primes below x of that form one has

$$\pi_c(x) = (1 + o(1))\frac{x^{1/c}}{\log x}. \tag{5.12}$$

Later the range for c in (5.12) was extended to $1 < c < 10/9 = 1.1111\ldots$ (G. Kolesnik [3452]), $c < 755/662 = 1.14048\ldots$ (D.R. Heath-Brown [2636]), $c < 1.14049\ldots$ (S.W. Graham [2303]), $c < 1.1470\ldots$ (G. Kolesnik [3459]), $c < 15/13 = 1.1538\ldots$ (H.Q. Liu, J. Rivat [3959]), $c < 1.1544$ (J. Rivat [5231]), $c < 1.1550$ (G.I. Arkhipov, V.N. Čubarikov [119]) and $c < 1.16177\ldots$ (J. Rivat, P. Sargos [5232]). It is also known that for $1 < c < 243/205 = 1.1853\ldots$ one has

$$\pi_c(x) \gg \frac{x^{1/c}}{\log x}$$

(J. Rivat, J. Wu [5233]). Earlier this had been known for $c < 13/11 = 1.1818\ldots$ (C.H. Jia [3135]) and for $c < 45/38 = 1.1842\ldots$ (A. Kumchev [3573]). In 1976 J.-M. Deshouillers [1484] showed that for every $c > 1$ one has

$$\pi_c(x) \ll \frac{x^{1/c}}{\log x},$$

and for almost all $c > 1$ the sequence $[n^c]$ contains infinitely many primes.

In 1992 A. Balog and J.B. Friedlander [303] proved that if $c \in [1, 21/20)$, then in Vinogradov's three primes theorem one can assume that the summands have the form $[n^c]$, and C.H. Jia [3140] extended the admissible interval for c to $[1, 16/15]$.

9. In 1957 D.A. Burgess [854–857] introduced a new method of evaluation of character sums which he applied to the study of quadratic residues, primitive roots and Dirichlet L-functions. His main result (proved in [856, 857]) showed that if k is square-free and χ is a primitive character mod k, then for every $H, \varepsilon > 0$ and $r = 1, 2, \ldots$ one has

$$\sum_{n=m+1}^{m+H} \chi(n) \le cH^{1-1/(r+1)}k^{1/4r+\varepsilon},$$

[4] Paul Joseph Cohen (1934–2007), professor at Stanford University. Fields Medal 1966 for proving the independence of the axiom of choice and the continuum hypothesis of the Zermelo–Fraenkel axioms of set theory.

with a certain constant $c = c(r, \varepsilon)$. For not square-free k this evaluation holds for $r = 2$. In [855] the case of prime k was treated, and the factor k^{ε} in the bound was replaced by $\log k$. This bound is better than that of Pólya–Vinogradov in the case when H is essentially smaller than $\sqrt{p}$.

Explicit values for the constants occurring in Burgess's inequalities were later provided by E. Grosswald [2364].

Burgess's bound has several applications. We have already mentioned in Sect. 3.1.4 its influence on bounds for the consecutive quadratic residues and non-residues, and it has been shown by D.A. Burgess [855] that the smallest primitive root mod p is $\ll p^{a}$ for every $a > 1/4$. In [856] he applied his result to improve the upper bound for the values of Dirichlet L-functions at the critical line, obtaining for primitive characters χ mod k and fixed t the bound

$$|L(1/2 + it, \chi)| \ll k^{7/32+\varepsilon},$$

and then, in [857], replacing the constant $7/32$ by $3/16$ (previously H. Davenport [1341] had $1/4$ in the exponent). Later he succeeded in eliminating the ε [859].

The best known bound,

$$|L(1/2 + it, \chi)| \ll (1 + |t|)^{A} k^{1/6+\varepsilon},$$

is due to J.B. Conrey and H. Iwaniec [1218].

An analogue of Burgess's bound for algebraic number fields was obtained by J. Hinz [2818].

10. In 1951 a new version of E.C. Titchmarsh's treatise [6170] appeared [6178], incorporating progress in the study of Riemann's zeta-function made since its first publication in 1930, and in 1957 K. Prachar published the book [5006] in which he gave a modern presentation of analytical methods in number theory.

Later more books devoted to Riemann's zeta-function appeared (A. Ivić [3038, 3041, 3042], A.A. Karatsuba, Y. Motohashi [4461], S.M. Voronin[5] [3256]). For prime number theory see the book by W.J. Ellison [1755].

11. The number $A(x)$ of distinct values of Euler's function $\varphi(n)$ for $n \leq x$ was considered by P. Erdős [1795] in 1945. He showed that $A(x) = (c + o(1))x$ with a certain $c > 0$, and it turned out later that $c = \zeta(2)\zeta(3)/\zeta(6)$ (R.E. Dressler [1625]).

The resulting error term was shown by P.T. Bateman [353] to be

$$O\left(x \exp\left(-c\sqrt{\log x \log\log x/2}\right)\right)$$

for every $c < 1/\sqrt{2}$ (see A. Smati [5827] for a fully effective bound), and this was improved in 1998 to

$$O\left(x \exp\left(-a \log^{3/5} x \,(\log\log x)^{-1/5}\right)\right)$$

by M. Balazard and G. Tenenbaum [289].

[5] Sergeĭ Mikhaĭlovič Voronin (1946–1997), worked at the Steklov Institute. See [117].

The same question for the divisor function $d(n)$ was treated by P. Erdős and L. Mirsky [1844]. They showed in 1952 that if $D(x)$ is the number of distinct values of $d(n)$ for $n \le x$, then

$$\log D(x) = \left(\frac{2\sqrt{2}}{\sqrt{3}}\pi + o(1)\right)\frac{\sqrt{\log x}}{\log\log x},$$

and conjectured that for infinitely many integers n the equality

$$d(n) = d(n+1) \tag{5.13}$$

holds.

This problem remained unresolved for a long time and the first result on it was obtained in 1973 by R.C. Vaughan [6345] who showed that if this conjecture fails, then there exist infinitely many primes p such that $8p+1$ is a product of two primes. In 1981 C. Spiro [5868] succeeded in showing that infinitely often one has $d(n) = d(n+5040)$ and her method was modified in 1984 by D.R. Heath-Brown [2637] to yield the solution of the Erdős–Mirsky problem. His result gave the lower bound $\gg x/\log^7 x$ for the number of $n \le x$ satisfying (5.13) and three years later this was improved to $\gg x/(\log\log x)^3$ by A. Hildebrand [2802]. On the other hand the upper bound $\ll x/\sqrt{\log\log x}$ was established by P. Erdős, C. Pomerance and A. Sárkőzy [1847]. In 1997 C. Pinner [4879] generalized the result of D.R. Heath-Brown showing that for every $a > 0$ one has $d(n) = d(n+a)$ infinitely often.

A similar question concerning consecutive values of the function $\omega(n)$ was answered by J.-C. Schlage-Puchta [5457] in 2003. Earlier it had been shown by P. Erdős, C. Pomerance and A. Sárkőzy [1846] that there exists a constant C such that the inequality $|\omega(n+1) - \omega(n)| \le C$ holds for at least $\gg x/\sqrt{\log\log x}$ integers $n \le x$. They also gave upper bounds for the number of $n \le x$ with either $\varphi(n+1) = \varphi(n)$ or $\sigma(n+1) = \sigma(n)$.

12. Another question related to Euler's function was answered in 1948 by P. Erdős [1797]. He obtained asymptotics for the number of $F(x)$ of integers $n \le x$ with $(\varphi(n), n) = 1$ by establishing the equality

$$F(x) = (e^{-\gamma} + o(1))\frac{x}{\log\log\log x}, \tag{5.14}$$

with γ being the Euler constant. Note that the condition $(\varphi(n), n) = 1$ characterizes integers n with the property that all groups of order n are isomorphic.

The same question was also considered for several other integer-valued functions. It was shown by P. Erdős and G.G. Lorentz [1842] that for a large class of functions f one has

$$\#\{n \le x : (f(n), n) = 1\} = \left(\frac{6}{\pi^2} + o(1)\right)x$$

(cf. R.R. Hall [2473–2475]), and E.J. Scourfield [5592] made a far reaching generalization of (5.14), showing that if f is integer-valued, multiplicative, and for $j = 1, 2, \dots$ and for every prime p one has $f(p^j) = V_j(p)$ with $V_j \in \mathbf{Z}[X]$, $V_1(0) \ne 0$ and $\deg V_1 \ge 1$, then

$$\#\{n \le x : (f(n), n) = 1\} = (C + o(1))\frac{x}{(\log\log x)^\lambda}$$

with $C > 0$ and rational $0 < \lambda \le 1$, depending only on V_1. A strong evaluation of the error term was provided by C. Spiro [5869] in the case of f multiplicative with $f(p)$ and $f(p^2)$

being independent of the prime p. Her result applied also to the asymptotics of the property $(f(n), P(n)) = 1$ with $P \in \mathbf{Z}[X]$. See also E.J. Scourfield [5593].

A corresponding result for a class of additive functions has been proved by A.S. Faĭnleĭb [1951].

5.2 Additive Problems

1. In 1942 H.B. Mann [4139] succeeded in proving the $\alpha + \beta$ *hypothesis* concerning the Schnirelman density $\delta(A)$, stating that the inequality (4.32) could be improved to

$$\delta(A+B) \geq \min\{1, \delta(A) + \delta(B)\}, \tag{5.15}$$

provided $0 \in A \cup B$.

Earlier A.J. Khintchine [3321] proved this in the case $\delta(A) = \delta(B)$; A. Brauer [679] obtained

$$\delta(A+B) \geq \frac{8}{9} \min\{1, \delta(A) + \delta(B)\},$$

and in [680] replaced 8/9 by 9/10. Other proofs of (5.15) were later given by E. Artin and P. Scherk [145], F.J. Dyson [1676], J.G. van der Corput [6293–6295], P. Scherk [5426] and B.K. Garrison [2197]. Dyson's paper also contains an extension to several summands.

Since the only set with Schnirelman's density equal to 1 is the set of all natural numbers, then Mann's theorem implies that if a set A of natural numbers containing 0 has positive Schnirelman density, then it is a basis of order not exceeding $1/\delta(A)$.

Mann's theorem is best possible, as B. Lepson [3833] proved that for any real numbers $0 \leq \alpha, \beta \leq 1$ there exist sets A, B containing 0 and having densities α and β, respectively, such that the set $A + B$ has density equal to $\max\{1, \alpha + \beta\}$.

The inequality (4.32) does not hold for the asymptotic density of sets, defined by

$$d(A) = \lim_{x\to\infty} \frac{A(x)}{x},$$

if this limit exists. Indeed, it fails in the case when A is the set of all even integers, when $d(A + A) = 1/2 = d(A)$. If however at least one of the sets A, B contains 1, then one has

$$d(A+B) \geq d(A) + d(B)/2,$$

as shown by P. Erdős [1791] (see H.N. Shapiro [5680] for a simple proof, covering also a case which Erdős excluded). For improvements see M. Kneser [3406], H.-H. Ostmann[6] [4703], H. Rohrbach, B. Volkmann [5267].

Cf. also Y.F. Bilu [509].

[6] Hans-Heinrich Ostmann (1913–1959), professor in Berlin. See [2907, 3243].

For a discussion of cases in which Mann's theorem fails for the asymptotic density see the paper [3406] of M. Kneser.

Note that Mann's theorem is an analogue of a result proved in 1813 by A. Cauchy [965], who showed that if p is a prime and $A = \{a_1, \dots, a_m\}$ and $B = \{b_1, \dots, b_n\}$ are two sets of residues mod p, then the set $A + B$ of all residue classes of the form $a_i + b_j$ has at least $\min\{m + n - 1, p\}$ elements. This result was rediscovered in 1935 by H. Davenport [1346, 1362] and is called the *Cauchy–Davenport theorem*. The cases when equality holds were determined by A.G. Vosper [6476] and S. Chowla, H.B. Mann, E.G. Straus [1098]. The last paper also contains applications to additive questions. See also G.A. Freiman [2075].

An analogue of the Cauchy–Davenport theorem for composite moduli was proved by I. Chowla [1063] (cf. S.S. Pillai [4866]).

For an extension of the Cauchy–Davenport theorem see J.M. Pollard [4943, 4945].

The Cauchy–Davenport theorem was later extended to all finite groups G by G. Károlyi [3258] who showed that if $A, B \subset G$, then

$$\#\{ab : a \in A, b \in B\} \geq \min\{p(G), \#A + \#B - 1\},$$

where $p(G)$ denotes the smallest prime factor of $\#G$.

Put

$$A \odot B = \{a + b : a \in A, b \in B, a \neq b\}.$$

It was conjectured in 1964 by P. Erdős and H. Heilbronn (the *Erdős–Heilbronn conjecture*) that if A is a set of residue classes mod p, then

$$\#(A \odot A) \geq \min\{p, 2\#A - 3\}.$$

This was established in 1994 by J.A. Dias da Silva and Y.O. Hamidoune [1530]. One year later N. Alon, M.B. Nathanson and I.Z. Ruzsa [58] provided a simpler proof, and established

$$\#(A \odot B) \geq \min\{p, \#A + \#B - 2\} \tag{5.16}$$

for $A, B \subset \mathbf{Z}/p\mathbf{Z}$ satisfying $\#A \neq \#B$. The last condition can actually be omitted (G. Károlyi [3261]; see also [4737]). Cf. N. Alon, M.B. Nathanson, I.Z. Ruzsa [59], G.A. Freiman, L. Low, J. Pitman [2083].

For the restricted addition $A \odot B$ in groups see G. Károlyi [3259], V.F. Lev [3842–3845] and Z.-W. Sun [5990]. The analogue of (5.16) for finite groups was obtained in 2009 by P. Ballister and J.P. Wheeler [293].

A book devoted to various addition theorems was published by H.B. Mann [4141] in 1965.

2. A difficult problem in the additive theory of numbers, which still awaits a solution, was proposed in 1941 by P. Erdős and P. Turán [1865]. It is usually stated in the following form.

If A is a basis of order 2 of the set of positive integers and $f_A(n)$ denotes the number of representations $n = a + b$ with $a, b \in A$, then $\limsup_{n\to\infty} f_A(n) = \infty$.

For a class of bases this was established in 2004 by J. Nešetřil and O. Serra [4560] but in the general case no essential progress has been achieved. It is known only that for every basis A of order 2 one has

$$\limsup_{n\to\infty} f_A(n) \geq 8.$$

This was proved by P. Borwein, S. Choi and F. Chu [650] in 2006 (earlier G. Grekos, L. Haddad, C. Helou and J. Pihko [2341] had 6 in place of 8).

This problem can also be reformulated in terms of bases of finite initial segments of the integers (M. Dowd [1618]).

Another problem concerning the function $f_A(n)$, posed by P. Erdős and P. Turán in [1865], asked for a proof that there is no sequence A such that with a certain positive c one has

$$\sum_{k=0}^{n} f_A(k) = cn + O(1). \tag{5.17}$$

This was established by P. Erdős and W.H.J. Fuchs [1829] in 1956 in a stronger form: they showed that the error term in the equality is $\Omega(n^{1/4}/\sqrt{\log n})$ (a simpler proof was given later by D.J. Newman[7] [4584]). For an analogue in real quadratic fields see W. Schaal [5420].

In 1990 H.L. Montgomery and R.C. Vaughan [4369] showed that $o(n^{1/4})$ is also impossible. This is not far from being optimal, as I.Z. Ruzsa [5348] proved in 1997 the existence of a sequence A for which the error term in (5.17) is $O(n^{1/4}\log n)$.

The question of whether there is a 2-basis A (i.e., a basis of order 2) with $f_A(n) = O(n^{\varepsilon})$ for every positive ε, was posed by S. Sidon[8] (see [1865]), and a positive answer is contained in the paper [1810] by P. Erdős from 1954. He showed that there are bases satisfying $f_A(n) = O(\log n)$, and actually almost all (in a certain sense) 2-bases have this property.

Erdős's proof was not effective, but an effective example was provided in 1995 by M.N. Kolountzakis [3460]. Another 2-basis A with small function f_A, satisfying

$$\sum_{n\le x} f_A^2(n) \le Bx$$

with some unspecified B was found in 1990 by I.Z. Ruzsa [5344], and M. Tang [6043] found a basis with explicit $B < 1.5 \cdot 10^9$. For a generalization to k-bases for $k \ge 3$ see P. Erdős and P. Tetali [1863].

Another question, attributed to S. Sidon, asked for the evaluation of the size $\Phi(n)$ of the largest set $A \subset [1, n]$ with all sums $a + b$ $(a, b \in A)$ distinct (such sets are called *Sidon sets*). The answer was provided in 1944 by S. Chowla [1084], who established

$$\lim_{n\to\infty} \frac{\Phi(n)}{\sqrt{n}} = 1,$$

confirming a conjecture by P. Erdős and P. Turán [1865].

---

[7]Donald J. Newman (1930–2007), professor at MIT, Brown University, Yeshiva University and Temple University. See [6766].

[8]Simon Sidon (1892–1941).

One also considers infinite sets having the Sidon property. Chowla's result implies the bound $A(x) = O(\sqrt{x})$ for such sets, and the largest known such set seems to be that constructed by I.Z. Ruzsa [5350] , for which one has

$$A(x) = (c + o(1))x^{\sqrt{2}-1+o(1)}$$

with $c > 0$.

The question of whether a Sidon set can be an asymptotical basis for the integers got a positive answer in 2009, when J.-M. Deshouillers and A. Plagne [1495] constructed such a basis of order 7. The question of whether there exists a Sidon set which is an asymptotical basis of order 3 (P. Erdős, A. Sárkőzy, V.T. Sós [1853]) is still open, but it is known that there is no such basis for all integers (G. Grekos, L. Haddad, C. Helou, J. Pihko [2342]).

It is not known whether there is a polynomial $P \in \mathbf{Z}[X]$ whose set of positive values is a Sidon set. There is an old conjecture that $P(X) = X^5$ has this property. It was shown in 2001 by I.Z. Ruzsa [5354] that for suitable $\alpha \in (0, 1)$ and n_0 the set $\{n^5 + [\alpha n^4] : n \geq n_0\}$ is a Sidon set.

3. Two sets A, B of positive integers are said to be *complementary* if every sufficiently large integer can be written in the form $a + b$ with $a \in A$ and $b \in B$. It was shown in 1954 by G.G. Lorentz [3990] that every infinite set A has a complementary set B with

$$B(x) = \sum_{\substack{b \leq x \\ b \in B}} 1 \ll \sum_{n \leq x} \frac{\log A(n)}{A(n)}.$$

In particular the set of primes has a complement B with $B(x) = O(\log^3 x)$, and this was improved to $B(x) = O(\log^2 x)$ by P. Erdős [1811], who asked whether one can even have $B(x) = O(\log x)$.

Much later D. Wolke [6717] showed that for every function $f(x)$ increasing to infinity there exists a set B of primes with $B(x) \ll f(x) \log x \log\log x$, such that almost every even integer is the sum of a prime and an element of B, and M.N. Kolountzakis [3461] proved that this holds also for $f(x) = 1$. In 1998 I. Ruzsa [5349] showed that in Wolke's result one can remove the factor $\log\log x$.

C. Hanani (see P. Erdős [1816]) asked whether there are complementary sets A, B with

$$\lim_{x\to\infty} \frac{A(x)B(x)}{x} = 1,$$

and such an example was provided in 1964 by L. Danzer [1327].

I.Z. Ruzsa [5347] showed in 1996 that this can be obtained with $A = \{a^k : k \geq 0\}$ for any integer $a \geq 3$. Later [5353] he proved that the same holds for any $A = \{a_1, a_2, \ldots\}$ satisfying $a_{n+1}/(na_n) \to \infty$.

4. It was proved in 1959 by E. Grosswald, A. and J. Calloway [2367] that integers which are sums of three squares but are not sums of three positive squares form a finite set, which conjecturally has 10 elements. This is related to the problem of determining the discriminants of binary quadratic forms with one class per genus. Cf. L.J. Mordell [4399], A. Schinzel [5432], F. Halter-Koch [2482].

Positive integers not divisible by 4 and having a unique representation as sums of three squares were considered in 1984 by P.T. Bateman and E. Grosswald [357], who showed that they form a finite set S, and related its determination to the solution of the class-number four problem for imaginary quadratic fields. Since this determination was later achieved by S. Arno [128] one knows now that S has 32 elements, the largest being 427. The corresponding problem for sums of four squares was solved by D.H. Lehmer [3786] in 1948, and for certain ternary quadratic forms by I. Kaplansky [3249].

5. The method used by H. Davenport [1354, 1356] for the treatment of the Waring problem for cubes and fourth powers enabled progress to be made for other small powers also. So the inequality $G(6) \leq 36$ was proved by K. Subba Rao [5982] and H. Davenport [1360], and in the same paper Davenport obtained $G(5) \leq 23$, improving upon the bounds 35 by R.D. James [3099], 29 by T. Estermann [1890] and 28 by L.K. Hua [2924]. This was followed by bounds for $G(k)$ in the cases $k = 8, 9$ and 10, equal to 73, 99 and 122, respectively, proved by V. Narasimhamurti [4534].

6. In 1944 C.L. Siegel returned to the problem of representing integers in an algebraic number field K as sums of kth powers in a more general setting. He showed in [5762] that every totally positive integer of a totally real field K which is the sum of kth powers of integers and has a sufficiently large norm is the sum of at most $kn(2^{k-1} + n) + 1$ such powers, with $n = [K : \mathbf{Q}]$, and in [5766] extended this result also to remaining fields. If $g(k, K)$ denotes the minimal number of summands needed for representing totally positive integers with sufficiently large norm, then Siegel expected that this number should depend only on k, established this in [5766] in the case $k = 2$, proving $g(2, K) \leq 5$, and conjectured that $g(2, K) = 4$. He proved also that if K is not totally real, then all totally positive integers are sums of squares if and only if the discriminant of K is odd, and in totally real fields this happens if and only if $K = \mathbf{Q}$ or $K = \mathbf{Q}(\sqrt{5})$. For higher powers this holds for totally real fields only in the case $K = \mathbf{Q}$, and in other fields if and only if every integer is the sum of kth powers.

Siegel's bound for $g(k, K)$ was reduced in 1958 by T. Tatuzawa [6070] to $g(k, K) \leq 8kn(n + k)$ using Vinogradov's method, and by O. Körner [3481], who in 1961 got

$$g(k, K) \leq kn\left(3\log k + 3\log\left(n^2 + 1\right) - 3\log 3 + 11\right).$$

The next step was in the sixties by B.J. Birch [522] and O. Körner [3482]. Birch reduced the question to a local problem by showing that if s exceeds 2^k, then every sufficiently large totally positive integer of K which is locally (i.e., in every $\mathfrak{p}$-adical completion of K) the sum of s kth integral powers is also the sum of s totally positive kth powers in K. He then used the investigations of the local question done by R.M. Stemmler [5922], who studied the easier Waring problem in algebraic number fields, to prove Siegel's conjecture for prime k, with the bound $1 + 2^k$ in this case. The final step was taken independently by B.J. Birch [524] and

C.P. Ramanujam[9] [5073]. Birch showed that every element of a local field, which is the sum of kth powers, is the sum of at most k^{16k^2} such powers, and Ramanujam got the bound $8k^5$ for the number of summands needed.

The best known upper bound for $g(k, K)$ was obtained in 1999 by M. Davidson [1409], who obtained

$$g(k, K) \leq kn(\log k + \log\log k + 6) + O\left(\frac{\log\log k}{\log k}\right).$$

This improved the earlier bounds of T. Tatuzawa [6071, 6072], Y. Eda [1683] and M. Davidson [1408].

Algebraic number fields K in which every integer is the sum of kth powers were described by M. Bhaskaran [502], who showed that this happens if and only if k and the discriminant of K are co-prime, and for every prime ideal $\mathfrak{p}$ with norm p^f there is no proper divisor r of f such that k is divisible by $(p^f - 1)/(p^r - 1)$. For prime k this was done earlier by P.T. Bateman and R.M. Stemmler [360]. A simpler proof of Bhaskaran's result was provided later by O. Körner [3483].

Integers of quadratic fields which are sums of kth powers were determined by J.H.E. Cohn [1161] in 1972.

A monograph dealing with the Waring problem in algebraic number fields was written by Y. Wang [6557].

7. In 1949 G.A. Freiman [2073] stated that if $2 \leq n_1 \leq n_2 \leq \cdots$ is a sequence of positive integers, then for every j there exists $r = r(n_j)$ such that every large integer n can be written in the form

$$x_0^{n_j} + x_1^{n_j+1} + \cdots + x_r^{n_j+r}$$

if and only if the series $\sum_{j=1}^{\infty} 1/n_j$ diverges. The first proof was published in 1960 by E.J. Scourfield [5589]. In the case $n_j = j+1$ her result gives $r(j+1) \ll j^5 \log j$.

Asymptotics for the number of representations were found by K. Thanigasalam [6126], and in 1995 K. Ford [2026] improved Scourfield's bound to $r(j+1) \ll j^2 \log j$. K. Thanigasalam [6125] showed in 1966 that in the Freiman–Scourfield theorem one can assume that all x_i's are prime powers.

8. A problem which can be regarded as an approximation of the Goldbach conjecture was attempted in 1951 by Yu.V. Linnik [3916]. He used the General Riemann Hypothesis to show the existence of a constant K such that every large even integer is the sum of two primes and K powers of 2. Two years later [3917] he was able to eliminate all unproven assumptions, but did not give any value for the constant K.

Such a value, alas rather large, ($K = 770$ under the General Riemann Hypothesis and $K = 54 \cdot 10^3$ unconditionally) was provided in 1998 (J.Y. and M.C. Liu, T.Z. Wang [3961, 3962], who quickly reduced it to $K = 200$ [3963]), and the best known results are $K = 7$ under the General Riemann Hypothesis (D.R. Heath-Brown, J.-C. Schlage-Puchta [2672], J. Pintz, I.Z. Ruzsa [4909]), and $K = 13$ [2672] without it. The development of this problem is presented in a survey paper by J. Pintz [4905].

[9] Chidambaran Padmanabhan Ramanujam (1938–1973), professor at the Tata Institute. See [5072].

9. The representation of integers as linear combinations of primes was studied in H.-E. Richert's dissertation [5203]. He showed that if $s \geq 3$ and the positive integers $a_1, \ldots, a_s$ are pairwise co-prime and $n > 0$ has the same parity as $\sum_{j=1}^{s} a_j$, then the equation

$$n = \sum_{j=1}^{s} a_j p_j$$

is solvable with prime p_j for n sufficiently large, and the asymptotics of the number of solutions can be determined. If the a_j's are of distinct signs, then this equation has for every n infinitely many solutions.

In the last case one can ask for solutions with small primes, and this was considered in 1967 by A. Baker [230] who showed that in the case $s = 3$ under certain additional assumptions one can have $|p_i| \leq C^m$ with $m = n^{\varepsilon}$ for every $\varepsilon > 0$. Later stronger bounds were found, the best being due to Li [3881], who showed that solutions exist with

$$\max |p_i| \ll \max\{n, |a_1|, |a_2|, |a_3|\}^B$$

with $B = 38$. It is known (K.K. Choi, M.C. Liu, K.M. Tsang [1056]) that under the General Riemann Hypothesis one can have $B = 4 + \varepsilon$ with any $\varepsilon > 0$.

5.3 Diophantine Equations and Congruences

1. The old conjecture asserting that the only solutions of the equation

$$\frac{x^n - 1}{x - 1} = y^q, \tag{5.18}$$

with $x, y, q \geq 2,\ n \geq 3$ are

$$(x, y, q, n) = (3, 11, 2, 5), (7, 20, 2, 4), (18, 7, 3, 3)$$

was considered in 1943 by W. Ljunggren [3975], who confirmed it in the case of even q (earlier T. Nagell [4501, 4502] proved the conjecture for n divisible by 3 or 4).

T.N. Shorey and R. Tijdeman showed in their book [5726] that (5.18) has finitely many solutions if x is fixed, or if one fixes a prime divisor of n or y.

In the case $x = 10$ the conjecture was settled in 1999 by Y. Bugeaud and M. Mignotte [817]. This implies that an integer > 1 all of whose decimal digits equal unity cannot be a proper power. It had been shown earlier by R. Obláth [4651] that the same assertion holds for integers having all their decimal digits equal to a fixed integer $\neq 1$.

Later, the case $p|n$ with odd prime $p \leq 11$ was settled by Y. Bugeaud, G. Hanrot and M. Mignotte [816]. Some other cases were treated by M.A. Bennett [409], Y. Bugeaud, M. Mignotte [818], Y. Bugeaud, M. Mignotte, Y. Roy [820], Y. Bugeaud, M. Mignotte, Y. Roy, T.N. Shorey [821], M.H. Le [3747], M. Mignotte [4293] and N. Saradha, T.N. Shorey [5396]. A survey was prepared in 2002 by Y. Bugeaud and M. Mignotte [819].

It was noted by T.N. Shorey in his survey [5723] of exponential Diophantine equations that the ABC conjecture implies that (5.18) has only finitely many solutions x, y, q, n (cf. [819]).

2. In 1950 P. Erdős [1803] considered the problem of representing a rational number $r = a/q > 0$ as the sum of unit fractions

$$\frac{a}{q} = \sum_{i=1}^{k} \frac{1}{n_i} \tag{5.19}$$

with $n_1 < n_2 < \cdots < n_k$, and presented an algorithm which accomplishes this with $k \le 8 \log q / \log\log q$ and $n_k \le 4q^2 \log q / \log\log q$. This improved essentially an old result by J.J. Sylvester [6007].

An algorithm quicker by a factor of 2 was proposed in 1986 by H. Yokota [6782, 6783], and its speed was doubled by G. Tenenbaum and Yokota [6110] in 1990.

Let $D(a, q)$ be the minimal number of terms in a representation (5.19) and put $D(q) = \max_{1 \le a < q}\{D(a, q)\}$. Erdős's bound $D(q) = O(\log q / \log\log q)$ was reduced in 1985 to $O(\sqrt{\log q})$ by M.D. Vose [6474].

Denote by $N(p, q)$ the minimum of the largest denominator in all possible representations (5.19) and put $N(q) = \max_{1 \le a < q}\{N(a, q)\}$. Erdős's bound for $N(q)$ was improved to $N(q) \le q(q - 1)$ by S.W. Golomb [2275], to $N(q) \le (1 + o(1))q \log^2 q$ by M.N. Bleicher and P. Erdős [555, 556], and to $N(q) \ll q \log^c q$ with any $c > 3/2$ by H. Yokota [6784], who in [6785] proved this for every $c > 1$. On the other hand, for suitable q one has $N(q) = \Omega(q \log q)$ [556].

It is still unknown whether for rational numbers with odd denominator the greedy algorithm, in which at every step one takes the smallest possible odd denominator, terminates. Even for rationals of small weight this algorithm may lead to denominators extremely large, for example for $r = 5/139$ one arrives after a few steps at denominators having more than 300 000 digits (see the review [4161] of the paper by J. Pihko [4855]).

It has been shown by H. Yokota [6786, 6787] that there are $\log n + O(\log\log n)$ integers which can be written as the sum of distinct unit fractions with denominators $\le n$ (see E. Croot [1282], H. Yokota [6787] for improvements of the error term). In [6788] asymptotics for the largest such integer were found.

3. In 1952 P. Erdős [1806], studying an additive problem, introduced the notion of *covering congruences*, i.e., finite sets of congruences

$$x \equiv a_i \pmod{n_i} \quad (i = 1, 2, \ldots, k)$$

with distinct moduli n_i such that every integer satisfies at least one of them.

There are several open problems concerning this notion, among them the following two proposed by P. Erdős and J.L. Selfridge.

(A): *Prove that for every N there exists a system of covering congruences with minimal modulus $\ge N$.*

An example with $N = 20$ was found by S.L.G. Choi [1057], and this was later superseded by R. Morikawa [4426] with $N = 24$, D.J. Gibson [2237] with $N = 25$ and P. Nielsen [4614] with $N = 40$.

(B): *Show that there is no system of covering congruences with odd moduli* > 1.

A necessary condition for the existence of such a system was given by M.A. Berger, A. Felzenbaum and A.S. Fraenkel [434, 435]. In [435] they showed that such a system must consist of at least six congruences. See also R.J. Simpson and D. Zeilberger [5798].

It was shown by J.H. Jordan [3157] that the analogue of (B) fails in the ring of integers of $\mathbf{Q}(\sqrt{-2})$.

In 1962 P. Erdős [1820] conjectured, strengthening an earlier conjecture of S.K. Stein [5918], that if a finite system of k congruences does not cover all integers, then it does not contain a positive integer $n \le 2^k$. This was shown to be true in 1970 by R.B. Crittenden and C.L. Vanden Eynden [1281].

Surveys on covering congruences were prepared by Š. Porubský [4995] in 1981, Š. Znám [6834] in 1982 and Š. Porubský, J. Schönheim [4997] in 2002.

4. In 1959 H. Davenport [1370] adapted the circle method to show that a cubic form over $\mathbf{Z}$ in more than 32 variables represents zero non-trivially.

Later he was able [1373, 1374] to replace 32 first by 29 and then by 16. It has been shown recently by D.R. Heath-Brown [2659] that 14 variables are sufficient, and it is conjectured that even 10 would do. This would be best possible.

Davenport's method was generalized by C.P. Ramanujam [5074] to cubic forms over algebraic number fields, and it was shown in 1975 by P.A.B. Pleasants[10] [4922] that 16 variables always suffice (earlier C. Ryavec [5355] had here 17). B.J. Birch [523] obtained analogues of Davenport's result for systems of arbitrary forms with rational coefficients.

Later W.M. Schmidt [5515] proved that a system of r cubic forms over $\mathbf{Q}$ in $\ge (10r)^5$ variables has a non-trivial rational zero. For a single non-singular cubic form F over the rationals (i.e., when the system of equations $\partial F/\partial x_i = 0$ has no non-trivial solutions) ten variables are sufficient, as shown in 1983 by D.R. Heath-Brown [2635], and C. Hooley [2900–2902] showed that in that case the Hasse principle holds for 9 variables.

5. In a letter to L. Moser, written around 1950, P. Erdős conjectured that for $k \ge 2$ the equality

$$1^k + 2^k + \cdots + (n-1)^k = n^k \tag{5.20}$$

is impossible, and L. Moser [4440] showed in 1953 that this is correct for all $n \le 10^{10^6}$ and gave a proof for odd k (another proof was later given by J. Urbanowicz [6253]).

Moser's bound was extended to $n \le 10^{9\,321\,155}$ by W. Butske, L.M. Jaje and D. Mayernik [869]. It was shown in 1994 by P. Moree, H.J.J. te Riele and J. Urbanowicz [4417] that the exponent k in (5.20) must be divisible by the least common multiple of all integers ≤ 200, which exceeds $10^{89.5}$. Later B.C. Kellner [3297] showed that k is divisible by every prime $p < 1000$, hence k must be a multiple of an integer $> 5.7 \cdot 10^{427}$. Recently Y. Gallot, P. Moree and V.V. Zudilin [2188], showed that the number of $n \le x$ for which (5.20) holds with some k is $O(\log x)$. They point out that (5.20) seems to be the only exponential equation in two variables for which it is not known whether it has only finitely many solutions.

---

[10] Peter Arthur Barry Pleasants (1939–2008), worked in Cambridge, Cardiff, Macquarie University, the University of the South Pacific and Brisbane. See [1036].

5.4 Elliptic Curves

1. In 1941 M. Deuring [1503] studied function fields of elliptic curves over fields of positive characteristic. His results give in particular a description of possible orders of elliptic curves over the field $\mathbf{F}_p$.

Such a description for curves over an arbitrary finite field was given in 1969 in W.C. Waterhouse's thesis [6571, Theorem 4.1]. Later H.-G. Rück [5336] used this result to describe the possible group structures of elliptic curves over finite fields (see also J. Miret, R. Moreno, A. Rio and M. Valls [4330], C. Wittmann [6704]). Waterhouse's thesis also brings important additions to the classification of Abelian varieties over finite fields, obtained by J. Tate [6062] and T. Honda [2853] (cf. W.C. Waterhouse and J.S. Milne [6575]).

2. If E is an elliptic curve, defined over a field K, and C is a non-singular curve over K on which the group $E(K)$ acts in a simply transitive way, then C is called a *principal homogeneous space* for E. The set $WC(E/K)$ of equivalence classes of such spaces under a natural equivalence was first considered by F. Châtelet [1003, 1004], who showed in 1941 that this set can be injected in the cohomology group $H^1(G_K, E)$, G_K being the Galois group of the extension $\overline{K}/K$. Later A.A. Weil [6622] noted that $WC(E/K)$ is actually a group (the *Weil–Châtelet group* for $E(K)$) isomorphic to $H^1(G_K, E)$ (see also S. Lang, J. Tate [3705]).

Recall that if G is a group acting on an Abelian group A, then the first cohomology group $H^1(G, A)$ is defined as the factor group of the group of all maps $f: G \to A$ satisfying $f(gh) = gf(h) + f(h)$ by the subgroup consisting of maps having the form $f(g) = ga - a$ for some $a \in A$.

For every completion $K_{\mathfrak{p}}$ of K there is a canonical homomorphism $WC(E/K) \to WC(E/K_{\mathfrak{p}})$, and the kernel of the resulting map

$$WC(E/K) \to \prod_{\mathfrak{p}} WC(E/K_{\mathfrak{p}}) \qquad (5.21)$$

is denoted by[11] $\text{Ш}(E/K)$ and called the *Tate–Šafarevič group of* $E(K)$. It has been conjectured that it is a finite group, and J.W.S. Cassels [939] has shown that $\#\text{Ш}(E/K)$ can be arbitrarily large. On the other hand I.R. Šafarevič [5364] proved in 1959 that for every n, $\text{Ш}(E/K)$ contains only finitely many elements of order $\leq n$.

The cardinality of $\text{Ш}(E/K)$ appears in the important conjecture of B.J. Birch and Swinnerton-Dyer, which will be described in Sect. 6.7.

The first examples of finite groups $\text{Ш}(E/K)$ were given by K. Rubin [5328] and V.A. Kolyvagin [3462–3464]. It is now known that the Birch–Swinnerton-Dyer conjecture implies that for infinitely many E over $\mathbf{Q}$ one has $\#\text{Ш}(E) \gg N^c$ for every $c < 1/4$, with $N = N_E$ denoting the conductor of E (L. Mai, M.R. Murty [4097]). Unconditionally one knows only that $\#\text{Ш}(E) \gg \exp(c \log N / \log\log N)$ holds infinitely often with a suitable $c > 0$ (K. Kramer [3510]). It is expected (B.M.M. de Weger [1514]) that the inequality

[11] The name of the letter Ш is "sha".

$\#\text{Ш}(E) \gg N^c$ holds for every $c < 1/2$ for infinitely many curves. Examples of curves with large ratio $\log \#\text{Ш}(E)/\log N$ were given by A. Nitaj [4619] and H.E. Rose [5284]. It has been conjectured by L. Mai and M.R. Murty that the cardinality of $\text{Ш}(E)$ is $O(N^{1/2+\varepsilon})$ for every $\varepsilon > 0$. For certain families of curves this has been shown to be true by D. Goldfeld and D. Lieman [2261].

The last conjecture is equivalent to the *Szpiro conjecture* ([6030]; cf. P. Vojta [6451]), which concerns the relation between conductor $N(E)$ and the minimal discriminant $\Delta(E)$ of an elliptic curve E over the rationals:

$$|\Delta(E)| \ll N(E)^c \tag{5.22}$$

for every $c > 6$.

It was shown by S. Lang in 1990 [3700] that the Szpiro conjecture is a consequence of the ABC conjecture. Later D. Goldfeld [2259] proved that it is equivalent to a weak form of the ABC conjecture. D.W. Masser [4175] showed that the exponent 6 cannot be improved.

One knows also that for modular curves satisfying the Birch–Swinnerton-Dyer conjecture, then the Szpiro conjecture is equivalent to the conjecture of L. Mai and M.R. Murty, quoted above (D. Goldfeld, L. Szpiro [2264]). Cf. É. Fouvry, M. Nair, G. Tenenbaum [2063].

The cokernel of the map in (5.21) was determined by M.I. Bašmakov [346].

The theory of WC-groups for Abelian varieties has been further developed by S. Lang and J. Tate [3705]. For an analogous construction for varieties over local fields see J. Tate [6058] and J.S. Milne [4312].

3. The *index* $\text{ind}(E)$ of an elliptic curve E defined over a field k is defined as the smallest degree of an extension K of k in which E has a point. It was conjectured by S. Lang and J. Tate [3705] that every positive integer is an index for a genus one curve over the rationals. I.R. Šafarevič [5363, 5364] proved that there is no upper bound for $\text{ind}(E(\mathbf{Q}))$. It had already been shown in 1931 by F.K. Schmidt [5479, p. 27] that over a finite field one has $\text{ind}(E) = 1$, and a purely algebraic proof was provided by E. Witt [6702]. The *period* $\text{per}(E)$ of E is the order of the image of E in the WC-group. In the same way one defines the period and index of every element of the Weil–Châtelet group. S. Lang and J. Tate proved $\text{per}(E)|\text{ind}(E)$, and J.W.S. Cassels showed that the equality $\text{per}(A) = \text{ind}(A)$ holds for $A \in \text{Ш}(E/K)$ [937]. He found also an elliptic curve $E(\mathbf{Q})$ with $\text{per}(E) \neq \text{ind}(E)$ [938].

It was shown by I.R. Šafarevič [5365] and A.P. Ogg [4664] that in the local case one has $\text{per}(E) = \text{ind}(E)$, provided $\text{per}(E)$ is not divisible by the field characteristic, and this proviso was later removed by S. Lichtenbaum [3885]. It was shown by P.L. Clark [1120] in 2006 that for any algebraic number field K every integer occurs as index of a genus one curve over K. A partial result was obtained earlier by W.A. Stein [5919].

5.5 Probabilistic Number Theory

1. We have noted already that an arithmetical result by G.H. Hardy and S. Ramanujan has a probabilistic interpretation. Namely, they showed in [2540] that the function $\log\log n$ is the normal order for the number $\omega(n)$ of prime divisors of n.

The *normal order* of a function $f(n)$ is defined as a function $g(n)$ such that for every $\varepsilon > 0$ and almost all n one has

$$|f(n) - g(n)| < \varepsilon g(n).$$

Usually one is interested in non-decreasing functions g.

It was P. Turán [6212] who, in 1934, provided a simple proof of this result, resembling the proof of Čebyšev's inequality[12] in probability theory. The first truly probabilistic approach to the study of the function $\omega(n)$ appeared in a paper by P. Erdős and M. Kac [1835]. They considered, more generally, real-valued additive functions f (i.e., satisfying $f(ab) = f(a) + f(b)$ for co-prime a, b) satisfying $f(p^n) = f(p)$ for prime p and $n = 1, 2, \ldots$, as well as the condition $|f(p)| \le 1$ for all primes p and showed that if we put

$$A_x = \sum_{p \le x} \frac{f(p)}{p}, \qquad B_x = \sqrt{\sum_{p \le x} \frac{f^2(p)}{p}}, \tag{5.23}$$

and B_x tends to infinity, then for every real t one has

$$\lim_{x \to \infty} \frac{1}{x} \# \left\{ n \le x : \frac{f(n) - A_x}{B_x} \le t \right\} = G(t), \tag{5.24}$$

where

$$G(t) = \frac{1}{\sqrt{2\pi}} \int_{-\infty}^{t} e^{-u^2/2} du.$$

This relation shows that the distribution of values of an additive function satisfying the assumptions of this theorem obeys the Gaussian distribution law. In the case of the function $\omega(n)$, considered by G.H. Hardy and S. Ramanujan, one gets for

$$N(x, t) = \# \left\{ n \le x : \frac{\omega(n) - \log\log x}{\sqrt{\log\log x}} \le t \right\}$$

the relation

$$\lim_{x \to \infty} \frac{N(x, t)}{x} = G(t). \tag{5.25}$$

The first proof used Brun's sieve, but it was later shown by H. Halberstam [2449] that this can be avoided, and a further simplification in the case $f(n) = \omega(n)$ was provided by P. Billingsley[13] [508]. Later the assumption $|f(p)| \le 1$ was replaced by a certain weaker condition (see, e.g., H. Halberstam [2449], J. Kubilius [3544], H.N. Shapiro [5681]).

The first bound for the resulting error term $R(x, t) = N(x, t)/x - G(t)$ in the Erdős–Kac theorem was evaluated in the case $f = \omega$ by W.J. LeVeque [3847], who showed

$$R(x, t) = O\left(\frac{\log\log\log x}{(\log\log x)^{1/4}} \right),$$

and conjectured $R(x, t) = O((\log\log x)^{-1/2})$.

---

[12] Turán himself was at that time unaware that his proof can be interpreted in that way.

[13] Patrick Billingsley (1925–2011), professor at the University of Chicago.

In 1956 J. Kubilius [3546] established

$$R(x,t) = O\left(\frac{\log\log\log x}{\sqrt{\log\log x}}\right)$$

and two years later A. Rényi and P. Turán [5168] confirmed LeVeque's conjecture. It was pointed out by G. Halász [2445] that this follows also from an earlier result by A. Selberg [5617].

Later H. Delange [1430] used Selberg's [5617] formula[14]

$$\sum_{n\le x} z^{\omega(n)} = \frac{x}{\Gamma(z)}\prod_p\left(1-\frac{1}{p}\right)^z\left(1+\frac{z}{p-1}\right)\log^{z-1}x + O(\log^{z-2}x),$$

valid for all complex numbers inside the disc $|z| < 2$, to obtain a simpler proof of (5.24) for $f = \omega$, and proved also an asymptotic expansion of $R(x)$.

For arbitrary additive functions the error term in the Erdős–Kac theorem has been evaluated by M.B. Barban and A.I. Vinogradov [328] (see also B.V. Levin and A.S. Faĭnleĭb [3857] and Chap. 20 of the book [1735] by P.D.T.A. Elliott).

The history of the discovery of the Erdős–Kac theorem is presented in the second volume of P.D.T.A. Elliot's book ([1736]).

The question of existence of a non-decreasing additive arithmetic function which is not a constant multiple of the logarithm was settled negatively in 1946 by P. Erdős [1796], who showed also that an additive function f satisfying

$$\lim_{n\to\infty}(f(n+1) - f(n)) = 0$$

equals $c\log n$.

For other characterizations of the logarithm among additive functions see, e.g., P.D.T.A. Elliott [1737–1739, 1748], A. Hildebrand [2794], I. Kátai [3270, 3271], J.-L. Mauclaire [4204, 4205], J.-L. Mauclaire, L. Murata [4208, 4209], B.M. Phong [4839], I.Z. Ruzsa [5342], Y.S. Tang, P.C. Shao [6044], E. Wirsing [6698].

P. Erdős also established in [1796] that the only monotonic multiplicative functions are powers. A simpler proof was found by L. Moser and J. Lambek [4442].

The last result was generalized in 1967 by B.J. Birch [526], who showed that the only multiplicative functions with a monotonic normal order are powers.

Additive functions having a non-decreasing normal order were characterized by P.D.T.A. Elliott [1733].

2. A new method in the study of additive functions was introduced by J. Kubilius [3546, 3547] based on consideration of independent random variables in a sequence of finite probability spaces. This approach enabled characterization of additive functions f for which there exist a function $\Phi(t)$ such that

$$\lim_{x\to\infty}\frac{1}{x}\#\left\{n \le x : \frac{f(n) - A_x}{B_x} \le t\right\} = \Phi(t)$$

holds at every point of continuity of Φ, with A_x, B_x defined by (5.23).

---

[14] A generalization of this formula was obtained in 1971 by H. Delange [1432].

J. Kubilius also established the important inequality (*Turán–Kubilius inequality*) which for additive functions f satisfying $f(p^k) = f(p)$ for primes p and every $k \geq 1$ acquires the form

$$\sum_{n \leq x} |f(n) - A_x|^2 \leq C(x) x B_x^2,$$

where $C(x)$ does not depend on f, and $C = \limsup_{x \to \infty} C(x)$ is an absolute constant.

The first effective bound, $C \leq 55$, was given by P.D.T.A. Elliott [1730] in 1970, and in 1975 J. Kubilius [3548] proved the inequalities

$$1.47 \leq C \leq 2.08.$$

Later P.D.T.A. Elliott [1734] obtained $C \leq 2$, noting that J. Kubilius had a proof of $C \leq 1.764$. Finally J. Kubilius [3549] and A. Hildebrand [2795] established $C = 1.5$, and in 1985 J. Kubilius [3550] showed

$$C(x) = 1.5 + O\left(\frac{1}{\log x}\right).$$

A broad exposition of probabilistic methods in number theory is contained in the book [1735, 1736] by P.D.T.A. Elliott.

5.6 Geometry of Numbers, Transcendence and Diophantine Approximations

1. In 1942 an old problem of H. Minkowski's found its solution. Minkowski asked in [4324] whether the following assertion is true.

Let $L_1, \dots, L_n$ be a system of real linear forms in n variables of determinant ± 1, and assume that the only integral solution of

$$|L_i(x_1, \dots, x_n)| < 1 \quad (i = 1, 2, \dots, n)$$

is $x_1 = \cdots = x_n = 0$. Then at least one of the forms L_i has integral coefficients.

The truth of this assertion was established by G. Hajós [2442], after B. Levi[15] [3854] in 1911 proved[16] it for $n \leq 4$ and O. Perron [4793, 4794] for $n \leq 9$. Later Hajós's proof was simplified by L. Rédei [5139, 5140]. Note that both Perron and Hajós used a reformulation of Minkowski's question, which appeared first in a paper by O.-H. Keller [3295].

[15] Beppo Levi (1875–1961), professor in Piacenza, Cagliari, Parma, Rosario and Bologna. See [5424].

[16] His argument in the general case turned out to be insufficient. See [3295].

2. An important result which later gained several applications was obtained in 1946 by K. Mahler [4083]. He proved that if Λ_n is a sequence of lattices in $\mathbf{R}^N$ with bounded discriminants (defined as the determinant formed by coefficients of a minimal set of generators) and there is a neighborhood U of the origin with $\Lambda_n \cap U = \{0\}$, then a subsequence of this sequence converges to a lattice (*Mahler compactness theorem*). Mahler himself applied his result to study lattices having no non-zero points in a fixed star body.

Mahler's theorem has been put in a much more general context by C. Chabauty [976], and an extension to algebraic number fields has been made by K. Rogers and H.P.F. Swinnerton-Dyer [5258].

3. One of the famous problems in the geometry of numbers is the *kissing problem*, in which one asks for the maximal number τ_n of non-overlapping unit spheres touching a fixed unit sphere in n-dimensional space. Its origin goes back to Newton who conjectured $\tau_3 = 12$. He was right, and this was confirmed by K. Schütte and B.L. van der Waerden [5579] in 1953 (a special case had been treated already in 1874 by C. Bender [402]).

For a long time no progress was made for dimensions larger than 3 until in 1979 A.M. Odlyzko and N.J.A. Sloane [4656] determined $\tau_8 = 240$ and $\tau_{24} = 196\,560$. Their paper also gives a method of proving upper bounds for τ_n in the general case (earlier such bounds had been given by H.S.M. Coxeter [1262]). The strongest known asymptotic bounds are due to G.A. Kabatiansky and V.I. Levenshtein [3176], who obtained

$$\tau_n \le 2^{an+o(n)}$$

with $a = 0.401$. On the other hand one has

$$\tau_n \ge 2^{bn+o(n)}$$

with $b = 1 - \log 3/(2\log 2) = 0.207518\ldots$ (A.D. Wyner [6765]). The lower and upper bounds for $n \le 23$ known in 1979 can be found in [4656]. Later O.R. Musin [4497] gave a new proof of $\tau_3 = 12$. Earlier he announced the equality $\tau_4 = 24$ [4496], and the proof appeared in 2008 [4498]. A recent paper by C. Bachoc and F. Vallentin [204] improves the upper bounds in the case of small n. For $32 \le n \le 128$ upper bounds were given by Y. Edel, E.M. Rains, N.J.A. Sloane [1685]. Earlier upper bounds were obtained for $n = 44$ (G. Nebe [4549]), $n = 64$ (G. Nebe [4550], N.D. Elkies [1718]), $n = 80$ (C. Bachoc, G. Nebe [202]) and $n = 128$ (N.D. Elkies [1720]).

For surveys see G. Nebe [4551] and F. Pfender, G.M. Ziegler [4832]. See also the book by J.H. Conway and N.J.A. Sloane [1230].

4. We have already mentioned K.F. Roth's result [5311] showing that if α is an algebraic number, then for every $\varepsilon > 0$ the inequality

$$\left|\alpha - \frac{p}{q}\right| < \frac{1}{q^{2+\varepsilon}}$$

can have at most finitely many solutions in integers $q > 0, p$. This result is not effective, as it does not give a bound for the possible values of q; however, an upper bound for the number of solutions of this inequality was given by H. Davenport and K.F. Roth [1401] in 1955. Roth's theorem was generalized by D. Ridout [5213, 5214] to the p-adic case (see A.S. Fraenkel [2066] for an analogue of the result of [1401] in this case).

Other proofs of Roth's theorem, some of them encompassing Ridout's generalization, were given by H. Esnault, E. Vieweg [1876], P. Vojta [6455] and E. Bombieri, A.J. van der Poorten [624]. The last paper also contains an improvement of the bounds given in [1401].

An effective version of Roth's theorem was deduced from the ABC conjecture by E. Bombieri [606] in 1994 (cf. M. van Frankenhuysen [6333]).

A generalization of Roth's theorem to algebraic number fields was given by W.J. LeVeque in his book [3850], published in 1956. He showed that if $k \subset K$ are distinct algebraic number fields, then for $\alpha \in K$ and $\varepsilon > 0$ the inequality

$$|\alpha - \beta| > \frac{1}{H(\beta)^{2+\varepsilon}}$$

can have only finitely many solutions $\beta \in k$. The question of approximating algebraic numbers by algebraic numbers of smaller degree was also considered in 1958 by E. Bombieri [592], who improved upon previous results by A. Brauer [675]. Later this subject was pursued by M. Cugiani [1291], K. Mahler [4090] and R. Güting [2397], who gave effective lower bounds for the difference of algebraic numbers.

In 1961 E. Wirsing [6693] showed that if α is a real number which is either transcendental or algebraic of degree $\geq n+1$, then for every $\varepsilon > 0$ there are infinitely many algebraic numbers β of degree $\leq n$, satisfying

$$|\alpha - \beta| \leq \frac{1}{H(\beta)^{A_n - \varepsilon}}, \tag{5.26}$$

with $A_n = n/2 + 3 - \varepsilon$. For a far reaching generalization see P. Vojta [6454].

In 1970 the following generalization of Roth's theorem was proved by W.M. Schmidt [5498].

If $a_1, \dots, a_n$ are real algebraic numbers such that $1, a_1, \dots, a_n$ are linearly independent over the rationals, then for every $\varepsilon > 0$ the system

$$\left| a_j - \frac{p_j}{q} \right| < \frac{1}{q^{1+1/n+\varepsilon}} \tag{5.27}$$

can have only finitely many solutions in integers $p_1, \dots, p_n$ and $q > 0$.

A p-adic analogue of Roth's theorem was established in 1978 by J.F. Morrison [4435]. Later V.I. Bernik and K.I. Tiščenko [464] obtained (5.26) with $A_n = n/2 + 3 + o(1)$ and K.I. Tiščenko (under the name Tsishchanka) improved this in [6207] to $A_n = n/2 + 4 + o(1)$. These results are still far from $A_n = n+1$, conjectured by W.M. Schmidt in his book [5513] (E. Wirsing [6693] conjectured this with an extra ε), except in the case $n = 2$, in which the conjecture was established by H. Davenport and W.M. Schmidt [1403]. It follows from Mahler's conjecture, proved by V.G. Sprindžuk (see [5875]), that Schmidt's conjecture holds for almost all real α.

After ten years E. Wirsing returned to this subject in [6699] and proved another generalization of Roth's theorem to algebraic number fields, showing that if α is algebraic and $\varepsilon > 0$, then there are at most finitely many algebraic numbers β of a fixed degree t such that

$$|\alpha - \beta| < \frac{1}{H(\beta)^{2t+\varepsilon}}. \tag{5.28}$$

An upper bound for the number of solutions of (5.28) was given later by J.-H. Evertse [1927] and H. Locher [3980]. Wirsing's result was improved by W.M. Schmidt [5498], who replaced $2t$ in the exponent by $t+1$. This result is best possible. An extension to simultaneous approximations was proved in 1975 by W.M. Schmidt [5506], and this was extended to the $\mathfrak{p}$-adic case by H.P. Schlickewei [5461].

5. In 1955 J.W.S. Cassels and H.P.F. Swinnerton-Dyer [953] considered a special case of a conjecture stated around 1930 by J.E. Littlewood (see [3947]). Littlewood conjectured that for every pair α, β of reals one has

$$\inf_{q>0} q\|q\alpha\| \cdot \|q\beta\| = 0, \tag{5.29}$$

$\|a\|$ denoting the distance of a to the nearest integer. Since for almost all real α one has

$$\inf_{q>0} q\|q\alpha\| = 0, \tag{5.30}$$

Littlewood's conjecture holds for almost all pairs α, β. In the paper [953] it was shown that (5.29) holds if α, β lie in the same cubic field.

An important step forward was taken in 2000 when A.D. Pollington and S.L. Velani [4952] showed that if α does not satisfy (5.30) then there exist non-denumerably many β violating (5.30) such that (5.29) holds and the triple $1, \alpha, \beta$ is linearly independent over the rationals. Effective examples of this situation were given later by B. de Mathan [4187]. It was recently proved by M. Einsiedler, A. Katok and E. Lindenstrauss [1707] that the set of exceptions to (5.29) is of zero Hausdorff dimension. For an exposition see A. Venkatesh [6380].

An analogous conjecture for the field of power series over an infinite field does not hold, as shown in 1963 by Davenport and Lewis [1394] (see also J.V. Armitage [127], A. Baker [222], R.T. Bumby [840], T.W. Cusick [1304], T. Komatsu [3468]).

6. In 1958 P. Erdős, P. Szüsz and P. Turán [1862] posed the following conjecture.

If $S(N, A, c)$ denotes the set of numbers $\xi \in (0, 1)$ for which there exists an integer $b \in [N, cN]$ such that

$$\|b\xi\| < \frac{A}{b},$$

then the limit

$$\lim_{N\to\infty} \mu(S(N, A, c)),$$

μ being the Lebesgue measure, exists.

They showed that for $0 < A < c/(1 + c^2)$ this is correct and found an explicit value for the limit (cf. P. Erdős [1817]).

H. Kesten [3308] extended this to $A \le 1/c$ and later, with V.T. Sós [3311], proved the existence of the limit in all cases. An explicit value was provided by F.P. Boca [566] in 2008.

7. It was shown in 1947 by M. Hall, Jr. [2471] that every real number can be written as the sum of two numbers whose continued fractions have partial quotients bounded by 4.

Generalizations and similar results were provided by T.W. Cusick and R.A. Lee [1311], J.L. Hlavka [2831], T.W. Cusick [1305, 1306], B. Diviš[17] [1595] and S. Astels [155–158].

8. In 1949 A.O. Gelfond [2228] confirmed in the case $d = 3$ the following conjecture proposed earlier by him and T. Schneider.

If $\alpha \neq 0, 1$ *is algebraic and* β *is an algebraic number of degree* d*, then the numbers*

$$\alpha, \alpha^{\beta}, \alpha^{\beta^2}, \dots, \alpha^{\beta^{d-1}} \tag{5.31}$$

are algebraically independent.

This conjecture is a consequence of the conjecture of Schanuel.

A quantitative form of Gelfond's result was proved in 1950 by Gelfond and N.I. Feldman [2232]. For improvements see W.D. Brownawell [758] and G.V. Chudnovsky [1107]. A generalization in which α is replaced by a complex number having good approximations by algebraic numbers of bounded degrees was later given by W.D. Brownawell and M. Waldschmidt [760].

In [2229] A.O. Gelfond showed that if $d \geq 4$, then at least two numbers (5.31) are algebraically independent (for a further development of his method see W.D. Brownawell [757] and for a survey see M. Waldschmidt [6503]).

The next step was taken in 1972 by A.A. Šmelev [5828], who showed that for $d \geq 19$ at least three of the numbers (5.31) are algebraically independent. Two years later G.V. Chudnovsky [1290] showed that this holds already for $d \geq 7$, and announced that for $d \geq 2$ at least $[\log_2(1+d)]$ numbers (5.31) are algebraically independent (for an exposition of his method see M. Waldschmidt [6499]). A proof of G.V. Chudnovsky's assertion was supplied by Yu.V. Nesterenko [4563, 4564], who in 1985 showed [4565] the algebraic independence of $[d/2]$ numbers (5.31). Later G. Diaz [1531] and Yu.V. Nesterenko [4567] replaced $[d/2]$ by $[(d+1)/2]$. Similar methods have been used to evaluate the transcendence degree of other sets of values of the exponential function. See A.O. Gelfond [2229], G.V. Chudnovsky [1105, 1290], W.D. Brownawell [757], G. Diaz [1531].

5.7 Other Questions

1. It was proved by M. Ward [6562] that there are infinitely many prime divisors of elements of a second order recurrence

$$u_{n+1} = au_n + bu_{n-1}$$

[17] Bohuslav Diviš (1942–1976), professor at Ohio State University. See [5585].

with $ab \neq 0$, provided the ratio of the roots of the associated polynomial $X^2 - AX - B$ is not a root of unity. The next year he studied [6563] a particular class of recurrences, namely *Lehmer numbers* v_n, defined by D.H. Lehmer [3773] in 1930 in the following way:

$$v_n = \begin{cases} (\alpha^n - \beta^n)/(\alpha - \beta) & \text{if } n \text{ is odd,} \\ (\alpha^n - \beta^n)/(\alpha^2 - \beta^2) & \text{if } n \text{ is even,} \end{cases}$$

where α, β are algebraic numbers such that $(\alpha + \beta)^2$ and $\alpha\beta$ are co-prime non-zero rational integers, and the ratio α/β is not a root of unity.

M. Ward proved that if the numbers α^2, β^2 are both real, and $n \geq 18$, then v_n has a primitive prime divisor, defined as a prime p with $p | v_n$ and $p \nmid (\alpha^2 - \beta^2)v_1 \cdots v_{n-1}$. All such sequences having members without primitive prime divisors were determined by L.K. Durst [1657].

The first result of this type goes back to K. Zsigmondy[18] [6836] who showed in 1892 that if $a > b$ are relatively prime positive integers and $u_n = (a^n - b^n)/(a - b)$, then for every n there is a prime p dividing u_n, but not dividing $u_1 u_2 \cdots u_{n-1}$, with the exception of the cases $n = 1$, $a = b + 1$, $n = 2$, $a + b = 2^k$ and $n = 6$, $a = 2$, $b = 1$. The case $b = 1$ was treated earlier by A.S. Bang [314]. This theorem has been rediscovered several times (see, e.g., G.D. Birkhoff[19], H.S. Vandiver[20] [541], R.D. Carmichael [913]), and is called commonly the *Birkhoff–Vandiver* theorem.

This result was extended in 1913 by R.D. Carmichael [916]. He considered *Lucas sequences* u_n defined by

$$u_n = \frac{\alpha^n - \beta^n}{\alpha - \beta},$$

where α, β are algebraic numbers such that $\alpha + \beta$ and $\alpha\beta$ are non-zero co-prime rational integers and α/β is not a root of unity, and showed that if α and β are real, then for $n \geq 13$ the element u_n has a *primitive prime divisor*, i.e., a prime dividing u_n which does not divide $(\alpha - \beta)^2 u_1 u_2 \cdots u_{n-1}$. This is best possible, as F_{12}, the twelfth Fibonacci number (which is the Lucas sequence corresponding to $\alpha = (1 + \sqrt{5})/2$, $\beta = (1 - \sqrt{5})/2$), equals $144 = 2^4 3^2$; hence since $F_2 = 2$, $F_3 = 3$ it does not have a primitive prime divisor.

This topic was treated later by C.G. Lekkerkerker [3811, 3812], L.K. Durst [1658] and A. Rotkiewicz [5316].

If one lifts the reality assumption on α^2 and β^2 in Ward's result, then, as shown by A. Schinzel [5434] in 1962, a similar result (which applies also to Lucas sequences) holds for n exceeding a value n_0 depending on α and β. In certain cases it can be shown that v_n has more than one primitive prime divisor (A. Schinzel [5436–5438]). Twelve years later A. Schinzel [5443] proved that for n_0 one can take an effective constant, not depending on

[18] Karl Zsigmondy (1867–1925), professor at the Technische Hochschule in Vienna. See [5472].

[19] George David Birkhoff (1884–1944), professor at the University of Wisconsin, Princeton and Harvard. See [4438, 6655].

[20] Harry Schultz Vandiver (1882–1973), professor at the University of Texas. See [1251].

α or β, and an explicit value for that constant, $n_0 = e^{452}4^{67}$ was provided by C.L. Stewart [5946]. This was later reduced to $n_0 = 2 \cdot 10^{10}$ and $c_0 = 30\,030$ by P.M. Voutier [6478–6480]. The final step was taken in 2001 by Y.F. Bilu, G. Hanrot and P.M. Voutier [518], who showed that one can take $n_0 = 30$. This allowed M. Abouzaid [7] to establish a complete list of all the elements of Lucas and Lehmer sequences without primitive prime divisors.

H. Hasse [2606] proved that the density of prime divisors of the sequence $2^n + 1$ equals $17/24$. Densities of the sets of prime divisors of certain other binary recurrences were found by J.C. Lagarias [3604], P. Moree [4410] and P. Moree and P. Stevenhagen [4416].

For prime divisors of recurrent sequences see also C. Ballot [294], J.-P. Bézivin [497, 498], H. Roskam [5290], I.E. Šparlinskiĭ [5858].

Asymptotics for the number of divisors $\leq x$ for certain such sequences were found by P. Moree [4411, 4412].

2. The Dirichlet convolution of arithmetical functions f, g is defined by

$$(f \star g)(n) = \sum_{d|n} f(d)g(n/d),$$

and the set of all complex-valued arithmetic functions forms an integral domain Ω with the convolution as multiplication. This ring is isomorphic to the ring of formal power series in countably many variables over the complex field. It was shown in 1959 by E.D. Cashwell and C.J. Everett [925, 926] that Ω is a unique factorization domain, and L. Carlitz [901] studied algebraic independence of various arithmetical functions in Ω.

This subject was later pursued by J. Popken [4991, 4992], H.N. Shapiro and G.H. Sparer [5683] and V. Laohakosol [3716].

The same question for the ring of arithmetical functions with usual multiplication was treated by R. Bellman and H.N. Shapiro [401].

In 1975 S.M. Voronin [6468] showed that every finite set $\{L_1, \dots, L_k\}$ of Dirichlet L-functions corresponding to inequivalent characters is independent in a much stronger sense: if $F_1, \dots, F_m$ are continuous functions in k variables, not all equal to 0, then the sum

$$\sum_{j=1}^{m} s^j F_j(L_1(s), \dots, L_k(s))$$

does not vanish identically.

Later F. Nicolae [4592] showed that for every normal extension of the rationals the Artin L-functions corresponding to irreducible characters of the Galois group are algebraically independent over $\mathbf{C}$.

3. The search for *amicable pairs*, i.e., pairs m, n of integers with

$$\sigma(m) = \sigma(n) = m + n$$

started a long time ago. Already the ancient Greeks knew that the pair 220, 284 has this property, and it was shown by Thabit ben Korrah (see [1545, p. 39]) in the ninth century that if the numbers $p = 3 \cdot 2^n - 1$, $q = 3 \cdot 2^{n-1} - 1$ and $r = 9 \cdot 2^{2n-1} - 1$ are primes, then $2^n pq$ and $2^n r$ form an amicable pair. Later more than 10^7 such pairs were found, but we still do not have a proof that their number is infinite.

The first result dealing with the number $A(x)$ of amicable pairs (a,b) with $a \leq x$ was established in 1955 by P. Erdős [1813], who proved $A(x) = o(x)$.

This was improved twenty years later to $A(x) = O(x/\log\log\log x)$ by P. Erdős and G.J. Rieger [1851]. The best known bound for $A(x)$ is due to C. Pomerance [4973], who in 1981 obtained $A(x) = O(x\exp(-\log^{1/3} x))$, improving upon his earlier bounds [4972]. In 1974 W. Borho [644] showed that there are only finitely many amicable pairs m,n with mn having a bounded number of prime divisors, and a new method of constructing amicable pairs was given in 1986 by Borho and H. Hoffmann [645]. A list of all known amicable pairs is provided on J.M. Pedersen's web page: http://amicable.homepage.dk/knwnc2.htm.

For a survey see M. Garcia, J.M. Pedersen, H.J.J. te Riele [2193].

Chapter 6
The Last Period

6.1 Analytic Number Theory

6.1.1 Sieves

1. Linnik's idea of the large sieve method was revived in 1964, when K.F. Roth [5312] presented his version of it at the Rogers–Roth seminar at University College London. Roth's result

$$\sum_{p\le x} pD(p) \ll Zx^2 \log x, \tag{6.1}$$

holds for $x \le N^{1/2}/\log^{1/2} N$ and essentially improves Rényi's earlier bounds. The proof was markedly distinct from Linnik's and Rényi's approach, as it was based on an evaluation of the integral of the square of an absolutely convergent Fourier series of the form

$$a_0 + 2\sum_{n=1}^{\infty} a_n \cos(2\pi nt),$$

where the first N coefficients a_n are close to 1.

An important improvement was obtained by E. Bombieri [595] who proved a result, known now as the *large sieve theorem.*

The large sieve *Let M, N be integers, and put*

$$S(t) = \sum_{M<n\le M+N} a_n \exp(2\pi i nt),$$

with complex coefficients a_n. Then one has

$$\sum_{q\le x} \sum_{\substack{1\le a<q\\(a,q)=1}} |S(a/q)|^2 \le B(x,N) \sum_{M<n\le M+N} |a_n|^2, \tag{6.2}$$

with $B(x,N) = 7\max\{N, x^2\}$.

W. Narkiewicz, *Rational Number Theory in the 20th Century*,
Springer Monographs in Mathematics,
DOI 10.1007/978-0-85729-532-3_6, © Springer-Verlag London Limited 2012

Bombieri also remarked that one cannot replace the coefficient 7 here by a number smaller that $3/\pi^2$.

The bound in (6.2) was reduced consecutively by H. Davenport and H. Halberstam [1383], P.X. Gallagher [2179], M.C. Liu [3966], E. Bombieri and H. Davenport [609, 610], E. Bombieri [599] and H.L. Montgomery and R.C. Vaughan [4364]. In the last paper inequality (6.2) was established with $B(x, N) = N + x^2$.

Gallagher's approach to the large sieve [2179] produced

$$\sum_{p\le x} pD(p) \le Z\left(\pi N + x^2\right)$$

for $x \ll \sqrt{N}$ and it was proved by P. Erdős and A. Rényi [1850] that the last inequality may fail if x is of larger order than $\sqrt{N \log N}$.

The paper [595] also contains an analogous result for character sums, which was later strengthened by P.X. Gallagher [2179] to the following form:

$$\sum_{q\le Q}\sum_{\chi \bmod q}\left|\sum_{n=M+1}^{M+N} a_n\chi(n)\right|^2 \le \left(N + Q^2\right)\sum_{n=M+1}^{M+N}|a_n|^2, \tag{6.3}$$

the second sum being taken over primitive characters.

2. It was shown by H.L. Montgomery [4353] that inequality (6.2) can be used to obtain upper bounds for the number of integers in an interval avoiding certain residue classes modulo primes.

Let A be a finite set of integers contained in an interval of length N and assume that for every prime $p \le Q$ there are $0 \le f(p) < p$ residue classes mod p *not containing any element of A. Then*

$$\#A \le \frac{Q^2 + \pi N}{L} \tag{6.4}$$

where

$$L = \sum_{q\le Q}\mu^2(q)\prod_{p|q}\frac{f(p)}{p - f(p)}.$$

Later H.L. Montgomery and R.C. Vaughan [4364] showed that the constant π in (6.4) can be replaced by 1.

This result was later extended by J. Johnsen [3146] and P.X. Gallagher [2183] to the case of residue classes modulo prime powers (cf. also [2182] for a variant which gives stronger results if $p - f(p)$ is small). See A. Selberg [5621] for an improvement. The case when the primes are replaced by numbers with a bounded number of prime factors has been dealt with by S. Salerno and C. Viola [5377]. I. Kobayashi [3421] and Y. Motohashi [4455] showed that inequality (6.4) can also be obtained with the use of Selberg's sieve.

A survey of the large sieve method was given in 1978 by H.L. Montgomery [4360].

3. Bombieri's paper [595] also brought important applications of the large sieve (6.2) to the theory of primes. The main application dealt with the mean value of the number $N(\alpha, T, \chi)$ of zeros $\sigma + it$ of $L(s, \chi)$ in the rectangle $\alpha \le \sigma \le 1$, $|t| \le T$ for $1/2 \le \alpha \le 1$.

Bombieri's density theorem *For every $\varepsilon > 0$ and $2 \le T \le \sqrt{Q}$ one has uniformly in α, T*

$$S(\alpha, Q, T) := \sum_{q \le Q} \sum_{\chi}{}^{*} N(\alpha, T, \chi) \ll Q^{1+2(1-\alpha)+\varepsilon} T^{1+\varepsilon},$$

the asterisk denoting summation over primitive characters. For the interval $2 \le T \le Q^2$ and $\alpha \ge 5/6$ one has

$$S(\alpha, Q, T) \ll Q^{1+\varepsilon} T^{2+\varepsilon}.$$

He also conjectured the bound

$$S(\alpha, Q, T) \ll Q^{4(1-\alpha)+\varepsilon} T^{1+\varepsilon}, \tag{6.5}$$

valid for every $\varepsilon > 0$, $1/2 \le \alpha \le 1$ and $T \ge 2$, uniformly in α. Later M.B. Barban [325] proposed a modification in the form

$$S(\alpha, Q, T) \ll Q^{2(1-\alpha)} T^{c},$$

with unspecified c, and now the following form of the conjecture is commonly accepted:

$$S(\alpha, Q, T) \ll (Q^2 T)^{2(1-\alpha)+\varepsilon}. \tag{6.6}$$

The inequalities (5.5), which we met in the preceding chapter, and (6.6) form the *density conjecture for L-functions.*

Bombieri's bound was soon improved by H.L. Montgomery [4355], who used his bounds for Dirichlet's polynomials [4354] to deduce

$$S(\alpha, Q, T) \ll \left(Q^2 T\right)^{c(\alpha)} \log^{13}(QT),$$

where

$$c(\alpha) = \begin{cases} 3(1-\alpha)/(2-\alpha) & \text{if } 1/2 \le \alpha \le 4/5, \\ 2(1-\alpha)/\alpha & \text{if } 4/5 \le \alpha \le 1. \end{cases} \tag{6.7}$$

This gives, for all $\sigma \in [1/2, 1]$, the bound (5.5), (6.6) with the coefficient 2.5 in place of 2 (cf. P.X. Gallagher [2181]). This coefficient has been reduced to 2.4 by M.N. Huxley [2978].

In 1972 M. Jutila [3168] and M.N. Huxley [2976] proved (6.6) for $\alpha > 5/6$ (cf. R. Balasubramanian, K. Ramachandra [281]), and this was improved to $\alpha \ge 21/26 = 0.8076\ldots$ by M. Jutila [3171], $\alpha > 0.8$ by M.N. Huxley and M. Jutila [2990], and to $\sigma > 11/14 = 0.7857\ldots$ by D.R. Heath-Brown [2629] (cf. J. Bourgain [662]).

The values of the exponent $c(\alpha)$ in (6.7) have subsequently been reduced by M. Jutila [3168, 3171] and Y. Motohashi [4451].

4. As a consequence of his density theorem E. Bombieri deduced a very strong bound for the error term in the Prime Number Theorem for progressions. This result was also obtained independently[1] by A.I. Vinogradov [6398] and is usually called the *Bombieri–Vinogradov theorem.* To state it recall that for co-prime integers k, l the function $\psi(x; k, l)$ is defined by

$$\psi(x; k, l) = \sum_{\substack{n \le x \\ n \equiv l \pmod k}} \Lambda(n),$$

where $\Lambda(n)$ is the von Mangoldt's function, defined by (2.9).

For any $A > 0$ *there exists* $B = B(A)$ *such that one has*

$$\sum_{k \le x^{1/2} \log^{-B} x} \max_{y \le x} \max_{(k,l)=1} \left| \psi(y; k, l) - \frac{y}{\varphi(k)} \right| \ll \frac{x}{\log^A x}. \tag{6.8}$$

Bombieri's proof gave $B(A) = 3A + 23$. Proofs of the Bombieri–Vinogradov theorem can be found in the books by H. Davenport [1377], M.N. Huxley (with $B = A + 10$, [2986, Chap. 24]) and H.L. Montgomery [4357, Chap. 15]. See also P.X. Gallagher [2180].

A modern proof with $B(A) = 2A + 6$ can be found in the book [3064] by H. Iwaniec and E. Kowalski. In 1975 R.C. Vaughan [6346] proved (6.8) with $B = A + 3.5$ (cf. B.V. Levin, N.M. Timofeev [3860]). Various extensions of the Bombieri–Vinogradov theorem were later obtained by É. Fouvry [2045, 2047, 2048], É. Fouvry and H. Iwaniec [2055, 2056] and E. Bombieri, J.B. Friedlander and H. Iwaniec [611–613].

The bound (6.8) implies a corresponding result for $\pi(x; k, l)$:

$$\sum_{k \le x^{1/2} \log^{-B} x} \max_{y \le x} \max_{(k,l)=1} \left| \pi(y; k, l) - \frac{\pi(y)}{\varphi(k)} \right| \ll \frac{x}{\log^A x}. \tag{6.9}$$

A weaker form of this assertion, with the summation extended over square-free $k \le x^{\delta}$ with some positive δ, had earlier been established by M.B. Barban [320]. As pointed out by P.D.T.A. Elliott [1735, p. 92] the presented proof works for $\delta < 3/23$. Later M.B. Barban [324] showed that one can take any $\delta < 3/8$ here, and k does not have to be square-free.

P.D.T.A. Elliott and H. Halberstam [1749] conjectured that the summation over k in (6.8) can be carried out up to $x^{1-\varepsilon}$ for every fixed $\varepsilon > 0$. The truth of his conjecture would have several deep consequences, e.g., for every positive δ almost every interval $(x, x + x^{\delta})$ would contain a prime (D.R. Heath-Brown [2633]).

The stronger form of this conjecture, in which the summation bound $x^{1-\varepsilon}$ is replaced by $x \log^{-c} x$ for sufficiently large c was refuted by J.B. Friedlander and A. Granville [2101–2103] in 1989. Later these authors in a joint paper with A. Hildebrand and H. Maier [2104] showed that even in the range $k \le x \exp(-\log^c x)$ with $c < 1/2$ the inequality (6.8) may fail (cf. T.Z. Xuan [6770]).

[1] Note however that in Vinogradov's result the summation covered the range $k \le x^{1/2-\varepsilon}$ with $\varepsilon > 0$.

An analogue of the Bombieri–Vinogradov theorem for a class of multiplicative functions f was proved in 1973 by D. Wolke [6713, 6714] (cf. Y. Motohashi [4452], D. Wolke [6716]), and this led to evaluations of the difference

$$\Delta_f(x;q,a) = \sum_{\substack{n\le x \\ n\equiv a \pmod q}} f(n) - \frac{1}{q}\sum_{\substack{n\le x \\ (n,q)=1}} f(n)$$

for $(a,q)=1$.

Important results on $\Delta_f(x;q,a)$ for a large class of functions were obtained in 1987 by P.D.T.A. Elliott [1740–1746]. He showed that if f is a complex-valued multiplicative function with $f(n)\le 1$, then one has uniformly for $x\le T$

$$\Delta_f(x;q,a) \ll x\left(\frac{\log\log T}{\log T}\right)^{1/8}\frac{\log T}{\log x},$$

except for q's which are multiples of a certain q_0, and pursued this topic in a series of papers [1741–1746]. See also G. Bachman [195, 196], A. Hildebrand [2806], P. Shiu [5713].

5. It was shown in 1966 by H. Davenport and H. Halberstam [1384] with the use of the large sieve that for $T\le x\log^{-A}x$ one has

$$\sum_{q\le T}\sum_{\substack{l \\ (k,l)=1}}\left|\psi(x;k,l)-\frac{x}{\varphi(k)}\right|^2 \ll \frac{x^2}{\log^{A+5}x}. \tag{6.10}$$

A similar result holds also with $\psi(\cdot)$ replaced by $\pi(\cdot)$ and $x/\varphi(k)$ by $\mathrm{li}(x)/\varphi(k)$. A slightly less precise result, with $A+5$ replaced by an unspecified $B(A)$ was obtained earlier by M.B. Barban [323, 324]. A few years later H.L. Montgomery [4356] replaced the inequality (6.10) by the equality

$$\sum_{q\le T}\sum_{\substack{l \\ (k,l)=1}}\left|\psi(x;k,l)-\frac{x}{\varphi(k)}\right|^2 = Tx\log T + O(Tx) + O\left(\frac{x^2}{\log^M x}\right),$$

valid for every $M>0$. He did not use the large sieve. In his paper he conjectured the bound

$$\sum_{k\le\sqrt{x}}\varphi(k)\sum_{\substack{l \\ (k,l)=1}}\left|\psi(x;k,l)-\frac{x}{\varphi(k)}\right|^2 \ll x^2\log^B x \tag{6.11}$$

for some $B>0$, quoting an observation by H. Halberstam that this bound follows from the large sieve in the case when k runs only over prime numbers. The truth of (6.11) would imply the non-vanishing of $\zeta(s)$ and L-functions in the half-plane $\Re s > 3/4$.

Further work on the Barban–Davenport–Halberstam theorem was done in a series of papers by C. Hooley [2869–2887] in which he also obtained results similar to (6.10) for other sequences. The seventeenth paper of that series [2885] contains a rather simple proof of (6.10).

For later research see E. Bombieri, J.B. Friedlander, H. Iwaniec [611–613], J.B. Friedlander, D.A. Goldston [2099, 2100], H.Q. Liu [3953].

See also R.C. Vaughan [6367–6369].

6. Another application of the large sieve was presented by R.C. Vaughan [6341] in 1970. P. Erdős and E.G. Straus conjectured in 1948 (see [1826]) that for every $n \geq 2$ the number $4/n$ can be written as the sum of three unit fractions, i.e., the equation

$$\frac{4}{n} = \frac{1}{x} + \frac{1}{y} + \frac{1}{z}$$

has positive integral solutions, and A. Schinzel [5450] extended this conjecture to fractions with arbitrary numerator $a > 0$ in place of 4, assuming that $n > n_0(a)$. A connection of the Erdős–Straus conjecture with covering congruences was pointed out by L. Bernstein [469], who also checked the Erdős–Straus conjecture for $n < 8000$. This was extended up to $n = 10^7$ by K. Yamamoto [6772], to 10^{10} by I. Kotsireas [3500], and to 10^{14} by A. Swett [5998].

R.C. Vaughan [6341] proved that the Erdős–Straus conjecture holds for almost all denominators n, more precisely, the number of integers $n \leq x$ for which it fails is $O(x \exp(-c(a) \log^{2/3} x))$, with certain $c(a) > 0$. An analogous result on the solvability of the equation

$$\frac{a}{N} = \sum_{j=1}^{k} \frac{1}{x_j}$$

for fixed a, k was established by C. Viola [6449]. For an improvement see Z. Shen [5686].

A further improvement was obtained by C. Elsholtz [1759] in 2001.

The possibility of representing every number $3/n$, with odd $n > 3$ as the sum of three unit fractions having odd denominators was established in 1956 by A. Schinzel [5431]. It was shown by G. Hofmeister and P. Stoll [2842] that for every fixed m and almost all n the number m/n can be written as the sum of two unit fractions.

7. In 1973 R.C. Vaughan [6344] used the large sieve to show that the number of integers $n \leq x$ such that all positive values of $n - 2^k \leq n - 2$ are primes is

$$O(x \exp(-c \log x \log\log\log x / \log\log x))$$

with a suitable $c > 0$. It was conjectured by Erdős [1801] in 1950 that the maximal n with this property is 105, and no larger integer has been found below 2^{44} (W.E. Mientka, R.C. Weitzenkamp [4289]). If the General Riemann Hypothesis is true, then Vaughan's bound can be replaced by $O(x^c)$ with some $c < 1/2$ (see [2868, Chap. 7] and [4536]).

It was shown by P.X. Gallagher [2184] that Linnik's theorem [3917] about representations of integers as the sum of two primes and powers of 2 can be proved with the use of the large sieve.

8. The large sieve method has been generalized to algebraic number fields by M.N. Huxley [2969–2971], R.J. Wilson [6679], W. Schaal [5419] (who used it to extend the Brun–Titchmarsh theorem to number fields), and P.D. Schumer [5572]. For applications see J. Hinz [2820]. Linnik's version of that sieve had been extended to the algebraic case earlier by G.J. Rieger [5220, 5222].

For analogues of the Bombieri–Vinogradov theorem for algebraic number fields see R.J. Wilson [6679], M.N. Huxley [2971], J. Hinz [2821] and M.D. Coleman and A. Swallow [1179].

9. A. Selberg published three fundamental papers on sieves [5619, 5620, 5622]. In the first the main ideas of his sieve method are explained, and in its last section and in the two next papers he presented the main ideas of the sieve invented by J.B. Rosser (not published), who essentially extended the method invented by A.A. Buhštab [831, 832] in the thirties (Selberg called this sieve the *Buchstab–Rosser sieve* (cf. also W.B. Jurkat, H.-E. Richert [3166]). This sieve was independently explored by H. Iwaniec [3057, 3058] and is called now the *Rosser–Iwaniec sieve*. For an exposition of the theory of this sieve see Selberg's lectures [5624]. The Rosser–Iwaniec sieve was generalized to algebraic number fields by M.D. Coleman [1178].

For expositions of the large sieve method and its applications see H.L. Montgomery [4357] and E. Bombieri [601] (see also the book by Y. Motohashi [4459]). A survey of the development of various sieves was presented by É. Fouvry [2052] in 2000.

6.1.2 Zeta-Functions and L-Functions

1. In 1973 H.L. Montgomery [4358] studied the difference between the imaginary parts $0 < \gamma_1 < \gamma_2 < \cdots$ of the consecutive zeros of $\zeta(s)$ on the critical line, and formulated the following conjecture.

Pair Correlation Conjecture (PCC) *If for $x, T \geq 1$ we put*

$$F(x,T) = 4 \sum_{0<\gamma,\gamma'\leq T} \frac{x^{i(\gamma-\gamma')}}{4+(\gamma-\gamma')^2},$$

γ, γ' running over roots of $\zeta(s)$ lying on the line $\Re s = 1/2$, then for every M and $T \leq x \leq T^M$ one has

$$F(x,T) = \left(\frac{1}{2\pi} + o(1)\right) T \log T.$$

He showed that the Riemann Hypothesis implies that for $1 \leq x = o(T)$ the function $F(x, T)$ is asymptotic to $(T \log x)/2\pi + (T \log^2 T / 2\pi x^2)$. He proved also under the same assumption that if we put

$$\mu = \liminf_{n\to\infty} \frac{(\gamma_{n+1} - \gamma_n)\log \gamma_n}{2\pi},$$

then $\mu \leq 0.68$, and the inequality $\mu < 1/4$ would imply an effective lower bound for the class-numbers of imaginary quadratic fields. One conjectures $\mu = 0$ and

$$\lambda = \limsup_{n\to\infty} \frac{(\gamma_{n+1} - \gamma_n)\log \gamma_n}{2\pi} = \infty.$$

Later H.L. Montgomery and A.M. Odlyzko [4363] deduced $\mu \leq 0.5179$ from the Riemann Hypothesis, and J.B. Conrey, A. Ghosh and S.M. Gonek [1213] proved $\mu \leq 0.5172$. The inequalities $\lambda \geq 1.9979$ and $\lambda \geq 2.337$ were deduced from the Riemann Hypothesis in [4363] and [1213], respectively. Stronger results can be obtained using the General Riemann Hypothesis: $\mu < 0.5172$, $\lambda > 2.68$ (J.B. Conrey, A. Ghosh, S.M. Gonek [1214]).

In 1999 R.R. Hall [2476] obtained $\lambda \geq 2.2635$ unconditionally, and later improved this to $\lambda \geq 2.3452$ [2477] and $\lambda \geq 2.6306$ [2478].

It was shown by J.B. Conrey and N.C. Snaith [1219] that the pair conjecture is a consequence of a conjecture by D.W. Farmer [1959], stating that for complex $\alpha, \beta, \gamma, \delta$ satisfying $\Re\gamma > 0$ and $\Re\delta > 0$ the integral

$$\int_0^T \frac{\zeta(s+\alpha)\zeta(s+\beta)}{\zeta(s+\gamma)\zeta(s+\delta)} dt$$

is for $\Re s = 1/2$ and $T \to \infty$ asymptotic to

$$\frac{T}{(\alpha+\beta)(\gamma+\delta)} \left((\alpha+\delta)(\beta+\gamma) - \frac{(\delta-\beta)(\gamma-\alpha)}{T^{\alpha+\beta}} \right).$$

H.L. Montgomery proved also, still assuming the Riemann Hypothesis, that at least $2/3$ of zeros of the zeta-function are simple, and it was shown later by P.X. Gallagher and J. Mueller [2185] that the Riemann Hypothesis and the Pair Correlation Conjecture imply that almost all zeros are simple. In 1998 J.B. Conrey, A. Ghosh and S.M. Gonek [1215] showed that one can replace the exponent $2/3$ by 0.6727, still under the Riemann Hypothesis.

The Pair Correlation Conjecture has many important consequences. P.X. Gallagher and J. Mueller [2185] showed that jointly with the Riemann Hypothesis it implies the equality

$$\psi(x) = x + o\left(\sqrt{x}\log^2 x\right),$$

and it was shown by J. Mueller [4466] that the same assumptions lead to

$$p_{n+1} - p_n \ll \sqrt{p_n}\log^{3/4} p_n.$$

The analogue of the conjecture for 3-tuples of zeta zeros was considered by D.A. Hejhal [2720], and for n-tuples by Z. Rudnick and P. Sarnak [5337, 5338], who also considered zeros of automorphic L-functions.

For a survey see P. Michel [4283].

Analogues of the Pair Correlation Conjecture for Dirichlet L-functions have been formulated by C.Y. Yildirim [6777], A.E. Özlük [4712] and A. Languasco and A. Perelli [3715], and, more generally, for functions of the Selberg class (see Sect. 6.1.4) by M.R. Murty and A. Perelli [4492].

H.L. Montgomery conjectured also that the zeros of $\zeta(s)$ behave asymptotically like eigenvalues of certain random matrices. This conjecture got numerical support by A.M. Odlyzko [4654, 4655]. For a heuristic approach see E.B. Bogomolny, J.P. Keating [572, 573].

In 1999 N.M. Katz and P. Sarnak [3281, 3282] extended Montgomery's conjecture to a broad class of L-functions. For further developments see H. Iwaniec, W. Luo and P. Sarnak [3066].

2. A rather striking property of the Riemann zeta-function, implying that it contains information about every decent function, was discovered in 1975 by S.M. Voronin [6467], who showed that if $f(s)$ is a function continuous and non-vanishing in a closed disc D of radius $r < 1/4$, centered at the origin and regular in the interior of D, then for every positive ε there exists $\tau > 0$ such that for $|s| \le r$ one has

$$\left| f(s) - \zeta\left(s + \frac{3}{4} + i\tau\right)\right| < \varepsilon.$$

The forerunner of this result is a theorem of G.D. Birkhoff [540] who in 1929 showed the existence of an entire function $F(z)$ having the property that for every entire function $f(z)$ there exists a sequence $z_1, z_2, \dots$ with the property that

$$\lim_{n\to\infty} F(z + z_n) = f(z),$$

the limit being uniform on every compact subset of the plane.

Soon it became clear that the disc in Voronin's theorem can be replaced by certain other compact sets and the zeta-function by several other functions having an Euler product (R. Bagchi [214, 215], J. Kaczorowski, M. Kulas [3195], A. Laurinčikas, K. Matsumoto, J. Steuding [3724–3726, 3728, 3730–3732] and A. Reich [5146]). A monograph on this subject was written by A. Laurinčikas [3727].

3. Explicit zero-free regions for L-functions were found by R.J. Miech [4287], T. Metsänkylä [4267] and T. Lepistö [3830]. A numerical search for small complex zeros was made first by D. Davies and C.B. Haselgrove [1410] in 1961 and then by R. Spira [5866]. Subsequent enlargements of the zero-free regions were given by K.S. McCurley [4224] and H. Li [3878]. For numerical results see R.S. Rumely [5340].

In a series of eight papers published in 1976–1977 J. Pintz [4883–4891] presented simple proofs of several important results of the theory of L-functions.

In 1981 M. Jutila [3173] proved that for infinitely many prime discriminants d one has $L_d(1/2) \neq 0$ and E. Stankus [5885] showed that this happens for infinitely many integers d from an arithmetic progression with odd difference. In his book [1095] S. Chowla conjectured that $L_d(1/2)$ never vanishes. Some numerical evidence supporting this conjecture was provided by R.S. Rumely [5340]. In 1999 A.E. Özlük and C. Snyder [4713] deduced from the General Riemann Hypothesis that for a positive proportion of d's this is true, and one year later K. Soundararajan [5851] showed unconditionally that for at least 87.5% of odd square-free positive integers k the L-function corresponding to the Kronecker character mod $8k$

does not vanish at $s = 1/2$. Similar results hold also for other characters. In 1992 R. Balasubramanian and V.K. Murty [279] showed that for large primes p, a positive proportion of characters χ mod p one has $L(1/2, \chi) \neq 0$. Later H. Iwaniec and P. Sarnak [3069] showed that this happens for almost 1/3 of primitive even characters χ mod k for large k.

4. In 1968 T. Mitsui [4344] and A.V. Sokolovskiĭ [5844] obtained the same result for the zero-free region of Dedekind's zeta-function as for Riemann's zeta. Namely, they proved that for large $|t|$ the function $\zeta_K(\sigma + it)$ does not vanish in the region

$$\sigma \geq 1 - \frac{A}{\log^{2/3}|t|(\log\log|t|)^{1/3}},$$

A being a positive constant depending on the field K. This result implies that the error term in the Prime Ideal Theorem has a bound of the same size as that obtained by I.M. Vinogradov in the case of the Prime Number Theorem.

The analogues of this result for Hecke zeta-functions and L-functions were obtained by J. Hinz [2816] and M.D. Coleman [1177]. The constants in Hinz's result were evaluated by K. Bartz [341].

A.V. Sokolovskiĭ [5844] sharpened also the upper bounds for ζ_K at the lines $\Re s = 1/2$ and $\Re s = 1$, by proving

$$\zeta_K\left(\frac{1}{2} + it\right) \ll |t|^a,$$

where

$$a = \frac{n}{4} - \frac{c}{n^2\log(n+2)},$$

with an absolute constant c, and n being the degree of K, and

$$\zeta_K(1 + it) \ll \log^{2/3}(|t| + 3).$$

He pointed also out that if K is totally real, then a stronger bound

$$\zeta_K(1/2 + it) \ll |t|^{(3n-1)/12}$$

can be obtained using the methods of H. Weyl and J.G. van der Corput.

5. In 1964 T. Kubota and H.-W. Leopoldt [3557] defined the analogues $L_p(s, \chi)$ of Dirichlet L-functions in the p-adic case, based on an earlier idea of Leopoldt [3827]. Another approach was presented by K. Iwasawa [3070, 3071], who also established relations between these functions and the class-number of $\mathbf{Z}_p$-extensions (cf. Y. Amice, J. Fresnel [64], J. Fresnel [2086], K. Shiratani [5711]). See also J. Coates, W. Sinnott [1127]. A precise conjecture on these relations proposed by K. Iwasawa (the *Iwasawa main conjecture*) was established in 1984 by B. Mazur and A. Wiles [4220]. For generalizations see A. Wiles [6667, 6669], K. Rubin [5330] (for an exposition see B. Perrin-Riou [4782]).

A survey of further generalizations of p-adic L-functions was given in 2000 by P. Colmez [1192].

The values of p-adic L-functions and/or their derivatives at integers have been studied by J. Diamond [1529], N. Koblitz [3425], B. Ferrero and R. Greenberg [1995], B.H. Gross [2354], K. Hatada [2612], H. Imai [3011], K. Shiratani [5712] and L.C. Washington [6570].

For an exposition of the theory of p-adic L-functions see the books of N. Koblitz [3423, 3426].

Later several p-adic analogues of classical functions were constructed. J.-P. Serre [5645] defined p-adic analogues of the Dedekind zeta-functions, and p-adic L-functions over totally real fields were defined by D. Barsky [339], P. Cassou-Noguès [956], P. Deligne, K. Ribet [1451], N.M. Katz [3280] and C. Queen [5021]. Y. Morita [4432] and J. Diamond [1528] defined the p-adic Γ-function. B.H. Gross and N. Koblitz [2355] found a formula relating this function to Gauss sums. Cf. N. Koblitz [3424].

An elementary proof of the Gross–Koblitz formula was found by A.M. Robert [5235].

The p-adic Γ-function was studied by D. Barsky [340], M. Boyarsky [669] and B. Dwork [1672]. For related functions see Y. Amice [63], F. Baldassari [290], P. Cassou-Noguès [955], B. Dwork [1673], M. Endo [1764] and H. Imai [3011].

Partial p-adic zeta-functions were studied by J. Coates [1126] and J. Coates and W. Sinnott [1128].

A formula for the residue of the p-adic analogues of Dedekind's zeta-functions at $s = 1$ was given by P. Colmez [1190], who later [1191] studied their behavior at $s = 0$.

p-adic L-functions associated with elliptic curves were defined by Yu.I. Manin [4137, 4138] and S. Lichtenbaum [3886]. They were studied by Y. Amice and J. Vélu [65], P. Cassou-Noguès [957], C. Goldstein [2265] and C. Goldstein and N. Schappacher [2266].

An analogue of the conjecture of Birch and Swinnerton-Dyer was formulated by B. Mazur, J. Tate and J. Teitelbaum [4219] who also gave numerical support for it (see also J. Teitelbaum [6100]). This conjecture was later proved by R. Greenberg and G. Stevens [2340].

See also P. Colmez [1193], M. Emerton [1762], L. Orton [4701], K. Kato [3275], S. Kobayashi [3422].

A similar conjecture was proposed by M. Bertolini and H. Darmon [471] for another kind of L-function associated with modular forms (see [472, 473]).

6.1.3 Prime Number Distribution

1. The Brun–Titchmarsh theorem was strengthened by N.I. Klimov [3361], who in 1961 established

$$\pi(x,k,l) < \frac{2x}{\varphi(k)\log(\sqrt{x/k})}\left(1 + O\left(\frac{\log\log(x/k)}{\log(x/k)}\right)\right)$$

in the case $x > 3k$ (see also [609]). Later J.H. van Lint and H.-E. Richert [6334] eliminated the iterated logarithm, and showed

$$\pi(x,k,l) < \frac{2x}{\varphi(k)\log(x/k)}\left(1+\frac{4}{\log(x/k)}\right)$$

for $k \leq x$ (cf. E. Bombieri and H. Davenport [599, 609]).

It was shown later by H.L. Montgomery and R.C. Vaughan [4364] that using the large sieve one can remove the bracketed factor in the last inequality. In 1982 H. Iwaniec [3059] proved that for $k \leq x^c$ $(c = 9/20 - \varepsilon)$ one has

$$\pi(x,k,l) \leq (2+\varepsilon)\frac{x}{\varphi(k)\log(xk^{-3/8})},$$

and showed that for cube-free k one can replace $3/8$ by $1/4$. For further improvements see R.C. Baker [256] (for $x^{3/7} \leq k \leq x^{9/20}$), and J.B. Friedlander and H. Iwaniec [2107] (for $x^c < k < x^d$ with $c > 6/11$, $d < 1$).

Bounds valid for fixed l and almost all k were obtained by C. Hooley [2865]. In 1974 Y. Motohashi [4450] showed that for $k \leq x^{1-\delta}$ (with $\delta > 0$) and almost all l mod k one has

$$\pi(x,k,l) \leq (2+\varepsilon)\frac{x}{\varphi(k)\log(xk^{-1/2})},$$

and later proved ([4456]) that if for large k one has

$$\pi(x,k,l) \leq \frac{ax}{\varphi(k)\log(x/k)}$$

with some fixed $a < 2$, then the L-functions mod$k$ do not have Siegel zeros (cf. also H. Siebert [5737]). This confirmed a conjecture posed by K.A. Rodosskiĭ (cf. N.I. Klimov [3361]).

For improvements see J.-M. Deshouillers, H. Iwaniec [1492], É. Fouvry [2050], H.H. Mikawa [4303], R.C. Baker, G. Harman [264].

2. The Bombieri–Vinogradov theorem was used in 1966 by E. Bombieri and H. Davenport [608] to obtain essential progress in the evaluation of

$$E = \liminf_{n\to\infty} \frac{p_{n+1}-p_n}{\log p_n}, \tag{6.12}$$

with p_n denoting the nth consecutive prime.

The inequality $E \leq 1$ is a simple consequence of the Prime Number Theorem, and G.H. Hardy and J.E. Littlewood deduced from the Riemann Hypothesis the inequality $E \leq 2/3$, but did not publish[2] it, and in 1940 R.A. Rankin [5106] deduced $E \leq 3/5$ from the General Riemann Hypothesis. The first unconditional result on E is due to P. Erdős who in 1940 obtained $E < 1$ [1789]. In 1947 R.A. Rankin [5107] improved Erdős's result establishing $E \leq 57/59$ with a method of A.A. Buhštab's. Three years later he improved this to $E < 42/43$ [5108]. In 1954 G. Ricci [5201]

[2]See the review of [1789] by S. Ikehara in Zentralblatt für Mathematik [3009].

got $E \leq 15/16$ (another proof was given by N.C. Ankeny and H. Onishi [105]), but P. Erdős and Rényi [1849] had already stated in 1950 that the inequality $E \leq 15/16$ had been shown by A. Selberg in an unpublished manuscript. In 1955 P. Erdős [1814] proved that the closure of the set $\{(p_{n+1} - p_n)/\log p_n\}$ is of positive measure and this result was made more precise by G. Ricci [5202].

The next improvement, $E \leq 29/32$, was made in 1965 by Y. Wang, S. Xie and K. Yu [6559], and in the next year a big step forward was made by E. Bombieri and H. Davenport [608], who showed

$$E \leq \frac{2+\sqrt{3}}{8} = 0.4665\ldots.$$

This was improved in 1972 by G.Z. Pilt'jai [4875] to $E \leq (2\sqrt{2} - 1)/4 = 0.4571\ldots$, and his method was modified by M.N. Huxley who first got $E \leq 0.4463$ [2974], and then achieved $E \leq 0.4442\ldots$ and $E \leq 0.4394$ [2975, 2980]. Later $E < 0.4342\ldots$ was shown by É. Fouvry and F. Grupp [2053], and in 1988 H. Maier [4099] proved $E < 0.2484\ldots$.

The final step was in the current century when D.A. Goldston, J. Pintz and C.Y. Yildirim [2273] showed $E = 0$ (cf. D.A. Goldston, Y. Motohashi, J. Pintz, C.Y. Yildirim [2272]; for a variant of the proof see D.A. Goldston, S.W. Graham, J. Pintz, C.Y. Yildirim [2270]). An even stronger result,

$$\liminf_{n\to\infty} \frac{p_{n+1} - p_n}{\sqrt{\log p_n}(\log\log p_n)^2} < \infty$$

was established in [2274].

This method has been used by D.A. Goldston, S.W. Graham, J. Pintz and C.Y. Yildirim [2270] to show that if q_n denotes the nth consecutive product of two primes, then

$$\liminf_{n\to\infty}(q_{n+1} - q_n) \leq 26.$$

In [2271] they were able to replace the bound 26 by 6. A similar result for numbers with a fixed number of prime divisors has been obtained by F. Thorne [6140].

3. H.-E. Richert [5209] used a variant of the Selberg sieve to show that for large x every interval $(x - x^\theta, x]$ with $\theta = 6/11 = 0.545\ldots$ contains a number P_2, i.e., an integer divisible by at most two primes, improving a previous result by Y. Wang [6552] who had $\theta = 10/17 = 0.588\ldots$. Later J.R. Chen [1023, 1024] proved that this holds with $\theta = 1/2$, and D.R. Heath-Brown [2626] pushed θ below $1/2$. The next reductions were made by M. Laborde ($\theta = 0.4867$, [3597]), J.R. Chen ($\theta = 0.477$, [1024]), H. Halberstam, D.R. Heath-Brown and H.-E. Richert ($\theta = 0.4548\ldots$, [2452]), H. Iwaniec and M. Laborde ($\theta = 0.45$, [3065]), H. Halberstam and H.-E. Richert ($\theta = 0.4476$, [2456]), É. Fouvry ($\theta = 0.4436$, [2051]), J. Wu ($\theta = 0.44$, [6756]), H. Li ($\theta = 0.4386$, [3875]), H.Q. Liu ($\theta = 0.436$, [3957]), and P. Sargos and J. Wu ($\theta = 0.43596$, [5398]).

Stronger results are obtainable if one looks for numbers P_2 in almost all short intervals. D.R. Heath-Brown [2626] showed in 1978 that for almost all integers n there is a P_2 in $(n, n + n^{1/11})$, the following year Y. Motohashi [4457] showed this for the intervals $(n, n + n^\varepsilon)$ (for every $\varepsilon > 0$) and in the same year D. Wolke [6715] proved this for $(n, n+\log^C n)$ with a large C. Later G. Harman [2548] showed that one can take for C any number larger than 7 and H. Mikawa [4302] replaced 7 by 5.

4. Although the first numerical comparisons between the values of $\pi(x;k,l)$ and $\pi(x;k,m)$ for $l \neq m$ had already been carried out in 1914 by A. Cunningham [1295, 1296], and at the same time J.E. Littlewood [3940] showed that the difference $\pi(x;3,2) - \pi(x;3,1)$ changes its sign infinitely often, this subject did not attract attention until the late fifties, when J. Leech [3759] and D. Shanks [5673] confirmed numerically Čebyšev's observations on primes in progressions mod 4. In particular D. Shanks observed that the majority of primes $p \le 3 \cdot 10^6$ are congruent to 3 mod 4 and made similar observations on primes mod k for $k = 8$, 10 and 12.

Computations of $\pi(x;4,\pm 1)$ for large x were made later by M. Deléglise, P. Dusart and X.-F. Roblot [1441], using a modification of the improvement of the old Meissel's [4236] method to compute $\pi(x)$, developed by J.C. Lagarias, V.S. Miller and A.M. Odlyzko [3607] and M. Deléglise and J. Rivat [1443].

Denote by $D(x)$ the number of $n \le x$ for which $\pi(x;4,1)$ exceeds $\pi(x;4,3)$. Čebyšev's assertion (2.15) led to the conjecture that $D(x)/x$ tends to zero [3381], but this was later shown by J. Kaczorowski [3188] to be incompatible with the General Riemann Hypothesis (see also [3190, 3192]).

The analogue of this question for arbitrary pairs of progressions with the same difference became the subject of two cycles of papers by P. Turán and S. Knapowski [3381–3396]. In [3381] a list of 60 related problems was given (a modified version of it appeared in [3389]). The first problem asked whether for every $l_1 \not\equiv l_2 \bmod k$ the difference $\pi(x;k,l_1) - \pi(x;k,l_2)$ changes sign infinitely often, and they proved [3381] that this holds in the case, when $l_1 = 1$ and no L-function mod k has a real zero (for $k = 4$, $l_1 = 1, l_2 = 3$ this had been shown already by J.E. Littlewood [2523]). Among the listed problems one finds the *Shanks–Rényi race problem.*

Shanks–Rényi race problem *For a given $k \ge 4$ let $l_1, l_2, \dots, l_r$ (with $r = \varphi(k)$) be a permutation of the residues* mod k, *prime to k. Show that for infinitely many integers n one has*

$$\pi(n;k,l_1) > \pi(n;k,l_2) > \cdots > \pi(n;k,l_r). \tag{6.13}$$

It was shown in 1994 by M. Rubinstein and P. Sarnak [5334] that the answer to this problem is positive under the assumption of the General Riemann Hypothesis and the simplicity and **Q**-linear independence of zeros of $L(s,\chi)$ on $\Re s = 1/2$, χ running over all primitive characters. In 1995 J. Kaczorowski [3191] deduced from the General Riemann Hypothesis a positive answer for $k = 5$, and in fact (6.13) holds in this case on a set of positive lower density. Earlier [3189] he deduced from the same hypothesis that for every k there are infinitely many integers m, n with

$$\pi(n;k,1) > \max_{a \not\equiv 1 \pmod k} \pi(n;k,a), \qquad \pi(m;k,1) < \min_{a \not\equiv 1 \pmod k} \pi(m;k,a).$$

For later results on this topic see C. Bays, K. Ford, R.H. Hudson and M. Rubinstein [366], C. Bays and R.H. Hudson [367], A. Feuerverger and G. Martin [1997], K. Ford and S.V. Konyagin [2035–2037], J. Kaczorowski [3193], J. Pintzand S. Salerno [4911–4913] and P. Moree [4414].

From the several results by S. Knapowski and P. Turán one should point out the deduction from the Haselgrove condition of the existence of infinitely many sign changes of the differences $\psi(x;k,l_1) - \psi(x;k,l_2)$ and $\Pi(x;k,l_1) - \Pi(x;k,l_2)$ [3385].

A modulus k is said to satisfy the *Haselgrove condition* if L-functions corresponding to characters mod k have no zeros in some rectangle

$$\{s : 1/2 < \Re s < 1, 0 \leq \Im s \leq A(k)\}$$

for some positive $A(k)$ (in certain cases they had to assume $A(k) \gg k^{10}$). The function Π occurring here is defined by

$$\Pi(x,k,l) = \sum_{n \equiv l \pmod{k}} \frac{\Lambda(n)}{\log n}.$$

They showed later that the same result holds for the difference $\pi(x;k,l_1) - \pi(x;k,l_2)$ provided l_1, l_2 are either both quadratic residues or quadratic non-residues mod k. In all considered cases they showed that the number of sign changes below T is $\gg \log\log T$. This was improved in the case of the function ψ to $\gg \sqrt{\log T}$ by J. Pintz and S. Salerno [4910] in 1984.

5. The question of prime polynomial values was considered by A. Schinzel and W. Sierpiński in [5452], where the following conjecture, called *conjecture H*, was stated.

Conjecture *H* *Let $f_1, f_2, \ldots, f_k$ be non-constant polynomials with integral coefficients, positive leading coefficient, irreducible over the rationals and without a fixed prime divisor. Then there are infinitely many integers n such that all numbers $f_i(n)$ are prime.*

A quantitative version of this conjecture was formulated by P.T. Bateman and R.A. Horn [358, 359]. It asserts that if the polynomials $f_1, \ldots, f_k$ satisfy the assumptions of that conjecture, then for the number $P(x)$ of integers $n \leq x$ for which all numbers $f_i(n)$ $(i = 1, 2, \ldots, k)$ are prime one has

$$P(x) = \left(\frac{C(f_1, \ldots, f_k)}{d_1 d_2 \cdots d_k} + o(1)\right) \int_2^x \frac{dt}{\log^k t},$$

with

$$C(f_1, \ldots, f_k) = \prod_p \left(1 - \frac{\omega(p)}{p}\right)\left(1 - \frac{1}{p}\right)^{-k},$$

where d_j is the degree of f_j, and $\omega(p)$ denotes the number of solutions of the congruence

$$\prod_{j=1}^{k} f_j(x) \equiv 0 \pmod{p}.$$

This forms a generalization of the conjectures **B**, **E** and **P** of G.H. Hardy and J.E. Littlewood [2531], covering the cases $k = 2$, $f_1(x)$, $f_2(x) = x + a$, $k = 1$, $f_1(x) = x^2 + 1$ and $k = 2$, $f_1(x) = x^2 + 1$, $f_2(x) = x^2 + 3$.

A kind of approximation to conjecture H was obtained by H. Halberstam and H.-E. Richert [2454] in 1972. They showed that under the assumptions of that conjecture the polynomials in question represent infinitely many integers with an explicitly bounded number of prime factors (see also [2455, Chap. 10]).

For computations of the constant $C(f_1, \dots, f_k)$ in the Bateman–Horn conjecture see H. Davenport, A. Schinzel [1402].

It was shown by J.B. Friedlander and A. Granville [2103] under the General Riemann Hypothesis that a uniform version of this conjecture cannot hold, and the General Riemann Hypothesis was eliminated from this assertion by M. Nair and A. Perelli [4521].

A heuristic approach to the Bateman–Horn conjecture was pursued by K. Conrad [1200] and M. Hindry, T. Rivoal [2813], who applied a method used earlier by S.W. Golomb [2276] to study the problem of twin primes. See also S. Baier [217].

6. The analogous question for general composite polynomials was first considered by R.J. Miech [4284] in 1964. He used Selberg's sieve method to show that if f is a product of k distinct irreducible polynomials, having their degrees equal to $d_1, \dots, d_k$ and does not have fixed divisors, then there are $\gg N/\log^k N$ integers $\le N$ represented by f with at most

$$\sum_{j=1}^{k} \left[\frac{9d_j}{5}\right] + \left[k \sum_{j=1}^{k} \frac{1}{j} + k\log(5/2)\right]$$

prime factors. In [4286] he gave a bound for the earliest occurrence of such a number and in [4285] showed that the sequence $f(p)$, p running over primes, contains infinitely many numbers with a bounded number of prime factors, but he did not give any bound for the number of these factors. In 1969 H.-E. Richert [5209] showed that if f is of degree d, then there are infinitely many primes p with $f(p)$ having at most $2d + 1$ prime factors (see [2455, Theorem 10.6]).

Richert's result has been extended to polynomials in algebraic number fields by J. Hinz [2823].

7. The situation looks much better in the case of polynomials in several variables. It was deduced from the General Riemann Hypothesis in 1957 by C. Hooley

[2854], that for every integer a the polynomial $x^2 + y^2 + a$ represents infinitely many primes. He also obtained an asymptotic formula for the sum

$$\sum_{p \le T} f_a(p),$$

where $f_a(p)$ denotes the number of representations of the prime p in the form $p = x^2 + y^2 + a$. These results were proved unconditionally two years later by B.M. Bredikhin [695] who applied Linnik's dispersion method (see below). He later used the same method in [697] to show that there are infinitely many primes of the form $p = F(x, y) + a$, where F is a positive definite quadratic form, whose discriminant is not a square.

It was pointed out in 1966 by P.D.T.A. Elliott and H. Halberstam [1749] that Bombieri's result can be used to make Hooley's argument in [2854] independent of the General Riemann Hypothesis.

The next step was made by Y. Motohashi [4447, 4448], who considered the case $a = 1$ and was able to show that if $S_a(N)$ denotes the number of primes $p \le N$ of the form $x^2 + y^2 + a$, then

$$\frac{N}{\log^2 N} \ll S_1(N) \ll \frac{N}{\log^{3/2} N}.$$

Moreover he conjectured that one has

$$S_1(N) = \left(\frac{3}{2}c + o(1)\right)\frac{N}{\log^{3/2} N},$$

with

$$c = \prod_{p \equiv 3 \pmod 4} \left(1 - \frac{1}{p^2}\right)^{-1/2} \left(1 - \frac{1}{p(p+1)}\right).$$

This conjecture had to be modified, as two years later H. Iwaniec [3052] used his sieve [3051] to prove the inequality

$$S_1(N) \le (c + o(1))\frac{N}{\log^{3/2} N}$$

as a particular case of his main theorem showing that for any quadratic form $f(x, y) = ax^2 + bxy + cy^2$ with co-prime integral coefficients and non-square discriminant the number $S_A(F; N)$ of primes $p \le N$ of the form $F(x, y) + A$ (with $A \ne 0$) satisfies

$$\frac{N}{\log^{3/2} N} \ll S_A(F; N) \ll \frac{N}{\log^{3/2} N}.$$

In [3054] he obtained an analogous result for irreducible quadratic polynomials in two variables which have no fixed factor and represent infinitely many positive odd integers.

In the case of higher degrees these conditions may be insufficient, as the example

$$F(x, y) = \left(y^2 + 15\right)\left(1 - \left(x^2 - 23y^2 - 1\right)^2\right) - 5$$

shows (D.R. Heath-Brown [2657]). This polynomial is irreducible, has no fixed factors, represents infinitely many positive integers, but nevertheless does not represent positive primes as all its positive values are divisible by 5.

The existence of primes of the form $x^2 + y^2 + 1$ in short intervals has been investigated by M.N. Huxley, H. Iwaniec [2989].

For improvements see J. Wu [6757] and K. Matomäki [4194]. The last author established the existence of such primes in every interval $(x, x + x^{10/11})$ for large x.

The case of cubic polynomials in several variables was considered by P.A.B. Pleasants [4920, 4921] who showed that every such polynomial having at least 10 variables and satisfying certain necessary conditions represents infinitely many primes. He showed also that if a quadratic polynomial F in $n \geq 3$ variables satisfies the necessary congruence conditions and its leading form is positive definite of rank ≥ 3, then F represents infinitely many primes.

In 1998 J.B. Friedlander and H. Iwaniec [2108] showed that there are

$$(c + o(1))\frac{N^{3/4}}{\log N}$$

primes $p \leq N$ of the form $x^4 + y^2$ with

$$c = \frac{4\Gamma(1/4)}{6\sqrt{2}\pi^{3/2}}.$$

A related result was achieved in 1997 by É. Fouvry and H. Iwaniec [2058], who showed the existence of infinitely many primes of the form $p^2 + x^2$ with prime p.

In 2001 D.R. Heath-Brown [2657] proved that the polynomial $x^2 + 2y^3$ represents infinitely many primes. He used an approach, axiomatized by J.B. Friedlander and H. Iwaniec in [2109], which makes it possible to detect prime numbers by a sieve method using strong analytical assumptions. It was observed earlier by E. Bombieri [602] that this cannot be done by a pure sieve.

The result in [2657] established the truth of the qualitative part of G.H. Hardy and J.E. Littlewood's conjecture **N** [2531]. The same assertion for a large class of cubic polynomials in two variables was later proved by D.R. Heath-Brown and B.Z. Moroz [2669, 2670].

6.1.4 Selberg Class

1. In 1989 A. Selberg [5623] introduced a large class S of functions, the *Selberg class*. It consists of functions $f(s)$, defined in the half-plane $\Re s > 1$ by an absolutely convergent Dirichlet series

$$f(s) = \sum_{n=1}^{\infty} \frac{a_f(n)}{n^s},$$

and satisfying the following four conditions.

(i) *$f(s)$ can be continued to a meromorphic function in the complex plane, its only singularity being a possible pole at $s = 1$.*
(ii) *There exist $r \geq 0$, $Q, \alpha_j > 0$, complex numbers μ_j with $\Re\mu_j \geq 0$, and ω on the unit circle such that for the function $\Phi(s)$ defined by*

$$\Phi(s) = Q^s \prod_{j=1}^{r} \Gamma(\alpha_j s + \mu_j) f(s)$$

the functional equation

$$\Phi(s) = \omega \overline{\Phi}(1-s)$$

holds.
(iii) *For every $\varepsilon > 0$ one has $a_f(n) = O(n^{\varepsilon})$.*
(iv) *In the half-plane $\Re s > 1$ one has $F(s) \neq 0$, and*

$$\log F(s) = \sum_{n=1}^{\infty} \frac{b_f(n)}{n^s},$$

where the coefficients $b_f(n)$ vanish, except when n is a prime power, and moreover one has $b_f(n) = O(n^{\theta})$ for some $\theta < 1/2$.

The Selberg class contains the Riemann zeta-function as well as Dedekind zeta-functions and the L-functions of Dirichlet and Hecke. Also the L-functions associated with holomorphic newforms of any level lie in S, and it has been conjectured that S also contains all of Artin's L-functions and automorphic L-functions associated with the matrix groups $GL_n(K)$ for algebraic number fields K.

The sum $d(f) = 2\sum_{j=1}^{r} \alpha_j$ is called the *degree* (or *dimension*) of f. Although the functional equation in (ii) is not unique, the sum $\sum_{j=1}^{r} \alpha_j$ depends only on f (S. Bochner [570], M.-F. Vignéras [6389]).

One denotes by S_d the set of all functions of degree d. In all known cases the degree is an integer and it is conjectured (the *degree conjecture*) that this is a general property.

One has $S_0 = \{1\}$ and $S_d = \emptyset$ if $0 < d < 1$ (this result appears first explicitly in the paper by J.B. Conrey and A. Ghosh [1212], but its essence can be found earlier: see S. Bochner [570] and H.-E. Richert [5206]). It was shown by J. Kaczorowski and A. Perelli [3197] that the class S_1 consists of functions $L(s + i\theta, \chi)$, where θ is real and χ is a primitive Dirichlet character.

It was shown later by J. Kaczorowski and A. Perelli [3201] that S_d is empty for $1 < d < 5/3$ (cf. K. Soundararajan [5853]).

Recently J. Kaczorowski and A. Perelli [3203] showed that S_d is empty for $1 < d < 2$.

2. A function $f \in S$ is called *primitive* if an equality $f = gh$ with $g, h \in S$ can hold only if g or h equals 1. A. Selberg conjectured that the factorization into primitives is unique. It was shown by J.B. Conrey and A. Ghosh [1212] and M.R. Murty [4488, 4489] that this is a consequence of the following conjecture, proposed by A. Selberg [5623].

Selberg orthonormality conjecture *If $f, g \in S$ are both primitive, then*

$$\sum_{p\le x} \frac{a_f(p)\overline{a_g(p)}}{p} = \begin{cases} c(f)\log\log x + O(1) & \text{if } f = g, \\ O(1) & \text{if } f \ne g. \end{cases}$$

This conjecture implies Artin's conjecture for his L-functions (M.R. Murty [4488, 4489]), as well as the truth of *Langlands conjecture* for solvable extensions $K/\mathbf{Q}$, which asserts that Artin L-functions are equal to L-functions of certain representations of the group of n-dimensional matrices over the ring of adeles of the rational number field. See also J. Kaczorowski, G. Molteni and A. Perelli [3196].

It was shown by M.R. Murty and A. Zaharescu [4494] that the orthonormality conjecture is essentially equivalent to an analogue of the Pair Correlation Conjecture for the Selberg class, formulated by M.R. Murty and A. Perelli [4492].

For surveys see the papers by J. Kaczorowski and A. Perelli [3194, 3204, 4776, 4777]. See also [3198–3200, 3202].

6.1.5 Other Questions

1. Exponential sums of the form

$$S_f(x,t) = \sum_{n\le x} f(n)\exp(2\pi i n t),$$

where f is a multiplicative function and t is real had already been considered in particular cases in 1937 by H. Davenport [1347, 1348], who obtained for every M the bound

$$\sum_{n\le x} \mu(n)\exp(2\pi i n t) \ll_M \frac{x}{\log^M x}, \tag{6.14}$$

uniform in t.

The general case was treated in 1974 by H. Daboussi and H. Delange in [1316]. They stated several results, one of them (which according to H. Delange [1435] is due to H. Daboussi) asserts that if $|f(n)| \le 1$, then for irrational t one has

$$S_f(x,t) = o(x). \tag{6.15}$$

Later [1317, 1435] H. Daboussi and H. Delange proved that it suffices to assume $\sum_{n\le x} |f(n)|^2 = O(x)$, or even $\sum_{n\le x} |f(n)|^m = O(x)$ for some $m > 1$.

For generalizations see L. Goubin [2290] and K.-H. Indlekofer, I. Kátai [3012, 3013], I. Kátai [3272].

See I. Kátai [3274] for a survey.

If the number t has good approximation by rationals, then the bound in (6.15) can be improved. This was achieved by H. Daboussi [1313] and H.L. Montgomery and R.C. Vaughan [4366].

Later improvements were made by G. Bachman [193, 194] and H. Maier and A. Sankaranarayanan [4103]. In the case considered by H. Davenport ($f = \mu$) it was shown by D. Hajela and B. Smith [2441] that the General Riemann Hypothesis implies in (6.14) the bound $O(x^c)$ for every $c > 5/6$. In 1991 R.C. Baker and G. Harman [262] proved this for all $c > 3/4$.

2. For an arithmetic function f its *mean value* $m(f)$ is defined by

$$m(f) = \lim_{x\to\infty} \frac{1}{x} \sum_{n\le x} f(n),$$

provided this limit exists. In his book [6690], published in 1943, A. Wintner considered the question of whether $m(f)$ exists for every multiplicative function f assuming only the values ± 1, and gave an insufficient argument supporting the affirmative answer. In 1959 Z. Ciesielski [1110] showed that Wintner's assertion holds for most such functions, and two years later E. Wirsing [6694] provided a correct proof for Wintner's assertion[3]. The question arose of whether a similar result holds also for complex-valued multiplicative functions f with $|f(n)| \le 1$. At the same time H. Delange [1428, 1429] gave a necessary and sufficient condition for functions of that class to have a non-zero mean value. The final step was in 1968 by G. Halász [2443] who showed that for these functions one has

$$\sum_{n\le x} f(n) = Cx^{1+ia} L(\log x) + o(x),$$

where C, a are constants, and the function $L(t)$ satisfies $|L(t)| = 1$ and is *slowly oscillating*, i.e., for every fixed $A > 0$ one has

$$\lim_{x\to\infty} \frac{L(Ax)}{L(x)} = 1.$$

(Such functions were used for the first time for arithmetical purposes by G. Pólya [4955] in 1917.) The results of E. Wirsing [6695] and G. Halász [2443] led to new proofs of the Prime Number Theorem.

Two elementary proofs of Halász's theorem were given by H. Daboussi and K.-H. Indlekofer [1318], and a generalization to mean values of multiplicative functions in arithmetic progressions appeared in H. Delange [1434].

Mean values of completely multiplicative functions f with $f(p)$ for prime p lying in a fixed subset of the unit disc were studied by A. Granville and K. Soundararajan [2324] in 2001.

3. In 1963 C. Hooley [2857, 2858] considered the problem of bounding sums of the form

$$V_\gamma(q) = \sum_{j=1}^{\varphi(q)} (a_{i+1} - a_i)^\gamma \quad (\gamma > 0),$$

[3] A simpler elementary proof was found in 1986 by A. Hildebrand [2798].

where $1 = a_1 < a_2 < \cdots < \cdots$ are integers prime to q and less than a fixed $q \geq 3$. In 1940 P. Erdős [1789] conjectured the bound

$$V_2(q) \ll \frac{q^2}{\varphi(q)}, \tag{6.16}$$

and Hooley established in [2857] the bounds

$$V_\gamma(q) \ll q^\gamma \varphi^{1-\gamma}(q)$$

for $1 < \gamma < 2$ and

$$V_2(q) \ll q \log\log q,$$

and in [2858] proved for $0 < \gamma < 2$ the equality

$$V_\gamma(q) = (\Gamma(1+\gamma) + o(1)) q^\gamma \varphi(q)^{1-\gamma}.$$

The next step was taken by M. Hausman and H.N. Shapiro [2618] in 1973, who showed

$$V_2(q) \ll \frac{q^2}{\varphi(q)} \cdot \max\left\{1, \sum_{p|q} \frac{\log p}{p-1}\right\}.$$

The final step was taken by H.L. Montgomery and R.C. Vaughan [4368], who established (6.16) and showed, more generally,

$$q^\gamma \varphi^{1-\gamma}(q) \ll V_\gamma(q) \ll q^\gamma \varphi^{1-\gamma}(q).$$

4. From the many books on the analytical theory of numbers published in the considered period we should mention here only few: H. Davenport wrote two books [1372, 1377], the first presenting applications of analytical methods in additive problems and the second bringing a short introduction into the contemporary problems of prime number theory. A. Walfisz presented an exposition [6535] of the theory of Weyl's sums and M.N. Huxley [2973] discussed modern sieve methods. From several textbooks let us mention here A. Blanchard [546] and K. Chandrasekharan [992, 993].

6.2 Additive Problems

1. In 1960 Yu.V. Linnik [3921–3923] established the truth of conjecture **J** by G.H. Hardy and J.E. Littlewood [2531], by showing that all large integers are sums of a prime and two squares. He found also an asymptotic formula for the number of such representations. This had earlier been shown to be a consequence of the General Riemann Hypothesis by C. Hooley [2854]. Yu.V. Linnik used his *dispersion method*, whose main idea can be described as follows: if we have a binary additive problem, consisting of estimating the number $f(n)$ of solutions of the equation $n = a + b$, where a, b are elements of certain sets A and B, respectively, then one first obtains heuristically an asymptotic formula for $f(n)$, say $f(n) = (1 + o(1))g(n)$, and then tries to obtain strong bounds for the moments

$$S_k(x) = \sum_{n \le x} |f(n) - g(n)|^k \quad (k = 1, 2, \ldots).$$

A simpler proof, based on the large sieve, was later given by P.D.T.A. Elliott and H. Halberstam [1749].

Linnik's method is also applicable to other problems. In [3920, 3925] Linnik used it to obtain asymptotics for the sum

$$\sum_{n \le x} d(n+1) d_k(n) \quad (k \ge 2),$$

generalizing C. Hooley's result [2855], which dealt with the case $k = 3$. This sum counts the number of solutions of the system

$$x_1 x_2 - y_1 \cdots y_k = 1, \quad y_1 \cdots y_k \le x,$$

so it can be regarded as a kind of binary additive problem.

The more general sums $\sum_{n \le x} d_k(n) d(n+a)$ were treated by B.M. Bredikhin [696], D. Wolke [6712], Y. Motohashi [4453, 4454], D. Redmond [5141–5143] and V.A. Bykovskiĭ and A.I. Vinogradov [877].

Yu.V. Linnik discussed this method in a book [3925] published in 1961 (see also B.M. Bredikhin's survey [698]).

2. A very simple proof of Waring's theorem was found in 1960 by D.J. Newman [4583], who melded the analytical approach and Schnirelman's method. Another proof of that type was presented by K. Thanigasalam [6129] in 1974.

A new way of proving asymptotics for the number of representations in the Waring problem was shown in 1982 by G. Lachaud [3598], who used adelic analysis.

The last open case of the Waring problem for small exponents, the determination of $g(4)$, the smallest integer k such that every positive integer is the sum of at most k fourth powers of integers, was finally settled in 1986, when R. Balasubramanian, J.-M. Deshouillers and F. Dress succeeded in establishing the long-awaited equality $g(4) = 19$ [276, 277, 1485, 1486]. Earlier the bounds 34, 30, 22, 21 and 20 for $g(4)$ were obtained by F. Dress [1620, 1622], H.E. Thomas, Jr. [6139] and R. Balasubramanian [274, 275], respectively, and it was known that every integer exceeding 10^{10^a} with $a = 88.39$ is the sum of at most 19 fourth powers (F.C. Auluck [172]).

The bounds for $G(k)$ in the case of small k have been quickly reduced due to the work of several authors. The old bound $G(5) \le 23$ of H. Davenport [1360] was reduced first to 22 (K. Thanigasalam [6134]) and then to 21 (R.C. Vaughan [6362] and K. Thanigasalam [6135]), 19 (R.C. Vaughan [6365, 6366]) 18 (J. Brüdern [771]), and finally to 17 (R.C. Vaughan, T.D. Wooley [6372]). Important new methods were introduced in the nineties by R.C. Vaughan and T.D. Wooley [6371–6375, 6724], which in particular led to $G(6) \le 24$, $G(7) \le 33$, $G(8) \le 42$, $G(9) \le 50$ and $G(10) \le 59$, which are the best known bounds at the time of writing.

Bounds for the number of $n \le x$ which are not sums of 15 and 16 fifth powers were given by K. Kawada and T.D. Wooley [3290], who also considered sums of 9 and 10 fourth powers (cf. [3290]).

3. For prime p and $k \geq 2$ denote by $\gamma(k, p)$ the smallest integer s such that for every $a \in [0, p-1]$ the congruence

$$\sum_{j=1}^{s} x_j^k \equiv a \pmod{p}$$

has a solution. The determination of the value of this constant forms an analogue of the Waring problem for the finite field $\mathbf{F}_p$. One sees easily that $\gamma(p-1, p) = p-1$, $\gamma((p-1)/2, p) = (p-1)/2$, and if $k \nmid p-1$, then $\gamma(k, p) = 1$, so one assumes usually that k is a divisor of $p-1$, and $t = (p-1)/k \geq 3$. The first result for $\gamma(k, p)$ goes back to G.H. Hardy and J.E. Littlewood, who in [2536] established $\gamma(k, p) \leq k$, and this bound was reduced in 1943 by I. Chowla [1065] to $k^{0.8771}$. One deduces from Weil's bound for exponential sums that if $k < p^2$, then $\gamma(k, P) = O(\log k)$, and for $k < p^{1/2-\delta}$ with $\delta > 0$ one has $\gamma(k, p) = O(1)$.

In 1964 H. Heilbronn conjectured in his book [2712], devoted to additive problems in $\mathbf{F}_p$, that for large t one has

$$\gamma(k, p) = O(k^{\varepsilon}) \tag{6.17}$$

for every positive ε, and for all $t \geq 3$ one has

$$\gamma(k, p) = O\left(\sqrt{k}\right). \tag{6.18}$$

He obtained also

$$\gamma(k, p) \ll_t p^{1/\varphi(t)}.$$

The first improvement upon Chowla's bound was obtained in 1971 by M. Dodson [1607], who reduced the exponent to $7/8$. The next steps were made by A. Tietäväinen [6157], who showed $\gamma(k, p) = O(k^{3/5+\varepsilon})$, and M. Dodson and A. Tietäväinen [1610], who in 1976 proved

$$\gamma(k, p) \leq 68\sqrt{k}\log^2 k.$$

The Waring problem for residue classes with regard to a composite modulus was considered in 1977 by C. Small [5822, 5823], who in particular confirmed a conjecture by I. Kaplansky showing that in a finite field of more than $(k-1)^4$ elements every element is the sum of at most two kth powers.

The truth of H. Heilbronn's conjecture (6.17) was established in 1992 by S.V. Konyagin [3470]. Later D.R. Heath-Brown and S.V. Konyagin [2667] showed that for $k \leq p^{2/3-\varepsilon}$, $\gamma(k, p)$ is bounded by a constant depending only on ε.

For the final step in conjecture (6.18) one had to wait till 2007, when it was established by J.A. Cipra, T. Cochrane and C. Pinner [1115]. The obtained bound cannot be further improved, since, as shown by M. Dodson and A. Tietäväinen [1610], if $p = 1 + 3k$ is a prime, then $\gamma(k, p) \gg \sqrt{k}$. For further progress see A. Alnaser, T. Cochrane [57].

Another additive problem concerning residues mod p was considered by P. Erdős and H. Heilbronn [1834] in 1964. They showed that if p is a prime and A is a set of residues mod p having at least $3\sqrt{6}\,p^{1/2}$ elements, then every residue class mod p can be written as the sum of distinct elements of A, and conjectured that the constant

$3\sqrt{6}$ can be replaced by 2. In the case $0 \notin A$ this was established four years later by J.E. Olson [4677]. In 1994 J.A. Dias da Silva and Y.O. Hamidoune [1530] removed from Olson's result the assumption $0 \notin A$.

The analogue for arbitrary Abelian groups was considered by G. Károlyi [3257, 3260] and G. Wang [6542].

4. At the beginning of the sixties G.A. Freiman wrote a series of papers [2074–2078], culminating in his book [2079], in which he studied the cardinality of sets $2A = \{a + b : a, b \in A\}$ for finite sets A of positive integers (and, more generally, finite subsets of $\mathbf{Z}^n$), as well as the inverse problem of the additive theory of numbers, in which one asks for properties of A knowing $2A$. One of his main results characterizes sets A of k elements with $\#(2A) \le ck$. For similar results in Abelian groups see I.Z. Ruzsa [5346, 5352].

For other proofs of the last result see Y.F. Bilu [510], M.-C. Chang [998], G.A. Freiman [2080], I.Z. Ruzsa [5345].

For a survey see G.A. Freiman [2081, 2082], and for an exposition of inverse problems in additive number theory see the book by M.B. Nathanson [4547].

6.3 Modular Forms

1. E. Hecke [2695] had already given sufficient conditions for a Dirichlet series to be an L-series of a modular form then in 1967 A. Weil [6627] extended Hecke's result to modular forms associated with the groups $\Gamma_0(N)$, assuming the existence of functional equations for twists of

$$L(s) = \sum_{n=1}^{\infty} \frac{a_n}{n^s},$$

defined by

$$L(s, \chi) = \sum_{n=1}^{\infty} \chi(n) \frac{a_n}{n^s},$$

χ being a primitive Dirichlet character (cf. R. Weissauer [6634]).

Further extensions of Hecke's theory, encompassing various kinds of automorphic functions, were made by H. Jacquet and R.P. Langlands [3090] (see also A. Weil [6628] for a more elementary treatment). In the nineties, J.B. Conrey and D.W. Farmer [1204] replaced the assumption on twists by some properties of the Euler factors of $L(s)$, provided the level is small ($N \le 17$, $N \ne 13$ or $N = 23$).

The case $N = 13$ was settled in 2007 by J.B. Conrey, D.W. Farmer, B.E. Odgers and N.C. Snaith [1207]. For later developments see A. Diaconu, A. Perelli, A. Zaharescu [1516], D.W. Farmer, K. Wilson [1960].

In 1964 it was shown by K. Wohlfahrt [6707] that zeros of Eisenstein series E_{2k} for certain small values of k lie on transforms of the unit circle, and conjectured that this holds for every k. (The Eisenstein series $E_{2k}(z)$ are defined by

$$E_{2k}(z) = 1 + (-1)^k \frac{4k}{B_{2k}} \sum_{n=1}^{\infty} \sigma_{2k-1}(n) z^n,$$

where B_k is the kth Bernoulli number.)

This was established in 1970 by F.K.C. Rankin and H.P.F. Swinnerton-Dyer [5104]. For a generalization see R.A. Rankin [5127].

2. In a paper dealing with congruences satisfied by the Ramanujan function $\tau(n)$ J.-P. Serre [5638] conjectured that for every prime ℓ there exists a continuous representation

$$\rho_\ell : Gal(K_\ell/\mathbf{Q}) \to GL_2(\mathbf{Q}_\ell)$$

(K_ℓ being the maximal extension of the rationals with ℓ being the only ramified prime) such that if for primes $p \neq \ell$, F_p denotes the corresponding Frobenius element, then

$$\det\left(1 - \rho_\ell\left(F_p\right) T\right) = 1 - \tau(p)T + p^{11}T^2,$$

thus the trace of $\rho_\ell(F_p)$ equals $\tau(p)$, and its determinant equals p^{11}.

Recall that the *Frobenius element* F_p is defined in the following way: if $K/\mathbf{Q}$ is a finite Galois extension unramified at p and $\mathfrak{p}$ is a prime ideal of K lying over p, then $F_{\mathfrak{p}}$ is the unique element of the Galois group of $K/\mathbf{Q}$ which satisfies

$$F_{\mathfrak{p}}(x) \equiv x^p \pmod{\mathfrak{p}}.$$

He mentioned also that a similar assertion is possibly true for all cusp forms, writing [5638, p. 14]: "Ce[4] qui a été dit pour τ peut l'être aussi pour les coefficients de toute forme parabolique de poids k $\Phi(X) = \sum_{n=1}^{\infty} a_n X^n$, $a_1 = 1$, qui est fonction propre des opérateurs de Hecke, et dont les coefficients appartient à $\mathbf{Z}$."

In the case of cusp forms f for $\Gamma_0(N)$ of weight 2 this result follows from the Eichler–Shimura theory (M. Eichler [1699], G. Shimura [5692]; see also the books [5698] by Shimura and [3399] by A.W. Knapp), which associates with every newform of weight k an elliptic curve defined over $\mathbf{Q}$ having the same L-function as f.

Serre's conjecture was established for weights $k \geq 2$ by P. Deligne [1444] in 1971. He constructed 2-dimensional ℓ-adic representations ρ_ℓ of the Galois group $G_{\mathbf{Q}}$ of $\overline{\mathbf{Q}}/\mathbf{Q}$, associated with cusp forms of weight $k \geq 2$

$$1 + \sum_{n=2}^{\infty} a_n q^n$$

[4] "What has been said for τ possibly also holds for coefficients of every cusp form of weight k $\Phi(X) = \sum_{n=1}^{\infty} a_n X^n$, $a_1 = 1$, which is an eigenfunction of Hecke operators and has coefficients in $\mathbf{Z}$."

which are eigenfunctions of Hecke operators and satisfy $a_n \in \mathbf{Z}$. This representation is unramified at primes $p \neq \ell$ and for the Frobenius element F_p (for $p \neq \ell$) the trace and the determinant of $\rho_\ell(F_p)$ are a_p and p^{k-1}, respectively.

For cusp forms of weight 1 for $\Gamma_0(N)$ the construction of a corresponding representation was given in 1974 by P. Deligne and J.-P. Serre [1452]. For an exposition see J.-P. Serre [5651].

The analogous problem of constructing λ-adic representations related to Hilbert modular forms associated with totally real number fields was solved by H. Carayol [896], A. Wiles and R. Taylor [6078–6080, 6668] (cf. D. Blasius, J. Rogawski [550]). The same problem in the case of imaginary quadratic fields has been considered by M. Harris, D. Soudry, R. Taylor [2571], D. Ramakrishnan [5068] and R. Taylor [6081–6084].

J.-P. Serre [5644] and H.P.F. Swinnerton-Dyer [6002, 6003] showed that for certain small weights the image of ρ_ℓ is for sufficiently large primes ℓ as large as possible. This has been generalized to all weights by K. Ribet [5184, 5185].

3. All the main results of the Hecke–Petersson theory of modular forms have been transferred to Siegel modular forms of degree $g = 2$ and any weight k by A.N. Andrianov [93, 94] in 1974 (for the case of congruence subgroups see S.A. Evdokimov [1910, 1911] and I. Matsuda [4196]). In particular A.N. Andrianov defined the zeta-function of a Siegel modular form f by an Euler product. The case of higher degrees has been studied by S.A. Evdokimov [1913] and M.-H. Kim [3334, 3335].

4. Several authors considered the possibility of lifting modular forms in one variable to Hilbert modular forms (see, e.g., H. Cohen [1139], M. Eichler [1703], S.S. Kudla [3558], H. Naganuma [4500], D. Zagier [6809]).

In 1977 H. Saito and N. Kurokawa [3585] conjectured the existence of a map $f \mapsto l_k(f)$ from $S_{2(k-1)}$ (for even $k \geq 4$) to the space of Siegel modular forms of weight k with respect to $Sp_2(\mathbf{Z})$, preserving the property of being an eigenfunction of Hecke operators. They also predicted that the ratio of the zeta-functions of f and $l_k(f)$ equals

$$\zeta(s-k+1)\zeta(s-k+2).$$

N. Kurokawa conjectured also (conjecture 2 in [3585]) the equality of $l_k(S_{2(k-1)})$ and the subspace M_k of the space of Siegel modular forms of dimension 2 and weight k, satisfying a certain condition on their Fourier coefficients, introduced by H. Maass (*Maass forms*). In a series of papers H. Maass [4046–4048] showed that the dimensions of M_k and $S_{2(k-1)}$ coincide and exhibited a map $M_k \to S_{2(k-1)}$ for which the condition on the ratio of zeta-functions was satisfied under some assumptions on the Fourier coefficients, which were soon removed by A.N. Andrianov [95].

The conjecture was finally established by D. Zagier [6810], who used an earlier result by W. Kohnen [3440]. One finds a proof of it in the book by M. Eichler and D. Zagier [1706] in which *Jacobi forms* were introduced which are functions of two complex variables, one of them lying in the upper half-plane. They behave in one variable like modular forms, whereas in the other variable they imitate elliptic functions. The Weierstrass $\wp$-function and Jacobi's theta-functions serve as examples of these forms. It follows from results by G. Shimura [5703] that Jacobi forms are related to modular forms of half-integral weight. For a generalization to higher degrees see C. Ziegler [6831]. A characterization of Jacobi forms by functional equations of corresponding L-functions, introduced by R. Berndt [458], has been given by Y. Martin [4162, 4163].

The analogue of the Saito–Kurokawa conjecture for forms of higher levels was established in 1993 by M. Manickam, B. Ramakrishnan and T.C. Vasudevan [4129].

For further developments on the theory of Maass forms see S. Breulmann [719] and S.A. Evdokimov [1912]. For properties of the Andrianov zeta-functions see T. Oda [4653].

5. In 1963 H. Shimizu [5690, 5691] considered automorphic functions on the product of n upper complex half-planes under the action of a discrete subgroup G of n copies of the group of analytic automorphisms of that plane. Under certain assumptions on the fundamental domain for the group G he determined the dimensions of the spaces of cusp forms, and applied this in particular in the case when G is the Hilbert modular group.

6. In 1972 the state the of art in the theory of modular functions was presented at a Summer School, held in Antwerp. It was followed by a conference, held in Bonn in 1976. The six volumes [531, 1450, 3567, 3568, 5663, 5664] of proceedings contain an overview of the theory.

Expositions of the theory of modular forms were given in the book by S. Lang [3696], R.A. Rankin [5125] and Schoeneberg [5549].

Later the books by H. Petersson [4821], F. Diamond and J. Shurman [1520] appeared. Presentations of the modern theory of modular and, more generally, automorphic forms were given by H. Iwaniec [3060, 3061].

6.4 Diophantine Approximations and Transcendence

6.4.1 Diophantine Approximations

1. A very important result concerning products of linear forms was obtained in 1971 by W.M. Schmidt [5500, 5501, 5503] (see also the books by W.M. Schmidt [5513, 5523] and E. Bombieri, W. Gubler [614]).

The subspace theorem *If $L_1, \dots, L_n$ is a system of linear forms in n variables with algebraic coefficients, then for every positive δ there is a finite set $\{T_1, \dots, T_m\}$ of proper subspaces of $\mathbf{Q}^n$ with the following property.*

If $\bar{x} \in \mathbf{Z}^n$ satisfies

$$\left| \prod_{j=1}^{n} L_j(\bar{x}) \right| < \frac{1}{|\bar{x}|^{\delta}},$$

then $\bar{x}$ lies in the union $\bigcup_{i=1}^{m} T_i$.

A p-adic version of this result was proved in 1976 by H.P. Schlickewei [5459–5461].

Another proof of Schmidt's theorem and Schlickewei's generalization was given by G. Faltings and G. Wüstholz [1956].

For quantitative versions of the subspace theorem in which a bound for the number of subspaces is given see W.M. Schmidt [5521, 5525], H.P. Schlickewei [5464], P. Vojta [6452]. For the p-adic version and in the case of normal extensions this has been done by H.P. Schlickewei [5464, 5467], who applied his result to generalize (5.27) to the p-adic case. An extension to arbitrary number fields was obtained in 1996 by J.-H. Evertse [1926].

A bound not depending on the discriminant of the field was obtained in 2002 by J.-H. Evertse and H.P. Schlickewei [1939]. Another extension was proved by M. Ru and P. Vojta [5327].

In 2007 B. Adamczewski and Y. Bugeaud [13] applied the subspace theorem to show that $q \geq 2$ is an integer and $\rho(n)$ denotes the number of distinct n-sequences appearing in the q-ary expansion of an irrational algebraic number, then

$$\lim_{n\to\infty} \frac{\rho(n)}{n} = \infty.$$

Earlier only the relation

$$\lim_{n\to\infty} (\rho(n) - n) = \infty$$

was known (S. Ferenczi, C. Mauduit [1987]). For an effective version of the last result see Y. Bugeaud [813]. It has been conjectured that for every irrational real algebraic number one has $\rho(n) = q^n$, hence in particular every such number is absolutely normal.

The subspace theorem has also been applied to prove the transcendence of various infinite sums (see, e.g., B. Adamczewski, Y. Bugeaud [14], P. Bundschuh, A. Pethő [847], P. Corvaja, U. Zannier [1253]). The last two authors used the subspace theorem in [1254] to give a proof of Siegel's finiteness theorem.

A survey of various applications of the subspace theorem is given by Y.F. Bilu [513].

The analogue of the subspace theorem for function fields was established by M. Ratliff [5131].

For effective versions see R.C. Mason [4169–4173], J.T.-Y. Wang [6543] and T.T.H. An and J.T.-Y. Wang [70].

A survey of the theory of approximations of algebraic numbers was published in 1971 by W.M. Schmidt [5502].

2. The length $l(r)$ of the continued fraction expansions of a rational number r was considered in 1969 by H. Heilbronn [2713], who proved

$$\frac{1}{\varphi(N)} \sum_{j<N,(j,N)=1} l(j/N) = \frac{12\log 2}{\pi^2} \log N + O\left(\log\log^4 N\right), \tag{6.19}$$

and J.D. Dixon [1597] showed that for every $\delta > 1/2$ and almost all pairs $j, N \leq x$ one has

$$\left| l(j, N) - \frac{12\log 2}{\pi^2} \log N \right| < \log^{\delta} N.$$

Note that $l(j/N)$ also counts the number of steps in the Euclidean algorithm for the pair j, N.

The error term in Heilbronn's result was later reduced by T. Tonkov [6193, 6194], and J.W. Porter [4994] replaced the error term in (6.19) by $c + O(N^{-a})$ with a certain constant c and any $a < 1/6$. An explicit form of c determined by J.W. Wrench, Jr. can be found in the paper by D. Knuth [3420].

See also V. Baladi, B. Vallée [271], V.A. Bykovskiĭ [876], D. Hensley [2753] on this topic.

3. In 1964 A. Baker [220] used hypergeometric series to show that in certain cases the constants in Thue–Siegel–Roth theorem can be made explicit, and later [221] applied the same method to obtain lower bounds for rational approximations of radicals. In particular he proved the inequality

$$\left| \sqrt[3]{2} - \frac{p}{q} \right| > \frac{1}{10^6 q^a}$$

with $a = 2.955$, and in [228, 229] he obtained the first effective improvement of Liouville's theorem by showing that if α is an algebraic number of degree $n \geq 3$, and $\varepsilon > 0$, then with an effective $c > 0$, depending on α and ε one has

$$\left| \alpha - \frac{p}{q} \right| > \frac{c \exp(\log^{1/(1+n+\varepsilon)} q)}{q^n}$$

for all integers $p, q > 0$.

Further studies of approximations of radicals were made by E. Bombieri [603], E. Bombieri and J. Mueller [617, 618] and G.V. Chudnovsky [1108]. It was pointed out in [617] that in certain cases, similar although weaker, results of this style had already been obtained by A. Thue [6145]. See also H. Davenport [1379], C.F. Osgood [4702].

4. In 1960 an important notion was introduced by K. Mahler [4089], now called the *Mahler measure* $M(f)$, defined for a complex polynomial $f(X)$ by

$$M(f) = \exp\left(\int_0^1 \log \left| f\left(e^{2\pi i t}\right) \right| dt \right).$$

It follows from Jensen's theorem (which is actually due to C.G.J. Jacobi [3074]) that $M(f)$ is related to the zeros of f by the formula

$$M(f) = |a_n| \prod_{i=1}^{n} \max\{1, |z_i|\}, \tag{6.20}$$

if $f(X) = a_n \prod_{i=1}^{n} (X - z_i)$.

An old result of L. Kronecker's [3525] implies that if f is an irreducible monic polynomial with integral coefficients and $M(f) = 1$, then either f is cyclotomic or $f(X) = X$. In the case of monic polynomials the product (6.20) had already been considered in 1933 by D.H. Lehmer [3778], who asked for non-cyclotomic monic polynomials $f \neq x$ having $M(f)$ arbitrarily close to 1. Now the assertion that there exists a positive constant c such that $M(f) \geq 1 + c$ holds for non-cyclotomic $f \in \mathbf{Z}[X]$, $f(X) \neq X$, assumed to be monic and irreducible, is called the *Lehmer conjecture*. This is now known to be true for f non-reciprocal[5] (R. Breusch [721], C.J. Smyth [5838]), or if it has many real roots (P.E. Blanksby [548]).

In the general case the inequality

$$M(f) \geq 1 + \frac{1}{52n \log(6n)}$$

(with $n = \deg f$) was established by P.E. Blanksby and H.L. Montgomery [549] in 1971, and this was improved in 1978 by E. Dobrowolski [1600] to

$$M(f) \geq 1 + c(n) \left(\frac{\log \log n}{\log n} \right)^3,$$

with $c(n) = 1 + o(1)$.

The value of $c(n)$ was later improved by D.C. Cantor, E.G. Straus [891], U. Rausch [5133] and R. Louboutin [3993], who obtained $c(n) = 2.25 + o(1)$. The best bound valid for all $n \geq 2$ is due to P.M. Voutier [6482], who showed that $c(n) \geq 1/4$. For numerical results see D.W. Boyd [670, 671], M.J. Mossinghoff [4444].

Later this topic was treated by G. Rhin and J.-M. Sac-Épée [5174], V. Flammang, G. Rhin and J.-M. Sac-Épée [2010] and M.J. Mossinghoff, G. Rhin and Q. Wu [4445].

Lower bounds for Mahler's measure $M(f)$ which are independent of the number of non-zero coefficients of f were obtained by E. Dobrowolski, W. Lawton and A. Schinzel [1603] and E. Dobrowolski [1601, 1602].

It was shown in 2007 by P. Borwein, E. Dobrowolski and M.J. Mossinghoff [651] that Lehmer's assertion holds for polynomials having all coefficients congruent to unity modulo an integer ≥ 2. Cf. N.C. Bonciocat [625], P. Borwein, K.G. Hare, M.J. Mossinghoff [652], A. Dubickas, M.J. Mossinghoff [1631].

A closely related question concerns the *house* $\overline{|a|}$ of an algebraic integer a, defined as the product of all conjugates of a, lying outside the unit circle. It was shown in 1965 by A. Schinzel and H. Zassenhaus [5455] that if a is not a root of unity and has $2s$ non-real conjugates, then

$$\overline{|a|} > 1 + \frac{1}{16 \cdot 4^s}.$$

They conjectured the existence of a constant $c > 0$ such that if $\deg a = n \geq 2$ and a is not a root of unity, then

$$\overline{|a|} \geq 1 + \frac{c}{n}. \tag{6.21}$$

(For a reformulation see C.G. Pinner and J.D. Vaaler [4880].)

[5] A polynomial f of degree n is said to be *reciprocal* if $f(X) = \pm X^n f(1/X)$.

The inequality (6.21) would follow from the truth of Lehmer's conjecture and it has been proved by J.W.S. Cassels [943] in the case when the minimal polynomial of a is non-reciprocal. P.E. Blanksby and H.L. Montgomery [549] proved

$$\overline{|a|} \geq 1 + \frac{1}{30n^2 \log(6n)},$$

and E. Dobrowolski improved this first [1599] to

$$\overline{|a|} \geq 1 + \frac{\log n}{6n^2},$$

and then [1600] to

$$\overline{|a|} \geq 1 + \frac{2 + o(1)}{n} \left(\frac{\log \log n}{\log n} \right)^3. \tag{6.22}$$

In 1991 E.M. Matveev [4201] showed

$$\overline{|a|} \geq (d + 1/2)^{1/d^2},$$

and in 1993 A. Dubickas [1628] replaced the coefficient 2 in (6.22) by $64/\pi^2 = 6.48\ldots$ (cf. A. Dubickas [1629, 1630]).

For further improvements see A. Dubickas, M.J. Mossinghoff [1631] and G. Rhin, Q. Wu [5176].

6.4.2 *Uniform Distribution*

1. In 1961 I. Niven [4621] made the first systematic study of uniform distribution of integer sequences in residue classes, which is a special case of uniform distribution of sequences in compact Abelian groups. The Weyl-type criterion derived by Niven was later applied by H. Delange [1431] to characterize integer-valued additive functions whose values are uniformly distributed in residue classes mod N.

In the case of polynomial sequences this question coincides with the problem of determining *permutation polynomials* mod N (i.e., polynomials inducing a permutation of the set of all residue classes mod N) considered in the case of prime N already by L.E. Dickson in his thesis [1535].

An important result concerning this class of polynomials was established in 1970 by M. Fried [2095] who proved a conjecture, stated in 1923 by I. Schur [5575]. I. Schur considered polynomials $f \in \mathbf{Z}[X]$ which are permutation polynomials for infinitely many primes p and showed that if the degree of f is a prime number, then f can be written as a composition of binomials $\alpha X^n + \beta$ and Čebyšev polynomials $T_n(X)$, defined by the relation $T_n(\cos t) = \cos(nt)$. He conjectured that the same assertion holds for polynomials of any degree, having coefficients in a fixed finite extension K of the rationals, in this case the primes being replaced by prime ideals of K. M. Fried proved this conjecture, using the Riemann Hypothesis for function fields and the theory of Riemann surfaces. Later elementary proofs were given by

G. Turnwald [6240] and P. Müller [4473]. See also the paper [6238] by G. Turnwald, where for a given polynomial f with integral coefficients in an algebraic number field K the set $S(f)$ of all ideals I of $\mathbf{Z}_K$ such that the residue classes mod I are permuted by f was described.

A classification of rational functions $f \in \mathbf{Q}[X]$ whose reductions mod p induce permutations of $\mathbf{Z}/p\mathbf{Z}$ for infinitely many primes p was obtained in 2003 by R.M. Guralnick, P. Müller and J. Saxl [2395].

It was conjectured in 1966 by L. Carlitz that if f is a polynomial over a finite field $\mathbf{F}_q$ (with $q = p^n$, $p \geq 3$, prime) which is a permutation polynomial in infinitely many finite extensions of $\mathbf{F}_q$, then $2 \nmid \deg f$. This was established in 1993 by Fried, R. Guralnick and J. Saxl [2098], who also showed that in the case $p \geq 5$ every such f is a composite of Chebyshev polynomials, cyclic polynomials and polynomials whose degree is a power of p.

For a survey of the theory of permutational polynomials see the book by R. Lidl and H. Niederreiter [3888, Chap. 7].

2. Criteria for the uniform distribution mod N of second order linear recurrences were given by R.T. Bumby [841] and W.A. Webb and C.T. Long [6597] (the case of prime N was also treated by M.B. Nathanson [4546] and of prime power N by P. Bundschuh and P.J.-S. Shiue [848]). In particular the sequence of Fibonacci numbers is uniformly distributed mod N if and only if N is a power of 5 (L. Kuipers, P.J.-S. Shiue [3571], H. Niedereiter [4598]).

These results have been extended to rings of integers in algebraic number fields and, more generally, to Dedekind domains by R.F. Tichy and G. Turnwald [6154] and G. Turnwald [6236, 6237].

The case of recurrences of orders 3 and 4 was considered by M.J. Knight, W.A. Webb [3410] and H. Niedereiter, P.J.-S. Shiue [4604, 4605].

3. A sequence of integers a_n is said to be *weakly uniformly distributed* mod N if it contains infinitely many integers prime to N, and for k prime to N all limits

$$\lim_{x\to\infty} \frac{\#\{n \leq x : a_n \equiv k \pmod{N}\}}{\#\{n \leq x : (a_n, N) = 1\}}$$

exist and do not depend on k. It was shown by J.-P. Serre [5648] that the sequence of values of Ramanujan's function $\tau(n)$ is weakly uniformly distributed mod N if and only if either N is odd and not divisible by 7 or N is even and $(N, 3 \cdot 23) = 1$ (see [4537], where also one finds similar results for a class of multiplicative functions).

Criteria for weak uniform distribution of second order recurrences were given by G. Turnwald [6239]. A criterion for recurrences of the form $u_{n+1} = au_n + b$ was found by R.F. Tichy and G. Turnwald [6155, 6239].

In the seventies two books dealing with uniform distribution appeared, authored by L. Kuipers and H. Niedereiter [3569] and E. Hlawka [2834]. The subsequent research was described in the book [1626] by M. Drmota and R.F. Tichy.

6.4.3 Transcendence and Rationality

1. It was proved by A.O. Gelfond [2226, 2230] in 1939 that if α_1, α_2 are algebraic numbers $\neq 0, 1$, whose logarithms are linearly independent over the rationals, then for every pair of non-zero algebraic numbers β_1, β_2 one can give an effective lower bound for the linear form $|\beta_1 \log \alpha_1 + \beta_2 \log \alpha_2|$. In particular Gelfond showed that if $H(\theta)$ denotes the height of an algebraic number θ, then for sufficiently large H one has for all θ with $H(\theta) \leq H$ the inequality

$$\left| \frac{\log \alpha_1}{\log \alpha_2} - \theta \right| > c_1 \exp(-\log^c H)$$

for every $c > 3$, with c_1 being a positive constant, depending on α_1, α_2. He conjectured also the existence of similar results for linear forms in logarithms in three and more variables. Important applications of Gelfond's result to linear recurrences, Diophantine equations and prime divisors of polynomials were given later by A. Schinzel [5439], who established in particular that if f is a quadratic polynomial in $\mathbf{Z}[X]$ having distinct roots, then the maximal prime divisor of $f(x)$ is $\gg \log \log x$.

In the case of more logarithms, A.O. Gelfond and Yu.V. Linnik [2233] showed in 1948 that if $\alpha_1, \alpha_2, \ldots, \alpha_n$ are algebraic and their logarithms are linearly independent over the rationals, then for every $\varepsilon > 0$ the inequality

$$\left| \sum_{j=1}^{n} x_j \log(\alpha_j) \right| < \exp\left(-\varepsilon \max_j |x_j| \right) \tag{6.23}$$

can have only finitely many integral solutions; however, their result was not effective.

In a sequence of four papers [223–226] in 1966 A. Baker confirmed Gelfond's conjecture providing effective lower bounds for non-zero linear combinations of logarithms of algebraic numbers, which found important applications in several problems.

He considered linear forms

$$\Lambda = \sum_{j=1}^{n} \beta_j \log \alpha_j \neq 0, \tag{6.24}$$

where $\alpha_i \neq 0, 1$ are algebraic numbers, $\beta_j \neq 0$ are algebraic and of degree $\leq d$ and $H = \max_j H(\beta_j)$, and proved the inequality

$$|\Lambda| > C \exp\left(-\log^k H \right), \tag{6.25}$$

with explicitly given $C = C(d, n, \alpha_1, \ldots, \alpha_n)$. In [223] he obtained this for every $k > n + 1$ and relaxed this condition to $k > n$ in [225]. In [226] the right-hand side of (6.25) was replaced in the case of rational β_i's by $C \exp(-\delta H)$ for every $0 < \delta < 1$, this time C depending also on δ. A. Baker also obtained a similar result in the inhomogeneous case, bounding effectively from below non-zero sums of the form

$$\beta_0 + \sum_{j=1}^{n} \beta_j \log \alpha_j. \tag{6.26}$$

All these bounds were later improved and this allowed important progress to be made in several problems. The first such results were given by N.I. Feldman [1977–1980], and the next improvements were obtained by A. Baker himself [234] and H.M. Stark [5902, 5903]. This led to progress in the class-number problem which will be presented in the next section. The next important sharpening came again from A. Baker [235–237], whose result in the case when the numbers a_i, β_i in (6.25) are rational, gives

$$|\Lambda| > \exp(-\log A \log H \log C),$$

where $A \geq \max\{4, H(\alpha_1), \dots, H(\alpha_n)\}$, and C depends on n and the maximal value of $H(\alpha_i)$ for $1 \leq i \leq n-1$. This was improved by Baker and H.M. Stark [244]. A survey of further improvements up to the year 1976 was given by A. Baker [238, 239]. Further progress was later made by A.J. van der Poorten [6301].

In the following years there were many improvements of the bounds for sums (6.24) and (6.26) (A. Baker, G. Wüstholz [245], C.D. Bennett, J. Blass, A.M.W. Glass, D.B. Meronk, R.P. Steiner [406], J. Blass, A.M.W. Glass, D.K. Manski, D.B. Meronk, R.P. Steiner [551, 552], E.M. Matveev [4202, 4203] and M. Waldschmidt [6501, 6506]).

In the case $n = 2$ various explicit bounds were given by M. Mignotte and M. Waldschmidt [4297–4299], M. Laurent, M. Mignotte and Yu.V. Nesterenko [3723] and N. Gouillon [2291].

2. The first important application of Baker's bounds appears in [223], where it is shown that if $\beta_1, \dots, \beta_n$ are real algebraic numbers, forming with 1 a set linearly independent over the rationals and $\alpha_1, \dots, \alpha_n$ are positive algebraic numbers distinct from 0 and 1, then the number

$$\prod_{i=1}^{n} \alpha_i^{\beta_i}$$

is transcendental. This generalized the Gelfond–Schneider theorem. An equivalent form of this theorem states that $\mathbf{Q}$-linearly independent logarithms of algebraic numbers are linearly independent over the field of algebraic numbers (a simpler proof of this result can be found in the lectures [6505] of M. Waldschmidt).

In [224] the assumption of reality of the β_i's was removed, and the results of [225] implied the transcendence of $\pi + \log \alpha$ (with algebraic $\alpha \neq 0$) and also of $\exp(\alpha\pi + \beta)$ for algebraic α and $\beta \neq 0$.

For applications of Baker's method to Diophantine equations see Sect. 6.6.

3. Around 1960 S.H. Schanuel (see [3691, pp. 30–31; 3694]) conjectured that if the complex numbers $a_1, a_2, \dots, a_n$ are linearly independent over the rationals, then the transcendence degree of the field

$$\mathbf{Q}(a_1, a_2, \dots, a_n, \exp(a_1), \dots, \exp(a_n))$$

is $\geq n$. If true, this would have many interesting corollaries, e.g., the algebraic independence of π and e.

A discussion of the various consequences of Schanuel's conjecture is contained in the survey paper by M. Waldschmidt [6509]. In 2001 Roy [5323] showed the equivalence of this conjecture with an algebraic statement regarding polynomials.

Similar conjectures for fields of power series and differential fields, also stated by D.H. Schanuel, were proved in 1971 by J. Ax [177, 178].

The question posed by T. Schneider in [5541] of whether at least one of the numbers e^e, e^{e^2} is transcendental was answered in the positive by M. Waldschmidt [6496, 6497] and W.D. Brownawell [756] (see [6498] for an exposition). Cf. also Šmelev [5829].

4. The following result, known as the *six exponentials theorem* was established by S. Lang [3692] and K. Ramachandra [5057].

Six exponentials theorem *If x_1, x_2 and y_1, y_2, y_3 are **Q**-linearly independent sets of complex numbers, then at least one of the numbers* $\exp(x_i y_j)$ $(i = 1, 2; j = 1, 2, 3)$ *is transcendental.*

This result also follows from T. Schneider's results in [5540] and a special case of it has been attributed to C.L. Siegel by L. Alaoglu and P. Erdős [41].

It was conjectured in [3691] and [5056, 5057] that the same assertion holds also in the case that the second set has only two elements (*four exponentials conjecture*). For extensions of the six exponentials theorem see D. Roy [5322], D. Roy and M. Waldschmidt [5326], M. Waldschmidt [6510].

In 1979 M. Waldschmidt [6500] considered values of analytical homomorphisms φ between two algebraic subgroups defined over the algebraic closure of **Q** and gave criteria for transcendence and algebraic independence of values of φ. This approach allowed several transcendence statements to be obtained on values of various special functions.

Later this idea was pursued by D.W. Masser and G. Wüstholz [4178, 4179], J.-C. Moreau [4408, 4409], P. Philippon [4835–4837], M. Waldschmidt [6502, 6504].

In particular it has been proved by P. Philippon [4835, 4836] and G. Wüstholz [6762] that if $a_1, \ldots, a_n$ are algebraic numbers linearly independent over an imaginary quadratic field k, and $\wp(z)$ is the Weierstrass elliptic function with algebraic invariance and complex multiplication by k, then the values $\wp(a_1), \ldots, \wp(a_n)$ are algebraically independent. This was known earlier for $n = 1$ (T. Schneider [5539]) and $n = 2, 3$ (G.V. Chudnovsky [1106]). A quantitative version of this result was obtained by Yu.V. Nesterenko [4565, 4568]. D.W. Masser and G. Wüstholz [4181] considered fields generated by values of $\wp(z)$ in the case, when there is no complex multiplication and proved lower bounds for the transcendence degrees. For an improvement see M. Takeuchi [6039].

An exposition of transcendence problems in the theory of elliptic functions was given by D.W. Masser [4174] in 1975.

5. In 1969 K. Mahler [4091] asked whether for all non-zero algebraic numbers q inside the unit disc the value $J(q)$, defined by (4.67) is transcendental. Earlier T. Schneider [5539] established the transcendence of numbers $j(z)$ for every z in the upper half-plane which does not lie in an imaginary quadratic field (if it does then it has long been known that $j(z)$ is an algebraic integer).

Mahler's conjecture can be also stated using properties of the Weierstrass $\wp$-function. Let ω_1, ω_2 be complex numbers, linearly independent over the reals, and let

$$\Lambda = \{x\omega_1 + y\omega_2 : x, y \in \mathbf{Z}\}$$

be the lattice generated by them. The Weierstrass function is defined by

$$\wp(z) = \frac{1}{z^2} + \sum_{\substack{\lambda\in\Lambda \\ \lambda\neq 0}} \left(\frac{1}{(z-\lambda)^2} - \frac{1}{\lambda_2} \right),$$

and it satisfies the differential equation

$$\wp'(z)^2 = 4\wp(z)^3 - g_2\wp(z) - g_3,$$

where g_2, g_3 are complex numbers, uniquely determined by the choice of ω_i's. The conjecture is equivalent to the statement that if the numbers g_2, g_3 are both algebraic, then

$$\exp\left(\frac{2\pi i\omega_1}{\omega_2}\right)$$

is transcendental.

This conjecture was shown to be true in 1996 by K. Barré-Sirieix, G. Diaz, F. Gramain and G. Philibert [337] (for an effective version see K. Barré [336]). They settled also a p-adic version of this conjecture proposed in 1971 by Yu.I. Manin [4135]. For an analogue in finite characteristics see J.F. Voloch [6466] and D.S. Thakur [6123].

The transcendence of values of modular forms has been studied by G.V. Čudnovsky [1105], who showed that for non-zero z inside the unit disc the transcendence degree of the field generated by $\{z, E_2(z), E_4(z), E_6(z)\}$ equals 2, hence at least two of these numbers are transcendental.

In 1996 Yu.V. Nesterenko [4569] succeeded in replacing 2 by 3 in G.V. Chudnovsky's result (see [4571] for a quantitative version). This implies in particular that $\pi, e^{\pi}, \Gamma(1/4)$ are algebraically independent, as well as $\pi, \exp(\pi\sqrt{D})$ for natural D. His paper contains also several other transcendence results.

See Yu.V. Nesterenko [4572] for a simplification of the proof. For expositions see V. Bosser [656] and M. Waldschmidt [6507].

It was shown by J.O. Shallit [5668, 5669] in 1979 that the values of the function[6]

$$F(x) = \sum_{k=0}^{\infty} \frac{1}{x^{2^k}} \tag{6.27}$$

at positive integers have bounded partial quotients. These were the first examples of simple-looking transcendental numbers (their transcendence was established in 1916 by A.J. Kempner [3299]) with the last property (cf. A. Pethő [4823], T. Wu [6761]). For connections with automata theory and paper-folding see J.-P. Allouche, A. Lubiw and M. Mendès France, A.J. van der Poorten and J.O. Shallit [56], A. Blanchard and M. Mendès France [547], H. Cohn [1158], M. Mendès France [4250, 4251], A.J. van der Poorten, J.O. Shallit [6307].

Surveys of the theory of transcendental numbers were presented by N.I. Feldman, A.B. Šidlovskiĭ [1985] in 1967, S. Lang [3694] in 1971 and R. Tijdeman [6160] in 1976. See also the book [1984] by N.I. Feldman and Yu.V. Nesterenko.

The development of the theory of transcendental numbers in the second half of the 20th century has been described by M. Waldschmidt [6507, 6508, 6512].

6. It has been known since Euler, that the values of the zeta-function at even integers are rational multiples of powers of π, which in view of F. Lindemann's theorem implies their transcendence. Much less is known about $\zeta(n)$ for n odd. In June 1978, at the *Journées Arithmétiques* held in Luminy, R. Apéry [110] announced a proof of the irrationality of $\zeta(3)$, and the first complete proofs were published by H. Cohen [1140], E. Reyssat [5172] and A.J. van der Poorten [6302]. Other proofs of the irrationality of $\zeta(3)$ were provided by F. Beukers [482, 484, 488], Yu.V. Nesterenko [4570] and V.N. Sorokin [5847].

Later K. Ball and T. Rivoal [292, 5234] established that $\zeta(2k+1)$ is irrational for infinitely many integers k, and for large n the dimension of the **Q**-linear space spanned by

$$\{1, \zeta(3), \zeta(5), \dots, \zeta(2n+1)\}$$

exceeds $c \log n$, with some positive c. Moreover, as shown by V.V. Zudilin [6842, 6844] at least one of the numbers $\zeta(5), \zeta(7), \zeta(9), \zeta(11)$ is irrational, and also for every m at least one of the values

$$\zeta(2m+1), \zeta(2m+3), \dots, \zeta(16m-9)$$

is irrational [6843]. For a survey see S. Fischler [2007]. It was shown in 2001 by G. Rhin and C. Viola [5175] that for rational p/q with large denominator one has

$$\left|\zeta(3) - \frac{p}{q}\right| > \frac{1}{q^{\alpha}}$$

for every $\alpha > 5.52$, which improved a previous result by M. Hata [2611].

[6] The series (6.27) has been repeatedly called the Fredholm series, but, as pointed out in [5670], this is due to a misunderstanding.

6.5 Gauss's Class-Number Problem

1. The sixties brought the solution of an old problem in the theory of quadratic forms, going back to C.F. Gauss, who in [2208, Sect. 303] conjectured that there are only finitely many negative determinants Δ with a given class-number counting the equivalence classes of quadratic forms $aX^2 + 2BXY + cY^2$ with determinant $\Delta = AC - B^2$ under the action of $SL_2(\mathbf{Z})$. Later this conjecture took the following more general form, in which the middle coefficient of the forms is allowed to be odd.

There are only finitely many negative discriminants d with a given class-number $h(d)$ of quadratic forms of discriminant d.

In this form Gauss's conjecture is closely related to the theory of quadratic fields. If D is a square-free rational integer $\neq 1$ and K_D denotes the field generated by $\sqrt{D}$, then one shows easily that its discriminant $d = d(K_D)$ equals D if D is congruent to unity mod 4 and equals $4D$ otherwise. There is a correspondence between ideals of $\mathbf{Z}_{K_D}$ and quadratic forms of discriminant d which can be used to show that the number $h(K_D)$ of ideal classes in $\mathbf{Z}_{K_D}$ is equal to the class-number of binary quadratic forms of discriminant d. Note that the discriminants of quadratic fields are always fundamental discriminants. It has long been known that the number of ideal classes of an algebraic number field K is equal to 1 if and only if there is a unique factorization law in the ring $\mathbf{Z}_K$. Thus the truth of Gauss's conjecture for discriminants $d < 0$ with $h(d) = 1$ would imply that there are only finitely many imaginary quadratic fields with unique factorization.

2. The first result dealing with Gauss's problem was obtained in 1903 by E. Landau [3621], who used an elementary approach to show that there are only finitely many even negative discriminants with class-number one, namely -4, -8, -12, -16 and -28. Actually this settled the original problem, since, as we have already noted, Gauss only considered forms with even middle coefficient, and this forced the discriminant to be even. Landau's argument consisted of an explicit construction of two inequivalent forms in the remaining cases. He pointed out in the summary of his paper in the *Jahrbuch* [3622] that the same result had already been obtained by P. Joubert [3163, p. 837] in 1860 with the use of elliptic functions. Joubert showed also that if $(h, 6) = 1$, then there are only finitely many even negative discriminants with class-number h. See D. Shanks [5675] for a modern proof.

A simpler proof of Landau's result was given by M. Lerch [3835], who based his argument on the classical formula relating the class-number of quadratic forms of discriminant $-d$ and the class-number of forms of discriminant $-a^2d$.

In 1911 L.E. Dickson [1542] checked that there are no discriminants with class-number one in the interval $[-1\,500\,000, -164]$.

The next step was taken by E. Hecke who in 1918 deduced the truth of Gauss's conjecture from the assumption that for every modulus d the Dirichlet L-functions corresponding to non-principal characters χ mod d do not have zeros in the interval

$(1 - c/\log|d|, 1)$ with a fixed positive c. More precisely, he deduced from this assumption the lower bound

$$h(d) \gg \frac{\sqrt{|d|}}{\log|d|}, \tag{6.28}$$

valid for all negative discriminants d. Hecke's proof was published by E. Landau [3659] who pointed out that a slightly weaker result is hidden in a paper by T.H. Gronwall [2350] published in 1913 (for another proof see L.J. Mordell [4384]). Landau used Hecke's result to show that if there exist infinitely many imaginary quadratic fields violating (6.28), then they must be rare.

3. In 1933 there was important progress due to M. Deuring [1497], who showed that the falsity of the Riemann Hypothesis implies that there are only finitely many discriminants $d < 0$ with $h(d) = 1$. He showed also that if there are infinitely many such discriminants, say $0 > d_1 > d_2 > \cdots > d_n > \cdots$, then

$$|d_{n+1}| > \exp\left(c\sqrt[4]{|d_n|}\right),$$

and S. Chowla [1073] replaced the quartic root by a square root.

Then L.J. Mordell [4385] proved that the falsity of the Riemann Hypothesis implies

$$\lim_{d\to-\infty} h(d) = \infty, \tag{6.29}$$

and finally H. Heilbronn [2708] established (6.29) unconditionally. This implied in particular that the equality $h(d) = 1$ can hold only for finitely many negative discriminants d, but to have a complete answer to Gauss's question one had to give a list of them. H. Heilbronn used his method in a joint paper with E.H. Linfoot [2718] to show that apart from the nine known discriminants with this property there can be at most one more, and its existence would contradict the General Riemann Hypothesis. E. Landau [3679] made Heilbronn's result more precise by showing that for every h there can be at most one negative discriminant $-d$ with $h(-d) = h$ and

$$d \le Bh^8 \log^6(3h).$$

This result was later improved by T. Tatuzawa who proved in 1951 [6066] that there can be at most one discriminant $d < 0$ with $|d| > 2100h^2\log^2(13h)$ and class-number h.

4. In 1931 B.A. Venkov [6383] found an elementary proof for Dirichlet formulas for the class-number of positive definite binary quadratic forms f in the case when the discriminant of f is of the form $4D$ with square-free $D < -3$, not congruent to unity mod 8. This also affects the corresponding formulas for the class-number of imaginary quadratic fields.

An elementary proof in the case of arbitrary imaginary positive definite binary quadratic forms was given[7] in 1978 by H.L.S. Orde [4698]. For further discussion see [4538, Chap. 5].

5. The final answer for discriminants with class-number one was given in 1967 by H.M. Stark [5897], who proved that the set of these discriminants equals $\{-3, -4, -7, -8, -11, -19, -43, -67, -163\}$. Earlier he showed [5895] that there are no other such discriminants of absolute value larger than $\exp(2.2 \cdot 10^7)$, improving upon the bound $5 \cdot 10^9$ obtained in 1933 by D.H. Lehmer [3777]. Actually the same result was published by K. Heegner [2704] in 1952, however his proof was for a long time regarded as incomplete, as it was based on certain unclear assertions in the book by H. Weber [6602]. The work of B.J. Birch [527], M. Deuring [1511] and H.M. Stark [5901] showed later that Heegner's arguments were sound.

The proof of Gauss's conjecture given by Stark used two series of the form

$$\sum_{(x,y)\neq(0,0)} \frac{\chi(Q(x,y))}{Q(x,y)^s},$$

where χ was a Dirichlet character mod 8 or mod 12, and Q was a positive definite binary quadratic form. A general theory of such functions, generalizing Epstein's zeta-function, and defined by

$$L(s,\chi,Q) = \sum_{\mathbf{x}\neq\mathbf{0}} \frac{\chi(Q(\mathbf{x}))}{Q(\mathbf{x})^s},$$

where χ is a Dirichlet character mod k, $\mathbf{x} = (x_1, \dots, x_n) \in \mathbf{Z}^n$ and Q a positive definite quadratic form in n variables and integral coefficients, was presented by Stark [5899, 5900] in 1968. This series converges in the half-plane $\Re s > n/2$ and if the discriminant of Q is prime to k and χ is primitive, then L can be extended to an entire function, satisfying a certain functional equation. Moreover $L(s,\chi,Q)$ can be represented as a quickly converging series. In 1971 Stark [5904] gave an explicit formula for $L(1,\chi,Q)$ in the case when χ is the Kronecker symbol, and stated a conjecture about a unit appearing in that formula. This conjecture was established by R. Schertz [5427, 5428] in 1973 (cf. K.H. Rosen [5285, 5286]). In [5905, 5906] H.M. Stark considered a similar question for a class of Artin L-functions.

Similar series with χ being an additive character were studied by T. Callahan and R.A. Smith [887].

An effective bound for negative discriminants with class-number one[8] also follows from Baker's theorem on lower bounds for linear forms in logarithms. The same approach led also to effective bounds for negative discriminants with class-number two, obtained independently by H.M. Stark [5902, 5903] and A. Baker

[7] The paper by Orde earned a hostile review in Math. Reviews (80a:10036), amended later by the editors.

[8] It had already been noted by A.O. Gelfond and Yu.V. Linnik [2233] in 1948 that effectivization of the inequality (6.23) leads to such bound.

[234]. Stark's paper gave the bound $|d| < 10^{4100}$ and Baker did not give any numerical bound. The complete list of such discriminants was given by H.M. Stark [5909] in 1975.

Later the same method was used to prove Stark's result on class-number one (P. Bundschuh, A. Hock [846]). Soon other proofs were found by J.M. Cherubini and R.V. Wallisser [1045], N.I. Feldman and N.G. Čudakov [1983], C. Meyer [4280] and C.L. Siegel [5772].

Earlier, in 1969, A. Baker [233] (see also M.A. Kenku [3300] and P.J. Weinberger [6632]) obtained a rather large effective bound for negative discriminants not congruent to 5 mod 8 having class-number two. It turned out two years later, that use of the theory of continued fractions can reduce the computational effort (W.J. Ellison, J. Pesek, D.S. Stall, W.F. Lunnon [1757]), and this led to a complete list of such fields (there are eight of them).

We have noted already, while considering idoneal numbers, that there are only finitely many negative quadratic discriminants with class-group being an elementary 2-group. The same assertion for elementary 3-groups was proved in 1973 by D.W. Boyd and H. Kisilevsky [672] and P.J. Weinberger [6633]. They showed also that the General Riemann Hypothesis implies that for the exponent $m(d)$ of the class-group of the field $\mathbf{Q}(\sqrt{-d})$ one has

$$m(d) \gg \frac{\log(|d|)}{\log\log(|d|)}.$$

F. Pappalardi [4740] proved later that the last inequality holds unconditionally for almost all d.

The finiteness of the set of discriminants $-d < 0$ with $m(d) = 2^k$ has been established for $k = 2$ by A.G. Earnest and D.R. Estes[9] [1679], and for any k by A.G. Earnest, and O. Körner [1680].

Recently D.R. Heath-Brown [2661] established the finiteness of the sets of discriminants with $m(d) = 5$ and $m(d) = 3 \cdot 2^k$.

An important step towards an effective determination of negative discriminants d with a given class-number $h(d)$ was taken by D. Goldfeld [2258], who related this problem to a question in the theory of elliptic curves. Specifically, he showed that if there is an elliptic curve whose L-function has at $s = 1$ a zero of order ≥ 3, then for every $\varepsilon > 0$ one has

$$h(-d) \leq B(\varepsilon) \log^{1-\varepsilon} d$$

with an effective constant $B(\varepsilon)$.

Ten years later the work of B.H. Gross and D. Zagier [2356, 2357] overcame this obstacle, and this led to an explicit inequality, which after amendments by J. Oesterlé [4662] took the form

$$h(d) \geq \frac{\log(|d|)}{55} \prod_{p|d,\, p<|d|} \left(1 - \frac{2\sqrt{p}}{p-1}\right).$$

[9] Dennis Ray Estes (1941–1999), professor at the University of South California.

This quickly led to the determination of discriminants with $h = 3$, and in the next twenty years all negative discriminants with small class-numbers were listed: the case $h = 4$ was treated in S. Arno's thesis [128], discriminants with $h = 5, 6, 7$ were listed by C. Wagner [6488] and the case $h \leq 23$ was resolved by S. Arno, M. Robinson and F. Wheeler [129].

All discriminants with $h \leq 100$ were given in M. Watkins's thesis [6577].

The question of whether there are infinitely many positive discriminants d with $h(d) = 1$ is still unanswered, however it has been shown by G. Lachaud [3600] that under the General Riemann Hypothesis one can list all such fields with discriminant of the form $4x^2 + 1$ (cf. R.A. Mollin, H.C. Williams [4349], R.A. Mollin [4347]).

In 2003 A. Biró [543, 544] proved unconditionally that for square-free $d = 4n^2 + 1$ and $d = n^2 + 4$ (with n square-free) the class-number exceeds one for $n > 13$ and $n > 17$, respectively, confirming the conjectures of S. Chowla [1097] and H. Yokoi [6781]. Later D. Byeon, M. Kim and J. Lee [874] proved that the same applies to $d = n^2 - 4$ for $n > 21$. In 2008 D. Byeon and J. Lee [875] determined all even square-free $d = n^2 + 1$ with $h(d) = 2$. This had been done earlier by R.A. Mollin and H.C. Williams [4350] under the assumption of the General Riemann Hypothesis. In 2009 J. Lee [3758] determined all square-free $d = n^2 \pm 2$ with $h(d) = 1$.

6.6 Diophantine Equations and Congruences

1. Baker's method turned out to be an extremely useful tool in the theory of Diophantine equations. The first such result was obtained by A. Baker [228], who gave an effective proof of the theorem of Thue by proving that if $F(X, Y)$ is an irreducible form with integral coefficients of degree $n \geq 3$, then all integral solutions x, y of the equation

$$F(x, y) = m \tag{6.30}$$

satisfy

$$\max\{|x|, |y|\} \leq C(F, \varepsilon) \exp\left(\log^{1+n+\varepsilon} m\right)$$

for every $\varepsilon > 0$, with an explicit constant $C(F, \varepsilon)$.

Later these bounds were improved several times, and the best known results were obtained by Y. Bugeaud [810]. Similar bounds for solutions of (6.30) which are prime to a given integer were given by J. Coates [1123–1125]. In the second of these papers he obtained for the first time a quantitative formulation of Mahler's result [4068–4070] on the maximal prime divisor of values of a form f of degree $n \geq 3$ with integral coefficients, by showing that if $(x, y) = 1$, then the maximal prime divisor of $f(x, y)$ exceeds

$$\left(\frac{\log\log X}{(10n)^8 \log h(F)}\right)^{1/4},$$

where

$$X = \max\{|x|, |y|\},$$

and $h(F)$ is the height of F, i.e., the maximal modulus of its coefficients. In the third paper he showed that the maximal prime divisor of the difference $x^3 - y^2$ for co-prime x, y exceeds $(\log\log X)^{1/4}/1000$.

Effective methods of solving Thue equations were given by A. Pethő and R. Schulenberg [4825], Y.F. Bilu and G. Hanrot [515, 517], G. Hanrot [2498], and N. Tzanakis and B.M.M. de Weger [6243]. See also R.J. Stroeker and N. Tzanakis [5973], who used T. Skolem's p-adic method [5807]. An algorithm in the case of rings of algebraic integers was given by N.P. Smart [5825].

The first complete solution of a parametrized family of cubic Thue equations was given by E. Thomas [6137], who showed that for $n \geq A = 1.365 \cdot 10^7$ the equation

$$x^3 - (n-1)x^2y - (n+1)xy^2 - y^3 = \pm 1$$

has only trivial solutions (i.e., with $y = \pm 1$), and M. Mignotte [4292] showed that for $4 \leq n < A$ there are no solutions. For the quartic case see M. Mignotte, A. Pethő and R. Roth [4295], A. Pethő [4824]. See also C. Heuberger and R.F. Tichy [2777].

Later all solutions for certain other families of cubic (A. Togbé [6183, 6184]), quartic (A. Dujella, B. Jadrijević [1647], B. Jadrijević [3092, 3093]), quintic (I. Gáal, G. Lettl [2170, 2171]) and sextic (A. Togbé [6185]) equations have been found. See also C. Heuberger [2774–2776].

Thue equations have been surveyed in the book [5523] by W.M. Schmidt.

2. In [232] A. Baker made C.L. Siegel's result [5746] effective by providing explicit bounds for solutions of the equation

$$y^n = f(x), \tag{6.31}$$

where n is fixed, and f is a polynomial with integral coefficients, having at least three simple zeros. Earlier W.J. LeVeque [3851] described the cases for which there are infinitely many solutions with bounded denominators in algebraic number fields.

Baker's bounds were later improved by S.A. Trelina [6198], B. Brindza [734], P.M. Voutier [6481] and Y. Bugeaud [807]. Bounds for the number of solutions were given by J.-H. Evertse, J.H. Silverman [1941]. For cubic f see J.H. Silverman [5792]. A quick way for finding solutions, based on Baker's method, was given by Y.F. Bilu and G. Hanrot [516] in 1998.

It was shown in 1976 by A. Schinzel and R. Tijdeman [5453] that equation (6.31) does not have integral solutions with $|y| \geq 2$ if m is sufficiently large, and f has at least two distinct zeros.

For quantitative versions of this result see J. Turk [6233], B. Brindza, J.-H. Evertse and K. Győry [736], Y. Bugeaud [806], A.Bérczes, B. Brindza and L. Hajdu [424].

The analogue of the theorem by A. Schinzel and R. Tijdeman [5453] in the case of function fields was established by B. Brindza, Á. Pintér and J.Végső [738].

The conjecture proposed in [5453], stating that a polynomial $f \in \mathbf{Z}[X]$ having at least three distinct zeros represents only finitely many square-full integers, is still open but it has been deduced from the ABC conjecture by Walsh [6538].

3. In [5501] (see also [5503]) W.M. Schmidt applied his subspace theorem to characterize $\mathbf{Z}$-submodules M of an algebraic number field K for which the equation $N_{K/\mathbf{Q}}(x) = a$ has infinitely many solutions $x \in M$. This generalizes both Thue's theorem ($n = 2$) and an earlier result by Schmidt [5489] who did it for $n = 3$ (partial results in that case were obtained earlier by T. Skolem [5807] and C. Chabauty [974]). The result in [5500, 5501] implies in particular that if θ is an algebraic number of degree N, and $k < n/2$, then for every rational r the $\mathbf{Z}$-module generated by $1, \theta, \dots, \theta^k$ contains only finitely many elements of norm r, and this ceases to be true if $k \geq n/2$ (M. Fujiwara [2142]).

An effective analogue of Schmidt's result in the case of forms over a function field was obtained by R.C. Mason [4169–4172].

In 1978 K. Győry and Z.Z. Papp [2414] applied Baker's method to get an effective version of results in [5501], giving explicit bounds for solutions of *norm form equations*, i.e., of equations

$$cN(a_1x_1 + \cdots + a_nx_n) = m, \tag{6.32}$$

where the a_i's lie in an algebraic number field K, $m \in \mathbf{Z}$, and c is an integer such that the left-hand side of (6.32) has integral coefficients.

For further improvements of bounds for the solutions of (6.32) see I. Gáal [2169] and K. Győry [2403, 2405, 2407, 2409].

See also A.Bérczes, K. Győry [426] and K. Győry, K. Yu [2416].

Effective bounds for the number of solutions of equation (6.32) were provided by W.M. Schmidt [5522]. See also M.A. Bean, J.L. Thunder [370], J.-H. Evertse [1929].

In 2002 J.L. Thunder [6150] considered forms of finite type, i.e., forms $F(\overline{\mathbf{x}})$ in n variables, which are products of linear forms, and satisfy the following condition.

The measure of the set X_m of real solutions of $|F(\overline{\mathbf{x}})| \leq m$ is finite for every m, and the same happens for the q-dimensional measure of the intersection of X_m with every q-dimensional linear subspace of $\mathbf{R}^n$, defined over $\mathbf{Q}$.

For such forms Thunder obtained asymptotics for the number of integral solutions of $|F(\overline{\mathbf{x}})| \leq t$, and showed also that this number is bounded by $c(n, d)t^{n/d}$ (with $d = \deg F$), as conjectured by W.M. Schmidt in [5522] (cf. Thunder [6150, 6152]). Earlier [6146, 6147] he treated the case $n = 2$. See also [6151].

4. In 1969 A. Baker and H. Davenport [242] applied the bounds given in [226] to show that the only solutions of the system

$$3x^2 - 2 = y^2, \qquad 8x^2 - 7 = z^2$$

occur for $x = 1$ and $x = 11$. This solved negatively the old problem (going back to Diophantus) of the existence of an integer $m \neq 120$ with the property that the product of any two of the numbers $1, 3, 8, m$ increased by 1 is a square.

The last problem can be generalized: call a sequence $a_1, a_2, \dots, a_m$ of positive integers a Diophantine m-tuple if each of the numbers $a_i a_j + 1$ $(i \neq j)$ is a square. Baker and H. Davenport's result was generalized in 1998 by Dujella and Pethő [1648] to state that the pair 1, 3 cannot be extended to a Diophantine 5-tuple. It is also known (A. Dujella [1640]) that if the numbers a_j are allowed to be rational, then most 4-tuples can be extended to a 5-tuple.

It is conjectured that no Diophantine 5-tuple exists and the non-existence of 9-tuples was established in 2001 by A. Dujella [1641], who showed three years later [1643] that there are no Diophantine 6-tuples and the number of 5-tuples is finite.

A further generalization was proposed by A. Dujella [1639] who in the definition of a Diophantine m-tuple replaced the sum $a_i a_j + 1$ by $a_i a_j + n$ for a fixed n (see A. Dujella [1642, 1644], A. Dujella, A. Filipin, C. Fuchs [1645], A. Dujella, C. Fuchs [1646]).

5. An important application of Baker's method to E. Catalan's[10] old problem was made by R. Tijdeman [6159] in 1976. Catalan [960] stated in 1842 that the equation

$$a^x - b^y = 1 \tag{6.33}$$

with $a, b \geq 2, x, y \geq 1$ only has the solution $x = b = 2, a = y = 3$, i.e., the numbers 8 and 9 are the only consecutive positive integers which are perfect powers. He repeated this assertion in [961]. It is clear that one can assume that the exponents a, b are prime numbers.

The case $y = 2$ had been dealt with already in the 19th century by V.A. Lebesgue [3753], and one had to wait till 1952 for further progress, when W.J. LeVeque [3848] showed that for fixed a, b the equation (6.33) has at most one solution, even admitting the exponent 1, with the only exception being $a = 3$, $b = 2$, when we have $3^2 - 2^3 = 3^1 - 2^1$, and in the next year his proof was simplified by J.W.S. Cassels [930, 931], who also obtained an explicit form of the possible solution. In 1965 the case $x = 2$ was dealt with by Chao Ko [1000], and a simpler proof was provided by E.Z. Chein [1008]. Tijdeman's result gave explicit upper bounds for the values a, b, x, y (with prime x, y) in (6.33). These bounds were subsequently reduced, so, for example, M. Langevin [3708] obtained $a^x < \exp_4(730)$ ($\exp_4$ denoting the fourth iteration of the exponential function).

Further research showed that the numbers a, b, x, y in (6.33) must be very large, apart from the known case. M. Aaltonen and K. Inkeri[11] [1] established $a, b \geq 10^{500}$, and M. Mignotte and Y. Roy [4296] showed $x, y > 30000$. On the other hand M. Mignotte [4291] established the bounds $x < 1.23 \cdot 10^{18}$, $y < 2.48 \cdot 10^{24}$.

The final step was after the end of the century when in 2002 P. Mihăilescu [4300] established Catalan's conjecture, making an ingenious application of the theory of cyclotomic fields. All details of the proof can be found in the papers by Y.F. Bilu [511, 512] and H. Cohen [1141], as well as in the recent book by R. Schoof [5566]. See also T. Metsänkylä [4274].

---

[10] Eugène Catalan (1814–1894), professor at l'École Polytechnique in Paris.

[11] Kuusta Adolf Inkeri (1908–1997), professor in Turku. See [4273].

The proof given in [4300] had to use Baker's method in one case but P. Mihăilescu [4301] showed later that this can be avoided.

The analogue of Tijdeman's result for algebraic number fields was established in 1986 by B. Brindza, K. Győry and R. Tijdeman [737].

For the history of Catalan's conjecture before Mihăilescu's success see the book by P. Ribenboim [5179] and the survey by M. Mignotte [4294]. Exponential Diophantine equations were surveyed in the book [5727] of T.N. Shorey and R. Tijdeman.

A survey of applications of Baker's method to Diophantine equations was given by T.N. Shorey, A.J. van der Poorten, R. Tijdeman and A. Schinzel [5731] in 1978.

For a later survey see K. Győry [2410].

6. It follows from Siegel's result in [5747] that the equation $f(x) = g(y)$, with $f, g \in \mathbf{Z}[X]$ can have infinitely many integral solutions only if the corresponding curve is of genus zero. Such equations were later considered by H. Davenport, D.J. Lewis and A. Schinzel [1398], M. Fried [2096], A. Schinzel [5445, 5448]. Effective bounds for the number of solutions have been provided by P.G. Walsh [6536].

For further effective results see S. Tengely [6111] and A. Sankaranarayanan and N. Saradha [5392]. In 2000 Y.F. Bilu and R.F. Tichy [519] provided a complete description of pairs f, g for which this equation has infinitely many rational solutions with bounded denominators.

7. In 1967 a counterexample to an old problem of Euler's was found by L.J. Lander and T.R. Parkin [3683]. Euler conjectured that for $n = 2, 3, \dots$ the equation

$$x_1^n + \cdots + x_{n-1}^n = y^n$$

has no positive solutions (for $n = 3$ this is Fermat's theorem in the cubic case). A computer search led to the counterexample

$$27^5 + 84^5 + 110^5 + 133^5 = 144^5,$$

and in 1988 a counterexample in the case $n = 4$ was found by N.D. Elkies [1715]:

$$2\,682\,400^4 + 15\,365\,639^4 + 18\,796\,760^4 = 20\,615\,673^4.$$

The minimal counterexample in this case, given later by R. Frye (see the review [2311] of [1715] in Math. Reviews), is

$$95\,800^4 + 217\,519^4 + 414\,560^4 = 422\,481^4.$$

8. The research around Hilbert's tenth problem culminated with its solution in 1970 by Yu.V. Matijasevič [4188, 4189], who established the non-existence of an algorithm determining whether a polynomial Diophantine equation has an integral solution.

Matijasevič's result formed the last step on a long path of research in mathematical logic starting with the famous paper of K. Gödel [2253], where it was shown that there exist undecidable statements of the form

$$S_1 S_2 \dots S_n (P = Q), \tag{6.34}$$

where P, Q are polynomials in several variables with non-negative integral coefficients, and each S_i is a quantifier. Note that the undecidability of (6.34) in the case when the S_i's are existential quantifiers and P, Q are suitably chosen would give the negative solution of Hilbert's tenth problem (cf. A. Church [1109]).

The next development consisted of reducing the size of the logical system in which one can find undecidable statements. The first step was in 1953 by M. Davis [1412], and it was followed by the results of M. Davis, H. Putnam and J. Robinson[12] [1415, 1416, 5019, 5241, 5242] on which Matijasevič's proof was based. In 1975 Yu.V. Matijasevič and J. Robinson [4193] showed that there exists a polynomial in 13 variables for which there is no algorithm for checking the existence of integral zeros.

In 1982 J.P. Jones [3155] replaced 13 by 9 in the last result, confirming a claim by Yu.V. Matijasevič [4192].

For expositions see J.-P. Azra [186], M. Davis [1414], H. Hermes[13] [2760] and Yu.V. Matijasevič [4190, 4191]. A simple presentation of the proof was provided by J.P. Jones and Yu.V. Matijasevič [3156] in 1991. Other proofs were later given by G.V. Čudnovskiĭ [1289], M. Davis [1413] and N.K. Kosovskiĭ [3496].

In the case of quadratic polynomials an algorithm for checking the existence of integral solutions was constructed by C.L. Siegel [5774] in 1972.

In 2004 F.J. Grunewald and D. Segal [2373] found an algorithm determining the existence of a solution of a quadratic equation in positive integers.

The non-existence of such algorithms for exponential Diophantine equations had already been established in 1961 (M. Davis, H. Putnam, J. Robinson [1416]). There exists, however, an algorithm for the corresponding question in p-adic fields, as shown in 1965 by J. Ax and S. Kochen [180] (for p-adic fields this had also been proved by J.L. Eršov [1872]), as well as for finite fields (J. Ax [176]).

It was shown in 2003 by B. Poonen [4986] that for certain sets S of density one of primes there is no algorithm for solving Diophantine equations in the set $\mathbf{Z}_S$ of rationals having all prime factors of denominators in S.

For several classes of algebraic number fields K a negative solution of the analogue of Hilbert's tenth problem for their rings of integers was proved by J. Denef [1464, 1465] (for K quadratic or totally real), J. Denef, L. Lipshitz [1469], T. Pheidas [4833] (K cubic, not totally real), H.N. Shapiro, A. Shlapentokh [5682] (for K Abelian), A. Shlapentokh [5717] (fields with one pair of complex embeddings). A survey was given in 1999 by A. Shlapentokh [5718], who also wrote a book on that subject [5719].

We have already noted in Sect. 3.4 that Hilbert's tenth problem has a positive solution in the ring of all algebraic integers (R. Rumely [5339]). There are however infinite extensions $K/\mathbf{Q}$ such that for the rings $\mathbf{Z}_K$ the solution is negative

[12] Julia Robinson (1919–1985), sister of C. Reid, professor at Berkeley. See [5152].

[13] Hans Hermes (1912–2003), professor in Münster and Freiburg. See [4647].

(A. Shlapentokh [5717]). See M. Jarden and C.R. Videla [3105] for this question in certain subrings of infinite extensions of the rationals.

For a survey see B. Poonen [4987].

9. The old question of whether the Fibonacci sequence F_n, defined by $F_1 = F_2 = 1$, $F_{n+1} = F_n + F_{n-1}$, contains squares distinct from 1 and $F_{12} = 12^2$ was settled in the negative in 1964 independently by J.H.E. Cohn [1160] and O. Wyler [6763]. In 1970 H. London and R. Finkelstein [3988] showed that $F_6 = 2^3$ is the only perfect cube in that sequence.

This result was generalized in 2006 by Y. Bugeaud, M. Mignotte and S. Siksek [822, 823], who proved that the only perfect powers ($\neq 0, 1$) in the Fibonacci sequence are F_6 and F_{12}. They showed also that the only Lucas number which is a power $\neq 1$ is $L_3 = 2^2$ (Lucas numbers are defined by $L_0 = 2$, $L_1 = 1$ and $L_{n+1} = L_n + L_{n-1}$). For earlier results see P. Ribenboim and W.L. McDaniel [5182, 5183], T. Kagawa and N. Terai [3214], A. Bremner and N. Tzanakis [703–705].

Now it is known (P. Corvaja, U. Zannier [1252]) that in general a linear recurrence of the form

$$u_n = \sum_{j=1}^{k} a_j b_j^n$$

with rational $a_j \neq 0$ and integral b_j can contain only finitely many perfect powers. A quantitative version of this result was obtained by C. Fuchs and R.F. Tichy [2123].

For second order recurrences this had been established earlier by A. Pethő [4822] and T.N. Shorey, C.L. Stewart [5724]. See also T.N. Shorey, C.L. Stewart [5725].

It was shown by J.H.E. Cohn [1165] in 1996 that the *Pell sequence*, defined by $P_0 = 0$, $P_1 = 1$, $P_{n+2} = 2P_{n+1} + P_n$ is never a non-trivial kth power for $k \geq 3$.

10. Let $a_1, \ldots, a_k$ be given co-prime positive integers. The problem of determining the smallest number $G = G(a_1, \ldots, a_k)$ with the property that every integer $n > G$ can be represented in the form

$$\sum_{j=1}^{k} a_j x_j = n$$

with non-negative integers x_j is usually associated with Frobenius and often called *the coin problem*. The solution in the case $k = 2$,

$$G(a_1, a_2) = a_1 a_2 - a_1 - a_2,$$

had been found already in 1884 by J.J. Sylvester [6011], but in the case $k = 3$ one had to wait until 1960 even for an algorithm, when it was found by S.M. Johnson [3147] (cf. A. Brauer, J.E. Shockley [686], Ö.J. Rödseth [5248, 5249], E.S. Selmer, Ö. Beyer [5636]).

For the general case such algorithms were given by B.R. Heap and M.S. Lynn [2624, 2625] and F. Aicardi [34].

The first polynomial time algorithm to find $G(a_1, \ldots, a_k)$ was found by R. Kannan [3238] in 1992, and another such algorithm, based on ideas from [5636] and [5248, 5249], was given by J.L. Davison [1417] in 1994.

See also D. Beihoffer, J. Hendry, A. Nijenhuis, S. Wagon [390], S. Böcker, Z. Lipták [571].

In many special cases formulas for G were obtained, e.g., when the a_i's either form an arithmetic progression, or differ from it slightly [3869, 5236, 5779]. It was shown in 1990 by F. Curtis [1303] that there is no simple polynomial formula for G in the case $k \geq 3$.

For the case $k = 3$ see also the papers of F. Aicardi [33], V.I. Arnold [133], L.G. Fel [1971, 1972], J.C. Rosales, P.A. García-Sanchéz [5283]

The first upper bound for G,

$$G(a_1, \ldots, a_k) \leq a_1 a_k + a_2 + \cdots + a_k \quad (a_1 \leq a_2 \leq \cdots \leq a_k)$$

was proved by I. Schur in his last lecture in Berlin in 1935 (see [682]). For later improvements see A. Brauer [682], A. Brauer, B.M. Seelbinder [685], P. Erdős, R.L. Graham [1830], E.S. Selmer [5635].

Asymptotic properties of $G(a_1, a_2, \ldots, a_k)$ were considered by V.I. Arnold[14] [130–132], J. Bourgain and Ya.G. Sinai [666], V. Ščur, Ya.G. Sinai and A.V. Ustinov [5595], A.V. Ustinov [6255].

There is a large literature concerned with the value of G under various assumptions on the sequence a_i (the paper [5635] gives a rather complete bibliography up to 1976). A monograph devoted to the Frobenius problem was written by J.L. Ramirez Alfonsin [5100].

11. In 1969 M. Hall, Jr. [2472] conjectured that for integral $x, y > 0$ one has

$$\left|x^3 - y^2\right| > C\sqrt{x}$$

with some positive constant C, and on the other hand for every $\varepsilon > 0$ one has

$$\left|x^3 - y^2\right| \leq B(\varepsilon) x^{1/2+\varepsilon}$$

infinitely often (in both cases one assumes $x^3 \neq y^2$). A few years earlier B.J. Birch, S. Chowla, M. Hall, Jr. and A. Schinzel [529] showed that the last inequality holds infinitely often with $\varepsilon = 0.1$. In 1982 L.V. Danilov [1325] established the truth of $|x^3 - y^2| < 0.97\sqrt{x}$ for infinitely many x, y, and on the other hand it was shown by S. Lang [3700] that the ABC conjecture implies $|x^3 - y^2| \gg x^{1/2-\varepsilon}$ for every $\varepsilon > 0$ and $x^3 \neq y^2$.

12. A new method of dealing with the number of solutions of congruences was discovered by S.A. Stepanov [5924] in 1969. He used completely elementary tools to show that if $f \in \mathbf{Z}[X]$ is of degree $n \geq 3$ and p is a large prime, then the number N_p of solutions of the congruence $y^2 \equiv f(x) \pmod{p}$ satisfies

$$\left|N_p - p\right| \leq c(n)\sqrt{p},$$

[14] Vladimir Igorevič Arnold (1937–2010), professor in Moscow.

with $c(n)$ being a constant depending only on n (the value of this constant was later reduced by H.M. Stark [5907]). In his next paper [5925] he obtained a similar result for the congruence $y^m \equiv f(x) \pmod{p}$ for $m \geq 3$, and in [5927] he did this for $m = 2$ in the case of an arbitrary finite field. In [5928, 5929] he proved a similar bound for the number of solutions of $f(x, y) = 0$ for polynomials over finite fields, and this led a new proof of the Riemann Hypothesis for curves, proved earlier by A. Weil with much stronger machinery (see E. Bombieri [600] and W.M. Schmidt [5505]). The same method led to elementary proofs of bounds for Kloosterman sums (S.A. Stepanov [5926]).

For an exposition of Stepanov's method see the book [5507] by W.M. Schmidt. Stepanov's lectures on the arithmetic of algebraic curves were published in 1991 [5930].

13. Let $f \in \mathbf{Z}[X]$ be irreducible, denote by m_k the number of solutions of the congruence

$$f(x) \equiv 0 \pmod{k},$$

and for $k \geq 2$ let $x_1^{(k)}, \ldots, x_{m_k}^{(k)}$ be these solutions. It was shown by C. Hooley [2861] that the sequence of the ratios $x_i^{(k)}/k$ is uniformly distributed mod 1. If f is quadratic, then the same holds when k runs over primes (W. Duke, J.B. Friedlander, H. Iwaniec [1649], A. Tóth [6196]).

14. The classical approach to Diophantine equations was presented in the book [4405] by L.J. Mordell, published in 1969, and the modern approach is represented by the books by M. Hindry and J.H. Silverman [2815], and S. Lang [3688, 3698, 3699, 3701]. Cf. also N.P. Smart [5826] and V.G. Sprindžuk [5879].

6.7 Elliptic Curves

1. In the sixties one became accustomed to the use of methods of algebraic geometry in the study of Diophantine problems. Such methods were developed in a precise fashion by A. Weil [6614–6616, 6619], and applied in Lang's books [3688, 3698] (see also his survey [3701]). The case of elliptic curves had been treated earlier by S. Lang in [3697] (the more traditional method was presented by him in [3695]). Lang's book [3688] earned a hostile review from L.J. Mordell [4401] in the Bulletin of AMS, and this was followed by a letter from Siegel to Mordell, made public several years later by S. Lang [3702].

Two important conjectures were stated at the Stockholm ICM Congress in 1962 by I.R. Šafarevič [5366]. The first (the *finiteness conjecture*) asserted that there are only finitely many (up to isomorphism) non-constant algebraic curves over an algebraic number field which are non-singular, irreducible, of fixed genus $g \geq 1$ (in the case $g = 1$ he assumed that the curve has a rational point), and have a good reduction outside a fixed finite set S of prime divisors. He gave a proof of the finiteness conjecture in the case of elliptic curves (in this case a simple proof due to J. Tate is given in the book [5639] by J.-P. Serre) and hyper-elliptic curves, i.e., curves defined by an equation of the form $y^2 = f(x)$, where f is a polynomial of degree ≥ 5 (cf. Yu.G. Zarhin [6814]). He stated also the analogue of the finiteness conjecture for algebraic function fields in one variable over algebraically closed fields.

The second conjecture by I.R. Šafarevič asserted that there are no curves over **Q** of genus ≥ 1 having good reduction everywhere.

In 2007 J.E. Cremona and M.P. Lingham [1279] presented an algorithm giving a list of all elliptic curves (up to isomorphism) over an algebraic number field having good reduction outside a given finite set of prime ideals.

2. The finiteness conjecture for curves over function fields over **C** was established by A.N. Paršin [4749], who dealt with the case $S = \emptyset$, and S.Yu. Arakelov [112] (cf. A.N. Paršin [4752]). The case of positive characteristics was later settled by L. Szpiro [6028]. For function fields having finite fields of constants see A.N. Paršin [4752, 4753].

In [4749] A.N. Paršin showed the strength of Šafarevič's conjecture by proving that it implies Mordell's conjecture, asserting that a curve of genus $g \geq 2$ can have at most finitely many rational points, and at the International Congress of Mathematicians in Nice in 1970 he presented conjectural extensions of the finiteness conjecture to Abelian varieties [4751].

3. In 1963 B.J. Birch and H.P.F. Swinnerton-Dyer [538] posed an important conjecture[15] (the *Birch–Swinnerton-Dyer conjecture*) concerning the L-function of an elliptic curve (see (7.6) for its definition). It asserts that if r denotes the rank of $E(\mathbf{Q})$, then

$$L_E(s) = a_r(s-1)^r + \sum_{j=r+1}^{\infty} a_j(s-1)^j$$

with $a_r = \lambda(E)\#\text{Ш}(E(\mathbf{Q})) \neq 0$, where

$$\lambda(E) = \frac{\Omega_E \cdot Reg_E \cdot \prod_p c_p}{(\#E_{tor}(\mathbf{Q}))^2}.$$

Here Ω_E denotes the real period, defined as the integral $\int_{E(\mathbf{R})} |\omega|$, ω being the invariant differential on a minimal Weierstrass equation of E, Reg_E (the regulator of E) is a discriminant related to the basis of the torsion-free part of E, and c_p is the cardinality of the factor group $E(\mathbf{F}_p)/E_0(\mathbf{F}_p)$, E_0 being the group of non-singular points of the reduction $E(\mathbf{F}_p)$ of E mod p (see, e.g., J. Tate [6063]).

The number c_p may also be defined as the Tamagawa number for primes p of bad reduction, whereas for primes of good reduction one has $c_p = 1$. The *Tamagawa number* of a connected semi-simple algebraic group G over a field K is defined as the Tamagawa measure (which is essentially the product of Haar measures

[15] It is one of the Millennium conjectures, with a prize of $\$10^6$ for its solution.

of the locally compact groups G_v, the local components of the adele group of G) of the factor group G_A/G_K, where G_A denotes the group of adeles of G, and G_K is its subgroup of principal adeles consisting of K-points of G embedded in G_A in a diagonal way. The principal properties of Tamagawa numbers have been studied by A. Weil [6623, 6624], who determined their values for many classical algebraic groups, conjectured that for simply connected classical groups it equals 1, and established this conjecture in many cases. For the remaining classical groups this was done later by J.G. Mars [4157]. For further work on this subject see A. Borel [628], A. Borel and Harish-Chandra [633], A. Weil [6626], R.P. Langlands [3711], G. Harder [2503]. See also K.F. Lai [3613], J. Oesterlé [4661].

One defines the analytical cardinality $\#Ш_{an}(E(\mathbf{Q}))$ of the Tate–Šafarevič group as the ratio $a_r/\lambda(E)$, and by the p-part of the conjecture of Birch–Swinnerton-Dyer ($BSD(E,p)$) one understands the equality of the p-parts of $\#Ш_{an}(E(\mathbf{Q}))$ and $Ш(E(\mathbf{Q}))$, provided the number $\#Ш_{an}(E(\mathbf{Q}))$ is rational.

It was established in the sixties by J.W.S. Cassels that if the p-part of $\#Ш(E(\mathbf{Q}))$ is finite, then it is a square [936], and the truth of $BSD(E,p)$ depends only on the isogeny class of E [940].

There exists ample evidence for the Birch–Swinnerton-Dyer conjecture (see, e.g., B.J. Birch, H.P.F. Swinnerton-Dyer [537, 538], J.A. Antoniadis, M. Bungert, G. Frey [107], J.P. Buhler, B.H. Gross, D. Zagier [830], A. Brumer, D. McGuiness [796], J.W.S. Cassels [940], J.E. Cremona [1278], C.D. Gonzales-Avilés [2278], G. Grigorov, A. Jorza, S. Patrikis, W.A. Stein, C. Tarniţă [2345], V.A. Kolyvagin [3463], A.R. Rajwade [5054, 5055], N.M. Stephens [5931]).

The first supporting result was proved in 1977 by J. Coates and A. Wiles [1130], who showed that if the curve E has complex multiplication by integers of an imaginary quadratic field K of class-number one and $L_E(1) \neq 0$, then $E(K)$ is finite, hence its rank equals 0. Another proof of this result was given by R. Gupta [2392] in 1985, and in 1987 K. Rubin [5328] showed that if E has complex multiplication and rank $E(\mathbf{Q}) \geq 2$, then the order of $L_E(s)$ at $s = 1$ is ≥ 2. This led to examples of elliptic curves for which the Birch–Swinnerton-Dyer conjecture holds.

The next step was made by B.H. Gross and D. Zagier [2357]. They proved that if $E(\mathbf{Q})$ is a modular elliptic curve and $L_E(s)$ has a simple zero at $s = 1$, then $E(\mathbf{Q})$ has a positive rank. An elliptic curve E is said to be *modular* if its L-function coincides with the Dirichlet series (as given in (4.62)) of a modular form of weight 2, having its level equal to the conductor of E. A few years later V.A. Kolyvagin [3462, 3464] showed that for modular curves E the non-vanishing of $L_E(s)$ at $s = 1$ implies $r(E) = 0$, and if $L_E(s)$ has a simple zero at 1, then $r(E) = 1$. (Cf. K. Rubin [5329], M.R. and V.K. Murty [4490], D. Bump, S. Friedberg, J. Hoffstein [842]. For an exposition see B. Perrin-Riou [4782].) Now one knows that these results apply to all elliptic curves over $\mathbf{Q}$, due to the proof of the Taniyama–Shimura conjecture (see Sect. 7.2). This result implies also the truth of the Birch–Swinnerton-Dyer conjecture in the case when the order of $L_E(s)$ at $s = 1$ equals 0 or 1.

In 1983 R. Greenberg [2339] proved that if the L-function of a CM-curve E has an odd order zero at $s = 1$, then $r(E) \geq 1$ (see also D.E. Rohrlich [5270, 5271]).

An algorithm for the determination of the rank of an elliptic curve over the rationals, depending on the truth of the Birch–Swinnerton-Dyer conjecture, was constructed in 1994 by J. Gebel and H.G. Zimmer [2217].

An analogue of this conjecture for Abelian varieties over function fields of positive characteristics was stated by J. Tate [6059–6061]. It was later established in certain cases by J. Tate [6062] and J.S. Milne [4311, 4313]. Cf. P. Schneider [5536] and W. Bauer [363]. In 2003 K. Kato and F. Trihan [3276] showed that the conjecture follows from the finiteness of ℓ-part of the Tate–Šafarevič group for some prime ℓ.

4. In 1965 J. Tate [6060, 6061] stated a conjecture (the *Sato–Tate conjecture*)[16] concerning the number of points in the reduction of elliptic curves modulo primes. (Note that isogenous curves over a finite field F have the same number of elements in F; this follows from results in M. Deuring's paper [1503], and can be also deduced from F.K. Schmidt's [5478, 5479] theory of his zeta-functions; cf. S. Lang [3686].) If the curve E is defined over $\mathbf{Q}$, has good reduction mod p and E mod p has N_p elements, then H. Hasse's proof of the Riemann conjecture for elliptic function fields [2593] implies that the ratio $ST(p) = (N_p - p - 1)/2\sqrt{p}$ lies in the interval $[-1, 1]$, hence one can write

$$ST(p) = \cos\vartheta(p)$$

with $0 \leq \vartheta \leq \pi$. The Sato–Tate conjecture asserts that for $0 \leq a \leq b \leq 1$ one has

$$\lim_{x\to\infty} \frac{\#\{a \leq \vartheta(p) \leq b\}}{\pi(x)} = \frac{2}{\pi}\int_a^b \sin^2 t\, dt.$$

In the case when the curve does not have complex multiplication this conjecture was reduced by J.-P. Serre [5639] to a question of non-vanishing at $s = 1$ of certain L-functions introduced by J. Tate [6060]. Later A.P. Ogg [4668] showed that it would be sufficient to have a continuation of these L-functions to the half-plane $\Re s > c$ for some $c < 1/2$ (cf. also V.K. Murty [4495]).

This approach was used by L. Clozel, M. Harris, N. Shepherd-Barron and R. Taylor [1122, 2570, 6088] to prove the Sato–Tate conjecture for all elliptic curves having multiplicative reduction at least one prime. The same result was obtained also in the case when E is defined over a totally real number field. For expositions see H. Carayol [898], L. Clozel [1121]. See also B. Mazur [4218] for the sketch of the proof in one particular case.

For the analogue of the Sato–Tate conjecture in the case of curves over function fields see H. Yoshida [6789].

5. Let E be an elliptic curve defined over a field K. For a prime ℓ not dividing the characteristics of K and $m = 1, 2, \ldots$ denote by $E[\ell^m]$ the subgroup of $E(\overline{K})$ consisting of elements with order dividing ℓ^m. The inverse limit

$$T_\ell(E) = \lim_{\leftarrow} E[\ell^m]$$

[16]Tate noted in [6061] that the computer calculations performed by M. Sato led him to formulate this conjecture.

is a free $\mathbf{Z}_\ell$-module, called the *Tate module*.

It was conjectured by J. Tate that curves with isomorphic Tate modules are isogenous, and this turned out to be a consequence of Faltings' theorem (see Sect. 7.2).

The Tate module becomes in a natural way a $Gal(\overline{K}/K)$ module, and the Galois action on it induces an *ℓ-adic representation* ρ_ℓ of $G = Gal(\overline{K}/K)$ in the group $GL_2(\mathbf{Z}_\ell)$ of automorphisms of the linear $\mathbf{Q}_\ell$-space $T_\ell(E) \otimes \mathbf{Q}_p$. The idea of associating with an elliptic curve (or, more generally, an Abelian variety) such a representation appeared first in the book [6616] by A. Weil, published in 1948.

The next important step was in 1957 by Y. Taniyama [6045] who showed that there is a finite set S of primes such that for $p \notin S$ the characteristic polynomials

$$P_{p,\rho_\ell}(T) = \det(1 - F_{p,\rho_\ell} T)$$

(where $F_{p,\rho_\ell} T$ is defined as the image under ρ_ℓ of the Frobenius element F_p) do not depend on ℓ. He considered, more generally, systems $\rho = (\rho_\ell)_\ell$ of ℓ-adic representations having the last property and the associated L-series.

The same approach works also for representations of the Galois group of $\overline{K}/K$ for algebraic number fields K.

J.-P. Serre [5639, 5643] studied the image of ρ_ℓ and showed in particular [5643, Theorem 2] that if for every finite extension L/K the curve $E(L)$ does not have complex multiplication, then for sufficiently large ℓ the image $\rho_p(G)$ coincides with $GL_2(\mathbf{Z}_\ell)$. This was applied (J.-P. Serre [5644], H.P.F. Swinnerton-Dyer [6002, 6003]) to the study of congruences for coefficients of modular forms, mentioned in Sect. 2.2.6.

6. The construction of Tate's module can be applied also to arbitrary Abelian varieties A, defined over a finite extension K of the rationals, leading to the Tate module $T_\ell(A)$. J. Tate [6061, 6062] posed two conjectures concerning this module. The first asserted that the representation of the Galois group $G(K)$ of $\overline{K}/K$ induced by the action of G on the tensor product

$$T_\ell(A) \otimes_{\mathbf{Z}_\ell} \mathbf{Q}_p$$

is semi-simple, and the second stated that the ring of Galois endomorphisms of $T_\ell(A)$ is isomorphic with the tensor product (over $\mathbf{Z}$) of the ring of K-endomorphisms of A and $\mathbf{Z}_\ell$. J. Tate [6062] himself proved the analogue of these conjectures for varieties over finite fields, A.N. Paršin [4752] did this for elliptic curves over function fields over finite fields of characteristics $\neq 2$, and Yu.G. Zarhin [6815–6817] established them for a large class of varieties in the case of function fields over finite fields. J.-P. Serre observed in [5639] that Tate's conjectures in the case of elliptic curves would follow from the conjecture that there are only finitely many Abelian varieties isogenous to a given variety. Since the last conjecture was later established by G. Faltings [1954], both of Tate's conjectures became theorems.

7. The second conjecture of I.R. Šafarevič [5366], stating that there are no elliptic curves defined over $\mathbf{Q}$ with good reduction at every prime, was established by J. Tate [6063] in 1974. He also gave an example of a curve over the field $\mathbf{Q}(\sqrt{29})$ with good reduction at every prime ideal of that field (cf. J.-P. Serre [5643, Sect. 5.10]).

Later other examples of curves defined on certain number fields and having everywhere good reduction were given (T. Kagawa [3211], D.E. Rohrlich [5269], B. Setzer [5666], R.J. Stroeker [5970] and S. Comalada [1194]). It was established by M. Hindry and J.H. Silverman [2814] that the torsion part of such curves has at most $cN \log N$ elements, with $N = [K : \mathbf{Q}]$ and $c = 1\,977\,408$. For fields over which no elliptic curve has good reduction everywhere see B. Setzer [5666], R.J. Stroeker [5971], H. Ishii [3032].

For later results see T. Kagawa [3212, 3213], M. Kida [3327–3329], M. Kida and T. Kagawa [3330].

It is still unknown whether there are infinitely many elliptic curves over $\mathbf{Q}$ having only one prime of bad reduction. It was shown in 2005 by J.B. Friedlander and H. Iwaniec [2112] that this would be a consequence of the existence of infinitely many D such that the Dirichlet L-function associated with the real character χ_D of conductor D satisfies

$$L(1, \chi_D) \leq \frac{1}{\log^{61} D}.$$

In 1985 J.M. Fontaine [2024] generalized Tate's result to all Abelian varieties over $\mathbf{Q}$, confirming a conjecture by I.R. Šafarevič [5366]. For dimensions two and three this was established earlier by V.A. Abraškin [9, 10].

Fontaine's result also holds for Abelian varieties over cyclotomic fields $\mathbf{Q}(\zeta_n)$ for $n = 3, 4, 5, 7, 8, 9, 11, 12$ and 15 (R. Schoof [5564]).

In 1972 J. Tate gave a Colloquium Lecture at a meeting of the American Mathematical Society, presenting a survey of the theory of elliptic curves [6063].

8. All possible finite groups which can form the torsion part $E_{tor}(\mathbf{Q})$ of an elliptic curve over the rationals were listed in 1977 by B. Mazur [4214–4216], who showed that $E(\mathbf{Q})$ cannot have points of prime order ≥ 17, and this in view of a preceding result by D. Kubert [3540] implied that $E_{tor}(\mathbf{Q})$ is either cyclic with $n \leq 10$ or $n = 12$ elements, or isomorphic to $C_{2n} \oplus C_2$ with $n \leq 4$ (for an exposition see J.-P. Serre [5653]). This confirmed a conjecture going back to B. Levi [3853], who formulated it at the ICM in Rome in 1908 in a geometric form (cf. also T. Nagell [4513] and A.P. Ogg [4670, 4671]). Already in 1908 B. Levi [3852] showed that each of these groups occurs infinitely often as $E_{tor}(\mathbf{Q})$. Several writers showed the non-existence of curves having points of a given order N (see, e.g., G. Billing, K. Mahler [507], C.-E. Lind [3891], T. Nagell [4514], A.P. Ogg [4669]).

It has long been conjectured that the cardinality of the torsion part $E_{tor}(K)$ of an elliptic curve is bounded by a constant depending only on the field K (J.W.S. Cassels in his survey [944] called this "part of the folklore"). In 1969 Yu.I. Manin [4133] proved that there can be only finitely many points in $E(K)$ whose order is a power of a given prime p, their number being bounded by a constant depending on K and p (for an exposition see J.-P. Serre [5642]). An effective version of Manin's result was given by V.G. Berkovič [441] in 1976. See also A.N. Paršin [4750].

It was proved in 1990 by M. Flexor and J. Oesterlé [2015] that Szpiro's conjecture implies an explicit bound for the cardinality of the torsion part of a curve $E(k)$, depending only on the degree of the field k.

In 1986 S. Kamienny [3223] proved that for a large class of curves over totally real fields the cardinalities of torsion subgroups are uniformly bounded. Ten years later L. Merel [4255] showed that there are only finitely many groups serving as the torsion part of $E(k)$ for fields k of a fixed degree n over the rationals. For $n = 2$ this had been established earlier by S. Kamienny [3224] (see S. Kamienny [3221, 3222], M.A. Kenku [3301, 3302], M.A. Kenku, F. Momose [3303], F. Momose [4351] for previous results), who in a joint paper with B. Mazur also proved this for $n \le 8$ [3225]. The case $n \le 14$ was settled by D. Abramovich [8]. Merel proved that if the field K is of degree $N \ge 2$, and $E(K)$ has a point of order p, with prime p, then $p < N^{3N^2}$, and his main theorem followed from the theorem of Yu.I. Manin [4133].

Merel's result was made fully effective by P. Parent [4741], who showed that if $E(K)$ has a point of order p^k, then

$$p^k \le \begin{cases} 4160(3^N - 1)N^6 & \text{if } p \ge 5, \\ 4160(5^N - 1)N^6 & \text{if } p = 3, \\ 94041(3^N - 1)N^6 & \text{if } p = 2, \end{cases}$$

N being the degree of K.

For a survey see B. Edixhoven [1688].

It has recently been shown by F. Breuer [717] that if E is an elliptic curve defined over a finitely generated field K of characteristics 0 and $[L : K] = N$, then the torsion group of $E(L)$ has $O(N(\log\log N)^c)$ elements with a certain $c = c(E)$, and this result is best possible. For an earlier bound see N. Ratazzi [5130].

Complete lists of possible torsion groups for curves with integral j-invariant were given by H.H. Müller, H. Ströher and H.G. Zimmer [4472] for quadratic fields, by G.W. Fung, H. Ströher, H.C. Williams and H.G. Zimmer [2145] for pure cubic fields, and A. Pethő, T. Weis and H.G. Zimmer [4826] in the general cubic case.

Cubic fields, without assumptions on the j-invariant, were treated by P. Parent [4742, 4743], who established that only primes ≤ 13 can divide the cardinality of the torsion group. This case has also been treated by D. Jeon, C.H. Kim and A. Schweizer [3125]. For quartic fields see C.S. Abel-Hollinger and H.G. Zimmer [4], D. Jeon, C.H. Kim and E. Park [3124] and T. Kishi [3341].

It was shown by K. Ribet [5186] in 1981 that for every elliptic curve (and, more generally, for every Abelian variety) defined over $\mathbf{Q}$ and for every infinite Abelian extension $K/\mathbf{Q}$ the torsion part of $E(K)$ is finite. In the case $K = \bigcup_{n=1}^{\infty} \mathbf{Q}(\zeta_{p^n})$ this had been proved earlier by H. Imai [3010]. Later the possible torsion groups of curves over the field $\mathbf{Q}(\{\sqrt{n} : n \in \mathbf{Z}\})$ were determined (M. Laska and M. Lorenz [3721] and Y. Fujita [2137, 2138]).

In 1997 V.A. Demyanenko [1461] obtained the bound 12 for the exponent of the torsion group of an elliptic curve over a cyclotomic field.

Bounds for the cardinality of the torsion part of curves defined over a global function field having transcendental invariant j were given in 1968 by M. Levin [3861].

9. In [229, 231], A. Baker used his method to give an effective upper bound for the number of rational integral points on an elliptic curve. These bounds were unsuitable for direct computations, hence modifications of Baker's approach were proposed to make feasible an effective computation of all such points. One of the first efforts in this direction was made in a joint paper by five mathematicians (F. and W.J. Ellison, J. Pesek, C.E. Stahl and D.S. Stall) from Ann Arbor[17] [1756], who used this method to find all solutions of the equation

$$x^3 - y^2 = k$$

in the case $k = 28$, which at that time was the smallest positive integer for which this equation had not been solved. Baker's bounds for solutions of such equations have been later improved by H.M. Stark [5908].

An effective bound for integral points on a curve of genus one was established by A. Baker and J. Coates [240]. For later improvements see W.M. Schmidt [5524], R. Gross and J.H. Silverman [2358], Á. Pintér [4881], Y. Bugeaud [811, 812], H.A. Helfgott and A. Venkatesh [2721].

Algorithms for finding all integral points on a curve of genus zero were presented by J. Gebel, A. Pethő and H.G. Zimmer [2215], R.J. Stroeker and N. Tzanakis [5974, 5975] and R.J. Stroeker and B.M.M. de Weger [5976], based on the results of S. David [1407] on linear forms of elliptic logarithms (cf. N. Tzanakis [6242]). This was generalized to the case of S-integral points by N.P. Smart [5824], and A. Pethő, H.G. Zimmer, J. Gebel and E. Herrmann [4827].

An algorithm for the number of points on an elliptic curve over a finite field was first given by R. Schoof [5562]. In [5563] he presented three such algorithms, due to J.-F. Mestre, G. Cornacchia and A.O.L. Atkin, N.D. Elkies. See also N.D. Elkies [1723].

For algorithms for other curves and varieties see L.M. Adleman and M.-D.A. Huang [19, 20], A. Chambert-Loir [985], J. Denef and F. Vercauteren [1470], M. Fouquet, P. Gaudry and R. Harley [2044], P. Gaudry and N. Gürel [2207], K.S. Kedlaya [3293], J. Pila [4856], T. Satoh [5415], T. Satoh, B. Skjernaa and Y. Taguchi [5416].

For procedures leading to rational points see, e.g., J.E. Cremona [1278], N.D. Elkies [1721] and J.H. Silverman [5795].

The results of J. Coates [1123–1125] provided an effective procedure to determine all elliptic curves with a given conductor (see also the book [3720] by M. Laska). For curves with small conductors see S. Akhtar [37], A. Brumer and K. Kramer [795], I. Miyawaki [4346], A.P. Ogg [4665, 4666] and B. Setzer [5665].

For later results see N.D. Elkies, M. Watkins [1726], W. Ivorra [3049].

There are $O_\varepsilon(N^{1/2+\varepsilon})$ elliptic curves over $\mathbf{Q}$ having conductor N, and it is expected that this number is actually $O(N^{c/\log\log N})$ for some c (A. Brumer, J.H. Silverman [797]). As an application of Baker's method, M.K. Agrawal, J. Coates, D.C. Hunt and A.J. van der Poorten [25] determined in 1980 all elliptic curves over $\mathbf{Q}$ of the minimal possible conductor 11.

The book [1278] by J.E. Cremona contains a list of all elliptic curves with conductor ≤ 1000.

---

[17] They tried, without success, to publish their paper under the name Anne Arbor.

10. There is an old problem of Euler's [1906] concerning quadratic forms, which in our times has been related to the structure of elliptic curves. Euler called two forms $x^2 + ay^2$ and $x^2 + by^2$ *concordant* if there exist integral arguments at which they both attain a square value, i.e., if the system

$$x^2 + ay^2 = t^2, \qquad x^2 + by^2 = u^2 \tag{6.35}$$

has an integral solution with $tu \neq 0$. Numbers a for which this equality holds with $b = -a$ are called *congruent numbers.* This definition goes back, according to L.E. Dickson [1545, p. 459], to an anonymous Arabian author from the tenth century. Congruent numbers can be also defined as integers a for which there exists a Pythagorean triangle with rational legs and area a.

Already Fermat knew that 1 is not a congruent number (this fact is equivalent to Fermat's Last Theorem for the exponent 4). In previous centuries several writers produced many examples of congruent and non-congruent numbers (see [1545, Chap. 16]) and showed also that certain classes of numbers are not congruent. The checking of numbers for the congruential property is not easy, as the example found by L. Bastien [350] in 1915 and quoted in R. Guy's book [2398, D27] shows where the numbers x, y, t, u in the smallest solution of (6.35) with $a = 101$, $b = -101$ exceed 10^{20}.

A list of all congruent numbers below 2000 was given in 1986 by G. Kramarz [3509], and it was extended to 10 000 by K. Noda and H. Wada [4623], and to 42 552 by F.R. Nemenzo [4557].

It was pointed out in 1939 by E.T. Bell [399] that the problem of concordant forms is related to elliptic curves (cf. E. Haentzschel [2428]), and he later gave [400] a condition for the solvability of (6.35) which is, however, of limited usefulness.

In 1983 Tunnell [6211] related the problem of congruent numbers to the theory of elliptic curves. He observed that n is a congruent number if and only if the curve $E_n : y^2 = x^3 - nx^2$ has infinitely many rational points, i.e., the rank of $E_n(\mathbf{Q})$ is non-zero, and used the result of J. Coates and A. Wiles [1130] to show that if n is congruent, then

$$\#\left\{n = x^2 + 2y^2 + 8z^2\right\} = 2\#\left\{n = x^2 + 2y^2 + 32z^2\right\},$$

and the converse is a consequence of the Birch–Swinnerton-Dyer conjecture. A highly readable exposition of this result is presented in the book by N. Koblitz [3427], published in 1984.

In 1972 R. Alter, T. Curtz and K.K. Kubota [60] conjectured that every $n \equiv 5, 6, 7 \pmod 8$ is congruent.

It was proved in 1984 by J.S. Chahal [978, 979] (with the use of an old identity, due to A. Desboves [1476]) that every residue class mod 8 contains infinitely many congruent numbers. This was extended to cover all arithmetic progressions by M.A. Bennett [412].

In 1996 K. Ono [4680] proved that if the elliptic curve

$$y^2 = x^3 + (a + b)x^2 + ab$$

has positive rank over $\mathbf{Q}$ then (6.35) has infinitely many solutions, and gave in the case of zero rank a criterion for the existence of a solution, which is unique if it exists.

11. An elliptic curve E over the rationals is called *supersingular* at a prime p, if $\#E(\mathbf{F}_p) = p + 1$, i.e., the coefficient $a_p(E)$ in Hasse's formula

$$\#E(\mathbf{F}_p) = p + 1 - a_p(E)$$

vanishes. M. Deuring's results [1503] imply that for a CM-curve E defined over the rationals the number $\pi_0(x; E)$ of primes $p \le x$ with vanishing a_p is asymptotically equal to $\pi(x)/2$. The behavior of $\pi_0(x; E)$ for non-CM-curves is the subject of a conjecture by S. Lang and H. Trotter [3706], formulated in 1976. It states that if E is defined over $\mathbf{Q}$ and is not a CM-curve, then one has

$$\#\{p \le x : a_p = 0\} = (C(E) + o(1))\frac{\sqrt{x}}{\log x}.$$

They also gave heuristic support for this conjecture based on a probabilistic model, pointing out that it had earlier been proposed by T.A. Tuškina [6241] on the basis of numerical calculations.

This conjecture was later extended to the case of several elliptic curves.

If $E_1, \dots, E_N$ are pairwise non-isogenous elliptic curves over $\mathbf{Q}$ and r is a given positive integer, then for the number $\pi^r_{E_1,\dots,E_N}(x)$ of primes $p \le x$ satisfying $a_p(E_i) = r$ for $i = 1, 2, \dots, N$ one has

$$\pi^r_{E_1,\dots,E_N}(x) = \begin{cases} c\sqrt{x}/\log x & \text{if } N = 1, \\ c\log\log x & \text{if } N = 2, \\ O(1) & \text{if } N \ge 3, \end{cases} \tag{6.36}$$

the value of $c \ge 0$ depending on $E_1, \dots, E_N$.

N.D. Elkies [1714] showed in 1987 that $\pi^0_E(x; E)$ tends to infinity and two years later obtained an analogue for curves over real algebraic number fields [1716]. The first lower bounds for $\pi^0_E(x; E)$ were obtained under the General Riemann Hypothesis (M.L. Brown [753], N.D. Elkies and M.R. Murty [1713]) and the first unconditional bound was given by É. Fouvry and M.R. Murty [2062] in 1996:

$$\pi^0_E(x; E) \ge \frac{\log\log\log x}{(\log\log\log\log x)^c}$$

for every $c > 1$ and large x. They showed also that for $N = 1$ and $r = 0$ the equality (6.36) holds on average. This is also true for $N = 1, 2$ and any r (É. Fouvry and M.R. Murty [2061], C. David and F. Pappalardi [1405]). For the case when the conjectured constant c in (6.36) vanishes see C. David, H. Kisilevsky and F. Pappalardi [1404].

See also A. Akbary, C. David and R. Juricevic [36] and C. David and F. Pappalardi [1406].

On the other hand J.-P. Serre [5654] obtained

$$\pi^0_E(x; E) \ll \frac{x}{\log^c x}$$

for every $c < 5/4$ and deduced the stronger bound $\pi^0_E(x; E) \ll x^{3/4}$ from the General Riemann Hypothesis. Later N.D. Elkies and M.R. Murty [1713] noted that the last bound follows unconditionally from the results of a paper by M. Kaneko [3231].

In 2005 A.C. Cojocaru and C. Hall [1172] reduced this to $O(x^{3/4}/\log x)$.

12. In [3706] one finds also another conjecture by S. Lang and H. Trotter.

Let E be an elliptic curve over $\mathbf{Q}$ without complex multiplication, and for every prime p for which E has good reduction denote by π_p a root of the polynomial

$$X^2 - a_p X + p,$$

where

$$a_p = p + 1 - \#E(\mathbf{F}_p).$$

Now if K is an imaginary quadratic field, then

$$A_{E,K}(x) := \#\{p \le x : \mathbf{Q}(\pi_p) = K\} = (c(K,E) + o(1))\frac{\sqrt{x}}{\log x}$$

holds with a suitable $c(K,E) > 0$.

It was shown by J.-P. Serre [5654] that the General Riemann Hypothesis implies

$$A_{E,K}(x) \ll x^{1-\theta}$$

with some positive θ, depending on K and E.

In 2005 A.C. Cojocaru, É. Fouvry and M.R. Murty [1171] proved that one can take for θ any number smaller than $1/18$. They also established the unconditional bound

$$A_{E,K}(x) \ll \omega(|D|)x\frac{(\log\log x)^{13/12}}{\log^{25/24} x},$$

with D being the discriminant of K.

There are more conjectural assertions in [3706], one of them stating that if two elliptic curves without CM have the same sets of supersingular primes, except for a finite number, then they are isogenous.

13. In 1975 I. Borosh, C.J. Moreno and H. Porta [647] conjectured that the set of primes for which the reduction of an elliptic curve is cyclic has a density (which may vanish). Two years later S. Lang and H. Trotter [3707] posed a similar conjecture analogous to Artin's conjecture on primitive roots.

If P is a point of infinite order of an elliptic curve E over the rationals, then for infinitely many p the reduction of E mod p is cyclic, generated by P mod p.

The first result dealing with these conjectures was obtained by J.-P. Serre, who stated in [5652] that for the number $C_E(x)$ of primes $p \le x$ for which E mod p is cyclic one has

$$C_E(x) = (c(E) + o(1))\,\mathrm{li}(x), \tag{6.37}$$

provided the General Riemann Hypothesis is true. Moreover, the constant $c(E)$ is positive if and only if E has an irrational 2-division point. In 1983 M.R. Murty

[4485] showed that for CM-curves (6.37) can be proved unconditionally, and in 1987 he established [4486] the existence of $\lim_{x\to\infty} C_E(x) = \infty$ for a class of non-CM-curves. Three years later, in a joint paper with R. Gupta [2394], he obtained the lower bound

$$C_E(x) \gg \frac{x}{\log^2 x}$$

for curves having an irrational 2-division point.

In 2002 A.C. Cojocaru [1168] gave another proof of J.-P. Serre's result for non-CM-curves with the error term $O(x \log\log x / \log^2 x)$ under a weaker form of the General Riemann Hypothesis. In 2004 A.C. Cojocaru and Murty [1173] gave a fresh proof of Serre's result with a stronger error term. See also A.C. Cojocaru [1169].

For further development see Y.-M. Chen and J. Yu [1040] and N. Nakazawa [4531, 4532].

For the case of curves over a function field see D.A. Clark and M. Kuwata [1119].

14. Still another conjecture was stated in 1988 by N. Koblitz [3428].

For an elliptic curve E over $\mathbf{Q}$ and prime p let N_p denote the number of points of the reduction of E mod p. *Then*

$$\#\{p \le x : N_p \text{ is prime}\} = (c_E + o(1)) \frac{x}{\log^2 x}.$$

In this direction there are results showing that for at least $cx/\log^2 x$ primes below x the number of prime divisors of N_p is bounded. They were achieved by S.A. Miri and V.K. Murty [4331] and J. Steuding and A. Weng [5941] under the General Riemann Hypothesis. A.C. Cojocaru [1170] and H. Iwaniec and J. Jiménez Urroz [3062] proved this unconditionally. In the latter paper the bound $\omega(N_p) \le 3$ for $\gg x/\log^2 x$ primes below x was obtained in the case of CM-curves.

15. Introductions to the theory of elliptic curves were published by J.W.S. Cassels [949], D.H. Husemöller, [2966], A.W. Knapp [3399], J.H. Silverman [5791, 5794], J.H. Silverman and J. Tate [5796].

Chapter 7
Fermat's Last Theorem

7.1 Classical Approach

1. *Fermat's Last Theorem* (*FLT*), stating that for $n \geq 3$ the equation

$$x^n + y^n = z^n \tag{7.1}$$

has no solution in positive integers x, y, z, was formulated by Fermat in the 17th century. The case $n = 4$ was settled by Fermat ([1989, nr. 45]; for an analysis of Fermat's proof see [1690, Sect. 1.6], cf. Euler [1898]); a simple proof was given in 1986 by Y. Suzuki [5997]. The case $n = 3$ was settled by Euler [1901]; for a simple proof in this case see H.B. Mann, W.A. Webb [4142]. There were some doubts about the validity of Euler's argument in the cubic case, since he used an assertion without providing a proof of it. It turned out later that this assertion can be found with a correct proof in another paper by Euler [1900]; see G. Bergmann [438] on this question. The next cases to be settled were $n = 5$ (A.M. Legendre in the third edition of [3767] and P.G. Dirichlet [1582]) and $n = 14$ (Dirichlet [1583]). An important step forward was made by E.E. Kummer, who in 1850 [3578] gave the following sufficient criterion for the truth of Fermat's assertion in the case of a prime exponent (to which the general case reduces easily).

If p is an odd prime with the property that none of the Bernoulli numbers B_{2k} with $k \leq (p-3)/2$ has its numerator divisible by p, then equation (7.1) *has no positive solutions for $n = p$.*

Recall that Bernoulli numbers B_n are defined by the identity

$$\sum_{n=0}^{\infty} \frac{B_n}{n!} z^n = \frac{z}{e^z - 1},$$

and this implies that they vanish for odd indices $n \geq 3$. These numbers are of importance in various parts of mathematics (see [5818] for a large bibliography covering the years 1713–1983). Kummer's condition is equivalent to the non-divisibility by p of the class-number h_p of the pth cyclotomic field. Such primes are called *regular*, and it is still unknown whether there are infinitely many of them. On the other hand,

W. Narkiewicz, *Rational Number Theory in the 20th Century*,
Springer Monographs in Mathematics,
DOI 10.1007/978-0-85729-532-3_7, © Springer-Verlag London Limited 2012

it was proved by K.L. Jensen in 1915 [3123] that the set of irregular primes is infinite (cf. L. Carlitz [902], H.S. Vandiver [6332]).

Kummer proved that his condition is satisfied by all primes $p < 100$ with the exception of $p = 37, 59$ and 67. Later [3580] he was able to establish *FLT* also for those three exponents. As noted by F. Mertens [4261] and H.S. Vandiver [6315] (see also A. Weil's comments in [3583]), not all of Kummer's arguments were correct, but since that time all inaccuracies in his proofs have been eliminated, mostly through the work of H.S. Vandiver [6315, 6317] and P. Dénes [1473].

Kummer [3580] showed also that for the truth of *FLT* for prime exponents in the so-called first case (in which the exponent n in (7.1) satisfies $(n, xyz) = 1$) it suffices to have in the sequence B_{2k} $(k \le (p-3)/2)$ at most one term with numerator divisible by p.

An exposition of Kummer's work on Fermat's theorem was given in the book by H.M. Edwards [1690]. A modern proof of Kummer's criterion was given in 1994 by W.G. McCallum [4222] using a method of bounding the number of rational points on curves of genus ≥ 2, developed by R.F. Coleman [1180].

In 1965 H.L. Montgomery [4352] proved that for every prime p there exist infinitely many irregular primes not congruent to 1 mod p, and this was generalized by T. Metsänkylä [4269, 4271], who showed that if H is a proper subgroup of the multiplicative group of residues $\bmod m$ then infinitely many irregular primes lie outside H. It was conjectured by Kummer that there are $(\frac{1}{2} + o(1))\pi(x)$ irregular primes below x, but this does not seem to correspond to computations performed by J.L. Selfridge, C.A. Nicol and H.S. Vandiver [5629], and C.L. Siegel [5771] suggested that the constant $1/2$ should be replaced by $\sqrt{e} - 1 = 0.6487\ldots$.

Several authors extended the list of irregular primes, and now all such primes below $12 \cdot 10^6$ are known (R. Ernvall and T. Metsänkylä [1870, 1871], J.P. Buhler et al. [827–829]).

2. The first progress in the 20th century was made by D. Mirimanoff[1] [4332] who strengthened Kummer's [3580] result for the first case of *FLT*. He showed that if $r(p)$ denotes the number of non-zero Bernoulli numbers B_k with $k \ge p - 3$, divisible by p ($r(p)$ is called the *index of irregularity of* p) and *FLT* for p fails in the first case, then $r(p) \ge 4$. This permitted him to establish *FLT* in this case for all prime exponents $p < 257$, the previous record being $p < 223$ (E. Maillet [4108]). A few years later L.E. Dickson extended this first to $p < 1700$ [1538, 1539], and then to $p < 6857$ [1539].

D. Mirimanoff's lower bound for $r(p)$ was replaced for very large p (exceeding $(45!)^{88} > 10^{4934}$) by $r(p) \ge 2[\log^{1/3} p]$ (M. Krasner [3511] in 1934), followed by $r(p) \ge 9$ (H. Wada [6487]), $r(p) \ge \max\{22, \log^{1/3} p\}$ (W. Keller, G. Löh [3296]), $r(p) > \log^{2/5} p$, $r(p) \ge (\log p / \log\log p)^{1/2}$ (A. Granville [2313]), and $r(p) > 2(\log p / \log\log p)^{1/2}$ for large p (V. Jha [3128]).

A survey of research on the index of irregularity was prepared in 1987 by T. Metsänkylä [4272].

[1] Dimitry Mirimanoff (1861–1945), professor in Geneva. See [6330].

3. In 1909 L.E. Dickson [1540] considered for fixed prime p and sufficiently large prime q the congruence

$$x^p + y^p + z^p \equiv 0 \pmod{q},$$

and gave a lower bound for the number of its solutions not divisible by q (cf. G. Cornacchia [1248]). It was shown earlier by A.E. Pellet [4766]), confirming an assertion stated in 1832 by G. Libri [3884, p. 275], that this congruence has non-trivial solutions for every large prime q, destroying the hope that one could prove Fermat's assertion using congruences.

Dickson's result was subsequently extended by A. Hurwitz [2964] (cf. W. Jänichen [3103]) to congruences of the form

$$ax^p + by^p + cz^p \equiv 0 \pmod{q}.$$

4. In 1909 a prize of 100 000 German Mark for the proof of Fermat's last theorem was announced [5009] as a result of the last will of P. Wolfskehl[2]. This brought an avalanche of fallacious proofs. For a long time the assistants in the Mathematical Seminar at Göttingen University had the duty of finding the errors, and the editors of the journal *Mathematische Annalen* announced that they would not consider papers containing alleged proofs of Fermat's theorem [6837].

E.T. Bell [397] recalled in 1923 a story that É. Lucas some day "in less than a quarter of hour" reduced the proof of Fermat's theorem to a problem of periodicity of certain symmetric functions of polynomial roots. Unfortunately, no trace of this could be found in Lucas' manuscripts.

5. Essential progress was achieved by A. Wieferich [6657] in 1909. Using the theory of algebraic numbers and utilizing certain congruences established by D. Mirimanoff [4332] four years earlier, he succeeded in showing that in the first case of *FLT* with exponent p one has

$$2^{p-1} \equiv 1 \pmod{p^2}.$$

This condition is equivalent to the statement that the numerator of the sum

$$\sum_{k=1}^{(p-1)/2} \frac{1}{k}$$

is divisible by p (see, e.g., P. Bachmann [201], M. Lerch [3837] and M.A. Stern [5934]). Primes satisfying this condition are called *Wieferich primes*. Even today only two Wieferich primes are known, namely, 1093 and 3511, found by W. Meissner [4237] in 1913 and N.G.W.H. Beeger [382] in 1922, respectively (the first examples of primes p for which the congruence $x^{p-1} \equiv 1 \pmod{p^2}$ is solvable with $x > 1$ had been given already in 1828 by C.G.J. Jacobi [3076]). It is not yet excluded that every large prime satisfies Wieferich's congruence.

---

[2]Paul Wolfskehl (1856–1906), studied medicine and mathematics, lectured in Darmstadt. See [329].

In 1939 N.G.W.H. Beeger [383] showed that there are no Wieferich primes other than 1093, 3511 below 16 000 and this search was later extended up to $3 \cdot 10^9$ (J. Brillhart, J. Tonascia and P.J. Weinberger [733]), $6 \cdot 10^9$ (D.H. Lehmer [3791]), $4 \cdot 10^{12}$ (R. Crandall, K. Dilcher and C. Pomerance [1276]) and $1.25 \cdot 10^{15}$ (J. Knauer and J. Richstein [3402]).

In 1988 J.H. Silverman [5793] showed that the ABC conjecture implies that there are infinitely many non-Wieferich primes. This follows also from a conjecture of P. Erdős's stating that every positive integer is the sum of a square-free number and a power of 2 (A. Granville and K. Soundararajan [2323]).

D. Mirimanoff [4333] simplified Wieferich's proof, and showed later [4334, 4335] that also the congruence $3^{p-1} \equiv 1 \pmod{p^2}$ gives a necessary condition. Simpler proofs of both results were given by G. Frobenius [2115, 2116], and P. Furtwängler [2162] made a common generalization, simpler proofs of which were found later by R. Fueter [2131] and H.S. Vandiver [6313, 6314].

These conditions were later extended to the form

$$q^{p-1} \equiv 1 \pmod{p^2} \tag{7.2}$$

for all primes $q \leq 43$, consecutively by D. Mirimanoff [4335], H.S. Vandiver [6312], G. Frobenius [2119], F. Pollaczek[3] [4942], T. Morishima[4] [4427], [4429], and J.B. Rosser [5297, 5298]. T. Morishima [4429] showed that these conditions for $q \leq 31$ apply also to Fermat's equation with exponent p in the pth cyclotomic field (for an extension see V.A. Kolyvagin [3466]).

Using T. Morishima's result [4429], J.B. Rosser [5296] proved in 1939 that Fermat's assertion is true in the first case for all prime exponents below 8 332 403, and the following year [5297] he extended this range to $p < 41 \cdot 10^6$. This record was beaten in 1941 by D.H. Lehmer and E. Lehmer [3793] who got up to 253 747 889.

However, some doubts were expressed by N.G. Gunderson [2380] about the correctness of arguments leading to this result for primes between 37 and 43. Fortunately, later work by A. Granville and M.B. Monagan [2319] confirmed the result, extended it to $q \leq 89$ and confirmed Fermat's statement in the first case for all primes $p \leq 7.1 \cdot 10^{14}$. This was improved to $p \leq 1.5 \cdot 10^{17}$ by J.W. Tanner and S.S. Wagstaff, Jr. [6047] and to $p < 7.568 \cdot 10^{17}$ by D. Coppersmith [1247]. In 1994 J. Suzuki [5996] showed that Wieferich's criterion works for all $q \leq 113$, and this allowed him to establish *FLT* in the first case for $p < 8.858 \cdot 10^{20}$.

6. P. Furtwängler's results [2153, 2156] in the class-field theory were used in 1910 by E. Hecke [2674] to prove Fermat's assertion in the first case for those primes p for which p^2 does not divide the ratio $h_p^- = h_p / h_p^+$, h_p^+ being the class-number of the maximal real subfield K_p^+ of K_p. The numbers h_p^-, h_p^+ were called the first and the second factors of h_p, but in recent times h_p^- is rather called the *relative*

---

[3] Felix Pollaczek (1892–1981), one of the pioneers of queueing theory, worked in Berlin and Paris. See [1149, 5567].

[4] Taro Morishima (1903–1989), professor in Tokyo.

class number or the *minus class-number*. Similar definitions apply also to the case of cyclotomic fields $\mathbf{Q}(\zeta_q)$ with prime power q.

The computation of the class-number h_p^+ presents difficulties and its value is known only for primes $p \leq 71$ ($p \leq 163$ under the General Riemann Hypothesis). In 2003 R. Schoof [5565] computed for $p < 10\,000$ the cardinality of a certain subgroup A_p of the class-group of K_p^+, giving thus a lower bound for h_p^+, and conjectured that actually A_p equals the full class-group.

It is easier to compute h_p^-, and this had already been done for $p < 100$ by E.E. Kummer [3582]. This was extended to $p < 257$ by G. Schrutka von Rechtenstamm [5570], to $p < 521$ by D.H. Lehmer and J.M. Masley [3799], to $p < 1097$ by S. Pajunen [4721, 4722], to $p < 3000$ by G. Fung, A. Granville and H.C. Williams [2144] and to $p < 10\,000$ by M.A. Shokrollahi [5720].

The following elementary formula for h_p^- was given by L. Carlitz and F.R. Olson [903]:

$$h_p^- = p^{-(p-3)/2}|\det D_p|,$$

where $D_p = [a_{ij}]$ denotes the $\frac{p-1}{2} \times \frac{p-1}{2}$ matrix, called the *Maillet matrix*, defined by

$$a_{ij} = ij' \mod p,$$

j' denoting the inverse of j mod p (E. Maillet [4111]). They stated also that the same formula was obtained independently by A. Weil and S. Chowla. Earlier, in 1914, E. Malo [4117] computed values of D_p for small p, and conjectured (incorrectly, as we now know) the equality $D_p = (-p)^{(p-3)/2}$ (cf. H.W. Turnbull[5] [6234]).

A similar formula for $h_{p^k}^-$ (defined as the ratio between the class-numbers of $\mathbf{Q}(\zeta_{p^k})$ and its maximal real subfield) was established by T. Metsänkylä [4266] in 1967.

7. In 1910 Hecke's result was strengthened by P. Furtwängler [2160] who proved that if Fermat's theorem fails in the first case for a prime p, then p^4 divides h_p^-. This line of research was later pursued by H.S. Vandiver [6314], who proved that under these circumstances one has $p^8|h_p^-$. Later T. Morishima [4428, 4429] and D.H. Lehmer [3775] replaced the exponent 8 by 12, and in 1952 T. Morishima [4430] stepped it up to 13.

In 1965 M. Eichler [1701] gave a remarkably simple argument, showing that if *FLT* fails in the first case, then h_p^- must be divisible by p^a with $a = [\sqrt{p}] - 1$.

A parallel result for the second case (i.e., when p divides xyz) was achieved by F. Bernstein [465], who proved Fermat's assertion in the case that p divides the class-number of the p^2th cyclotomic field in the first power.

---

[5] Herbert Westren Turnbull (1885–1961), professor in Oxford. See [3757].

8. It was conjectured in 1851 by E.E. Kummer [3579] that h_p^- is asymptotically equivalent to

$$L(p) = 2p\left(\frac{\sqrt{p}}{2\pi}\right)^{(p-1)/2}.$$

The first step in this problem was in 1949 by N.C. Ankeny and S. Chowla [103, 104], who obtained

$$\log h_p^- = \log L(p) + o(\log p), \tag{7.3}$$

and in 1953 T. Tatuzawa [6069] proved

$$h_p^- - \log L(p) \leq \log^c p \tag{7.4}$$

for some c.

The equality (7.3) was extended to the fields $\mathbf{Q}(\zeta_n)$ with composite n by T. Lepistö [3828, 3829]. Later he showed [3831] that in (7.4) one can put $c = 2$ and got also an explicit lower bound for the difference $h_p^- - \log L(p)$, which implied that for $p \geq 230$ the class-number of $\mathbf{Q}(\zeta_p)$ exceeds 10^{46}.

In 1964 C.L. Siegel [5771] proved

$$\lim_{p\to\infty} \frac{\log h_p^-}{\log L(p)} = \frac{1}{4},$$

and in 1976 J.M. Masley and H.L. Montgomery [4167] established for large p the inequalities

$$0.199p^{5.5} \leq \frac{h_p^-}{L(p)} \leq 0.2p^7.$$

Later, in 1990, A. Granville [2314] proved that for a positive fraction of primes the ratio $h_p^-/L(p)$ lies between two positive constants, and in 2001 M.R. Murty and Y.N. Petridis [4493] established the last result for all primes.

9. Siegel's result in [5771] implied that there are only finitely many cyclotomic fields $\mathbf{Q}(\zeta_p)$ (with prime p) with a given class-number and he conjectured that class-number one occurs only for $p \leq 19$. It was shown in 1970 by T. Metsänkylä [4268] that the factor h_p^- of h_p exceeds $2p\exp((p-3)/1000)$, which implies $h_p^- > 1$ for $p > 5 \cdot 10^7$. K. Uchida showed that this happens already for $p > 2400$ [6246] and the final step in proving Siegel's conjecture was made by K. Uchida [6247] in 1971. Five years later J.M. Masley and H.L. Montgomery [4167] determined all cyclotomic fields with class-number one. For fixed p the sequence $h_{p^k}^-$ is for $p^{k+1} > 100$ strictly increasing, as shown by T. Metsänkylä [4270] in 1972.

In [6245] K. Uchida proved that there are only finitely many complex Abelian fields with a given class-number. Among them there are 172 fields with class-number one, as shown in 1992 by K. Yamamura [6774, 6775]. All complex Abelian fields K with $h(K) = h(K^+)$ were found in 2000 by K.-Y. Chang and S.-H. Kwon [997].

10. In 1929 H.S. Vandiver [6321] first extended Kummer's list of prime exponents for which Fermat's equation does not have positive solutions by showing that this holds for all primes below 211. Later he extended this bound, first to $p < 269$ [6322], then to $p < 307$ [6323] and in 1939 he reached 619 [6326]. He published more than 40 papers on *FLT*, discovering several new criteria for the truth of *FLT* (see, e.g., [6316, 6318, 6320] and [6314]). In 1940 he presented a criterion in the first case, using Euler's numbers [6327], and in 1946 he published a broad survey [6329] of work done on Fermat's Last Theorem, expressing the view that it is certainly true in the first case, and hoping that it will be found to be false in general. He wrote: "I can think of nothing more interesting from the standpoint of the development of number theory, than to have it turn out that the Fermat relation has solutions, for a finite number > 0, of primes l."

11. In 1932 H.S. Vandiver [6324, 6325] gave a short proof of Kummer's theorem about Fermat's Last Theorem in the case of regular prime exponents, and in the following year M. Moriya [4433] gave a simple proof of E. Maillet's result [4108] on the insolvability of Fermat's equation

$$x^{p^k} + y^{p^k} = z^{p^k}$$

for prime p and sufficiently large k in the case when $p \nmid xyz$, showing that Fermat's assertion holds in the first case for infinitely many pairwise co-prime integers.

See also H. Kapferer [3247], Y. Hellegouarch [2723], L.C. Washington [6569], S. Sitaraman [5800].

12. In the general case, H.S. Vandiver's record ($p \leq 619$) was surpassed in 1954 by D.H. Lehmer, E. Lehmer and H.S. Vandiver, who went up to $p < 2000$ [3798] and H.S. Vandiver [6331] extended this to $p \leq 2521$. In 1955 J.L. Selfridge, C.A. Nicol and H.S. Vandiver [5629] reached $p \leq 4002$, and in 1964 J.L. Selfridge and B.W. Pollack [5630] got up to 25 000. Later the development acquired more speed. In 1975 W. Johnson [3148] reached 30 000, three years later S.S. Wagstaff, Jr. [6489] got up to $125 \cdot 10^3$, and in 1987 J.W. Tanner and S.S. Wagstaff, Jr. [6046] covered the range $[125 \cdot 10^3, 150 \cdot 10^3]$.

13. In 1951 P. Dénes [1471] returned to the idea of Sophie Germain who in the early 19th century showed that if $p \geq 3$ and $2p + 1$ are both primes, then Fermat's assertion holds in the first case for the exponent p. He showed that in Germain's result one can replace the prime $2p + 1$ by any prime of the form $2kp + 1$ with $k \leq 55$ and $3 \nmid k$.

In the last result one can take any $k \leq 89$, as shown in 1977 by A.V. Tolstikov [6190]. It is still not known whether the assumptions of Dénes's or Tolstikov's theorems are satisfied by infinitely many prime numbers, but a development of this idea was utilized in 1985 by L.M. Adleman and D.R. Heath-Brown [18] to prove that the first case of Fermat's theorem holds for infinitely many prime exponents. The crucial point in their proof was provided by a theorem of É. Fouvry [2050] who showed that for a positive proportion of primes $p \equiv 2 \pmod 3$ the maximal prime divisor of $p-1$ exceeds p^δ with some $\delta > 2/3$. This was known earlier for some smaller values of δ (C. Hooley [2866], J.-M. Deshouillers, H. Iwaniec [1492], É. Fouvry [2046]). In the same year A. Granville [2312] and D.R. Heath-Brown [2639] deduced from Faltings' theorem that Fermat's assertion holds for almost all exponents. In the first case this had been shown earlier by N.C. Ankeny [101].

For an improvement of Dénes' result see A. Simalarides [5797], who also presented an elementary criterion for the truth of Fermat's assertion in terms of Lucas numbers.

The bound $O(N \log^{2+1/(p-1)} N)$ for the number of solutions of $x^p + y^p = z^p$ with a fixed prime p and $1 \le x, y, z \le N$ was established in 1951 by P. Turán [6222], and in a joint paper with P. Dénes [1474] he improved this to $O(N^{2/p}/\log^{2(1-1/p)})$.

In 1965 D. Mumford [4476] obtained the bound $O(\log\log N)$ as a special case of his result dealing with rational points on curves of genus $g \ge 2$.

Lower bounds for the unknowns in Fermat's equation were given by R. Obláth [4650] (cf. H.J.A. Duparc and A. van Wijngaarden [1654]), K. Inkeri [3026], M.H. Le [3744], A. Grytczuk [2376].

14. A rather surprising result was achieved in 1977 by G. Terjanian [6119], who gave a completely elementary argument to establish the remarkable fact that if p is an odd prime, and x, y, z are positive integers satisfying

$$x^{2p} + y^{2p} = z^{2p},$$

then at least one of the numbers x, y is divisible by $2p$. Under the extra assumption $p \not\equiv 1 \pmod 8$ this had already been proved in 1837 by E.E. Kummer [3575] (a simple proof of Kummer's result was given in 1943 by F. Niedermeier [4597]).

7.2 Finale

1. The first application of modern methods of algebraic geometry to Fermat's problem was made in 1983, when G. Faltings [1954] established in a strong form the finiteness conjecture of I.R. Šafarevič. He showed that there are only finitely many isogeny classes of Abelian varieties of genus $g \ge 2$ having good reduction outside a fixed finite set. This achievement brought him the Fields Medal in 1986. In that way L.J. Mordell's conjecture about rational points on curves of genus ≥ 2 became a theorem, and it resulted, as a particular case, that for every exponent $n > 3$ Fermat's equation can have at most finitely many solutions in co-prime integers.

For expositions of Faltings' proof see P. Deligne [1449] and L. Szpiro [6029]. Cf. also the books by G. Faltings and G. Wüstholz [1955] and M. Hindry and J.H. Silverman [2815].

Earlier the analogue of Mordell's conjecture for curves over function fields with an algebraically closed field of constants of characteristics 0 was established by Yu.I. Manin [4131] (for other proofs see H. Grauert [2329] and A.N. Paršin [4749]). P. Samuel[6] [5383] showed that the same is true also in positive characteristics (for another proof see L. Szpiro [6028]). An exposition of Manin's proof was given by R.F. Coleman [1181], eliminating a gap in it (see also C.-L. Chai [980]). A uniform version of Manin's theorem was given by L. Caporaso [894] and G. Heier [2705].

Other proofs of Mordell's conjecture were later given by E. Bombieri [604], D.W. Masser, G. Wüstholz [4184] and P. Vojta [6453, 6454]. Bombieri's proof gave effective bounds for the number of rational points on a curve of genus ≥ 2. There are still no bounds known for the size of these points in the general case (for Thue equations such bounds in the function field case were provided by W.M. Schmidt [5509]).

See also B. Farhi [1958]. In the function field case this result was obtained by G. Heier [2705].

Faltings' result implied also the existence of a bound $c(E)$ for the degrees of isogenies $E \to E'$. This bound was made effective by D.V. and G.V. Chudnovsky [1104] in the case of the rational field, and by D.W. Masser and G. Wüstholz [4183] in the general case (for an exposition see D. Bertrand [474]). An effective form of this result was given by S. David [1407] (cf. F. Pellarin [4765]). For a generalization to Abelian varieties see D.W. Masser and G. Wüstholz [4184].

2. The prehistory of the Taniyama–Shimura conjecture, which played a great role in the solution of Fermat's problem, had already begun at the end of the thirties, when H. Hasse asked whether for a fixed elliptic curve E its L-function, defined as the product of congruence L-functions of function fields, defined by its reduction E_p, is entire[7].

In 1953 M. Deuring [1506–1509] associated with every algebraic curve defined over an algebraic number field k a zeta-function. In the case when k is the field of rational numbers and E is an elliptic curve, defined by the equation

$$Y^2 = f(X), \tag{7.5}$$

where

$$f(X) = aX^3 + bX^2 + cX + d \in \mathbf{Z}[X],$$

this zeta-function has the form

$$\zeta(s, E) = \prod_p \zeta(s, E_p),$$

where $\zeta(s, E_p)$ is the congruence zeta-function of the function field $\mathbf{F}_p(x, y)$, corresponding to the reduction $\bmod p$ of the curve (7.5). It turned out that if

[6] Pierre Samuel (1921–2009), professor in Paris.

[7] I was informed by Peter Roquette that Hasse gave this problem to one of his students before 1939.

for every prime p not dividing the discriminant $\Delta(E)$ of the polynomial we put $t_p = p + 1 - A_p$, A_p being the number of solutions of the congruence

$$Y^2 \equiv f(X) \pmod{p},$$

and for primes $p|\Delta(E)$ one puts $t_p = 0, -1$ or 1, depending on the geometric properties of the reduction of E mod p, then

$$\zeta(E,s) = \zeta(s)\zeta(s-1)L_E(s),$$

with

$$L_E(s) = \prod_{p|\Delta} \frac{1}{1 - t_p p^{-s}} \prod_{p\nmid\Delta} \frac{1}{1 - t_p p^{-s} + p^{1-2s}}. \tag{7.6}$$

The L-function so defined is regular for $\Re s > 3/2$, and M. Deuring proved in [1506–1509] that for all elliptic curves admitting complex multiplication it can be extended to an entire function. In certain special cases the function $L_E(s)$ occurred in earlier papers. Wiman [6687] showed that the L-function of the elliptic curve

$$y^2 = x^3 + (3 \cdot 7 \cdot 11 \cdot 17 \cdot 41)^2 x$$

can be continued across the line $\Re s = 3/2$ and has a zero of order ≥ 4 at $s = 1$ and A. Weil [6620] (cf. Y. Taniyama [6045]) showed in 1952 that in certain cases the L-function equals a product of Hecke L-functions with shifted argument, hence it can be continued to an entire function, and conjectured that this is true for all elliptic curves. This was established by M. Deuring [1506–1509] for all elliptic curves admitting complex multiplication, and in 1966 G. Shimura [5695] did this for some particular curves without *CM* (see also M. Eichler [1699], G. Shimura [5692, 5693]).

3. An elliptic curve E is called *modular* if its L-function coincides with the Dirichlet series associated to a modular form[8]. In 1955, during the Tokyo symposium on algebraic number theory, Taniyama formulated certain problems which later, in hands of G. Shimura [5692, 5693, 5696, 5698] and A. Weil [6627] acquired the following form.

Every elliptic curve over **Q** *is modular.*

For elliptic curves with complex multiplication this conjecture was established by G. Shimura [5699, 5700] in 1971, and in 1987 J.-P. Serre [5656] deduced it in full generality from his conjecture on representations of the Galois group of the field of all algebraic numbers.

The end of the century brought the expected proof of the Taniyama–Shimura conjecture in the general case. The last steps were achieved through the efforts of B. Conrad and F. Diamond [1518], F. Diamond and K. Kramer [1519], F. Diamond and R. Taylor [1199] and C. Breuil, B. Conrad, F. Diamond and R. Taylor [718].

[8] For the history of this conjecture see S. Lang [3703].

4. A more general conjecture, relating odd representations of $G = Gal(\overline{\mathbf{Q}}/\mathbf{Q})$ and modular forms was formulated in 1973 by J.-P. Serre (first published in [5655, 5656], for an earlier special case see [5648, Sect. 3]).

Let ρ be an irreducible 2-dimensional continuous representation of the Galois group $G_{\mathbf{Q}}$ of $\overline{\mathbf{Q}}/\mathbf{Q}$ over the algebraic closure of the finite field $\mathbf{F}_\ell$. It is called an *odd representation* if for the complex conjugation τ one has $\det\rho(\tau) = -1$. Serre's conjecture asserts that if ρ is odd, then there exists a cusp form f, which is an eigenfunction of all Hecke operators and has the property that ρ coincides with the reduction of the representation ρ_f, associated with f, modulo a prime ideal $\mathfrak{p}$, dividing p, of the ring of integers of the field containing the coefficients of f. Moreover J.-P. Serre predicted the values of the level $N(\rho)$, weight k_ρ and character $\varepsilon_\rho(\ell)$ for f. If this happens, then one says that the representation ρ is modular. (Note that the Taniyama–Shimura conjecture is a consequence of Serre's conjecture, and so is Fermat's Last Theorem.)

Some modifications were later introduced (J.-P. Serre [5657]; cf. B. Edixhoven [1689]), and this led to a modified definition of k_ρ. B. Edixhoven [1687] used it to show that if a representation ρ is modular of some height, then it is also modular of weight k_ρ. The same assertion for the level $N(\rho)$ was established by H. Carayol [897] for $l \geq 5$, with the use of K. Ribet's [5188] method of level-lowering.

The status of the conjecture at the end of the century was described by K. Ribet and W.A. Stein [5192] (cf. B. Edixhoven [1689]).

In 2006 C. Khare [3312] established Serre's conjecture in the case $k = 1$. Surveys of the proof were given by C. Khare [3313] and J.-P. Wintenberger [6689]. L. Dieulefait [1573] established the conjecture for $k = 2$, $N = 1$, and C. Khare and J.-P. Wintenberger [3314] did this for certain small levels and weights. Finally they presented in [3315, 3316] the proof of the complete conjecture (cf. M. Kisin [3344]).

For an analogue of Serre's conjecture for n-dimensional representations see A. Ash [149], A. Ash and W. Sinnott [151], and A. Ash, D. Doud and D. Pollack [150].

It was conjectured by R.P. Langlands (see [3713, 3714], A. Borel [629]) that every representation of $Gal(\mathbf{Q}/Q)$ in $GL_2(\mathbf{C})$ is induced by a modular form (see [631]). The local form of this conjecture was established by M. Harris, R. Taylor [2572] and G. G. Henniart [2729].

See also G. Henniart [2730]. An analogue of Serre's conjecture for representations of the Galois group of the extension $\overline{\mathbf{Q}}/K$, where K is an imaginary quadratic field, was formulated in 1999 by L.M. Figueredo [1999]. It was confirmed in certain cases by M.H. Şengün [5637].

5. The first relation between elliptic curves and Fermat's problem appeared in a paper by Y. Hellegouarch [2722] in 1970. See also V.A. Demyanenko [1460], G. Frey [2088], Y. Hellegouarch [2724] and J. Vélu [6379].

In 1986 G. Frey [2089, 2090] associated with a presumed counterexample $a^p - b^p = c^p$ to Fermat's assertion for a prime exponent $p \geq 5$ an elliptic curve E having several remarkable properties, and wrote: "... the properties of E are so excellent that one suspects that such a curve cannot exist" [2089, p. 2]. The first version of E was

$$E: \quad y^2 + xy = x^3 + \frac{a^p + b^p - 1}{4}x^2 + \frac{(ab)^p}{4}x, \tag{7.7}$$

(with $2|a$ and $b \equiv 1 \pmod 4$), but it turned out later that it is advisable to consider the curve

$$y^2 = x(x - a^p)(x - b^p). \tag{7.8}$$

In particular the curve E would be a counterexample to the conjecture of Szpiro, at least for large p. Frey also gave reasons to believe that the existence of E would contradict two important conjectures: the Taniyama–Shimura conjecture and J.-P. Serre's conjecture [5648, 5655, 5656] concerning modular representations of weight 2.

6. It was established later by K. Ribet [5188, 5189] that if $p \geq 5$ and $a^p + b^p = c^p$, then the curve (7.8) is not modular, hence Fermat's theorem is a consequence of the Taniyama–Shimura conjecture.

Finally the path-breaking work of A. Wiles and R. Taylor [6089, 6670], who established the modularity of a large class of curves, led in 1995 to a complete solution of Fermat's problem. Expositions of Wiles's proof can be found in [1249], H. Darmon, F. Diamond and R. Taylor [1333], K. Ribet [5190], K. Rubin and A. Silverberg [5331], (see also Y. Hellegouarch [2725]). A simplification was given by N.D. Elkies [1722].

For a presentation of Wiles' approach see H. Darmon [1332], J.-P. Serre [5660], J. Oesterlé [4659].

Fermat's last theorem in cyclotomic fields was considered by V.A. Kolyvagin [3465–3467].

For the history of Fermat's last theorem see the books of P. Ribenboim [5178] and A.J. van der Poorten [6303]. The story of Wiles' proof was described by C.J. Mozzochi [4463].

7. Methods created in the proof of Fermat's Last Theorem turned out to be very useful in dealing with related Diophantine equations.

Important progress was made towards the proof of the following conjecture, formulated by A. Beal (see R.D. Mauldin [4211]).

If the integers r, s, t exceed 2, then the equation

$$x^r + y^s = z^t \tag{7.9}$$

has no solutions in positive integers x, y, z satisfying $(x, y, z) = 1$.

It was shown in 1995 by H. Darmon and A. Granville [1334] with the use of Faltings' theorem that if $1/r + 1/s + 1/t < 1$, then equation (7.9) can have at most finitely many primitive solutions, i.e., satisfying $(x, y, z) = 1$. Since from

$r, s, t \geq 3$ the inequality $1/r + 1/s + 1/t < 1$ follows with the exception of the case $r = s = t = 3$, hence Beal's equations have finitely many solutions, the exceptional case being covered by the cubic case of Fermat's Last Theorem. They showed also that in the case $1/r + 1/s + 1/t > 1$ the equation (7.9) has infinitely many primitive solutions, and F. Beukers [489] proved that all these solutions are contained in a finite set of parametrized solutions. In some cases with $\min\{r, s, t\} = 2$ the corresponding parameterizations were explicitly given in N. Bruin's thesis [790].

For certain triples (r, s, t) Beal's conjecture was established (see M.A. Bennett [414], N. Bruin [791], H. Darmon [1330, 1331], H. Darmon, L. Merel [1335], A. Kraus [3520], B. Poonen [4984]). Some of these results were achieved assuming the truth of the Taniyama–Shimura conjecture, which is now a theorem.

If one allows one of the numbers r, s, t to be equal 2, then (7.9) may have primitive solutions, even if $1/r + 1/s + 1/t < 1$ holds, as the examples $1^3 + 2^3 = 3^2$ or $2^5 + 7^2 = 3^4$ show. Eight other examples are listed by H. Darmon and A. Granville [1334] and R.D. Mauldin [4211], and it was conjectured in [1334] that this list is complete. For several cases for which all such solutions can be listed see N. Bruin [789, 790, 792, 793] and B. Poonen, E.F. Schaefer and M. Stoll [4988]. In 1915 L. Holzer [2850] had already proved that there are no primitive solutions for $(r, s, t) = (2, 6, 3)$.

All polynomial solutions of (7.9) had already been determined in 1904 by V.P. Velmin[9] [6378]. His result was rediscovered in 1917 by A. Korselt [3492].

8. The same approach was used by K. Ribet [5191] to prove that for prime $p \equiv 1 \pmod 4$ the equation

$$x^p + 2^\alpha y^p + z^p = 0 \quad (\alpha \geq 1)$$

has no non-zero solutions. This confirmed an old conjecture by P. Dénes [1472]. Ribet noted also that for primes $p \geq 11$ and $q \in \{3, 5, 7, 11, 13, 17, 19, 23, 29, 53, 59\}$ the results of [5188, 5656, 6089, 6670] can be used to show that the equation

$$x^p + q^\alpha y^p + z^p = 0 \quad (\alpha \geq 0)$$

has no non-trivial solutions.

The more general equation

$$Ax^p + By^p = Cz^p \tag{7.10}$$

was considered by A. Kraus [3519], who established the non-existence of non-trivial solutions in several cases.

For further studies of (7.10) see E. Halberstadt and A. Kraus [2448].

This method has also been applied by M.A. Bennett and C.M. Skinner [419] and W. Ivorra and A. Kraus [3050] to treat various equations of the form $ax^m + by^n = cz^2$.

[9] Vladimir Petrovič Velmin (1885–1974), professor in Warsaw, Rostov and Kiev. See [3220].

References[1]

1. Aaltonen, M., Inkeri, K.: Catalan's equation $x^p - y^q = 1$ and related congruences. Math. Comput. **56**, 359–370 (1991)
2. Aas, H.-F.: Congruences for the coefficients of the modular invariant $j(\tau)$. Math. Scand. **14**, 185–192 (1964)
3. Aas, H.-F.: Congruences for the coefficients of the modular invariant $j(\tau)$. Math. Scand. **15**, 64–68 (1964)
4. Abel-Hollinger, C.S., Zimmer, H.G.: Torsion groups of elliptic curves with integral j-invariant over multiquadratic fields. In: Number-Theoretic and Algebraic Methods in Computer Science, Moscow, 1993, pp. 69–87. World Scientific, Singapore (1995)
5. Abikoff, W., Birman, J.S., Kuiken, K. (eds.): The Mathematical Legacy of Wilhelm Magnus: Groups, Geometry and Special Functions. Contemp. Math., vol. 169. Am. Math. Soc., Providence (1994)
6. Ablialimov, S.B.: The number of lattice points in ovals. Izv. Vysš. Učebn. Zaved., Mat. **3**, 30–37 (1970) (in Russian)
7. Abouzaid, M.: Les nombres de Lucas et Lehmer sans diviseurs primitif. J. Théor. Nr. Bordx. **18**, 299–313 (2006)
8. Abramovich, D.: Formal finiteness and the torsion conjecture on elliptic curves. A footnote to the paper "Rational torsion of prime order in elliptic curves over number fields" by S. Kamienny and B. Mazur. Astérisque **228**(3), 5–17 (1995)
9. Abraškin, V.A.: Good reduction of two-dimensional Abelian varieties. Izv. Akad. Nauk SSSR, Ser. Mat. **40**, 262–272 (1976) (in Russian)
10. Abraškin, V.A.: p-divisible groups over Z. Izv. Akad. Nauk SSSR, Ser. Mat. **41**, 987–1007 (1977) (in Russian)
11. Acerbi, F.: A reference to perfect numbers in Plato's Theaetetus. Arch. Hist. Exact Sci. **59**, 319–348 (2005)
12. Acreman, D., Loxton, J.H.: Asymptotic analysis of Ramanujan pairs. Aequ. Math. **30**, 106–117 (1986)
13. Adamczewski, B., Bugeaud, Y.: On the complexity of algebraic numbers I. Expansions in integer bases. Ann. Math. **165**, 547–565 (2007)
14. Adamczewski, B., Bugeaud, Y.: On the complexity of algebraic numbers II. Continued fractions. Acta Math. **195**, 1–20 (2005)
15. Adams, W.W.: Transcendental numbers in the P-adic domain. Am. J. Math. **88**, 279–308 (1966)
16. Adhikari, S.D., Balasubramanian, R., Sankaranarayanan, A.: An Ω-result related to $r_4(n)$. Hardy-Ramanujan J. **12**, 20–30 (1989)

[1] Abbreviations are listed in accordance with ISSN standards.

W. Narkiewicz, *Rational Number Theory in the 20th Century*,
Springer Monographs in Mathematics,
DOI 10.1007/978-0-85729-532-3,

17. Adhikari, S.D., Pétermann, Y.-F.S.: Lattice points in ellipsoids. Acta Arith. **59**, 329–338 (1991)
18. Adleman, L.M., Heath-Brown, D.R.: The first case of Fermat's last theorem. Invent. Math. **79**, 409–416 (1985)
19. Adleman, L.M., Huang, M.-D.A.: Counting rational points on curves and Abelian varieties over finite fields. In: Lecture Notes in Comput. Sci., vol. 1122, pp. 1–16. Springer, Berlin (1996)
20. Adleman, L.M., Huang, M.-D.A.: Counting points on curves and Abelian varieties over finite fields. J. Symb. Comput. **32**, 171–189 (2001)
21. Adleman, L.M., Pomerance, C., Rumely, R.: On distinguishing prime numbers from composite numbers. Ann. Math. **117**, 173–206 (1983)
22. Adolphine, D.: Solution de question 386. Nouv. Ann. Math. **16**, 288–290 (1857)
23. Agapito, J.: Weighted Brianchon-Gram decomposition. Can. Math. Bull. **49**, 161–169 (2006)
24. Agrawal, M., Kayal, N., Saxena, N.: Primes is in P. Ann. Math. **160**, 781–793 (2004)
25. Agrawal, M.K., Coates, J.H., Hunt, D.C., van der Poorten, A.J.: Elliptic curves of conductor 11. Math. Comput. **35**, 991–1002 (1980)
26. Ahlfors, L., Carleson, L.: Arne Beurling in memoriam. Acta Math. **161**, 1–9 (1988)
27. Ahlgren, S.: Distribution of the partition function modulo composite integers M. Math. Ann. **318**, 795–803 (2000)
28. Ahlgren, S., Boylan, M.: Arithmetic properties of the partition function. Invent. Math. **153**, 487–502 (2003)
29. Ahlgren, S., Boylan, M.: Coefficients of half-integral weight modular forms modulo l^j. Math. Ann. **331**, 219–239 (2005); add.: 241–242
30. Ahlgren, S., Ono, K.: Congruence properties for the partition function. Proc. Natl. Acad. Sci. USA **98**, 12882–12884 (2001)
31. Ahlswede, R., Khachatrian, L.H., Sárkőzy, A.: On the counting function of primitive sets of integers. J. Number Theory **79**, 330–344 (1999)
32. Ahlswede, R., Khachatrian, L.[H.], Sárkőzy, A.: On the density of primitive sets. J. Number Theory **109**, 319–361 (2004)
33. Aicardi, F.: A short proof of Fel's theorems on 3-D Frobenius problem. Funct. Anal. Other Math. **2**, 241–246 (2009)
34. Aicardi, F.: On the geometry of the Frobenius problem. Funct. Anal. Other Math. **2**, 111–127 (2009)
35. Aistleitner, C., Berkes, I.: On the law of the iterated logarithm for the discrepancy of $\langle n_k x \rangle$. Monatshefte Math. **156**, 103–121 (2009)
36. Akbary, A., David, C., Juricevic, R.: Average distributions and products of special values of L-series. Acta Arith. **111**, 239–268 (2004)
37. Akhtar, S.: Elliptic curves of prime power conductor. Punjab Univ. J. Math. **9**, 1–12 (1976)
38. Akhtari, S.: The Diophantine equation $aX^4 - bY^2 = 1$. J. Reine Angew. Math. **630**, 33–57 (2009)
39. Akhtari, S., Togbé, A., Walsh, P.G.: On the equation $aX^4 - bY^2 = 2$. Acta Arith. **131**, 145–169 (2008)
40. Akhtari, S., Togbé, A., Walsh, P.G.: Addendum on the equation $aX^4 - bY^2 = 2$. Acta Arith. **137**, 199–202 (2009)
41. Alaoglu, L., Erdős, P.: On highly composite and similar numbers. Trans. Am. Math. Soc. **56**, 448–469 (1944)
42. Alasia, C.: Ernest Cesàro. 1859–1916. Enseign. Math. **9**, 5–24 (1907)
43. Albert, A.A.: Leonard Eugene Dickson. Bull. Am. Math. Soc. **61**, 331–345 (1955)
44. Alder, H.L.: Generalizations of the Rogers-Ramanujan identities. Pac. J. Math. **4**, 161–168 (1954)
45. Alder, H.L.: Partition identities—from Euler to the present. Am. Math. Mon. **76**, 733–746 (1969)
46. Aleksandrov, P.S. (ed.): Hilbert's Problems. Nauka, Moscow (1969) (in Russian) [German translation: Harri Deutsch, Frankfurt 1998]

47. Aleksandrov, P.S.: To the memory of Lazar Aronovič Lusternik. Usp. Mat. Nauk **37**(1), 125–126 (1982) (in Russian)
48. Aleksandrov, P.S., Oleĭnik, O.A.: In memoriam Richard Courant. Usp. Mat. Nauk **30**(4), 205–226 (1975) (in Russian)
49. Aleksentsev, Yu.M.: On the measure of approximation of π by algebraic numbers. Mat. Zametki **66**, 483–493 (1999) (in Russian)
50. Alemu, Y.: On zeros of forms over local fields. Acta Arith. **45**, 163–171 (1985)
51. Alexander, R.: Principles of a new method in the study of irregularities of distribution. Invent. Math. **103**, 279–296 (1991)
52. Alexanderson, G.L., Lange, L.H.: Obituary: George Pólya. Bull. Lond. Math. Soc. **19**, 559–608 (1987)
53. Alford, W.R., Granville, A., Pomerance, C.: There are infinitely many Carmichael numbers. Ann. Math. **139**, 703–722 (1994)
54. Alladi, K., Gordon, B.: Generalizations of Schur's partition theorem. Manuscr. Math. **79**, 113–126 (1993)
55. Allen, P.B.: On the multiplicity of linear recurrence sequences. J. Number Theory **126**, 212–216 (2007)
56. Allouche, J.-P., Lubiw, A., Mendès France, M., van der Poorten, A.J., Shallit, J.: Convergents of folded continued fractions. Acta Arith. **77**, 77–96 (1996)
57. Alnaser, A., Cochrane, T.: Waring's number mod m. J. Number Theory **128**, 2582–2590 (2008)
58. Alon, N., Nathanson, M.B., Ruzsa, I.: Adding distinct congruence classes modulo a prime. Am. Math. Mon. **102**, 250–255 (1995)
59. Alon, N., Nathanson, M.B., Ruzsa, I.: The polynomial method and restricted sums of congruence classes. J. Number Theory **56**, 404–417 (1996)
60. Alter, R., Curtz, T., Kubota, K.K.: Remarks and results on congruent numbers. In: Proceedings of the Third Southeastern Conference on Combinatorics, Graph Theory and Computing, Boca Raton, pp. 27–35 (1972)
61. Alter, R., Kubota, K.K.: Multiplicities of second order linear recurrences. Trans. Am. Math. Soc. **178**, 271–284 (1973)
62. Alzer, H.: On rational approximation to e. J. Number Theory **68**, 57–62 (1998)
63. Amice, Y.: Fonction Γ p-adique associée à un caractére de Dirichlet. Groupe de travail d'analyse ultramétrique **7–8**(exp. 17), 1–11 (1978/1981)
64. Amice, Y., Fresnel, J.: Fonctions zêta p-adiques des corps de nombres abeliens réels. Acta Arith. **20**, 353–384 (1972)
65. Amice, Y., Vélu, J.: Distributions p-adiques associées aux séries de Hecke. Astérisque **24/25**, 119–131 (1975)
66. Amice, Y., et al.: Charles Pisot. Acta Arith. **51**, 1–4 (1988)
67. Amitsur, S.A.: Arithmetic linear transformations and abstract prime number theorems. Can. J. Math. **13**, 83–109 (1961)
68. Amoroso, F., Viada, E.: Small points on subvarieties of a torus. Duke Math. J. **150**, 407–442 (2009)
69. Amou, M.: On Sprindžuk's classification of transcendental numbers. J. Reine Angew. Math. **470**, 27–50 (1996)
70. An, T.T.H., Wang, J.T.-Y.: An effective Schmidt's subspace theorem for non-linear forms over function fields. J. Number Theory **125**, 210–228 (2007)
71. Anderson, I.: An application of a theorem of de Bruijn, Tengbergen, and Kruyswijk. J. Comb. Theory **3**, 43–47 (1967)
72. Anderson, R.J.: Simple zeros of the Riemann zeta function. J. Number Theory **17**, 176–182 (1983)
73. Anderson, R.J., Stark, H.M.: Oscillation theorems. In: Lecture Notes in Math. vol. 899, pp. 79–106. Springer, Berlin (1981)
74. Andersson, J.: Summation formulae and zeta functions. Ph.D. thesis, Stockholm Univ. (2006)

75. Andersson, J.: Disproof of some conjectures of P. Turán. Acta Math. Acad. Sci. Hung. **117**, 245–250 (2007)
76. Andrews, G.E.: An asymptotic expression for the number of solutions of a general class of Diophantine equations. Trans. Am. Math. Soc. **99**, 272–277 (1961)
77. Andrews, G.E.: A lower bound for the volume of strictly convex bodies with many boundary lattice points. Trans. Am. Math. Soc. **106**, 270–279 (1963)
78. Andrews, G.E.: On the general Rogers-Ramanujan theorem. Mem. Am. Math. Soc. **152**, 1–86 (1974)
79. Andrews, G.E.: On the Alder polynomials and a new generalization of the Rogers-Ramanujan identities. Trans. Am. Math. Soc. **204**, 40–64 (1975)
80. Andrews, G.E.: The Theory of Partitions. Addison-Wesley, Reading (1976). [Reprint: Cambridge, 1998]
81. Andrews, G.E.: An incredible formula of Ramanujan. Aust. Math. Soc. Gaz. **6**, 80–89 (1979)
82. Andrews, G.E.: q-Series: Their Development and Application in Analysis, Number Theory, Combinatorics, Physics, and Computer Algebra. Am. Math. Soc., Providence (1986)
83. Andrews, G.E.: The Rogers-Ramanujan identities without Jacobi's triple product. Rocky Mt. J. Math. **17**, 659–672 (1987)
84. Andrews, G.E., Askey, R.A., Berndt, B.C., Ramanathan, K.G., Rankin, R.A. (eds.): Ramanujan Revisited (Proceedings of the Centenary Conference). Academic Press, San Diego (1988)
85. Andrews, G.E., Baxter, R.J.: A motivated proof of the Rogers-Ramanujan identities. Am. Math. Mon. **96**, 401–409 (1989)
86. Andrews, G.E., Berndt, B.C.: Ramanujan's Lost Notebook, I. Springer, Berlin (2005)
87. Andrews, G.E., Berndt, B.C.: Ramanujan's Lost Notebook, II. Springer, Berlin (2009)
88. Andrews, G.E., Berndt, B.C.: Ramanujan's Lost Notebook, III, IV. To appear
89. Andrews, G.E., Bressoud, D.M., Parson, L.A. (eds.): The Rademacher Legacy to Mathematics. Contemp. Math., vol. 166. Am. Math. Soc., Providence (1994)
90. Andrews, G.E., Ekhad, S.B., Zeilberger, D.: A short proof of Jacobi's formula for the number of representations of an integer as a sum of four squares. Am. Math. Mon. **100**, 274–276 (1993)
91. Andrews, G.E., Garvan, F.G.: Dyson's crank of a partition. Bull. Am. Math. Soc. **18**, 167–171 (1988)
92. Andrews, G.E., Ono, K.: Ramanujan's congruences and Dyson's crank. Proc. Natl. Acad. Sci. USA **102**, 15277 (2005)
93. Andrianov, A.N.: Dirichlet series with Euler product in the theory of Siegel modular forms of genus two. Tr. Mat. Inst. Steklova **112**, 73–94 (1971) (in Russian)
94. Andrianov, A.N.: Euler products corresponding to Siegel's modular forms of genus 2. Usp. Mat. Nauk **29**(3), 43–110 (1974) (in Russian)
95. Andrianov, A.N.: Modular descent and the Saito-Kurokawa conjecture. Invent. Math. **53**, 267–280 (1979)
96. Andrianov, A.N., et al.: Boris F. Skubenko. An essay on his life and work. Zap. Nauč. Semin. POMI **212**, 5–14 (1995) (in Russian)
97. Andriyasyan, A.K., Ilin, I.V., Malyšev, A.V.: Application of computers to estimates of Čebotarev type in the nonhomogeneous Minkowski conjecture. Zap. Nauč. Semin. LOMI **151**, 7–25 (1986) (in Russian)
98. Anglin, W.S.: Simultaneous Pell equations. Math. Comput. **65**, 355–359 (1996)
99. Ankeny, N.C.: The least quadratic non-residue. Ann. Math. **55**, 65–71 (1952)
100. Ankeny, N.C.: Representations of primes by quadratic forms. Am. J. Math. **74**, 913–919 (1952)
101. Ankeny, N.C.: The insolubility of sets of Diophantine equations in the rational numbers. Proc. Natl. Acad. Sci. USA **38**, 880–884 (1952)
102. Ankeny, N.C.: Sums of three squares. Proc. Am. Math. Soc. **8**, 316–319 (1957)
103. Ankeny, N.C., Chowla, S.: The class number of the cyclotomic field. Proc. Natl. Acad. Sci. USA **35**, 529–532 (1949)

104. Ankeny, N.C., Chowla, S.: The class number of the cyclotomic field. Can. J. Math. **3**, 486–491 (1951)
105. Ankeny, N.C., Onishi, H.: The general sieve. Acta Arith. **10**, 31–62 (1964)
106. Ankeny, N.C., Rogers, C.A.: A conjecture of Chowla. Ann. Math. **53**, 541–550 (1951); add.: **58**, 1953, p. 591
107. Antoniadis, J.A., Bungert, M., Frey, G.: Properties of twists of elliptic curves. J. Reine Angew. Math. **405**, 1–28 (1990)
108. Apéry, F.: Roger Apéry, 1916–1999: A radical mathematician. Math. Intell. **18**(2), 54–61 (1996)
109. Apéry, R.: Sur une équation diophantienne. C. R. Acad. Sci. Paris **251**, 1263–1264 (1960), 1451–1452
110. Apéry, R.: Irrationalité de $\zeta(2)$ et $\zeta(3)$. Astérisque **61**, 11–13 (1979)
111. Apostol, T.M.: Generalized Dedekind sums and transformation formulae of certain Lambert series. Duke Math. J. **17**, 147–157 (1950)
112. Arakelov, S.Yu.: Families of algebraic curves with fixed degeneracies. Izv. Akad. Nauk SSSR, Ser. Mat. **35**, 1269–1293 (1971)
113. Aramata, H.: Über die Teilbarkeit der Zetafunktionen gewisser algebraischer Zahlkörper. Proc. Imp. Acad. (Tokyo) **7**, 334–336 (1931)
114. Aramata, H.: Über die Teilbarkeit der Dedekindschen Zetafunktionen. Proc. Imp. Acad. (Tokyo) **9**, 31–34 (1933)
115. Archibald, R.C.: 881.F [Review of [4582]]. Math. Tables Other Aids Comput. **5**, 135–138 (1951)
116. Arkhipov, G.I.: The Hilbert-Kamke problem. Izv. Akad. Nauk SSSR, Ser. Mat. **48**, 3–52 (1984) (in Russian)
117. Arkhipov, G.I., et al.: Sergeĭ Mikhaĭlovič Voronin. Usp. Mat. Nauk **53**(4), 125–128 (1998) (in Russian)
118. Arkhipov, G.I., et al.: Nikolaĭ Mihaĭlovič Timofeev. Usp. Mat. Nauk **58**(4), 135–138 (2003) (in Russian)
119. Arkhipov, G.I., Čubarikov, V.N.: On the distribution of prime numbers in a sequence of the form $[n^c]$. Vestn. Mosk. Univ. Ser. I, Mat. Mekh. **6**, 25–35 (1999) (in Russian)
120. Arkhipov, G.I., Čubarikov, V.N.: Anatoliĭ Alekseevič Karatsuba (on the occasion of his sixtieth birthday). Usp. Mat. Nauk **53**(2), 173–176 (1998) (in Russian)
121. Arkhipov, G.I., Karatsuba, A.A.: On a local representation of zero by a form. Izv. Akad. Nauk SSSR, Ser. Mat. **45**, 948–961 (1981) (in Russian)
122. Arkhipov, G.I., Karatsuba, A.A.: On a problem in the theory of congruences. Usp. Mat. Nauk **37**(5), 161–162 (1982) (in Russian)
123. Arkhipov, G.I., Karatsuba, A.A., Čubarikov, V.N.: Multiple trigonometric sums. Tr. Mat. Inst. Steklova **151**, 1–128 (1980) (in Russian)
124. Arkhipov, G.I., Karatsuba, A.A., Čubarikov, V.N.: Theory of Multiple Trigonometric Sums. Nauka, Moscow (1987) (in Russian) [English translation: Trigonometric Sums in Number Theory and Analysis, de Gruyter, 2004]
125. Arkhipov, G.I., Zhitkov, A.N.: Waring's problem with nonintegral exponents. Izv. Akad. Nauk SSSR, Ser. Mat. **48**, 1138–1150 (1984) (in Russian)
126. Arkhontova, R.A., et al.: Kirill Andreevič Rodosskiĭ. Usp. Mat. Nauk **61**(5), 173–175 (2006) (in Russian)
127. Armitage, J.V.: An analogue of a problem of Littlewood. Mathematika **16**, 101–105 (1969); corr.: **17**, 173–178 (1970)
128. Arno, S.: The imaginary quadratic fields of class number 4. Acta Arith. **60**, 321–334 (1992)
129. Arno, S., Robinson, M., Wheeler, F.: Imaginary quadratic fields with small odd class number. Acta Arith. **83**, 295–330 (1998)
130. Arnold, V.I.: Weak asymptotics of the numbers of solutions of Diophantine equations. Funkc. Anal. Prilozh. **33**, 65–66 (1999) (in Russian)
131. Arnold, V.I.: Geometry and growth rate of Frobenius numbers of additive semigroups. Math. Phys. Anal. Geom. **9**, 95–108 (2006)

132. Arnold, V.I.: Arithmetical turbulence of self-similar fluctuations statistics of large Frobenius numbers of additive semigroups of integers. Mosc. Math. J. **7**, 173–193 (2007)
133. Arnold, V.I.: Geometry of continued fractions associated with Frobenius numbers. Funct. Anal. Other Math. **2**, 129–138 (2009)
134. Arthur, J., et al.: Armand Borel (1923–2003). Not. Am. Math. Soc. **51**, 498–524 (2004)
135. Artin, E.: Über die Zetafunktionen gewisser algebraischer Zahlkörper. Math. Ann. **89**, 147–156 (1923) [[144], pp. 95–104]
136. Artin, E.: Quadratische Körper im Gebiet der höheren Kongruenzen, I. Math. Z. **19**, 153–206 (1924) [[144], pp. 1–54]
137. Artin, E.: Quadratische Körper im Gebiet der höheren Kongruenzen, II. Math. Z. **19**, 207–246 (1924) [[144], pp. 55–94]
138. Artin, E.: Über eine neue Art von L-Reihen. Abh. Math. Semin. Univ. Hamb. **3**, 89–108 (1924) [[144], pp. 105–124]
139. Artin, E.: Beweis des allgemeinen Reziprozitätsgesetzes. Abh. Math. Semin. Univ. Hamb. **5**, 353–363 (1927) [[144], pp. 131–141]
140. Artin, E.: Über die Zerlegung definiter Funktionen in Quadrate. Abh. Math. Semin. Univ. Hamb. **5**, 100–115 (1927) [[144], pp. 273–288]
141. Artin, E.: Idealklassen in Oberkörpern und allgemeines Reziprozitätsgesetz. Abh. Math. Semin. Univ. Hamb. **7**, 46–51 (1930) [[144], pp. 159–164]
142. Artin, E.: Zur Theorie der L-Reihen mit allgemeinen Gruppencharakteren. Abh. Math. Semin. Univ. Hamb. **8**, 292–306 (1930) [[144], pp. 165–179]
143. Artin, E.: Über die Bewertungen algebraischer Zahlkörper. J. Reine Angew. Math. **167**, 157–159 (1932) [[144], pp. 199–201]
144. Artin, E.: Collected Papers. Addison-Wesley, Reading (1963)
145. Artin, E., Scherk, P.: On the sum of two sets of integers. Ann. Math. **44**, 138–142 (1943) [[144], pp. 346–350]
146. Artin, E., Schreier, O.: Algebraische Konstruktion reeller Körper. Abh. Math. Semin. Univ. Hamb. **5**, 85–99 (1927) [[144], pp. 258–272]
147. Artin, E., Tate, J.: Class Field Theory. Benjamin, Elmsford (1968); 2nd ed. Addison-Wesley, 1990 [Reprint: Chelsea (2009)]
148. Arwin, A.: Common solutions of two simultaneous Pell equations. Ann. Math. **23**, 307–312 (1923)
149. Ash, A.: Galois representations attached to mod p cohomology of $GL(n, Z)$. Duke Math. J. **65**, 235–255 (1992)
150. Ash, A., Doud, D., Pollack, D.: Galois representations with conjectural connections to arithmetic cohomology. Duke Math. J. **112**, 521–579 (2002)
151. Ash, A., Sinnott, W.: An analogue of Serre's conjecture for Galois representations and Hecke eigenclasses in the mod p cohomology of $GL(n, Z)$. Duke Math. J. **105**, 1–24 (2000)
152. Ashworth, M.H.: Congruence and identical properties of modular forms. Ph.D. thesis, Oxford (1968)
153. Askey, R., de Boor, C.: In memoriam: I.J. Schoenberg (1903–1990). J. Approx. Theory **63**, 1–2 (1990)
154. Askey, R., Nevai, P.: Gabor Szegő: 1895–1985. Math. Intell. **18**, 10–22 (1996)
155. Astels, S.: Cantor sets and numbers with restricted partial quotients. Trans. Am. Math. Soc. **352**, 133–170 (2000)
156. Astels, S.: Sums of numbers with small partial quotients. Proc. Am. Math. Soc. **130**, 637–642 (2002)
157. Astels, S.: Sums of numbers with small partial quotients, II. J. Number Theory **91**, 187–205 (2001)
158. Astels, S.: Products and quotients of numbers with small partial quotients. J. Théor. Nr. Bordx. **14**, 387–402 (2002)
159. Athanasiadis, C.A.: Ehrhart polynomials, simplicial polytopes, magic squares and a conjecture of Stanley. J. Reine Angew. Math. **583**, 163–174 (2005)
160. Atiyah, M.F.: John Arthur Todd. Bull. Lond. Math. Soc. **30**, 305–316 (1998)

161. Atkin, A.O.L.: Proof of a conjecture of Ramanujan. Glasg. Math. J. **8**, 14–32 (1967)
162. Atkin, A.O.L.: Multiplicative congruence properties and density problems for $p(n)$. Proc. Lond. Math. Soc. **18**, 563–576 (1968)
163. Atkin, A.O.L.: Ramanujan congruences for $p_{-k}(n)$. Can. J. Math. **20**, 67–78 (1968); corr. **21**, 256 (1968)
164. Atkin, A.O.L., Lehner, J.: Hecke operators on $\Gamma_0(m)$. Math. Ann. **185**, 134–160 (1970)
165. Atkin, A.O.L., Morain, F.: Elliptic curves and primality proving. Math. Comput. **61**, 29–68 (1993)
166. Atkin, A.O.L., O'Brien, J.N.: Some properties of $p(n)$ and $c(n)$ modulo powers of 13. Trans. Am. Math. Soc. **126**, 442–459 (1967)
167. Atkin, A.O.L., Swinnerton-Dyer, H.P.F.: Some properties of partitions. Proc. Lond. Math. Soc. **4**, 84–106 (1954)
168. Atkinson, F.V.: A divisor problem. Q. J. Math. **12**, 193–200 (1941)
169. Atkinson, F.V.: The mean-value of the Riemann zeta-function. Acta Math. **81**, 353–376 (1949)
170. Atkinson, O.D., Brüdern, J., Cook, R.J.: Simultaneous additive congruences to a large prime modulus. Mathematika **39**, 1–9 (1992)
171. Atkinson, O.D., Cook, R.J.: Pairs of additive congruences to a large prime modulus. J. Aust. Math. Soc. A **46**, 438–455 (1989)
172. Auluck, F.C.: On Waring's problem for biquadrates. Proc. Indian Acad. Sci., Sect. A, Phys. Sci. **11**, 437–450 (1940)
173. Auluck, F.C., Chowla, S.: The representation of a large number as a sum of "almost equal" squares. Proc. Indian Acad. Sci. Math. Sci. **6**, 81–82 (1937)
174. Avidon, M.R.: On the distribution of primitive abundant numbers. Acta Arith. **77**, 195–205 (1996)
175. Ax, J.: Zeroes of polynomials over finite fields. Am. J. Math. **86**, 255–261 (1964)
176. Ax, J.: The elementary theory of finite fields. Ann. Math. **88**, 239–271 (1968)
177. Ax, J.: On Schanuel's conjectures and Skolem's method. In: Proc. Symposia Pure Math., vol. 20, pp. 206–212. Am. Math. Soc., Providence (1971)
178. Ax, J.: On Schanuel's conjectures. Ann. Math. **93**, 252–268 (1971)
179. Ax, J., Kochen, S.: Diophantine problems over local fields, I. Am. J. Math. **87**, 605–630 (1965)
180. Ax, J., Kochen, S.: Diophantine problems over local fields, II. A complete set of axioms for p-adic number theory. Am. J. Math. **87**, 631–648 (1965)
181. Axer, A.: Über einige Grenzwertsätze. Sitz.ber. Philos.-Hist. Kl. Kais. Akad. Wiss. **120**, 1253–1298 (1911)
182. Ayad, M.: Sur le théorème de Runge. Acta Arith. **58**, 203–209 (1991)
183. Ayoub, R.[G.]: On Rademacher's extension of the Goldbach-Vinogradoff theorem. Trans. Am. Math. Soc. **74**, 482–491 (1953)
184. Ayoub, R.[G.]: An Introduction to the Analytic Theory of Numbers. Am. Math. Soc., Providence (1963)
185. Ayoub, R.G., Huard, J.G., Williams, K.S.: Sarvadaman Chowla (1907–1995). Not. Am. Math. Soc. **45**, 594–598 (1998)
186. Azra, J.-P.: Relations diophantiennes et la solution négative du 10e problème de Hilbert (d'après M. Davis, H. Putnam, J. Robinson et I. Matiasevitch). In: Lecture Notes in Math., vol. 244, pp. 11–28. Springer, Berlin (1971)
187. Baas, N.A., Skau, F.C.: The lord of the number, Atle Selberg. On his life and mathematics. Bull. Am. Math. Soc. **45**, 617–649 (2008)
188. Babai, L., Pomerance, C., Vértesi, P.: The mathematics of Paul Erdös. Not. Am. Math. Soc. **45**, 19–31 (1998)
189. Babai, L., Spencer, J.: Paul Erdős (1913–1996). Not. Am. Math. Soc. **45**, 64–73 (1998)
190. Bach, E.: Explicit bounds for primality testing and related problems. Math. Comput. **55**, 355–380 (1990)
191. Bach, E., Sorenson, J.: Explicit bounds for primes in residue classes. Math. Comput. **65**, 1717–1735 (1996)

192. Bacher, R., Venkov, B.: Réseaux entiers unimodulaires sans racines en dimensions 27 et 28. In: Réseaux euclidiens, designs sphériques et formes modulaires, pp. 212–267. Enseignement Math., Genéve (2001)
193. Bachman, G.: On exponential sums with multiplicative coefficients. In: Number Theory, Halifax, NS, 1994, pp. 29–38. Am. Math. Soc., Providence (1995)
194. Bachman, G.: On exponential sums with multiplicative coefficients, II. Acta Arith. **106**, 41–57 (2003)
195. Bachman, G.: Some remarks on nonnegative multiplicative functions on arithmetic progressions. J. Number Theory **73**, 72–91 (1998)
196. Bachman, G.: On a Brun-Titchmarsh inequality for multiplicative functions. Acta Arith. **106**, 1–25 (2003)
197. Bachmann, P.: Zur Theorie von Jacobis Kettenbruch-Algorithmen. J. Reine Angew. Math. **75**, 25–34 (1873)
198. Bachmann, P.: Zahlentheorie, I–V. Leipzig, 1872–1905
199. Bachmann, P.: Niedere Zahlentheorie, I–II. Leipzig, 1902–1910
200. Bachmann, P.: Über Gauß' zahlentheoretische Arbeiten. Nachr. Ges. Wiss. Göttingen, 1911, 455–508
201. Bachmann, P.: Über den Rest von $\frac{2^{p-1}-1}{p}$ (mod. p). J. Reine Angew. Math. **142**, 41–50 (1913)
202. Bachoc, C., Nebe, G.: Extremal lattices of minimum 8 related to the Mathieu group M_{22}. J. Reine Angew. Math. **494**, 155–171 (1998)
203. Bachoc, C., Nebe, G., Venkov, B.: Odd unimodular lattices of minimum 4. Acta Arith. **101**, 151–158 (2002)
204. Bachoc, C., Vallentin, F.: New upper bounds for kissing numbers from semidefinite programming. J. Am. Math. Soc. **21**, 909–924 (2008)
205. Backlund, R.J.: Einige numerische Rechnungen, die Nullpunkte der Riemannschen ζ-Funktion betreffend. Öfversigt af Finska Vetenskaps. Soc. Förh. **54a**(3), 1–7 (1911/1912)
206. Backlund, R.J.: Sur les zéros de la fonction $\zeta(s)$ de Riemann. C. R. Acad. Sci. Paris **158**, 1979–1981 (1914)
207. Backlund, R.J.: Über die Beziehung zwischen Anwachsen und Nullstellen der Zetafunktion. Öfversigt af Finska Vetenskaps. Soc. Förh. **61a**(9), 1–8 (1918/1919)
208. Backlund, R.J.: Über die Nullstellen der Riemannschen Zetafunktion. Acta Math. **41**, 345–375 (1918)
209. Backlund, R.J.: Über die Differenzen zwischen den Zahlen, die zu den n ersten Primzahlen teilerfremd sind. Ann. Acad. Sci. Fenn., Ser. A 1 Math. **32**(2), 1–9 (1929)
210. Baer, W.S.: Über die Zerlegung der ganzen Zahlen in sieben Kuben. Math. Ann. **74**, 511–514 (1913)
211. Baer, W.S.: Beiträge zum Waringschen Problem. Dissertation, Göttingen (1913)
212. Baeza, R., Coulangeon, R., Icaza, M.I., O'Ryan, M.: Hermite's constant for quadratic number fields. Exp. Math. **10**, 543–551 (2001)
213. Báez-Duarte, L.: On Beurling's real variable reformulation of the Riemann hypothesis. Adv. Math. **101**, 10–30 (1993)
214. Bagchi, B.: The statistical behaviour and universality properties of the Riemann zeta-function and other allied Dirichlet series. Ph.D. thesis, Calcutta (1981)
215. Bagchi, B.: A joint universality theorem for Dirichlet L-functions. Math. Z. **181**, 319–334 (1982)
216. Bagemihl, F., McLaughlin, R.C.: Generalization of some classical theorems concerning triples of consecutive convergents to simple continued fractions. J. Reine Angew. Math. **221**, 146–149 (1966)
217. Baier, S.: On the Bateman-Horn conjecture. J. Number Theory **96**, 432–448 (2002)
218. Bailey, W.N.: Some identities in combinatory analysis. Proc. Lond. Math. Soc. **49**, 421–425 (1947)
219. Bailey, W.N.: Ernst William Barnes. J. Lond. Math. Soc. **29**, 498–503 (1954)
220. Baker, A.: Rational approximations to certain algebraic numbers. Proc. Lond. Math. Soc. **14**, 385–398 (1964)

221. Baker, A.: Rational approximations to $\sqrt[3]{2}$ and other algebraic numbers. Q. J. Math. **15**, 375–383 (1964)
222. Baker, A.: On an analogue of Littlewood's Diophantine approximation problem. Mich. Math. J. **11**, 247–250 (1964)
223. Baker, A.: Linear forms in the logarithms of algebraic numbers, I. Mathematika **13**, 204–216 (1966)
224. Baker, A.: Linear forms in the logarithms of algebraic numbers, II. Mathematika **14**, 102–107 (1967)
225. Baker, A.: Linear forms in the logarithms of algebraic numbers, III. Mathematika **14**, 220–228 (1968)
226. Baker, A.: Linear forms in the logarithms of algebraic numbers, IV. Mathematika **15**, 204–216 (1968)
227. Baker, A.: On a theorem of Sprindžuk. Proc. R. Soc. Lond. Ser. A, Math. Phys. Sci. **292**, 92–104 (1966)
228. Baker, A.: Contributions to the theory of Diophantine equations, I. On the representation of integers by binary forms. Philos. Trans. R. Soc. Lond. Ser. A, Math. Phys. Sci. **263**, 173–191 (1967)
229. Baker, A.: Contributions to the theory of Diophantine equations, II. The Diophantine equation $y^2 = x^3 + k$. Philos. Trans. R. Soc. Lond. Ser. A, Math. Phys. Sci. **263**, 193–208 (1968)
230. Baker, A.: On some diophantine inequalities involving primes. J. Reine Angew. Math. **228**, 166–181 (1967)
231. Baker, A.: The diophantine equation $y^2 = ax^3 + bx^2 + cx + d$. J. Lond. Math. Soc. **43**, 1–9 (1968)
232. Baker, A.: Bounds for the solution of the hyperelliptic equation. Proc. Camb. Philos. Soc. **65**, 439–444 (1968)
233. Baker, A.: A remark on the class number of quadratic fields. Bull. Lond. Math. Soc. **1**, 98–102 (1969)
234. Baker, A.: Imaginary quadratic fields with class number 2. Ann. Math. **94**, 139–152 (1971)
235. Baker, A.: A sharpening of the bounds for linear forms in logarithms, I. Acta Arith. **21**, 117–129 (1972)
236. Baker, A.: A sharpening of the bounds for linear forms in logarithms, II. Acta Arith. **24**, 33–36 (1973)
237. Baker, A.: A sharpening of the bounds for linear forms in logarithms, III. Acta Arith. **27**, 247–252 (1975)
238. Baker, A.: Transcendental Number Theory. Cambridge University Press, Cambridge (1975); 2nd ed. 1990
239. Baker, A.: The theory of linear forms in logarithms. In: Transcendence Theory: Advances and Applications, pp. 1–27. Academic Press, San Diego (1977)
240. Baker, A., Coates, J.: Integer points on curves of genus 1. Proc. Camb. Philos. Soc. **67**, 595–602 (1970)
241. Baker, A., Coates, J.: Fractional parts of powers of rationals. Math. Proc. Camb. Philos. Soc. **77**, 269–279 (1975)
242. Baker, A., Davenport, H.: The equations $3x^2 - 2 = y^2$ and $8x^2 - 7 = z^2$. Q. J. Math. **20**, 129–137 (1969) [[1380], vol. 4, pp. 1748–1756]
243. Baker, A., Schmidt, W.M.: Diophantine approximation and Hausdorff dimension. Proc. Lond. Math. Soc. **21**, 1–11 (1970)
244. Baker, A., Stark, H.M.: On a fundamental inequality in number theory. Ann. Math. **94**, 190–199 (1971)
245. Baker, A., Wüstholz, G.: Logarithmic forms and group varieties. J. Reine Angew. Math. **442**, 19–62 (1993)
246. Baker, H.F.: Percy Alexander MacMahon. J. London Math. Soc., **5**, 307–318
247. Baker, R.C.: Sprindzuk's theorem and Hausdorff dimension. Mathematika **23**, 184–197 (1976)
248. Baker, R.C.: Fractional parts of several polynomials. Q. J. Math. **28**, 453–471 (1977)

249. Baker, R.C.: Fractional parts of several polynomials, II. Mathematika **25**, 76–93 (1978)
250. Baker, R.C.: Metric number theory and the large sieve. J. Lond. Math. Soc. **24**, 34–40 (1981)
251. Baker, R.C.: On the fractional parts of αn^2 and βn. Glasg. Math. J. **22**, 181–183 (1981)
252. Baker, R.C.: Weyl sums and Diophantine approximation. J. Lond. Math. Soc. **25**, 25–34 (1982)
253. Baker, R.C.: On the fractional parts of αn^3, βn^2 and γn. In: Journées Arithmétiques 1980, Exeter, 1980, pp. 226–231. Cambridge University Press, Cambridge (1982)
254. Baker, R.C.: Diophantine Inequalities. Clarendon, Oxford (1986)
255. Baker, R.C.: The greatest prime factor of the integers in an interval. Acta Arith. **47**, 193–231 (1986)
256. Baker, R.C.: The Brun-Titchmarsh theorem. J. Number Theory **56**, 343–365 (1996)
257. Baker, R.C.: On irregularities of distribution, II. J. Lond. Math. Soc. **59**, 50–64 (1999)
258. Baker, R.C., Brüdern, J.: On pairs of additive cubic equations. J. Reine Angew. Math. **391**, 157–180 (1988)
259. Baker, R.C., Brüdern, J., Harman, G.: The fractional part of αn^k for square-free n. Q. J. Math. **42**, 421–431 (1991)
260. Baker, R.C., Brüdern, J., Harman, G.: Simultaneous Diophantine approximation with square-free numbers. Acta Arith. **63**, 51–60 (1993)
261. Baker, R.C., Gajraj, J.: Some non-linear Diophantine approximations. Acta Arith. **31**, 325–341 (1976)
262. Baker, R.C., Harman, G.: Exponential sums formed with the Möbius function. J. Lond. Math. Soc. **43**, 193–198 (1991)
263. Baker, R.C., Harman, G.: Numbers with a large prime factor. Acta Arith. **73**, 119–145 (1995)
264. Baker, R.C., Harman, G.: The Brun-Titchmarsh theorem on average. Prog. Math. **138**, 39–103 (1996)
265. Baker, R.C., Harman, G.: The difference between consecutive primes. Proc. Lond. Math. Soc. **72**, 261–280 (1996)
266. Baker, R.C., Harman, G.: Shifted primes without large prime factors. Acta Arith. **83**, 331–361 (1998)
267. Baker, R.C., Harman, G., Pintz, J.: The difference between consecutive primes, II. Proc. Lond. Math. Soc. **83**, 532–562 (2001)
268. Baker, R.C., Montgomery, H.L.: Oscillations of quadratic L-functions. Prog. Math. **85**, 23–40 (1990)
269. Baker, R.C., Pintz, J.: The distribution of squarefree numbers. Acta Arith. **46**, 73–79 (1985)
270. Baker, R.C., Schlickewei, H.P.: Indefinite quadratic forms. Proc. Lond. Math. Soc. **54**, 385–411 (1987)
271. Baladi, V., Vallée, B.: Euclidean algorithms are Gaussian. J. Number Theory **110**, 331–386 (2005)
272. Balanzario, E.P.: An example in Beurling's theory of primes. Acta Arith. **87**, 121–139 (1998)
273. Balasubramanian, R.: An improvement on a theorem of Titchmarsh on the mean square of $|\zeta\frac{1}{2}+it)|$. Proc. Lond. Math. Soc. **36**, 540–576 (1978)
274. Balasubramanian, R.: On Waring's problem: $g(4) \leq 21$. Hardy-Ramanujan J. **2**, 1–31 (1979)
275. Balasubramanian, R.: On Waring's problem: $g(4) \leq 20$. Hardy-Ramanujan J. **8**, 1–40 (1985)
276. Balasubramanian, R., Deshouillers, J.-M., Dress, F.: Problème de Waring pour les bicarrés. Schéma de la solution. C. R. Acad. Sci. Paris **303**, 85–88 (1986)
277. Balasubramanian, R., Deshouillers, J.-M., Dress, F.: Problème de Waring pour les bicarrés. Résultats auxiliaires pour le théorème asymptotique. C. R. Acad. Sci. Paris **303**, 161–163 (1986)
278. Balasubramanian, R., Murty, M.R.: An Ω-theorem for Ramanujan's τ-function. Invent. Math. **68**, 241–252 (1982)

279. Balasubramanian, R., Murty, V.K.: Zeros of Dirichlet L-functions. Ann. Sci. Éc. Norm. Super. **25**, 567–615 (1992)
280. Balasubramanian, R., Nagaraj, S.V.: Density of Carmichael numbers with three prime factors. Math. Comput. **66**, 1705–1708 (1997)
281. Balasubramanian, R., Ramachandra, K.: Two remarks on a result of Ramachandra. J. Indian Math. Soc. (N.S.) **38**, 395–397 (1974)
282. Balasubramanian, R., Ramachandra, K.: On the frequency of Titchmarsh's phenomenon for $\zeta(s)$. III. Proc. Indian Acad. Sci. Math. Sci. **86**, 341–351 (1977)
283. Balasubramanian, R., Ramachandra, K.: Some problems of analytic number theory, II. Studia Sci. Math. Hung. **14**, 193–202 (1979)
284. Balasubramanian, R., Ramachandra, K.: On square-free numbers. In: Proceedings of the Ramanujan Centennial International Conference, Annamalainagar, 1987, pp. 27–30 (1988)
285. Balasubramanian, R., Ramachandra, K.: Proof of some conjectures on the mean-value of Titchmarsh series, I, Hardy. Ramanujan J. **13**, 1–20 (1990)
286. Balasubramanian, R., Ramachandra, K., Subbarao, M.V.: On the error function in the asymptotic formula for the counting function of k-full numbers. Acta Arith. **50**, 107–118 (1988)
287. Balasubramanian, R., Shorey, T.N.: On the equation $a(x^m-1)/(x-1)=b(y^n-1)/(y-1)$. Math. Scand. **46**, 177–182 (1980)
288. Balazard, M., de Roton, A.: Notes de lecture de l'article "Partial sums of the Möbius function" de Kannan Soundararajan, to appear. arXiv:0810.3587
289. Balazard, M., Tenenbaum, G.: Sur la répartition des valeurs de la fonction d'Euler. Compos. Math. **110**, 239–250 (1998)
290. Baldassarri, F.: Higher p-adic gamma functions and Dwork cohomology. Astérisque **119/120**, 111–127 (1984)
291. Ball, K.: A lower bound for the optimal density of lattice packings. Internat. Math. Res. Notices 217–221 (1992)
292. Ball, K., Rivoal, T.: Irrationalité d'une infinité de valeurs de la fonction zeta aux entiers impairs. Invent. Math. **146**, 193–207 (2001)
293. Ballister, P., Wheeler, J.P.: The Erdős-Heilbronn problem for finite groups. Acta Arith. **140**, 105–118 (2009)
294. Ballot, C.: Density of prime divisors of linear recurrences. Mem. Am. Math. Soc. **115**, 1–102 (1995)
295. Balog, A.: Numbers with a large prime factor. Studia Sci. Math. Hung. **15**, 139–146 (1980)
296. Balog, A.: Numbers with a large prime factor, II. In: Topics in Classical Number Theory, pp. 49–67. North-Holland, Amsterdam (1984)
297. Balog, A.: On the fractional part of p^θ. Arch. Math. **40**, 434–440 (1983)
298. Balog, A.: On the distribution of p^θ mod 1. Acta Math. Acad. Sci. Hung. **45**, 179–199 (1985)
299. Balog, A.: A remark on the distribution of αp modulo one. In: Analytic and Elementary Number Theory, Marseille, 1983, pp. 6–24. University Paris XI, Orsay (1986)
300. Balog, A.: Linear equations in primes. Mathematika **39**, 367–378 (1992)
301. Balog, A.: Six primes and an almost prime in four linear equations. Can. J. Math. **50**, 465–486 (1998)
302. Balog, A., Friedlander, J.B.: Simultaneous Diophantine approximation using primes. Bull. Lond. Math. Soc. **20**, 289–292 (1988)
303. Balog, A., Friedlander, J.B.: A hybrid of theorems of Vinogradov and Piatetski-Shapiro. Pac. J. Math. **156**, 45–62 (1992)
304. Balog, A., Harman, G., Pintz, J.: Numbers with a large prime factor, IV. J. Lond. Math. Soc. **28**, 218–226 (1983)
305. Balog, A., Pelikán, J., Pintz, J., Szemerédi, E.: Difference sets without κ-th powers. Acta Math. Acad. Sci. Hung. **65**, 165–187 (1994)
306. Balog, A., Perelli, A.: Diophantine approximation by square-free numbers. Ann. Sc. Norm. Super. Pisa, Cl. Sci. **11**, 353–359 (1984)

307. Bambah, R.P.: Two congruence properties of Ramanujan's function $\tau(n)$. J. Lond. Math. Soc. **21**, 91–93 (1946)
308. Bambah, R.P.: On lattice coverings by spheres. Proc. Natl. Inst. Sci. India **20**, 25–52 (1954)
309. Bambah, R.P., Chowla, S.: A note on Ramanujan's function $\tau(n)$. Q. J. Math. **18**, 122–123 (1947)
310. Bambah, R.P., Davenport, H.: The covering of n-dimensional space by spheres. J. Lond. Math. Soc. **27**, 224–229 (1952) [[1380], vol. 2, pp. 603–608]
311. Bambah, R.P., Woods, A.C.: Minkowski's conjecture for $n = 5$; a theorem of Skubenko. J. Number Theory **12**, 27–48 (1980)
312. Bambah, R.P., Woods, A.C., Zassenhaus, H.: Three proofs of Minkowski's second inequality in the geometry of numbers. J. Aust. Math. Soc. **5**, 453–462 (1965)
313. Banerjee, D.P.: On the solution of the "easier" Waring problem. Bull. Calcutta Math. Soc. **34**, 197–199 (1942)
314. Bang, A.S.: Taltheoritiske undersøgelser. Mat. Tidsskr. **4**, 70–80, 130–137 (1886)
315. Banks, W.D., Luca, F.: Composite integers n for which $\varphi(n)|n-1$. Acta Math. Sin. **23**, 2007 (1915–1918)
316. Baragar, A.: Integral solutions of Markoff-Hurwitz equations. J. Number Theory **49**, 27–44 (1994)
317. Baragar, A.: Asymptotic growth of Markoff-Hurwitz numbers. Compos. Math. **94**, 1–18 (1994)
318. Baragar, A.: On the unicity conjecture for Markoff numbers. Can. Math. Bull. **39**, 3–9 (1996)
319. Baragar, A.: The exponent for the Markoff-Hurwitz equations. Pac. J. Math. **182**, 1–21 (1998)
320. Barban, M.B.: New applications of the "large sieve" of Yu.V. Linnik. Tr. Inst. Mat. Akad. Nauk Uzbek. SSR **22**, 1–20 (1961); Corr. in: Theory of Probability and Mathematical Statistics, pp. 130–133, Tashkent (1964) (in Russian)
321. Barban, M.B.: Linnik's "large sieve" and a limit theorem for the class numbers of ideals of an imaginary quadratic field. Izv. Akad. Nauk SSSR, Ser. Mat. **26**, 573–580 (1962) (in Russian)
322. Barban, M.B.: The "density" of the zeros of Dirichlet L-series and the problem of the sum of primes and "almost" primes. Mat. Sb. **61**, 418–425 (1963) (in Russian)
323. Barban, M.B.: Analogues of the divisor problem of Titchmarsh. Vestn. Leningr. Univ., Ser. Mat. Mekh. Astronom. **18**(4), 5–13 (1963) (in Russian)
324. Barban, M.B.: The "large sieve" method and its applications in number theory. Usp. Mat. Nauk **21**(1), 51–102 (1966) (in Russian)
325. Barban, M.B.: The density hypothesis of E. Bombieri. Dokl. Akad. Nauk SSSR **172**, 999–1000 (1967) (in Russian)
326. Barban, M.B., Gordover, G.: On the moments of the class-numbers of purely radical quadratic forms of negative determinant. Dokl. Akad. Nauk SSSR **167**, 267–269 (1966) (in Russian)
327. Barban, M.B., Linnik, Yu.V., Čudakov, N.G.: On prime numbers in an arithmetic progression with a prime-power difference. Acta Arith. **9**, 375–390 (1964)
328. Barban, M.B., Vinogradov, A.I.: On the number-theoretic basis of probabilistic number theory. Dokl. Akad. Nauk SSSR **154**, 495–496 (1964) (in Russian)
329. Barner, K.: Paul Wolfskehl and the Wolfskehl prize. Not. Am. Math. Soc. **44**, 1294–1303 (1997)
330. Barnes, E.S.: The complete enumeration of extreme senary forms. Philos. Trans. R. Soc. Lond. Ser. A, Math. Phys. Sci. **249**, 461–506 (1957)
331. Barnes, E.S.: The construction of perfect and extreme forms, I. Acta Arith. **5**, 57–79 (1958)
332. Barnes, E.S.: The construction of perfect and extreme forms, II. Acta Arith. **5**, 205–222 (1959)
333. Barnes, E.S.: Criteria for extreme forms. J. Aust. Math. Soc. **1**, 17–20 (1959/1960)
334. Barnes, E.S., Wall, G.E.: Some extreme forms defined in terms of Abelian groups. J. Aust. Math. Soc. **1**, 47–63 (1959/1960)

335. Barreira, L., Saussol, B., Schmeling, J.: Distribution of frequencies of digits via multifractal analysis. J. Number Theory **97**, 410–438 (2002)
336. Barré[-Sirieix], K.: Mesure d'approximation simultanée de q et $J(q)$. J. Number Theory **66**, 102–128 (1997)
337. Barré-Sirieix, K., Diaz, G., Gramain, F., Philibert, G.: Une preuve de la conjecture de Mahler-Manin. Invent. Math. **124**, 1–9 (1996)
338. Barrucand, P., Louboutin, S.: Minoration au point 1 des fonctions L attachés à des caractères de Dirichlet. Colloq. Math. **65**, 301–306 (1993)
339. Barsky, D.: Fonctions zeta p-adiques d'une classe de rayon de corps totalement réels. Groupe de travail d'analyse ultramétrique **5**(exp. 23), 1–16 (1977/1978)
340. Barsky, D.: On Morita's p-adic gamma function. Math. Proc. Camb. Philos. Soc. **89**, 23–27 (1981)
341. Bartz, K.: On zero-free regions for the Hecke-Landau zeta functions. Funct. Approx. Comment. Math. **14**, 101–107 (1984)
342. Bartz, K.: An effective order of Hecke-Landau zeta functions near the line $\sigma = 1$, I. Acta Arith. **50**, 183–193 (1988); corr. **58** 211 (1991)
343. Bartz, K.: An effective order of Hecke-Landau zeta functions near the line $\sigma = 1$, II. (Some applications). Acta Arith. **52**, 163–170 (1989)
344. Bartz, K.: On an effective order estimate of the Riemann zeta function in the critical strip. Monatshefte Math. **109**, 267–270 (1990)
345. Barvinok, A.I.: Computing the Ehrhart polynomial of a convex lattice polytope. Discrete Comput. Geom. **12**, 35–48 (1994)
346. Bašmakov, M.I.: Cohomology of Abelian varieties over a number field. Usp. Mat. Nauk **27**, 25–66 (1972) (in Russian)
347. Bass, H., Lam, T.Y.: Irving Kaplansky 1917–2006. Not. Am. Math. Soc. **54**, 1477–1499 (2007)
348. Bass, H., Lazard, M., Serre, J.-P.: Sous-groupes d'indice fini dans $SL(n, \mathbf{Z})$. Bull. Am. Math. Soc. **70**, 385–392 (1964) [[5661], vol. 2, pp. 222–229]
349. Bass, H., Milnor, J., Serre, J.-P.: Solution of the congruence subgroup problem for SL_n $(n \geq 3)$ and Sp_{2n} $(n \geq 2)$. Publ. Math. Inst. Hautes Études Sci. **33**, 59–137 (1967); corr. **44**, 241–244 (1974) [[5661], vol. 2, pp. 342–420]
350. Bastien, L.: Nombres congruents. L'Intermédiaire Math. **22**, 231–232 (1915)
351. Bateman, P.T.: A multiplicative function, Problem E-2051. Am. Math. Mon. **76**, 190–191 (1969)
352. Bateman, P.T.: On the representation of a number as the sum of three squares. Trans. Am. Math. Soc. **71**, 70–101 (1951)
353. Bateman, P.T.: The distribution of values of the Euler function. Acta Arith. **21**, 329–345 (1972)
354. Bateman, P.T., Chowla, S., Erdős, P.: Remarks on the size of $L(1, \chi)$. Publ. Math. (Debr.) **1**, 165–182 (1950)
355. Bateman, P.T., Grosswald, E.: On a theorem of Erdös and Szekeres. Ill. J. Math. **2**, 88–98 (1958)
356. Bateman, P.T., Grosswald, E.: On Epstein's zeta function. Acta Arith. **9**, 365–373 (1964)
357. Bateman, P.T., Grosswald, E.: Positive integers expressible as a sum of three squares in essentially only one way. J. Number Theory **19**, 301–308 (1984)
358. Bateman, P.T., Horn, R.A.: A heuristic asymptotic formula concerning the distribution of prime numbers. Math. Comput. **16**, 363–367 (1962)
359. Bateman, P.T., Horn, R.A.: Primes represented by irreducible polynomials in one variable. In: Proc. Symposia Pure Math., vol. 8, pp. 119–132. Am. Math. Soc., Providence (1965)
360. Bateman, P.T., Stemmler, R.M.: Waring's problem for algebraic number fields and primes of the form $(p^r - 1)/(p^d - 1)$. Ill. J. Math. **6**, 142–156 (1962)
361. Bauer, M.L., Bennett, M.A.: Applications of the hypergeometric method to the generalized Ramanujan-Nagell equation. Ramanujan J. **6**, 209–270 (2000)
362. Bauer, P.J.: Zeros of Dirichlet L-series on the critical line. Acta Arith. **93**, 37–52 (2000)

363. Bauer, W.: On the conjecture of Birch and Swinnerton-Dyer for Abelian varieties over function fields in characteristic $p > 0$. Invent. Math. **108**, 263–287 (1992)
364. Baxa, C.: On the growth of the denominators of convergents. Acta Math. Acad. Sci. Hung. **83**, 125–130 (1999)
365. Baxa, C.: Lévy constants of transcendental numbers. Proc. Am. Math. Soc. **137**, 2243–2249 (2009)
366. Bays, C., Ford, K., Hudson, R.H., Rubinstein, M.: Zeros of Dirichlet L-functions near the real axis and Chebyshev's bias. J. Number Theory **87**, 54–76 (2001)
367. Bays, C., Hudson, R.H.: Zeroes of Dirichlet L-functions and irregularities in the distribution of primes. Math. Comput. **69**, 861–866 (2000)
368. Bays, C., Hudson, R.H.: A new bound for the smallest x with $\pi(x) > \mathrm{li}(x)$. Math. Comput. **69**, 1285–1296 (2000)
369. Beach, B.D., Williams, H.C.: Some computer results on periodic continued fractions. In: Proc. Second Louisiana Conf. on Combinatorics, Graph Theory and Computing, pp. 133–146. Louisiana State University, Baton Rouge (1971)
370. Bean, M.A., Thunder, J.L.: Isoperimetric inequalities for volumes associated with decomposable forms. J. Lond. Math. Soc. **54**, 39–49 (1996)
371. Becher, V., Figueira, S., Picchi, R.: Turing's unpublished algorithm for normal numbers. Theor. Comput. Sci. **377**, 126–138 (2007)
372. Beck, J.: On a problem of K.F. Roth concerning irregularities of point distribution. Invent. Math. **74**, 477–487 (1983)
373. Beck, J.: Some upper bounds in the theory of irregularities of distribution. Acta Arith. **43**, 115–130 (1984)
374. Beck, J.: Irregularities of distribution. Acta Math. **159**, 1–49 (1987)
375. Beck, J.: Irregularities of distribution, II. Proc. Lond. Math. Soc. **56**, 1–50 (1988)
376. Beck, J.: A two-dimensional van Aardenne-Ehrenfest theorem in irregularities of distribution. Compos. Math. **72**, 269–339 (1989)
377. Beck, J.: Probabilistic Diophantine approximation, I. Kronecker sequences. Ann. of Math., (2) **140**, 109–160, 449–502 (1994)
378. Beck, J., Chen, W.W.L.: Irregularities of Distribution. Cambridge University Press, Cambridge (1987). [Reprint: 2008]
379. Beck, J., Sós, V.T.: Discrepancy theory. In: Handbook of Combinatorics, pp. 1405–1446. Elsevier, Amsterdam (1995)
380. Beck, M., Pine, E., Tarrant, W., Jensen, K.Y.: New integer representations as the sum of three cubes. Math. Comput. **76**, 1683–1690 (2007)
381. Becker, P.-G., Töpfer, T.: Transcendency results for sums of reciprocals of linear recurrences. Math. Nachr. **168**, 5–17 (1994)
382. Beeger, N.G.W.H.: On a new case of the congruence $2^{p-1} \equiv 1 \pmod{p^2}$. Messenger Math. **51**, 149–150 (1921/1922)
383. Beeger, N.G.W.H.: On the congruence $2^{p-1} \equiv 1 \pmod{p^2}$ and Fermat's last theorem. Nieuw Arch. Wiskd. **20**, 51–54 (1939)
384. Behnke, H.: Zur Theorie der diophantischen Approximationen. Abh. Math. Semin. Univ. Hamb. **3**, 261–318 (1924)
385. Behnke, H.: Otto Blumenhal zum Gedächtnis. Math. Ann. **136**, 387–392 (1958)
386. Behr, H.: Eine endliche Präsentation der symplektischen Gruppe $Sp_4(Z)$. Math. Z. **141**, 47–56 (1975)
387. Behrend, F.: On sequences of numbers not divisible one by another. J. Lond. Math. Soc. **10**, 42–44 (1935)
388. Behrend, F.: On sequences of integers containing no arithmetic progression. Čas. Mat. Fys. **67**, 235–238 (1938)
389. Behrend, F.: On sets of integers which contain no three terms in arithmetic progression. Proc. Natl. Acad. Sci. USA **32**, 331–332 (1946)
390. Beihoffer, D., Hendry, J., Nijenhuis, A., Wagon, S.: Faster algorithms for Frobenius numbers. Electron. J. Comb. **12**(27), 1–38 (2005)

391. Béjian, R.: Minoration de la discrépance a l'origine d'une suite quelconque. Ann. Fac. Sci. Toulouse **1**, 201–213 (1979)
392. Béjian, R.: Minoration de la discrépance d'une suite quelconque sur T. Acta Arith. **41**, 185–202 (1982)
393. Belhoste, B.: Cauchy, 1789–1857. Paris (1985). [English translation: Augustin-Louis Cauchy. A biography. Springer, 1991]
394. Bell, E.T.: On the number of representations of $2n$ as a sum of $2r$ squares. Bull. Am. Math. Soc. **26**, 19–25 (1919)
395. Bell, E.T.: On the representations of numbers as sums of 3, 5, 7, 9, 11 and 13 squares. Am. J. Math. **42**, 168–188 (1920)
396. Bell, E.T.: Class numbers and the form $xy + yz + zx$. Tohoku Math. J. **19**, 105–116 (1921)
397. Bell, E.T.: Analogies between the u_n, v_n of Lucas and elliptic functions. Bull. Am. Math. Soc. **29**, 401–406 (1923)
398. Bell, E.T.: Note on an easier Waring problem. J. Lond. Math. Soc. **11**, 277–278 (1936)
399. Bell, E.T.: Euler's concordant forms. Proc. Natl. Acad. Sci. USA **25**, 46–48 (1939)
400. Bell, E.T.: The problems of congruent numbers and concordant forms. Proc. Natl. Acad. Sci. USA **33**, 326–328 (1947)
401. Bellman, R., Shapiro, H.N.: The algebraic independence of arithmetic functions. I. Multiplicative functions. Duke Math. J. **15**, 229–235 (1948)
402. Bender, C.: Bestimmung der grössten Anzahl gleich Kugeln, welche sich auf eine Kugel von demselben Radius, wie die übrigen, aufliegen lassen. Arch. Math. **6**, 302–306 (1874)
403. Bennett, A.A.: On sets of three consecutive integers which are quadratic residues of primes. Bull. Am. Math. Soc. **31**, 411–412 (1925)
404. Bennett, A.A.: Large primes have four consecutive quadratic residues. Tohoku Math. J. **27**, 53–57 (1926)
405. Bennett, A.A.: Large primes have at least five consecutive quadratic residues. Bull. Am. Math. Soc. **32**, 33 (1926)
406. Bennett, C.D., Blass, J., Glass, A.M.W., Meronk, D.B., Steiner, R.P.: Linear forms in the logarithms of three positive rational numbers. J. Théor. Nr. Bordx. **9**, 97–136 (1997)
407. Bennett, M.A.: Solving families of simultaneous Pell equations. J. Number Theory **67**, 246–251 (1997)
408. Bennett, M.A.: On the number of solutions of simultaneous Pell equations. J. Reine Angew. Math. **498**, 173–199 (1998)
409. Bennett, M.A.: Rational approximation to algebraic number of small height; the Diophantine equation $|ax^n - by^n| = 1$. J. Reine Angew. Math. **535**, 1–49 (2001)
410. Bennett, M.A.: On some exponential equations of S.S. Pillai. Can. J. Math. **53**, 897–922 (2001)
411. Bennett, M.A.: On the representation of unity by binary cubic forms. Trans. Am. Math. Soc. **353**, 1507–1534 (2001)
412. Bennett, M.A.: Lucas' square pyramid problem revisited. Acta Arith. **105**, 341–347 (2002)
413. Bennett, M.A.: Pillai's conjecture revisited. J. Number Theory **98**, 228–235 (2003)
414. Bennett, M.A.: The equation $x^{2n} + y^{2n} = z^5$. J. Théor. Nr. Bordx. **18**, 315–321 (2006)
415. Bennett, M.A., Bruin, N., Győry, K., Hajdu, L.: Powers from products of consecutive terms in arithmetic progression. Proc. Lond. Math. Soc. **92**, 273–306 (2006)
416. Bennett, M.A., Cipu, M., Mignotte, M., Okazaki, R.: On the number of solutions of simultaneous Pell equations, II. Acta Arith. **122**, 407–417 (2006)
417. Bennett, M.A., de Weger, B.M.M.: On the Diophantine equation $|ax^n - by^n| = 1$. Math. Comput. **67**, 413–438 (1998)
418. Bennett, M.A., Filaseta, M., Trifonov, O.: Yet another generalization of the Ramanujan-Nagell equation. Acta Arith. **134**, 211–217 (2008)
419. Bennett, M.A., Skinner, C.M.: Ternary Diophantine equations via Galois representations and modular forms. Can. J. Math. **56**, 23–54 (2004)
420. Bennett, M.A., Togbé, A., Walsh, P.G.: A generalization of a theorem of Bumby on quartic Diophantine equations, Internat. J. Number Theory **2**, 195–206 (2006)

421. Bennett, M.A., Walsh, [P.]G.: The Diophantine equation $b^2X^4 - dY^2 = 1$. Proc. Am. Math. Soc. **127**, 3481–3491 (1999)
422. Bentkus, V., Götze, F.: Lattice point problems and distribution of values of quadratic forms. Ann. Math. **150**, 977–1027 (1999)
423. Bercovici, H., Foias, C.: A real variable restatement of Riemann's hypothesis. Isr. J. Math. **48**, 57–68 (1984)
424. Bérczes, A., Brindza, B., Hajdu, L.: On the power values of polynomials. Publ. Math. (Debr.) **53**, 375–381 (1988)
425. Bérczes, A., Evertse, J.-H., Győry, K.: On the number of equivalence classes of binary forms of given degree and given discriminant. Acta Arith. **113**, 363–399 (2004)
426. Bérczes, A., Győry, K.: On the number of solutions of decomposable polynomial equations. Acta Arith. **101**, 171–187 (2002)
427. Bérczes, A., Pink, I.: On the Diophantine equation $x^2 + p^{2k} = y^n$. Arch. Math. **91**, 505–517 (2008)
428. Beresnevich, V.: On approximation of real numbers by real algebraic numbers. Acta Arith. **90**, 97–112 (1999)
429. Beresnevich, V.: On a theorem of V. Bernik in the metric theory of Diophantine approximation. Acta Arith. **117**, 71–80 (2005)
430. Beresnevich, V., Bernik, V.[I.]: On a metrical theorem of W. Schmidt. Acta Arith. **75**, 219–233 (1996)
431. Berg, C.: Børge Jessen, 19.6.1907–20.3.1993. J. Électron. Hist. Probab. Stat. **5**(1), 1–15 (2009)
432. Bergelson, V., Leibman, A.: Polynomial extensions of van der Waerden's and Szemerédi's theorems. J. Am. Math. Soc. **9**, 725–753 (1996)
433. Bergelson, V., McCutcheon, R.: An ergodic IP polynomial Szemerédi theorem. Mem. Am. Math. Soc. **146**, 1–106 (2000)
434. Berger, M.A., Felzenbaum, A., Fraenkel, A.[S.]: Necessary condition for the existence of an incongruent covering system with odd moduli. Acta Arith. **45**, 375–379 (1986)
435. Berger, M.A., Felzenbaum, A., Fraenkel, A.[S.]: Necessary condition for the existence of an incongruent covering system with odd moduli, II. Acta Arith. **48**, 73–79 (1987)
436. Bergman, G.: A generalization of a theorem of Nagell. Ark. Mat. **2**, 299–305 (1952)
437. Bergman, G.: On the exceptional group of a Weierstrass curve in an algebraic number field. Acta Math. **91**, 113–124 (1954)
438. Bergmann, G.: Über Eulers Beweis des grossen Fermatschen Satzes für den Exponenten 3. Math. Ann. **164**, 159–175 (1966)
439. Berkes, I., Philipp, W.: The size of trigonometric and Walsh series and uniform distribution mod 1. J. Lond. Math. Soc. **50**, 454–464 (1994)
440. Berkes, I., Philipp, W., Tichy, R.F.: Empirical processes in probabilistic number theory: the LIL or the discrepancy of $(n_k\omega)$ mod 1. Ill. J. Math. **50**, 107–145 (2006)
441. Berkovič, V.G.: Rational points on the Jacobians of modular curves. Mat. Sb. **101**, 542–567 (1976) (in Russian)
442. Berlekamp, E.R.: A construction for partitions which avoid long arithmetic progressions. Can. Math. Bull. **11**, 409–414 (1968)
443. Bernays, P.: Über die Darstellung von positiven, ganzen Zahlen durch die primitiven, binären quadratischen Formen einer nicht-quadratischen Diskriminante. Dissertation, Göttingen (1912)
444. Berndt, B.C.: Arithmetical identities and Hecke's functional equation. Proc. Edinb. Math. Soc. **16**, 221–226 (1969)
445. Berndt, B.C.: Identities involving the coefficients of a class of Dirichlet series, V. Trans. Am. Math. Soc. **160**, 139–156 (1971)
446. Berndt, B.C.: The Voronoï summation formula. In: Lecture Notes in Math., vol. 251, pp. 21–36. Springer, Berlin (1972)
447. Berndt, B.C.: Ramanujan's Notebooks, I–V. Springer, Berlin (1985–1998)
448. Berndt, B.C.: Bruno Schoeneberg (1906–1995). Jahresber. Dtsch. Math.-Ver. **99**, 83–89 (1997)

449. Berndt, B.C.: Number Theory in the Spirit of Ramanujan. Am. Math. Soc., Providence (2006)
450. Berndt, B.C., Evans, R.: Sums of Gauss, Eisenstein, Jacobi, Jacobsthal, and Brewer. Ill. J. Math. **23**, 374–437 (1979)
451. Berndt, B.C., Evans, R.: Sums of Gauss, Jacobi and Jacobsthal. J. Number Theory **11**, 349–398 (1979)
452. Berndt, B.C., Evans, R.: The determination of Gauss sums. Bull. Am. Math. Soc. **5**, 107–129 (1981)
453. Berndt, B.C., Evans, R., Williams, K.S.: Gauss and Jacobi Sums. Wiley, New York (1998)
454. Berndt, B.C., Kohnen, W., Ono, K.: The life and work of R.A. Rankin (1915–2001). Ramanujan J. **7**, 11–40 (2003)
455. Berndt, B.C., Ono, K.: Ramanujan's unpublished manuscript on the partition and tau functions with proofs and commentary. In: The Andrews Festschrift, Maratea, 1998. Sém. Lothar. Combin., vol. 42, pp. 1–63 (1999), Art. B42c
456. Berndt, B.C., Rankin, R.A. (eds.): Ramanujan: Letters and Commentary. Am. Math. Soc./Lond. Math. Soc., Providence/London (1995)
457. Berndt, B.C., Rankin, R.A. (eds.): Ramanujan: Essays and Surveys. Am. Math. Soc./Lond. Math. Soc., Providence/London (2001)
458. Berndt, R.: L-functions for Jacobi forms à la Hecke. Manuscr. Math. **84**, 101–112 (1994)
459. Bernik, V.I.: The Baker-Schmidt conjecture. Dokl. Akad. Nauk Belorus. **23**, 392–395 (1979) (in Russian)
460. Bernik, V.I.: Application of the Hausdorff dimension in the theory of Diophantine approximations. Acta Arith. **42**, 219–253 (1983) (in Russian)
461. Bernik, V.I.: The exact order of approximating zero by values of integral polynomials. Acta Arith. **53**, 17–28 (1989) (in Russian)
462. Bernik, V.I., Dodson, M.M.: Metric Diophantine Approximation on Manifolds. Cambridge University Press, Cambridge (1999)
463. Bernik, V.I., Melničuk, Yu.V.: Diophantine approximations and Hausdorff dimension. Nauka i Tekhnika, Minsk (1988) (in Russian)
464. Bernik, V.I., Tiščenko, K.I.: Integral polynomials with difference in the values of coefficients and Wirsing's theorem. Dokl. Akad. Nauk Belorus. **37**(5), 9–11 (1993) (in Russian)
465. Bernstein, F.: Ueber den zweiten Fall des letzten Fermatschen Lehrsatzes. Nachr. Ges. Wiss. Göttingen, 1910, 507–516
466. Bernstein, F.: Über eine Anwendung der Mengenlehre auf ein aus der Theorie der säkularen Störungen herrührendes Problem. Math. Ann. **71**, 417–439 (1912)
467. Bernstein, L.: The Jacobi-Perron Algorithm—Its Theory and Application. Lecture Notes in Math., vol. 207. Springer, Berlin (1971)
468. Bernstein, L., Hasse, H.: Einheitenberechnung mittels des Jacobi-Perronschen Algorithmus. J. Reine Angew. Math. **218**, 51–69 (1965)
469. Bernstein, L.: Zur Lösung der diophantischen Gleichung $m/n = 1/x + 1/y + 1/z$, insbesondere im Fall $m = 4$. J. Reine Angew. Math. **211**, 1–10 (1962)
470. Bertin, M.J., Decomps-Guilloux, A., Grandet-Hugot, M., Pathiaux-Delefosse, M., Schreiber, J.-P.: Pisot and Salem Numbers. Birkhäuser, Basel (1992)
471. Bertolini, M., Darmon, H.: Heegner points on Mumford-Tate curves. Invent. Math. **126**, 413–456 (1996)
472. Bertolini, M., Darmon, H.: A rigid analytic Gross-Zagier formula and arithmetic applications. Ann. Math. **146**, 111–147 (1997)
473. Bertolini, M., Darmon, H.: Non-triviality of families of Heegner points and ranks of Selmer groups over anticyclotomic towers. J. Ramanujan Math. Soc. **13**, 15–24 (1998)
474. Bertrand, D.: Hauteurs et isogénies. Astérisque **183**, 107–125 (1990)
475. Besicovitch, A.S.: On linear sets of points of fractional dimension. Math. Ann. **101**, 161–193 (1929)
476. Besicovitch, A.S.: On the sum of digits of real numbers represented in the dyadic system. On sets of fractional dimensions. II. Math. Ann. **110**, 321–330 (1934)

477. Besicovitch, A.S.: Sets of fractional dimension. IV. On rational approximation to real numbers. J. Lond. Math. Soc. **9**, 126–131 (1934)
478. Besicovitch, A.S.: On the sum of digits of real numbers represented in the dyadic system. Math. Ann. **110**, 321–330 (1935)
479. Besicovitch, A.S.: On the density of the sum of two sequences of integers. J. Lond. Math. Soc. **10**, 246–248 (1935)
480. Besov, O.V., et al.: Sergeĭ Borisovič Stečkin. Usp. Mat. Nauk **51**(6), 3–10 (1996) (in Russian)
481. Beukers, F.: The lattice-points of n-dimensional tetrahedra. Indag. Math. **37**, 365–372 (1975)
482. Beukers, F.: A note on the irrationality of $\zeta(2)$ and $\zeta(3)$. Bull. Lond. Math. Soc. **11**, 268–272 (1979)
483. Beukers, F.: The multiplicity of binary recurrences. Compos. Math. **40**, 251–267 (1980)
484. Beukers, F.: Padé-approximations in number theory. In: Lecture Notes in Math., vol. 888, pp. 90–99. Springer, Berlin (1981)
485. Beukers, F.: On the generalized Ramanujan-Nagell equation, I. Acta Arith. **38**, 389–410 (1980/1981)
486. Beukers, F.: On the generalized Ramanujan-Nagell equation, II. Acta Arith. **39**, 113–123 (1980/1981)
487. Beukers, F.: Fractional part of powers of rationals. Math. Proc. Camb. Philos. Soc. **90**, 13–20 (1981)
488. Beukers, F.: Irrationality proofs using modular forms. Astérisque **147/148**, 271–283 (1987)
489. Beukers, F.: The diophantine equation $Ax^p + By^q = Cz^r$. Duke Math. J. **91**, 61–88 (1998)
490. Beukers, F., Bézivin, J.-P., Robba, P.: An alternative proof of the Lindemann-Weierstrass theorem. Am. Math. Mon. **97**, 193–197 (1990)
491. Beukers, F., Brownawell, W.D., Heckman, G.: Siegel normality. Ann. Math. **127**, 279–308 (1988)
492. Beukers, F., Schlickewei, H.P.: The equation $x + y = 1$ in finitely generated groups. Acta Arith. **78**, 189–199 (1996)
493. Beukers, F., Tijdeman, R.: On the multiplicities of binary complex recurrences. Compos. Math. **51**, 193–213 (1984)
494. Beurling, A.: Analyse de la loi asymptotique de la distribution des nombres premiers généralisés, I. Acta Math. **68**, 255–291 (1937)
495. Beurling, A.: A closure problem related to the Riemann zeta-function. Proc. Natl. Acad. Sci. USA **41**, 312–314 (1955)
496. Bezdek, K.: Improving Roger's upper bound for the density of unit ball packings via estimating the surface area of Voronoi cells from below in Euclidean d-space for all $d \geq 8$. Discrete Comput. Geom. **28**, 75–106 (2002)
497. Bézivin, J.-P.: Diviseurs premiers de suites récurrentes linéaires. Electron. J. Comb. **7**, 199–204 (1986)
498. Bézivin, J.-P.: Sur les diviseurs premiers des suites récurrentes linéaires. Ann. Fac. Sci. Toulouse **8**, 61–73 (1986/1987)
499. Bhargava, M.: On the Conway-Schneeberger Fifteen Theorem. In: Contemp. Math., vol. 272, pp. 27–37. Am. Math. Soc., Providence (2000)
500. Bhargava, M.: Finiteness theorem for quadratic forms, to appear
501. Bhargava, M., Hanke, J.: Universal quadratic forms and the 290-theorem. Invent. Math., to appear
502. Bhaskaran, M.: Sums of mth powers in algebraic and Abelian number fields. Arch. Math. **17**, 497–504 (1966)
503. Bieberbach, L., Schur, I.: Über die Minkowskische Reduktionstheorie der positiven quadratischen Formen. SBer. Preuß. Akad. Wiss. Berlin, 510–535, 1928
504. Bierstedt, R.G., Mills, W.H.: On the bound for a pair of consecutive quartic residues of a prime. Proc. Am. Math. Soc. **14**, 628–632 (1963)
505. Bilharz, H.: Primdivisorem mit vorgegebener Primitivwurzel. Math. Ann. **114**, 476–492 (1937)

506. Billing, G.: Beiträge zur arithmetischen Theorie der ebenen kubischen Kurven vom Geschlechte Eins. Nova Acta R. Soc. Sci. Ups. **11**(1), 1–165 (1938)
507. Billing, G., Mahler, K.: On exceptional points on cubic curves. J. Lond. Math. Soc. **15**, 32–43 (1940)
508. Billingsley, P.: On the central limit theorem for the prime divisor function. Am. Math. Mon. **76**, 132–139 (1969)
509. Bilu, Y.[F.]: Addition of sets of integers of positive density. J. Number Theory **64**, 233–275 (1997)
510. Bilu, Y.[F.]: Structure of sets with small sumset. Astérisque **258**, 77–108 (1999)
511. Bilu, Y.F.: Catalan's conjecture (after Mihăilescu). Astérisque **294**, 1–26 (2004)
512. Bilu, Y.F.: Catalan without logarithmic forms. J. Théor. Nr. Bordx. **17**, 69–85 (2005)
513. Bilu, Y.F.: The many faces of the subspace theorem after Adamczewski, Bugeaud, Corvaja, Zannier ... Astérisque **317**, 1–38 (2008)
514. Bilu, Y.[F.], Bugeaud, Y.: Démonstration du théorème de Baker-Feldman via les formes linéaires en deux logarithmes. J. Théor. Nr. Bordx. **12**, 13–23 (2000)
515. Bilu, Y.[F.], Hanrot, G.: Solving Thue equations of high degree. J. Number Theory **60**, 373–392 (1996)
516. Bilu, Y.[F.], Hanrot, G.: Solving superelliptic Diophantine equations by Baker's method. Compos. Math. **112**, 273–312 (1998)
517. Bilu, Y.[F.], Hanrot, G.: Thue equations with composite fields. Acta Arith. **88**, 311–326 (1999)
518. Bilu, Y.[F.], Hanrot, G., Voutier, P.: Existence of a primitive divisor of Lucas and Lehmer numbers. J. Reine Angew. Math. **539**, 75–122 (2001)
519. Bilu, Y.[F.], Tichy, R.: The Diophantine equation $f(x) = g(y)$. Acta Arith. **95**, 261–288 (2000)
520. Bingham, N.H.: The work of Alfréd Rényi: some aspects in probability and number theory. Studia Sci. Math. Hung. **26**, 165–183 (1991)
521. Birch, B.J.: Homogeneous forms of odd degree in a large number of variables. Mathematika **4**, 102–105 (1957)
522. Birch, B.J.: Waring's problem in algebraic number fields. Proc. Camb. Philos. Soc. **57**, 449–459 (1961)
523. Birch, B.J.: Forms in many variables. Proc. R. Soc. Lond. Ser. A, Math. Phys. Sci. **265**, 245–263 (1961/1962)
524. Birch, B.J.: Waring's problem for $\mathfrak{p}$-adic number fields. Acta Arith. **9**, 169–176 (1964)
525. Birch, B.J.: Diagonal equations over $\mathfrak{p}$-adic fields. Acta Arith. **9**, 291–300 (1964)
526. Birch, B.J.: Multiplicative functions with non-decreasing normal order. J. Lond. Math. Soc. **32**, 149–151 (1967)
527. Birch, B.J.: Weber's class invariants. Mathematika **16**, 283–294 (1969)
528. Birch, B.J.: Small zeros of diagonal forms of odd degree in many variables. Proc. Lond. Math. Soc. **21**, 12–18 (1970)
529. Birch, B.J., Chowla, S., Hall, M. Jr., Schinzel, A.: On the difference $x^3 - y^2$. Norske Vid. Selsk. Forh., Trondheim **38**, 65–69 (1965)
530. Birch, B.J., Davenport, H.: Indefinite quadratic forms in many variables. Mathematika **5**, 8–12 (1958) [[1380], vol. 3, pp. 1077–1081]
531. Birch, B.J., Kuyk, W. (eds.): Modular Functions of One Variable IV. Lecture Notes in Math., vol. 476. Springer, Berlin (1975)
532. Birch, B.J., Lewis, D.J.: $\mathfrak{p}$-adic forms. J. Indian Math. Soc. (N.S.) **23**, 11–32 (1960)
533. Birch, B.J., Lewis, D.J.: Systems of three quadratic forms. Acta Arith. **10**, 423–442 (1964/1965)
534. Birch, B.J., Lewis, D.J., Murphy, T.G.: Simultaneous quadratic forms. Am. J. Math. **84**, 110–115 (1962)
535. Birch, B.J., Merriman, J.R.: Finiteness theorems for binary forms with given discriminant. Proc. Lond. Math. Soc. **24**, 385–394 (1972)
536. Birch, B.J., Swinnerton-Dyer, H.P.F.: On the inhomogeneous minimum of the product of n linear forms. Mathematika **3**, 25–39 (1956)

537. Birch, B.J., Swinnerton-Dyer, H.P.F.: Notes on elliptic curves, I. J. Reine Angew. Math. **212**, 7–25 (1963)
538. Birch, B.J., Swinnerton-Dyer, H.P.F.: Notes on elliptic curves, II. J. Reine Angew. Math. **218**, 79–108 (1965)
539. Birch, B.J., Swinnerton-Dyer, H.P.F.: The Hasse problem for rational surfaces. J. Reine Angew. Math. **274/275**, 164–174 (1975)
540. Birkhoff, G.D.: Démonstration d'un théorème élémentaire sur les fonctions entières. C. R. Acad. Sci. Paris **189**, 473–475 (1929)
541. Birkhoff, G.D., Vandiver, H.S.: On the integral divisors of $a^n - b^n$. Ann. Math. **5**, 173–180 (1904)
542. Birman, J.S.: On Siegel's modular group. Math. Ann. **191**, 59–68 (1971)
543. Biró, A.: Yokoi's conjecture. Acta Arith. **106**, 85–104 (2003)
544. Biró, A.: Chowla's conjecture. Acta Arith. **106**, 179–194 (2003)
545. Blake, I.F.: The Leech lattice as a code for the Gaussian channel. Inf. Control **19**, 66–74 (1971)
546. Blanchard, A.: Initiation à la théorie analytique des nombres premiers. Dunod, Paris (1969)
547. Blanchard, A., Mendès France, M.: Symétrie et transcendance. Bull. Sci. Math. **106**, 325–335 (1982)
548. Blanksby, P.E.: A note on algebraic integers. J. Number Theory **1**, 155–160 (1969)
549. Blanksby, P.E., Montgomery, H.L.: Algebraic integers near the unit circle. Acta Arith. **18**, 355–369 (1971)
550. Blasius, D., Rogawski, J.: Galois representations for Hilbert modular forms. Bull. Am. Math. Soc. **21**, 65–69 (1989)
551. Blass, J., Glass, A.M.W., Manski, D.K., Meronk, D.B., Steiner, R.P.: Constants for lower bounds for linear forms in the logarithms of algebraic numbers, I. The general case. Acta Arith. **55**, 1–14 (1990)
552. Blass, J., Glass, A.M.W., Manski, D.K., Meronk, D.B., Steiner, R.P.: Constants for lower bounds for linear forms in the logarithms of algebraic numbers, II. The homogeneous rational case. Acta Arith. **55**, 15–22 (1993); corr., **65**, 83 (1993)
553. Blecksmith, R., Brillhart, J., Gerst, I.: A computer-assisted investigation of Ramanujan pairs. Math. Comput. **46**, 731–749 (1986)
554. Bleicher, M.N.: Lattice coverings of n-space by spheres. Can. J. Math. **14**, 632–650 (1962)
555. Bleicher, M.N., Erdős, P.: Denominators of Egyptian fractions. J. Number Theory **8**, 157–168 (1976)
556. Bleicher, M.N., Erdős, P.: Denominators of Egyptian fractions, II. Ill. J. Math. **20**, 598–613 (1976)
557. Blichfeldt, H.F.: A new principle in the geometry of numbers with some applications. Trans. Am. Math. Soc. **15**, 227–235 (1914)
558. Blichfeldt, H.F.: Notes on diophantine approximations. Bull. Am. Math. Soc. **28**, 284–285 (1922)
559. Blichfeldt, H.F.: On the minimum value of positive real quadratic forms in 6 variables. Bull. Am. Math. Soc. **31**, 386 (1925)
560. Blichfeldt, H.F.: The minimum value of positive quadratic forms in seven variables. Bull. Am. Math. Soc. **32**, 99 (1926)
561. Blichfeldt, H.F.: The minimum value of quadratic forms, and the closest packing of spheres. Math. Ann. **101**, 605–608 (1929)
562. Blichfeldt, H.F.: The minimum values of positive quadratic forms in six, seven and eight variables. Math. Z. **39**, 1–15 (1935)
563. Blichfeldt, H.F.: A new upper bound to the minimum value of the sum of linear homogeneous forms. Monatshefte Math. Phys. **43**, 410–414 (1936)
564. Blichfeldt, H.F.: Note on the minimum value of the discriminant of an algebraic field. Monatshefte Math. Phys. **48**, 531–533 (1939)
565. Blumenthal, O.: Über Modulfunktionen von mehreren Veränderlichen. Math. Ann. **56**, 509–548 (1903); **58**, 1904, 497–527

566. Boca, F.P.: A problem of Erdős, Szűsz and Turán concerning Diophantine approximations. Int. J. Number Theory **4**, 691–708 (2008)
567. Bochnak, J., Oh, B.-K.: Almost regular quaternary quadratic forms. Ann. Inst. Fourier **58**, 1499–1549 (2008)
568. Bochnak, J., Oh, B.-K.: Almost-universal quadratic forms: an effective solution of a problem of Ramanujan. Duke Math. J. **147**, 131–156 (2009)
569. Bochner, S.: Ein Satz von Landau und Ikehara. Math. Z. **37**, 1–9 (1933)
570. Bochner, S.: On Riemann's functional equation with multiple Gamma factors. Ann. Math. **67**, 29–41 (1958)
571. Böcker, S., Lipták, Z.: A fast and simple algorithm for the money changing problem. Algorithmica **48**, 413–432 (2007)
572. Bogomolny, E.B., Keating, J.P.: Random matrix theory and the Riemann zeros. I. Three- and four-point correlations. Nonlinearity **8**, 1115–1131 (1995)
573. Bogomolny, E.B., Keating, J.P.: Random matrix theory and the Riemann zeros. II. n-point correlations. Nonlinearity **9**, 911–935 (1996)
574. Bogomolov, F., et al.: Vassiliĭ Alekseevič Iskovskih. Usp. Mat. Nauk **64**(5), 167–174 (2009) (in Russian)
575. Bohl, P.: Über ein in der Theorie der säkularen Störungen vorkommendes Problem. J. Reine Angew. Math. **135**, 189–283 (1909)
576. Bohr, H.: Sur l'existence de valeurs arbitrairement petites de la fonction $\zeta(s) = \zeta(\sigma + it)$ de Riemann pour $\sigma > 1$. Oversigt Kgl. Danske Videnskab. Selsk, Forh. **3**, 201–208 (1911) [[578], vol. 1, B3]
577. Bohr, H.: Zur Theorie der Riemannschen Zetafunktion im kritischen Streifen. Acta Math. **40**, 67–100 (1916) [[578], vol. 1, B19]
578. Bohr, H.: Collected Mathematical Works. Dansk Matematish Forening, København (1952)
579. Bohr, H., Courant, R.: Neue Anwendungen der Theorie der Diophantischen Approximationen auf die Riemannsche Zetafunktion. J. Reine Angew. Math. **144**, 249–274 (1914) [[578], vol. 1, B15]
580. Bohr, H., Cramér, H.: Die neuere Entwicklung der analytischen Zahlentheorie. In: Enzyklopädie der Mathematischen Wissenschaften, vol. 2, C8, pp. 722–849 (1923) [[578], vol. 3, H; [1274], vol. 1, pp. 289–416]
581. Bohr, H., Jessen, B.: Über die Werteverteilung der Riemannscher Zetafunktion. Acta Math. **54**, 1–35 (1930)
582. Bohr, H., Jessen, B.: Über die Werteverteilung der Riemannscher Zetafunktion, II. Acta Math. **58**, 1–55 (1932) [[578], vol. 1, B23, B24]
583. Bohr, H., Jessen, B.: Mean-value theorems for the Riemann zeta-function. Q. J. Math. **5**, 43–47 (1934) [[578], B25]
584. Bohr, H., Landau, E.: Über das Verhalten von $\zeta(s)$ und $\zeta_\kappa(s)$ in der Nähe der Geraden $\sigma = 1$. Nachr. Ges. Wiss. Göttingen, 1910, 303–330. [[578], vol. 3, B1; [3680], vol. 4, pp. 221–248]
585. Bohr, H., Landau, E.: Beiträge zur Theorie der Riemannschen Zetafunktion. Math. Ann. **74**, 3–30 (1913) [[578], vol. 3, B11; [3680], vol. 5, pp. 454–481]
586. Bohr, H., Landau, E.: Sur les zéros de la fonction $\zeta(s)$ de Riemann. C. R. Acad. Sci. Paris **158**, 106–110 (1914) [[578], vol. 3, B14; [3680], vol. 6, pp. 56–60]
587. Bohr, H., Landau, E.: Ein Satz über Dirichletsche Reihen mit Anwendung an die ζ-Funktion und die L-Funktionen. Rend. Circ. Mat. Palermo **37**, 269–272 (1914) [[578], vol. 3, B13; [3680], vol. 6, pp. 45–48]
588. Bohr, H., Landau, E., Littlewood, J.E.: Sur la fonction $\zeta(s)$ dans le voisinage de la droite $\sigma = 1/2$. Bull. Acad. Roy. Belg., 1913, 3–35. [[578], vol. 3, B12; [3680], vol. 6, pp. 56–93]
589. Boklan, K.D.: The asymptotic formula in Waring's problem. Mathematika **41**, 329–347 (1994)
590. Bokowski, J., Hadwiger, H., Wills, J.M.: Eine Ungleichung zwischen Volumen, Oberfläche und Gitterpunktanzahl konvexer Körper im n-dimensionalen euklidischen Raum. Math. Z. **127**, 363–364 (1972)

591. Bokowski, J., Wills, J.M.: Eine Ungleichung zwischen Volumen, Oberfläche und Gitterpunktanzahl konvexer Mengen im R^3. Acta Math. Acad. Sci. Hung. **25**, 7–13 (1974)
592. Bombieri, E.: Sull'approssimazione di numeri algebrici mediante numeri algebrici. Boll. Unione Mat. Ital. **13**, 351–354 (1958)
593. Bombieri, E.: Maggiorazione del resto nel "Primzahlsatz" col metodo di Erdős-Selberg. Rend. - Ist. Lomb. Sci. Lett., Accad. Sci. Lett. **96**, 343–350 (1962)
594. Bombieri, E.: Sul theorema di Tschebotarev. Acta Arith. **8**, 273–281 (1963)
595. Bombieri, E.: On the large sieve. Mathematika **12**, 201–225 (1965)
596. Bombieri, E.: On exponential sums in finite fields. Am. J. Math. **88**, 71–105 (1966)
597. Bombieri, E.: On exponential sums in finite fields, II. Invent. Math. **47**, 29–39 (1978)
598. Bombieri, E.: On a theorem of van Lint and Richert. In: Symposia Mathematica, vol. 4, pp. 175–180. Academic Press, London (1970)
599. Bombieri, E.: A note on the large sieve. Acta Arith. **18**, 401–404 (1971)
600. Bombieri, E.: Counting points on curves over finite fields (d'après Stepanov). In: Lecture Notes in Math., vol. 383, pp. 234–241. Springer, Berlin (1974)
601. Bombieri, E.: Le grand crible dans la théorie analytique des nombres. Astérisque **18**, 2–87 (1974) [2nd ed. 1987]
602. Bombieri, E.: The asymptotic sieve. Rend. Accad. Naz. dei XL, (5), 1975/1976, 243–269
603. Bombieri, E.: On the Thue-Siegel-Dyson theorem. Acta Math. **148**, 255–296 (1982)
604. Bombieri, E.: The Mordell conjecture revisited. Ann. Sc. Norm. Super. Pisa, Cl. Sci. **17**, 615–640 (1990); corr. **18**, 473 (1991)
605. Bombieri, E.: Effective Diophantine approximation on G_m. Ann. Sc. Norm. Super. Pisa, Cl. Sci. **20**, 61–89 (1993)
606. Bombieri, E.: Roth's theorem and the *abc* conjecture. Preprint, ETH Zürich (1994)
607. Bombieri, E.: On the Thue-Mahler equation, II. Acta Arith. **67**, 69–96 (1994)
608. Bombieri, E., Davenport, H.: Small differences between prime numbers. Proc. R. Soc. Lond. Ser. A, Math. Phys. Sci. **293**, 1–18 (1966) [[1380], vol. 4, pp. 1639–1656]
609. Bombieri, E., Davenport, H.: On the large sieve method. In: Abhandlungen aus Zahlentheorie und Analysis. Zur Erinnerung an Edmund Landau, pp. 11–22. VEB Deutscher Verlag der Wissenschaften, Berlin (1969) [[1380], vol. 4, pp. 1673–1684]
610. Bombieri, E., Davenport, H.: Some inequalities involving trigonometric polynomials. Ann. Sc. Norm. Super. Pisa, Cl. Sci. **23**, 223–241 (1969) [[1380], vol. 4, pp. 1685–1703]
611. Bombieri, E., Friedlander, J.B., Iwaniec, H.: Primes in arithmetic progressions to large moduli. Acta Math. **156**, 203–251 (1986)
612. Bombieri, E., Friedlander, J.B., Iwaniec, H.: Primes in arithmetic progressions to large moduli, II. Math. Ann. **277**, 361–393 (1987)
613. Bombieri, E., Friedlander, J.B., Iwaniec, H.: Primes in arithmetic progressions to large moduli, III. J. Am. Math. Soc. **2**, 215–224 (1989)
614. Bombieri, E., Gubler, W.: Heights in Diophantine Geometry. Cambridge University Press, Cambridge (2006)
615. Bombieri, E., Iwaniec, H.: On the order of $\zeta(\frac{1}{2}+it)$. Ann. Sc. Norm. Super. Pisa, Cl. Sci. **13**, 449–472 (1986)
616. Bombieri, E., Iwaniec, H.: Some mean-value theorems for exponential sums. Ann. Sc. Norm. Super. Pisa, Cl. Sci. **13**, 473–486 (1986)
617. Bombieri, E., Mueller, J.: On effective measures of irrationality for $\sqrt[r]{a/b}$ and related numbers. J. Reine Angew. Math. **342**, 173–196 (1983)
618. Bombieri, E., Mueller, J.: Remarks on the approximation to an algebraic number by algebraic numbers. Mich. Math. J. **33**, 83–93 (1986)
619. Bombieri, E., Mueller, J., Poe, M.: The unit equation and the cluster principle. Acta Arith. **79**, 361–389 (1997)
620. Bombieri, E., Pila, J.: The number of integral points on arcs and ovals. Duke Math. J. **59**, 337–357 (1989)
621. Bombieri, E., Schmidt, W.M.: On Thue's equation. Invent. Math. **88**, 69–81 (1987)
622. Bombieri, E., Swinnerton-Dyer, H.P.F.: On the local zeta function of a cubic threefold. Ann. Sc. Norm. Super. Pisa, Cl. Sci. **21**, 1–29 (1967)

623. Bombieri, E., Vaaler, J.: On Siegel's lemma. Invent. Math. **73**, 11–32 (1983); add.: **75**, 377 (1984)
624. Bombieri, E., van der Poorten, A.J.: Some quantitative results related to Roth's theorem. J. Aust. Math. Soc. A **45**, 233–248 (1988); corr., **A48**, 154–155 (1990)
625. Bonciocat, N.C.: Congruences and Lehmer's problem. Int. J. Number Theory **4**, 587–596 (2008)
626. Borcherds, R.E.: Monstrous moonshine and monstrous Lie superalgebras. Invent. Math. **109**, 405–444 (1992)
627. Borcherds, R.E.: Classification of positive definite lattices. Duke Math. J. **105**, 525–567 (2000)
628. Borel, A.: Some finiteness properties of adele groups over number fields. Publ. Math. Inst. Hautes Études Sci. **16**, 5–30 (1963)
629. Borel, A.: Formes automorphes et séries de Dirichlet (d'aprés R.P. Langlands). In: Lecture Notes in Math., vol. 514, pp. 185–222. Springer, Berlin (1976)
630. Borel, A.: Values of indefinite quadratic forms at integral points and flows on spaces of lattices. Bull. Am. Math. Soc. **32**, 184–204 (1995)
631. Borel, A., Casselman, W. (eds.): Automorphic Forms, Representations and L-functions, I, II. Proc. Symposia Pure Math., vol. 33. Am. Math. Soc., Providence (1979)
632. Borel, A., Chowla, S., Herz, C.S., Iwasawa, K., Serre, J.-P.: Seminar on Complex Multiplication. Lecture Notes in Math., vol. 21. Springer, Berlin (1966)
633. Borel, A., Harish-Chandra: Arithmetic subgroups of algebraic groups. Ann. Math. **75**, 485–535 (1962)
634. Borel, É.: Sur la nature arithmétique du nombre e. C. R. Acad. Sci. Paris **128**, 596–599 (1899)
635. Borel, É.: Sur l'approximation des nombres par des nombres rationnels. C. R. Acad. Sci. Paris **136**, 1054–1055 (1903)
636. Borel, É.: Contribution à l'analyse arithmétique du continu. J. Math. Pures Appl. **9**, 329–375 (1903)
637. Borel, É.: Les probabilités dénombrables et leurs applications arithmétiques. Rend. Circ. Mat. Palermo **27**, 247–271 (1909)
638. Borel, É.: Leçons sur la théorie de la croissance. Gauthier-Villars, Paris (1910)
639. Borel, É.: Sur un problème de probabilités relatif aux fractions continues. Math. Ann. **72**, 578–584 (1912)
640. Borel, É.: Leçons sur la théorie des fonctions. Gauthier-Villars, Paris (1914)
641. Borevič, Z.I.: On the proof of the principal ideal theorem. Vestn. Leningr. Univ. **12**(13), 5–8 (1957) (in Russian)
642. Borevič, Z.I., Šafarevič, I.R.: Number Theory. Nauka, Moscow (1964); 2nd ed. 1972; 3rd ed. 1985 (in Russian) [German translation: Zahlentheorie, Birkhäuser, 1966; English translation: Academic Press, 1966; French translation: Théorie des nombres, Gauthier-Villars, 1967; reprint: Jacques Gabay, 1993]
643. Borevič, Z.I., et al.: A.V. Malyšev (1928–1993). Scientist and teacher. Zap. Nauč. Semin. POMI **211**, 7–29 (1994) (in Russian)
644. Borho, W.: Befreundete Zahlen mit gegebener Primteileranzahl. Math. Ann. **209**, 183–193 (1974)
645. Borho, W., Hoffmann, H.: Breeding amicable pairs in abundance. Math. Comput. **46**, 281–293 (1986)
646. Bőrőczky, K. Jr.: Finite Packing and Covering. Cambridge University Press, Cambridge (2004)
647. Borosh, I., Moreno, C.J., Porta, H.: Elliptic curves over finite fields, II. Math. Comput. **29**, 951–964 (1975)
648. Borozdkin, K.G.: On the question of I.M. Vinogradov's constant. In: Trudy III Vsesoj. Mat. S'ezda, Moscow, vol. 1, p. 3 (1956) (in Russian)
649. Borwein, J., Choi, K.K.S.: On the representations of $xy + yz + zx$. Exp. Math. **9**, 153–158 (2000)

650. Borwein, P., Choi, S.[K.K.S.], Chu, F.: An old conjecture of Erdős-Turán on additive bases. Math. Comput. **75**, 475–484 (2006)
651. Borwein, P., Dobrowolski, E., Mossinghoff, M.J.: Lehmer's problem for polynomials with odd coefficients. Ann. Math. **166**, 347–366 (2007)
652. Borwein, P., Hare, K.G., Mossinghoff, M.J.: The Mahler measure of polynomials with odd coefficients. Bull. Lond. Math. Soc. **36**, 332–338 (2004)
653. Borwein, P., Ingalls, C.: The Prouhet-Tarry-Escott problem revisited. Enseign. Math. **40**, 3–27 (1994)
654. Borwein, P., Lisoněk, P., Percival, C.: Computational investigation of the Prouhet-Tarry-Escott problem. Math. Comput. **72**, 2063–2070 (2003)
655. Bosman, J.: Explicit computations with modular Galois representations. Ph.D. thesis, Leiden Univ. (2008)
656. Bosser, V.: Indépendance algébrique de valeurs de séries d'Eisenstein (théorème de Nesterenko). In: Formes modulaires et transcendance, pp. 119–178. Soc. Math. France, Paris (2005)
657. Boulet, C., Pak, I.: A combinatorial proof of the Rogers-Ramanujan and Schur identities. J. Comb. Theory, Ser. A **113**, 1019–1030 (2006)
658. Boulyguine, B.[2]: Sur la représentation d'un nombre entier par une somme de carrés. C.R. Acad. Sci. Paris **158**, 328–330 (1914); **161**, 28–30 (1915)
659. Bouniakowsky, V.: Nouveaux théorèmes relatifs à la distinction des nombres premiers et à la décomposition des entiers en facteurs. Mém. Acad. Imp. Sci. St.-Pétersbg., Sci. Math. Phys. Nat. **6**, 305–329 (1857)
660. Bourgain, J.: On triples in arithmetic progression. Geom. Funct. Anal. **9**, 968–984 (1999)
661. Bourgain, J.: On large values estimates for Dirichlet polynomials and the density hypothesis for the Riemann zeta function. Internat. Math. Res. Notices, 2000, 133–146
662. Bourgain, J.: On the distribution of Dirichlet sums, II. In: Number Theory for the Millennium, I, pp. 87–109. Peters, Wellesley (2002)
663. Bourgain, J.: Mordell's exponential sum estimate revisited. J. Am. Math. Soc. **18**, 477–499 (2005)
664. Bourgain, J.: Roth's theorem on progressions revisited. J. Anal. Math. **104**, 155–192 (2008)
665. Bourgain, J.: Estimates on polynomial exponential sums. Isr. J. Math. **176**, 221–240 (2010)
666. Bourgain, J., Sinai, Ya.G.: Limit behavior of large Frobenius numbers. Usp. Mat. Nauk **62**(4), 77–90 (2007) (in Russian)
667. Boutin, A., Teilhet, P.F.: L'Intermédiaire des Math., **11**, 68, 182 (1904)
668. Bovey, J.D.: $\Gamma^*(8)$. Acta Arith. **25**, 145–150 (1974)
669. Boyarsky, M.: p-adic gamma functions and Dwork cohomology. Trans. Am. Math. Soc. **257**, 359–369 (1980)
670. Boyd, D.W.: Reciprocal polynomials having small measure. Math. Comput. **35**, 1361–1377 (1980), S1–S5
671. Boyd, D.W.: Reciprocal polynomials having small measure, II. Math. Comput. **53**, 355–357 (1989), S1–S5
672. Boyd, D.W., Kisilevsky, H.: On the exponent of the ideal class groups of complex quadratic fields. Proc. Am. Math. Soc. **31**, 433–436 (1972)
673. Brauer, A.: Über einige spezielle diophantische Gleichungen. Math. Z. **25**, 499–504 (1926)
674. Brauer, A.: Über Sequenzen von Potenzresten. SBer. Preuß. Akad. Wiss. Berlin, 1928, 9–16
675. Brauer, A.: Über diophantische Gleichungen mit endlich vielen Lösungen. J. Reine Angew. Math. **160**, 70–99 (1929)
676. Brauer, A.: Über den kleinsten quadratischen Nichtrest. Math. Z. **33**, 161–176 (1931)
677. Brauer, A.: Über die Verteilung der Potenzreste. Math. Z. **35**, 39–50 (1932)
678. Brauer, A.: Nachruf auf Hermann Zeitz. SBer. Berliner Math. Ges. **33**, 2–6 (1934)
679. Brauer, A.: Über die Dichte von Mengen positiver ganzer Zahlen. Ann. Math. **39**, 322–340 (1938)

[2] See also Bulygin, V.V.

680. Brauer, A.: Über die Dichte von Mengen positiver ganzer Zahlen, II. Ann. Math. **42**, 959–988 (1941)
681. Brauer, A.: On addition chains. Bull. Am. Math. Soc. **45**, 736–739 (1939)
682. Brauer, A.: On a problem on partitions. Am. J. Math. **64**, 299–312 (1942)
683. Brauer, A., Macon, N.: On the approximation of irrational numbers by the convergents of their continued fractions. Am. J. Math. **71**, 349–361 (1949)
684. Brauer, A., Macon, N.: On the approximation of irrational numbers by the convergents of their continued fractions, II. Am. J. Math. **72**, 419–424 (1950)
685. Brauer, A., Seelbinder, B.M.: On a problem on partitions, II. Am. J. Math. **76**, 343–346 (1954)
686. Brauer, A., Shockley, J.E.: On a problem of Frobenius. J. Reine Angew. Math. **211**, 215–220 (1962)
687. Brauer, A., Zeitz, H.: Über eine zahlentheoretische Behauptung von Legendre. SBer. Berliner Math. Ges. **29**, 116–125 (1930)
688. Brauer, R.: A note on systems of homogeneous algebraic equations. Bull. Am. Math. Soc. **51**, 749–755 (1945)
689. Brauer, R.: On the zeta-functions of algebraic number fields. Am. J. Math. **69**, 243–250 (1947)
690. Brauer, R.: On the zeta-functions of algebraic number fields, II. Am. J. Math. **72**, 739–746 (1950)
691. Brauer, R.: On Artin's L-series with general group characters. Ann. Math. **48**, 502–514 (1947)
692. Braun, H.: Konvergenz verallgemeinerter Eisensteinscher Reihen. Math. Z. **44**, 387–397 (1938)
693. Bravais, A.: Mémoire sur les systémes formées par des points distribués réguliérement sur un plan ou dans l'espace. J. Éc. Polytech. **19**, 1–128 (1850) [German translation, Ostwald, 1897]
694. Bredikhin, B.M.: Binary additive problems of indeterminate type. Izv. Akad. Nauk SSSR, Ser. Mat. **27**, 439–462 (1963) (in Russian)
695. Bredikhin, B.M.: Binary additive problems of indeterminate type, II. Analogue of the problem of Hardy and Littlewood. Izv. Akad. Nauk SSSR, Ser. Mat. **27**, 577–612 (1963) (in Russian)
696. Bredikhin, B.M.: Binary additive problems of indeterminate type, III. The additive problem of divisors. Izv. Akad. Nauk SSSR, Ser. Mat. **27**, 777–794 (1963) (in Russian)
697. Bredikhin, B.M.: Binary additive problems of indeterminate type, IV. The analogue of the generalized Hardy–Littlewood problem. Izv. Akad. Nauk SSSR, Ser. Mat. **28**, 1409–1440 (1964) (in Russian)
698. Bredikhin, B.M.: The dispersion method and definite binary additive problems. Usp. Mat. Nauk **20**(2), 89–130 (1965) (in Russian)
699. Bredikhin, B.M.: On the ternary Goldbach's theorem. Issled. Teor. Čisel, Sarat. **6**, 5–18 (1975) (in Russian)
700. Bredikhin, B.M., Grišina, T.I.: An elementary evaluation of $G(n)$ in Waring's problem. Mat. Zametki **24**, 7–18 (1978) (in Russian)
701. Bredikhin, B.M.: Applications of the dispersion method in the Goldbach problem. Acta Arith. **27**, 253–263 (1975) (in Russian)
702. Bremner, A.: On sums of three cubes. In: Number Theory, Halifax, NS, 1994, pp. 87–91. Am. Math. Soc., Providence (1995)
703. Bremner, A., Tzanakis, N.: Lucas sequences whose 12th or 9th term is a square. J. Number Theory **107**, 215–227 (2004)
704. Bremner, A., Tzanakis, N.: On squares in Lucas sequences. J. Number Theory **124**, 511–520 (2007)
705. Bremner, A., Tzanakis, N.: Lucas sequences whose nth term is a square or an almost square. Acta Arith. **126**, 261–280 (2007)
706. Brent, R.P.: The first occurrence of large gaps between successive primes. Math. Comput. **27**, 959–963 (1973)

707. Brent, R.P.: On the zeros of the Riemann zeta-function in the critical strip. Math. Comput. **33**, 1361–1372 (1979)
708. Brent, R.P.: The first occurrence of certain large prime gaps. Math. Comput. **35**, 1435–1436 (1980)
709. Brent, R.P.: Factorization of the eleventh Fermat number. Abstr. Pap. Present. Am. Math. Soc. **10**, 89T-11-73 (1989)
710. Brent, R.P.: Factorization of the tenth Fermat number. Math. Comput. **68**, 429–451 (1999)
711. Brent, R.P., Cohen, G.L., te Riele, H.J.J.: Improved techniques for lower bounds for odd perfect numbers. Math. Comput. **57**, 857–868 (1991)
712. Brent, R.P., Pollard, J.M.: Factorization of the eighth Fermat number. Math. Comput. **36**, 627–630 (1981)
713. Brent, R.P., van de Lune, J., te Riele, H.J.J., Winter, D.T.: On the zeros of the Riemann zeta-function in the critical strip, II. Math. Comput. **39**, 681–688 (1982); corr.: **46**, 771 (1986)
714. Bressoud, D.M.: An easy proof of the Rogers-Ramanujan identities. J. Number Theory **16**, 235–241 (1983)
715. Bressoud, D.M., Zeilberger, D.: Generalized Rogers-Ramanujan bijections. Adv. Math. **78**, 42–75 (1989)
716. Bretschneider, C.A.: Tafeln zur Zerlegung der Zahlen bis 4100 in Biquadrate. J. Reine Angew. Math. **46**, 1–23 (1853)
717. Breuer, F.: Torsion bounds for elliptic curves and Drinfeld modules. J. Number Theory **130**, 1241–1250 (2010)
718. Breuil, C., Conrad, B., Diamond, F., Taylor, R.: On the modularity of elliptic curves over **Q**: wild 3-adic exercises. J. Am. Math. Soc. **14**, 843–939 (2001)
719. Breulmann, S.: On Hecke eigenforms in the Maaß space. Math. Z. **232**, 527–530 (1999)
720. Breusch, R.: Zur Verallgemeinerung des Bertrandschen Postulates, daß zwischen x und $2x$ stets Primzahlen liegen. Math. Z. **34**, 505–526 (1932)
721. Breusch, R.: On the distribution of the roots of a polynomial with integral coefficients. Proc. Am. Math. Soc. **3**, 939–941 (1951)
722. Breusch, R.: An elementary proof of the prime number theorem with remainder term. Pac. J. Math. **10**, 487–497 (1960)
723. Brewer, B.W.: On certain character sums. Trans. Am. Math. Soc. **99**, 241–245 (1961)
724. Brewer, B.W.: On primes of the form $u^2 + 5v^2$. Proc. Am. Math. Soc. **17**, 502–509 (1966)
725. Brezinski, C.: History of Continued Fractions and Padé Approximants. Springer, Berlin (1991)
726. Brezinski, C.: A Bibliography on Continued Fractions, Padé Approximation, Sequence Transformation and Related Subjects. Prensas Universitarias de Zaragoza, Zaragoza (1991)
727. Brieskorn, E. (ed.): Felix Hausdorff zum Gedächtnis, I. Vieweg, Braunschweig (1996)
728. Briggs, W.E.: An elementary proof of a theorem about the representation of primes by quadratic forms. Can. J. Math. **6**, 353–363 (1954)
729. Brillhart, J.: Derrick Henry Lehmer. Acta Arith. **62**, 207–213 (1992)
730. Brillhart, J.: Commentary on Lucas' test. In: High Primes and Misdemeanors, pp. 103–109. Am. Math. Soc., Providence (2004)
731. Brillhart, J.: Emma Lehmer (1906–2007). Not. Am. Math. Soc. **54**, 1500–1501 (2007)
732. Brillhart, J., Lehmer, D.H., Lehmer, E.: Bounds for pairs of consecutive seventh and higher power residues. Math. Comput. **18**, 397–407 (1964)
733. Brillhart, J., Tonascia, J., Weinberger, P.: On the Fermat quotient. In: Computers in Number Theory, pp. 213–222. Academic Press, San Diego (1971)
734. Brindza, B.: On S-integral solutions of the equation $y^m = f(x)$. Acta Math. Acad. Sci. Hung. **44**, 133–139 (1984)
735. Brindza, B.: On large values of binary forms. Rocky Mt. J. Math. **26**, 839–845 (1996)
736. Brindza, B., Evertse, J.-H., Győry, K.: Bounds for the solutions of some Diophantine equations in terms of discriminants. J. Aust. Math. Soc. A **51**, 8–26 (1991)
737. Brindza, B., Győry, K., Tijdeman, R.: On the Catalan equation over algebraic number fields. J. Reine Angew. Math. **367**, 90–102 (1986)

738. Brindza, B., Pintér, Á., Végső, J.: The Schinzel-Tijdeman theorem over function fields. C. R. Acad. Sci. Paris Acad. Sci. Soc. R. Can. **16**, 53–57 (1994)
739. Brion, M.: Points entiers dans les polyèdres convexes. Ann. Sci. Éc. Norm. Super. **21**, 653–663 (1988)
740. Brion, M.: Polyèdres et réseaux. Enseign. Math. **38**, 71–88 (1992)
741. Brion, M.: Points entiers dans les polytopes convexes. Astérisque **227**, 145–169 (1995)
742. Brion, M., Vergne, M.: Residue formulae, vector partition functions and lattice points in rational polytopes. J. Am. Math. Soc. **10**, 797–833 (1997)
743. Brisebarre, N., Philibert, G.: Effective lower and upper bounds for the Fourier coefficients of powers of the modular invariant j. J. Ramanujan Math. Soc. **20**, 255–282 (2005)
744. Browder, F. (ed.): Mathematical Developments Arising from Hilbert Problems, I, II. Am. Math. Soc., Providence (1976)
745. Browkin, J.: Forms over p-adic fields. Bull. Acad. Pol. Sci., Sér. Sci. Math. Astron. Phys. **14**, 489–492 (1966)
746. Browkin, J.: On zeros of forms. Bull. Acad. Pol. Sci., Sér. Sci. Math. Astron. Phys. **17**, 611–616 (1969)
747. Browkin, J.: The abc-conjecture. In: Number Theory, pp. 75–105. Birkhäuser, Basel (2000)
748. Browkin, J.: The abc-conjecture for algebraic numbers. Acta Math. Sin. **22**, 211–222 (2006)
749. Browkin, G.[J.], Schinzel, A.: Sur les nombres de Mersenne qui sont triangulaires. C. R. Acad. Sci. Paris **242**, 1780–1781 (1956) [[5449], vol. 1, pp. 11–12]
750. Browkin, G.[J.], Schinzel, A.: On the equation $2^n - D = y^2$. Bull. Acad. Pol. Sci., Sér. Sci. Math. Astron. Phys. **3**, 311–318 (1960)
751. Brown, G., Moran, W., Pollington, A.D.: Normality to noninteger bases. C. R. Acad. Sci. Paris **316**, 1241–1244 (1993)
752. Brown, G., Moran, W., Pollington, A.D.: Normality with respect to powers of a base. Duke Math. J. **88**, 247–265 (1997)
753. Brown, M.L.: Note on supersingular primes of elliptic curves over Q. Bull. Lond. Math. Soc. **20**, 293–296 (1988)
754. Brown, S.: Bounds on transfer principle for algebraically closed and complete discretely valued fields. Mem. Am. Math. Soc. **15**, 1–92 (1978)
755. Brown, T.C., Shiue, P.J.-S.: Sums of fractional parts of integer multiples of an irrational. J. Number Theory **50**, 181–192 (1995)
756. Brownawell, W.D.: The algebraic independence of certain numbers related by the exponential function. J. Number Theory **6**, 22–31 (1974)
757. Brownawell, W.D.: Gelfond's method for algebraic independence. Trans. Am. Math. Soc. **210**, 1–26 (1975)
758. Brownawell, W.D.: On the Gel'fond-Fel'dman measure of algebraic independence. Compos. Math. **38**, 355–368 (1979)
759. Brownawell, W.D.: On p-adic zeros of forms. J. Number Theory **18**, 342–349 (1984)
760. Brownawell, W.D., Waldschmidt, M.: The algebraic independence of certain numbers to algebraic powers. Acta Arith. **32**, 63–71 (1977)
761. Browning, T.D.: Equal sums of two kth powers. J. Number Theory **96**, 293–318 (2002)
762. Browning, T.D., Heath-Brown, D.R.: Rational points on quartic hypersurfaces. J. Reine Angew. Math. **629**, 37–88 (2009)
763. Bruce, J.W.: A really trivial proof of the Lucas-Lehmer test. Am. Math. Mon. **100**, 370–371 (1993)
764. Brüdern, J.: Sums of squares and higher powers. J. Lond. Math. Soc. **35**, 233–243 (1987)
765. Brüdern, J.: Sums of squares and higher powers, II. J. Lond. Math. Soc. **35**, 244–250 (1987)
766. Brüdern, J.: On Waring's problem for cubes and biquadrates. J. Lond. Math. Soc. **37**, 25–42 (1988)
767. Brüdern, J.: On Waring's problem for cubes and biquadrates, II. Math. Proc. Camb. Philos. Soc. **104**, 199–206 (1988)
768. Brüdern, J.: A problem in additive number theory. Math. Proc. Camb. Philos. Soc. **103**, 27–33 (1988)

769. Brüdern, J.: Sums of four cubes. Monatshefte Math. **107**, 179–188 (1989)
770. Brüdern, J.: On pairs of diagonal cubic forms. Proc. Lond. Math. Soc. **61**, 273–343 (1990)
771. Brüdern, J.: On Waring's problem for fifth powers and some related topics. Proc. Lond. Math. Soc. **61**, 457–479 (1990)
772. Brüdern, J.: On Waring's problem for cubes. Math. Proc. Camb. Philos. Soc. **109**, 229–256 (1991)
773. Brüdern, J.: A sieve approach to the Waring-Goldbach problem. II: On the seven cubes theorem. Acta Arith. **72**, 211–227 (1995)
774. Brüdern, J., Godinho, H.: On Artin's conjecture, I: Systems of diagonal forms. Bull. Lond. Math. Soc. **150**, 305–313 (1999)
775. Brüdern, J., Godinho, H.: On Artin's conjecture, II. Pairs of additive forms. Proc. Lond. Math. Soc. **84**, 513–538 (2002)
776. Brüdern, J., Perelli, A.: The addition of primes and power. Can. J. Math. **48**, 512–526 (1996)
777. Brüdern, J., Perelli, A.: Exponential sums of additive problems involving square-free numbers. Ann. Sc. Norm. Super. Pisa, Cl. Sci. **28**, 591–613 (1999)
778. Brüdern, J., Wooley, T.D.: On Waring's problem: a square, four cubes and a biquadrate. Math. Proc. Camb. Philos. Soc. **127**, 193–200 (1999)
779. Brüdern, J., Wooley, T.D.: On Waring's problem: three cubes and a sixth power. Nagoya Math. J. **163**, 13–53 (2001)
780. Brüdern, J., Wooley, T.D.: The Hasse principle for pairs of diagonal cubic forms. Ann. Math. **166**(2), 865–895 (2007)
781. Bruggeman, R.W.: Fourier coefficients of cusp forms. Invent. Math. **45**, 1–18 (1978)
782. Bruggeman, R.W.: Fourier Coefficients of Automorphic Forms. Lecture Notes in Math., vol. 865. Springer, Berlin (1981)
783. de Bruijn, N.G.: Over Modulaire Vormen van Meer Veranderlijken. Thesis, Free University of Amsterdam (1943)
784. de Bruijn, N.G.: The asymptotic behaviour of a function occurring in the theory of primes. J. Indian Math. Soc. (N.S.) **15**, 25–32 (1951)
785. de Bruijn, N.G.: On the number of positive integers $\le x$ and free of prime factors $> y$. Indag. Math. **15**, 50–60 (1951)
786. de Bruijn, N.G.: On the number of positive integers $\le x$ and free of prime factors $> y$, II. Indag. Math. **28**, 239–247 (1966)
787. de Bruijn, N.G.: Johannes G. van der Corput (1890–1973). A biographical note. Acta Arith. **32**, 207–208 (1977)
788. de Bruijn, N.G.: In memoriam Jack van Lint (1932–2004). Nieuw Arch. Wiskd. **6**, 105–109 (2005)
789. Bruin, N.: The diophantine equations $x^2 \pm y^4 = \pm z^6$ and $x^2 + y^8 = z^3$. Compos. Math. **118**, 305–321 (1999)
790. Bruin, N.: Chabauty methods and covering techniques applied to generalized Fermat equations. Thesis, Univ. Leiden (1999). [CWI Tracts **133**, 1–77 (2002)]
791. Bruin, N.: On powers as sums of two cubes. In: Lecture Notes in Comput. Sci., vol. 1838, pp. 169–184. Springer, Berlin (2000)
792. Bruin, N.: Chabauty methods using elliptic curves. J. Reine Angew. Math. **562**, 27–49 (2003)
793. Bruin, N.: The primitive solutions to $x^3 + y^9 = z^2$. J. Number Theory **111**, 179–189 (2005)
794. Bruinier, J.H., Ono, K.: Coefficients of half-integral weight modular forms. J. Number Theory **99**, 164–179 (2003)
795. Brumer, A., Kramer, K.: The rank of elliptic curves. Duke Math. J. **44**, 715–743 (1977)
796. Brumer, A., McGuiness, D.: The behavior of the Mordell-Weil group of elliptic curves. Bull. Am. Math. Soc. **23**, 375–382 (1990)
797. Brumer, A., Silverman, J.H.: The number of elliptic curves over Q with conductor N. Manuscr. Math. **91**, 95–102 (1996)
798. Brun, V.: Über das Goldbachsche Gesetz und die Anzahl der Primzahlpaare. Arch. Math. Naturvidensk. **34**(8), 1–19 (1915)

799. Brun, V.: Om fordelingen av primtallene i forskjellige talklasser. En övre begraensning. Nyt Tidsskr. Math. B **27**, 47–58 (1916)
800. Brun, V.: Le crible d'Eratosthène et le théorème de Goldbach. C. R. Acad. Sci. Paris **168**, 544–546 (1919)
801. Brun, V.: La série $\frac{1}{5}+\frac{1}{7}+\frac{1}{11}+\frac{1}{13}+\ldots$ où les dénominateurs sont "nombres premiers jumeaux" est convergente ou finie. Bull. Sci. Math., (2) 43, 100–104, 124–128 (1919)
802. Brun, V.: Le crible d'Eratosthène et le théorème de Goldbach. Videnselsk. Skr. **1**(3), 1–36 (1920)
803. Brun, V.: Untersuchungen über das Siebverfahren des Eratosthenes. Jahresber. Dtsch. Math.-Ver. **33**, 81–96 (1925)
804. Brun, V.: Carl Størmer in memoriam. Nordisk Mat. Tidsskr. **5**, 169–175 (1957)
805. Brünner, R., Perelli, A., Pintz, J.: The exceptional set for the sum of a prime and a square. Acta Math. Acad. Sci. Hung. **53**, 347–365 (1989)
806. Bugeaud, Y.: Sur la distance entre deux puissances pures. C. R. Acad. Sci. Paris **322**, 1119–1121 (1996)
807. Bugeaud, Y.: Bounds for the solutions of superelliptic equations. Compos. Math. **107**, 187–219 (1997)
808. Bugeaud, Y.: On the Diophantine equation $x^2-2^m=\pm y^n$. Proc. Am. Math. Soc. **125**, 3203–3208 (1997)
809. Bugeaud, Y.: On the Diophantine equation $x^2-p^m=\pm y^n$. Acta Arith. **80**, 213–223 (1997)
810. Bugeaud, Y.: Bornes effectives pour les solutions des équations en S-unités et des équations de Thue-Mahler. J. Number Theory **71**, 227–244 (1998)
811. Bugeaud, Y.: On the size of integer solutions of elliptic equations. Bull. Aust. Math. Soc. **57**, 199–206 (1998)
812. Bugeaud, Y.: On the size of integer solutions of elliptic equations, II. Bull. Soc. Math. Grèce **43**, 125–130 (2000)
813. Bugeaud, Y.: An explicit lower bound for the block complexity of an algebraic number. Atti Accad. Naz. Lincei, (9) **19**, 229–235 (2008)
814. Bugeaud, Y., Dubickas, A.: On a problem of Mahler and Szekeres on approximation by roots of integers. Mich. Math. J. **56**, 703–715 (2008)
815. Bugeaud, Y., Győry, K.: Bounds for the solutions of Thue-Mahler equations and norm form equations. Acta Arith. **74**, 273–292 (1996)
816. Bugeaud, Y., Hanrot, G., Mignotte, M.: Sur l'équation diophantienne $(x^n-1)/(x-1)=y^q$, III. Proc. Lond. Math. Soc. **84**, 59–78 (2002)
817. Bugeaud, Y., Mignotte, M.: On integers with identical digits. Mathematika **46**, 411–417 (1999)
818. Bugeaud, Y., Mignotte, M.: Sur l'équation diophantienne $(x^n-1)/(x-1)=y^q$, II. C. R. Acad. Sci. Paris **328**, 741–744 (1999)
819. Bugeaud, Y., Mignotte, M.: L'équation de Nagell-Ljunggren $\frac{x^n-1}{x-1}=y^q$. Enseign. Math. **48**, 47–168 (2002)
820. Bugeaud, Y., Mignotte, M., Roy, Y.: On the Diophantine equation $(x^n-1)/(x-1)=y^q$. Pac. J. Math. **193**, 257–268 (2000)
821. Bugeaud, Y., Mignotte, M., Roy, Y., Shorey, T.N.: The equation $(x^n-1)/(x-1)=y^q$ has no solution with x square. Math. Proc. Camb. Philos. Soc. **127**, 353–372 (1999)
822. Bugeaud, Y., Mignotte, M., Siksek, S.: Classical and modular approaches to exponential Diophantine equations, I. Fibonacci and Lucas perfect powers. Ann. Math. **163**, 969–1018 (2006)
823. Bugeaud, Y., Mignotte, M., Siksek, S.: Classical and modular approaches to exponential Diophantine equations, II. The Lebesgue-Nagell equation. Compos. Math. **142**, 31–62 (2006)
824. Bugeaud, Y., Shorey, T.N.: On the Diophantine equation $\frac{x^m-1}{x-1}=\frac{y^n-1}{y-1}$. Pacific J. Math., 2007, 61–75
825. Bugulov, E.A.: On the question of the existence of odd hyperperfect numbers. Uč. Zap. Kabardino-Balkarsk. Gos. Univ. **30**, 9–19 (1966) (in Russian)

826. Buhler, J.P.: Icosahedral Galois Representations. Lecture Notes in Math., vol. 654. Springer, Berlin (1978)
827. Buhler, J.[P.], Crandall, R., Ernvall, R., Metsänkylä, T.: Irregular primes and cyclotomic invariants to four million. Math. Comput. **61**, 151–153 (1993)
828. Buhler, J.[P.], Crandall, R., Ernvall, R., Metsänkylä, T., Shokrollahi, M.A.: Irregular primes and cyclotomic invariants to 12 million. J. Symb. Comput. **31**, 89–96 (2001)
829. Buhler, J.P., Crandall, R., Sompolski, R.W.: Irregular primes to one million. Math. Comput. **59**, 717–722 (1992)
830. Buhler, J.P., Gross, B.H., Zagier, D.B.: On the conjecture of Birch and Swinnerton-Dyer for an elliptic curve of rank 3. Math. Comput. **44**, 473–481 (1985)
831. Buhštab, A.A.: Asymptotical evaluation of a general number-theoretical function. Mat. Sb. **2**, 1239–1246 (1937) (in Russian)
832. Buhštab, A.A.: New improvements in the method of the sieve of Eratosthenes. Mat. Sb. **4**, 375–387 (1938) (in Russian)
833. Buhštab, A.A.: Sur la décomposition des nombres pairs en somme de deux composantes dont chacune est formée d'un nombre borné de facteurs premiers. Dokl. Akad. Nauk SSSR **29**, 544–548 (1940)
834. Buhštab, A.A.: On those numbers in an arithmetic progression all prime factors of which are small in order of magnitude. Dokl. Akad. Nauk SSSR **67**, 5–8 (1949) (in Russian)
835. Buhštab, A.A.: On an asymptotic estimate of the number of numbers of an arithmetic progression which are not divisible by "relatively" small prime numbers. Mat. Sb. **28**, 165–184 (1951) (in Russian)
836. Buhštab, A.A.: New results in the investigation of the Goldbach-Euler problem and the problem of prime pairs. Dokl. Akad. Nauk SSSR **162**, 735–738 (1965) (in Russian)
837. Buhštab, A.A.: A combinatorial strengthening of the Eratosthenes sieve. Usp. Mat. Nauk **22**(3), 199–256 (1967) (in Russian)
838. Bulota, K.: On Hecke Z-functions and the distribution of the prime numbers of an imaginary quadratic field. Liet. Mat. Rink. **4**, 309–328 (1964)
839. Bulygin, V.V.[3]: Sur une application des fonctions elliptiques au probleme de représentation des nombres entiers par une somme de carrés. Bull. Acad. Sci. St. Petersbourg, (6) **8**, 389–404 (1914)
840. Bumby, R.T.: On the analog of Littlewood's problem in power series fields. Proc. Am. Math. Soc. **18**, 1125–1127 (1967)
841. Bumby, R.T.: A distribution property for linear recurrence of the second order. Proc. Am. Math. Soc. **50**, 101–106 (1975)
842. Bump, D., Friedberg, S., Hoffstein, J.: Nonvanishing theorems for L-functions of modular forms and their derivatives. Invent. Math. **102**, 543–618 (1990)
843. Bundschuh, P.: Irrationalitätsmaße für e^a, $a \neq 0$ rational oder Liouville-Zahl. Math. Ann. **192**, 229–242 (1971)
844. Bundschuh, P.: Generalization of a recent irrationality result of Mahler. J. Number Theory **19**, 248–253 (1984)
845. Bundschuh, P.: Einführung in die Zahlentheorie. Springer, Berlin (1988)
846. Bundschuh, P., Hock, A.: Bestimmung aller imaginär-quadratischen Zahlkörper der Klassenzahl Eins mit Hilfe eines Satzes von Baker. Math. Z. **111**, 191–204 (1969)
847. Bundschuh, P., Pethő, A.: Zur Transzendenz gewisser Reihen. Monatshefte Math. **104**, 199–223 (1987)
848. Bundschuh, P., Shiue, [P.]J.-S.: Solution of a problem on the uniform distribution of integers. Atti Accad. Naz. Lincei **55**, 172–177 (1973)
849. Bundschuh, P., Shiue, P.J.-S., Yu, X.: Transcendence and algebraic independence connected with Mahler type numbers. Publ. Math. (Debr.) **56**, 121–120 (2000)
850. Bundschuh, P., Zassenhaus, H.: Theodor Schneider (1911–1988). J. Number Theory **39**, 129–143 (1991); corr. **41**, 129 (1992)

[3]See also Boulyguine, B.

851. Burckhardt, J.J.: Rudolf Fueter. Vierteljahr. Naturforsch. Ges. Zürich **95**, 284–287 (1950)
852. Burckhardt, J.J., Fellman, E.A., Habicht, W. (eds.): Leonhard Euler 1707–1783: Beiträge zu Leben und Werk. Birkhäuser, Basel (1983)
853. Bureau, J., Morales, J.: A note on indefinite ternary quadratic forms representing all odd integers. Bol. Soc. Parana. Mat. **23**, 85–92 (2005)
854. Burgess, D.A.: The distribution of quadratic residues and non-residues. Mathematika **4**, 106–112 (1957)
855. Burgess, D.A.: On character sums and primitive roots. Proc. Lond. Math. Soc. **12**, 179–192 (1962)
856. Burgess, D.A.: On character sums and L-series. Proc. Lond. Math. Soc. **12**, 193–206 (1962)
857. Burgess, D.A.: On character sums and L-series. Proc. Lond. Math. Soc. **13**, 524–536 (1963)
858. Burgess, D.A.: A note on the distribution of residues and non-residues. J. Lond. Math. Soc. **38**, 253–256 (1963)
859. Burgess, D.A.: A note on L-functions. J. Lond. Math. Soc. **39**, 103–108 (1964)
860. Burgess, D.A.: Estimating $L_x(1)$. Norske Vid. Selsk. Forh., Trondheim **39**, 101–108 (1967)
861. Burgess, D.A.: The average of the least primitive root. In: Number Theory, Debrecen, 1968. Colloq. János Bolyai Math. Soc., pp. 11–14. North-Holland, Amsterdam (1970)
862. Burgess, D.A.: The average of the least primitive root modulo p^2. Acta Arith. **18**, 263–271 (1971)
863. Burgess, D.A., Elliott, P.D.T.A.: The average of the least primitive root. Mathematika **15**, 39–50 (1968)
864. Burkill, J.C.: Albert Edward Ingham. Bull. Lond. Math. Soc. **1**, 109–124 (1969)
865. Burkill, J.C.: John Edensor Littlewood. Bull. Lond. Math. Soc. **11**, 59–103 (1979)
866. Burkill, H., Hooley, C., Ledermann, W., Perfect, H.: Leon Mirsky. Bull. Lond. Math. Soc. **18**, 195–206 (1986)
867. Busam, R., Freitag, E.: Hans Maaß. Jahresber. Dtsch. Math.-Ver. **101**, 135–150 (1999)
868. Busche, E.: Lösung einer Aufgabe über Teileranzahlen. Mitt. Math. Ges. Hamb. **4**, 229–237 (1906)
869. Butske, W., Jaje, L.M., Mayernik, D.: On the equation $\sum_{p|N}(1/p) + (1/N) = 1$, pseudoperfect numbers, and perfectly weighted graphs. Math. Comput. **69**, 407–420 (2000)
870. Button, J.O.: The uniqueness of the prime Markoff numbers. J. Lond. Math. Soc. **58**, 9–17 (1998)
871. Buzzard, K., Dickinson, M., Shepherd-Barron, N., Taylor, R.: On icosahedral Artin representations. Duke Math. J. **109**, 283–318 (2001)
872. Buzzard, K., Stein, W.A.: A mod five approach to modularity of icosahedral Galois representations. Pac. J. Math. **203**, 265–282 (2002)
873. Buzzard, K., Taylor, R.: Companion forms and weight one forms. Ann. Math. **149**, 905–919 (1999)
874. Byeon, D., Kim, M., Lee, J.: Mollin's conjecture. Acta Arith. **126**, 99–114 (2007)
875. Byeon, D., Lee, J.: Class number 2 problem for certain real quadratic fields of Richaud-Degert type. J. Number Theory **128**, 865–883 (2008)
876. Bykovskiĭ, V.A.: An estimate for the dispersion of lengths of finite continued fractions. Fundam. Prikl. Mat. **11**(6), 15–26 (2005) (in Russian)
877. Bykovskiĭ, V.A., Vinogradov, A.I.: Inhomogeneous convolutions. Zap. Nauč. Semin. LOMI **160**, 16–30 (1987) (in Russian)
878. Cahen, E.: Sur la fonction $\zeta(s)$ de Riemann et sur des fonctions analogues. Ann. Sci. Éc. Norm. Super. **11**, 75–164 (1894)
879. Cahen, E.: Éléments de la théorie des nombres. Gauthier-Villars, Paris (1900)
880. Cahen, P.-J., Chabert, J.-L.: Integer-Valued Polynomials. Am. Math. Soc., Providence (1997)
881. Cai, Y.: On the distribution of square-full integers. Acta Math. Sin. **13**, 269–280 (1997)
882. Cai, Y.: A remark on Chen's theorem. Acta Arith. **102**, 339–352 (2002)
883. Cai, Y.: A remark on Chen's theorem, II. Chin. Ann. Math., Ser. B **29**, 687–698 (2008)
884. Cai, Y.: On Chen's theorem, II. J. Number Theory **128**, 1336–1357 (2008)

885. Cai, Y., Lu, M.G.: On Chen's theorem. In: Analytic Number Theory, Beijing/Kyoto, 1999, pp. 99–119. Kluwer Academic, Dordrecht (2002)
886. Cai, Y., Lu, M.G.: On the upper bound for $\pi_2(x)$. Acta Arith. **110**, 275–298 (2003)
887. Callahan, T., Smith, R.A.: L-functions of a quadratic form. Trans. Am. Math. Soc. **217**, 297–309 (1976)
888. Canfield, E.R., Erdős, P., Pomerance, C.: On a problem of Oppenheim concerning "factorisatio numerorum". J. Number Theory **17**, 1–28 (1983)
889. Cantor, D., Gordon, B., Hales, A., Schacher, M.: Biography—Ernst G. Straus 1922–1983. Pac. J. Math. **118**(2), i–xx (1985)
890. Cantor, D.C., Roquette, P.: On Diophantine equations over the ring of all algebraic integers. J. Number Theory **18**, 1–26 (1984)
891. Cantor, D.C., Straus, E.G.: On a conjecture of D.H. Lehmer. Acta Arith. **42**, 97–100 (1982/1983); corr. p. 327
892. Cao, X.: The distribution of square-full integers. Period. Math. Hung. **28**, 43–54 (1994)
893. Cao, X.: On the distribution of square-full integers. Period. Math. Hung. **34**(3), 169–175 (1997)
894. Caporaso, L.: On certain uniformity properties of curves over function fields. Compos. Math. **130**, 1–19 (2002)
895. Cappell, S.E., Shaneson, J.L.: Genera of algebraic varieties and counting of lattice points. Bull. Am. Math. Soc. **30**, 62–69 (1994)
896. Carayol, H.: Sur les représentations l-adiques associées aux formes modulaires de Hilbert. Ann. Sci. Éc. Norm. Super. **19**, 409–468 (1986)
897. Carayol, H.: Sur les représentations galoisiennes modulo ℓ attachées aux formes modulaires. Duke Math. J. **59**, 785–801 (1989)
898. Carayol, H.: The Sato-Tate conjecture (after Clozel, Harris, Shepherd-Barron, Taylor). Astérisque **317**, 345–392 (2008)
899. Carey, J.C.: The Riemann hypothesis as a sequence of surface to volume ratios. Linear Algebra Appl. **165**, 131–151 (1992)
900. Carleman, T.: L.E. Phragmén in memoriam. Acta Math. **69**, xxxi–xxxiii (1938)
901. Carlitz, L.: Independence of arithmetic functions. Duke Math. J. **19**, 65–70 (1952)
902. Carlitz, L.: Note on irregular primes. Proc. Am. Math. Soc. **5**, 329–331 (1954)
903. Carlitz, L., Olson, F.R.: Maillet's determinant. Proc. Am. Math. Soc. **6**, 265–269 (1955)
904. Carlitz, L., Uchiyama, S.: Bounds for exponential sums. Duke Math. J. **24**, 37–41 (1957)
905. Carlson, F.: Über die Nullstellen der Dirichletschen Reihen und der Riemannscher ζ-Funktion. Ark. Mat. Astron. Fys. **15**(20), 1–28 (1920)
906. Carlson, F.: Contributions á la théorie des séries de Dirichlet, I. Ark. Mat. Astron. Fys. **16**(18), 1–19 (1922)
907. Carlson, F.: Contributions á la théorie des séries de Dirichlet, II. Ark. Mat. Astron. Fys. **19**(25), 1–17 (1926)
908. Carmichael, R.D.: Multiply perfect numbers of three different primes. Ann. Math. **8**, 49–56 (1906)
909. Carmichael, R.D.: On Euler's φ-function. Bull. Am. Math. Soc. **13**, 241–243 (1906/1907)
910. Carmichael, R.D.: A table of multiply perfect numbers. Bull. Am. Math. Soc. **13**, 383–386 (1906/1907)
911. Carmichael, R.D.: Multiply perfect numbers of four different primes. Ann. Math. **8**, 149–158 (1907)
912. Carmichael, R.D.: Even multiply perfect numbers of five different prime factors. Bull. Am. Math. Soc. **15**, 7–8 (1908/1909)
913. Carmichael, R.D.: On the numerical factors of certain arithmetic forms. Am. Math. Mon. **16**, 153–159 (1909)
914. Carmichael, R.D.: Note on a new number theory function. Bull. Am. Math. Soc. **16**, 232–238 (1909/1910)
915. Carmichael, R.D.: On composite numbers P, which satisfy the Fermat congruence $a^{P-1} \equiv 1 \pmod{P}$. Am. Math. Mon. **19**, 22–27 (1912)

916. Carmichael, R.D.: On the numerical factors of the arithmetic forms $\alpha^n \pm \beta^n$. Ann. Math. **15**, 30–70 (1913)
917. Carmichael, R.D.: The Theory of Numbers. Wiley, New York (1914)
918. Carmichael, R.D.: Diophantine Analysis. Wiley, New York (1915)
919. Carmichael, R.D.: On sequences of integers defined by recurrence relations. Q. J. Math. **48**, 343–372 (1920)
920. Carmichael, R.D.: Note on Euler's φ-function. Bull. Am. Math. Soc. **28**, 109–110 (1922)
921. Carmichael, R.D.: A simple principle of unification in the elementary theory of numbers. Am. Math. Mon. **36**, 132–143 (1929)
922. Carmichael, R.D.: Expansions of arithmetical functions in infinite series. Proc. Lond. Math. Soc. **34**, 1–26 (1932)
923. Carslaw, H.S., Hardy, G.H.: John Raymond Wilton. J. Lond. Math. Soc. **20**, 58–64 (1945)
924. Cartwright, M.L.: Edward Charles Titchmarsh. J. Lond. Math. Soc. **39**, 544–565 (1964)
925. Cashwell, E.D., Everett, C.J.: The ring of number theoretic functions. Pac. J. Math. **9**, 975–985 (1959)
926. Cashwell, E.D., Everett, C.J.: Formal power series. Pac. J. Math. **13**, 45–64 (1963)
927. Cassels, J.W.S.: The Markoff chain. Ann. Math. **50**, 676–685 (1949)
928. Cassels, J.W.S.: A note on the division values of $\wp(u)$. Proc. Camb. Philos. Soc. **45**, 167–172 (1949)
929. Cassels, J.W.S.: The rational solutions of the diophantine equation $Y^2 = X^3 - D$. Acta Math. **82**, 243–273 (1950); corr. **84**, 299 (1951)
930. Cassels, J.W.S.: On the equation $a^x - b^y = 1$. Am. J. Math. **75**, 159–162 (1953); corr. **57**, 187 (1961)
931. Cassels, J.W.S.: On the equation $a^x - b^y = 1$, II. Proc. Camb. Philos. Soc. **56**, 97–103 (1960); corr. **57**, 187 (1961)
932. Cassels, J.W.S.: A new inequality with application to the theory of diophantine approximation. Math. Ann. **126**, 108–118 (1953)
933. Cassels, J.W.S.: An Introduction to Diophantine Approximation. Cambridge University Press, Cambridge (1957) [Reprint: Hafner, 1972]
934. Cassels, J.W.S.: An Introduction to the Geometry of Numbers. Springer, Berlin (1959); 2nd ed. 1971. [Reprint: 1997]
935. Cassels, J.W.S.: Arithmetic on curves of genus 1, I. On a conjecture of Selmer. J. Reine Angew. Math. **202**, 52–99 (1959)
936. Cassels, J.W.S.: Arithmetic on curves of genus 1, III. The Tate-Šafarevič and Selmer groups. Proc. Lond. Math. Soc. **12**, 259–296 (1962); corr.: **13**, 768 (1963)
937. Cassels, J.W.S.: Arithmetic on curves of genus 1, IV. Proof of the Hauptvermutung. J. Reine Angew. Math. **211**, 95–112 (1962)
938. Cassels, J.W.S.: Arithmetic on curves of genus 1, V. Two counterexamples. J. Lond. Math. Soc. **38**, 244–248 (1963)
939. Cassels, J.W.S.: Arithmetic on curves of genus 1, VI. The Tate-Šafarevič group can be arbitrarily large. J. Reine Angew. Math. **214/215**, 65–70 (1964)
940. Cassels, J.W.S.: Arithmetic on curves of genus 1, VIII. On conjectures of Birch and Swinnerton-Dyer. J. Reine Angew. Math. **217**, 180–199 (1965)
941. Cassels, J.W.S.: On a problem of Steinhaus about normal numbers. Colloq. Math. **7**, 95–101 (1959)
942. Cassels, J.W.S.: On a class of exponential equations. Ark. Mat. **4**, 231–233 (1961)
943. Cassels, J.W.S.: On a problem of Schinzel and Zassenhaus. J. Math. Sci. **1**, 1–8 (1966)
944. Cassels, J.W.S.: Diophantine equations with special reference to elliptic curves. J. Lond. Math. Soc. **41**, 193–291 (1966)
945. Cassels, J.W.S.: On Kummer sums. Proc. Lond. Math. Soc. **21**, 19–27 (1970)
946. Cassels, J.W.S.: Trygve Nagell. Acta Arith. **55**, 109–118 (1991)
947. Cassels, J.W.S.: Obituary of Kurt Mahler. Acta Arith. **58**, 215–228 (1991)
948. Cassels, J.W.S.: Mordell's finite basis theorem revisited. Math. Proc. Camb. Philos. Soc. **100**, 31–41 (1986)

949. Cassels, J.W.S.: Lectures on Elliptic Curves. Cambridge University Press, Cambridge (1991)
950. Cassels, J.W.S., Fröhlich, A. (eds.): Algebraic Number Theory. Academic Press, San Diego (1967) [Reprint: 1986]
951. Cassels, J.W.S., Fröhlich, A.: Hans Arnold Heilbronn. Bull. Lond. Math. Soc. **9**, 219–232 (1977)
952. Cassels, J.W.S., Guy, M.J.T.: On the Hasse principle for cubic surfaces. Mathematika **13**, 111–120 (1966)
953. Cassels, J.W.S., Swinnerton-Dyer, H.P.F.: On the product of three homogeneous linear forms and indefinite ternary quadratic forms. Philos. Trans. R. Soc. Lond. Ser. A, Math. Phys. Sci. **248**, 73–96 (1955)
954. Cassels, J.W.S., Vaughan, R.C.: Obituary: Ivan Matveevich Vinogradov. Bull. Lond. Math. Soc. **17**, 584–600 (1985)
955. Cassou-Noguès, P.: Analogues p-adiques des fonctions Γ-multiples. Astérisque **61**, 43–55 (1979)
956. Cassou-Noguès, P.: Valeurs aux entiers négatifs des fonctions zêta et fonctions zêta p-adiques. Invent. Math. **51**, 29–59 (1979)
957. Cassou-Noguès, P.: p-adic L-functions for elliptic curves with complex multiplication, I. Compos. Math. **42**, 31–56 (1980/1981)
958. Cassou-Noguès, P.: Prolongement de certaines séries de Dirichlet. Am. J. Math. **105**, 13–58 (1983)
959. Cassou-Noguès, Ph., Chinburg, T., Fröhlich, A., Taylor, M.J.: L-functions and Galois modules. In: L-functions and Arithmetic, Durham, 1989, pp. 75–139. Cambridge University Press, Cambridge (1991)
960. Catalan, E.: Théorèmes et problèmes. 48. Théorème. Nouv. Ann. Math. **1**, 520 (1842)
961. Catalan, E.: Note extraite d'une lettre adressée à l'éditeur. J. Reine Angew. Math. **27**, 192 (1844)
962. Catalan, E.: Savin Realis †. Nouv. Ann. Math. **5**, 200–203 (1886)
963. Catlin, P.A.: Two problems in metric Diophantine approximation, I. J. Number Theory **8**, 282–288 (1976)
964. Catlin, P.A.: Two problems in metric Diophantine approximation, II. J. Number Theory **8**, 289–297 (1976)
965. Cauchy, A.: Recherches sur les nombres. J. Éc. Polytech. **9**, 99–123 (1813) [Oeuvres, (2), vol. 1, pp. 39–63, Paris, 1905]
966. Cauchy, A.: Démonstration du théorème genéral de Fermat sur les nombres polygones. Mem. Sci. Math. Phys. Inst. Fr. **14**, 177–220 (1813–1815) [Oeuvres, (2), vol. 6, pp. 320–353, Paris, 1887]
967. Cauer, D.: Neue Anwendungen der Pfeifferschen Methode zur Abschätzung zahlentheoretischer Funktionen. Dissertation, Göttingen (1914)
968. Čebotarev, N.G.[4]: On a theorem of Minkowski. Učen. Zap. Univ. Kazan **94**(7), 3–16 (1934) (in Russian)
969. Čebyšev, P.L.: The Theory of Congruences, St. Petersburg, 1849 (in Russian) [2nd ed. 1879, 3rd ed. 1901; [973], vol. 1, pp. 10–172. German translation: Theorie der Congruenzen, Berlin, 1889, 1902; Italian translation: Teoria delle congruenze, Rome 1895]
970. Čebyšev, P.L.: Mémoire sur nombres premiers. Mémoires des savants étrangers de l'Acad. Sci. St. Pétersbourg **7**, 17–33 (1850) [J. math. pures appl., **17**, 366–390 (1852); Oeuvres, vol. 1, pp. 49–70, S. Pétersbourg 1899, reprint: Chelsea, 1962; [973], vol. 1, pp. 191–207 (in Russian)]
971. Čebyšev, P.L.: On an arithmetic question. Zap. Imp. Akad. Nauk. **10**(4), 1–54 (1868) (in Russian) [[973], vol. 1, pp. 237–275; French translation: Oeuvres, vol. 1, pp. 639–684, St. Pétersbourg 1899]

[4]See also Tschebotareff.

972. Čebyšev, P.L.: Lettre de M. le professeur Tchébychev à M. Fuss, sur un nouveau théorème relatif aux nombres premiers contenus dans les formes $4n + 1$ et $4n + 3$. Bull. Cl. Phys.-Math. Acad. Imp. St. Pétersbourg, **11**, p. 208 [[973], vol. 1, p. 276 (in Russian)]
973. Čebyšev, P.L.: Polnoe Sobranie Sočineniĭ. Nauka, Moscow (1946–1955)
974. Chabauty, C.: Sur les équations diophantiennes liées aux unités d'un corps de nombres algébriques fini. Ann. Mat. Pura Appl. **117**, 127–168 (1938)
975. Chabauty, C.: Démonstration nouvelle d'un théorème de Thue et Mahler sur les formes binaires. Bull. Sci. Math. **65**, 112–130 (1941)
976. Chabauty, C.: Limite d'ensembles et géométrie des nombres. Bull. Soc. Math. Fr. **78**, 143–151 (1950)
977. Chace, C.E.: The divisor problem for arithmetic progressions with small modulus. Acta Arith. **61**, 35–50 (1992)
978. Chahal, J.S.: On an identity of Desboves. Proc. Jpn. Acad. Sci. **60**, 105–108 (1984)
979. Chahal, J.S.: Congruent numbers and elliptic curves. Am. Math. Mon. **113**, 308–317 (2006)
980. Chai, C.-L.: A note on Manin's theorem of the kernel. Am. J. Math. **113**, 387–389 (1991)
981. Chaix, H.: Démonstration élémentaire d'un théorème de Van der Corput. C. R. Acad. Sci. Paris **275**, 883–885 (1972)
982. Chakri, L., El Hanine, M.: Anisotropic forms modulo p^2. Acta Arith. **108**, 147–151 (2003)
983. Chalk, J.H.H.: On Hua's estimates for exponential sums. Mathematika **34**, 115–123 (1987)
984. Chalk, J.H.H., Smith, R.A.: On Bombieri's estimate for exponential sums. Acta Arith. **18**, 191–212 (1971)
985. Chambert-Loir, A.: Compter (rapidement) le nombre de solutions d'équations dans les corps finis. Astérisque **317**, 39–90 (2008)
986. Chamizo, F., Ubis, A.: An average formula for the class number. Acta Arith. **122**, 75–90 (2006)
987. Chamizo, F., Iwaniec, H.: On the sphere problem. Rev. Mat. Iberoam. **11**, 417–429 (1995)
988. Chamizo, F., Iwaniec, H.: On the Gauss mean-value formula for class number. Nagoya Math. J. **151**, 199–208 (1998)
989. Champernowne, D.G.: The construction of decimals normal in the scale of ten. J. Lond. Math. Soc. **8**, 254–260 (1933)
990. Chan, W.K., Oh, B.-K.: Positive ternary quadratic forms with finitely many exceptions. Proc. Am. Math. Soc. **132**, 1567–1573 (2004)
991. Chandrasekharan, K.: Obituary: S.S. Pillai. J. Indian Math. Soc. (N.S.), Ser. A **15**, 1–10 (1951)
992. Chandrasekharan, K.: Introduction to Analytic Number Theory. Springer, Berlin (1968)
993. Chandrasekharan, K.: Arithmetical Functions. Springer, Berlin (1970)
994. Chandrasekharan, K., Narasimhan, R.: Hecke's functional equation and the average order of arithmetic functions. Acta Arith. **6**, 487–503 (1960/1961)
995. Chandrasekharan, K., Narasimhan, R.: The approximate functional equation for a class of zeta-functions. Math. Ann. **152**, 30–64 (1963)
996. Chang, K.L.: On some Diophantine equations $y^2 = x^3 + k$ with no rational solutions. Q. J. Math. **19**, 181–188 (1948)
997. Chang, K.-Y., Kwon, S.-H.: Class numbers of imaginary Abelian number fields. Proc. Am. Math. Soc. **128**, 2517–2528 (2000)
998. Chang, M.-C.: A polynomial bound in Freiman's theorem. Duke Math. J. **113**, 399–419 (2002)
999. Chao, K.F., Plymen, R.: A new bound for the smallest x with $\pi(x) > \mathrm{li}(x)$, Internat. J. Number Theory **6**(3), 1–10 (2010)
1000. Chao, K.: On the diophantine equation $x^2 = y^n + 1, xy \neq 0$. Sci. Sin. **14**, 457–460 (1965)
1001. Châtelet, A.: Sur certains ensembles de tableaux et leur application à la théorie des nombres. Ann. Sci. Éc. Norm. Super. **28**, 105–202 (1911)
1002. Châtelet, F.: Points exceptionnels d'une cubique de Weierstrass. C. R. Acad. Sci. Paris **210**, 90–92 (1940)
1003. Châtelet, F.: Courbes réduites dans les classes de courbes de genre 1. C. R. Acad. Sci. Paris **212**, 320–322 (1941)

1004. Châtelet, F.: Variations sur un théme de H. Poincaré. Ann. Sci. Éc. Norm. Super. **61**, 249–300 (1944)
1005. Châtelet, F.: Points exceptionnels des cubiques. In: Algèbre et Théorie des Nombres. Colloq. Internat. du CNRS, vol. 24, pp. 71–72. CNRS, Paris (1950)
1006. Chaundy, T.W.: The arithmetic minima of positive quadratic forms. Q. J. Math. **17**, 166–192 (1946)
1007. Cheer, A.Y., Goldston, D.A.: Simple zeros of the Riemann zeta-function. Proc. Am. Math. Soc. **118**, 365–372 (1993)
1008. Chein, E.Z.: A note on the equation $x^2 = y^q + 1$. Proc. Am. Math. Soc. **56**, 83–84 (1976)
1009. Chein, E.Z.: An odd perfect number has at least 8 prime factors. Ph.D. thesis, Pennsylvania State Univ. (1979)
1010. Chen, B.: Lattice points, Dedekind sums, and Ehrhart polynomials of lattice polyhedra. Discrete Comput. Geom. **28**, 175–199 (2002)
1011. Chen, J.H., Voutier, P.: Complete solution of the Diophantine equation $X^2 + 1 = dY^4$ and a related family of quartic Thue equations. J. Number Theory **62**, 71–99 (1997)
1012. Chen, J.R.: On Waring's problem for n-th powers. Acta Math. Sin. **8**, 253–257 (1958) (in Chinese) [English translation: Chinese Math. Acta, **8**, 1966, 849–853]
1013. Chen, J.R.: On the representation of a natural number as a sum of terms of the form $x(x + 1)\cdots(x + k - 1)/k!$. Acta Math. Sin. **9**, 264–270 (1959) (in Chinese)
1014. Chen, J.R.: The number of lattice points in a given region. Acta Math. Sin. **12**, 408–420 (1962) (in Chinese) [English translation: Chinese Math., **3**, 439–452 (1963)]
1015. Chen, J.R.: The lattice points in a circle. Sci. Sin. **12**, 633–649 (1963)
1016. Chen, J.R.: Improvements on [the] asymptotic formulas for the number of lattice points in a region of three dimensions. Sci. Sin. **12**, 151–161 (1963)
1017. Chen, J.R.: Improvements on [the] asymptotic formulas for the number of lattice points in a region of three dimensions, II. Sci. Sin. **12**, 751–764 (1963)
1018. Chen, J.R.: Waring's problem for $g(5)$. Sci. Sin. **1**, 1547–1568 (1964)
1019. Chen, J.R.: On the divisor problem for $d_3(n)$. Sci. Sin. **14**, 19–29 (1965)
1020. Chen, J.R.: On the least prime in an arithmetic progression. Sci. Sin. **14**, 1868–1871 (1965)
1021. Chen, J.R.: On the representation of a large even integer as the sum of a prime and the product of at most two primes. Sci. Sin. **16**, 157–176 (1973)
1022. Chen, J.R.: On the representation of a large even integer as the sum of a prime and the product of at most two primes, II. Sci. Sin. **21**, 421–430 (1978)
1023. Chen, J.R.: On the distribution of almost primes in an interval. Sci. Sin. **18**, 611–627 (1975)
1024. Chen, J.R.: On the distribution of almost primes in an interval, II. Sci. Sin. **22**, 253–275 (1979)
1025. Chen, J.R.: On Professors Hua estimate of exponential sums. Sci. Sin. **20**, 711–719 (1977)
1026. Chen, J.R.: On the Goldbach's problem and the sieve methods. Sci. Sin. **21**, 701–739 (1978)
1027. Chen, J.R.: On the least prime in an arithmetical progression and theorems concerning the zeros of Dirichlet's L-functions, II. Sci. Sin. **22**, 859–889 (1979)
1028. Chen, J.R.: The exceptional set of Goldbach numbers, II. Sci. Sin. **26**, 714–731 (1983)
1029. Chen, J.R., Liu, J.: On the least prime in an arithmetical progression, III. Sci. China Ser. A **32**, 654–673 (1989)
1030. Chen, J.R., Liu, J.: On the least prime in an arithmetical progression, IV. Sci. China Ser. A **32**, 792–807 (1989)
1031. Chen, J.R., Liu, J.: On the least prime in an arithmetical progression and theorems concerning the zeros of Dirichlet's L-functions, V. In: International Symposium in Memory of Hua Loo Keng, I, pp. 19–42. Springer, Berlin (1991)
1032. Chen, J.R., Liu, J.: The exceptional set of Goldbach numbers, III. Chin. Q. J. Math. **4**, 1–15 (1989)
1033. Chen, J.R., Pan, C.D.: The exceptional set of Goldbach numbers, I. Sci. Sin. **23**, 416–430 (1980)
1034. Chen, J.R., Wang, T.Z.: A theorem on exceptional zeros of L-functions. Acta Math. Sin. **32**, 841–858 (1989)

1035. Chen, J.R., Wang, T.Z.: The Goldbach problem for odd numbers. Acta Math. Sin. **39**, 169–174 (1996)
1036. Chen, W., Matthews, K., Street, R.: Obituary: Peter A.B. Pleasants. Aust. Math. Soc. Gaz. **35**, 162–165 (2008)
1037. Chen, W.L. (ed.): Proceedings of the Symposium on Analytic Number Theory. Imperial College, London (1985)
1038. Chen, Y.G.: On the Siegel-Tatuzawa-Hoffstein theorem. Acta Arith. **130**, 361–367 (2007)
1039. Chen, Y.G., Sun, X.G.: On Romanoff's constant. J. Number Theory **106**, 275–284 (2004)
1040. Chen, Y.-M., Yu, J.: On primitive points of elliptic curves with complex multiplication. J. Number Theory **114**, 66–87 (2005)
1041. Cheng, Y.[F.]: An explicit zero-free region for the Riemann zeta-function. Rocky Mt. J. Math. **30**, 135–148 (2000)
1042. Cheng, Y.F.: Explicit estimate on primes between consecutive cubes. Rocky Mt. J. Math. **40**, 117–153 (2010)
1043. Cheng, Y.F., Graham, S.W.: Explicit estimates for the Riemann zeta-function. Rocky Mt. J. Math. **34**, 1261–1289 (2004)
1044. Chernick, J.: Ideal solutions of the Tarry-Escott problem. Am. Math. Mon. **44**, 626–633 (1937)
1045. Cherubini, J.M., Wallisser, R.V.: On the computation of all imaginary quadratic fields of class number one. Math. Comput. **49**, 295–299 (1987)
1046. Chevalley, C.: La théorie du symbole de restes normiques. J. Reine Angew. Math. **169**, 140–157 (1933)
1047. Chevalley, C.: Sur la théorie du corps de classes dans les corps finis et les corps locaux. J. Fac. Sci. Univ. Tokyo **2**, 365–476 (1933)
1048. Chevalley, C.: Démonstration d'une hypothèse de M. Artin. Abh. Math. Semin. Univ. Hamb. **11**, 73–75 (1935)
1049. Chevalley, C.: Généralisation de la théorie du corps de classes pour les extensions infinies. J. Math. Pures Appl. **15**, 359–371 (1936)
1050. Chevalley, C.: La théorie du corps de classes. Ann. Math. **41**, 394–418 (1940)
1051. Chevalley, C.: Emil Artin (1898–1962). Bull. Soc. Math. Fr. **92**, 1–10 (1964)
1052. Chevalley, C., Nehrkorn, H.: Sur les démonstrations arithmétiques dans la théorie du corps de classes. Math. Ann. **111**, 364–371 (1935)
1053. Chevalley, C., Weil, A.: Hermann Weyl (1885–1955). Enseign. Math. **3**, 157–187 (1957)
1054. Chih, T.: The Dirichlet's divisor problem. Sci. Rep. Nat. Tsing-Hua Univ., Ser. A **5**, 402–427 (1950)
1055. Chinburg, T.: "Easier" Waring problem for commutative rings. Acta Arith. **35**, 303–331 (1979)
1056. Choi, K.K., Liu, M.C., Tsang, K.M.: Conditional bounds for small prime solutions of linear equations. Manuscr. Math. **74**, 321–340 (1992)
1057. Choi, S.L.G.: Covering the set of integers by congruence classes of distinct moduli. Math. Comput. **25**, 885–895 (1971)
1058. Choie, Y.J., Lichiardopol, N., Moree, P., Solé, P.: On Robin's criterion for the Riemann hypothesis. J. Théor. Nr. Bordx. **19**, 357–372 (2007)
1059. Chopra, S.D.: Hansraj Gupta 1902–1988: a biographical sketch. J. Indian Math. Soc. **57**, 1–10 (1991)
1060. Chor, B., Lemke, P., Mador, Z.: On the number of ordered factorizations of n. Discrete Math. **214**, 123–133 (2000)
1061. Choudhry, A.: On sums of seventh powers. J. Number Theory **81**, 266–269 (2000)
1062. Choudhry, A., Wróblewski, J.: Ideal solutions of the Tarry-Escott problem of degree eleven with applications to sums of thirteenth powers. Hardy-Ramanujan J. **31**, 1–13 (2008)
1063. Chowla, I.: A theorem on the addition of residue classes: application to the number $\Gamma(k)$ in Waring's problem. Q. J. Math. **8**, 99–102 (1937)
1064. Chowla, I.: A new evaluation of the number $\Gamma(k)$ in Waring's problem. Proc. Indian Acad. Sci. Math. Sci. **6**, 97–103 (1937)

1065. Chowla, I.: On Waring's problem mod p. Proc. Natl. Acad. Sci., India **13**, 195–220 (1943)
1066. Chowla, P., Chowla, S.: Problems on periodic simple continued fractions. Proc. Natl. Acad. Sci. USA **69**, 37–45 (1972)
1067. Chowla, S.: A property of positive odd integers. J. Indian Math. Soc. **18**, 129–130 (1929)
1068. Chowla, S.D.: On the greatest prime factor of a certain product. J. Indian Math. Soc. **18**, 135–137 (1929)
1069. Chowla, S.D.: Remarks on Waring's problem. J. Lond. Math. Soc. **5**, 155–158 (1930)
1070. Chowla, S.D.: A generalization of a theorem of Wolstenholme. J. Lond. Math. Soc. **5**, 158–160 (1930)
1071. Chowla, S.: Contributions to the analytic theory of numbers. Math. Z. **35**, 279–299 (1932)
1072. Chowla, S.: Congruence properties of partitions. J. Lond. Math. Soc. **9**, 247 (1934)
1073. Chowla, S.: The class-number of binary quadratic forms. Q. J. Math. **5**, 302–303 (1934)
1074. Chowla, S.: An extension of Heilbronn's theorem. Q. J. Math. **5**, 304–307 (1934)
1075. Chowla, S.: A theorem on irrational indefinite quadratic forms. J. Lond. Math. Soc. **9**, 162–163 (1934)
1076. Chowla, S.: On the least prime in an arithmetical progression. J. Indian Math. Soc. **1**, 1–3 (1934)
1077. Chowla, S.: On abundant numbers. J. Indian Math. Soc. **1**, 41–44 (1934)
1078. Chowla, S.: On the k-analogue of a result in the theory of the Riemann zeta function. Math. Z. **38**, 483–487 (1934)
1079. Chowla, S.: Primes in an arithmetical progression. Indian Phys.-Math. J. **5**, 35–43 (1934)
1080. Chowla, S.: The representation of a number as a sum of four squares and a prime. Acta Arith. **1**, 115–122 (1935)
1081. Chowla, S.: A theorem on sums of powers with application to the additive theory of numbers. Proc. Natl. Acad. Sci., India **1**, 698–700 (1935)
1082. Chowla, S.: A theorem of Erdös. Proc. Indian Acad. Sci. Math. Sci. **5**, 37–39 (1937)
1083. Chowla, S.: A remark on $g(n)$. Proc. Natl. Acad. Sci., India **9**, 20–21 (1939)
1084. Chowla, S.: Solution of a problem of Erdös and Turan in additive number theory. Proc. Natl. Acad. Sci., India **14**, 1–2 (1944)
1085. Chowla, S.: A formula similar to Jacobsthal's for the explicit value of x in $p = x^2 + y^2$ where p is a prime of the form $4k + 1$. Proc. Lahore Philos. Soc. **7**, 1–2 (1945)
1086. Chowla, S.: On an unsuspected real zero of Epstein's zeta function. Proc. Natl. Inst. Sci. India **13**(4), 1 (1947)
1087. Chowla, S.: On the class-number of the corpus $P(\sqrt{-k})$. Proc. Natl. Inst. Sci. India **13**, 197–200 (1947)
1088. Chowla, S.: On a theorem of Walfisz. J. Lond. Math. Soc. **22**, 136–140 (1947)
1089. Chowla, S.: Improvement of a theorem of Linnik and Walfisz. Proc. Lond. Math. Soc. **50**, 423–429 (1949)
1090. Chowla, S.: A new proof of a theorem of Siegel. Ann. Math. **51**, 120–122 (1950)
1091. Chowla, S.: On a conjecture of J.F. Gray. Norske Vid. Selsk. Forh., Trondheim **33**, 58–59 (1960)
1092. Chowla, S.: Review of [2497]. Math. Rev. **28**, #1179
1093. Chowla, S.: Bounds for the fundamental unit of a real quadratic field. Norske Vid. Selsk. Forh., Trondheim **37**, 88–90 (1964)
1094. Chowla, S.: A remark on a conjecture of C.L. Siegel. Proc. Natl. Acad. Sci. USA **51**, 774–775 (1964)
1095. Chowla, S.: The Riemann Hypothesis and Hilbert's Tenth Problem. Gordon & Breach, New York (1965)
1096. Chowla, S., Briggs, W.E.: On discriminants of binary quadratic forms with a single class in each genus. Can. J. Math. **6**, 463–470 (1954)
1097. Chowla, S., Friedlander, J.B.: Class numbers and quadratic residues. Glasg. Math. J. **17**, 47–52 (1976)
1098. Chowla, S., Mann, H.B., Straus, E.G.: Some applications of the Cauchy-Davenport theorem. Norske Vid. Selsk. Forh., Trondheim **32**, 74–80 (1959)

1099. Chowla, S., Pillai, S.S.: Hypothesis K of Hardy and Littlewood. Math. Z. **41**, 537–540 (1936)
1100. Chowla, S., Selberg, A.: On Epstein's zeta function, I. Proc. Natl. Acad. Sci. USA **35**, 371–374 (1949) [[5625], vol. 1, pp. 367–370]
1101. Chowla, S., Shimura, G.: On the representation of zero by a linear combination of k-th powers. Norske Vid. Selsk. Forh., Trondheim **36**, 169–176 (1963)
1102. Chowla, S., Vijayaraghavan, T.: On the largest prime divisors of numbers. J. Indian Math. Soc. **11**, 31–37 (1947)
1103. Chua, K.S.: Real zeros of Dirichlet zeta-functions of real quadratic fields. Math. Comput. **74**, 1457–1470 (2005)
1104. Chudnovsky, D.V., Chudnovsky, G.V.: Padé approximations and Diophantine geometry. Proc. Natl. Acad. Sci. USA **82**, 2212–2216 (1985)
1105. Chudnovsky, G.V.[5]: Algebraic independence of values of exponential and elliptic functions. In: Proc. of ICM, Helsinki, 1978, pp. 339–350 (1980)
1106. Chudnovsky, G.[V.]: Algebraic independence of the values of elliptic function at algebraic points. Elliptic analogue of the Lindemann-Weierstrass theorem. Invent. Math. **61**, 267–290 (1980)
1107. Chudnovsky, G.V.: Indépendance algébrique dans la méthode de Gel'fond-Schneider. C. R. Acad. Sci. Paris **291**, A365–A368 (1980)
1108. Chudnovsky, G.V.: On the method of Thue-Siegel. Ann. Math. **117**, 325–383 (1983)
1109. Church, A.: An unsolvable problem of elementary number theory. Am. J. Math. **58**, 345–363 (1936)
1110. Ciesielski, Z.: On multiplicative sequences. Colloq. Math. **7**, 265–268 (1959/1960)
1111. Cigler, J.: Hausdorffsche Dimensionen spezieller Punktmengen. Math. Z. **76**, 22–30 (1961)
1112. Cigler, J.: Über eine Verallgemeinerung des Hauptsatzes der Theorie der Gleichverteilung. J. Reine Angew. Math. **210**, 141–147 (1962)
1113. Cijsouw, P.L.: A transcendence measure for π. In: Transcendence Theory: Advances and Applications, pp. 93–100. Academic Press, San Diego (1977)
1114. Cipolla, M.: Sui numeri compositi P, che verificano la congruenza di Fermat $a^{P-1} \equiv 1 \pmod{= P}$. Ann. Mat. Pura Appl. **9**, 139–160 (1904)
1115. Cipra, J.A., Cochrane, T., Pinner, C.: Heilbronn's conjecture on Waring's number mod p. J. Number Theory **125**, 289–297 (2007)
1116. Cipu, M., Mignotte, M.: On the number of solutions to simultaneous hyperbolic Diophantine equations. J. Number Theory **125**, 365–392 (2007)
1117. Cipu, M., Mignotte, M.: On a conjecture on exponential Diophantine equation. Acta Arith. **140**, 251–270 (2009)
1118. Clark, D.: An arithmetical function associated with the rank of elliptic curves. Can. Math. Bull. **34**, 181–185 (1991)
1119. Clark, D.A., Kuwata, M.: Generalized Artin's conjecture for primitive roots and cyclicity mod p of elliptic curves over function fields. Can. Math. Bull. **38**, 167–173 (1995)
1120. Clark, P.L.: There are genus one curves of every index over every number field. J. Reine Angew. Math. **594**, 201–206 (2006)
1121. Clozel, L.: The Sato-Tate conjecture. In: Current Development in Mathematics, Somerville, 2006, pp. 1–34 (2008)
1122. Clozel, L., Harris, M., Taylor, R.: Automorphy for some ℓ-adic lifts of automorphic mod ℓ representations. Publ. Math. Inst. Hautes Études Sci. **108**, 1–181 (2008)
1123. Coates, J.: An effective p-adic analogue of a theorem of Thue. Acta Arith. **15**, 279–305 (1968/1969)
1124. Coates, J.: An effective p-adic analogue of a theorem of Thue, II. The greatest prime factor of a binary form. Acta Arith. **16**, 399–412 (1969/1970)
1125. Coates, J.: An effective p-adic analogue of a theorem of Thue, III. The diophantine equation $y^2 = x^3 + k$. Acta Arith. **16**, 425–435 (1969/1970)

[5] See also Čudnovskiĭ, G.V.

1126. Coates, J.: Fonctions zêta partielles d'un corps de nombres totalement réel. Sémin. Delange–Pisot–Poitou **16** (exp. 1), 1–9 (1974/75)
1127. Coates, J., Sinnott, W.: On p-adic L-functions over real quadratic fields. Invent. Math. **25**, 253–279 (1974)
1128. Coates, J., Sinnott, W.: Integrality properties of the values of partial zeta functions. Proc. Lond. Math. Soc. **34**, 365–384 (1977)
1129. Coates, J.H., van der Poorten, A.J.: Kurt Mahler (1903–1988). Number Theory Research Reports 92-118, Macquarie Univ. (1992)
1130. Coates, J., Wiles, A.: On the conjecture of Birch and Swinnerton-Dyer. Invent. Math. **39**, 223–251 (1977)
1131. Cochrane, T.: Trigonometric approximation and uniform distribution modulo one. Proc. Am. Math. Soc. **103**, 695–702 (1988)
1132. Cochrane, T., Pinner, C., Rosenhouse, J.: Sparse polynomial exponential sums. Acta Arith. **108**, 37–52 (2003)
1133. Cochrane, T., Zheng, Z.: On upper bounds of Chalk and Hua for exponential sums. Proc. Am. Math. Soc. **129**, 2505–2516 (2001)
1134. Codecá, P.: A note on Euler's φ-function. Ark. Mat. **19**, 261–263 (1981)
1135. Cohen, G.L.: On the largest component of an odd perfect number. J. Aust. Math. Soc. A **42**, 280–286 (1987)
1136. Cohen, G.L., Hagis, P. Jr.: On the number of prime factors of n if $\varphi(n)|(n-1)$. Nieuw Arch. Wiskd. **28**, 177–185 (1980)
1137. Cohen, G.L., Hagis, P. Jr.: Results concerning odd multiperfect numbers. Bull. Malays. Math. Soc. **8**, 23–26 (1985)
1138. Cohen, G.L., Hendy, M.D.: On odd multiperfect numbers. Addendum to: "Polygonal supports for sequences of primes". Math. Chron. **9**, 120–136 (1980)
1139. Cohen, H.: A lifting of modular forms in one variable to Hilbert modular forms in two variables. In: Lecture Notes in Math., vol. 627, pp. 175–196. Springer, Berlin (1977)
1140. Cohen, H.: Démonstration de l'irrationalité de $\zeta(3)$. Sém. Th. Nombres, Grenoble, 1978
1141. Cohen, H.: Démonstration de la conjecture de Catalan. In: Théorie algorithmique des nombres et équations diophantiennes, Palaiseau, 2005, pp. 1–83. Springer, Berlin (2005)
1142. Cohen, H.: Number Theory, vols. I–II. Springer, Berlin (2007)
1143. Cohen, H., Lenstra, H.W. Jr.: Primality testing and Jacobi sums. Math. Comput. **42**, 297–330 (1984)
1144. Cohen, H., Lenstra, H.W. Jr.: Implementation of a new primality test. Math. Comput. **48**, 103–121, S1–S4 (1987)
1145. Cohen, H., Oesterlé, J.: Dimensions des espaces de formes modulaires. In: Lecture Notes in Math., vol. 627, pp. 69–78. Springer, Berlin (1977)
1146. Cohen, I.B. (ed.): Benjamin Peirce: "Father of Pure Mathematics" in America. Arno Press, New York (1980)
1147. Cohen, J.: Primitive roots in quadratic fields. Int. J. Number Theory **2**, 7–23 (2006)
1148. Cohen, J.: Primitive roots in quadratic fields, II. J. Number Theory **124**, 429–441 (2007)
1149. Cohen, J.W.: Felix Pollaczek, 1892–1981. J. Appl. Probab. **18**, 958–963 (1981)
1150. Cohen, P.B.: Interactions between number theory and operator algebras in the study of the Riemann zeta function (d'après Bost-Connes and Connes). In: Number Theory, New York, 2003, pp. 87–103. Springer, Berlin (2004)
1151. Cohen, P.J.: On a conjecture of Littlewood and idempotent measures. Am. J. Math. **82**, 191–212 (1960)
1152. Cohen, S.D.: Obituary: Robert Winston Keith Odoni (1947–2002). Glasg. Math. J. **45**, 565–575 (2003)
1153. Cohen, S.D., Odoni, R.W.K., Stothers, W.W.: On the least primitive root modulo p^2. Bull. Lond. Math. Soc. **6**, 42–46 (1974)
1154. Cohn, H.: Minkowski's conjecture on critical lattices in the metric $(|\xi|^p + |\eta|^p)^{1/p}$. Ann. Math. **51**, 734–738 (1950)
1155. Cohn, H.: Approach to Markoff's minimal forms through modular functions. Ann. Math. **61**, 1–12 (1955)

1156. Cohn, H.: Representation of Markoff's binary quadratic forms by geodesics on a perforated torus. Acta Arith. **18**, 125–136 (1971)
1157. Cohn, H.: Markoff forms and primitive words. Math. Ann. **196**, 8–22 (1972)
1158. Cohn, H.: Symmetry and specializability in continued fractions. Acta Arith. **75**, 297–320 (1996)
1159. Cohn, H., Elkies, N.[D.]: New upper bounds on sphere packings, I. Ann. Math. **157**, 689–714 (2003)
1160. Cohn, J.H.E.: On square Fibonacci numbers. J. Lond. Math. Soc. **39**, 537–540 (1964)
1161. Cohn, J.H.E.: Waring's problem in quadratic number fields. Acta Arith. **20**, 1–16 (1972)
1162. Cohn, J.H.E.: The Diophantine equation $x^4 - Dy^2 = 1$, II. Acta Arith. **78**, 401–403 (1997)
1163. Cohn, J.H.E.: The length of the period of the simple continued fraction of $\sqrt{d}$. Pac. J. Math. **71**, 21–32 (1977)
1164. Cohn, J.H.E.: The Diophantine equation $x^2 + C = y^n$. Acta Arith. **65**, 367–381 (1993)
1165. Cohn, J.H.E.: Perfect Pell powers. Glasg. Math. J. **38**, 19–20 (1996)
1166. Cohn, J.H.E.: The Diophantine equation $x^4 + 1 = Dy^2$. Math. Comput. **66**, 1347–1351 (1997)
1167. Cohn, P.M.: Shimshon Avraham Amitsur. Bull. Lond. Math. Soc. **28**, 433–439 (1996)
1168. Cojocaru, A.C.: On the cyclicity of the group of F_p-rational points of non-CM elliptic curves. J. Number Theory **96**, 335–350 (2002)
1169. Cojocaru, A.C.: Cyclicity of CM elliptic curves modulo p. Trans. Am. Math. Soc. **355**, 2651–2662 (2003)
1170. Cojocaru, A.C.: Reductions of an elliptic curve with almost prime orders. Acta Arith. **119**, 265–289 (2005)
1171. Cojocaru, A.C., Fouvry, É., Murty, M.R.: The square sieve and the Lang-Trotter conjecture. Can. J. Math. **57**, 1155–1177 (2005)
1172. Cojocaru, A.C., Hall, C.: Uniform results for Serre's theorem for elliptic curves. Internat. Math. Res. Notices, 2005, 3065–3080
1173. Cojocaru, A.C., Murty, M.R.: Cyclicity of elliptic curves modulo p and elliptic curve analogues of Linnik's problem. Math. Ann. **330**, 601–625 (2004)
1174. Cole, F.N.: On the factoring of large numbers. Bull. Am. Math. Soc. **10**, 134–137 (1903/1904)
1175. Colebrook, C.M., Kemperman, J.H.B.: On non-normal numbers. Indag. Math. **30**, 1–11 (1968)
1176. Coleman, M.D.: On the equation $b_1 p - b_2 P_2 = b_3$. J. Reine Angew. Math. **403**, 1–66 (1990)
1177. Coleman, M.D.: A zero-free region for the Hecke L-functions. Mathematika **37**, 287–304 (1990)
1178. Coleman, M.D.: The Rosser-Iwaniec sieve in number fields, with an application. Acta Arith. **65**, 53–83 (1993)
1179. Coleman, M.[D.], Swallow, A.: Localised Bombieri-Vinogradov theorems in imaginary quadratic fields. Acta Arith. **120**, 349–377 (2005)
1180. Coleman, R.F.: Effective Chabauty. Duke Math. J. **52**, 765–770 (1985)
1181. Coleman, R.F.: Manin's proof of the Mordell conjecture over function fields. Enseign. Math. **36**, 393–427 (1990)
1182. Colin de Verdière, Y.: Nombre de points entiers dans une famille homothétique de domains de R. Ann. Sci. Éc. Norm. Super. **10**, 559–575 (1977)
1183. Collingwood, E.F.: Emile Borel. J. Lond. Math. Soc. **34**, 488–512 (1959); add.: **35**, 384 (1960)
1184. Colliot-Thélène, J.-L.: L'arithmétique des variétés rationnelles. Ann. Fac. Sci. Toulouse **1**, 295–336 (1992)
1185. Colliot-Thélène, J.-L.: Les grands thèmes de François Châtelet. Enseign. Math. **34**, 387–405 (1988)
1186. Colliot-Thélène, J.-L., Kanevsky, D., Sansuc, J.-J.: Arithmétique des surfaces cubiques diagonales. In: Lecture Notes in Math., vol. 1290, pp. 1–108. Springer, Berlin (1987)

1187. Colliot-Thélène, J.-L., Poonen, B.: Algebraic families of nonzero elements of Shafarevich-Tate groups. J. Am. Math. Soc. **13**, 83–99 (2000)
1188. Colliot-Thélène, J.-L., Sansuc, J.-J.: Sur le principle de Hasse et l'approximation faible, et sur une hypothèse de Schinzel. Acta Arith. **41**, 34–53 (1982)
1189. Colliot-Thélène, J.-L., Swinnerton-Dyer, P.: Hasse principle and weak approximation for pencils of Severi-Brauer and similar varieties. J. Reine Angew. Math. **453**, 49–112 (1994)
1190. Colmez, P.: Résidu en $s = 1$ des fonctions zêta p-adiques. Invent. Math. **91**, 371–389 (1988)
1191. Colmez, P.: Fonctions zêta p-adiques en $s = 0$. J. Reine Angew. Math. **467**, 89–107 (1995)
1192. Colmez, P.: Fonctions L p-adiques. Astérisque **266**, 21–58 (2000)
1193. Colmez, P.: La conjecture de Birch et Swinnerton-Dyer p-adique. Astérisque **294**, 251–319 (2004)
1194. Comalada, S.: Elliptic curves with trivial conductor over quadratic fields. Pac. J. Math. **144**, 237–258 (1990)
1195. Conn, W., Vaserstein, L.N.: On sums of three integral cubes. In: Contemp. Math., vol. 166, 285–294. Am. Math. Soc., Providence (1994)
1196. Connes, A.: Formule de trace en géométrie non-commutative et hypothèse de Riemann. C. R. Acad. Sci. Paris **323**, 1231–1236 (1996)
1197. Connes, A.: Trace formula in noncommutative geometry and the zeros of the Riemann zeta function. Sel. Math. New Ser. **5**, 29–106 (1999)
1198. Connor, W.G.: Partition theorems related to some identities of Rogers and Watson. Trans. Am. Math. Soc. **214**, 95–111 (1997)
1199. Conrad, B., Diamond, F., Taylor, R.: Modularity of certain potentially Barsotti-Tate Galois representations. J. Am. Math. Soc. **12**, 521–567 (1999)
1200. Conrad, K.: Hardy–Littlewood constants. In: Mathematical Properties of Sequences and Other Combinatorial Structures, Los Angeles, CA, 2002, pp. 133–154. Kluwer Academic, Dordrecht (2003)
1201. Conrey, J.B.: Zeros of derivatives of Riemann's ζ-function on the critical line. J. Number Theory **16**, 49–74 (1983)
1202. Conrey, J.B.: More than two fifth of the zeros of the Riemann zeta function are on the critical line. J. Reine Angew. Math. **399**, 1–26 (1989)
1203. Conrey, J.B.: The Riemann Hypothesis. Not. Am. Math. Soc. **50**, 341–353 (2003)
1204. Conrey, J.B., Farmer, D.W.: An extension of Hecke's converse theorem. Internat. Math. Res. Notices, 1995, 445–463
1205. Conrey, J.B., Farmer, D.W., Keating, J.P., Rubinstein, M., Snaith, N.C.: Integral moments of L-functions. Proc. Lond. Math. Soc. **91**, 33–104 (2005)
1206. Conrey, J.B., Farmer, D.W., Keating, J.P., Rubinstein, M.O., Snaith, N.C.: Lower order terms in the full moment conjecture for the Riemann zeta function. J. Number Theory **128**, 1516–1554 (2008)
1207. Conrey, J.B., Farmer, D.W., Odgers, B.E., Snaith, N.C.: A converse theorem for $\Gamma_0(13)$. J. Number Theory **122**, 314–323 (2007)
1208. Conrey, J.B., Ghosh, A.: On mean values of the zeta-function. Mathematika **31**, 159–161 (1984)
1209. Conrey, J.B., Ghosh, A.: On mean values of the zeta-function, II. Acta Arith. **52**, 367–371 (1989)
1210. Conrey, J.B., Ghosh, A.: On mean values of the zeta-function, III. In: Proceedings of the Conference on Analytic Number Theory, Maiori, 1989, pp. 35–59. Universit'a di Salerno, Salerno (1992)
1211. Conrey, J.B., Ghosh, A.: A simpler proof of Levinson's theorem. Math. Proc. Camb. Philos. Soc. **97**, 385–395 (1985)
1212. Conrey, J.B., Ghosh, A.: On the Selberg class of Dirichlet series: small degrees. Duke Math. J. **72**, 673–693 (1993)
1213. Conrey, J.B., Ghosh, A., Gonek, S.M.: A note on gaps between zeros of the zeta function. Bull. Lond. Math. Soc. **16**, 421–424 (1984)

1214. Conrey, J.B., Ghosh, A., Gonek, S.M.: Large on gaps between zeros of the zeta function. Mathematika **33**, 212–238 (1986)
1215. Conrey, J.B., Ghosh, A., Gonek, S.M.: Simple zeros of the Riemann zeta-function. Proc. Lond. Math. Soc. **76**, 497–522 (1998)
1216. Conrey, J.B., Gonek, S.M.: High moments of the Riemann zeta-function. Duke Math. J. **107**, 577–604 (2001)
1217. Conrey, J.B., Granville, A., Poonen, B., Soundararajan, K.: Zeros of Fekete polynomials. Ann. Inst. Fourier **50**, 865–889 (2000)
1218. Conrey, J.B., Iwaniec, H.: The cubic moment of central values of automorphic L-functions. Ann. Math. **151**, 1175–1216 (2000)
1219. Conrey, J.B., Snaith, N.C.: Applications of the L-functions ratios conjectures. Proc. Lond. Math. Soc. **94**, 594–646 (2007)
1220. Conrey, J.B., Soundararajan, K.: Real zeros of quadratic Dirichlet L-functions. Invent. Math. **150**, 1–44 (2002)
1221. Contini, S., Croot, E., Shparlinski, I.E.: Complexity of inverting the Euler function. Math. Comput. **75**, 983–996 (2006)
1222. Conway, J.H.: A group of order 8,315,553,613,086,720,000. Bull. Lond. Math. Soc. **1**, 79–88 (1969)
1223. Conway, J.H.: A characterisation of Leech's lattice. Invent. Math. **7**, 137–142 (1969)
1224. Conway, J.H.: Universal quadratic forms and the fifteen theorem. In: Contemp. Math., vol. 272, pp. 23–26. Am. Math. Soc., Providence (2000)
1225. Conway, J.H., Norton, S.P.: Monstrous Moonshine. Bull. Lond. Math. Soc. **11**, 308–339 (1979)
1226. Conway, J.H., Sloane, N.J.A.: Lorentzian forms for the Leech lattice. Bull. Am. Math. Soc. **6**, 215–217 (1982)
1227. Conway, J.H., Sloane, N.J.A.: Twenty-three constructions for the Leech lattice. Proc. R. Soc. Lond. Ser. A, Math. Phys. Sci. **381**, 275–283 (1982)
1228. Conway, J.H., Sloane, N.J.A.: On the enumeration of lattices of determinant one. J. Number Theory **15**, 83–94 (1982)
1229. Conway, J.H., Sloane, N.J.A.: The unimodular lattices of dimension up to 23 and the Minkowski-Siegel mass constants. Electron. J. Comb. **3**, 219–231 (1982)
1230. Conway, J.H., Sloane, N.J.A.: Sphere Packings, Lattices and Groups. Springer, Berlin (1988); 2nd ed. 1993; 3rd ed. 1998
1231. Cook, R.J.: Pairs of additive equations. Mich. Math. J. **19**, 325–331 (1972)
1232. Cook, R.J.: Pairs of additive equations, II. Large odd degree. J. Number Theory **17**, 80–92 (1983)
1233. Cook, R.J.: Pairs of additive equations, III. Quintic equations. Proc. Edinb. Math. Soc. **26**, 191–211 (1983)
1234. Cook, R.J.: Pairs of additive equations, IV. Sextic equations. Acta Arith. **43**, 227–243 (1984)
1235. Cook, R.J.: On the fractional parts of a set of points. Mathematika **19**, 63–68 (1972)
1236. Cook, R.J.: On the fractional parts of a set of points, II. Pac. J. Math. **45**, 81–85 (1973)
1237. Cook, R.J.: On the fractional parts of a set of points, III. J. Lond. Math. Soc. **9**, 490–494 (1974/1975)
1238. Cook, R.J.: On the fractional parts of a set of points, IV. Indian J. Math. **19**, 7–23 (1977)
1239. Cook, R.J.: On the fractional parts of a polynomial. Can. J. Math. **28**, 168–173 (1976)
1240. Cook, R.[J.]: On sums of powers of integers. J. Number Theory **11**, 516–528 (1979)
1241. Cook, R.J.: Small fractional parts of quadratic forms in many variables. Mathematika **27**, 25–29 (1980)
1242. Cook, R.J.: An effective seven cube theorem. Bull. Aust. Math. Soc. **30**, 381–385 (1984)
1243. Cook, R.J.: Pairs of additive congruences: cubic congruences. Mathematika **32**, 286–300 (1985)
1244. Cook, R.J.: Pairs of additive congruences: quintic congruences. Indian J. Pure Appl. Math. **17**, 786–799 (1986)

1245. Cook, R.J.: Computations for additive Diophantine equations: pairs of quintic congruences, II. In: Computers in Mathematical Research, Cardiff, 1986, pp. 93–117. Clarendon Press, Oxford (1988)
1246. Copeland, A.H., Erdős, P.: Note on normal numbers. Bull. Am. Math. Soc. **52**, 857–860 (1946)
1247. Coppersmith, D.: Fermat's last theorem (case 1) and the Wieferich criterion. Math. Comput. **54**, 895–902 (1990)
1248. Cornacchia, G.: Sulla congruenza $x^n + y^n \equiv z^n \pmod{p}$. Giornale Mat. **16**, 219–268 (1909)
1249. Cornell, G., Silverman, J.H., Stevens, S. (eds.): Modular Forms and Fermat's Last Theorem. Springer, Berlin (1997)
1250. Corrádi, K., Kátai, I.: A note on a paper of K.S. Gangadharan. Mag. Tud. Akad. Mat. Fiz. Oszt. Kőzl. **17**, 89–97 (1967) (in Hungarian)
1251. Corry, L.: Fermat comes to America: Harry Schultz Vandiver and FLT (1914–1963). Math. Intell. **29**, 30–40 (2007)
1252. Corvaja, P., Zannier, U.: Diophantine equations with power sums and universal Hilbert sets. Indag. Math. **9**, 317–332 (1998)
1253. Corvaja, P., Zannier, U.: Some new applications of the subspace theorem. Compos. Math. **131**, 319–340 (2002)
1254. Corvaja, P., Zannier, U.: A subspace theorem approach to integral points on curves. C. R. Acad. Sci. Paris **334**, 267–271 (2002)
1255. Corvaja, P., Zannier, U.: On the rational approximations to the powers of an algebraic number: solution of two problems of Mahler and Mendès France. Acta Math. **193**, 175–191 (2004)
1256. Costa Pereira, N.: Elementary estimates for the Chebyshev function $\psi(x)$ and for the Möbius function $M(x)$. Acta Arith. **52**, 307–337 (1989)
1257. Cougnard, J.: Les travaux de A. Fröhlich, Ph. Cassou-Noguès et M.J. Taylor sur les bases normales. Astérisque **105/106**, 25–38 (1983)
1258. Coulangeon, R., Icaza, M.I., O'Ryan, M.: Lenstra's constant and extreme forms in number fields. Exp. Math. **16**, 455–462 (2007)
1259. Courant, R.: Felix Klein. Jahresber. Dtsch. Math.-Ver. **34**, 197–213 (1926)
1260. Courant, R.: Carl Runge als Mathematiker. Naturwissenschaften **15**, 229–231 (1927)
1261. Coxeter, H.S.M.: Extreme forms. Can. J. Math. **3**, 391–441 (1951)
1262. Coxeter, H.S.M.: An upper bound for the number of equal nonoverlapping spheres that can touch another of the same size. In: Proc. Symposia Pure Math., vol. 7, pp. 53–71. Am. Math. Soc., Providence (1963)
1263. Coxeter, H.S.M., Few, L., Rogers, C.A.: Covering space with equal spheres. Mathematika **6**, 147–157 (1959)
1264. Coxeter, H.S.M., Todd, J.A.: An extreme duodenary form. Can. J. Math. **5**, 384–392 (1953)
1265. Craig, M.: A cyclotomic construction for Leech's lattice. Mathematika **25**, 236–241 (1978)
1266. Cramér, H.: Some theorems concerning prime numbers. Ark. Math. Astron. Fys. **15**(5), 1–33 (1921) [[1274], vol. 1, pp. 138–170]
1267. Cramér, H.: On the distribution of primes. Proc. Camb. Philos. Soc. **20**, 272–280 (1921) [[1274], vol. 1, pp. 171–179]
1268. Cramér, H.: Ein Mittelwertsatz in der Primzahltheorie. Math. Z. **12**, 147–153 (1922) [[1274], vol. 1, pp. 229–235]
1269. Cramér, H.: Über zwei Sätze von Herrn G.H. Hardy. Math. Z. **15**, 201–210 (1922) [[1274], vol. 1, pp. 236–245]
1270. Cramér, H.: Über das Teilerproblem von Piltz. Ark. Math. Astron. Fys. **16**(21), 1–40 (1922) [[1274], vol. 1, pp. 184–223]
1271. Cramér, H.: Contributions to the analytic theory of numbers. In: Verhandl. V Skand. Math. Kongress, Helsingfors, pp. 266–272 (1923) [[1274], vol. 1, pp. 246–252]
1272. Cramér, H.: Prime numbers and probability. In: Verhandl. VIII Skand. Math. Kongress, Stockholm, pp. 107–115 (1934) [[1274], vol. 2, pp. 827–835]

1273. Cramér, H.: On the order of magnitude of the difference between consecutive prime numbers. Acta Arith. **2**, 23–46 (1936) [[1274], vol. 2, pp. 871–894]
1274. Cramér, H.: Collected Works, vols. 1, 2. Springer, Berlin (1994)
1275. Crandall, R.E.: New representations for the Madelung constant. Exp. Math. **8**, 367–374 (1999)
1276. Crandall, R.[E.], Dilcher, K., Pomerance, C.: A search for Wieferich and Wilson primes. Math. Comput. **66**, 433–449 (1997)
1277. Crandall, R.[E.], Pomerance, C.: Prime Numbers. A Computational Perspective, Springer, Berlin (2001); 2nd ed. 2005
1278. Cremona, J.E.: Algorithms for Modular Elliptic Curves. Cambridge University Press, Cambridge (1992); 2nd ed. 1997
1279. Cremona, J.E., Lingham, M.P.: Finding all elliptic curves with good reduction outside a given set of primes. Exp. Math. **16**, 303–312 (2007)
1280. Cremona, J.E., Odoni, R.W.K.: Some density results for negative Pell equations; an application of graph theory. J. Lond. Math. Soc. **39**, 16–28 (1989)
1281. Crittenden, R.B., Vanden Eynden, C.L.: Any n arithmetic progressions covering the first 2^n integers cover all integers. Proc. Am. Math. Soc. **24**, 475–481 (1970)
1282. Croot, E.S. III: On some questions of Erdős and Graham about Egyptian fractions. Mathematika **46**, 359–372 (1999)
1283. Croot, E.S. III: On a coloring conjecture about unit fractions. Ann. Math. **157**, 545–556 (2003)
1284. Čubarikov, V.N.: Simultaneous representation of natural numbers by sums of powers of primes. Dokl. Akad. Nauk SSSR **286**, 828–831 (1986) (in Russian)
1285. Čudakov, N.G.[6]: On the functions $\zeta(s)$ and $\pi(x)$. Dokl. Akad. Nauk SSSR **21**, 425–426 (1938) (in Russian)
1286. Čudakov, N.G.: On the density of the set of even numbers which are not representable as a sum of two primes. Mat. Sb. **2**, 25–40 (1938) (in Russian)
1287. Čudakov, N.G.: On Siegel's theorem. Izv. Akad. Nauk SSSR, Ser. Mat. **6**, 135–142 (1942) (in Russian)
1288. Čudakov, N.G.: On zeros of Dirichlet's L-functions. Mat. Sb. **19**, 47–56 (1946) (in Russian)
1289. Čudnovskiĭ, G.V.[7]: Diophantine predicates. Usp. Mat. Nauk **25**(4), 185–186 (1970) [Errata: Math. Rev. **44**, #1632] (in Russian)
1290. Čudnovskiĭ, G.V.: Algebraic independence of several values of the exponential function. Mat. Zametki **15**, 661–672 (1974) (in Russian)
1291. Cugiani, M.: Sull'approssimabilità di un numero algebrico mediante numeri algebrici di un corpo assegnato. Boll. Unione Mat. Ital. **14**, 151–162 (1959)
1292. Cugiani, M.: Giovanni Ricci (1904–1973). Acta Arith. **46**, 303–311 (1986)
1293. Cummins, C.J., Gannon, T.: Modular equations and the genus zero property of moonshine functions. Invent. Math. **129**, 413–443 (1997)
1294. Cunningham, A.: On finding factors. Messenger Math. **20**, 37–45 (1890)
1295. Cunningham, A.: Number of primes of given linear forms. Proc. Lond. Math. Soc. **10**, 249–253 (1912)
1296. Cunningham, A.: On the number of primes of same residuacity. Proc. Lond. Math. Soc. **13**, 258–272 (1914)
1297. Cunningham, A.: Binomial Factorisations. vols. 1–7. Hodgson, London (1924–1925)
1298. Cunningham, A., Western, A.E.: On Fermat's numbers. Proc. Lond. Math. Soc. **1**, 175 (1903)
1299. Cunningham, A., Woodall, H.J.: Determination of successive high primes. Messenger Math. **31**, 165–176 (1901/1902); **34**, 72–89, 184–192 (1904/1905)

[6] See also Tchudakoff, N.

[7] See also Chudnovsky, G.V.

1300. Cunningham, A., Woodall, H.J., Creak, T.G.: On least primitive roots. Proc. Lond. Math. Soc. **21**, 343–358 (1922)
1301. Cunningham, A., Woodall, H.J.: Factorisation of $y^n \pm 1$, $y = 2, 3, 5, 6, 7, 10, 11, 12$ up to High Powers (n). Hodgson, London (1925)
1302. Cupr, K.: Prof. Mathias Lerch. Časopis Mat. Fys. **52**, 301–313 (1923)
1303. Curtis, F.: On formulas for the Frobenius number of a numerical semigroup. Math. Scand. **67**, 190–192 (1990)
1304. Cusick, T.W.: Littlewood's Diophantine approximation problem for series. Proc. Am. Math. Soc. **18**, 920–924 (1967)
1305. Cusick, T.W.: Sums and products of continued fractions. Proc. Am. Math. Soc. **27**, 35–38 (1971)
1306. Cusick, T.W.: On M. Hall's continued fraction theorem. Proc. Am. Math. Soc. **38**, 253–254 (1973)
1307. Cusick, T.W.: Continuants with bounded digits, I. Mathematika **24**, 166–172 (1977)
1308. Cusick, T.W.: Continuants with bounded digits, II. Mathematika **25**, 107–109 (1978)
1309. Cusick, T.W.: Continuants with bounded digits, III. Monatshefte Math. **99**, 105–109 (1985)
1310. Cusick, T.W., Flahive, M.E.: The Markoff and Lagrange Spectra. Am. Math. Soc., Providence (1989)
1311. Cusick, T.W., Lee, R.A.: Sums of sets of continued fractions. Proc. Am. Math. Soc. **30**, 241–246 (1971)
1312. Cutter, P.A.: Finding prime pairs with particular gaps. Math. Comput. **70**, 1737–1744 (2001)
1313. Daboussi, H.: Fonctions multiplicatives presque périodiques B. Astérisque **24–25**, 321–324 (1975)
1314. Daboussi, H.: Sur le théorème des nombres premiers. C. R. Acad. Sci. Paris **298**, 161–164 (1984)
1315. Daboussi, H.: On the prime number theorem for arithmetic progressions. J. Number Theory **31**, 243–254 (1989)
1316. Daboussi, H., Delange, H.: Quelques propriétés des fonctions multiplicatives de module au plus égal à 1. C. R. Acad. Sci. Paris **278**, 657–660 (1974)
1317. Daboussi, H., Delange, H.: On multiplicative arithmetical functions whose modulus does not exceed one. J. Lond. Math. Soc. **26**, 245–264 (1982)
1318. Daboussi, H., Indlekofer, K.-H.: Two elementary proofs of Halász's theorem. Math. Z. **209**, 43–52 (1992)
1319. Damgård, I., Landrock, P., Pomerance, C.: Average case error estimates for the strong probable prime test. Math. Comput. **61**, 177–194 (1993)
1320. Dani, S.G.: A proof of Margulis' theorem on values of quadratic forms, independent of the axiom of choice. Enseign. Math. **40**, 49–58 (1994)
1321. Dani, S.G., Margulis, G.A.: Values of quadratic forms at primitive integral points. Invent. Math. **98**, 405–424 (1989)
1322. Dani, S.G., Margulis, G.A.: Values of quadratic forms at integral points: An elementary approach. Enseign. Math. **36**, 143–174 (1990)
1323. Danicic, I.: An extension of a theorem of Heilbronn. Mathematika **5**, 30–37 (1958)
1324. Danicic, I.: On the fractional parts of θx^2 and ϕx^2. J. Lond. Math. Soc. **34**, 353–357 (1959)
1325. Danilov, L.V.: The diophantine equation $x^3 - y^2 = k$ and the conjecture of M. Hall. Mat. Zametki **32**, 273–275 (1982); corr. **36**, 457–458 (1984) (in Russian)
1326. Danset, R.: Méthode du cercle adélique et principe de Hasse fin pour certains systémes de formes. Enseign. Math. **31**, 1–66 (1985)
1327. Danzer, L.: Über eine Frage von G. Hanani aus der additiven Zahlentheorie. J. Reine Angew. Math. **214/215**, 392–394 (1964)
1328. Darling, H.B.C.: On Mr. Ramanujan's congruence properties of $p(n)$. Proc. Camb. Philos. Soc. **19**, 217–218 (1917–1920)
1329. Darling, H.B.C.: Proofs of certain identities and congruences enunciated by S. Ramanujan. Proc. Lond. Math. Soc. **19**, 350–372 (1921)

1330. Darmon, H.: The equation $x^4 - y^4 = z^p$. C. R. Acad. Sci. Can. **15**, 286–290 (1993)
1331. Darmon, H.: The equations $x^n + y^n = z^2$ and $x^n + y^n = z^3$. Internat. Math. Res. Notices, 1993, 263–274
1332. Darmon, H.: The Shimura-Taniyama conjecture. Usp. Mat. Nauk **50**(3), 503–548 (1995) (in Russian)
1333. Darmon, H., Diamond, F., Taylor, R.: Fermat's last theorem. In: Current Developments in Mathematics, 1995, pp. 1–154. International Press, Cambridge (1994) [Updated version in: Elliptic Curves, Modular Forms & Fermat's Last Theorem (Hong Kong, 1993), pp. 2–140. Cambridge, MA, 1997]
1334. Darmon, H., Granville, A.: On the equations $z^m = F(x, y)$ and $Ax^p + By^q = Cz^r$. Bull. Lond. Math. Soc. **27**, 513–543 (1995)
1335. Darmon, H., Merel, L.: Winding quotients and some variants of Fermat's Last Theorem. J. Reine Angew. Math. **490**, 81–100 (1997)
1336. Dartyge, C.: Entiers de la forme $n^2 + 1$ sans grand facteur premier. Acta Math. Acad. Sci. Hung. **72**, 1–34 (1996)
1337. Dartyge, C., Martin, G., Tenenbaum, G.: Polynomial values free of large prime factors. Period. Math. Hung. **43**, 111–119 (2001)
1338. Datskovsky, B.A.: A mean-value theorem for class numbers of quadratic extensions. In: Contemp. Math., vol. 143, pp. 179–242. Am. Math. Soc., Providence (1993)
1339. Davenport, H.: On the distribution of quadratic residues (mod p). J. Lond. Math. Soc. **6**, 49–54 (1931) [[1380], vol. 4, pp. 1451–1456]
1340. Davenport, H.: On the distribution of quadratic residues (mod p) II. J. Lond. Math. Soc. **8**, 46–52 (1933) [[1380], vol. 4, pp. 1481–1487]
1341. Davenport, H.: On Dirichlet's L-functions. J. Lond. Math. Soc. **6**, 198–202 (1931) [[1380], vol. 4, pp. 1759–1763]
1342. Davenport, H.: On the distribution of l-th power residues (mod p). J. Lond. Math. Soc. **7**, 117–121 (1932) [[1380], vol. 4, pp. 1457–1461]
1343. Davenport, H.: On certain exponential sums. J. Reine Angew. Math. **169**, 158–176 (1933) [[1380], vol. 4, pp. 1462–1480]
1344. Davenport, H.: Über numeri abundantes. SBer. Preuß. Akad. Wiss. Berlin **27**, 830–837 (1933) [[1380], vol. 4, pp. 1834–1841]
1345. Davenport, H.: Note on mean-value theorems for the Riemann zeta-function. J. Lond. Math. Soc. **10**, 136–138 (1935) [[1380], vol. 4, pp. 1764–1766]
1346. Davenport, H.: On the addition of residue classes. J. Lond. Math. Soc. **10**, 30–32 (1935) [[1380], vol. 4, pp. 1854–1856]
1347. Davenport, H.: On some infinite series involving arithmetical functions, I. Q. J. Math. **8**, 8–13 (1937) [[1380], vol. 4, pp. 1781–1786]
1348. Davenport, H.: On some infinite series involving arithmetical functions, II. Q. J. Math. **8**, 313–320 (1937) [[1380], vol. 4, pp. 1787–1794]
1349. Davenport, H.: Sur les sommes de puissances entiéres. C. R. Acad. Sci. Paris **207**, 1366–1368 (1938) [[1380], vol. 3, pp. 972–974]
1350. Davenport, H.: On the product of three homogeneous linear forms, II. Proc. Lond. Math. Soc. **44**, 412–431 (1938) [[1380], vol. 1, pp. 14–33]
1351. Davenport, H.: On the product of three homogeneous linear forms, III. Proc. Lond. Math. Soc. **45**, 98–125 (1939) [[1380], vol. 1, pp. 34–61]
1352. Davenport, H.: On the product of three homogeneous linear forms, IV. Proc. Camb. Philos. Soc. **39**, 1–21 (1943) [[1380], vol. 1, pp. 66–86]
1353. Davenport, H.: A simple proof of Remak's theorem on the product of three linear forms. J. Lond. Math. Soc. **14**, 47–51 (1939) [[1380], vol. 1, pp. 92–96]
1354. Davenport, H.: On Waring's problem for fourth powers. Ann. Math. **40**, 731–747 (1939) [[1380], vol. 3, pp. 946–962]
1355. Davenport, H.: On character sums in finite fields. Acta Math. **71**, 99–121 (1939) [[1380], vol. 4, pp. 1529–1551]
1356. Davenport, H.: On Waring's problem for cubes. Acta Math. **71**, 123–143 (1939) [[1380], vol. 3, pp. 925–945]

1357. Davenport, H.: Minkowski's inequality for the minima associated with a convex body. Q. J. Math. **10**, 119–121 (1939)
1358. Davenport, H.: Note on sums of fourth powers. J. Lond. Math. Soc. **16**, 3–4 (1941) [[1380], vol. 3, pp. 997–998]
1359. Davenport, H.: Note on the product of three homogeneous linear forms. J. Lond. Math. Soc. **16**, 98–101 (1941) [[1380], vol. 1, pp. 62–65]
1360. Davenport, H.: On Waring's problem for fifth and sixth powers. Am. J. Math. **64**, 199–207 (1942) [[1380], vol. 3, pp. 963–971]
1361. Davenport, H.: On a theorem of Tschebotareff. J. Lond. Math. Soc. **21**, 28–34 (1946); corr. **24**, 316 (1949) [[1380], vol. 1, pp. 151–157]
1362. Davenport, H.: A historical note. J. Lond. Math. Soc. **22**, 100–101 (1947) [[1380], vol. 4, pp. 1857–1858]
1363. Davenport, H.: Sums of three positive cubes. J. Lond. Math. Soc. **25**, 339–343 (1950) [[1380], vol. 3, pp. 999–1003]
1364. Davenport, H.: On the class-number of binary cubic forms, I. J. Lond. Math. Soc. **26**, 183–192 (1951) [[1380], vol. 2, pp. 509–518]
1365. Davenport, H.: On the class-number of binary cubic forms, II. J. Lond. Math. Soc. **26**, 192–198 (1951) [[1380], vol. 2, pp. 519–525]
1366. Davenport, H.: The covering of space by spheres. Rend. Circ. Mat. Palermo **1**, 92–107 (1952) [[1380], vol. 2, pp. 609–624]
1367. Davenport, H.: Indefinite quadratic forms in many variables. Mathematika **3**, 81–101 (1956) [[1380], vol. 3, pp. 1035–1055]
1368. Davenport, H.: Indefinite quadratic forms in many variables, II. Proc. Lond. Math. Soc. **8**, 109–126 (1958) [[1380], vol. 3, pp. 1058–1075]
1369. Davenport, H.: T. Vijayaraghavan. J. Lond. Math. Soc. **33**, 252–255 (1958)
1370. Davenport, H.: Cubic forms in thirty-two variables. Philos. Trans. R. Soc. Lond. Ser. A, Math. Phys. Sci. **251**, 193–232 (1959) [[1380], vol. 3, pp. 1149–1187]
1371. Davenport, H.: On a theorem of P.J. Cohen. Mathematika **7**, 93–97 (1960) [[1380], vol. 4, pp. 1602–1606]
1372. Davenport, H.: Analytic Methods for Diophantine Equations and Diophantine Inequalities. University of Michigan Press, Ann Arbor (1962); 2nd ed. Cambridge, 2005
1373. Davenport, H.: Cubic forms in 29 variables. Philos. Trans. R. Soc. Lond. Ser. A, Math. Phys. Sci. **266**, 287–298 (1962) [[1380], vol. 3, pp. 1197–1208]
1374. Davenport, H.: Cubic forms in sixteen variables. Proc. R. Soc. Lond. Ser. A, Math. Phys. Sci. **272**, 285–303 (1963) [[1380], vol. 3, pp. 1209–1227]
1375. Davenport, H.: L.J. Mordell. Acta Arith. **9**, 4–12 (1964)
1376. Davenport, H.: Eine Bemerkung über Dirichlet's L-Funktionen. Nachr. Ges. Wiss. Göttingen, 1966, 203–212 [[1380], vol. 4, pp. 1816–1825]
1377. Davenport, H.: Multiplicative Number Theory. Markham, Chicago (1967); 2nd ed. Springer, 1980; 3rd ed. 2000
1378. Davenport, H.: On a theorem of Heilbronn. Q. J. Math. **18**, 339–344 (1967) [[1380], vol. 3, pp. 1307–1312]
1379. Davenport, H.: A note on Thue's theorem. Mathematika **15**, 76–87 (1968) [[1380], vol. 2, pp. 757–768]
1380. Davenport, H.: The Collected Works, vols. 1–4. Academic Press, San Diego (1977)
1381. Davenport, H., Erdős, P.: The distribution of quadratic and higher residues. Publ. Math. (Debr.) **2**, 252–265 (1952) [[1380], vol. 4, pp. 1562–1575]
1382. Davenport, H., Erdős, P.: Note on normal decimals. Can. J. Math. **4**, 58–63 (1952) [[1380], vol. 4, pp. 1859–1864]
1383. Davenport, H., Halberstam, H.: The values of a trigonometric polynomial at well spaced points. Mathematika **13**, 91–96 (1966); corr., **14**, 229–232 (1967) [[1380], vol. 4, pp. 1657–1666]
1384. Davenport, H., Halberstam, H.: Primes in arithmetic progressions. Mich. Math. J. **13**, 485–489 (1966); corr., **15**, 505 (1968) [[1380], vol. 4, pp. 1667–1672]

1385. Davenport, H., Hasse, H.: Die Nullstellen der Kongruenzzetafunktionen in gewissen zyklischen Fällen. J. Reine Angew. Math. **172**, 151–182 (1935) [[1380], vol. 4, pp. 1488–1519]
1386. Davenport, H., Heilbronn, H.: On the zeros of certain Dirichlet series. J. Lond. Math. Soc. **11**, 181–185 (1936) [[1380], vol. 4, pp. 1769–1773, 1774–1779]
1387. Davenport, H., Heilbronn, H.: On the zeros of certain Dirichlet series, II. J. Lond. Math. Soc. **11**, 307–313 (1936) [[2715], pp. 272–276, 277–282]
1388. Davenport, H., Heilbronn, H.: On Waring's problem for fourth powers. Proc. Lond. Math. Soc. **41**, 143–150 (1936) [[1380], vol. 3, pp. 875–882; [2715], pp. 283–290]
1389. Davenport, H., Heilbronn, H.: On Waring's problem: two cubes and one square. Proc. Lond. Math. Soc. **43**, 73–104 (1937) [[1380], vol. 43, pp. 883–914; [2715], pp. 296–327]
1390. Davenport, H., Heilbronn, H.: Note on a result in the additive theory of numbers. Proc. Lond. Math. Soc. **43**, 142–151 (1937) [[1380], vol. 3, pp. 915–924; [2715], pp. 328–337]
1391. Davenport, H., Heilbronn, H.: On indefinite quadratic forms in five variables. J. Lond. Math. Soc. **21**, 185–193 (1946) [[1380], vol. 3, pp. 1010–1018; [2715], pp. 360–368]
1392. Davenport, H., Landau, E.: On the representation of positive integers as sums of three cubes of positive rational numbers. In: Number Theory and Analysis, pp. 49–53. Plenum, New York (1969) [[1380], vol. 4, pp. 1894–1896; [3680], vol. 9, pp. 445–447]
1393. Davenport, H., Lewis, D.J.: Exponential sums in many variables. Am. J. Math. **84**, 649–665 (1962) [[1380], vol. 3, pp. 1228–1244]
1394. Davenport, H., Lewis, D.J.: An analogue of a problem of Littlewood. Mich. Math. J. **10**, 157–160 (1963) [[1380], vol. 4, pp. 1866–1869]
1395. Davenport, H., Lewis, D.J.: Homogeneous additive equations. Proc. R. Soc. Lond. Ser. A, Math. Phys. Sci. **274**, 443–460 (1963) [[1380], vol. 3, pp. 1313–1330]
1396. Davenport, H., Lewis, D.J.: Cubic equations of additive type. Philos. Trans. R. Soc. Lond. Ser. A, Math. Phys. Sci. **261**, 97–136 (1966) [[1380], vol. 3, pp. 1331–1370]
1397. Davenport, H., Lewis, D.J.: Simultaneous equations of additive type. Philos. Trans. R. Soc. Lond. Ser. A, Math. Phys. Sci. **264**, 557–595 (1969) [[1380], vol. 3, pp. 1402–1440]
1398. Davenport, H., Lewis, D.J., Schinzel, A.: Equations of the form $f(x) = g(y)$. Q. J. Math. **12**, 304–312 (1961) [[1380], vol. 4, pp. 1711–1719]
1399. Davenport, H., Ridout, D.: Indefinite quadratic forms. Proc. Lond. Math. Soc. **9**, 544–555 (1959) [[1380], vol. 3, pp. 1105–1116]
1400. Davenport, H., Rogers, C.A.: Hlawka's theorem in the geometry of numbers. Duke Math. J. **14**, 367–375 (1947) [[1380], vol. 1, pp. 318–326]
1401. Davenport, H., Roth, K.F.: Rational approximations to algebraic numbers. Mathematika **2**, 160–167 (1955) [[1380], vol. 2, pp. 689–696]
1402. Davenport, H., Schinzel, A.: A note on certain arithmetical constants. Ill. J. Math. **10**, 181–185 (1966) [[1380], vol. 4, pp. 1811–1815]
1403. Davenport, H., Schmidt, W.M.: Approximation to real numbers by quadratic irrationals. Acta Arith. **13**, 169–176 (1967/1968) [[1380], vol. 2, pp. 770–777]
1404. David, C., Kisilevsky, H., Pappalardi, F.: Galois representations with non-surjective traces. Can. J. Math. **51**, 936–951 (1999)
1405. David, C., Pappalardi, F.: Average Frobenius distributions of elliptic curves. Internat. Math. Res. Notices, 1999, nr. 4, 165–183
1406. David, C., Pappalardi, F.: Average Frobenius distribution for inerts in $Q(i)$. J. Ramanujan Math. Soc. **19**, 181–201 (2004)
1407. David, S.: Minorations de formes linéaires de logarithmes elliptiques. Mém. Soc. Math. Fr. **62**, 1–143 (1995)
1408. Davidson, M.: Sums of k-th powers in number fields. Mathematika **45**, 359–370 (1998)
1409. Davidson, M.: On Waring's problem in number fields. J. Lond. Math. Soc. **59**, 435–447 (1999)
1410. Davies, D., Haselgrove, C.B.: The evaluation of Dirichlet L-functions. Proc. R. Soc. Lond. Ser. A, Math. Phys. Sci. **264**, 122–132 (1961)
1411. Davis, C.S.: Note on a conjecture of Minkowski. J. Lond. Math. Soc. **23**, 172–175 (1948)
1412. Davis, M.: Arithmetical problems and recursively enumerable predicates. J. Symb. Comput. **18**, 33–41 (1953)

1413. Davis, M.: An explicit diophantine definition of the exponential function. Commun. Pure Appl. Math. **24**, 137–145 (1971)
1414. Davis, M.: Hilbert's tenth problem is unsolvable. Am. Math. Mon. **80**, 233–269 (1973)
1415. Davis, M., Putnam, H.: Reductions of Hilbert's tenth problem. J. Symb. Comput. **23**, 183–187 (1958)
1416. Davis, M., Putnam, H., Robinson, J.: The decision problem for exponential diophantine equations. Ann. Math. **74**, 425–436 (1961)
1417. Davison, J.L.: On the linear diophantine problem of Frobenius. J. Number Theory **48**, 353–363 (1994)
1418. Deaux, R.: René Goormaghtigh. Mathesis **69**, 257–273 (1961)
1419. Decaillot-Laulagnet, A.-M.: Édouard Lucas (1842–1891): le parcours original d'un scientifique français dans la deuxième moitié du XIXe siècle. Thése, Univ. Paris V (1999)
1420. Dedekind, R.: Erläuterungen zu den vorstehenden Fragmenten. In: Riemann, B. (ed.) Gesammelte mathematische Werke, pp. 438–447. Teubner, Leipzig (1876); 2nd ed. 1892; [Reprint: Dover, 1953]; 3rd ed. Springer, Teubner, 1990. [English translation: Collected Papers, Kendrick Press, 2004]
1421. Dedekind, R.: Schreiben an Herrn Borchardt über die Theorie der elliptischen Modulfunktionen. J. Reine Angew. Math. **83**, 265–292 (1877) [[1423], vol. 1, pp. 174–201]
1422. Dedekind, R.: Über die Anzahl von Idealklassen in reinen kubischen Zahlkörpern. J. Reine Angew. Math. **121**, 40–123 (1900) [[1423], vol. 2, pp. 148–233]
1423. Dedekind, R.: Gesammelte mathematische Werke. Vieweg, Wiesbaden (1931). [Reprint: Chelsea, 1968]
1424. de la Bretèche, R., Tenenbaum, G.: Entiers friables: inégalité de Turán-Kubilius et applications. Invent. Math. **159**, 531–588 (2005)
1425. de la Bretèche, R., Tenenbaum, G.: Propriétés statistiques des entiers friables. Ramanujan J. **9**, 139–202 (2005)
1426. Delange, H.: Généralisation du théorème de Ikehara. Ann. Sci. Éc. Norm. Sup. **71**, 213–242 (1954)
1427. Delange, H.: Sur la distribution des entiers ayant certaines propriétés. Ann. Sci. Éc. Norm. Sup. **73**, 15–74 (1956)
1428. Delange, H.: Un théorème sur les fonctions arithmétiques multiplicatives et ses applications. Ann. Sci. Éc. Norm. Sup. **78**, 1–29 (1961)
1429. Delange, H.: Sur les fonctions arithmétiques multiplicatives. Ann. Sci. Éc. Norm. Sup. **78**, 273–304 (1961)
1430. Delange, H.: Sur le nombre des diviseurs premiers de n. Acta Arith. **7**, 191–215 (1962)
1431. Delange, H.: On integral-valued additive functions. J. Number Theory **1**, 419–430 (1969)
1432. Delange, H.: Sur des formules de Atle Selberg. Acta Arith. **19**, 105–146 (1971)
1433. Delange, H.: Sur la distribution des valeurs des fonctions arithmétiques additives. Journées Arithmétiques 1971, exp. 2, 1–13, Marseille 1971
1434. Delange, H.: Sur les fonctions arithmétiques multiplicatives de module ≤ 1. Acta Arith. **42**, 121–151 (1983)
1435. Delange, H.: Generalization of Daboussi's theorem. In: Topics in Classical Number Theory, pp. 305–318. North-Holland, Amsterdam (1984)
1436. Delaunay, B.[8]: La solution générale de l'équation $X^3\varrho + Y^3 = 1$. C. R. Acad. Sci. Paris **162**, 150–151 (1916)
1437. Delaunay, B.: Sur la représentation des nombres par les formes binaires. C. R. Acad. Sci. Paris **178**, 1458–1461 (1924)
1438. Delaunay, B.: Vollständige Lösung der unbestimmten Gleichung $X^3q + Y^3 = 1$ in ganzen Zahlen. Math. Z. **28**, 1–9 (1928)
1439. Delaunay, B.: Über die Darstellung der Zahlen durch die binären kubischen Formen von negativer Diskriminante. Math. Z. **31**, 1–26 (1930).

[8]See also Delone, B.N.

1440. Deléglise, M.: Bounds for the density of abundant integers. Exp. Math. **7**, 137–143 (1998)
1441. Deléglise, M., Dusart, P., Roblot, X.-F.: Counting primes in residue classes. Math. Comput. **73**, 1565–1575 (2004)
1442. Deléglise, M., Hernane, M.O., Nicolas, J.-L.: Grandes valeurs et nombres champions de la fonction arithmétique de Kalmár. J. Number Theory **128**, 1676–1716 (2008)
1443. Deléglise, M., Rivat, J.: Computing $\pi(x)$: the Meissel, Lehmer, Lagarias, Miller, Odlyzko method. Math. Comput. **65**, 235–245 (1996)
1444. Deligne, P.: Formes modulaires et représentations ℓ-adiques. In: Lecture Notes in Math., vol. 179, pp. 139–172. Springer, Berlin (1971)
1445. Deligne, P.: La conjecture de Weil pour les surfaces $K3$. Invent. Math. **15**, 206–226 (1972)
1446. Deligne, P.: Les constantes des équations fonctionnelles des fonctions L. In: Lecture Notes in Math., vol. 349, pp. 501–597. Springer, Berlin (1973)
1447. Deligne, P.: La conjecture de Weil, I. Publ. Math. Inst. Hautes Études Sci. **43**, 273–307 (1974)
1448. Deligne, P.: La conjecture de Weil, II. Publ. Math. Inst. Hautes Études Sci. **52**, 137–252 (1980)
1449. Deligne, P.: Preuve des conjectures de Tate et de Shafarevitch (d'aprés G. Faltings). Astérisque **121–122**, 25–41 (1985)
1450. Deligne, P., Kuijk, W. (eds.): Modular Functions of One Variable II. Lecture Notes in Math., vol. 349. Springer, Berlin (1973)
1451. Deligne, P., Ribet, K.A.: Values of Abelian L-functions at negative integers over totally real fields. Invent. Math. **59**, 227–286 (1980)
1452. Deligne, P., Serre, J.-P.: Formes modulaires de poids 1. Ann. Sci. Éc. Norm. Super. **7**, 507–530 (1974) [[5661], vol. 3, pp. 193–216]
1453. Delmer, F.: Sur la somme de diviseurs $\sum_{k\le x}\{d[f(k)]\}^s$. C. R. Acad. Sci. Paris **272**, A849–A852 (1971)
1454. Delone, B.N.[9]: Solving the indeterminate equation $X^3\varrho + Y^3 = 1$. Comm. Soc. Math. Charkov, (2) **15**, 1915/1916, 1–16, 46–48, 75–76 (in Russian)
1455. Delone, B.N., Faddeev, D.K.: The theory of irrationalities of third degree. Tr. Mat. Inst. Steklova **11**, 1–340 (1940) (in Russian). [English translation, Am. Math. Soc., 1964]
1456. Delone, B.N., Ryškov, S.S.: Solution of the problem on the least dense lattice covering of a 4-dimensional space by equal spheres. Dokl. Akad. Nauk SSSR **152**, 523–524 (1963) (in Russian)
1457. Demyanenko, V.A.: Sums of four cubes. Izv. Vysš. Učeb. Zaved. Mat., 1966, nr. 5, 64–69 (in Russian)
1458. Demyanenko, V.A.: Rational points of a class of algebraic curves. Izv. Akad. Nauk SSSR, Ser. Mat. **30**, 1373–1396 (1966) (in Russian)
1459. Demyanenko, V.A.: Rational points of certain curves of higher genus. Acta Arith. **12**, 333–354 (1966/1967)
1460. Demyanenko, V.A.: The points of finite order of elliptic curves. Acta Arith. **19**, 185–194 (1971)
1461. Demyanenko, V.A.: Torsion of elliptic curves over cyclotomic fields. Algebra Anal. **9**, 51–64 (1997) (in Russian)
1462. Demyanov, V.B.: On cubic forms in discretely normed fields. Dokl. Akad. Nauk SSSR **74**, 889–891 (1950) (in Russian)
1463. Demyanov, V.B.: Pairs of quadratic forms over a complete field with a discrete norm and a finite field of residue classes. Izv. Akad. Nauk SSSR, Ser. Mat. **20**, 307–324 (1956) (in Russian)
1464. Denef, J.: Hilbert's tenth problem for quadratic rings. Proc. Am. Math. Soc. **48**, 214–220 (1975)
1465. Denef, J.: Diophantine sets over algebraic integer rings. II. Trans. Am. Math. Soc. **257**, 227–236 (1980)

[9]See also Delaunay, B.

1466. Denef, J.: The rationality of the Poincaré series associated to p-adic points on a variety. Invent. Math. **77**, 1–23 (1984)
1467. Denef, J.: p-adic semi-algebraic sets and cell decomposition. J. Reine Angew. Math. **369**, 154–166 (1986)
1468. Denef, J.: On the degree of Igusa's local zeta function. Am. J. Math. **109**, 991–1008 (1987)
1469. Denef, J., Lipshitz, L.: Diophantine sets over some rings of algebraic integers. J. Lond. Math. Soc. **18**, 385–391 (1978)
1470. Denef, J., Vercauteren, F.: An extension of Kedlaya's algorithm to hyperelliptic curves in characteristic 2. J. Cryptol. **19**, 1–25 (2006)
1471. Dénes, P.: An extension of Legendre's criterion in connection with the first case of Fermat's last theorem. Publ. Math. (Debr.) **2**, 115–120 (1951)
1472. Dénes, P.: Über die Diophantische Gleichung $x^l + y^l = cz^l$. Acta Math. **88**, 241–251 (1952)
1473. Dénes, P.: Proof of a conjecture of Kummer. Publ. Math. (Debr.) **2**, 206–214 (1952)
1474. Dénes, P., Turán, P.: A second note on Fermat's conjecture. Publ. Math. (Debr.) **4**, 28–32 (1955) [[6227], vol. 1, pp. 760–764]
1475. Derksen, H.: A Skolem-Mahler-Lech theorem in positive characteristic and finite automata. Invent. Math. **168**, 175–224 (2007)
1476. Desboves, A.: Sur l'emploi des identités algébriques dans la résolution, en nombres entiers, des équations d'un degré supérieur au second. C. R. Acad. Sci. Paris **87**, 159–161 (1878)
1477. Descartes, R.: Letter to Mersenne, 27.07.1638, Oeuvres, vol. 2, p. 256. Paris, 1898
1478. Descartes, R.: Letter to Mersenne, 23.08.1638, Oeuvres, vol. 2, pp. 337–338. Paris, 1898
1479. Descombes, R., Poitou, G.: Sur l'approximation dans $R(i\sqrt{11})$. C. R. Acad. Sci. Paris **231**, 264–266 (1950)
1480. Deshouillers, J.-M.: Amélioration de la constante de Šnirelman dans le problème de Goldbach. Sém. Delange–Pisot–Poitou **14**(exp. 14), 1–4 (1975/1976)
1481. Deshouillers, J.-M.: Problème de Waring avec exposants non entiers. Bull. Soc. Math. Fr. **101**, 285–295 (1973)
1482. Deshouillers, J.-M.: Un problème binaire en théorie additive. Acta Arith. **25**, 393–403 (1973/1974)
1483. Deshouillers, J.-M.: Sur la constante de Šnirelman. Sém. Delange–Pisot–Poitou, **17**, 1975/1976, exp. 16
1484. Deshouillers, J.-M.: Nombres premiers de la forme $[n^c]$. C. R. Acad. Sci. Paris **282**, 131–133 (1976)
1485. Deshouillers, J.-M., Dress, F.: Sums of 19 biquadrates: on the representation of large integers. Ann. Sc. Norm. Super. Pisa, Cl. Sci. **19**, 113–153 (1992)
1486. Deshouillers, J.-M., Dress, F.: Numerical results for sums of five and seven biquadrates and consequences for the sums of 19 biquadrates. Math. Comput. **61**, 195–207 (1993)
1487. Deshouillers, J.-M., Effinger, G., te Riele, H., Zinoviev, D.: A complete Vinogradov 3-primes theorem under the Riemann Hypothesis. Electron. Res. Announc. Am. Math. Soc. **3**, 99–104 (1997)
1488. Deshouillers, J.-M., Hennecart, F., Landreau, B.: Waring's problem for sixteen biquadrates—numerical results. J. Théor. Nr. Bordx. **212**, 411–422 (2000)
1489. Deshouillers, J.-M., Iwaniec, H.: An additive divisor problem. J. Lond. Math. Soc. **26**, 1–14 (1982)
1490. Deshouillers, J.-M., Iwaniec, H.: Kloosterman sums and Fourier coefficients of cusp forms. Invent. Math. **70**, 219–288 (1982/1983)
1491. Deshouillers, J.-M., Iwaniec, H.: On the greatest prime factor of $n^2 + 1$. Ann. Inst. Fourier **32**(4), 1–11 (1982)
1492. Deshouillers, J.-M., Iwaniec, H.: On the Brun-Titchmarsh theorem on average. In: Topics in Classical Number Theory, pp. 319–333. North-Holland, Amsterdam (1984)
1493. Deshouillers, J.-M., Hennecart, F., Landreau, B.: 7 373 170 279 850. Math. Comput. **69**, 421–439 (2000)
1494. Deshouillers, J.-M., Kawada, K., Wooley, T.D.: On sums of sixteen biquadrates. Mém. Soc. Math. Fr. **100**, 1–120 (2005)

1495. Deshouillers, J.-M., Plagne, A.: A Sidon basis. Acta Math. Acad. Sci. Hung. **123**, 233–238 (2009)
1496. Deuring, M.: Zur arithmetischen Theorie der algebraischen Funktionen. Math. Ann. **106**, 77–102 (1932)
1497. Deuring, M.: Imaginäre quadratische Zahlkörper mit der Klassenzahl 1. Math. Z. **37**, 405–415 (1933)
1498. Deuring, M.: Zetafunktionen quadratischen Formen. J. Reine Angew. Math. **172**, 226–252 (1935)
1499. Deuring, M.: Über den Tschebotareffschen Dichtigkeitssatz. Math. Ann. **110**, 414–415 (1935)
1500. Deuring, M.: Arithmetische Theorie der Korrespondenzen algebraischer Funktionenkörper, I. J. Reine Angew. Math. **177**, 161–191 (1937)
1501. Deuring, M.: Arithmetische Theorie der Korrespondenzen algebraischer Funktionenkörper, II. J. Reine Angew. Math. **183**, 25–36 (1940)
1502. Deuring, M.: On Epstein's zeta function. Ann. Math. **38**, 584–593 (1937)
1503. Deuring, M.: Die Typen der Multiplikatorenringe elliptischer Funktionenkörper. Abh. Math. Semin. Hansischen Univ. **14**, 197–272 (1941)
1504. Deuring, M.: Algebraische Begründung der komplexen Multiplikation. Abh. Math. Semin. Univ. Hamb. **16**, 32–47 (1949)
1505. Deuring, M.: Die Struktur der elliptischen Funktionenkörper und die Klassenkörper der imaginären quadratischer Zahlkörper. Math. Ann. **124**, 393–426 (1952)
1506. Deuring, M.: Die Zetafunktion einer algebraischen Kurve vom Geschlechte Eins, I. Nachr. Ges. Wiss. Göttingen, 1953, 85–94
1507. Deuring, M.: Die Zetafunktion einer algebraischen Kurve vom Geschlechte Eins, II. Nachr. Ges. Wiss. Göttingen, 1955, 13–42
1508. Deuring, M.: Die Zetafunktion einer algebraischen Kurve vom Geschlechte Eins, III. Nachr. Ges. Wiss. Göttingen, 1956, 37–76
1509. Deuring, M.: Die Zetafunktion einer algebraischen Kurve vom Geschlechte Eins, IV. Nachr. Ges. Wiss. Göttingen, 1957, 55–80
1510. Deuring, M.: Die Klassenkörper der komplexen Multiplikation. In: Enzyklopädie der Mathematischen Wissenschaften, vol. I_2, Heft 10. Teubner, Leipzig (1958)
1511. Deuring, M.: Imaginäre quadratische Zahlkörper mit der Klassenzahl Eins. Invent. Math. **5**, 169–179 (1968)
1512. Deuring, M.: Lectures on the Theory of Algebraic Functions of One Variable. Lecture Notes in Math., vol. 314. Springer, Berlin (1973)
1513. de Weger, B.M.M.: Equal binomial coefficients: some elementary considerations. J. Number Theory **63**, 373–386 (1997)
1514. de Weger, B.M.M.: $A+B=C$ and big III's. Q. J. Math. **49**, 105–128 (1998)
1515. Diaconu, A., Goldfeld, D., Hoffstein, J.: Multiple Dirichlet series and moments of zeta and L-functions. Compos. Math. **139**, 197–360 (2003)
1516. Diaconu, A., Perelli, A., Zaharescu, A.: A note on GL_2 converse theorems. C. R. Acad. Sci. Paris **334**, 621–624 (2002)
1517. Diamond, F.: Congruence primes for cusp forms of weight $k \geq 2$. Astérisque **196/197**, 205–213 (1991)
1518. Diamond, F.: On deformation rings and Hecke rings. Ann. Math. **144**, 137–166 (1996)
1519. Diamond, F., Kramer, K.: Modularity of a family of elliptic curves. Math. Res. Lett. **2**, 299–304 (1995)
1520. Diamond, F., Shurman, J.: A First Course in Modular Forms. Springer, Berlin (2005)
1521. Diamond, H.G.: The prime number theorem for Beurling's generalized numbers. J. Number Theory **1**, 200–207 (1969)
1522. Diamond, H.G.: A set of generalized numbers showing Beurling's theorem to be sharp. Ill. J. Math. **14**, 29–34 (1970)
1523. Diamond, H.G.: Changes of sign of $\pi(x) - \mathrm{li}(x)$. Enseign. Math. **21**, 1–14 (1975)
1524. Diamond, H.G.: When do Beurling generalized integers have a density? J. Reine Angew. Math. **295**, 22–39 (1977)

1525. Diamond, H.G.: Elementary methods in the study of the distribution of prime numbers. Bull. Am. Math. Soc. **7**, 553–589 (1982)
1526. Diamond, H.G., Erdős, P.: On sharp elementary prime number estimates. Enseign. Math. **26**, 313–321 (1980)
1527. Diamond, H.G., Steinig, J.: An elementary proof of the prime number theorem with a remainder term. Invent. Math. **11**, 199–258 (1970)
1528. Diamond, J.: The p-adic log gamma function and p-adic Euler constants. Trans. Am. Math. Soc. **233**, 321–337 (1977)
1529. Diamond, J.: On the values of p-adic L-functions at positive integers. Acta Arith. **35**, 223–237 (1979)
1530. Dias da Silva, J.A., Hamidoune, Y.O.: Cyclic spaces for Grassmann derivatives and additive theory. Bull. Lond. Math. Soc. **26**, 140–146 (1994)
1531. Diaz, G.: Grands degrés de transcendance pour des familles d'exponentielles. J. Number Theory **31**, 1–23 (1989)
1532. Diaz, R., Robins, S.: The Ehrhart polynomial of a lattice polytope. Ann. Math. **145**, 503–518 (1997); err.: vol. 146, 1997, p. 237
1533. Dick, A.: Franz Mertens, 1840–1927. Forschungszentrum Graz, Graz (1981)
1534. Dickmann, K.: On the frequency of numbers containing prime factors of a certain relative magnitude. Ark. Mat. **22**(10), 1–14 (1930)
1535. Dickson, L.E.: The analytic representation of substitutions on a power of a prime number of letters with a discussion of the linear group. Ph.D. thesis, Chicago (1897). Also Ann. Math., **11**, 65–120, 161–183 (1897)
1536. Dickson, L.E.: On higher congruences and modular invariants. Bull. Am. Math. Soc. **14**, 313–318 (1907/1908)
1537. Dickson, L.E.: A new extension of Dirichlet's theorem on prime numbers. Messenger Math. **33**, 155–161 (1903/1904)
1538. Dickson, L.E.: On the last theorem of Fermat. Messenger Math. **38**, 14–32 (1908/1909)
1539. Dickson, L.E.: On the last theorem of Fermat, II. Q. J. Math. **40**, 27–45 (1908)
1540. Dickson, L.E.: Lower limit for the number of sets of solutions of $x^e + y^e + z^e \equiv 0 \pmod{p}$. J. Reine Angew. Math. **135**, 181–188 (1909)
1541. Dickson, L.E.: On the factorization of integral functions with p-adic coefficients. Bull. Am. Math. Soc. **17**, 19–23 (1910/1911)
1542. Dickson, L.E.: On the negative discriminants for which there is a single class of positive primitive binary quadratic forms. Bull. Am. Math. Soc. **17**, 534–537 (1910/1911)
1543. Dickson, L.E.: Finiteness of the odd perfect and primitive abundant numbers with n distinct prime factors. Am. J. Math. **35**, 413–422 (1913)
1544. Dickson, L.E.: Even abundant numbers. Am. J. Math. **35**, 423–426 (1913)
1545. Dickson, L.E.: History of the Theory of Numbers. Carnegie Institution of Washington, Washington (1919) [Reprints: Chelsea, 1952, 1966]
1546. Dickson, L.E.: Integers represented by positive ternary quadratic forms. Bull. Am. Math. Soc. **33**, 63–70 (1927)
1547. Dickson, L.E.: Quaternary quadratic forms representing all integers. Am. J. Math. **49**, 39–56 (1927)
1548. Dickson, L.E.: Generalizations of Waring's theorem on fourth, sixth and eighth powers. Am. J. Math. **49**, 241–250 (1927)
1549. Dickson, L.E.: All positive integers are sums of values of a quadratic function of x. Bull. Am. Math. Soc. **33**, 713–720 (1927)
1550. Dickson, L.E.: Ternary quadratic forms and congruences. Ann. Math. **28**, 333–341 (1927)
1551. Dickson, L.E.: Simpler proofs of Waring's theorem for cubes, with various generalizations. Trans. Am. Math. Soc. **30**, 1–18 (1928)
1552. Dickson, L.E.: Additive number theory for all quadratic functions. Am. J. Math. **50**, 1–48 (1928)
1553. Dickson, L.E.: Quadratic functions of forms, sums of whose values give all positive integers. J. Math. Pures Appl. **7**, 319–336 (1928)

1554. Dickson, L.E.: Studies in the Theory of Numbers. University of Chicago Press, Chicago (1930)
1555. Dickson, L.E.: Proof of a Waring theorem on fifth powers. Bull. Am. Math. Soc. **37**, 549–553 (1931)
1556. Dickson, L.E.: Recent progress on Waring's theorem and its generalizations. Bull. Am. Math. Soc. **39**, 701–727 (1933)
1557. Dickson, L.E.: A new method for universal Waring theorems with details for seventh powers. Am. Math. Mon. **41**, 547–555 (1934)
1558. Dickson, L.E.: Polygonal numbers and related Waring problems. Q. J. Math. **5**, 283–290 (1934)
1559. Dickson, L.E.: Waring's problem for ninth powers. Bull. Am. Math. Soc. **40**, 487–493 (1934)
1560. Dickson, L.E.: Universal Waring theorem for eleventh powers. J. Lond. Math. Soc. **9**, 201–206 (1934)
1561. Dickson, L.E.: Cyclotomy, higher congruences, and Waring's problem, II. Am. J. Math. **57**, 463–474 (1935)
1562. Dickson, L.E.: On Waring's problem and its generalization. Ann. Math. **37**, 293–316 (1936)
1563. Dickson, L.E.: The Ideal Waring Theorem for twelfth powers. Duke Math. J. **2**, 192–204 (1936)
1564. Dickson, L.E.: Proof of the Ideal Waring Theorem for exponents 7–180. Am. J. Math. **58**, 521–529 (1936)
1565. Dickson, L.E.: Solution of Waring's problem. Am. J. Math. **58**, 530–535 (1936)
1566. Dickson, L.E.: A generalization of Waring's problem. Bull. Am. Math. Soc. **42**, 525–529 (1936)
1567. Dickson, L.E.: All integers except 23 and 239 are sums of eight cubes. Bull. Am. Math. Soc. **45**, 588–591 (1939)
1568. Dickson, L.E.: Hans Frederik Blichfeldt, 1873–1945. Bull. Am. Math. Soc. **53**, 882–883 (1947)
1569. Dickson, L.E., Mitchell, H.H., Vandiver, H.S., Wahlin, G.E.: Algebraic Numbers, Report of the Committee on Algebraic Numbers. National Research Council, Washington (1923)
1570. Dietmann, R.: Simultaneous Diophantine approximation by square-free numbers. Q. J. Math. **59**, 311–319 (2008)
1571. Dieudonné, J.: Jacques Herbrand et la théorie des nombres. In: Proceedings of the Herbrand Symposium, Marseilles, 1981, pp. 3–7. North-Holland, Amsterdam (1982)
1572. Dieudonné, J., Tits, J.: Claude Chevalley (1909–1984). Bull. Am. Math. Soc. **17**, 1–7 (1987)
1573. Dieulefait, L.: The level 1 weight 2 case of Serre's conjecture. Rev. Math. Iberoam. **23**, 1115–1124 (2007)
1574. Ding, P.: An improvement to Chalk's estimation of exponential sums. Acta Arith. **59**, 149–155 (1991)
1575. Ding, P.: On a conjecture of Chalk. J. Number Theory **65**, 116–129 (1997)
1576. Ding, P.: On an exponential sum. Zap. Nauč. Semin. POMI **322**, 63–75 (2005)
1577. Ding, P., Qi, M.G.: Further estimate of complete trigonometric sums. J. Tsinghua Univ. **29**, 74–85 (1989)
1578. Ding, S., Kang, M.-C., Tan, E.-T.: Chiungtze C. Tsen (1898–1940) and Tsen's theorems. Rocky Mt. J. Math. **29**, 1237–1269 (1999)
1579. Dinghas, A.: Erhard Schmidt (Erinnerungen und Werk). Jahresber. Dtsch. Math.-Ver. **72**, 3–17 (1970/1971)
1580. Dipert, R.R.: The life and logical contribution of O.H. Mitchell, Peirce's gifted student. Trans. C.S. Peirce's Soc. **30**, 515–542 (1994)
1581. Dirac, G.A.: Percy John Heawood. J. Lond. Math. Soc. **38**, 263–277 (1963)
1582. Dirichlet, P.G.L.: Mémoire sur l'impossibilité de quelques équations indéterminées du cinquiéme degré. J. Reine Angew. Math. **3**, 354–375 (1828) [[1593], pp. 21–46]
1583. Dirichlet, P.G.L.: Démonstration du théorème de Fermat pour le cas de 14iémes puissances. J. Reine Angew. Math. **9**, 390–393 (1832) [[1593], pp. 189–194]

1584. Dirichlet, P.G.L.: Beweis eines Satzes ueber die arithmetische Progression. Ber. Verhandl. Kgl. Preuß. Akad. Wiss., 1837, 108–110 [[1593], pp. 307–312]
1585. Dirichlet, P.G.L.: Beweis des Satzes, dass jede unbegrenzte arithmetische Progression, deren erstes Glied und Differenz ganze Zahlen ohne gemeinschaftlichen Factor sind, unendlich viele Primzahlen enthält. Abhandl. Kgl. Preuß. Akad. Wiss., 1837, 45–81 [[1593], pp. 313–342; French translation: J. Math. Pures Appl. **4**, 393–422 (1839)]
1586. Dirichlet, P.G.L.: Sur l'usage des séries infinies dans la théorie des nombres. J. Reine Angew. Math. **18**, 259–274 (1838) [[1593], pp. 357–374]
1587. Dirichlet, P.G.L.: Recherches sur diverses applications de l'analyse infinitésimale à la théorie des nombres. J. Reine Angew. Math. **19**, 324–369 (1839), **21**, 1–12, 134–155 (1840) [[1593], pp. 411–496]
1588. Dirichlet, P.G.L.: Über eine Eigenschaft der quadratischen Formen. Ber. Verhandl. Kgl. Preuß. Akad. Wiss., 1840, 49–52 [[1593], pp. 497–502]
1589. Dirichlet, P.G.L.: Verallgemeinerung eines Satzes aus der Lehre von den Kettenbrüchen nebst einigen Anwendungen auf die Theorie der Zahlen. SBer. Kgl. Preuß. Akad. Wiss. Berlin, 1842, 93–95 [[1593], pp. 633–638]
1590. Dirichlet, P.G.L.: Über die Bestimmung der mittleren Werte in der Zahlentheorie. Abh. Kgl. Preuß. Akad. Wiss. Berlin, 1849, 69–83 [[1594], pp. 49–66; French translation: J. Math. Pures Appl., (2) **1**, 353–370 (1856)]
1591. Dirichlet, P.G.L.: Letter to Kronecker (23.07.1858) [[1594], pp. 406–408]
1592. Dirichlet, P.G.L.: Vorlesungen über Zahlentheorie. Springer, Berlin (1863), 2nd ed. 1871, 3rd ed. 1879, 4th ed. 1894 [Reprint: Chelsea, 1968; English translation: Am. Math. Soc., 1999]
1593. Dirichlet, P.G.L.: Werke, vol. 1, Springer, Berlin (1889) [Reprint: Chelsea, 1969]
1594. Dirichlet, P.G.L.: Werke, vol. 2, Springer, Berlin (1897) [Reprint: Chelsea, 1969]
1595. Diviš, B.: On the sums of continued fractions. Acta Arith. **22**, 157–173 (1973)
1596. Dixon, A.L., Ferrar, W.L.: Lattice points summation formulae. Q. J. Math. **2**, 31–54 (1931)
1597. Dixon, J.D.: The number of steps in the Euclidean algorithm. J. Number Theory **2**, 414–422 (1970)
1598. Dixon, J.D.: Asymptotically fast factorization of integers. Math. Comput. **36**, 255–260 (1981)
1599. Dobrowolski, E.: On the maximal modulus of conjugates of an algebraic integer. Bull. Acad. Pol. Sci., Sér. Sci. Math. Astr. Phys. **26**, 291–292 (1978)
1600. Dobrowolski, E.: On a question of Lehmer and the number of irreducible factors of a polynomial. Acta Arith. **34**, 391–401 (1979)
1601. Dobrowolski, E.: Mahler's measure of a polynomial in function of the number of its coefficients. Can. Math. Bull. **34**, 186–195 (1991)
1602. Dobrowolski, E.: Mahler's measure of a polynomial in terms of the number of its monomials. Acta Arith. **123**, 201–231 (2006)
1603. Dobrowolski, E., Lawton, W., Schinzel, A.: On a problem of Lehmer. In: Studies in Pure Mathematics, pp. 135–144. Birkhäuser, Basel (1983)
1604. Dobrowolski, E., Williams, K.S.: An upper bound for the sum $\sum_{n=a+1}^{a+H} f(n)$ for a certain class of functions f. Proc. Am. Math. Soc. **114**, 29–35 (1992)
1605. Dodson, M.: Homogeneous additive congruences. Philos. Trans. R. Soc. Lond. Ser. A, Math. Phys. Sci. **261**, 163–210 (1967)
1606. Dodson, M.: The average order of two arithmetical functions. Acta Arith. **16**, 71–84 (1969/1970)
1607. Dodson, M.: On Waring's problem in $GF[p]$. Acta Arith. **19**, 147–173 (1971)
1608. Dodson, M.: On Waring's problem in p-adic fields. Acta Arith. **22**, 315–327 (1972/1973)
1609. Dodson, M.: Some estimates for diagonal equations over $\mathfrak{p}$-adic fields. Acta Arith. **40**, 117–124 (1981/1982)
1610. Dodson, M., Tietäväinen, A.: A note on Waring's problem in $GF(p)$. Acta Arith. **30**, 159–167 (1976)
1611. Domar, Y.: On the Diophantine equation $|Ax^n - By^n| = 1$, $n \geq 5$. Math. Scand. **2**, 29–32 (1954)

1612. Dong, X., Shiu, W.C., Chu, C.I., Cao, Z.: The simultaneous Pell equations $y^2 - Dz^2 = 1$ and $x^2 - 2Dz^2 = 1$. Acta Arith. **126**, 115–123 (2007)
1613. Dörge, K.: Zur Verteilung der quadratischen Reste. Jahresber. Dtsch. Math.-Ver. **38**, 41–49 (1929)
1614. Dörner, E.: Simultaneous diagonal equations over certain p-adic fields. J. Number Theory **36**, 1–11 (1990)
1615. Dorwart, H.L.: Concerning certain reducible polynomials. Duke Math. J. **1**, 70–73 (1935)
1616. Dorwart, H.L., Brown, O.E.: The Tarry-Escott problem. Am. Math. Mon. **44**, 613–626 (1937)
1617. Doud, D.: A procedure to calculate torsion of elliptic curves over **Q**. Manuscr. Math. **95**, 463–469 (1998)
1618. Dowd, M.: Questions related to the Erdös-Turán conjecture. SIAM J. Discrete Math. **1**, 142–150 (1988)
1619. Dransfield, M.R., Liu, L., Marek, V.W., Truszczyński, M.: Satisfiability and computing van der Waerden numbers. Electron. J. Comb. **11**, #R41 (2004)
1620. Dress, F.: Amélioration de la majoration de $g(4)$ dans le probleme de Waring: $g(4) \leq 34$. Sém. Delange–Pisot–Poitou, 1969/70, exp. 15, 1–23
1621. Dress, F.: Intersections d'ensembles normaux. J. Number Theory **2**, 352–362 (1970)
1622. Dress, F.: Amélioration de la majoration de $g(4)$ dans le probleme de Waring: $g(4) \leq 30$. Acta Arith. **22**, 137–147 (1973)
1623. Dress, F., Mendès France, M.: Caractérisation des ensembles normaux dans Z. Acta Arith. **17**, 115–120 (1970)
1624. Dress, F., Olivier, M.: Polynômes prenant des valeurs premières. Exp. Math. **8**, 319–338 (1999)
1625. Dressler, R.E.: A density which counts multiplicity. Pac. J. Math. **34**, 371–378 (1970)
1626. Drmota, M., Tichy, R.F.: Sequences, Discrepancies and Applications. In: Lecture Notes in Math., vol. 1651, Springer, Berlin (1997)
1627. Dubickas, A.: A lower bound on the value of $\|(3/2)^k\|$. Usp. Mat. Nauk **45**(4), 153–154 (1990)
1628. Dubickas, A.: On a conjecture of A. Schinzel and H. Zassenhaus. Acta Arith. **63**, 15–20 (1993)
1629. Dubickas, A.: On the maximal conjugate of a totally real algebraic integer. Liet. Mat. Rink. **37**, 13–19 (1997)
1630. Dubickas, A.: The maximal conjugate of a non-reciprocal algebraic integer. Liet. Mat. Rink. **37**, 168–174 (1997)
1631. Dubickas, A., Mossinghoff, M.J.: Auxiliary polynomials for some problems regarding Mahler's measure. Acta Arith. **119**, 65–79 (2005)
1632. Dubois, E., Rhin, G.: Sur la majoration de formes linéaires à coefficients algébriques réels et p-adiques. Démonstration d'une conjecture de K. Mahler. C. R. Acad. Sci. Paris **282**, 1211–1214 (1976)
1633. Dubouis, E.: Solution d'un problémè de J. Tannery. L'Intermédiaire des Math. **18**, 55–56, 224–225 (1911)
1634. Duffin, R.J., Schaeffer, A.C.: Khintchine's problem in metric Diophantine approximation. Duke Math. J. **8**, 243–255 (1941)
1635. Dufresnoy, J., Pisot, C.: Sur les dérivés successifs d'un ensemble fermé d'entiers algébriques. Bull. Sci. Math. **77**, 129–136 (1953)
1636. Dufresnoy, J., Pisot, C.: Sur un ensemble fermé d'entiers algébriques. Ann. Sci. Éc. Norm. Sup. **70**, 105–133 (1953)
1637. Dufresnoy, J., Pisot, C.: Sur les petits éléments d'un ensemble remarquable d'entiers algébriques. C. R. Acad. Sci. Paris **238**, 1551–1553 (1954)
1638. Dufresnoy, J., Pisot, C.: Étude de certaines fonctions méromorphes bornées sur le cercle unité. Application à un ensemble fermé d'entiers algébriques. Ann. Sci. Éc. Norm. Sup. **72**, 69–92 (1955)
1639. Dujella, A.: Generalization of a problem of Diophantus. Acta Arith. **65**, 15–27 (1993)

1640. Dujella, A.: On Diophantine quintuples. Acta Arith. **81**, 69–79 (1997)
1641. Dujella, A.: An absolute bound for the size of Diophantine m-tuples. J. Number Theory **89**, 126–150 (2001)
1642. Dujella, A.: On the size of Diophantine m-tuples. Math. Proc. Camb. Philos. Soc. **132**, 23–33 (2002)
1643. Dujella, A.: There are only finitely many Diophantine quintuples. J. Reine Angew. Math. **566**, 183–214 (2004)
1644. Dujella, A.: Bounds for the size of sets with the property $D(n)$. Glas. Mat. **39**, 199–205 (2004)
1645. Dujella, A., Filipin, A., Fuchs, C.: Effective solution of the $D(-1)$-quadruple conjecture. Acta Arith. **128**, 319–338 (2007)
1646. Dujella, A., Fuchs, C.: Complete solution of a problem of Diophantus and Euler. J. Lond. Math. Soc. **71**, 33–52 (2005)
1647. Dujella, A., Jadrijević, B.: A parametric family of quartic Thue equations. Acta Arith. **101**, 159–170 (2002)
1648. Dujella, A., Pethő, A.: A generalization of a theorem of Baker and Davenport. Q. J. Math. **49**, 291–306 (1998)
1649. Duke, W., Friedlander, J.B., Iwaniec, H.: Equidistribution of roots of a quadratic congruence to prime moduli. Ann. Math. **141**, 423–441 (1995)
1650. Duke, W., Schulze-Pillot, R.: Representation of integers by positive ternary quadratic forms and equidistribution of lattice points on ellipsoids. Invent. Math. **99**, 49–57 (1990)
1651. Dumir, V.C., Hans-Gill, R.J.: Mathematical contributions of Professor Hansraj Gupta. J. Indian Math. Soc. **57**, 11–16 (1991)
1652. Dunford, N.: Einar Hille (June 28, 1894–February 12, 1980). Bull. Am. Math. Soc. **4**, 303–319 (1981)
1653. Dunnington, G.W.: Carl Friedrich Gauss. Titan of Science. Exposition Press, New York (1955) [Reprint: Math. Assoc. of America, 2004]
1654. Duparc, H.J.A., van Wijngaarden, A.: A remark on Fermat's last theorem. Nieuw Arch. Wiskd. **1**, 123–128 (1953)
1655. Dupré, A.: Examen d'une proposition de Legendre relative à la théorie des nombres. Mallet-Bachelier, Paris (1859)
1656. Durand, A.: Quatre problèmes de Mahler sur la fonction ordre d'un nombre transcendant. Bull. Soc. Math. Fr. **102**, 365–377 (1974)
1657. Durst, L.K.: Exceptional real Lehmer sequences. Pac. J. Math. **9**, 437–441 (1959)
1658. Durst, L.K.: Exceptional real Lucas sequences. Pac. J. Math. **11**, 489–494 (1961)
1659. Dusart, P.: The kth prime is greater than $k(\ln k + \ln\ln k - 1)$ for $k \geq 2$. Math. Comput. **68**, 411–415 (1999)
1660. Dusart, P.: Inégalités explicites pour $\psi(X)$, $\theta(X)$, $\pi(X)$ et les nombres premiers. C. R. Acad. Sci. Paris **21**, 53–59 (1999)
1661. Dusart, P.: Estimates of $\theta(x; k, l)$ for large values of x. Math. Comput. **71**, 1137–1168 (2002)
1662. Dusart, P.: Sur la conjecture $\pi(x + y) \leq \pi(x) + \pi(y)$. Acta Arith. **102**, 295–308 (2002)
1663. Dvornicich, R., Zannier, U.: Local-global divisibility of rational points in some commutative algebraic groups. Bull. Soc. Math. Fr. **129**, 317–338 (2001)
1664. Dvornicich, R., Zannier, U.: An analogue for elliptic curves of the Grunwald-Wang example. C. R. Acad. Sci. Paris **338**, 47–50 (2004)
1665. Dvornicich, R., Zannier, U.: On a local-global principle for the divisibility of a rational point by a positive integer. Bull. Lond. Math. Soc. **39**, 27–34 (2007)
1666. Dwork, B.: On the Artin root number. Am. J. Math. **78**, 444–472 (1956)
1667. Dwork, B.: On the rationality of the zeta function of an algebraic variety. Am. J. Math. **82**, 631–648 (1960)
1668. Dwork, B.: On the zeta function of a hypersurface. Publ. Math. Inst. Hautes Études Sci. **12**, 5–68 (1962)
1669. Dwork, B.: On the zeta function of a hypersurface, II. Ann. Math. **80**, 227–299 (1964)

1670. Dwork, B.: On the zeta function of a hypersurface, III. Ann. Math. **83**, 457–519 (1966)
1671. Dwork, B.: On the zeta function of a hypersurface, IV. Ann. Math. **90**, 335–352 (1969)
1672. Dwork, B.: A note on the p-adic gamma function. Groupe de travail d'analyse ultramétrique **9**(exp. 5), 1–10 (1981/1982)
1673. Dwork, B.: Generalized Hypergeometric Functions. Oxford University Press, Oxford (1990)
1674. Dyson, F.J.: Three identities in combinatory analysis. J. Lond. Math. Soc. **18**, 35–39 (1943)
1675. Dyson, F.J.: Some guesses in the theory of partitions. Eureka **8**, 10–15 (1944)
1676. Dyson, F.J.: A theorem on the densities of sets of integers. J. Lond. Math. Soc. **20**, 8–14 (1945)
1677. Dyson, F.J.: The approximation to algebraic numbers by rationals. Acta Math. **79**, 225–240 (1947)
1678. Dyson, F.J.: On the product of four non-homogeneous linear forms. Ann. Math. **49**, 82–109 (1948)
1679. Earnest, A.G., Estes, D.R.: An algebraic approach to the growth of class numbers of binary quadratic lattices. Mathematika **28**, 160–168 (1981)
1680. Earnest, A.G., Körner, O.H.: On ideal class groups of 2-power exponent. Proc. Am. Math. Soc. **86**, 196–198 (1982)
1681. Ebeling, W.: Lattices and Codes. Vieweg, Leipzig (1994); 2nd ed. 2002
1682. Ecklund, E.F. Jr., Erdős, P., Selfridge, J.L.: A new function associated with the prime factors of $\binom{n}{k}$. Math. Comput. **28**, 647–649 (1974)
1683. Eda, Y.: On the Waring problem in an algebraic number field. In: Seminar on Modern Methods in Number Theory, Tokyo (1971), 11 pp.
1684. Eda, Y., Nakagoshi, N.: An elementary proof of the prime ideal theorem with remainder term. Sci. Rep. Kanazawa Univ. **12**, 1–12 (1967)
1685. Edel, Y., Rains, E.M., Sloane, N.J.A.: On kissing numbers in dimensions 32 to 128. Electron. J. Comb. **5**(22), 1–5 (1998)
1686. Edgorov, Ž.: The divisor problem in special arithmetic progressions. Izv. Akad. Nauk Uzb. SSR., Ser. Fiz.-Mat. Nauk, 1977, nr. 2, 9–13 (in Russian)
1687. Edixhoven, B.: The weight in Serre's conjectures on modular forms. Invent. Math. **109**, 563–594 (1992)
1688. Edixhoven, B.: Rational torsion points on elliptic curves over number fields (after Kamienny and Mazur). Astérisque **227**, 209–227 (1995)
1689. Edixhoven, B.: Serre's conjecture. In: Cornell, G., Silverman, J.H., Stevens, S. (eds.) Modular Forms and Fermat's Last Theorem, pp. 209–242. Springer, Berlin (1997)
1690. Edwards, H.M.: Fermat's Last Theorem, A Genetic Introduction to Algebraic Number Theory. Springer, Berlin (1977) [Reprint: 1996]
1691. Edwards, H.M.: Riemann's Zeta Function. Academic Press, San Diego (1974) [Reprint: Dover 2001]
1692. Eggleston, H.G.: The fractional dimension of a set defined by decimal properties. Q. J. Math. **20**, 31–36 (1949)
1693. Ehlich, H.: Ein elementarer Beweis des Primzahlsatzes für binäre quadratische Formen. J. Reine Angew. Math. **201**, 1–36 (1959)
1694. Ehrhart, E.: Sur les polyèdres rationnels homothétiques à n dimensions. C. R. Acad. Sci. Paris **254**, 616–618 (1962)
1695. Ehrhart, E.: Sur un problème de géométrie diophantienne linéaire. I. Polyèdres et réseaux. J. Reine Angew. Math. **226**, 1–29 (1967)
1696. Ehrhart, E.: Sur un problème de géométrie diophantienne linéaire. II, Systèmes diophantiens linéaires. J. Reine Angew. Math. **227**, 25–49 (1967)
1697. Ehrhart, E.: Polynômes arithmétiques et méthode des polyèdres en combinatoire. Institut de recherche mathématique avancée, Strasbourg (1974); 2nd ed. Birkhäuser, 1977.
1698. Eichhorn, D., Ono, K.: Congruences for partition functions. In: Contemp. Math., vol. 138, pp. 309–321. Am. Math. Soc., Providence (1996)
1699. Eichler, M.: Quaternäre quadratische Formen und die Riemannsche Vermutung für die Kongruenzzetafunktion. Arch. Math. **5**, 355–366 (1954)

1700. Eichler, M.: Über die Darstellbarkeit von Modulformen durch Thetareihen. J. Reine Angew. Math. **195**, 156–171 (1955)
1701. Eichler, M.: Eine Bemerkung zur Fermatschen Vermutung. Acta Arith. **11**, 129–131 (1965); corr. p. 261
1702. Eichler, M.: The basis problem for modular forms and the traces of the Hecke operators. In: Lecture Notes in Math., vol. 320, pp. 75–151. Springer, Berlin (1973); corr. vol. 476, pp. 145–147 (1975)
1703. Eichler, M.: Theta functions over Q and over $Q(\sqrt{q})$. In: Lecture Notes in Math., vol. 627, pp. 197–225. Springer, Berlin (1977)
1704. Eichler, M.: Das wissenschaftliche Werk von Max Deuring. Acta Arith. **47**, 187–192 (1986)
1705. Eichler, M.: Alexander Ostrowski. Über sein Leben und Werk. Acta Arith. **51**, 295–298 (1988)
1706. Eichler, M., Zagier, D.: The Theory of Jacobi Forms. Progr. Math., **55**, 1985
1707. Einsiedler, M., Katok, A., Lindenstrauss, E.: Invariant measures and the set of exceptions to Littlewood's conjecture. Ann. Math. **164**, 513–560 (2006)
1708. Eisenstein, G.: Beweis des Reciprocitätssatzes für die cubischen Reste in der Theorie der aus dritten Wurzeln der Einheit zusammengesetzten complexen Zahlen. J. Reine Angew. Math. **27**, 289–310 (1844)
1709. Eisenstein, G.: Neue Theoreme der höheren Arithmetik. J. Reine Angew. Math. **35**, 117–136 (1847)
1710. Eisenstein, G.: Note sur la représentation d'un nombre par la somme de cinq carrés. J. Reine Angew. Math. **35**, 368 (1847)
1711. Eisenstein, G.: Über ein einfaches Mittel zur Auffindung der höheren Reziprozitätsgesetze und der mit ihnen verbundenen Ergänzungssätzen. J. Reine Angew. Math. **39**, 351–364 (1850)
1712. Eliahou, S., et al.: Claude Michel Kervaire (26 avril 1927–19 novembre 2007). Math. Gaz. **116**, 77–82 (2009)
1713. Elkies, N.[D.]: Supersingular primes of a given elliptic curve over a number field. Ph.D. thesis, Harvard Univ. (1987)
1714. Elkies, N.D.: The existence of infinitely many supersingular primes for every elliptic curve over Q. Invent. Math. **89**, 561–567 (1987)
1715. Elkies, N.D.: On $A^4 + B^4 + C^4 = D^4$. Math. Comput. **51**, 825–835 (1988)
1716. Elkies, N.D.: Supersingular primes for elliptic curves over real number fields. Compos. Math. **72**, 165–172 (1989)
1717. Elkies, N.D.: ABC implies Mordell. Int. Math. Res. Not. **7**, 99–109 (1991)
1718. Elkies, N.D.: Mordell-Weil lattices in characteristic 2. I. Construction and first properties. Internat. Math. Res. Notices, 1994, 343–361
1719. Elkies, N.D.: Mordell-Weil lattices in characteristic 2. II. The Leech lattice as a Mordell-Weil lattice. Invent. Math. **128**, 1–8 (1997)
1720. Elkies, N.D.: Mordell-Weil lattices in characteristic 2. III. A Mordell-Weil lattice of rank 128. Exp. Math. **10**, 467–473 (2001)
1721. Elkies, N.D.: Heegner point computation. In: Lecture Notes in Comput. Sci., vol. 877, pp. 122–133. Springer, Berlin (1994)
1722. Elkies, N.D.: Wiles minus epsilon implies Fermat. In: Elliptic Curves, Modular Forms, & Fermat's Last Theorem, Hong Kong, 1993, pp. 38–40. Cambridge University Press, Cambridge (1995)
1723. Elkies, N.D.: Elliptic and modular curves over finite fields and related computational issues. In: Computational Perspectives on Number Theory, Chicago, IL, 1995, pp. 21–76. Am. Math. Soc., Providence (1998)
1724. Elkies, N.D.: Three lectures on elliptic surfaces and curves of high rank. Preprint, arXiv:0709.2908
1725. Elkies, N.D., Gross, B.H.: The exceptional cone and the Leech lattice. Internat. Math. Res. Notices, 1996, 665–698
1726. Elkies, N.D., Watkins, M.: Elliptic curves of large rank and small conductor. In: Lecture Notes in Comput. Sci., vol. 3076, pp. 42–56. Springer, Berlin (2004)

1727. Elliott, P.D.T.A.: A note on a recent paper of U.V. Linnik and A.I. Vinogradov. Acta Arith. **13**, 103–105 (1967/1968)
1728. Elliott, P.D.T.A.: The distribution of primitive roots. Can. J. Math. **21**, 822–841 (1969)
1729. Elliott, P.D.T.A.: On the size of $L(1,\chi)$. J. Reine Angew. Math. **236**, 26–36 (1969)
1730. Elliott, P.D.T.A.: The Turán-Kubilius inequality, and a limitation theorem for the large sieve. Am. J. Math. **92**, 293–300 (1970)
1731. Elliott, P.D.T.A.: On the mean value of $f(p)$. Proc. Lond. Math. Soc. **21**, 28–96 (1970)
1732. Elliott, P.D.T.A.: On the least pair of consecutive quadratic non-residues (mod p). In: Proceedings of the Number Theory Conference, Boulder, CO, 1972, pp. 75–79. University of Colorado Press, Boulder (1972)
1733. Elliott, P.D.T.A.: On a problem of Hardy and Ramanujan. Mathematika **23**, 10–17 (1976)
1734. Elliott, P.D.T.A.: The Turán-Kubilius inequality. Proc. Am. Math. Soc. **65**, 8–10 (1977)
1735. Elliott, P.D.T.A.: Probabilistic Number Theory, I. Springer, Berlin (1979)
1736. Elliott, P.D.T.A.: Probabilistic Number Theory, II. Springer, Berlin (1980)
1737. Elliott, P.D.T.A.: On sums of an additive arithmetic function with shifted arguments. J. Lond. Math. Soc. **22**, 25–38 (1980)
1738. Elliott, P.D.T.A.: On additive arithmetic functions $f(n)$ for which $f(an+b)-f(cn+d)$ is bounded. J. Number Theory **16**, 285–310 (1983)
1739. Elliott, P.D.T.A.: Arithmetic Functions and Integer Products. Springer, Berlin (1985)
1740. Elliott, P.D.T.A.: Multiplicative functions on arithmetic progressions. Mathematika **34**, 199–206 (1987)
1741. Elliott, P.D.T.A.: Multiplicative functions on arithmetic progressions, II. Mathematika **35**, 38–50 (1988)
1742. Elliott, P.D.T.A.: Multiplicative functions on arithmetic progressions, III. The large moduli. In: A tribute to Paul Erdős, pp. 177–194. Cambridge University Press, Cambridge (1990)
1743. Elliott, P.D.T.A.: Multiplicative functions on arithmetic progressions, IV. The middle moduli. J. Lond. Math. Soc. **41**, 201–216 (1990)
1744. Elliott, P.D.T.A.: Multiplicative functions on arithmetic progressions, V. Composite moduli. J. Lond. Math. Soc. **41**, 408–424 (1990)
1745. Elliott, P.D.T.A.: Multiplicative functions on arithmetic progressions, VI. More middle moduli. J. Number Theory **44**, 178–208 (1993)
1746. Elliott, P.D.T.A.: Multiplicative functions on arithmetic progressions, VII. Large moduli. J. Lond. Math. Soc. **66**, 14–28 (2002)
1747. Elliott, P.D.T.A.: The least prime primitive root and Linnik's theorem. In: Number Theory for the Millennium, vol. 1, pp. 393–418. AK Peters, Wellesley (2002)
1748. Elliott, P.D.T.A.: Monotone additive functions on shifted primes. Ramanujan J. **15**, 87–102 (2008)
1749. Elliott, P.D.T.A., Halberstam, H.: Some applications of Bombieri's theorem. Mathematika **13**, 196–203 (1966)
1750. Elliott, P.D.T.A., Murata, L.: On the average of the least primitive root modulo p. J. Lond. Math. Soc. **56**, 435–454 (1997)
1751. Elliott, P.D.T.A., Ryavec, C.: The distribution of the values of additive arithmetical functions. Acta Math. **126**, 143–164 (1971)
1752. Ellison, F.: Three diagonal quadratic forms. Acta Arith. **23**, 137–151 (1973)
1753. Ellison, W.J.: On a theorem of Srivasankanarayama Pillai. Sém. Théor. Nombres Bordeaux, 1970/1971, nr. 12, 1–10
1754. Ellison, W.J.: Waring's problem. Am. Math. Mon. **78**, 10–36 (1971)
1755. Ellison, W.J. (in collaboration with M. Mendès France): Les nombres premiers. Hermann, 1975 [English version (with F. Ellison): Prime Numbers, Wiley & Hermann, 1985]
1756. Ellison, W.J., Ellison, F., Pesek, J., Stahl, C.E., Stall, D.S.: The diophantine equation $y^2+k=x^3$. J. Number Theory **4**, 107–117 (1972)
1757. Ellison, W.J., Pesek, J., Stall, D.S., Lunnon, W.F.: A postscript to a paper of A. Baker. Bull. Lond. Math. Soc. **3**, 75–78 (1971)
1758. Elsenhans, A.-S., Jahnel, J.: New sums of three cubes. Math. Comput. **78**, 1227–1230 (2009)

1759. Elsholtz, C.: Sums of k unit fractions. Trans. Am. Math. Soc. **353**, 3209–3227 (2001)
1760. Elsholtz, C.: The number $\Gamma(k)$ in Waring's problem. Acta Arith. **131**, 43–49 (2008)
1761. Elstrodt, J., Grunewald, F.: The Petersson scalar product. Jahresber. Dtsch. Math.-Ver. **100**, 253–283 (1998)
1762. Emerton, M.: p-adic L-functions and unitary completions of representations of p-adic reductive groups. Duke Math. J. **130**, 353–392 (2005)
1763. Endler, O.: Valuation Theory. Springer, Berlin (1972)
1764. Endo, M.: p-adic multiple gamma functions. Comment. Math. Univ. St. Pauli **43**, 35–54 (1994)
1765. Engstrom, T.: On sequences defined by linear recurrence relations. Trans. Am. Math. Soc. **33**, 210–218 (1931)
1766. Ennola, V.: On the lattice constant of a symmetric convex domain. J. Lond. Math. Soc. **36**, 135–138 (1961)
1767. Ennola, V.: A note on a divisor problem. Ann. Univ. Turku **118**, 1–11 (1968)
1768. Epstein, P.: Zur Theorie allgemeiner Zetafunctionen. Math. Ann. **56**, 615–644 (1903)
1769. Epstein, P.: Zur Theorie allgemeiner Zetafunctionen, II. Math. Ann. **63**, 205–216 (1907)
1770. Epstein, P.: Zur Auflösbarkeit der Gleichung $x^2 - Dy^2 = -1$. J. Reine Angew. Math. **171**, 243–252 (1934)
1771. Erdős, P.: Beweis eines Satzes von Tschebyschef. Acta Litt. Sci. Reg. Univ. Hung. Franc.-Jos., Sect. Math. **5**, 194–198 (1930/1932)
1772. Erdős, P.: On the density of abundant numbers. J. Lond. Math. Soc. **9**, 278–282 (1934)
1773. Erdős, P.: A theorem of Sylvester and Schur. J. Lond. Math. Soc. **9**, 282–288 (1934)
1774. Erdős, P.: Über die Primzahlen gewisser arithmetischer Reihen. Math. Z. **39**, 473–491 (1935)
1775. Erdős, P.: On the arithmetical density of the sum of two sequences one of which forms a basis for the integers. Acta Arith. **1**, 197–200 (1935)
1776. Erdős, P.: On the difference of consecutive primes. Q. J. Math. **6**, 124–128 (1935)
1777. Erdős, P.: On primitive abundant numbers. J. Lond. Math. Soc. **10**, 49–58 (1935)
1778. Erdős, P.: On the density of some sequences of numbers. J. Lond. Math. Soc. **10**, 120–125 (1935)
1779. Erdős, P.: On the density of some sequences of numbers, III. J. Lond. Math. Soc. **13**, 119–127 (1938)
1780. Erdős, P.: Note on sequences of integers no one of which is divisible by any other. J. Lond. Math. Soc. **10**, 126–128 (1935)
1781. Erdős, P.: On the normal number of prime factors of $p - 1$ and some related problems concerning Euler's φ-function. Q. J. Math. **6**, 205–213 (1935)
1782. Erdős, P.: On the representation of an integer as the sum of k k-th powers. J. Lond. Math. Soc. **11**, 133–136 (1936)
1783. Erdős, P.: On the integers which are the totient of a product of three primes. Q. J. Math. **7**, 16–19 (1936)
1784. Erdős, P.: On the integers which are the totient of a product of two primes. Q. J. Math. **7**, 227–229 (1936)
1785. Erdős, P.: On the easier Waring problem for powers of primes, I. Proc. Camb. Philos. Soc. **33**, 6–12 (1937)
1786. Erdős, P.: On the easier Waring problem for powers of primes, II. Proc. Camb. Philos. Soc. **35**, 149–165 (1939)
1787. Erdős, P.: Note on products of consecutive integers. J. Lond. Math. Soc. **14**, 194–198, 245–249 (1939)
1788. Erdős, P.: On the smoothness of the asymptotic distribution of additive arithmetical functions. Am. J. Math. **61**, 722–725 (1939)
1789. Erdős, P.: The difference of consecutive primes. Duke Math. J. **6**, 438–441 (1940)
1790. Erdős, P.: On some asymptotic formulas in the theory of the "factorisation numerorum". Ann. Math. **42**, 989–993 (1941)
1791. Erdős, P.: On the asymptotic density of the sum of two sequences. Ann. Math. **43**, 65–68 (1942)

1792. Erdős, P.: On an elementary proof of some asymptotic formulas in the theory of partitions. Ann. Math. **43**, 437–450 (1942)
1793. Erdős, P.: On highly composite numbers. J. Lond. Math. Soc. **19**, 130–133 (1944)
1794. Erdős, P.: On the least primitive root of a prime p. Bull. Am. Math. Soc. **51**, 131–132 (1945)
1795. Erdős, P.: Some remarks on Euler's ϕ-function and some related problems. Bull. Am. Math. Soc. **51**, 540–544 (1945)
1796. Erdős, P.: On the distribution function of additive functions. Ann. Math. **47**, 1–20 (1946)
1797. Erdős, P.: Some asymptotic formulas in number theory. J. Indian Math. Soc. **12**, 75–78 (1948)
1798. Erdős, P.: On the density of some sequences of integers. Bull. Am. Math. Soc. **54**, 685–692 (1948)
1799. Erdős, P.: On some applications of Brun's method. Acta Sci. Math. **13**, 57–63 (1949)
1800. Erdős, P.: On a new method in elementary number theory which leads to an elementary proof of the prime number theorem. Proc. Natl. Acad. Sci. USA **35**, 374–384 (1949)
1801. Erdős, P.: On integers of the form $2^k + p$ and some related problems. Summa Bras. Math. **2**, 113–123 (1950)
1802. Erdős, P.: On almost primes. Am. Math. Mon. **57**, 404–407 (1950)
1803. Erdős, P.: On a diophantine equation. Mat. Lapok **1**, 192–210 (1950) (in Hungarian)
1804. Erdős, P.: On a Diophantine equation. J. Lond. Math. Soc. **26**, 176–178 (1951)
1805. Erdős, P.: Some problems and results in elementary number theory. Publ. Math. (Debr.) **2**, 103–109 (1951)
1806. Erdős, P.: On a problem concerning congruence systems. Mat. Lapok **3**, 122–128 (1952) (in Hungarian)
1807. Erdős, P.: On the sum $\sum_{k=1}^{x} d(f(k))$. J. Lond. Math. Soc. **27**, 7–15 (1952)
1808. Erdős, P.: On the greatest prime factor of $\prod_{k=1}^{x} f(k)$. J. Lond. Math. Soc. **27**, 379–384 (1952)
1809. Erdős, P.: Arithmetical properties of polynomials. J. Lond. Math. Soc. **28**, 416–425 (1953)
1810. Erdős, P.: On a problem of Sidon in additive number theory. Acta Sci. Math. **15**, 255–259 (1954)
1811. Erdős, P.: Some results on additive number theory. Proc. Am. Math. Soc. **5**, 847–853 (1954)
1812. Erdős, P.: On the products of consecutive integers, III. Indag. Math. **17**, 85–90 (1955)
1813. Erdős, P.: On amicable numbers. Publ. Math. (Debr.) **4**, 108–112 (1955)
1814. Erdős, P.: Some problems on the distribution of prime numbers. C.I.M.E., Teoria dei numeri, 1955, 8 pp.
1815. Erdős, P.: On pseudoprimes and Carmichael numbers. Publ. Math. (Debr.) **4**, 201–206 (1956)
1816. Erdős, P.: Some unsolved problems. Mich. Math. J. **4**, 291–300 (1957)
1817. Erdős, P.: Some results on diophantine approximation. Acta Arith. **5**, 359–369 (1959)
1818. Erdős, P.: Remarks on number theory. III. On adition chains. Acta Arith. **6**, 77–81 (1960)
1819. Erdős, P.: An asymptotic inequality in the theory of numbers. Vestnik Leningr. Univ., Ser. Mat. Mekh. Astronom., **15**(13), 41–49 (1960) (in Russian)
1820. Erdős, P.: Remarks on number theory, IV. Extremal problems in number theory, I. Mat. Lapok **13**, 228–255 (1962) (in Hungarian)
1821. Erdős, P.: On the integers relatively prime to n and on a number-theoretic function considered by Jacobsthal. Math. Scand. **10**, 163–170 (1962)
1822. Erdős, P.: Problems and results on diophantine approximations. Compos. Math. **16**, 52–65 (1964)
1823. Erdős, P.: On the distribution of the convergents of almost all real numbers. J. Number Theory **2**, 425–441 (1970)
1824. Erdős, P.: Problems and results on Diophantine approximations, II. In: Lecture Notes in Math., vol. 475, pp. 89–99. Springer, Berlin (1975)
1825. Erdős, P.: Problems and results on consecutive integers. Publ. Math. (Debr.) **23**, 271–282 (1976)

1826. Erdős, P.: E. Straus (1921–1983). Rocky Mt. J. Math. **15**, 331–341 (1985)
1827. Erdős, P.: Some personal reminiscences of the mathematical work of Paul Turán. Acta Arith. **37**, 3–8 (1980)
1828. Erdős, P., Few, L., Rogers, C.A.: The amount of overlapping in partial coverings of space by equal spheres. Mathematika **11**, 171–184 (1964)
1829. Erdős, P., Fuchs, W.H.J.: On a problem of additive number theory. J. Lond. Math. Soc. **31**, 67–73 (1956)
1830. Erdős, P., Graham, R.L.: On a linear diophantine problem of Frobenius. Acta Arith. **21**, 399–408 (1972)
1831. Erdős, P., Graham, R.L.: Old and New Problems and Results in Combinatorial Number Theory. Université de Genève, Genève (1980)
1832. Erdős, P., Hall, R.R.: On the values of Euler's ϕ-function. Acta Arith. **22**, 201–206 (1973)
1833. Erdős, P., Hall, R.R.: Distinct values of Euler's ϕ-function. Mathematika **23**, 1–3 (1976)
1834. Erdős, P., Heilbronn, H.: On the addition of residue classes mod p. Acta Arith. **9**, 149–159 (1964) [[2715], pp. 455–465]
1835. Erdős, P., Kac, M.: The Gaussian law of errors in the theory of additive arithmetical functions. Am. J. Math. **62**, 738–742 (1940)
1836. Erdős, P., Kac, M., van Kampen, E.R., Wintner, A.: Ramanujan sums and almost periodic functions. Stud. Math. **9**, 43–53 (1940)
1837. Erdős, P., Koksma, J.F.: On the uniform distribution modulo 1 of lacunary sequences. Indag. Math. **11**, 79–88 (1949)
1838. Erdős, P., Koksma, J.F.: On the uniform distribution modulo 1 of sequences $(f(n,\theta))$. Indag. Math. **11**, 299–302 (1949)
1839. Erdős, P., Kolesnik, G.: Prime power divisors of binomial coefficients. Discrete Math. **200**, 101–117 (1999)
1840. Erdős, P., Lacampagne, C.B., Selfridge, J.L.: Estimates of the least prime factor of a binomial coefficient. Math. Comput. **61**, 215–224 (1993)
1841. Erdős, P., Lehner, J.: The distribution of the number of summands in the partitions of a positive integer. Duke Math. J. **8**, 335–345 (1941)
1842. Erdős, P., Lorentz, G.G.: On the probability that n and $g(n)$ are relatively prime. Acta Arith. **5**, 35–44 (1958)
1843. Erdős, P., Mahler, K.: On the number of integers which can be represented by a binary form. J. Lond. Math. Soc. **13**, 136–139 (1939)
1844. Erdős, P., Mirsky, L.: The distribution of values of the divisor function $d(n)$. Proc. Lond. Math. Soc. **2**, 257–271 (1952)
1845. Erdős, P., Nicolas, J.-L.: Répartition des nombres superabondants. Bull. Soc. Math. Fr. **103**, 65–90 (1975)
1846. Erdős, P., Pomerance, C., Sárkőzy, A.: On locally repeated values of certain arithmetic functions. Acta Math. Acad. Sci. Hung. **49**, 251–259 (1987)
1847. Erdős, P., Pomerance, C., Sárkőzy, A.: On locally repeated values of certain arithmetic functions, III. Proc. Am. Math. Soc. **101**, 1–7 (1987)
1848. Erdős, P., Rado, R.: Combinatorial theorems on classifications of subsets of a given set. Proc. Lond. Math. Soc. **2**, 417–439 (1952)
1849. Erdős, P., Rényi, A.: Some problems and results on consecutive primes. Simon Stevin **27**, 115–125 (1950)
1850. Erdős, P., Rényi, A.: Some remarks on the large sieve of Yu.V. Linnik. Ann. Univ. Sci. Bp. **11**, 3–13 (1968)
1851. Erdős, P., Rieger, G.J.: Ein Nachtrag über befreundete Zahlen. J. Reine Angew. Math. **273**, 220 (1975)
1852. Erdős, P., Rogers, C.A.: The covering of n-dimensional space by spheres. J. Lond. Math. Soc. **28**, 287–293 (1953)
1853. Erdős, P., Sárkőzy, A., Sós, V.T.: On additive properties of general sequences. Discrete Math. **136**, 75–99 (1994)
1854. Erdős, P., Sárkőzy, A., Szeméredi, E.: On a theorem of Behrend. J. Aust. Math. Soc. **7**, 9–16 (1967)

1855. Erdős, P., Sárkőzy, A., Szemérdi, E.: On an extremal problem concerning primitive sequences. J. Lond. Math. Soc. **42**, 484–488 (1967)
1856. Erdős, P., Schinzel, A.: On the greatest prime factor of $\prod_{k=1}^{x} f(k)$. Acta Arith. **55**, 191–200 (1990)
1857. Erdős, P., Selfridge, J.L.: The product of consecutive integers is never a power. Ill. J. Math. **19**, 292–301 (1975)
1858. Erdős, P., Shapiro, H.N.: On the changes of sign of a certain error function. Can. J. Math. **3**, 375–385 (1951)
1859. Erdős, P., Shapiro, H.N.: On the least primitive root of a prime. Pac. J. Math. **7**, 861–865 (1957)
1860. Erdős, P., Stewart, C.L., Tijdeman, R.: Some diophantine equations with many solutions. Compos. Math. **66**, 37–56 (1988)
1861. Erdős, P., Szekeres, G.: Über die Anzahl der Abelschen Gruppen gegebener Ordnung und über ein verwandtes zahlentheoretisches Problem. Acta Litt. Sci. Reg. Univ. Hung. Franc.-Jos., Sect. Math. **7**, 95–102 (1934/1935)
1862. Erdős, P., Szűsz, P., Turán, P.: Remarks on the theory of diophantine approximation. Colloq. Math. **6**, 119–126 (1958) [[6227], vol. 2, pp. 1038–1045]
1863. Erdős, P., Tetali, P.: Representation of integers as the sum of k terms. Random Struct. Algorithms **1**, 245–261 (1990)
1864. Erdős, P., Turán, P.: On some sequences of integers. J. Lond. Math. Soc. **11**, 261–264 (1936) [[6227], vol. 1, pp. 45–48]
1865. Erdős, P., Turán, P.: On a problem of Sidon in additive number theory and on some related problems. J. Lond. Math. Soc. **16**, 212–215 (1941). Addendum: **19**, 208 (1944) [[6227], vol. 1, pp. 257–260]
1866. Erdős, P., Turán, P.: On a problem in the theory of uniform distribution, I. Indag. Math. **10**, 370–378 (1948)
1867. Erdős, P., Turán, P.: On a problem in the theory of uniform distribution, II. Indag. Math. **10**, 406–413 (1948)
1868. Erdős, P., Turán, P.: On some general problems in the theory of partitions, I. Acta Arith. **18**, 53–62 (1971) [[6227], vol. 3, pp. 2100–2109]
1869. Erdős, P., Wintner, A.: Additive arithmetical functions and statistical independence. Am. J. Math. **61**, 713–721 (1939)
1870. Ernvall, R., Metsänkylä, T.: Cyclotomic invariants for primes between 125 000 and 150 000. Math. Comput. **56**, 851–858 (1991)
1871. Ernvall, R., Metsänkylä, T.: Cyclotomic invariants for primes to one million. Math. Comput. **59**, 249–250 (1992)
1872. Eršov, J.L.: On elementary theory of local fields. Algebra Log. Semin. **4**, 5–30 (1965) (in Russian)
1873. Escott, E.B.: L'Intermédiaire des Math., **14**, 1908, 109–111
1874. Eskin, A., Margulis, G.A., Mozes, S.: On a quantitative version of the Oppenheim conjecture. Electron. Res. Announc. Am. Math. Soc. **1**, 124–130 (1995)
1875. Eskin, A., Margulis, G.A., Mozes, S.: Upper bounds and asymptotics in a quantitative version of the Oppenheim conjecture. Ann. Math. **147**, 93–141 (1998)
1876. Esnault, H., Vieweg, E.: Dyson's lemma for polynomials in several variables (and the theorem of Roth). Invent. Math. **78**, 445–490 (1984)
1877. Estermann, T.: On the divisor problem in a class of residues. J. Lond. Math. Soc. **3**, 247–250 (1928)
1878. Estermann, T.: Vereinfachter Beweis eines Satzes von Kloosterman. Abh. Math. Semin. Univ. Hamb. **7**, 82–98 (1929)
1879. Estermann, T.: On the representations of a number as the sum of three products. Proc. Lond. Math. Soc. **29**, 453–478 (1929)
1880. Estermann, T.: On the representations of a number as the sum of two products, I. Proc. Lond. Math. Soc. **31**, 123–133 (1930)
1881. Estermann, T.: On the representations of a number as the sum of two products, II. J. Lond. Math. Soc. **5**, 131–137 (1930)

1882. Estermann, T.: Über die Darstellung einer Zahl als Differenz von zwei Produkten. J. Reine Angew. Math. **164**, 173–182 (1931)
1883. Estermann, T.: On the representation of a number as the sum of two numbers not divisible by k-th powers. J. Lond. Math. Soc. **6**, 37–40 (1931)
1884. Estermann, T.: On the representations of a number as the sum of a prime and a quadratfrei number. J. Lond. Math. Soc. **6**, 219–221 (1931)
1885. Estermann, T.: Einige Sätze über quadratfreie Zahlen. Math. Ann. **105**, 653–662 (1931)
1886. Estermann, T.: Eine neue Darstellung und neue Anwendungen der Viggo Brunschen Methode. J. Reine Angew. Math. **168**, 106–116 (1932)
1887. Estermann, T.: Proof that every large integer is a sum of seventeen biquadrates. Proc. Lond. Math. Soc. **41**, 126–142 (1936)
1888. Estermann, T.: Proof that every large integer is a sum of two primes and a square. Proc. Lond. Math. Soc. **42**, 501–516 (1937)
1889. Estermann, T.: A new result in the additive theory of numbers. Q. J. Math. **8**, 32–38 (1937)
1890. Estermann, T.: On Waring's problem for fourth and higher powers. Acta Arith. **2**, 197–211 (1937)
1891. Estermann, T.: On Goldbach's problem: proof that almost all even positive integers are sums of two primes. Proc. Lond. Math. Soc. **44**, 307–314 (1938)
1892. Estermann, T.: Note on a theorem of Minkowski. J. Lond. Math. Soc. **21**, 179–182 (1946)
1893. Estermann, T.: On Dirichlet's L-functions. J. Lond. Math. Soc. **23**, 275–279 (1948)
1894. Estermann, T.: On the representation of a number as a sum of three squares. Proc. Lond. Math. Soc. **9**, 575–594 (1959)
1895. Estermann, T., Barham, C.L.: On the representations of a number as the sum of four or more N-numbers. Proc. Lond. Math. Soc. **38**, 340–353 (1934)
1896. Euler, J.A.: Item 36 in L. Euler's Fragmenta arithmetica ex adversariis mathematicis deprompta. In: Opera postuma, vol. 1, pp. 203–204. A pund Eggers et Socios, Petropoli (1862)
1897. Euler, L.: Observationes de theoremate quodam Fermatiano aliisque ad numeros primos spectantibus. Comment. Acad. Sci. Petropol. **6**(1732/1733), 103–107 (1738) [[1908], ser. 1, vol. 2, pp. 1–5]
1898. Euler, L.: Theorematum quorundam arithmeticorum demonstrationes. Comment. Acad. Sci. Petropol. **10**(1738), 125–146 (1747) [[1908], ser. 1, vol. 2, pp. 38–58]
1899. Euler, L.: Introductio in analysin infinitorum, I. Bousquet, Lausannae (1748) [Opera Omnia, ser. 1, vol. 8, Leipzig, 1915. Translations: French: Paris, 1796; German: Springer, 1885; English: Springer, 1988]
1900. Euler, L.: Supplementum quorundam theorematum arithmeticorum quae in nonnullis demonstrationibus supponuntur. Novi Comment. Acad. Sci. Petropol. **8**(1760/1761), 105–128 (1763) [[1908], ser. 1, vol. 2, pp. 556–575]
1901. Euler, L.: Vollständige Anleitung zur Algebra. Akademie der Wissenschaften, St. Petersbourg (1770) [[1908], ser. 1, vol. 1]
1902. Euler, L.: Extrait d'une lettre de M. Euler le pére à M. Bernoulli concernant la mémoire imprimé parmi ceux de 1771, p. 318. Nouv. Mém. Acad. Berlin, 1772/1774, 35–36 [[1908], ser. 1, vol. 3, pp. 335–337]
1903. Euler, L.: De formulis speciei $mxx + nyy$ ad numeros primos explorandos idoneis, eorumque mirabilius proprietatibus. Nova Acta Acad. Sci. Petropol. **12**(1794), 22–46 (1801) [[1908], ser. 1, vol. 4, pp. 269–289]
1904. Euler, L.: De variis modis numeros praegrandes examinandi, utrum sint primi necne? Nova Acta Acad. Sci. Petropol. **13**(1795/1796), 14–44 (1802) [[1908], ser. 1, vol. 4, pp. 303–328]
1905. Euler, L.: Illustratio paradoxi circa progressionem numerorum idoneorum sive congruorum. Nova Acta Acad. Sci. Petropol. **15**(1799–1802), 29–32 (1806) [[1908], ser. 1, vol. 4, pp. 395–398]
1906. Euler, L.: De binis formuli speciei $xx + myy$ et $xx + nyy$ inter se concordibus et disconcordibus. Mém. Acad. Sci. St. Petersbourg **8**(1817–1818), 3–16 (1826) [Commentat. Arithm., **2**, 406–413 (1849); [1908], ser. 1, vol. 5, pp. 48–60]

1907. Euler, L.: De numeris amicabilibus. Comment. Arith. **2**, 627–636 (1849) [[1908], ser. 1, vol. 5, pp. 353–365]

1908. Euler, L.: Opera Omnia. Started Leipzig, 1911, still continued by Birkhäuser

1909. Euler, L., Goldbach, C.: Briefwechsel 1729–1764. Akademie Verlag, Berlin (1965)

1910. Evdokimov, S.A.: Euler products for congruence subgroups of the Siegel group of genus 2. Mat. Sb. **99**, 483–513 (1976) (in Russian)

1911. Evdokimov, S.A.: Analytic properties of Euler products for congruence-subgroups of $Sp_2(Z)$. Mat. Sb. **110**, 369–398 (1979) (in Russian)

1912. Evdokimov, S.A.: Characterization of the Maass space of Siegel modular cusp forms of genus 2. Mat. Sb. **112**, 133–142 (1980) (in Russian)

1913. Evdokimov, S.A.: A basis composed of eigenfunctions of Hecke operators in the theory of modular forms of genus n. Mat. Sb. **115**, 337–363 (1981); corr. **116**, 603 (1981) (in Russian)

1914. Evelyn, C.J.A., Linfoot, E.H.: On a problem in the additive theory of numbers. Math. Z. **30**, 433–448 (1929)

1915. Evelyn, C.J.A., Linfoot, E.H.: On a problem in the additive theory of numbers, II. J. Reine Angew. Math. **164**, 131–140 (1931)

1916. Evelyn, C.J.A., Linfoot, E.H.: On a problem in the additive theory of numbers, III. Math. Z. **34**, 637–644 (1932)

1917. Evelyn, C.J.A., Linfoot, E.H.: On a problem in the additive theory of numbers, IV. Ann. Math. **32**, 261–270 (1931)

1918. Evelyn, C.J.A., Linfoot, E.H.: On a problem in the additive theory of numbers, V. Q. J. Math. **3**, 152–160 (1932)

1919. Everest, G., van der Poorten, A.J., Shparlinski, I., Ward, T.: Recurrence Sequences. Am. Math. Soc., Providence (2003)

1920. Evertse, J.-H.: On the equation $ax^n - by^n = c$. Compos. Math. **47**, 289–315 (1982)

1921. Evertse, J.-H.: Upper bounds for the number of solutions of Diophantine equations. In: Math. Centre Tracts, vol. 168, pp. 1–125. Mathematisch Centrum, Amsterdam (1983)

1922. Evertse, J.-H.: On the representation of integers by binary cubic forms of positive discriminant. Invent. Math. **73**, 117–138 (1983); corr. **75**, 379 (1984)

1923. Evertse, J.-H.: On sums of S-units and linear recurrences. Compos. Math. **53**, 225–244 (1984)

1924. Evertse, J.-H.: On equations in S-units and the Thue-Mahler equation. Invent. Math. **75**, 561–584 (1984)

1925. Evertse, J.-H.: Estimates for reduced binary forms. J. Reine Angew. Math. **434**, 159–190 (1993)

1926. Evertse, J.-H.: An improvement of the quantitative subspace theorem. Compos. Math. **101**, 225–311 (1996)

1927. Evertse, J.-H.: The number of algebraic numbers of given degree approximating a given algebraic number. In: Analytic Number Theory, pp. 53–83. Cambridge University Press, Cambridge (1997)

1928. Evertse, J.-H.: The number of solutions of the Thue-Mahler equation. J. Reine Angew. Math. **482**, 121–149 (1997)

1929. Evertse, J.-H.: On the norm form inequality $|F(x)| \le h$. Publ. Math. (Debr.) **56**, 337–374 (2000)

1930. Evertse, J.-H., Győry, K.: On unit equations and decomposable form equations. J. Reine Angew. Math. **358**, 6–19 (1985)

1931. Evertse, J.-H., Győry, K.: On the number of solutions of weighted unit equations. Compos. Math. **66**, 329–354 (1988)

1932. Evertse, J.-H., Győry, K.: Finiteness criteria for decomposable form equations. Acta Arith. **50**, 357–379 (1988)

1933. Evertse, J.-H., Győry, K.: Thue-Mahler equations with a small number of solutions. J. Reine Angew. Math. **399**, 60–80 (1989)

1934. Evertse, J.-H., Győry, K.: Effective finiteness results for binary forms with given discriminant. Compos. Math. **79**, 169–204 (1991)

1935. Evertse, J.-H., Győry, K.: Effective finiteness theorems for decomposable forms of given discriminant. Acta Arith. **60**, 233–277 (1992)
1936. Evertse, J.-H., Győry, K., Stewart, C.L., Tijdeman, R.: S-unit equations and their applications. In: New Advances in Transcendence Theory, Durham, 1986, pp. 110–174. Cambridge University Press, Cambridge (1988)
1937. Evertse, J.-H., Győry, K., Stewart, C.L., Tijdeman, R.: On S-units equations in two unknowns. Invent. Math. **92**, 461–477 (1988)
1938. Evertse, J.-H., Schlickewei, H.P.: The absolute subspace theorem and linear equations with unknowns from a multiplicative group. In: Number Theory in Progress, I, pp. 121–142. de Gruyter, Berlin (1999)
1939. Evertse, J.-H., Schlickewei, H.P.: A quantitative version of the absolute subspace theorem. J. Reine Angew. Math. **548**, 21–127 (2002)
1940. Evertse, J.-H., Schlickewei, H.P., Schmidt, W.M.: Linear equations in variables which lie in a multiplicative group. Ann. Math. **155**, 807–836 (2002)
1941. Evertse, J.-H., Silverman, J.H.: Uniform bounds for the number of solutions to $Y^n = f(X)$. Math. Proc. Camb. Philos. Soc. **100**, 237–248 (1986)
1942. Ewell, J.A.: A simple derivation of Jacobi's four-square formula. Proc. Am. Math. Soc. **85**, 323–326 (1982)
1943. Ewell, J.A.: A formula for Ramanujan's tau function. Proc. Am. Math. Soc. **91**, 37–40 (1984)
1944. Ewell, J.A.: On sums of sixteen squares. Rocky Mt. J. Math. **17**, 295–299 (1987)
1945. Ewell, J.A.: On necessary conditions for the existence of odd perfect numbers. Rocky Mt. J. Math. **29**, 165–175 (1999)
1946. Ewing, H., Gehring, F.W. (eds.): Paul Halmos. Celebrating 50 Years of Mathematics. Springer, Berlin (1991)
1947. Faber, G.: Über arithmetische Eigenschaften analytischer Funktionen. Math. Ann. **58**, 545–557 (1904)
1948. Faber, G.: Über stetige Funktionen. Math. Ann. **69**, 372–443 (1910)
1949. Faddeev, D.K.: Über die Gleichung $x^4 - Ay^4 = \pm 1$. Tr. Mat. Inst. Steklova **5**, 41–52 (1934)
1950. Faddeev, D.K., et al.: Boris Nikolaevič Delone (on his life and creativity). Tr. Mat. Inst. Steklova **196**, 3–10 (1991) (in Russian)
1951. Faĭnleĭb, A.S.: On the mutual primality of n and $f(n)$. Mat. Zametki **11**, 259–268 (1972) (in Russian)
1952. Faĭnleĭb, A.S.: On the distribution of Beurling integers. J. Number Theory **111**, 227–247 (2005)
1953. Faivre, C.: Distribution of Lévy constants for quadratic numbers. Acta Arith. **61**, 13–34 (1992)
1954. Faltings, G.: Endlichkeitssätze für abelsche Varietäten über Zahlkörpern. Invent. Math. **73**, 349–366 (1983); corr. vol. 75, 1984, p. 381
1955. Faltings, G., Wüstholz, G. (eds.): Rational Points. Vieweg, Braunschweig (1984); 2nd ed., 1986, 3rd ed. 1992
1956. Faltings, G., Wüstholz, G.: Diophantine approximations on projective spaces. Invent. Math. **116**, 109–138 (1994)
1957. Farey, J.: On a curious property of vulgar fractions. Philos. Mag. **47**, 385–386 (1816)
1958. Farhi, B.: Une approche polynomiale du théorème de Faltings. C. R. Acad. Sci. Paris **340**, 103–106 (2005)
1959. Farmer, D.W.: Long mollifiers of the Riemann zeta-function. Mathematika **40**, 71–87 (1993)
1960. Farmer, D.W., Wilson, K.: Converse theorems assuming a partial Euler product. Ramanujan J. **15**, 205–218 (2008)
1961. Faure, H.: Discrépances de suites associées a un systeme de numération (en dimension un). Bull. Soc. Math. Fr. **109**, 143–182 (1981)
1962. Faure, H.: Discrépance de suites associées a un systeme de numération (en dimension s). Acta Arith. **41**, 337–351 (1982)

1963. Fay, J.D.: Theta Functions on Riemann Surfaces. In: Lecture Notes in Math., vol. 352. Springer, Berlin (1973)
1964. Fejes Tóth, G.: Recent progress on packing and covering. In: Contemp. Math., vol. 223, pp. 145–162. Am. Math. Soc., Providence (1999)
1965. Fejes Tóth, G., Kuperberg, W.: Blichfeldt's density bound revisited. Math. Ann. **295**, 721–727 (1993)
1966. Fejes Tóth, L.: Über die dichteste Kugellagerung. Math. Z. **48**, 676–684 (1943)
1967. Fejes Tóth, L.: Lagerungen in der Ebene, auf der Kugel und im Raum. Springer, Berlin (1953); 2nd ed. 1972
1968. Fejes Tóth, L.: Regular Figures. Macmillan, New York (1964). [German version: Reguläre Figuren, Budapest, 1965]
1969. Fekete, M.: The zeros of Riemann's zeta-function on the critical line. J. Lond. Math. Soc. **1**, 15–19 (1926)
1970. Fekete, M., Pólya, G.: Über ein Problem von Laguerre. Rend. Circ. Mat. Palermo **34**, 89–120 (1912)
1971. Fel, L.G.: Frobenius problem for semigroups $S(d_1, d_2, d_3)$. Funct. Anal. Other Math. **1**, 119–157 (2006)
1972. Fel, L.G.: Analytic representations in the three-dimensional Frobenius problem. Funct. Anal. Other Math. **2**, 27–44 (2008)
1973. Feldman, N.I.: Approximation of certain transcendental numbers, I, Approximation of logarithms of algebraic numbers. Izv. Akad. Nauk SSSR, Ser. Mat. **15**, 53–74 (1951) (in Russian)
1974. Feldman, N.I.: Approximation of certain transcendental numbers, II. The approximation of certain numbers connected with the Weierstrass function $\wp(z)$. Izv. Akad. Nauk SSSR, Ser. Mat. **15**, 153–176 (1951) (in Russian)
1975. Feldman, N.I.: The measure of transcendency of the number π. Izv. Akad. Nauk SSSR, Ser. Mat. **24**, 357–368 (1960) (in Russian)
1976. Feldman, N.I.: On the measure of transcendence of the number e. Usp. Mat. Nauk **18**(3), 207–213 (1963) (in Russian)
1977. Feldman, N.I.: Estimation of a linear form in the logarithms of algebraic numbers. Mat. Sb. **76**, 304–319 (1968) (in Russian)
1978. Feldman, N.I.: An improvement of the estimate of a linear form in the logarithms of algebraic numbers. Mat. Sb. **77**, 423–436 (1968) (in Russian)
1979. Feldman, N.I.: An inequality for a linear form in the logarithms of algebraic numbers. Mat. Zametki **5**, 681–689 (1969) (in Russian)
1980. Feldman, N.I.: Refinement of two effective inequalities of A. Baker. Mat. Zametki **6**, 767–769 (1969) (in Russian)
1981. Feldman, N.I.: An effective strengthening of Liouville's theorem. Izv. Akad. Nauk SSSR, Ser. Mat. **35**, 973–990 (1971) (in Russian)
1982. Feldman, N.I.: The Seventh Problem of Hilbert. Moscow State University Press, Moscow (1982) (in Russian)
1983. Feldman, N.I., Čudakov, N.G.: On a theorem of Stark. Mat. Zametki **11**, 329–340 (1972) (in Russian)
1984. Feldman, N.I., Nesterenko, Yu.V.: Number Theory IV. Transcendental Numbers. Encyclopaedia of Mathematical Sciences, vol. 44. Springer, Berlin (1998)
1985. Feldman, N.I., Šidlovskiĭ, A.B.: The development and contemporary state of the theory of transcendental numbers. Usp. Mat. Nauk **22**(3), 3–81 (1967) (in Russian)
1986. Fellman, E.A.: Leonhard Euler. Rowohlt, Hamburg (1995). [English translation: Birkhäuser, 2007]
1987. Ferenczi, S., Mauduit, C.: Transcendence of numbers with a low complexity expansion. J. Number Theory **67**, 146–161 (1997)
1988. Ferguson, S.P.: Sphere packings, V, Pentahedral prisms. Discrete Comput. Geom. **36**, 167–204 (2006)
1989. Fermat, P.: Observationes Domini Petri de Fermat. In: [2040], vol. 1, pp. 292–342 [French translation: [1991], vol. 3, pp. 241–274]

1990. Fermat, P.: Letter to Mersenne, September 1636. In: [2040], vol. 2, p. 65
1991. Fermat, P.: Oeuvres, vols. 1–4. Gauthier-Villars, Paris (1891–1912)
1992. Ferrar, W.L.: Summation formulae and their relation to Dirichlet's series. Compos. Math. **1**, 344–360 (1935)
1993. Ferrar, W.L.: Summation formulae and their relation to Dirichlet's series, II. Compos. Math. **4**, 394–405 (1937)
1994. Ferrar, W.L.: Arthur Lee Dixon. J. Lond. Math. Soc. **31**, 126–128 (1956)
1995. Ferrero, B., Greenberg, R.: On the behavior of p-adic L-functions at $s = 0$. Invent. Math. **50**, 91–102 (1978/1979)
1996. Ferrers, N.M.: Series for $\frac{\pi}{\sqrt{7}}, \frac{\pi}{\sqrt{11}}, \frac{\pi}{\sqrt{19}}$. Messenger Math. **31**, 92–94 (1901/1902)
1997. Feuerverger, A., Martin, G.: Biases in the Shanks-Rényi prime number race. Exp. Math. **9**, 535–570 (2000)
1998. Fields, J.C.: A simple statement of proof of reciprocal theorem. Am. J. Math. **13**, 189–190 (1891)
1999. Figueredo, L.M.: Serre's conjecture for imaginary quadratic fields. Compos. Math. **118**, 103–122 (1999)
2000. Filakovszky, P., Hajdu, L.: The resolution of the Diophantine equation $x(x + d)\cdots(x + (k - 1)d) = by^2$ for fixed d. Acta Arith. **98**, 151–154 (2001)
2001. Filaseta, M.: Short interval results for squarefree numbers. J. Number Theory **35**, 128–149 (1990)
2002. Filaseta, M., Trifonov, O.: On gaps between squarefree numbers. Prog. Math. **85**, 235–253 (1990)
2003. Filaseta, M., Trifonov, O.: On gaps between squarefree numbers, II. J. Lond. Math. Soc. **45**, 215–221 (1992)
2004. Filaseta, M., Trifonov, O.: The distribution of squarefull numbers in short intervals. Acta Arith. **67**, 323–333 (1994)
2005. Filaseta, M., Trifonov, O.: The distribution of fractional parts with applications to gap results in number theory. Proc. Lond. Math. Soc. **73**, 241–278 (1996)
2006. Finkelstein, R., London, H.: Completion of a table of O. Hemer. In: Proceedings of the Washington State University Conference on Number Theory, pp. 148–159. Washington State Univ., Pullman (1971) (Finkelstein, R. = Steiner, R.P.)
2007. Fischler, S.: Irrationalité de valeurs de zêta (d'aprés Apèry, Rivoal, ...). Astérisque **294**, 27–62 (2004)
2008. Fisher, T.: A counterexample to a conjecture of Selmer. In: Number Theory and Algebraic Geometry, pp. 119–131. Cambridge University Press, Cambridge (2003)
2009. Fiske, T.S.: Frank Nelson Cole. Bull. Am. Math. Soc. **33**, 773–777 (1927)
2010. Flammang, V., Rhin, G., Sac-Épée, J.-M.: Integer transfinite diameter and polynomials with small Mahler measure. Math. Comput. **75**, 1527–1540 (2006)
2011. Flanders, H.: Generalization of a theorem of Ankeny and Rogers. Ann. Math. **57**, 392–400 (1953)
2012. Fleck, A.: Über die Darstellung ganzer Zahlen als Summen von positiven Kuben und als Summen von Biquadraten ganzer Zahlen. SBer. Berliner Math. Ges. **5**, 2–9 (1906)
2013. Fleck, A.: Über die Darstellung ganzer Zahlen als Summen von sechsten Potenzen ganzer Zahlen. Math. Ann. **64**, 561–566 (1909)
2014. Flett, T.M.: On the function $\sum_{n=1}^{\infty} \frac{1}{n} \sin \frac{t}{n}$. J. Lond. Math. Soc. **25**, 5–19 (1950)
2015. Flexor, M., Oesterlé, J.: Sur les points de torsion des courbes elliptiques. Astérisque **183**, 25–36 (1990)
2016. Fluch, W.: Verwendung der Zeta-Function beim Sieb von Selberg. Acta Arith. **5**, 381–405 (1959)
2017. Fogels, E.: On average values of arithmetic functions. Proc. Camb. Philos. Soc. **37**, 358–372 (1941)
2018. Fogels, E.: On the zeros of Hecke's L-function, I. Acta Arith. **7**, 87–106 (1962)
2019. Fogels, E.: On the zeros of Hecke's L-function, II. Acta Arith. **7**, 131–147 (1962)
2020. Fogels, E.: On the zeros of Hecke's L-function, III. Acta Arith. **7**, 225–240 (1962)

2021. Fogels, E.: On the distribution of prime ideals. Acta Arith. **7**, 255–269 (1962)
2022. Fogels, E.: On the abstract theory of primes, I. Acta Arith. **10**, 136–182 (1964)
2023. Fomenko, O.M.: On the mean value of solutions of certain congruences. Zap. Nauč. Semin. POMI **254**, 192–206 (1998) (in Russian)
2024. Fontaine, J.M.: Il n'y a pas de variété abélienne sur **Z**. Invent. Math. **81**, 515–538 (1985)
2025. Ford, K.B.: On the representation of numbers as sums of unlike powers. J. Lond. Math. Soc. **51**, 14–26 (1995)
2026. Ford, K.B.: On the representation of numbers as sums of unlike powers, II. J. Am. Math. Soc. **9**, 919–940 (1996); add.: **12**, 1213 (1999)
2027. Ford, K.B.: New estimates for mean values of Weyl sums. Internat. Math. Res. Notices, 1995, nr. 3, 155–171
2028. Ford, K.: The distribution of totients. Ramanujan J. **2**, 67–151 (1998)
2029. Ford, K.: The number of solutions of $\varphi(x) = m$. Ann. Math. **150**, 283–311 (1999)
2030. Ford, K.: Waring's problem with polynomial summands. J. Lond. Math. Soc. **61**, 671–680 (2000)
2031. Ford, K.: Zero-free regions for the Riemann zeta function. In: Number Theory for the Millennium, II, pp. 25–56. AK Peters, Wellesley (2002)
2032. Ford, K.: Vinogradov's integral and bounds for the Riemann zeta function. Proc. Lond. Math. Soc. **85**, 565–633 (2002)
2033. Ford, K.: The distribution of integers with a divisor in a given interval. Ann. Math. **168**, 367–433 (2008)
2034. Ford, K., Konyagin, S.: On two conjectures of Sierpiński concerning the arithmetic functions σ and φ. In: Number Theory in Progress, vol. 2, pp. 795–803. de Gruyter, Berlin (1999)
2035. Ford, K., Konyagin, S.: Chebyshev's conjecture and the prime number race. In: IV International Conference "Modern Problems of Number Theory and Its Applications". Current Problems, vol. 2, pp. 67–91. Moscow State University Press, Moscow (2002)
2036. Ford, K., Konyagin, S.: The prime number race and zeros of L-functions off the critical line. Duke Math. J. **113**, 313–330 (2002)
2037. Ford, K., Konyagin, S.: The prime number race and zeros of L-functions off the critical line, II. Bonner Math. Schr. **360**, 1–40 (2003)
2038. Ford, L.: Rational approximations to irrational complex numbers. Trans. Am. Math. Soc. **19**, 1–42 (1918)
2039. Ford, L.R.: On the closeness of approach of complex rational fractions to a complex irrational number. Trans. Am. Math. Soc. **27**, 146–154 (1925)
2040. Forder, H.G.: A simple proof of a result on diophantine approximation. Math. Gaz. **47**, 237–238 (1963)
2041. Forman, W., Shapiro, H.N.: Abstract prime number theorems. Commun. Pure Appl. Math. **7**, 587–619 (1954)
2042. Forsyth, A.R.: James Whitbread Lee Glaisher. J. Lond. Math. Soc. **4**, 101–112 (1929)
2043. Forti, M., Viola, C.: Density estimates for the zeros of L-functions. Acta Arith. **23**, 379–391 (1973)
2044. Fouquet, M., Gaudry, P., Harley, R.: An extension of Satoh's algorithm and its implementation. J. Ramanujan Math. Soc. **15**, 281–318 (2000)
2045. Fouvry, É.: Répartition des suites dans les progressions arithmétiques. Acta Arith. **41**, 359–382 (1982)
2046. Fouvry, É.: Sur le théorème de Brun-Titchmarsh. Acta Arith. **43**, 417–424 (1984)
2047. Fouvry, É.: Autour du théorème de Bombieri-Vinogradov. Acta Math. **152**, 219–244 (1984)
2048. Fouvry, É.: Autour du théorème de Bombieri-Vinogradov, II. Ann. Sci. Éc. Norm. Super. **20**, 617–640 (1987)
2049. Fouvry, É.: Sur le problème des diviseurs de Titchmarsh. J. Reine Angew. Math. **357**, 51–76 (1985)
2050. Fouvry, É.: Théorème de Brun-Titchmarsh: application au théorème de Fermat. Invent. Math. **79**, 383–407 (1985)

2051. Fouvry, É.: Nombres presque premiers dans les petits intervalles. In: Lecture Notes in Math., vol. 1434, pp. 65–85. Springer, Berlin (1990)
2052. Fouvry, É.: Cinquante ans de théorie analytique des nombres. Un point de vue parmi d'autres: celui des méthodes de crible. In: Development of Mathematics 1950–2000, pp. 485–514. Birkhäuser, Basel (2000)
2053. Fouvry, É., Grupp, F.: On the switching principle in sieve theory. J. Reine Angew. Math. **370**, 101–126 (1986)
2054. Fouvry, É., Grupp, F.: Weighted sieves and twin prime type equations. Duke Math. J. **58**, 731–748 (1989)
2055. Fouvry, É., Iwaniec, H.: On a theorem of Bombieri-Vinogradov type. Mathematika **27**, 135–152 (1980)
2056. Fouvry, É., Iwaniec, H.: Primes in arithmetic progressions. Acta Arith. **42**, 197–218 (1983)
2057. Fouvry, É., Iwaniec, H.: The divisor function over arithmetic progressions. Acta Arith. **61**, 271–287 (1992)
2058. Fouvry, É., Iwaniec, H.: Gaussian primes. Acta Arith. **79**, 249–287 (1997)
2059. Fouvry, É., Katz, N.[M.]: A general stratification theorem for exponential sums, and applications. J. Reine Angew. Math. **540**, 115–166 (2001)
2060. Fouvry, É., Michel, P.: Sur certaines sommes d'exponentielles sur les nombres premiers. Ann. Sci. Éc. Norm. Super. **31**, 93–130 (1998)
2061. Fouvry, É., Murty, M.R.: Supersingular primes common to two elliptic curves. In: Number Theory, Paris, 1992–1993, pp. 91–102. Cambridge University Press, Cambridge (1995)
2062. Fouvry, É., Murty, M.R.: On the distribution of supersingular primes. Can. J. Math. **48**, 81–104 (1996)
2063. Fouvry, É., Nair, M., Tenenbaum, G.: L'ensemble exceptionnel dans la conjecture de Szpiro. Bull. Soc. Math. Fr. **120**, 485–506 (1992)
2064. Fouvry, É., Tenenbaum, G.: Entiers sans grand facteur premier en progressions arithmetiques. Proc. Lond. Math. Soc. **63**, 449–494 (1991)
2065. Fraenkel, A.: Axiomatische Begründung von Hensel's p-adischen Zahlen. J. Reine Angew. Math. **141**, 43–76 (1912)
2066. Fraenkel, A.S.: On a theorem of D. Ridout in the theory of Diophantine approximations. Trans. Am. Math. Soc. **105**, 84–101 (1962)
2067. Franel, J.: Les suites de Farey et le problème des nombres premiers. Nachr. Ges. Wiss. Göttingen 198–201 (1924)
2068. Frank, E.: Oskar Perron (1880–1975). J. Number Theory **14**, 281–291 (1982)
2069. Frasch, H.: Die Erzeugenden der Hauptkongruenzgruppen für Primzahlstufen. Math. Ann. **108**, 229–252 (1933)
2070. Fréchet, M.: La vie et l'oeuvre d'Émile Borel. Enseign. Math. **11**, 1–94 (1965)
2071. Frei, G.: Helmut Hasse (1898–1979). Expo. Math. **3**, 55–69 (1985)
2072. Frei, G.: Heinrich Weber and the emergence of class field theory. In: The History of Modern Mathematics, vol. 1, pp. 425–450. Academic Press, San Diego (1989)
2073. Freiman, G.A.: The solution of Waring's problem in a new setting. Usp. Mat. Nauk **4**(1), 193 (1949) (in Russian)
2074. Freiman, G.A.: The addition of finite sets. I, Izv. Vysš. Učeb. Zaved. Mat., 1959, nr. 6, 202–213 (in Russian)
2075. Freiman, G.A.: Inverse problems in additive number theory. Addition of sets of residues modulo a prime. Dokl. Akad. Nauk SSSR **141**, 571–573 (1961) (in Russian)
2076. Freiman, G.A.: Inverse problems of additive number theory, VII. The addition of finite sets, IV. The method of trigonometric sums. Izv. Vysš. Učeb. Zaved. Mat., 1962, nr. 6, 131–144. (in Russian)
2077. Freiman, G.A.: On the addition of finite sets. Dokl. Akad. Nauk SSSR **158**, 1038–1041 (1964) (in Russian)
2078. Freiman, G.A.: Inverse problems in additive number theory, IX. The addition of finite sets, V. Izv. Vysš. Učeb. Zaved. Mat., 1964, nr. 6, 168–178 (in Russian)

2079. Freiman, G.A.: Foundations of a Structural Theory of Set Addition. Kazan. Gos. Ped. Inst. & Elabuz. Gos. Ped. Inst., Kazan (1966) (in Russian) [English translation: Am. Math. Soc., 1973]
2080. Freiman, G.A.: What is the structure of K if $K + K$ is small? In: Lecture Notes in Math., vol. 1240, pp. 109–134. Springer, Berlin (1987)
2081. Freiman, G.A.: Structure theory of set addition. Astérisque **258**, 1–33 (1999)
2082. Freiman, G.A.: Structure theory of set addition, II, Results and problems. In: Paul Erdős and His Mathematics, I, Budapest, 1999, pp. 243–260. Janos Bolyai Math. Soc., Budapest (2002)
2083. Freiman, G.A., Low, L., Pitman, J.: Sumsets with distinct summands and the Erdős-Heilbronn conjecture on sums of residues. Astérisque **258**, 163–172 (1999)
2084. Freitag, E.: Siegelsche Modulfunktionen. Springer, Berlin (1983)
2085. Freitag, E.: Hilbert Modular Forms. Springer, Berlin (1990)
2086. Fresnel, J.: Nombres de Bernoulli et fonctions L p-adiques. Ann. Inst. Fourier **17**(2), 281–333 (1967)
2087. Frewer, M.: Felix Bernstein. Jahresber. Dtsch. Math.-Ver. **83**, 84–95 (1981)
2088. Frey, G.: Rationale Punkte auf Fermatkurven und getwisteten Modulkurven. J. Reine Angew. Math. **331**, 185–191 (1982)
2089. Frey, G.: Links between stable elliptic curves and certain Diophantine equations. Ann. Univ. Sarav. Ser. Math. **1**, 1–40 (1986)
2090. Frey, G.: Links between elliptic curves and solutions of $A - B = C$. J. Indian Math. Soc. **51**, 117–145 (1987)
2091. Fricke, R.: Ueber die Substitutionsgruppen, welche zu den aus dem Legrendre'schen Integralmodul $k^2(\omega)$ gezogenen Wurzeln gehören. Math. Ann. **29**, 99–118 (1886)
2092. Fricke, R.: Automorphe Funktionen mit Einschluss der elliptischen Modulfunktionen. In: Enzyklopädie der Mathematischen Wissenschaften, vol. 2_2, pp. 349–470. Teubner, Leipzig (1913)
2093. Fricker, F.: Einführung in die Gitterpunktlehre. Birkhäuser, Basel (1982)
2094. Fridlender, V.R.: On the least n-th power non-residues. Dokl. Akad. Nauk SSSR **66**, 351–352 (1949) (in Russian)
2095. Fried, M.: On a conjecture of Schur. Mich. Math. J. **17**, 41–55 (1970)
2096. Fried, M.: On a theorem of Ritt and related Diophantine problems. J. Reine Angew. Math. **264**, 40–55 (1973)
2097. Fried, M.: Constructions arising from Néron's high rank curves. Trans. Am. Math. Soc. **281**, 615–631 (1984)
2098. Fried, M.D., Guralnick, R., Saxl, J.: Schur covers and Carlitz's conjecture. Isr. J. Math. **82**, 157–225 (1993)
2099. Friedlander, J.B., Goldston, D.A.: Variance of distribution of primes in residue classes. Q. J. Math. **47**, 313–336 (1996)
2100. Friedlander, J.B., Goldston, D.A.: Note on a variance in the distribution of primes. In: Number Theory in Progress, vol. 2, pp. 841–848. de Gruyter, Berlin (1999)
2101. Friedlander, J.B., Granville, A.: Limitations to the equi-distribution of primes, I. Ann. Math. **129**, 369–382 (1989)
2102. Friedlander, J.B., Granville, A.: Limitations to the equi-distribution of primes, III. Compos. Math. **81**, 19–32 (1992)
2103. Friedlander, J.B., Granville, A.: Limitations to the equi-distribution of primes, IV. Proc. R. Soc. Lond. Ser. A, Math. Phys. Sci. **435**, 197–204 (1991)
2104. Friedlander, J.B., Granville, A., Hildebrand, A., Maier, H.: Oscillation theorems for primes in arithmetic progressions and for sifting functions. J. Am. Math. Soc. **4**, 25–86 (1991)
2105. Friedlander, J.B., Iwaniec, H.: Incomplete Kloosterman sum and a divisor problem. Ann. Math. **121**, 319–350 (1985)
2106. Friedlander, J.B., Iwaniec, H.: The divisor problem for arithmetic progressions. Acta Arith. **45**, 273–277 (1985)
2107. Friedlander, J.B., Iwaniec, H.: The Brun-Titchmarsh theorem. In: Analytic Number Theory, Kyoto, 1996, pp. 85–93. Cambridge University Press, Cambridge (1997)

2108. Friedlander, J.B., Iwaniec, H.: The polynomial $X^2 + Y^4$ captures its primes. Ann. Math. **148**, 945–1040 (1998)
2109. Friedlander, J.B., Iwaniec, H.: Asymptotic sieve for primes. Ann. Math. **148**, 1041–1065 (1998)
2110. Friedlander, J.B., Iwaniec, H.: Exceptional characters and prime numbers in arithmetic progressions. Int. Math. Res. Not. **37**, 2033–2050 (2003)
2111. Friedlander, J.B., Iwaniec, H.: Exceptional characters and prime numbers in short intervals. Sel. Math. New Ser. **10**, 61–69 (2004)
2112. Friedlander, J.B., Iwaniec, H.: The illusory sieve. Int. J. Number Theory **1**, 459–494 (2005)
2113. Friesen, C.: On continued fractions of given period. Proc. Am. Math. Soc. **103**, 9–14 (1988)
2114. Frobenius, G.: Über Beziehungen zwischen den Primidealen eines algebraischen Körpers und den Substitutionen seiner Gruppe. SBer. Preuß. Akad. Wiss. Berlin, 1896, 689–703. [[2120], vol. 2, pp. 719–733]
2115. Frobenius, G.: Über den Fermatschen Satz. J. Reine Angew. Math. **137**, 314–316 (1910) [[2120], vol. 3, pp. 428–430]
2116. Frobenius, G.: Über den Fermatschen Satz, II. SBer. Kgl. Preuß. Akad. Wiss. Berlin, 1910, 200–208 [[2120], vol. 3, pp. 431–439]
2117. Frobenius, G.: Über den Stridbergschen Beweis des Waringschen Satzes. SBer. Preuß. Akad. Wiss. Berlin, 1912, 666–670 [[2120], vol. 3, pp. 568–572]
2118. Frobenius, G.: Über die Markoffsche Zahlen. SBer. Kgl. Preuß. Akad. Wiss. Berlin, 1913, 458–487 [[2120], vol. 3, pp. 598–627]
2119. Frobenius, G.: Über den Fermatschen Satz. III, SBer. Kgl. Preuß. Akad. Wiss. Berlin, 1914, 653–681 [[2120], vol. 3, pp. 648–676]
2120. Frobenius, G.: Gesammelte Abhandlungen, vols. 1–3. Springer, Berlin (1968)
2121. Fröhlich, A.: Galois Module Structure of Algebraic Integers. Springer, Berlin (1983)
2122. Frostman, O.: Fritz Carlson in memoriam. Acta Math. **90**, ix–xii (1953)
2123. Fuchs, C., Tichy, R.F.: Perfect powers in linear recurring sequences. Acta Arith. **107**, 9–25 (2003)
2124. Fuchs, W.H.J., Wright, E.M.: The 'easier' Waring problem. Q. J. Math. **10**, 190–209 (1939)
2125. Fueter, R.: Die Theorie der Zahlstrahlen. J. Reine Angew. Math. **130**, 197–237 (1905)
2126. Fueter, R.: Die Theorie der Zahlstrahlen, II. J. Reine Angew. Math. **132**, 255–269 (1907)
2127. Fueter, R.: Die Klassenanzahl der Körper der komplexen Multiplikation. Nachr. Ges. Wiss. Göttingen, 1907, 288–298
2128. Fueter, R.: Die verallgemeinerte Kroneckersche Grenzformel und ihre Anwendung auf die Berechnung der Klassenzahlen. Rend. Circ. Mat. Palermo **29**, 380–395 (1910)
2129. Fueter, R.: Die Klassenkörper der komplexen Multiplikation und ihr Einfluss auf die Entwicklung der Zahlentheorie. Jahresber. Dtsch. Math.-Ver. **20**, 1–47 (1911) [Reprint: Teubner, 1958]
2130. Fueter, R.: Abelsche Gleichungen in quadratisch-imaginären Zahlkörpern. Math. Ann. **75**, 177–255 (1914)
2131. Fueter, R.: Kummers Kriterium zum letzten Theorem von Fermat. Math. Ann. **85**, 11–20 (1922)
2132. Fueter, R.: Vorlesungen über die singulären Moduln und die komplexe Multiplikation der elliptischen Funktionen. Teubner, Leipzig (1924–1927)
2133. Fueter, R.: Ueber kubische diophantische Gleichungen. Comment. Math. Helv. **2**, 69–89 (1930)
2134. Fujii, A.: Some remarks on Goldbach's problem. Acta Arith. **32**, 27–35 (1977)
2135. Fujisaki, G.: On the zeta-function of the simple algebra over the field of rational numbers. J. Fac. Sci. Univ. Tokyo **I7**, 567–604 (1958)
2136. Fujisaki, G.: On L-functions of simple algebras over the field of rational numbers. J. Fac. Sci. Univ. Tokyo **I9**, 293–311 (1962)
2137. Fujita, Y.: Torsion subgroups of elliptic curves with non-cyclic torsion over Q in elementary abelian 2-extensions of Q. Acta Arith. **115**, 29–45 (2004)
2138. Fujita, Y.: Torsion subgroups of elliptic curves in elementary abelian 2-extensions of Q. J. Number Theory **114**, 124–134 (2005)

2139. Fujiwara, M.: Bemerkung zur Theorie der Approximation der irrationalen Zahlen durch rationale Zahlen. Tohoku Math. J. **11**, 239–242 (1917)
2140. Fujiwara, M.: Bemerkung zur Theorie der Approximation der irrationalen Zahlen durch rationale Zahlen. Tohoku Math. J. **14**, 109–115 (1918)
2141. Fujiwara, M.: Hasse principle in algebraic equations. Acta Arith. **22**, 267–276 (1972)
2142. Fujiwara, M.: Some applications of a theorem of W.M. Schmidt. Mich. Math. J. **19**, 315–319 (1972)
2143. Fujiwara, M., Sudo, M.: Some forms of odd degree for which the Hasse principle fails. Pac. J. Math. **67**, 161–169 (1976)
2144. Fung, G.[W.], Granville, A., Williams, H.C.: Computation of the first factor of the class number of cyclotomic fields. J. Number Theory **42**, 297–312 (1992)
2145. Fung, G.W., Ströher, H., Williams, H.C., Zimmer, H.G.: Torsion groups of elliptic curves with integral j-invariant over pure cubic fields. J. Number Theory **36**, 12–45 (1990)
2146. Furstenberg, H.: Ergodic behaviour of diagonal measures and a theorem of Szemerédi. J. Anal. Math. **31**, 204–256 (1977)
2147. Furstenberg, H., Katznelson, Y.: An ergodic Szemerédi theorem for commuting transformations. J. Anal. Math. **34**, 275–291 (1978)
2148. Furstenberg, H., Katznelson, Y., Ornstein, D.: The ergodic theoretical proof of Szemerédi's theorem. Bull. Am. Math. Soc. **7**, 527–552 (1982)
2149. Furstenberg, H., Weiss, B.: Markov processes and Ramsey theory for trees. Comb. Probab. Comput. **12**, 547–563 (2003)
2150. Furtwängler, Ph.: Über das Reciprocitätsgesetz der l^{ten} Potenzreste in algebraischen Zahlkörpern, wenn l eine ungerade Primzahl bedeutet. Abhandl. Ges. Wiss. Göttingen Math. Phys. Kl. **2**(3), 3–82 (1902)
2151. Furtwängler, Ph.: Die Konstruktion des Klassenkörpers für solche algebraische Zahlkörper, die eine $l$te Einheitswurzel enthalten und deren Idealklassen eine cyclische Gruppe vom Grade l^h bilden. Nachr. Ges. Wiss. Göttingen, 1903, 203–217
2152. Furtwängler, Ph.: Ueber die Konstruktion des Klassenkörpers für beliebige algebraische Zahlkörper, die eine $l$te Einheitswurzel enthalten. Nachr. Ges. Wiss. Göttingen, 1903, 282–303
2153. Furtwängler, Ph.: Die Konstruktion des Klassenkörpers für beliebige algebraische Zahlkörper. Nachr. Ges. Wiss. Göttingen, 1904, 173–195
2154. Furtwängler, Ph.: Über die Reziprozitätsgesetze zwischen l-ten Potenzresten in algebraischen Zahlkörpern, wenn l eine ungerade Primzahl bedeutet. Math. Ann. **58**, 1–50 (1904)
2155. Furtwängler, Ph.: Eine charakteristische Eigenschaft des Klassenkörpers. Nachr. Ges. Wiss. Göttingen, 1906, 417–434; 1907, 1–24
2156. Furtwängler, P.: Allgemeiner Existenzbeweis für den Klassenkörper eines beliebigen algebraischen Zahlkörpers. Math. Ann. **63**, 1–37 (1907)
2157. Furtwängler, P.: Die Reziprozitätsgesetze für Potenzreste mit Primzahlexponenten in algebraischen Zahlkörpern, I. Math. Ann. **67**, 1–31 (1909)
2158. Furtwängler, P.: Die Reziprozitätsgesetze für Potenzreste mit Primzahlexponenten in algebraischen Zahlkörpern, II. Math. Ann. **72**, 346–386 (1912)
2159. Furtwängler, P.: Die Reziprozitätsgesetze für Potenzreste mit Primzahlexponenten in algebraischen Zahlkörpern, III. Math. Ann. **74**, 413–429 (1913)
2160. Furtwängler, P.: Untersuchungen über die Kreisteilungskörper und den letzten Fermatschen Satz. Nachr. Ges. Wiss. Göttingen, 1910, 554–562
2161. Furtwängler, P.: Allgemeiner Beweis des Zerlegungssatzes für den Klassenkörper. Nachr. Ges. Wiss. Göttingen, 1911, 293–317
2162. Furtwängler, P.: Letzter Fermatscher Satz und Eisensteinsches Reziprozitätsprinzip. SBer. Kais. Akad. Wissensch. Wien **121**, 589–592 (1912)
2163. Furtwängler, P.: Über das Verhalten der Ideale des Grundkörpers im Klassenkörper. Monatshefte Math. Phys. **27**, 1–15 (1916)
2164. Furtwängler, P.: Über die Reziprozitätgesetze für Primzahlpotenzexponente. J. Reine Angew. Math. **157**, 15–25 (1927)

2165. Furtwängler, P.: Über die simultane Approximation von Irrationalzahlen. Math. Ann. **96**, 169–175 (1927)
2166. Furtwängler, P.: Über die simultane Approximation von Irrationalzahlen, II. Math. Ann. **99**, 71–83 (1928)
2167. Furtwängler, P.: Beweis des Hauptidealsatzes für die Klassenkörper algebraischer Zahlkörper. Abh. Math. Semin. Univ. Hamb. **7**, 14–36 (1930)
2168. Fuss, P.H.: Correspondance mathématique et physique de quelques célèbres géomètres du XVIIIéme siécle. St. Pétersbourg (1843) [Reprint: Johnson Reprint Co. 1968]
2169. Gaál, I.: Inhomogeneous discriminant form equations and integral elements with given discriminant over finitely generated integral domains. Publ. Math. (Debr.) **34**, 109–122 (1987)
2170. Gaál, I., Lettl, G.: A parametric family of quintic Thue equations. Math. Comput. **69**, 851–859 (2000)
2171. Gaál, I., Lettl, G.: A parametric family of quintic Thue equations, II. Monatshefte Math. **131**, 29–35 (2000)
2172. Gaborit, P.: Construction of new extremal unimodular lattices. Electron. J. Comb. **25**, 549–564 (2004)
2173. Gabriel Oltramare. Enseign. Math. **8**, 378–382 (1906)
2174. Gafurov, N.: An improvement of a certain asymptotic formula of C. Hooley. Dokl. Akad. Nauk Tadjik. SSR **15**(10), 6–10 (1972) (in Russian)
2175. Gage, W.G.: Representations in the form $xy + yz + zx$. Am. J. Math. **51**, 345–348 (1929)
2176. Gajda, W., Górnisiewicz, K.: Linear dependence in Mordell-Weil groups. J. Reine Angew. Math. **630**, 219–233 (2009)
2177. Gajraj, J.: Simultaneous approximation to certain polynomials. J. Lond. Math. Soc. **14**, 527–534 (1976)
2178. Gallagher, P.[X.]: Approximation by reduced fractions. J. Math. Soc. Jpn. **13**, 342–345 (1961)
2179. Gallagher, P.X.: The large sieve. Mathematika **14**, 14–20 (1967)
2180. Gallagher, P.X.: Bombieri's mean value theorem. Mathematika **15**, 1–6 (1968)
2181. Gallagher, P.X.: A large sieve density estimate near $\sigma = 1$. Invent. Math. **11**, 329–339 (1970)
2182. Gallagher, P.X.: The larger sieve. Acta Arith. **18**, 77–81 (1971)
2183. Gallagher, P.X.: Sieving by prime powers. Acta Arith. **24**, 491–497 (1974)
2184. Gallagher, P.X.: Primes and powers of 2. Invent. Math. **29**, 125–142 (1975)
2185. Gallagher, P.X., Mueller, J.H.: Primes and zeros in short intervals. J. Reine Angew. Math. **303/304**, 205–220 (1978)
2186. Gallardo, L.H.: Congruences for odd perfect numbers modulo some powers of 2. Int. J. Contemp. Math. Sci. **3**, 999–1016 (2008)
2187. Gallardo, L.H., Rahavandrainy, O.: New congruences for odd perfect numbers. Rocky Mt. J. Math. **36**, 225–235 (2006)
2188. Gallot, Y., Moree, P., Zudilin, W.[V.]: The Erdős-Moser equation $1^k + 2^k + \cdots + (m-1)^k = m^k$ revisited using continued fractions. Preprint (2009). arxiv:0907.1356
2189. Ganelius, T.H.: Tauberian Remainder Theorems. In: Lecture Notes in Math., vol. 232, pp. 1–75. Springer, Berlin (1971)
2190. Gangadharan, K.S.: Two classical lattice point problems. Proc. Camb. Philos. Soc. **57**, 699–721 (1961)
2191. Gannon, T.: Monstrous Moonshine: The first twenty years. Bull. Lond. Math. Soc. **38**, 1–33 (2006)
2192. Gannon, T.: Moonshine Beyond the Monster. The Bridge Connecting Algebra, Modular Forms and Physics. Cambridge University Press, Cambridge (2006)
2193. Garcia, M., Pedersen, J.M., Riele, H.: Amicable pairs, a survey. In: High Primes and Misdemeanors, pp. 179–196. Am. Math. Soc., Providence (2004)
2194. Gardiner, V.L., Lazarus, R.B., Stein, P.R.: Solutions of the diophantine equation $x^3 + y^3 = z^3 - d$. Math. Comput. **18**, 408–413 (1964)
2195. Gårding, L.: Marcel Riesz in memoriam. Acta Math. **124**, x–xi (1970)

2196. Gardner, R.J.: The Brunn–Minkowski inequality. Bull. Am. Math. Soc. **39**, 355–405 (2002)
2197. Garrison, B.K.: A nontransformational proof of Mann's density theorem. J. Reine Angew. Math. **245**, 41–46 (1970)
2198. Garsia, A.M., Milne, S.C.: Method for constructing bijections for classical partition identities. Proc. Natl. Acad. Sci. USA **78**, 2026–2028 (1981)
2199. Garsia, A.M., Milne, S.C.: A Rogers-Ramanujan bijection. J. Comb. Theory **A31**, 289–339 (1981)
2200. Garvan, F.G.: A simple proof of Watson's partition congruences for powers of 7. J. Aust. Math. Soc. **A36**, 316–334 (1984)
2201. Garvan, F.G.: Combinatorial interpretation of Ramanujan's partition congruences. In: Andrews, G.E., Askey, R.A., Berndt, B.C., Ramanathan, K.G., Rankin, R.A. (eds.) Ramanujan Revisited (Proceedings of the Centenary Conference), pp. 29–45. Academic Press, New York (1988)
2202. Garvan, F.G.: New combinatorial interpretations of Ramanujan's partition congruences mod 5, 7 and 11. Trans. Am. Math. Soc. **305**, 47–77 (1988)
2203. Garvan, F.[G.], Kim, D., Stanton, D.: Cranks and t-cores. Invent. Math. **101**, 1–17 (1990)
2204. Garvan, F.[G.], Stanton, D.: Sieved partition functions and q-binomial coefficients. Math. Comput. **55**, 299–311 (1990)
2205. Gasbarri, C.: The strong abc conjecture over function fields [after McQuillan and Yamanoi]. Astérisque **326**, 219–256 (2009)
2206. Gatteschi, L.: Un perfezionamento di un teorema di I. Schur sulla frequenza dei numeri primi. Boll. Unione Mat. Ital. **2**, 123–125 (1947)
2207. Gaudry, P., Gürel, N.: Counting points in medium characteristic using Kedlaya's algorithm. Exp. Math. **12**, 395–402 (2003)
2208. Gauss, C.F.: Disquisitiones Arithmeticae. Fleischer, Leipzig (1801) [[2214], vol. 1, pp. 1–474; German translation: Untersuchungen über höhere Arithmetik, Springer, 1889; reprint: Chelsea, 1965; English translation: Yale, 1966; Springer, 1986]
2209. Gauss, C.F.: Summatio quarumdam serierum singularium. Comm. Soc. Reg. Sci. Gottingensis **1** (1811) [[2214], vol. 2, pp. 9–45]
2210. Gauss, C.F.: Letter to Laplace, 30.01.1812. In: [2263], vol. 10_1, pp. 371–374
2211. Gauss, C.F.: Theoria residuorum biquadraticorum. Comm. Soc. Reg. Sci. Gottingensis, **6** (1828); **7** (1832) [[2214], vol. 2, pp. 65–92, 93–148]
2212. Gauss, C.F.: Anzeige: "Untersuchungen über die Eigenschaften der positiven ternären quadratischen Formen" von Ludwig August Seeber. Göttingische gelehrte Anzeigen **7** (1831). Also J. Reine Angew. Math., **20**, 1840, 312–320 [[2214], vol. 2, pp. 188–196]
2213. Gauss, C.F.: De nexu inter multitudinem classium, in quae formae binariae secundi gradus distribuuntur, earumque determinantem, I, II. Göttinger Abhandl., 1834, 1837. [[2214], vol. 2, pp. 269–291]
2214. Gauss, C.F.: Werke. vols. 1–12. Teubner/Springer, Leipzig/Berlin (1863–1929) [Reprint: G. Olms, 1973]
2215. Gebel, J., Pethő, A., Zimmer, H.G.: Computing integral points on elliptic curves. Acta Arith. **68**, 171–192 (1994)
2216. Gebel, J., Pethő, A., Zimmer, H.G.: On Mordell's equation. Compos. Math. **110**, 335–367 (1998)
2217. Gebel, J., Zimmer, H.G.: Computing the Mordell-Weil group of an elliptic curve over Q. In: Elliptic Curves and Related Topics, pp. 61–83. Am. Math. Soc., Providence (1994)
2218. Gegenbauer, L.: Arithmetische Sätze. SBer. Kais. Akad. Wissensch. Wien **112**, 1055–1078 (1885)
2219. Gegenbauer, L.: Asymptotische Gesetze der Zahlentheorie. Denkschr. Akad. Wiss. Wien **49**, 37–80 (1885)
2220. Gelbart, S.: Automorphic forms and Artin's conjecture. In: Lecture Notes in Math., vol. 627, pp. 241–276. Springer, Berlin (1977)
2221. Gelbcke, M.: Zum Waringschen Problem. Math. Ann. **105**, 637–652 (1931)
2222. Gelfond, A.O.: Sur les nombres transcendants. C. R. Acad. Sci. Paris **189**, 1224–1228 (1929)

2223. Gelfond, A.O.: On transcendental numbers. In: Pervyi Vsesojuznyi S'ezd Matematikov v Kharkove (1930) (in Russian) [[2231], pp. 14–15]

2224. Gelfond, A.O.: On the seventh problem of Hilbert. Dokl. Akad. Nauk SSSR **2**, 1–6 (1934) (in Russian) [[2231], pp. 48–50]

2225. Gelfond, A.O.: Sur la septième problem de Hilbert. Izv. Akad. Nauk SSSR, Ser. Mat. **7**, 623–634 (1934)

2226. Gelfond, A.O.: On the approximation of the ratio of logarithms of two algebraic numbers by algebraic numbers. Izv. Akad. Nauk SSSR, Ser. Mat. **3**, 509–518 (1939) (in Russian)

2227. Gelfond, A.O.: Approximation of algebraic irrationalities and their logarithms. Vestn. Mosk. Univ. **9**, 3–25 (1948) (in Russian) [[2231], pp. 129–150]

2228. Gelfond, A.O.: On the algebraic independence of algebraic powers of algebraic numbers. Dokl. Akad. Nauk SSSR **64**, 277–280 (1949) (in Russian)

2229. Gelfond, A.O.: On the algebraic independence of transcendental numbers of certain classes. Usp. Mat. Nauk **4**, 14–48 (1949) (in Russian)

2230. Gelfond, A.O.: Transcendental and algebraic numbers. GITTL, Moscow (1952) (in Russian) [English translation: Dover, 1960]

2231. Gelfond, A.O.: Selected Papers. Nauka, Moscow (1973) (in Russian)

2232. Gelfond, A.O., Feldman, N.I.: On the measure of relative transcendentality of certain numbers. Izv. Akad. Nauk SSSR, Ser. Mat. **14**, 493–500 (1950)

2233. Gelfond, A.O., Linnik, Yu.V.: On Thue's method and the problem of effectivization in quadratic fields. Dokl. Akad. Nauk SSSR **61**, 773–776 (1948) (in Russian) [[3929], vol. 2, pp. 40–44]

2234. Gelfond, A.O., Linnik, Yu.V.: Elementary Methods in the Analytic Theory of Numbers. Fizmatgiz, Moscow (1962) (in Russian) [English translations: Rand McNally & Co., 1965; Pergamon Press, 1966. French translation: Méthodes élémentaires dans la théorie analytique des nombres, Gauthier-Villars, 1965]

2235. Gerst, I.: On the theory of *n*th power residues and a conjecture of Kronecker. Acta Arith. **17**, 121–139 (1970)

2236. Getz, J.: On congruence properties of the partition function. Int. J. Math. Math. Sci. **23**, 493–496 (2000)

2237. Gibson, D.J.: A covering system with least modulus 25. Math. Comput. **78**, 1127–1146 (2009)

2238. Giles, J.R., Wallis, J.S.: George Szekeres. With affection and respect. J. Aust. Math. Soc. A **21**, 385–392 (1976)

2239. Gimbel, S., Jaroma, J.H.: Sylvester: ushering in the modern era of research on odd perfect numbers. Integers **3**(#A16), 1–26 (2003)

2240. Girstmair, K., Kühleitner, M., Müller, W., Nowak, W.G.: The Piltz divisor problem in number fields: an improved lower bound by Soundararajan's method. Acta Arith. **117**, 187–206 (2005)

2241. Giudici, R.E., Muskat, J.B., Robinson, S.F.: On the evaluation of Brewer's character sums. Trans. Am. Math. Soc. **171**, 317–347 (1972)

2242. Glaisher, J.W.L.: Classes of recurring formulae involving Bernoullian numbers. Messenger Math. **28**, 36–79 (1898/1899)

2243. Glaisher, J.W.L.: On a set of coefficients analogous to the Eulerian numbers. Proc. Lond. Math. Soc. **31**, 216–235 (1899)

2244. Glaisher, J.W.L.: A congruence theorem relating to Eulerian numbers and other coefficients. Proc. Lond. Math. Soc. **32**, 171–198 (1900)

2245. Glaisher, J.W.L.: A series for $\frac{\pi}{\sqrt{7}}$. Messenger Math. **31**, 50–52 (1901/1902)

2246. Glaisher, J.W.L.: On series for $\frac{\pi}{\sqrt{p}}$. Messenger Math. **31**, 98–115 (1901/1902)

2247. Glaisher, J.W.L.: A general congruence theorem relating to the Bernoullian function. Proc. Lond. Math. Soc. **33**, 27–56 (1901)

2248. Glaisher, J.W.L.: On the number of representations of a number as a sum of $2r$ squares, where $2r$ does not exceed eighteen. Proc. Lond. Math. Soc. **5**, 479–490 (1907)

2249. Glimm, J., et al. (ed.): The Legacy of John von Neumann. Proceedings of the Summer Research Institute Held at Hofstra University, Hempstead, New York, May 29–June 4, 1988. Proc. Symposia Pure Math., vol. 50. Am. Math. Soc., Providence (1990)
2250. Gloden, A.: Sur les égalités multigrades. Luxembourg 1938 [German translation: Mehrgradige Gleichungen, Noordhoff, 1944]
2251. Gluchoff, A.: Pure mathematics applied in early twentieth-century America: the case of T.H. Gronwall, consulting mathematician. Hist. Mat. **32**, 312–357 (2005)
2252. Gnedenko, B.V., Kolmogorov, A.N.: Aleksandr Jakovlevič Hinčin. Usp. Mat. Nauk **15**(4), 97–110 (1960) (in Russian)
2253. Gödel, K.: Über formal unentscheidbare Sätze der Principia Mathematica und verwandter Systeme. I. Monatshefte Math. Phys. **38**, 173–198 (1931)
2254. Gogišvili, G.P.: A relation between the number of representation of numbers by quadratic forms and the corresponding singular series. Tr. Tbil. Mat. Inst. **57**, 40–62 (1977)
2255. Gogišvili, G.P.: A type of formula for the number of representations of numbers by positive quadratic forms. Tr. Tbil. Mat. Inst. **63**, 25–35 (1980)
2256. Goldberg, M.: Morris Newman—a discrete mathematician for all seasons. Linear Algebra Appl. **254**, 7–18 (1997)
2257. Goldfeld, D.M.: A simple proof of Siegel's theorem. Proc. Natl. Acad. Sci. USA **71**, 1055 (1974)
2258. Goldfeld, D.: The class number of quadratic fields and the conjectures of Birch and Swinnerton-Dyer. Ann. Sc. Norm. Super. Pisa, Cl. Sci. **3**, 624–663 (1976)
2259. Goldfeld, D.: Modular forms, elliptic curves and the ABC-conjecture. In: A Panorama of Number Theory or the View from Baker's Garden, Zürich, 1999, pp. 128–147. Cambridge University Press, Cambridge (2002)
2260. Goldfeld, D., Hoffstein, J.: Eisenstein series of $1/2$-integral weight and the mean value of real Dirichlet L-series. Invent. Math. **80**, 185–208 (1985)
2261. Goldfeld, D., Lieman, D.: Effective bounds on the size of the Tate-Shafarevich group. Math. Res. Lett. **3**, 309–318 (1996)
2262. Goldfeld, D., Sarnak, P.: Sums of Kloosterman sums. Invent. Math. **71**, 143–250 (1983)
2263. Goldfeld, D., Schinzel, A.: On Siegel's zero. Ann. Sc. Norm. Super. Pisa, Cl. Sci. **2**, 571–583 (1975) [[5449], vol. 2, pp. 1199–1210]
2264. Goldfeld, D., Szpiro, L.: Bounds for the order of the Tate-Shafarevich group. Compos. Math. **97**, 71–87 (1995)
2265. Goldstein, C.: Courbes elliptiques et théorie d'Iwasawa. Publ. Math. d'Orsay **82**, 1–41 (1982)
2266. Goldstein, C., Schappacher, N.: Séries d'Eisenstein et fonctions L de courbes elliptiques à multiplication complexe. J. Reine Angew. Math. **327**, 184–218 (1981)
2267. Goldstein, L.J.: On a conjecture of Hecke concerning elementary class number formulas. Manuscr. Math. **9**, 245–305 (1973)
2268. Goldston, D.A.: On a result of Littlewood concerning prime numbers. Acta Arith. **40**, 263–271 (1981/1982)
2269. Goldston, D.A.: On a result of Littlewood concerning prime numbers, II. Acta Arith. **43**, 49–51 (1983)
2270. Goldston, D.A., Graham, S.W., Pintz, J., Yildirim, C.Y.: Small gaps between primes or almost primes. Trans. Am. Math. Soc. **361**, 5285–5330 (2009)
2271. Goldston, D.A., Graham, S.W., Pintz, J., Yildirim, C.Y.: Small gaps between products of two primes. Proc. Lond. Math. Soc. **98**, 741–774 (2009)
2272. Goldston, D.A., Motohashi, Y., Pintz, J., Yildirim, C.Y.: Small gaps between primes exist. Proc. Jpn. Acad. Sci. **82**, 61–65 (2006)
2273. Goldston, D.A., Pintz, J., Yildirim, C.Y.: Primes in tuples, I. Ann. Math. **170**, 819–862 (2009)
2274. Goldston, D.A., Pintz, J., Yildirim, C.Y.: Primes in tuples, II. Acta Math. **204**, 1–47 (2010)
2275. Golomb, S.W.: An algebraic algorithm for the representation problems of the Ahmes papyrus. Am. Math. Mon. **69**, 785–786 (1962)

2276. Golomb, S.W.: The lambda method in prime number theory. J. Number Theory **2**, 193–198 (1970)
2277. Golubeva, E.P.: Representation of large numbers by ternary quadratic forms. Mat. Sb. **129**, 40–54 (1986) (in Russian)
2278. Gonzalez-Avilés, C.D.: On the conjecture of Birch and Swinnerton-Dyer. Trans. Am. Math. Soc. **349**, 4181–4200 (1997)
2279. Good, A.: Ein Ω-Resultat für das quadratische Mittel der Riemannscher Zetafunktion auf der kritischen Linie. Invent. Math. **41**, 233–251 (1977)
2280. Good, I.J.: The fractional dimensional theory of continued fractions. Proc. Camb. Philos. Soc. **37**, 199–228 (1941)
2281. Goormaghtigh, R.: L'Intermédiaire Math. **24**, 88 (1917)
2282. Gordon, B.: A combinatorial generalization of the Rogers-Ramanujan identities. Am. J. Math. **83**, 393–399 (1961)
2283. Gordon, B.: Some continued fractions of the Rogers-Ramanujan type. Duke Math. J. **32**, 741–748 (1965)
2284. Gordon, B.: Ramanujan congruences for p_{-k} mod 11^r. Glasg. Math. J. **24**, 107–123 (1983)
2285. Gorškov, D.S.: On the distance from zero on the interval [0, 1] of polynomials with integral coefficients. In: Trudy III Vsesoj. Mat. S'ezda, Moscow, vol. 4 (1956) (in Russian)
2286. Goto, T., Ohno, Y.: Odd perfect numbers have a prime factor exceeding 10^8. Math. Comput. **77**, 2008 (1859–1868)
2287. Gottschling, E.: Explizite Bestimmung der Randflächen des Fundamentalbereiches der Modulgruppe zweiten Grades. Math. Ann. **138**, 103–124 (1959)
2288. Gottschling, E.: Über die Fixpunkte der Siegelschen Modulgruppe. Math. Ann. **143**, 111–149 (1961)
2289. Götzky, F.: Über eine zahlentheoretische Anwendung von Modulfunktionen zweier Veränderlichen. Math. Ann. **100**, 411–437 (1928)
2290. Goubin, L.: Sommes d'exponentielles et principe de l'hyperbole. Acta Arith. **73**, 303–324 (1995)
2291. Gouillon, N.: Explicit lower bounds for linear forms in two logarithms. J. Théor. Nr. Bordx. **18**, 125–146 (2006)
2292. Gourdon, X.: The 10^{13} first zeros of the Riemann zeta function, and zeros computation at very large height. Preprint (2004)
2293. Gowers, W.T.: A new proof of Szemerédi's theorem for arithmetic progressions of length four. Geom. Funct. Anal. **8**, 529–551 (1998)
2294. Gowers, W.T.: A new proof of Szemerédi's theorem. Geom. Funct. Anal. **11**, 465–588 (2001); err.: p. 869
2295. Grabner, P.J., Tichy, R.F.: Remark on an inequality of Erdős-Turán-Koksma. Anz. Österreich. AW, Math.-Natur. Kl. **127**, 15–22 (1990)
2296. Grace, J.H.: Note on a diophantine approximation. Proc. Lond. Math. Soc. **17**, 316–319 (1919)
2297. Gradštein, I.S.: On odd perfect numbers. Mat. Sb. **32**, 476–510 (1925) (in Russian)
2298. Graham, R.L.: On quadruples of consecutive kth power residues. Proc. Am. Math. Soc. **15**, 196–197 (1964)
2299. Graham, R.L., Rothschild, B.L.: A short proof of van der Waerden's theorem on arithmetic progression. Proc. Am. Math. Soc. **42**, 385–386 (1974)
2300. Graham, S.[W.]: Applications of sieve methods. Ph.D. thesis, Univ. of Michigan, Ann Arbor (1979)
2301. Graham, S.[W.]: On Linnik's constant. Acta Arith. **39**, 163–179 (1981)
2302. Graham, S.W.: The distribution of square-free numbers. J. Lond. Math. Soc. **24**, 54–64 (1981)
2303. Graham, S.W.: An algorithm for computing optimal exponent pairs. J. Lond. Math. Soc. **33**, 203–218 (1986)
2304. Graham, S.W., Kolesnik, G.: On the difference between consecutive squarefree integers. Acta Arith. **49**, 435–447 (1988)

2305. Graham, S.W., Kolesnik, G.: Van der Corput's Method of Exponential Sums. Cambridge University Press, Cambridge (1991)
2306. Graham, S.W., Pintz, J.: The distribution of r-free numbers. Acta Math. Acad. Sci. Hung. **53**, 213–236 (1989)
2307. Graham, S.W., Ringrose, C.J.: Lower bounds for least quadratic non-residues. Prog. Math. **85**, 269–309 (1990)
2308. Gram, J.P.: Ludvig Henrik Ferdinand Oppermann. Mat. Tidsskr. **1**, 137–144 (1883) (in Danish)
2309. Gram, J.P.: Note sur le calcul de la fonction $\zeta(s)$ de Riemann. Oversigt Kong. Dansk. Vid. Selsk. Forh., 1895, 303–308
2310. Gram, J.P.: Note sur les zéros de la fonction $\zeta(s)$. Acta Math. **27**, 289–304 (1903)
2311. Grant, D.: Review of [1715]. Math. Rev. 89h:11012
2312. Granville, A.: The set of exponents, for which Fermat's last theorem is true, has density one. C. R. Acad. Sci. Paris Acad. Sci. Soc. R. Can. **7**, 55–60 (1985)
2313. Granville, A.: On Krasner's criteria for the first case of Fermat's last theorem. Manuscr. Math. **56**, 67–70 (1986)
2314. Granville, A.: On the size of the first factor of the class number of a cyclotomic field. Invent. Math. **100**, 321–338 (1990)
2315. Granville, A.: Integers, without large prime factors, in arithmetic progressions, I. Acta Math. **170**, 255–273 (1993)
2316. Granville, A.: Integers, without large prime factors, in arithmetic progressions, II. Philos. Trans. R. Soc. Lond. Ser. A, Math. Phys. Sci. **345**, 349–362 (1993)
2317. Granville, A.: ABC allows us to count squarefrees. Internat. Math. Res. Notices, 1998, 991–1009
2318. Granville, A.: It is easy to determine whether a given integer is prime. Bull. Am. Math. Soc. **42**, 3–38 (2005)
2319. Granville, A., Monagan, M.B.: The first case of Fermat's last theorem is true for all prime exponents up to 714 591 416 091 389. Trans. Am. Math. Soc. **306**, 329–359 (1988)
2320. Granville, A., Pomerance, C.: On the least prime in certain arithmetic progressions. J. Lond. Math. Soc. **41**, 193–200 (1990)
2321. Granville, A., Pomerance, C.: Two contradictory conjectures concerning Carmichael numbers. Math. Comput. **71**, 883–908 (2002)
2322. Granville, A., Ramaré, O.: Explicit bounds on exponential sums and the scarcity of squarefree binomial coefficients. Mathematika **43**, 73–107 (1996)
2323. Granville, A., Soundararajan, K.: A binary additive problem of Erdős and the order of 2 mod p^2. Ramanujan J. **2**, 283–298 (1998)
2324. Granville, A., Soundararajan, K.: The spectrum of multiplicative functions. Ann. Math. **153**, 407–470 (2001)
2325. Granville, A., Soundararajan, K.: Upper bounds for $|L(1,\chi)|$. Q. J. Math. **53**, 265–284 (2002)
2326. Granville, A., Soundararajan, K.: Large character sums: pretentious characters and the Pólya-Vinogradov theorem. J. Am. Math. Soc. **20**, 357–384 (2007)
2327. Granville, A., Stark, H.M.: ABC implies no "Siegel zeros" for L-functions of characters with negative discriminant. Invent. Math. **139**, 509–523 (2000)
2328. Granville, A., van de Lune, J., ter Riele, H.J.J.: Checking the Goldbach conjecture on a vector computer. In: Number Theory and Applications, Banff, AB, 1988, pp. 423–433. Kluwer Academic, Dordrecht (1989)
2329. Grauert, H.: Mordell's Vermutung über rationale Punkte auf algebraischen Kurven und Funktionenkörper. Publ. Math. Inst. Hautes Études Sci. **25**, 131–149 (1965)
2330. Greaves, G.: On the representation of a number as a sum of two fourth powers. Math. Z. **94**, 223–234 (1966)
2331. Greaves, G.: Large prime factors of binary forms. J. Number Theory **3**, 35–59 (1971); corr., vol. 9, 1977, pp. 561–562
2332. Greaves, G.: An application of a theorem of Barban, Davenport and Halberstam. Bull. Lond. Math. Soc. **6**, 1–9 (1974)

2333. Green, B.: Roth's theorem in the primes. Ann. Math. **161**, 1609–1636 (2005)
2334. Green, B., Tao, T.: The primes contain arbitrarily long arithmetic progressions. Ann. Math. **167**, 481–547 (2008)
2335. Green, B., Tao, T.: Linear equations in primes. Ann. Math. **171**, 1753–1850 (2010)
2336. Green, B., Tao, T.: Restriction theory of the Selberg sieve, with applications. J. Théor. Nr. Bordx. **18**, 147–182 (2006)
2337. Green, B., Tao, T.: New bounds for Szemerédi's theorem. II. A new bound for $r_4(N)$. In: Analytic Number Theory, pp. 180–204. Cambridge University Press, Cambridge (2009)
2338. Greenberg, M.J.: Lectures on Forms in Many Variables. Benjamin, Elmsford (1969)
2339. Greenberg, R.: On the Birch and Swinnerton-Dyer conjecture. Invent. Math. **72**, 241–265 (1983)
2340. Greenberg, R., Stevens, G.: p-adic L-functions and p-adic periods of modular forms. Invent. Math. **111**, 407–447 (1993)
2341. Grekos, G., Haddad, L., Helou, C., Pihko, J.: On the Erdős-Turán conjecture. J. Number Theory **102**, 339–352 (2003)
2342. Grekos, G., Haddad, L., Helou, C., Pihko, J.: Representation functions, Sidon sets and bases. Acta Arith. **130**, 149–156 (2007)
2343. Greminger, H.: Sur le nombre e. Enseign. Math. **29**, 255–259 (1930)
2344. Griess, R.: The friendly giant. Invent. Math. **68**, 1–102 (1982)
2345. Grigorov, G., Jorza, A., Patrikis, S., Stein, W.A., Tarniţa, C.: Computational verification of the Birch and Swinnerton-Dyer conjecture for individual elliptic curves. Math. Comput. **78**, 2397–2425 (2009)
2346. Grimshaw, M.E.: Hans Ludwig Hamburger. J. Lond. Math. Soc. **33**, 377–383 (1958)
2347. Gritsenko, S.A.: Three additive problems. Izv. Ross. Akad. Nauk, Ser. Mat. **56**, 1198–1206 (1992) (in Russian)
2348. Gritzmann, P., Sturmfels, B.: Victor L. Klee 1925–2007. Not. Am. Math. Soc. **55**, 467–475 (2008)
2349. Grodzenskiĭ, S.Ya.: Andreĭ Andreevič Markov. 1856–1922. Nauka, Moscow (1987) (in Russian)
2350. Gronwall, T.H.: Sur les séries de Dirichlet correspondant à des caractéres complexes. Rend. Circ. Mat. Palermo **35**, 145–159 (1913)
2351. Gronwall, T.H.: Theory of prime numbers. Bull. Am. Math. Soc. **20**, 368–376 (1913/1914)
2352. Grošev, A.V.: Un théorème sur les systemes de formes linéaires. Dokl. Akad. Nauk SSSR **19**, 151–152 (1938)
2353. Gross, B.H.: Arithmetic on Elliptic Curves with Complex Multiplication. Lecture Notes in Math., vol. 776. Springer, Berlin (1980)
2354. Gross, B.H.: p-adic L-series at $s=0$. J. Fac. Sci. Univ. Tokyo **28**, 979–994 (1981)
2355. Gross, B.H., Koblitz, N.: Gauss sums and the p-adic Γ-function. Ann. Math. **109**, 569–581 (1979)
2356. Gross, B.H., Zagier, D.: Points de Heegner et dérivées de fonctions L. C. R. Acad. Sci. Paris **297**, 85–87 (1983)
2357. Gross, B.H., Zagier, D.: Heegner points and derivatives of L-series. Invent. Math. **84**, 225–320 (1986)
2358. Gross, R., Silverman, J.: S-integer points on elliptic curves. Pac. J. Math. **167**, 263–288 (1995)
2359. Großmann, J.: Über die Nullstellen der Riemannscher ζ-Funktion und der Dirichletschen L-Funktionen. Dissertation, Univ. Göttingen (1913)
2360. Grosswald, E.: On the structure of some subgroups of the modular group. Am. J. Math. **72**, 809–834 (1950)
2361. Grosswald, E.: Negative discriminants of binary quadratic forms with one class in each genus. Acta Arith. **8**, 295–306 (1963)
2362. Grosswald, E.: On some generalizations of theorems by Landau and Pólya. Isr. J. Math. **3**, 211–220 (1965)
2363. Grosswald, E.: Oscillation theorems of arithmetical functions. Trans. Am. Math. Soc. **126**, 1–28 (1967)

2364. Grosswald, E.: On Burgess' bound for primitive roots modulo primes and an application to $\Gamma(p)$. Am. J. Math. **103**, 1171–1183 (1981)
2365. Grosswald, E.: Arithmetic progressions that consist only of primes. J. Number Theory **14**, 9–31 (1982)
2366. Grosswald, E.: Representations of Integers as Sums of Squares. Springer, Berlin (1985)
2367. Grosswald, E., Calloway, A., Calloway, J.: The representation of integers by three positive squares. Proc. Am. Math. Soc. **10**, 451–455 (1959)
2368. Grothendieck, A.: Formule de Lefschetz et rationalité des fonctions L. In: Sém. Bourbaki, vol. 279, pp. 41–55. Benjamin, Elmsford (1966)
2369. Grube, F.: Ueber einige Euler'sche Sätze aus der Theorie der quadratischen Formen. Z. Math. Phys. **19**, 492–519 (1874)
2370. Gruber, P.M.: Convex and Discrete Geometry. Springer, Berlin (2007)
2371. Gruber, P.M., Wills, J.M. (eds.): Handbook of Convex Geometry, vols. I–II. North-Holland, Amsterdam (1993)
2372. Gruenberger, F., Armerding, G.: Statistics of the first six million primes. Report P-2460, The Rand Corporation (1961)
2373. Grunewald, F., Segal, D.: On the integer solutions of quadratic equations. J. Reine Angew. Math. **569**, 13–45 (2004)
2374. Grunewald, F.J., Zimmert, R.: Über einige rationale elliptische Kurven mit freiem Rang ≥ 8. J. Reine Angew. Math. **296**, 100–107 (1977)
2375. Grunsky, H.: Ludwig Bieberbach zum Gedächtnis. Jahresber. Dtsch. Math.-Ver. **88**, 190–205 (1986)
2376. Grytczuk, A.: On the equation of Fermat in the second case. C. R. Acad. Sci. Paris **317**, 25–28 (1993)
2377. Grytczuk, A., Schinzel, A.: On Runge's theorem about Diophantine equations. In: Sets, Graphs and Numbers, Budapest, 1991, pp. 329–356. North-Holland, Amsterdam (1992) [[5449], vol. 1, pp. 93–115]
2378. Guenther, R.B.: Obituary: Lothar Collatz, 1910–1990. Aequ. Math. **43**, 117–119 (1992)
2379. Guinand, A.P.: A class of self-reciprocal functions connected with summation formulae. Proc. Lond. Math. Soc. **43**, 439–448 (1937)
2380. Gunderson, N.G.: Derivation of criteria for the first case of Fermat's last theorem and the combination of these criteria to produce a new lower bound for the exponent. Ph.D. thesis, Cornell Univ. (1949)
2381. Gundlach, K.-B.: Über die Darstellung der ganzen Spitzenformen zu den Idealstufen der Hilbertschen Modulgruppe und die Abschätzung ihrer Fourierkoeffizienten. Acta Math. **92**, 309–345 (1954)
2382. Gunning, R.C.: Lectures on Modular Forms. Princeton University Press, Princeton (1962)
2383. Günther, A.: Über transzendente p-adische Zahlen. I. Ein Satz über algebraische Abhängigkeit p-adischer Funktionen als Prinzip für Transzendenzbeweise p-adischer Zahlen. J. Reine Angew. Math. **192**, 155–166 (1953)
2384. Guo, Y., Le, M.H.: A note on the exponential Diophantine equation $x^2 - 2^m = y^n$. Proc. Am. Math. Soc. **123**, 3627–3629 (1995)
2385. Gupta, H.: A table of partitions. Proc. Lond. Math. Soc. **39**, 142–149 (1935)
2386. Gupta, H.: A table of partitions, II. Proc. Lond. Math. Soc. **42**, 546–549 (1937)
2387. Gupta, H.: On a table of values of $L(n)$. J. Indian Math. Soc. **12**, 407–409 (1940)
2388. Gupta, H.: Some idiosyncratic numbers of Ramanujan. Proc. Indian Acad. Sci. Math. Sci. **13**, 519–520 (1941)
2389. Gupta, H.: Congruence properties of $\tau(n)$. Proc. Benares Math. Soc. **5**, 17–22 (1943)
2390. Gupta, H.: A table of values of $\tau(n)$. Proc. Natl. Inst. Sci. India **13**, 201–206 (1947)
2391. Gupta, H.: The vanishing of Ramanujan's function. Curr. Sci. **17**, 180 (1948)
2392. Gupta, R.: Ramification in the Coates-Wiles tower. Invent. Math. **81**, 59–69 (1985)
2393. Gupta, R., Murty, M.R.: A remark on Artin's conjecture. Invent. Math. **78**, 127–130 (1984)
2394. Gupta, R., Murty, M.R.: Cyclicity and generation of points mod$p$ on elliptic curves. Invent. Math. **101**, 225–235 (1990)

2395. Guralnick, R.M., Müller, P., Saxl, J.: The rational function analogue of a question of Schur and exceptionality of permutation representations. Mem. Am. Math. Soc. **162**, 1–79 (2003)
2396. Gurwood, C.: Diophantine approximations and the Markov chains. Ph.D. thesis, New York Univ. (1976)
2397. Güting, R.: Approximation of algebraic numbers by algebraic numbers. Mich. Math. J. **8**, 149–159 (1961)
2398. Guy, R.K.: Unsolved Problems in Number Theory. Springer, Berlin (1981); 2nd ed. 1994, 3rd ed. 2004
2399. Gyires, B.: Alfréd Rényi (1920–1970). Publ. Math. (Debr.) **17**, 1–17 (1970)
2400. Győry, K.: Sur les polynômes à coefficients entiers et de discriminant donné. Acta Arith. **23**, 419–426 (1973)
2401. Győry, K.: On the number of solutions of linear equations in units of an algebraic number field. Comment. Math. Helv. **54**, 583–600 (1979)
2402. Győry, K.: On S-integral solutions of norm form, discriminant form and index form equations. Studia Sci. Math. Hung. **16**, 149–161 (1981)
2403. Győry, K.: Bounds for the solutions of norm form, discriminant form and index form equations in finitely generated integral domains. Acta Math. Acad. Sci. Hung. **42**, 45–80 (1983)
2404. Győry, K.: Some recent applications of S-unit equations. Astérisque **209**, 17–38 (1992)
2405. Győry, K.: On the numbers of families of solutions of systems of decomposable form equations. Publ. Math. (Debr.) **42**, 65–101 (1993)
2406. Győry, K.: On the Diophantine equation $\binom{n}{k} = x^l$. Acta Arith. **80**, 289–295 (1997)
2407. Győry, K.: Bounds for the solutions of decomposable form equations. Publ. Math. (Debr.) **52**, 1–31 (1998)
2408. Győry, K.: Power values of products of consecutive integers and binomial coefficients. In: Number Theory and Applications, pp. 145–156. Kluwer Academic, Dordrecht (1999)
2409. Győry, K.: On the distribution of solutions of decomposable form equations. In: Number Theory in Progress, I, pp. 237–265. de Gruyter, Berlin (1999)
2410. Győry, K.: Solving Diophantine equations by Baker's theory. In: A Panorama in Number Theory or The View from Baker's Garden, pp. 38–72. Cambridge University Press, Cambridge (2002)
2411. Győry, K.: On the *abc* conjecture in algebraic number fields. Acta Arith. **133**, 281–295 (2008)
2412. Győry, K., Hajdu, L., Pintér, Á.: Perfect powers from products of consecutive terms in arithmetic progression. Compos. Math. **145**, 845–864 (2009)
2413. Győry, K., Hajdu, L., Saradha, N.: On the Diophantine equation $n(n+d)\cdots(n+(k-1)d = by^l)$. Can. Math. Bull. **47**, 373–388 (2004)
2414. Győry, K., Papp, Z.Z.: Effective estimates for the integer solutions of norm form and discriminant form equations. Publ. Math. (Debr.) **25**, 311–325 (1978)
2415. Győry, K., Pethő, A., Pintér, Á.: Béla Brindza (1958–2003). Publ. Math. (Debr.) **65**, 1–11 (2004)
2416. Győry, K., Yu, K.: Bounds for the solutions of S-unit equations and decomposable form equations. Acta Arith. **123**, 9–41 (2006)
2417. Haas, A.: The geometry of Markoff forms. In: Number Theory, New York, 1984–1985. Lecture Notes in Math., vol. 1240, pp. 135–144. Springer, Berlin (1987)
2418. Haberland, K.: Perioden von Modulformen einer Variablen und Gruppencohomologie, I. Math. Nachr. **112**, 245–282 (1983)
2419. Haberland, K.: Perioden von Modulformen einer Variablen und Gruppencohomologie, II. Math. Nachr. **112**, 283–295 (1983)
2420. Haberland, K.: Perioden von Modulformen einer Variablen und Gruppencohomologie, III. Math. Nachr. **112**, 297–315 (1983)
2421. Haberzetle, M.: The Waring problem with summands x^m, $m \geq n$. Duke Math. J. **5**, 49–57 (1939)
2422. Habsieger, L.: Explicit lower bounds for $\|(3/2)^k\|$. Acta Arith. **106**, 299–309 (2003)
2423. Habsieger, L., Roblot, X.-F.: On integers of the form $p+2^k$. Acta Arith. **122**, 45–50 (2006)

2424. Hadamard, J.: Étude sur le propriétés des fonctions entières et en particulier d'une fonction considérée par Riemann. J. Math. Pures Appl. **9**, 171–215 (1893) [Selecta, pp. 52–93, Gauthier-Villars, Paris, 1935]
2425. Hadamard, J.: Sur les zéros de la fonction $\zeta(s)$ de Riemann. C. R. Acad. Sci. Paris **122**, 1470–1473 (1896)
2426. Hadamard, J.: Sur la distribution des zéros de la fonction $\zeta(s)$ et ses conséquences arithmétiques. Bull. Soc. Math. Fr. **24**, 199–200 (1896) [Selecta, pp. 111–132, Gauthier-Villars, Paris, 1935]
2427. Hadamard, J.: Obituary: Émile Picard. J. Lond. Math. Soc. **18**, 114–128 (1943)
2428. Haentzschel, E.: Bedingungen für die Lösung des Fermatschen Problem $y^2 = a_0x^4 + 4a_1x^3 + 6a_2x^2 + 4a_3x + a_4$. J. Reine Angew. Math. **144**, 275–283 (1914)
2429. Hafner, J.L.: New omega theorems for two classical lattice point problems. Invent. Math. **63**, 181–186 (1981)
2430. Hafner, J.L.: On the average order of a class of arithmetic functions. J. Number Theory **15**, 36–76 (1982)
2431. Hafner, J.L.: Zeros on the critical line for Dirichlet series attached to certain cusp forms. Math. Ann. **264**, 21–37 (1983)
2432. Hafner, J.L., Ivić, A.: On the mean-square of the Riemann zeta-function on the critical line. J. Number Theory **32**, 151–191 (1989)
2433. Hafner, J.L., Ivić, A.: On sums of Fourier coefficients of cusp forms. Enseign. Math. **35**, 375–382 (1989)
2434. Hagedorn, T.: Computation of Jacobsthal's function $h(n)$ for $n < 50$. Math. Comput. **78**, 1073–1087 (2009)
2435. Hagis, P. Jr.: Outline of a proof that every odd perfect number has at least eight prime factors. Math. Comput. **35**, 1027–1032 (1980)
2436. Hagis, P. Jr.: Sketch of a proof that an odd perfect number relatively prime to 3 as at least eleven prime factors. Math. Comput. **40**, 399–404 (1983)
2437. Hagis, P. Jr.: On the equation $M\varphi(n) = n - 1$. Nieuw Arch. Wiskd. **6**, 255–261 (1988)
2438. Hagis, P. Jr., Cohen, G.L.: Every odd perfect number has a prime factor which exceeds 10^6. Math. Comput. **67**, 1323–1330 (1998)
2439. Hagis, P. Jr., McDaniel, W.L.: On the largest prime divisor of an odd perfect number. Math. Comput. **27**, 955–957 (1973)
2440. Hagis, P. Jr., McDaniel, W.L.: On the largest prime divisor of an odd perfect number, II. Math. Comput. **29**, 922–924 (1975)
2441. Hajela, D., Smith, B.: On the maximum of an exponential sum of the Möbius function. In: Lecture Notes in Math., vol. 1240, pp. 145–164. Springer, Berlin (1987)
2442. Hajós, G.: Über einfache und mehrfache Bedeckung des n-dimensionalen Raums mit einem Würfelgitter. Math. Z. **47**, 427–467 (1942)
2443. Halász, G.: Über die Mittelwerte multiplikativer zahlentheoretischer Funktionen. Acta Math. Acad. Sci. Hung. **19**, 365–403 (1968)
2444. Halász, G.: The number-theoretic work of Paul Turán. Acta Arith. **37**, 9–19 (1980)
2445. Halász, G.: Note to paper 95. In: Turán, P. (ed.) Collected Papers, vol. 2, pp. 1005–1007. Akad. Kiadó, Budapest (1990)
2446. Halász, G., Lovász, L., Simonovits, M., Sós, V.T. (eds.): Paul Erdős and His Mathematics, vols. I, II. Springer, Berlin (2002)
2447. Halász, G., Turán, P.: On the distribution of roots of Riemann zeta and allied functions, I. J. Number Theory **1**, 122–137 (1969) [[6227], vol. 3, pp. 1919–1935]
2448. Halberstadt, E., Kraus, A.: Courbes de Fermat: résultats et problèmes. J. Reine Angew. Math. **548**, 167–234 (2002)
2449. Halberstam, H.: On the distribution of additive number-theoretic functions. J. Lond. Math. Soc. **30**, 43–53 (1955)
2450. Halberstam, H.: A proof of Chen's theorem. Astérisque **24/25**, 281–293 (1975)
2451. Halberstam, H.: Loo-keng Hua: obituary. Acta Arith. **51**, 99–117 (1988)

2452. Halberstam, H., Heath-Brown, D.R., Richert, H.-E.: Almost-primes in short intervals. In: Recent Progress in Analytic Number Theory, Durham, 1979, vol. 1, pp. 69–101. Academic Press, San Diego (1981)
2453. Halberstam, H., Jurkat, W., Richert, H.-E.: Un nouveau résultat de la méthode du crible. C. R. Acad. Sci. Paris **264**, 920–923 (1967)
2454. Halberstam, H., Richert, H.-E.: The distribution of polynomial sequences. Mathematika **19**, 25–50 (1972)
2455. Halberstam, H., Richert, H.-E.: Sieve Methods. Academic Press, San Diego (1974); 2nd ed. Springer, 1983
2456. Halberstam, H., Richert, H.-E.: A weighted sieve of Greaves type, II. In: Elementary and Analytic Theory of Numbers, Warsaw, 1982. Banach Center Publ., vol. 17, pp. 183–215. Polish Acad. Sci., Warsaw (1985)
2457. Halberstam, H., Roth, K.F.: Sequences, I. Oxford University Press, Oxford (1966); 2nd. ed. Springer, 1983
2458. Halberstam, H., Rotkiewicz, A.: A gap theorem for pseudoprimes in arithmetic progressions. Acta Arith. **13**, 395–404 (1968)
2459. Hales, T.: The sphere packing problem. J. Comput. Appl. Math. **44**, 41–76 (1992)
2460. Hales, T.: Sphere packings, I. Discrete Comput. Geom. **17**, 1–51 (1997)
2461. Hales, T.: Sphere packings, II. Discrete Comput. Geom. **18**, 135–149 (1997)
2462. Hales, T.: Sphere packings, III. Extremal cases. Discrete Comput. Geom. **36**, 71–110 (2006)
2463. Hales, T.: Sphere packings, IV. Detailed bounds. Discrete Comput. Geom. **36**, 111–116 (2006)
2464. Hales, T.: Sphere packings, VI. Tame graphs and linear programs. Discrete Comput. Geom. **36**, 205–265 (2006)
2465. Hales, T.: A proof of the Kepler conjecture. Ann. Math. **162**, 1065–1185 (2005)
2466. Hales, T.: Historical overview of the Kepler conjecture. Discrete Comput. Geom. **36**, 5–20 (2006)
2467. Hales, T., Ferguson, S.P.: A formulation of the Kepler conjecture. Discrete Comput. Geom. **36**, 21–69 (2006)
2468. Hales, T., Harrison, J., McLaughlin, S., Nipkow, T., Obua, S., Zumkeller, R.: A revision of the proof of the Kepler conjecture. Discrete Comput. Geom. **44**, 1–34 (2010)
2469. Hales, T., McLaughlin, S.: The dodecahedral conjecture. J. Am. Math. Soc. **23**, 299–344 (2010)
2470. Hall, M. [Jr.]: An isomorphism between linear recurring sequences and algebraic rings. Trans. Am. Math. Soc. **44**, 196–218 (1938)
2471. Hall, M. [Jr.]: On the sum and product of continued fractions. Ann. Math. **48**, 966–993 (1947)
2472. Hall, M. Jr.: The Diophantine equation $x^3 - y^2 = k$. In: Computers in Number Theory, pp. 173–198. Academic Press, San Diego (1971)
2473. Hall, R.R.: On the probability that n and $f(n)$ are relatively prime. Acta Arith. **17**, 169–183 (1970); corr. vol. 19, 1971, pp. 203–204
2474. Hall, R.R.: On the probability that n and $f(n)$ are relatively prime, II. Acta Arith. **19**, 175–184 (1971)
2475. Hall, R.R.: On the probability that n and $f(n)$ are relatively prime, III. Acta Arith. **20**, 267–289 (1972)
2476. Hall, R.R.: The behaviour of the Riemann zeta-function on the critical line. Mathematika **46**, 281–313 (1999)
2477. Hall, R.R.: A Wirtinger type inequality and the spacing of the zeros of the Riemann zeta-function. J. Number Theory **93**, 235–245 (2002)
2478. Hall, R.R.: A new unconditional result about large spaces between zeta zeros. Mathematika **52**, 101–113 (2005)
2479. Hall, R.R., Tenenbaum, G.: Divisors. Cambridge University Press, Cambridge (1988)
2480. Hall, R.S.: The prime number theorem for generalized primes. J. Number Theory **4**, 313–320 (1972)

2481. Halmos, P.: Note on almost-universal forms. Bull. Am. Math. Soc. **44**, 141–144 (1938)
2482. Halter-Koch, F.: Darstellung natürlicher Zahlen als Summen von Quadraten. Acta Arith. **42**, 11–20 (1982)
2483. Halter-Koch, F.: Reell-quadratische Zahlkörper mit großer Grundeinheit. Abh. Math. Semin. Univ. Hamb. **59**, 171–181 (1989)
2484. Halter-Koch, F.: Continued fractions of given symmetric period. Fibonacci Q. **29**, 298–303 (1991)
2485. Halter-Koch, F., Warlimont, R.: Analytic number theory in formations based on additive arithmetical semigroups. Math. Z. **215**, 99–128 (1994)
2486. Halton, J.H.: On the efficiency of certain quasi-random sequences of points in evaluating multidimensional integrals. Numer. Math. **2**, 84–90 (1960)
2487. Halupczok, K.: On the number of representations in the ternary Goldbach problem with one prime number in a given residue class. J. Number Theory **117**, 292–300 (2006)
2488. Halupczok, K.: Zum ternären Goldbachproblem mit Kongruenzbedingungen an die Primzahlen. In: Elementary and Analytic Number Theory. Proceedings of the ELAZ conference, Stuttgart, May 24–28, 2004, pp. 57–64. Franz Steiner Verlag, Stuttgart (2006)
2489. Hamburger, H.: Über die Riemannsche Funktionalgleichung der ζ-Funktion. SBer. Berl. Math. Ges. **20**, 4–9 (1921)
2490. Hamburger, H.: Über die Funktionalgleichung der L-Reihen. SBer. Berl. Math. Ges. **20**, 10–13 (1921)
2491. Hamburger, H.: Über die Riemannsche Funktionalgleichung der ζ-Funktion, I. Math. Z. **10**, 240–254 (1921)
2492. Hamburger, H.: Über die Riemannsche Funktionalgleichung der ζ-Funktion, II. Math. Z. **11**, 224–245 (1922)
2493. Hamburger, H.: Über die Riemannsche Funktionalgleichung der ζ-Funktion, III. Math. Z. **13**, 283–311 (1922)
2494. Hamburger, H.: Bemerkungen zu einem Satze über die Riemannsche ζ Funktion. SBer. Bayer. Akad. Wiss., 1922, 151–156
2495. Hančl, J.: Two proofs of transcendency of π and e. Czechoslov. Math. J. **35**, 543–549 (1985)
2496. Hancock, H.: Development of the Minkowski Geometry of Numbers. Macmillan, New York (1939) [Reprint: Dover, 1964]
2497. Haneke, W.: Verschärfung der Abschätzung von $\zeta(1/2 + it)$. Acta Arith. **8**, 357–430 (1962/1963)
2498. Hanrot, G.: Solving Thue equations without the full unit group. Math. Comput. **129**, 1011–1033 (2000)
2499. Hans-Gill, R.J., Madhu, R., Ranjeet, S.: On conjectures of Minkowski and Woods for $n = 7$. J. Number Theory **129**, 1011–1033 (2009)
2500. Hanson, D.: On a theorem of Sylvester and Schur. Can. Math. Bull. **16**, 195–199 (1973)
2501. Harald Cramér Symposium. Proceedings of the Symposium held in Stockholm, September 24–25, 1993. Scand. Actuar. J. 1995, 1–152
2502. Harder, G.: Eine Bemerkung zu einer Arbeit von P.E. Newstead. J. Reine Angew. Math. **242**, 16–25 (1970)
2503. Harder, G.: Chevalley groups over function fields and automorphic forms. Ann. Math. **100**, 249–306 (1974)
2504. Hardy, G.H.: A formula for the prime factors of any numbers. Messenger Math. **35**, 145–146 (1905/1906)
2505. Hardy, G.H.: Sur les zéros de la fonction $\zeta(s)$ de Riemann. C. R. Acad. Sci. Paris **158**, 1012–1014 (1915)
2506. Hardy, G.H.: On the expression of a number as the sum of two squares. Q. J. Math. **46**, 263–283 (1915)
2507. Hardy, G.H.: Sur le problème de diviseurs de Dirichlet. C. R. Acad. Sci. Paris **160**, 617–619 (1915)
2508. Hardy, G.H.: On Dirichlet's divisor problem. Proc. Lond. Math. Soc. **15**, 1–25 (1916)

2509. Hardy, G.H.: The average order of the arithmetical functions $P(x)$ and $\Delta(x)$. Proc. Lond. Math. Soc. **15**, 192–213 (1916)
2510. Hardy, G.H.: A problem of diophantine approximation. J. Indian Math. Soc. **11**, 162–166 (1919)
2511. Hardy, G.H.: On the representation of a number as the sum of any number of squares, and in particular of five. Trans. Am. Math. Soc. **21**, 255–284 (1920)
2512. Hardy, G.H.: Srinivasa Ramanujan (1887–1920). Proc. Lond. Math. Soc. **19**, xl–lviii (1921) [[5088], xxi–xxxvi]
2513. Hardy, G.H.: Note on Ramanujan's trigonometrical function $c_q(n)$, and certain series of arithmetical functions. Proc. Camb. Philos. Soc. **20**, 263–271 (1921)
2514. Hardy, G.H.: Note on Ramanujan's arithmetical function $\tau(h)$. Proc. Camb. Philos. Soc. **23**, 675–680 (1927)
2515. Hardy, G.H.: An introduction to the theory of numbers. Bull. Am. Math. Soc. **35**, 778–818 (1929)
2516. Hardy, G.H.: Ramanujan. Cambridge University Press, Cambridge (1940)
2517. Hardy, G.H.: Divergent Series. Oxford University Press, Oxford (1949)
2518. Hardy, G.H., Heilbronn, H.: Edmund Landau. J. Lond. Math. Soc. **13**, 302–310 (1938) [[3680], vol. 1, pp. 15–23; [2715], pp. 351–359]
2519. Hardy, G.H., Littlewood, J.E.: Some problems of diophantine approximation. In: Proceedings of the 5th ICM, pp. 223–229. Cambridge University Press, Cambridge (1912)
2520. Hardy, G.H., Littlewood, J.E.: Some problems of diophantine approximation, I. The fractional part of $n^k\Theta$. Acta Math. **37**, 155–191 (1914)
2521. Hardy, G.H., Littlewood, J.E.: Some problems of diophantine approximation, II. The trigonometrical series associated with the elliptic ϑ-functions. Acta Math. **37**, 193–238 (1914)
2522. Hardy, G.H., Littlewood, J.E.: Some theorems concerning power series and Dirichlet's series. Messenger Math. **43**, 134–147 (1914)
2523. Hardy, G.H., Littlewood, J.E.: Contributions to the theory of Riemann zeta-function and the theory of distribution of primes. Acta Math. **41**, 119–196 (1917)
2524. Hardy, G.H., Littlewood, J.E.: A new solution of Waring's problem. Q. J. Math. **48**, 272–293 (1919)
2525. Hardy, G.H., Littlewood, J.E.: Some problems of 'Partitio Numerorum'; I: A new solution of Waring's problem. Nachr. Ges. Wiss. Göttingen, 1920, 33–54
2526. Hardy, G.H., Littlewood, J.E.: Some problems of 'Partitio Numerorum': II. Proof that every large number is a sum of at most 21 biquadrates. Math. Z. **9**, 14–27 (1921)
2527. Hardy, G.H., Littlewood, J.E.: The zeros of Riemann's zeta-function on the critical line. Math. Z. **10**, 283–317 (1921)
2528. Hardy, G.H., Littlewood, J.E.: Some problems of diophantine approximation: the lattice-points in a right-angled triangle. Proc. Lond. Math. Soc. **20**, 15–36 (1922)
2529. Hardy, G.H., Littlewood, J.E.: Some problems of diophantine approximation: the lattice-points in a right-angled triangle, II. Abh. Math. Semin. Univ. Hamburg **1**, 212–2496 (1922)
2530. Hardy, G.H., Littlewood, J.E.: The approximate functional equation in the theory of zeta-function, with applications to the divisor-problems of Dirichlet and Piltz. Proc. Lond. Math. Soc. **21**, 39–74 (1923)
2531. Hardy, G.H., Littlewood, J.E.: Some problems of 'Partitio Numerorum': III. On the expression of a number as a sum of primes. Acta Math. **44**, 1–70 (1923)
2532. Hardy, G.H., Littlewood, J.E.: Some problems of 'Partitio Numerorum': IV. The singular series in Waring's problem and the value of the number $G(k)$. Math. Z. **12**, 161–188 (1922)
2533. Hardy, G.H., Littlewood, J.E.: On Lindelöf's hypothesis concerning the Riemann zeta-function. Proc. R. Soc. Lond. Ser. A, Math. Phys. Sci. **103**, 403–412 (1923)
2534. Hardy, G.H., Littlewood, J.E.: Some problems of 'Partitio Numerorum' (V): A further contribution to the study of Goldbach's problem. Proc. Lond. Math. Soc. **22**, 46–56 (1923)
2535. Hardy, G.H., Littlewood, J.E.: Some problems of 'Partitio Numerorum' (VI): Further researches in Waring's problem. Math. Z. **23**, 1–37 (1925)

2536. Hardy, G.H., Littlewood, J.E.: Some problems of 'Partitio Numerorum' (VIII): The number $\Gamma(k)$ in Waring's problem. Proc. Lond. Math. Soc. **28**, 518–542 (1928)
2537. Hardy, G.H., Littlewood, J.E.: The approximate functional equations for $\zeta(s)$ and $\zeta^2(s)$. Proc. Lond. Math. Soc. **29**, 81–97 (1929)
2538. Hardy, G.H., Littlewood, J.E.: A new proof of a theorem on rearrangements. J. Lond. Math. Soc. **23**, 163–168 (1949)
2539. Hardy, G.H., Ramanujan, S.: Asymptotic formulae for the distribution of integers of various types. (A problem in the analytic theory of numbers). Proc. Lond. Math. Soc. **16**, 112–132 (1916) [[5088], pp. 245–261]
2540. Hardy, G.H., Ramanujan, S.: The normal number of prime factors of a number n. Q. J. Math. **48**, 76–92 (1917) [[5088], pp. 262–275]
2541. Hardy, G.H., Ramanujan, S.: Asymptotic formulae in combinatory analysis. Proc. Lond. Math. Soc. **17**, 75–115 (1918) [[5088], pp. 276–309]
2542. Hardy, G.H., Ramanujan, S.: On the coefficients in the expansion of certain modular functions. Proc. R. Soc. Lond. Ser. A, Math. Phys. Sci. **95**, 144–155 (1919) [[5088], pp. 310–321]
2543. Hardy, G.H., Riesz, M.: The General Theory of Dirichlet's series. Cambridge University Press, Cambridge (1915)
2544. Hardy, G.H., Wright, E.M.: Leudesdorf's extension of Wolstenholme's theorem. J. Lond. Math. Soc. **9**, 38–41 (1934); corr. p. 240
2545. Hardy, G.H., Wright, E.M.: An Introduction to the Theory of Numbers. Oxford University Press, Oxford (1938). [2nd ed. 1945, 3rd ed. 1954, 4th ed. 1960, 5th ed. 1979, 6th ed. 2008]
2546. Hare, K.G.: More on the total number of prime factors of an odd perfect number. Math. Comput. **74**, 1003–1008 (2005)
2547. Hare, K.G.: New techniques for bounds on the total number of prime factors of an odd perfect number. Math. Comput. **76**, 2241–2248 (2007)
2548. Harman, G.: Almost-primes in short intervals. Math. Ann. **258**, 107–112 (1981/1982)
2549. Harman, G.: Trigonometric sums over primes, I. Mathematika **28**, 249–254 (1981)
2550. Harman, G.: Trigonometric sums over primes, II. Glasg. Math. J. **24**, 23–37 (1983)
2551. Harman, G.: Trigonometric sums over primes, III. J. Théor. Nr. Bordx. **15**, 727–740 (2003)
2552. Harman, G.: Primes in short intervals. Math. Z. **180**, 335–348 (1982)
2553. Harman, G.: On the distribution of $\sqrt{p}$ modulo one. Mathematika **30**, 104–116 (1983)
2554. Harman, G.: On the distribution of αp modulo one. J. Lond. Math. Soc. **27**, 9–18 (1983)
2555. Harman, G.: On the distribution of αp modulo one, II. Proc. Lond. Math. Soc. **72**, 241–260 (1996)
2556. Harman, G.: Diophantine approximation with square-free integers. Math. Proc. Camb. Philos. Soc. **95**, 381–388 (1984)
2557. Harman, G.: Metric Diophantine approximation with two restricted variables. I. Two square-free integers, or integers in arithmetic progressions. Proc. Camb. Philos. Soc. **103**, 197–206 (1988)
2558. Harman, G.: Simultaneous Diophantine approximation with primes. J. Lond. Math. Soc. **39**, 405–413 (1989)
2559. Harman, G.: Some cases of the Duffin and Schaeffer conjecture. Q. J. Math. **41**, 395–404 (1990)
2560. Harman, G.: Numbers badly approximable by fractions with prime denominator. Math. Proc. Camb. Philos. Soc. **118**, 1–5 (1995)
2561. Harman, G.: Metric Number Theory. Oxford University Press, Oxford (1998)
2562. Harman, G.: Integers without large prime factors in short intervals and arithmetic progressions. Acta Arith. **91**, 279–289 (1999)
2563. Harman, G.: One hundred years of normal numbers. In: Number Theory for the Millennium, vol. II, pp. 149–166. AK Peters, Berlin (2002)
2564. Harman, G.: On the number of Carmichael numbers up to x. Bull. Lond. Math. Soc. **37**, 641–650 (2005)
2565. Harman, G.: Prime-Detecting Sieves. Princeton University Press, Princeton (2007)

2566. Harman, G.: Watt's mean value theorem and Carmichael numbers. Int. J. Number Theory **4**, 241–248 (2008)
2567. Harman, G., Kumchev, A.: On sums of squares of primes. Math. Proc. Camb. Philos. Soc. **140**, 1–13 (2006)
2568. Harman, G., Lewis, P.: Gaussian primes in narrow sectors. Mathematika **48**, 119–135 (2001)
2569. Haros, C^{en} [Citoyen]: Tables pour évaluer une fraction ordinaire avec autant de décimales qu'on voudra; et por trouver la fraction ordinaire la plus simple, et qui approche sensiblement d'une fraction décimale. J. Éc. Polytech. **4**, 364–368 (1802)
2570. Harris, M., Shepherd-Barron, N., Taylor, R.: Ihara's lemma and potential automorphy. Preprint, http://people.math.jussieu.fr/~preprints/pdf/399.pdf
2571. Harris, M., Soudry, D., Taylor, R.: l-adic representations associated to modular forms over imaginary quadratic fields, I, Lifting to $GSp_4(Q)$. Invent. Math. **112**, 377–411 (1993)
2572. Harris, M., Taylor, R.: The Geometry and Cohomology of Some Simple Shimura Varieties. Ann. of Math. Stud., vol. 151. Princeton University Press, Princeton (2001)
2573. Hartman, P.: Aurel Wintner. J. Lond. Math. Soc. **37**, 483–503 (1962)
2574. Haselgrove, C.B.: A disproof of a conjecture of Pólya. Mathematika **5**, 141–145 (1958)
2575. Haselgrove, C.B., Miller, J.C.P.: Tables of the Riemann Zeta Function. Cambridge University Press, Cambridge (1960)
2576. Haslam-Jones, U.S.: Bertram Martin Wilson. Proc. Edinb. Math. Soc. **4**, 268–269 (1936)
2577. Hasse, H.: Über die Darstellbarkeit von Zahlen durch quadratische Formen im Körper der rationalen Zahlen. J. Reine Angew. Math. **152**, 129–148 (1923) [[2607], vol. 1, pp. 3–42]
2578. Hasse, H.: Über die Äquivalenz quadratischer Formen im Körper der rationalen Zahlen. J. Reine Angew. Math. **152**, 205–224 (1923) [[2607], vol. 1, pp. 23–42]
2579. Hasse, H.: Zur Theorie des quadratischen Hilbertschen Normenrestsymbols in algebraischen Körpern. J. Reine Angew. Math. **153**, 76–93 (1923)
2580. Hasse, H.: Darstellbarkeit von Zahlen durch quadratische Formen in einem beliebigen algebraischen Zahlkörper. J. Reine Angew. Math. **153**, 113–130 (1924) [[2607], vol. 1, pp. 75–92]
2581. Hasse, H.: Äquivalenz quadratischer Formen in einem beliebigen algebraischen Zahlkörper. J. Reine Angew. Math. **153**, 158–162 (1924) [[2607], vol. 1, pp. 93–97]
2582. Hasse, H.: Bericht über neuere Untersuchungen und Probleme aus der Theorie der algebraischen Zahlkörper, I. Jahresber. Dtsch. Math.-Ver. **35**, 1–55 (1926) [2nd ed. 1965, 3rd ed. 1970, Physica Verlag]
2583. Hasse, H.: Bericht über neuere Untersuchungen und Probleme aus der Theorie der algebraischen Zahlkörper, Ia. Jahresber. Dtsch. Math.-Ver. **36**, 233–311 (1927) [2nd ed. 1965, 3rd ed. 1970, Physica Verlag]
2584. Hasse, H.: Bericht über neuere Untersuchungen und Probleme aus der Theorie der algebraischen Zahlkörper, II. In: Jber. Deutsch. Math.-Verein. Ergänzungsband, vol. 6, pp. 1–204. Teubner, Leipzig (1930) [2nd ed. 1965, 3rd ed. 1970, Physica Verlag]
2585. Hasse, H.: Das Eisensteinsche Reziprozitätsgesetz der n-ten Potenzreste. Math. Ann. **97**, 599–623 (1927) [[2607], vol. 1, pp. 269–293]
2586. Hasse, H.: Über das Reziprozitätsgesetz der m-ten Potenzreste. J. Reine Angew. Math. **158**, 228–259 (1927) [[2607], vol. 1, pp. 294–325]
2587. Hasse, H.: Neue Begründung der komplexen Multiplikation, I. J. Reine Angew. Math. **157**, 115–139 (1927) [[2607], vol. 2, pp. 3–27]
2588. Hasse, H.: Neue Begründung der komplexen Multiplikation, II. J. Reine Angew. Math. **165**, 64–88 (1931) [[2607], vol. 2, pp. 28–52]
2589. Hasse, H.: Zum expliziten Reziprozitätsgesetz. Abh. Math. Semin. Univ. Hamb. **7**, 52–63 (1929) [[2607], vol. 1, pp. 343–354]
2590. Hasse, H.: Neue Begründung und Verallgemeinerung der Theorie des Normenrestsymbols. J. Reine Angew. Math. **162**, 134–144 (1930) [[2607], vol. 1, pp. 134–144[10]]

[10]The same pagination occurring in two places is not a result of a printing error, but occurred really.

2591. Hasse, H.: Die Normenresttheorie relativ-abelscher Zahlkörper als Klassenkörpertheorie im Kleinen. J. Reine Angew. Math. **162**, 145–154 (1930) [[2607], vol. 1, pp. 145–154]
2592. Hasse, H.: Beweis eines Satzes und Widerlegung einer Vermutung über das allgemeine Normenrestsymbol. Nachr. Ges. Wiss. Göttingen, 64–69 (1931) [[2607], vol. 1, pp. 155–160]
2593. Hasse, H.: Beweis des Analogons der Riemannscher Vermutung für die Artinschen und F.K. Schmidtschen Kongruenzzetafunktionen in gewissen elliptischen Fällen. Nachr. Ges. Wiss. Göttingen, 253–262 (1933) [[2607], vol. 2, pp. 85–94]
2594. Hasse, H.: Die Struktur der R. Brauerschen Algebrenklassengruppe über einem algebraischen Zahlkörper. Math. Ann. **107**, 731–760 (1933) [[2607], vol. 1, pp. 501–530]
2595. Hasse, H.: Über die Kongruenzzetafunktionen. Unter Benutzung von Mitteilungen von Prof. Dr. F.K. Schmidt und Prof. Dr. E. Artin. SBer. Preuß. Akad. Wiss. Berl. **17**, 250–263 (1934) [[2607], vol. 2, pp. 95–108]
2596. Hasse, H.: Abstrakte Begründung der komplexen Multiplikation und Riemannsche Vermutung in Funktionenkörper. Abh. Math. Semin. Univ. Hamb. **10**, 325–348 (1934) [[2607], vol. 2, pp. 109–132]
2597. Hasse, H.: Theorie der relativ-zyklischen algebraischen Funktionenkörper, insbesondere bei endlichem Konstantenkörper. J. Reine Angew. Math. **172**, 37–54 (1935) [[2607], vol. 2, pp. 133–150]
2598. Hasse, H.: Zur Theorie der abstrakten elliptischen Funktionenkörper, I. Die Struktur der Gruppe der Divisorenklassen endlicher Ordnung. J. Reine Angew. Math. **175**, 55–62 (1936) [[2607], vol. 2, pp. 223–230]
2599. Hasse, H.: Zur Theorie der abstrakten elliptischen Funktionenkörper, II. Automorphismen und Meromorphismen. J. Reine Angew. Math. **175**, 69–88 (1936) [[2607], vol. 2, pp. 231–250]
2600. Hasse, H.: Zur Theorie der abstrakten elliptischen Funktionenkörper, III. Die Struktur des Meromorphismenringes. Die Riemannsche Vermutung. J. Reine Angew. Math. **175**, 193–208 (1936) [[2607], vol. 2, pp. 251–266]
2601. Hasse, H.: Der n-Teilungskörper eines abstrakten elliptischen Funktionenkörpers als Klassenkörper, nebst Anwendung auf den Mordell-Weilschen Endlichkeitssatz. Math. Z. **48**, 48–66 (1942) [[2607], vol. 2, pp. 403–421]
2602. Hasse, H.: Punti razionali sopra curve algebriche a congruenze. Atti Conv., Reale Accad. d'Italia **9**, 85–140 (1943) [[2607], vol. 2, pp. 295–350]
2603. Hasse, H.: Kurt Hensel zum Gedächtnis. J. Reine Angew. Math. **187**, 1–13 (1949)
2604. Hasse, H.: Rein-arithmetischer Beweis des Siegelschen Endlichkeitssatzes für binäre diophantische Gleichungen im Spezialfall des Geschlechts 1. Abhandl. Deutsch. Akad. Wiss., 1951, nr. 2, 1–19
2605. Hasse, H.: Wissenschaftlicher Nachruf auf Hermann Ludwig Schmid. Math. Nachr. **18**, 1–18 (1958)
2606. Hasse, H.: Über die Dichte der Primzahlen p, für die eine vorgegebene ganzrationale Zahl $\neq 0$ von gerader bzw. ungerader Ordnung mod p ist. Math. Ann. **166**, 19–23 (1966)
2607. Hasse, H.: Mathematische Abhandlungen, vols. 1–3. de Gruyter, Berlin (1975)
2608. Hasse, H., Hensel, K.: Über die Normenreste eines relativ-zyklischen Körpers vom Primzahlgrad l nach einem Primteiler $\mathfrak{l}$ von l. Math. Ann. **90**, 262–278 (1923) [[2607], vol. 1, pp. 101–117]
2609. Hasse, H., Schmidt, F.K.: Die Struktur diskret bewerteter Körper. J. Reine Angew. Math. **170**, 4–63 (1934) [[2607], vol. 3, pp. 460–520]
2610. Hata, M.: Rational approximations to π and some other numbers. Acta Arith., **1993**, 335–349
2611. Hata, M.: A new irrationality measure for $\zeta(3)$. Acta Arith. **92**, 47–57 (2000)
2612. Hatada, K.: On the values at rational integers of the p-adic Dirichlet L functions. J. Math. Soc. Jpn. **31**, 7–27 (1979)
2613. Hathaway, A.S.: A memoir in the theory of numbers. Am. J. Math. **9**, 162–179 (1887)
2614. Hausdorff, F.: Ueber die Erzeugung der Invarianten durch Integration. Nachr. Ges. Wiss. Göttingen, 1897, 71–90

2615. Hausdorff, F.: Zur Hilbertschen Lösung des Waringschen Problems. Math. Ann. **67**, 301–302 (1909)
2616. Hausdorff, F.: Grundzüge der Mengenlehre. Teubner, Leipzig (1914)
2617. Hausdorff, F.: Dimension und äusseres Mass. Math. Ann. **79**, 157–179 (1918)
2618. Hausman, M., Shapiro, H.N.: On the mean square distribution of primitive roots of unity. Commun. Pure Appl. Math. **26**, 539–547 (1973)
2619. Haussner, R.: Tafeln für das Goldbach'sche Gesetz. Abhandlungen der Kais. Leopold. Carol. Deutsch. Akad. der Naturforscher, Halle **1**, 1–214 (1899)
2620. Hayes, D.: Leonard Carlitz (1907–1999). Not. Am. Math. Soc. **48**, 1322–1324 (2001)
2621. Hayman, W.K.: Rolf Nevanlinna. Bull. Lond. Math. Soc. **14**, 419–436 (1982)
2622. Hazewinkel, M.: Local class field theory is easy. Adv. Math. **18**, 148–181 (1975)
2623. He, B., Togbé, A.: On the number of solutions of Goormaghtigh equation for given x and y. Indag. Math. **19**, 65–72 (2008)
2624. Heap, B.R., Lynn, M.S.: A graph-theoretic algorithm for the solution of a linear Diophantine problem of Frobenius. Numer. Math. **6**, 346–354 (1964)
2625. Heap, B.R., Lynn, M.S.: On a linear Diophantine problem of Frobenius: An improved algorithm. Numer. Math. **7**, 226–231 (1965)
2626. Heath-Brown, D.R.: Almost-primes in arithmetic progressions and short intervals. Proc. Camb. Philos. Soc. **83**, 357–375 (1978)
2627. Heath-Brown, D.R.: Zero density estimates for the Riemann zeta-function and Dirichlet L-functions. J. Lond. Math. Soc. **19**, 221–232 (1979)
2628. Heath-Brown, D.R.: The fourth power moment of the Riemann zeta function. Proc. Lond. Math. Soc. **38**, 385–422 (1979)
2629. Heath-Brown, D.R.: The density of zeros of Dirichlet's L-functions. Can. J. Math. **31**, 231–240 (1979)
2630. Heath-Brown, D.R.: Fractional moments of the Riemann zeta-function. J. Lond. Math. Soc. **24**(2), 65–78 (1981)
2631. Heath-Brown, D.R.: Mean values of the zeta functions and divisor problems. In: Recent Progress in Analytic Number Theory, Durham, 1979. vol. 1, pp. 115–119. Academic Press, San Diego (1981)
2632. Heath-Brown, D.R.: Gaps between primes, and the pair correlation of zeros of the zeta function. Acta Arith. **41**, 85–99 (1982)
2633. Heath-Brown, D.R.: Primes in "almost all" short intervals. J. Lond. Math. Soc. **26**, 385–396 (1982)
2634. Heath-Brown, D.R.: Prime twins and Siegel zeros. Proc. Lond. Math. Soc. **47**, 193–224 (1983)
2635. Heath-Brown, D.R.: Cubic forms in ten variables. Proc. Lond. Math. Soc. **47**, 225–257 (1983)
2636. Heath-Brown, D.R.: The Pjateckiĭ Šapiro prime number theorem. J. Number Theory **16**, 242–266 (1983)
2637. Heath-Brown, D.R.: The divisor function at consecutive integers. Mathematika **31**, 141–149 (1994)
2638. Heath-Brown, D.R.: Diophantine approximation with square-free numbers. Math. Z. **187**, 335–344 (1984)
2639. Heath-Brown, D.R.: Fermat's Last Theorem for "almost all" exponents. Bull. Lond. Math. Soc. **17**, 15–16 (1985)
2640. Heath-Brown, D.R.: Artin's conjecture on primitive roots. Q. J. Math. **37**, 27–38 (1986)
2641. Heath-Brown, D.R.: The divisor function $d_3(n)$ in arithmetic progressions. Acta Arith. **47**, 29–56 (1986)
2642. Heath-Brown, D.R.: Integer sets containing no arithmetic progressions. J. Lond. Math. Soc. **35**, 385–394 (1987)
2643. Heath-Brown, D.R.: Weyl's inequality, Hua's inequality and Waring's problem. J. Lond. Math. Soc. **38**, 216–230 (1988)
2644. Heath-Brown, D.R.: Ternary quadratic forms and sums of three square-full numbers. Prog. Math. **75**, 137–163 (1988)

2645. Heath-Brown, D.R.: The fractional part of αn^k. Mathematatika **35**, 28–37 (1988)
2646. Heath-Brown, D.R.: Siegel zeros and the least prime in an arithmetic progression. Q. J. Math. **41**, 405–418 (1990)
2647. Heath-Brown, D.R.: Square-full numbers in short intervals. Math. Proc. Camb. Philos. Soc. **110**, 1–3 (1991)
2648. Heath-Brown, D.R.: The distribution and moments of the error term in the Dirichlet divisor problem. Acta Arith. **60**, 389–415 (1992)
2649. Heath-Brown, D.R.: Searching for solutions of $x^3 + y^3 + z^3 = k$. In: Séminaire de Théorie des Nombres, Paris 1989–1990, Progr. Math., vol. 102, pp. 71–76 (1992)
2650. Heath-Brown, D.R.: Zero-free regions for Dirichlet L-functions and the least prime in an arithmetic progression. Proc. Lond. Math. Soc. **64**, 265–338 (1992)
2651. Heath-Brown, D.R.: Odd perfect numbers. Math. Proc. Camb. Philos. Soc. **115**, 191–196 (1994)
2652. Heath-Brown, D.R.: The largest prime factor of the integers in an interval. Sci. China Ser. A **39**, 449–476 (1996)
2653. Heath-Brown, D.R.: The density of rational points on cubic surfaces. Acta Arith. **79**, 17–30 (1997)
2654. Heath-Brown, D.R.: Almost-prime k-tuples. Mathematika **44**, 245–266 (1997)
2655. Heath-Brown, D.R.: Lattice points in the sphere. In: Number Theory in Progress, vol. 2, pp. 883–892. de Gruyter, Berlin (1999)
2656. Heath-Brown, D.R.: The largest prime factor of $X^3 + 2$. Proc. Lond. Math. Soc. **82**, 554–596 (2001)
2657. Heath-Brown, D.R.: Primes represented by $x^3 + 2y^3$. Acta Math. **186**, 1–84 (2001)
2658. Heath-Brown, D.R.: The density of rational points on curves and surfaces. Ann. Math. **155**, 553–595 (2002)
2659. Heath-Brown, D.R.: Cubic forms in 14 variables. Invent. Math. **170**, 199–230 (2007)
2660. Heath-Brown, D.R.: Carmichael numbers with three prime factors. Hardy-Ramanujan J. **30**, 6–12 (2007)
2661. Heath-Brown, D.R.: Imaginary quadratic fields with class group exponent 5. Forum Math. **20**, 275–283 (2008)
2662. Heath-Brown, D.R., Goldston, D.A.: A note on the differences between consecutive primes. Math. Ann. **266**, 317–320 (1984)
2663. Heath-Brown, D.R., Huxley, M.N.: Exponential sums with a difference. Proc. Lond. Math. Soc. **61**, 227–250 (1990)
2664. Heath-Brown, D.R., Iwaniec, H.: On the difference between consecutive prime numbers. Invent. Math., **55**, 49–69
2665. Heath-Brown, D.R., Jia, C.H.: The largest prime factor of the integers in an interval, II. J. Reine Angew. Math. **498**, 39–59 (1998)
2666. Heath-Brown, D.R., Jia, C.H.: The distribution of αp modulo one. Proc. Lond. Math. Soc. **84**, 79–104 (2002)
2667. Heath-Brown, D.R., Konyagin, S.: New bounds for Gauss sums derived from kth powers, and for Heilbronn's exponential sum. Q. J. Math. **51**, 221–235 (2000)
2668. Heath-Brown, D.R., Lioen, W.M., te Riele, H.J.J.: On solving the diophantine equation $x^3 + y^3 + z^3 = k$ on a vector computer. Math. Comput. **61**, 235–244 (1993)
2669. Heath-Brown, D.R., Moroz, B.Z.: Primes represented by binary cubic forms. Proc. Lond. Math. Soc. **84**, 257–288 (2002)
2670. Heath-Brown, D.R., Moroz, B.Z.: On the representation of primes by cubic polynomials in two variables. Proc. Lond. Math. Soc. **88**, 289–312 (2004)
2671. Heath-Brown, D.R., Patterson, S.J.: The distribution of Kummer sums at prime arguments. J. Reine Angew. Math. **310**, 111–130 (1979)
2672. Heath-Brown, D.R., Schlage-Puchta, J.-C.: Integers represented as a sum of primes and powers of two. Asian J. Math. **6**, 535–565 (2002)
2673. Heawood, P.J.: The classification of rational approximations. Proc. Lond. Math. Soc. **20**, 233–250 (1921)

2674. Hecke, E.: Über nicht-reguläre Primzahlen und den Fermatschen Satz. Nachr. Ges. Wiss. Göttingen, 1910, 420–424 [[2703], pp. 59–63]
2675. Hecke, E.: Höhere Modulfunktionen und ihre Anwendungen auf die Zahlentheorie. Math. Ann. **71**, 1–37 (1912) [[2703], pp. 21–57]
2676. Hecke, E.: Über die Konstruktion relativ-Abelscher Zahlkörper durch Modulfunktionen von zwei Variabeln. Math. Ann. **74**, 465–510 (1913) [[2703], pp. 64–68]
2677. Hecke, E.: Über die Zetafunktion beliebiger algebraischer Zahlkörper. Nachr. Ges. Wiss. Göttingen, 1917, 77–89 [[2703], pp. 159–171]
2678. Hecke, E.: Über die L-Funktionen und den Dirichletschen Primzahlsatz für einen beliebigen Zahlkörper. Nachr. Ges. Wiss. Göttingen, 1917, 299–318 [[2703], pp. 178–197]
2679. Hecke, E.: Eine neue Art von Zetafunktionen und ihre Beziehung zur Verteilung der Primzahlen. Math. Z. **1**, 357–376 (1918) [[2703], pp. 215–234]
2680. Hecke, E.: Eine neue Art von Zetafunktionen und ihre Beziehung zur Verteilung der Primzahlen, II. Math. Z. **6**, 11–51 (1920) [[2703], pp. 249–289]
2681. Hecke, E.: Reziprozitätsgesetz und Gauss'sche Summen in quadratischen Zahlkörpern. Nachr. Ges. Wiss. Göttingen, 1919, 265–278 [[2703], pp. 235–248]
2682. Hecke, E.: Über analytische Funktionen und die Verteilung von Zahlen mod. Abh. Math. Semin. Univ. Hamb. **1**, 54–76 (1921) [[2703], pp. 313–335]
2683. Hecke, E.: Über die Lösung der Riemannscher Funktionalgleichung. Math. Z. **16**, 301–307 (1923) [[2703], pp. 374–380]
2684. Hecke, E.: Analytische Funktionen und algebraische Zahlen, II. Abh. Math. Semin. Univ. Hamb. **3**, 213–236 (1924) [[2703], pp. 381–404]
2685. Hecke, E.: Vorlesungen über die Theorie der algebraischen Zahlen. Akademische Verlag, Leipzig (1923) [Reprint: Chelsea, 1948; 2nd ed., Leipzig, 1954; reprint: Chelsea, 1970; English translation: Lectures on the Theory of Algebraic Numbers, Springer, 1981]
2686. Hecke, E.: Zur Theorie der elliptischen Modulfunktionen. Math. Ann. **97**, 210–242 (1927) [[2703], pp. 428–460]
2687. Hecke, E.: Theorie der Eisensteinschen Reihen höheren Stufe und ihre Anwendung auf Funktionentheorie und Arithmetik. Abh. Math. Semin. Univ. Hamb. **5**, 199–224 (1927) [[2703], pp. 461–486]
2688. Hecke, E.: Über das Verhalten von $\sum_{m,n} e^{2\pi i \tau \frac{|m^2-2n^2|}{8}}$ und ähnlichen Funktionen bei Modulsubstitutionen. J. Reine Angew. Math. **157**, 159–170 (1927) [[2703], pp. 487–498]
2689. Hecke, E.: Bestimmung der Perioden gewisser Integrale durch die Theorie der Klassenkörper. Math. Z. **28**, 708–727 (1928) [[2703], pp. 505–524]
2690. Hecke, E.: Über ein Fundamentalproblem aus der Theorie der elliptischen Modulfunktionen. Abh. Math. Semin. Univ. Hamb. **6**, 235–257 (1928) [[2703], pp. 525–547]
2691. Hecke, E.: Über das Verhalten der Integrale 1. Gattung bei Abbildungen, insbesondere in der Theorie der elliptischen Modulformen. Abh. Math. Semin. Univ. Hamb. **8**, 271–281 (1930) [[2703], pp. 548–558]
2692. Hecke, E.: Die Riemannschen Periodenrelationen für die elliptischen Modulfunktionen. J. Reine Angew. Math. **167**, 337–345 (1932) [[2703], pp. 559–567]
2693. Hecke, E.: Die eindeutige Bestimmung der Modulfunktionen q-ter Stufe durch algebraische Eigenschaften. Math. Ann. **111**, 293–301 (1935) [[2703], pp. 568–576]
2694. Hecke, E.: Die Primzahlen in der Theorie der elliptischen Modulfunktionen. Kgl. Danske Vid. Selsk. Mat.-Fys. Medd. **13**(10), 1–16 (1935) [[2703], pp. 577–590]
2695. Hecke, E.: Über die Bestimmung Dirichletscher Reihen durch ihre Funktionalgleichungen. Math. Ann. **112**, 664–699 (1936) [[2703], pp. 591–626]
2696. Hecke, E.: Neuere Fortschritte in der Theorie der elliptischen Modulfunktionen. In: Comptes Rendus du Congrès International des Mathématiciens, Oslo, 1936, pp. 140–156 (1937)
2697. Hecke, E.: Über Modulfunktionen und die Dirichletschen Reihen mit Eulerschen Produktentwicklung, I. Math. Ann. **114**, 1–28 (1937) [[2703], pp. 644–707]
2698. Hecke, E.: Über Modulfunktionen und die Dirichletschen Reihen mit Eulerschen Produktentwicklung, II. Math. Ann. **114**, 316–351 (1937) [[2703], pp. 644–707]

2699. Hecke, E.: Über Dirichletreihen mit Funktionalgleichung und ihre Nullstellen auf der Mittelgeraden. SBer. Bayer. Akad. Wiss., 1937, 73–95 [[2703], pp. 708–730]
2700. Hecke, E.: Über die Darstellung der Determinante einer positiven quadratischen Form durch die Form. Vierteljahr. Naturforsch. Ges. Zürich **85**, 64–70 (1940) [[2703], pp. 782–788]
2701. Hecke, E.: Analytische Arithmetik der positiven quadratischen Formen. Kgl. Danske Vid. Selsk. Mat.-Fys. Medd. **17**(12), 1–134 (1940) [[2703], pp. 789–918]
2702. Hecke, E.: Herleitung des Euler-Produktes dr Zetafunktion und einiger L-Reihen aus ihrer Funktionalgleichung. Math. Ann. **119**, 266–287 (1944) [[2703], pp. 919–940]
2703. Hecke, E.: Mathematische Werke. Vandenhoeck & Ruprecht, Göttingen (1959); 2nd. ed. 1970; 3rd ed. 1983
2704. Heegner, H.: Diophantische Analysis und Modulfunktionen. Math. Z. **56**, 227–253 (1952)
2705. Heier, G.: Uniformly effective Shafarevich conjecture on families of hyperbolic curves over a curve with prescribed degeneracy locus. J. Math. Pures Appl. **83**, 845–867 (2004)
2706. Heilbronn, H.: Über die Verteilung der Primzahlen in Polynomen. Math. Ann. **104**, 794–799 (1931) [[2715], pp. 57–62]
2707. Heilbronn, H.: Über den Primzahlsatz von Herrn Hoheisel. Math. Z., **36**, 394–423. [[2715], pp. 70–99]
2708. Heilbronn, H.: On the class-number in imaginary quadratic fields. Q. J. Math. **5**, 150–160 (1934) [[2715], pp. 177–187]
2709. Heilbronn, H.: Über das Waringsche Problem. Acta Arith. **1**, 212–221 (1935) [[2715], pp. 252–271]
2710. Heilbronn, H.: On Dirichlet series which satisfy a certain functional equation. Q. J. Math. **9**, 194–195 (1938) [[2715], pp. 343–344]
2711. Heilbronn, H.: On the distribution of the sequence $n^2\theta$ mod 1. Q. J. Math. **19**, 249–256 (1948)
2712. Heilbronn, H.: Lecture Notes on Additive Number Theory mod p. California Institute of Technology, Pasadena (1964)
2713. Heilbronn, H.: On the average length of a class of finite continued fractions. In: Number Theory and Analysis, pp. 87–96. Plenum, New York (1969) [[2715], pp. 518–525]
2714. Heilbronn, H.: On real zeros of Dedekind ζ-functions. Can. J. Math. **25**, 870–873 (1973) [[2715], pp. 554–557]
2715. Heilbronn, H.: Collected Papers. Wiley, New York (1988)
2716. Heilbronn, H., Landau, E.: Bemerkungen zur vorstehender Arbeit von Herrn Bochner. Math. Z. **37**, 10–16 (1933) [[2715], pp. 143–149; [3680], vol. 9, pp. 215–221]
2717. Heilbronn, H., Landau, E., Scherk, P.: Alle grossen ganzen Zahlen lassen sich als Summe von höchstens 71 Primzahlen darstellen. Časopis Mat. Fys. **65**, 117–141 (1936) [[2715], pp. 197–221; [3680], vol. 9, pp. 351–375]
2718. Heilbronn, H., Linfoot, E.H.: On the imaginary quadratic corpora of class-number one. Q. J. Math. **5**, 293–301 (1934) [[2715], pp. 188–196]
2719. Heinhold, J.: Oskar Perron. Jahresber. Dtsch. Math.-Ver. **90**, 184–199 (1988)
2720. Hejhal, D.A.: On the triple correlation of zeros of the zeta function. Internat. Math. Res. Notices, 1994, 293–302
2721. Helfgott, H.A., Venkatesh, A.: Integral points on elliptic curves and 3-torsion in class groups. J. Am. Math. Soc. **19**, 527–550 (2006)
2722. Hellegouarch, Y.: Étude des points d'ordre fini des variétés abéliennes de dimension un définies sur un anneau principal. J. Reine Angew. Math. **244**, 20–36 (1970)
2723. Hellegouarch, Y.: Sur un théorème de Maillet. C. R. Acad. Sci. Paris **273**, 477–478 (1971)
2724. Hellegouarch, Y.: Points d'ordre $2p^h$ sur les courbes elliptiques. Acta Arith. **26**, 253–263 (1974/1975)
2725. Hellegouarch, Y.: Invitation aux mathématiques de Fermat-Wiles. Masson, Paris (1997); 2nd ed. 2001. [English translation: Invitation to the Mathematics of Fermat-Wiles, Academic Press, 2002]
2726. Hemer, O.: On the Diophantine equation $y^2 - k = x^3$. Dissertation, Univ. of Uppsala (1952)
2727. Hemer, O.: Notes on the Diophantine equation $y^2 - k = x^3$. Ark. Mat. **3**, 67–77 (1954)

2728. Henkin, L.: In memoriam: Raphael Mitchell Robinson. Bull. Symb. Log. **1**, 340–343 (1995)
2729. Henniart, G.: Une preuve simple des conjectures de Langlands pour $GL(n)$ sur un corps p-adique. Invent. Math. **139**, 439–455 (2000)
2730. Henniart, G.: Une caractérisation de la correspondance de Langlands locale pour $GL(n)$. Bull. Soc. Math. Fr. **130**, 587–602 (2002)
2731. Hensel, K.: Über eine neue Begründung der Theorie der algebraischen Zahlen. Jahresber. Dtsch. Math.-Ver. **6**, 83–88 (1897)
2732. Hensel, K.: Über die Entwicklung der algebraischen Zahlen in Potenzreihen. Math. Ann. **55**, 301–336 (1902)
2733. Hensel, K.: Neue Grundlagen der Arithmetik. J. Reine Angew. Math. **127**, 51–84 (1904)
2734. Hensel, K.: Über eine neue Begründung der Theorie der algebraischen Zahlen. J. Reine Angew. Math. **128**, 1–32 (1905)
2735. Hensel, K.: Über die zu einem algebraischem Körper gehörigen Invarianten. J. Reine Angew. Math. **129**, 68–85 (1905)
2736. Hensel, K.: Theorie der algebraischen Zahlen. Teubner, Leipzig (1908)
2737. Hensel, K.: Über die zu einer algebraischen Gleichung gehörigen Auflösungskörper. J. Reine Angew. Math. **136**, 183–209 (1909)
2738. Hensel, K.: Zahlentheorie. Göschen, Berlin (1913)
2739. Hensel, K.: Über die Grundlagen einer neuen Theorie der quadratischer Zahlkörper. J. Reine Angew. Math. **144**, 57–70 (1914)
2740. Hensel, K.: Die Exponentialdarstellung der Zahlen eines algebraischen Zahlkörpers für den Bereich eines Primdivisors. In: Festschrift H.A. Schwarz, Berlin, pp. 61–75 (1914)
2741. Hensel, K.: Untersuchungen der Zahlen eines algebraischen Körpers für den Bereich eines beliebigen Primteilers. J. Reine Angew. Math. **145**, 92–113 (1915)
2742. Hensel, K.: Die multiplikative Darstellung der algebraischen Zahlen für den Bereich eines beliebigen Primteilers. J. Reine Angew. Math. **146**, 189–215 (1916)
2743. Hensel, K.: Untersuchung der Zahlen eines algebraischen Körpers für eine beliebige Primteilerpotenz als Modul. J. Reine Angew. Math. **146**, 216–228 (1916)
2744. Hensel, K.: Allgemeine Theorie der Kongruenzklassgruppen und ihrer Invarianten in algebraischen Körpern. J. Reine Angew. Math. **147**, 1–15 (1917)
2745. Hensel, K.: Eine neue Theorie der algebraischen Zahlen. Math. Z. **2**, 433–452 (1918)
2746. Hensel, K.: Über die Zerlegung der Primteiler in relativ cyklischen Körpern, nebst einer Anwendung auf die Kummerschen Körper. J. Reine Angew. Math. **151**, 112–120 (1921)
2747. Hensel, K.: Über die Zerlegung der Primteiler eines beliebigen Zahlkörpers in einem auflösbaren Oberkörper. J. Reine Angew. Math. **151**, 200–209 (1921)
2748. Hensel, K.: Zur multiplikativen Darstellung der algebraischen Zahlen für den Bereich eines Primteilers. J. Reine Angew. Math. **151**, 210–212 (1921)
2749. Hensel, K.: Paul Bachmann und sein Lebenswerk. Jahresber. Dtsch. Math.-Ver. **36**, 31–73 (1927)
2750. Hensley, D.: The distribution of round numbers. Proc. Lond. Math. Soc. **54**, 412–444 (1987)
2751. Hensley, D.: The Hausdorff dimensions of some continued fraction Cantor sets. J. Number Theory **33**, 182–198 (1989)
2752. Hensley, D.: Continued fraction Cantor sets, Hausdorff dimension, and functional analysis. J. Number Theory **40**, 336–358 (1992)
2753. Hensley, D.: The number of steps in the Euclidean algorithm. J. Number Theory **49**, 142–182 (1994)
2754. Hensley, D.: A polynomial time algorithm for the Hausdorff dimension of continued fraction Cantor sets. J. Number Theory **58**, 9–45 (1996); corr. **59**, 419 (1996)
2755. Hensley, D., Richards, I.: On the incompatibility of two conjectures concerning primes. In: Proc. Symposia Pure Math., vol. 24, pp. 123–127. Am. Math. Soc., Providence (1973)
2756. Hensley, D., Richards, I.: Primes in intervals. Acta Arith. **25**, 375–391 (1974)
2757. Herbrand, J.: Théorie arithmétique des corps de nombres à degré infini. Math. Ann. **106**, 473–501 (1932)

2758. Herbrand, J.: Théorie arithmétique des corps de nombres à degré infini, II. Math. Ann. **108**, 699–717 (1933)
2759. Herglotz, G.: Zur letzten Eintragung im Gaussschen Tagebuch. Ber. Sächs. Akad. Wiss. **73**, 271–276 (1921)
2760. Hermes, H.: Unlösbarkeit des zehnten Hilbertschen Problems. Enseign. Math. **18**, 47–56 (1972)
2761. Hermite, C.: Lettres de M. Ch. Hermite à M. Jacobi sur différents objets de la théorie des nombres. J. Reine Angew. Math. **40**, 261–315 (1850) [[2766], vol. 1, pp. 100–163]
2762. Hermite, C.: Sur la théorie des formes quadratiques, II. J. Reine Angew. Math. **47**, 343–368 (1854) [[2766], vol. 1, pp. 234–263]
2763. Hermite, C.: Extrait d'une lettre de M.C.Hermite à M. Borchardt sur le nombre limité d'irrationalités auxquelles se réduisent les racines des équations à coefficients entiers complexes d'un degré et d'un discriminant donnés. J. Reine Angew. Math. **53**, 182–192 (1857) [[2766], vol. 1, pp. 414–428]
2764. Hermite, C.: Sur la fonction exponentielle. C. R. Acad. Sci. Paris **77**, 18–24, 74–79, 221–233, 285–293 (1873) [[2766], vol. 3, pp. 150–181]
2765. Hermite, C.: Sur une extension donnée a la théorie des fractions continues par M. Tchebychef. (Extrait d'une lettre de M.Ch. Hermite a M. Borchardt). J. Reine Angew. Math. **88**, 10–15 (1879) [[2766], vol. 3, pp. 513–519]
2766. Hermite, C.: Oeuvres, vols. 1–4. Gauthier-Villars, Paris (1905–1917)
2767. Hernane, M.-O., Nicolas, J.-L.: Grandes valeurs du nombre de factorisations d'un entier en produit ordonné de facteurs premiers. Ramanujan J. **14**, 277–304 (2007)
2768. Herrmann, O.: Eine metrische Charakterisierung eines Fundamentalbereichs der Hilbertschen Modulgruppen. Math. Z. **60**, 148–155 (1954)
2769. Herrmann, O.: Über Hilbertsche Modulfunktionen und die Dirichletschen Reihen mit Eulerscher Produktentwicklung. Math. Ann. **127**, 357–400 (1954)
2770. Herrmann, O.: Über die Berechnung der Fourierkoeffizienten der Funktion $j(\tau)$. J. Reine Angew. Math. **274/275**, 187–195 (1975)
2771. Herschfeld, A.: The equation $2^x - 3^y = d$. Bull. Am. Math. Soc. **42**, 231–234 (1936)
2772. Herzog, E.: Note zum Wieferichschen Beweis der Darstellbarkeit der ganzen Zahlen durch neun Kuben. Acta Arith. **3**, 86–88 (1938)
2773. Herzog, J., Smith, P.R.: Lower bounds for a certain class of error functions. Acta Arith. **60**, 289–305 (1992)
2774. Heuberger, C.: On families of parametrized Thue equations. J. Number Theory **76**, 45–61 (1999)
2775. Heuberger, C.: On a conjecture of E. Thomas concerning parametrized Thue equations. Acta Arith. **98**, 375–394 (2001)
2776. Heuberger, C.: On explicit bounds for the solutions of a class of parametrized Thue equations of arbitrary degree. Monatshefte Math. **132**, 325–339 (2001)
2777. Heuberger, C., Tichy, R.: Effective solution of families of Thue equations containing several parameters. Acta Arith. **91**, 147–163 (1999)
2778. Hickerson, D.R.: Length of period of simple continued fraction expansion of $\sqrt{d}$. Pac. J. Math. **46**, 429–431 (1973)
2779. Hijikata, H., Pizer, A., Shemanske, T.: The basis problem for modular forms on $\Gamma_0(N)$. Proc. Jpn. Acad. Sci. **56**, 280–284 (1980)
2780. Hilano, T.: On the distribution of zeros of Dirichlet's L-functions on the line $\sigma = 1/2$, II. Tokyo J. Math. **1**, 285–304 (1978)
2781. Hilbert, D.: Ueber die Darstellung definiter Formen als Summe von Formenquadraten. Math. Ann. **32**, 342–350 (1888) [[2792], vol. 2, pp. 154–161]
2782. Hilbert, D.: Ein neuer Beweis des Kronecker'schen Fundamentalsatzes über Abel'sche Zahlkörper. Nachr. Ges. Wiss. Göttingen, 1896, 29–39
2783. Hilbert, D.: Die Theorie der algebraischer Zahlkörper. Jahresber. Dtsch. Math.-Ver. **4**, 175–546 (1897) [[2792], **1**, 63–363; English translation: The Theory of Algebraic Number Fields, Springer, 1998; French translation: Ann. Fac. Sci. Toulouse **1**, 257–328 (1909); **2**, 225–456 (1910)]

2784. Hilbert, D.: Über Diophantische Gleichungen. Nachr. Ges. Wiss. Göttingen, 1897, 48–54 [[2792], vol. 2, pp. 384–389]
2785. Hilbert, D.: Ueber die Theorie der relativ-Abelschen Zahlkörper Nachr. Ges. Wiss. Göttingen, 1898, 377–399. Acta Math. **26**, 1902, 99–132 [[2792], vol. 1, pp. 483–509]
2786. Hilbert, D.: Ueber die Theorie des relativ-quadratischen Zahlkörpers. Math. Ann. **51**, 1–127 (1899) [[2792], vol. 1, pp. 370–482]
2787. Hilbert, D.: Grundlagen der Geometrie. Teubner, Leipzig (1899); 2nd ed. 1903, 3rd ed. 1909, ..., 14th ed. 1999
2788. Hilbert, D.: Mathematische Probleme, Arch. Math. **1**, 44–63, 213–237 (1901). [[2792], vol. 3, pp. 290–329; English translation: Mathematical problems. Bull. Am. Math. Soc. **8**, 437–479 (1901/1902); reprint, (N.S.), **37**, 407–436 (2000)]
2789. Hilbert, D.: Beweis für die Darstellbarkeit der ganzen Zahlen durch eine feste Anzahl n-ter Potenzen (Waringsches Problem). Math. Ann. **67**, 281–300 (1909) [[2792], vol. 1, pp. 510–535]
2790. Hilbert, D.: Herrmann Minkowski. Math. Ann. **68**, 445–471 (1910) [[2792], vol. 3, pp. 339–364]
2791. Hilbert, D.: Adolf Hurwitz. Math. Ann. **83**, 161–172 (1921) [[2792], vol. 3, pp. 370–377]
2792. Hilbert, D.: Gesammelte Abhandlungen, vols. 1–3. Springer, Berlin (1932–1935); 2nd ed. 1970. [Reprint: Chelsea, 1965]
2793. Hilbert, D., Hurwitz, A.: Über die diophantischen Gleichungen vom Geschlecht Null. Acta Math. **14**, 217–224 (1891) [[2792], vol. 2, pp. 258–263]
2794. Hildebrand, A.: A characterization of the logarithm as an additive arithmetic function. Arch. Math. **38**, 535–539 (1982)
2795. Hildebrand, A.: An asymptotic formula for the variance of an additive function. Math. Z. **183**, 145–170 (1983)
2796. Hildebrand, A.: Über die punktweise Konvergenz von Ramanujan-Entwicklungen zahlentheoretischer Funktionen. Acta Arith. **44**, 110–140 (1984)
2797. Hildebrand, A.: The prime number theorem via the large sieve. Mathematika **33**, 23–30 (1986)
2798. Hildebrand, A.: On Wirsing's mean value theorem for multiplicative functions. Bull. Lond. Math. Soc. **18**, 147–152 (1986)
2799. Hildebrand, A.: On the number of positive integers $\leq x$ and free of prime factors $> y$. J. Number Theory **22**, 289–307 (1986)
2800. Hildebrand, A.: On consecutive kth power residues. Monatshefte Math. **102**, 103–114 (1986)
2801. Hildebrand, A.: On consecutive kth power residues, II. Mich. Math. J. **39**, 241–253 (1991)
2802. Hildebrand, A.: The divisor function at consecutive integers. Pac. J. Math. **129**, 307–319 (1987)
2803. Hildebrand, A.: On the least pair of consecutive quadratic nonresidues. Mich. Math. J. **34**, 57–62 (1987)
2804. Hildebrand, A.: Large values of character sums. J. Number Theory **29**, 271–296 (1988)
2805. Hildebrand, A.: Additive and multiplicative functions on shifted primes. Proc. Lond. Math. Soc. **59**, 209–232 (1989)
2806. Hildebrand, A.: Multiplicative functions on arithmetic progressions. Proc. Am. Math. Soc. **108**, 307–318 (1990)
2807. Hildebrand, A., Tenenbaum, G.: On integers free of large prime factors. Trans. Am. Math. Soc. **296**, 265–290 (1986)
2808. Hildebrand, A., Tenenbaum, G.: On the number of prime factors of an integer. Duke Math. J. **56**, 471–501 (1988)
2809. Hildebrand, A., Tenenbaum, G.: Integers without large prime factors. J. Théor. Nr. Bordx. **5**, 411–484 (1993)
2810. Hille, E.: A problem in "Factorisatio Numerorum". Acta Arith. **2**, 134–144 (1936)
2811. Hilliker, D.L., Straus, E.G.: Determination of bounds for the solutions to those binary Diophantine equations that satisfy the hypotheses of Runge's theorem. Trans. Am. Math. Soc. **280**, 637–657 (1983)

2812. Hilton, P.J.: Heinz Hopf. Bull. Lond. Math. Soc. **4**, 202–217 (1972)
2813. Hindry, M., Rivoal, T.: Le λ-calcul de Golomb et la conjecture de Bateman-Horn. Enseign. Math. **51**, 265–318 (2005)
2814. Hindry, M., Silverman, J.: Sur le nombre de points de torsion rationnels sur une courbe elliptique. C. R. Acad. Sci. Paris **329**, 97–100 (1999)
2815. Hindry, M., Silverman, J.: Diophantine Geometry. Springer, Berlin (2000)
2816. Hinz, J.: Eine Erweiterung des nullstellenfreien Bereiches der Heckeschen Zetafunktion und Primideale in Idealklassen. Acta Arith. **38**, 209–254 (1980/1981)
2817. Hinz, J.: Character sums in algebraic number fields. J. Number Theory **17**, 52–70 (1983)
2818. Hinz, J.: The average order of magnitude of least primitive roots in algebraic number fields. Mathematika **30**, 11–25 (1983)
2819. Hinz, J.: A note on Artin's conjecture in algebraic number fields. J. Number Theory **22**, 334–349 (1986)
2820. Hinz, J.: Methoden des grossen Siebes in algebraischen Zahlkörpern. Manuscr. Math. **57**, 181–194 (1987)
2821. Hinz, J.: A generalization of Bombieri's prime number theorem to algebraic number fields. Acta Arith. **51**, 173–193 (1988)
2822. Hinz, J.: On the prime ideal theorem. J. Indian Math. Soc. **59**, 243–260 (1993)
2823. Hinz, J.: An application of algebraic sieve theory. Arch. Math. **80**, 586–599 (2003)
2824. Hirschhorn, M.D.: Two further Ramanujan pairs. J. Aust. Math. Soc. **30**, 1–4 (1980/1981)
2825. Hirschhorn, M.D.: A simple proof of Jacobi's four-square theorem. Proc. Am. Math. Soc. **101**, 436–438 (1987)
2826. Hirschhorn, M.D.: On the parity of $p(n)$, II. J. Comb. Theory, Ser. A **62**, 128–138 (1993)
2827. Hirschhorn, M.D.: Jacobi's two-square theorem and related identities. Ramanujan J. **3**, 153–158 (1999)
2828. Hirschhorn, M.D., Subbarao, M.V.: On the parity of $p(n)$. Acta Arith. **50**, 355–356 (1988)
2829. Hirst, K.E.: The length of periodic continued fractions. Monatshefte Math. **76**, 428–435 (1972)
2830. Hirzebruch, F., Zagier, D.: The Atiyah-Singer Theorem and Elementary Number Theory. Birkhäuser, Boston (1974)
2831. Hlavka, J.L.: Results on sums of continued fractions. Trans. Am. Math. Soc. **211**, 123–134 (1975)
2832. Hlawka, E.: Zur Geometrie der Zahlen. Math. Z. **49**, 285–312 (1943/1944)
2833. Hlawka, E.: Zur formalen Theorie der Gleichverteilung in kompakten Gruppen. Rend. Circ. Mat. Palermo **4**, 33–47 (1955)
2834. Hlawka, E.: Theorie der Gleichverteilung. Bibliogr. Inst., Mannheim (1979) [English translation: The Theory of Uniform Distribution, Academic Publishers, 1984]
2835. Hlawka, E.: Carl Ludwig Siegel (31/12/1896–4/4/1981). J. Number Theory **20**, 373–404 (1985)
2836. Hlawka, E.: Nachruf auf Nikolaus Hofreiter. Monatshefte Math. **116**, 263–273 (1993)
2837. Hlawka, E., Firneis, F., Zinterhof, P.: Zahlentheoretische Methoden in der numerischen Mathematik. Oldenbourg, Wien (1981)
2838. Ho, K.-H., Tsang, K.-M.: On almost prime k-tuples. J. Number Theory **120**, 33–46 (2006)
2839. Hodges, A.: Alan Turing. In: Cambridge Scientific Minds, pp. 253–268. Cambridge University Press, Cambridge (2002)
2840. Hoffstein, J.: On the Siegel-Tatuzawa theorem. Acta Arith. **38**, 167–174 (1980)
2841. Hofmeister, G.: Thin bases of order two. J. Number Theory **86**, 118–132 (2001)
2842. Hofmeister, G., Stoll, P.: Note on Egyptian fractions. J. Reine Angew. Math. **362**, 141–145 (1985)
2843. Hofreiter, N.: Über Extremformen. Monatshefte Math. Phys. **40**, 129–152 (1933)
2844. Hofreiter, N.: Diophantische Approximationen in imaginär quadratischen Zahlkörpern. Monatshefte Math. Phys. **45**, 175–190 (1937)
2845. Hofreiter, N.: Über das Produkt von Linearformen. Monatshefte Math. Phys. **49**, 295–298 (1940)

2846. Hofreiter, N.: Nachruf auf Philipp Furtwängler. Monatshefte Math. Phys. **49**, 219–227 (1940)
2847. Hoheisel, G.: Primzahlprobleme in der Analysis. SBer. Preuß. Akad. Wiss. Berlin, 1930, 580–588
2848. Hölder, O.: Über einen asymptotischen Ausdruck. Acta Math. **59**, 89–97 (1932)
2849. Holzapfel, R.-P.: The Ball and some Hilbert Problems. Birkhäuser, Basel (1995)
2850. Holzer, L.: Über die Gleichung $(x + y)(x^2 + y^2) = 4Cz^3$. Monatshefte Math. Phys. **26**, 289–294 (1915)
2851. Holzer, L.: Minimal solutions of diophantine equations. Can. J. Math. **2**, 238–244 (1950)
2852. Honda, K.: Teiji Takagi: a biography—on the 100th anniversary of his birth. Comment. Math. Univ. St. Pauli **24**, 141–167 (1975/1976)
2853. Honda, T.: Isogeny classes of abelian varieties over finite fields. J. Math. Soc. Jpn. **20**, 83–95 (1968)
2854. Hooley, C.: On the representation of a number as the sum of two squares and a prime. Acta Math. **97**, 189–210 (1957)
2855. Hooley, C.: An asymptotic formula in the theory of numbers. Proc. Lond. Math. Soc. **7**, 396–413 (1957)
2856. Hooley, C.: On the representations of a number as the sum of two cubes. Math. Z. **82**, 259–266 (1963)
2857. Hooley, C.: On the difference of consecutive numbers prime to n. Acta Arith. **8**, 343–347 (1962/1963)
2858. Hooley, C.: On the difference of consecutive numbers prime to n, II. Publ. Math. (Debr.) **12**, 39–49 (1965)
2859. Hooley, C.: On the number of divisors of a quadratic polynomial. Acta Math., **110**, 97–114
2860. Hooley, C.: On the representations of a number as the sum of two h-th powers. Math. Z. **84**, 126–136 (1964)
2861. Hooley, C.: On the distribution of the roots of polynomial congruences. Mathematika **11**, 39–49 (1964)
2862. Hooley, C.: On Artin's conjecture. J. Reine Angew. Math. **225**, 209–220 (1967)
2863. Hooley, C.: On the greatest prime factor of a quadratic polynomial. Acta Math. **117**, 281–299 (1967)
2864. Hooley, C.: On the power free values of polynomials. Mathematika **14**, 21–26 (1967)
2865. Hooley, C.: On the Brun-Titchmarsh theorem. J. Reine Angew. Math. **255**, 60–79 (1972); Proc. Lond. Math. Soc. **30**, 114–128 (1975)
2866. Hooley, C.: On the largest prime factor of $p + a$. Mathematika **20**, 135–143 (1973)
2867. Hooley, C.: On the distribution of square-free numbers. Can. J. Math. **25**, 1216–1223 (1973)
2868. Hooley, C.: Applications of Sieve Methods. Academic Press, San Diego (1974)
2869. Hooley, C.: On the Barban-Davenport-Halberstam theorem, I. J. Reine Angew. Math. **274/275**, 206–223 (1975)
2870. Hooley, C.: On the Barban-Davenport-Halberstam theorem, II. J. Lond. Math. Soc. **9**, 625–636 (1974/1975)
2871. Hooley, C.: On the Barban-Davenport-Halberstam theorem, III. J. Lond. Math. Soc. **10**, 249–256 (1975)
2872. Hooley, C.: On the Barban-Davenport-Halberstam theorem, IV. J. Lond. Math. Soc. **11**, 399–407 (1975)
2873. Hooley, C.: On the Barban-Davenport-Halberstam theorem, V. Proc. Lond. Math. Soc. **33**, 535–548 (1976)
2874. Hooley, C.: On the Barban-Davenport-Halberstam theorem, VI. J. Lond. Math. Soc. **13**, 57–64 (1976)
2875. Hooley, C.: On the Barban-Davenport-Halberstam theorem, VII. J. Lond. Math. Soc. **16**, 1–8 (1977)
2876. Hooley, C.: On the Barban-Davenport-Halberstam theorem, VIII. J. Reine Angew. Math. **499**, 1–46 (1998)

2877. Hooley, C.: On the Barban-Davenport-Halberstam theorem, IX. Acta Arith. **83**, 17–30 (1998)
2878. Hooley, C.: On the Barban-Davenport-Halberstam theorem, X. Hardy-Ramanujan J. **21**, 2–11 (1998)
2879. Hooley, C.: On the Barban-Davenport-Halberstam theorem, XI. Acta Arith. **91**, 1–41 (1999)
2880. Hooley, C.: On the Barban-Davenport-Halberstam theorem, XII. In: Number Theory in Progress, vol. 2, pp. 893–910. de Gruyter, Berlin (1999)
2881. Hooley, C.: On the Barban-Davenport-Halberstam theorem, XIII. Acta Arith. **94**, 53–86 (2000)
2882. Hooley, C.: On the Barban-Davenport-Halberstam theorem, XIV. Acta Arith. **101**, 247–292 (2002)
2883. Hooley, C.: On the Barban-Davenport-Halberstam theorem, XV. Acta Arith. **111**, 205–224 (2004)
2884. Hooley, C.: On the Barban-Davenport-Halberstam theorem, XVI. Bonner Math. Schr., **360**, 2003, 18 pp.
2885. Hooley, C.: On the Barban-Davenport-Halberstam theorem, XVII. Bonner Math. Schr., **360**, 2003, 5 pp.
2886. Hooley, C.: On the Barban-Davenport-Halberstam theorem, XVIII. Ill. J. Math. **49**, 581–643 (2005)
2887. Hooley, C.: On the Barban-Davenport-Halberstam theorem, XIX. Hardy-Ramanujan J. **30**, 56–67 (2007)
2888. Hooley, C.: On the representations of a number as the sum of four cubes. Proc. Lond. Math. Soc. **36**, 117–140 (1978)
2889. Hooley, C.: On the representations of a number as the sum of four cubes, II. J. Lond. Math. Soc. **16**, 424–428 (1977)
2890. Hooley, C.: On power-free numbers and polynomials, I. J. Reine Angew. Math. **293/294**, 67–85 (1977)
2891. Hooley, C.: On power-free numbers and polynomials, II. J. Reine Angew. Math. **295**, 1–21 (1977)
2892. Hooley, C.: On the greatest prime factor of a cubic polynomial. J. Reine Angew. Math. **303/304**, 21–50 (1978)
2893. Hooley, C.: On the numbers that are representable as the sum of two cubes. J. Reine Angew. Math. **314**, 146–173 (1980)
2894. Hooley, C.: On Waring's problem for two squares and three cubes. J. Reine Angew. Math. **328**, 161–207 (1981)
2895. Hooley, C.: On another sieve method and the numbers that are a sum of two h-th powers. Proc. Lond. Math. Soc. **43**, 73–109 (1981)
2896. Hooley, C.: On another sieve method and the numbers that are a sum of two h-th powers, II. J. Reine Angew. Math. **475**, 55–75 (1996)
2897. Hooley, C.: On a new approach to various problems of Waring's type. In: Recent Progress in Analytic Number Theory, vol. 1, pp. 127–191. Academic Press, San Diego (1981)
2898. Hooley, C.: On the Pellian equation and the class number of indefinite binary quadratic forms. J. Reine Angew. Math. **353**, 98–131 (1984)
2899. Hooley, C.: On Waring's problem. Acta Math. **157**, 48–97 (1986)
2900. Hooley, C.: On nonary cubic forms. J. Reine Angew. Math. **386**, 32–98 (1988)
2901. Hooley, C.: On nonary cubic forms, II. J. Reine Angew. Math. **415**, 95–165 (1991)
2902. Hooley, C.: On nonary cubic forms, III. J. Reine Angew. Math. **456**, 53–63 (1994)
2903. Hooley, C.: On a problem of Hardy and Littlewood. Acta Arith. **79**, 289–311 (1997)
2904. Hooley, C.: On hypothesis K for biquadrates. Proc. Indian Acad. Sci. Math. Sci. **109**, 333–343 (1999)
2905. Hopf, H.: Über die Verteilung quadratischer Reste. Math. Z. **32**, 222–231 (1930)
2906. Hornfeck, B.: Zur Dichte der Menge der vollkommenen Zahlen. Arch. Math. **6**, 442–443 (1955)

2907. Hornfeck, B.: Hans-Heinrich Ostmann: 1913–1959. Jahresber. Dtsch. Math.-Ver. **86**, 31–35 (1984)
2908. Hornfeck, B., Wirsing, E.: Über die Häufigkeit vollkommener Zahlen. Math. Ann. **133**, 431–438 (1957)
2909. Horváth, J.: L'oeuvre mathématique de Marcel Riesz, I. In: Proceedings of the Seminar on the History of Mathematics, vol. 3, pp. 83–121. Institut H. Poincaré, Paris (1982)
2910. Horváth, J.: L'oeuvre mathématique de Marcel Riesz, II. In: Proceedings of the Seminar on the History of Mathematics, vol. 4, pp. 1–59. Institut H. Poincaré, Paris (1983)
2911. Hsia, J.S., Icaza, M.I.: Effective version of Tartakowsky's theorem. Acta Arith. **89**, 235–253 (1999)
2912. Hsiao, H.J., Lee, H.C.: Hecke operators on modular forms. Arch. Math. **32**, 158–165 (1979)
2913. Hu, P.-C., Yang, C.-C.: The "abc" conjecture over function fields. Proc. Jpn. Acad. Sci. **76**, 118–120 (2000)
2914. Hu, P.-C., Yang, C.-C.: A generalized abc-conjecture over function fields. J. Number Theory **94**, 286–298 (2002)
2915. Hua, L.K.: Waring's problem for cubes. Bull. Calcutta Math. Soc. **26**, 139–140 (1935)
2916. Hua, L.K.: On Waring's problem with polynomial summands. Am. J. Math. **58**, 553–562 (1936)
2917. Hua, L.K.: An easier Waring-Kamke problem. J. Lond. Math. Soc. **11**, 4–5 (1936)
2918. Hua, L.K.: On a generalized Waring problem. Proc. Lond. Math. Soc. **43**, 161–182 (1937) [[2937], pp. 17–38]
2919. Hua, L.K.: On a generalized Waring problem, II. J. Chin. Math. Soc. **2**, 175–191 (1940) [[2937], pp. 61–73]
2920. Hua, L.K.: Some results in the additive prime-number theory. Q. J. Math. **9**, 68–80 (1938)
2921. Hua, L.K.: On Waring's problem. Q. J. Math. **9**, 199–202 (1938) [[2937], pp. 39–42]
2922. Hua, L.K.: On Tarry's problem. Q. J. Math. **9**, 315–320 (1938) [[2937], pp. 43–48]
2923. Hua, L.K.: On the representation of numbers as the sums of the powers of primes. Math. Z. **44**, 335–346 (1939) [[2937], pp. 49–60]
2924. Hua, L.K.: On Waring's problem for fifth powers. Proc. Lond. Math. Soc. **45**, 144–160 (1939)
2925. Hua, L.K.: On an exponential sum. J. Chin. Math. Soc. **2**, 301–312 (1940) [[2937], pp. 101–109]
2926. Hua, L.K.: Sur le problème de Waring relatif à un polynome du troisième degré. C. R. Acad. Sci. Paris **210**, 650–652 (1940)
2927. Hua, L.K.: On a system of diophantine equations. Dokl. Akad. Nauk SSSR **27**, 312–313 (1940)
2928. Hua, L.K.: The lattice-points in a circle. Q. J. Math. **13**, 18–29 (1942) [[2937], pp. 124–135]
2929. Hua, L.K.: On the least primitive root of a prime. Bull. Am. Math. Soc. **48**, 726–730 (1942)
2930. Hua, L.K.: On the number of partitions of a number into unequal parts. Trans. Am. Math. Soc. **51**, 194–201 (1942) [[2937], pp. 110–117]
2931. Hua, L.K.: On the least solution of Pell's equation. Bull. Am. Math. Soc. **48**, 731–735 (1942) [[2937], pp. 119–123]
2932. Hua, L.K.: Additive Theory of Prime Numbers. Tr. Mat. Inst. Steklova **22**, 1–197 (1947) (in Russian) [English translation: Am. Math. Soc., 1965; German translation: Additive Zahlentheorie, Leipzig, 1959]
2933. Hua, L.K.: Some results on additive theory of numbers. Proc. Natl. Acad. Sci. USA **33**, 136–137 (1947)
2934. Hua, L.K.: An improvement of Vinogradov's mean-value theorem and several applications. Q. J. Math. **20**, 48–61 (1949) [[2937], pp. 178–191]
2935. Hua, L.K.: Improvement of a result of Wright. J. Lond. Math. Soc. **24**, 17–159 (1949) [[2937], 175–177]
2936. Hua, L.K.: On exponential sums over an algebraic field. Can. J. Math. **3**, 44–51 (1951) [[2937], 192–199]
2937. Hua, L.K.: Selected Papers. Springer, Berlin (1983)

2938. Hua, L.K.: Die Abschätzung von Exponentialsummen und ihre Anwendung in der Zahlentheorie. In: Enzyklopädie der mathematischen Wissenschaften, vol. 1_2, part 1 of Heft 13
2939. Hua, L.K., Reiner, I.: On the generators of the symplectic modular group. Trans. Am. Math. Soc. **65**, 415–426 (1949) [[2937], pp. 491–502]
2940. Hua, L.K., Wang, Y.: Applications of Number Theory to Numerical Analysis. Springer, Berlin (1981)
2941. Huard, J.G., Williams, K.S.: Sums of twelve squares. Acta Arith. **109**, 195–204 (2003)
2942. Huber, A.: Philipp Furtwängler. Jahresber. Dtsch. Math.-Ver. **50**, 167–178 (1940)
2943. Hudson, R.H.: On sequences of consecutive quadratic non-residues. J. Number Theory **3**, 178–181 (1971)
2944. Hudson, R.H.: On a conjecture of Issai Schur. J. Reine Angew. Math. **289**, 215–220 (1977)
2945. Hughes, J.F., Shallit, J.O.: On the number of multiplicative partitions. Am. Math. Mon. **90**, 468–471 (1983)
2946. Hughes, K.: Ramanujan congruences for $p_{-k}(n)$ modulo powers of 17. Can. J. Math. **43**, 506–525 (1991)
2947. Humbert, G.: Sur la représentation d'un entier par une somme de dix ou de douze carrés. C. R. Acad. Sci. Paris **144**, 874–878 (1907)
2948. Humbert, G.: Remarques sur certaines suites d'approximation. J. Math. Pures Appl. **2**, 155–167 (1916)
2949. Humbert, P.: Théorie de la réduction des formes quadratiques définies positives dans un corps algébrique K fini. Comment. Math. Helv. **12**, 263–306 (1940)
2950. Humbert, P.: Réduction de formes quadratiques dans un corps algébrique fini. Comment. Math. Helv. **23**, 50–63 (1949)
2951. Hummel, P.: On consecutive quadratic non-residues: a conjecture of Issai Schur. J. Number Theory **103**, 257–266 (2003)
2952. Humphreys, M.G.: On the Waring problem with polynomial summands. Duke Math. J. **1**, 361–375 (1935)
2953. Hunter, W.: The representation of numbers by sums of fourth powers. J. Lond. Math. Soc. **16**, 177–179 (1941)
2954. Hurwitz, A.: Grundlagen einer independenten Theorie der elliptischen Modulfunktionen und Theorie der Multiplicatorgleichungen erster Stufe. Math. Ann. **18**, 528–592 (1881) [[2965], vol. 1, pp. 1–66]
2955. Hurwitz, A.: Einige Eigenschaften der Dirichlet'schen Functionen $F(s) = \sum(\frac{D}{n}) \cdot \frac{1}{n^s}$, die bei der Bestimmung der Classenanzahlen binärer quadratischer Formen auftreten. Z. Angew. Math. Phys. **27**, 86–101 (1882) [[2965], vol. 1, pp. 72–88]
2956. Hurwitz, A.: Sur la décomposition des nombres en cinq carrés. C. R. Acad. Sci. Paris **98**, 504–507 (1884) [[2965], vol. 2, pp. 5–7]
2957. Hurwitz, A.: Über die Entwicklung complexer Grössen in Kettenbrüche. Acta Math. **11**, 187–200 (1887/1888) [[2965], vol. 2, pp. 72–83]
2958. Hurwitz, A.: Über die angenäherte Darstellung der Irrationalzahlen durch rationale Brüche. Math. Ann. **39**, 279–284 (1891) [[2965], vol. 2, pp. 122–128]
2959. Hurwitz, A.: Über höhere Kongruenzen. Arch. Math. **5**, 17–27 (1903) [[2965], vol. 2, pp. 374–384]
2960. Hurwitz, A.: Über die Theorie der elliptischen Modulfunktionen. Math. Ann. **58**, 343–360 (1904) [[2965], vol. 2, pp. 577–595]
2961. Hurwitz, A.: Über eine Aufgabe der unbestimmten Analyse. Arch. Math. **11**, 185–196 (1907) [[2965], vol. 2, pp. 410–421]
2962. Hurwitz, A.: Somme de trois carrés. L'Intermédiaire Math. **14**, 106–107 (1907) [[2965], vol. 2, p. 751]
2963. Hurwitz, A.: Über die Darstellung der ganzen Zahlen als Summen von n^{ten} Potenzen ganzer Zahlen. Math. Ann. **65**, 424–427 (1908) [[2965], vol. 2, pp. 422–426]
2964. Hurwitz, A.: Über die Kongruenz $ax^e + by^e + cz^e \equiv 0 (\bmod p)$. J. Reine Angew. Math. **136**, 272–292 (1909) [[2965], vol. 2, pp. 430–445.]
2965. Hurwitz, A.: Mathematische Werke, vols. 1, 2. Birkhäuser, Basel (1932)

2966. Husemöller, D.H.: Elliptic Curves. Springer, Berlin (1987); 2nd ed. 2004
2967. Huston, R.E.: Asymptotic generalizations of Waring's theorem. Proc. Lond. Math. Soc. **39**, 82–115 (1935)
2968. Hutchinson, J.I.: On the roots of the Riemann zeta function. Trans. Am. Math. Soc. **27**, 49–60 (1925)
2969. Huxley, M.N.: The large sieve inequality for algebraic number fields. Mathematika **15**, 178–187 (1968)
2970. Huxley, M.N.: The large sieve inequality for algebraic number fields, II. Proc. Lond. Math. Soc. **21**, 108–128 (1970)
2971. Huxley, M.N.: The large sieve inequality for algebraic number fields, III. J. Lond. Math. Soc. **3**, 233–240 (1971)
2972. Huxley, M.N.: On the difference between consecutive primes. Invent. Math. **15**, 164–170 (1972)
2973. Huxley, M.N.: The Distribution of Prime Numbers. Large Sieves and Zero-Density Theorems. Clarendon Press, Oxford (1972)
2974. Huxley, M.N.: Small differences between consecutive primes. Mathematika **20**, 229–232 (1973)
2975. Huxley, M.N.: Small differences between consecutive primes, II. Mathematika **24**, 142–152 (1977)
2976. Huxley, M.N.: Large values of Dirichlet polynomials. Acta Arith. **24**, 329–346 (1973)
2977. Huxley, M.N.: Large values of Dirichlet polynomials, II. Acta Arith. **27**, 159–169 (1974)
2978. Huxley, M.N.: Large values of Dirichlet polynomials, III. Acta Arith. **26**, 435–444 (1974)
2979. Huxley, M.N.: A note on polynomial congruences. In: Recent Progress in Analytic Number Theory, vol. 1, pp. 193–196. Academic Press, San Diego (1981)
2980. Huxley, M.N.: An application of the Fouvry-Iwaniec theorem. Acta Arith. **43**, 441–443 (1984)
2981. Huxley, M.N.: Exponential sums and lattice points. Proc. Lond. Math. Soc. **60**, 471–502 (1990)
2982. Huxley, M.N.: Exponential sums and lattice points, II. Proc. Lond. Math. Soc. **66**, 279–301 (1993); corr. **68**, 264 (1994)
2983. Huxley, M.N.: Exponential sums and lattice points, III. Proc. Lond. Math. Soc. **87**, 591–609 (2003)
2984. Huxley, M.N.: Exponential sums and the Riemann zeta function, IV. Proc. Lond. Math. Soc. **66**, 1–40 (1993)
2985. Huxley, M.N.: Exponential sums and the Riemann zeta function, V. Proc. Lond. Math. Soc. **90**, 1–41 (2005)
2986. Huxley, M.N.: Area, Lattice Points and Exponential Sums. Oxford University Press, Oxford (1996)
2987. Huxley, M.N.: Moments of differences between square-free numbers. In: Sieve Methods, Exponential Sums, and Their Applications in Number Theory, Cardiff, 1995, pp. 187–204. Cambridge University Press, Cambridge (1997)
2988. Huxley, M.N.: The rational points close to a curve, II. Acta Arith. **93**, 201–219 (2000)
2989. Huxley, M.N., Iwaniec, H.: Bombieri's theorem in short intervals. Mathematika **22**, 188–194 (1975)
2990. Huxley, M.N., Jutila, M.: Large values of Dirichlet polynomials, IV. Acta Arith. **32**, 297–312 (1977)
2991. Huxley, M.N., Kolesnik, G.: Exponential sums and the Riemann zeta function, III. Proc. Lond. Math. Soc. **62**, 449–468 (1991); corr. **66**, 302 (1993)
2992. Huxley, M.N., Nair, M.: Power free values of polynomials, III. Proc. Lond. Math. Soc. **41**, 66–82 (1980)
2993. Huxley, M.N., Trifonov, O.: The square-full numbers in an interval. Math. Proc. Camb. Philos. Soc. **119**, 201–208 (1996)
2994. Huxley, M.N., Watt, N.: Exponential sums and the Riemann zeta function. Proc. Lond. Math. Soc. **57**, 1–24 (1988)

2995. Hwang, J.S.: Small zeros of additive forms in algebraic number fields. Bull. Lond. Math. Soc. **29**, 295–298 (1997)
2996. Hwang, J.S.: Small zeros of additive forms in several variables. Acta Math. Sin. New Ser. **14**, 57–66 (1998)
2997. Hykšová, M.: Karel Rychlík (1885–1968). Dějiny Matematiky, vol. 22. CVUT Praha, Praha (2003)
2998. Hyyrö, S.: Über die Gleichung $ax^n - by^n = z$ und das Catalansche Problem. Ann. Acad. Sci. Fenn., Ser. A 1 Math. **355**, 1–50 (1964)
2999. Iannucci, D.E., Sorli, M.: On the total number of prime factors of an odd perfect number. Math. Comput. **72**, 2077–2084 (2003)
3000. Ibragimov, I.A., et al.: Juriĭ Vladimirovič Linnik: obituary. Usp. Mat. Nauk **28**(2), 197–213 (1973) (in Russian)
3001. Icaza, M.I.: Hermite constant and extreme forms for algebraic number fields. J. Lond. Math. Soc. **55**, 11–22 (1997)
3002. Igusa, J.-I.: On the theory of algebraic correspondences and its application to the Riemann hypothesis in function-fields. J. Math. Soc. Jpn. **1**, 147–197 (1949)
3003. Igusa, J.-I.: Theta Functions. Springer, Berlin (1972)
3004. Igusa, J.-I.: Complex powers and asymptotic expansions, II, Asymptotic expansions. J. Reine Angew. Math. **278/279**, 307–321 (1975)
3005. Igusa, J.-I.: Some observations of higher degree characters. Am. J. Math. **99**, 393–417 (1977)
3006. Ihara, Y.: Hecke polynomials as congruence ζ functions in elliptic modular case. Ann. Math. **85**, 267–295 (1967)
3007. Ihara, T.: Takuro Shintani (1943–1980). J. Fac. Sci. Univ. Tokyo **28**(3), iii–vi (1981)
3008. Ikehara, S.: An extension of Landau's theorem in the analytical theory of numbers. J. Math. Phys. MIT **10**, 1–12 (1931)
3009. Ikehara, S.: Review of [1789]. Zent.bl. Math. 0023.2980
3010. Imai, H.: A remark on the rational points of Abelian varieties with values in cyclotomic $\mathbb{Z}_p$ extensions. Proc. Jpn. Acad. Sci. **51**, 12–16 (1975)
3011. Imai, H.: Values of p-adic L-functions at positive integers and p-adic log multiple gamma functions. Tohoku Math. J. **45**, 505–510 (1993)
3012. Indlekofer, K.-H., Kátai, I.: Exponential sums with multiplicative coefficients. Acta Math. Acad. Sci. Hung. **54**, 263–268 (1989)
3013. Indlekofer, K.-H., Kátai, I.: On a theorem of H. Daboussi. Publ. Math. (Debr.) **57**, 145–152 (2000)
3014. Indlekofer, K.-H., Manstavičius, E.: Additive and multiplicative functions on arithmetical semigroups. Publ. Math. (Debr.) **45**, 1–17 (1994)
3015. Indlekofer, K.-H., Manstavičius, E.: Distribution of multiplicative functions defined on semigroups. QMS. Quest. Maths Sci. **24**, 335–347 (2001)
3016. Ingham, A.E.: Some asymptotic formulae in the theory of numbers. J. Lond. Math. Soc. **2**, 202–208 (1927)
3017. Ingham, A.E.: Mean-value theorems in the theory of Riemann zeta-function. Proc. Lond. Math. Soc. **27**, 273–300 (1928)
3018. Ingham, A.E.: The Distribution of Prime Numbers. Cambridge University Press, Cambridge (1932) [Reprints: Stechert-Hafner 1964; Hafner 1971; Cambridge, 1990]
3019. Ingham, A.E.: Mean-value theorems and the Riemann zeta-function. Q. J. Math. **4**, 278–280 (1933)
3020. Ingham, A.E.: A note on the distribution of primes. Acta Arith. **1**, 201–211 (1935)
3021. Ingham, A.E.: On the difference between consecutive primes. Q. J. Math. **8**, 255–266 (1937)
3022. Ingham, A.E.: On two classical lattice point problems. Proc. Camb. Philos. Soc. **36**, 131–138 (1940)
3023. Ingham, A.E.: On the estimation of $N(\sigma, T)$. Q. J. Math. **11**, 291–292 (1940)
3024. Ingham, A.E.: On two conjectures in the theory of numbers. Am. J. Math. **64**, 313–319 (1942)

3025. Ingham, A.E.: Some Tauberian theorems connected with the prime number theorem. J. Lond. Math. Soc. **20**, 171–180 (1945)
3026. Inkeri, K.: Abschätzungen für eventuelle Lösungen der Gleichung im Fermatschen Problem. Ann. Univ. Turku A **16**(1), 1–9 (1953)
3027. Iosevich, A.: Lattice points and generalized Diophantine conditions. J. Number Theory **90**, 19–30 (2001)
3028. Iseki, K.: A remark on the Goldbach-Vinogradov theorem. Proc. Jpn. Acad. Sci. **25**, 185–187 (1949)
3029. Ishibashi, M., Kanemitsu, S.: Some asymptotic formulas of Ramanujan. In: Lecture Notes in Math., vol. 1434, pp. 149–167. Springer, Berlin (1990)
3030. Ishida, M.: On the divisibility of Dedekind's zeta-functions. Proc. Jpn. Acad. Sci. **33**, 293–297 (1957)
3031. Ishida, M.-N.: Polyhedral Laurent series and Brion's equalities. Int. J. Math. **1**, 251–265 (1990)
3032. Ishii, H.: The non-existence of elliptic curves with everywhere good reduction over certain quadratic fields. Jpn. J. Math. **12**, 45–52 (1986)
3033. Iskovskih, V.A.: A counterexample to the Hasse principle for systems of two quadratic forms in five variables. Mat. Zametki **10**, 253–257 (1971) (in Russian)
3034. Iskovskih, V.A.: Verification of the Riemann hypothesis for certain local zeta-functions. Usp. Mat. Nauk **28**(3), 181–182 (1973) (in Russian)
3035. Itard, J.: Pierre Fermat. Birkhäuser, Basel (1950) [2nd ed. 1979]
3036. Ivić, A.: On the asymptotic formulas for powerful numbers. Publ. Inst. Math. (Belgr.) **23**, 85–94 (1978)
3037. Ivić, A.: Exponent pairs and the zeta function of Riemann. Studia Sci. Math. Hung. **15**, 157–181 (1980)
3038. Ivić, A.: Topics in Recent Zeta Function Theory. Université de Paris-Sud, Orsay (1983)
3039. Ivić, A.: Exponent pairs and power moments of the zeta-function. In: Topics in Classical Number Theory, Colloq. Math. Soc. J. Bólyai, vol. 34, pp. 749–768. North-Holland, Amsterdam (1984)
3040. Ivić, A.: The distribution of primitive abundant numbers. Studia Sci. Math. Hung. **20**, 183–187 (1985)
3041. Ivić, A.: The Riemann Zeta-function. Wiley, New York (1985) [Reprint: Dover 2003]
3042. Ivić, A.: Lectures on Mean Values of the Riemann Zeta Function. Springer, Berlin (1991)
3043. Ivić, A., Krätzel, E., Kühleitner, M., Nowak, W.G.: Lattice points in large regions and related arithmetic functions: Recent developments in a very classical topic. In: Elementare und analytische Zahlentheorie, Proc. ELAZ-Conference, Stuttgart, May 24–28, 2004, pp. 89–128 (2006)
3044. Ivić, A., Motohashi, Y.: The mean square of the error term for the fourth power moment of the zeta-function. Proc. Lond. Math. Soc. **69**, 309–329 (1994)
3045. Ivić, A., Motohashi, Y.: On the fourth power moment of the Riemann zeta-function. J. Number Theory **51**, 16–45 (1995)
3046. Ivić, A., Ouellet, M.: Some new estimates in the Dirichlet divisor problem. Acta Arith. **52**, 241–253 (1989)
3047. Ivić, A., Sargos, P.: On the higher moments of the error term in the divisor problem. Ill. J. Math. **51**, 353–377 (2007)
3048. Ivić, A., Shiu, P.: The distribution of powerful integers. Ill. J. Math. **26**, 576–690 (1982)
3049. Ivorra, W.: Courbes elliptiques sur Q, ayant un point d'ordre 2 rationnel sur Q, de conducteur $2^N p$. Diss. Math. **429**, 1–55 (2004)
3050. Ivorra, W., Kraus, A.: Quelques résultats sur les équations $ax^p + by^p = cz^2$. Can. J. Math. **58**, 115–153 (2006)
3051. Iwaniec, H.: On the error term in the linear sieve. Acta Arith. **19**, 1–30 (1971)
3052. Iwaniec, H.: Primes of the type $\phi(x, y) + A$ where ϕ is a quadratic form. Acta Arith. **21**, 203–234 (1972)
3053. Iwaniec, H.: On zeros of Dirichlet's L series. Invent. Math. **23**, 435–459 (1974)

3054. Iwaniec, H.: Primes represented by quadratic polynomials in two variables. Acta Arith. **24**, 435–459 (1974)
3055. Iwaniec, H.: Almost-primes represented by quadratic polynomials. Invent. Math. **47**, 171–188 (1978)
3056. Iwaniec, H.: On the problem of Jacobsthal. Demonstr. Math. **11**, 225–231 (1978)
3057. Iwaniec, H.: Rosser's sieve. Acta Arith. **36**, 171–202 (1980)
3058. Iwaniec, H.: A new form of the error term in the linear sieve. Acta Arith. **37**, 307–320 (1980)
3059. Iwaniec, H.: On the Brun-Titchmarsh theorem. J. Math. Soc. Jpn. **34**, 95–123 (1982)
3060. Iwaniec, H.: Promenade along modular forms and analytic number theory. In: Topics in Analytic Number Theory, Austin, TX, 1982, pp. 221–303 (1985)
3061. Iwaniec, H.: Topics in Classical Automorphic Forms. Am. Math. Soc., Providence (1997)
3062. Iwaniec, H., Jiménez Urroz, J.: Orders of CM elliptic curves modulo p with at most two primes. Preprint (2006)
3063. Iwaniec, H., Jutila, M.: Primes in short intervals. Ark. Mat. **17**, 167–176 (1979)
3064. Iwaniec, H., Kowalski, E.: Analytic Number Theory. Am. Math. Soc., Providence (2004)
3065. Iwaniec, H., Laborde, M.: P_2 in short intervals. Ann. Inst. Fourier **31**(4), 37–56 (1981)
3066. Iwaniec, H., Luo, W., Sarnak, P.: Low lying zeros of families of L-functions. Publ. Math. Inst. Hautes Études Sci. **91**, 55–131 (2000)
3067. Iwaniec, H., Mozzochi, C.J.: On the divisor and circle problems. J. Number Theory **29**, 60–93 (1988)
3068. Iwaniec, H., Pintz, J.: Primes in short intervals. Monatshefte Math. **98**, 115–143 (1984)
3069. Iwaniec, H., Sarnak, P.: L-functions at the central point. In: Number Theory in Progress, vol. 2, pp. 941–952. de Gruyter, Berlin (1999)
3070. Iwasawa, K.: On p-adic L-functions. Ann. Math. **89**, 198–205 (1969)
3071. Iwasawa, K.: Lectures on p-adic L-functions. Ann. of Math. Stud. **74**, 1972
3072. Iyanaga, S.: Zum Beweis des Hauptidealsatzes. Abh. Math. Semin. Univ. Hamb. **10**, 349–357 (1934)
3073. Iyanaga, S.: Travaux de Claude Chevalley sur la théorie du corps de classes: introduction. Jpn. J. Math. **1**, 25–85 (2006)
3074. Jacobi, C.G.J.: Ueber den Ausdruck der verschiedenen Wurzeln einer Gleichung durch bestimmte Integrale. J. Reine Angew. Math. **2**, 1–8 (1827) [[3082], vol. 6, pp. 12–20]
3075. Jacobi, C.G.J.: Note sur la décomposition d'un nombre donné en quatre carrés. J. Reine Angew. Math. **3**, 191 (1828) [[3082], vol. 1, pp. 245–247]
3076. Jacobi, C.G.J.: Beantwortung der Aufgabe Seite 212 des 3ten Bandes des Crelleschen Journals: "Kann $\alpha^{\mu-1}$, wenn μ eine Primzahl und α eine ganze Zahl und kleiner als μ und grösser als 1 ist, durch $\mu\mu$ teilbar sein?" J. Reine Angew. Math. **3**, 238–239 (1828) [[3082], vol. 6, pp. 238–239]
3077. Jacobi, C.G.J.: Fundamenta nova theoriae functionum ellipticarum. Bornträger, Regiomonti (1829) [[3082], vol. 1, pp. 49–239]
3078. Jacobi, C.G.J.: De usu legitimo formulae summatoriae Maclaurinianae. J. Reine Angew. Math. **12**, 263–272 (1834) [[3082], vol. 6, pp. 64–75]
3079. Jacobi, C.G.J.: De usu theoriae integralium ellipticorum et integralium abelianorum in analysi diophantea. J. Reine Angew. Math. **13**, 353–355 (1835) [[3082], vol. 2, pp. 51–55]
3080. Jacobi, C.G.J.: Über die Zusammensetzung der Zahlen aus ganzen positiven Cuben; nebst einer Tabelle für die kleinste Cubenanzahl aus welcher jede Zahl bis 12000 zusammengesetzt werden kann. J. Reine Angew. Math. **42**, 41–69 (1851) [[3082], vol. 6, pp. 322–354]
3081. Jacobi, C.G.J.: Allgemeine Theorie de kettenbruchähnlichen Algorithmen, in welchen jede Zahl aus drei vorhergehenden gebildet wird. J. Reine Angew. Math. **69**, 29–64 (1868) [[3082], vol. 6, pp. 385–426]
3082. Jacobi, C.G.J.: Gesammelte Werke, vols. 1–6. Reiner (1881–1891)
3083. Jacobsthal, E.: Anwendungen einer Formel aus der Theorie der quadratischen Reste. Dissertation, Univ. Berlin (1906)
3084. Jacobsthal, E.: Über die Darstellung der Primzahlen der Form $4n + 1$ als Summe zweier Quadrate. J. Reine Angew. Math. **132**, 238–245 (1907)

3085. Jacobsthal, E.: Über Sequenzen ganzer Zahlen, von denen keine zu n teilerfremd ist, I. Norske Vid. Selsk. Forh., Trondheim **33**, 117–124 (1961)
3086. Jacobsthal, E.: Über Sequenzen ganzer Zahlen, von denen keine zu n teilerfremd ist, II. Norske Vid. Selsk. Forh., Trondheim **33**, 125–131 (1961)
3087. Jacobsthal, E.: Über Sequenzen ganzer Zahlen, von denen keine zu n teilerfremd ist, III. Norske Vid. Selsk. Forh., Trondheim **33**, 132–139 (1961)
3088. Jacobsthal, E.: Über Sequenzen ganzer Zahlen, von denen keine zu n teilerfremd ist, IV. Norske Vid. Selsk. Forh., Trondheim **34**, 1–7 (1961)
3089. Jacobsthal, E.: Über Sequenzen ganzer Zahlen, von denen keine zu n teilerfremd ist, V. Norske Vid. Selsk. Forh., Trondheim **34**, 110–115 (1961)
3090. Jacquet, H., Langlands, R.: Automorphic Forms on $GL(2)$. Lecture Notes in Math., vol. 113. Springer, Berlin (1970)
3091. Jacquet-Chiffelle, D.-O.: Énumeration complète des classes de formes parfaites en dimension 7. Ann. Inst. Fourier **43**, 21–55 (1993)
3092. Jadrijević, B.: A system of Pellian equations and related two-parametric family of quartic Thue equations. Rocky Mt. J. Math. **35**, 547–571 (2005)
3093. Jadrijević, B.: On two-parametric family of quartic Thue equations. J. Théor. Nr. Bordx. **17**, 161–167 (2005)
3094. Jager, H., Lekkerkerker, C.G.: In memoriam Prof. Dr. J. Popken. Nieuw Arch. Wiskd. **19**, 1–9 (1971)
3095. Jager, H., Liardet, P.: Distributions arithmétiques des dénominateurs de convergents de fractions continues. Indag. Math. **50**, 181–197 (1988)
3096. Jagy, W.C.: Five regular or nearly-regular ternary quadratic forms. Acta Arith. **77**, 361–367 (1996)
3097. Jagy, W.C., Kaplansky, I., Schiemann, A.: There are 913 regular ternary forms. Mathematika **44**, 332–341 (1997)
3098. James, R.D.: The representation of integers as sums of pyramidal numbers. Math. Ann. **109**, 196–199 (1933)
3099. James, R.D.: On Waring's problem for odd powers. Proc. Lond. Math. Soc. **37**, 257–291 (1934)
3100. James, R.D.: The value of the number $g(k)$ in Waring's problem. Trans. Am. Math. Soc. **35**, 395–444 (1934)
3101. James, R.D.: The representation of integers as sums of values of cubic polynomials. Am. J. Math. **56**, 303–315 (1934)
3102. James, R.D.: The representation of integers as sums of values of cubic polynomials, II. Am. J. Math. **59**, 393–398 (1937)
3103. Jänichen, W.: Über einen zahlentheoretischen Satz von Hurwitz. Math. Z. **17**, 277–292 (1923)
3104. Janusz, G.J.: Irving Reiner 1924–1986. Ill. J. Math. **32**, 315–328 (1988)
3105. Jarden, M., Videla, C.R.: Undecidability of families of rings of totally real integers. Int. J. Number Theory **4**, 835–850 (2008)
3106. Jarník, V.: O mřížových bodech v rovině. Rozpravy České Akad. Věd a Umění, **33**(36), 1–23 (1924)
3107. Jarník, V.: Über die Gitterpunkte auf konvexen Kurven. Math. Z. **24**, 500–518 (1925)
3108. Jarník, V.: Über Gitterpunkte in mehrdimensionalen Ellipsoiden. Math. Ann. **100**, 699–721 (1928)
3109. Jarník, V.: Über Gitterpunkte in mehrdimensionalen Ellipsoiden, II. Math. Ann. **101**, 136–146 (1929)
3110. Jarník, V.: Diophantische Approximationen und Hausdorffsches Mass. Mat. Sb. **36**, 371–382 (1929)
3111. Jarník, V.: Przyczynek do metrycznej teorji przybliżeń diofantowych. Pr. Mat.-Fiz. **36**, 91–106 (1929)
3112. Jarník, V.: Über die Mittelwertsätze der Gitterpunktlehre, I. Math. Z. **33**, 62–84 (1931)
3113. Jarník, V.: Über die Mittelwertsätze der Gitterpunktlehre, II. Math. Z. **33**, 85–97 (1931)

3114. Jarník, V.: Über die Mittelwertsätze der Gitterpunktlehre, III. Math. Z. **36**, 518–617 (1933)
3115. Jarník, V.: Über Gitterpunkte in mehrdimensionalen Ellipsoden: eine Anwendung des Hausdorffschen Maßbegriffes. Math. Z. **38**, 217–256 (1934)
3116. Jarník, V., Landau, E.: Untersuchungen über einen van der Corputschen Satz. Math. Z. **39**, 745–767 (1935) [[3680], vol. 9, pp. 327–349]
3117. Jarník, V., Walfisz, A.: Über Gitterpunkte in mehrdimensionalen Ellipsoiden. Math. Z. **32**, 152–160 (1930)
3118. Jehne, W.: On knots in algebraic number theory. J. Reine Angew. Math. **311/312**, 215–254 (1979)
3119. Jeltsch-Fricker, R.: In memoriam: Alexander M. Ostrowski (1893 bis 1986). Elem. Math. **43**, 33–38 (1988)
3120. Jenkins, P.M.: Odd perfect numbers have a prime factor exceeding 10^7. Math. Comput. **72**, 1549–1554 (2003)
3121. Jenkinson, O.: On the density of Hausdorff dimensions of bounded type continued fraction sets: the Texan conjecture. Stoch. Dyn. **4**, 63–76 (2004)
3122. Jensen, C.U.: On the solvability of a certain class of non-Pellian equations. Math. Scand. **10**, 71–84 (1962)
3123. Jensen, K.L.: Om talteoretiske Egenskaber ved de Bernoulliske tal. Nyt Tidsskr. Mat. B **26**, 73–83 (1915)
3124. Jeon, D., Kim, C.H., Park, E.: On the torsion of elliptic curves over quartic number fields. J. Lond. Math. Soc. **74**, 1–12 (2006)
3125. Jeon, D., Kim, C.H., Schweizer, A.: On the torsion of elliptic curves over cubic number fields. Acta Arith. **113**, 291–301 (2004)
3126. Jeśmanowicz, L.: Kilka uwag o liczbach pitagorejskich. Wiadom. Mat. **1**, 196–202 (1955/1956)
3127. Jessen, B.: Harald Bohr, 22 April 1887–22 January 1951. Acta Math. **86**, I–XXIII (1951)
3128. Jha, V.: On Krasnér's theorem for the first case of Fermat's last theorem. Colloq. Math. **67**, 25–31 (1994)
3129. Ji, C.G., Lu, H.W.: Lower bound of real primitive L-function at $s = 1$. Acta Arith. **111**, 405–409 (2004)
3130. Jia, C.H.: The distribution of square-free numbers. Beijing Daxue Xuebao, 3, 21–27 (1987) (in Chinese)
3131. Jia, C.H.: The square-full integers in short intervals. Acta Math. Sin. **30**, 614–621 (1987) (in Chinese)
3132. Jia, C.H.: The greatest prime factor of integers in short intervals, II. Acta Math. Sin. **32**, 188–199 (1989) (in Chinese)
3133. Jia, C.H.: Three primes theorem in a short interval, V. Acta Math. Sin. New Ser. **7**, 135–170 (1991)
3134. Jia, C.H.: The distribution of square-free numbers. Sci. China Ser. A **36**, 154–169 (1993)
3135. Jia, C.H.: On Pjateckiĭ-Šapiro prime number theorem, II. Sci. China Ser. A **36**, 913–926 (1993)
3136. Jia, C.H.: On the distribution of αp modulo one. J. Number Theory **45**, 241–253 (1993)
3137. Jia, C.H.: On the distribution of αp modulo one, II. Sci. China Ser. A **43**, 703–721 (2000)
3138. Jia, C.H.: The greatest prime factor of integers in a short interval, III. Acta Math. Sin. New Ser. **9**, 321–336 (1993)
3139. Jia, C.H.: The greatest prime factor of integers in a short interval, IV. Acta Math. Sin. New Ser. **12**, 433–445 (1996)
3140. Jia, C.H.: On the Piatetski-Shapiro-Vinogradov theorem. Acta Arith. **73**, 1–28 (1995)
3141. Jia, C.H.: Difference between consecutive primes. Sci. China Ser. A **38**, 1163–1186 (1995)
3142. Jia, C.H.: Goldbach numbers in short interval, I. Sci. China Ser. A **38**, 385–406 (1995)
3143. Jia, C.H.: Goldbach numbers in short interval, II. Sci. China Ser. A **38**, 513–523 (1995)
3144. Jia, C.H.: Almost all short intervals containing prime numbers. Acta Arith. **76**, 21–84 (1996)
3145. Jia, C.H., Liu, M.-C.: On the largest prime factor of integers. Acta Arith. **95**, 17–48 (2000)

3146. Johnsen, J.: On the large sieve method in $GF(q,x)$. Mathematika **18**, 172–184 (1971)
3147. Johnson, S.M.: A linear diophantine problem. Can. J. Math. **12**, 390–398 (1960)
3148. Johnson, W.: Irregular primes and cyclotomic invariants. Math. Comput. **29**, 113–120 (1975)
3149. Joly, J.-R.: Sommes de puissances d-iémes dans un anneau commutatif. Acta Arith. **17**, 37–115 (1970)
3150. Joly, J.-R.: Équations et variétés algébriques sur un corps fini. Enseign. Math. **19**, 1–117 (1973)
3151. Jones, B.W.: The regularity of a genus of positive ternary quadratic forms. Trans. Am. Math. Soc. **33**, 111–124 (1931)
3152. Jones, B.W., Pall, G.: Regular and semi-regular positive ternary quadratic forms. Acta Math. **70**, 165–191 (1939)
3153. Jones, H.: Khinchin's theorem in k dimensions with prime numerator and denominator. Acta Arith. **99**, 205–225 (2001)
3154. Jones, J.P.: Diophantine representation of Mersenne and Fermat primes. Acta Arith. **35**, 209–221 (1979)
3155. Jones, J.P.: Universal Diophantine equation. J. Symb. Comput. **47**, 549–571 (1982)
3156. Jones, J.P., Matijasevič, Y.V.: Proof of recursive unsolvability of Hilbert's tenth problem. Am. Math. Mon. **98**, 689–709 (1991)
3157. Jordan, J.H.: A covering class of residues with odd moduli. Acta Arith. **13**, 335–338 (1967/1968)
3158. Jorgenson, J., Krantz, S.G.: Serge Lang, 1927–2005. Not. Am. Math. Soc. **53**, 536–553 (2006)
3159. Jorgenson, J., Krantz, S.G.: The mathematical contributions of Serge Lang. Not. Am. Math. Soc. **54**, 476–497 (2007)
3160. Joris, H.: Ω-Sätze für zwei arithmetische Funktionen. Comment. Math. Helv. **47**, 220–248 (1972)
3161. Joris, H.: An Ω result for coefficients of cusp forms. Mathematika **22**, 12–19 (1975)
3162. Joseph, A., Melnkov, A., Rentschler, R. (eds.): Studies in Memory of Issai Schur. Progr. Math., vol. 210. Birkhäuser Boston, Boston (2003)
3163. Joubert, P.: Sur la théorie des fonctions elliptiques et son application à la théorie des nombres. C. R. Acad. Sci. Paris **50**, 832–837 (1860)
3164. Juel, C.S.: Ueber die Parameterbestimmung von Punkten auf Curven zweiter und dritter Ordnung. Eine geometrische Einleitung in die Theorie der logarithmischen und elliptischen Funktionen. Math. Ann. **47**, 72–104 (1896)
3165. Jurkat, W.B.: On the Martens[11] conjecture and related general Ω-theorem. In: Proc. Symposia Pure Math., vol. 24, pp. 147–158. Am. Math. Soc., Providence (1972)
3166. Jurkat, W.B., Richert, H.-E.: An improvement of Selberg's sieve method, I. Acta Arith. **11**, 217–240 (1965)
3167. Jutila, M.: A new estimate for Linnik's constant. Ann. Acad. Sci. Fenn. Ser. A1 **471**, 1–8 (1970)
3168. Jutila, M.: On a density theorem of H.L. Montgomery for L-functions. Ann. Acad. Sci. Fenn. Ser. A1 **520**, 1–13 (1972)
3169. Jutila, M.: On numbers with a large prime factor. J. Indian Math. Soc. **37**, 43–53 (1973)
3170. Jutila, M.: On large values of Dirichlet polynomials. In: Topics in Number Theory, Proc. Colloq. Debrecen, 1974, pp. 129–140. North-Holland, Amsterdam (1976)
3171. Jutila, M.: Zero-density estimates for L-functions. Acta Arith. **32**, 55–62 (1977)
3172. Jutila, M.: On Linnik's constant. Math. Scand. **41**, 45–62 (1977)
3173. Jutila, M.: On the mean value of $L(1/2,\chi)$ for real characters. Analysis **1**, 149–161 (1981)
3174. Jutila, M.: Riemann's zeta function and the divisor problem. Ark. Mat. **21**, 75–96 (1983)
3175. Jutila, M.: Riemann's zeta function and the divisor problem, II. Ark. Mat. **31**, 61–70 (1993)

[11] It should be 'Mertens'.

3176. Kabatiansky, G.A., Levenštein, V.I.: Bounds on packing on a sphere and in space. Probl. Inf. Transm. **14**, 1–17 (1978)
3177. Kac, M.: Hugo Steinhaus—a reminiscence and a tribute. Am. Math. Mon. **81**, 572–581 (1974)
3178. Kac, M., van Kampen, E.R., Wintner, A.: Ramanujan sums and almost periodic functions. Am. J. Math. **62**, 107–114 (1940)
3179. Kaczorowski, J.: On sign-changes in the remainder-term of the prime-number formula. Acta Arith. **44**, 365–377 (1984)
3180. Kaczorowski, J.: On sign-changes in the remainder-term of the prime-number formula, II. Acta Arith. **45**, 65–74 (1985)
3181. Kaczorowski, J.: On sign-changes in the remainder-term of the prime-number formula, III. Acta Arith. **48**, 347–371 (1987)
3182. Kaczorowski, J.: On sign-changes in the remainder-term of the prime-number formula, IV. Acta Arith. **50**, 15–21 (1988)
3183. Kaczorowski, J.: The k-functions in multiplicative number theory, I. On complex explicit formulae. Acta Arith. **56**, 195–211 (1990)
3184. Kaczorowski, J.: The k-functions in multiplicative number theory, II. Uniform distribution of zeta zeros. Acta Arith. **56**, 213–224 (1990)
3185. Kaczorowski, J.: The k-functions in multiplicative number theory, III. Uniform distribution of zeta zeros; discrepancy. Acta Arith. **57**, 199–210 (1991)
3186. Kaczorowski, J.: The k-functions in multiplicative number theory, IV. On a method of A.E. Ingham. Acta Arith. **57**, 231–244 (1991)
3187. Kaczorowski, J.: The k-functions in multiplicative number theory, V. Changes of sign of some arithmetical error terms. Acta Arith. **59**, 37–58 (1991)
3188. Kaczorowski, J.: The boundary values of generalized Dirichlet series and a problem of Chebyshev. Astérisque **209**, 227–235 (1992)
3189. Kaczorowski, J.: A contribution to the Shanks-Rényi race problem. Q. J. Math. **44**, 451–458 (1993)
3190. Kaczorowski, J.: Results on the distributions of primes. J. Reine Angew. Math. **446**, 89–113 (1994)
3191. Kaczorowski, J.: On the Shanks-Rényi race problem mod 5. J. Number Theory **50**, 106–118 (1995)
3192. Kaczorowski, J.: On the distribution of primes (mod 4). Analysis **15**, 159–171 (1995)
3193. Kaczorowski, J.: On the Shanks-Rényi race problem. Acta Arith. **74**, 31–46 (1996)
3194. Kaczorowski, J.: Axiomatic theory of L-functions: the Selberg class. In: Lecture Notes in Math., vol. 1891, pp. 133–209. Springer, Berlin (2006)
3195. Kaczorowski, J., Kulas, M.: On the non-trivial zeros off the critical line for L-functions from the extended Selberg class. Monatshefte Math. **150**, 217–232 (2007)
3196. Kaczorowski, J., Molteni, G., Perelli, A.: Linear independence in the Selberg class. C.R. Acad. Sci. Can. **21**, 28–32 (1999)
3197. Kaczorowski, J., Perelli, A.: On the structure of the Selberg class. I. $0 \leq d \leq 1$. Acta Math. **182**, 207–241 (1999)
3198. Kaczorowski, J., Perelli, A.: On the structure of the Selberg class. II. Invariants and conjectures. J. Reine Angew. Math. **524**, 73–96 (2000)
3199. Kaczorowski, J., Perelli, A.: On the structure of the Selberg class. III. Sarnak's rigidity conjecture. Duke Math. J. **101**, 529–554 (2000)
3200. Kaczorowski, J., Perelli, A.: On the structure of the Selberg class. IV. Basic invariants. Acta Arith. **104**, 97–116 (2002)
3201. Kaczorowski, J., Perelli, A.: On the structure of the Selberg class. V. $1 < d < 5/3$. Invent. Math. **150**, 485–516 (2002)
3202. Kaczorowski, J., Perelli, A.: On the structure of the Selberg class. VI. Non-linear twists. Acta Arith. **116**, 315–341 (2005)
3203. Kaczorowski, J., Perelli, A.: On the structure of the Selberg class. VII. $1 < d < 2$. Ann. Math. **173**, 1397–1441 (2011)

3204. Kaczorowski, J., Perelli, A.: The Selberg class: a survey. In: Number Theory in Progress, vol. 2, pp. 953–992. de Gruyter, Berlin (1999)
3205. Kaczorowski, J., Perelli, A., Pintz, J.: A note on the exceptional set for Goldbach's problem in short intervals. Monatshefte Math. **116**, 275–282 (1993)
3206. Kaczorowski, J., Pintz, J.: Oscillatory properties of arithmetical functions, I. Acta Math. Acad. Sci. Hung. **48**, 173–185 (1986)
3207. Kaczorowski, J., Pintz, J.: Oscillatory properties of arithmetical functions, II. Acta Math. Acad. Sci. Hung. **49**, 441–453 (1987)
3208. Kaczorowski, J., Staś, W.: On the number of sign-changes in the remainder-term of the prime-ideal theorem. Discuss. Math. **9**, 83–102 (1988)
3209. Kaczorowski, J., Staś, W.: On the number of sign changes in the remainder-term of the prime-ideal theorem. Colloq. Math. **56**, 185–197 (1988)
3210. Kadiri, H.: Short effective intervals containing primes in arithmetic progressions and the seven cubes problem. Math. Comput. **77**, 1733–1748 (2008)
3211. Kagawa, T.: Determination of elliptic curves with everywhere good reduction over real quadratic fields. Arch. Math. **73**, 25–32 (1999)
3212. Kagawa, T.: Nonexistence of elliptic curves having everywhere good reduction and cubic discriminant. Proc. Jpn. Acad. Sci. **76**, 141–142 (2000)
3213. Kagawa, T.: Determination of elliptic curves with everywhere good reduction over real quadratic fields $Q(\sqrt{3p})$. Acta Arith. **96**, 231–245 (2001)
3214. Kagawa, T., Terai, N.: Squares in Lucas sequences and some Diophantine equations. Manuscr. Math. **96**, 195–202 (1998)
3215. Kahane, J.-P.: Quelques aspects de la vie et de l'oeuvre de Georges Poitou. Math. Gaz. **44**, 3–8 (1990)
3216. Kahane, J.-P.: Sur les nombres premiers généralisés de Beurling. Preuve d'une conjecture de Bateman et Diamond. J. Théor. Nr. Bordx. **9**, 251–266 (1997)
3217. Kahane, J.-P.: Un théorème de Littlewood pour les nombres premiers de Beurling. Bull. Lond. Math. Soc. **31**, 424–430 (1999)
3218. Kallies, J.: Verallgemeinerte Dedekindsche Summen und ein Gitterpunktproblem im n-dimensionalen Raum. J. Reine Angew. Math. **344**, 22–37 (1983)
3219. Kalmár, L.: Über die mittlere Anzahl der Produktdarstellung der Zahlen. Acta Sci. Math. **5**, 95–107 (1930/1932)
3220. Kalužnin, L.A., et al.: Vladimir Petrovič Velmin. Usp. Mat. Nauk **31**(1), 229–235 (1976) (in Russian)
3221. Kamienny, S.: Torsion points on elliptic curves over all quadratic fields. Duke Math. J. **53**, 157–162 (1986)
3222. Kamienny, S.: Torsion points on elliptic curves over all quadratic fields, II. Bull. Soc. Math. Fr. **114**, 119–122 (1986)
3223. Kamienny, S.: On the torsion subgroups of elliptic curves over totally real fields. Invent. Math. **83**, 545–551 (1986)
3224. Kamienny, S.: Torsion points on elliptic curves and q-coefficients of modular forms. Invent. Math. **109**, 221–229 (1992)
3225. Kamienny, S., Mazur, B.: Rational torsion of prime order in elliptic curves over number fields. Astérisque **228**, 81–100 (1995)
3226. Kamke, E.: Verallgemeinerung des Waring-Hilbertschen Satzes. Math. Ann. **83**, 85–112 (1921)
3227. Kamke, E.: Zum Waringschen Problem für rationale Zahlen und Polynome. Math. Ann. **87**, 238–245 (1922)
3228. Kamke, E.: Über die Zerfällung rationaler Zahlen in rationale Polynomwerte. Math. Z. **12**, 323–328 (1922)
3229. Kamke, E.: Bemerkung zum allgemeinen Waringschen Problem. Math. Z. **15**, 188–194 (1922)
3230. Kamke, E.: Zur Arithmetik der Polynome. Math. Z. **19**, 247–264 (1924)
3231. Kaneko, M.: Supersingular j-invariants as singular moduli mod p. Osaka Math. J. **26**, 849–855 (1989)

3232. Kanemitsu, S.: Some asymptotic formulas of Ramanujan, II. Rep. Fac. Sci. Engrg. Saga Univ. **19**(1), 1–16 (1991)
3233. Kanemitsu, S., Yoshimoto, M.: Farey series and the Riemann Hypothesis. Acta Arith. **75**, 351–374 (1996)
3234. Kanemitsu, S., Yoshimoto, M.: Farey series and the Riemann Hypothesis, II. Acta Math. Acad. Sci. Hung. **78**, 287–304 (1998)
3235. Kanemitsu, S., Yoshimoto, M.: Farey series and the Riemann Hypothesis, III. Ramanujan J. **1**, 363–378 (1997)
3236. Kaniecki, L.: On Šnirelman's constant under the Riemann hypothesis. Acta Arith. **72**, 361–374 (1995)
3237. Kanigel, R.: The Man Who Knew Infinity. Charles Scribner's Sons, New York (1991)
3238. Kannan, R.: Lattice translates of a polytope and the Frobenius problem. Combinatorica **12**, 161–177 (1992)
3239. Kanold, H.-J.: Folgerungen aus dem Vorkommen einer Gauss'schen Primzahl in der Primfaktorenzerlegung einer ungeraden vollkommenen Zahl. J. Reine Angew. Math. **186**, 25–29 (1944)
3240. Kanold, H.-J.: Eine Bemerkung über die Menge der vollkommenen Zahlen. Math. Ann. **131**, 390–392 (1956)
3241. Kanold, H.-J.: Über einen Satz von L.E. Dickson, II. Math. Ann. **132**, 246–255 (1956)
3242. Kanold, H.-J.: Über die Verteilung der vollkommenen Zahlen und allgemeineren Zahlenmengen. Math. Ann. **132**, 442–450 (1957)
3243. Kanold, H.-J.: Das wissenschaftliche Werk von Hans-Heinrich Ostmann. SBer. Berliner Math. Ges., 1961–1964, 51–59
3244. Kanold, H.-J.: Über eine zahlentheoretische Funktion von Jacobsthal. Math. Ann. **170**, 314–326 (1967)
3245. Kantor, J.-M.: Sur le polynôme associé à un polytope à sommets entiers dans R^n. C. R. Acad. Sci. Paris **314**, 669–672 (1992)
3246. Kantor, J.-M., Khovanskii, A.: Une application du théorème de Riemann-Roch combinatoire au polynôme d'Ehrhart des polytopes entiers de R^d. C. R. Acad. Sci. Paris **317**, 501–507 (1993)
3247. Kapferer, H.: Verifizierung des symmetrischen Teils der Fermatschen Vermutung für unendlich viele paarweise teilerfremde Exponenten. J. Reine Angew. Math. **214/215**, 360–372 (1964)
3248. Kaplansky, I.: Ternary positive quadratic forms that represent all odd positive integers. Acta Arith. **70**, 209–214 (1995)
3249. Kaplansky, I.: Integers uniquely represented by certain ternary forms. In: The Mathematics of Paul Erdős, I, pp. 86–94. Springer, Berlin (1997)
3250. Kaplansky, I.: Hilbert's Problems. University of Chicago Press, Chicago (1977)
3251. Karatsuba, A.A.: A uniform valuation of the error term in Dirichlet's divisor problem. Izv. Akad. Nauk SSSR, Ser. Mat. **36**, 475–483 (1972) (in Russian)
3252. Karatsuba, A.A.: Mean value of the modulus of a trigonometric sum. Izv. Akad. Nauk SSSR, Ser. Mat. **37**, 1203–1227 (1973) (in Russian)
3253. Karatsuba, A.A.: The function $G(n)$ in Waring's problem. Izv. Akad. Nauk SSSR, Ser. Mat. **49**, 935–947 (1985) (in Russian)
3254. Karatsuba, A.A., Korolev, M.A.: The argument of the Riemann zeta function. Usp. Mat. Nauk **60**(3), 41–96 (2005) (in Russian)
3255. Karatsuba, A.A., Kovalevskaya, E.I., Kubilyus, I.P. [Kubilius, J.]: Biography and scientific work of V.G. Sprindzhuk. Acta Arith. **53**, 5–16 (1989) (in Russian)
3256. Karatsuba, A.A., Voronin, S.M.: The Riemann Zeta-Function. Phys.-Mat. Lit., Moscow (1994) (in Russian) [English translation: de Gruyter, 1992]
3257. Károlyi, G.: The Erdős-Heilbronn problem in abelian groups. Isr. J. Math. **139**, 349–359 (2004)
3258. Károlyi, G.: Cauchy-Davenport theorem in group extensions. Enseign. Math. **51**, 239–254 (2005)

3259. Károlyi, G.: A compactness argument in the additive theory and the polynomial method. Discrete Math. **302**, 124–144 (2005)
3260. Károlyi, G.: An inverse theorem for the restricted set addition in abelian groups. J. Algebra **290**, 557–593 (2005)
3261. Károlyi, G.: Restricted set addition: the exceptional case of the Erdős-Heilbronn conjecture. J. Comb. Theory, Ser. A **116**, 741–746 (2009)
3262. Kasch, F.: Abschätzung der Dichte von Summenmengen, I. Math. Z. **62**, 368–387 (1955)
3263. Kasch, F.: Abschätzung der Dichte von Summenmengen, II. Math. Z. **64**, 243–257 (1956)
3264. Kasch, F.: Abschätzung der Dichte von Summenmengen, III. Math. Z. **66**, 164–172 (1956)
3265. Kasch, F.: Ein metrischer Beitrag über Mahlersche Vermutung über S-Zahlen, II. J. Reine Angew. Math. **203**, 157–159 (1960)
3266. Kasch, F., Volkmann, B.: Zur Mahlerschen Vermutung über S-Zahlen. Math. Ann. **136**, 442–453
3267. Kasimov, A.M.: On the I.M. Vinogradov constant in the Goldbach ternary problem. Uzbek Mat. Zh., 1992, nr. 3–4, 55–64 (in Russian)
3268. Kátai, I.: The number of lattice points in a circle. Ann. Univ. Sci. Bp. **8**, 39–60 (1965) (in Russian)
3269. Kátai, I.: On distribution of arithmetical functions on the set prime plus one. Compos. Math. **19**, 278–289 (1968)
3270. Kátai, I.: On a problem of P. Erdős. J. Number Theory **2**, 1–6 (1970)
3271. Kátai, I.: On additive functions. Publ. Math. (Debr.) **25**, 251–257 (1978)
3272. Kátai, I.: A remark on a theorem of H. Daboussi. Acta Math. Acad. Sci. Hung. **47**, 223–225 (1986)
3273. Kátai, I.: M.V. Subbarao in memoriam, 1921–2006. Ann. Univ. Sci. Bp. Rolando Eötvös Nomin., Sect. Comput. **26**, 3–4 (2006)
3274. Kátai, I.: Some remarks on a theorem of H. Daboussi. Math. Pannon. **19**, 71–80 (2008)
3275. Kato, K.: p-adic Hodge theory and values of zeta functions of modular forms. Astérisque **295**, 117–290 (2004)
3276. Kato, K., Trihan, F.: On the conjectures of Birch and Swinnerton-Dyer in characteristic $p > 0$. Invent. Math. **153**, 537–592 (2003)
3277. Katz, N.M.: On a theorem of Ax. Am. J. Math. **93**, 485–499 (1971)
3278. Katz, N.M.: Higher congruences between modular forms. Ann. Math. **101**, 332–367 (1975)
3279. Katz, N.M.: An overview of Deligne's proof of the Riemann hypothesis for varieties over finite fields. In: Proc. Symposia Pure Math., vol. 28, pp. 275–305. Am. Math. Soc., Providence (1976)
3280. Katz, N.M.: p-adic L-functions for CM fields. Invent. Math. **49**, 199–297 (1978)
3281. Katz, N.M., Sarnak, P.: Zeroes of zeta functions and symmetry. Bull. Am. Math. Soc. **36**, 1–26 (1999)
3282. Katz, N.M., Sarnak, P.: Random Matrices, Frobenius Eigenvalues, and Monodromy. Am. Math. Soc., Providence (1999)
3283. Katz, N.M., Tate, J.: Bernard Dwork (1923–1998). Not. Am. Math. Soc. **46**, 338–343 (1999)
3284. Kaufman, R.M.: The distribution of $\{\sqrt{p}\}$. Mat. Zametki **26**, 497–504 (1979) (in Russian)
3285. Kaufmann-Bühler, W.: Gauss. A Biographical Study. Springer, Berlin (1981)
3286. Kawada, K.: On sums of seven cubes of almost primes. Acta Arith. **117**, 213–245 (2005)
3287. Kawada, K., Wooley, T.D.: Sums of fifth powers and related topics. Acta Arith. **87**, 27–65 (1998)
3288. Kawada, K., Wooley, T.D.: Sums of fourth powers and related topics. J. Reine Angew. Math. **512**, 173–223 (1999)
3289. Kawada, K., Wooley, T.D.: On the Waring-Goldbach problem for fourth and fifth powers. Proc. Lond. Math. Soc. **83**, 1–50 (2001)
3290. Kawada, K., Wooley, T.D.: Slim exceptional sets for sums of fourth and fifth powers. Acta Arith. **103**, 225–248 (2002)
3291. Keates, M.: On the greatest prime factor of a polynomial. Proc. Edinb. Math. Soc. **16**, 301–303 (1968/1969)

3292. Keating, J.P., Snaith, N.C.: Random matrix theory and $\zeta(\frac{1}{2}+it)$. Commun. Math. Phys. **214**, 57–89 (2000)
3293. Kedlaya, K.S.: Counting points on hyperelliptic curves using Monsky–Washnitzer cohomology. J. Ramanujan Math. Soc. **16**, 323–338 (2001)
3294. Keiper, J.B.: On the zeros of the Ramanujan τ-Dirichlet series in the critical strip. Math. Comput. **65**, 1613–1619 (1996)
3295. Keller, O.-H.: Über die lückenlose Erfüllung des Raumes mit Würfeln. J. Reine Angew. Math. **163**, 231–248 (1930)
3296. Keller, W., Löh, G.: The criteria of Kummer and Mirimanoff extended to include 22 consecutive irregular pairs. Tokyo J. Math. **6**, 397–402 (1983)
3297. Kellner, B.C.: Über irreguläre Paare höherer Ordnung. Diplomarbeit, Univ. Göttingen (2002)
3298. Kempner, A.: Bemerkungen zum Waringschen Problem. Math. Ann. **72**, 387–399 (1912)
3299. Kempner, A.: On transcendental numbers. Trans. Am. Math. Soc. **17**, 476–482 (1916)
3300. Kenku, M.A.: Determination of the even discriminants of complex quadratic fields with class-number 2. Proc. Lond. Math. Soc. **22**, 734–746 (1970)
3301. Kenku, M.A.: Certain torsion points on elliptic curves defined over quadratic fields. J. Lond. Math. Soc. **19**, 233–240 (1979)
3302. Kenku, M.A.: Rational torsion points on elliptic curves defined over quadratic fields. J. Niger. Math. Soc. **2**, 1–16 (1983)
3303. Kenku, M.A., Momose, F.: Torsion points on elliptic curves defined over quadratic fields. Nagoya Math. J. **109**, 125–149 (1988)
3304. Kershner, R.: The number of circles covering a set. Am. J. Math. **61**, 665–671 (1939)
3305. Kersten, I.: Ernst Witt 1911–1991. Jahresber. Dtsch. Math.-Ver. **95**, 166–180 (1993)
3306. Kervaire, M.: Unimodular lattices with a complete root system. Enseign. Math. **40**, 59–104 (1994)
3307. Kesseböhmer, M., Zhu, S.: Dimension sets for infinite IFSs: the Texan conjecture. J. Number Theory **116**, 230–246 (2006)
3308. Kesten, H.: Some probabilistic theorems on Diophantine approximations. Trans. Am. Math. Soc. **103**, 189–217 (1962)
3309. Kesten, H.: The discrepancy of random sequences $\{kx\}$. Acta Arith. **10**, 183–213 (1964)
3310. Kesten, H.: The influence of Mark Kac on probability theory. Ann. Probab. **14**, 1103–1128 (1986)
3311. Kesten, H., Sós, V.T.: On two problems of Erdös, Szüsz and Turán concerning diophantine approximations. Acta Arith. **12**, 183–192 (1966/1967)
3312. Khare, C.: Serre's modularity conjecture: the level one case. Duke Math. J. **134**, 557–589 (2006)
3313. Khare, C.: Serre's modularity conjecture: a survey of the level one case. In: L-functions and Galois Representations, pp. 270–299. Cambridge University Press, Cambridge (2007)
3314. Khare, C., Wintenberger, J.-P.: On Serre's conjecture for 2-dimensional mod p representations of $\mathrm{Gal}(\overline{Q}/Q)$. Ann. Math. **169**, 229–253 (2009)
3315. Khare, C., Wintenberger, J.-P.: Serre's modularity conjecture, I. Invent. Math. **178**, 485–504 (2009)
3316. Khare, C., Wintenberger, J.-P.: Serre's modularity conjecture, II. Invent. Math. **178**, 505–586 (2009)
3317. Khintchine, A.J.: Über dyadische Brüche. Math. Z. **18**, 109–116 (1923)
3318. Khintchine, A.J.: Ein Satz über Kettenbrüche, mit arithmetischen Anwendungen. Math. Z. **18**, 289–306 (1923)
3319. Khintchine, A.J.: Einige Sätze über Kettenbrüche, mit Anwendungen auf die Theorie der Diophantischen Approximationen. Math. Ann. **92**, 115–125 (1924)
3320. Khintchine, A.J.: Zur metrischen Theorie der diophantischen Approximationen. Math. Z. **24**, 706–714 (1926)
3321. Khintchine, A.J.: Zur additiven Zahlentheorie. Mat. Sb. **39**(3), 27–34 (1932)
3322. Khintchine, A.J.: Über ein metrisches Problem der additiven Zahlentheorie. Mat. Sb. **40**, 180–189 (1933)

3323. Khintchine, A.J.: Metrische Kettenbruchprobleme. Compos. Math. **1**, 361–382 (1935)
3324. Khintchine, A.J.: Zur metrischen Kettenbruchtheorie. Compos. Math. **3**, 276–285 (1936)
3325. Khintchine, A.J.: Ein Satz über lineare diophantische Approximationen. Math. Ann. **113**, 398–415 (1937)
3326. Khintchine, A.J.: On a problem of Čebyšev. Izv. Akad. Nauk SSSR, Ser. Mat. **10**, 281–294 (1946) (in Russian)
3327. Kida, M.: Reduction of elliptic curves over certain real quadratic number fields. Math. Comput. **68**, 1679–1685 (1999)
3328. Kida, M.: Nonexistence of elliptic curves having good reduction everywhere over certain quadratic fields. Arch. Math. **76**, 436–440 (2001)
3329. Kida, M.: Good reduction of elliptic curves over imaginary quadratic fields. J. Théor. Nr. Bordx. **13**, 201–209 (2001)
3330. Kida, M., Kagawa, T.: Nonexistence of elliptic curves with good reduction everywhere over real quadratic fields. J. Number Theory **66**, 201–210 (1997)
3331. Kienast, A.: Über die asymptotische Darstellung der summatorischen Funktion von Dirichletreihen mit positiven Koeffizienten. Math. Z. **44**, 115–126 (1939); corr. vol. 45, 1939, pp. 554–558
3332. Kifer, Y., Peres, Y., Weiss, B.: A dimension gap for continued fractions with independent digits. Isr. J. Math. **124**, 61–76 (2001)
3333. Kim, H.H.: Functoriality and number of solutions of congruences. Acta Arith. **128**, 235–243 (2007)
3334. Kim, M.-H.: The canonical decomposition of Siegel modular forms, I. J. Korean Math. Soc. **26**, 57–65 (1989)
3335. Kim, M.-H.: The canonical decomposition of Siegel modular forms, II. J. Korean Math. Soc. **29**, 209–223 (1992)
3336. Kiming, I., Olsson, J.B.: Congruences like Ramanujan's for powers of the partition function. Arch. Math. **59**, 348–360 (1992)
3337. Kiming, I., Wang, X.D.: Examples of 2-dimensional, odd Galois representations of A_5-type over **Q** satisfying the Artin conjecture. In: Lecture Notes in Math., vol. 1585, pp. 109–121. Springer, Berlin (1994)
3338. Kinkelin, H.: Allgemeine Theorie der harmonischen Reihen, mit Anwendungen auf die Zahlentheorie. Programm der Gewerbeschule Basel, 1861/62, 1–32
3339. Kinney, J.R., Pitcher, T.S.: The dimension of some sets defined in terms of f-expansions. Z. Wahrscheinlichkeitstheor. Verw. Geb. **4**, 293–315 (1965/1966)
3340. Kirmse, J.: Zur Darstellung total positiver Zahlen als Summen von vier Quadraten. Math. Z. **21**, 195–202 (1924)
3341. Kishi, T.: On torsion subgroups of elliptic curves with integral j-invariant over imaginary cyclic quartic fields. Tokyo J. Math. **20**, 315–329 (1997)
3342. Kishore, M.: On the number of distinct prime factors of n for which $\varphi(n)|n-1$. Nieuw Arch. Wiskd. **25**, 48–53 (1977)
3343. Kishore, M.: Odd perfect numbers not divisible by 3, II. Math. Comput. **40**, 405–411 (1983)
3344. Kisin, M.: Modularity of 2-adic Barsotti-Tate representations. Invent. Math. **178**, 587–634 (2009)
3345. Kitaoka, Y.: On a space of some theta functions. Nagoya Math. J. **42**, 89–93 (1971)
3346. Kitaoka, Y.: Arithmetic of Quadratic Forms. Cambridge University Press, Cambridge (1993)
3347. Kiuchi, I.: On an exponential sum involving the arithmetic function $\sigma_a(n)$. Math. J. Okayama Univ. **29**, 193–205 (1987)
3348. Klazar, M., Luca, F.: On the maximal order of numbers in the "factorisatio numerorum" problem. J. Number Theory **124**, 470–490 (2007)
3349. Klee, V.L. Jr.: On a conjecture of Carmichael. Bull. Am. Math. Soc. **53**, 1183–1186 (1947)
3350. Klein, F.: Ueber die Transformation der elliptischen Functionen und die Auflösung der Gleichungen fünften Grades. Math. Ann. **14**, 111–172 (1879)
3351. Klein, F.: Zur Theorie der elliptischen Modulfunctionen. Math. Ann. **17**, 62–70 (1880)

3352. Klein, F.: Neue Beiträge zur Riemann'schen Functionentheorie. Math. Ann. **21**, 141–218 (1883)
3353. Klein, F.: Bericht über den Stand der Herausgabe von Gauß' Werken, Neunter Bericht. Nachr. Ges. Wiss. Göttingen, 1911, 26–32
3354. Klein, F.: Vorlesungen über die Entwicklung der Mathematik im 19. Jahrhundert, I. Springer, Berlin (1926)
3355. Klein, F.: Vorlesungen über die Entwicklung der Mathematik im 19. Jahrhundert, II. Springer, Berlin (1927)
3356. Klein, F., Fricke, R.: Vorlesungen über die Theorie der elliptischen Modulfunktionen, I. Teubner, Leipzig (1890)
3357. Klein, F., Fricke, R.: Vorlesungen über die Theorie der elliptischen Modulfunktionen, II. Teubner, Leipzig (1892)
3358. Kleinbock, D.: Extremal subspaces and their submanifolds. Geom. Funct. Anal. **13**, 437–466 (2003)
3359. Kleinbock, D.: An extension of quantitative nondivergence and applications to Diophantine exponents. Trans. Am. Math. Soc. **360**, 6497–6523 (2008)
3360. Kleinbock, D.Y., Margulis, G.A.: Flows on homogeneous spaces and Diophantine approximation on manifolds. Ann. Math. **148**, 339–360 (1998)
3361. Klimov, N.I.: Almost prime numbers. Usp. Mat. Nauk **16**(3), 181–188 (1961) (in Russian)
3362. Klimov, N.I., Pilt'jai, G.Z., Šeptickaĭa, T.A.: An estimate for the absolute constant in the Goldbach–Šnirelman problem. Issled. Teor. Čisel, Sarat. **4**, 35–51 (1972) (in Russian)
3363. Klimov, N.I.: On primes in an arithmetic progression. Issled. Teor. Čisel, Sarat. **5**, 44–54 (1975) (in Russian)
3364. Klingen, H.: Charakterisierung der Siegelschen Modulgruppe durch ein endliches System definierender Relationen. Math. Ann. **144**, 64–82 (1961)
3365. Klingen, H.: Zur Struktur der Siegelschen Modulgruppe. Math. Z. **136**, 169–178 (1974)
3366. Klingen, H.: Introductory Lectures on Siegel Modular Forms. Cambridge University Press, Cambridge (1990)
3367. Kloosterman, H.D.: Over het splitsen van geheele postieve getallen in een som van kwadraten. Dissertation, Univ. of Groningen (1923)
3368. Kloosterman, H.D.: On the representation of numbers in the form $ax^2+by^2+cz^2+dt^2$. Proc. Lond. Math. Soc. **25**, 143–173 (1926)
3369. Kloosterman, H.D.: On the representation of numbers in the form $ax^2+by^2+cz^2+dt^2$. Acta Math. **49**, 407–464 (1926)
3370. Kloosterman, H.D.: Über Gitterpunkte in vierdimensionalen Ellipsoiden. Math. Z. **24**, 514–529 (1926)
3371. Kloosterman, H.D.: Asymptotische Formeln für die Fourierkoeffizienten ganzer Modulformen. Abh. Math. Semin. Univ. Hamb. **5**, 337–352 (1927)
3372. Kloosterman, H.D.: Theorie der Eisensteinschen Reihen von mehreren Veränderlichen. Abh. Math. Semin. Univ. Hamb. **6**, 163–188 (1928)
3373. Kloosterman, H.D.: Thetareihen in total-reellen algebraischen Zahlkörpern. Math. Ann. **103**, 279–299 (1930)
3374. Klotz, W.: Eine obere Schranke für die Reichweite einer Extremalbasis zweiter Ordnung. J. Reine Angew. Math. **238**, 161–168 (1969)
3375. Kløve, T.: Recurrence formulae for the coefficients of modular forms and congruences for the partition function and for the coefficients of $j(\tau)$, $(j(\tau)-1728)^{1/2}$ and $(j(\tau))^{1/3}$. Math. Scand. **23**, 133–159 (1968)
3376. Kløve, T.: Density problems for $p(n)$. J. Lond. Math. Soc. **2**, 504–508 (1970)
3377. Knapowski, S.: On sign-changes in the prime-number formula. J. Lond. Math. Soc. **36**, 451–460 (1961)
3378. Knapowski, S.: On sign-changes of the difference $\pi(x)-\mathrm{li}(x)$. Acta Arith. **7**, 107–119 (1961/1962)
3379. Knapowski, S.: On oscillations of certain means formed from the Möbius series, I. Acta Arith. **8**, 311–320 (1962/1963)

3380. Knapowski, S.: On oscillations of certain means formed from the Möbius series, II. Acta Arith. **10**, 377–386 (1964/1965)
3381. Knapowski, S., Turán, P.: Comparative prime-number theory, I. Acta Math. Acad. Sci. Hung. **13**, 299–314 (1962) [[6227], vol. 2, pp. 1329–1343]
3382. Knapowski, S., Turán, P.: Comparative prime-number theory, II. Acta Math. Acad. Sci. Hung. **13**, 315–342 (1962) [[6227], vol. 2, pp. 1344–1371]
3383. Knapowski, S., Turán, P.: Comparative prime-number theory, III. Acta Math. Acad. Sci. Hung. **13**, 343–364 (1962) [[6227], vol. 2, pp. 1372–1393]
3384. Knapowski, S., Turán, P.: Comparative prime-number theory, IV. Acta Math. Acad. Sci. Hung. **14**, 31–42 (1963) [[6227], vol. 2, pp. 1408–1419]
3385. Knapowski, S., Turán, P.: Comparative prime-number theory, V. Acta Math. Acad. Sci. Hung. **14**, 43–63 (1963) [[6227], vol. 2, pp. 1420–1440]
3386. Knapowski, S., Turán, P.: Comparative prime-number theory, VI. Acta Math. Acad. Sci. Hung. **14**, 65–78 (1963) [[6227], vol. 2, pp. 1441–1454]
3387. Knapowski, S., Turán, P.: Comparative prime-number theory, VII. Acta Math. Acad. Sci. Hung. **14**, 241–250 (1963) [[6227], vol. 2, pp. 1465–1474]
3388. Knapowski, S., Turán, P.: Comparative prime-number theory, VIII. Acta Math. Acad. Sci. Hung. **14**, 251–268 (1963) [[6227], vol. 2, pp. 1475–1492]
3389. Knapowski, S., Turán, P.: Further developments in the comparative prime-number theory, I. Acta Arith. **9**, 23–40 (1964) [[6227], vol. 2, pp. 1534–1551]
3390. Knapowski, S., Turán, P.: Further developments in the comparative prime-number theory, II. Acta Arith. **10**, 293–313 (1964) [[6227], vol. 2, pp. 1572–1592]
3391. Knapowski, S., Turán, P.: Further developments in the comparative prime-number theory, III. Acta Arith. **11**, 115–127 (1965) [[6227], vol. 2, pp. 1619–1631]
3392. Knapowski, S., Turán, P.: Further developments in the comparative prime-number theory, IV. Acta Arith. **11**, 147–161 (1965) [[6227], vol. 2, pp. 1656–1670]
3393. Knapowski, S., Turán, P.: Further developments in the comparative prime-number theory, V. Acta Arith. **11**, 193–202 (1965) [[6227], vol. 2, pp. 1671–1680]
3394. Knapowski, S., Turán, P.: Further developments in the comparative prime-number theory, VI. Acta Arith. **12**, 85–96 (1966) [[6227], vol. 2, pp. 1738–1749]
3395. Knapowski, S., Turán, P.: Further developments in the comparative prime-number theory, VII. Acta Arith. **21**, 193–201 (1972) [[6227], vol. 2, pp. 2230–2238]
3396. Knapowski, S., Turán, P.: On an assertion of Chebyshev. J. Anal. Math. **14**, 267–274 (1965) [[6227], vol. 2, pp. 1605–1612]
3397. Knapowski, S., Turán, P.: On the sign changes of $\pi(x) - \mathrm{li}(x)$, I. Coll. Math. Soc. János Bólyai **13**, 153–165 (1974) [[6227], vol. 3, pp. 2378–2394]
3398. Knapowski, S., Turán, P.: On the sign changes of $\pi(x) - \mathrm{li}(x)$, II. Monatshefte Math. **82**, 163–175 (1976) [[6227], vol. 3, pp. 2404–2416]
3399. Knapp, A.W.: Elliptic Curves. Princeton University Press, Princeton (1992)
3400. Knapp, M.P.: Systems of diagonal equations over p-adic fields. J. Lond. Math. Soc. **63**, 257–267 (2001)
3401. Knapp, M.P.: Artin's conjecture for forms of degree 7 and 11. J. Lond. Math. Soc. **63**, 268–274 (2001)
3402. Knauer, J., Richstein, J.: The continuing search for Wieferich primes. Math. Comput. **74**, 1559–1563 (2005)
3403. Kneser, A.: Leopold Kronecker. Jahresber. Dtsch. Math.-Ver. **33**, 210–228 (1925)
3404. Kneser, A., Meder, A.: Piers Bohl zum Gedächtnis. Jahresber. Dtsch. Math.-Ver. **33**, 25–32 (1925)
3405. Kneser, M.: Zum expliziten Reziprozitätsgesetz von I.R. Shafarevič. Math. Nachr. **6**, 89–96 (1951)
3406. Kneser, M.: Abschätzung der asymptotischen Dichte von Summenmengen. Math. Z. **58**, 459–484 (1953)
3407. Kneser, M.: Martin Eichler (1912–1992). Acta Arith. **65**, 293–296 (1993)
3408. Knichal, V.: Dyadische Entwicklungen und Hausdorffsches Maß. Vestnik ČSAV, 1933, nr. 14, 1–19

3409. Knichal, V.: Dyadische Entwicklungen und Hausdorffsches Mass. Časopis Mat. Fys. **65**, 195–210 (1936)
3410. Knight, M.J., Webb, W.A.: Uniform distribution of third-order linear recurrence sequences. Acta Arith. **36**, 7–20 (1980)
3411. Knödel, W.: Carmichaelsche Zahlen. Math. Nachr. **9**, 343–350 (1953)
3412. Knödel, W.: Eine obere Schranke für die Anzahl der Carmichaelschen Zahlen kleiner als x. Arch. Math. **4**, 282–284 (1953)
3413. Knopfmacher, A., Mays, M.E.: A survey of factorization counting functions. Int. J. Number Theory **1**, 563–581 (2005)
3414. Knopfmacher, J.: Abstract Analytic Number Theory. North-Holland, Amsterdam (1975); 2nd ed. Dover 1990
3415. Knopfmacher, J.: Analytic Arithmetic of Algebraic Function Fields. Dekker, New York (1979)
3416. Knopfmacher, J., Zhang, W.-B.: Number Theory Arising from Finite Fields. Dekker, New York (2001)
3417. Knopp, K.: Hans von Mangoldt. Jahresber. Dtsch. Math.-Ver. **36**, 332–348 (1927)
3418. Knopp, K.: Edmund Landau. Jahresber. Dtsch. Math.-Ver. **54**, 55–62 (1951)
3419. Knopp, M., Sheingorn, M. (eds.): A Tribute to Emil Grosswald: Number Theory and Related Analysis. Contemp. Math., vol. 143. Am. Math. Soc., Providence (1993)
3420. Knuth, D.E.: Evaluation of Porter's constant. Comput. Math. Appl. **2**, 137–139 (1976)
3421. Kobayashi, I.: A note on the Selberg sieve and the large sieve. Proc. Jpn. Acad. Sci. **49**, 1–5 (1973)
3422. Kobayashi, S.: An elementary proof of the Mazur-Tate-Teitelbaum conjecture for elliptic curves. Doc. Math., Extra Volume, 567–575 (2006)
3423. Koblitz, N.: P-adic Numbers, p-adic Analysis and Zeta-Functions. Springer, Berlin (1977); 2nd ed. Springer 1984
3424. Koblitz, N.: Interpretation of the p-adic log gamma function and Euler constants using the Bernoulli measure. Trans. Am. Math. Soc. **242**, 261–269 (1978)
3425. Koblitz, N.: A new proof of certain formulas for p-adic L-functions. Duke Math. J. **46**, 455–468 (1979)
3426. Koblitz, N.: p-adic Analysis: A Short Course on Recent Work. Cambridge University Press, Cambridge (1980)
3427. Koblitz, N.: Introduction to Elliptic Curves and Modular Forms. Springer, Berlin (1984); 2nd ed. 1993
3428. Koblitz, N.: Primality of the number of points of an elliptic curve over a finite field. Pac. J. Math. **131**, 157–165 (1988)
3429. Koch, H.: Nachruf auf Hans Reichardt. Jahresber. Dtsch. Math.-Ver. **95**, 135–140 (1993)
3430. Koch, H., Nebe, G.: Extremal even unimodular lattices of rank 32 and related codes. Math. Nachr. **161**, 309–319 (1993)
3431. Koch, H., Venkov, B.B.: Über ganzzahlige unimodulare euklidische Gitter. J. Reine Angew. Math. **398**, 144–168 (1989)
3432. Koch, H., Venkov, B.B.: Über gerade unimodulare Gitter der Dimension 32, III. Math. Nachr. **152**, 191–213 (1991)
3433. von Koch, H.: Sur la distribution des nombres premiers. Acta Math. **24**, 159–182 (1901)
3434. von Koch, H.: Ueber die Riemannsche Primzahlfunction. Math. Ann. **55**, 440–464 (1902)
3435. von Koch, H.: Contribution à la théorie des nombres premiers. Acta Math. **33**, 293–320 (1910)
3436. Koecher, M.: Zur Theorie der Modulformen n-ten Grades. I. Math. Z. **59**, 399–416 (1954)
3437. Koecher, M.: Zur Theorie der Modulformen n-ten Grades. II. Math. Z. **61**, 455–466 (1955)
3438. Koenigsberger, L.: Carl Gustav Jacob Jacobi. Jahresber. Dtsch. Math.-Ver. **13**, 405–435 (1904)
3439. Koenigsberger, L.: Carl Gustav Jacob Jacobi. Teubner, Leipzig (1904)
3440. Kohnen, W.: Modular forms of half-integral weight on $\Gamma_0(4)$. Math. Ann. **248**, 249–266 (1980)

3441. Kohnen, W.: Newforms of half-integral weight. J. Reine Angew. Math. **333**, 32–72 (1982)
3442. Kohnen, W.: A remark on the Shimura correspondence. Glasg. Math. J. **30**, 285–291 (1988)
3443. Koksma, J.F.: Ein mengentheoretischer Satz über die Gleichverteilung modulo Eins. Compos. Math. **2**, 250–258 (1935)
3444. Koksma, J.F.: Diophantische Approximationen. Springer, Berlin (1936) [Reprint: Springer, 1974]
3445. Koksma, J.F.: Über die Mahlersche Klasseneinteilung der transzendenten Zahlen und die Approximation komplexer Zahlen durch algebraische Zahlen. Monatshefte Math. Phys. **48**, 176–189 (1939)
3446. Koksma, J.F.: Some theorems on Diophantine inequalities. Math. Centrum Amsterdam **5**, 1–51 (1950)
3447. Koksma, J.F., Popken, J.: Zur Transzendenz von e^{π}. J. Reine Angew. Math. **168**, 211–230 (1932)
3448. Kolberg, O.: Note on the parity of the partition function. Math. Scand. **7**, 377–378 (1959)
3449. Kolberg, O.: Congruences for Ramanujan's τ-function. Årbok Univ. Bergen (Mat.-Naturv. Ser.), 1962, nr. 12
3450. Kolberg, O.: Note on Ramanujan's function $\tau(n)$. Math. Scand. **10**, 171–172 (1962)
3451. Kolberg, O.: Congruences for the coefficients of the modular invariant $j(\tau)$. Math. Scand. **10**, 173–181 (1962)
3452. Kolesnik, G.A.: On the distribution of primes in sequences of the form $[n^c]$. Mat. Zametki **2**, 117–128 (1967) (in Russian)
3453. Kolesnik, G.A.: An improvement of the remainder term in the divisor problem. Mat. Zametki **6**, 545–554 (1969) (in Russian)
3454. Kolesnik, G.A.: On the estimation of certain trigonometric sums. Acta Arith. **25**, 7–30 (1973) (in Russian)
3455. Kolesnik, G.: On the estimation of multiple exponential sums. In: Recent Progress in Analytic Number Theory, Durham, 1979, vol. 1, pp. 231–246. Academic Press, San Diego (1981)
3456. Kolesnik, G.: On the order of $\zeta(\frac{1}{2}+it)$ and $\Delta(R)$. Pac. J. Math. **98**, 107–122 (1982)
3457. Kolesnik, G.: An improvement of the method of exponent pairs. In: Topics in Classical Number Theory. Colloq. Math. Soc. J. Bólyai, vol. 34, pp. 907–926. North-Holland, Amsterdam (1984)
3458. Kolesnik, G.: On the method of exponent pairs. Acta Arith. **45**, 115–143 (1985)
3459. Kolesnik, G.: Primes of the form $[n^c]$. Pac. J. Math. **118**, 437–447 (1985)
3460. Kolountzakis, M.N.: An effective basis for the integers. Discrete Math. **145**, 307–313 (1995)
3461. Kolountzakis, M.N.: On the additive complements of the primes and sets of similar growth. Acta Arith. **77**, 1–8 (1996)
3462. Kolyvagin, V.A.: Finiteness of $E(\mathbf{Q})$ and $Ш(E,\mathbf{Q})$ for a subclass of Weil curves. Izv. Akad. Nauk SSSR, Ser. Mat. **52**, 522–540 (1988) (in Russian)
3463. Kolyvagin, V.A.: On the Mordell-Weil and Shafarevich-Tate groups for elliptic Weil curves. Izv. Akad. Nauk SSSR, Ser. Mat. **52**, 1154–1180 (1988) (in Russian)
3464. Kolyvagin, V.A.: Euler systems. Prog. Math. **87**, 435–483 (1990)
3465. Kolyvagin, V.A.: The Fermat equations over cyclotomic fields. Tr. Mat. Inst. Steklova **208**, 163–185 (1995) (in Russian)
3466. Kolyvagin, V.A.: On the first case of the Fermat theorem for cyclotomic fields. J. Math. Sci. **106**, 3302–3311 (2001)
3467. Kolyvagin, V.A.: The Fermat equation over a tower of cyclotomic fields. Izv. Akad. Nauk SSSR, Ser. Mat. **65**, 85–122 (2001) (in Russian)
3468. Komatsu, T.: Extension of Baker's analogue of Littlewood's Diophantine approximation problem. Kodai Math. Semin. Rep. **14**, 335–340 (1991)
3469. Konyagin, S.V.: On the Littlewood problem. Izv. Akad. Nauk SSSR, Ser. Mat. **45**, 243–265 (1981) (in Russian)
3470. Konyagin, S.V.: On estimates of Gaussian sums and Waring's problem for a prime modulus. Tr. Mat. Inst. Steklova **198**, 111–124 (1992) (in Russian)

3471. Konyagin, S.V.: Estimates of the least prime factor of a binomial coefficient. Mathematika **46**, 41–55 (1999)
3472. Konyagin, S.V.: An additive problem with fractional powers. Mat. Zametki **73**, 633–636 (2003) (in Russian)
3473. Konyagin, S.V., Stečkin, S.B.: An estimate for the number of solutions of nth degree congruence with one unknown. Tr. Mat. Inst. Steklova **219**, 249–257 (1997) (in Russian)
3474. Kopetzky, H.G.: Über die Grössenordnung der Teilerfunktion in Restklassen. Monatshefte Math. **82**, 287–295 (1976)
3475. Kopřiva, J.: O jednom vztahu Fareyovy řady k Riemannově domněnce o nulových bodach funkce ζ. Časopis Pěst. Math. **78**, 49–55 (1953)
3476. Kopřiva, J.: Přispěvek k vztahu Fareyovy řady a Riemannovy domněnky. Časopis Pěst. Math. **79**, 77–82 (1954)
3477. Korevaar, J.: Tauberian Theory. Springer, Berlin (2004)
3478. Korkine, A., Zolotarev, E.: Sur les formes quadratiques positives quaternaires. Math. Ann. **5**, 581–583 (1872)
3479. Korkine, A., Zolotarev, E.: Sur les formes quadratiques. Math. Ann. **6**, 360–389 (1873)
3480. Korkine, A., Zolotarev, E.: Sur les formes quadratiques positives. Math. Ann. **11**, 242–292 (1877)
3481. Körner, O.: Über das Waringsche Problem in algebraischen Zahlkörpern. Math. Ann. **144**, 224–238 (1961)
3482. Körner, O.: Über Mittelwerte trigonometrischer Summen und ihre Anwendung in algebraischen Zahlkörpern. Math. Ann. **147**, 205–239 (1962)
3483. Körner, O.: Über durch Potenzen erzeugte Ringe und Gruppen in algebraischen Zahlkörpern. Manuscr. Math. **3**, 157–174 (1970)
3484. Körner, O., Stähle, H.: Remarks on Hua's estimate of complete trigonometrical sums. Acta Arith. **35**, 353–359 (1979)
3485. Korobov, N.M.: Concerning some questions of uniform distribution. Izv. Akad. Nauk SSSR, Ser. Mat. **14**, 215–238 (1950) (in Russian)
3486. Korobov, N.M.: On zeros of the function $\zeta(s)$. Dokl. Akad. Nauk SSSR **118**, 431–432 (1958) (in Russian)
3487. Korobov, N.M.: Estimates of exponential sums and their applications. Usp. Mat. Nauk **13**(4), 185–192 (1958) (in Russian)
3488. Korobov, N.M.: Weyl's estimates of sums and the distribution of primes. Dokl. Akad. Nauk SSSR **123**, 28–31 (1958) (in Russian)
3489. Korobov, N.M., Nesterenko, Yu.V., Shidlovskii, A.B.: Naum Il'ič Feldman. Usp. Mat. Nauk **50**(6), 157–162 (1995) (in Russian)
3490. Korobov, N.M., Postnikov, A.G.: Some general theorems on the uniform distribution of fractional parts. Dokl. Akad. Nauk SSSR **84**, 217–220 (1952) (in Russian)
3491. Korselt, A.: Problème chinois. L'Intermédiaire Math. **6**, 142–143 (1899)
3492. Korselt, A.: Veränderliche Lösung von $x^m + y^n + z^r = 0$. Arch. Math. Phys. **25**, 89–91 (1917)
3493. Kortum, H.: Rudolf Lipschitz. Jahresber. Dtsch. Math.-Ver. **15**, 56–59 (1906)
3494. Koshliakov [Košliakov], N.S.: On Voronoï's sum-formula. Messenger Math. **58**, 30–32 (1928/1929)
3495. Košliakov, N.S.: Application of Mellin's formula to the deduction of certain summation formulas. Izv. Akad. Nauk SSSR, Ser. Mat. **5**, 43–56 (1941) (in Russian)
3496. Kosovskiĭ, N.K.: The diophantine representations of a sequence of solutions of the Pell equation. Zap. Nauč. Semin. LOMI **20**, 49–59 (1971) (in Russian)
3497. Kotnik, T.: The prime-counting function and its analytic approximations: $\pi(x)$ and its approximations. Adv. Comput. Math. **29**, 55–70 (2008)
3498. Kotnik, T., te Riele, H.J.J.: The Mertens conjecture revisited. In: Lecture Notes in Comput. Sci., vol. 4076, pp. 156–167. Springer, Berlin (2006)
3499. Kotov, S.V.: On the largest prime divisor of a polynomial. Mat. Zametki **13**, 515–522 (1973) (in Russian)

3500. Kotsireas, I.: The Erdős-Straus conjecture on Egyptian fractions. In: Paul Erdős and His Mathematics, Budapest, 1999, pp. 140–144. János Bolyai Math. Soc., Budapest (1999)
3501. Kovalčik, F.B.: Density theorems and the distribution of primes in sectors and progressions. Dokl. Akad. Nauk SSSR **219**, 31–34 (1974) (in Russian)
3502. Kovalevskaya, E.I.: A metric theorem on the exact order of approximation of zero by values of integer polynomials in Q_p. Dokl. Akad. Nauk Belorus. **43**, 34–36 (1999) (in Russian)
3503. Kowalski, E.: Some local-global applications of Kummer theory. Manuscr. Math. **111**, 105–139 (2003)
3504. Koyama, K.: Tables of solutions of the Diophantine equation $x^3 + y^3 + z^3 = n$. Math. Comput. **62**, 941–942 (1994)
3505. Koyama, K.: On searching for solutions of the Diophantine equation $x^3 + y^3 + 2z^3 = n$. Math. Comput. **69**, 1735–1742 (2000)
3506. Koyama, K., Tsuruoka, Y., Sekigawa, H.: On searching for solutions of the Diophantine equation $x^3 + y^3 + z^3 = n$. Math. Comput. **66**, 841–851 (1997)
3507. Kozlov, V.Ya., Mardzanišvili, K.K.: Andreĭ Borisovič Šidlovskiĭ (on the occasion of his sixtieth birthday). Usp. Mat. Nauk **31**(3), 225–232 (1976) (in Russian)
3508. Kra, B.: The Green-Tao theorem on arithmetic progressions in the primes: an ergodic point of view. Bull. Am. Math. Soc. **43**, 3–23 (2006)
3509. Kramarz, G.: All congruent numbers less than 2000. Math. Ann. **273**, 337–340 (1986)
3510. Kramer, K.: A family of semistable elliptic curves with large Tate-Shafarevitch groups. Proc. Am. Math. Soc. **89**, 379–386 (1983)
3511. Krasner, M.: Sur le premier cas du théoreme de Fermat. C. R. Acad. Sci. Paris **199**, 256–258 (1934)
3512. Krasner, M.: Nombre des extensions d'un degré donné d'un corps $\mathfrak{p}$-adique. C. R. Acad. Sci. Paris **254**, 3470–3472 (1962); **255**, 224–226, 1682–1684, 2342–2344, 3095–3097 (1962)
3513. Krass, S.: Estimates for n-dimensional Diophantine approximation constants for $n \geq 4$. J. Number Theory **20**, 172–176 (1985)
3514. Krätzel, E.: Bemerkungen zu einem Gitterpunktproblem. Math. Ann. **179**, 90–96 (1969)
3515. Krätzel, E.: Zahlen k-ter Art. Am. J. Math. **94**, 309–328 (1972)
3516. Krätzel, E.: Divisor problems and powerful numbers. Math. Nachr. **114**, 97–104 (1983)
3517. Krätzel, E.: Lattice Points. Kluwer Academic, Dordrecht (1988)
3518. Krätzel, E.: The distribution of powerful integers of type 4. Acta Arith. **52**, 141–145 (1989)
3519. Kraus, A.: Majorations effectives pour l'équation de Fermat généralisée. Can. J. Math. **49**, 1139–1161 (1997)
3520. Kraus, A.: Sur l'équation $a^3 + b^3 = c^p$. Exp. Math. **7**, 1–13 (1998)
3521. Krečmar, V.: Sur les propriétés de la divisibilité d'une fonction additive. Izv. Akad. Nauk SSSR, Ser. Mat., 1933, 763–780
3522. Kretschmer, T.J.: Construction of elliptic curves with large rank. Math. Comput. **46**, 627–635 (1986)
3523. Křížek, M., Luca, F., Somer, L.: 17 Lectures on Fermat Numbers. Springer, Berlin (2001)
3524. Kronecker, L.: Über die algebraisch auflösbaren Gleichungen. Mon. Ber. Kgl. Preuß. Akad. Wiss., 1853, 365–374; 1856, 203–215. [[3532], vol. 4, pp. 1–11, 25–37]
3525. Kronecker, L.: Zwei Sätze über Gleichungen mit ganzzahligen Coefficienten. J. Reine Angew. Math. **53**, 173–175 (1857) [[3532], vol. 1, pp. 103–108]
3526. Kronecker, L.: Ueber die Anzahl der verschiedenen Classen quadratischer Formen von negativer Determinante. J. Reine Angew. Math. **57**, 248–255 (1860) [[3532], vol. 4, pp. 185–195]
3527. Kronecker, L.: Über die Irreductibilität der Gleichungen. Mon. Ber. Kgl. Preuß. Akad. Wiss., 1880, 155–163 [[3532], vol. 2, pp. 85–93]
3528. Kronecker, L.: Grundzüge einer arithmetischen Theorie der algebraischen Grössen. J. Reine Angew. Math. **92**, 1–122 (1882) [[3532], vol. 2, pp. 237–387]
3529. Kronecker, L.: Über bilineare Formen mit vier Variabeln. Abh. Kgl. Preuß. Akad. Wiss. Berl. **2**, 1–60 (1883) [[3532], vol. 2, pp. 425–495]
3530. Kronecker, L.: Die Periodensysteme von Funktionen reeller Variablen. SBer. Kgl. Preuß. Akad. Wiss, Berlin 1884, 1071–1080 [[3532], vol. 3_1, pp. 33–46]

3531. Kronecker, L.: Ueber einige Anwendungen der Modulsysteme auf elementare algebraische Fragen. J. Reine Angew. Math. **99**, 329–371 (1886) [[3532], vol. 3_1, pp. 145–208]
3532. Kronecker, L.: Werke, vols. 1–5. Teubner, Leipzig (1895–1931) [Reprint: Chelsea, 1968]
3533. Krull, W.: Galoissche Theorie der unendlichen algebraischen Erweiterungen. Math. Ann. **100**, 687–698 (1928)
3534. Krull, W.: Idealtheorie in unendlichen Zahlkörpern. Math. Z. **29**, 42–54 (1928)
3535. Krull, W.: Idealtheorie in unendlichen Zahlkörpern, II. Math. Z. **31**, 527–557 (1930)
3536. Krull, W.: Allgemeine Bewertungstheorie. J. Reine Angew. Math. **167**, 160–196 (1932)
3537. Krummhiebel, B., Amthor, A.: Das Problema bovinum des Archimedes. Zeitschr. Math. Phys. **25**, 121–136, 153–171 (1880)
3538. Kuba, G.: A mean-value theorem on sums of two kth powers of numbers in residue classes. Abh. Math. Semin. Univ. Hamb. **60**, 249–256 (1990)
3539. Kuba, G.: A mean-value theorem on sums of two kth powers of numbers in residue classes, II. Abh. Math. Semin. Univ. Hamb. **63**, 87–95 (1993)
3540. Kubert, D.S.: Universal bounds on the torsion of elliptic curves. Proc. Lond. Math. Soc. **33**, 193–237 (1976)
3541. Kubilis, I.P. [Kubilius, J.], et al.: Aleksandr Adolfovič Buhštab (obituary). Usp. Mat. Nauk **46**(1), 201–202 (1991) (in Russian)
3542. Kubilius, I.P. [J.]: The decomposition of prime numbers into two squares. Dokl. Akad. Nauk SSSR **77**, 791–794 (1951) (in Russian)
3543. Kubilius, I.P. [J.]: On some problems in the geometry of numbers. Mat. Sb. **31**, 507–542 (1952) (in Russian)
3544. Kubilius, I.P. [J.]: On the distribution of values of additive arithmetic functions. Dokl. Akad. Nauk SSSR **100**, 623–626 (1955) (in Russian)
3545. Kubilius, I.P. [J.]: On a problem in the n-dimensional analytic theory of numbers. Vilniaus Valst. Univ. Mokslo Darbai **4**, 5–43 (1955) (in Lithuanian)
3546. Kubilius, I.P. [J.]: Probabilistic methods in the theory of numbers. Usp. Mat. Nauk **11**(2), 31–66 (1956) (in Russian)
3547. Kubilius, J.: Probability Methods in Number Theory. Gos. Izd. Polit. Nauchn. Lit. SSR, Vilnius (1959); 2nd ed. 1962 (in Russian) [English translation: Probabilistic Methods in the Theory of Numbers, Am. Math. Soc., 1964]
3548. Kubilius, J.: On an inequality for additive arithmetic functions. Acta Arith. **27**, 371–383 (1975)
3549. Kubilius, J.: Estimation of the second central moment for strongly additive arithmetic functions. Liet. Mat. Rink. **23**, 122–133 (1983) (in Russian)
3550. Kubilius, J.: Sharpening of the estimate of the second central moment for additive arithmetical functions. Liet. Mat. Rink. **25**, 104–110 (1985) (in Russian)
3551. Kubilius, J., Reiziņš, L., Riekstiņš, R.: Ernests Fogels (1910–1985). Acta Arith. **57**, 179–184 (1991)
3552. Kubina, J.M., Wunderlich, M.C.: Extending Waring's conjecture to $471,600,000$. Math. Comput. **55**, 815–820 (1990)
3553. Kubota, K.K.: On a conjecture of Morgan Ward, I. Acta Arith. **33**, 11–28 (1977)
3554. Kubota, K.K.: On a conjecture of Morgan Ward, II. Acta Arith. **33**, 29–48 (1977)
3555. Kubota, K.K.: On a conjecture of Morgan Ward, III. Acta Arith. **33**, 99–109 (1977)
3556. Kubota, T.: Obituary note: Matsusaburô Fujiwara (1881–1946). Tohoku Math. J. **1**, 1–2 (1949)
3557. Kubota, T., Leopoldt, H.W.: Eine p-adische Theorie der Zetawerte, I, Einführung der p-adischen Dirichletschen L-Funktionen. J. Reine Angew. Math. **214/215**, 328–339 (1964)
3558. Kudla, S.S.: Theta-functions and Hilbert modular forms. Nagoya Math. J. **69**, 97–106 (1978)
3559. Kudryavtsev, M.V.: Estimation of a complete rational trigonometric sum, I. Mat. Zametki **53**, 59–67 (1993) (in Russian)
3560. Kuharev, V.G.: The critical determinant of the region $|x|^p + |y|^p \leq 1$. Dokl. Akad. Nauk SSSR **169**, 1273–1275 (1966) (in Russian)

3561. Kühleitner, M.: On a question of A. Schinzel concerning the sum $\sum_{n\le x}(r(n))^2$. In: Österreichisch-Ungarisch-Slowakisches Kolloquium über Zahlentheorie, Maria Trost, 1992, pp. 63–67. Karl-Franzens-Universität Graz, Graz (1993)
3562. Kühleitner, M., Nowak, W.G., Schoissengeier, J., Wooley, T.D.: On sums of two cubes: an Ω_+-estimate for the error term. Acta Arith. **85**, 179–195 (1998)
3563. Kuhn, P.: Neue Abschätzungen auf Grund der Viggo Brunschen Siebmethode. In: 12 Skand. Mat. Kongr., Lund, 1953, pp. 160–168 (1954)
3564. Kuhn, P.: Über die Primteiler eines Polynoms. In: Proc. ICM Amsterdam, vol. 2, pp. 35–37. North-Holland, Amsterdam (1956)
3565. Kuhn, P.: Eine Verbesserung des Restgliedes beim elementaren Beweis des Primzahlsatzes. Math. Scand. **3**, 75–89 (1955)
3566. Kühnel, U.: Verschärfung der notwendigen Bedingungen für die Existenz von ungeraden vollkommenen Zahlen. Math. Z. **52**, 202–211 (1949)
3567. Kuijk, W. (ed.): Modular Functions of One Variable I. Lecture Notes in Math., vol. 320. Springer, Berlin (1973)
3568. Kuijk, W., Serre, J.-P. (eds.): Modular Functions of One Variable III. Lecture Notes in Math., vol. 350. Springer, Berlin (1973)
3569. Kuipers, L., Niederreiter, H.: Uniform Distribution of Sequences. Wiley, New York (1974)
3570. Kuipers, L., Popken, J.: In memoriam J.F. Koksma (1904–1964). Nieuw Arch. Wiskd. **13**, 1–18 (1965)
3571. Kuipers, L., Shiue, [P.]J.-S.: A distribution property of the sequence of Fibonacci numbers. Fibonacci Q. **10**, 375–376 (1972)
3572. Kulas, M.: Refinement of an estimate for the Hurwitz zeta function in a neighbourhood of the line $\sigma = 1$. Acta Arith. **89**, 301–309 (1999)
3573. Kumchev, A.: On the distribution of prime numbers of the form $[n^c]$. Glasg. Math. J. **41**, 85–102 (1999)
3574. Kumchev, A.V.: Review of [5013]. Math. Rev. 2005a:11121
3575. Kummer, E.E.: De aequatione $x^{2\lambda} + y^{2\lambda} = z^{2\lambda}$ per numeros integros resolvenda. J. Reine Angew. Math. **17**, 203–209 (1837) [[3583], vol. 1, pp. 135–141]
3576. Kummer, E.E.: Eine Aufgabe betreffend die Theorie der cubischen Reste. J. Reine Angew. Math. **23**, 285–286 (1842) [[3583], vol. 1, pp. 143–144]
3577. Kummer, E.E.: De residuis cubicis disquisitiones nonnullae analyticae. J. Reine Angew. Math. **32**, 341–359 (1846) [[3583], vol. 1, pp. 145–163]
3578. Kummer, E.E.: Allgemeiner Beweis des Fermat'schen Satzes, dass die Gleichung $x^\lambda + y^\lambda = z^\lambda$ durch ganze Zahlen unlösbar ist, für alle diejenigen Potenz-Exponenten λ, welche ungerade Primzahlen sind und in den Zählern der ersten $\frac{1}{2}(\lambda - 3)$ Bernoulli'schen Zahlen als Factoren nicht vorkommen. J. Reine Angew. Math. **40**, 130–138 (1850) [[3583], vol. 1, pp. 336–344]
3579. Kummer, E.E.: Mémoire sur la théorie des nombres complexes composés de racines de l'unité et des nombres entiers. J. Math. Pures Appl. **16**, 377–498 (1851) [[3583], vol. 1, pp. 363–484]
3580. Kummer, E.E.: Einige Sätze über die aus den Wurzeln der Gleichung $\alpha^\lambda = 1$ gebildeten complexen Zahlen, für den Fall, dass die Klassenzahl durch λ theilbar ist, nebst Anwendung derselben auf einen weiteren Beweis des letzten Fermat'schen Lehrsatzes. Abhandl. Kgl. Preuss. Akad. Wiss. Berlin, 1857, 41–74. [[3583], vol. 1, pp. 639–672]
3581. Kummer, E.E.: Über die allgemeinen Reziprozitätsgesetze unter den Resten und Nichtresten der Potenzen, deren Grad eine Primzahl ist. Abhandl. Kgl. Preuss. Akad. Wiss. Berlin, 1859, 19–159. [[3583], vol. 1, pp. 699–839]
3582. Kummer, E.E.: Über die Classenanzahl der aus n-ten Einheitswurzeln gebildeten complexen Zahlen. Mon. Ber. Kgl. Preuß. Akad. Wiss., 1861, 1051–1053 [[3583], pp. 883–885]
3583. Kummer, E.E.: Collected Papers, vols. 1, 2, Springer, Berlin (1975)
3584. Kunz, E., Nastold, H.-J.: In memoriam Friedrich Karl Schmidt. Jahresber. Dtsch. Math.-Ver. **83**, 169–181 (1981)
3585. Kurokawa, N.: Examples of eigenvalues of Hecke operators on Siegel cusp forms of degree two. Invent. Math. **49**, 149–165 (1978)

3586. Kűrschak, J.: Über Limesbildung und allgemeine Körpertheorie. J. Reine Angew. Math. **142**, 211–253 (1913)
3587. Kurzweil, J.: O životě a díle člena korespondenta ČSAV Prof. Vladimíra Knichala. Časopis Pěst. Mat. **100**, 314–324 (1975)
3588. Kuzel, A.V.: Elementary solution of Waring's problem for polynomials using Yu.V. Linnik's method. Usp. Mat. Nauk **11**(3), 165–168 (1956) (in Russian)
3589. Kuzmin, R.O.: On a problem of Gauss. Dokl. Akad. Nauk SSSR, 1928, 375–380 (in Russian)
3590. Kuzmin, R.O.: On a new class of transcendental numbers. Izv. Akad. Nauk SSSR, Ser. Mat., 1930, 585–597
3591. Kuzmin, R.O.: Sur un problème de Gauss. In: Atti congresso int. Mat., Bologna, 1928, vol. 6, pp. 83–89 (1932)
3592. Kuzmin, R.O.: On the zeros of Riemann's function $\zeta(s)$. Dokl. Akad. Nauk SSSR, 1934, nr. 2, 398–400 (in Russian)
3593. Kuzmin, R.O.: On the theory of Dirichlet series. Dokl. Akad. Nauk SSSR, 1934, nr. 2, 1470–1491 (in Russian)
3594. Kuzmin, R.O.: The life and scientific activity of Egor Ivanovič Zolotarev. Usp. Mat. Nauk **2**(6), 21–51 (1947) (in Russian)
3595. Kuznetsov, N.V.: The Petersson conjecture for cusp forms of weight zero and the Linnik conjecture. Sums of Kloosterman sums. Mat. Sb. **111**, 334–383 (1980) (in Russian)
3596. Laborde, M.: Equirépartition des solutions du problème de Waring. Sém. Delange–Pisot–Poitou **18**(exp. 20), 1–16 (1976/1977)
3597. Laborde, M.: Nombres presque-premiers dans de petits intervalles. Sém. Théor. Nombres Bordeaux, 1977/1978, exp. 15, 1–17
3598. Lachaud, G.: Une présentation adélique de la série singulière et du problème de Waring. Enseign. Math. **28**, 139–169 (1982)
3599. Lachaud, G.: Variations sur un thème de Mahler. Invent. Math. **52**, 149–162 (1979)
3600. Lachaud, G.: On real quadratic fields. Bull. Am. Math. Soc. **17**, 307–311 (1987)
3601. Lagarias, J.C.: On the computational complexity of determining the solvability or unsolvability of the equation $X^2 - DY^2 = -1$. Trans. Am. Math. Soc. **260**, 485–508 (1980)
3602. Lagarias, J.C.: Sets of primes determined by systems of polynomial congruences. Ill. J. Math. **27**, 224–239 (1983)
3603. Lagarias, J.C.: The $3x + 1$ problem and its generalizations. Am. Math. Mon. **92**, 3–23 (1985)
3604. Lagarias, J.C.: The set of primes dividing the Lucas numbers has density $2/3$. Pac. J. Math. **118**, 449–461 (1985). Errata: **162**, 393–397 (1994)
3605. Lagarias, J.C.: An elementary problem equivalent to the Riemann Hypothesis. Am. Math. Mon. **109**, 534–543 (2002)
3606. Lagarias, J.C.: The $3x + 1$ problem: an annotated bibliography (1963–2000). arXiv: math/0309224v9
3607. Lagarias, J.C., Miller, V.S., Odlyzko, A.M.: Computing $\pi(x)$: the Meissel-Lehmer method. Math. Comput. **44**, 537–560 (1985)
3608. Lagarias, J.C., Montgomery, H.L., Odlyzko, A.M.: A bound for the least ideal in the Chebotarev density theorem. Invent. Math. **54**, 271–296 (1979)
3609. Lagrange, J.: Décomposition d'un entier en somme de carrés et fonction multiplicative. Sém. Delange–Pisot–Poitou **14**(exp. 1), 1–5 (1972/1973)
3610. Lagrange, J.L.: Démonstration d'un théorème d'arithmétique. Nouveaux Mémoires de l'Acad. Royale des Sciences et Belles Lettres de Berlin, 1770 [Oeuvres, vol. 3, pp. 189–201, Paris, 1869]
3611. Lagrange, J.L.: Recherches d'arithmétique, Nouv. Mém. Acad. Roy. Sci. Bell. Lettr. de Berlin, 1773, 1775 [Oeuvres, vol. 3, pp. 695–795, Paris, 1869]
3612. Lahiri, D.B.: On Ramanujan's function $\tau(n)$ and the divisor function $\sigma_k(n)$, II. Bull. Calcutta Math. Soc. **39**, 33–52 (1947)
3613. Lai, K.F.: Tamagawa number of reductive algebraic groups. Compos. Math. **41**, 153–188 (1980)

3614. Laishram, L., Shorey, T.N.: The greatest prime divisor of a product of consecutive integers. Acta Arith. **120**, 299–306 (2005)
3615. Landau, E.: Neuer Beweis der Gleichung $\sum_1^\infty \mu(k)/k = 0$. Dissertation, Berlin (1899) [[3680], vol. 1, pp. 69–83]
3616. Landau, E.: Ueber die zahlentheoretische Funktion $\varphi(n)$ und ihre Beziehung zum Goldbachschen Satz. Nachr. Ges. Wiss. Göttingen, 1900, 177–186. [[3680], vol. 1, pp. 106–115]
3617. Landau, E.: Sur quelques problèmes rélatifs à la distribution des nombres premiers. Bull. Soc. Math. Fr. **28**, 25–38 (1900) [[3680], vol. 1, pp. 92–105]
3618. Landau, E.: Ueber die asymptotischen Werte einiger zahlentheoretischer Functionen. Math. Ann. **54**, 570–591 (1901) [[3680], vol. 1, pp. 141–162]
3619. Landau, E.: Über die zu einem algebraischen Zahlkörper gehörige Zetafunction und die Ausdehnung der Tchebyschefschen Primzahltheorie auf das Problem der Verteilung der Primideale. J. Reine Angew. Math. **125**, 64–188 (1902) [[3680], vol. 1, pp. 201–325]
3620. Landau, E.: Neuer Beweis des Primzahlsatzes und Beweis des Primidealsatzes. Math. Ann. **56**, 645–670 (1903) [[3680], vol. 1, pp. 327–332]
3621. Landau, E.: Über die Klassenzahl der binären quadratischen Formen von negativer Diskriminante. Math. Ann. **56**, 671–676 (1903) [[3680], vol. 1, pp. 354–359]
3622. Landau, E.: Summary of [3621]. Jahrbuch f. d. Fortschr. Mathematik 34.0241.09
3623. Landau, E.: Über die Primzahlen einer arithmetischer Progression. SBer. Kais. Akad. Wissensch. Wien **112**, 493–535 (1903) [[3680], vol. 2, pp. 17–59]
3624. Landau, E.: Über die Darstellung der Anzahl der Idealklassen eines algebraischen Körpers durch eine unendliche Reihe. J. Reine Angew. Math. **127**, 167–174 (1904) [[3680], vol. 2, pp. 145–152]
3625. Landau, E.: Über einen Satz von Tschebyschef. Math. Ann. **61**, 527–550 (1905) [[3680], vol. 2, pp. 206–229]
3626. Landau, E.: Sur quelques inégalités dans la théorie de la fonction $\zeta(s)$ de Riemann. Bull. Soc. Math. Fr. **33**, 229–241 (1905) [[3680], vol. 2, pp. 167–179]
3627. Landau, E.: Über die Verteilung der Primideale in den Idealklassen eines algebraischen Zahlkörpers. Math. Ann. **63**, 145–204 (1906) [[3680], vol. 3, pp. 181–204]
3628. Landau, E.: Über die Darstellung einer ganzen Zahl als Summe von Biquadraten. Rend. Circ. Mat. Palermo **23**, 91–96 (1907) [[3680], vol. 3, pp. 269–272]
3629. Landau, E.: Zwei neue Herleitungen für die asymptotische Anzahl der Primzahlen unter einer gegebener Grenze. SBer. Berlin, 1908, 746–764. [[3680], vol. 4, pp. 21–39]
3630. Landau, E.: Beiträge zur analytischen Zahlentheorie. Rend. Circ. Mat. Palermo **26**, 169–302 (1908) [[3680], vol. 3, pp. 411–544]
3631. Landau, E.: Über die Primzahlen in einer arithmetischen Progression und die Primideale in einer Idealklasse. SBer. Kais. Akad. Wissensch. Wien **117**, 1095–1107 (1908) [[3680], vol. 4, pp. 73–85]
3632. Landau, E.: Über die Verteilung der Nullstellen der Riemannschen Zetafunktion und einer Klasse verwandter Funktionen. Math. Ann. **66**, 419–445 (1908) [[3680], vol. 4, pp. 149–175]
3633. Landau, E.: Über die Einteilung der positiven ganzen Zahlen in vier Klasen nach der Mindestzahl der zu ihrer additiven Zusammensetzung erforderlichen Quadrate. Arch. Math. **13**, 305–312 (1908) [[3680], vol. 4, pp. 59–66]
3634. Landau, E.: Neue Beiträge zur analytischen Zahlentheorie. Rend. Circ. Mat. Palermo **27**, 46–58 (1909) [[3680], vol. 4, pp. 41–53]
3635. Landau, E.: Über eine Anwendung der Primzahltheorie auf das Waringsche Problem in der elementaren Zahlentheorie. Math. Ann. **66**, 102–105 (1909) [[3680], vol. 4, pp. 55–58]
3636. Landau, E.: Handbuch der Lehre von der Verteilung der Primzahlen, Teubner, Leipzig (1909) [Reprint: Chelsea, 1953]
3637. Landau, E.: Zur Theorie der Riemannschen Zetafunktion. Vierteljahrsschr. Naturf. Ges. Zürich **56**, 125–148 (1911) [[3680], vol. 5, pp. 22–45]
3638. Landau, E.: Über die Äquivalenz zweier Hauptsätze der analytischen Zahlentheorie. SBer. Kais. Akad. Wissensch. Wien **120**, 973–988 (1911) [[3680], vol. 5, pp. 46–61]

3639. Landau, E.: Über die Verteilung der Zahlen welche aus ν Primfaktoren zusammengesetzt sind. Nachr. Ges. Wiss. Göttingen, 1911, 361–381. [[3680], vol. 4, pp. 443–463]
3640. Landau, E.: Über das Nichtverschwinden der Dirichletschen Reihen, welche komplexen Charakteren entsprechen. Math. Ann. **70**, 69–78 (1911) [[3680], vol. 4, pp. 249–258]
3641. Landau, E.: Gelöste und ungelöste Probleme aus der Theorie der Primzahlverteilung und der Riemannschen Zetafunktion. In: Proc. 5th. ICM, vol. 1, pp. 93–108. Cambridge University Press, Cambridge (1913) [Jahresber. Dtsch. Math.-Ver., vol. 21, 1912, pp. 208–228; [3680], vol. 5, pp. 240–255]
3642. Landau, E.: Über die Anzahl der Gitterpunkte in gewissen Bereichen. Nachr. Ges. Wiss. Göttingen, 1912, 687–770 [[3680], vol. 5, pp. 156–239]
3643. Landau, E.: Über die Anzahl der Gitterpunkte in gewissen Bereichen, II. Nachr. Ges. Wiss. Göttingen, 1915, 209–243 [[3680], vol. 6, pp. 308–342]
3644. Landau, E.: Die Bedeutung der Pfeifferschen Methode für die analytische Zahlentheorie. SBer. Kais. Akad. Wissensch. Wien **121**, 2195–2332 (1912) [[3680], vol. 5, pp. 284–419]
3645. Landau, E.: Über die Zerlegung der Zahlen in zwei Quadrate. Ann. Mat. Pura Appl. **20**, 1–28 (1912) [[3680], vol. 5, pp. 256–283]
3646. Landau, E.: Über einige Summen, die von den Nullstellen der Riemannschen Zetafunktion abhängen. Acta Math. **35**, 271–294 (1912) [[3680], vol. 5, pp. 62–85]
3647. Landau, E.: Über eine idealtheoretische Funktion. Trans. Am. Math. Soc. **13**, 1–21 (1912) [[3680], vol. 5, pp. 107–127]
3648. Landau, E.: Über die Primzahlen in definiten quadratischen Formen und die Zetafunktion reiner kubischen Körper. In: H.A. Schwarz Festschrift, Berlin, pp. 244–273 (1914) [[3680], vol. 6, pp. 105–134]
3649. Landau, E.: Über Dirichlet's Teilerproblem. SBer. Bayer. Akad. Wiss., 1915, 317–328 [[3680], vol. 6, pp. 343–354]
3650. Landau, E.: Über die Hardysche Entdeckung unendlich vieler Nullstellen der Zetafunktion mit reellem Teil 1/2. Math. Ann. **76**, 212–243 (1915) [[3680], vol. 6, pp. 135–166]
3651. Landau, E.: Zur analytischen Zahlentheorie der definiten quadratischen Formen. (Über die Gitterpunkte in einem mehrdimensionalen Ellipsoid). SBer. Kgl. Preuß. Akad. Wiss. Berl. **31**, 458–476 (1915) [[3680], vol. 6, pp. 200–218]
3652. Landau, E.: Über die Gitterpunkte in einem Kreise. Nachr. Ges. Wiss. Göttingen, 1915, 148–160, 209–243 [[3680], vol. 6, pp. 187–199, 308–342]
3653. Landau, E.: Über den Mellinschen Satz. Arch. Math. **24**, 97–107 (1915) [[3680], vol. 6, pp. 230–240]
3654. Landau, E.: Einführung in die elementare und analytische Theorie der algebraischen Zahlen und Ideale. Teubner, Leipzig (1918) [2nd ed. 1929; reprint: Chelsea, 1949]
3655. Landau, E.: Über einige ältere Vermutungen und Behauptungen in der Primzahltheorie, I. Math. Z. **1**, 1–24 (1918) [[3680], vol. 6, pp. 469–492]
3656. Landau, E.: Über einige ältere Vermutungen und Behauptungen in der Primzahltheorie, II. Math. Z. **1**, 213–219 (1918) [[3680], vol. 6, pp. 469–492]
3657. Landau, E.: Über Ideale und Primideale in Idealklassen. Math. Z. **2**, 52–154 (1918) [[3680], vol. 7, pp. 11–154]
3658. Landau, E.: Abschätzungen von Charaktersummen, Einheiten und Klassenzahlen. Nachr. Ges. Wiss. Göttingen, 1918, 79–97 [[3680], vol. 7, pp. 114–132]
3659. Landau, E.: Über die Klassenzahl imaginärquadratischer Zahlkörper. Nachr. Ges. Wiss. Göttingen, 1918, 285–295 [[3680], vol. 7, pp. 150–160]
3660. Landau, E.: Über die Zerlegung total positiver Zahlen in Quadrate. Nachr. Ges. Wiss. Göttingen, 1919, 392–396 [[3680], vol. 7, pp. 201–204]
3661. Landau, E.: Über Dirichlet's Teilerproblem. Nachr. Ges. Wiss. Göttingen, 1920, 13–32 [[3680], vol. 7, pp. 232–251]
3662. Landau, E.: Zur Hardy-Littlewoodschen Lösung des Waringschen Problems. Nachr. Ges. Wiss. Göttingen, 1921, 88–92 [[3680], vol. 7, pp. 327–331]
3663. Landau, E.: Zum Waringschen Problem. Math. Z. **12**, 219–247 (1922) [[3680], vol. 7, pp. 383–411]

3664. Landau, E.: Zur additiven Primzahltheorie. Rend. Circ. Mat. Palermo **46**, 349–356 (1922) [[3680], vol. 7, pp. 436–443]
3665. Landau, E.: Über die Gitterpunkte in einem Kreise, IV. Nachr. Ges. Wiss. Göttingen, 1923, 58–65 [[3680], vol. 8, pp. 59–66]
3666. Landau, E.: Über die ζ-Funktion und die L-Funktionen. Math. Z. **20**, 105–125 (1924) [[3680], vol. 8, pp. 77–97]
3667. Landau, E.: Über Gitterpunkte in mehrdimensionalen Ellipsoiden. Math. Z. **21**, 126–132 (1924) [[3680], vol. 8, pp. 137–143]
3668. Landau, E.: Über Gitterpunkte in mehrdimensionalen Ellipsoiden, II. Math. Z. **24**, 299–310 (1925) [[3680], vol. 8, pp. 277–288]
3669. Landau, E.: Bemerkungen zu der vorstehenden Abhandlung von Herrn Franel. Nachr. Ges. Wiss. Göttingen, 1924, 202–206 [[3680], vol. 8, pp. 166–170]
3670. Landau, E.: Über die Möbiussche Funktion. Rend. Circ. Mat. Palermo **48**, 277–280 (1924) [[3680], vol. 8, pp. 171–174]
3671. Landau, E.: Die Winogradowsche Methode zum Beweise des Waring-Hilbert-Kamkeschen Satzes. Acta Math. **48**, 217–253 (1926) [[3680], vol. 8, pp. 307–343]
3672. Landau, E.: On the representation of a number as the sum of two k-th powers. J. Lond. Math. Soc. **1**, 72–74 (1926) [[3680], vol. 8, pp. 373–375]
3673. Landau, E.: Über Dirichletsche Reihen mit komplexen Charakteren. J. Reine Angew. Math. **157**, 26–32 (1926) [[3680], vol. 8, pp. 377–383]
3674. Landau, E.: Vorlesungen über Zahlentheorie, vols. I–III. Hirzel, Leipzig (1927) [Reprint: Chelsea, 1950, 1969; English translation of vol. I: Elementary Number Theory, Chelsea, 1958]
3675. Landau, E.: Über die neue Winogradoffsche Behandlung des Waringschen Problems. Math. Z. **31**, 319–338 (1929) [[3680], vol. 9, pp. 137–156]
3676. Landau, E.: Die Goldbachsche Vermutung und der Schnirelmannsche Satz. Nachr. Ges. Wiss. Göttingen, 1930, 255–276 [[3680], vol. 9, pp. 167–188]
3677. Landau, E.: Über die Fareyreihe und die Riemannsche Vermutung. Nachr. Ges. Wiss. Göttingen, 1932, 347–352 [[3680], vol. 9, pp. 197–202]
3678. Landau, E.: Der Paleysche Satz über Charaktere. Math. Z. **37**, 28–32 (1933) [[3680], vol. 9, pp. 233–237]
3679. Landau, E.: Bemerkungen zum Heilbronnschen Satz. Acta Arith. **1**, 1–18 (1935) [[3680], vol. 9, pp. 265–282]
3680. Landau, E.: Collected Works, vols. 1–9. Thales Verlag, Essen (1985–1987)
3681. Landau, E., Ostrowski, A.: On the diophantine equation $ay^2 + by + c = dx^n$. Proc. Lond. Math. Soc. **19**, 276–280 (1920); add.: **20**, xxxix (1922) [[3680], vol. 7, pp. 322–326, 364]
3682. Landau, E., van der Corput, J.G.: Über Gitterpunkte in ebenen Bereichen. Nachr. Ges. Wiss. Göttingen, 1920, 135–171 [[3680], vol. 7, pp. 285–321]
3683. Lander, L.J., Parkin, T.R.: A counterexample to Euler's sum of powers conjecture. Math. Comput. **21**, 101–103 (1967)
3684. Lander, L.J., Parkin, T.R.: On first appearance of prime differences. Math. Comput. **21**, 483–488 (1967)
3685. Lang, S.: On quasi algebraic closure. Ann. Math. **55**, 373–390 (1952)
3686. Lang, S.: Algebraic groups over finite fields. Am. J. Math. **78**, 555–563 (1956)
3687. Lang, S.: Integral points on curves. Publ. Math. Inst. Hautes Études Sci. **6**, 27–43 (1960)
3688. Lang, S.: Diophantine Geometry. Interscience, New York (1962)
3689. Lang, S.: A transcendence measure for E-functions. Mathematika **9**, 157–161 (1962)
3690. Lang, S.: Algebraic Numbers. Addison-Wesley, Reading (1964)
3691. Lang, S.: Introduction to Transcendental Numbers. Addison-Wesley, Reading (1966)
3692. Lang, S.: Algebraic values of meromorphic functions, II. Topology **5**, 363–370 (1966)
3693. Lang, S.: Algebraic Number Theory. Addison-Wesley, Reading (1970) [2nd ed. Springer, 1994]
3694. Lang, S.: Transcendental numbers and Diophantine approximations. Bull. Am. Math. Soc. **77**, 635–677 (1971)

3695. Lang, S.: Elliptic Functions. Addison-Wesley, Reading (1973); 2nd ed. 1987, Springer, 1987
3696. Lang, S.: Introduction to Modular Forms. Springer, Berlin (1976); reprint 1995
3697. Lang, S.: Elliptic Curves: Diophantine Analysis. Springer, Berlin (1978)
3698. Lang, S.: Fundamentals of Diophantine Geometry. Springer, Berlin (1983)
3699. Lang, S.: Hyperbolic and Diophantine analysis. Bull. Am. Math. Soc. **14**, 159–205 (1986)
3700. Lang, S.: Old and new conjectured Diophantine inequalities. Bull. Am. Math. Soc. **23**, 37–75 (1990)
3701. Lang, S.: Number Theory. III. Diophantine Geometry. Encyclopaedia of Mathematical Sciences, vol. 60. Springer, Berlin (1991)
3702. Lang, S.: Mordell's review, Siegel's letter to Mordell, Diophantine geometry, and 20th century mathematics. Not. Am. Math. Soc. **42**, 339–350 (1995)
3703. Lang, S.: Some history of the Shimura-Taniyama conjecture. Not. Am. Math. Soc. **42**, 1301–1307 (1995)
3704. Lang, S., Néron, A.: Rational points of abelian varieties over function fields. Am. J. Math. **81**, 95–118 (1959)
3705. Lang, S., Tate, J.: Principal homogeneous spaces over abelian varieties. Am. J. Math. **80**, 659–684 (1958)
3706. Lang, S., Trotter, H.: Frobenius Distributions in GL_2-extensions. Lecture Notes in Math., vol. 504. Springer, Berlin (1976)
3707. Lang, S., Trotter, H.: Primitive points on elliptic curves. Bull. Am. Math. Soc. **83**, 289–292 (1977)
3708. Langevin, M.: Quelques applications de nouveaux résultats de van der Poorten. Sém. Delange–Pisot–Poitou **17**(12), 1–11 (1975/1976)
3709. Langevin, M.: Plus grand facteur premier d'entiers en progression arithmétique. Sém. Delange–Pisot–Poitou **18**(3), 1–7 (1976/1977)
3710. Langevin, M.: Facteurs premiers d'entiers en progression arithmétique. Acta Arith. **39**, 241–249 (1981)
3711. Langlands, R.P.: The volume of the fundamental domain for some arithmetical subgroups of Chevalley groups. In: Proc. Symposia Pure Math., vol. 9, pp. 143–148. Am. Math. Soc., Providence (1965)
3712. Langlands, R.P.: On Artin's L-functions. In: Complex Analysis, Rice University Studies, vol. 56, pp. 23–28. Rice University, Houston (1970)
3713. Langlands, R.P.: Problems in the theory of automorphic forms. In: Lecture Notes in Math., vol. 170, pp. 18–61. Springer, Berlin (1970)
3714. Langlands, R.P.: Automorphic representations, Shimura varieties, and motives. Ein Märchen. In: Proc. Symposia Pure Math., vol. 33, part 2, pp. 205–246. Am. Math. Soc., Providence (1977)
3715. Languasco, A., Perelli, A.: A pair correlation hypothesis and the exceptional set in Goldbach's problem. Mathematika **43**, 349–361 (1996)
3716. Laohakosol, V.: Dependence of arithmetic functions and Dirichlet series. Proc. Am. Math. Soc. **115**, 637–645 (1992)
3717. Lapin, A.I.: Subfields of hyperelliptic fields. I. Izv. Akad. Nauk SSSR, Ser. Mat. **28**, 953–988 (1964) (in Russian)
3718. Lapin, A.I.: On the rational points of an elliptic curve. Izv. Akad. Nauk SSSR, Ser. Mat. **29**, 701–716 (1965) (in Russian)
3719. Lapin, A.I.: The integral points on curves of genus $p > 1$. Izv. Akad. Nauk SSSR, Ser. Mat. **35**, 754–761 (1971) (in Russian)
3720. Laska, M.: Elliptic Curves over Number Fields with Prescribed Reduction Type. Vieweg, Wiesbaden (1983)
3721. Laska, M., Lorenz, M.: Rational points on elliptic curves over Q in elementary abelian 2-extensions of Q. J. Reine Angew. Math. **355**, 163–172 (1985)
3722. Laugwitz, D.: Bernhard Riemann 1826–1866. Wendepunkte in der Auffassung der Mathematik. Birkhäuser, Basel (1996) [English translation: Birkhäuser, 1999, reprint: 2008]

3723. Laurent, M., Mignotte, M., Nesterenko, Y. [Yu.V.]: Formes linéaires en deux logarithmes et déterminants d'interpolation. J. Number Theory **55**, 285–321 (1995)
3724. Laurinčikas, A.: Distribution of values of generating Dirichlet series of multiplicative functions. Liet. Mat. Rink. **22**, 56–63 (1982) (in Russian)
3725. Laurinčikas, A.: The universality theorem. Liet. Mat. Rink. **23**, 53–62 (1983) (in Russian)
3726. Laurinčikas, A.: The universality theorem, II. Liet. Mat. Rink. **24**, 113–121 (1984) (in Russian)
3727. Laurinčikas, A.: Limit Theorems for the Riemann Zeta-Function. Kluwer Academic, Dordrecht (1996)
3728. Laurinčikas, A.: Universality of the Lerch zeta function. Liet. Mat. Rink. **37**, 367–375 (1997) (in Russian)
3729. Laurinčikas, A., Garunkštis, R.: The Lerch Zeta-Function. Springer, Berlin (2003)
3730. Laurinčikas, A., Matsumoto, K.: The joint universality and the functional independence for Lerch zeta-functions. Nagoya Math. J. **157**, 211–227 (2000)
3731. Laurinčikas, A., Matsumoto, K.: The universality of zeta-functions attached to certain cusp forms. Acta Arith. **98**, 345–359 (2001)
3732. Laurinčikas, A., Matsumoto, K., Steuding, J.: The universality of L-functions associated with newforms. Izv. Ross. Akad. Nauk, Ser. Mat. **67**, 77–90 (2003) (in Russian)
3733. Lau, Y.K., Tsang, K.M.: Mean square of the remainder term in the Dirichlet divisor problem. J. Théor. Nr. Bordx. **7**, 75–92 (1995)
3734. Lavrik, A.F.: The functional equation for Dirichlet L-functions and the problem of divisors in arithmetic progressions. Izv. Akad. Nauk SSSR, Ser. Mat. **30**, 433–448 (1966) (in Russian)
3735. Lavrik, A.F.: Approximate functional equations of Dirichlet functions. Izv. Akad. Nauk SSSR, Ser. Mat. **32**, 134–185 (1968) (in Russian)
3736. Lavrik, A.F.: The moments of the number of classes of primitive quadratic forms with negative discriminant. Dokl. Akad. Nauk SSSR **197**, 32–35 (1971) (in Russian)
3737. Lavrik, A.F.: Methods of studying the distribution laws of prime numbers. Tr. Mat. Inst. Steklova **163**, 118–142 (1984) (in Russian)
3738. Lavrik, A.F., Edgorov, Ž.: The divisor problem in arithmetic progressions. Izv. Akad. Nauk Uzb. SSR., Ser. Fiz.-Mat. Nauk **17**(5), 14–18 (1973) (in Russian)
3739. Lavrik, A.F., Sobirov, A.Š.: On the remainder term in the elementary proof of the theorem about the number of primes. Dokl. Akad. Nauk SSSR **211**, 534–536 (1973) (in Russian)
3740. Laxton, R.R., Lewis, D.J.: Forms of degrees 7 and 11 over p-adic fields. In: Proc. Symposia Pure Math., vol. 8, pp. 16–21. Am. Math. Soc., Providence (1965)
3741. Le, M.H.: A note on the Diophantine equation $x^{2p} - Dy^2 = 1$. Proc. Am. Math. Soc. **107**, 27–34 (1989)
3742. Le, M.H.: On the generalized Ramanujan-Nagell equation $x^2 - D = p^n$. Acta Arith. **58**, 289–298 (1991)
3743. Le, M.H.: On the number of solutions of the generalized Ramanujan-Nagell equation $x^2 - D = 2^{n+2}$. Acta Arith. **60**, 149–167 (1991)
3744. Le, M.H.: Lower bounds for the solutions in the second case of Fermat's last theorem. Proc. Am. Math. Soc. **111**, 921–923 (1991)
3745. Le, M.H.: On the generalized Ramanujan-Nagell equation $x^2 - D = 2^{n+2}$. Trans. Am. Math. Soc. **334**, 805–825 (1992)
3746. Le, M.H.: On the number of solutions of the Diophantine equation $x^2 + D = p^n$. C. R. Acad. Sci. Paris **317**, 135–138 (1993)
3747. Le, M.H.: A note on the Diophantine equation $(x^m - 1)/(x - 1) = y^n + 1$. Proc. Camb. Philos. Soc. **116**, 385–389 (1994)
3748. Le, M.H.: The Diophantine equation $x^2 + D^m = 2^{n+2}$. Comment. Math. Univ. St. Pauli **43**, 127–133 (1994)
3749. Le, M.H.: On the Diophantine equation $(x^3 - 1)/(x - 1) = (y^n - 1)/(y - 1)$. Trans. Am. Math. Soc. **351**, 1063–1074 (1999)
3750. Le, M.H.: A conjecture concerning the exponential Diophantine equation $a^x + b^y = c^z$. Acta Arith. **106**, 345–353 (2003)

3751. Le, M.H.: A conjecture concerning the pure exponential Diophantine equation $a^x + b^y = c^z$. Acta Math. Sin. **21**, 943–948 (2005)
3752. Le, M.H.: A note on Jeśmanowicz' conjecture concerning primitive Pythagorean triplets. Acta Arith. **138**, 137–144 (2009)
3753. Lebesgue, V.A.: Sur l'impossibilié en nombres entiers de l'équation $x^m = y^2 + 1$. Nouv. Ann. Math. **9**, 178–181 (1850)
3754. Lebesgue, V.A.: Exercises d'analyse numerique. Leiber et Faraguet, Paris (1859)
3755. Lebesgue, V.A.: Introduction à la théorie des nombres. Mallet-Bachelier, Paris (1862)
3756. Lech, C.: A note on recurring series. Ark. Mat. **2**, 417–421 (1953)
3757. Ledermann, W.: Herbert Westren Turnbull. J. Lond. Math. Soc. **38**, 123–128 (1963)
3758. Lee, J.: The complete determination of wide Richaud-Degert types which are not 5 modulo 8 with class number one. Acta Arith. **140**, 1–29 (2009)
3759. Leech, J.: Note on the distribution of prime numbers. J. Lond. Math. Soc. **32**, 56–58 (1957)
3760. Leech, J.: Some sphere packings in higher space. Can. J. Math. **16**, 657–682 (1964)
3761. Leech, J.: Notes on sphere packings. Can. J. Math. **19**, 251–267 (1967)
3762. Leep, D.B.: Systems of quadratic forms. J. Reine Angew. Math. **350**, 109–116 (1984)
3763. Leep, D.B., Schmidt, W.M.: Systems of homogeneous equations. Invent. Math. **71**, 539–549 (1983)
3764. Leep, D.B., Schueller, L.M.: Classification of pairs of symmetric and alternating bilinear forms. Expo. Math. **17**, 385–414 (1999)
3765. Leep, D.B., Yeomans, C.C.: Quintic forms over p-adic fields. J. Number Theory **57**, 231–241 (1996)
3766. Legendre, A.M.: Recherches d'analyse indéterminée. Histoire de l'Academie Royale des Sciences de Paris (1785), 465–559, Paris, 1788
3767. Legendre, A.M.: Essai sur la théorie des nombres, Duprat, Paris (1798) [2nd ed. 1808, 3rd ed.: Théorie des nombres, Paris, 1830; German translation: Teubner, 1886, 1894]
3768. Lehman, R.S.: On the difference $\pi(x) - \mathrm{li}(x)$. Acta Arith. **11**, 397–410 (1966)
3769. Lehman, R.S.: Separation of zeros of the Riemann zeta-function. Math. Comput. **20**, 523–541 (1966)
3770. Lehman, R.S.: Factoring large integers. Math. Comput. **28**, 637–646 (1974)
3771. Lehmer, D.H.: Tests for primality by the converse of Fermat's theorem. Bull. Am. Math. Soc. **33**, 327–340 (1927)
3772. Lehmer, D.H.: A further note on the converse of Fermat's theorem. Bull. Am. Math. Soc. **34**, 54–56 (1928)
3773. Lehmer, D.H.: An extended theory of Lucas' functions. Ann. Math. **31**, 419–448 (1930)
3774. Lehmer, D.H.: Note on Mersenne numbers. Bull. Am. Math. Soc. **38**, 383–384 (1932)
3775. Lehmer, D.H.: A note on Fermat's last theorem. Bull. Am. Math. Soc. **38**, 723–724 (1932)
3776. Lehmer, D.H.: On Euler's totient function. Bull. Am. Math. Soc. **38**, 745–751 (1932)
3777. Lehmer, D.H.: On imaginary quadratic fields whose class-number is unity. Bull. Am. Math. Soc. **39**, 360 (1933)
3778. Lehmer, D.H.: Factorization of certain cyclotomic functions. Ann. Math. **34**, 461–479 (1933)
3779. Lehmer, D.H.: On Lucas's test for the primality of Mersenne's numbers. J. Lond. Math. Soc. **10**, 162–165 (1935)
3780. Lehmer, D.H.: On a conjecture of Ramanujan. J. Lond. Math. Soc. **11**, 114–118 (1936)
3781. Lehmer, D.H.: The lattice points of an n-dimensional tetrahedron. Duke Math. J. **7**, 341–353 (1940)
3782. Lehmer, D.H.: Properties of the coefficients of the modular invariant $J(\tau)$. Am. J. Math. **64**, 488–502 (1942)
3783. Lehmer, D.H.: Ramanujan's function $\tau(n)$. Duke Math. J. **10**, 483–492 (1943)
3784. Lehmer, D.H.: The Tarry-Escott problem. Scr. Math. **13**, 37–41 (1947)
3785. Lehmer, D.H.: The vanishing of Ramanujan's function $\tau(n)$. Duke Math. J. **14**, 429–433 (1947)
3786. Lehmer, D.H.: On the partition of numbers into squares. Am. Math. Mon. **55**, 476–481 (1948)

3787. Lehmer, D.H.: On the roots of the Riemann zeta-function. Acta Math. **95**, 291–298 (1956)
3788. Lehmer, D.H.: Extended computation of the Riemann zeta-function. Mathematika **3**, 102–108 (1956)
3789. Lehmer, D.H.: On the exact number of primes less than a given limit. Ill. J. Math. **3**, 381–388 (1959)
3790. Lehmer, D.H.: The primality of Ramanujan's tau-function. Am. Math. Mon. **72**, 15–18 (1965)
3791. Lehmer, D.H.: On Fermat's quotient, base two. Math. Comput. **36**, 289–290 (1981)
3792. Lehmer, D.H.: The mathematical work of Morgan Ward. Math. Comput. **61**, 307–311 (1993)
3793. Lehmer, D.H., Lehmer, E.: On the first case of Fermat's last theorem. Bull. Am. Math. Soc. **47**, 139–142 (1941)
3794. Lehmer, D.H., Lehmer, E.: Heuristic anyone. In: Studies in Mathematical Analysis and Related Topics, pp. 202–210. Stanford University Press, Stanford (1962)
3795. Lehmer, D.H., Lehmer, E.: On runs of residues. Proc. Am. Math. Soc. **13**, 102–106 (1962)
3796. Lehmer, D.H., Lehmer, E., Mills, W.H.: Pairs of consecutive power residues. Can. J. Math. **15**, 172–177 (1963)
3797. Lehmer, D.H., Lehmer, E., Mills, W.H., Selfridge, J.L.: Machine proof of a theorem on cubic residues. Math. Comput. **16**, 407–415 (1962)
3798. Lehmer, D.H., Lehmer, E., Vandiver, H.S.: An application of high-speed computing to Fermat's last theorem. Proc. Natl. Acad. Sci. USA **40**, 25–33 (1954)
3799. Lehmer, D.H., Masley, J.M.: Table of the cyclotomic class numbers $h^*(p)$ and their factors for $200 < p < 521$. Math. Comput. **32**, 577–582 (1978)
3800. Lehmer, D.H., Powers, R.E.: On factoring large numbers. Bull. Am. Math. Soc. **37**, 770–776 (1931)
3801. Lehmer, D.N.: Multiply perfect numbers. Ann. Math. **2**, 103–104 (1901)
3802. Lehmer, D.N.: A theorem in the theory of numbers. Bull. Am. Math. Soc. **13**, 501–502 (1907)
3803. Lehmer, D.N.: Factor Table for the First Ten Millions, Containing the Smallest Factor of Every Number not Divisible by 2, 3, 5, or 7 Between the Limits 0 and 10 017 000. Carnegie Institution of Washington, Washington (1909)
3804. Lehmer, D.N.: List of Prime Numbers from 1 to 10 006 721. Carnegie Institution of Washington, Washington (1914)
3805. Lehmer, D.N.: A theorem on factorization. Bull. Am. Math. Soc. **33**, 35–36 (1927)
3806. Lehmer, E.: On the location of Gauss sums. Math. Tables Other Aids Comput. **10**, 194–202 (1956)
3807. Lehner, J.: Further congruence properties of the Fourier coefficients of the modular invariant $j(\tau)$. Am. J. Math. **71**, 373–386 (1949)
3808. Lehner, J.: Proof of Ramanujan's partition conjecture for the modulus 11^α. Proc. Am. Math. Soc. **1**, 172–181 (1950)
3809. Lehner, J.: Discontinuous Groups and Automorphic Functions. Am. Math. Soc., Providence (1964)
3810. Lehner, J.: A Short Course in Automorphic Functions. Holt, Rinehart & Winston, New York (1966)
3811. Lekkerkerker, C.G.: Prime factors of the elements of certain sequences of integers, I. Indag. Math. **15**, 265–276 (1953)
3812. Lekkerkerker, C.G.: Prime factors of the elements of certain sequences of integers, II. Indag. Math. **15**, 277–280 (1953)
3813. Lekkerkerker, C.G.: On the Minkowski-Hlawka theorem. Indag. Math. **18**, 426–434 (1956)
3814. Lekkerkerker, C.G.: Geometry of Numbers. Wolters-Noordhoff, Groningen (1969); 2nd ed. (with P.M. Gruber), North-Holland, 1987
3815. Lemmermeyer, F.: A note on Pépin's counter-examples to Hasse principle for curves of genus 1. Abh. Math. Semin. Univ. Hamb. **69**, 335–345 (1999)
3816. Lemmermeyer, F.: Reciprocity Laws. Springer, Berlin (2000)

3817. Lenskoĭ, D.N., Linnik, Yu.V.: Nikolai Grigorevič Čudakov (on his sixtieth birthday). Usp. Mat. Nauk **20**(2), 237–240 (1965) (in Russian)
3818. Lenstra, A.K., Lenstra, H.W. Jr. (eds.): The Development of the Number Field Sieve. Lecture Notes in Math., vol. 1554. Springer, Berlin (1993)
3819. Lenstra, A.K., Lenstra, H.W. Jr., Lovász, L.: Factoring polynomials with integral coefficients. Math. Ann. **261**, 515–534 (1982)
3820. Lenstra, A.K., Lenstra, H.W. Jr., Manasse, M.S., Pollard, J.M.: Factorization of the ninth Fermat number. Math. Comput. **61**, 319–349 (1993); add.: **64**, 1357 (1995)
3821. Lenstra, H.W. Jr.: On Artin's conjecture and Euclid's algorithm in global fields. Invent. Math. **42**, 201–224 (1977)
3822. Lenstra, H.W. Jr.: Factoring integers with elliptic curves. Ann. Math. **126**, 649–673 (1987)
3823. Lenstra, H.W. Jr.: Solving the Pell equation. Not. Am. Math. Soc. **49**, 182–192 (2002)
3824. Lenstra, H.W. Jr., Pomerance, C.: A rigorous time bound for factoring integers. J. Am. Math. Soc. **5**, 483–516 (1992)
3825. Lenstra, H.W. Jr., Pomerance, C.: Primality testing with Gaussian periods. Preprint (2005)
3826. Lenz, H., et al.: Richard Rado, 1906–1989. Jahresber. Dtsch. Math.-Ver. **93**, 127–145 (1991)
3827. Leopoldt, H.-W.: Zur Arithmetik in abelschen Zahlkörpern. J. Reine Angew. Math. **209**, 54–71 (1962)
3828. Lepistö, T.: The first factor of the class number of the cyclotomic field $k(e^{2\pi i/p^n})$. Ann. Univ. Turku **70**, 1–7 (1963)
3829. Lepistö, T.: On the first factor of the class number of the cyclotomic field and Dirichlet's L-functions. Ann. Univ. Turku **387**, 1–53 (1966)
3830. Lepistö, T.: A zero-free region for certain L-functions. Ann. Acad. Sci. Fenn., Ser. A 1 Math. **576**, 1–13 (1974)
3831. Lepistö, T.: On the growth of the first factor of the class number of the prime cyclotomic field. Ann. Acad. Sci. Fenn., Ser. A 1 Math. **577**, 1–21 (1974)
3832. Lepowsky, J., Meurman, A.: An E_8-approach to the Leech lattice and the Conway group. J. Algebra **77**, 484–504 (1982)
3833. Lepson, B.: Certain best possible results in the theory of Schnirelman density. Proc. Am. Math. Soc. **1**, 592–594 (1950)
3834. Lerch, M.: Sur la fonction $\mathfrak{K}(w,x,s)=\sum_{k=0}^{\infty}\frac{\exp(2k\pi ix)}{(w+k)^s}$. Acta Math. **11**, 19–24 (1887/1888)
3835. Lerch, M.: Über die arithmetische Gleichung $\mathrm{Cl}(-\Delta)=1$. Math. Ann. **57**, 568–571 (1903)
3836. Lerch, M.: Question 1547. L'Intermédiaire Math. **11**, 145–146 (1904)
3837. Lerch, M.: Zur Theorie des Fermatschen Quotienten $\frac{a^{p-1}-1}{p}=q(a)$. Math. Ann. **60**, 471–490 (1905)
3838. Lesage, J.-L.: Différence entre puissances et carrés d'entiers. J. Number Theory **73**, 390–425 (1998)
3839. Lesca, J., Mendès France, M.: Ensembles normaux. Acta Arith. **17**, 273–282 (1970)
3840. Leudersdorf, C.: Some results in the elementary theory of numbers. Proc. Lond. Math. Soc. **20**, 199–212 (1888)
3841. Leung, M.-C., Liu, M.C.: On generalized quadratic equations in three prime variables. Monatshefte Math. **115**, 133–169 (1993)
3842. Lev, V.F.: Restricted set addition in groups, I. The classical setting. J. Lond. Math. Soc. **62**, 27–40 (2000)
3843. Lev, V.F.: Restricted set addition in groups, II. A generalization of the Erdős-Heilbronn conjecture. Electron. J. Comb. **7**(4), 1–10 (2000)
3844. Lev, V.F.: Restricted set addition in groups, III. Integer sumsets with generic restrictions. Period. Math. Hung. **42**, 89–98 (2001)
3845. Lev, V.F.: Restricted set addition in abelian groups: results and conjectures. J. Théor. Nr. Bordx. **17**, 181–193 (2005)
3846. Levenštein, V.I.: Bounds for packings in n-dimensional Euclidean space. Dokl. Akad. Nauk SSSR **245**, 1299–1303 (1979) (in Russian)

3847. LeVeque, W.J.: On the size of certain number-theoretic functions. Trans. Am. Math. Soc. **66**, 440–463 (1948)
3848. LeVeque, W.J.: On the equation $a^x - b^y = 1$. Am. J. Math. **74**, 325–331 (1952)
3849. LeVeque, W.J.: Mahler's U-numbers. J. Lond. Math. Soc. **28**, 220–229 (1953)
3850. LeVeque, W.J.: Topics in Number Theory. Addison-Wesley, Reading (1956)
3851. LeVeque, W.J.: On the equation $y^m = f(x)$. Acta Arith. **9**, 209–219 (1964)
3852. Levi, B.: Saggio per una teoria aritmetica della forme cubiche ternarie. Atti della Reale Accademia delle Scienze di Torino **43**, 99–120, 413–434, 672–681 (1908)
3853. Levi, B.: Sull'equazione indeterminata del 3[e] ordine. In: Atti del IV Congresso Internazionale di matematici, Roma, 6–11 Aprile 1908, vol. 2, pp. 173–177 (1908)
3854. Levi, B.: Un teorema del Minkowski sui sistemi di forme lineari a variabili intere. Rend. Circ. Mat. Palermo **31**, 318–340 (1911)
3855. Levin, B.V.: Distribution of 'almost primes' in polynomial sequences. Mat. Sb. **61**, 389–407 (1963) (in Russian)
3856. Levin, B.V.: The one-dimensional sieve. Acta Arith. **10**, 387–397 (1964) (in Russian)
3857. Levin, B.V., Faĭnleĭb, A.S.: Application of certain integral equations to questions of the theory of numbers. Usp. Mat. Nauk **22**(3), 119–197 (1967) (in Russian)
3858. Levin, B.V., Feldman, N.I., Šidlovskiĭ, A.B.: Alexander O. Gelfond. Acta Arith. **17**, 315–336 (1971)
3859. Levin, B.V., Timofeev, N.M.: An analytic method in probabilistic number theory. Učen. Zap. Vladimir. Gos. Ped. Inst. **38**, 57–150 (1971) (in Russian)
3860. Levin, B.V., Timofeev, N.M.: Distribution of arithmetic functions in mean in progressions (theorems of Vinogradov-Bombieri type). Mat. Sb. **125**, 558–572 (1984) (in Russian)
3861. Levin, M.: On the group of rational points on elliptic curves over function fields. Am. J. Math. **90**, 456–462 (1968)
3862. Levinson, N.: Wiener's life. Bull. Am. Math. Soc. **72**(1, pt. 2), 1–32 (1966)
3863. Levinson, N.: Ω-theorems for the Riemann zeta-function. Acta Arith. **20**, 317–330 (1972)
3864. Levinson, N.: More than one third of zeros of Riemann's zeta-function are on $\sigma = 1/2$. Adv. Math. **13**, 383–436 (1974)
3865. Levinson, N.: On the number of sign-changes of $\pi(x) - \mathrm{li}(x)$. Colloq. Math. Soc. János Bolyai **13**, 171–177 (1974)
3866. Lévy, P.: Sur les lois de probabilité dont dépend les quotients complets et incomplets d'une fraction continue. Bull. Soc. Math. Fr. **57**, 178–194 (1929)
3867. Lévy, P.: Sur le développement en fraction continue d'un nombre choisi au hasard. Compos. Math. **3**, 286–306 (1936)
3868. Lévy, P., Mandelbrojt, S., Malgrange, B., Malliavin, P.: La vie et l'oeuvre de Jacques Hadamard (1865–1963). L'Enseignement Mathematique Universite, Genéve (1967)
3869. Lewin, M.: An algorithm for a solution of a problem of Frobenius. J. Reine Angew. Math. **276**, 68–82 (1975)
3870. Lewis, D.J.: Cubic homogeneous polynomials over p-adic fields. Ann. Math. **56**, 473–478 (1952)
3871. Lewis, D.J.: Cubic forms over algebraic number fields. Mathematika **4**, 97–101 (1957)
3872. Lewis, D.J.: Cubic congruences. Mich. Math. J. **4**, 85–95 (1957)
3873. Lewis, D.J., Mahler, K.: On the representation of integers by binary forms. Acta Arith. **6**, 333–363 (1960/1961)
3874. Lewis, D.J., Montgomery, H.L.: On zeros of p-adic forms. Mich. Math. J. **30**, 83–87 (1983)
3875. Li, H.: Almost primes in short intervals. Sci. China **A37**, 1428–1441 (1994)
3876. Li, H.: The divisor problem for arithmetic progressions. Chin. Sci. Bull. **40**, 265–267 (1995)
3877. Li, H.: Goldbach numbers in short intervals. Sci. China Ser. A **38**, 641–652 (1995)
3878. Li, H.: Zero-free regions for Dirichlet L-functions. Q. J. Math. **50**, 13–23 (1999)
3879. Li, H.: The exceptional set of Goldbach numbers. Q. J. Math. **50**, 471–482 (1999)
3880. Li, H.: The exceptional set of Goldbach numbers, II. Acta Arith. **92**, 71–88 (2000)
3881. Li, H.: Small prime solutions of linear ternary equations. Acta Arith. **98**, 293–309 (2001)

3882. Li, H.: The exceptional set for the sum of a prime and a square. Acta Math. Acad. Sci. Hung. **99**, 123–141 (2003)
3883. Li, W.-C.W.: Newforms and functional equations. Math. Ann. **212**, 285–315 (1975)
3884. Libri, G.: Mémoire sur la théorie des nombres. J. Reine Angew. Math. **9**, 54–80, 169–188, 261–276 (1832)
3885. Lichtenbaum, S.: The period-index problem for elliptic curves. Am. J. Math. **90**, 1209–1223 (1968)
3886. Lichtenbaum, S.: On p-adic L-functions associated to elliptic curves. Invent. Math. **56**, 19–55 (1980)
3887. Lidl, R., Mullen, G.L., Turnwald, G.: Dickson Polynomials. Longman/Wiley, New York (1993)
3888. Lidl, R., Niederreiter, H.: Finite Fields. Addison–Wesley, Reading (1983); 2nd ed. 1986 [Reprint: 2008]
3889. Lieuwens, E.: Do there exist composite numbers M for which $k\varphi(M) = M - 1$ holds? Nieuw Arch. Wiskd. **18**, 165–169 (1970)
3890. Lin, K.-P., Yau, S.S.-T.: A sharp upper estimate of the number of integral points in a 5-dimensional tetrahedra. J. Number Theory **93**, 207–234 (2002)
3891. Lind, C.-E.: Untersuchungen über die rationalen Punkte der ebenen kubischen Kurven vom Geschlecht Eins. Dissertation, Uppsala (1940)
3892. Lindelöf, E.: Sur une formule sommatoire générale. Acta Math. **27**, 305–312 (1903)
3893. Lindelöf, E.: Quelques remarques sur la croissance de la fonction $\zeta(s)$. Bull. Sci. Math. **32**, 341–356 (1908)
3894. Lindelöf, E.: Robert Hjalmar Mellin. Acta Math. **61**, I–VI (1932)
3895. Lindemann, F.: Über die Zahl π. Math. Ann. **20**, 213–225 (1882)
3896. Lindemann, F.: Über den Fermatschen Satz betreffend die Unmöglichkeit der Gleichung $x^n = y^n + z^n$. Sitzungsber. - Bayer. Akad. Wiss., Philos.- Hist. Kl. **31**, 185–202 (1901)
3897. Lindemann, F.: Über das sogenannte letzte Fermatsche Theorem. Sitzungsber. - Bayer. Akad. Wiss., Philos.- Hist. Kl. **37**, 287–305 (1907)
3898. Lindemann, F.: Über den sogenannten letzten Fermatschen Satz. Teubner, Leipzig (1909)
3899. Lindsey, J.H.: Sphere packing in R^3. Mathematika **33**, 137–147 (1986)
3900. Linnik, Yu.V.: On certain results relating to positive ternary quadratic forms. Mat. Sb. **5**, 453–471 (1939)
3901. Linnik, Yu.V.: On the representations of large numbers by positive ternary quadratic forms. Izv. Akad. Nauk SSSR, Ser. Mat. **4**, 363–402 (1940) (in Russian) [[3929], vol. 1, pp. 84–122]
3902. Linnik, Yu.V.: The large sieve. Dokl. Akad. Nauk SSSR **30**, 290–292 (1941) (in Russian) [[3929], vol. 1, pp. 293–296]
3903. Linnik, Yu.V.: Remark on the smallest quadratic non-residue. Dokl. Akad. Nauk SSSR **36**, 131–132 (1941) (in Russian) [[3929], vol. 1, pp. 296–297]
3904. Linnik, Yu.V.: On Erdős's theorem on the addition of numerical sequences. Mat. Sb. **10**, 67–78 (1942) (in Russian)
3905. Linnik, Yu.V.: On a conditional theorem by J.E. Littlewood. Dokl. Akad. Nauk SSSR **37**, 142–144 (1942) (in Russian)
3906. Linnik, Yu.V.: On the representation of large numbers as a sum of seven cubes. Mat. Sb. **13**, 218–224 (1943) (in Russian) [[3929], vol. 1, pp. 122–128]
3907. Linnik, Yu.V.: An elementary solution of the problem of Waring by Schnirelman's method. Mat. Sb. **12**, 225–230 (1943) (in Russian) [[3929] vol. 1, pp. 297–303]
3908. Linnik, Yu.V.: On Weyl's sums. Mat. Sb. **12**, 28–39 (1943) (in Russian) [[3929], vol. 1, pp. 302–314]
3909. Linnik, Yu.V.: On the possibility to avoid Riemann's hypothesis in the investigation of primes in progressions. Dokl. Akad. Nauk SSSR **44**, 147–150 (1944) [[3929], vol. 1, pp. 323–327, Leningrad 1979]
3910. Linnik, Yu.V.: On the least prime in an arithmetic progression. I. The basic theorem. Mat. Sb. **15**, 139–178 (1944) (in Russian) [[3929], vol. 1, pp. 336–378]

3911. Linnik, Yu.V.: On the least prime in an arithmetic progression. II. The Deuring-Heilbronn phenomenon. Mat. Sb. **15**, 347–368 (1944) (in Russian) [[3929], vol. 1, pp. 378–399]
3912. Linnik, Yu.V.: On the possibility of a unique method in certain problems of "additive" and "distributive" prime number theorem. Dokl. Akad. Nauk SSSR **49**, 3–7 (1945) (in Russian) [[3929], vol. 2, pp. 7–13]
3913. Linnik, Yu.V.: On the density of zeros of L-series. Izv. Akad. Nauk SSSR, Ser. Mat. **10**, 35–46 (1946) (in Russian) [[3929], vol. 2, pp. 13–22]
3914. Linnik, Yu.V.: A new proof of the Goldbach–Vinogradov theorem. Mat. Sb. **19**, 3–8 (1946) (in Russian) [[3929], vol. 2, pp. 23–27]
3915. Linnik, Yu.V.: An elementary proof of Siegel's theorem based on the method of I.M. Vinogradov. Izv. Akad. Nauk SSSR, Ser. Mat. **14**, 327–342 (1950) [[3929], vol. 2, pp. 44–59]
3916. Linnik, Yu.V.: Prime numbers and powers of two. Tr. Mat. Inst. Steklova **38**, 152–169 (1951) (in Russian)
3917. Linnik, Yu.V.: Addition of prime numbers and powers of the same number. Mat. Sb. **32**, 3–60 (1953) (in Russian) [[3929], vol. 2, pp. 76–131]
3918. Linnik, Yu.V.: Asymptotical distribution of integral points on a sphere. Dokl. Akad. Nauk SSSR **96**, 909–912 (1954) (in Russian) [[3929], vol. 2, pp. 134–138]
3919. Linnik, Yu.V.: Asymptotical-geometric and ergodic properties of the set of integral points on a sphere. Mat. Sb. **43**, 257–276 (1957) (in Russian) [[3929], vol. 2, pp. 209–228]
3920. Linnik, Yu.V.: Dispersion of divisors and quadratic forms in progressions, and certain binary additive problems. Dokl. Akad. Nauk SSSR **120**, 960–962 (1958) (in Russian) [[3929], vol. 2, pp. 175–179]
3921. Linnik, Yu.V.: All large numbers are sums of a prime and two squares (On a problem of Hardy and Littlewood), I. Mat. Sb. **52**, 661–700 (1960) (in Russian) [[3929], vol. 2, pp. 217–289]
3922. Linnik, Yu.V.: All large numbers are sums of a prime and two squares (On a problem of Hardy and Littlewood), II. Mat. Sb. **53**, 3–38 (1961) (in Russian) [[3929], vol. 2, pp. 217–289]
3923. Linnik, Yu.V.: An asymptotic formula in an additive problem of Hardy-Littlewood. Izv. Akad. Nauk SSSR, Ser. Mat. **24**, 629–706 (1960) (in Russian)
3924. Linnik, Yu.V.: New variants and applications of the dispersion method in binary additive problems. Dokl. Akad. Nauk SSSR **137**, 1299–1302 (1961) (in Russian) [[3929], vol. 2, pp. 291–294]
3925. Linnik, Yu.V.: The Dispersion Method in Binary Additive Problems. Nauka, Leningrad (1961) (in Russian) [English translation: Am. Math. Soc., 1963]
3926. Linnik, Yu.V.: Additive problems and eigenvalues of the modular operators. In: Proc. ICM, Stockholm, 1962, pp. 270–284. Inst. Mittag-Leffler, Djursholm (1963)
3927. Linnik, Yu.V.: Ergodic Properties of Algebraic Fields. Izdat. Leningrad. Univ., Leningrad (1967) (in Russian) [English translation: Springer, 1968]
3928. Linnik, Yu.V.: Additive problems involving squares, cubes and almost primes. Acta Arith. **21**, 413–422 (1972) [[3929], vol. 1, pp. 284–292]
3929. Linnik, Yu.V.: Selected Works, vols. 1, 2. Nauka, Leningrad (1979–1980) (in Russian)
3930. Linnik, Yu.V., Malyšev, A.V.: On integral points on a sphere. Dokl. Akad. Nauk SSSR **89**, 209–211 (1953) (in Russian) [[3929], vol. 1, pp. 128–133]
3931. Linnik, Yu.V., Malyšev, A.V.: Applications of the arithmetic of quaternions to the theory of ternary quadratic forms and to the decomposition of numbers into cubes. Usp. Mat. Nauk **8**(5), 3–71 (1953); corr.: **10**(1), 243–244 (1955) (in Russian)
3932. Linnik, Yu.V., Vinogradov, A.I.: Hyperelliptic curves and the smallest prime quadratic residue. Dokl. Akad. Nauk SSSR **168**, 259–261 (1966) (in Russian)
3933. Liouville, J.: Sur de classes très eténdues de quantités dont la valeur n'est ni algébrique, ni même reductible à des irrationelles algébriques. C. R. Acad. Sci. Paris **18**, 883–885 (1844)
3934. Liouville, J.: Sur de classes très eténdues de quantités dont la valeur n'est ni algébrique, ni même reductible à des irrationelles algébriques, II. J. Math. Pures Appl. **16**, 133–142 (1851)

3935. Liouville, J.: Note de M. Liouville. J. Math. Pures Appl. **7**, 44–48 (1862)
3936. Liouville, J.: Nombre des représentations d'un entier quelconque sous la forme d'une somme de dix carrés. J. Math. Pures Appl. **11**, 1–8 (1866)
3937. Lipschitz, R.: Untersuchungen der Eigenschaften einer Gattung von unendlichen Reihen. J. Reine Angew. Math. **105**, 127–156 (1889)
3938. Listratenko, Yu.M.: Estimation of the number of summands in Waring's problem for the integer parts of fixed powers of nonnegative numbers. Chebyshev. Sb. **3**, 71–77 (2002)
3939. Littlewood, J.E.: Quelques conséquences de l'hypothése que la fonction $\zeta(s)$ de Riemann n'a pas de zéros dans la demi-plan $R(s) > 1/2$. C. R. Acad. Sci. Paris **154**, 263–266 (1912)
3940. Littlewood, J.E.: Sur la distribution des nombres premiers. C. R. Acad. Sci. Paris **158**, 1869–1872 (1914)
3941. Littlewood, J.E.: Researches in the theory of the Riemann ζ-function. Proc. Lond. Math. Soc. **20**, xxii–xxviii (1922)
3942. Littlewood, J.E.: On the zeros of the Riemann zeta-function. Proc. Camb. Philos. Soc. **22**, 295–318 (1924)
3943. Littlewood, J.E.: On the Riemann zeta-function. Proc. Lond. Math. Soc. **24**, 175–201 (1925)
3944. Littlewood, J.E.: Mathematical notes: 3; On a theorem concerning the distribution of prime numbers. J. Lond. Math. Soc. **2**, 41–45 (1927)
3945. Littlewood, J.E.: On the class-number of the corpus $P(\sqrt{-k})$. Proc. Lond. Math. Soc. **27**, 358–372 (1928)
3946. Littlewood, J.E.: A Mathematician's Miscellany. Methuen, London (1953); 2nd ed. ("Littlewood's miscellany"), Cambridge, 1986
3947. Littlewood, J.E.: Some Problems in Real and Complex Analysis. Heath, Lexington (1968)
3948. Littlewood, J.E., Walfisz, A.: The lattice points in a circle. Proc. R. Soc. Lond. Ser. A, Math. Phys. Sci. **106**, 478–488 (1924)
3949. Litver, E.L., Yudina, G.E.: Primitive roots for the first million primes and their powers. In: Matematičeskiĭanaliz i ego priloženiya, vol. 3, pp. 106–109. Izdat. Rostov. Univ., Rostov na Donu (1971) (in Russian)
3950. Liu, H.Q.: On the prime twins problem. Sci. China **A33**, 281–298 (1990)
3951. Liu, H.Q.: On square-full numbers in short intervals. Acta Math. Sin. New Ser. **6**, 148–164 (1990)
3952. Liu, H.Q.: The greatest prime factor of the integers in an interval. Acta Arith. **65**, 301–328 (1993)
3953. Liu, H.Q.: Lower bounds for sums of Barban-Davenport-Halberstam type. J. Reine Angew. Math. **438**, 163–174 (1993); suppl.: Manuscr. Math., **87**, 1995, 159–166
3954. Liu, H.Q.: The number of squarefull numbers in an interval. Acta Arith. **64**, 129–149 (1993)
3955. Liu, H.Q.: The distribution of square-full integers. Ark. Mat. **32**, 449–454 (1994)
3956. Liu, H.Q.: The distribution of 4-full numbers. Acta Arith. **67**, 165–176 (1994)
3957. Liu, H.Q.: Almost primes in short intervals. J. Number Theory **57**, 303–322 (1996)
3958. Liu, H.Q.: On a fundamental result in van der Corput's method of estimating exponential sums. Acta Arith. **90**, 357–370 (1999)
3959. Liu, H.Q., Rivat, J.: On the Pjateckiĭ-Šapiro prime number theorem. Bull. Lond. Math. Soc. **24**, 143–147 (1992)
3960. Liu, H.-Q., Wu, J.: Numbers with a large prime factor. Acta Arith. **89**, 163–187 (1999)
3961. Liu, J., Liu, M.C., Wang, T.Z.: The number of powers of 2 in the representation of large even integer, I. Sci. China **41**, 386–398 (1998)
3962. Liu, J., Liu, M.C., Wang, T.Z.: The number of powers of 2 in the representation of large even integer, II. Sci. China **41**, 1255–1271 (1998)
3963. Liu, J., Liu, M.C., Wang, T.Z.: On the almost Goldbach problem of Linnik. J. Théor. Nr. Bordx. **11**, 133–147 (1999)
3964. Liu, J., Zhan, T.: Distribution of integers that are sums of three squares of primes. Acta Arith. **98**, 207–228 (2001)
3965. Liu, J., Zhan, T.: The exceptional set in Hua's theorem for three squares of primes. Acta Math. Sin. **21**, 335–350 (2005)

3966. Liu, M.C.: On a result of Davenport and Halberstam. J. Number Theory **1**, 385–389 (1969)
3967. Liu, M.C.: On a theorem of Heilbronn concerning the fractional part of θn^2. Can. J. Math. **22**, 784–788 (1970)
3968. Liu, M.C.: Simultaneous approximation of additive forms. Trans. Am. Math. Soc. **206**, 361–373 (1975)
3969. Liu, M.C., Wang, T.Z.: On the Vinogradov bound in the three primes Goldbach conjecture. Acta Arith. **105**, 133–175 (2002)
3970. Ljunggren, W.: Einige Eigenschaften der Einheiten reeller quadratischer und rein-biquadratischer Zahlkörper. Mit Anwendung auf die Lösung einer Klasse unbestimmter Gleichungen 4. Grades. Skr. Norske Vidensk. Akad. Oslo I 1936, Nr. 12, 1–73
3971. Ljunggren, W.: A note on simultaneous Pell equations. Norsk Mat. Tidsskr. **23**, 132–138 (1941) (in Norwegian)
3972. Ljunggren, W.: Einige Bemerkungen über die Darstellung ganzer Zahlen durch binäre kubische Formen mit positiver Diskriminante. Acta Math. **75**, 1–21 (1942)
3973. Ljunggren, W.: Über die Gleichung $x^4 - Dy^2 = 1$. Arch. Math. Naturvidensk. **45**, 61–70 (1942)
3974. Ljunggren, W.: Zur Theorie der Gleichung $x^2 + 1 = Dy^4$. Avh. Norske Vid. Akad. Oslo, I, 1942, nr. 5, 1–27
3975. Ljunggren, W.: Some theorems on indeterminate equations of the form $(x^n - 1)/(x - 1) = y^q$. Norsk Mat. Tidsskr. **25**, 17–20 (1943) (in Norwegian)
3976. Ljunggren, W.: Einige Sätze über unbestimmte Gleichungen von der Form $Ax^4 + Bx^2 + C = Dy^2$. Skr. Norske Vid. Akad., 1943, nr. 9, 1–53
3977. Ljunggren, W.: Some remarks on the diophantine equations $x^2 - Dy^4 = 1$ and $x^4 - Dy^2 = 1$. J. Lond. Math. Soc. **41**, 542–544 (1966)
3978. Ljunggren, W.: On the diophantine equation $Ax^4 - By^2 = C$ $(C = 1, 4)$. Math. Scand. **21**, 149–158 (1967)
3979. Ljunggren, W.: A remark concerning the paper "On the diophantine equation $Ax^4 - By^2 = C$ $(C = 1, 4)$". Math. Scand. **22**, 282 (1968)
3980. Locher, H.: On the number of good approximations of algebraic numbers by algebraic numbers of bounded degree. Acta Arith. **89**, 97–122 (1999)
3981. Lochs, G.: Über die Lösungszahl einer linearen, diophantischen Ungleichung. Jahresber. Dtsch. Math.-Ver. **54**, 41–51 (1950)
3982. Lochs, G.: Über die Anzahl der Gitterpunkte in einem Tetraeder. Monatshefte Math. **56**, 233–239 (1952)
3983. Loh, W.K.A.: On Hua's lemma. Bull. Aust. Math. Soc. **50**, 451–458 (1994)
3984. Lomadze, G.A.: On the representation of numbers by sums of squares. Tr. Tbil. Inst. Razmadze **16**, 231–275 (1948) (in Russian)
3985. Lomadze, G.A.: On the representation of numbers by sums of squares. Tr. Tbil. Inst. Razmadze **20**, 47–87 (1954) (in Russian)
3986. Lomadse, G.A.: The scientific work of Arnold Walfisz. Acta Arith. **10**, 227–237 (1964)
3987. London, F.: Ueber Doppelfolgen und Doppelreihen. Math. Ann. **53**, 322–370 (1900)
3988. London, H., Finkelstein, R.: On Fibonacci and Lucas numbers which are perfect powers. Fibonacci Q. **7**, 476–481 (1969); corr. vol. 8, 1970, p. 248
3989. London, H., Finkelstein, R.: On Mordell's Equation $y^2 - k = x^3$. Bowling Green State University Press, Bowling Green (1973)
3990. Lorentz, G.G.: On a problem of additive number theory. Proc. Am. Math. Soc. **5**, 838–841 (1954)
3991. Lou, S.T., Yao, Q.: A Chebychev's type of prime number theorem in a short interval, II. Hardy-Ramanujan J. **15**, 1–33 (1992)
3992. Lou, S.T., Yao, Q.: The number of primes in a short interval. Hardy-Ramanujan J. **16**, 21–43 (1993)
3993. Louboutin, R.: Sur la mesure de Mahler d'un nombre algébrique. C. R. Acad. Sci. Paris **296**, 707–708 (1083)

3994. Louboutin, S.: Majoration au point 1 des fonctions L associées aux caractères de Dirichlet primitifs, ou au caractère d'une extension quadratique d'un corps quadratique imaginaire principal. J. Reine Angew. Math. **419**, 213–219 (1991)
3995. Louboutin, S.: Minoration au point 1 des fonctions L et détermination des corps sextiques abéliens totalement imaginaires principaux. Acta Arith. **62**, 109–124 (1992)
3996. Louboutin, S.: Majorations explicites de $|L(1, \chi)|$. C. R. Acad. Sci. Paris **316**, 11–14 (1993)
3997. Louboutin, S.: Majorations explicites de $|L(1, \chi)|$, II. C. R. Acad. Sci. Paris **323**, 443–446 (1996)
3998. Louboutin, S.: Majorations explicites de $|L(1, \chi)|$, III. C. R. Acad. Sci. Paris **332**, 95–98 (2001)
3999. Louboutin, S.: Majorations explicites de $|L(1, \chi)|$, IV. C. R. Acad. Sci. Paris **334**, 625–628 (2002)
4000. Louboutin, S.: Explicit upper bounds for $|L(1, \chi)|$ for primitive even Dirichlet characters. Acta Arith. **101**, 1–18 (2002)
4001. Louboutin, S.: Explicit upper bounds for values at $s = 1$ of Dirichlet L-series associated with primitive even characters. J. Number Theory **104**, 118–131 (2004)
4002. Louboutin, S.: Explicit upper bounds for $|L(1, \chi)|$ for primitive even Dirichlet characters χ. Q. J. Math. **55**, 57–68 (2004)
4003. Louboutin, S., Mollin, R.A., Williams, H.C.: Class numbers of real quadratic fields, continued fractions, reduced ideals, prime-producing quadratic polynomials and quadratic residue covers. Can. J. Math. **44**, 824–842 (1992)
4004. Low, L., Pitman, J., Wolff, A.: Simultaneous diagonal congruences. J. Number Theory **29**, 31–59 (1988)
4005. Low, M.: Real zeros of Dedekind zeta function of an imaginary quadratic field. Acta Arith. **14**, 117–140 (1968)
4006. Loxton, J.H.: On the determination of Gauss sums. Sém. Delange–Pisot–Poitou **18**(Exp. 27), 1–12 (1976/1977)
4007. Loxton, J.H.: Some conjectures concerning Gauss sums. J. Reine Angew. Math. **297**, 153–158 (1978)
4008. Loxton, J.H.: Estimates for complete multiple exponential sums. Acta Arith. **92**, 277–290 (2000)
4009. Loxton, J.H., Smith, R.A.: On Hua's estimate for exponential sums. J. Lond. Math. Soc. **26**, 15–20 (1982)
4010. Loxton, J.H., van der Poorten, A.J.: Transcendence and algebraic independence by a method of Mahler. In: Boker, A., Mosser, D.W. (eds.) Transcendence Theory: Advances and Applications, pp. 211–226. Academic Press, London (1977)
4011. Loxton, J.H., Vaughan, R.C.: The estimation of complete exponential sums. Can. Math. Bull. **28**, 440–454 (1985)
4012. Lü, G.: On Romanoff's constant and its generalized problem. Adv. Math. (China) **36**, 94–100 (2007)
4013. Lü, G.: Number of solutions of certain congruences. Acta Arith. **140**, 317–328 (2009)
4014. Lu, H.W.: The length of the period of the simple continued fraction of a real quadratic irrational number. Acta Math. Sin. **29**, 433–443 (1986) (in Chinese)
4015. Lu, H.W.: A new application of Davenport's method. Sci. China **A34**, 385–394 (1991)
4016. Lu, M.G.: Improvement on a theorem of Roth's. J. China Univ. Sci. Tech., 1982, suppl. 1, 13–18
4017. Lu, M.G.: Estimate of a complete trigonometric sum. Sci. Sin. **A28**, 561–578 (1985)
4018. Lu, M.G.: On Waring's problem for cubes and fifth power. Sci. China Ser. A **36**, 641–662 (1993)
4019. Lu, M.G., Yu, H.B.: Waring's problem for cubes and biquadrates. Sci. China **38**, 257–266 (1995)
4020. Lu, W.C.: On the elementary proof of the prime number theorem with a remainder term. Rocky Mt. J. Math. **29**, 979–1053 (1999)
4021. Lu, W.C.: Exceptional set of Goldbach number. J. Number Theory **130**, 2359–2392 (2010)

4022. Lubkin, S.: On a conjecture of André Weil. Am. J. Math. **89**, 443–548 (1967)
4023. Lubkin, S.: A p-adic proof of Weil' conjectures. Ann. Math. **87**, 105–194, 195–255 (1968)
4024. Luca, F., Mukhopadhyay, A., Srinivas, K.: Some results on Oppenheim's "Factorisatio Numerorum" function. Acta Arith. **142**, 41–50 (2010)
4025. Lucas, É.: Note sur l'application des séries récurrentes á la recherche de la loi de distribution des nombres premiers. C. R. Acad. Sci. Paris **82**, 165–167 (1876)
4026. Lucas, É.: Sur la décomposition des nombres en bicarrés. Nouv. Corresp. Math. **4**, 323–325 (1878)
4027. Lucas, É.: Sur un théoréme de M. Liouville, concernant la décomposition des nombres en bicarrés. Nouv. Ann. Math. **17**, 536–537 (1878)
4028. Lucas, É.: Théorie des fonctions numériques simplement périodiques. Am. J. Math. **1**, 184–240, 289–321 (1878)
4029. Lucas, É.: Sur le neuviéme nombre parfait. Mathesis **7**, 45–46 (1887)
4030. Lucas, É.: Théorie des Nombres. Gauthier-Villars, Paris (1891) [Reprint: Blanchard, Paris, 1961]
4031. Lucht, L.[G.]: Weighted relationship theorems and Ramanujan expansions. Acta Arith. **70**, 25–42 (1995)
4032. Lucht, L.G., Reifenrath, K.: Mean-value theorems in arithmetic semigroups. Acta Math. Acad. Sci. Hung. **93**, 27–57 (2001)
4033. Lusternik, L.A.: Die Brunn-Minkowskische Ungleichung für beliebige meßbare Mengen. Dokl. Akad. Nauk SSSR **3**, 55–58 (1935)
4034. Lutz, É.: Sur l'équation $y^2 = x^3 - Ax - B$ dans les corps p-adic. J. Reine Angew. Math. **177**, 238–247 (1937)
4035. Lützen, J.: Joseph Liouville 1809–1882: Master of Pure and Applied Mathematics. Springer, Berlin (1990)
4036. Lygeros, N., Rozier, O.: A new solution to the equation $\tau(p) \equiv 0 \pmod{p}$. J. Integer Seq. **13**, art. 10.7.4 (2010)
4037. Maass, H.: Über Gruppen von hyperabelschen Transformationen. SBer. Heidelberg. Akad. Wiss., 1940, nr. 2, 1–26
4038. Maass, H.: Über die Darstellung total positiver Zahlen des Körpers $R(\sqrt{5})$ als Summe von drei Quadraten. Abh. Math. Semin. Hansischen Univ. **14**, 185–191 (1941)
4039. Maass, H.: Über eine neue Art von nichtanalytischen automorphen Funktionen und die Bestimmung Dirichletscher Reihen durch Funktionalgleichungen. Math. Ann. **121**, 141–183 (1949)
4040. Maass, H.: Modulformen zweiten Grades und Dirichletreihen. Math. Ann. **122**, 90–108 (1950)
4041. Maass, H.: Über die Darstellung der Modulformen n-ten Grades durch Poincarésche Reihen. Math. Ann. **123**, 125–151 (1951)
4042. Maass, H.: Die Primzahlen in der Theorie der Siegelschen Modulfunktionen. Math. Ann. **124**, 87–122 (1951)
4043. Maass, H.: Die Bestimmung der Dirichletreihen mit Grössencharakteren zu den Modulformen n-ten Grades. J. Indian Math. Soc. **19**, 1–23 (1955)
4044. Maass, H.: Lectures on Modular Functions of One Complex Variable. Tata, Bombay (1964)
4045. Maass, H.: Siegel's modular forms and Dirichlet series. In: Lecture Notes in Math., vol. 216. Springer, Berlin (1971)
4046. Maass, H.: Über eine Spezialschar von Modulformen zweiten Grades. Invent. Math. **52**, 95–104 (1979)
4047. Maass, H.: Über eine Spezialschar von Modulformen zweiten Grades, II. Invent. Math. **53**, 249–253 (1979)
4048. Maass, H.: Über eine Spezialschar von Modulformen zweiten Grades, III. Invent. Math. **53**, 255–265 (1979)
4049. Macdonald, I.G.: The volume of a lattice polyhedron. Proc. Camb. Philos. Soc. **59**, 719–726 (1963)
4050. Macdonald, I.G.: Polynomials associated with finite cell-complexes. J. Lond. Math. Soc. **4**, 181–192 (1971)

4051. Macintyre, A.: Rationality of p-adic Poincaré series: uniformity in p. Ann. Pure Appl. Log. **49**, 31–74 (1990)
4052. Maclaurin, C.: Treatise of Fluxions. Ruddimans, Edinburgh (1742)
4053. MacLeod, R.A.: An extremal result for divisor functions. J. Number Theory **23**, 365–366 (1986)
4054. MacLeod, R.A.: Extreme values for divisor functions. Bull. Aust. Math. Soc. **37**, 447–465 (1988)
4055. MacMahon, P.A.: Combinatory Analysis. Cambridge University Press, Cambridge (1915) [Reprint: Chelsea, 1960]
4056. MacMahon, P.A.: Dirichlet series and the theory of partitions. Proc. Lond. Math. Soc. **22**, 404–411 (1923)
4057. Madritsch, M.G., Thuswaldner, J., Tichy, R.F.: Normality of numbers generated by the values of entire functions. J. Number Theory **128**, 1127–1145 (2008)
4058. Magnus, W.: Über den Beweis des Hauptidealsatzes. J. Reine Angew. Math. **170**, 235–240 (1934)
4059. Mahlburg, K.: Partition congruences and the Andrews-Garvan-Dyson crank. Proc. Natl. Acad. Sci. USA **102**, 15373–15376 (2005)
4060. Mahler, K.: Über einen Satz von Mellin. Math. Ann. **100**, 384–398 (1928)
4061. Mahler, K.: Arithmetische Eigenschaften der Lösungen einer Klasse der Funktionalgleichungen. Math. Ann. **101**, 342–366 (1929); corr. **103**, 532 (1930)
4062. Mahler, K.: Über das Verschwinden von Potenzreihen mehrerer Veränderlichen in speziellen Punktfolgen. Math. Ann. **103**, 573–587 (1930)
4063. Mahler, K.: Arithmetische Eigenschaften einer Klasse der transzendental-transzendenter Funktionen. Math. Z. **32**, 545–585 (1930)
4064. Mahler, K.: Zur Approximation der Exponentialfunktion und des Logarithmus, I. J. Reine Angew. Math. **166**, 118–136 (1932)
4065. Mahler, K.: Zur Approximation der Exponentialfunktion und des Logarithmus, II. J. Reine Angew. Math. **166**, 137–150 (1932)
4066. Mahler, K.: Über das Maßder Menge aller S-Zahlen. Math. Ann. **106**, 131–139 (1932)
4067. Mahler, K.: Ein Beweis der Transzendenz der P-adischen Exponentialfunktion. J. Reine Angew. Math. **169**, 61–66 (1932)
4068. Mahler, K.: Zur Approximation algebraischer Zahlen, I. Über den grössten Primteiler binärer Formen. Math. Ann. **107**, 691–730 (1933)
4069. Mahler, K.: Zur Approximation algebraischer Zahlen, II. Über die Anzahl der Darstellungen ganzer Zahlen durch Binärformen. Math. Ann. **108**, 37–55 (1933)
4070. Mahler, K.: Zur Approximation algebraischer Zahlen, III. Acta Math. **62**, 91–166 (1934)
4071. Mahler, K.: Über den größten Primteiler der Polynome $x^2 \pm 1$. Arch. Math. Naturvidensk. **41**(1), 1–8 (1933)
4072. Mahler, K.: Über Diophantische Approximationen im Gebiet der p-adischen Zahlen. Jahresber. Dtsch. Math.-Ver. **44**, 250–255 (1934)
4073. Mahler, K.: Zur Approximation p-adischer Irrationalzahlen. Nieuw Arch. Wiskd. **18**, 22–34 (1934)
4074. Mahler, K.: Über rationalen Punkte auf Kurven vom Geschlecht Eins. J. Reine Angew. Math. **170**, 168–178 (1934)
4075. Mahler, K.: Über eine Klasseneinteilung der P-adischen Zahlen. Mathematica B **3**, 177–185 (1935)
4076. Mahler, K.: Über transzendente P-adische Zahlen. Compos. Math. **2**, 259–275 (1935); corr. **8**, 112 (1951)
4077. Mahler, K.: Über den größten Primteiler spezieller Polynomen zweiten Grades. Arch. Math. Naturvidensk. **41**(6), 1–26 (1935)
4078. Mahler, K.: Eine arithmetische Eigenschaft der Taylor-Koeffizienten rationaler Funktionen. Proc. Akad. Wet. Amst. **38**, 50–60 (1935)
4079. Mahler, K.: On the lattice points on curves of genus 1. Proc. Lond. Math. Soc. **39**, 431–466 (1935)

4080. Mahler, K.: Note on hypothesis K of Hardy and Littlewood. J. Lond. Math. Soc. **11**, 136–138 (1936)
4081. Mahler, K.: Arithmetische Eigenschaften einer Klasse von Dezimalbrüchen. Proc. Akad. Wet. Amst. **40**, 421–428 (1937)
4082. Mahler, K.: Ein P-adisches Analogon zu einem Satz von Tchebycheff. Mathematica **7**, 2–6 (1938)
4083. Mahler, K.: On lattice points in n-dimensional star bodies, I. Existence theorems. Proc. R. Soc. Lond. Ser. A, Math. Phys. Sci. **187**, 151–187 (1946)
4084. Mahler, K.: The theorem of Minkowski-Hlawka. Duke Math. J. **13**, 611–621 (1946)
4085. Mahler, K.: On the minimum determinant and the circumscribed hexagons of a convex domain. Indag. Math. **9**, 326–337 (1947)
4086. Mahler, K.: On the approximation of π. Indag. Math. **15**, 30–42 (1953)
4087. Mahler, K.: On the approximation of logarithms of algebraic numbers. Philos. Trans. R. Soc. Lond. Ser. A, Math. Phys. Sci. **245**, 371–398 (1953)
4088. Mahler, K.: On the fractional part of the powers of a rational number, II. Mathematika **4**, 122–124 (1957)
4089. Mahler, K.: An application of Jensen's formula to polynomials. Mathematika **7**, 98–100 (1960)
4090. Mahler, K.: On a theorem of Bombieri. Indag. Math. **23**, 245–253 (1960); corr. **23**, 141
4091. Mahler, K.: Remarks on a paper by W. Schwarz. J. Number Theory **1**, 512–521 (1969)
4092. Mahler, K.: On the order function of a transcendental number. Acta Arith. **18**, 63–76 (1971)
4093. Mahler, K.: Lectures on Transcendental Numbers. Lecture Notes in Math., vol. 546. Springer, Berlin (1976)
4094. Mahler, K.: On some irrational decimal fractions. J. Number Theory **13**, 268–269 (1981)
4095. Mahler, K., Szekeres, G.: On the approximation of real numbers by roots of integers. Acta Arith. **12**, 315–320 (1966/1967)
4096. Mahoney, M.S.: The Mathematical Career of Pierre de Fermat. Princeton University Press, Princeton (1973); 2nd ed. 1994
4097. Mai, L., Murty, M.R.: A note on quadratic twists of an elliptic curve. In: Elliptic Curves and Related Topics. CRM Proc. Lect. Notes, vol. 4, pp. 121–124. Am. Math. Soc., Providence (1994)
4098. Maier, H.: Primes in short intervals. Mich. Math. J. **32**, 221–225 (1985)
4099. Maier, H.: Small differences between prime numbers. Mich. Math. J. **35**, 323–344 (1988)
4100. Maier, H., Montgomery, H.L.: The sum of the Möbius function. Bull. Lond. Math. Soc. **41**, 213–226 (2009)
4101. Maier, H., Pomerance, C.: On the number of distinct values of Euler's φ-function. Acta Arith. **49**, 263–275 (1988)
4102. Maier, H., Pomerance, C.: Unusually large gaps between consecutive primes. Trans. Am. Math. Soc. **322**, 201–237 (1990)
4103. Maier, H., Sankaranarayanan, A.: On a certain general exponential sum. Int. J. Number Theory **1**, 183–192 (2005)
4104. Maier, H., Sankaranarayanan, A.: On certain exponential sums over primes. J. Number Theory **129**, 1669–1677 (2009)
4105. Maier, H., Tenenbaum, G.: On the set of divisors of an integer. Invent. Math. **76**, 121–128 (1984)
4106. Maillet, E.: Sur la décomposition d'un nombre entier en une somme de cubes d'entiers positifs. In: C.R. 24 Session Assoc. Française pour l'avancement des Sciences, Paris, vol. 2, pp. 242–247 (1895)
4107. Maillet, E.: Quelques extensions du théorème de Fermat sur les nombres polygones. J. Math. Pures Appl. **2**, 363–380 (1896)
4108. Maillet, E.: Sur l'équation indéterminée $ax^{\lambda^t} + by^{\lambda^t} = cz^{\lambda^t}$. In: C.R. 26 Session Assoc. Française pour l'avancement des Sciences, Paris, vol. 2, pp. 156–168 (1898)
4109. Maillet, E.: Introduction à la théorie des nombres transcendants et des propriétes arithmétiques des fonctions. Gauthier-Villars, Paris (1906)

4110. Maillet, E.: Sur la décomposition d'un entier en une somme de puissances huitièmes d'entiers (Problème de Waring). Bull. Soc. Math. Fr. **36**, 69–77 (1908)
4111. Maillet, E.: Question 4269. L'Intermédiaire Math. **20**, 218 (1913)
4112. Maillet, E.: Sur un théorème de M. Axel Thue. Nouv. Ann. Math. **16**, 338–345 (1916)
4113. Maillet, E.: Détermination des points entiers des courbes algébriques unicursales à coefficients entiers. C. R. Acad. Sci. Paris **168**, 217–220 (1919)
4114. Maillet, E.: Détermination des points entiers des courbes algébriques unicursales à coefficients entiers dans l'espace à k dimensions. J. Éc. Polytech. **20**, 115–156 (1920)
4115. Malliavin, P.: Sur le reste de la loi asymptotique de répartition des nombres premiers généralisés de Beurling. Acta Math. **106**, 281–298 (1961)
4116. Malmstén, C.J.: De integralibus quibusdam definitis, seriebusque infinitis. J. Reine Angew. Math. **38**, 1–39 (1849)
4117. Malo, E.: L'Intermédiaire Math., **21**, 1914, 173–176
4118. Malyšev, A.V.: An asymptotic law for the representation of numbers by some positive ternary quadratic forms. Dokl. Akad. Nauk SSSR **93**, 771–774 (1953); corr.: **95**, 700 (1954) (in Russian)
4119. Malyšev, A.V.: On integral points on ellipsoids. Vestnik Leningrad. Univ. Ser. Mat. Mekh. Astronom. **11**(19), 18–34 (1956) (in Russian)
4120. Malyšev, A.V.: Theory of ternary quadratic forms. I. Arithmetic of hermitions. Vestnik Leningrad. Univ. Ser. Mat. Mekh. Astronom. **14**, 55–71 (1959) (in Russian)
4121. Malyšev, A.V.: On the theory of ternary quadratic forms. III. On the representation of large numbers by positive quadratic forms of odd coprime invariants. Vestnik Leningrad. Univ. Ser. Mat. Mekh. Astronom. **15**(1), 70–84 (1960) (in Russian)
4122. Malyšev, A.V.: On the representation of integers by positive quadratic forms. Tr. Mat. Inst. Steklova **65**, 1–212 (1962) (in Russian)
4123. Malyshev, A.V.: Yu.V. Linnik's works in number theory. Acta Arith. **27**, 3–10 (1975)
4124. Malyšev, A.V.: Estimates for the inhomogeneous arithmetic minimum of a product of linear forms. Zap. Nauč. Semin. LOMI **160**, 138–150 (1986) (in Russian)
4125. von Mangoldt, H.: Zu Riemann's Abhandlung "Ueber die Anzahl der Primzahlen unter einer gegebener Größe". J. Reine Angew. Math. **114**, 255–305 (1895)
4126. von Mangoldt, H.: Beweis der Gleichung $\sum_{k=1}^{\infty} \frac{\mu(k)}{k} = 0$. SBer. Kgl. Preuß. Akad. Wiss. Berlin, 1897, 835–852 [French translation: Ann. Sci. Éc. Norm. Sup. **15**, 431–454 (1898)]
4127. von Mangoldt, H.: Zur Verteilung der Nullstellen der Riemannscher Funktion $\xi(t)$. Math. Ann. **60**, 1–19 (1905)
4128. Manickam, M., Ramakrishnan, B., Vasudevan, T.C.: On the theory of newforms of half-integral weight. J. Number Theory **34**, 210–224 (1990)
4129. Manickam, M., Ramakrishnan, B., Vasudevan, T.C.: On Saito-Kurokawa descent for congruence subgroups. Manuscr. Math. **81**, 161–182 (1993)
4130. Manin, Yu.I.: On cubic congruences to a prime modulus. Izv. Akad. Nauk SSSR, Ser. Mat. **20**, 673–678 (1956) (in Russian)
4131. Manin, Yu.I.: Rational points on algebraic curves over function fields. Izv. Akad. Nauk SSSR, Ser. Mat. **27**, 1395–1440 (1963) (in Russian)
4132. Manin, Yu.I.: Correspondences, motifs and monoidal transformations. Mat. Sb. **77**, 475–507 (1968) (in Russian)
4133. Manin, Yu.I.: The p-torsion of elliptic curves is uniformly bounded. Izv. Akad. Nauk SSSR, Ser. Mat. **33**, 459–465 (1969) (in Russian)
4134. Manin, Yu.I.: Le groupe de Brauer-Grothendieck en géométrie diophantienne. Actes ICM Nice **1**, 401–411 (1970)
4135. Manin, Yu.I.: Cyclotomic fields and modular curves. Usp. Mat. Nauk **26**(6), 7–71 (1971) (in Russian)
4136. Manin, Yu.I.: Cubic Forms. Nauka, Moscow (1972) (in Russian) [English translation: North-Holland, 1974, 2nd ed. 1986]
4137. Manin, Yu.I.: Periods of cusp forms, and p-adic Hecke series. Mat. Sb. **92**, 378–401 (1973)

4138. Manin, Yu.I.: Values of p-adic Hecke series at lattice points of the critical strip. Mat. Sb. **93**, 621–626 (1974)
4139. Mann, H.B.: A proof of the fundamental theorem on the density of sums of sets of positive integers. Ann. Math. **43**, 523–527 (1942)
4140. Mann, H.B.: A generalization of a theorem of Ankeny and Rogers. Rend. Circ. Mat. Palermo **3**, 106–108 (1954)
4141. Mann, H.B.: Addition Theorems: the Addition Theorems of Group Theory and Number Theory. Interscience, New York (1965) [Reprint: Robert E. Krieger Publ., 1976]
4142. Mann, H.B., Webb, W.A.: A short proof of Fermat's theorem for $n = 3$. Math. Stud. **46**, 103–104 (1982)
4143. Mardžanišvili, K.K.: On the simultaneous representations of n integers by sums of complete first, second, n-th powers. Izv. Akad. Nauk SSSR, Ser. Mat. **1**, 609–631 (1937) (in Russian)
4144. Mardžanišvili, K.K.: Sur un système d'équations de Diophante. Dokl. Akad. Nauk SSSR **22**, 467–470 (1939)
4145. Mardžanišvili, K.K.: On an additive problem in number theory. Izv. Akad. Nauk SSSR, Ser. Mat. **4**, 193–213 (1940) (in Russian)
4146. Mardžanišvili, K.K.: Sur la démonstration du théorème de Goldbach-Vinogradoff. Dokl. Akad. Nauk SSSR **30**, 687–689 (1941)
4147. Mardžanišvili, K.K.: On an asymptotic formula of the additive theory of prime numbers. Soobšč. AN Gruz. SSR **8**, 597–604 (1997) (in Russian)
4148. Margulis, G.A.: Formes quadratiques indéfinies et flots unipotents sur les espaces homogènes. C. R. Acad. Sci. Paris **304**, 249–253 (1987)
4149. Margulis, G.A.: Discrete subgroups and ergodic theory. In: Number Theory, Trace Formulas and Discrete Groups, Oslo, 1989, pp. 377–398 (1989)
4150. Margulis, G.A.: Indefinite quadratic forms and unipotent flows on homogeneous spaces. Banach Cent. Publ. **23**, 399–409 (1989)
4151. Marke, P.W.: Über die Bestimmung Dirichletscher Reihen durch ihre Funktionalgleichung. Math. Ann. **114**, 29–56 (1937)
4152. Márki, L., Steinfeld, O., Szép, J.: Short review of the work of László Rédei. Studia Sci. Math. Hung. **16**, 3–14 (1981)
4153. Markoff, A.A.: Sur les formes quadratiques binaires indéfinies. Math. Ann. **15**, 381–406 (1879)
4154. Markoff, A.A.: Sur les formes quadratiques binaires indéfinies, II. Math. Ann. **17**, 379–399 (1884)
4155. Markoff, A.A.: Dèmonstration d'un théorème de Tchébycheff. C. R. Acad. Sci. Paris **120**, 1032–1034 (1895)
4156. Markoff, A.A.: Sur les formes quadratiques ternaires indéfinies. Math. Ann. **56**, 233–251 (1903)
4157. Mars, J.G.: The Tamagawa number of ${}^2A_n$. Ann. Math. **89**, 557–574 (1969)
4158. Marstrand, J.M.: On Khinchin's conjecture about strong uniform distribution. Proc. Lond. Math. Soc. **21**, 540–556 (1970)
4159. Marszałek, R.: On the product of consecutive elements of an arithmetic progression. Monatshefte Math. **100**, 215–222 (1985)
4160. Martin, G.: The least prime primitive root and the shifted sieve. Acta Arith. **80**, 277–288 (1997)
4161. Martin, G.: Review of [4855]. Math. Rev. 2002h:11029
4162. Martin, Y.: A converse theorem for Jacobi forms. J. Number Theory **61**, 181–193 (1996)
4163. Martin, Y.: L-functions for Jacobi forms of arbitrary degree. Abh. Math. Semin. Univ. Hamb. **58**, 45–63 (1998)
4164. Martinet, J.: Les réseaux parfaits des espaces euclidiens. Masson, Paris (1996) [English version: Perfect Lattices in Euclidean Spaces, Springer, 2003]
4165. Masai, P., Valette, A.: A lower bound for a counterexample to Carmichael's conjecture. Boll. Unione Mat. Ital., A **1**, 313–316 (1982)
4166. Masani, P.R.: Norbert Wiener: 1894–1964. Birkhäuser, Basel (1990)

4167. Masley, J.M.: Cyclotomic fields with unique factorization. J. Reine Angew. Math. **286/287**, 248–256 (1976)
4168. Mason, R.C.: Diophantine Equations over Function Fields. Cambridge University Press, Cambridge (1984)
4169. Mason, R.C.: Norm form equations, I. J. Number Theory **22**, 190–207 (1986)
4170. Mason, R.C.: Norm form equations, III. Positive characteristic. Math. Proc. Camb. Philos. Soc. **99**, 409–423 (1986)
4171. Mason, R.C.: Norm form equations, IV. Rational functions. Mathematika **33**, 204–211 (1986)
4172. Mason, R.C.: Norm form equations, V. Degenerate modules. J. Number Theory **25**, 239–248 (1987)
4173. Mason, R.C.: The study of Diophantine equations over function fields. In: New Advances in Transcendence Theory, Durham, 1986, pp. 229–247. Cambridge University Press, Cambridge (1988)
4174. Masser, D.[W.]: Elliptic Functions and Transcendence. In: Lecture Notes in Math., vol. 437. Springer, Berlin (1975)
4175. Masser, D.W.: Note on a conjecture of Szpiro. Astérisque **183**, 19–23 (1990)
4176. Masser, D.W.: A note on Siegel's lemma. Rocky Mt. J. Math. **26**, 1057–1068 (1996)
4177. Masser, D.W., Rickert, J.H.: Simultaneous Pell equations. J. Number Theory **61**, 52–66 (1996)
4178. Masser, D.W., Wüstholz, G.: Zero estimates on group varieties, I. Invent. Math. **64**, 489–516 (1981)
4179. Masser, D.W., Wüstholz, G.: Zero estimates on group varieties, II. Invent. Math. **80**, 233–267 (1985)
4180. Masser, D.W., Wüstholz, G.: Algebraic independence properties of values of elliptic functions. In: Journées Arithmétiques 1980, pp. 360–363. Cambridge University Press, Cambridge (1982)
4181. Masser, D.W., Wüstholz, G.: Fields of large transcendence degree generated by values of elliptic functions. Invent. Math. **72**, 407–464 (1983)
4182. Masser, D.W., Wüstholz, G.: Algebraic independence of values of elliptic functions. Math. Ann. **276**, 1–17 (1986)
4183. Masser, D.W., Wüstholz, G.: Estimating isogenies on elliptic curves. Invent. Math. **100**, 1–24 (1990)
4184. Masser, D.W., Wüstholz, G.: Isogeny estimates for abelian varieties, and finiteness theorems. Ann. Math. **137**, 459–472 (1993)
4185. Massias, J.-P., Robin, G.: Bornes effectives pour certaines fonctions concernant les nombres premiers. J. Théor. Nr. Bordx. **8**, 215–242 (1996)
4186. de Mathan, B.: Numbers contravening a condition in density modulo 1. Acta Math. Acad. Sci. Hung. **36**, 237–241 (1981)
4187. de Mathan, B.: Conjecture de Littlewood et récurrences linéaires. J. Théor. Nombres Bordeaux **15**, 249–266 (2003)
4188. Matijasevič, Yu.V.: The diophantineness of enumerable sets. Dokl. Akad. Nauk SSSR **191**, 279–282 (1970) (in Russian)
4189. Matijasevič, Yu.V.: Diophantine representation of enumerable predicates. Izv. Akad. Nauk SSSR, Ser. Mat. **35**, 3–30 (1971) (in Russian)
4190. Matijasevič, Yu.V.: Diophantine representation of enumerable predicates. Mat. Zametki **12**, 115–120 (1972) (in Russian)
4191. Matijasevič, Yu.V.: Diophantine sets. Usp. Mat. Nauk **27**(5), 185–222 (1972) (in Russian)
4192. Matijasevič, Yu.V.: Primes are enumerated by a polynomial in 10 variables. Zap. Nauč. Semin. LOMI **68**, 62–82 (1977) (in Russian)
4193. Matijasevič, Yu.V.: Reduction of an arbitrary Diophantine equation to one in 13 unknowns. Acta Arith. **27**, 521–553 (1975)
4194. Matomäki, T.: Prime numbers of the form $p = m^2 + n^2 + 1$ in short intervals. Acta Arith. **128**, 193–200 (2007)

4195. Matomäki, T.: The distribution of αp modulo one. Math. Proc. Camb. Philos. Soc. **147**, 267–283 (2009)
4196. Matsuda, I.: Dirichlet series corresponding to Siegel modular forms of degree 2, level N. Sci. Papers College Gen. Ed. Univ. Tokyo **28**, 21–49 (1978)
4197. Matsumoto, H.: Sur les sous-groupes arithmétiques des groupes semi-simples déployés. Ann. Sci. Éc. Norm. Super. **2**, 1–62 (1969)
4198. Matsumoto, K.: A remark on Smith's result on a divisor problem in arithmetic progressions. Nagoya Math. J. **98**, 37–42 (1985)
4199. Matthews, C.: Gauss sums and elliptic functions, I. The Kummer sum. Invent. Math. **52**, 163–185 (1979)
4200. Matthews, C.: Gauss sums and elliptic functions, II. The quartic sum. Invent. Math. **54**, 23–52 (1979)
4201. Matveev, E.M.: On the size of algebraic integers. Mat. Zametki **49**, 152–154 (1991)
4202. Matveev, E.M.: Explicit lower bound for a homogeneous rational linear form in logarithms of algebraic numbers, I. Izv. Ross. Akad. Nauk, Ser. Mat. **62**(4), 81–136 (1998) (in Russian)
4203. Matveev, E.M.: Explicit lower bound for a homogeneous rational linear form in logarithms of algebraic numbers, II. Izv. Ross. Akad. Nauk, Ser. Mat. **64**(6), 125–180 (2000) (in Russian)
4204. Mauclaire, J.-L.: Sur la régularité des fonctions additives. Enseign. Math. **18**, 167–174 (1972)
4205. Mauclaire, J.-L.: On a problem of Kátai. Acta Sci. Math. **36**, 205–207 (1974)
4206. Mauclaire, J.-L.: Sommes de Gauss modulo p^α, I. Proc. Jpn. Acad. Sci. **59**, 109–112 (1983)
4207. Mauclaire, J.-L.: Sommes de Gauss modulo p^α, II. Proc. Jpn. Acad. Sci. **59**, 161–163 (1983)
4208. Mauclaire, J.-L., Murata, L.: On the regularity of arithmetic multiplicative functions, I. Proc. Jpn. Acad. Sci. **56**, 438–441 (1980)
4209. Mauclaire, J.-L., Murata, L.: On the regularity of arithmetic multiplicative functions, II. Proc. Jpn. Acad. Sci. **57**, 130–133 (1981)
4210. Mauduit, C.: Automates finis et ensembles normaux. Ann. Inst. Fourier **36**(2), 1–25 (1986)
4211. Mauldin, R.D.: A generalization of Fermat's last theorem: the Beal conjecture and prize problem. Not. Am. Math. Soc. **44**, 1436–1437 (1997)
4212. Maxwell, J.W.: William J. LeVeque (1923–2007). Not. Am. Math. Soc. **55**, 1261–1262 (2008)
4213. Mazur, B.: Rational points of abelian varieties with values in towers of number fields. Invent. Math. **18**, 183–226 (1972)
4214. Mazur, B.: Rational points on modular curves. In: Lecture Notes in Math., vol. 601, pp. 107–148. Springer, Berlin (1977)
4215. Mazur, B.: Modular curves and the Eisenstein ideal. Publ. Math. Inst. Hautes Études Sci. **47**, 33–186 (1977)
4216. Mazur, B.: Rational isogenies of prime degree (with an appendix by D. Goldfeld). Invent. Math. **44**, 129–162 (1978)
4217. Mazur, B.: On the passage from local to global in number theory. Bull. Am. Math. Soc. **29**, 14–50 (1993)
4218. Mazur, B.: Finding meaning in error terms. Bull. Am. Math. Soc. **45**, 185–228 (2008)
4219. Mazur, B., Tate, J., Teitelbaum, J.: On p-adic analogues of the conjectures of Birch and Swinnerton-Dyer. Invent. Math. **84**, 1–48 (1986)
4220. Mazur, B., Wiles, A.: Class fields of abelian extensions of Q. Invent. Math. **76**, 179–330 (1984)
4221. Maz'ya, V., Shaposhnikova, T.: Jacques Hadamard, a Universal Mathematician. Am. Math. Soc., Providence (1998)
4222. McCallum, W.G.: On the method of Coleman and Chabauty. Math. Ann. **299**, 565–596 (1994)
4223. McCurley, K.S.: An effective seven cube theorem. J. Number Theory **19**, 381–385 (1984)
4224. McCurley, K.S.: Explicit zero-free regions for Dirichlet L-functions. J. Number Theory **19**, 7–32 (1984)

4225. McCurley, K.S.: Explicit estimates for the error term in the prime number theorem for arithmetic progressions. Math. Comput. **42**, 265–285 (1984)
4226. McCurley, K.S.: Explicit estimates for $\theta(x; 3, l)$ and $\psi(x; 3, l)$. Math. Comput. **42**, 287–296 (1984)
4227. McCurley, K.S.: Prime values of polynomials and irreducible testing. Bull. Am. Math. Soc. **11**, 155–158 (1984)
4228. McCurley, K.S.: The smallest prime value of $x^n + a$. Can. J. Math. **38**, 925–936 (1986)
4229. McCurley, K.S.: Polynomials with no small prime values. Proc. Am. Math. Soc. **97**, 393–395 (1986)
4230. McGehee, O.C., Pigno, L., Smith, B.: Hardy's inequality and the L_1 norm of exponential sums. Ann. Math. **113**, 613–618 (1981)
4231. McGettrick, A.D.: On the biquadratic Gauss sum. Proc. Camb. Philos. Soc. **71**, 79–83 (1972)
4232. McKee, J.: On the average number of divisors of quadratic polynomials. Math. Proc. Camb. Philos. Soc. **117**, 389–392 (1995)
4233. McKee, J.: The average number of divisors of an irreducible quadratic polynomial. Math. Proc. Camb. Philos. Soc. **126**, 17–22 (1999)
4234. McKee, J.: Speeding Fermat's factoring method. Math. Comput. **68**, 1729–1737 (1999)
4235. McMullen, C.T.: Minkowski's conjecture, well-rounded lattices and topological dimension. J. Am. Math. Soc. **18**, 711–734 (2005)
4236. Meissel, E.: Ueber die Bestimmung der Primzahlmenge innerhalb gegebener Grenzen. Math. Ann. **2**, 636–642 (1870)
4237. Meissner, W.: Über die Teilbarkeit von $2^p - 2$ durch das Quadrat der Primzahl $p = 1093$. SBer. Preuß. Akad. Wiss. Berlin, 1913, 663–667
4238. Mellin, H.: Über eine Verallgemeinerung der Riemannscher Funktion $\zeta(s)$. Acta Soc. Sci. Fenn. **24**(10), 1–50 (1899)
4239. Mellin, H.: Über den Zusammenhang zwischen den linearen Differential- und Differenzengleichungen. Acta Math. **25**, 139–164 (1901)
4240. Mellin, H.: Eine Formel für den Logarithmus transcendenter Funktionen von endlichen Geschlecht. Acta Soc. Sci. Fenn. **29**(4), 1–50 (1900) [A shortened version: Acta Math. **25**, 165–184 (1901)]
4241. Mellin, H.: Die Dirichletschen Reihen, die zahlentheoretischen Funktionen und die unendlichen Produkte von endlichem Geschlecht. Acta Math. **28**, 37–64 (1904)
4242. Mellin, H.: Abriß einer einheitlichen Theorie der Gamma- und hypergeometrischen Funktionen. Math. Ann. **68**, 305–337 (1910)
4243. Mellin, H.: Remarks concerning the proof of a theorem of Hardy on the zeta-function. Acta Soc. Fenn. A **11**, 1917, nr. 3 (in Finnish)
4244. Melzak, Z.A.: A note on the Tarry-Escott problem. Can. Math. Bull. **4**, 233–237 (1961)
4245. Mendès France, M.: Nombres normaux et fonctions pseudo-aléatoires. Ann. Inst. Fourier **13**(2), 91–104 (1963)
4246. Mendès France, M.: Deux remarques concernant l'équirépartition des suites. Acta Arith. **14**, 163–167 (1967/1968)
4247. Mendès France, M.: Nombres transcendants et ensembles normaux. Acta Arith. **15**, 189–192 (1968/1969)
4248. Mendès France, M.: La réunion des ensembles normaux. J. Number Theory **2**, 345–351 (1970)
4249. Mendès France, M.: Les ensembles de Bésineau. Sém. Delange–Pisot–Poitou **15**(exp. 7), 1–6 (1973/1974)
4250. Mendès France, M.: Principe de la symétrie perturbée. Prog. Math. **12**, 77–98 (1981)
4251. Mendès France, M.: Some applications of the theory of automata. In: Prospects of Mathematical Science, Tokyo, 1986, pp. 127–140. World Scientific, Singapore (1988)
4252. Mennicke, J.L.: Finite factor groups of the unimodular group. Ann. Math. **81**, 31–37 (1965)
4253. Menzer, H.: The distribution of powerful integers of type 4. Monatshefte Math. **107**, 69–75 (1989)

4254. Menzer, H.: On the distribution of powerful numbers. Abh. Math. Semin. Univ. Hamb. **67**, 221–237 (1997)
4255. Merel, L.: Bornes pour la torsion des courbes elliptiques sur le corps de nombres. Invent. Math. **124**, 437–449 (1996)
4256. Merlin, J.: Un travail de Jean Merlin sur les nombres premiers. Bull. Sci. Math. **39**, 121–136 (1915)
4257. Mertens, F.: Ueber einige asymptotische Gesetze der Zahlentheorie. J. Reine Angew. Math. **77**, 289–338 (1874)
4258. Mertens, F.: Ueber eine zahlentheoretische Function. SBer. Kais. Akad. Wissensch. Wien **106**, 761–830 (1897)
4259. Mertens, F.: Über eine Eigenschaft der Riemannscher ζ-Funktion. SBer. Kais. Akad. Wissensch. Wien **107**, 1429–1434 (1898)
4260. Mertens, F.: Beweis, dass jede lineare Function mit ganzen complexen teilerfremden Coefficienten unendlich viele complexe Primzahlen darstellt. SBer. Kais. Akad. Wissensch. Wien **108**, 517–556 (1899)
4261. Mertens, F.: Über den Kummerschen Logarithmus einer komplexen Zahl des Bereichs einer primitiven λ-ten Einheitswurzel in bezug auf den Modul λ^{1+n}, wo λ eine ungerade Primzahl bezeichnet. SBer. Kais. Akad. Wissensch. Wien **126**, 1337–1343 (1917)
4262. Merzbach, U.C.: Robert Remak and the estimation of units and regulators. In: Amphora, pp. 481–522. Birkhäuser, Basel (1992)
4263. Mestre, J.F.: Construction d'une courbe elliptique de rang $\geq$ 12. C. R. Acad. Sci. Paris **295**, 643–644 (1982)
4264. Mestre, J.F.: Formules explicites et minorations de conducteurs de variétés algébriques. Compos. Math. **58**, 209–232 (1986)
4265. Mestre, J.F.: Un exemple de courbe elliptique sur Q de rang $\geq$ 15. C. R. Acad. Sci. Paris **314**, 453–455 (1992)
4266. Metsänkylä, T.: Bemerkungen über den ersten Faktor der Klassenzahl des Kreiskörpers. Ann. Univ. Turku, Ser. AI **105**, 1–15 (1967)
4267. Metsänkylä, T.: Zero-free regions of Dirichlet's L-functions near the point 1. Ann. Univ. Turku, Ser. AI **139**, 1–11 (1970)
4268. Metsänkylä, T.: Estimations for L-functions and the class numbers of certain imaginary cyclic fields. Ann. Univ. Turku, Ser. AI **140**, 1–11 (1970)
4269. Metsänkylä, T.: Note on the distribution of irregular primes. Ann. Acad. Sci. Fenn., Ser. A 1 Math. **492**, 1–7 (1971)
4270. Metsänkylä, T.: On the growth of the first factor of the cyclotomic class number. Ann. Univ. Turku **155**, 1–12 (1972)
4271. Metsänkylä, T.: Distribution of irregular prime numbers. J. Reine Angew. Math. **282**, 126–130 (1976)
4272. Metsänkylä, T.: The index of irregularity of primes. Expo. Math. **5**, 143–156 (1987)
4273. Metsänkylä, T.: Kustaa Inkeri. Portrait of a mathematician. In: Collected Papers of Kustaa Inkeri, pp. 1–8. Queen's University Press, Kingston (1992)
4274. Metsänkylä, T.: Catalan's conjecture: another old diophantine problem solved. Bull. Am. Math. Soc. **41**, 43–57 (2004)
4275. Meurman, T.: A simple proof of Voronoï's identity. Astérisque **209**, 265–274 (1992)
4276. Meurman, T.: The mean square of the error term in a generalization of Dirichlet's divisor problem. Acta Arith. **74**, 351–364 (1996)
4277. Meuser, D.: On the rationality of certain generating functions. Math. Ann. **256**, 303–310 (1981)
4278. Meuser, D.: The meromorphic continuation of a zeta function of Weil and Igusa type. Invent. Math. **85**, 493–514 (1986)
4279. Meyer, A.: Ueber die Kriterien für die Auflösbarkeit der Gleichung $ax^2+by^2+cz^2+du^2=0$ in ganzen Zahlen. Vierteljahr. Naturforsch. Ges. Zürich **29**, 209–222 (1884)
4280. Meyer, C.: Bemerkungen zum Satz von Heegner-Stark über die imaginär-quadratischen Zahlkörper mit der Klassenzahl Eins. J. Reine Angew. Math. **242**, 179–214 (1970)

4281. Meyer, Y.: Nombres algébriques, nombres transcendants et équirépartition modulo 1. Acta Arith. **16**, 347–350 (1969/1970)
4282. Meyer, Y.: Nombres de Pisot, nombres de Salem et analyse harmonique. In: Lecture Notes in Math., vol. 117. Springer, Berlin (1970)
4283. Michel, P.: Répartition des zéros des fonctions L et matrices aléatoires. Astérisque **282**, 211–248 (2002)
4284. Miech, R.J.: Almost primes generated by a polynomial. Acta Arith. **10**, 9–30 (1964)
4285. Miech, R.J.: Primes, polynomials and almost primes. Acta Arith. **11**, 35–56 (1965)
4286. Miech, R.J.: A uniform result on almost primes. Acta Arith. **11**, 371–391 (1966)
4287. Miech, R.J.: A number-theoretic constant. Acta Arith. **15**, 119–137 (1968/1969)
4288. Mientka, W.E.: An application of the Selberg sieve method. J. Indian Math. Soc. **25**, 129–138 (1961)
4289. Mientka, W.E., Weitzenkamp, R.C.: On f-plentiful numbers. J. Comb. Theory **7**, 374–377 (1969)
4290. Mignosi, G.: Michele Cipolla. Ann. Mat. Pura Appl. **26**, 217–220 (1947)
4291. Mignotte, M.: Sur l'équation de Catalan, II. Theor. Comput. Sci. **123**, 145–149 (1994)
4292. Mignotte, M.: Verification of a conjecture of E. Thomas. J. Number Theory **44**, 172–177 (1993)
4293. Mignotte, M.: On the diophantine equation $\frac{x^n-1}{x-1} = y^q$. In: Algebraic Number Theory and Diophantine Analysis, pp. 305–310. de Gruyter, Berlin (2000)
4294. Mignotte, M.: Catalan's equation just before 2000. In: Number Theory, Turku 1999, pp. 247–254. de Gruyter, Berlin (2001)
4295. Mignotte, M., Pethő, A., Roth, R.: Complete solutions of a family of quartic Thue and index form equations. Math. Comput. **65**, 341–354 (1996)
4296. Mignotte, M., Roy, Y.: Lower bounds for Catalan's equation. Ramanujan J. **1**, 351–356 (1997)
4297. Mignotte, M., Waldschmidt, M.: Linear forms in two logarithms and Schneider's method. Math. Ann. **231**, 241–267 (1977/1978)
4298. Mignotte, M., Waldschmidt, M.: Linear forms in two logarithms and Schneider's method, II. Acta Arith. **53**, 251–287 (1989)
4299. Mignotte, M., Waldschmidt, M.: Linear forms in two logarithms and Schneider's method, III. Ann. Fac. Sci. Toulouse **97**, 43–75 (1989)
4300. Mihăilescu, P.: Primary cyclotomic units and a proof of Catalan's conjecture. J. Reine Angew. Math. **572**, 167–195 (2004)
4301. Mihăilescu, P.: On the class groups of cyclotomic extensions in presence of a solution to Catalan's equation. J. Number Theory **118**, 123–144 (2006)
4302. Mikawa, H.: Almost-primes in arithmetic progressions and short intervals. Tsukuba J. Math. **13**, 387–401 (1989)
4303. Mikawa, H.: On the Brun-Titchmarsh theorem. Tsukuba J. Math. **15**, 31–40 (1991)
4304. Mikawa, H.: On the exceptional set in Goldbach's problem. Tsukuba J. Math. **16**, 513–543 (1992)
4305. Mikolás, M.: Farey series and their connection with the prime number problem, I. Acta Sci. Math. **13**, 93–117 (1949)
4306. Mikolás, M.: Farey series and their connection with the prime number problem, II. Acta Sci. Math. **14**, 5–21 (1951)
4307. Mikolás, M., Sato, K.I.: On the asymptotic behaviour of Franel's sum and the Riemann Hypothesis. Results Math. **21**, 368–378 (1992)
4308. Miller, J.C.P.: Alfred Edward Western. J. Lond. Math. Soc. **38**, 278–281 (1963)
4309. Miller, J.C.P., Wheeler, D.J.: Large prime numbers. Nature **168**, 838 (1951)
4310. Miller, J.C.P., Woollett, M.F.C.: Solutions of the Diophantine equation $x^3 + y^3 + z^3 = k$. J. Lond. Math. Soc. **30**, 101–110 (1955)
4311. Milne, J.S.: The Tate-Šafarevič group of a constant abelian variety. Invent. Math. **6**, 91–105 (1968)
4312. Milne, J.S.: Weil-Châtelet groups over local fields. Ann. Sci. Éc. Norm. Super. **3**, 273–284 (1970); add.: **5**, 261–264 (1972)

4313. Milne, J.S.: On a conjecture of Artin and Tate. Ann. Math. **102**, 517–533 (1975)
4314. Milne, J.S.: New infinite families of exact sums of squares formulas, Jacobi elliptic functions, and Ramanujan's tau function. Proc. Natl. Acad. Sci. USA **93**, 15004–15008 (1996)
4315. Milne, S.C.: Infinite families of exact sums of squares formulas, Jacobi elliptic functions, continued fractions, and Schur functions. Ramanujan J. **6**, 7–149 (2002)
4316. Min, S.-H.: On the order of $\zeta(1/2+it)$. Trans. Am. Math. Soc. **65**, 448–472 (1949)
4317. Mingarelli, A.B.: A glimpse into the life and times of F.V. Atkinson. Math. Nachr. **278**, 1364–1387 (2005)
4318. Minkowski, H.: Mémoire sur la théorie des formes quadratiques à coefficients entières. Mémoires présentés par divers savants à l'Académie **29**(2), 1–178 (1887) [[4329], vol. 1, pp. 3–144]
4319. Minkowski, H.: Zur Theorie der positiven quadratischen Formen. J. Reine Angew. Math. **101**, 196–202 (1887) [[4329], vol. 1, pp. 212–239]
4320. Minkowski, H.: Über die Bedingungen, unter welchen zwei quadratische Formen mit rationalen Koeffizienten ineinander transformiert werden können. J. Reine Angew. Math. **106**, 5–26 (1890) [[4329], vol. 1, pp. 219–239]
4321. Minkowski, H.: Ueber Geometrie der Zahlen. Jahresber. Dtsch. Math.-Ver. **1**, 64–65 (1890/1891) [[4329], vol. 1, pp. 264–265]
4322. Minkowski, H.: Über die positiven quadratischen Formen und über kettenbruchähnliche Algorithmen. J. Reine Angew. Math. **107**, 278–297 (1891) [[4329], vol. 1, pp. 243–260]
4323. Minkowski, H.: Généralisation de la théorie des fractions continues. Ann. Sci. Éc. Norm. Super. **13**, 41–60 (1896) [German translation: [4329], vol. 1, pp. 278–285]
4324. Minkowski, H.: Geometrie der Zahlen. Teubner, Leipzig (1896). 1910 [Reprints: Teubner, 1925; Chelsea, 1953; Johnson 1968; English translation with commentaries: [2496]]
4325. Minkowski, H.: Dichteste gitterförmige Lagerung kongruenter Körper. Nachr. Ges. Wiss. Göttingen, 1904, 311–355 [[4329], vol. 2, pp. 3–42]
4326. Minkowski, H.: Diskontinuitätsbereich für arithmetische Äquivalenz. J. Reine Angew. Math. **129**, 220–274 (1905) [[4329], vol. 2, pp. 53–100]
4327. Minkowski, H.: Peter Gustav Lejeune Dirichlet und seine Bedeutung für die heutige Mathematik. Jahresber. Dtsch. Math.-Ver. **14**, 149–163 (1905) [[4329], vol. 2, pp. 447–461]
4328. Minkowski, H.: Diophantische Approximationen. Teubner, Leipzig (1907) [Reprints: Teubner, 1927; Chelsea, 1957; Physica Verlag, 1961]
4329. Minkowski, H.: Gesammelte Abhandlungen, vols. 1, 2. Teubner, Leipzig (1911)
4330. Miret, J., Moreno, R., Rio, A., Valls, M.: Computing the l-power torsion of an elliptic curve over a finite field. Math. Comput. **78**, 1767–1786 (2009)
4331. Miri, S.A., Murty, V.K.: An application of sieve methods to elliptic curves. In: Lecture Notes in Comput. Sci., vol. 2247, pp. 91–98. Springer, Berlin (2001)
4332. Mirimanoff, D.: L'équation indéterminée $x^l + y^l + z^l = 0$ et le criterion de Kummer. J. Reine Angew. Math. **128**, 45–68 (1905)
4333. Mirimanoff, D.: Sur le dernier théorème de Fermat et le critérium de M.A. Wieferich. Enseign. Math. **11**, 455–459 (1909)
4334. Mirimanoff, D.: Sur le dernier théorème de Fermat. C. R. Acad. Sci. Paris **150**, 204–206 (1910)
4335. Mirimanoff, D.: Sur le dernier théorème de Fermat. J. Reine Angew. Math. **139**, 309–324 (1911)
4336. Mirsky, L.: On the number of representations of an integer as the sum of three r-free integers. Proc. Camb. Philos. Soc. **43**, 433–441 (1947)
4337. Mirsky, L.: On a theorem in the additive theory of numbers due to Evelyn and Linfoot. Proc. Camb. Philos. Soc. **44**, 305–312 (1948)
4338. Mitchell, O.H.: On binomial congruences; comprising an extension of Fermat's and Wilson's theorems, and a theorem on which both are special cases. Am. J. Math. **3**, 294–315 (1880)
4339. Mitchell, O.H.: Some theorems in numbers. Am. J. Math. **4**, 25–38 (1881)
4340. Mitkin, D.A.: Estimate for the number of summands in the Hilbert-Kamke problem. Mat. Sb. **129**, 549–577 (1986) (in Russian)

4341. Mitkin, D.A.: Estimate for the number of summands in the Hilbert-Kamke problem, II. Mat. Sb. **132**, 345–351 (1987) (in Russian)
4342. Mitkin, D.A.: An elementary proof of A. Weil's estimate for rational trigonometric sums with a prime denominator. Izv. Vysš. Učeb. Zaved. Mat., 1986, nr. 6, 14–17 (in Russian)
4343. Mitkin, D.A.: The Hilbert-Kamke problem in prime numbers. Usp. Mat. Nauk **42**(5), 205–206 (1987) (in Russian)
4344. Mitsui, T.: On the prime ideal theorem. Jpn. J. Math. **20**, 233–247 (1968)
4345. Miyake, K.: Teiji Takagi, founder of the Japanese School of Modern Mathematics. Jpn. J. Math. **2**, 151–164 (2007)
4346. Miyawaki, I.: Elliptic curves of prime power conductor with Q-rational points of finite order. Osaka Math. J. **10**, 309–323 (1973)
4347. Mollin, R.A.: Class number one criteria for real quadratic fields, I. Proc. Jpn. Acad. Sci. **63**, 121–125 (1987)
4348. Mollin, R.A.: Quadratics. CRC Press, Boca Raton (1996)
4349. Mollin, R.A., Williams, H.C.: A conjecture of S. Chowla via the generalized Riemann hypothesis. Proc. Am. Math. Soc. **102**, 794–796 (1988)
4350. Mollin, R.A., Williams, H.C.: On a solution of a class number two problem for a family of real quadratic fields. In: Computational Number Theory, Debrecen, 1989, pp. 95–101. de Gruyter, Berlin (1991)
4351. Momose, F.: p-torsion points on elliptic curves defined over quadratic fields. Nagoya Math. J. **96**, 139–165 (1984)
4352. Montgomery, H.L.: Distribution of irregular primes. Ill. J. Math. **9**, 553–558 (1965)
4353. Montgomery, H.L.: A note on the large sieve. J. Lond. Math. Soc. **43**, 93–98 (1968)
4354. Montgomery, H.L.: Mean and large values of Dirichlet polynomials. Invent. Math. **8**, 334–345 (1969)
4355. Montgomery, H.L.: Zeros of L-functions. Invent. Math. **8**, 346–354 (1969)
4356. Montgomery, H.L.: Primes in arithmetic progressions. Mich. Math. J. **17**, 33–39 (1970)
4357. Montgomery, H.L.: Topics in Multiplicative Number Theory. Lecture Notes in Math., vol. 227. Springer, Berlin (1971)
4358. Montgomery, H.L.: The pair correlation of zeros of the zeta function. In: Proc. Symposia Pure Math., vol. 24, pp. 181–193. Am. Math. Soc., Providence (1973)
4359. Montgomery, H.L.: Distribution of the zeros of the Riemann zeta function. In: Proceedings of the International Congress of Mathematicians, Vancouver, B.C., 1974. vol. 1, pp. 379–381 (1975)
4360. Montgomery, H.L.: The analytic principle of the large sieve. Bull. Am. Math. Soc. **84**, 547–567 (1978)
4361. Montgomery, H.L.: Fluctuations in the mean of Euler's phi function. Proc. Indian Acad. Sci. Math. Sci. **97**, 239–245 (1987)
4362. Montgomery, H.L.: Ten Lectures on the Interface Between Analytic Number Theory and Harmonic Analysis. Am. Math. Soc., Providence (1994)
4363. Montgomery, H.L., Odlyzko, A.M.: Gaps between zeros of the zeta function. In: Topics in Classical Number Theory, pp. 1079–1106. North-Holland, Amsterdam (1984)
4364. Montgomery, H.L., Vaughan, R.C.: The large sieve. Mathematika **20**, 119–134 (1973)
4365. Montgomery, H.L., Vaughan, R.C.: The exceptional set in Goldbach's problem. Acta Arith. **27**, 353–370 (1975)
4366. Montgomery, H.L., Vaughan, R.C.: Exponential sums with multiplicative coefficients. Invent. Math. **43**, 69–82 (1977)
4367. Montgomery, H.L., Vaughan, R.C.: The distribution of squarefree numbers. In: Recent Progress in Analytic Number Theory, Durham, 1979. vol. 1, pp. 247–256. Academic Press, San Diego (1981)
4368. Montgomery, H.L., Vaughan, R.C.: On the distribution of reduced residues. Ann. Math. **123**, 311–333 (1986)
4369. Montgomery, H.L., Vaughan, R.C.: On the Erdős-Fuchs theorem. In: A Tribute to Paul Erdős, pp. 331–338. Cambridge University Press, Cambridge (1990)

4370. Montgomery, H.L., Vaughan, R.C., Wooley, T.D.: Some remarks on Gauss sums associated with kth powers. Math. Proc. Camb. Philos. Soc. **118**, 21–33 (1995)
4371. Montgomery, H.L., Vorhauer, U.: Changes of sign of the error term in the prime number theorem. Funct. Approx. Comment. Math. **35**, 235–247 (2006)
4372. Morain, F.: La primalité en temps polynomial (d'aprés Adleman, Huang; Agrawal, Kayal, Saxena). Astérisque **294**, 205–230 (2004)
4373. Moran, W., Pollington, A.D.: The discrimination theorem for normality to non-integer bases. Isr. J. Math. **100**, 339–347 (1997)
4374. Mordell, L.J.: The diophantine equation $y^2 - k = x^3$. Proc. Lond. Math. Soc. **13**, 60–80 (1913)
4375. Mordell, L.J.: On the representations of numbers as a sum of 2^r squares. Q. J. Math. **48**, 93–104 (1917)
4376. Mordell, L.J.: On Mr. Ramanujan's empirical expansions of modular functions. Proc. Camb. Philos. Soc. **19**, 117–124 (1917–1920)
4377. Mordell, L.J.: On the representations of numbers as a sum of an odd number of squares. Trans. Camb. Philos. Soc. **22**, 361–372 (1919)
4378. Mordell, L.J.: A statement by Fermat. Proc. Lond. Math. Soc. **18**, v (1919)
4379. Mordell, L.J.: On the rational solutions of the indeterminate equations of the third and fourth degrees. Proc. Camb. Philos. Soc. **21**, 179–192 (1922)
4380. Mordell, L.J.: On the integer solutions of the equation $ey^2 = ax^3 + bx^2 + cx + d$. Proc. Lond. Math. Soc. **21**, 415–419 (1923)
4381. Mordell, L.J.: On the number of solutions in positive integers of the equation $yz+zx+xy=n$. Am. J. Math. **45**, 1–4 (1923)
4382. Mordell, L.J.: On a sum analogous to a Gauss's sum. Q. J. Math. **3**, 161–167 (1932)
4383. Mordell, L.J.: The number of solutions of some congruences in two variables. Math. Z. **37**, 193–209 (1933)
4384. Mordell, L.J.: Some arithmetical results in the geometry of numbers. Compos. Math. **1**, 248–253 (1934)
4385. Mordell, L.J.: On the Riemann hypothesis and imaginary quadratic fields with a given class number. J. Lond. Math. Soc. **9**, 289–298 (1934)
4386. Mordell, L.J.: On the four integer cubes problem. J. Lond. Math. Soc. **11**, 208–218 (1936); add. **12**, 80 (1937)
4387. Mordell, L.J.: A remark on indeterminate equations in several variables. J. Lond. Math. Soc. **12**, 127–129 (1937)
4388. Mordell, L.J.: An application of quaternions to the representation of a binary quadratic form as a sum of four linear squares. Q. J. Math. **8**, 58–61 (1937)
4389. Mordell, L.J.: Tschebotareff's theorem on the product of non-homogeneous linear forms. Vierteljschr. Naturforsch. Ges. Zürich **85**, 47–50 (1940)
4390. Mordell, L.J.: Lattice points in the region $|Ax^4+By^4| \leq 1$. J. Lond. Math. Soc. **16**, 152–156 (1941)
4391. Mordell, L.J.: The product of three homogeneous linear ternary forms. J. Lond. Math. Soc. **17**, 107–115 (1942)
4392. Mordell, L.J.: Observation on the minimum of a positive quadratic form in eight variables. J. Lond. Math. Soc. **19**, 3–6 (1944)
4393. Mordell, L.J.: A Chapter in the Theory of Numbers. Cambridge University Press, Cambridge (1947)
4394. Mordell, L.J.: On some Diophantine equations $y^2 = x^3 + k$ with no rational solutions. Arch. Math. Naturvidensk. **49**, 143–150 (1947)
4395. Mordell, L.J.: On some Diophantine equations $y^2 = x^3 + k$ with no rational solutions, II. In: 1969 Number Theory and Analysis, pp. 223–232. Plenum, New York (1969)
4396. Mordell, L.J.: Rational points on cubic surfaces. Publ. Math. (Debr.) **1**, 1–6 (1949)
4397. Mordell, L.J.: Lattice points in a tetrahedron and generalized Dedekind sums. J. Indian Math. Soc. **15**, 41–46 (1951)
4398. Mordell, L.J.: On the representation of a number as a sum of three squares. Rev. Roum. Math. Pures Appl. **3**, 25–27 (1958)

4399. Mordell, L.J.: The representation of integers by three positive squares. Mich. Math. J. **7**, 289–290 (1960)
4400. Mordell, L.J.: On a cubic exponential sum in three variables. Am. J. Math. **85**, 49–52 (1963)
4401. Mordell, L.J.: Book Review: Diophantine Geometry. Bull. Am. Math. Soc. **70**, 491–498 (1964)
4402. Mordell, L.J.: On the conjecture for the rational points on a cubic surface. J. Lond. Math. Soc. **40**, 149–158 (1965)
4403. Mordell, L.J.: The infinity of rational solutions of $y^2 = x^3 + k$. J. Lond. Math. Soc. **41**, 523–525 (1966)
4404. Mordell, L.J.: On the magnitude of the integer solutions of the equation $ax^2 + by^2 + cz^2 = 0$. J. Number Theory **1**, 1–3 (1969)
4405. Mordell, L.J.: Diophantine Equations. Academic Press, San Diego (1969)
4406. Mordell, L.J.: Harold Davenport (1907–1969). Acta Arith. **18**, 1–4 (1971)
4407. Mordell, L.J.: Some aspects of Davenport's work. Acta Arith. **18**, 5–11 (1971)
4408. Moreau, J.-C.: Démonstrations géométriques de lemmes de zéros, I. Prog. Math. **38**, 201–205 (1983)
4409. Moreau, J.-C.: Démonstrations géométriques de lemmes de zéros, II. Prog. Math. **31**, 191–197 (1983)
4410. Moree, P.: On the prime density of Lucas sequences. J. Théor. Nr. Bordx. **8**, 449–459 (1996)
4411. Moree, P.: On the divisors of $a^k + b^k$. Acta Arith. **80**, 197–212 (1997)
4412. Moree, P.: Counting divisors of Lucas numbers. Pac. J. Math. **186**, 267–284 (1998)
4413. Moree, P.: Approximation of singular series and automata. Manuscr. Math. **101**, 385–399 (2000)
4414. Moree, P.: Chebyshev's bias for composite numbers with restricted prime divisors. Math. Comput. **73**, 425–449 (2004)
4415. Moree, P.: On some claims in Ramanujan's 'unpublished' manuscript on the partition and tau functions. Ramanujan J. **8**, 317–330 (2004)
4416. Moree, P., Stevenhagen, P.: Prime divisors of Lucas sequences. Acta Arith. **82**, 403–410 (1997)
4417. Moree, P., te Riele, H.J.J., Urbanowicz, J.: Divisibility properties of integers x, k satisfying $1^k + \cdots + (x-1)^k = x^k$. Math. Comput. **63**, 799–815 (1994)
4418. Morehead, J.C.: Note on Fermat's numbers. Bull. Am. Math. Soc. **11**, 543–545 (1904/1905)
4419. Morehead, J.C.: Note on the factors of Fermat's numbers. Bull. Am. Math. Soc. **12**, 449–451 (1905/1906)
4420. Morehead, J.C.: Extension of the sieve of Eratosthenes to arithmetical progressions and application. Ann. Math. **10**, 88–104 (1909)
4421. Morehead, J.C., Western, A.E.: Note on Fermat's numbers. Bull. Am. Math. Soc. **16**, 1–6 (1909/1910)
4422. Morelli, R.: Pick's theorem and the Todd class of a toric variety. Adv. Math. **100**, 183–231 (1993)
4423. Moreno, C.J.: Prime number theorems for the coefficients of modular forms. Bull. Am. Math. Soc. **78**, 796–798 (1972)
4424. Moreno, C.J., Wagstaff, S.S. Jr.: Sums of Squares of Integers. Chapman & Hall/CRC, London/Boca Raton (2006)
4425. Moret-Bailly, L.: Hauteurs et classes de Chern sur les surfaces arithmétiques. Astérisque **183**, 37–58 (1990)
4426. Morikawa, R.: Some examples of covering sets. Bull. Fac. Lib. Arts, Nagasaki Univ., Nat. Sci. **21**(2), 1–4 (1981)
4427. Morishima, T.: Über den Fermatschen Quotienten. Jpn. J. Math. **8**, 159–173 (1931)
4428. Morishima, T.: Über die Fermatsche Vermutung, VII. Proc. Jpn. Acad. Sci. **8**, 63–66 (1932)
4429. Morishima, T.: Über die Fermatsche Vermutung, XI. Jpn. J. Math. **11**, 241–252 (1935)
4430. Morishima, T.: On Fermat's last theorem (thirteenth paper). Trans. Am. Math. Soc. **72**, 67–81 (1952)
4431. Morita, Y.: Hecke polynomials $H_k^{(p)}(u)$ ($p = 2$ or 3). J. Fac. Sci. Univ. Tokyo **15**, 99–105 (1968)

4432. Morita, Y.: A p-adic analogue of the Γ-function. J. Fac. Sci. Univ. Tokyo **22**, 255–266 (1975)
4433. Moriya, M.: Über die Fermatsche Vermutung. J. Reine Angew. Math. **169**, 92–97 (1933)
4434. Moriya, M.: Rein arithmetisch-algebraischer Aufbau der Klassenkörpertheorie über algebraischen Funktionenkörpern einer Unbestimmten mit endlichem Konstantenkörper. Jpn. J. Math. **14**, 67–84 (1938)
4435. Morrison, J.F.: Approximation of p-adic numbers by algebraic numbers of bounded degree. J. Number Theory **10**, 334–350 (1978)
4436. Morrison, M.A., Brillhart, J.: The factorization of F_7. Bull. Am. Math. Soc. **77**, 264 (1971)
4437. Morrison, M.A., Brillhart, J.: A method of factoring and the factorization of F_7. Math. Comput. **29**, 183–205 (1975)
4438. Morse, M.: George David Birkhoff and his mathematical work. Bull. Am. Math. Soc. **52**, 357–391 (1946)
4439. Morton, P.: On the nonexistence of abelian conditions governing solvability of the -1 Pell equation. J. Reine Angew. Math. **405**, 147–155 (1990)
4440. Moser, L.: On the diophantine equation $1^n + 2^n + 3^n + \cdots + (m-1)^n = m^n$. Scr. Math. **19**, 84–88 (1953)
4441. Moser, L.: Notes on number theory. II. On a theorem of van der Waerden. Can. Math. Bull. **3**, 23–25 (1960)
4442. Moser, L., Lambek, J.: On monotone multiplicative functions. Proc. Am. Math. Soc. **4**, 544–545 (1953)
4443. Moshchevitin, N.G.: On small fractional parts of polynomials. J. Number Theory **129**, 349–357 (2009)
4444. Mossinghoff, M.J.: Polynomials with small Mahler measure. Math. Comput. **67**, 1697–1705, S11–S14 (1998)
4445. Mossinghoff, M.J., Rhin, G., Wu, Q.: Minimal Mahler measures. Exp. Math. **17**, 451–458 (2008)
4446. Mostow, G.D.: Abraham Robinson, 1918–1974. Isr. J. Math. **25**, 4–14 (1976)
4447. Motohashi, Y.: On the distribution of prime numbers which are of the form $x^2 + y^2 + 1$. Acta Arith. **16**, 351–363 (1969/1970)
4448. Motohashi, Y.: On the distribution of prime numbers which are of the form $x^2 + y^2 + 1$, II. Acta Math. Acad. Sci. Hung. **22**, 207–210 (1971/1972)
4449. Motohashi, Y.: A note on the least prime in an arithmetic progression with a prime difference. Acta Arith. **17**, 283–285 (1970)
4450. Motohashi, Y.: On some improvements of the Brun-Titchmarsh theorem. J. Math. Soc. Jpn. **26**, 306–323 (1974)
4451. Motohashi, Y.: On a density theorem of Linnik. Proc. Jpn. Acad. Sci. **51**, 815–817 (1975)
4452. Motohashi, Y.: An induction principle for the generalization of Bombieri's prime number theorem. Proc. Jpn. Acad. Sci. **52**, 273–275 (1976)
4453. Motohashi, Y.: On some additive divisor problems. J. Math. Soc. Jpn. **28**, 772–784 (1976)
4454. Motohashi, Y.: On some additive divisor problems, II. Proc. Jpn. Acad. Sci. **52**, 279–281 (1976)
4455. Motohashi, Y.: A note on the large sieve, II. Proc. Jpn. Acad. Sci. **53**, 122–124 (1977)
4456. Motohashi, Y.: A note on Siegel's zeros. Proc. Jpn. Acad. Sci. **55**, 190–192 (1979)
4457. Motohashi, Y.: A note on almost-primes in short intervals. Proc. Jpn. Acad. Sci. **55**, 225–226 (1979)
4458. Motohashi, Y.: An elementary proof of Vinogradov's zero-free region for the Riemann zeta function. In: Recent Progress in Analytic Number Theory, Durham, 1979. vol. 1, pp. 257–267. Academic Press, San Diego (1981)
4459. Motohashi, Y.: Lectures on Sieve Methods and Prime Number Theory. Tata Institute, Bombay (1983)
4460. Motohashi, Y.: The binary additive divisor problem. Ann. Sci. Éc. Norm. Super. **27**, 529–572 (1994)
4461. Motohashi, Y.: Spectral Theory of the Riemann Zeta-function. Cambridge University Press, Cambridge (1997)

4462. Mozzochi, C.J.: On the difference between consecutive primes. J. Number Theory **24**, 181–187 (1986)
4463. Mozzochi, C.J.: The Fermat Diary. Am. Math. Soc., Providence (2000)
4464. Muder, D.J.: Putting the best face on a Voronoï polyhedron. Proc. Lond. Math. Soc. **56**, 329–348 (1988)
4465. Muder, D.J.: A new bound on the local density of sphere packings. Discrete Comput. Geom. **10**, 351–375 (1993)
4466. Mueller, J.: On the difference between consecutive primes. In: Recent Progress in Analytic Number Theory, vol. 1, pp. 269–273. Academic Press, San Diego (1981)
4467. Mueller, J., Schmidt, W.M.: Trinomial Thue equations and inequalities. J. Reine Angew. Math. **379**, 76–99 (1987)
4468. Mueller, J., Schmidt, W.M.: Thue's equation and a conjecture of Siegel. Acta Math. **160**, 207–247 (1988)
4469. Mulholland, H.P.: On the product of n complex homogeneous linear forms. J. Lond. Math. Soc. **35**, 241–250 (1960)
4470. Müller, G.H.: Paul J. Bernays (1888–1977). Math. Intell. **1**, 27–28 (1978/1979)
4471. Müller, H.: Über die asymptotische Verteilung von Beurlingschen Zahlen. J. Reine Angew. Math. **289**, 181–187 (1977)
4472. Müller, H.H., Ströher, H., Zimmer, H.G.: Torsion groups of elliptic curves with integral j-invariant over quadratic fields. J. Reine Angew. Math. **397**, 100–161 (1989)
4473. Müller, P.: A Weil-bound free proof of Schur's conjecture. Finite Fields Appl. **3**, 25–32 (1997)
4474. Müller, W., Nowak, W.G.: Lattice points in domains $|x|^p + |y|^p \leq R^p$. Arch. Math. **51**, 55–59 (1988)
4475. Müller, W., Nowak, W.G.: Lattice points in planar domains: applications of Huxley's "discrete Hardy-Littlewood method". In: Lecture Notes in Math., vol. 1452, pp. 139–164. Springer, Berlin (1990)
4476. Mumford, D.: A remark on Mordell's conjecture. Am. J. Math. **87**, 1007–1016 (1965)
4477. Mumford, D.: Tata Lectures in Theta, vol. 1. Birkhäuser, Basel (1983) [Reprint: 2007]
4478. Mumford, D.: Tata Lectures in Theta, vol. 2. Birkhäuser, Basel (1984) [Reprint: 2007]
4479. Mumford, D.: Tata Lectures in Theta, vol. 3. Birkhäuser, Basel (1991) [Reprint: 2007]
4480. Müntz, Ch.H.: Allgemeine Begründung der Theorie der höheren ζ-Funktionen. Abh. Math. Semin. Univ. Hamb. **3**, 1–11 (1924)
4481. Müntz, Ch.H.: Zur Gittertheorie n-dimensionalen Ellipsoiden. Math. Z. **25**, 150–165 (1926)
4482. Murata, L.: On the magnitude of the least prime primitive root. J. Number Theory **37**, 47–66 (1991)
4483. Murty, M.R.: Some Ω-results for Ramanujan's τ-function. In: Lecture Notes in Math., vol. 938, pp. 123–137. Springer, Berlin (1982)
4484. Murty, M.R.: Oscillations of Fourier coefficients of modular forms. Math. Ann. **262**, 431–446 (1983)
4485. Murty, M.R.: On Artin's conjecture. J. Number Theory **16**, 147–168 (1983)
4486. Murty, M.R.: On the supersingular reduction of elliptic curves. Proc. Indian Acad. Sci. Math. Sci. **97**, 247–250 (1987)
4487. Murty, M.R.: The Ramanujan τ function. In: Andrews, G.E., Askey, R.A., Berndt, B.C., Ramanathan, K.G., Rankin, R.A. (eds.) Ramanujan Revisited (Proceedings of the Centenary Conference), pp. 269–288. Academic Press, San Diego (1988)
4488. Murty, M.R.: Selberg's conjectures and Artin L-functions. Bull. Am. Math. Soc. **31**, 1–14 (1994)
4489. Murty, M.R.: Selberg's conjectures and Artin L-functions, II. In: Current Trends in Mathematics and Physics, pp. 154–168. Narosa Publishing House, New Delhi (1995)
4490. Murty, M.R., Murty, V.K.: Mean values of derivatives of modular L-series. Ann. Math. **133**, 447–475 (1991)
4491. Murty, M.R., Murty, V.K., Shorey, T.N.: Odd values of the Ramanujan function. Bull. Soc. Math. Fr. **115**, 391–395 (1987)

4492. Murty, M.R., Perelli, A.: The pair correlation of zeros of functions in the Selberg class. Internat. Math. Res. Notices, 1999, 531–545
4493. Murty, M.R., Petridis, Y.N.: On Kummer's conjecture. J. Number Theory **80**, 294–303 (2001)
4494. Murty, [M.]R., Zaharescu, A.: Explicit formulas for the pair correlation of zeros of functions in the Selberg class. Forum Math. **14**, 65–83 (2002)
4495. Murty, V.K.: Explicit formulae and the Lang-Trotter conjecture. Rocky Mt. J. Math. **15**, 535–551 (1985)
4496. Musin, O.R.: The problem of the twenty-five spheres. Usp. Mat. Nauk **58**(4), 153–154 (2003) (in Russian)
4497. Musin, O.R.: The kissing problem in three dimensions. Discrete Comput. Geom. **35**, 375–384 (2006)
4498. Musin, O.R.: The kissing number in four dimensions. Ann. Math. **168**, 1–32 (2008)
4499. Myrberg, P.J.: Ernst Lindelöf in memoriam. Acta Math. **79**, I–IV (1947)
4500. Naganuma, H.: On the coincidence of two Dirichlet series associated with cusp forms of Hecke's "Neben"-type and Hilbert modular forms over a real quadratic field. J. Math. Soc. Jpn. **25**, 547–555 (1973)
4501. Nagel[l], T.: The indeterminate equations $x^2+x+1=y^n$ et $x^2+x+1=3y^n$. Norsk. Mat. Forenings Skr. **2**, 1–14 (1920) (in Norwegian)
4502. Nagel[l], T.: A note on the indeterminate equation $\frac{x^n-1}{x-1}=y^q$. Norsk Mat. Tidsskr. **2**, 75–78 (1920) (in Norwegian)
4503. Nagel[l], T.: Généralisation d'un théorème de Tchebycheff. J. Math. Pures Appl. **4**, 343–356 (1921)
4504. Nagel[l], T.: Zur Arithmetik der Polynome. Abh. Math. Semin. Univ. Hamb. **1**, 179–194 (1922)
4505. Nagell, T.: Über Einheiten in reinen kubischen Körpern. Christiania Vid. Selsk, Skr., 1923, nr. 11, 1–34
4506. Nagell, T.: Über die rationalen Punkte auf einigen kubischen Kurven. Tohoku Math. J. **24**, 48–53 (1924/1925)
4507. Nagell, T.: Über einige kubische Gleichungen mit zwei Unbestimmten. Math. Z. **24**, 422–447 (1925)
4508. Nagell, T.: Solution complète de quelques équations cubiques à deux indéterminées. J. Math. Pures Appl. **4**, 209–270 (1925)
4509. Nagell, T.: Darstellung ganzer Zahlen durch binäre kubische Formen mit negativer Diskriminante. Math. Z. **28**, 10–29 (1928)
4510. Nagell, T.: Sur les propriétés arithmétiques des cubiques planes du premier genre. Acta Math. **52**, 93–126 (1928)
4511. Nagell, T.: Solution de quelques problèmes dans la théorie arithmétique des cubiques planes de premier genre. Vid. Akad. Skr. Oslo **1**(1), 1–25 (1935)
4512. Nagell, T.: Über den grössten Primteiler gewisser Polynome dritten Grades. Math. Ann. **114**, 284–292 (1937)
4513. Nagell, T.: Problems in the theory of exceptional points on plane cubics of genus one. In: Den 11te Skandinaviske Matematikerkongres, Trondheim, 1949. pp. 71–76. J. Grundt. Tanums Forlag, Oslo (1952)
4514. Nagell, T.: Sur la division des périodes de la fonction $\wp(u)$ et les points exceptionnels des cubiques. Nova Acta Soc. Sci. Uppsaliensis **15**(8), 1–73 (1953)
4515. Nagell, T.: Les points exceptionnels rationnels sur certaines cubiques du premier genre. Acta Arith. **5**, 333–357 (1959)
4516. Nagell, T.: Anders Wiman in memoriam. Acta Math. **103**, i–vi (1960)
4517. Nagell, T.: The diophantine equation $x^2+7=2^n$. Ark. Mat. **4**, 185–187 (1961)
4518. Nagell, T.: Thoralf Skolem in memoriam. Acta Math. **110**, i–xi (1963)
4519. Nair, M.: Power free values of polynomials. Mathematika **23**, 159–183 (1976)
4520. Nair, M.: Power free values of polynomials, II. Proc. Lond. Math. Soc. **38**, 353–368 (1979)
4521. Nair, M., Perelli, A.: On the prime ideal theorem and irregularities in the distribution of primes. Duke Math. J. **77**, 1–20 (1995)

4522. Nair, R.: On strong uniform distribution. Acta Arith. **56**, 183–193 (1990)
4523. Nair, R.: On strong uniform distribution, II. Monatshefte Math. **132**, 341–348 (2001)
4524. Nair, R.: On strong uniform distribution, III. Indag. Math. **14**, 233–240 (2003)
4525. Nair, R.: On strong uniform distribution, IV. J. Inequal. Appl., 2005, 319–327
4526. Nakagawa, J.: Binary forms and orders of algebraic number fields. Invent. Math. **97**, 219–235 (1989)
4527. Nakagawa, J., Horie, K.: Elliptic curves with no rational points. Proc. Am. Math. Soc. **104**, 20–24 (1988)
4528. Nakai, Y., Shiokawa, I.: A class of normal numbers. Jpn. J. Math. **16**, 17–29 (1990)
4529. Nakai, Y., Shiokawa, I.: A class of normal numbers, II. In: Number Theory and Cryptography, Sydney, 1989, pp. 204–210. Cambridge University Press, Cambridge (1990)
4530. Nakai, Y., Shiokawa, I.: Normality of numbers generated by the values of polynomials at primes. Acta Arith. **81**, 345–356 (1997)
4531. Nakazawa, N.: Parametric families of elliptic curves with cyclic F_p-rational points groups. Tokyo J. Math. **28**, 381–392 (2005)
4532. Nakazawa, N.: Construction of elliptic curves with cyclic groups over prime fields. Bull. Aust. Math. Soc. **73**, 245–254 (2006)
4533. Narasinga Rao, A., Ganapathy Iyer, V.: R. Vaidyanathaswamy (1894–1960). Math. Stud. **29**, 1–14 (1961) [Reprint: J. Indian Math. Soc. **64**, 1997, 1–13]
4534. Narasimhamurti, V.: On Waring's problem for 8th, 9th, and 10th powers. J. Indian Math. Soc. **5**, 122 (1941)
4535. Narkiewicz, W.: Divisibility properties of a class of multiplicative functions. Colloq. Math. **18**, 219–232 (1967)
4536. Narkiewicz, W.: On a conjecture of Erdös. Colloq. Math. **37**, 313–315 (1977)
4537. Narkiewicz, W.: Uniform Distribution of Sequences of Integers in Residue Classes. Lecture Notes in Math., vol. 1087. Springer, Berlin (1984)
4538. Narkiewicz, W.: Classical Problems in Number Theory. PWN, Warsaw (1986)
4539. Narkiewicz, W.: A note on Artin's conjecture in algebraic number fields. J. Reine Angew. Math. **381**, 110–115 (1987)
4540. Narkiewicz, W.: Hermann Weyl and the theory of numbers. In: Exact Sciences and Their Philosophical Foundations, pp. 51–60. Peter Lang, Oxford (1988)
4541. Narkiewicz, W.: Polynomial Mappings. Lecture Notes in Math., vol. 1600. Springer, Berlin (1995)
4542. Narkiewicz, W.: The Development of Prime Number Theory. Springer, Berlin (2000)
4543. Narkiewicz, W.: Elementary and Analytic Theory of Algebraic Numbers, 3rd ed. Springer, Berlin (2004)
4544. Narumi, S.: An extension of a theorem of Liouville's. Tohoku Math. J. **11**, 128–142 (1917)
4545. Narzullaev, Kh.N., Skubenko, B.F.: A refinement of an estimation of the arithmetical minimum of the product of inhomogeneous linear forms (on the inhomogeneous Minkowski conjecture). Zap. Nauč. Semin. LOMI **82**, 88–94 (1979) (in Russian)
4546. Nathanson, M.B.: Linear recurrences and uniform distribution. Proc. Am. Math. Soc. **48**, 289–291 (1975)
4547. Nathanson, M.B.: Additive Number Theory. Inverse Problems and the Geometry of Sumsets. Springer, Berlin (1996)
4548. Nathanson, M.B.: Elementary Methods in Number Theory. Springer, Berlin (2000)
4549. Nebe, G.: Some cyclo-quaternionic lattices. J. Algebra **199**, 472–498 (1998)
4550. Nebe, G.: Construction and investigation of lattices with matrix groups. In: Contemp. Math., vol. 249, pp. 205–219. Am. Math. Soc., Providence (1999)
4551. Nebe, G.: Gitter und Modulformen. Jahresber. Dtsch. Math.-Ver. **104**, 125–144 (2002)
4552. Nečaev, V.I.: Waring's problem for polynomials. Tr. Mat. Inst. Steklova **38**, 190–243 (1951) (in Russian)
4553. Nečaev, V.I.: On the representation of natural numbers by summands of the form $\frac{x(x+1)\cdots(x+k-1)}{k!}$. Izv. Akad. Nauk SSSR, Ser. Mat. **17**, 485–498 (1953) (in Russian)
4554. Nečaev, V.I.: Estimate of the complete trigonometric sum. Mat. Zametki **17**, 839–843 (1975) (in Russian)

4555. Nečaev, V.I., Topunov, V.L.: Estimation of the modulus of complete rational trigonometric sums of degree three and four. Tr. Mat. Inst. Steklova **158**, 125–129 (1981) (in Russian)
4556. Nelson, H.L.: A solution to Archimedes' cattle problem. J. Recreat. Math. **13**, 162–176 (1980/1981)
4557. Nemenzo, F.R.: All congruent numbers less than 40000. Proc. Jpn. Acad. Sci. **74**, 29–31 (1998)
4558. Néron, A.: Problèmes arithmétiques et géométriques rattachés à la notion de rang d'une courbe algébrique dans un corps. Bull. Soc. Math. Fr. **80**, 101–166 (1952)
4559. Néron, A.: Propriétés arithmétiques de certaines familles de courbes algébriques. In: Proc. ICM, Amsterdam, 1954, vol. 3, pp. 481–488. North-Holland, Amsterdam (1956)
4560. Nešetřil, J., Serra, O.: The Erdős-Turán property for a class of bases. Acta Arith. **115**, 245–254 (2004)
4561. Nesterenko, Yu.V.: An order function for almost all numbers. Mat. Zametki **15**, 405–414 (1974) (in Russian)
4562. Nesterenko, Yu.V.: Estimate of the orders of the zeroes of functions of a certain class, and their application in the theory of transcendental numbers. Izv. Akad. Nauk SSSR, Ser. Mat. **41**, 253–284 (1977) (in Russian)
4563. Nesterenko, Yu.V.: On the algebraical independence of algebraic numbers to algebraic powers. Prog. Math. **31**, 199–220 (1983)
4564. Nesterenko, Yu.V.: Algebraic independence of algebraic powers of algebraic numbers. Mat. Sb. **123**, 435–459 (1984) (in Russian)
4565. Nesterenko, Yu.V.: Measure of algebraic independence of values of an elliptic function at algebraic points. Usp. Mat. Nauk **40**(4), 221–222 (1985) (in Russian)
4566. Nesterenko, Yu.V.: On the number π. Vestnik Moskov. Univ. Ser. I. Mat. Mekh., 1987, nr. 3, 7–10 (in Russian)
4567. Nesterenko, Yu.V.: Degrees of transcendence of some fields that are generated by values of an exponential function. Mat. Zametki **45**, 40–49 (1989) (in Russian)
4568. Nesterenko, Yu.V.: On the measure of algebraic independence of values of an elliptic function. Izv. Ross. Akad. Nauk, Ser. Mat. **59**, 155–178 (1995) (in Russian)
4569. Nesterenko, Yu.V.: Modular functions and transcendence questions. Mat. Sb. **187**(9), 65–96 (1996) (in Russian)
4570. Nesterenko, Yu.V.: Some remarks on $\zeta(3)$. Mat. Zametki **59**, 865–880 (1996) (in Russian)
4571. Nesterenko, Yu.V.: On the measure of algebraic independence of values of Ramanujan functions. Tr. Mat. Inst. Steklova **218**, 299–334 (1997) (in Russian)
4572. Nesterenko, Yu.V.: On the algebraic independence of values of Ramanujan functions. Vestnik Moskov. Univ. Ser. I. Mat. Mekh., 2001, nr. 2, 6–10 (in Russian)
4573. Nesterenko, Yu.V., Shorey, T.N.: On an equation of Goormaghtigh. Acta Arith. **83**, 381–389 (1999)
4574. Neubauer, G.: Eine empirische Untersuchung zur Mertensscher Funktion. Numer. Math. **5**, 1–13 (1963)
4575. Neukirch, J.: Class Field Theory. Springer, Berlin (1986)
4576. Neukirch, J.: Algebraische Zahlentheorie. Springer, Berlin (1992) [English translation: Algebraic Number Theory, Springer, 1999]
4577. Neumann, B.H.: Felix Adalbert Behrend. J. Lond. Math. Soc. **38**, 308–310 (1963)
4578. von Neumann, J., Goldstine, H.H.: A numerical study of a conjecture of Kummer. Math. Tables Other Aids Comput. **7**, 133–134 (1953)
4579. Neumann, O.: Two proofs of the Kronecker-Weber theorem "according to Kronecker, and Weber". J. Reine Angew. Math. **323**, 105–126 (1981)
4580. Neumann, P.M., Rayner, M.E.: Obituary: William Leonard Ferrar. Bull. Lond. Math. Soc. **26**, 395–401 (1994)
4581. Nevanlinna, F., Nevanlinna, R.: Über die Nullstellen der Riemannschen Zetafunktion. Math. Z. **20**, 253–263 (1923); **23**, 159–160 (1925)
4582. Neville, E.H. (ed.): The Farey Series of Order 1025. Cambridge University Press, Cambridge (1950)

4583. Newman, D.J.: A simplified proof of Waring's conjecture. Mich. Math. J. **7**, 291–295 (1960)
4584. Newman, D.J.: A simplified proof of the Erdős-Fuchs theorem. Proc. Am. Math. Soc. **75**, 209–210 (1979)
4585. Newman, M.: Congruences for the coefficients of modular forms and some new congruences for the partition function. Can. J. Math. **9**, 549–552 (1957)
4586. Newman, M.: Congruences for the coefficients of modular forms and for the coefficients of $j(\tau)$. Proc. Am. Math. Soc. **9**, 609–612 (1958)
4587. Newman, M.: Periodicity modulo m and divisibility properties of the partition function. Trans. Am. Math. Soc. **97**, 225–236 (1960)
4588. Newman, M.: Maximal normal subgroups of the modular group. Proc. Am. Math. Soc. **19**, 1138–1144 (1968)
4589. Newman, M.H.A.: Hermann Weyl. J. Lond. Math. Soc. **33**, 500–511 (1958)
4590. Ng, N.: The distribution of the summatory function of the Möbius function. Proc. Lond. Math. Soc. **89**, 361–389 (2004)
4591. Nicely, T.R.: New maximal prime gaps and first occurrences. Math. Comput. **68**, 1311–1315 (1999)
4592. Nicolae, F.: On Artin's L-functions, I. J. Reine Angew. Math. **539**, 179–184 (2001)
4593. Nicolas, J.-L.: Ordre maximal d'un élément du groupe S_n des permutations et "highly composite numbers". Bull. Soc. Math. Fr. **97**, 129–191 (1969)
4594. Nicolas, J.-L.: Répartition des nombres hautement composés de Ramanujan. Can. J. Math. **23**, 116–130 (1971)
4595. Nicolas, J.-L.: Nombres hautement composés. Acta Arith. **49**, 395–412 (1988)
4596. Nicolas, J.-L.: On highly composite numbers. In: Andrews, G.E., Askey, R.A., Berndt, B.C., Ramanathan, K.G., Rankin, R.A. (eds.) Ramanujan Revisited (Proceedings of the Centenary Conference), pp. 215–244. Academic Press, San Diego (1988)
4597. Niedermeier, F.: Ein elementarer Beitrag zu Fermatschen Vermutung. J. Reine Angew. Math. **185**, 111–112 (1943)
4598. Niederreiter, H.: Distribution of Fibonacci numbers mod 5^k. Fibonacci Q. **10**, 373–374 (1972)
4599. Niederreiter, H.: Low-discrepancy point sets. Monatshefte Math. **102**, 155–167 (1986)
4600. Niederreiter, H.: On an irrationality theorem of Mahler and Bundschuh. J. Number Theory **24**, 197–199 (1986)
4601. Niederreiter, H.: Point sets and sequences with small discrepancy. Monatshefte Math. **104**, 273–337 (1987)
4602. Niederreiter, H.: Low-discrepancy and low-dispersion sequences. J. Number Theory **30**, 51–70 (1988)
4603. Niederreiter, H.: Random Number Generation and quasi-Monte Carlo Methods. SIAM, Philadelphia (1992)
4604. Niederreiter, H., Shiue, [P.]J.-S.: Equidistribution of linear recurring sequences in finite fields. Indag. Math. **39**, 397–405 (1977)
4605. Niederreiter, H., Shiue, [P.]J.-S.: Equidistribution of linear recurring sequences in finite fields, II. Acta Arith. **38**, 197–207 (1980/1981)
4606. Niederreiter, H., Xing, C.P.: Low-discrepancy sequences obtained from algebraic function fields over finite fields. Acta Arith. **72**, 281–298 (1995)
4607. Niederreiter, H., Xing, C.P.: Explicit global function fields over the binary field with many rational places. Acta Arith. **75**, 383–396 (1996)
4608. Niederreiter, H., Xing, C.P.: Low-discrepancy sequences and global function fields with many rational places. Finite Fields Appl. **2**, 241–273 (1996)
4609. Niederreiter, H., Xing, C.P.: Quasirandom points and global function fields. In: Finite Fields and Applications, Glasgow, 1995, pp. 269–296. Cambridge University Press, Cambridge (1996)
4610. Niederreiter, H., Xing, C.P.: Cyclotomic function fields, Hilbert class fields, and global function fields with many rational places. Acta Arith. **79**, 59–76 (1997)

4611. Nieland, L.W.: Zum Kreisproblem. Math. Ann. **98**, 717–736 (1928); corr., **100**, p. 480
4612. Nielsen, P.: An upper bound for odd perfect numbers. Integers **3**, #A14 (2003)
4613. Nielsen, P.: Odd perfect numbers have at least nine distinct prime factors. Math. Comput. **76**, 2109–2126 (2007)
4614. Nielsen, P.: A covering system whose smallest modulus is 40. J. Number Theory **129**, 640–666 (2009)
4615. Niemeier, H.-V.: Definite quadratische Formen der Dimension 24 und Diskriminante 1. J. Number Theory **5**, 142–178 (1973)
4616. Nikolaĭ Mikhaĭlovič Korobov (November 23, 1917–October 25, 2004). Chebyshev. Sb. **6**, 224–230 (2005) (in Russian)
4617. Nikolskiĭ, S.M., Zaharov, V.K.: Konstantin Konstantinovič Mardžanišvili. Usp. Mat. Nauk **38**(5), 107–109 (1983) (in Russian)
4618. Nishioka, K.: Mahler functions and transcendence. Lecture Notes in Math., vol. 1631. Springer, Berlin (1996)
4619. Nitaj, A.: Invariants des courbes de Frey-Hellegouarch et grands groupes de Tate-Shafarevich. Acta Arith. **93**, 303–327 (2000)
4620. Niven, I.: An unsolved case of the Waring problem. Am. J. Math. **66**, 137–143 (1944)
4621. Niven, I.: Uniform distribution of sequences of integers. Trans. Am. Math. Soc. **98**, 52–61 (1961)
4622. Niwa, S.: Modular forms of half integral weight and the integral of certain theta-functions. Nagoya Math. J. **56**, 147–161 (1975)
4623. Noda, K., Wada, H.: All congruent numbers less than 10 000. Proc. Jpn. Acad. Sci. **69**, 175–178 (1993)
4624. Noether, M.: Charles Hermite. Math. Ann. **55**, 337–385 (1902)
4625. Nongkynrih, A.: On prime primitive roots. Acta Arith. **72**, 45–53 (1995)
4626. Norton, K.K.: Numbers with small prime factors, and the least kth power non-residue. Mem. Am. Math. Soc. **106**, 1–106 (1971)
4627. Nosarzewska, M.: Évaluation de la différence entre l'aire d'une région plane convexe et le nombre des points aux coordonnées entieres couverts par elle. Colloq. Math. **1**, 305–311 (1948)
4628. Novák, B., Schwarz, Š.: Vojtěch Jarník. Acta Arith. **20**, 107–115 (1972)
4629. Nowak, W.G.: Ein kurzer Beweis eines Satzes von Sierpiński. Ann. Univ. Sci. Bp. **24**, 153–156 (1981)
4630. Nowak, W.G.: Eine Bemerkung zum Kreisproblem in der p-Norm. SBer. Österr. Akad. Wiss. Math. Natur. Kl. II **191**, 125–132 (1982)
4631. Nowak, W.G.: Lattice points in a circle and divisors in arithmetic progressions. Manuscr. Math. **49**, 195–205 (1984)
4632. Nowak, W.G.: Zur Gitterpunktlehre der euklidischen Ebene. Indag. Math. **46**, 209–223 (1984)
4633. Nowak, W.G.: Zur Gitterpunktlehre der euklidischen Ebene, II. SBer. Österr. Akad. Wiss. Math. Natur. Kl. **194**, 31–37 (1985)
4634. Nowak, W.G.: On the divisor problem in an arithmetic progression. Comment. Math. Univ. St. Pauli **33**, 209–217 (1984)
4635. Nowak, W.G.: An Omega-estimate for the lattice rest of a convex planar domain. Proc. R. Soc. Edinb., Sect. A, Math. **100**, 295–299 (1985)
4636. Nowak, W.G.: An Ω_+-estimate for the number of lattice points in a sphere. Rend. Semin. Mat. Univ. Padova **73**, 31–40 (1985)
4637. Nowak, W.G.: Einige Beiträge zur Theorie der Gitterpunkte. In: Lecture Notes in Math., vol. 1114, pp. 98–117. Springer, Berlin (1985)
4638. Nowak, W.G.: On the lattice rest of a convex body in R^s. Arch. Math. **45**, 284–288 (1985)
4639. Nowak, W.G.: On the lattice rest of a convex body in R^s, II. Arch. Math. **47**, 232–237 (1986)
4640. Nowak, W.G.: On the lattice rest of a convex body in R^s, III. Czechoslov. Math. J. **41**, 359–367 (1991)

4641. Nowak, W.G.: Zum Kreisproblem. SBer. Österreich. AW SBer. Math.-Natur. Kl. **194**, 265–271 (1985)
4642. Nowak, W.G.: On the distribution of integer ideals in algebraic number fields. Math. Nachr. **161**, 59–74 (1993)
4643. Nowak, W.G.: Sums of two kth powers: an omega estimate for the error term. Arch. Math. **68**, 27–35 (1997)
4644. Nowak, W.G.: Lattice points in a circle: an improved mean-square asymptotics. Acta Arith. **113**, 259–272 (2004)
4645. Nyman, B.: A general prime number theorem. Acta Math. **81**, 299–307 (1949)
4646. Nyman, B.: On the one-dimensional translation group and semi-group in certain function spaces. Dissertation, Univ. Uppsala (1950)
4647. Oberschelp, W.: Hans Hermes 12.2.1912 bis 10.11.2003. Jahresber. Dtsch. Math.-Ver. **109**, 99–109 (2007)
4648. Obláth, R.: Über Produkte aufeinanderfolgenden Zahlen. Tohoku Math. J. **38**, 73–92 (1933)
4649. Obláth, R.: Note on the binomial coefficients. J. Lond. Math. Soc. **23**, 252–253 (1948)
4650. Obláth, R.: Untere Schranken für Lösungen der Fermatschen Gleichung. Port. Math. **11**, 129–132 (1952)
4651. Obláth, R.: Une propriété des puissances parfaites. Mathesis **65**, 356–364 (1956)
4652. Obreškov, N.: On the approximation of irrational numbers by rational numbers. C.R. Acad. Bolgare Sci. **3**(1), 1–4 (1950) (in Russian)
4653. Oda, T.: On the poles of Andrianov L-functions. Math. Ann. **256**, 323–340 (1981)
4654. Odlyzko, A.M.: On the distribution of spacings between zeros of the zeta function. Math. Comput. **48**, 273–308 (1987)
4655. Odlyzko, A.M.: The 10^{22}-nd zero of the Riemann zeta function. In: Contemp. Math., vol. 290, pp. 139–144. Am. Math. Soc., Providence (2001)
4656. Odlyzko, A.M., Sloane, N.J.A.: New bounds on the number of unit spheres that can touch a unit sphere in n dimensions. J. Comb. Theory, Ser. A **26**, 210–214 (1979)
4657. Odlyzko, A., te Riele, H.J.J.: Disproof of the Mertens conjecture. J. Reine Angew. Math. **357**, 138–160 (1985)
4658. Odoni, R.W.K.: On Gauss sums $(\bmod p^n)$, $n \geq 2$. Bull. Lond. Math. Soc. **5**, 325–327 (1973)
4659. Oesterlé, J.: Travaux de Wiles (et Taylor, ...), II. Astérisque **237**, 333–355 (1996)
4660. Oesterlé, J.: Effective versions of the Chebotarev density theorem. Astérisque **61**, 165–167 (1979)
4661. Oesterlé, J.: Nombres de Tamagawa et groupes unipotents en caractéristique p. Invent. Math. **78**, 13–88 (1984)
4662. Oesterlé, J.: Nombre de classes des corps quadratiques imaginaires. Astérisque **121/122**, 309–323 (1985)
4663. Oesterlé, J.: Nouvelles approches du "théorème" de Fermat. Astérisque **161/162**, 165–186 (1988)
4664. Ogg, A.P.: Cohomology of abelian varieties over function fields. Ann. Math. **76**, 185–212 (1962)
4665. Ogg, A.P.: Abelian curves of 2-power conductor. Proc. Camb. Philos. Soc. **62**, 143–148 (1966)
4666. Ogg, A.P.: Abelian curves of small conductor. J. Reine Angew. Math. **226**, 204–215 (1967)
4667. Ogg, A.[P.]: Modular Forms and Dirichlet Series. Benjamin, Elmsford (1969)
4668. Ogg, A.P.: A remark on the Sato-Tate conjecture. Invent. Math. **9**, 198–200 (1970)
4669. Ogg, A.[P.]: Rational points of finite order on elliptic curves. Invent. Math. **12**, 105–111 (1971)
4670. Ogg, A.[P.]: Rational points on certain elliptic modular curves. In: Proc. Symposia Pure Math., vol. 24, pp. 221–231. Am. Math. Soc., Providence (1973)
4671. Ogg, A.[P.]: Diophantine equations and modular forms. Bull. Am. Math. Soc. **81**, 14–27 (1975)
4672. Ohta, M.: On theta series mod p. J. Fac. Sci. Univ. Tokyo **28**, 679–686 (1981)
4673. Okano, T.: A note on the rational approximations to e. Tokyo J. Math. **15**, 129–133 (1992)

4674. Okazaki, R.: Geometry of a cubic Thue equation. Publ. Math. (Debr.) **61**, 267–314 (2002)
4675. Oliveira e Silva, T.: Maximum excursion and stopping time record-holders for the $3x+1$ problem: computational results. Math. Comput. **68**, 371–384 (1999)
4676. Olsen, L.: Applications of multifractal divergence points to sets of numbers defined by their N-adic expansion. Math. Proc. Camb. Philos. Soc. **136**, 139–165 (2004)
4677. Olson, J.E.: An addition theorem modulo p. J. Comb. Theory **5**, 45–52 (1968)
4678. Olson, J.[E.]: Henry B. Mann. In: Number Theory and Algebra, pp. xx–xxv. Academic Press, San Diego (1977)
4679. Oltramare, G.: L'Intermédiaire Math. **1**, 25 (1894)
4680. Ono, K.: Euler's concordant forms. Acta Arith. **78**, 101–123 (1996)
4681. Ono, K.: Parity of the partition function in arithmetic progressions. J. Reine Angew. Math. **472**, 1–15 (1996)
4682. Ono, K.: The partition function in arithmetic progressions. Math. Ann. **213**, 251–260 (1998)
4683. Ono, K.: Distribution of the partition function modulo m. Ann. Math. **151**, 293–307 (2000)
4684. Ono, K.: Representations of integers as sums of squares. J. Number Theory **95**, 253–258 (2002)
4685. Ono, K., Soundararajan, K.: Ramanujan's ternary quadratic form. Invent. Math. **130**, 415–454 (1997)
4686. Opolka, H., Scharlau, W.: From Fermat to Minkowski. Springer, Berlin (1985)
4687. Oppenheim, A.: On an arithmetic function. J. Lond. Math. Soc. **1**, 205–211 (1926)
4688. Oppenheim, A.: On an arithmetic function, II. J. Lond. Math. Soc. **2**, 123–130 (1927)
4689. Oppenheim, A.: Some identities in the theory of numbers. Proc. Lond. Math. Soc. **26**, 295–350 (1927)
4690. Oppenheim, A.: The minima of indefinite quaternary quadratic forms. Proc. Natl. Acad. Sci. USA **15**, 724–727 (1929)
4691. Oppenheim, A.: The determination of all universal ternary quadratic forms. Q. J. Math. **1**, 179–185 (1930)
4692. Oppenheim, A.: The minima of indefinite quaternary quadratic forms. Ann. Math. **32**, 271–298 (1931)
4693. Oppenheim, A.: The minima of quaternary quadratic forms of signature zero. Proc. Lond. Math. Soc. **37**, 63–81 (1934)
4694. Oppenheim, A.: Remark on the minimum of quadratic forms. J. Lond. Math. Soc. **21**, 251–252 (1946)
4695. Oppenheim, A.: Values of quadratic forms, I. Q. J. Math. **4**, 54–59 (1953)
4696. Oppenheim, A.: Values of quadratic forms, II. Q. J. Math. **4**, 60–66 (1953)
4697. Oppermann, L.: Om vor Kundskab om Primtallenes Moengde mellem givne Groenser. Overs. Dansk. Vidensk. Selsk. Forh., 1882, 169–179
4698. Orde, H.L.S.: On Dirichlet's class number formula. J. Lond. Math. Soc. **18**, 409–420 (1978)
4699. Ore, O.: Anzahl der Wurzeln höherer Kongruenzen. Norsk Mat. Tidsskr. **3**, 63–66 (1921)
4700. Ortiz, E.L., Pinkus, A.: Herman Müntz: a mathematician's odyssey. Math. Intell. **27**, 22–31 (2005)
4701. Orton, L.: An elementary proof of a weak exceptional zero conjecture. Can. J. Math. **56**, 373–405 (2004)
4702. Osgood, C.F.: The diophantine approximation of roots of positive integers. J. Res. Natl. Bur. Stand. **74B**, 241–244 (1970)
4703. Ostmann, H.-H.: Verfeinerte Lösung der asymptotischen Dichtenaufgabe. J. Reine Angew. Math. **187**, 183–188 (1950)
4704. Ostrowski, A.: Über einige Fragen der allgemeinen Körpertheorie. J. Reine Angew. Math. **143**, 255–284 (1913)
4705. Ostrowski, A.: Über sogenannte perfekte Körper. J. Reine Angew. Math. **147**, 191–204 (1917)
4706. Ostrowski, A.: Über einige Lösungen der Funktionalgleichung $\varphi(x)\varphi(y) = \varphi(xy)$. Acta Math. **41**, 271–284 (1918)

4707. Ostrowski, A.: Über ganzwertige Polynome in algebraischen Zahlkörpern. J. Reine Angew. Math. **149**, 117–124 (1919)
4708. Ostrowski, A.: Bemerkungen zur Theorie der Diophantischen Approximationen. Abh. Math. Semin. Univ. Hamb. **1**, 77–98 (1922); corr.: 250–251; **4**, 224 (1922)
4709. Ostrowski, A.: Untersuchungen zur arithmetischen Theorie der Körper. (Die Theorie der Teilbarkeit in allgemeinen Körpern). Math. Z. **39**, 269–404, 321–404 (1935)
4710. Oystein Ore (1899–1968). J. Comb. Theory, **8**, i–iii (1970)
4711. Ožigova, E.P.: The development of number theory in Russia. Nauka, Leningrad (1972) (in Russian)
4712. Özlük, A.E.: On the q-analogue of the pair correlation conjecture. J. Number Theory **56**, 319–351 (1996)
4713. Özlük, A.E., Snyder, C.: On the distribution of the non-trivial zeros of quadratic L-functions close to the real axis. Acta Arith. **91**, 209–228 (1999)
4714. Padé, H.: Sur la généralisation des fractions continues algébriques. J. Math. Pures Appl. **10**, 291–329 (1894)
4715. Page, A.: An asymptotic formula in the theory of numbers. J. Lond. Math. Soc. **7**, 24–27 (1932)
4716. Page, A.: A statement by Ramanujan. J. Lond. Math. Soc. **7**, 105–112 (1932)
4717. Page, A.: On the representations of a number as a sum of squares and products, I. Proc. Lond. Math. Soc. **36**, 241–256 (1933)
4718. Page, A.: On the representations of a number as a sum of squares and products, II. Proc. Lond. Math. Soc. **36**, 470–484 (1933)
4719. Page, A.: On the representations of a number as a sum of squares and products, III. Proc. Lond. Math. Soc. **37**, 1–16 (1934)
4720. Page, A.: On the number of primes in an arithmetic progression. Proc. Lond. Math. Soc. **39**, 116–141 (1935)
4721. Pajunen, S.: Computations on the growth of the first factor for prime cyclotomic fields. Nor-disk Tidskr. Inf. **16**, 85–87 (1976)
4722. Pajunen, S.: Computations on the growth of the first factor for prime cyclotomic fields, II. Nor-disk Tidskr. Inf. **17**, 113–114 (1977)
4723. Pak, I.: Partition bijections, a survey. Ramanujan J. **12**, 5–75 (2006)
4724. Paley, R.E.A.C.: On the k-analogues of some theorems in the theory of the Riemann zeta function. Proc. Lond. Math. Soc. **32**, 273–311 (1931)
4725. Paley, R.E.A.C.: A theorem on characters. J. Lond. Math. Soc. **7**, 28–32 (1932)
4726. Pall, G.: Large positive integers are sums of four or five values of a quadratic function. Am. J. Math. **54**, 66–78 (1932)
4727. Pall, G.: On sums of squares. Am. Math. Mon. **40**, 10–18 (1933)
4728. Pall, G.: An almost universal form. Bull. Am. Math. Soc. **46**, 291 (1940)
4729. Pall, G.: Review of [3901]. Math. Reviews, 2, 348f
4730. Pall, G.: The completion of a problem of Kloosterman. Am. J. Math. **68**, 47–58 (1946)
4731. Pan, C.D.: On the least prime in an arithmetical progression. Sci. Rec. **1**, 311–313 (1957)
4732. Pan, C.D.: On the representation of even numbers as the sum of a prime and an almost prime. Sci. Sin. **11**, 873–888 (1962) (in Russian)
4733. Pan, C.D.: On the representation of even numbers as the sum of a prime and a product of no more than 4 primes. Sci. Sin. **12**, 455–473 (1963) (in Russian)
4734. Pan, C.D.: A new application of the Yu.V. Linnik large sieve method. Acta Math. Sin. **14**, 597–606 (1964) (in Chinese) [English translation: Chinese Math. Acta, **5**, 1964, 642–652]
4735. Pan, C.D., Ding, X.X., Wang, Y.: On the representation of every large even integer as a sum of a prime and an almost prime. Sci. Sin. **18**, 599–610 (1975)
4736. Pan, C.D., Wang, Y.: Chen Jingrun: a brief outline of his life and works. Acta Math. Sin. New Ser. **12**, 225–233 (1996)
4737. Pan, G.: Review of [3261]. Math. Rev. 2010b:11011
4738. Panaitopol, L.: A special case of the Hardy-Littlewood conjecture. Math. Rep. (Bucuresti) **4**, 265–268 (2002)

4739. Papier, E.: Représentations l-adiques. Compos. Math. **71**, 303–362 (1989)
4740. Pappalardi, F.: On the exponent of the ideal class group of $Q(\sqrt{-d})$. Proc. Am. Math. Soc. **123**, 663–671 (1995)
4741. Parent, P.: Bornes effectives pour la torsion des courbes elliptiques sur les corps de nombres. J. Reine Angew. Math. **506**, 85–116 (1999)
4742. Parent, P.: Torsion des courbes elliptiques sur les corps cubiques. Ann. Inst. Fourier **50**(3), 723–749 (2000)
4743. Parent, P.: No 17-torsion on elliptic curves over cubic number fields. J. Théor. Nr. Bordx. **15**, 831–838 (2003)
4744. Parmankulov, Š.S.: The set of zeros of the Dirichlet L-function on the line $\sigma = 1/2$. Izv. Akad. Nauk Uzb. SSR, Ser. Fiz.-Mat. Nauk **16**, 32–37 (1972) (in Russian)
4745. Parry, C.J.: The p-adic generalization of the Thue-Siegel theorem. J. Lond. Math. Soc. **15**, 293–305 (1940)
4746. Parry, C.J.: The $\mathfrak{p}$-adic generalization of the Thue-Siegel theorem. Acta Math. **83**, 1–100 (1950)
4747. Parsell, S.T., Wooley, T.D.: On pairs of diagonal quintic forms. Compos. Math. **131**, 61–96 (2002)
4748. Parshall, K.H.: James Joseph Sylvester. Jewish Mathematician in a Victorian World. Johns Hopkins University Press, Baltimore (2006)
4749. Paršin, A.N.: Algebraic curves over function fields. I. Izv. Akad. Nauk SSSR, Ser. Mat. **32**, 1191–1219 (1968) (in Russian)
4750. Paršin, A.N.: Isogenies and torsion of elliptic curves. Izv. Akad. Nauk SSSR, Ser. Mat. **34**, 409–424 (1970) (in Russian)
4751. Paršin, A.N.: Quelques conjectures de finitude en géométrie diophantienne. In: Actes du Congrès International des Mathématiciens, Nice, 1970, vol. 1, pp. 467–471. Gauthier-Villars, Paris (1971)
4752. Paršin, A.N.: Minimal models of curves of genus 2, and homomorphisms of abelian varieties defined over a field of finite characteristic. Izv. Akad. Nauk SSSR, Ser. Mat. **36**, 67–109 (1972) (in Russian)
4753. Paršin, A.N.: Algebraic curves over function fields with a finite field of constants. Mat. Zametki **15**, 561–570 (1974) (in Russian)
4754. Pas, J.: Uniform p-adic cell decomposition and local zeta functions. J. Reine Angew. Math. **399**, 137–172 (1989)
4755. Pas, J.: Cell decomposition and local zeta functions in a tower of unramified extensions of a p-adic field. Proc. Lond. Math. Soc. **60**, 37–67 (1990)
4756. Pas, J.: Local zeta functions and Meuser's invariant functions. J. Number Theory **38**, 287–299 (1991)
4757. Paszkiewicz, A.: A new prime p for which the least primitive root (mod$p$) and the least primitive root (mod p^2) are not equal. Math. Comput. **78**, 1193–1195 (2009)
4758. Paszkiewicz, A., Schinzel, A.: Numerical calculation of the density of prime numbers with a given least primitive root. Math. Comput. **71**, 1781–1797 (2002)
4759. Patterson, S.J.: The distribution of general Gauss sums and similar arithmetic functions at prime arguments. Proc. Lond. Math. Soc. **54**, 193–215 (1987)
4760. Paule, P.: Über das Involutionsprinzip von Garsia und Milne. Bayreuth. Math. Schr. **21**, 295–319 (1986)
4761. Pears, A.R.: Edward Maitland Wright, 1906–2005. Bull. Lond. Math. Soc. **39**, 857–865 (2007)
4762. Peck, L.G.: Diophantine equations in algebraic number fields. Am. J. Math. **71**, 387–402 (1949)
4763. Peetre, J.: Outline of a scientific biography of Ernst Meissel (1826–1895). Hist. Math. **22**, 154–178 (1995)
4764. Peirce, B.: Math. Diary **2**, 267–277 (1832)
4765. Pellarin, F.: Sur une majoration explicite pour un degré d'isogénie liant deux courbes elliptiques. Acta Arith. **100**, 203–243 (2001)

4766. Pellet, A.E.: Mémoire sur la théorie algébrique des équations. Bull. Soc. Math. Fr. **15**, 61–103 (1887)
4767. Pen, A.S., Skubenko, B.F.: Upper bound for the period of a quadratic irrationality. Mat. Zametki **5**, 413–418 (1969) (in Russian)
4768. Penney, D.E., Pomerance, C.: A search for elliptic curves with large rank. Math. Comput. **28**, 851–853 (1974)
4769. Penney, D.E., Pomerance, C.: Three elliptic curves with rank at least seven. Math. Comput. **29**, 965–967 (1975)
4770. Pennington, W.B.: On the order of magnitude or Ramanujan's arithmetical function $\tau(n)$. Proc. Camb. Philos. Soc. **47**, 668–678 (1951)
4771. Pépin, T.: Sur certains nombres complexes compris dans la formule $a+b\sqrt{-c}$. J. Math. Pures Appl. **1**, 317–372 (1875)
4772. Pépin, T.: Sur la formule $2^{2^n}+1$. C. R. Acad. Sci. Paris **85**, 329–331 (1877)
4773. Pépin, T.: Théorèmes d'analyse indéterminée. C. R. Acad. Sci. Paris **88**, 1255–1257 (1879)
4774. Pépin, T.: Démonstration du théorème de Fermat sur les nombres polygones. Atti Accad. Naz. Lincei **46**, 119–131 (1893)
4775. Pépin, T.: Solution de l'équation $X^4+35Y^4=Z^2$. J. Math. Pures Appl. **1**, 351–358 (1895)
4776. Perelli, A.: A survey of the Selberg class of L-functions, I. Milan J. Math. **73**, 19–52 (2005)
4777. Perelli, A.: A survey of the Selberg class of L-functions, II. Riv. Mat. Univ. Parma Ser. 7 **3***, 83–118 (2004)
4778. Perelli, A., Pintz, J.: On the exceptional set for the $2k$-twin primes problem. Compos. Math. **82**, 355–372 (1992)
4779. Perelli, A., Pintz, J.: On the exceptional set for Goldbach's problem in short intervals. J. Lond. Math. Soc. **47**, 41–49 (1993)
4780. Perelli, A., Zaccagnini, A.: On the sum of a prime and a kth power. Izv. Ross. Akad. Nauk, Ser. Mat. **59**, 185–200 (1995)
4781. Perelmuter, G.I.: The Z-function of a class of cubic surfaces. In: Studies in Number Theory, Saratov, vol. 1, pp. 49–58 (1966) (in Russian)
4782. Perrin-Riou, B.: Travaux de Kolyvagin et Rubin. Astérisque **189–190**, 69–106 (1990)
4783. Perron, O.: Grundlagen für eine Theorie des Jacobischen Kettenbruchalgorithmus. Math. Ann. **64**, 1–76 (1907)
4784. Perron, O.: Zur Theorie der Dirichletschen Reihen. J. Reine Angew. Math. **134**, 95–143 (1908)
4785. Perron, O.: Die Lehre von den Kettenbrüchen. Teubner, Leipzig (1913); 2nd ed. 1929, 3rd ed., vols. 1, 2, Stuttgart 1954, 1957
4786. Perron, O.: Abschätzung der Lösung der Pellschen Gleichung. J. Reine Angew. Math. **144**, 71–73 (1914)
4787. Perron, O.: Über diophantische Approximationen. Math. Ann. **83**, 77–84 (1921)
4788. Perron, O.: Über die Approximation irrationaler Zahlen durch rationale, I, II. SBer. Heidelberg. Akad. Wiss., 1921
4789. Perron, O.: Über die Approximation einer komplexen Zahl durch Zahlen des Körpers $\mathfrak{K}(i)$. Math. Ann. **103**, 533–544 (1930)
4790. Perron, O.: Über die Approximation einer komplexen Zahl durch Zahlen des Körpers $\mathfrak{K}(i)$, II. Math. Ann. **105**, 160–164 (1931)
4791. Perron, O.: Über einen Approximationssatz von Hurwitz und über die Approximation einer komplexen Zahl des Körpers der dritten Einheitswurzeln. SBer. Bayer. Akad. Wiss., 1931, 129–154
4792. Perron, O.: Diophantische Approximationen in imaginären quadratischen Körpern, insbesondere im Körper $\mathfrak{K}(i\sqrt{2})$. Math. Z. **37**, 749–767 (1933)
4793. Perron, O.: Modulartige lückenlose Ausfüllung des R_n mit kongruenten Würfeln, I. Math. Ann. **117**, 415–447 (1940)
4794. Perron, O.: Modulartige lückenlose Ausfüllung des R_n mit kongruenten Würfeln, II. Math. Ann. **117**, 609–658 (1940)
4795. Perron, O.: Alfred Pringsheim. Jahresber. Dtsch. Math.-Ver. **56**, 1–6 (1952)

4796. Petečuk, M.M.: The sum of values of a function of divisors in arithmetic progressions with a difference equal to a power of an odd prime number. Izv. Akad. Nauk SSSR, Ser. Mat. **43**, 892–908 (1979) (in Russian)
4797. Pétermann, Y.-F.S.: Changes of sign of error terms related to Euler's function and to divisor functions. Comment. Math. Helv. **61**, 84–101 (1986)
4798. Pétermann, Y.-F.S.: Changes of sign of error terms related to Euler's function and to divisor functions, II. Acta Arith. **51**, 321–333 (1988)
4799. Pétermann, Y.-F.S.: An Ω-theorem for an error term related to the sum-of-divisors function. Monatshefte Math. **103**, 145–157 (1987); Add.: **105**, 1988, 145–153
4800. Pétermann, Y.-F.S.: About a theorem of Paolo Codecà's and omega estimates for arithmetical convolutions, II. Ann. Sc. Norm. Super. Pisa, Cl. Sci. **17**, 343–353 (1990)
4801. Pétermann, Y.-F.S., Wu, J.: On some asymptotic formulae of Ramanujan. Math. Nachr. **242**, 165–178 (2002)
4802. Peters, M.: Darstellungen durch definite ternäre quadratische Formen. Acta Arith. **34**, 57–80 (1977/1978)
4803. Peters, M.: The diophantine equation $xy + yx + zx = n$ and indecomposable binary quadratic forms. Exp. Math. **13**, 273–274 (2004)
4804. Petersson, H.: Über die Anzahl der Gitterpunkte in mehrdimensionalen Ellipsoiden. Abh. Math. Semin. Univ. Hamb. **5**, 116–150 (1927)
4805. Petersson, H.: Theorie der automorphen Formen beliebiger reeller Dimension und ihre Darstellung durch eine neue Art Poincaréscher Reihen. Math. Ann. **103**, 369–436 (1930)
4806. Petersson, H.: Über die Entwicklungskoeffizienten der automorphen Funktionen. Acta Math. **58**, 169–215 (1932)
4807. Petersson, H.: Über die Entwicklungskoeffizienten einer allgemeinen Klasse automorpher Formen. Math. Ann. **108**, 370–377 (1933)
4808. Petersson, H.: Zur analytischen Theorie der Grenzkreisgruppen. Math. Ann. **115**, 23–67 (1937)
4809. Petersson, H.: Zur analytischen Theorie der Grenzkreisgruppen, II. Divisorentheorie der automorphen Formen komplexer Dimension. Math. Ann. **115**, 175–204 (1937)
4810. Petersson, H.: Zur analytischen Theorie der Grenzkreisgruppen, III. Divisorentheorie und automorphe Primformen; Summandensysteme; arithmetisch ausgezeichnete Multiplikatorsysteme. Math. Ann. **115**, 518–572 (1937)
4811. Petersson, H.: Zur analytischen Theorie der Grenzkreisgruppen, IV. Anwendungen der Formen- und Divisorentheorie. Math. Ann. **115**, 670–709 (1937)
4812. Petersson, H.: Zur analytischen Theorie der Grenzkreisgruppen, V. Theorie der Poincaréschen Reihen zu den hyperbolischen Fixpunktpaaren bei beliebigen Grenzkreisgruppen. Math. Z. **44**, 127–155 (1939)
4813. Petersson, H.: Konstruktion der sämtlicher Lösungen einer Riemannscher Funktionalgleichung durch Dirichletreihen mit Eulerschen Produktentwicklung, I. Math. Ann. **116**, 401–412 (1939)
4814. Petersson, H.: Konstruktion der sämtlicher Lösungen einer Riemannscher Funktionalgleichung durch Dirichletreihen mit Eulerschen Produktentwicklung, II. Math. Ann. **117**, 39–64 (1939/1940)
4815. Petersson, H.: Konstruktion der sämtlicher Lösungen einer Riemannscher Funktionalgleichung durch Dirichletreihen mit Eulerschen Produktentwicklung, III. Math. Ann. **117**, 277–300 (1939/1940)
4816. Petersson, H.: Über eine Metrisierung der ganzen Modulformen. Jahresber. Dtsch. Math.-Ver. **49**, 49–75 (1939)
4817. Petersson, H.: Über eine Metrisierung der automorphen Formen und die Theorie der Poincaréschen Reihen. Math. Ann. **117**, 453–537 (1939/1940)
4818. Petersson, H.: Das wissenschaftliche Werk von E. Hecke. Abh. Math. Semin. Univ. Hamb. **16**, 7–31 (1949)
4819. Petersson, H.: Konstruktion der Modulformen und der zu gewissen Grenzkreisgruppen gehörigen automorphen Formen von positiver reeller Dimension und die vollständige Bestimmung ihrer Fourierkoeffizienten. SBer. Heidelberg. Akad. Wiss., 1950, 417–494

4820. Petersson, H.: Über automorphe Orthogonalfunktionen und die Konstruktion der automorphen Formen von positiver reeller Dimension. Math. Ann. **127**, 33–81 (1954)
4821. Petersson, H.: Modulfunktionen und quadratische Formen. Springer, Berlin (1982)
4822. Pethő, A.: Perfect powers in second order linear recurrences. J. Number Theory **15**, 5–13 (1982)
4823. Pethő, A.: Simple continued fractions for the Fredholm numbers. J. Number Theory **14**, 232–236 (1982)
4824. Pethő, A.: Complete solutions to families of quartic Thue equations. Math. Comput. **57**, 777–798 (1991)
4825. Pethő, A., Schulenberg, R.: Effektives Lösen von Thue Gleichungen. Publ. Math. (Debr.) **34**, 189–196 (1987)
4826. Pethő, A., Weis, T., Zimmer, H.G.: Torsion groups of elliptic curves with integral j-invariant over general cubic number fields. Int. J. Algebra Comput. **7**, 353–413 (1997)
4827. Pethő, A., Zimmer, H.G., Gebel, J., Herrmann, E.: Computing all S-integral points on elliptic curves. Math. Proc. Camb. Philos. Soc. **127**, 383–402 (1999)
4828. Petr, K.: Über die Anzahl der Darstellungen einer Zahl als Summe von zehn und zwölf Quadraten. Arch. Math. **11**, 83–85 (1907)
4829. Petrov, F.V.: On the number of rational points on a strictly convex curve. Funkc. Anal. Prilozh. **40**, 30–42 (2006) (in Russian)
4830. Peyre, E.: Obstructions au principe de Hasse et à l'approximation faible. Astérisque **299**, 165–193 (2005)
4831. Pezda, T.: Cycles of polynomial mappings in several variables over rings of integers in finite extensions of the rationals. Acta Arith. **108**, 127–146 (2003)
4832. Pfender, F., Ziegler, G.M.: Kissing number, sphere packings and some unexpected proofs. Not. Am. Math. Soc. **51**, 873–883 (2004)
4833. Pheidas, T.: Hilbert's tenth problem for a class of rings of algebraic integers. Proc. Am. Math. Soc. **104**, 611–620 (1988)
4834. Philipp, W.: Limit theorems for lacunary series and uniform distribution mod 1. Acta Arith. **26**, 241–251 (1974/1975)
4835. Philippon, P.: Variétés abéliennes et indépendance algébrique, I. Invent. Math. **70**, 289–318 (1982/1983)
4836. Philippon, P.: Variétés abéliennes et indépendance algébrique, II. Un analogue abélien du théoréme de Lindemann-Weierstraß. Invent. Math. **72**, 389–405 (1983)
4837. Philippon, P.: Lemmes de zéros dans les groupes algébriques commutatifs. Bull. Soc. Math. Fr. **114**, 355–383 (1986)
4838. Phillips, E.: The zeta-function of Riemann: Further developments of van der Corput's method. Q. J. Math. **4**, 209–225 (1933)
4839. Phong, B.M.: Characterizations of the logarithm as an additive function. Ann. Univ. Sci. Bp., Sect. Comput. **16**, 45–57 (1996)
4840. Phragmén, E.: Sur le logarithme intégral et la fonction $f(x)$ de Riemann. Öfversigt Kongl. Vet.-Akad. Förhandl., Stockholm **48**, 599–616 (1891)
4841. Phragmén, E., Lindelöf, E.: Sur une extension d'un principe classique d'analyse et sur quelques propriétés du fonctions monogènes dans le voisinage d'un point singulier. Acta Math. **31**, 381–406 (1908)
4842. Piateckiĭ-Šapiro, I.I.: On the distribution of prime numbers in sequences of the form $[f(n)]$. Mat. Sb. **33**, 559–566 (1953) (in Russian)
4843. Piateckiĭ-Šapiro, I.I.: On the theory of abelian modular functions. Dokl. Akad. Nauk SSSR **106**, 973–976 (1956) (in Russian)
4844. Piateckiĭ-Šapiro, I.I.: Theory of modular functions and related problems in the theory of discrete groups. Usp. Mat. Nauk **15**(1), 99–136 (1960) (in Russian)
4845. Piateckiĭ-Šapiro, I.I.: Geometry of Classical Domains and Theory of Automorphic Functions, Fizmatqiz, Moscow (1961) (in Russian) [French translation: Géométrie des domaines classiques et théorie des fonctions automorphes, Dunod, 1966; English translation: Automorphic Functions and the Geometry of Classical Domains, Gordon & Breach, 1969]

4846. Picard, É.: Sur une classe de groupes discontinus de substitutions linéaires et sur les fonctions de deux variables indépendantes restant invariables par ces substitutions. Acta Math. **1**, 297–321 (1882)
4847. Picard, É.: Sur les formes quadratiques ternaires indéfinies à indéterminées conjuguées et sur les fonctions hyperfuchsiennes correspondantes. Acta Math. **5**, 121–182 (1884)
4848. Picard, É.: Sur les fonctions hyperabéliennes. J. Math. Pures Appl. **1**, 87–128 (1885)
4849. Picard, É.: L'oeuvre scientifique de Charles Hermite. Ann. Sci. Éc. Norm. Super. **18**, 9–34 (1901)
4850. Picard, É.: L'oeuvre de Henri Poincaré. Ann. Sci. Éc. Norm. Super. **30**, 463–482 (1913)
4851. Pichorides, S.K.: A lower bound for the L_1 norm of exponential sums. Mathematika **21**, 155–159 (1974)
4852. Pichorides, S.K.: On a conjecture of Littlewood concerning exponential sums, I. Bull. Soc. Math. Grèce, (N.S.) **18**(1), 8–16 (1977)
4853. Pichorides, S.K.: On the L_1 norm of exponential sums. Ann. Inst. Fourier **30**(2), 79–89 (1980)
4854. Pick, G.: Ueber gewisse ganzzahlige lineare Substitutionen, welche sich nicht durch algebraische Congruenzen erklären lassen. Math. Ann. **28**, 119–124 (1886)
4855. Pihko, J.: Remarks on the "greedy odd" Egyptian fraction algorithm. Fibonacci Q. **39**, 221–227 (2001)
4856. Pila, J.: Frobenius maps of abelian varieties and finding roots of unity in finite fields. Math. Comput. **55**, 745–763 (1990)
4857. Pila, J.: Geometric postulation of a smooth function and the number of rational points. Duke Math. J. **63**, 449–463 (1991)
4858. Pila, J.: Density of integer points on plane algebraic curves. Internat. Math. Res. Notices, 1996, 903–912
4859. Pillai, S.S.: On the representation of a number as the sum of two positive k-th powers. J. Lond. Math. Soc. **3**, 56–62 (1928)
4860. Pillai, S.S.: On some functions connected with $\phi(n)$. Bull. Am. Math. Soc. **35**, 832–836 (1929)
4861. Pillai, S.S.: On the inequality $0 < a^x - b^y \leq n$. J. Indian Math. Soc. **19**, 1–11 (1931)
4862. Pillai, S.S.: On $a^x - b^y = c$. J. Indian Math. Soc. **2**, 119–122 (1936/1937); corr. 215
4863. Pillai, S.S.: On Waring's problem. J. Annamalai Univ. **5**, 145–166 (1936)
4864. Pillai, S.S.: On Waring's problem. J. Indian Math. Soc. **2**, 16–44 (1936)
4865. Pillai, S.S.: On Waring's problem, V. J. Indian Math. Soc. **2**, 213–214 (1937)
4866. Pillai, S.S.: On the addition of residue classes. Proc. Indian Acad. Sci. Math. Sci. **7**, 1–4 (1938)
4867. Pillai, S.S.: On numbers which are not multiples of any other in the set. Proc. Indian Acad. Sci. Math. Sci. **10**, 392–394 (1939)
4868. Pillai, S.S.: On Waring's problem $g(6) = 73$. Proc. Natl. Acad. Sci. India, Ser. A **12**, 30–40 (1940)
4869. Pillai, S.S.: Waring's problem with indices $\geq n$. Proc. Natl. Acad. Sci. India, Ser. A **12**, 41–45 (1940)
4870. Pillai, S.S.: Highly abundant numbers. Bull. Calcutta Math. Soc. **35**, 141–156 (1943)
4871. Pillai, S.S.: On the smallest primitive root of a prime. J. Indian Math. Soc. **8**, 14–17 (1944)
4872. Pillai, S.S.: On the equation $2^x - 3^y = 2^X + 3^Y$. Bull. Calcutta Math. Soc. **37**, 18–20 (1945)
4873. Pillai, S.S., Chowla, S.D.: On the error term in some asymptotic formulae in the theory of numbers, I. J. Lond. Math. Soc. **5**, 95–101 (1930)
4874. Pillai, S.S., Chowla, S.D.: On the error term in some asymptotic formulae in the theory of numbers, II. J. Indian Math. Soc. **18**, 181–184 (1930)
4875. Pilt'jai, G.Z.: The magnitude of the difference between consecutive primes. Issled. teor. čisel, Sarat. **4**, 73–79 (1974) (in Russian)
4876. Piltz, A.: Über das Gesetz, nach welchem die mittlere Darstellbarkeit der natürlichen Zahlen als Produkte einer gegebenen Anzahl Faktoren mit der Größe der Zahlen wächst. Dissertation, Berlin (1881)

4877. Piltz, A.: Über die Häufigkeit der Primzahlen in arithmetischen Progressionen und über verwandte Gesetze. Habilitationsschrift, Jena (1884)
4878. Pinch, R.G.E.: The pseudoprimes up to 10^{13}. In: Lecture Notes in Comput. Sci., vol. 1838, pp. 459–473. Springer, Berlin (2000)
4879. Pinner, C.G.: Repeated values of the divisor function. Q. J. Math. **48**, 499–502 (1997)
4880. Pinner, C.G., Vaaler, J.D.: The number of irreducible factors of a polynomial, III. In: Number Theory in Progress, I, pp. 395–405. de Gruyter, Berlin (1999)
4881. Pintér, Á.: On the magnitude of integer points on elliptic curves. Bull. Aust. Math. Soc. **52**, 195–199 (1995)
4882. Pintz, J.: On Siegel's theorem. Acta Arith. **24**, 543–551 (1973/1974)
4883. Pintz, J.: Elementary methods in the theory of L-functions, I. Hecke's theorem. Acta Arith. **31**, 53–60 (1976)
4884. Pintz, J.: Elementary methods in the theory of L-functions, II. On the greatest real zero of a real L-function. Acta Arith. **31**, 273–289 (1976)
4885. Pintz, J.: Elementary methods in the theory of L-functions, III. The Deuring-phenomenon. Acta Arith. **31**, 295–306 (1976)
4886. Pintz, J.: Elementary methods in the theory of L-functions, IV. The Heilbronn phenomenon. Acta Arith. **31**, 419–429 (1976)
4887. Pintz, J.: Elementary methods in the theory of L-functions, V. The theorems of Landau and Page. Acta Arith. **32**, 163–171 (1977)
4888. Pintz, J.: Elementary methods in the theory of L-functions, VI. On the least prime quadratic residue (mod ϱ). Acta Arith. **32**, 173–178 (1977)
4889. Pintz, J.: Elementary methods in the theory of L-functions, VII. Upper bound for $L(1,\chi)$. Acta Arith. **32**, 397–406 (1977); Corr.: **33**, 1977, 293–295
4890. Pintz, J.: Elementary methods in the theory of L-functions, VIII. Real zeros of real L-functions. Acta Arith. **33**, 89–98 (1977)
4891. Pintz, J.: Elementary methods in the theory of L-functions, IX. Density theorems. Acta Arith. **49**, 387–394 (1988)
4892. Pintz, J.: Bemerkungen zur Arbeit von S. Knapowski und P. Turán. Monatshefte Math. **82**, 199–206 (1976)
4893. Pintz, J.: On the remainder term of the prime number formula, II. On a theorem of Ingham. Acta Arith. **37**, 209–220 (1980)
4894. Pintz, J.: On the remainder term of the prime number formula, III. Sign changes of $\pi(x) - \mathrm{li}\, x$. Studia Sci. Math. Hung. **12**, 345–369 (1977)
4895. Pintz, J.: On the remainder term of the prime number formula, IV. Sign changes of $\pi(x) - \mathrm{li}\, x$. Studia Sci. Math. Hung. **13**, 29–42 (1978)
4896. Pintz, J.: On the remainder term of the prime number formula, V. Effective mean value theorems. Studia Sci. Math. Hung. **15**, 215–223 (1980)
4897. Pintz, J.: Oscillatory properties of $M(x) = \sum_{n\le x} \mu(n)$, I. Acta Arith. **42**, 49–55 (1982/1983)
4898. Pintz, J.: Oscillatory properties of $M(x) = \sum_{n\le x} \mu(n)$, II. Studia Sci. Math. Hung. **15**, 491–496 (1980)
4899. Pintz, J.: Oscillatory properties of $M(x) = \sum_{n\le x} \mu(n)$, III. Acta Arith. **43**, 105–113 (1984)
4900. Pintz, J.: On primes in short intervals, I. Studia Sci. Math. Hung. **16**, 395–414 (1981)
4901. Pintz, J.: On primes in short intervals, II. Studia Sci. Math. Hung. **19**, 89–96 (1984)
4902. Pintz, J.: An effective disproof of the Mertens conjecture. Astérisque **147/148**, 325–333 (1987)
4903. Pintz, J.: On an assertion of Riemann concerning the distribution of prime numbers. Acta Math. Acad. Sci. Hung. **58**, 383–387 (1991)
4904. Pintz, J.: Very large gaps between consecutive primes. J. Number Theory **63**, 286–301 (1997)
4905. Pintz, J.: Recent results on the Goldbach problem. In: Elementary and Analytic Number Theory. Proceedings of the ELAZ Conference, Stuttgart, May 24–28, 2004, pp. 220–254 (2006)

4906. Pintz, J.: A note on Romanov's constant. Acta Math. Acad. Sci. Hung. **112**, 1–14 (2006)
4907. Pintz, J.: Cramér vs. Cramér. On Cramérs probabilistic model for primes. Funct. Approx. Comment. Math. **37**, 361–376 (2007)
4908. Pintz, J.: Landau's problems on primes. J. Théor. Nr. Bordx. **21**, 357–404 (2009)
4909. Pintz, J., Ruzsa, I.Z.: On Linnik's approximation to Goldbach's problem, I. Acta Arith. **109**, 169–194 (2003)
4910. Pintz, J., Salerno, S.: On the comparative theory of primes. Ann. Sc. Norm. Super. Pisa, Cl. Sci. **11**, 245–260 (1984)
4911. Pintz, J., Salerno, S.: Irregularities in the distribution of primes in arithmetic progressions, I. Arch. Math. **42**, 439–447 (1984)
4912. Pintz, J., Salerno, S.: Irregularities in the distribution of primes in arithmetic progressions, II. Arch. Math. **43**, 351–357 (1984)
4913. Pintz, J., Salerno, S.: Some consequences of the general Riemann hypothesis in the comparative theory of primes. J. Number Theory **23**, 183–194 (1986)
4914. Pipping, N.: Die Goldbachsche Vermutung und der Goldbach-Vinogradowsche Satz. Acta Acad Aboensis, M.Ph. **11**(4), 1–25 (1938)
4915. Pipping, N.: Goldbachsche Spaltungen der geraden Zahlen x für $x = 60\,000$–$99\,998$. Acta Acad Aboensis, M.Ph. **12**(11), 1–18 (1940)
4916. Pisot, Ch.: Sur une propriété de certains nombres algébriques. C. R. Acad. Sci. Paris **202**, 892–894 (1936)
4917. Pisot, Ch.: Répartition (mod 1) des puissances successives des nombres réels. Comment. Math. Helv. **19**, 153–160 (1946)
4918. Pitt, H.R.: Tauberian Theorems. Oxford University Press, Oxford (1958)
4919. Plagne, A.: A uniform version of Jarník's theorem. Acta Arith. **87**, 255–267 (1999)
4920. Pleasants, P.A.B.: The representation of primes by cubic polynomials. Acta Arith. **12**, 23–45 (1966)
4921. Pleasants, P.A.B.: The representation of primes by quadratic and cubic polynomials. Acta Arith. **12**, 131–163 (1966)
4922. Pleasants, P.A.B.: Cubic polynomials over algebraic number fields. J. Number Theory **7**, 310–344 (1975)
4923. Plesken, W.: Hans Zassenhaus: 1912–1991. Jahresber. Dtsch. Math.-Ver. **96**, 1–20 (1994)
4924. Plünnecke, H.: Eine zahlentheoretische Anwendung der Graphentheorie. J. Reine Angew. Math. **243**, 171–183 (1970)
4925. Podsypanin, E.V.: The length of the period of quadratic irrationalities. Zap. Nauč. Semin. LOMI **82**, 95–99 (1979) (in Russian)
4926. Podsypanin, V.G.: On the indeterminate equation $x^3 = y^2 + Az^6$. Mat. Sb. **24**, 391–403 (1949) (in Russian)
4927. Poe, M.: On the distribution of solutions of S-unit equations. J. Number Theory **62**, 221–241 (1997)
4928. Pohst, M.: In memoriam: Hans Zassenhaus (1912–1991). J. Number Theory **47**, 1–19 (1994)
4929. Pohst, M.E., Wagner, M.: On the computation of Hermite-Humbert constants for real quadratic number fields. J. Théor. Nr. Bordx. **17**, 905–920 (2005)
4930. Poincaré, H.: Sur les fonctions uniformes qui se reproduisent par des substitutions linéaires. Math. Ann. **19**, 553–565 (1881)
4931. Poincaré, H.: Théorie des groupes fuchsiens. Acta Math. **1**, 1–62 (1882)
4932. Poincaré, H.: Mémoire sur les fonctions fuchsiennes. Acta Math. **1**, 193–294 (1883)
4933. Poincaré, H.: Mémoire sur les groupes kleinéens. Acta Math. **3**, 49–92 (1883)
4934. Poincaré, H.: Sur les groupes des équations linéaires. Acta Math. **4**, 201–312 (1884)
4935. Poincaré, H.: Extension aux nombres premiers complexes des théorèmes de M. Tchebicheff. J. Math. Pures Appl. **8**, 25–68 (1892)
4936. Poincaré, H.: Sur les propriétés arithmétiques des courbes algébriques. J. Math. Pures Appl. **7**, 161–233 (1901)
4937. Poisson, S.D.: Suite de mémoire sur les intégrales définies et sur la sommation des séries. J. Éc. Polytech. **12**, 404–509 (1823)

4938. Poisson, S.D.: Mémoire sur le calcul numérique des intégrales définies. Mém. de l'Institut **6**, 571–602 (1823)

4939. Poitou, G.: Sur l'approximation des nombres complexes par les nombres des corps imaginaires quadratiques dénués d'idéaux non principaux, particuliérement lorsque vaut l'algorithme d'Euclide. Ann. Sci. Éc. Norm. Super. **70**, 199–265 (1953)

4940. de Polignac, A.: Recherches nouvelles sur les nombres premiers. C. R. Acad. Sci. Paris **29**, 397–401 (1849). Corr.: 738–739

4941. de Polignac, A.: Six propositions arithmologiques déduites du crible d'Ératosthène. Nouv. Ann. Math. **8**, 423–429 (1849)

4942. Pollaczek, F.: Über den grossen Fermatschen Satz. SBer. Kais. Akad. Wissensch. Wien **126**, 45–59 (1917)

4943. Pollard, J.M.: A generalisation of the theorem of Cauchy and Davenport. J. Lond. Math. Soc. **8**, 460–462 (1974)

4944. Pollard, J.M.: Theorems on factorization and primality testing. Proc. Camb. Philos. Soc. **76**, 521–528 (1974)

4945. Pollard, J.M.: Addition properties of residue classes. J. Lond. Math. Soc. **11**, 147–152 (1975)

4946. Pollard, J.M.: A Monte Carlo method for factorization. Nor-disk Tidskr. Inf. **15**, 331–334 (1975)

4947. Pollington, A.D.: On the density of sequence $\{n_k\xi\}$. Ill. J. Math. **23**, 511–515 (1979)

4948. Pollington, A.D.: The Hausdorff dimension of a set of normal numbers. Pac. J. Math. **95**, 193–204 (1981)

4949. Pollington, A.D.: The Hausdorff dimension of a set of normal numbers, II. J. Aust. Math. Soc. **44**, 259–264 (1988)

4950. Pollington, A.D., Vaughan, R.C.: The k-dimensional Duffin and Schaeffer conjecture. J. Théor. Nr. Bordx. **1**, 81–88 (1989)

4951. Pollington, A.D., Vaughan, R.C.: The k-dimensional Duffin and Schaeffer conjecture. Mathematika **37**, 190–200 (1990)

4952. Pollington, A.D., Velani, S.L.: On a problem in simultaneous Diophantine approximation: Littlewood's conjecture. Acta Math. **185**, 287–306 (2000)

4953. Pólya, G.: Ueber ganzwertige ganze Funktionen. Rend. Circ. Mat. Palermo **40**, 1–16 (1915) [[4963], vol. 1, pp. 1–16]

4954. Pólya, G.: Généralisation d'un théorème de M. Störmer. Arch. Math. Naturvidensk. **35**(5), 1–8 (1917) (in Swedish)

4955. Pólya, G.: Über eine neue Weise bestimmte Integrale in der analytischen Zahlentheorie zu gebrauchen. Nachr. Ges. Wiss. Göttingen, 1917, 149–159

4956. Pólya, G.: Zur arithmetischen Untersuchungen der Polynome. Math. Z. **1**, 143–148 (1918)

4957. Pólya, G.: Über die Verteilung der quadratischen Reste und Nichtreste. Nachr. Ges. Wiss. Göttingen, 1918, 21–29

4958. Pólya, G.: Über ganzwertige Polynome in algebraischen Zahlkörpern. J. Reine Angew. Math. **149**, 97–116 (1919)

4959. Pólya, G.: Verschiedene Bemerkungen zur Zahlentheorie. Jahresber. Dtsch. Math.-Ver. **28**, 31–40 (1919)

4960. Pólya, G.: Untersuchungen über Lücken und Singularitäten von Potenzreihen. Math. Z. **29**, 549–640 (1929) [[4963], vol. 1, pp. 363–454]

4961. Pólya, G.: Über das Vorzeichen des Restgliedes im Primzahlsatz. Nachr. Ges. Wiss. Göttingen, 1930, 19–27 [Corrected reprint in: Number Theory and Analysis 233–244, Plenum, 1969; [4963], vol. 1, pp. 459–467]

4962. Pólya, G.: On polar singularities of power series and of Dirichlet series. Proc. Lond. Math. Soc. **33**, 85–101 (1932) [[4963], vol. 1, pp. 521–537]

4963. Pólya, G.: Collected Papers. MIT Press, Cambridge (1975)

4964. Polyakov, I.V.: On the exceptional set of sums of a prime and a square of an integer. Izv. Akad. Nauk SSSR, Ser. Mat. **45**, 1391–1423 (1981) (in Russian)

4965. Polyakov, I.V.: Addition of a prime number and the square of an integer. Mat. Zametki **47**, 90–99 (1990) (in Russian)

4966. Pomerance, C.: Odd perfect numbers are divisible by at least seven distinct primes. Acta Arith. **25**, 265–300 (1974)
4967. Pomerance, C.: Multiply perfect numbers, Mersenne primes and effective computability. Math. Ann. **226**, 195–206 (1977)
4968. Pomerance, C.: On composite n for which $\varphi(n)|n-1$. Acta Arith. **28**, 387–389 (1976)
4969. Pomerance, C.: On composite n for which $\varphi(n)|n-1$, II. Pac. J. Math. **69**, 177–186 (1977)
4970. Pomerance, C.: A note on the least prime in an arithmetic progression. J. Number Theory **12**, 218–223 (1980)
4971. Pomerance, C.: Popular values of Euler's function. Mathematika **27**, 84–89 (1980)
4972. Pomerance, C.: On the distribution of amicable numbers. J. Reine Angew. Math. **293/294**, 217–222 (1977)
4973. Pomerance, C.: On the distribution of amicable numbers, II. J. Reine Angew. Math. **325**, 183–188 (1981)
4974. Pomerance, C.: On the distribution of pseudoprimes. Math. Comput. **37**, 587–593 (1981)
4975. Pomerance, C.: A new lower bound for the pseudoprime counting function. Ill. J. Math. **26**, 4–9 (1982)
4976. Pomerance, C.: Analysis and comparison of some integer factoring algorithms. In: Computational Methods in Number Theory, 1. Math. Centre Tracts, vol. 154, pp. 89–139. Mathematisch Centrum, Amsterdam (1982)
4977. Pomerance, C.: On the distribution of round numbers. In: Lecture Notes in Math., vol. 1122, pp. 173–200. Springer, Berlin (1985)
4978. Pomerance, C.: The quadratic sieve factoring algorithm. In: Lecture Notes in Comput. Sci., vol. 209, pp. 169–182. Springer, Berlin (1985)
4979. Pomerance, C.: On the distribution of the values of Euler's function. Acta Arith. **47**, 63–70 (1986)
4980. Pomerance, C.: Carmichael numbers. Nieuw Arch. Wiskd. **11**, 199–209 (1993)
4981. Pomerance, C., Selfridge, J.L., Wagstaff, S.S. Jr.: The pseudoprimes to $25\cdot 10^9$. Math. Comput. **35**, 1003–1026 (1980)
4982. Pommerenke, C.: Über die Gleichverteilung von Gitterpunkten auf m-dimensionalen Ellipsoiden. Acta Arith. **5**, 227–257 (1959); corr., **7**, 279 (1961/1962)
4983. Pommersheim, J.E.: Toric varieties, lattice points and Dedekind sums. Math. Ann. **295**, 1–24 (1993)
4984. Poonen, B.: Some Diophantine equations of the form $x^n+y^n=z^m$. Acta Arith. **86**, 193–205 (1998)
4985. Poonen, B.: An explicit algebraic family of genus-one curves violating the Hasse principle. J. Théor. Nr. Bordx. **13**, 263–274 (2001)
4986. Poonen, B.: Hilbert's tenth problem and Mazur's conjecture for large subrings of Q. J. Am. Math. Soc. **16**, 981–990 (2003)
4987. Poonen, B.: Undecidability in Number Theory. Not. Am. Math. Soc. **55**, 344–350 (2008)
4988. Poonen, B., Schaefer, E.F., Stoll, M.: Twists of $X(7)$ and primitive solutions to $x^2+y^3=z^7$. Duke Math. J. **137**, 103–158 (2007)
4989. Popken, J.: Zur Transzendenz von e. Math. Z. **29**, 525–541 (1929)
4990. Popken, J.: Zur Transzendenz von π. Math. Z. **29**, 542–548 (1929)
4991. Popken, J.: Algebraic dependence of arithmetic functions. Indag. Math. **24**, 155–168 (1962)
4992. Popken, J.: On multiplicative arithmetic functions. In: Studies in Mathematical Analysis, pp. 285–293. Stanford University Press, Stanford (1962)
4993. Porter, J.W.: The generalized Titchmarsh-Linnik divisor problem. Proc. Lond. Math. Soc. **24**, 15–26 (1972)
4994. Porter, J.W.: On a theorem of Heilbronn. Mathematika **22**, 20–28 (1975)
4995. Porubský, Š.: Results and problems on covering systems of residue classes. Mitt. Math. Semin. Giessen **150**, 1–85 (1981)
4996. Porubský, Š.: Leben und Werk von Matyáš Lerch [1860–1922]. In: Mathematik im Wandel, pp. 347–373. Franzbecker, Hildesheim (2001)

4997. Porubský, Š., Schönheim, J.: Covering systems of Paul Erdős: past, present and future. In: Halász, G., et al. (ed.) Paul Erdős and His Mathematics, 1. Bólyai Soc. Math. Stud., vol. 11, pp. 581–627. Springer, Berlin (2002)
4998. Posse, K.A.: A.N. Korkin. Mat. Sb. **27**, 1–27 (1909) (in Russian)
4999. Potter, H.S.A.: The mean values of certain Dirichlet series, I. Proc. Lond. Math. Soc. **46**, 467–478 (1940)
5000. Potter, H.S.A.: The mean values of certain Dirichlet series, II. Proc. Lond. Math. Soc. **47**, 1–19 (1940)
5001. Potter, H.S.A., Titchmarsh, E.C.: The zeros of Epstein's zeta-functions. Proc. Lond. Math. Soc. **39**, 372–384 (1935)
5002. Poulakis, D.: Bounds for the minimal solutions of genus zero diophantine equations. Acta Arith. **86**, 51–90 (1998)
5003. Powers, R.E.: The tenth perfect number. Am. Math. Mon. **18**, 195–197 (1911)
5004. Powers, R.E.: A Mersenne prime. Bull. Am. Math. Soc. **20**, 531 (1913/1914)
5005. Prachar, K.: Über ein Problem vom Waring-Goldbachschen Typ. Monatshefte Math. **57**, 66–74 (1953)
5006. Prachar, K.: Primzahlverteilung. Springer, Berlin (1957) [Reprint: Springer, 1978]
5007. Prachar, K.: Über die kleinste Primzahl einer arithmetischen Reihe. J. Reine Angew. Math. **206**, 3–4 (1961)
5008. Prasad, D., Yogananda, C.S.: A report on Artin's holomorphy conjecture. In: Bambah, R.P., et al. (ed.) Number Theory, pp. 301–314. Birkhäuser, Basel (2000)
5009. Preisausschreibung der Kgl. Gesellschaft der Wissenschaften zu Göttingen für den Beweis des Fermatschen Satzes. Math. Ann. **66**, 143 (1909)
5010. Preissmann, E.: Sur la moyenne quadratique du terme de reste du problème du cercle. C. R. Acad. Sci. Paris **306**, 151–154 (1988)
5011. Pringsheim, A.: Zur Theorie der zweifach unendlichen Zahlenfolgen. Math. Ann. **53**, 289–321 (1900)
5012. Pritsker, I.E.: The Gelfond-Schnirelman method in prime number theory. Can. J. Math. **57**, 1080–1101 (2005)
5013. Proskurin, N.V.: On general Kloosterman sums. Zap. Naučn. Semin. POMI **302**, 107–134 (2003) (in Russian)
5014. Prouhet, E.: Mémoire sur les quelques relations entre les puissances des nombres. C. R. Acad. Sci. Paris **33**, 225 (1851)
5015. Prudnikov, V.E.: Pafnutiĭ Lvovič Čebyšev (1821–1894). Nauka, Moscow (1976) (in Russian)
5016. Pupyrev, Yu.A.: Effectivization of a lower bound for $\|(4/3)^k\|$. Mat. Zametki **85**, 927–935 (2009) (in Russian)
5017. Purdy, G.: The real zeros of the Epstein zeta function. Ph.D. thesis, University of Illinois (1972)
5018. Pustylnikov, L.D.: New estimates of Weyl sums and the remainder term in the law of distribution of the fractional part of a polynomial. Ergod. Theory Dyn. Syst. **11**, 515–534 (1991)
5019. Putnam, H.: An unsolvable problem in number theory. J. Symb. Comput. **25**, 220–232 (1960)
5020. Qi, M.G.: A result concerning the divisor function. J. Tsinghua Univ., **29**(3), 20–29 (1989) (in Chinese)
5021. Queen, C.: The existence of p-adic Abelian L-functions. In: Number Theory and Algebra, pp. 263–288. Academic Press, San Diego (1977)
5022. Questions proposées. Ann. Math. Pures Appl. **9**, 320 (1818/1819)
5023. Rabinowitsch, G.: Eindeutigkeit der Zerlegung in Primzahlfaktoren in quadratischen Zahlkörpern. J. Reine Angew. Math. **142**, 153–164 (1913)
5024. Rademacher, H.: Beiträge zur Viggo Brunschen Methode in der Zahlentheorie. Abh. Math. Semin. Univ. Hamb. **3**, 12–30 (1923) [[5040], vol. 1, pp. 259–277]
5025. Rademacher, H.: Über die Anwendung der Viggo Brunschen Methode auf die Theorie der algebraischen Zahlkörper. SBer. Preuß. Akad. Wiss. Berlin, 1923, 211–218 [[5040], vol. 1, pp. 280–287]

5026. Rademacher, H.: Zur additiven Primzahltheorie algebraischer Zahlkörper, I. Über die Darstellung totalpositiver Zahlen als Summe von totalpositiven Primzahlen im reellquadratischen Zahlkörper. Abh. Math. Semin. Univ. Hamb. **3**, 109–163 (1924) [[5040], vol. 1, pp. 306–360]
5027. Rademacher, H.: Zur additiven Primzahltheorie algebraischer Zahlkörper, II. Über die Darstellung von Körperzahlen als Summe von Primzahlen im imaginär-quadratischen Zahlkörper. Abh. Math. Semin. Univ. Hamb. **3**, 331–378 (1924) [[5040], vol. 1, pp. 362–409]
5028. Rademacher, H.: Zur additiven Primzahltheorie algebraischer Zahlkörper, III. Über die Darstellung totalpositiver Zahlen als Summen von totalpositiven Primzahlen in einem beliebigen Zahlkörper. Math. Z. **27**, 321–426 (1928) [[5040], vol. 1, pp. 445–550]
5029. Rademacher, H.: Über eine Erweiterung des Goldbachschen Problems. Math. Z. **25**, 627–657 (1926) [[5040], vol. 1, pp. 413–443]
5030. Rademacher, H.: Über die Erzeugenden von Kongruenzuntergruppe der Modulgruppe. Abh. Math. Semin. Univ. Hamb. **7**, 134–148 (1929) [[5040], vol. 1, pp. 630–644]
5031. Rademacher, H.: Zur Theorie der Modulfunktionen. J. Reine Angew. Math. **167**, 312–336 (1932) [[5040], vol. 1, pp. 652–676]
5032. Rademacher, H.: Bestimmung einer gewisser Einheitswurzel in der Theorie der Modulfunktionen. J. Lond. Math. Soc. **7**, 14–19 (1932) [[5040], vol. 2, pp. 3–9]
5033. Rademacher, H.: Primzahlen reell-quadratischer Zahlkörper in Winkelräumen. Math. Ann. **111**, 209–228 (1935) [[5040], vol. 2, pp. 38–58]
5034. Rademacher, H.: Über die Anzahl der Primzahlen eines reell-quadratischen Zahlkörpers, deren Konjugierten unterhalb gegebener Grenzen liegen. Acta Arith. **1**, 67–77 (1935) [[5040], vol. 2, pp. 59–70]
5035. Rademacher, H.: On prime numbers of real quadratic fields in rectangles. Trans. Am. Math. Soc. **39**, 380–398 (1936) [[5040], vol. 2, pp. 80–99]
5036. Rademacher, H.: A convergent series for the partition function $p(n)$. Proc. Natl. Acad. Sci. USA **23**, 78–84 (1937) [[5040], vol. 2, pp. 10–107]
5037. Rademacher, H.: On the partition function $p(n)$. Proc. Lond. Math. Soc. **43**, 241–254 (1937) [[5040], vol. 2, pp. 108–122]
5038. Rademacher, H.: The Fourier coefficients of the modular invariant $j(\tau)$. Am. J. Math. **60**, 501–512 (1938) [[5040], vol. 2, pp. 123–134]
5039. Rademacher, H.: Eine Bemerkung über die Heckeschen Operatoren $T(n)$. Abh. Math. Semin. Univ. Hamb. **31**, 149–151 (1967) [[5040], vol. 2, pp. 586–588]
5040. Rademacher, H.: Collected Papers, vols. 1, 2. MIT Press, Cambridge (1974)
5041. Rademacher, H., Grosswald, E.: Dedekind Sums. Math. Assoc. of America, Washington (1972)
5042. Rademacher, H., Whiteman, A.: Theorems on Dedekind sums. Am. J. Math. **63**, 377–407 (1941) [[5040], vol. 2, pp. 220–250]
5043. Rademacher, H., Zuckerman, H.S.: On the Fourier coefficients of certain modular forms of positive dimension. Ann. Math. **39**, 433–462 (1938); corr. **64**, 456 (1942) [[5040], vol. 2, pp. 136–165]
5044. Rado, R.: Verallgemeinerung eines Satzes von van der Waerden mit Anwendungen auf ein Problem der Zahlentheorie. SBer. Preuß. Akad. Wiss. Berlin, 1933, 589–596
5045. Rados, G.: Zur Theorie der Congruenzen höheren Grades. J. Reine Angew. Math. **99**, 258–260 (1886)
5046. Rados, G.: Über Kongruenzbedingung der rationalen Lösbarkeit von algebraischen Gleichungen. Math. Ann. **87**, 78–83 (1922)
5047. Radziejewski, M.: Oscillations of error terms associated with certain arithmetical functions. Monatshefte Math. **144**, 113–130 (2005)
5048. Raghavan, S.: Professor K.G. Ramanathan (1920–1992). Acta Arith. **64**, 1–6 (1993)
5049. Raghavan, S., Rangachari, S.S.: On zeta functions of quadratic forms. Ann. Math. **85**, 46–57 (1967)
5050. Raghunathan, M.S.: The congruence subgroup problem. Proc. Indian Acad. Sci. Math. Sci. **114**, 299–308 (2004); corr. 369

5051. Rai, T.: Easier Waring problem. J. Sci. Res. Benares Hindu Univ. **1**, 5–12 (1950/1951)
5052. Raĭkov, D.: On bases of the natural numbers. Mat. Sb. **2**, 595–597 (1937) (in Russian)
5053. Raĭkov, D.: On multiplicative bases for the natural series. Mat. Sb. **3**, 569–576 (1938) (in Russian)
5054. Rajwade, A.R.: Arithmetic on curves with complex multiplication by $\sqrt{-2}$. Proc. Camb. Philos. Soc. **64**, 659–672 (1968)
5055. Rajwade, A.R.: Arithmetic on curves with complex multiplication by Eisenstein integers. Proc. Camb. Philos. Soc. **65**, 59–73 (1969)
5056. Ramachandra, K.: Contributions to the theory of transcendental numbers, I. Acta Arith. **14**, 65–72 (1967/1968)
5057. Ramachandra, K.: Contributions to the theory of transcendental numbers, II. Acta Arith. **14**, 73–88 (1967/1968)
5058. Ramachandra, K.: A note on numbers with a large prime factor. J. Lond. Math. Soc. **1**, 303–306 (1970)
5059. Ramachandra, K.: A note on numbers with a large prime factor, II. J. Indian Math. Soc. **34**, 39–48 (1970)
5060. Ramachandra, K.: On the number of Goldbach numbers in small intervals. J. Indian Math. Soc. **37**, 157–170 (1973)
5061. Ramachandra, K.: Application of a theorem of Montgomery and Vaughan to the zeta-function. J. Lond. Math. Soc. **10**, 482–486 (1975)
5062. Ramachandra, K.: Some new density estimates for the zeros of the Riemann zeta-function. Ann. Acad. Sci. Fenn., Ser. A 1 Math. **1**, 177–182 (1975)
5063. Ramachandra, K.: Some remarks on the mean value of the Riemann zeta-function and other Dirichlet series, I. Hardy-Ramanujan J. **1**, 1–15 (1978)
5064. Ramachandra, K.: Some remarks on the mean value of the Riemann zeta-function and other Dirichlet series, II. Hardy-Ramanujan J. **3**, 1–24 (1980)
5065. Ramachandra, K.: One more proof of Siegel's theorem. Hardy-Ramanujan J. **3**, 25–40 (1980)
5066. Ramachandra, K.: On the Mean-Value and Omega-Theorems for the Riemann Zeta-function. Springer, Berlin (1995)
5067. Ramachandra, K., Sankaranarayanan, A.: On some theorems of Littlewood and Selberg, I. J. Number Theory **44**, 281–291 (1993)
5068. Ramakrishnan, D.: A refinement of the strong multiplicity one theorem for $GL(2)$. Invent. Math. **116**, 645–649 (1994)
5069. Ramanathan, K.: The theory of units of quadratic and Hermitian forms. Am. J. Math. **73**, 233–255 (1951)
5070. Ramanathan, K.: Units of quadratic forms. Ann. Math. **56**, 1–10 (1952)
5071. Ramanathan, K.: Zeta functions of quadratic forms. Acta Arith. **7**, 39–69 (1961/1962)
5072. Ramanathan, K. (ed.): C.P. Ramanujam—a Tribute. Springer, Berlin (1978)
5073. Ramanujam, C.P.: Sums of m-th powers in p-adic rings. Mathematika **10**, 137–146 (1963)
5074. Ramanujam, C.P.: Cubic form over algebraic number fields. Proc. Camb. Philos. Soc. **59**, 683–705 (1963)
5075. Ramanujan, S.: Question 464. J. Indian Math. Soc. **5**, 120 (1913) [[5088], p. 327]
5076. Ramanujan, S.: Highly composite numbers. Proc. Lond. Math. Soc. **14**, 347–409 (1915) [[5088], pp. 78–128]
5077. Ramanujan, S.: Some formulae in the analytic theory of numbers. Messenger Math. **45**, 81–84 (1915/1916) [[5088], pp. 133–135]
5078. Ramanujan, S.: On certain arithmetical functions. Trans. Camb. Philos. Soc. **22**, 159–184 (1916) [[5088], pp. 136–162]
5079. Ramanujan, S.: On the expression of a number in the form $ax^2 + by^2 + cz^2 + du^2$. Proc. Camb. Philos. Soc. **19**, 11–21 (1917–1920) [[5088], pp. 169–178]
5080. Ramanujan, S.: On certain trigonometrical sums and their applications in the theory of numbers. Trans. Camb. Philos. Soc. **22**, 259–276 (1918) [[5088], pp. 179–199]
5081. Ramanujan, S.: Some properties of $p(n)$, the number of partitions of n. Proc. Camb. Philos. Soc. **19**, 207–210 (1917–1920) [[5088], pp. 210–213]

5082. Ramanujan, S.: Proof of certain identities in combinatory analysis. Proc. Camb. Philos. Soc. **19**, 214–216 (1919) [[5088], pp. 214–215]

5083. Ramanujan, S.: Congruence properties of partitions. Proc. Lond. Math. Soc., (2) **18**, 1920, Records for 13 March 1919 [[5088], p. 230]

5084. Ramanujan, S.: Congruence properties of partitions. Math. Z. **9**, 147–153 (1921) [[5088], pp. 232–238]

5085. Ramanujan, S.: Notebooks. vols. I, II. Tata, Bombay (1957)

5086. Ramanujan, S.: The Lost Notebook and Other Unpublished Papers. Narosa, New Delphi (1988); Springer, 1988

5087. Ramanujan, S.: Highly composite numbers. Ramanujan J. **1**, 119–153 (1997)

5088. Ramanujan, S.: Collected Papers. Cambridge University Press, Cambridge (1927) [Reprint: Chelsea, 2000]

5089. Rao Rama, N.: An extension of Leudesdorf's theorem. J. Lond. Math. Soc. **12**, 247–250 (1937)

5090. Ramaré, O.: On Šnirelman's constant. Ann. Sc. Norm. Super. Pisa, Cl. Sci. **22**, 645–706 (1995)

5091. Ramaré, O.: Approximate formulae for $L(1,\chi)$. Acta Arith. **100**, 245–266 (2001)

5092. Ramaré, O.: Approximate formulae for $L(1,\chi)$, II. Acta Arith. **112**, 141–149 (2004)

5093. Ramaré, O.: An explicit seven cube theorem. Acta Arith. **118**, 375–382 (2005)

5094. Ramaré, O.: An explicit result of the sum of seven cubes. Manuscr. Math. **124**, 59–75 (2007)

5095. Ramaré, O., Rumely, R.: Primes in arithmetic progressions. Math. Comput. **65**, 397–425 (1996)

5096. Ramaré, O., Saouter, Y.: Short effective intervals containing primes. J. Number Theory **98**, 10–33 (2003)

5097. Ramaré, O., Schlage-Puchta, J.-C.: Improving on the Brun-Titchmarsh theorem. Acta Arith. **131**, 351–366 (2008)

5098. Ramaswami, V.: The number of positive integers $\leq x$ and free of prime divisors $> x^c$, and a problem of S.S. Pillai. Duke Math. J. **16**, 99–109 (1949)

5099. Ramharter, G.: Some metrical properties of continued fractions. Mathematika **30**, 117–132 (1983)

5100. Ramirez Alfonsin, J.L.: The Diophantine Frobenius Problem. Oxford University Press, Oxford (2005)

5101. Ramshaw, L.: On the discrepancy of the sequence formed by the multiples of an irrational number. J. Number Theory **13**, 138–175 (1981)

5102. Randol, B.: A lattice-point problem. Trans. Am. Math. Soc. **121**, 257–268 (1966)

5103. Rane, V.V.: On an approximate functional equation for Dirichlet L-series. Math. Ann. **264**, 137–145 (1983)

5104. Rankin, F.K.C., Swinnerton-Dyer, H.P.F.: On the zeros of Eisenstein series. Bull. Lond. Math. Soc. **2**, 169–170 (1970)

5105. Rankin, R.A.: The difference between consecutive prime numbers, I. J. Lond. Math. Soc. **13**, 242–247 (1938)

5106. Rankin, R.A.: The difference between consecutive prime numbers, II. Proc. Camb. Philos. Soc. **36**, 255–266 (1940)

5107. Rankin, R.A.: The difference between consecutive prime numbers, III. J. Lond. Math. Soc. **22**, 226–230 (1947)

5108. Rankin, R.A.: The difference between consecutive prime numbers, IV. Proc. Am. Math. Soc. **1**, 143–150 (1950)

5109. Rankin, R.A.: The difference between consecutive prime numbers, V. Proc. Edinb. Math. Soc. **13**, 331–332 (1962/1963)

5110. Rankin, R.A.: Contributions to the theory of Ramanujan's function $\tau(n)$ and similar arithmetic functions. I. The zeros of the function $\sum_{n=1}^{\infty}\tau(n)/n^s$ on the line $\Re s = 13/2$. Proc. Camb. Philos. Soc. **35**, 351–356 (1939)

5111. Rankin, R.A.: Contributions to the theory of Ramanujan's function $\tau(n)$ and similar arithmetic functions. II. The order of the Fourier coefficients of integral modular forms. Proc. Camb. Philos. Soc. **35**, 357–372 (1939)
5112. Rankin, R.A.: Contributions to the theory of Ramanujan's function $\tau(n)$ and similar arithmetic functions. III. A note on the sum function of the Fourier coefficients of integral modular forms. Proc. Camb. Philos. Soc. **36**, 150–151 (1940)
5113. Rankin, R.A.: On the closest packing of spheres in n dimensions. Ann. Math. **48**, 1062–1081
5114. Rankin, R.A.: The scalar product of modular forms. Proc. Lond. Math. Soc. **2**, 198–217 (1952)
5115. Rankin, R.A.: Van der Corput's method and the theory of exponent pairs. Q. J. Math. **6**, 147–153 (1955)
5116. Rankin, R.A.: Diophantine approximation and horocyclic groups. Can. J. Math. **9**, 277–290 (1957)
5117. Rankin, R.A.: Sets of integers containing not more than a given number of terms in arithmetical progression. Proc. R. Soc. Edinb., Sect. A, Math. **65**, 332–344 (1960/1961)
5118. Rankin, R.A.: The divisibility of divisor function. Glasg. Math. J. **5**, 35–40 (1961)
5119. Rankin, R.A.: On the representation of a number as the sum of any number of squares, and in particular of twenty. Acta Arith. **7**, 399–407 (1961/1962)
5120. Rankin, R.A.: Sums of squares and cusp forms. Am. J. Math. **87**, 857–860 (1965)
5121. Rankin, R.A.: George Neville Watson. J. Lond. Math. Soc. **41**, 551–565 (1966)
5122. Rankin, R.A.: William Barry Pennington. Bull. Lond. Math. Soc. **1**, 382–385 (1969)
5123. Rankin, R.A.: An Ω result for the coefficients of cusp forms. Math. Ann. **203**, 239–250 (1973)
5124. Rankin, R.A.: Ramanujan's unpublished work on congruences. In: Lecture Notes in Math., vol. 601, pp. 3–15. Springer, Berlin (1977)
5125. Rankin, R.A.: Modular Forms and Functions. Cambridge University Press, Cambridge (1977)
5126. Rankin, R.A.: Gertrude Katherine Stanley. Bull. Lond. Math. Soc. **14**, 554–555 (1982)
5127. Rankin, R.A.: The zeros of certain Poincaré series. Compos. Math. **46**, 255–272 (1982)
5128. Rankin, R.A.: Fourier coefficients of cusp forms. Math. Proc. Camb. Philos. Soc. **100**, 5–29 (1986)
5129. Rapinchuk, A.S.: Congruence subgroup problem for algebraic groups: old and new. Astérisque **209**, 73–84 (1992)
5130. Ratazzi, N.: Borne sur la torsion dans les variétés abéliennes de type CM. Ann. Sci. Éc. Norm. Super. **40**, 951–983 (2007)
5131. Ratliff, M.: The Thue-Siegel-Roth-Schmidt theorem for algebraic functions. J. Number Theory **10**, 99–126 (1978)
5132. Raulf, N.: Asymptotics of class-numbers for fundamental discriminants. Oberwolfach Rep. **4**, 1963–1964 (2007)
5133. Rausch, U.: On a theorem of Dobrowolski about the product of conjugate numbers. Colloq. Math. **50**, 137–142 (1985)
5134. Rauzy, G.: Caractérisation des ensembles normaux. Bull. Soc. Math. Fr. **98**, 401–414 (1970)
5135. Rauzy, G.: Nombres normaux et processus d'eterministes. Acta Arith. **29**, 211–225 (1976)
5136. Realis, S.: Note sur un théorème d'arithmétique. Nouv. Corresp. Math. **4**, 209–210 (1878)
5137. Recknagel, W.: Varianten des Gaußschen Kreisproblems. Abh. Math. Semin. Univ. Hamb. **59**, 183–189 (1989)
5138. Rédei, L.: Über die Pellsche Gleichung $t^2 - du^2 = -1$. J. Reine Angew. Math. **173**, 193–221 (1935)
5139. Rédei, L.: Vereinfachter Beweis des Satzes von Minkowski-Hajós. Acta Sci. Math. **13**, 21–35 (1949)
5140. Rédei, L.: Kurzer Beweis des gruppentheoretischen Satzes von Hajós. Comment. Math. Helv. **23**, 272–282 (1949)

5141. Redmond, D.: An asymptotic formula in the theory of numbers. Math. Ann. **224**, 247–268 (1976)
5142. Redmond, D.: An asymptotic formula in the theory of numbers, II. Math. Ann. **234**, 221–238 (1978)
5143. Redmond, D.: An asymptotic formula in the theory of numbers, III. Math. Ann. **243**, 143–151 (1979)
5144. Redouaby, M.: Sur la méthode de van der Corput pour des sommes d'exponentielles. J. Théor. Nr. Bordx. **13**, 583–607 (2001)
5145. Redouaby, M., Sargos, P.: Sur la transformation B de van der Corput. Expo. Math. **17**, 207–232 (1999)
5146. Reich, A.: Universelle Werteverteilung von Eulerprodukten. Nachr. Ges. Wiss. Göttingen, 1977, 1–17
5147. Reichardt, H.: Über die Diophantische Gleichung $ax^4 + bx^2y^2 + cy^4 = ez^2$. Math. Ann. **117**, 235–276 (1940/1941)
5148. Reichardt, H.: Einige im Kleinem überall lösbare, im Großen unlösbare diophantische Gleichungen. J. Reine Angew. Math. **184**, 12–18 (1942)
5149. Reid, C.: Hilbert. Springer, Berlin (1970) [Reprint: Hilbert–Courant, Springer, 1986]
5150. Reid, C.: Courant in Göttingen and New York. The Story of an Improbable Mathematician. Springer, Berlin (1976) [Reprints: Hilbert–Courant, Springer, 1986 and Copernicus, 1996; German translation: Richard Courant: 1888–1972. Der Mathematiker als Zeitgenosse, Springer, 1979]
5151. Reid, C.: The Search for E.T. Bell. Math. Assoc. of America, Washington (1993)
5152. Reid, C.: Julia: A Life in Mathematics. Math. Assoc. of America, Washington (1996)
5153. Reidlinger, H.: Über ungerade mehrfach vollkommene Zahlen. SBer. Österreich. AW SBer. Math.-Natur. Kl., II **192**, 237–266 (1983)
5154. Reiner, I.: Normal subgroups of the unimodular group. Ill. J. Math. **2**, 142–144 (1958)
5155. Reinhardt, K.: Über die dichteste gitterförmige Lagerung kongruenter Bereiche in der Ebene und eine besondere Art konvexer Kurven. Abh. Math. Semin. Univ. Hamb. **10**, 216–230 (1934)
5156. Reiter, C.: Effective lower bounds on large fundamental units of real quadratic fields. Osaka Math. J. **22**, 755–765 (1985)
5157. Remak, R.: Bemerkung zu Herrn Strindberg's Beweis des Waringschen Theorems. Math. Ann. **72**, 153–156 (1912)
5158. Remak, R.: Abschätzung der Lösung der Pellschen Gleichung im Anschluss an den Dirichletschen Existenzsatz. J. Reine Angew. Math. **143**, 250–254 (1913)
5159. Remak, R.: Verallgemeinerung eines Minkowskischen Satzes. Math. Z. **17**, 1–34 (1923); **18**, 173–200 (1923)
5160. Remak, R.: Über indefinite binäre quadratische Minimalformen. Math. Ann. **92**, 155–182 (1924)
5161. Remak, R.: Vereinfachung eines Blichfeldtschen Beweises aus der Geometrie der Zahlen. Math. Z. **26**, 694–699 (1927)
5162. Ren, X.M., Tsang, K.M.: Waring–Goldbach problem for unlike primes. Acta Math. Sin. **23**, 265–280 (2007)
5163. Rényi, A.: On the representation of even numbers as the sum of a prime and an almost prime number. Izv. Akad. Nauk SSSR, Ser. Mat. **12**, 57–78 (1948) (in Russian)
5164. Rényi, A.: Un nouveau théorème concernant les fonctions indépendantes et ses applications à la théorie des nombres. J. Math. Pures Appl. **28**, 137–149 (1949)
5165. Rényi, A.: On a general theorem of probability theory and its application in number theory. Časopis Mat. Fys. **74**, 167–175 (1949) (in Russian)
5166. Rényi, A.: On the large sieve of Yu.V. Linnik. Compos. Math. **8**, 68–75 (1951)
5167. Rényi, A.: Probabilistic methods in number theory. Prog. Math. **4**, 465–510 (1958) (in Chinese)
5168. Rényi, A., Turán, P.: On a theorem of Erdős-Kac. Acta Arith. **4**, 71–84 (1958) [[6227], vol. 2, pp. 991–1004]

5169. Révész, S.G.: Effective oscillation theorems for a general class of real-valued remainder terms. Acta Arith. **49**, 481–505 (1988)
5170. Revoy, P.: Sur les sommes de quatre cubes. Enseign. Math. **29**, 209–220 (1983)
5171. Revoy, P.: Le problème facile de Waring. Enseign. Math. **37**, 223–234 (1991)
5172. Reyssat, E.: Irrationalité de $\zeta(3)$ selon Apéry. Sém. Delange–Pisot–Poitou **20**(exp. 6), 1–6 (1978/1979)
5173. Rhin, G.: Sur les mesures d'irrationalité de certains nombres transcendants. Groupe d'étude en théorie analytique des nombres **1**(exp. 28), 1–6 (1984–1985)
5174. Rhin, G., Sac-Épée, J.-M.: New methods providing high degree polynomials with small Mahler measure. Exp. Math. **12**, 457–461 (2003)
5175. Rhin, G., Viola, C.: The group structure for $\zeta(3)$. Acta Arith. **97**, 269–293 (2001)
5176. Rhin, G., Wu, Q.: On the smallest value of the maximal modulus of an algebraic integer. Math. Comput. **76**, 1025–1038 (2007)
5177. Ribenboim, P.: Théorie des valuations. Press de l'Université de Montréal, Montreal (1968)
5178. Ribenboim, P.: 13 Lectures on Fermat's Last Theorem. Springer, Berlin (1979)
5179. Ribenboim, P.: Catalan's Conjecture. Academic Press, San Diego (1994)
5180. Ribenboim, P.: The Theory of Classical Valuations. Springer, Berlin (1999)
5181. Ribenboim, P.: The Little Book on Bigger Primes, 2nd ed. Springer, Berlin (2004)
5182. Ribenboim, P., McDaniel, W.L.: The square terms in Lucas sequences. J. Number Theory **58**, 104–123 (1996)
5183. Ribenboim, P., McDaniel, W.L.: Squares in Lucas sequences having an even first parameter. Colloq. Math. **78**, 29–34 (1998)
5184. Ribet, K.A.: An l-adic representations attached to modular forms. Invent. Math. **28**, 245–275 (1975)
5185. Ribet, K.A.: On l-adic representations attached to modular forms, II. Glasg. Math. J. **27**, 185–194 (1985)
5186. Ribet, K.A.: Torsion points of Abelian varieties in cyclotomic extensions. Enseign. Math. **27**, 315–319 (1981)
5187. Ribet, K.A.: Congruence relations between modular forms. In: Proceedings of the International Congress of Mathematicians, Warsaw, 1983, pp. 503–514 (1984)
5188. Ribet, K.A.: On modular representations of Gal$(\overline{Q}/Q)$ arising from modular forms. Invent. Math. **100**, 431–476 (1990)
5189. Ribet, K.A.: From the Taniyama-Shimura conjecture to Fermat's last theorem. Ann. Fac. Sci. Toulouse **11**, 116–139 (1990)
5190. Ribet, K.A.: Galois representations and modular forms. Bull. Am. Math. Soc. **32**, 375–402 (1995)
5191. Ribet, K.A.: On the equation $a^p + 2^\alpha b^p + c^p = 0$. Acta Arith. **79**, 7–16 (1997)
5192. Ribet, K.A., Stein, W.A.: Lectures on Serre's conjectures. In: Arithmetic Algebraic Geometry, Park City, UT, 1999, pp. 143–232. Am. Math. Soc., Providence (2001)
5193. Ricci, G.: Sul teorema di Dirichlet relativo alla progressione aritmetica. Boll. Unione Mat. Ital. **12**, 304–309 (1933)
5194. Ricci, G.: Ricerche aritmetiche sui polinomi. Rend. Circ. Mat. Palermo **57**, 433–475 (1933)
5195. Ricci, G.: Ricerche aritmetiche sui polinomi, II. Rend. Circ. Mat. Palermo **58**, 190–208 (1934)
5196. Ricci, G.: Sui teoremi di Dirichlet e di Bertrand-Tchebychef relativi alla progressione aritmetica. Boll. Unione Mat. Ital. **13**, 1–11 (1934)
5197. Ricci, G.: Su un teorema di Tchebychef-Nagel. Ann. Mat. Pura Appl. **12**, 295–303 (1934)
5198. Ricci, G.: Sul settimo problema di Hilbert. Ann. Sc. Norm. Super. Pisa, Cl. Sci. **4**, 341–372 (1935)
5199. Ricci, G.: Su la congettura di Goldbach e la costante di Schnirelmann, I. Ann. Sc. Norm. Super. Pisa, Cl. Sci. **6**, 71–90 (1937)
5200. Ricci, G.: Su la congettura di Goldbach e la costante di Schnirelmann, II. Ann. Sc. Norm. Super. Pisa, Cl. Sci. **6**, 91–116 (1937)
5201. Ricci, G.: Sull'andamento della differenza di numeri primi consecutivi. Riv. Mat. Univ. Parma **5**, 3–54 (1954)

5202. Ricci, G.: Recherches sur l'allure de la suite $\{\frac{p_{n+1}-p_n}{\log p_n}\}$. In: Colloque sur la Théorie des Nombres, Bruxelles, 1955, pp. 92–106. G. Thone, Liège (1956)
5203. Richert, H.-E.: Aus der additiven Primzahltheorie. J. Reine Angew. Math. **191**, 179–198 (1953)
5204. Richert, H.-E.: Verschärfung der Abschätzung beim Dirichletschen Teilerproblem. Math. Z. **58**, 204–218 (1953)
5205. Richert, H.-E.: On the difference between consecutive squarefree numbers. J. Lond. Math. Soc. **29**, 16–20 (1954)
5206. Richert, H.-E.: Über Dirichletreihen mit Funktionalgleichung. Publ. Inst. Math. Acad. Serbe **11**, 73–124 (1957)
5207. Richert, H.-E.: Einführung in die Theorie der starken Rieszschen Summierbarkeit von Dirichletreihen. Nachr. Ges. Wiss. Göttingen, 1960, 17–75
5208. Richert, H.-E.: Zur Abschätzung der Riemannschen Zetafunktion in der Nähe der Vertikalen $\sigma = 1$. Math. Ann. **169**, 97–101 (1967)
5209. Richert, H.-E.: Selberg's sieve with weights. Mathematika **16**, 1–22 (1969)
5210. Richmond, H.W.: An elementary note upon Waring's problem for cubes, positive and negative. Messenger Math. **51**, 177–186 (1921/1922)
5211. Richmond, H.W.: On analogues of Waring's problem for rational numbers. Proc. Lond. Math. Soc. **21**, 401–409 (1922)
5212. Richstein, J.: Verifying the Goldbach conjecture up to $4 \cdot 10^{14}$. Math. Comput. **70**, 1745–1749 (2001)
5213. Ridout, D.: Rational approximations to algebraic numbers. Mathematika **4**, 125–131 (1957)
5214. Ridout, D.: The p-adic generalization of the Thue-Siegel-Roth theorem. Mathematika **5**, 40–48 (1958)
5215. Ridout, D.: Indefinite quadratic forms. Mathematika **5**, 122–124 (1958)
5216. Rieger, G.J.: Zur Hilbertschen Lösung des Waringschen Problems, Abschätzung von $g(n)$. Arch. Math. **4**, 275–281 (1953)
5217. Rieger, G.J.: Verallgemeinerung der Siebmethode von A. Selberg auf algebraische Zahlkörper, I. J. Reine Angew. Math. **199**, 208–214 (1958)
5218. Rieger, G.J.: Verallgemeinerung der Siebmethode von A. Selberg auf algebraische Zahlkörper, II. J. Reine Angew. Math. **201**, 157–171 (1959)
5219. Rieger, G.J.: Verallgemeinerung der Siebmethode von A. Selberg auf algebraische Zahlkörper, III. J. Reine Angew. Math. **208**, 79–90 (1961)
5220. Rieger, G.J.: Zum Sieb von Linnik. Arch. Math. **11**, 14–22 (1960)
5221. Rieger, G.J.: Dedekindsche Summen in algebraischen Zahlkörpern. Math. Ann. **141**, 377–383 (1960)
5222. Rieger, G.J.: Das große Sieb von Linnik für algebraische Zahlen. Arch. Math. **12**, 184–187 (1961)
5223. Rieger, G.J.: Die metrische Theorie der Kettenbrüche seit Gauß. Abh. Braunschw. Wiss. Ges. **27**, 103–117 (1977)
5224. Riemann, B.: Ueber die Anzahl der Primzahlen unter einer gegebener Grösse. Monatsber. Kgl. Preuß. Akad. Wiss. Berlin, 1860, 671–680 [Gesammelte mathematische Werke, 3–47, Leipzig, 1876; Reprint: Dover 1953; Springer, Teubner, 1990; English translation: Kendrick Press, 2004; French translation: Blanchard, 1968]
5225. Riesel, H., Vaughan, R.C.: On sums of primes. Ark. Mat. **21**, 45–74 (1983)
5226. Riesz, M.: Sur l'hypothèse de Riemann. Acta Math. **40**, 185–190 (1916)
5227. Rigge, O.: Über ein diophantisches Problem. In: IX Congress Math. Scandinav., Helsingfors, pp. 155–160 (1938)
5228. Rigge, O.: On a diophantine problem. Ark. Math. Astr. Fys. A **27**(3), 1–10 (1940)
5229. Rignaux, M.: L'Intermédiaire Math. **25**, 94–95 (1918)
5230. Ringrose, C.J.: Sums of three cubes. J. Lond. Math. Soc. **33**, 407–413 (1986)
5231. Rivat, J.: Autour d'un théorème de Piatetski-Shapiro (nombres premiers dans la suite $[n^c]$). Thése, Univ. Paris-Sud, Orsay (1992)
5232. Rivat, J., Sargos, P.: Nombres premiers de la forme $\lfloor n^c \rfloor$. Can. J. Math. **53**, 414–433 (2001)

5233. Rivat, J., Wu, J.: Prime numbers of the form $[n^c]$. Glasg. Math. J. **43**, 237–254 (2001)
5234. Rivoal, T.: La fonction zêta de Riemann prend une infinité de valeurs irrationnelles aux entiers impairs. C. R. Acad. Sci. Paris **331**, 267–270 (2000)
5235. Robert, A.M.: The Gross-Koblitz formula revisited. Rend. Semin. Mat. Univ. Padova **105**, 157–170 (2001)
5236. Roberts, J.A.: Note on linear forms. Proc. Am. Math. Soc. **7**, 465–469 (1956)
5237. Roberts, S.: King of Infinite Space: Donald Coxeter, the Man Who Saved Geometry. Walker, New York (2006)
5238. Robin, G.: Estimation de la fonction de Tchebychef θ sur le k-ième nombre premier et grandes valeurs de la fonction $\omega(n)$ nombre de diviseurs premiers de n. Acta Arith. **42**, 367–389 (1983)
5239. Robin, G.: Grandes valeurs de la somme des diviseurs et hypothèse de Riemann. J. Math. Pures Appl. **63**, 187–213 (1984)
5240. Robinson, A., Roquette, P.: On the finiteness theorem of Siegel and Mahler concerning Diophantine equations. J. Number Theory **7**, 121–176 (1975)
5241. Robinson, J.: Existential definability in arithmetic. Trans. Am. Math. Soc. **72**, 437–449 (1952)
5242. Robinson, J.: Unsolvable diophantine problems. Proc. Am. Math. Soc. **22**, 534–538 (1969)
5243. Robinson, R.M.: Mersenne and Fermat numbers. Proc. Am. Math. Soc. **5**, 842–846 (1954)
5244. Rodosskiĭ, K.A.: On the least prime number in an arithmetic progression. Mat. Sb. **34**, 331–356 (1954) (in Russian)
5245. Rodosskiĭ, K.A.: On non-residues and zeros of L-functions. Izv. Akad. Nauk SSSR, Ser. Mat. **20**, 303–306 (1956) (in Russian)
5246. Rodosskiĭ, K.A.: On the exceptional zero. Izv. Akad. Nauk SSSR, Ser. Mat. **20**, 667–672 (1956) (in Russian)
5247. Rodriquez, G.: Sul problema dei divisori di Titchmarsh. Boll. Unione Mat. Ital. **20**, 358–366 (1965)
5248. Rödseth, Ö.J.: On a linear diophantine problem of Frobenius. J. Reine Angew. Math. **301**, 171–178 (1978)
5249. Rödseth, Ö.J.: On a linear diophantine problem of Frobenius, II. J. Reine Angew. Math. **307/308**, 431–440 (1979)
5250. Rogers, C.A.: Existence theorems in the geometry of numbers. Ann. Math. **48**, 994–1002
5251. Rogers, C.A.: The product of n real homogeneous forms. Acta Math. **82**, 185–208 (1950)
5252. Rogers, C.A.: The number of lattice points in a set. Proc. Lond. Math. Soc. **6**, 305–320 (1956)
5253. Rogers, C.A.: The packing of equal spheres. Proc. Lond. Math. Soc. **8**, 609–620 (1958)
5254. Rogers, C.A.: Packing and Covering. Cambridge University Press, Cambridge (1964)
5255. Rogers, C.A.: Hausdorff Measures. Cambridge University Press, Cambridge (1970); reprint: 1998
5256. Rogers, C.A.: A brief survey of the work of Harold Davenport. Acta Arith. **18**, 13–17 (1971)
5257. Rogers, C.A.: Richard Rado. Bull. Lond. Math. Soc. **30**, 185–195 (1988)
5258. Rogers, K., Swinnerton-Dyer, H.P.F.: The geometry of numbers over algebraic number fields. Trans. Am. Math. Soc. **88**, 227–242 (1958)
5259. Rogers, L.J.: On the expansion of some infinite products. Proc. Lond. Math. Soc. **24**, 337–352 (1892)
5260. Rogers, L.J.: Second memoir on the expansion of some infinite products. Proc. Lond. Math. Soc. **25**, 318–343 (1893)
5261. Rogosinski, W.: Neue Anwendung der Pfeifferschen Methode bei Dirichlets Teilerproblem. Dissertation, Göttingen (1922)
5262. Rogosinski, W.W.: Obituary: Michael Fekete. J. Lond. Math. Soc. **33**, 496–500 (1958)
5263. Rohrbach, H.: Ein Beitrag zur additiven Zahlentheorie. Math. Z. **42**, 1–30 (1937)
5264. Rohrbach, H.: Erhard Schmidt. Ein Lebensbild. Jahresber. Dtsch. Math.-Ver. **69**, 209–224 (1967/1968)

5265. Rohrbach, H.: Richard Brauer zum Gedächtnis. Jahresber. Dtsch. Math.-Ver. **83**, 125–134 (1981)
5266. Rohrbach, H.: Alfred Brauer zum Gedächtnis. Jahresber. Dtsch. Math.-Ver. **90**, 145–154 (1988)
5267. Rohrbach, H., Volkmann, B.: Zur Theorie der asymptotischen Dichtenaufgabe. J. Reine Angew. Math. **192**, 102–112 (1953)
5268. Rohrbach, H., Weis, J.: Zum finiten Fall des Bertrandschen Postulats. J. Reine Angew. Math. **214/215**, 432–440 (1964)
5269. Rohrlich, D.E.: Elliptic curves with good reduction everywhere. J. Lond. Math. Soc. **25**, 216–222 (1982)
5270. Rohrlich, D.E.: On L-functions of elliptic curves and anticyclotomic towers. Invent. Math. **75**, 383–408 (1984)
5271. Rohrlich, D.E.: On L-functions of elliptic curves and cyclotomic towers. Invent. Math. **75**, 409–423 (1984)
5272. Romanoff, N.P.: Über einige Sätze der additiven Zahlentheorie. Math. Ann. **109**, 668–678 (1934)
5273. Romanov, N.P.: On Goldbach's problem. Izv. NII Mat. Tech. Univ. Tomsk **1**, 34–38 (1935) (in Russian)
5274. Rooney, P.G.: Peter Scherk, 1910–1985. Trans. R. Soc. Can. **1**, 343–346 (1986)
5275. Roquette, P.: Riemannsche Vermutung in Funktionenkörpern. Arch. Math. **4**, 6–16 (1953)
5276. Roquette, P.: Arithmetischer Beweis der Riemannschen Vermutung in Kongruenzfunktionenkörpern beliebigen Geschlechts. J. Reine Angew. Math. **191**, 199–252 (1953)
5277. Roquette, P.: Über die algebraisch-zahlentheoretischen Arbeiten von Max Deuring. Jahresber. Dtsch. Math.-Ver. **91**, 109–125 (1989)
5278. Roquette, P.: History of Valuation Theory, Part I. In: Valuation Theory and Its Applications, Fields Int. Comm. Ser., vol. 32, pp. 291–355. Am. Math. Soc., Providence (2002)
5279. Roquette, P.: The Riemann hypothesis in characteristic p, its origin and development, I. Mitt. Math. Ges. Hamb. **21**, 79–157 (2002)
5280. Roquette, P.: The Riemann hypothesis in characteristic p, its origin and development, II. Mitt. Math. Ges. Hamb. **23**, 5–74 (2004)
5281. Roquette, P.: The Riemann hypothesis in characteristic p, its origin and development, III. Mitt. Math. Ges. Hamb. **25**, 103–176 (2006)
5282. Roquette, P.: Artin's Reciprocity Law and Hasse's Local-Global Principle in Historical Perspective, to be published
5283. Rosales, J.C., García-Sánchez, P.A.: Numerical semigroups with embedding dimension three. Arch. Math. **83**, 488–496 (2004)
5284. Rose, H.E.: On some elliptic curves with large sha. Exp. Math. **9**, 85–89 (2000)
5285. Rosen, K.H.: L-series for quadratic forms at $s = 1$, I. Applications of a theorem of Söhngen. Am. J. Math. **104**, 905–917 (1982)
5286. Rosen, K.H.: L-series for quadratic forms at $s = 1$, II. The modular functions that arise. Am. J. Math. **104**, 919–933 (1982)
5287. Rosen, K.H., Snyder, W.M.: p-adic Dedekind sums. J. Reine Angew. Math. **361**, 23–26 (1985)
5288. Rosen, M.I.: A proof of the Lucas-Lehmer test. Am. Math. Mon. **95**, 855–856 (1988)
5289. Roskam, H.: A quadratic analogue of Artin's conjecture on primitive roots. J. Number Theory **81**, 93–109 (2000); corr. **85**, 108 (2000)
5290. Roskam, H.: Prime divisors of linear recurrences and Artin's primitive root conjecture for number fields. J. Théor. Nr. Bordx. **13**, 303–314 (2001)
5291. Roskam, H.: Artin's primitive root conjecture for quadratic fields. J. Théor. Nr. Bordx. **14**, 287–324 (2002)
5292. Ross, A.E.: On criteria for universality of ternary quadratic forms. Q. J. Math. **4**, 147–158 (1933)
5293. Ross, A.E.: On a problem of Ramanujan. Am. J. Math. **68**, 29–46 (1946)
5294. Ross, A.E., Pall, G.: An extension of a problem of Kloosterman. Am. J. Math. **68**, 59–65 (1946)

5295. Ross, P.M.: On Chen's theorem that each large even number has the form $p_1 + p_2$ or $p_1 + p_2p_3$. J. Lond. Math. Soc. **10**, 500–506 (1975)
5296. Rosser, [J.]B.: On the first case of Fermat's last theorem. Bull. Am. Math. Soc. **45**, 636–640 (1939)
5297. Rosser, [J.]B.: A new lower bound for the exponent in the first case of Fermat's last theorem. Bull. Am. Math. Soc. **46**, 299–304 (1940)
5298. Rosser, [J.]B.: An additional criterion for the first case of Fermat's last theorem. Bull. Am. Math. Soc. **47**, 109–110 (1941)
5299. Rosser, [J.]B.: Explicit bounds for some functions of prime numbers. Am. J. Math. **63**, 211–232 (1941)
5300. Rosser, J.B.: Real roots of Dirichlet L-series. Bull. Am. Math. Soc. **55**, 906–913 (1949)
5301. Rosser, J.B.: Real roots of real Dirichlet L-series. J. Res.Natl. Bur. Stand. **45**, 505–514 (1950)
5302. Rosser, J.B., Schoenfeld, L.: Approximate formulas for some functions of prime numbers. Ill. J. Math. **6**, 64–94 (1962)
5303. Rosser, J.B., Schoenfeld, L.: Sharper bounds for the Chebyshev functions $\theta(x)$ and $\psi(x)$. Math. Comput. **29**, 243–269 (1975)
5304. Rosser, J.B., Yohe, J.M., Schoenfeld, L.: Rigorous computation and the zeros of the Riemann zeta-function. In: Information Processing 68, vol. 1, pp. 70–76. North-Holland, Amsterdam (1969)
5305. Roth, K.F.: Proof that almost all positive integers are sums of a square, a positive cube and a fourth power. J. Lond. Math. Soc. **24**, 4–13 (1949)
5306. Roth, K.F.: A problem in additive number theory. Proc. Lond. Math. Soc. **53**, 381–395 (1951)
5307. Roth, K.F.: On the gaps between squarefree numbers. J. Lond. Math. Soc. **26**, 263–268 (1951)
5308. Roth, K.F.: Sur quelques ensembles d'entiers. C. R. Acad. Sci. Paris **234**, 388–390 (1952)
5309. Roth, K.F.: On certain sets of integers. J. Lond. Math. Soc. **28**, 104–109 (1953)
5310. Roth, K.F.: On irregularities of distribution. Mathematika **1**, 73–79 (1954)
5311. Roth, K.F.: Rational approximations to algebraic numbers. Mathematika **2**, 1–20 (1955); corr. p. 168
5312. Roth, K.F.: On the large sieves of Linnik and Rényi. Mathematika **12**, 1–9 (1965)
5313. Roth, K.F.: Irregularities of sequences relative to arithmetic progressions, III. J. Number Theory, **2** (1970)
5314. Roth, K.F.: Irregularities of sequences relative to arithmetic progressions, IV. Period. Math. Hung. **2**, 301–326 (1972)
5315. Roth, K.F., Vaughan, R.C.: Theodor Estermann. Bull. Lond. Math. Soc. **26**, 593–606 (1994)
5316. Rotkiewicz, A.: On Lucas numbers with two intrinsic prime divisors. Bull. Acad. Pol. Sci., Aér. Sci. Math. Astron. Phys. **10**, 229–232 (1962)
5317. Rotkiewicz, A.: Sur les nombres pseudopremiers de la forme $ax + b$. C. R. Acad. Sci. Paris **257**, 2601–2604 (1963)
5318. Rotkiewicz, A.: Sur les nombres composés de la forme $cx + d$ pour lesquels $n|a^{n-1} - b^{n-1}$. Bull. Soc. R. Sci. Liège **32**, 823–829 (1963)
5319. Rotkiewicz, A.: On the pseudoprimes of the form $ax + b$. Proc. Camb. Philos. Soc. **63**, 389–392 (1967)
5320. Rotkiewicz, A.: Pseudoprime Numbers and their Generalizations. University of Novi Sad, Novi Sad (1972)
5321. Rotkiewicz, A., Schinzel, A.: Sur les nombres pseudopremiers de la forme $ax^2 + bxy + cy^2$. C. R. Acad. Sci. Paris **258**, 3617–3620 (1964)
5322. Roy, D.: Matrices whose coefficients are linear forms in logarithms. J. Number Theory **41**, 22–47 (1992)
5323. Roy, D.: An arithmetic criterion for the values of the exponential function. Acta Arith. **97**, 183–194 (2001)
5324. Roy, D., Thunder, J.L.: A note on Siegel's lemma over number fields. Monatshefte Math. **120**, 307–318 (1995)

5325. Roy, D., Thunder, J.L.: An absolute Siegel's lemma. J. Reine Angew. Math. **476**, 1–26 (1996)
5326. Roy, D., Waldschmidt, M.: Autour du théorème du sous-groupe algébrique. Can. Math. Bull. **36**, 358–367 (1993)
5327. Ru, M., Vojta, P.: Schmidt's subspace theorem with moving targets. Invent. Math. **127**, 51–65 (1997)
5328. Rubin, K.: Tate-Shafarevich groups and L-functions of elliptic curves with complex multiplication. Invent. Math. **89**, 527–559 (1987)
5329. Rubin, K.: The work of Kolyvagin on the arithmetic of elliptic curves. In: Lecture Notes in Math., vol. 1399, pp. 128–136. Springer, Berlin (1989)
5330. Rubin, K.: The "main conjectures" of Iwasawa theory for imaginary quadratic fields. Invent. Math. **103**, 25–68 (1991)
5331. Rubin, K., Silverberg, A.: A report on Wiles' Cambridge lectures. Bull. Am. Math. Soc. **31**, 15–38 (1994)
5332. Rubin, K., Silverberg, A.: Ranks of elliptic curves. Bull. Am. Math. Soc. **39**, 455–474 (2002)
5333. Rubin, K., Wiles, A.: Mordell-Weil groups of elliptic curves over cyclotomic fields. In: Number Theory Related to Fermat's Last Theorem, pp. 237–254. Birkhäuser, Basel (1982)
5334. Rubinstein, M., Sarnak, P.: Chebyshev's bias. Exp. Math. **3**, 173–197 (1994)
5335. Rubugunday, R.: On $g(k)$ in Waring's problem. J. Indian Math. Soc. **6**, 192–198 (1942)
5336. Rück, H.-G.: A note on elliptic curves over finite fields. Math. Comput. **49**, 301–304 (1987)
5337. Rudnick, Z., Sarnak, P.: n-level correlations of zeros of the zeta function. C. R. Acad. Sci. Paris **319**, 1027–1032 (1994)
5338. Rudnick, Z., Sarnak, P.: Zeros of principal L-functions and random matrix theory. Duke Math. J. **81**, 269–322 (1996)
5339. Rumely, R.S.: Arithmetic over the ring of all algebraic integers. J. Reine Angew. Math. **368**, 127–133 (1986)
5340. Rumely, R.[S.]: Numerical computations concerning the ERH. Math. Comput. **61**, 415–440 (1993)
5341. Runge, C.: Ueber ganzzahlige Lösungen von Gleichungen zwischen zwei Veränderlichen. J. Reine Angew. Math. **100**, 425–435 (1887)
5342. Ruzsa, I.Z.: Additive functions with bounded difference. Period. Math. Hung. **10**, 67–70 (1979)
5343. Ruzsa, I.Z.: Essential components. Proc. Lond. Math. Soc. **54**, 38–56 (1987)
5344. Ruzsa, I.Z.: A just basis. Monatshefte Math. **109**, 145–151 (1990)
5345. Ruzsa, I.Z.: Arithmetic progressions in sumsets. Acta Arith. **60**, 191–202 (1991)
5346. Ruzsa, I.Z.: Generalized arithmetical progressions and sumsets. Acta Math. Acad. Sci. Hung. **65**, 379–388 (1994)
5347. Ruzsa, I.Z.: An asymptotically exact additive completion. Studia Sci. Math. Hung. **32**, 51–57 (1996)
5348. Ruzsa, I.Z.: A converse to a theorem of Erdős and Fuchs. J. Number Theory **62**, 397–402 (1997)
5349. Ruzsa, I.Z.: On the additive completion of primes. Acta Arith. **86**, 269–275 (1998)
5350. Ruzsa, I.Z.: An infinite Sidon sequence. J. Number Theory **68**, 63–71 (1998)
5351. Ruzsa, I.Z.: Erdős and the integers. J. Number Theory **79**, 115–163 (1999)
5352. Ruzsa, I.Z.: An analog of Freiman's theorem in groups. Astérisque **258**, 323–326 (1999)
5353. Ruzsa, I.Z.: Additive completion of lacunary sequences. Combinatorica **21**, 279–291 (2001)
5354. Ruzsa, I.Z.: An almost polynomial Sidon sequence. Studia Sci. Math. Hung. **38**, 367–375 (2001)
5355. Ryavec, C.: Cubic forms over algebraic number fields. Proc. Camb. Philos. Soc. **66**, 323–333 (1969)
5356. Rychlik, K.: Přispěvek k theorii těles. Časopis Mat. Fys. **48**, 145–165 (1919)
5357. Ryll-Nardzewski, C.: On the ergodic theorem, II. Stud. Math. **12**, 74–79 (1951)

5358. Ryškov, S.S.: Effective realization of a method of Davenport in the theory of coverings. Dokl. Akad. Nauk SSSR **175**, 303–305 (1967) (in Russian)
5359. Ryškov, S.S., Baranovskiĭ, E.P.: Solution of the problem of the least dense lattice covering of five-dimensional space by equal spheres. Dokl. Akad. Nauk SSSR **222**, 39–42 (1975) (in Russian)
5360. Ryškov, S.S., Baranovskiĭ, E.P.: S-types of n-dimensional lattices and five-dimensional primitive parallelohedra (with an application to covering theory). Tr. Mat. Inst. Steklova **137**, 1–131 (1976) (in Russian)
5361. Sacks, G.E.: John Barkley Rosser (1907–1989). Not. Am. Math. Soc. **36**, 1367 (1989)
5362. Šafarevič, I.R.: A general reciprocity law. Mat. Sb. **26**, 113–146 (1950) (in Russian)
5363. Šafarevič, I.R.: Exponents of elliptic curves. Dokl. Akad. Nauk SSSR **114**, 714–716 (1957) (in Russian)
5364. Šafarevič, I.R.: The group of principal homogeneous algebraic manifolds. Dokl. Akad. Nauk SSSR **124**, 42–43 (1959) (in Russian)
5365. Šafarevič, I.R.: Principal homogeneous spaces, defined over a function field. Tr. Mat. Inst. Steklova **64**, 316–346 (1961) (in Russian)
5366. Šafarevič, I.R.: Algebraic number fields. In: Proc. ICM Stockholm, Djursholm, 1962, pp. 163–176 (1963) (in Russian)
5367. Šafarevič, I.R.: Dmitriĭ Konstantinovič Faddeev. Algebra Anal. **2**, 3–9 (1990) (in Russian)
5368. Saffari, B.: Ω-théorèmes sur le terme résiduel dans la loi de répartition des entiers non divisibles par une puisance rième, $r > 1$ ("r-free"). C. R. Acad. Sci. Paris **272**, A95–A97 (1971)
5369. Saias, E.: Sur le nombre des entiers sans grand facteur premier. J. Number Theory **32**, 78–99 (1989)
5370. Salberger, P.: On obstructions to the Hasse principle. In: Number Theory and Algebraic Geometry, pp. 251–277. Cambridge University Press, Cambridge (2003)
5371. Salem, R.: Sets of uniqueness and sets of multiplicity. Trans. Am. Math. Soc. **54**, 218–228 (1943)
5372. Salem, R.: Sets of uniqueness and sets of multiplicity, II. Trans. Am. Math. Soc. **56**, 32–49 (1944); corr.: **63**, 595–598 (1948)
5373. Salem, R.: A remarkable class of algebraic integers. Proof of a conjecture of Vijayaraghavan. Duke Math. J. **11**, 103–108 (1944)
5374. Salem, R.: On a Problem of Littlewood. Am. J. Math. **77**, 535–540 (1955)
5375. Salem, R., Spencer, D.C.: On sets of integers which contain no three terms in arithmetic progression. Proc. Natl. Acad. Sci. USA **28**, 561–563 (1942)
5376. Salem, R., Spencer, D.C.: On sets which do not contain a given number of terms in arithmetic progression. Nieuw Arch. Wiskd. **23**, 33–143 (1950)
5377. Salerno, S., Viola, C.: Sieving by almost-primes. J. Lond. Math. Soc. **14**, 221–234 (1976)
5378. Salié, H.: Über die Kloostermansche Summen. Math. Z. **34**, 91–109 (1932)
5379. Salié, H.: Zur Abschätzung der Fourierkoeffizienten ganzer Modulformen. Math. Z. **36**, 263–278 (1933)
5380. Salié, H.: Über den kleinsten positiven quadratischen Nichtrest nach einer Primzahl. Math. Nachr. **3**, 7–8 (1949)
5381. Salikhov, V.Kh.: Irreducibility of hypergeometric equations, and algebraic independence of values of E-functions. Acta Arith. **53**, 453–471 (1990) (in Russian)
5382. Salikhov, V.Kh.: On the irrationality measure of π. Usp. Mat. Nauk **63**(3), 163–164 (2008) (in Russian)
5383. Samuel, P.: Compléments à un article de Hans Grauert sur la conjecture de Mordell. Publ. Math. Inst. Hautes Études Sci. **29**, 55–62 (1966)
5384. Sander, J.W.: Prime power divisors of $\binom{2n}{n}$. J. Number Theory **39**, 65–74 (1991)
5385. Sander, J.W.: On prime divisors of binomial coefficients. Bull. Lond. Math. Soc. **24**, 140–142 (1992)
5386. Sander, J.W.: Irrationality criteria for Mahler's numbers. J. Number Theory **52**, 145–156 (1995)

5387. Sandham, H.F.: A square as the sum of 7 squares. Q. J. Math. **4**, 230–236 (1953)
5388. Sandham, H.F.: A square as the sum of 9, 11 and 13 squares. J. Lond. Math. Soc. **29**, 31–38 (1954)
5389. Sándor, G.: Über die Anzahl der Lösungen einer Kongruenz. Acta Math. **87**, 13–16 (1952)
5390. Šanin, A.A.: Determination of constants in the method of Brun–Šnirelman. Volž. Mat. Sb. **2**, 261–265 (1964) (in Russian)
5391. Sankaranarayanan, A.: On the sign changes in the remainder term of an asymptotic formula for the number of square-free numbers. Arch. Math. **60**, 51–57 (1993)
5392. Sankaranarayanan, A., Saradha, N.: Estimates for the solutions of certain Diophantine equations by Runge's method. Int. J. Number Theory **4**, 475–493 (2008)
5393. Saouter, Y.: Checking the odd Goldbach conjecture up to 10^{20}. Math. Comput. **67**, 863–866 (1998)
5394. Saouter, Y., Demichel, P.: A sharp region where $\pi(x) - \mathrm{li}(x)$ is positive. Math. Comput. **79**, 2395–2405 (2010)
5395. Saradha, N.: Squares in products of terms in an arithmetical progression. Acta Arith. **86**, 27–43 (1998)
5396. Saradha, N., Shorey, T.N.: The equation $(x^n - 1)/(x - 1) = y^q$ with x square. Math. Proc. Camb. Philos. Soc. **125**, 1–19 (1999)
5397. Sarges, H.: Eine Anwendung des Selbergschen Siebes auf Zahlkörper. Acta Arith. **28**, 433–455 (1976)
5398. Sargos, P., Wu, J.: Multiple exponential sums with monomials and their applications in number theory. Acta Math. Acad. Sci. Hung. **87**, 333–354 (2000)
5399. Sárkőzy, A.: On difference sets of sequences of integers, I. Acta Math. Acad. Sci. Hung. **31**, 125–149 (1978)
5400. Sárkőzy, A.: On difference sets of sequences of integers, II. Ann. Univ. Sci. Bp. **21**, 45–53 (1978)
5401. Sárkőzy, A.: On difference sets of sequences of integers, III. Acta Math. Acad. Sci. Hung. **31**, 355–386 (1978)
5402. Sárkőzy, A.: On divisors of binomial coefficients, I. J. Number Theory **20**, 70–80 (1985)
5403. Sárkőzy, A.: On divisibility properties of sequences of integers. In: The Mathematics of Paul Erdős, vol. I, pp. 241–250. Springer, Berlin (1997)
5404. Sarma, M.L.N.: On the error term in a certain sum. Proc. Natl. Acad. Sci. India, Ser. A **3**, 338 (1936)
5405. Sarnak, P.: Class numbers of indefinite binary quadratic forms. J. Number Theory **15**, 229–247 (1983); corr. vol. 16, 1983, p. 284
5406. Sarnak, P.: Class numbers of indefinite binary quadratic forms. J. Number Theory **21**, 333–346 (1985)
5407. Sarnak, P.: Some Applications of Modular Forms. Cambridge University Press, Cambridge (1990)
5408. Sarrus, F.: Questions résolues. Démonstration de la fausseté du théorème énoncé à la page 320 du IX.e volume de ce recueil. Ann. Math. Pures Appl. **10**, 184–187 (1819/1820)
5409. Sastry, S., Singh, R.: A problem in additive number theory. J. Sci. Res. Banaras Hindu Univ. **9**, 261–265 (1955/1956)
5410. Satgé, Ph.: Quelques résultats sur les entiers qui sont somme des cubes de deux rationnels. Astérisque **147/148**, 335–341 (1987)
5411. Sathe, L.G.: On a problem of Hardy on the distribution of integers having a given number of prime factors. J. Indian Math. Soc. **17**, 63–82 (1953)
5412. Sathe, L.G.: On a problem of Hardy on the distribution of integers having a given number of prime factors, II. J. Indian Math. Soc. **17**, 83–141 (1953)
5413. Sathe, L.G.: On a problem of Hardy on the distribution of integers having a given number of prime factors, III. J. Indian Math. Soc. **18**, 27–42 (1954)
5414. Sathe, L.G.: On a problem of Hardy on the distribution of integers having a given number of prime factors, IV. J. Indian Math. Soc. **18**, 43–81 (1954)
5415. Satoh, T.: The canonical lift of an ordinary elliptic curve over a finite field and its point counting. J. Ramanujan Math. Soc. **15**, 247–270 (2000)

5416. Satoh, T., Skjernaa, B., Taguchi, Y.: Fast computation of canonical lifts of elliptic curves and its application to point counting. Finite Fields Appl. **9**, 89–101 (2003)
5417. du Sautoy, M.P.F.: Mersenne primes, irrationality and counting subgroups. Bull. London Math. Soc. **29**, 185–294 (1997)
5418. Sayers, M.: An improved lower bound for the total number of prime factors of an odd perfect number. M.A. thesis, New South Wales Institute of Technology (1986)
5419. Schaal, W.: On the large sieve method in algebraic number fields. J. Number Theory **2**, 249–270 (1970)
5420. Schaal, W.: Der Satz von Erdös und Fuchs in reell-quadratischen Zahlkörpern. Acta Arith. **32**, 147–156 (1977)
5421. Schanuel, S.H.: An extension of Chevalley's theorem to congruences modulo prime powers. J. Number Theory **6**, 284–290 (1974)
5422. Schappacher, N.: On the history of Hilbert's twelfth problem: a comedy of errors. In: Matériaux pour l'histoire des mathématiques au XXe siècle, Nice, 1996, pp. 243–273. Soc. Math. France, Paris (1998)
5423. Schappacher, N., Scholz, E.: Oswald Teichmüller—Leben und Werk. Jahresber. Dtsch. Math.-Ver., vol. 94, pp. 1–35. Springer, Berlin (1992)
5424. Schappacher, N., Schoof, R.: Beppo Levi and the arithmetic of elliptic curves. Math. Intell. **18**, 57–69 (1996)
5425. Schärtlin, G.: Hermann Kinkelin. Mitt. Schweiz. Versicherungsmath. **28**, 1–17 (1933)
5426. Scherk, P.: On a theorem of Mann and Dyson. J. Reine Angew. Math. **217**, 49–63 (1965)
5427. Schertz, R.: L-Reihen in imaginär-quadratischen Zahlkörpern und ihre Anwendung auf Klassenzahlprobleme bei quadratischen und biquadratischen Zahlkörpern, I. J. Reine Angew. Math. **262/263**, 120–133 (1973)
5428. Schertz, R.: L-Reihen in imaginär-quadratischen Zahlkörpern und ihre Anwendung auf Klassenzahlprobleme bei quadratischen und biquadratischen Zahlkörpern, II. J. Reine Angew. Math. **270**, 195–212 (1974)
5429. Schilling, O.F.G.: The Theory of Valuations. Am. Math. Soc., Providence (1950)
5430. Schinzel, A.: Sur l'équation $\varphi(x) = m$. Elem. Math. **11**, 75–78 (1956) [[5449], vol. 2, pp. 871–874]
5431. Schinzel, A.: Sur quelques propriétés des nombres $3/n$ et $4/n$, où n est un nombre impair. Mathesis **65**, 219–222 (1956) [[5449], vol. 1, pp. 13–16]
5432. Schinzel, A.: Sur les sommes de trois carrés. Bull. Acad. Pol. Sci., Aér. Sci. Math. Astron. Phys. **7**, 307–310 (1959) [[5449], vol. 2, pp. 871–874]
5433. Schinzel, A.: Remarks on the paper "Sur certaines hypothéses concernant les nombres premiers". Acta Arith. **7**, 1–8 (1961/1962) [[5449], vol. 2, pp. 871–874]
5434. Schinzel, A.: The intrinsic divisors of Lehmer numbers in the case of negative discriminant. Ark. Mat. **4**, 413–416 (1962)
5435. Schinzel, A.: Remark on the paper of K. Prachar "Über die kleinste Primzahl einer arithmetischen Reihe". J. Reine Angew. Math. **210**, 121–122 (1962)
5436. Schinzel, A.: On primitive prime factors of Lehmer numbers, I. Acta Arith. **8**, 213–223 (1962/1963); [[5449], vol. 2, pp. 1046–1058]
5437. Schinzel, A.: On primitive prime factors of Lehmer numbers, II. Acta Arith. **8**, 251–257 (1962/1963); [[5449], vol. 2, pp. 1059–1065]
5438. Schinzel, A.: On primitive prime factors of Lehmer numbers, III. Acta Arith. **15**, 49–70 (1968); corr. **16**, 101 (1969/1970) [[5449], vol. 2, pp. 1066–1089]
5439. Schinzel, A.: On two theorems of Gelfond and some of their applications. Acta Arith. **13**, 177–236 (1967/1968)
5440. Schinzel, A.: An improvement of Runge's theorem on Diophantine equations. Comment. Pontificia Acad. Sci. **2**(20), 1–9 (1969) [[5449], vol. 1, pp. 36–40]
5441. Schinzel, A.: A refinement of a theorem of Gerst on power residues. Acta Arith. **17**, 161–168 (1970)
5442. Schinzel, A.: Wacław Sierpiński's papers on the theory of numbers. Acta Arith. **21**, 7–13 (1972)

5443. Schinzel, A.: Primitive divisors of the expression $A^n - B^n$ in algebraic number fields. J. Reine Angew. Math. **268/269**, 27–33 (1974) [[5449], vol. 2, pp. 1090–1097]
5444. Schinzel, A.: On power residues and exponential congruences. Acta Arith. **27**, 397–420 (1975) [[5449], vol. 2, pp. 915–938]
5445. Schinzel, A.: Selected Topics on Polynomials. University of Michigan Press, Ann Arbor (1982)
5446. Schinzel, A.: Hasse's principle for systems of ternary quadratic forms and for one biquadratic form. Stud. Math. **77**, 103–109 (1983) [[5449], vol. 1, pp. 87–92]
5447. Schinzel, A.: On an analytic problem considered by Sierpiński and Ramanujan. In: New Trends in Probability and Statistics, Palanga, 1991, vol. 2, pp. 165–171. VSP, Utrecht (1992) [[5449], vol. 2, pp. 1217–1223]
5448. Schinzel, A.: Polynomials with Special Regard to Reducibility. Cambridge University Press, Cambridge (2000)
5449. Schinzel, A.: Selecta, vols. 1, 2. Eur. Math. Soc., Zurich (2007)
5450. Schinzel, A., Sierpiński, W.: Sur l'équation $x^2 + y^2 + 1 = xyz$. Matematiche **10**, 30–36 (1955)
5451. Schinzel, A., Sierpiński, W.: Sur les sommes de quatre cubes. Acta Arith. **4**, 20–30 (1958)
5452. Schinzel, A., Sierpiński, W.: Sur certaines hypothèses concernant les nombres premiers. Acta Arith. **4**, 185–208 (1958) [[5449], vol. 2, pp. 1113–1133]
5453. Schinzel, A., Tijdeman, R.: On the equation $y^m = P(x)$. Acta Arith. **31**, 199–204 (1976) [[5449], vol. 1, pp. 41–46]
5454. Schinzel, A., Wirsing, E.: Multiplicative properties of the partition function. Proc. Indian Acad. Sci. Math. Sci. **97**, 297–303 (1987) [[5449], vol. 2, pp. 1211–1216]
5455. Schinzel, A., Zassenhaus, H.: A refinement of two theorems of Kronecker. Mich. Math. J. **12**, 81–85 (1965) [[5449], vol. 1, pp. 175–178]
5456. Schlafly, A., Wagon, S.: Carmichael's conjecture on the Euler function is valid below $10^{10,000,000}$. Math. Comput. **63**, 415–419 (1994)
5457. Schlage-Puchta, J.-C.: The equation $\omega(n) = \omega(n+1)$. Mathematika **50**, 99–101 (2003)
5458. Schlage-Puchta, J.-C.: The exponential sum over squarefree integers. Acta Arith. **115**, 265–268 (2004)
5459. Schlickewei, H.P.: Linearformen mit algebraischen Koeffizienten. Manuscr. Math. **18**, 147–185 (1976)
5460. Schlickewei, H.P.: Die p-adische Verallgemeinerung des Satzes von Thue-Siegel-Roth-Schmidt. J. Reine Angew. Math. **288**, 86–105 (1976)
5461. Schlickewei, H.P.: The $\mathfrak{p}$-adic Thue-Siegel-Roth-Schmidt theorem. Arch. Math. **29**, 267–270 (1977)
5462. Schlickewei, H.P.: Über die diophantische Gleichung $x_1 + x_2 + \cdots + x_n = 0$. Acta Arith. **33**, 183–185 (1977)
5463. Schlickewei, H.P.: p-adic T-numbers do exist. Acta Arith. **39**, 181–191 (1981)
5464. Schlickewei, H.P.: The number of subspaces occurring in the p-adic subspace theorem in Diophantine approximation. J. Reine Angew. Math. **406**, 44–108 (1990)
5465. Schlickewei, H.P.: S-unit equations over number fields. Invent. Math. **102**, 95–107 (1990)
5466. Schlickewei, H.P.: An explicit upper bound for the number of solutions of the S-unit equation. J. Reine Angew. Math. **406**, 109–120 (1990)
5467. Schlickewei, H.P.: The quantitative subspace theorem for number fields. Compos. Math. **82**, 245–273 (1992)
5468. Schlickewei, H.P.: Multiplicities of algebraic linear recurrences. Acta Math. **170**, 151–180 (1993)
5469. Schlickewei, H.P.: Multiplicities of recurrence sequences. Acta Math. **176**, 171–243 (1996)
5470. Schlickewei, H.P.: The multiplicity of binary recurrences. Invent. Math. **129**, 11–36 (1997)
5471. Schmid, H.L., Teichmüller, O.: Ein neuer Beweis für die Funktionalgleichung der L-Reihen. Abh. Math. Semin. Univ. Hamb. **15**, 85–96 (1947)
5472. Schmid, Th.: Karl Zsigmondy. Jahresber. Dtsch. Math.-Ver. **36**, 167–169 (1927)
5473. Schmidt, A.L.: Approximations theorems of Borel and Fujiwara. Math. Scand. **14**, 35–38 (1964)

5474. Schmidt, E.: Über die Anzahl der Primzahlen unter gegebener Grenze. Math. Ann. **57**, 195–204 (1903)
5475. Schmidt, E.: Zum Hilbertschen Beweise des Waringschen Theorems. Math. Ann. **74**, 271–274 (1913)
5476. Schmidt, F.K.: Zur Zahlentheorie in Körpern von der Charakteristik p. SBer. Erlangen **58/59**, 159–172 (1926/1927)
5477. Schmidt, F.K.: Zur Klassenkörpertheorie im Kleinen. J. Reine Angew. Math. **162**, 155–168 (1930)
5478. Schmidt, F.K.: Die Theorie der Klassenkörper über einem Körper algebraischer Funktionen in einer Unbestimmten und mit endlichen Koeffizientenbereich. SBer. Erlangen **62**, 267–284 (1931)
5479. Schmidt, F.K.: Analytische Zahlentheorie in Körpern der Charakteristik p. Math. Z. **33**, 1–32 (1931)
5480. Schmidt, F.K.: Mehrfach perfekte Körper. Math. Ann. **108**, 1–25 (1933)
5481. Schmidt, H.: Hermann Ludwig Schmid. Jahresber. Dtsch. Math.-Ver. **61**, 7–11 (1958)
5482. Schmidt, P.G.: Zur Anzahl quadratvoller Zahlen in kurzen Intervallen. Acta Arith. **46**, 159–164 (1986)
5483. Schmidt, W.[M.]: Eine neue Abschätzung der kritischen Determinante von Sternkörpern. Monatshefte Math. **60**, 1–10 (1956)
5484. Schmidt, W.M.: On normal numbers. Pac. J. Math. **10**, 661–672 (1960)
5485. Schmidt, W.M.: Über die Normalität von Zahlen zu verschiedenen Basen. Acta Arith. **7**, 299–309 (1961–1962)
5486. Schmidt, W.M.: Two combinatorial theorems on arithmetic progressions. Duke Math. J. **29**, 129–140 (1962)
5487. Schmidt, W.M.: Metrical theorems on fractional parts of sequences. Trans. Am. Math. Soc. **110**, 493–518 (1964)
5488. Schmidt, W.M.: Metrische Sätze über simultane Approximation abhängiger Grössen. Monatshefte Math. **68**, 154–166 (1964)
5489. Schmidt, W.M.: Some diophantine equations in three variables with only finitely many solutions. Mathematika **14**, 113–120 (1967)
5490. Schmidt, W.M.: Irregularities of distribution. Q. J. Math. **19**, 181–191 (1968)
5491. Schmidt, W.M.: Irregularities of distribution, II. Trans. Am. Math. Soc. **136**, 347–360 (1969)
5492. Schmidt, W.M.: Irregularities of distribution, III. Pac. J. Math. **29**, 225–234 (1969)
5493. Schmidt, W.M.: Irregularities of distribution, IV. Invent. Math. **7**, 55–82 (1969)
5494. Schmidt, W.M.: Irregularities of distribution, V. Proc. Am. Math. Soc. **25**, 608–614 (1970)
5495. Schmidt, W.M.: Irregularities of distribution, VI. Compos. Math. **24**, 63–74 (1972)
5496. Schmidt, W.M.: Irregularities of distribution, VII. Acta Arith. **21**, 45–50 (1972)
5497. Schmidt, W.M.: T-numbers do exist. Symp. Math. INDAM **4**, 3–26 (1970)
5498. Schmidt, W.M.: Simultaneous approximations to algebraic numbers by rationals. Acta Math. **125**, 189–201 (1970)
5499. Schmidt, W.M.: Mahler's T-numbers. In: Proc. Symposia Pure Math., vol. 20, pp. 275–286. Am. Math. Soc., Providence (1971)
5500. Schmidt, W.M.: Linear forms with algebraic coefficients, I. J. Number Theory **3**, 253–277 (1971)
5501. Schmidt, W.M.: Linear forms with algebraic coefficients, II. Math. Ann. **191**, 1–20 (1971)
5502. Schmidt, W.M.: Approximation to algebraic numbers. Enseign. Math. **17**, 187–253 (1971)
5503. Schmidt, W.M.: Norm form equations. Ann. Math. **96**, 526–551 (1972)
5504. Schmidt, W.M.: Volume, surface area and the number of integer points covered by a convex set. Arch. Math. **23**, 527–543 (1972)
5505. Schmidt, W.M.: Zur Methode von Stepanov. Acta Arith. **24**, 347–368 (1973)
5506. Schmidt, W.M.: Simultaneous approximation to algebraic numbers by elements of a number field. Monatshefte Math. **79**, 55–66 (1975)
5507. Schmidt, W.M.: Equations over Finite Fields. An Elementary Approach. Lecture Notes in Math., vol. 536. Springer, Berlin (1976); 2nd ed., Kendrick Press, 2004

5508. Schmidt, W.M.: Small Fractional Parts of Polynomials. Am. Math. Soc., Providence (1977)
5509. Schmidt, W.M.: Thue's equation over function fields. J. Aust. Math. Soc. A **25**, 385–422 (1978)
5510. Schmidt, W.M.: Small zeros of additive forms in many variables. Trans. Am. Math. Soc. **248**, 121–133 (1979)
5511. Schmidt, W.M.: Small zeros of additive forms in many variables, II. Acta Math. **143**, 219–232 (1979)
5512. Schmidt, W.M.: Simultaneous p-adic zeros of quadratic forms. Monatshefte Math. **90**, 45–65 (1980)
5513. Schmidt, W.M.: Diophantine Approximation. Lecture Notes in Math., vol. 785. Springer, Berlin (1980)
5514. Schmidt, W.M.: On cubic polynomials, III. Systems of p-adic equations. Monatshefte Math. **93**, 211–223 (1982)
5515. Schmidt, W.M.: On cubic polynomials, IV. Systems of rational equations. Monatshefte Math. **93**, 329–348 (1982)
5516. Schmidt, W.M.: The solubility of certain p-adic equations. J. Number Theory **19**, 63–80 (1984)
5517. Schmidt, W.M.: Integer points on curves and surfaces. Monatshefte Math. **99**, 45–72 (1985)
5518. Schmidt, W.M.: The density of integer points on homogeneous varieties. Acta Math. **154**, 243–296 (1985)
5519. Schmidt, W.M.: Integer points on hypersurfaces. Monatshefte Math. **102**, 27–58 (1986)
5520. Schmidt, W.M.: Thue equations with few coefficients. Trans. Am. Math. Soc. **303**, 241–255 (1987)
5521. Schmidt, W.M.: The subspace theorem in diophantine approximations. Compos. Math. **69**, 121–173 (1989)
5522. Schmidt, W.M.: The number of solutions of norm form equations. Trans. Am. Math. Soc. **317**, 197–227 (1990)
5523. Schmidt, W.M.: Diophantine Approximations and Diophantine Equations. Lecture Notes in Math., vol. 1467. Springer, Berlin (1991)
5524. Schmidt, W.M.: Integer points on curves of genus 1. Compos. Math. **81**, 33–59 (1992)
5525. Schmidt, W.M.: Vojta's refinement of the subspace theorem. Trans. Am. Math. Soc. **340**, 705–731 (1993)
5526. Schmidt, W.M.: The zero multiplicity of linear recurrence sequences. Acta Math. **182**, 189–201 (1999)
5527. Schmidt, W.M.: Zeros of linear recurrence sequences. Publ. Math. (Debr.) **56**, 609–630 (2000)
5528. Schmidt, W.M.: Mahler and Koksma classification of points in R^n and C^n. Funct. Approx. Comment. Math. **35**, 307–319 (2006)
5529. Schmidt, W.M.: Edmund Hlawka (1916–2009). Acta Arith. **139**, 303–320 (2009)
5530. Schmitz, T.: Abschätzung der Lösung der Pellschen Gleichungen. Arch. Math. **24**, 87–89 (1916)
5531. Schmutz, P.: Systoles of arithmetic surfaces and the Markoff spectrum. Math. Ann. **305**, 191–203 (1996)
5532. Schnabel, L.: Über eine Verallgemeinerung des Kreisproblems. Wiss. Z. F.-S. Univ. Jena **31**, 667–681 (1982)
5533. Schnack, I. (ed.): Lebensbilder aus Hessen: Marburger Gelehrte in der ersten Hälfte des 20. Jahrhunderts. Elwert, Marburg (1977)
5534. Schnee, W.: Über die Koeffizientendarstellungsformel in der Theorie der Dirichletschen Reihen. Nachr. Ges. Wiss. Göttingen, 1910, 1–42
5535. Schneeberger, W.A.: Arithmetic and geometry of integral lattices. Ph.D. thesis, Princeton Univ. (1995)
5536. Schneider, P.: Zur Vermutung von Birch und Swinnerton-Dyer über globalen Funktionenkörpern. Math. Ann. **260**, 495–510 (1982)
5537. Schneider, T.: Transzendenzuntersuchungen periodischer Funktionen, I. Transzendenz von Potenzen. J. Reine Angew. Math. **172**, 65–69 (1935)

5538. Schneider, T.: Transzendenzuntersuchungen periodischer Funktionen, II. Tranzendenzeigenschaften elliptischer Funktionen. J. Reine Angew. Math. **172**, 70–74 (1935)
5539. Schneider, T.: Arithmetische Untersuchungen elliptischer Integrale. Math. Ann. **113**, 1–13 (1937)
5540. Schneider, T.: Zur Theorie der Abelschen Funktionen und Integrale. J. Reine Angew. Math. **183**, 110–128 (1941)
5541. Schneider, T.: Einführung in die transzendenten Zahlen. Springer, Berlin (1957). [French translation: Introduction aux nombres transcendantes, Gauthier-Villars, 1959]
5542. Schneider, T.: Anwendung eines abgeänderten Roth-Ridoutschen Satzes auf diophantische Gleichungen. Math. Ann. **169**, 177–182 (1967)
5543. Schneider, T.: Das Werk C.L. Siegels in der Zahlentheorie. Jahresber. Dtsch. Math.-Ver. **85**, 147–157 (1983)
5544. Schnirelmann, L.G.[12]: Über additive Eigenschaften von Zahlen. Math. Ann. **107**, 649–690 (1933)
5545. Schnorr, C.-P.: Refined analysis and improvements on some factoring algorithms. J. Algorithms **3**, 101–127 (1982)
5546. Schoenberg, I.: Über die asymptotische Verteilung reeller Zahlen mod 1. Math. Z. **28**, 171–199 (1928)
5547. Schoenberg, I.: On asymptotic distributions of arithmetical functions. Trans. Am. Math. Soc. **39**, 315–330 (1936)
5548. Schoeneberg, B.: Indefinite Quarternionen und Modulfunktionen. Math. Ann. **113**, 382–391 (1937)
5549. Schoeneberg, B.: Elliptic Modular Functions. Springer, Berlin (1974)
5550. Schoeneberg, B.: Erich Hecke 1887–1947. Jahresber. Dtsch. Math.-Ver. **91**, 168–190 (1989)
5551. Schoenfeld, L.: The order of the zeta function near the line $\sigma = 1$. Duke Math. J. **24**, 601–609 (1957)
5552. Schoenfeld, L.: Sharper bounds for the Chebyshev functions $\theta(x)$ and $\psi(x)$, II. Math. Comput. **30**, 337–360 (1976); corr., 900
5553. Schoißengeier, J.: The integral mean of the discrepancy of the sequence $(n\alpha)$. Monatshefte Math. **131**, 227–234 (2000)
5554. Scholz, A.: Die Abgrenzungssätze für Kreiskörper und Klassenkörper. SBer. Preuß. Akad. Wiss. Berl. **20**, 417–426 (1931)
5555. Scholz, A.: Totale Normenreste, die keine Normen sind, als Erzeuger nichtabelscher Körpererweiterungen, I. J. Reine Angew. Math. **175**, 100–107 (1936)
5556. Scholz, A.: Totale Normenreste, die keine Normen sind, als Erzeuger nichtabelscher Körpererweiterungen, II. J. Reine Angew. Math. **182**, 217–234 (1940)
5557. Scholz, A.: Aufgabe 253. Jahresber. Dtsch. Math.-Ver. **47**, 41–42 (1937)
5558. Scholz, A.: Zur Idealtheorie in unendlichen algebraischen Zahlkörpern. J. Reine Angew. Math. **185**, 113–126 (1943)
5559. Scholz, B.: Bemerkungen zu einem Beweis von Wieferich. Jahresber. Dtsch. Math.-Ver. **58**, 45–48 (1955)
5560. Schöneborn, H.: In memoriam Wolfgang Krull. Jahresber. Dtsch. Math.-Ver. **82**, 51–62 (1980)
5561. Schönhage, A.: Eine Bemerkung zur Konstruktion grosser Primzahldifferenzen. Arch. Math. **14**, 29–30 (1936)
5562. Schoof, R.: Elliptic curves over finite fields and the computation of square roots mod p. Math. Comput. **44**, 483–494 (1985)
5563. Schoof, R.: Counting points on elliptic curves over finite fields. J. Théor. Nr. Bordx. **7**, 219–254 (1995)
5564. Schoof, R.: Abelian varieties over cyclotomic fields with good reduction everywhere. Math. Ann. **325**, 413–448 (2003)

[12]See also Šnirelman, L.G.

5565. Schoof, R.: Class numbers of real cyclotomic fields of prime conductor. Math. Comput. **72**, 913–937 (2003)
5566. Schoof, R.: Catalan's Conjecture. Springer, Berlin (2008)
5567. Schreiber, F.: Emigration in Europe—the life and work of Félix Pollaczek. In: Charlemagne and His Heritage. 1200 Years of Civilization and Science in Europe, vol. II, pp. 179–194. Brepols, Turnhout (1998)
5568. Schreier, O.: Über eine Arbeit von Herrn Tschebotareff. Abh. Math. Semin. Univ. Hamb. **5**, 1–6 (1927)
5569. von Schrutka, L.: Ein Beweis für die Zerlegbarkeit der Primzahlen von der Form $6n + 1$ in ein einfaches und ein dreifaches Quadrat. J. Reine Angew. Math. **140**, 252–265 (1911)
5570. Schrutka von Rechtenstamm, G.: Tabelle der (Relativ)-Klassenzahlen der Kreiskörper, deren φ-Funktion des Wurzelexponenten (Grad) nicht grösser als 256 ist. Abhandl. Deutsch. Akad. Wiss., 1964, nr. 2, 1–64
5571. Schumann, H.G.: Zum Beweis des Hauptidealsatzes. Abh. Math. Semin. Univ. Hamb. **12**, 42–47 (1937)
5572. Schumer, P.D.: On the large sieve inequality in an algebraic number field. Mathematika **33**, 31–54 (1986)
5573. Schur, I.: Ein Beitrag zur additiven Zahlentheorie und zur Theorie der Kettenbrüche. SBer. Preuß. Akad. Wiss. Berlin, 1917, 302–321 [[5578], vol. 2, pp. 117–136]
5574. Schur, I.: Einige Bemerkungen zu der vorstehender Arbeit des Herrn G. Pólya: Über die Verteilung der quadratischen Reste und Nichtreste. Nachr. Ges. Wiss. Göttingen, 1918, 30–36 [[5578], vol. 2, pp. 239–245]
5575. Schur, I.: Über den Zusammenhang zwischen einem Problem der Zahlentheorie und einem Satz über algebraische Funktionen. SBer. Preuß. Akad. Wiss. Berlin, 1923, 123–134. [[5578], vol. 2, pp. 428–439]
5576. Schur, I.: Einige Sätze über Primzahlen mit Anwendungen auf Irreduzibilitätsfragen, I. SBer. Preuß. Akad. Wiss. Berl. **23**, 125–136 (1929) [[5578], vol. 3, pp. 140–151]
5577. Schur, I.: Einige Sätze über Primzahlen mit Anwendungen auf Irreduzibilitätsfragen, II. SBer. Preuß. Akad. Wiss. Berl. **23**, 370–391 (1929) [[5578], vol. 3, 152–173]
5578. Schur, I.: Gesammelte Abhandlungen, vols. 1–3. Springer, Berlin (1973)
5579. Schütte, K., van der Waerden, B.L.: Das Problem der dreizehn Kugeln. Math. Ann. **125**, 325–334 (1953)
5580. Schwarz, W.: Zur Darstellung von Zahlen durch Summen von Primzahlpotenzen. J. Reine Angew. Math. **206**, 78–112 (1961)
5581. Schwarz, W.: Über die Ramanujan-Entwicklung multiplikativer Funktionen. Acta Arith. **27**, 269–279 (1975)
5582. Schwarz, W.: Ramanujan expansions of arithmetical functions. In: Andrews, G.E., Askey, R.A., Berndt, B.C., Ramanathan, K.G., Rankin, R.A. (eds.) Ramanujan Revisited (Proceedings of the Centenary Conference), pp. 187–214. Academic Press, San Diego (1988)
5583. Schwarz, W.: John Knopfmacher, analytic number theory, and the theory of arithmetical functions. Quaest. Math. **24**, 273–290 (2001)
5584. Schwarz, W., Volkmann, B.: Hans Rohrbach zum Gedächtnis. 27.2.1903–19.12.1993. Jahresber. Dtsch. Math.-Ver. **105**, 89–99 (2003)
5585. Schwarz, W., Zassenhaus, H.: In memoriam Bohuslav Diviš: December 20, 1942–July 26, 1976. J. Number Theory **9**, iv–vii (1977)
5586. Schweiger, F.: Abschätzung der Hausdorffdimension für Mengen mit vorgeschriebenen Häufigkeiten der Ziffern. Monatshefte Math. **76**, 138–142 (1972)
5587. Scott, J.F.: The Mathematical Work of John Wallis, D.D., F.R.S. (1616–1703). Taylor & Francis, London (1938) [Reprint: Chelsea, 1981]
5588. Scott, S.J.: On the number of zeros of a cubic recurrence. Am. Math. Mon. **67**, 169–170 (1960)
5589. Scourfield, E.J.: A generalization of Waring's problem. J. Lond. Math. Soc. **35**, 98–116 (1960)
5590. Scourfield, E.J.: The divisors of a quadratic polynomial. Proc. Glasg. Math. Assoc. **5**, 8–20 (1961)

5591. Scourfield, E.J.: On the divisibility of $\sigma_\nu(n)$. Acta Arith. **10**, 245–288 (1964)
5592. Scourfield, E.J.: An asymptotic formula for the property $(n, f(n)) = 1$ for a class of multiplicative functions. Acta Arith. **29**, 401–423 (1976)
5593. Scourfield, E.J.: A uniform coprimality result for some arithmetic functions. J. Number Theory **20**, 315–353 (1985)
5594. Scriba, C.J.: Zur Erinnerung an Viggo Brun. Mitt. Math. Ges. Hamb. **11**, 271–290 (1985)
5595. Ščur, V., Sinai, Ya.[G.], Ustinov, A.[V.]: Limiting distribution of Frobenius numbers for $n = 3$. J. Number Theory **129**, 2778–2789 (2009)
5596. Seeber, L.A.: Untersuchungen über die Eigenschaften der positiven ternären quadratischen Formen. Freiburg im Breisgau (1831)
5597. Seelhoff, P.: Die neunte vollkommene Zahl. Z. Angew. Math. Phys. **31**, 174–178 (1886)
5598. Segal, B.I.: Generalisation of Brun's theorem. Dokl. Akad. Nauk SSSR, 1930, 501–507 (in Russian)
5599. Segal, B.I.: On a theorem analogous to Waring's theorem. Dokl. Akad. Nauk SSSR **1**, 47–49 (1933) (in Russian)
5600. Segal, B.I.: Waring's theorem for powers with rational or irrational exponents. Tr. Mat. Inst. Steklova **5**, 79–86 (1934) (in Russian)
5601. Segal, B.I.: On some problems of the additive theory of numbers. Ann. Math. **36**, 507–520 (1935)
5602. Segal, S.L.: On $\pi(x+y) \le \pi(x) + \pi(y)$. Trans. Am. Math. Soc. **104**, 523–527 (1962)
5603. Segre, B.: Sulle irrazionalita da cui puo farsi dipendere la determinazioni di S_k appartenenti a arieta intersezioni complete di forme. Atti Accad. Naz. Lincei **4**, 149–154 (1948)
5604. Sekigawa, H., Koyama, K.: Nonexistence conditions of a solution for the congruence $x_1^k + \cdots + x_s^k \equiv N \pmod{p^n}$. Math. Comput. **68**, 1283–1297 (1999)
5605. Selberg, A.: Über einige arithmetische Identitäten. Avh. Norske Vidensk. I 1936, Nr. 8, 1–23 [[5625], pp. 1–21]
5606. Selberg, A.: On the zeros of Riemann's zeta-function on the critical line. Arch. Math. Naturvidensk. **45**, 101–114 (1942) [[5625], pp. 142–155]
5607. Selberg, A.: On the zeros of Riemann's zeta-function. Skr. Norske Vid. Akad. Oslo, 1942, nr. 10, 1–59 [[5625], pp. 85–141]
5608. Selberg, A.: On the normal density of primes in small intervals, and the difference between consecutive primes. Arch. Math. Naturvidensk. **47**, 87–105 (1943) [[5625], pp. 160–178]
5609. Selberg, A.: On the remainder in the formula for $N(T)$, the number of zeros of $\zeta(s)$ in the strip $0 < t < T$. Avh. Norske Vidensk. I 1944, nr. 1, 1–27 [[5625], pp. 179–203]
5610. Selberg, A.: Contributions to the theory of the Riemann zeta-function. Arch. Math. Naturvidensk. **48**, 89–155 (1946) [[5625], pp. 214–280]
5611. Selberg, A.: On an elementary method in the theory of primes. Norske Vid. Selsk. Forh., Trondheim **19**, 64–67 (1947) [[5625], pp. 363–366]
5612. Selberg, A.: An elementary proof of Dirichlet's theorem about primes in an arithmetic progressions. Ann. Math. **50**, 297–304 (1949) [[5625], pp. 371–378]
5613. Selberg, A.: An elementary proof of the prime number theorem. Ann. Math. **50**, 305–313 (1949) [[5625], pp. 379–387]
5614. Selberg, A.: On elementary methods in prime number theory and their limitations. In: 11 Skandinaviske Matematiker Kongress, Trondheim, 1949, pp. 13–22. Johan Grundt Tonums Forlag, Oslo (1952) [[5625], pp. 388–397]
5615. Selberg, A.: An elementary proof of the prime number theorem for arithmetic progressions. Can. J. Math. **2**, 66–78 (1950) [[5625], pp. 398–410]
5616. Selberg, A.: The general sieve method and its place in prime number theory. In: Proc. ICM 1950, vol. 1, pp. 286–292. Am. Math. Soc., Providence (1952) [[5625], pp. 411–417]
5617. Selberg, A.: Note on a paper by L.G. Sathe. J. Indian Math. Soc. **18**, 83–87 (1954) [[5625], pp. 418–422]
5618. Selberg, A.: On the estimation of Fourier coefficients of modular forms. In: Proc. Symposia Pure Math., vol. 8, pp. 1–15. Am. Math. Soc., Providence (1965) [[5625], pp. 506–520]
5619. Selberg, A.: Sieve methods. In: Proc. Symposia Pure Math., vol. 20, pp. 311–351. Am. Math. Soc., Providence (1971) [[5625], pp. 568–608]

5620. Selberg, A.: Remarks on sieves. In: Proceedings of the Number Theory Conference, Univ. Colorado, Boulder, 1972, pp. 205–216 (1972) [[5625], pp. 609–625]
5621. Selberg, A.: Remarks on multiplicative functions. In: Lecture Notes in Math., vol. 626, pp. 232–241. Springer, Berlin (1977) [[5625], pp. 616–625]
5622. Selberg, A.: Sifting problems, sifting density, and sieves. In: Number Theory, Trace Formulas and Discrete Groups, Oslo, 1987, pp. 467–484. Academic Press, Boston (1989) [[5625], pp. 675–694]
5623. Selberg, A.: Old and new conjectures and results about a class of Dirichlet series. In: Proceedings of the Amalfi Conference on Analytic Number Theory, Maiori, 1989, pp. 367–385. Univ. de Salerno, Salerno (1992). [[5626], pp. 47–63]
5624. Selberg, A.: Lectures on sieves. In: Selberg, A. (ed.) Collected Papers, vol. 2, pp. 65–257. Springer, Berlin (1991)
5625. Selberg, A.: Collected Papers, vol. 1. Springer, Berlin (1989)
5626. Selberg, A.: Collected Papers, vol. 2. Springer, Berlin (1991)
5627. Selberg, A., Chowla, S.: On Epstein's zeta function. J. Reine Angew. Math. **227**, 86–110 (1967) [[5625], pp. 521–545]
5628. Selberg, S.: Ernst Jacobsthal. Norske Vid. Selsk. Forh., Trondheim **38**, 70–73 (1965)
5629. Selfridge, J.L., Nicol, C.A., Vandiver, H.S.: Proof of Fermat's last theorem for all prime exponents less than 4002. Proc. Natl. Acad. Sci. USA **41**, 970–973 (1955)
5630. Selfridge, J.L., Pollack, B.W.: Fermat's last theorem is true for any exponent up to 25, 000. Not. Am. Math. Soc. **11**, 97 (1964)
5631. Sellers, J.A., Sills, A.V., Mullen, G.L.: Bijections and congruences for generalizations of partition identities of Euler and Guy. Electron. J. Combin. **11**, Res. Paper 43, 1–19 (2004)
5632. Selmer, E.S.: The diophantine equation $ax^3 + by^3 + cz^3 = 0$. Acta Math. **85**, 203–362 (1951)
5633. Selmer, E.S.: Sufficient congruence conditions for the existence of rational points on certain cubic surfaces. Math. Scand. **1**, 113–119 (1953)
5634. Selmer, E.S.: The rational solutions of the Diophantine equation $\eta^2 = \xi^3 - D$ for $|D| \leq 100$. Math. Scand. **4**, 281–286 (1956)
5635. Selmer, E.S.: On the linear Diophantine problem of Frobenius. J. Reine Angew. Math. **293/294**, 1–17 (1977)
5636. Selmer, E.S., Beyer, Ö.: On the linear diophantine problem of Frobenius in three variables. J. Reine Angew. Math. **301**, 161–170 (1978)
5637. Şengün, M.H.: The nonexistence of certain representations of the absolute Galois group of quadratic fields. Proc. Am. Math. Soc. **137**, 27–35 (2009)
5638. Serre, J.-P.: Une interprétation des congruences relatives à la fonction τ de Ramanujan. Sém. Delange–Pisot–Poitou **9**(14), 1–17 (1967/1968) [[5661], vol. 2, pp. 498–511]
5639. Serre, J.-P.: Abelian l-adic Representations and Elliptic Curves. Benjamin, Elmsford (1968) [Reprints: Addison-Wesley, 1989; AK Peters, 1998]
5640. Serre, J.-P.: Le problème des groupes de congruence pour SL_2. Ann. Math. **92**, 489–527 (1970) [[5661], vol. 2, pp. 498–511]
5641. Serre, J.-P.: Cours d'arithmétique. Presses Universitaires de France, Paris (1970); 2nd ed. 1977 [English translation: A Course in Arithmetic, Springer, 1973; 2nd ed. 1978]
5642. Serre, J.-P.: p-torsion des courbes elliptiques (d'après Y. Manin). In: Lecture Notes in Math., vol. 180, pp. 281–294. Springer, Berlin (1971)
5643. Serre, J.-P.: Propriétés galoisiennes des points d'ordre fini des courbes elliptiques. Invent. Math. **15**, 259–331 (1972) [[5661], vol. 3, pp. 1–73]
5644. Serre, J.-P.: Congruences et formes modulaires. Sém. Bourbaki, **24**, 1971/1972, nr. 416 [[5661], vol. 3, pp. 74–78]
5645. Serre, J.-P.: Formes modulaires et fonctions zêta p-adiques. In: Modular Functions of One Variable, III. Lecture Notes in Math., vol. 350, pp. 191–268. Springer, Berlin (1973); corr. ibid, IV, Lecture Notes in Math., vol. 476, pp. 149–150 (1975) [[5661], vol. 3, pp. 95–172]
5646. Serre, J.-P.: Divisibilité de certaines fonctions arithmétiques. Sém. Delange–Pisot–Poitou **16**(exp. 20), 1–28 (1974/1975) [[5661], vol. 3, pp. 250–283]

5647. Serre, J.-P.: Divisibilité des coefficients des formes modulaires de poids entier. C. R. Acad. Sci. Paris **279**, 679–682 (1974) [[5661], vol. 3, pp. 189–192]
5648. Serre, J.-P.: Valeurs propres des opérateurs de Hecke modulo l. Astérisque **24/25**, 109–117 (1975) [[5661], vol. 3, pp. 226–234]
5649. Serre, J.-P.: Valeurs propres des endomorphismes de Frobenius (d'aprés P. Deligne). In: Lecture Notes in Math., vol. 431, pp. 190–204. Springer, Berlin (1975) [[5661], vol. 3, pp. 179–188]
5650. Serre, J.-P.: Divisibilité de certains fonctions arithmétiques. Enseign. Math. **22**, 227–260 (1976) [[5661], vol. 3, pp. 250–283]
5651. Serre, J.-P.: Modular forms of weight one and Galois representations. In: Algebraic Number Fields (L-functions and Galois Properties), pp. 193–268. Academic Press, London (1977) [[5661], vol. 3, pp. 292–367]
5652. Serre, J.-P.: Résumé des cours 1977–1978. In: Oeuvres, vol. 3, pp. 465–468. Springer, Berlin (1986–2000)
5653. Serre, J.-P.: Points rationnels des courbes modulaires $X_0(N)$ [d'aprés Barry Mazur]. In: Lecture Notes in Math., vol. 710, pp. 89–100. Springer, Berlin (1979)
5654. Serre, J.-P.: Quelques applications du théoreme de densité de Chebotarev. Publ. Math. Inst. Hautes Études Sci. **54**, 123–202 (1981) [[5661], vol. 3, pp. 563–641]
5655. Serre, J.-P.: Lettre à J.-F. Mestre. In: Contemp. Math., vol. 67, pp. 263–268. Am. Math. Soc., Providence (1987) [[5661], vol. 4, pp. 101–106]
5656. Serre, J.-P.: Sur les représentations modulaires de degré 2 de $\mathrm{Gal}(\overline{Q}/Q)$. Duke Math. J. **54**, 179–230 (1987) [[5661], vol. 4, pp. 107–158]
5657. Serre, J.-P.: Résumé des cours au Collège de France, 1987–1988. In: Oeuvres, vol. 4, pp. 163–166. Springer, Berlin (1986–2000)
5658. Serre, J.-P.: Lectures on the Mordell-Weil Theorem, 3rd ed. Vieweg, Wiesbaden (1989) 1997
5659. Serre, J.-P.: Smith, Minkowski et l'Académie des Sciences. Math. Gaz. **56**, 3–9 (1993)
5660. Serre, J.-P.: Travaux de Wiles (et Taylor, ...), I. Astérisque **237**, 319–332 (1996) [[5661], vol. 4, pp. 524–542]
5661. Serre, J.-P.: Oeuvres, vols. 1–4, Springer, Berlin (1986–2000)
5662. Serre, J.-P., Stark, H.M.: Modular forms of weight 1/2. In: Lecture Notes in Math., vol. 627, pp. 27–67. Springer, Berlin (1977) [[5661], vol. 3, pp. 401–440]
5663. Serre, J.-P., Zagier, D.B.: Modular Functions of One Variable V. Lecture Notes in Math., vol. 601. Springer, Berlin (1977)
5664. Serre, J.-P., Zagier, D.B.: Modular Functions of One Variable VI. Lecture Notes in Math., vol. 627. Springer, Berlin (1977)
5665. Setzer, B.: Elliptic curves of prime conductor. J. Lond. Math. Soc. **10**, 367–378 (1975)
5666. Setzer, B.: Elliptic curves over complex quadratic fields. Pac. J. Math. **74**, 235–250 (1978)
5667. Ševelev, F.J.: A short summary of the scientific work of N.V. Bugaev. Istor.-Mat. Issled. **12**, 525–558 (1959) (in Russian)
5668. Shallit, J.: Simple continued fractions for some irrational numbers. J. Number Theory **11**, 209–217 (1979)
5669. Shallit, J.: Simple continued fractions for some irrational numbers, II. J. Number Theory **14**, 228–231 (1982)
5670. Shallit, J.: Real numbers with bounded partial quotients: a survey. Enseign. Math. **38**, 151–187 (1992)
5671. Shan, Z.: On composite n for which $\varphi(n)|n-1$. J. China Univ. Sci. Tech. **15**, 109–112 (1985)
5672. Shan, Z., Wang, E.T.H.: Generalization of a theorem of Mahler. J. Number Theory **32**, 111–113 (1989)
5673. Shanks, D.: Quadratic residues and the distribution of primes. Math. Tables Other Aids Comput. **13**, 272–284 (1959)
5674. Shanks, D.: The second-order term in the asymptotic expansion of $B(x)$. Math. Comput. **18**, 75–86 (1964)

5675. Shanks, D.: In the Gaussian formulation the class number problem is easier. Math. Comput. **23**, 151–163 (1969)
5676. Shapiro, H.N.: Note on a theorem of Dickson. Bull. Am. Math. Soc. **55**, 450–452 (1949)
5677. Shapiro, H.N.: An elementary proof of the prime ideal theorem. Commun. Pure Appl. Math. **2**, 309–323 (1949)
5678. Shapiro, H.N.: On primes in arithmetic progressions, I. Ann. Math. **52**, 217–230 (1950)
5679. Shapiro, H.N.: On primes in arithmetic progressions, II. Ann. Math. **52**, 231–243 (1950)
5680. Shapiro, H.N.: Some remarks on a theorem of Erdös concerning asymptotic density. Proc. Am. Math. Soc. **1**, 590–592 (1950)
5681. Shapiro, H.N.: Distribution functions of additive arithmetic functions. Proc. Natl. Acad. Sci. USA **42**, 426–430 (1956)
5682. Shapiro, H.N., Shlapentokh, A.: Diophantine relationships between algebraic number fields. Commun. Pure Appl. Math. **42**, 1113–1122 (1989)
5683. Shapiro, H.N., Sparer, G.H.: On algebraic independence of Dirichlet series. Commun. Pure Appl. Math. **39**, 695–745 (1986)
5684. Shapiro, H.N., Warga, J.: On representation of large integers as sum of primes. Commun. Pure Appl. Math. **3**, 153–176 (1950)
5685. Shelah, S.: Primitive recursive bounds for van der Waerden numbers. J. Am. Math. Soc. **1**, 683–697 (1988)
5686. Shen, Z.: On the Diophantine equation $\sum_{i=0}^{k} 1/x_i = a/n$. Chin. Ann. Math., Ser. B **7**, 213–220 (1986)
5687. Shepherd-Barron, N.I., Taylor, R.: mod 2 and mod 5 icosahedral representations. J. Am. Math. Soc. **10**, 283–298 (1997)
5688. Shibata, K.: On the approximation of irrational numbers by rational numbers. Tohoku Math. J. **23**, 328–337 (1924)
5689. Shields, P.: Representation of integers as a sum of two h-th powers of primes. Proc. Lond. Math. Soc. **38**, 369–384 (1979)
5690. Shimizu, H.: On discontinuous groups operating on the product of the upper half planes. Ann. Math. **77**, 33–71 (1963)
5691. Shimizu, H.: On traces of Hecke operators. J. Fac. Sci. Univ. Tokyo **10**, 1–19 (1963)
5692. Shimura, G.: Correspondances modulaires et les fonctions ζ de courbes algébriques. J. Math. Soc. Jpn. **10**, 1–28 (1958)
5693. Shimura, G.: On the zeta-functions of the algebraic curves uniformized by certain automorphic functions. J. Math. Soc. Jpn. **13**, 275–331 (1961)
5694. Shimura, G.: On Dirichlet series and abelian varieties attached to automorphic forms. Ann. Math. **76**, 237–294 (1962)
5695. Shimura, G.: A reciprocity law in non-solvable extensions. J. Reine Angew. Math. **221**, 209–220 (1966)
5696. Shimura, G.: Construction of class fields and zeta functions of algebraic curves. Ann. Math. **85**, 58–159 (1967)
5697. Shimura, G.: Automorphic Functions and Number Theory. Lecture Notes in Math., vol. 54. Springer, Berlin (1968)
5698. Shimura, G.: Introduction to the Arithmetic Theory of Automorphic Functions. Princeton University Press, Princeton (1971); reprint 1994
5699. Shimura, G.: On elliptic curves with complex multiplication as factors of the Jacobians of modular function fields. Nagoya Math. J. **43**, 199–208 (1971)
5700. Shimura, G.: On the zeta-function of an abelian variety with complex multiplication. Ann. Math. **94**, 504–533 (1971)
5701. Shimura, G.: On modular forms of half integral weight. Ann. Math. **97**, 440–481 (1973)
5702. Shimura, G.: Modular forms of half integral weight. In: Modular forms of One Variable I. Lecture Notes in Math., vol. 320, pp. 57–74. Springer, Berlin (1973)
5703. Shimura, G.: On certain reciprocity-laws for theta functions and modular forms. Acta Math. **141**, 35–71 (1978)
5704. Shimura, G.: Yutaka Taniyama and his time. Bull. Lond. Math. Soc. **21**, 186–196 (1989)

5705. Shimura, G., Taniyama, Y.: Complex Multiplication of Abelian Varieties and Its Applications to Number Theory. Math. Soc. Jpn., Tokyo (1961) [Expanded version: Shimura, G., Abelian Varieties with Complex Multiplication and Modular Functions, Princeton (1998)]
5706. Shintani, T.: On zeta-functions associated with the vector space of quadratic forms. J. Fac. Sci. Univ. Tokyo **22**, 25–65 (1975)
5707. Shintani, T.: On construction of holomorphic cusp forms of half integral weight. Nagoya Math. J. **58**, 83–126 (1975)
5708. Shintani, T.: On evaluation of zeta functions of totally real algebraic number fields at non-positive integers. J. Fac. Sci. Univ. Tokyo **23**, 393–417 (1976)
5709. Shintani, T.: On certain ray class invariants of real quadratic fields. J. Math. Soc. Jpn. **30**, 139–167 (1978)
5710. Shiokawa, I.: A remark on a theorem of Copeland-Erdős. Proc. Jpn. Acad. Sci. **50**, 273–276 (1974)
5711. Shiratani, K.: On certain values of p-adic L-functions. Mem. Fac. Sci. Kyushu Univ. **28**, 59–82 (1974)
5712. Shiratani, K.: On a formula for p-adic L-functions. J. Fac. Sci. Univ. Tokyo **24**, 45–53 (1977)
5713. Shiu, P.: A Brun-Titchmarsh theorem for multiplicative functions. J. Reine Angew. Math. **313**, 161–170 (1980)
5714. Shiu, P.: On square-full integers in a short interval. Glasg. Math. J. **25**, 127–134 (1984)
5715. Shiu, P.: Counting sums of two squares: The Meissel-Lehmer method. Math. Comput. **47**, 351–360 (1986)
5716. Shiu, P.: Counting sums of three squares. Bull. Lond. Math. Soc. **20**, 203–208 (1988)
5717. Shlapentokh, A.: Extension of Hilbert's tenth problem to some algebraic number fields. Commun. Pure Appl. Math. **42**, 939–962 (1989)
5718. Shlapentokh, A.: Hilbert's Tenth Problem over number fields, a survey. In: Contemp. Math., vol. 270, pp. 107–137. Am. Math. Soc., Providence (2000)
5719. Shlapentokh, A.: Hilbert's Tenth Problem. Diophantine Classes and Extensions to Global Fields. Cambridge University Press, Cambridge (2007)
5720. Shokrollahi, M.A.: Relative class number of imaginary abelian fields of prime conductor below 10000. Math. Comput. **68**, 1717–1728 (1999)
5721. Shorey, T.N.: On gaps between numbers with a large prime factor, II. Acta Arith. **25**, 365–373 (1973)
5722. Shorey, T.N.: On the equation $ax^m - by^n = k$. Indag. Math. **48**, 353–358 (1986)
5723. Shorey, T.N.: Exponential Diophantine equations involving products of consecutive integers and related equations. In: Number Theory, pp. 463–495. Birkhäuser, Basel (2000)
5724. Shorey, T.N., Stewart, C.L.: On the Diophantine equation $ax^{2t} + bx^t y + cy^2 = d$ and pure powers in recurrence sequences. Math. Scand. **52**, 24–36 (1983)
5725. Shorey, T.N., Stewart, C.L.: Pure powers in recurrence sequences and some related Diophantine equations. J. Number Theory **27**, 324–352 (1987)
5726. Shorey, T.N., Tijdeman, R.: New applications of Diophantine approximations to Diophantine equations. Math. Scand. **39**, 5–18 (1976)
5727. Shorey, T.N., Tijdeman, R.: Exponential Diophantine Equations. Cambridge University Press, Cambridge (1986)
5728. Shorey, T.N., Tijdeman, R.: Perfect powers in products of terms in an arithmetical progression. Compos. Math. **75**, 307–344 (1990)
5729. Shorey, T.N., Tijdeman, R.: Perfect powers in products of terms in an arithmetical progression, II. Compos. Math. **82**, 107–117 (1992)
5730. Shorey, T.N., Tijdeman, R.: Perfect powers in products of terms in an arithmetical progression, III. Acta Arith. **61**, 391–398 (1992)
5731. Shorey, T.N., van der Poorten, A.J., Tijdeman, R., Schinzel, A.: Applications of the Gel'fond-Baker method to Diophantine equations. In: Transcendence Theory: Advances and Applications, pp. 59–77. Academic Press, San Diego (1977)

5732. Shparlinski, I.[13]: On exponential sums with sparse polynomials and rational functions. J. Number Theory **60**, 233–244 (1996)
5733. Sidelnikov, V.M.: New bounds for density of sphere packing in an n-dimensional Euclidean space. Mat. Sb. **93**, 148–158 (1974) (in Russian)
5734. Šidlovskiĭ, A.B.: On a criterion for algebraic independence of values of a class of entire functions. Dokl. Akad. Nauk SSSR **100**, 221–224 (1955) (in Russian)
5735. Šidlovskiĭ, A.B.: On a criterion for algebraic independence of values of a class of entire functions. Izv. Akad. Nauk SSSR, Ser. Mat. **23**, 35–66 (1959) (in Russian)
5736. Šidlovskiĭ, A.B.: Transcendental Numbers. Nauka, Moscow (1987) (in Russian) [English translation: de Gruyter, 1989]
5737. Siebert, H.: Sieve methods and Siegel's zeros. In: Studies in Pure Mathematics, pp. 659–668. Birkhäuser, Basel (1983)
5738. Siegel, C.L.: Approximation algebraischer Zahlen. Math. Z. **10**, 173–213 (1921) [[5778], vol. 1, pp. 6–46]
5739. Siegel, C.L.: Ueber den Thueschen Satz. Christiania Vid. Selsk, Skr., 1921, nr. 16 [[5778], vol. 1, pp. 103–112]
5740. Siegel, C.L.: Darstellung total positiver Zahlen durch Quadrate. Math. Z. **11**, 246–275 (1921) [[5778], vol. 1, pp. 47–76]
5741. Siegel, C.L.: Über die Koeffizienten in der Taylorschen Entwicklung rationeller Funktionen. Tohoku Math. J. **20**, 26–31 (1921) [[5778], vol. 1, pp. 97–102]
5742. Siegel, C.L.: Additive Theorie der Zahlkörper, I. Math. Ann. **87**, 1–35 (1922) [[5778], vol. 1, pp. 119–153]
5743. Siegel, C.L.: Additive Theorie der Zahlkörper, II. Math. Ann. **88**, 184–210 (1923) [[5778], vol. 1, pp. 180–206]
5744. Siegel, C.L.: Neuer Beweis für die Funktionalgleichung der Dedekindschen Zetafunktion. Math. Ann. **85**, 123–128 (1923) [[5778], vol. 1, pp. 113–118]
5745. Siegel, C.L.: Neuer Beweis für die Funktionalgleichung der Dedekindschen Zetafunktion, II. Nachr. Ges. Wiss. Göttingen, 1922, 25–31 [[5778], vol. 1, pp. 173–179]
5746. Siegel, C.L. (under the pseudonym X): The integer solutions of $y^2 = ax^n + bx^{n-1} + \cdots + k$. J. London Math. Soc. **1**, 66–68 (1926) [[5778], vol. 1, pp. 207–208]
5747. Siegel, C.L.: Über einige Anwendungen diophantischer Approximationen. Abh. Kgl. Preuß. Akad. Wiss. Berlin, 1929, 1–70 [[5778], vol. 1, pp. 209–266]
5748. Siegel, C.L.: Über die Perioden elliptischer Funktionen. J. Reine Angew. Math. **167**, 62–69 (1932) [[5778], vol. 1, pp. 267–274]
5749. Siegel, C.L.: Über Riemanns Nachlaßzur analytischen Zahlentheorie. Quellen Stud. Gesch. Math., Astron. Phys. **2**, 45–80 (1932) [[5778], vol. 1, pp. 275–310]
5750. Siegel, C.L.: Über die analytische Theorie der quadratischen Formen. Ann. Math. **36**, 527–606 (1935) [[5778], vol. 1, pp. 326–405]
5751. Siegel, C.L.: Über die analytische Theorie der quadratischen Formen, II. Ann. Math. **37**, 230–263 (1936) [[5778], vol. 1, pp. 410–443]
5752. Siegel, C.L.: Über die analytische Theorie der quadratischen Formen, III. Ann. Math. **38**, 212–291 (1937) [[5778], vol. 1, pp. 469–548]
5753. Siegel, C.L.: Über die Classenzahl quadratischer Zahlkörper. Acta Arith. **1**, 83–86 (1935) [[5778], vol. 1, pp. 406–409]
5754. Siegel, C.L.: Die Gleichung $ax^n - by^n = c$. Math. Ann. **114**, 57–68 (1937) [[5778], vol. 2, pp. 8–19]
5755. Siegel, C.L.: Über die Zetafunktionen indefiniter quadratischer Formen. Math. Z. **43**, 682–708 (1938) [[5778], vol. 2, pp. 41–67]
5756. Siegel, C.L.: Über die Zetafunktionen indefiniter quadratischer Formen, II. Math. Z. **44**, 398–426 (1939) [[5778], vol. 2, pp. 68–96]
5757. Siegel, C.L.: Einführung in die Theorie der Modulfunktionen n-ten Grades. Math. Ann. **116**, 617–657 (1939) [[5778], vol. 2, pp. 97–137]

[13] See also Šparlinskiĭ, I.E.

5758. Siegel, C.L.: Einheiten quadratischer Formen. Abh. Math. Semin. Hansischen Univ. **13**, 209–239 (1940) [[5778], vol. 2, pp. 138–168]
5759. Siegel, C.L.: Equivalence of quadratic forms. Am. J. Math. **63**, 658–680 (1941) [[5778], vol. 2, pp. 217–239]
5760. Siegel, C.L.: Contributions to the theory of the Dirichlet L-series and the Epstein zeta-functions. Ann. Math. **44**, 143–172 (1943) [[5778], vol. 2, pp. 360–389]
5761. Siegel, C.L.: Symplectic geometry. Am. J. Math. **65**, 1–86 (1943) [[5778], vol. 2, pp. 274–359]
5762. Siegel, C.L.: Generalization of Waring's problem to algebraic number fields. Am. J. Math. **66**, 122–136 (1944) [[5778], vol. 2, pp. 406–420]
5763. Siegel, C.L.: On the theory of indefinite quadratic forms. Ann. Math. **45**, 577–622 (1944) [[5778], vol. 2, pp. 421–466]
5764. Siegel, C.L.: The average measure of quadratic forms with given discriminant and signature. Ann. Math. **45**, 667–685 (1944) [[5778], vol. 2, pp. 473–491]
5765. Siegel, C.L.: Algebraic integers whose conjugates lie in the unit circle. Duke Math. J. **11**, 597–602 (1944)
5766. Siegel, C.L.: Sums of m-th powers of algebraic integers. Ann. Math. **46**, 313–339 (1945) [[5778], vol. 3, pp. 12–38]
5767. Siegel, C.L.: A mean value theorem in geometry of numbers. Ann. Math. **46**, 340–347 (1945) [[5778], vol. 3, pp. 39–46]
5768. Siegel, C.L.: Transcendental Numbers. Ann. Math. Stud., vol. 16. Princeton University Press, Princeton (1949) [German translation: Transzendente Zahlen, Mannheim (1967)]
5769. Siegel, C.L.: Indefinite quadratische Formen und Funktionentheorie. I. Math. Ann. **124**, 17–54 (1951) [[5778], vol. 3, pp. 85–91]
5770. Siegel, C.L.: Indefinite quadratische Formen und Funktionentheorie. II. Math. Ann. **124**, 364–387 (1951) [[5778], vol. 3, pp. 85–91]
5771. Siegel, C.L.: Zu zwei Bemerkungen Kummers. Nachr. Ges. Wiss. Göttingen, 1964, 51–57 [[5778], vol. 3, pp. 436–442]
5772. Siegel, C.L.: Zum Beweise des Starkschen Satzes. Invent. Math. **5**, 180–191 (1968) [[5778], vol. 4, pp. 41–52]
5773. Siegel, C.L.: Einige Erläuterungen zu Thues Untersuchungen über Annäherungswerte algebraischer Zahlen und diophantische Gleichungen. Nachr. Ges. Wiss. Göttingen, 1970, 169–195 [[5778], vol. 4, pp. 140–166]
5774. Siegel, C.L.: Zur Theorie der quadratischen Formen. Nachr. Ges. Wiss. Göttingen, 1972, 21–46 [[5778], vol. 4, pp. 224–249]
5775. Siegel, C.L.: Normen algebraischer Zahlen. Nachr. Ges. Wiss. Göttingen, 1973, nr. 11, 197–215 [[5778], vol. 4, pp. 250–268]
5776. Siegel, C.L.: Advanced Analytic Number Theory. Tata, Bombay (1980)
5777. Siegel, C.L.: Lectures on the Geometry of Numbers. Springer, Berlin (1989)
5778. Siegel, C.L.: Gesammelte Abhandlungen, vols. 1–4. Springer, Berlin (1966–1979)
5779. Siering, E.: Über lineare Formen und ein Problem von Frobenius. J. Reine Angew. Math. **271**, 177–202 (1994)
5780. Sierpiński, W.: O pewnem zagadnieniu z rachunku funkcyj asymptotycznych. Pr. Mat.-Fiz. **17**, 77–118 (1906) [French translation: [5785], vol. 1, pp. 73–108]
5781. Sierpiński, W.: O sumowaniu szeregu $\sum_{n>a}^{n\le b}\tau(n)f(n)$, gdzie $\tau(n)$ oznacza liczbę rozkładów liczby n na sumę kwadratów dwóch liczb całkowitych. Pr. Mat.-Fiz. **18**, 1–60 (1907) [French translation: [5785], vol. 1, pp. 109–154]
5782. Sierpiński, W.: Sur la valeur asymptotique d'une certaine somme. Bull. Acad. Sci. Cracovie, A 1910, 9–11. [[5785], vol. 1, pp. 158–160]
5783. Sierpiński, W.: Démonstration élementaire du théorème de M. Borel sur les nombres absolument normaux et détermination effective d'un tel nombre. Bull. Soc. Math. Fr. **45**, 125–132 (1917) [[5785], vol. 1, pp. 161–166]
5784. Sierpiński, W.: Elementary Theory of Numbers. PWN, Warsaw (1964); 2nd ed. North-Holland & PWN (1988)

5785. Sierpiński, W.: Oeuvres Choisies, vols. 1, 2. PWN, Warsaw (1974)
5786. Sikirič, M.D., Schürmann, A., Vallentin, F.: Classification of eight-dimensional perfect forms. Electron. Res. Announc. Am. Math. Soc. **13**, 21–32 (2007)
5787. Sikorav, J.-C.: Valeurs des formes quadratiques indéfinies irrationnelles (d'après G.A. Margulis). Prog. Math. **81**, 307–315 (1990)
5788. Siksek, S., Cremona, J.E.: On the Diophantine equation $x^2 + 7 = y^m$. Acta Arith. **109**, 143–149 (2003)
5789. Silverman, J.H.: Integer points on curves of genus 1. J. Lond. Math. Soc. **28**, 1–7 (1983)
5790. Silverman, J.H.: Representations of integers by binary forms and the rank of the Mordell-Weil group. Invent. Math. **74**, 281–292 (1983)
5791. Silverman, J.H.: The Arithmetic of Elliptic Curves. Springer, Berlin (1986); 2nd ed. 2009 [Reprint: 1992]
5792. Silverman, J.H.: A quantitative version of Siegel's theorem: integral points on elliptic curves and Catalan curves. J. Reine Angew. Math. **378**, 60–100 (1987)
5793. Silverman, J.H.: Wieferich's criterion and the *abc*-conjecture. J. Number Theory **30**, 226–237 (1988)
5794. Silverman, J.H.: Advanced Topics in the Arithmetic of Elliptic Curves. Springer, Berlin (1994)
5795. Silverman, J.H.: Computing rational points on rank 1 elliptic curves via L-series and canonical heights. Math. Comput. **68**, 835–858 (1999)
5796. Silverman, J.H., Tate, J.: Rational Points on Elliptic Curves. Springer, Berlin (1992)
5797. Simalarides, A.: Sophie Germain's principle and Lucas numbers. Math. Scand. **67**, 167–176 (1990)
5798. Simpson, R.J., Zeilberger, D.: Necessary conditions for distinct covering systems with square-free moduli. Acta Arith. **59**, 59–70 (1991)
5799. Sinisalo, M.: Checking the Goldbach conjecture up to $4 \cdot 10^{11}$. Math. Comput. **61**, 931–934 (1993)
5800. Sitaraman, S.: Vandiver revisited. J. Number Theory **57**, 122–129 (1996)
5801. Skewes, S.: On the difference $\pi(x) - \mathrm{li}(x)$. J. Lond. Math. Soc. **8**, 277–283 (1933)
5802. Skewes, S.: On the difference $\pi(x) - \mathrm{li}(x)$, II. Proc. Lond. Math. Soc. **5**, 48–70 (1955)
5803. Skinner, C.M., Wooley, T.D.: Sums of two *k*th powers. J. Reine Angew. Math. **462**, 57–68 (1995)
5804. Skolem, Th.: Über ganzzahlige Lösungen einer Klasse unbestimmter Gleichungen. Norsk mat. for. Skr., Ser. 1, 1922, nr. 10, 1–12
5805. Skolem, Th.: Einige Sätze über gewisse Reihenentwicklungen und exponentiale Beziehungen mit Anwendung auf diophantische Gleichungen. Skr. Norske Vid. Akad. Oslo, 1933, nr. 6, 1–61
5806. Skolem, Th.: A procedure for treating certain exponential equations and Diophantine equations. In: 8th Skand. Mat. Kongr., Stockholm, pp. 163–188 (1934) (in Swedish)
5807. Skolem, Th.: Einige Sätze über $\mathfrak{p}$-adische Potenzreihen mit Anwendung auf gewisse exponentielle Gleichungen. Math. Ann. **111**, 399–424 (1935)
5808. Skolem, Th.: Unlösbarkeit von Gleichungen, deren entsprechende Kongruenz für jeden Modul lösbar ist. Avh. Norske Vid. Akad. Oslo, 1942, nr. 4, 1–28
5809. Skolem, Th.: Einige Bemerkungen über die Auffindung der rationalen Punkte auf gewissen algebraischen Gebilden. Math. Z. **63**, 295–312 (1955)
5810. Skolem, T., Chowla, S., Lewis, D.J.: The diophantine equation $2^{n+2} - 7 = x^2$ and related problems. Proc. Am. Math. Soc. **10**, 663–669 (1959)
5811. Skorobogatov, A.N.: Beyond the Manin obstruction. Invent. Math. **135**, 399–424 (1999)
5812. Skorobogatov, A.[N.]: Torsors and Rational Points. Cambridge University Press, Cambridge (2001)
5813. Skriganov, M.M.: On integer points in polygons. Ann. Inst. Fourier **43**, 313–323 (1993)
5814. Skriganov, M.M.: Ergodic theory on $SL(n)$, Diophantine approximations and anomalies in the lattice point problem. Invent. Math. **132**, 1–72 (1998)
5815. Skriganov, M.M., Starkov, A.N.: On logarithmically small errors in the lattice point problem. Ergod. Theory Dyn. Syst. **20**, 1469–1476 (2000)

5816. Skubenko, B.F.: A proof of Minkowski's conjecture on the product of n linear inhomogeneous forms in n variables for $n \leq 5$. Tr. Mat. Inst. Steklova **133**, 4–36 (1973) (in Russian)
5817. Skubenko, B.F.: On Minkowski's conjecture for large n. Tr. Mat. Inst. Steklova **148**, 218–224 (1978) (in Russian)
5818. Skula, L., Slavutskii, I.Sh.: Bernoulli Numbers: Bibliography (1713–1983). J.E. Purkyne University, Brno (1987)
5819. Slater, L.J.: A new proof of Rogers's transformations of infinite series. Proc. Lond. Math. Soc. **53**, 460–475 (1951)
5820. Slater, L.J.: Further identities of the Rogers-Ramanujan type. Proc. Lond. Math. Soc. **54**, 147–167 (1952)
5821. Slijepčević, S.: A polynomial Sárkőzy-Furstenberg theorem with upper bounds. Acta Math. Acad. Sci. Hung. **98**, 111–128 (2003)
5822. Small, C.: Waring's problem mod n. Am. Math. Mon. **84**, 12–25 (1977)
5823. Small, C.: Solution of Waring's problem mod n. Am. Math. Mon. **84**, 356–359 (1977)
5824. Smart, N.P.: S-integral points on elliptic curves. Math. Proc. Camb. Philos. Soc. **116**, 391–399 (1994)
5825. Smart, N.P.: Thue and Thue-Mahler equations over rings of integers. J. Lond. Math. Soc. **56**, 455–462 (1997)
5826. Smart, N.P.: The Algorithmic Resolution of Diophantine Equations. Cambridge University Press, Cambridge (1998)
5827. Smati, A.: Evaluation effective du nombre d'entiers n tels que $\varphi(n) \leq x$. Acta Arith. **61**, 143–159 (1992)
5828. Šmelev, A.A.: On the question of the algebraic independence of algebraic powers of algebraic numbers. Mat. Zametki **11**, 635–644 (1972) (in Russian)
5829. Šmelev, A.A.: Analogue of the Brownawell-Waldschmidt theorem on transcendental numbers. Mat. Zametki **32**, 765–775 (1982) (in Russian)
5830. Smiley, M.F.: On the zeros of a cubic recurrence. Am. Math. Mon. **63**, 171–172 (1956)
5831. Smith, H.J.S.: Report on the theory of numbers. Rep. British Association, 1859, 228–267; 1860, 120–169; 1861, 292–340; 1862, 503–526; 1863, 768–786; 1865, 322–375. [[5834], vol. 1, pp. 38–364]
5832. Smith, H.J.S.: On the orders and genera of ternary quadratic forms. Philos. Trans. R. Soc. Lond. Ser. A, Math. Phys. Sci. **157**, 255–298 (1867) [[5834], vol. 1, pp. 455–506]
5833. Smith, H.J.S.: Mémoire sur la représentation des nombres par des sommes de cinq carrés. Mémoires présentés par divers savants à l'Académie **29**(2), 1–72 (1887) [[5834], vol. 2, pp. 623–680]
5834. Smith, H.J.S.: The Collected Mathematical Papers. vols. 1, 2. Oxford University Press, Oxford (1894)
5835. Smith, R.A.: An error term of Ramanujan. J. Number Theory **2**, 91–96 (1970)
5836. Smith, R.A.: Estimates for exponential sums. Proc. Am. Math. Soc. **79**, 365–368 (1980)
5837. Smith, R.A.: The generalized divisor problem over arithmetic progressions. Math. Ann. **260**, 255–268 (1982)
5838. Smyth, C.J.: On the product of the conjugates outside the unit circle of an algebraic integer. Bull. Lond. Math. Soc. **3**, 169–175 (1971)
5839. Smyth, C.J.: Ideal 9th-order multigrades and Letac's elliptic curve. Math. Comput. **57**, 817–823 (1991)
5840. Šnirelman, L.G.:[14] On additive properties of integers. Izv. Donsk. Politehn. Inst. (Novočerkask) **14**, 3–28 (1930) (in Russian)
5841. Sobol, I.M.: Distribution of points in a cube and approximate evaluation of integrals. Ž. Vyčisl. Mat. Mat. Fiz. **7**, 784–802 (1967) (in Russian)
5842. Söhne, P.: The Pólya-Vinogradov inequality for totally real algebraic number fields. Acta Arith. **65**, 197–212 (1993)

[14]See also Schnirelmann, L.G.

5843. Soifer, A.: Pierre Joseph Henry Baudet: Ramsey theory before Ramsey. Geombinatorics Q. **6**, 60–70 (1996)
5844. Sokolovskiĭ, A.V.: A theorem on the zeros of Dedekind's zeta function and the distance between "neighbouring" prime ideals. Acta Arith. **13**, 321–334 (1967/1968) (in Russian)
5845. Solé, P.: Sato-Tate conjectures and Chebyshev polynomials. Ramanujan J. **1**, 211–220 (1997)
5846. Sommer, J.: Vorlesungen über Zahlentheorie. Einführung in die Theorie der algebraischen Zahlkörper. Teubner, Leipzig (1907)
5847. Sorokin, V.N.: On Apéry's theorem. Vestnik Moskov. Univ. Ser. I. Mat. Mekh., 1998, nr. 3, 48–53
5848. Sós, V.T.: On the theory of diophantine approximations. I. Acta Math. Acad. Sci. Hung. **8**, 461–472 (1957)
5849. Sós, V.T.: On strong irregularities of the distribution of $\{n\alpha\}$ sequences. In: Studies in Pure Mathematics, pp. 685–700. Birkhäuser, Basel (1983)
5850. Soundararajan, K.: Mean-values of the Riemann zeta-function. Mathematika **42**, 158–174 (1995)
5851. Soundararajan, K.: Nonvanishing of quadratic Dirichlet L-functions at $s = 1/2$. Ann. Math. **152**, 447–488 (2000)
5852. Soundararajan, K.: Omega results for the divisor and circle problem. Int. Math. Res. Not. **36**, 2003 (1987–1998)
5853. Soundararajan, K.: Degree 1 elements of the Selberg class. Expo. Math. **23**, 65–70 (2005)
5854. Soundararajan, K.: Extreme values of zeta and L-functions. Math. Ann. **342**, 467–486 (2008)
5855. Soundararajan, K.: The distribution of smooth numbers in arithmetic progressions. In: Anatomy of Integers, pp. 115–128. Am. Math. Soc., Providence (2008)
5856. Soundararajan, K.: Moments of the Riemann zeta-function. Ann. Math. **179**, 981–993 (2009)
5857. Soundararajan, K.: Partial sums of the Möbius function. J. Reine Angew. Math. **631**, 141–152 (2009)
5858. Šparlinskiĭ, I.E.[15]: The number of prime divisors of recurrence sequences. Mat. Zametki **38**, 29–34 (1985) (in Russian)
5859. Šparlinskiĭ, I.E.: Estimates for Gauss sums. Mat. Zametki **50**, 122–130 (1991) (in Russian)
5860. Spears, J.L., Maxfield, J.E.: Further examples of normal numbers. Publ. Math. (Debr.) **16**, 119–127 (1969)
5861. Speiser, A.: Die Theorie der Gruppen der endlichen Ordnung, 2nd ed. Springer, Berlin (1923) 1927, 3rd ed. 1937
5862. Speiser, A.: Rudolf Fueter. Elem. Math. **5**, 98–99 (1950)
5863. Spencer, D.C.: On a Hardy-Littlewood problem of diophantine approximation. Proc. Camb. Philos. Soc. **35**, 527–547 (1939)
5864. Spencer, D.C.: The lattice points of tetrahedra. J. Math. Phys. MIT **21**, 189–197 (1942)
5865. Spira, R.: Zeros of sections of the zeta function, II. Math. Comput. **22**, 163–173 (1968)
5866. Spira, R.: Calculation of Dirichlet L-functions. Math. Comput. **23**, 489–497 (1969)
5867. Spira, R.: The lowest zero of sections of the zeta function. J. Reine Angew. Math. **255**, 170–189 (1972)
5868. Spiro, C.: The frequency with which an integral-valued, prime-independent, multiplicative or additive function of n divides a polynomial function of n. Ph.D. thesis, Univ. Illinois (1981)
5869. Spiro, C.: Frequency of coprimality of the values of a polynomial and a prime-independent multiplicative function. Rocky Mt. J. Math. **16**, 385–393 (1986)
5870. Sprindžuk, V.G.: On a classification of transcendental numbers, Lit. Mat. Sb. **2**, 215–219 (1962) (in Russian)

[15] See also Shparlinski, I.

5871. Sprindžuk, V.G.: On the conjecture of K. Mahler on the measure of the set of S-numbers. Liet. Mat. Rink. **2**(2), 221–226 (1963) (in Russian)
5872. Sprindžuk, V.G.: On Mahler's conjecture. Dokl. Akad. Nauk SSSR **154**, 783–786 (1964) (in Russian)
5873. Sprindžuk, V.G.: More on Mahler's conjecture. Dokl. Akad. Nauk SSSR **155**, 54–56 (1964) (in Russian)
5874. Sprindžuk, V.G.: Proof of Mahler's conjecture on the measure of the set of S-numbers. Izv. Akad. Nauk SSSR, Ser. Mat. **29**, 379–436 (1965) (in Russian)
5875. Sprindžuk, V.G.: Mahler's Problem in Metric Number Theory. Nauka i Tekhnika, Minsk (1967) (in Russian) [English translation: Am. Math. Soc., 1969]
5876. Sprindžuk, V.G.: On the largest prime divisor of a binary form. Dokl. Akad. Nauk Belorus. **15**, 389–391 (1971) (in Russian)
5877. Sprindžuk, V.G.: Metric Theory of Diophantine Approximations. Nauka, Moscow (1977) (in Russian) [English translation: Wiley, 1979]
5878. Sprindžuk, V.G.: Achievements and problems of the theory of Diophantine approximations. Usp. Mat. Nauk **35**(4), 3–68 (1980) (in Russian)
5879. Sprindžuk, V.G.: Classical Diophantine Equations in Two Unknowns. Nauka, Moscow (1982) (in Russian) [English translation: Classical Diophantine Equations, Lecture Notes in Math., vol. 1559, pp. 1–228 (1993)]
5880. Springer, T.A.: H.D. Kloosterman and his work. Not. Am. Math. Soc. **47**, 862–867 (2000)
5881. Srinivasan, A.: Markoff numbers and ambiguous classes. J. Théor. Nr. Bordx. **21**, 755–768 (2009)
5882. Srinivasan, B.R., Sampath, A.: An elementary proof of the prime number theorem with a remainder term. J. Indian Math. Soc. **53**, 1–50 (1988)
5883. Stacey, K.C.: The enumeration of perfect septenary forms. J. Lond. Math. Soc. **10**, 97–104 (1975)
5884. Stäckel, P.: Ueber Goldbachs empirisches Theorem: Jede grade Zahl kann als Summe von zwei Primzahlen dargestellt werden. Nachr. Ges. Wiss. Göttingen, 1896, 292–299
5885. Stankus, E.: The mean value of $L(1/2, \chi_d)$ for $d \equiv l \pmod D$. Liet. Mat. Rink. **23**, 169–177 (1983) (in Russian)
5886. Stanley, G.K.: On the representation of a number as the sum of seven squares. J. Lond. Math. Soc. **2**, 91–96 (1927)
5887. Stanley, G.K.: On the representations of a number as the sum of squares and primes. J. Lond. Math. Soc. **3**, 62–64 (1928)
5888. Stanley, G.K.: Two assertions made by Ramanujan. J. Lond. Math. Soc. **3**, 232–237 (1928); corr. **4**, 32 (1929)
5889. Stanley, G.K.: On the representation of a number as a sum of squares and primes. Proc. Lond. Math. Soc. **29**, 122–144 (1929)
5890. Stanley, G.K.: The representation of a number as the sum of one square and a number of k-th powers. Proc. Lond. Math. Soc. **31**, 512–553 (1930)
5891. Stanley, G.K.: The representation of a number as a sum of squares and cubes. J. Lond. Math. Soc. **6**, 194–197 (1931)
5892. Stanley, R.P.: Decompositions of rational convex polytopes. Ann. Discrete Math. **6**, 333–342 (1980)
5893. Stanley, R.P.: Decompositions of rational convex polytopes. Ann. Discrete Math. **6**, 333–342 (1980)
5894. Stanton, R.G., Sudler, C. Jr., Williams, H.C.: An upper bound for the period of the simple continued fraction for $\sqrt{D}$. Pac. J. Math. **67**, 525–536 (1976)
5895. Stark, H.M.: On complex quadratic fields with class number equal to one. Trans. Am. Math. Soc. **122**, 112–119 (1966)
5896. Stark, H.M.: On the asymptotic density of the k-free integers. Proc. Am. Math. Soc. **17**, 1211–1214 (1966)
5897. Stark, H.M.: A complete determination of the complex quadratic fields of class-number one. Mich. Math. J. **14**, 1–27 (1967)

5898. Stark, H.M.: On the zeros of Epstein's zeta function. Mathematika **14**, 47–55 (1967)
5899. Stark, H.M.: L-functions and character sums for quadratic form, I. Acta Arith. **14**, 35–50 (1968)
5900. Stark, H.M.: L-functions and character sums for quadratic form, II. Acta Arith. **15**, 307–317 (1969)
5901. Stark, H.M.: On the "gap" in a theorem of Heegner. J. Number Theory **1**, 16–27 (1969)
5902. Stark, H.M.: A transcendence theorem for class-number problems. Ann. Math. **94**, 153–173 (1971)
5903. Stark, H.M.: A transcendence theorem for class-number problems, II. Ann. Math. **96**, 174–209 (1973)
5904. Stark, H.M.: Values of L-functions at $s = 1$. I. L-functions for quadratic forms. Adv. Math. **7**, 301–343 (1971)
5905. Stark, H.M.: Values of L-functions at $s = 1$. II. Artin L-functions with rational characters. Adv. Math. **17**, 60–92 (1975)
5906. Stark, H.M.: Values of L-functions at $s = 1$. III. Totally real fields and Hilbert twelfth problem. Adv. Math. **22**, 64–84 (1976)
5907. Stark, H.M.: On the Riemann Hypothesis in hyperelliptic function fields. In: Proc. Symposia Pure Math., vol. 24, pp. 285–302. Am. Math. Soc., Providence (1973)
5908. Stark, H.M.: Effective estimates of solutions of some diophantine equations. Acta Arith. **24**, 251–259 (1973)
5909. Stark, H.M.: On complex quadratic fields with class-number two. Math. Comput. **29**, 289–302 (1975)
5910. Staś, W.: Über eine Abschätzung des Restgliedes in Primzahlsatz. Acta Arith. **5**, 427–434 (1959)
5911. Staś, W.: Über die Umkehrung eines Satzes von Ingham. Acta Arith. **6**, 435–446 (1960/1961)
5912. Staś, W.: Über das Verhalten der Riemannschen ζ-Funktion und einiger verwandter Funktionen, in der Nähe der Geraden $\sigma = 1$. Acta Arith. **7**, 217–224 (1961/1962)
5913. Staś, W.: On the order of Dedekind zeta-functions in the critical strip. Funct. Approx. Comment. Math. **4**, 19–26 (1976)
5914. Staś, W.: On sign-changes in the remainder term of the prime ideal formula. Funct. Approx. Comment. Math. **13**, 159–166 (1982)
5915. Stečkin, S.B.: A bound for Gauss's sums. Mat. Zametki **17**, 579–588 (1975) (in Russian)
5916. Stečkin, S.B.: Mean values of the modulus of a trigonometric sum. Tr. Mat. Inst. Steklova **134**, 283–309 (1975) (in Russian)
5917. Stečkin, S.B.: An estimate of a complete rational trigonometric sum. Tr. Mat. Inst. Steklova **143**, 188–207 (1977) (in Russian)
5918. Stein, S.K.: Unions of arithmetic sequences. Math. Ann. **134**, 289–294 (1958)
5919. Stein, W.A.: There are genus one curves over Q of every odd index. J. Reine Angew. Math. **547**, 139–147 (2002)
5920. Steinhaus, H.: Sur un thèoréme de M.V. Jarnik. Colloq. Math. **1**, 1–5 (1947)
5921. Steinig, J.: The changes of sign of certain arithmetical error-terms. Comment. Math. Helv. **44**, 385–400 (1969)
5922. Stemmler, R.M.: The easier Waring problem in algebraic number fields. Acta Arith. **6**, 447–468 (1961)
5923. Stemmler, R.M.: The ideal Waring theorem for exponents 401–200 000. Math. Comput. **18**, 144–146 (1964)
5924. Stepanov, S.A.: On the number of points of a hyperelliptic curve over a finite prime field. Izv. Akad. Nauk SSSR, Ser. Mat. **33**, 1103–1114 (1969) (in Russian)
5925. Stepanov, S.A.: Elementary method in the theory of congruences for a prime modulus. Acta Arith. **17**, 231–247 (1970)
5926. Stepanov, S.A.: Estimation of Kloosterman sums. Izv. Akad. Nauk SSSR, Ser. Mat. **35**, 308–323 (1971) (in Russian)
5927. Stepanov, S.A.: An elementary proof of the Hasse-Weil theorem for hyperelliptic curves. J. Number Theory **4**, 118–143 (1972)

5928. Stepanov, S.A.: Congruences with two unknowns. Izv. Akad. Nauk SSSR, Ser. Mat. **36**, 683–711 (1972) (in Russian)
5929. Stepanov, S.A.: A constructive method in the theory of equations over finite fields. Tr. Mat. Inst. Steklova **132**, 237–246 (1973) (in Russian)
5930. Stepanov, S.A.: Arithmetic of Algebraic Curves. Nauka, Moscow (1991) (in Russian) [English translation: New York, 1994]
5931. Stephens, N.M.: The diophantine equation $X^3 + Y^3 = DZ^3$ and the conjectures of Birch and Swinnerton-Dyer. J. Reine Angew. Math. **231**, 121–162 (1968)
5932. Stephens, P.J.: An average result for Artin's conjecture. Mathematika **16**, 178–188 (1969)
5933. Stephens, P.J.: Optimizing the size of $L(1, \chi)$. Proc. Lond. Math. Soc. **24**, 1–14 (1972)
5934. Stern, M.A.: Einige Bemerkungen über die Congruenz $\frac{r^p - r}{p} \equiv a \pmod{p}$. J. Reine Angew. Math. **100**, 182–188 (1887)
5935. von Sterneck, R.D.: Empirische Untersuchung über den Verlauf der zahlentheoretischen Function $\sigma(n) = \sum_{x=1}^{x=n} \mu(x)$ im Intervalle von 0 bis 150 000. SBer. Kais. Akad. Wissensch. Wien **106**, 835–1024 (1897)
5936. von Sterneck, R.D.: Bemerkung über die Summirung einiger zahlentheoretischen Functionen. Monatshefte Math. Phys. **9**, 43–45 (1898)
5937. von Sterneck, R.D.: On the distribution of quadratic residues and non-residues of a prime number. Mat. Sb. **20**, 267–284 (1898) (in Russian)
5938. von Sterneck, R.D.: Empirische Untersuchung über den Verlauf der zahlentheoretischen Funktion $\sigma(n) = \sum_{x=1}^{x=n} \mu(x)$ im Intervalle von 150 000 bis 500 000. SBer. Kais. Akad. Wissensch. Wien **110**, 1053–1102 (1901)
5939. von Sterneck, R.D.: Über die kleinste Anzahl von Kuben, aus welchen jede Zahl bis 40 000 zusammengesetzt werden kann. SBer. Kais. Akad. Wissensch. Wien **112**, 1627–1666 (1903)
5940. Steuding, J.: Voronoïs contribution to modern number theory. Šiauliai Math. Semin. **2**, 67–106 (2007)
5941. Steuding, J., Weng, A.: On the number of prime divisors of the order of elliptic curves mod p. Acta Arith. **117**, 341–352 (2005)
5942. Stevenhagen, P.: A density conjecture for the negative Pell equation. In: Computational Algebra and Number Theory, Sydney, 1992, pp. 187–200. Kluwer Academic, Dordrecht (1995)
5943. Stevenhagen, P.: The correction factor in Artin's primitive root conjecture. J. Théor. Nr. Bordx. **15**, 383–391 (2003)
5944. Stevens, H.: On Jacobsthal's $g(n)$-function. Math. Ann. **226**, 95–97 (1977)
5945. Stevenson, E.: The Artin conjecture for three diagonal cubic forms. J. Number Theory **14**, 374–390 (1982)
5946. Stewart, C.L.: Primitive divisors of Lucas and Lehmer numbers. In: Transcendence Theory: Advances and Applications, pp. 79–92. Academic Press, San Diego (1977)
5947. Stewart, C.L.: Cubic Thue equations with many solutions. Internat. Math. Res. Notices, 2008, nr. 13, art. 40, 1–11
5948. Stewart, C.L.: Integer points on cubic Thue equations. C. R. Acad. Sci. Paris **347**, 715–718 (2009)
5949. Stewart, C.L., Tijdeman, R.: On the Oesterlé-Masser conjecture. Monatshefte Math. **102**, 251–257 (1986)
5950. Stewart, C.L., Yu, K.: On the *abc* conjecture. Math. Ann. **291**, 225–230 (1991)
5951. Stewart, C.L., Yu, K.: On the *abc* conjecture, II. Duke Math. J. **108**, 169–181 (2003)
5952. Stieltjes, T.J.: Letter to Hermite, July 11th 1885. In: Baillaud, B., Bourget, H. (eds.) Correspondance d'Hermite et de Stieltjes. Gauthier-Villars, Paris (1905), letter 79
5953. Stiemke, E.: Über unendliche algebraische Zahlkörper. Math. Z. **25**, 9–39 (1926)
5954. Stockhofe, D.: Bijektive Abbildungen auf der Menge der Partitionen einer natürlichen Zahl. Bayreuth. Math. Schr. **10**, 1–59 (1982)
5955. Stöhr, A.: Eine Basis h-ter Ordnung für die Menge aller natürlichen Zahlen. Math. Z. **42**, 739–743 (1937)

5956. Stöhr, A.: Gelöste und ungelöste Fragen über Basen der natürlichen Zahlenreihe, I. J. Reine Angew. Math. **194**, 40–65 (1955)
5957. Stöhr, A.: Gelöste und ungelöste Fragen über Basen der natürlichen Zahlenreihe, II. J. Reine Angew. Math. **194**, 111–140 (1955)
5958. Stöhr, A., Wirsing, E.: Beispiele von wesentlichen Komponenten, die keine Basis sind. J. Reine Angew. Math. **196**, 96–98 (1956)
5959. Stöhr, K.-O., Voloch, J.F.: Weierstrass points and curves over finite fields. Proc. Lond. Math. Soc. **52**, 1–19 (1986)
5960. Stolz, O.: Leopold Gegenbauer. Monatshefte Math. Phys. **15**, 3–10, 129–136 (1904)
5961. Størmer, C.: Sur une équation indéterminée. C. R. Acad. Sci. Paris **127**, 752–754 (1898)
5962. Størmer, C.: Quelques propriétés arithmétiques des intégrales elliptiques et leurs applications à des fonctions entières transcendantes. Acta Math. **27**, 185–208 (1903)
5963. Stothers, W.W.: Polynomial identities and Hauptmoduln. Q. J. Math. **32**, 349–370 (1981)
5964. Strade, H.: Hel Braun: 1914–1986. Mitt. Math. Ges. Hamb. **11**, 373–376 (1987)
5965. Straßmann, R.: Über den Wertevorrat von Potenzreihen im Gebiet der $\mathfrak{p}$-adischen Zahlen. J. Reine Angew. Math. **159**, 13–28 (1928); add. **159**, 65–66 (1928)
5966. Strauch, O.: Some new criterions for sequences which satisfy Duffin–Schaeffer conjecture, I. Acta Math. Univ. Comen. **42/43**, 87–95 (1983) (in Russian)
5967. Strauch, O.: Some new criterions for sequences which satisfy Duffin–Schaeffer conjecture, II. Acta Math. Univ. Comen. **44/45**, 55–65 (1984) (in Russian)
5968. Strauch, O.: Some new criterions for sequences which satisfy Duffin–Schaeffer conjecture, III. Acta Math. Univ. Comen. **48/49**, 37–50 (1986) (in Russian)
5969. Stridsberg, E.: Sur la démonstration de M. Hilbert du théorème de Waring. Math. Ann. **72**, 145–152 (1912)
5970. Stroeker, R.J.: Elliptic curves over complex quadratic fields. A diophantine approach. Dissertation, Amsterdam (1975)
5971. Stroeker, R.J.: Reduction of elliptic curves over imaginary quadratic fields. Pac. J. Math. **108**, 451–463 (1983)
5972. Stroeker, R.J., Tijdeman, R.: Diophantine equations. In: Computational Methods in Number Theory, pp. 321–369. Mathematisch Centrum, Amsterdam (1982)
5973. Stroeker, R.J., Tzanakis, N.: On the application of Skolem's p-adic method to the solution of Thue equations. J. Number Theory **29**, 166–195 (1988)
5974. Stroeker, R.J., Tzanakis, N.: Solving elliptic Diophantine equations by estimating linear forms in elliptic logarithms. Acta Arith. **67**, 177–196 (1994)
5975. Stroeker, R.J., Tzanakis, N.: Computing all integer solutions of a genus 1 equation. Math. Comput. **72**, 1917–1933 (2003)
5976. Stroeker, R.J., de Weger, B.M.M.: Solving elliptic Diophantine equations: the general cubic case. Acta Arith. **87**, 339–365 (1999)
5977. Stroeker, R.J., de Weger, B.M.M.: Elliptic binomial Diophantine equations. Math. Comput. **68**, 1257–1281 (1999)
5978. Strzelecki, E.: On sequences $\{\xi t_n \pmod 1\}$. Can. Math. Bull. **18**, 727–738 (1975)
5979. Stuhler, U.: Martin Kneser (21.1.1928–16.2.2004). Jahresber. Dtsch. Math.-Ver. **108**, 45–61 (2006)
5980. Stürzbecher, M.: Dr. med. Albert Fleck und die Suche nach seiner Fermat-Klinik. Acta Hist. Leopold. **27**, 339–346 (1997)
5981. Subba Rao, K.: Some easier Waring's problems. Math. Z. **40**, 477–483 (1935)
5982. Subba Rao, K.: On Waring's problem for smaller powers. J. Indian Math. Soc. **5**, 117–121 (1941)
5983. Subbarao, M.V.: Some remarks on the partition function. Am. Math. Mon. **73**, 851–854 (1966)
5984. Subkhankulov, M.A.: Tauberian Theorems with Remainder. Nauka, Moscow (1976) (in Russian)
5985. Suetuna, Z.: On the mean value of L-functions. Jpn. J. Math. **1**, 69–82 (1924)
5986. Suetuna, Z.: The zeros of L-functions on the critical line. Tohoku Math. J. **24**, 313–331 (1925)

5987. Suetuna, Z.: Über die approximative Funktionalgleichung für Dirichletsche L-Funktionen. Jpn. J. Math. **9**, 111–116 (1932)
5988. Sugawara, M.: An invariant property of Siegel's modular function. Proc. Jpn. Acad. Sci. **14**, 1–3 (1938)
5989. Sun, Q., Yuan, P.Z.: A note on the Diophantine equation $x^4 - Dy^2 = 1$. Sichuan Daxue Xuebao **34**, 265–268 (1997) (in Chinese)
5990. Sun, Z.W.: A survey of problems and results on restricted sumsets. In: Number Theory, Sailing on the Sea of Number Theory, pp. 190–213. World Scientific, Singapore (2007)
5991. Surroca, A.: Siegel's theorem and the *abc* conjecture. Riv. Mat. Univ. Parma Ser. 7 **3***, 323–332 (2004)
5992. Surroca, A.: Sur l'effectivité du théorème de Siegel et la conjecture *abc*. J. Number Theory **124**, 267–290 (2007)
5993. Suryanarayana, D.: On the order of the error function of the square-full integers, II. Period. Math. Hung. **14**, 69–75 (1983)
5994. Suryanarayana, D., Sitaramachandra Rao, R.: The distribution of square-full integers. Ark. Mat. **11**, 195–201 (1973)
5995. Suryanarayana, D., Sitaramachandra Rao, R.: On the average order of the function $E(x) = \sum_{n\leq x}\varphi(n) - \frac{3x^2}{\pi^2}$. Ark. Mat. **10**, 99–106 (1972)
5996. Suzuki, J.: On the generalized Wieferich criteria. Proc. Jpn. Acad. Sci. **70**, 230–234 (1994)
5997. Suzuki, Y.: Simple proof of Fermat's last theorem for $n = 4$. Proc. Jpn. Acad. Sci. **62**, 209–210 (1986)
5998. Swett, A.: The Erdos-Strauss conjecture. http://math.uindy.edu/swett/esc.htm
5999. Swift, J.D.: Note on discriminants of binary quadratic forms with a single class in each genus. Bull. Am. Math. Soc. **54**, 560–561 (1948)
6000. Swinnerton-Dyer, H.P.F.: Two special cubic surfaces. Mathematika **9**, 54–56 (1962)
6001. Swinnerton-Dyer, H.P.F.: On the product of three homogeneous linear forms. Acta Arith. **18**, 371–385 (1971)
6002. Swinnerton-Dyer, H.P.F.: On l-adic representations and congruences for coefficients of modular forms, I. In: Lecture Notes in Math., vol. 350, pp. 1–5. Springer, Berlin (1973); corr: vol. 476, p. 149 (1975)
6003. Swinnerton-Dyer, H.P.F.: On l-adic representations and congruences for coefficients of modular forms, II. In: Lecture Notes in Math., vol. 601, pp. 63–90. Springer, Berlin (1977)
6004. Swinnerton-Dyer, H.P.F.: The number of lattice points on a convex curve. J. Number Theory **6**, 128–135 (1974)
6005. Swinnerton-Dyer, H.P.F.: The solubility of diagonal cubic surfaces. Ann. Sci. Éc. Norm. Super. **34**, 891–912 (2001)
6006. Sylvester, J.J.: On the partition of an even number into two primes. Proc. Lond. Math. Soc. **4**, 4–6 (1871) [[6014], vol. 2, pp. 709–711]
6007. Sylvester, J.J.: On a point in the theory of vulgar fractions. Am. J. Math. **3**, 332–335, 388–389 (1880) [[6014], vol. 3, pp. 440–445]
6008. Sylvester, J.J.: On Tchebycheff's theory of the totality of the prime numbers comprised within given limits. Am. J. Math. **4**, 230–247 (1881) [[6014], vol. 3, pp. 230–247]
6009. Sylvester, J.J.: A constructive theory of partitions, arranged in three acts, an interact and an exodion. Am. J. Math. **5**, 251–330 (1882) [[6014], vol. 4, pp. 1–83]
6010. Sylvester, J.J.: Sur les nombres de fractions ordinaires inégales qu'on peut exprimer en se servant de chiffres qui n'excedent pas un nombre donné. C. R. Acad. Sci. Paris **96**, 409–413 (1883) [[6014], vol. 4, pp. 84–87]
6011. Sylvester, J.J.: Question 7382. Math. Quest. Educ. Times **41**, 21 (1884)
6012. Sylvester, J.J.: Sur l'impossibilité de l'existence d'un nombre parfait impair qui ne contient pas au moins 5 diviseurs premiers distinct. C. R. Acad. Sci. Paris **106**, 522–526 (1888) [[6014], vol. 4, pp. 611–614]
6013. Sylvester, J.J.: On arithmetical series. Messenger of Math., (2) **21**, 1–19, 87–120 (1891/1892) [[6014], vol. 4, pp. 686–731]

6014. Sylvester, J.J.: Collected Mathematical Papers, vols. 1–4. Cambridge University Press, Cambridge (1908–1912)
6015. Synge, J.L.: John Charles Fields. J. Lond. Math. Soc. **8**, 153–160 (1933)
6016. Szalay, L.: On the resolution of simultaneous Pell equations. Ann. Math. Inst. Inform. **34**, 77–87 (2007)
6017. Szegö, G.: Beiträge zur Theorie der Laguerreschen Polynome. II. Zahlentheoretische Anwendungen. Math. Z. **25**, 388–404 (1926)
6018. Szegö, G., Walfisz, A.: Über das Piltzsche Teilerproblem in algebraischen Zahlkörpern, I. Math. Z. **26**, 138–156 (1927)
6019. Szegö, G., Walfisz, A.: Über das Piltzsche Teilerproblem in algebraischen Zahlkörpern, II. Math. Z. **26**, 467–486 (1927)
6020. Szekeres, G.: Search for the three-dimensional approximation constant. In: Diophantine Analysis, Kensington, 1985, pp. 139–146. Cambridge University Press, Cambridge (1986)
6021. Szekeres, G., Turán, P.: Über das zweite Hauptproblem der "Factorisatio Numerorum". Acta Sci. Math. **6**, 143–154 (1933) [[6227], vol. 1, pp. 1–12]
6022. Szemerédi, E.: On sets of integers containing no four elements in arithmetic progression. Acta Math. Acad. Sci. Hung. **20**, 89–104 (1969)
6023. Szemerédi, E.: On sets of integers containing no k elements in arithmetic progression. Acta Arith. **27**, 199–245 (1975)
6024. Szemerédi, E.: Integer sets containing no arithmetic progressions. Acta Math. Acad. Sci. Hung. **56**, 155–158 (1990)
6025. Szemerédi, E., Vu, V.H.: Long arithmetic progressions in sum-sets and the number of x-sum-free sets. Proc. Lond. Math. Soc. **90**, 273–296 (2005)
6026. Szemerédi, E., Vu, V.H.: Finite and infinite arithmetic progressions in sumsets. Ann. Math. **163**, 1–35 (2006)
6027. Szemerédi, E., Vu, V.H.: Long arithmetic progressions in sumsets: thresholds and bounds. J. Am. Math. Soc. **19**, 119–169 (2006)
6028. Szpiro, L.: Sur le théorème de rigidité de Parsin et Arakelov. Astérisque **64**, 169–202 (1979)
6029. Szpiro, L.: La conjecture (d'aprés G. Faltings). Astérisque **121/122**, 83–103 (1985)
6030. Szpiro, L.: Discriminant et conducteur des courbes elliptiques. Astérisque **183**, 7–18 (1990)
6031. Szűsz, P.: Über einen Kusminschen Satz. Acta Math. Acad. Sci. Hung. **12**, 447–453 (1961)
6032. Szűsz, P., Volkmann, B.: A combinatorial method for constructing normal numbers. Forum Math. **6**, 399–414 (1994)
6033. Szydło, B.: Über Vorzeichenwechsel einiger arithmetischer Funktionen, I. Math. Ann. **283**, 139–149 (1989)
6034. Szydło, B.: Über Vorzeichenwechsel einiger arithmetischer Funktionen, II. Math. Ann. **283**, 151–163 (1989)
6035. Szydło, B.: Über Vorzeichenwechsel einiger arithmetischer Funktionen, III. Monatshefte Math. **109**, 325–336 (1989)
6036. Szydło, B.: Über eine Methode, die Anzahl der Vorzeichenwechsel arithmetischer Funktionen von unten abzuschätzen. Arch. Math. **57**, 267–272 (1991)
6037. Takagi, T.: Über eine Theorie des relativ-Abelschen Zahlkörpers. J. Fac. Sci. Univ. Tokyo **41**(9), 1–133 (1920)
6038. Taketa, K.: Neuer Beweis eines Satzes von Herrn Furtwängler über die metabelschen Gruppen. Jpn. J. Math. **9**, 199–218 (1932)
6039. Takeuchi, M.: Quantitative results of algebraic independence and Baker's method. Acta Arith. **119**, 211–241 (2005)
6040. Tallini, G.: Beniamino Segre. Acta Arith. **45**, 1–18 (1985)
6041. Tamagawa, T.: On Hilbert's modular group. J. Math. Soc. Jpn. **11**, 241–246 (1959)
6042. Tammela, P.P.: An estimate of the critical determinant of a two-dimensional convex symmetric domain. Izv. Vysš. Učeb. Zaved. Mat., 1970, nr. 12, 103–107
6043. Tang, M.: A note on a result of Ruzsa. Bull. Aust. Math. Soc. **77**, 91–98 (2008)
6044. Tang, Y.S., Shao, P.C.: Proof of a conjecture of Kátai. Acta Math. Sin. **39**, 190–195 (1996)
6045. Taniyama, Y.: L-functions of number fields and zeta functions of abelian varieties. J. Math. Soc. Jpn. **9**, 330–366 (1957)

6046. Tanner, J.W., Wagstaff, S.S. Jr.: New congruences for the Bernoulli numbers. Math. Comput. **48**, 341–350 (1987)
6047. Tanner, J.W., Wagstaff, S.S. Jr.: New bound for the first case of Fermat's last theorem. Math. Comput. **53**, 743–750 (1989)
6048. Tao, T.: What is good mathematics? Bull. Am. Math. Soc. **44**, 623–634 (2007)
6049. Tao, T., Vu, V.[H.]: Additive Combinatorics. Cambridge University Press, Cambridge (2006) [2nd ed. 2010]
6050. Tao, T., Ziegler, T.: The primes contain arbitrarily long polynomial progressions. Acta Math. **201**, 213–305 (2008)
6051. Tardy, P.: Trasformazione di un prodotto di n fattori. Ann. Sc. Mat. Fis. Roma **2**, 287–291 (1851)
6052. Tarry, G.: L'Intermédiaire Math., **20**, 1913, 68–70
6053. Tartakovskiĭ, V.A.: Auflösung der Gleichung $x^4 - \varrho y^4 = 1$. Bull. Acad. Sci. Leningrad **20**, 301–324 (1926)
6054. Tartakovskiĭ, V.A.: Die Gesamtheit der Zahlen, die durch eine positive quadratische Form $F(x_1, x_2, \dots, x_s)$ $(s \geq 4)$ darstellbar sind, I. Izv. Akad. Nauk SSSR, Ser. Mat. **2**, 111–122 (1929)
6055. Tartakovskiĭ, V.A.: Die Gesamtheit der Zahlen, die durch eine positive quadratische Form $F(x_1, x_2, \dots, x_s)$ $(s \geq 4)$ darstellbar sind, II. Izv. Akad. Nauk SSSR, Ser. Mat. **2**, 165–196 (1929)
6056. Tartakovskiĭ, V.A.: La méthode du crible approximatif "électif". Dokl. Akad. Nauk SSSR **23**, 126–129 (1939)
6057. Tate, J.: Fourier analysis in number fields and Hecke's zeta functions. Ph.D. thesis, Princeton Univ. (1950) [Reprint: [950], pp. 305–347]
6058. Tate, J.: WC-groups over p-adic fields. Sém. Bourbaki **10**(exp. 156), 1–13 (1957/1958)
6059. Tate, J.: On the conjectures of Birch and Swinnerton-Dyer and a geometric analogue. Sém. Bourbaki, 1965/1966, nr. 306, 415–440
6060. Tate, J.: Algebraic cycles and poles of zeta functions. In: Arithmetical Algebraic Geometry, pp. 93–110. Harper&Row, New York (1965)
6061. Tate, J.: Algebraic cohomology classes. Usp. Mat. Nauk **20**(6), 27–40 (1965) (in Russian)
6062. Tate, J.: Endomorphisms of abelian varieties over finite fields. Invent. Math. **2**, 134–144 (1966)
6063. Tate, J.: The arithmetic of elliptic curves. Invent. Math. **23**, 179–206 (1974)
6064. Tate, J.: Local constants. In: Algebraic Number Fields: L-functions and Galois Properties, pp. 89–131. Academic Press, San Diego (1977)
6065. Tate, J., Šafarevič, I.R.: The rank of elliptic curves. Dokl. Akad. Nauk SSSR **175**, 770–773 (1967) (in Russian)
6066. Tatuzawa, T.: On a theorem of Siegel. Jpn. J. Math. **21**, 163–178 (1951)
6067. Tatuzawa, T.: On the number of the primes in an arithmetic progression. Jpn. J. Math. **21**, 93–111 (1951)
6068. Tatuzawa, T.: The approximate functional equation for Dirichlet's L-series. Jpn. J. Math. **22**, 19–25 (1952)
6069. Tatuzawa, T.: On the product of $L(1, \chi)$. Nagoya Math. J. **5**, 105–111 (1953)
6070. Tatuzawa, T.: On the Waring problem in an algebraic number field. J. Math. Soc. Jpn. **10**, 322–341 (1958)
6071. Tatuzawa, T.: On Waring's problem in an algebraic number field. Acta Arith. **24**, 37–60 (1973)
6072. Tatuzawa, T.: Fourier analysis used in analytic number theory. Acta Arith. **28**, 263–272 (1975)
6073. Tatuzawa, T.: On a proof of Siegel's theorem. In: Number Theory and Combinatorics, Japan, 1984, pp. 383–416. World Scientific, Singapore (1985)
6074. Tatuzawa, T., Iseki, K.: On Selberg's elementary proof of prime number theorem. Proc. Jpn. Acad. Sci. **27**, 340–342 (1951)
6075. Taussky-Todd, O.: Arnold Scholz zum Gedächtnis. Math. Nachr. **7**, 379–386 (1952)

6076. Taylor, A.D.: A note on van der Waerden's theorem. J. Comb. Theory, Ser. A **33**, 215–219 (1982)
6077. Taylor, M.J.: Albrecht Fröhlich, 1916–2001. Bull. Lond. Math. Soc. **38**, 329–350 (2006)
6078. Taylor, R.: On Galois representations associated to Hilbert modular forms. Invent. Math. **98**, 265–280 (1989)
6079. Taylor, R.: On Galois representations associated to Hilbert modular forms, II. In: Elliptic Curves, Modular Forms, & Fermat's Last Theorem, pp. 185–191. International Press, Cambridge (1995)
6080. Taylor, R.: Representations of Galois groups associated to Hilbert modular forms. In: Automorphic Forms, Shimura Varieties, and L-functions, vol. 2, pp. 323–336. Academic Press, San Diego (1990)
6081. Taylor, R.: Galois representations associated to Siegel modular forms of low weight. Duke Math. J. **63**, 281–332 (1991)
6082. Taylor, R.: On the l-adic cohomology of Siegel threefolds. Invent. Math. **114**, 289–310 (1993)
6083. Taylor, R.: l-adic representations associated to modular forms over imaginary quadratic fields, II. Invent. Math. **116**, 619–643 (1994)
6084. Taylor, R.: Representations of Galois groups associated to modular forms. In: Proceedings of the International Congress of Mathematicians, Zürich, 1994, pp. 435–442. Birkhäuser, Basel (1995)
6085. Taylor, R.: Icosahedral Galois representations. Pac. J. Math., 1997, Special Issue, 337–347
6086. Taylor, R.: On icosahedral Artin representations, II. Am. J. Math. **125**, 549–566 (2003)
6087. Taylor, R.: Galois representations. Ann. Fac. Sci. Toulouse **13**, 73–119 (2004)
6088. Taylor, R.: Automorphy for some ℓ-adic lifts of automorphic mod ℓ Galois representations, II. Publ. Math. Inst. Hautes Études Sci. **108**, 183–239 (2008)
6089. Taylor, R., Wiles, A.: Ring theoretic properties of certain Hecke algebras. Ann. Math. **141**, 553–572 (1995)
6090. Taylor, S.J.: Abram Samoilovitch Besicovitch. Bull. Lond. Math. Soc. **7**, 191–210 (1975)
6091. Taylor, S.J.: Paul Lévy. Bull. Lond. Math. Soc. **7**, 300–320 (1975)
6092. Taylor, S.J.: Thomas Muirhead Flett. Bull. Lond. Math. Soc. **9**, 330–339 (1977)
6093. Tchudakoff, N.[16]: On zeros of Dirichlet's L-functions. Mat. Sb. **1**, 591–602 (1936)
6094. Tchudakoff, N.: On zeros of the function $\zeta(s)$. Dokl. Akad. Nauk SSSR, 1936, nr. 1, 201–204
6095. Tchudakoff, N.: On the difference between two neighbouring prime numbers. Mat. Sb. **1**, 799–814 (1936)
6096. Tchudakoff, N.: Sur le problème de Goldbach. Dokl. Akad. Nauk SSSR **17**, 335–338 (1937)
6097. Tchudakoff, N.: On Goldbach-Vinogradov theorem. Ann. Math. **48**, 515–545 (1947)
6098. Teichmüller, O.: Über die Struktur diskret bewerteter perfekten Körper. Nachr. Ges. Wiss. Gött. **1**, 151–161 (1936)
6099. Teichmüller, O.: Diskret bewertete perfekte Körper mit unvollkommenen Restklassenkörper. J. Reine Angew. Math. **176**, 141–152 (1937)
6100. Teitelbaum, J.: p-adic periods of genus two Mumford-Schottky curves. J. Reine Angew. Math. **385**, 117–151 (1988)
6101. Teitelbaum, J.: A.O.L. Atkin (1923–2008). Not. Am. Math. Soc. **56**, 505 (2009)
6102. Tenenbaum, G.: Sur la répartition des diviseurs. Sém. Delange–Pisot–Poitou **17**(exp. G14), 1–5 (1975/1976). Paris, 1977
6103. Tenenbaum, G.: Lois de répartition des diviseurs, III. Acta Arith. **39**, 19–31 (1981)
6104. Tenenbaum, G.: Sur la probabilité qu'un entier posséde un diviseur dans un intervalle donné. Compos. Math. **51**, 243–263 (1984)
6105. Tenenbaum, G.: Théorèmes taubériens avec restes. In: Analytic and Elementary Number Theory, Marseille, 1983, pp. 159–169. Univ. Paris XI, Orsay (1986)

[16]See also Čudakov, N.G.

6106. Tenenbaum, G.: Un problème de probabilité conditionnelle en arithmétique. Acta Arith. **49**, 165–187 (1987)
6107. Tenenbaum, G.: Introduction à la théorie analytique et probabiliste des nombres. Institut Élie Cartan, Nancy (1990); 2nd ed., Paris, 1995 [English translation: Introduction to Analytic and Probabilistic Number Theory, Cambridge (1995)]
6108. Tenenbaum, G.: Sur une question d'Erdős et Schinzel, II. Invent. Math. **99**, 215–224 (1990)
6109. Tenenbaum, G.: Cribler les entiers sans grand facteur premier. Philos. Trans. R. Soc. Lond. Ser. A, Math. Phys. Sci. **345**, 377–384 (1993)
6110. Tenenbaum, G., Yokota, H.: Length and denominators of Egyptian fractions, III. J. Number Theory **35**, 150–156 (1990)
6111. Tengely, Sz.: On the Diophantine equation $F(x) = G(y)$. Acta Arith. **110**, 185–200 (2003)
6112. Tennison, B.R.: On the quartic threefold. Proc. Lond. Math. Soc. **29**, 714–734 (1974)
6113. Terai, N.: The Diophantine equation $a^x + b^y = c^z$. Proc. Jpn. Acad. Sci. **70**, 22–26 (1994)
6114. Terquem, O.: Footnote on p. 183 in Nouv. Ann. Math., **16**, 1857
6115. te Riele, H.J.J.: Some historical and other notes about the Mertens conjecture and its recent disproof. Nieuw Arch. Wiskd. **3**, 237–243 (1985)
6116. te Riele, H.J.J.: On the sign of the difference $\pi(x) - \mathrm{li}(x)$. Math. Comput. **48**, 323–328 (1987)
6117. Terjanian, G.: Un contre-exemple à une conjecture d'Artin. C. R. Acad. Sci. Paris **262**, 612 (1966)
6118. Terjanian, G.: Dimension arithmétique d'un corps. J. Algebra **22**, 517–545 (1972)
6119. Terjanian, G.: Sur l'équation $x^{2p} + y^{2p} = z^{2p}$. C. R. Acad. Sci. Paris **285**, 973–975 (1977)
6120. Terjanian, G.: Sur la dimension diophantienne des corps p-adiques. Acta Arith. **34**, 127–130 (1977/1978)
6121. Terjanian, G.: Formes p-adiques anisotropes. J. Reine Angew. Math. **313**, 217–220 (1980)
6122. Teterin, Yu.G.: Asymptotic formula for the number of representations by completely positive ternary quadratic forms. Izv. Akad. Nauk SSSR, Ser. Mat. **49**, 393–426 (1985) (in Russian)
6123. Thakur, D.S.: Automata-style proof of Voloch's result on transcendence. J. Number Theory **58**, 60–63 (1996)
6124. Thang, N.Q.: A note on the Hasse principle. Acta Arith. **54**, 171–184 (1990)
6125. Thanigasalam, K.: A generalization of Waring's problem for prime powers. Proc. Lond. Math. Soc. **16**, 193–212 (1966)
6126. Thanigasalam, K.: Asymptotic formula in a generalized Waring's problem. Proc. Camb. Philos. Soc. **63**, 87–98 (1967)
6127. Thanigasalam, K.: On additive number theory. Acta Arith. **13**, 237–258 (1967/1968)
6128. Thanigasalam, K.: Note on the representation of integers as sums of certain powers. Proc. Camb. Philos. Soc. **65**, 445–446 (1969)
6129. Thanigasalam, K.: Note on Waring's problem. Port. Math. **33**, 163–165 (1974)
6130. Thanigasalam, K.: On sums of powers and a related problem. Acta Arith. **36**, 125–141 (1980); corr. **42**, 425 (1983)
6131. Thanigasalam, K.: On Waring's problem. Acta Arith. **38**, 141–155 (1980)
6132. Thanigasalam, K.: On certain additive representations of integers. Port. Math. **42**, 447–465 (1983/1984)
6133. Thanigasalam, K.: Improvement on Davenport's iterative method and new results in additive number theory, I. Acta Arith. **46**, 1–31 (1986)
6134. Thanigasalam, K.: Improvement on Davenport's iterative method and new results in additive number theory, II. Proof that $G(5) \leq 22$. Acta Arith. **46**, 91–112 (1986)
6135. Thanigasalam, K.: Improvement on Davenport's iterative method and new results in additive number theory, III. Acta Arith. **48**, 97–116 (1987)
6136. Thomas, A.: Discrépance en dimension un. Ann. Fac. Sci. Toulouse **10**, 369–399 (1989)
6137. Thomas, E.: Complete solutions to a family of cubic Diophantine equations. J. Number Theory **34**, 235–250 (1990)
6138. Thomas, E.: Counting solutions to trinomial Thue equations: a different approach. Trans. Am. Math. Soc. **352**, 3595–3622 (2000)

6139. Thomas, H.E. Jr.: Waring's problem for twenty-two biquadrates. Trans. Am. Math. Soc. **193**, 427–430 (1974)
6140. Thorne, F.: Bounded gaps between products of primes with applications to ideal class groups and elliptic curves. Internat. Math. Res. Notices, 2008, nr. 5, art. 156, 1–41
6141. Thue, A.: Bemerkungen über gewisse Näherungsbrüche algebraischer Zahlen. Christiania Vid. Selsk, Skr., 1908, nr. 3, 1–34
6142. Thue, A.: Über Annäherungswerte algebraischer Zahlen. J. Reine Angew. Math. **135**, 284–305 (1909)
6143. Thue, A.: Über eine Eigenschaft, die keine transzendente Grösse haben kann. Christiania Vid. Selsk, Skr., 1912, nr. 20
6144. Thue, A.: Über die Unlösbarkeit der Gleichung $ax^2 + bx + c = dy^n$ in grossen ganzen Zahlen x und y. Ark. Math. Astron. Fys. **34**(16), 1–6 (1917)
6145. Thue, A.: Computation of all solutions of certain equations of the form $ax^r - by^r = f$. Christiania Vid. Selsk, Skr., 1918, nr. 4, 1–9 (in Norwegian)
6146. Thunder, J.L.: The number of solutions to cubic Thue inequalities. Acta Arith. **66**, 237–243 (1994)
6147. Thunder, J.L.: On Thue inequalities and a conjecture of Schmidt. J. Number Theory **52**, 319–328 (1995)
6148. Thunder, J.L.: Siegel's lemma for function fields. Mich. Math. J. **42**, 147–162 (1995)
6149. Thunder, J.L.: On cubic Thue inequalities and a result of Mahler. Acta Arith. **83**, 31–44 (1998)
6150. Thunder, J.L.: Decomposable form inequalities. Ann. Math. **153**, 767–804 (2001)
6151. Thunder, J.L.: Inequalities for decomposable forms of degree $n + 1$ in n variables. Trans. Am. Math. Soc. **354**, 3855–3868 (2002)
6152. Thunder, J.L.: Asymptotic estimates for the number of integer solutions to decomposable form inequalities. Compos. Math. **141**, 271–292 (2005)
6153. Tichy, R.F.: Nachruf auf Edmund Hlawka. Monatshefte Math. **158**, 107–120 (2009)
6154. Tichy, R.F., Turnwald, G.: Uniform distribution of recurrences in Dedekind domains. Acta Arith. **46**, 81–89 (1985)
6155. Tichy, R.F., Turnwald, G.: Weak uniform distribution of $u_n + 1 = au_n + b$ in Dedekind domains. Manuscr. Math. **61**, 11–22 (1988)
6156. Tietäväinen, A.: On a problem of Chowla and Shimura. J. Number Theory **3**, 247–252 (1971)
6157. Tietäväinen, A.: Note on Waring's problem mod p. Ann. Acad. Sci. Fenn., Ser. A 1 Math. **554**, 1–7 (1973)
6158. Tietze, H.: Nachruf: Gustav Herglotz. Jahrbuch Bayer. Akad. Wiss., 1953, 188–194
6159. Tijdeman, R.: On the equation of Catalan. Acta Arith. **29**, 197–209 (1976)
6160. Tijdeman, R.: Hilbert's seventh problem: on the Gel'fond-Baker method and its applications. In: Proc. Symposia Pure Math., vol. 28, pp. 241–268. Am. Math. Soc., Providence (1976)
6161. Tijdeman, R.: Diophantine equations and diophantine approximations. In: Number Theory and Applications, pp. 215–243. Kluwer Academic, Dordrecht (1989)
6162. Tijdeman, R., Wagner, G.: A sequence has almost nowhere small discrepancy. Monatshefte Math. **90**, 315–329 (1980)
6163. Tikhomirov, V.T., Uspenskii, V.: Lev Genrikhovič Šnirelman. Kvant, 1996, nr. 2, 2–4 (in Russian)
6164. Timofeev, N.M.: The Erdös-Kubilius conjecture on the distribution of the values of additive functions on the sequence of shifted primes. Acta Arith. **58**, 113–131 (1991) (in Russian)
6165. Titchmarsh, E.C.: A consequence of the Riemann Hypothesis. J. Lond. Math. Soc. **2**, 247–254 (1927)
6166. Titchmarsh, E.C.: The mean-value of the zeta-function on the critical line. Proc. Lond. Math. Soc. **27**, 137–150 (1927)
6167. Titchmarsh, E.C.: On an inequality satisfied by the zeta-function of Riemann. Proc. Lond. Math. Soc. **28**, 70–80 (1928)

6168. Titchmarsh, E.C.: On the zeros of Riemann's zeta function. Proc. Lond. Math. Soc. **30**, 319–321 (1929)
6169. Titchmarsh, E.C.: A divisor problem. Rend. Circ. Mat. Palermo **54**, 414–429 (1930); corr., **57**, 478–479 (1933)
6170. Titchmarsh, E.C.: The Zeta-Function of Riemann. Oxford University Press, Oxford (1930)
6171. Titchmarsh, E.C.: On van der Corput's method and the Zeta function of Riemann. Q. J. Math. **2**, 161–173 (1931)
6172. Titchmarsh, E.C.: On van der Corput's method and the Zeta function of Riemann, II. Q. J. Math. **2**, 313–320 (1931)
6173. Titchmarsh, E.C.: On van der Corput's method and the Zeta function of Riemann, III. Q. J. Math. **3**, 133–141 (1932)
6174. Titchmarsh, E.C.: On van der Corput's method and the Zeta function of Riemann, V. Q. J. Math. **5**, 195–210 (1934)
6175. Titchmarsh, E.C.: The lattice-points in a circle. Proc. Lond. Math. Soc. **38**, 96–115 (1935); corr. **38**, 555 (1935)
6176. Titchmarsh, E.C.: The zeros of the Riemann zeta-function. Proc. R. Soc. Lond. Ser. A, Math. Phys. Sci. **151**, 234–255 (1935)
6177. Titchmarsh, E.C.: On $\zeta(s)$ and $\pi(x)$. Q. J. Math. **9**, 97–108 (1938)
6178. Titchmarsh, E.C.: The Theory of Riemann Zeta-function. Oxford University Press, Oxford (1951); 2nd ed. 1986 [Reprints: 1986, 1994]
6179. Titchmarsh, E.C.: On the order of $\zeta(\frac{1}{2}+it)$. Q. J. Math. **13**, 11–17 (1942)
6180. Titchmarsh, E.C.: On the zeros of the Riemann zeta function. Q. J. Math. **18**, 4–16 (1947)
6181. Titchmarsh, E.C.: Godfrey Harold Hardy. J. Lond. Math. Soc. **25**, 82–101 (1950)
6182. Todd, J.A.: John Hilton Grace. J. Lond. Math. Soc. **34**, 113–117 (1959)
6183. Togbé, A.: A parametric family of cubic Thue equations. J. Number Theory **107**, 63–79 (2004)
6184. Togbé, A.: Complete solutions of a family of cubic Thue equations. J. Théor. Nr. Bordx. **18**, 285–298 (2006)
6185. Togbé, A.: A parametric family of sextic Thue equations. Acta Arith. **125**, 347–361 (2006)
6186. Togbé, A., Voutier, P.M., Walsh, P.G.: Solving a family of Thue equations with an application to the equation $x^2 - Dy^4 = 1$. Acta Arith. **120**, 39–58 (2005)
6187. Tolev, D.I.: On a theorem of Bombieri-Vinogradov type for prime numbers from a thin set. Acta Arith. **81**, 57–68 (1997)
6188. Tolev, D.I.: On the number of representations of an odd integer as a sum of three primes, one of which belongs to an arithmetic progression. Tr. Mat. Inst. Steklova **218**, 415–432 (1997) (in Russian)
6189. Tolev, D.I.: On the exponential sum with square-free numbers. Bull. Lond. Math. Soc. **37**, 827–834 (2005)
6190. Tolstikov, A.V.: A remark on Fermat's last theorem. Zap. Nauč. Semin. LOMI **67**, 223–224 (1977) (in Russian)
6191. Tong, K.C.: On divisor problems, II. Acta Math. Sin. **6**, 139–152 (1956) (in Chinese)
6192. Tong, K.C.: On divisor problems, III. Acta Math. Sin. **6**, 515–541 (1956) (in Chinese)
6193. Tonkov, T.: On the average length of finite continued fractions. Acta Arith. **26**, 47–57 (1974/1975)
6194. Tonkov, T.: The mean length of finite continued fractions. Math. Balk. **4**, 617–629 (1974) (in Russian)
6195. Tonkov, T.: The contribution of Nikola Obreshkoff to the theory of Diophantine approximation. Ann. Univ. Sofia Fac. Math. Inform. **89**, 47–58 (1995)
6196. Tóth, A.: Roots of quadratic congruences. Internat. Math. Res. Notices, 2000, 719–739
6197. Touibi, C., Zargouni, H.S.: Sur le théorème des idéaux premiers. Colloq. Math. **57**, 157–172 (1989)
6198. Trelina, L.A.: S-integral solutions of Diophantine equations of hyperbolic type. Dokl. Akad. Nauk Belorus. **22**, 881–884 (1978) (in Russian)
6199. Trifonov, O.: On the squarefree problem, II. Math. Balk. **3**, 284–295 (1989)

6200. Trifonov, O.: Lattice points close to a smooth curve and squarefull numbers in short intervals. J. Lond. Math. Soc. **65**, 303–319 (2002)
6201. Trost, E.: Zur Theorie der Potenzreste. Nieuw Arch. Wiskd. **18**, 58–61 (1934)
6202. Tsang, K.-M.: Higher power moments of $\Delta(x)$, $E(t)$ and $P(x)$. Proc. Lond. Math. Soc. **65**, 65–84 (1992)
6203. Tsang, K.-M.: Mean square of the remainder term in the Dirichlet divisor problem, II. Acta Arith. **71**, 279–299 (1995)
6204. Tsang, K.-M.: Counting lattice points in the sphere. Bull. Lond. Math. Soc. **32**, 679–688 (2000)
6205. Tschebotareff, N.G.: Die Bestimmung der Dichtigkeit einer Menge von Primzahlen, welche zu einer gegebener Substitutionsklasse gehören. Math. Ann. **95**, 191–228 (1926) (see also Čebotarev)
6206. Tsen, C.C.: Zur Stufentheorie und der Quasialgebraisch-Abgeschlossenheit kommutativer Körper. J. Chin. Math. Soc. **1**, 81–92 (1936)
6207. Tsishchanka, K.I.: On approximation of real numbers by algebraic numbers of bounded degree. J. Number Theory **123**, 290–314 (2007)
6208. Tubbs, R.: On the measure of algebraic independence of certain values of elliptic functions. J. Number Theory **23**, 60–79 (1986)
6209. Tubbs, R.: Diophantine problem on elliptic curves. Trans. Am. Math. Soc. **309**, 325–338 (1988)
6210. Tunnell, J.: Artin's conjecture for representations of octahedral type. Bull. Am. Math. Soc. **5**, 173–175 (1981)
6211. Tunnell, J.: A classical Diophantine problem and modular forms of weight 3/2. Invent. Math. **72**, 323–334 (1983)
6212. Turán, P.: On a theorem of Hardy and Ramanujan. J. Lond. Math. Soc. **9**, 274–276 (1934) [[6227], vol. 1, pp. 18–20]
6213. Turán, P.: Über die Primzahlen der arithmetischen Progression. Acta Sci. Math. **8**, 226–235 (1937) [[6227], vol. 1, pp. 64–73]
6214. Turán, P.: Über die Primzahlen der arithmetischen Progression, II. Acta Sci. Math. **9**, 187–192 (1939) [[6227], vol. 1, pp. 144–149]
6215. Turán, P.: Über die Verteilung der Primzahlen, I. Acta Sci. Math. **10**, 81–104 (1941) [[6227], vol. 1, pp. 207–230]
6216. Turán, P.: On Riemann's hypothesis. Izv. Akad. Nauk SSSR, Ser. Mat. **11**, 197–262 (1947) [[6227], vol. 1, pp. 306–362]
6217. Turán, P.: On some approximative Dirichlet polynomials in the theory of the zeta function of Riemann. Kgl. Danske Vid. Selsk. Mat.-Fys. Medd. **24**(17), 1–36 (1948) [[6227], vol. 1, pp. 369–402]
6218. Turán, P.: On the remainder-term of the prime-number formula, I. Acta Math. Acad. Sci. Hung. **1**, 48–63 (1950) [[6227], vol. 1, pp. 515–530]
6219. Turán, P.: On the remainder-term of the prime-number formula, II. Acta Math. Acad. Sci. Hung. **1**, 155–166 (1950) [[6227], vol. 1, pp. 541–551]
6220. Turán, P.: Results of number theory in Soviet Union. Mat. Lapok **1**, 243–266 (1950) (in Hungarian)
6221. Turán, P.: On Carlson's theorem in the theory of the zeta-function of Riemann. Acta Math. Acad. Sci. Hung. **2**, 39–73 (1951) [[6227], vol. 1, pp. 584–617]
6222. Turán, P.: A note on Fermat's conjecture. J. Indian Math. Soc. **15**, 47–50 (1951) [[6227], vol. 1, pp. 569–572]
6223. Turán, P.: Eine neue Methode in der Analysis und deren Anwendungen. Akad. Kiadó, Budapest (1953) [English translation of an expanded version: On a New Method of Analysis and Its Applications, Wiley, 1984]
6224. Turán, P.: On the zeros of the zeta-function of Riemann. J. Indian Math. Soc. **20**, 17–36 (1956) [[6227], vol. 2, pp. 890–909]
6225. Turán, P.: On a density theorem of Yu.V. Linnik. Magyar Tud. Akad. Mat. Kutató Int. Kőzl. **6**, 165–179 (1961) [[6227], vol. 2, pp. 1242–1255]

6226. Turán, P.: Commemoration on Stanisław Knapowski. Colloq. Math. **23**, 310–318 (1971) [[6227], vol. 3, pp. 2149–2160]
6227. Turán, P.: Collected Papers, vols. 1, 2. Akad. Kiadó, Budapest (1990)
6228. Turganaliev, R.T.: An asymptotic formula for the mean values of the fractional power of the Riemann zeta function. Tr. Mat. Inst. Steklova **158**, 203–226 (1981) (in Russian)
6229. Turing, A.M.: Some calculations of the Riemann zeta-function. Proc. Lond. Math. Soc. **3**, 99–117 (1953)
6230. Turing, S.: Alan M. Turing. Heffer, Cambridge (1959)
6231. Turjányi, S.: Eine Bemerkung zum Hilbert-Kamke-Problem. Publ. Math. (Debr.) **21**, 89–93 (1974)
6232. Turk, J.: The product of two or more neighboring integers is never a power. Ill. J. Math. **27**, 392–403 (1983)
6233. Turk, J.: On the difference between perfect powers. Acta Arith. **45**, 289–307 (1986)
6234. Turnbull, H.W.: On certain modular determinants. Edinb. Math. Notes **32**, 23–30 (1941)
6235. Turnbull, H.W.: Colin Maclaurin. Am. Math. Mon. **54**, 318–322 (1947)
6236. Turnwald, G.: Gleichverteilung von linearen rekursiven Folgen. SBer. Österr. Akad. Wiss. Math. Natur. Kl. **193**, 201–245 (1984)
6237. Turnwald, G.: Uniform distribution of second-order linear recurring sequences. Proc. Am. Math. Soc. **96**, 189–198 (1986)
6238. Turnwald, G.: On a problem concerning permutation polynomials. Trans. Am. Math. Soc. **302**, 251–257 (1987)
6239. Turnwald, G.: Weak uniform distribution of second-order linear recurring sequences. In: Lecture Notes in Math., vol. 1380, pp. 242–253. Springer, Berlin (1989)
6240. Turnwald, G.: On Schur's conjecture. J. Aust. Math. Soc. **58**, 312–357 (1995)
6241. Tuškina, T.A.: A numerical experiment on the calculation of the Hasse invariant for certain curves. Izv. Akad. Nauk SSSR, Ser. Mat. **29**, 1203–1204 (1965) (in Russian)
6242. Tzanakis, N.: Solving elliptic Diophantine equations by estimating linear forms in elliptic logarithms. The case of quartic equations. Acta Arith., 1996, 175–190
6243. Tzanakis, N., de Weger, B.M.M.: On the practical solution of the Thue equation. J. Number Theory **31**, 99–132 (1989)
6244. Tzanakis, N., de Weger, B.M.M.: How to explicitly solve a Thue-Mahler equation. Compos. Math. **84**, 223–288 (1992); corr.: **89**, 241–242 (1993)
6245. Uchida, K.: Class numbers of imaginary abelian number fields, I. Tohoku Math. J. **23**, 97–104 (1971)
6246. Uchida, K.: Class numbers of imaginary abelian number fields, II. Tohoku Math. J. **23**, 335–348 (1971)
6247. Uchida, K.: Class numbers of imaginary abelian number fields, III. Tohoku Math. J. **23**, 573–580 (1971)
6248. Uchida, K.: On Artin L-functions. Tohoku Math. J. **27**, 75–81 (1975)
6249. Uchiyama, S.: On a theorem concerning the distribution of almost primes. J. Fac. Sci. Hokkaido Univ., I **17**, 152–159 (1963)
6250. Uchiyama, S.: On the distribution of almost primes in an arithmetic progression. J. Fac. Sci. Hokkaido Univ., I **18**, 1–22 (1964)
6251. Ulam, S.: John von Neumann, 1903–1957. Bull. Am. Math. Soc. **64**, 1–49 (1958)
6252. Ul'ĭanov, P.L.: Remembering Sergeĭ Borisovič Stečkin. Usp. Mat. Nauk **51**(6), 11–20 (1996) (in Russian)
6253. Urbanowicz, J.: Remarks on the equation $1^k + 2^k + \cdots + (x-1)^k = x^k$. Indag. Math. **50**, 343–348 (1988)
6254. Uspensky, J.V.: Asymptotic expressions for arithmetical functions occurring in questions concerning partitions of integers into summands. Izv. Russ. Akad. Nauk **14**, 199–213 (1920) (in Russian)
6255. Ustinov, A.V.: Solution of the Arnold problem on weak asymptotics for Frobenius numbers with three arguments. Mat. Sb. **200**(4), 131–160 (2009) (in Russian)
6256. Vaaler, J.D.: Some extremal functions in Fourier analysis. Bull. Am. Math. Soc. **12**, 183–216 (1985)

6257. Vahlen, T.: Über Näherungswerte und Kettenbrüche. J. Reine Angew. Math. **115**, 221–233 (1895)
6258. Vaidyanathaswamy, R.: The theory of multiplicative arithmetic functions. Trans. Am. Math. Soc. **33**, 579–662 (1931)
6259. Valfiš, A.Z.[17]: On the class-number of binary quadratic forms. Trudy Tbil. Mat. Inst. **11**, 57–71 (1942) (in Russian)
6260. Valfiš, A.Z.: On the representation of integers by sums of squares. Asymptotical formulas. Usp. Mat. Nauk **7**(6), 97–178 (1951) (in Russian)
6261. Valfiš, A.Z.: On Euler's function. Tr. Tbil. Mat. Inst. **19**, 1–31 (1953) (in Russian)
6262. Vallée, B.: Generation of elements with small modular squares and provably fast integer factoring algorithms. Math. Comput. **56**, 823–849 (1991)
6263. de la Vallée-Poussin, C.J.: Recherches analytiques sur la théorie des nombres premiers. Ann. Soc. Sci. Bruxelles **20**, 183–256, 281–397 (1896)
6264. de la Vallée-Poussin, C.J.: Sur la fonction $\zeta(s)$ de Riemann et le nombre des nombres premiers inférieurs à une limite donnée. Mem. Couronnés de l'Acad. Roy. Sci. Bruxelles **59**, 1–74 (1899) [Reprint: Colloque Théorie des Nombres, Bruxelles 1955, 9–66, Liége, 1956]
6265. de la Vallée-Poussin, C.J.: Sur les zéros de $\zeta(s)$ de Riemann. C. R. Acad. Sci. Paris **163**, 418–421 (1916)
6266. Valson, C.A.: La vie et les travaux du baron Cauchy, vols. I–II, Gauthier-Villars, Paris (1868)
6267. van Aardenne-Ehrenfest, T.: Proof of the impossibility of a just distribution of an infinite sequence of points over an interval. Indag. Math. **7**, 71–76 (1945)
6268. van Aardenne-Ehrenfest, T.: Impossibility of a just distribution. Indag. Math. **12**, 264–269 (1949)
6269. van de Lune, J., te Riele, H.J.J.: On the zeros of the Riemann zeta function in the critical strip, III. Math. Comput. **41**, 759–767 (1983); corr. **46**, 771 (1986)
6270. van de Lune, J., te Riele, H.J.J.: Recent progress on the numerical verification of the Riemann hypothesis. CWI Newsletter, 1984, nr. 2, 35–37
6271. van de Lune, J., te Riele, H.J.J., Winter, D.T.: On the zeros of the Riemann zeta function in the critical strip, IV. Math. Comput. **46**, 667–681 (1986)
6272. van den Dries, L.: Elimination theory for the ring of algebraic integers. J. Reine Angew. Math. **388**, 189–205 (1988)
6273. van den Dries, L., Macintyre, A.: The logic of Rumely's local-global principle. J. Reine Angew. Math. **407**, 33–56 (1990)
6274. van der Corput, J.G.: Over roosterpunten in het platte vlak. (De beteekenis van de methoden van Voronoï en Pfeiffer.), pp. 1–128. Leiden, Groningen (1919)
6275. van der Corput, J.G.: Über Gitterpunkte in der Ebene. Math. Ann. **81**, 1–10 (1920)
6276. van der Corput, J.G.: Zahlentheoretische Abschätzungen. Math. Ann. **84**, 53–79 (1921)
6277. van der Corput, J.G.: Verschärfung der Abschätzung beim Teilerproblem. Math. Ann. **87**, 39–65 (1922); corr. **89**, 160 (1923)
6278. van der Corput, J.G.: Neue zahlentheoretische Abschätzungen. Math. Ann. **89**, 215–254 (1923); corr. **100**, 480
6279. van der Corput, J.G.: Neue zahlentheoretische Abschätzungen, II. Math. Z. **29**, 397–426 (1929)
6280. van der Corput, J.G.: Méthodes d'approximation dans le calcul du nombre des points a coordonnées entiers. Enseign. Math. **23**, 5–29 (1923)
6281. van der Corput, J.G.: Zahlentheoretische Abschätzungen mit Anwendung auf Gitterpunktprobleme. Math. Z. **17**, 250–259 (1923)
6282. van der Corput, J.G.: Zum Teilerproblem. Math. Ann. **98**, 697–716 (1928)
6283. van der Corput, J.G.: Diophantische Ungleichungen. I: Zur Gleichverteilung Modulo Eins. Acta Math. **56**, 373–456 (1931)

[17] See also Walfisz, A.

6284. van der Corput, J.G.: Diophantische Ungleichungen. II: Rhythmische Systeme. Acta Math. **59**, 209–328 (1932)
6285. van der Corput, J.G.: Verteilungsfunktionen, I. Proc. Akad. Wet. Amst. **38**, 813–821 (1935)
6286. van der Corput, J.G.: Verteilungsfunktionen, II. Proc. Akad. Wet. Amst. **38**, 1058–1066 (1935)
6287. van der Corput, J.G.: Sur l'hypothése de Goldbach pour presque tous les nombres pairs. Acta Arith. **2**, 266–290 (1937)
6288. van der Corput, J.G.: Contribution à la théorie additive des nombres, I. Proc. Akad. Wet. Amst. **41**, 227–237 (1938)
6289. van der Corput, J.G.: Contribution à la théorie additive des nombres, II. Proc. Akad. Wet. Amst. **41**, 350–361 (1938)
6290. van der Corput, J.G.: Contribution à la théorie additive des nombres, III. Proc. Akad. Wet. Amst. **41**, 442–453 (1938)
6291. van der Corput, J.G.: Contribution à la théorie additive des nombres, IV. Proc. Akad. Wet. Amst. **41**, 556–557 (1938)
6292. van der Corput, J.G.: Über Summen von Primzahlen und Primzahlquadraten. Math. Ann. **116**, 1–50 (1939)
6293. van der Corput, J.G.: On sets of integers, I. Indag. Math. **9**, 159–168 (1947)
6294. van der Corput, J.G.: On sets of integers, II. Indag. Math. **9**, 198–208 (1947)
6295. van der Corput, J.G.: On sets of integers, III. Indag. Math. **9**, 257–263 (1947)
6296. van der Corput, J.G.: On de Polignac's conjecture. Simon Stevin **27**, 99–105 (1950) (in Dutch)
6297. van der Corput, J.G., Koksma, J.F.: Sur l'ordre de grandeur de la fonction $\zeta(s)$ de Riemann dans la bande critique. Ann. Fac. Sci. Toulouse **22**, 1–39 (1930)
6298. van der Corput, J.G., Pisot, C.: Sur la discrépance modulo un, I. Indag. Math. **1**, 143–153 (1939)
6299. van der Corput, J.G., Pisot, C.: Sur la discrépance modulo un, II. Indag. Math. **1**, 184–195 (1939)
6300. van der Corput, J.G., Pisot, C.: Sur la discrépance modulo un, III. Indag. Math. **1**, 260–269 (1939)
6301. van der Poorten, A.J.: On Baker's inequality for linear forms in logarithms. Math. Proc. Camb. Philos. Soc. **80**, 233–248 (1976)
6302. van der Poorten, A.J.: A proof that Euler missed. Apéry's proof of the irrationality of $\zeta(3)$. An informal report. Math. Intell. **1**, 195–203 (1978/1979)
6303. van der Poorten, A.J.: Notes on Fermat's Last Theorem. Wiley, New York (1996)
6304. van der Poorten, A.J., Schlickewei, H.P.: The growth condition for recurrence sequences. Macquarie Univ. Math. Rep. 82-0041, North Ryde (1984)
6305. van der Poorten, A.J., Schlickewei, H.P.: Zeros of recurrence sequences. Bull. Aust. Math. Soc. **44**, 215–223 (1991)
6306. van der Poorten, A.J., Schlickewei, H.P.: Additive relations in fields. J. Aust. Math. Soc. **A51**, 154–170 (1991)
6307. van der Poorten, A.J., Shallit, J.: Folded continued fractions. J. Number Theory **40**, 237–250 (1992)
6308. van der Waal, R.W.: On a conjecture of Dedekind on zeta-functions. Indag. Math. **37**, 83–86 (1975)
6309. van der Waerden, B.L.: Beweis einer Baudet'schen Vermutung. Nieuw Arch. Wiskd. **15**, 212–216 (1927)
6310. van der Waerden, B.L.: Nachruf auf Otto Hölder. Math. Ann. **116**, 157–165 (1939)
6311. van der Waerden, B.L.: How the proof of Baudet's conjecture was found. In: Studies in Pure Mathematics, pp. 251–260. Academic Press, San Diego (1971)
6312. Vandiver, H.S.: Extensions of the criteria of Wieferich and Mirimanoff in connection with Fermat's last theorem. J. Reine Angew. Math. **144**, 314–317 (1914)
6313. Vandiver, H.S.: A property of cyclotomic integers and its relation to Fermat's last theorem, I. Ann. Math. **21**, 73–80 (1919/1920)

6314. Vandiver, H.S.: A property of cyclotomic integers and its relation to Fermat's last theorem, II. Ann. Math. **26**, 217–232 (1925)
6315. Vandiver, H.S.: On Kummer's memoir of 1857, concerning Fermat's last theorem. Proc. Natl. Acad. Sci. USA **6**, 266–269 (1920)
6316. Vandiver, H.S.: On the class number of the field $\Omega(e^{2i\pi/p^n})$ and the second case of Fermat's last theorem. Proc. Natl. Acad. Sci. USA **6**, 416–421 (1920)
6317. Vandiver, H.S.: On Kummer's memoir of 1857, concerning Fermat's last theorem. Bull. Am. Math. Soc. **28**, 400–407 (1922)
6318. Vandiver, H.S.: A new type of criteria for the first case of Fermat's last theorem. Ann. Math. **26**, 88–94 (1924)
6319. Vandiver, H.S.: On sets of three consecutive integers which are quadratic or cubic residues of primes. Bull. Am. Math. Soc. **31**, 33–38 (1925)
6320. Vandiver, H.S.: Note on trinomial congruences and the first case of Fermat's last theorem. Ann. Math. **27**, 54–56 (1925)
6321. Vandiver, H.S.: On Fermat's last theorem. Trans. Am. Math. Soc. **31**, 613–642 (1929)
6322. Vandiver, H.S.: Summary of results and proofs on Fermat's last theorem, V. Proc. Natl. Acad. Sci. USA **16**, 298–304 (1930)
6323. Vandiver, H.S.: Summary of results and proofs on Fermat's last theorem, VI. Proc. Natl. Acad. Sci. USA **17**, 661–673 (1931)
6324. Vandiver, H.S.: On the method of infinite descent in connection with Fermat's Last Theorem. Comment. Math. Helv. **4**, 1–8 (1932)
6325. Vandiver, H.S.: On Bernoulli's numbers and Fermat's last theorem. Duke Math. J. **3**, 569–584 (1937)
6326. Vandiver, H.S.: On Bernoulli's numbers and Fermat's last theorem, II. Duke Math. J. **5**, 418–427 (1939)
6327. Vandiver, H.S.: Note on Euler number criteria for the first case of Fermat's last theorem. Am. J. Math. **62**, 79–82 (1940)
6328. Vandiver, H.S.: Some theorems in finite field theory with applications to Fermat's Last Theorem. Proc. Natl. Acad. Sci. USA **30**, 362–367 (1944)
6329. Vandiver, H.S.: Fermat's last theorem. Am. Math. Mon. **53**, 555–578 (1946); suppl. **60**, 164–167 (1953)
6330. Vandiver, H.S.: Les travaux mathématiques de Dmitry Mirimanoff. Enseign. Math. **39**, 169–179 (1942/1950)
6331. Vandiver, H.S.: Examination of methods of attack on the second case of Fermat's last theorem. Proc. Natl. Acad. Sci. USA **40**, 732–735 (1954)
6332. Vandiver, H.S.: Is there an infinity of regular primes? Scr. Math. **21**, 306–309 (1955)
6333. van Frankenhuysen, M.: The ABC conjecture implies Roth's theorem and Mordell's conjecture. Mat. Contemp. **16**, 45–72 (1999)
6334. van Lint, J.H., Richert, H.-E.: On primes in arithmetic progressions. Acta Arith. **11**, 209–216 (1965)
6335. van Veen, S.C.: Thomas Jan Stieltjes (1856–1894). Nieuw Arch. Wiskd. **26**, 84–95 (1978)
6336. van Wijngaarden, A.: On the coefficients of the modular invariant $J(\tau)$. Indag. Math. **15**, 389–400 (1953)
6337. Vaserstein, L.N.: Waring's problem for commutative rings. J. Number Theory **26**, 299–307 (1987)
6338. Vaserstein, L.N.: Every integer is a sum or difference of 28 integral eighth powers. J. Number Theory **28**, 66–68 (1988)
6339. Vaserstein, L.N.: Quantum (abc)-theorems. J. Number Theory **81**, 351–358 (2000)
6340. Vaughan, R.C.: On the representation of numbers as sums of powers of natural numbers. Proc. Lond. Math. Soc. **21**, 160–180 (1970)
6341. Vaughan, R.C.: On a problem of Erdős, Straus and Schinzel. Mathematika **17**, 193–198 (1970)
6342. Vaughan, R.C.: On sums of mixed powers. J. Lond. Math. Soc. **3**, 677–688 (1971)
6343. Vaughan, R.C.: On Goldbach's problem. Acta Arith. **22**, 21–48 (1972)

6344. Vaughan, R.C.: Some applications of Montgomery's sieve. J. Number Theory **5**, 64–79 (1973)
6345. Vaughan, R.C.: A remark on the divisor function $d(n)$. Glasg. Math. J. **14**, 54–55 (1973)
6346. Vaughan, R.C.: Mean value theorems in prime number theory. J. Lond. Math. Soc. **10**, 153–162 (1975)
6347. Vaughan, R.C.: A note on Šnirel'man's approach to Goldbach's problem. Bull. Lond. Math. Soc. **8**, 245–250 (1976)
6348. Vaughan, R.C.: On the order of magnitude of Jacobsthal's function. Proc. Edinb. Math. Soc. **20**, 329–331 (1976/1977)
6349. Vaughan, R.C.: On the estimation of Schnirelman's constant. J. Reine Angew. Math. **290**, 93–108 (1977)
6350. Vaughan, R.C.: Sommes trigonométriques sur les nombres premiers. C. R. Acad. Sci. Paris **285**, 981–983 (1977)
6351. Vaughan, R.C.: On the distribution of αp modulo 1. Mathematika **24**, 135–141 (1977)
6352. Vaughan, R.C.: Homogeneous additive equations and Waring's problem. Acta Arith. **33**, 231–253 (1977)
6353. Vaughan, R.C.: On pairs of additive cubic equations. Proc. Lond. Math. Soc. **34**, 354–364 (1977)
6354. Vaughan, R.C.: A ternary additive problem. Proc. Lond. Math. Soc. **41**, 516–532 (1980)
6355. Vaughan, R.C.: An elementary method in prime number theory. Acta Arith. **37**, 111–115 (1980)
6356. Vaughan, R.C.: An elementary method in prime number theory. In: Recent Progress in Analytic Number Theory, Durham, 1979, vol. 1, pp. 341–348. Academic Press, San Diego (1981)
6357. Vaughan, R.C.: The Hardy-Littlewood Method. Cambridge University Press, Cambridge (1981); 2nd ed., 1997
6358. Vaughan, R.C.: Sur le problème de Waring pour les cubes. C. R. Acad. Sci. Paris **301**, 253–255 (1985)
6359. Vaughan, R.C.: Sums of three cubes. Bull. Lond. Math. Soc. **17**, 17–20 (1985)
6360. Vaughan, R.C.: On Waring's problem for cubes. J. Reine Angew. Math. **365**, 122–170 (1986)
6361. Vaughan, R.C.: On Waring's problem for cubes, II. J. Lond. Math. Soc. **39**, 205–218 (1989)
6362. Vaughan, R.C.: On Waring's problem for smaller exponents. Proc. Lond. Math. Soc. **52**, 445–463 (1986)
6363. Vaughan, R.C.: On Waring's problem for smaller exponents, II. Mathematika **33**, 6–22 (1986)
6364. Vaughan, R.C.: On Waring's problem: one square and five cubes. Q. J. Math. **37**, 117–127 (1986)
6365. Vaughan, R.C.: A new iterative method in Waring's problem. Acta Math. **162**, 1–71 (1989)
6366. Vaughan, R.C.: A new iterative method in Waring's problem, II. J. Lond. Math. Soc. **39**, 219–230 (1989)
6367. Vaughan, R.C.: On a variance associated with the distribution of primes in arithmetic progressions. Proc. Lond. Math. Soc. **82**, 533–553 (2001)
6368. Vaughan, R.C.: Moments for primes in arithmetic progressions, I. Duke Math. J. **120**, 371–383 (2003)
6369. Vaughan, R.C.: Moments for primes in arithmetic progressions, II. Duke Math. J. **120**, 385–403 (2003)
6370. Vaughan, R.C.: Hardy's legacy to number theory. J. Aust. Math. Soc. A **65**, 238–266 (1998)
6371. Vaughan, R.C., Wooley, T.D.: On Waring's problem: some refinements. Proc. Lond. Math. Soc. **63**, 35–68 (1991)
6372. Vaughan, R.C., Wooley, T.D.: Further improvements in Waring's problem. Acta Math. **174**, 147–240 (1995)
6373. Vaughan, R.C., Wooley, T.D.: Further improvements in Waring's problem, II. Sixth powers. Duke Math. J. **76**, 683–710 (1994)

6374. Vaughan, R.C., Wooley, T.D.: Further improvements in Waring's problem, III. Eighth powers. Philos. Trans. R. Soc. Lond. Ser. A, Math. Phys. Sci. **345**, 385–396 (1993)
6375. Vaughan, R.C., Wooley, T.D.: Further improvements in Waring's problem, IV. Higher powers. Acta Arith. **94**, 203–285 (2000)
6376. Velammal, G.: Is the binomial coefficient $\binom{2n}{n}$ squarefree? Hardy-Ramanujan J. **18**, 23–45 (1995)
6377. Veldkamp, G.R.: Ein Transzendenz-Satz für p-adische Zahlen. J. Lond. Math. Soc. **15**, 183–192 (1940)
6378. Velmin, V.P.: Solution of the indeterminate equation $u^m + v^n = w^k$. Mat. Sb. **24**, 633–661 (1904) (in Russian)
6379. Vélu, J.: Courbes modulaires et courbes de Fermat. Sém. Théor. Nombres Bordeaux, 1975/1976, exp. 16, 1–10
6380. Venkatesh, A.: The work of Einsiedler, Katok and Lindenstrauss on the Littlewood conjecture. Bull. Am. Math. Soc. **45**, 117–134 (2008)
6381. Venkov, A.B., Proskurin, N.V.: Automorphic functions and Kummer's problem. Usp. Mat. Nauk **37**(3), 143–165 (1982) (in Russian)
6382. Venkov, B.A.[18]: On the arithmetic of quaternions. Izv. Akad. Nauk SSSR, Ser. Mat., 1922, 205–220, 221–246; 1929, 489–504, 535–562, 607–622
6383. Venkov, B.A.: Über die Klassenzahl positiver binärer quadratischer Formen. Math. Z. **33**, 350–374 (1931)
6384. Venkov, B.A.: On an extremal problem of Markov for indefinite ternary quadratic forms. Dokl. Akad. Nauk SSSR **9**, 429–494 (1945) (in Russian)
6385. Venkov, B.A., Natanson, I.P.: Obituary: Rodion Osievič Kuzmin (1891–1949). Usp. Mat. Nauk **4**(4), 148–155 (1949) (in Russian)
6386. Venkov, B.B., Malyšev, A.V.: Boris Alekseevič Venkov (1900–1962). In: B.A. Venkov, Izbrannye Trudy, pp. 435–445. Nauka, Leningrad (1981) (in Russian)
6387. Verebrusov, A.S.: On the number of solutions of cubic equations in two variables. Mat. Sb. **26**, 115–129 (1907) (in Russian)
6388. Veselý, V.: O jedné obdobě Waringova problému. Časopis Mat. Fys. **62**, 123–127 (1933)
6389. Vignéras, M.-F.: Facteurs gamma et équations fonctionnelles. In: Lecture Notes in Math., vol. 627, pp. 79–103. Springer, Berlin (1977)
6390. Vijayaraghavan, T.: Periodic simple continued fractions. Proc. Lond. Math. Soc. **26**, 403–414 (1927)
6391. Vijayaraghavan, T.: On the fractional parts of powers of numbers. J. Lond. Math. Soc. **15**, 159–160 (1940)
6392. Vijayaraghavan, T.: On the fractional parts of powers of numbers, II. Proc. Camb. Philos. Soc. **37**, 349–357 (1941)
6393. Vijayaraghavan, T.: On the fractional parts of powers of numbers, III. J. Lond. Math. Soc. **17**, 137–138 (1942)
6394. Vijayaraghavan, T.: On the fractional parts of powers of numbers, IV. J. Indian Math. Soc. **12**, 33–39 (1948)
6395. Vilčinskiĭ, V.T.: Rational approximations to almost all real numbers. Vesti AN BSSR, 1979, nr. 6, 20–24 (in Russian)
6396. Vilčinskiĭ, V.T.: Calculation of the diophantine dimension of $\mathcal{P}$-adic fields. Dokl. Akad. Nauk Belorus. SSR **29**, 972–975 (1985) (in Russian)
6397. Vinogradov, A.I.: Application of $\zeta(s)$ to the sieve of Eratosthenes. Mat. Sb. **41**, 49–80 (1957); corr. 415–416 (in Russian)
6398. Vinogradov, A.I.: The density hypothesis for Dirichlet L-series. Izv. Akad. Nauk SSSR, Ser. Mat. **29**, 903–934 (1965); corr. **30**, 719–720 (1966) (in Russian)
6399. Vinogradov, A.I.: Artin L-series and his conjectures. Tr. Mat. Inst. Steklova **112**, 123–140 (1971) (in Russian)

[18] [Wenkov, B.A.]

6400. Vinogradov, A.I.: On a binary problem of Hardy-Littlewood. Acta Arith. **46**, 33–56 (1985) (in Russian)
6401. Vinogradov, A.I., Linnik, Yu.V.: Estimate of the sum of the number of divisors in a short segment of an arithmetic progression. Usp. Mat. Nauk **12**(4), 277–280 (1957) (in Russian)
6402. Vinogradov, A.I., Levin, B.V., Malyšev, A.V., Romanov, N.P., Čudakov, N.G.: Mark Borisovič Barban. Usp. Mat. Nauk **24**(2), 213–216 (1969) (in Russian)
6403. Vinogradov, I.M.: A new method to find asymptotical expression for arithmetical functions. Izv. Ross. Akad. Nauk, Ser. Mat. **11**, 1347–1378 (1917) (in Russian)
6404. Vinogradov, I.M.: On the mean value of the class-number of primitive forms of negative discriminant. Commun. Soc. Math. Charkov **16**, 10–38 (1918) (in Russian)
6405. Vinogradov, I.M.: On an asymptotical equality of the theory of quadratic forms. Ž. Fiz.-Mat. Obšč. Univ. Perm **1**, 18–28 (1918) (in Russian)
6406. Vinogradov, I.M.: Sur la distribution des résidus et des nonrésidus des puissances. Ž. Fiz.-Mat. Obšč. Univ. Perm **1**, 94–98 (1918)
6407. Vinogradov, I.M.: On the distribution of quadratic residues and non-residues. Ž. Fiz.-Mat. Obšč. Univ. Perm **2**, 1–16 (1919) (in Russian)
6408. Vinogradov, I.M.: On a general theorem of Waring. Mat. Sb. **31**, 490–507 (1924) (in Russian)
6409. Vinogradov, I.M.: On a bound for the smallest non-residue of n-th power. Izv. Akad. Nauk SSSR, Ser. Mat. **20**, 47–58 (1926) (in Russian)
6410. Vinogradov, I.M.: On a general theorem concerning the distribution of the residues and non-residues of powers. Trans. Am. Math. Soc. **29**, 209–217 (1927)
6411. Vinogradov, I.M.: On the bound of the least non-residue of n-th powers. Trans. Am. Math. Soc. **29**, 218–226 (1927)
6412. Vinogradov, I.M.: Analytischer Beweis des Satzes über die Verteilung der Bruchteile eines Polynoms. Izv. Akad. Nauk SSSR, Ser. Mat. **21**, 567–578 (1927)
6413. Vinogradov, I.M.: On Waring's theorem. Izv. Akad. Nauk SSSR, Ser. Mat. **1**, 393–400 (1928) (in Russian)
6414. Vinogradov, I.M.: On representations of a number by an integral polynomial in several variables. Izv. Akad. Nauk SSSR, Ser. Mat. **1**, 401–414 (1928) (in Russian)
6415. Vinogradov, I.M.: On the smallest primitive root. Dokl. Akad. Nauk SSSR, 1930, nr. 1, 7–11 (in Russian)
6416. Vinogradov, I.M.: A new solution of Waring's problem. Dokl. Akad. Nauk SSSR, 1934, nr. 2, 337–341 (in Russian)
6417. Vinogradov, I.M.: On the upper bound for $G(n)$ in Waring's problem. Izv. Akad. Nauk SSSR, Ser. Mat. **7**, 1455–1470 (1934) (in Russian)
6418. Vinogradov, I.M.: A new evaluation of $g(k)$ in Waring's problem. Dokl. Akad. Nauk SSSR, 1934, nr. 4, 249–253
6419. Vinogradov, I.M.: On Waring's problem. Ann. Math. **36**, 395–405 (1935)
6420. Vinogradov, I.M.: On Weyl's sums. Mat. Sb. **42**, 521–530 (1935)
6421. Vinogradov, I.M.: An asymptotic formula for the number of representations in Waring's problem. Mat. Sb. **42**, 531–534 (1935)
6422. Vinogradov, I.M.: On fractional parts of certain functions. Ann. Math. **37**, 448–455 (1936)
6423. Vinogradov, I.M.: On the number of fractional parts of a polynom lying in a given interval. Izv. Akad. Nauk SSSR, Ser. Mat. **1**, 3–8 (1936); add. 405–408
6424. Vinogradov, I.M.: A new method of resolving of certain general questions of the theory of numbers. Mat. Sb. **1**, 9–19 (1936)
6425. Vinogradov, I.M.: On asymptotic formula in Warings problem. Mat. Sb. **1**, 169–174 (1936)
6426. Vinogradov, I.M.: Sur quelques inégalités nouvelles de la théorie des nombres. C. R. Acad. Sci. Paris **202**, 1361–1362 (1936)
6427. Vinogradov, I.M.: A new method of estimation of trigonometric sums. Mat. Sb. **1**, 175–188 (1936) (in Russian)
6428. Vinogradov, I.M.: A new method in analytical number theory. Tr. Mat. Inst. Steklova **10**, 1–122 (1937) (in Russian)

6429. Vinogradov, I.M.: Representation of an odd number as a sum of three prime numbers. Dokl. Akad. Nauk SSSR **15**, 169–172 (1937)
6430. Vinogradov, I.M.: Some theorems concerning the theory of primes. Mat. Sb. **2**, 179–194 (1937)
6431. Vinogradov, I.M.: A new evaluation of a sum containing prime numbers. Mat. Sb. **2**, 783–791 (1937) (in Russian)
6432. Vinogradov, I.M.: Distribution of fractional parts of polynomial values under the condition that the argument runs over prime numbers of an arithmetic progression. Izv. Akad. Nauk SSSR, Ser. Mat. **1**, 505–514 (1937) (in Russian)
6433. Vinogradov, I.M.: Two theorems of the analytical number theory. Tr. Tbil. Mat. Inst. **5**, 167–180 (1938) (in Russian)
6434. Vinogradov, I.M.: Two theorems relating to the theory of distribution of prime numbers. Dokl. Akad. Nauk SSSR **30**, 287–288 (1941)
6435. Vinogradov, I.M.: On the estimation of trigonometrical sums. Dokl. Akad. Nauk SSSR **34**, 182–183 (1942) (in Russian)
6436. Vinogradov, I.M.: The method of trigonometrical sums in the theory of numbers. Tr. Mat. Inst. Steklova **23**, 1–109 (1947) (in Russian) [English translation: London, 1954]
6437. Vinogradov, I.M.: On an estimate of trigonometric sums with prime numbers. Izv. Akad. Nauk SSSR, Ser. Mat. **12**, 225–248 (1948) (in Russian)
6438. Vinogradov, I.M.: Improvement of the remainder term in an asymptotic formula. Izv. Akad. Nauk SSSR, Ser. Mat. **13**, 97–110 (1949) (in Russian)
6439. Vinogradov, I.M.: Improvement of asymptotic formulas for the number of lattice points in a three-dimensional region. Izv. Akad. Nauk SSSR, Ser. Mat. **19**, 3–10 (1955) (in Russian)
6440. Vinogradov, I.M.: Special cases of estimations of trigonometric sums. Izv. Akad. Nauk SSSR, Ser. Mat. **20**, 289–302 (1956) (in Russian)
6441. Vinogradov, I.M.: On the function $\zeta(s)$. Dokl. Akad. Nauk SSSR **118**, 631 (1958) (in Russian)
6442. Vinogradov, I.M.: A new estimate of the function $\zeta(1+it)$. Izv. Akad. Nauk SSSR, Ser. Mat. **22**, 161–164 (1958) (in Russian)
6443. Vinogradov, I.M.: On the question about an upper bound for $G(m)$. Izv. Akad. Nauk SSSR, Ser. Mat. **23**, 637–642 (1959) (in Russian)
6444. Vinogradov, I.M.: On the number of integral points in a given domain. Izv. Akad. Nauk SSSR, Ser. Mat. **24**, 777–786 (1960) (in Russian)
6445. Vinogradov, I.M.: On the number of integral points in a three-dimensional domain. Izv. Akad. Nauk SSSR, Ser. Mat. **27**, 3–8 (1963) (in Russian)
6446. Vinogradov, I.M.: On the number of integral points in a sphere. Izv. Akad. Nauk SSSR, Ser. Mat. **27**, 957–968 (1963) (in Russian)
6447. Vinogradov, I.M.: The Method of Trigonometrical Sums in the Theory of Numbers. Nauka, Moscow (1971); 2nd ed. 1980 (in Russian) [English translation: Calcutta, 1978]
6448. Vinogradov, I.M.: Special Variants of the Method of Trigonometrical Sums. Nauka, Moscow (1976) (in Russian)
6449. Viola, C.: On the diophantine equations $\prod_{i=0}^k x_i - \sum_{i=0}^k x_i = n$ and $\sum_{i=0}^r 1/x_i = a/n$. Acta Arith. **22**, 339–352 (1973)
6450. Vlăduţ, S.G.: Kronecker's Jugendtraum and Modular Functions. Gordon & Breach, New York (1991)
6451. Vojta, P.: Diophantine Approximations and Value Distribution Theory. Lecture Notes in Math., vol. 1239. Springer, Berlin (1987)
6452. Vojta, P.: A refinement of Schmidt's subspace theorem. Am. J. Math. **111**, 489–518 (1989)
6453. Vojta, P.: Siegel's theorem in the compact case. Ann. Math. **133**, 509–548 (1991)
6454. Vojta, P.: A generalization of theorems of Faltings and Thue-Siegel-Roth-Wirsing. J. Am. Math. Soc. **5**, 763–804 (1992)
6455. Vojta, P.: Roth's theorem with moving targets. Internat. Math. Res. Notices, 1996, 109–114
6456. Vojta, P.: A more general *abc* conjecture. Internat. Math. Res. Notices, 1998, 1103–1116
6457. Volkmann, B.: Über Hausdorffsche Dimensionen von Mengen die durch Zifferneigenschaften charakterisiert sind, I. Math. Z. **58**, 284–287 (1953)

6458. Volkmann, B.: Über Hausdorffsche Dimensionen von Mengen die durch Zifferneigenschaften charakterisiert sind, II. Math. Z. **59**, 247–254 (1953)
6459. Volkmann, B.: Über Hausdorffsche Dimensionen von Mengen die durch Zifferneigenschaften charakterisiert sind, III. Math. Z. **59**, 259–270 (1953)
6460. Volkmann, B.: Über Hausdorffsche Dimensionen von Mengen die durch Zifferneigenschaften charakterisiert sind, IV. Math. Z. **59**, 425–433 (1953)
6461. Volkmann, B.: Über Hausdorffsche Dimensionen von Mengen die durch Zifferneigenschaften charakterisiert sind, V. Math. Z. **65**, 389–413 (1956)
6462. Volkmann, B.: Über Hausdorffsche Dimensionen von Mengen die durch Zifferneigenschaften charakterisiert sind, VI. Math. Z. **68**, 439–449 (1958)
6463. Volkmann, B.: Ein metrischer Beitrag über Mahlersche Vermutung über S-Zahlen, I. J. Reine Angew. Math. **203**, 154–156 (1960)
6464. Volkmann, B.: Zur metrischen Theorie der S-Zahlen. J. Reine Angew. Math. **209**, 201–210 (1962)
6465. Volkmann, B.: Zur metrischen Theorie der S-Zahlen, II. J. Reine Angew. Math. **213**, 58–65 (1963)
6466. Voloch, J.F.: Transcendence of elliptic modular functions in characteristic p. J. Number Theory **58**, 55–59 (1996)
6467. Voronin, S.M.: A theorem on the "universality" of the Riemann zeta-function. Izv. Ross. Akad. Nauk, Ser. Mat. **39**, 475–486 (1975) (in Russian)
6468. Voronin, S.M.: The functional independence of Dirichlet L-functions. Acta Arith. **27**, 493–503 (1975) (in Russian)
6469. Voronoï, G.F.: Sur un problème du calcul des fonctions asymptotiques. J. Reine Angew. Math. **126**, 241–282 (1903)
6470. Voronoï, G.F.: Sur une fonction transcendante et ses applications a la sommation de quelques séries. Ann. Sci. École Norm. Sup., (3) **21**, 207–267, 459–533 (1904)
6471. Voronoï, G.F.: Nouvelles applications des paramétres continus à la théorie des formes quadratiques. Premier mémoire. Sur quelques propriétes des formes quadratiques positives parfaites. J. Reine Angew. Math. **133**, 97–178 (1908)
6472. Voronoï, G.F.: Nouvelles applications des paramétres continus à la théorie des formes quadratiques. Deuxième mémoire. Recherches sur les paralléloèdres primitifs. J. Reine Angew. Math. **134**, 198–287 (1908)
6473. Voronoï, G.F.: Nouvelles applications des paramétres continus à la théorie des formes quadratiques. Deuxième mémoire, Seconde partie. J. Reine Angew. Math. **136**, 67–181 (1909)
6474. Vose, M.D.: Egyptian fractions. Bull. Lond. Math. Soc. **17**, 21–24 (1985)
6475. Voskresenskiĭ, V.E., Malyšev, A.V., Perelmuter, G.I.: Nikolai Grigorevič Čudakov (on the occasion of his seventieth birthday). Usp. Mat. Nauk **30**(3), 195–197 (1975) (in Russian)
6476. Vosper, A.G.: The critical pairs of subsets of a group of prime order. J. Lond. Math. Soc. **31**, 200–205 (1956); add.: 280–282
6477. Voss, A.: Heinrich Weber. Jahresber. Dtsch. Math.-Ver. **23**, 431–444 (1914)
6478. Voutier, P.M.: Primitive divisors of Lucas and Lehmer sequences. Math. Comput. **64**, 869–888 (1995)
6479. Voutier, P.M.: Primitive divisors of Lucas and Lehmer sequences, II. J. Théor. Nr. Bordx. Math. Comput. **8**, 251–274 (1996)
6480. Voutier, P.M.: Primitive divisors of Lucas and Lehmer sequences, III. Math. Proc. Camb. Philos. Soc. **123**, 407–419 (1998)
6481. Voutier, P.M.: An upper bound for the size of integral solutions to $Y^m = f(X)$. J. Number Theory **53**, 247–271 (1995)
6482. Voutier, P.M.: An effective lower bound for the height of algebraic numbers. Acta Arith. **74**, 81–95 (1996)
6483. Voutier, P.M.: A further note on "On the equation $aX^4 - bY^2 = 2$". Acta Arith. **137**, 203–206 (2009)
6484. Vulakh, L.J.: Hurwitz constants. Izv. Vysš. Učeb. Zaved. Mat., 1977, nr. 2, 21–23 (in Russian)

6485. Vulakh, L.J.: Diophantine approximation on Bianchi groups. J. Number Theory **54**, 73–80 (1995)
6486. Vulakh, L.J.: Farey polytopes and continued fractions associated with discrete hyperbolic groups. Trans. Am. Math. Soc. **351**, 2295–2323 (1999)
6487. Wada, H.: Some computations on the criteria of Kummer. Tokyo J. Math. **3**, 173–176 (1980)
6488. Wagner, C.: Class numbers 5, 6 and 7. Math. Comput. **65**, 785–800 (1996)
6489. Wagstaff, S.S. Jr.: The irregular primes to 125 000. Math. Comput. **32**, 583–591 (1978)
6490. Wagstaff, S.S. Jr.: The Cunningham Project. In: High Primes and Misdemeanours, pp. 367–378. Am. Math. Soc., Providence (2004)
6491. Wahlin, G.E.: A new development of the theory of algebraic numbers. Trans. Am. Math. Soc. **16**, 502–508 (1915)
6492. Wahlin, G.E.: The equation $x^l - A \equiv 0 \pmod{p}$. J. Reine Angew. Math. **145**, 114–138 (1915)
6493. Wahlin, G.E.: On the principal units of an algebraic domain $k(\mathfrak{p}, \alpha)$. Bull. Am. Math. Soc. **23**, 450–455 (1916/1917)
6494. Wahlin, G.E.: The factorization of rational primes in a cubic domain. Am. J. Math. **44**, 191–203 (1922)
6495. Wakabayashi, I.: Number of solutions for cubic Thue equations with automorphisms. Ramanujan J. **14**, 131–154 (2007)
6496. Waldschmidt, M.: Solution d'un problème de Schneider sur les nombres transcendants. C. R. Acad. Sci. Paris **271**, A697–A700 (1970)
6497. Waldschmidt, M.: Solution du huitième problème de Schneider. J. Number Theory **5**, 191–202 (1973)
6498. Waldschmidt, M.: Nombres transcendants. Lecture Notes in Math., vol. 402. Springer, Berlin (1974)
6499. Waldschmidt, M.: Indépendance algébrique par la méthode de G.V. Čudnovskij. Sém. Delange–Pisot–Poitou, **16**, 1974/1975, fasc. 2, exp. G8, 1–18
6500. Waldschmidt, M.: Nombres transcendants et groupes algébriques. Astérisque **69–70**, 1–218 (1979)
6501. Waldschmidt, M.: A lower bound for linear forms in logarithms. Acta Arith. **37**, 257–283 (1980)
6502. Waldschmidt, M.: Transcendance et exponentielles en plusieurs variables. Invent. Math. **63**, 97–127 (1981)
6503. Waldschmidt, M.: Algebraic independence of transcendental numbers. Gel'fond's method and its developments. In: Perspectives in Mathematics, pp. 551–571. Birkhäuser, Basel (1984)
6504. Waldschmidt, M.: Groupes algébriques et grands degrés de transcendance. Acta Math. **156**, 253–302 (1986)
6505. Waldschmidt, M.: Linear independence of logarithms of algebraic numbers. IMS Report, 116, Institute of Math. Sciences, Madras (1992)
6506. Waldschmidt, M.: Minorations de combinaisons linéaires de logarithmes de nombres algébriques. Can. J. Math. **45**, 176–224 (1993)
6507. Waldschmidt, M.: Sur la nature arithmétique des valeurs de fonctions modulaires. Astérisque **245**, 105–140 (1997)
6508. Waldschmidt, M.: Un demi-siécle de transcendance. In: Development of Mathematics 1950–2000, pp. 1121–1186. Birkhäuser, Basel (2000)
6509. Waldschmidt, M.: Open Diophantine problems. Mosc. Math. J. **4**, 245–305 (2004)
6510. Waldschmidt, M.: Further variations on the six exponentials theorem. Hardy-Ramanujan J. **28**, 1–9 (2005)
6511. Waldschmidt, M.: Les contributions de Serge Lang à la théorie des nombres transcendants. Math. Gaz. **108**, 35–46 (2006)
6512. Waldschmidt, M.: Elliptic functions and transcendence. In: Surveys in Number Theory, pp. 143–188. Springer, Berlin (2008)
6513. Waldspurger, J.-L.: Engendrement par des séries thêta de certains espaces de formes modulaires. Invent. Math. **50**, 135–168 (1978/1979)

6514. Waldspurger, J.-L.: Correspondance de Shimura. J. Math. Pures Appl. **59**, 1–132 (1980)
6515. Walfisz, A.: Über die summatorischen Funktionen einiger Dirichletscher Reihen. Dissertation, Göttingen (1922)
6516. Walfisz, A.: Über Gitterpunkte in mehrdimensionalen Ellipsoiden. Math. Z. **19**, 300–307 (1924)
6517. Walfisz, A.: Über Gitterpunkte in mehrdimensionalen Ellipsoiden, II. Math. Z. **26**, 106–124 (1927)
6518. Walfisz, A.: Über Gitterpunkte in mehrdimensionalen Ellipsoiden, III. Math. Z. **27**, 245–268 (1927)
6519. Walfisz, A.: Über Gitterpunkte in mehrdimensionalen Ellipsoiden, IV. Math. Z. **35**, 212–229 (1932)
6520. Walfisz, A.: Über Gitterpunkte in mehrdimensionalen Ellipsoiden, VII. Tr. Tbil. Mat. Inst. **5**, 1–68 (1939)
6521. Walfisz, A.: Über Gitterpunkte in mehrdimensionalen Ellipsoiden, VIII. Tr. Tbil. Mat. Inst. **5**, 181–196 (1939)
6522. Walfisz, A.: Zur Abschätzung von $\zeta(\frac{1}{2}+it)$. Nachr. Ges. Wiss. Göttingen, 1924, 155–158
6523. Walfisz, A.: Über das Piltzsche Teilerproblem in algebraischen Zahlkörpern. Math. Z. **22**, 153–188 (1925)
6524. Walfisz, A.: Über das Piltzsche Teilerproblem in algebraischen Zahlkörpern, II. Math. Z. **26**, 487–494 (1927)
6525. Walfisz, A.: Über zwei Gitterpunktprobleme. Math. Ann. **95**, 69–83 (1926)
6526. Walfisz, A.: Teilerprobleme. Math. Z. **26**, 66–88 (1927)
6527. Walfisz, A.: Teilerprobleme, II. Math. Z. **34**, 448–472 (1931)
6528. Walfisz, A.: Teilerprobleme, III. J. Reine Angew. Math. **169**, 111–130 (1933)
6529. Walfisz, A.: Über die Koeffizientensummen einiger Modulformen. Math. Ann. **108**, 75–90 (1933)
6530. Walfisz, A.: Zur additiven Zahlentheorie. Acta Arith. **1**, 123–160 (1935)
6531. Walfisz, A.: Zur additiven Zahlentheorie, II. Math. Z. **40**, 592–607 (1936)
6532. Walfisz, A.: Gitterpunkte in mehrdimensionalen Kugeln. PWN, Warsaw (1957)
6533. Walfisz, A.: Über die Wirksamkeit einiger Abschätzungen trigonometrischer Summen. Acta Arith. **4**, 108–180 (1958)
6534. Walfisz, A.: Über Gitterpunkte in vierdimensionalen Ellipsoiden. Math. Z. **72**, 259–278 (1959/1960)
6535. Walfisz, A.: Weylsche Exponentialsummen in der neueren Zahlentheorie. Deutscher Verlag der Wissenschaften, Berlin (1963)
6536. Walsh, P.G.: A quantitative version of Runge's theorem on Diophantine equations. Acta Arith. **62**, 157–172 (1992); corr. **73**, 397–398 (1995)
6537. Walsh, P.G.: The Diophantine equation $X^2 - db^2Y^4 = 1$. Acta Arith. **87**, 179–188 (1998)
6538. Walsh, P.G.: On a conjecture of Schinzel and Tijdeman. In: Number Theory in Progress, vol. 1, pp. 577–582. de Gruyter, Berlin (1999)
6539. Walsh, P.G.: Sharp bounds for the number of solutions to simultaneous Pell equations. Acta Arith. **126**, 125–137 (2007)
6540. Walter, W.: Das wissenschafliche Werk von Erich Kamke. Jahresber. Dtsch. Math.-Ver. **69**, 193–205 (1968)
6541. Wan, D.Q.: An elementary proof of a theorem of Katz. Am. J. Math. **111**, 1–8 (1989)
6542. Wang, G.: On restricted sumsets in abelian groups of odd order. Integers **8**(A22), 1–8 (2008)
6543. Wang, J.T.-Y.: An effective Schmidt's subspace theorem over function fields. Math. Z. **246**, 811–844 (2004)
6544. Wang, T.Z.: On the exceptional set for the equation $n = p + k^2$. Acta Math. Sin. **11**, 156–167 (1995)
6545. Wang, T.Z.: The Goldbach problem for odd numbers under the generalized Riemann hypothesis, II. Adv. Math. (China) **25**, 339–346 (1996) (in Chinese)
6546. Wang, T.Z., Chen, J.R.: On odd Goldbach problem under general Riemann hypothesis. Sci. China **36**, 682–691 (1993)

6547. Wang, W.: On the least prime in an arithmetic progression. Acta Math. Sin. **29**, 826–836 (1986) (in Chinese)
6548. Wang, W.: On the least prime in an arithmetic progression. Acta Math. Sin. **7**, 279–289 (1991)
6549. Wang, Y.: On the representation of large even integer as a sum of a prime and a product of at most four primes. Acta Math. Sin. **6**, 565–582 (1956) (in Chinese)
6550. Wang, Y.: On sieve methods and some of their applications. Sci. Rec. **1**, 1–3 (1957)
6551. Wang, Y.: On sieve methods and some of the related problems. Sci. Rec. **1**, 9–12 (1957)
6552. Wang, Y.: On sieve methods and some of their applications. Sci. Sin. **8**, 357–381 (1959)
6553. Wang, Y.: A note on the least primitive root of a prime. Sci. Rec. **3**, 174–179 (1959)
6554. Wang, Y.: On the representation of large integer as a sum of a prime and an almost prime. Sci. Sin. **11**, 1033–1054 (1962)
6555. Wang, Y.: Bounds for solutions of additive equations in an algebraic number field, I. Acta Arith. **48**, 117–144 (1987)
6556. Wang, Y.: Bounds for solutions of additive equations in an algebraic number field, II. Acta Arith. **48**, 307–323 (1987)
6557. Wang, Y.: Diophantine Equations and Inequalities in Algebraic Number Fields. Springer, Berlin (1991)
6558. Wang, Y.: Hua LooKeng: a brief outline of his life and works. In: International Symposium in Memory of Hua Loo Keng, vol. 1, pp. 1–14. Springer, Berlin (1991)
6559. Wang, Y., Xie, S., Yu, K.: Remarks on the difference of consecutive primes. Sci. Sin. **14**, 786–788 (1965)
6560. Ward, M.: The arithmetical theory of linear recurring series. Trans. Am. Math. Soc. **35**, 600–628 (1933)
6561. Ward, M.: Note on an arithmetical property of recurring series. Math. Z. **39**, 211–214 (1935)
6562. Ward, M.: Prime divisors of second order recurring sequences. Duke Math. J. **21**, 607–614 (1954)
6563. Ward, M.: The intrinsic divisors of Lehmer numbers. Ann. Math. **62**, 230–236 (1955)
6564. Ward, M.: On the number of vanishing terms in an integral cubic recurrence. Am. Math. Mon. **62**, 155–160 (1955)
6565. Waring, E.: Meditationes Algebraicae. Cambridge University Press, Cambridge (1770); 2nd ed. 1782 [English translation: Am. Math. Soc., 1991]
6566. Warlimont, R.: Arithmetical semigroups. V. Multiplicative functions. Manuscr. Math. **77**, 361–383 (1992)
6567. Warlimont, R.: Richard John Knopfmacher and additive arithmetical semigroups. Quaest. Math. **24**, 269–272 (2001)
6568. Warning, E.: Bemerkung zur vorstehenden Arbeit von Herrn Chevalley. Abh. Math. Semin. Univ. Hamb. **11**, 76–83 (1935)
6569. Washington, L.C.: On Fermat's last theorem. J. Reine Angew. Math. **289**, 115–117 (1977)
6570. Washington, L.C.: The calculation of $L_p(1,\chi)$. J. Number Theory **9**, 175–178 (1977)
6571. Waterhouse, W.C.: Abelian varieties over finite fields. Ann. Sci. Éc. Norm. Super. **2**, 521–560 (1969)
6572. Waterhouse, W.C.: Pairs of quadratic forms. Invent. Math. **37**, 157–164 (1976)
6573. Waterhouse, W.C.: Pairs of symmetric bilinear forms in characteristic 2. Pac. J. Math. **69**, 275–283 (1977)
6574. Waterhouse, W.C.: A nonsymmetric Hasse-Minkowski theorem. Am. J. Math. **99**, 755–759 (1977)
6575. Waterhouse, W.C., Milne, J.S.: Abelian varieties over finite fields. In: Proc. Symposia Pure Math., vol. 20, pp. 53–64. Am. Math. Soc., Providence (1971)
6576. Watkins, M.: Real zeros of real odd Dirichlet L-functions. Math. Comput. **73**, 415–423 (2004)
6577. Watkins, M.: Class numbers of imaginary quadratic fields. Math. Comput. **73**, 907–938 (2004)

6578. Watson, G.L.: A proof of the seven cubes theorem. J. Lond. Math. Soc. **26**, 153–156 (1951)
6579. Watson, G.L.: Some problems in the theory of numbers. Ph.D. thesis, University College London (1953)
6580. Watson, G.L.: Minkowski's conjectures on the critical lattices of the region $|x|^p + |y|^p \leq 1$, I. J. Lond. Math. Soc. **28**, 305–309 (1953)
6581. Watson, G.L.: Minkowski's conjectures on the critical lattices of the region $|x|^p + |y|^p \leq 1$, II. J. Lond. Math. Soc. **28**, 402–410 (1953)
6582. Watson, G.L.: On indefinite quadratic forms in three and four variables. J. Lond. Math. Soc. **28**, 239–242 (1953)
6583. Watson, G.L.: Representation of integers by indefinite quadratic forms. Mathematika **2**, 32–38 (1955)
6584. Watson, G.L.: Distinct small values of quadratic forms. Mathematika **7**, 36–40 (1960)
6585. Watson, G.L.: Indefinite quadratic polynomials. Mathematika **7**, 141–144 (1960)
6586. Watson, G.L.: Indefinite quadratic Diophantine equations. Mathematika **8**, 32–38 (1961)
6587. Watson, G.L.: Quadratic Diophantine equations. Philos. Trans. R. Soc. Lond. Ser. A, Math. Phys. Sci. **253**, 227–254 (1960/1961)
6588. Watson, G.L.: On sums of a square and five cubes. J. Lond. Math. Soc. **5**, 215–218 (1972)
6589. Watson, G.N.: Some properties of the extended zeta function. Proc. Lond. Math. Soc. **12**, 288–296 (1913)
6590. Watson, G.N.: Über Ramanujansche Kongruenzeigenschaften der Zefällungszahlen, I. Math. Z. **39**, 713–731 (1935)
6591. Watson, G.N.: Ramanujan's Vermutung über Zerfällungszahlen. J. Reine Angew. Math. **179**, 97–128 (1938)
6592. Watson, G.N.: A table of Ramanujan's function $\tau(n)$. Proc. Lond. Math. Soc. **51**, 1–13 (1949)
6593. Watson, T.C.: An improved characterization of normal sets and some counter-examples. Isr. J. Math. **109**, 173–179 (1999)
6594. Watt, N.: Exponential sums and the Riemann zeta function, II. J. Lond. Math. Soc. **39**, 385–404 (1989)
6595. Watt, N.: Short intervals almost all containing primes. Acta Arith. **72**, 131–167 (1995)
6596. Weaver, R.L.: New congruences for the partition function. Ramanujan J. **5**, 53–63 (2001)
6597. Webb, W.A., Long, C.T.: Distribution modulo p^h of the general linear second order recurrence. Atti Accad. Naz. Lincei **58**, 92–100 (1975)
6598. Weber, H.: Beweis des Satzes, dass jede eigentlich primitive quadratische Form unendlich viele Primzahlen darzustellen fähig ist. Math. Ann. **20**, 301–329 (1882)
6599. Weber, H.: Theorie der Abelscher Zahlkörper. Acta Math. **8**, 193–263 (1886)
6600. Weber, H.: Theorie der Abelscher Zahlkörper, II. Acta Math. **9**, 105–130 (1887)
6601. Weber, H.: Elliptische Funktionen und algebraische Zahlen. Vieweg, Braunschweig (1894)
6602. Weber, H.: Lehrbuch der Algebra. vols. I–III. Vieweg, Braunschweig (1894–1908)
6603. Weber, H.: Über Zahlengruppen in algebraischen Körpern. Math. Ann. **48**, 433–437 (1897)
6604. Weber, H.: Über Zahlengruppen in algebraischen Körpern, II. Math. Ann. **49**, 83–100 (1897)
6605. Weber, H.: Über Zahlengruppen in algebraischen Körpern, III. Math. Ann. **50**, 1–26 (1898)
6606. Weber, H.: Über komplexe Primzahlen in Linearformen. J. Reine Angew. Math. **129**, 35–62 (1905)
6607. Wedeniwski, S.: Results connected with the first 100 billion zeros of the Riemann zeta function. http://www.zetagrid.net/zeta/math/zeta.result.100billion.zeros.html
6608. Weierstrass, C.: Zu Lindemann's Abhandlung "Über die Ludolphsche Zahl". SBer. Kgl. Preuß. Akad. Wiss. Berlin, 1885, 1067–1085. [Mathematische Werke, vol. 2, pp. 341–362, Berlin (1895)]
6609. Weil, A.: L'arithmétique sur les courbes algébriques. Acta Math. **52**, 281–315 (1928) [[6631], vol. 1, pp. 11–35]
6610. Weil, A.: Sur un théorème de Mordell. Bull. Sci. Math. **54**, 182–191 (1929) [[6631], vol. 1, pp. 47–56]

6611. Weil, A.: Remarques sur des résultats récents de C. Chevalley. C. R. Acad. Sci. Paris **203**, 1208–1210 (1936) [[6631], vol. 1, pp. 145–146]
6612. Weil, A.: Sur les fonctions algébriques à corps de constantes fini. C. R. Acad. Sci. Paris **210**, 592–594 (1940) [[6631], vol. 1, pp. 257–259]
6613. Weil, A.: On the Riemann hypothesis in function-fields. Proc. Natl. Acad. Sci. USA **27**, 345–347 (1941) [[6631], vol. 1, pp. 277–279]
6614. Weil, A.: Foundations of Algebraic Geometry. Am. Math. Soc., Providence (1946), 2nd. ed. 1962
6615. Weil, A.: Sur les courbes algébriques et les variétés qui s'en déduisent. Publ. Inst. Mat. Strasbourg **7**, 1–85 (1945) [Reprint: Hermann, 1948]
6616. Weil, A.: Variétés abéliennes et courbes algébriques. Publ. Inst. Mat. Strasbourg **8**, 1–165 (1946) [Reprint: Herrmann, 1948]
6617. Weil, A.: On some exponential sums. Proc. Natl. Acad. Sci. USA **34**, 204–207 (1948) [[6631], vol. 1, pp. 386–389]
6618. Weil, A.: Numbers of solutions of equations in finite fields. Bull. Am. Math. Soc. **55**, 497–508 (1949) [[6631], vol. 1, pp. 399–410]
6619. Weil, A.: Arithmetic on algebraic varieties. Ann. Math. **53**, 412–444 (1951) [[6631], vol. 1, pp. 450–482]
6620. Weil, A.: Jacobi sums as "Grössencharaktere". Trans. Am. Math. Soc. **73**, 487–495 (1952) [[6631], vol. 2, pp. 63–71]
6621. Weil, A.: Footnote to a recent paper. Am. J. Math. **76**, 347–350 (1954) [[6631], vol. 2, pp. 161–164]
6622. Weil, A.: On algebraic groups and homogeneous spaces. Am. J. Math. **77**, 493–512 (1955) [[6631], vol. 2, pp. 235–254]
6623. Weil, A.: Adèles et groupes algébriques. Sém. Bourbaki, **11**, 1958/1959, exp. 186 [[6631], vol. 2, pp. 398–404]
6624. Weil, A.: Adeles and Algebraic Groups. Princeton University Press, Princeton (1961); 2nd ed. Prog. Math., **23**, 1982
6625. Weil, A.: Sur la théorie des formes quadratiques. In: Colloq. Théorie des Groupes Algébriques, Bruxelles, 1962, pp. 9–22. Gauthier-Villars, Louvain–Paris (1962) [[6631], vol. 2, pp. 471–485]
6626. Weil, A.: Sur la formule de Siegel dans la théorie des groupes classiques. Acta Math. **113**, 1–87 (1965) [[6631], vol. 3, pp. 71–157]
6627. Weil, A.: Über die Bestimmung Dirichletscher Reihen durch Funktionalgleichungen. Math. Ann. **168**, 149–156 (1967) [[6631], vol. 3, pp. 165–172]
6628. Weil, A.: Dirichlet Series and Automorphic Functions. Lecture Notes in Math., vol. 189. Springer, Berlin (1971)
6629. Weil, A.: Basic Number Theory. Springer, Berlin (1967); 2nd ed. 1973 [Reprint: 1975], 3rd ed. 1974
6630. Weil, A.: Two lectures on number theory, past and present. Enseign. Math. **20**, 87–110 (1974) [[6631], vol. 3, pp. 279–302]
6631. Weil, A.: Oeuvres Scientifiques, vols. 1–3. Springer, Berlin (1979)
6632. Weinberger, P.J.: A proof of a conjecture of Gauss on class-number two. Ph.D. thesis, Berkeley (1969)
6633. Weinberger, P.J.: Exponents of the class groups of complex quadratic fields. Acta Arith. **22**, 117–124 (1973)
6634. Weissauer, R.: Der Heckesche Umkehrsatz. Abh. Math. Semin. Univ. Hamb. **61**, 83–119 (1991)
6635. Weissinger, J.: Theorie der Divisorenkongruenzen. Abh. Math. Semin. Univ. Hamb. **12**, 115–126 (1938)
6636. Wennberg, S.: On the theory of Dirichlet series. Dissertation, Uppsala (1920) (in Swedish)
6637. Western, A.E.: Computations concerning numbers representable by four or five cubes. J. Lond. Math. Soc. **1**, 244–250 (1926)
6638. Western, A.E.: Allan Joseph Cunningham. J. Lond. Math. Soc. **3**, 317–318 (1928)

6639. Western, A.E.: On Lucas's and Pepin's tests for the primeness of Mersenne's numbers. J. Lond. Math. Soc. **7**, 130–137 (1932)
6640. Western, A.E.: Note on the magnitude of the difference between successive primes. J. Lond. Math. Soc. **9**, 276–278 (1934)
6641. Westlund, J.: Note on multiply perfect numbers. Ann. Math. **2**, 172–174 (1901)
6642. Weston, T.: Kummer theory of abelian varieties and reductions of Mordell-Weil groups. Acta Arith. **110**, 77–88 (2003)
6643. Westzynthius, E.: Sur la distribution des entiers qui ne sont divisibles par aucun parmi les n plus petits nombres premiers. C. R. Acad. Sci. Paris **193**, 805–807 (1931)
6644. Westzynthius, E.: Über die Verteilung der Zahlen, die zu der n ersten Primzahlen teilerfremd sind. Commun. Phys. Math. Helsingfors, (5) **25**, 1–37 (1931)
6645. Weyl, H.: Über die Gibbs'sche Erscheinung und verwandte Kongruenzphänomene. Rend. Circ. Mat. Palermo **30**, 377–407 (1910)
6646. Weyl, H.: Über ein Problem aus dem Gebiet der Diophantischen Approximationen. Nachr. Ges. Wiss. Göttingen, 1914, 234–244
6647. Weyl, H.: Über die Gleichverteilung von Zahlen mod. Eins. Math. Ann. **77**, 313–352 (1916)
6648. Weyl, H.: Zur Abschätzung von $\zeta(1+it)$. Math. Z. **10**, 88–101 (1921)
6649. Weyl, H.: Bemerkung zur Hardy-Littlewoodschen Lösung des Waringschen Problems. Nachr. Ges. Wiss. Göttingen, 1922, 189–192
6650. Whiteman, A.L.: A note on Kloosterman sums. Bull. Am. Math. Soc. **51**, 373–377 (1945)
6651. Whiteman, A.L.: Theorems analogous to Jacobsthal's theorem. Duke Math. J. **16**, 619–626 (1949)
6652. Whiteman, A.L.: Theorems on quadratic partitions. Proc. Natl. Acad. Sci. USA **36**, 60–65 (1950)
6653. Whiteman, A.L.: Theorems on Brewer and Jacobsthal sums, I. In: Proc. Symposia Pure Math., vol. 8, pp. 44–55. Am. Math. Soc., Providence (1965)
6654. Whiteman, A.L.: Theorems on Brewer and Jacobsthal sums, II. Mich. Math. J. **12**, 65–80 (1965)
6655. Whittaker, E.T.: George David Birkhoff. J. Lond. Math. Soc. **20**, 121–128 (1945)
6656. Wieferich, A.: Beweis des Satzes, dass sich jede ganze Zahl als Summe von höchstens neun Kuben darstellen lässt. Math. Ann. **66**, 95–101 (1909)
6657. Wieferich, A.: Zum letzten Fermatschen Theorem. J. Reine Angew. Math. **136**, 293–302 (1909)
6658. Wieferich, A.: Über die Darstellung der Zahlen als Summen von Biquadraten. Math. Ann. **66**, 106–108 (1909)
6659. Wieferich, A.: Zur Darstellung der Zahlen als Summen von 5-ten und 7-ten Potenzen positiver ganzer Zahlen. Math. Ann. **67**, 61–75 (1909)
6660. Wiener, N.: A new method in Tauberian theorems. J. Math. Phys. MIT **7**, 161–184 (1928)
6661. Wiener, N.: Tauberian theorems. Ann. Math. **33**, 1–100 (1932)
6662. Wiener, N.: R.E.A.C. Paley—in memoriam. Bull. Am. Math. Soc. **39**, 476 (1933)
6663. Wiener, N.: Godfrey Harold Hardy (1877–1947). Bull. Am. Math. Soc. **55**, 72–77 (1949)
6664. Wiertelak, K.: On the application of Turán's method to the theory of Dirichlet L-functions. Acta Arith. **19**, 249–259 (1971)
6665. Wigert, S.: Sur l'ordre de grandeur du nombre des diviseurs d'un entier. Ark. Math. Astron. Fys. **3**(18), 1–9 (1906/1907)
6666. Wigert, S.: Sur quelques fonctions arithmétiques. Acta Math. **37**, 113–140 (1914)
6667. Wiles, A.: On p-adic representations for totally real fields. Ann. Math. **123**, 407–456 (1986)
6668. Wiles, A.: On ordinary λ-adic representations associated to modular forms. Invent. Math. **94**, 529–573 (1988)
6669. Wiles, A.: The Iwasawa conjecture for totally real fields. Ann. Math. **131**, 493–540 (1990)
6670. Wiles, A.: Modular elliptic curves and Fermat's Last Theorem. Ann. Math. **141**, 443–551 (1995)
6671. Willerding, M.F.: Determination of all classes of positive quaternary quadratic forms which represent all (positive) integers. Bull. Am. Math. Soc. **54**, 334–337 (1948)

6672. Williams, H.C.: Daniel Shanks (1917–1996). Not. Am. Math. Soc. **44**, 812–816 (1997)
6673. Williams, H.C.: Édouard Lucas and Primality Testing. Wiley, New York (1998)
6674. Williams, H.C.: Solving the Pell equation. In: Number Theory for the Millennium, vol. III, pp. 397–435. AK Peters, Wellesley (2002)
6675. Williams, K.S.: Note on the Kloosterman sum. Proc. Am. Math. Soc. **30**, 61–62 (1971)
6676. Wills, J.M.: Gitterzahlen und innere Volumina. Comment. Math. Helv. **53**, 508–524 (1978)
6677. Wilson, B.M.: Proofs of some formulae enunciated by Ramanujan. Proc. Lond. Math. Soc. **21**, 235–255 (1922)
6678. Wilson, B.M.: An application of Pfeiffer's method to a problem of Hardy and Littlewood. Proc. Lond. Math. Soc. **22**, 248–253 (1923)
6679. Wilson, R.J.: The large sieve in algebraic number fields. Mathematika **16**, 189–204 (1969)
6680. Wilton, J.R.: The lattice points of a circle. Proc. R. Soc. Edinb. **48**, 191–200 (1928)
6681. Wilton, J.R.: A note on Ramanujan's arithmetical function $\tau(n)$. Proc. Camb. Philos. Soc. **25**, 121–129 (1929)
6682. Wilton, J.R.: On Ramanujan's arithmetical function $\Sigma_{r,s}(n)$. Proc. Camb. Philos. Soc. **25**, 255–264 (1929)
6683. Wilton, J.R.: Congruence properties of Ramanujan's function $\tau(n)$. Proc. Lond. Math. Soc. **31**, 1–10 (1930)
6684. Wilton, J.R.: Voronoi's summation formula. Q. J. Math. **3**, 26–32 (1932)
6685. Wilton, J.R.: An extended form of Dirichlet's divisor problem. Proc. Lond. Math. Soc. **36**, 391–426 (1933)
6686. Wiman, A.: Über den Rang von Kurven $y^2 = x(x+a)(x+b)$. Acta Math. **76**, 225–251 (1945)
6687. Wiman, A.: Über rationale Punkte auf Kurven $y^2 = x(x^2 - c^2)$. Acta Math. **77**, 281–320 (1945)
6688. Wiman, A.: Über rationale Punkte auf Kurven dritter Ordnung vom Geschlechte Eins. Acta Math. **80**, 223–257 (1948)
6689. Wintenberger, J.-P.: La conjecture de modularité de Serre: le cas de conducteur 1 (d'aprés C. Khare). Astérisque **311**, 99–121 (2007)
6690. Wintner, A.: Eratosthenian Averages. Waverly Press, Baltimore (1943)
6691. Wirsching, G.J.: The dynamical system generated by the $3n + 1$ function. In: Lecture Notes in Math., vol. 1681, pp. 1–158. Springer, Berlin (1998)
6692. Wirsing, E.: Bemerkung zu der Arbeit über vollkommene Zahlen. Math. Ann. **137**, 316–318 (1959)
6693. Wirsing, E.: Approximation mit algebraischen Zahlen beschränktes Grades. J. Reine Angew. Math. **206**, 67–77 (1961)
6694. Wirsing, E.: Das asymptotische Verhalten von Summen über multiplikative Funktionen. Math. Ann. **143**, 75–102 (1961)
6695. Wirsing, E.: Das asymptotische Verhalten von Summen über multiplikative Funktionen, II. Acta Math. Acad. Sci. Hung. **18**, 411–467 (1967)
6696. Wirsing, E.: Elementare Beweise des Primzahlsatzes mit Restglied, I. J. Reine Angew. Math. **211**, 205–214 (1962)
6697. Wirsing, E.: Elementare Beweise des Primzahlsatzes mit Restglied, II. J. Reine Angew. Math. **214/215**, 1–8 (1964)
6698. Wirsing, E.: A characterization of $\log n$ as an additive arithmetic function. In: Symposia Mathematica, vol. 4, pp. 45–57. Academic Press, London (1970)
6699. Wirsing, E.: On approximations of algebraic numbers by algebraic numbers of bounded degree. In: Proc. Symposia Pure Math., vol. 20, pp. 213–247. Am. Math. Soc., Providence (1971)
6700. Wirsing, E.: On the theorem of Gauss-Kusmin-Levy and a Frobenius-type theorem for function spaces. Acta Arith. **24**, 507–528 (1974)
6701. Wisdom, J.M.: On the representations of a number as the sum of four fifth powers. J. Lond. Math. Soc. **60**, 399–419 (1999)
6702. Witt, E.: Über ein Gegenbeispiel zum Normensatz. Math. Z. **39**, 462–467 (1935)

6703. Witt, E.: Theorie der quadratischen Formen in beliebigen Körpern. J. Reine Angew. Math. **176**, 31–44 (1937)
6704. Wittmann, C.: Group structure of elliptic curves over finite fields. J. Number Theory **88**, 335–344 (2001)
6705. Wohlfahrt, K.: Über Operatoren Heckescher Art bei Modulformen reeller Dimension. Math. Nachr. **16**, 233–156 (1957)
6706. Wohlfahrt, K.: Über Dedekindsche Summen und Untergruppen der Modulgruppe. Abh. Math. Semin. Univ. Hamb. **23**, 5–10 (1959)
6707. Wohlfahrt, K.: Über die Nullstellen einiger Eisensteinreihen. Math. Nachr. **26**, 381–383 (1963/1964)
6708. Wohlfahrt, K.: Zur Struktur der rationalen Modulgruppe. Math. Ann. **174**, 79–99 (1967)
6709. Wohlfahrt, K.: Hans Petersson zum Gedächtnis. Jahresber. Dtsch. Math.-Ver. **96**, 117–129 (1994)
6710. Wójtowicz, M.: Robin's inequality and the Riemann hypothesis. Proc. Jpn. Acad. Sci. **83**, 47–49 (2007)
6711. Wolke, D.: Moments of the number of classes of primitive quadratic forms with negative discriminant. J. Number Theory **1**, 502–511 (1969)
6712. Wolke, D.: Über das summatorische Verhalten zahlentheoretischer Funktionen. Math. Ann. **194**, 147–166 (1971)
6713. Wolke, D.: Über die mittlere Verteilung der Werte zahlentheoretischer Funktionen auf Restklassen, I. Math. Ann. **202**, 1–25 (1973)
6714. Wolke, D.: Über die mittlere Verteilung der Werte zahlentheoretischer Funktionen auf Restklassen, II. Math. Ann. **204**, 145–153 (1973)
6715. Wolke, D.: Fast-Primzahlen in kurzen Intervallen. Math. Ann. **244**, 233–242 (1979)
6716. Wolke, D.: Ein Problem im Zusammenhang mit dem Bombierischen Primzahlsatz. Arch. Math. **46**, 46–51 (1986)
6717. Wolke, D.: On a problem of Erdős in additive number theory. J. Number Theory **59**, 209–213 (1996)
6718. Wolstenholme, J.: On certain properties of prime numbers. Q. J. Math. **5**, 35–39 (1862)
6719. Wong, S.: Power residues on abelian varieties. Manuscr. Math. **102**, 129–138 (2000)
6720. Wooldridge, K.: Values taken many times by Euler's phi-function. Proc. Am. Math. Soc. **76**, 229–234 (1979)
6721. Wooley, T.D.: On simultaneous additive equations, I. Proc. Lond. Math. Soc. **63**, 1–34 (1991)
6722. Wooley, T.D.: On simultaneous additive equations, II. J. Reine Angew. Math. **419**, 141–198 (1991)
6723. Wooley, T.D.: On simultaneous additive equations, III. Mathematika **37**, 85–96 (1990)
6724. Wooley, T.D.: Large improvements in Waring's problem. Ann. Math. **135**, 131–164 (1992)
6725. Wooley, T.D.: On Vinogradov's mean value theorem. Mathematika **39**, 379–399 (1992)
6726. Wooley, T.D.: On Vinogradov's mean value theorem, II. Mich. Math. J. **40**, 175–180 (1993)
6727. Wooley, T.D.: The application of a new mean value theorem to the fractional parts of polynomials. Acta Arith. **65**, 163–179 (1993)
6728. Wooley, T.D.: Breaking classical convexity in Waring's problem: sums of cubes and quasi-diagonal behaviour. Invent. Math. **122**, 421–451 (1995)
6729. Wooley, T.D.: Sums of two cubes. Internat. Math. Res. Notices, 1995, 181–184
6730. Wooley, T.D.: Some remarks on Vinogradov's mean value theorem and Tarry's problem. Monatshefte Math. **122**, 265–273 (1996)
6731. Wooley, T.D.: On exponential sums over smooth numbers. J. Reine Angew. Math. **488**, 79–140 (1997)
6732. Wooley, T.D.: Forms in many variables. In: Analytic Number Theory, Kyoto, 1996, pp. 361–376. Cambridge University Press, Cambridge (1997)
6733. Wooley, T.D.: An explicit version of Birch's theorem. Acta Arith. **85**, 79–96 (1998)
6734. Wooley, T.D.: On the local solubility of Diophantine systems. Compos. Math. **111**, 149–165 (1998)

6735. Wooley, T.D.: Sums of three cubes. Mathematika **47**, 53–61 (2000)
6736. Wooley, T.D.: Slim exceptional sets in Waring's problem: one square and five cubes. Q. J. Math. **53**, 111–118 (2002)
6737. Wooley, T.D.: Artin's conjecture for septic and unidecic forms. Acta Arith. **133**, 25–35 (2008)
6738. Wooley, T.D.: Vinogradov's mean value theorem via efficient congruencing, to appear. arXiv:1101.0574
6739. Wright, E.M.: Asymptotic partition formulae, I. Plane partitions. Q. J. Math. **2**, 177–189 (1931)
6740. Wright, E.M.: Asymptotic partition formulae, II. Weighted partitions. Proc. Lond. Math. Soc. **36**, 117–141 (1932)
6741. Wright, E.M.: Asymptotic partition formulae, III. Partitions into kth powers. Acta Math. **63**, 143–191 (1934)
6742. Wright, E.M.: The representation of a number as a sum of five or more squares. Q. J. Math. **4**, 37–51, 228–232 (1933)
6743. Wright, E.M.: An extension of Waring's problem. Philos. Trans. R. Soc. Lond. Ser. A, Math. Phys. Sci. **232**, 1–26 (1933)
6744. Wright, E.M.: Proportionality conditions in Waring's problem. Math. Z. **38**, 730–746 (1934)
6745. Wright, E.M.: An easier Waring's problem. J. Lond. Math. Soc. **9**, 267–272 (1934)
6746. Wright, E.M.: On Tarry's problem, I. Q. J. Math. **7**, 43–48 (1936)
6747. Wright, E.M.: On Tarry's problem, II. Q. J. Math. **6**, 261–267 (1935)
6748. Wright, E.M.: On Tarry's problem, III. Q. J. Math. **8**, 48–50 (1937)
6749. Wright, E.M.: The representation of a number as a sum of three or four squares. Proc. Lond. Math. Soc. **42**, 481–500 (1937)
6750. Wright, E.M.: The representation of a number as a sum of four "almost equal" squares. Q. J. Math. **8**, 278–279 (1937)
6751. Wright, E.M.: Equal sums of like powers. Bull. Am. Math. Soc. **54**, 755–757 (1948)
6752. Wright, E.M.: On the Prouhet-Lehmer problem. J. Lond. Math. Soc. **23**, 279–285 (1948)
6753. Wright, E.M.: Equal sums of like powers. Proc. Edinb. Math. Soc. **8**, 138–142 (1949)
6754. Wright, E.M.: Prouhet's 1851 solution of the Tarry-Escott problem of 1919. Bull. Am. Math. Soc. **66**, 199–201 (1959)
6755. Wu, J.: Sur la suite des nombres premiers jumeaux. Acta Arith. **55**, 365–394, 365–394 (1990)
6756. Wu, J.: P_2 dans les petits intervalles. Prog. Math. **102**, 233–267 (1992)
6757. Wu, J.: Primes of the form $p = 1 + m^2 + n^2$ in short intervals. Proc. Am. Math. Soc. **126**, 1–8 (1998)
6758. Wu, J.: On the distribution of square-full integers. Arch. Math. **77**, 233–240 (2001)
6759. Wu, J.: Chen's double sieve, Goldbach's conjecture and the twin prime problem. Acta Arith. **114**, 215–273 (2004)
6760. Wu, J.: Chen's double sieve, Goldbach's conjecture and the twin prime problem, II. Acta Arith. **131**, 367–387 (2008)
6761. Wu, T.: On the proof of continued fraction expansions for irrationals. J. Number Theory **23**, 55–59 (1986)
6762. Wüstholz, G.: Über das Abelsche Analogon des Lindemannschen Satzes, I. Invent. Math. **72**, 363–388 (1983)
6763. Wyler, O.: Solution of Problem 5080. Am. Math. Mon. **71**, 220–222 (1964)
6764. Wyman, M.: Leo Moser (1921–1970). Can. Math. Bull. **15**, 1–4 (1972)
6765. Wyner, A.D.: On coding and information theory. SIAM Rev. **11**, 317–346 (1969)
6766. Xu, Y.: In memoriam Donald J. Newman (1930–2007). J. Approx. Theory **154**, 37–58 (2008)
6767. Xu, Y., Yau, S.S.-T.: A sharp estimate of the number of integral points in a tetrahedron. J. Reine Angew. Math. **423**, 199–219 (1992)
6768. Xu, Y., Yau, S.S.-T.: A sharp estimate of the number of integral points in a 4-dimensional tetrahedra. J. Reine Angew. Math. **473**, 1–23 (1996)

6769. Xuan, T.Z.: Integers with no large prime factors. Acta Arith. **69**, 303–327 (1995)
6770. Xuan, T.Z.: Irregularities in the distribution of primes in an arithmetic progression. Acta Arith. **77**, 173–177 (1996)
6771. Yamamoto, K.: Theory of arithmetic linear transformations and its application to an elementary proof of Dirichlet's theorem. J. Math. Soc. Jpn. **7**, 424–434 (1955)
6772. Yamamoto, K.: On the diophantine equation $4/n = 1/x + 1/y + 1/z$. Mem. Fac. Sci. Kyushu Univ. **19**, 37–47 (1965)
6773. Yamamoto, Y.: Real quadratic number fields with large fundamental units. Osaka Math. J. **8**, 261–270 (1971)
6774. Yamamura, K.: The determination of the imaginary abelian number fields with class number one. Proc. Jpn. Acad. Sci. **68**, 21–24 (1992)
6775. Yamamura, K.: The determination of the imaginary abelian number fields with class number one. Math. Comput. **62**, 899–921 (1994)
6776. Yamanoi, K.: The second main theorem for small functions and related problems. Acta Math. **192**, 225–294 (2004)
6777. Yildirim, C.Y.: The pair correlation of zeros of Dirichlet L-functions and primes in arithmetic progressions. Manuscr. Math. **72**, 325–334 (1991)
6778. Yin, W.L.: On Dirichlet's divisor problem. Sci. Rec. **3**, 6–8 (1959)
6779. Yin, W.L.: Piltz's divisor problem for $k = 3$. Sci. Rec. **3**, 169–173 (1959)
6780. Yin, W.L.: The lattice point in a circle. Sci. Sin. **11**, 10–15 (1962)
6781. Yokoi, H.: Class number one problem for certain kind of real quadratic fields. In: Proc. Internat. Conf., Katata, 1986, pp. 125–137. Nagoya University Press, Nagoya (1986)
6782. Yokota, H.: Length and denominators of Egyptian fractions. J. Number Theory **24**, 249–258 (1986)
6783. Yokota, H.: Length and denominators of Egyptian fractions, II. J. Number Theory **28**, 272–282 (1988)
6784. Yokota, H.: Denominators of Egyptian fractions. J. Number Theory **28**, 258–271 (1988)
6785. Yokota, H.: On a problem of Bleicher and Erdős. J. Number Theory **30**, 198–207 (1988)
6786. Yokota, H.: On number of integers representable as a sum of unit fractions, II. J. Number Theory **67**, 162–169 (1997)
6787. Yokota, H.: On number of integers representable as a sum of unit fractions, III. J. Number Theory **96**, 351–372 (2002)
6788. Yokota, H.: The largest integer expressible as a sum of reciprocal of integers. J. Number Theory **76**, 206–216 (1999)
6789. Yoshida, H.: On an analogue of the Sato conjecture. Invent. Math. **19**, 261–277 (1973)
6790. Yoshimoto, M.: Farey series and the Riemann Hypothesis, II. Acta Math. Acad. Sci. Hung. **78**, 287–304 (1998)
6791. Yoshimoto, M.: Farey series and the Riemann Hypothesis, IV. Acta Math. Acad. Sci. Hung. **87**, 109–119 (2000)
6792. Yoshimoto, M.: Abelian theorems, Farey series and the Riemann hypothesis. Ramanujan J. **8**, 131–145 (2004)
6793. Young, J., Potler, A.: First occurrence prime gaps. Math. Comput. **52**, 221–224 (1989)
6794. Young, W.H.: Adolf Hurwitz. Proc. Lond. Math. Soc. **20**, xlviii–liv (1922)
6795. Yu, G.: The distribution of 4-full numbers. Monatshefte Math. **118**, 145–152 (1994)
6796. Yu, H.B.: On Waring's problem with polynomial summands. Acta Arith. **76**, 131–144 (1996)
6797. Yu, H.B.: On Waring's problem with polynomial summands, II. Acta Arith. **86**, 245–254 (1998)
6798. Yu, H.B.: On Waring's problem with quartic polynomial summands. Acta Arith. **80**, 77–82 (1997)
6799. Yu, H.B.: On Waring's problem with quartic polynomial summands, II. J. China Univ. Sci. Tech. **28**, 629–635 (1998)
6800. Yu, K.R.: A generalization of Mahler's classification to several variables. J. Reine Angew. Math. **377**, 113–126 (1987)

6801. Yuan, P.Z.: On the number of solutions of simultaneous Pell equations. Acta Arith. **101**, 215–221 (2002)
6802. Yuan, P.Z.: On the number of solutions of $x^2 - 4m(m+1)y^2 = y^2 - bz^2 - 1$. Proc. Am. Math. Soc. **132**, 1561–1566 (2004)
6803. Yuan, P.Z.: On the Diophantine equation $\frac{x^3-1}{x-1} = \frac{y^n-1}{y-1}$. J. Number Theory **112**, 20–25 (2005)
6804. Yüh, M.I.: A divisor problem. Sci. Rec. **2**, 326–328 (1958)
6805. Yüh, M.I., Wu, F.: On the divisor problem for $d_3(n)$. Sci. Sin. **11**, 1055–1060 (1962)
6806. Zaccagnini, A.: On the exceptional set for the sum of a prime and a kth power. Mathematika **39**, 400–421 (1992)
6807. Zaccagnini, A.: A note on the sum of a prime and a polynomial. Q. J. Math. **52**, 519–524 (2001)
6808. Zagier, D.: Higher dimensional Dedekind sums. Math. Ann. **202**, 149–172 (1973)
6809. Zagier, D.: Modular forms associated to real quadratic fields. Invent. Math. **30**, 1–46 (1975)
6810. Zagier, D.: Sur la conjecture de Saito-Kurokawa (d'après H. Maass). Prog. Math. **12**, 371–394 (1981)
6811. Zagier, D.: On the number on Markoff numbers below a given bound. Math. Comput. **39**, 709–723 (1982)
6812. Zagier, D.: A proof of the Kac-Wakimoto affine denominator formula for the strange series. Math. Res. Lett. **7**, 597–604 (2000)
6813. Zaharescu, A.: Small values of $n^2\alpha$ (mod 1). Invent. Math. **121**, 379–388 (1995)
6814. Zarhin, Yu.G.: Minimal models of curves of genus 2 and homomorphisms of abelian varieties defined over a field of positive characteristic. Izv. Akad. Nauk SSSR, Ser. Mat. **36**, 67–109 (1972) (in Russian)
6815. Zarhin, Yu.G.: A finiteness theorem for isogenies of abelian varieties over function fields of finite characteristic. Funkc. Anal. Prilozh. **8**, 31–34 (1974) (in Russian)
6816. Zarhin, Yu.G.: Isogenies of abelian varieties over fields of finite characteristic. Mat. Sb. **95**, 461–470 (1974) (in Russian)
6817. Zarhin, Yu.G.: Endomorphisms of Abelian varieties over fields of finite characteristic. Izv. Akad. Nauk SSSR, Ser. Mat. **39**, 272–277 (1975) (in Russian)
6818. Zassenhaus, H.: Über die Existenz von Primzahlen in arithmetischen Progressionen. Comment. Math. Helv. **22**, 232–259 (1949)
6819. Zassenhaus, H.: Marshall Hall, Jr., 1910–1990. Not. Am. Math. Soc. **37**, 1033 (1990)
6820. Zavorotnyĭ, N.I.: On the fourth moment of the Riemann zeta function. In: Automorphic Functions and Number Theory, Vladivostok, pp. 69–124 (1989) (in Russian)
6821. Zhai, W.: On higher-power moments of $\Delta(x)$. Acta Arith. **112**, 367–395 (2004)
6822. Zhai, W.: On higher-power moments of $\Delta(x)$, II. Acta Arith. **114**, 35–54 (2004)
6823. Zhai, W.: On higher-power moments of $\Delta(x)$, III. Acta Arith. **118**, 263–281 (2005)
6824. Zhan, T.: Distribution of k-full integers. Sci. China **A32**, 20–37 (1989)
6825. Zhang, M.Y., Ding, P.: An improvement to the Schnirelman constant. J. China Univ. Sci. Techn. **13**, 31–53 (1983)
6826. Zhang, W.-B.: Mean-value theorems of multiplicative functions on additive arithmetic semigroups. Math. Z. **229**, 195–233 (1998)
6827. Zhang, W.-B.: Probabilistic number theory in additive arithmetic semigroups, III. Ramanujan J. **6**, 387–428 (2002)
6828. Zhang, W.-B.: Mean-value theorems of multiplicative functions on additive arithmetic semigroups via Halász's method. Monatshefte Math. **138**, 319–353 (2003)
6829. Zhang, Y.: Congruence and uniqueness of certain Markoff numbers. Acta Arith. **128**, 295–301 (2007)
6830. Zhu, W., Yu, K.: The distribution of square-full integers. Pure Appl. Math. **12**, 113–122 (1996)
6831. Ziegler, C.: Jacobi forms of higher degree. Abh. Math. Semin. Univ. Hamb. **59**, 191–224 (1989)

6832. Zimmer, H.G.: An elementary proof of the Riemann hypothesis for an elliptic curve over a finite field. Pac. J. Math. **36**, 267–278 (1971)
6833. Zinoview, D.: On Vinogradov's constant in Goldbach's ternary problem. J. Number Theory **65**, 334–358 (1997)
6834. Znám, Š.: A survey of covering systems of congruences. Acta Math. Univ. Comen. **40**, 59–79 (1982)
6835. Zornow, A.: De compositione numerorum e cubis integris positivis. J. Reine Angew. Math. **14**, 276–280 (1835)
6836. Zsigmondy, K.: Zur Theorie der Potenzreste. Monatshefte Math. Phys. **3**, 265–284 (1892)
6837. Zu dem Preisausschreiben der Kgl. Gesellschaft der Wissenschaften zu Göttingen für den Beweis des Fermatschen Satzes. Math. Ann. **66**, 7 (1909)
6838. Zuckerman, H.S.: New results for the number $g(n)$ in Waring's problem. Am. J. Math. **58**, 545–552 (1936)
6839. Zuckerman, H.S.: The computation of smaller coefficients of $j(\tau)$. Bull. Am. Math. Soc. **45**, 917–919 (1939)
6840. Zuckerman, H.S.: On the coefficients of certain modular forms belonging to subgroups of the modular group. Trans. Am. Math. Soc. **45**, 298–321 (1939)
6841. Zuckerman, H.S.: On the expansions of certain modular forms of positive dimension. Am. J. Math. **62**, 127–152 (1940)
6842. Zudilin, V.V.: One of the numbers $\zeta(5), \zeta(7), \zeta(9), \zeta(11)$ is irrational. Usp. Mat. Nauk **56**(4), 149–150 (2001) (in Russian)
6843. Zudilin, V.V.: On the irrationality of the values of the Riemann zeta-function. Izv. Ross. Akad. Nauk, Ser. Mat. **66**, 49–102 (2002) (in Russian)
6844. Zudilin, W. [V.V.]: Arithmetic of linear forms involving odd zeta values. J. Théor. Nr. Bordx. **16**, 251–291 (2004)
6845. Zudilin, W. [V.V.]: A new lower bound for $\|(3/2)^k\|$. J. Théor. Nr. Bordx. **19**, 311–323 (2007)
6846. Zulauf, A.: Beweis einer Erweiterung des Satzes von Goldbach-Vinogradov. J. Reine Angew. Math. **190**, 169–198 (1952)
6847. Zulauf, A.: Über die Darstellung natürlichen Zahlen als Summen von Primzahlen aus gegebenen Restklassen und Quadraten mit gegebenen Koeffizienten, I. J. Reine Angew. Math. **192**, 210–229 (1953)
6848. Zulauf, A.: Über die Darstellung natürlichen Zahlen als Summen von Primzahlen aus gegebenen Restklassen und Quadraten mit gegebenen Koeffizienten, II. J. Reine Angew. Math. **193**, 39–53 (1954)
6849. Zulauf, A.: The distribution of Farey numbers. J. Reine Angew. Math. **289**, 209–213 (1977)

Author Index

W. Narkiewicz, *Rational Number Theory in the 20th Century*, Springer Monographs in Mathematics,
DOI 10.1007/978-0-85729-532-3, © Springer-Verlag London Limited 2012

I

M

Q

R

S

T

Subject Index

W. Narkiewicz, *Rational Number Theory in the 20th Century*,
Springer Monographs in Mathematics,
DOI 10.1007/978-0-85729-532-3,

MIX
Papier aus verantwortungsvollen Quellen
Paper from responsible sources
FSC® C105338

If you have any concerns about our products,
you can contact us on
ProductSafety@springernature.com

In case Publisher is established outside the EU,
the EU authorized representative is:
Springer Nature Customer Service Center GmbH
Europaplatz 3, 69115 Heidelberg, Germany

Printed by Libri Plureos GmbH
in Hamburg, Germany